Prealgebra

Prealgebra

Seventh Edition

Elayn Martin-Gay

University of New Orleans

PEARSON

Boston   Columbus   Indianapolis   New York   San Francisco   Upper Saddle River
Amsterdam   Cape Town   Dubai   London   Madrid   Milan   Munich   Paris   Montréal   Toronto
Delhi   Mexico City   São Paulo   Sydney   Hong Kong   Seoul   Singapore   Taipei   Tokyo

Editorial Director, Mathematics: *Christine Hoag*
Editor-in-Chief: *Michael Hirsch*
Acquisitions Editor: *Mary Beckwith*
Senior Content Editor: *Lauren Morse*
Editorial Assistant: *Matthew Summers*
Development Editor: *Dawn Nuttall*
Senior Managing Editor: *Karen Wernholm*
Production Project Manager: *Patty Bergin*
Cover and Illustration Design: *Tamara Newnam*
Program Design Lead: *Heather Scott*
Interior Design: *Integra*
Digital Assets Manager: *Marianne Groth*
Supplements Production Project Manager: *Katherine Roz*
Executive Content Manager, MathXL: *Rebecca Williams*
Senior Content Developer, TestGen: *John Flanagan*
Executive Manager, Course Production: *Peter Silvia*
Media Producer: *Audra Walsh*
Executive Marketing Manager: *Michelle Renda*
Marketing Assistant: *Caitlin Ghegan*
Senior Author Support/Technology Specialist: *Joe Vetere*
Procurement Specialist: *Debbie Rossi*
Production Management and Composition: *Integra*
Text Art: *Scientific Illustrators*
Answer Art: *Integra*

Library of Congress Cataloging-in-Publication Data
Martin-Gay, K. Elayn
  Prealgebra / Elayn Martin-Gay, University of New Orleans. – 7th edition.
    pages cm
  Includes index.
  ISBN 978-0-321-95504-3
  1. Arithmetic–Textbooks.   I. Title.
  QA107.2.M37 2015
  510–dc23   2013023122

1 2 3 4 5 6 7 8 9 10—CRK—17 16 15 14

www.pearsonhighered.com

ISBN-10: 0-321-95504-8 (Student Edition)
ISBN-13: 978-0-321-95504-3

In loving memory of my son
Bryan Jackson Gay

His two favorite quotes:

I can do everything through Him who gives me strength.
—Philippians 4:13

When one man, for whatever reason, has the opportunity to lead
an extraordinary life, he has no right to keep it to himself.
—Jacques-Yves Cousteau

Contents

Appendices

Preface

Prealgebra, **Seventh Edition**, was written to help students make the transition from arithmetic to algebra. To help them reach this goal, I introduce algebraic concepts early and repeat them as I cover traditional arithmetic topics, thus laying the groundwork for the next algebra course your students will take. A second goal was to show students the relevancy of mathematics in everyday life and in the workplace.

The many factors that contributed to the success of the previous editions have been retained. In preparing the Seventh Edition, I considered comments and suggestions of colleagues, students, and many users of the prior edition throughout the country.

What's New in the Seventh Edition?

- **The Martin-Gay Program** has been revised and enhanced with a new design in the text and MyMathLab® to actively encourage students to use the text, video program, Video Organizer, and Student Organizer as an integrated learning system.

- **The new Video Organizer** is designed to help students take notes and work practice exercises while watching the Interactive Lecture Series videos (available in MyMathLab and on DVD). All content in the Video Organizer is presented in the same order as it is presented in the videos, making it easy for students to create a course notebook and build good study habits.
  - Covers all of the video examples in order.
  - Provides ample space for students to write down key definitions and properties.
  - Includes "Play" and "Pause" button icons to prompt students to follow along with the author for some exercises while they try others on their own.

The Video Organizer is available in a loose-leaf, notebook-ready format. It is also available for download in MyMathLab.

- **Vocabulary, Readiness & Video Check** questions have been added prior to every section exercise set. These exercises quickly check a student's understanding of new vocabulary words. The **readiness** exercises center on a student's understanding of a concept that is necessary in order to continue to the exercise set. **New Video check questions for the Martin-Gay Interactive Lecture videos** are now included in every section for each learning objective. **These exercises are all available for assignment in MyMathLab** and are a great way to assess whether students have viewed and understood the key concepts presented in the videos.

- **New Student Success Tips Videos** are 3- to -5 minute video segments designed to be daily reminders to students to continue practicing and maintaining good organizational and study habits. They are organized in three categories and are available in MyMathLab and the Interactive Lecture Series. The categories are:
  1. Success Tips that apply to any course in college in general, such as Time Management.
  2. Success Tips that apply to any mathematics course. One example is based on understanding that mathematics is a course that requires homework to be completed in a timely fashion.
  3. Section- or Content-specific Success Tips to help students avoid common mistakes or to better understand concepts that often prove challenging. One example of this type of tip is how to apply the order of operations to simplify an expression such as $5 - 3(x + 2)$.

- **Interactive DVD Lecture Series**, featuring your text author (Elayn Martin-Gay), provides students with active learning at their own pace. The videos offer the following resources and more:

  A complete lecture for each section of the text highlights key examples and exercises from the text. "Pop-ups" reinforce key terms, definitions, and concepts.

  An interface with menu navigation features allows students to quickly find and focus on the examples and exercises they need to review.

  Interactive Concept Check exercises measure students' understanding of key concepts and common trouble spots.

  New Student Success Tips Videos.

- **The Interactive DVD Lecture Series** also includes the following resources for test prep:

  The Chapter Test Prep Videos help students during their most teachable moment—when they are preparing for a test. This innovation provides step-by-step solutions for the exercises found in each Chapter Test. For the Seventh Edition, the chapter test prep videos are also available on YouTube™. The videos are captioned in English and Spanish.

  The Practice Final Exam Videos help students prepare for an end-of-course final. Students can watch full video solutions to each exercise in the Practice Final Exam at the end of this text.

- **The Martin-Gay MyMathLab** course has been updated and revised to provide more exercise coverage, including assignable video check questions and an expanded video program. There are section lecture videos for every section, which students can also access at the specific objective level; Student Success Tips videos; and an increased number of watch clips at the exercise level to help students while doing homework in MathXL. Suggested homework assignments have been premade for assignment at the instructor's discretion.

- **New MyMathLab Ready to Go courses** (access code required) provide students with all the same great MyMathLab features that you're used to, but make it easier for instructors to get started. Each course includes preassigned homework and quizzes to make creating your course even simpler. Ask your Pearson representative about the details for this particular course or to see a copy of this course.

Key Pedagogical Features

The following key features have been retained and/or updated for the Seventh Edition of the text:

Problem-Solving Process This is formally introduced in Chapter 3 with a four-step process that is integrated throughout the text. The four steps are **Understand, Translate, Solve,** and **Interpret**. The repeated use of these steps in a variety of examples shows their wide applicability. Reinforcing the steps can increase students' comfort level and confidence in tackling problems.

Exercise Sets Revised and Updated The exercise sets have been carefully examined and extensively revised. Special focus was placed on making sure that even- and odd-numbered exercises are paired and that real-life applications were updated.

Examples Detailed, step-by-step examples were added, deleted, replaced, or updated as needed. Many examples reflect real life. Additional instructional support is provided in the annotated examples.

Practice Exercises   Throughout the text, each worked-out example has a parallel Practice exercise. These invite students to be actively involved in the learning process. Students should try each Practice exercise after finishing the corresponding example. Learning by doing will help students grasp ideas before moving on to other concepts. Answers to the Practice exercises are provided at the bottom of each page.

Helpful Hints   Helpful Hints contain practical advice on applying mathematical concepts. Strategically placed where students are most likely to need immediate reinforcement, Helpful Hints help students avoid common trouble areas and mistakes.

Concept Checks   This feature allows students to gauge their grasp of an idea as it is being presented in the text. Concept Checks stress conceptual understanding at the point-of-use and help suppress misconceived notions before they start. Answers appear at the bottom of the page. Exercises related to Concept Checks are included in the exercise sets.

Mixed Practice Exercises   In the section exercise sets, these exercises require students to determine the problem type and strategy needed to solve it just as they would need to do on a test.

Integrated Reviews   This unique, mid-chapter exercise set helps students assimilate new skills and concepts that they have learned separately over several sections. These reviews provide yet another opportunity for students to work with "mixed" exercises as they master the topics.

Vocabulary Check   This feature provides an opportunity for students to become more familiar with the use of mathematical terms as they strengthen their verbal skills. These appear at the end of each chapter before the Chapter Highlights. Vocabulary, Readiness & Video exercises provide practice at the section level.

Chapter Highlights   Found at the end of every chapter, these contain key definitions and concepts with examples to help students understand and retain what they have learned and help them organize their notes and study for tests.

Chapter Review   The end of every chapter contains a comprehensive review of topics introduced in the chapter. The Chapter Review offers exercises keyed to every section in the chapter, as well as Mixed Review exercises that are not keyed to sections.

Chapter Test and Chapter Test Prep Videos   The Chapter Test is structured to include those problems that involve common student errors. The **Chapter Test Prep Videos** gives students instant access to a step-by-step video solution of each exercise in the Chapter Test.

Cumulative Review   This review follows every chapter in the text (except Chapter 1). Each odd-numbered exercise contained in the Cumulative Review is an earlier worked example in the text that is referenced in the back of the book along with the answer.

Writing Exercises   ✎ These exercises occur in almost every exercise set and require students to provide a written response to explain concepts or justify their thinking.

Applications   Real-world and real-data applications have been thoroughly updated, and many new applications are included. These exercises occur in almost every exercise set and show the relevance of mathematics and help students gradually and continuously develop their problem-solving skills.

Review Exercises   These exercises occur in each exercise set (except in Chapter 1) and are keyed to earlier sections. They review concepts learned earlier in the text that will be needed in the next section or chapter.

Exercise Set Resource Icons   Located at the opening of each exercise set, these icons remind students of the resources available for extra practice and support:

See Student Resources descriptions on page xv for details on the individual resources available.

Exercise Icons   These icons facilitate the assignment of specialized exercises and let students know what resources can support them.

- DVD Video icon: exercise worked on the Interactive DVD Lecture Series.
- △ Triangle icon: identifies exercises involving geometric concepts.
- Pencil icon: indicates a written response is needed.
- Calculator icon: optional exercises intended to be solved using a scientific or graphing calculator.

Group Activities   Found at the end of each chapter, these activities are for individual or group completion, and are usually hands-on or data-based activities that extend the concepts found in the chapter, allowing students to make decisions and interpretations and to think and write about algebra.

Optional: Calculator Exploration Boxes and Calculator Exercises   The optional Calculator Explorations provide keystrokes and exercises at appropriate points to give students an opportunity to become familiar with these tools. Section exercises that are best completed by using a calculator are identified by ▦ for ease of assignment.

Student and Instructor Resources

STUDENT RESOURCES

Student Organizer	Student Solutions Manual
Guides students through the 3 main components of studying effectively—notetaking, practice, and homework. The Organizer includes before-class preparation exercises, notetaking pages in a 2-column format for use in class, and examples paired with exercises for practice for each section. Includes an outline and questions for use with the Student Success Tip Videos. It is 3-hole-punched. Available in loose-leaf, notebook-ready format and in MyMathLab.	Provides completely worked-out solutions to the odd-numbered section exercises; all exercises in the Integrated Reviews, Chapter Reviews, Chapter Tests, and Cumulative Reviews
Interactive DVD Lecture Series Videos	**Video Organizer**
Provides students with active learning at their pace. The videos offer: ● A complete lecture for each text section. The interface allows easy navigation to examples and exercises students need to review. ● Interactive Concept Check exercises ● Student Success Tips Videos ● Practice Final Exam ● Chapter Test Prep Videos	Designed to help students take notes and work practice exercises while watching the Interactive Lecture Series videos. ● Covers all of the video examples in order. ● Provides ample space for students to write down key definitions and rules. ● Includes "Play" and "Pause" button icons to prompt students to follow along with the author for some exercises while they try others on their own. ● Includes Student Success Tips Outline and Questions Available in loose-leaf, notebook-ready format and in MyMathLab.

INSTRUCTOR RESOURCES

Annotated Instructor's Edition	Instructor's Resource Manual with Tests and Mini-Lectures
Contains all the content found in the student edition, plus the following: ● Answers to exercises on the same text page ● Teaching Tips throughout the text placed at key points	● Mini-lectures for each text section ● Additional practice worksheets for each section ● Several forms of test per chapter—free response and multiple choice ● Answers to all items **Instructor's Solutions Manual** **TestGen**® (Available for download from the IRC)
Instructor-to-Instructor Videos—available in the Instructor Resources section of the MyMathLab course.	**Online Resources** **MyMathLab**® (access code required) **MathXL**® (access code required)

Acknowledgments

There are many people who helped me develop this text, and I will attempt to thank some of them here. Courtney Slade and Cindy Trimble were *invaluable* for contributing to the overall accuracy of the text. Dawn Nuttall was *invaluable* for her many suggestions and contributions during the development and writing of this Seventh Edition. Debbie Meyer and Patty Bergin provided guidance throughout the production process.

A very special thank you goes to my editor, Mary Beckwith, for being there 24/7/365, as my students say. And, my thanks to the staff at Pearson for all their support: Heather Scott, Lauren Morse, Matt Summers, Michelle Renda, Michael Hirsch, Chris Hoag, and Greg Tobin.

I would like to thank the following reviewers for their input and suggestions:

Lisa Angelo, *Bucks Community College*
Victoria Baker, *Nicholls State College*
Teri Barnes, *McLennan Community College*
Laurel Berry, *Bryant & Stratton*
Thomas Blackburn, *Northeastern Illinois University*
Gail Burkett, *Palm Beach Community College*
Anita Collins, *Mesa Community College*
Lois Colpo, *Harrisburg Area Community College*
Fay Dang, *Joliet Junior. College*
Robert Diaz, *Fullerton College*
Tamie Dickson, *Reading Area Community College*
Latonya Ellis, *Gulf Coast Community College*
Sonia Ford, *Midland College*
Cheryl Gibby, *Cypress College*
Kathryn Gunderson, *Three Rivers Community College*
Elizabeth Hamman, *Cypress College*
Craig Hardesty, *Hillsborough Community College*
Lloyd Harris, *Gulf Coast Community College*

Teresa Hasenauer, *Indian River College*
Julia Hassett, *Oakton Community College*
Jeff Koleno, *Lorain County Community College*
Judy Langer, *Westchester Community College*
Sandy Lofstock, *St. Petersburg College*
Stan Mattoon, *Merced College*
Dr. Kris Mudunuri, *Long Beach City College*
Carol Murphy, *San Diego Miramar College*
Greg Nguyen, *Fullerton College*
Jean Olsen, *Pikes Peak Community College*
Darlene Ornelas, *Fullerton College*
Warren Powell, *Tyler Junior College*
Jeanette Shea, *Central Texas College*
Katerina Vishnyakova, *Collin County Community College*
Corey Wadlington, *West Kentucky Community and Technical College*
Edward Wagner, *Central Texas College*
Jenny Wilson, *Tyler Junior College*

I would also like to thank the following dedicated group of instructors who participated in our focus groups, Martin-Gay Summits, and our design review for the series. Their feedback and insights have helped to strengthen this edition of the text. These instructors include:

Billie Anderson, *Tyler Junior College*
Cedric Atkins, *Mott Community College*
Lois Beardon, *Schoolcraft College*
Laurel Berry, *Bryant & Stratton*
John Beyers, *University of Maryland*
Bob Brown, *Community College of Baltimore County–Essex*
Lisa Brown, *Community College of Baltimore County–Essex*
NeKeith Brown, *Richland College*
Gail Burkett, *Palm Beach Community College*

Cheryl Cantwell, *Seminole Community College*
Jackie Cohen, *Augusta State College*
Julie Dewan, *Mohawk Valley Community College*
Janice Ervin, *Central Piedmont Community College*
Richard Fielding, *Southwestern College*
Cindy Gaddis, *Tyler Junior College*
Nita Graham, *St. Louis Community College*
Pauline Hall, *Iowa State College*
Pat Hussey, *Triton College*

Dorothy Johnson, *Lorain County Community College*

Sonya Johnson, *Central Piedmont Community College*

Irene Jones, *Fullerton College*

Paul Jones, *University of Cincinnati*

Kathy Kopelousous, *Lewis and Clark Community College*

Nancy Lange, *Inver Hills Community College*

Judy Langer, *Westchester Community College*

Lisa Lindloff, *McLinnan Community College*

Sandy Lofstock, *St. Petersburg College*

Kathy Lovelle, *Westchester Community College*

Jean McArthur, *Joliet Junior College*

Kevin McCandless, *Evergreen Valley College*

Daniel Miller, *Niagra County Community College*

Marica Molle, *Metropolitan Community College*

Carol Murphy, *San Diego Miramar College*

Greg Nguyen, *Fullerton College*

Eric Oilila, *Jackson Community College*

Linda Padilla, *Joliet Junior College*

Davidson Pierre, *State College of Florida*

Marilyn Platt, *Gaston College*

Ena Salter, *Manatee Community College*

Carole Shapero, *Oakton Community College*

Janet Sibol, *Hillsborough Community College*

Anne Smallen, *Mohawk Valley Community College*

Barbara Stoner, *Reading Area Community College*

Jennifer Strehler, *Oakton Community College*

Ellen Stutes, *Louisiana State University Elinice*

Tanomo Taguchi, *Fullerton College*

MaryAnn Tuerk, *Elsin Community College*

Walter Wang, *Baruch College*

Leigh Ann Wheeler, *Greenville Technical Community College*

Valerie Wright, *Central Piedmont Community College*

A special thank you to those students who participated in our design review: Katherine Browne, Mike Bulfin, Nancy Canipe, Ashley Carpenter, Jeff Chojnachi, Roxanne Davis, Mike Dieter, Amy Dombrowski, Kay Herring, Todd Jaycox, Kaleena Levan, Matt Montgomery, Tony Plese, Abigail Polkinghorn, Harley Price, Eli Robinson, Avery Rosen, Robyn Schott, Cynthia Thomas, and Sherry Ward.

Elayn Martin-Gay

About the Author

Elayn Martin-Gay has taught mathematics at the University of New Orleans for more than 25 years. Her numerous teaching awards include the local University Alumni Association's Award for Excellence in Teaching, and Outstanding Developmental Educator at University of New Orleans, presented by the Louisiana Association of Developmental Educators.

Prior to writing textbooks, Elayn Martin-Gay developed an acclaimed series of lecture videos to support developmental mathematics students in their quest for success. These highly successful videos originally served as the foundation material for her texts. Today, the videos are specific to each book in the Martin-Gay series. The author has also created Chapter Test Prep Videos to help students during their most "teachable moment"—as they prepare for a test—along with Instructor-to-Instructor videos that provide teaching tips, hints, and suggestions for each developmental mathematics course, including basic mathematics, prealgebra, beginning algebra, and intermediate algebra.

Elayn is the author of 12 published textbooks as well as multimedia, interactive mathematics, all specializing in developmental mathematics courses. She has also published series in Algebra 1, Algebra 2, and Geometry. She has participated as an author across the broadest range of educational materials: textbooks, videos, tutorial software, and courseware. This provides an opportunity of various combinations for an integrated teaching and learning package offering great consistency for the student.

Applications Index

The Whole Numbers

A Selection of Resources for Success in this Mathematics Course

Text book

Instructor

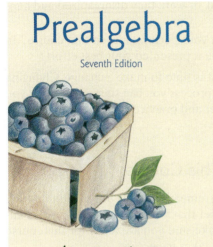

MyMathLab and MathXL

Video Organizer

Student Organizer

Interactive Lecture Series

For more information about the resources illustrated above, read Section 1.1.

Whole numbers are the basic building blocks of mathematics. The whole numbers answer the question "How many?"

This chapter covers basic operations on whole numbers. Knowledge of these operations provides a good foundation on which to build further mathematical skills.

1.1 Study Skill Tips for Success in Mathematics ▶

Objectives

A Get Ready for This Course. ▶

B Understand Some General Tips for Success. ▶

C Know How to Use This Text. ▶

D Know How to Use Text Resources. ▶

E Get Help as Soon as You Need It. ▶

F Learn How to Prepare for and Take an Exam. ▶

G Develop Good Time Management. ▶

Before reading Section 1.1, you might want to ask yourself a few questions.

1. When you took your last math course, were you organized? Were your notes and materials from that course easy to find, or were they disorganized and hard to find—if you saved them at all?

2. Were you satisfied—really satisfied—with your performance in that course? In other words, do you feel that your outcome represented your best effort?

If the answer is "no" to these questions, then it is time to make a change. Changing to or resuming good study skill habits is not a process you can start and stop as you please. It is something that you must remember and practice each and every day. To begin, continue reading this section.

Objective A Getting Ready for This Course ▶

Now that you have decided to take this course, remember that a *positive attitude* will make all the difference in the world. Your belief that you can succeed is just as important as your commitment to this course. Make sure you are ready for this course by having the time and positive attitude that it takes to succeed.

Make sure that you are familiar with the way that this course is being taught. Is it a traditional course, in which you have a printed textbook and meet with an instructor? Is it taught totally online, and your textbook is electronic and you e-mail your instructor? Or is your course structured somewhere in between these two methods? (Not all of the tips that follow will apply to all forms of instruction.)

Also make sure that you have scheduled your math course for a time that will give you the best chance for success. For example, if you are also working, you may want to check with your employer to make sure that your work hours will not conflict with your course schedule.

On the day of your first class period, double-check your schedule and allow yourself extra time to arrive on time in case of traffic problems or difficulty locating your classroom. Make sure that you are aware of and bring all necessary class materials.

Objective B General Tips for Success ▶

Below are some general tips that will increase your chance for success in a mathematics class. Many of these tips will also help you in other courses you may be taking.

Most important! Organize your class materials. In the next couple pages, many ideas will be presented to help you organize your class materials—notes, any handouts, completed homework, previous tests, etc. In general, you MUST have these materials organized. All of them will be valuable references throughout your course and when studying for upcoming tests and the final exam. One way to make sure you can locate these materials when you need them is to use a three-ring binder. This binder should be used solely for your mathematics class and should be brought to each and every class or lab. This way, any material can be immediately inserted in a section of this binder and will be there when you need it.

Form study groups and/or exchange names and e-mail addresses. Depending on how your course is taught, you may want to keep in contact with your fellow students. Some ways of doing this are to form a study group—whether in person or through the Internet. Also, you may want to ask if anyone is interested in exchanging e-mail addresses or any other form of contact.

> **Helpful Hint**
>
> **MyMathLab® and MathXL®**
> When assignments are turned in online, keep a hard copy of your complete written work. You will need to refer to your written work to be able to ask questions and to study for tests later.

Choose to attend all class periods. If possible, sit near the front of the classroom. This way, you will see and hear the presentation better. It may also be easier for you to participate in classroom activities.

Do your homework. You've probably heard the phrase "practice makes perfect" in relation to music and sports. It also applies to mathematics. You will find that the more time you spend solving mathematics exercises, the easier the process becomes. Be sure to schedule enough time to complete your assignments before the due date assigned by your instructor.

Check your work. Review the steps you took while working a problem. Learn to check your answers in the original exercises. You may also compare your answers with the "Answers to Selected Exercises" section in the back of the book. If you have made a mistake, try to figure out what went wrong. Then correct your mistake. If you can't find what went wrong, **don't** erase your work or throw it away. Show your work to your instructor, a tutor in a math lab, or a classmate. It is easier for someone to find where you had trouble if he or she looks at your original work.

Learn from your mistakes and be patient with yourself. Everyone, even your instructor, makes mistakes. (That definitely includes me—Elayn Martin-Gay.) Use your errors to learn and to become a better math student. The key is finding and understanding your errors.

Was your mistake a careless one, or did you make it because you can't read your own math writing? If so, try to work more slowly or write more neatly and make a conscious effort to carefully check your work.

Did you make a mistake because you don't understand a concept? Take the time to review the concept or ask questions to better understand it.

Did you skip too many steps? Skipping steps or trying to do too many steps mentally may lead to preventable mistakes.

Know how to get help if you need it. It's all right to ask for help. In fact, it's a good idea to ask for help whenever there is something that you don't understand. Make sure you know when your instructor has office hours and how to find his or her office. Find out whether math tutoring services are available on your campus. Check on the hours, location, and requirements of the tutoring service.

Don't be afraid to ask questions. You are not the only person in class with questions. Other students are normally grateful that someone has spoken up.

Turn in assignments on time. This way, you can be sure that you will not lose points for being late. Show every step of a problem and be neat and organized. Also be sure that you understand which problems are assigned for homework. If allowed, you can always double-check the assignment with another student in your class.

Objective C Knowing and Using Your Text

Flip through the pages of this text or view the e-text pages on a computer screen. Start noticing examples, exercise sets, end-of-chapter material, and so on. Every text is organized in some manner. Learn the way this text is organized by reading about and then finding an example in your text of each type of resource listed below. Finding and using these resources throughout your course will increase your chance of success.

- *Practice Exercises.* Each example in every section has a parallel Practice exercise. As you read a section, try each Practice exercise after you've finished the corresponding example. Answers are at the bottom of the page. This "learn-by-doing" approach will help you grasp ideas before you move on to other concepts.

- *Symbols at the Beginning of an Exercise Set.* If you need help with a particular section, the symbols listed at the beginning of each exercise set will remind you of the resources available.

Helpful Hint

MyMathLab® and MathXL®
If you are doing your homework online, you can work and re-work those exercises that you struggle with until you master them. Try working through all the assigned exercises twice before the due date.

Helpful Hint

MyMathLab® and MathXL®
If you are completing your homework online, it's important to work each exercise on paper before submitting the answer. That way, you can check your work and follow your steps to find and correct any mistakes.

Helpful Hint

MyMathLab® and MathXL®
Be aware of assignments and due dates set by your instructor. Don't wait until the last minute to submit work online.

- *Objectives.* The main section of exercises in each exercise set is referenced by an objective, such as **A** or **B**, and also an example(s). There is also often a section of exercises entitled "Mixed Practice," which is referenced by two or more objectives or sections. These are mixed exercises written to prepare you for your next exam. Use all of this referencing if you have trouble completing an assignment from the exercise set.

- *Icons (Symbols).* Make sure that you understand the meaning of the icons that are beside many exercises. ▶ tells you that the corresponding exercise may be viewed on the video Lecture Series that corresponds to that section. ＼ tells you that this exercise is a writing exercise in which you should answer in complete sentences. △ tells you that the exercise involves geometry.

- *Integrated Reviews.* Found in the middle of each chapter, these reviews offer you a chance to practice—in one place—the many concepts that you have learned separately over several sections.

- *End-of-Chapter Opportunities.* There are many opportunities at the end of each chapter to help you understand the concepts of the chapter.

   Vocabulary Checks contain key vocabulary terms introduced in the chapter.

   Chapter Highlights contain chapter summaries and examples.

   Chapter Reviews contain review problems. The first part is organized section by section and the second part contains a set of mixed exercises.

   Chapter Tests are sample tests to help you prepare for an exam. The Chapter Test Prep Videos found in the Interactive Lecture Series, MyMathLab, and YouTube provide the video solution to each question on each Chapter Test.

   Cumulative Reviews start at Chapter 2 and are reviews consisting of material from the beginning of the book to the end of that particular chapter.

- *Student Resources in Your Textbook.* You will find a **Student Resources** section at the back of this textbook. It contains the following to help you study and prepare for tests:

   Study Skill Builders contain study skills advice. To increase your chance for success in the course, read these study tips, and answer the questions.

   Bigger Picture—Study Guide Outline provides you with a study guide outline of the course, with examples.

   Practice Final provides you with a Practice Final Exam to help you prepare for a final.

- *Resources to Check Your Work.* The **Answers to Selected Exercises** section provides answers to all odd-numbered section exercises and to all integrated review, chapter test, and cumulative review exercises. Use the **Solutions to Selected Exercises** to see the worked-out solution to every other odd-numbered exercise.

Helpful Hint

MyMathLab®

In MyMathLab, you have access to the following video resources:

- Lecture Videos for each section
- Chapter Test Prep Videos

Use these videos provided by the author to prepare for class, review, and study for tests.

Objective **D** Knowing and Using Video and Notebook Organizer Resources ▶

Video Resources

Below is a list of video resources that are all made by me—the author of your text, Elayn Martin-Gay. By making these videos, I can be sure that the methods presented are consistent with those in the text.

- *Interactive DVD Lecture Series.* Exercises marked with a ▶ are fully worked out by the author on the DVDs and within MyMathLab. The lecture series provides approximately 20 minutes of instruction per section and is organized by Objective.

- *Chapter Test Prep Videos.* These videos provide solutions to all of the Chapter Test exercises worked out by the author. They can be found in MyMathLab, the Interactive Lecture series, and You Tube. This supplement is very helpful before a test or exam.
- *Student Success Tips.* These video segments are about 3 minutes long and are daily reminders to help you continue practicing and maintaining good organizational and study habits.
- *Final Exam Videos.* These video segments provide solutions to each question. These videos can be found within MyMathLab and the Interactive Lecture Series.

Notebook Organizer Resources

The resources below are in three-ring notebook ready form. They are to be inserted in a three-ring binder and completed. Both resources are numbered according to the sections in your text to which they refer.

- *Video Organizer.* This organizer is closely tied to the Interactive Lecture (Video) Series. Each section should be completed while watching the lecture video on the same section. Once completed, you will have a set of notes to accompany the Lecture (Video) Series section by section.
- *Student Organizer.* This organizer helps you study effectively through note-taking hints, practice, and homework while referencing examples in the text and examples in the Lecture Series.

Objective E Getting Help ▶

If you have trouble completing assignments or understanding the mathematics, get help as soon as you need it! This tip is presented as an objective on its own because it is so important. In mathematics, usually the material presented in one section builds on your understanding of the previous section. This means that if you don't understand the concepts covered during a class period, there is a good chance that you will not understand the concepts covered during the next class period. If this happens to you, get help as soon as you can.

Where can you get help? Many suggestions have been made in this section on where to get help, and now it is up to you to get it. Try your instructor, a tutoring center, or a math lab, or you may want to form a study group with fellow classmates. If you do decide to see your instructor or go to a tutoring center, make sure that you have a neat notebook and are ready with your questions.

Objective F Preparing for and Taking an Exam ▶

Make sure that you allow yourself plenty of time to prepare for a test. If you think that you are a little "math anxious," it may be that you are not preparing for a test in a way that will ensure success. The way that you prepare for a test in mathematics is important. To prepare for a test:

1. Review your previous homework assignments.
2. Review any notes from class and section-level quizzes you have taken. (If this is a final exam, also review chapter tests you have taken.)
3. Review concepts and definitions by reading the Chapter Highlights at the end of each chapter.
4. Practice working out exercises by completing the Chapter Review found at the end of each chapter. (If this is a final exam, go through a Cumulative Review. There is one found at the end of each chapter except Chapter 1. Choose the review found at the end of the latest chapter that you have covered in your course.) *Don't stop here!*

Helpful Hint

MyMathLab® and MathXL®

- Use the **Help Me Solve This** button to get step-by-step help for the exercise you are working. You will need to work an additional exercise of the same type before you can get credit for having worked it correctly.
- Use the **Video** button to view a video clip of the author working a similar exercise.

Helpful Hint

MyMathLab® and MathXL® Review your written work for previous assignments. Then, go back and re-work previous assignments. Open a previous assignment, and click **Similar Exercise** to generate new exercises. Re-work the exercises until you fully understand them and can work them without help features.

5. It is important that you place yourself in conditions similar to test conditions to find out how you will perform. In other words, as soon as you feel that you know the material, get a few blank sheets of paper and take a sample test. There is a Chapter Test available at the end of each chapter, or you can work selected problems from the Chapter Review. Your instructor may also provide you with a review sheet. During this sample test, do not use your notes or your textbook. Then check your sample test. If your sample test is the Chapter Test in the text, don't forget that the video solutions are in MyMathLab, the Interactive Lecture Series, and YouTube. If you are not satisfied with the results, study the areas that you are weak in and try again.

6. On the day of the test, allow yourself plenty of time to arrive at where you will be taking your exam.

When taking your test:

1. Read the directions on the test carefully.

2. Read each problem carefully as you take the test. Make sure that you answer the question asked.

3. Watch your time and pace yourself so that you can attempt each problem on your test.

4. If you have time, check your work and answers.

5. Do not turn your test in early. If you have extra time, spend it double-checking your work.

Objective G Managing Your Time ▶

As a college student, you know the demands that classes, homework, work, and family place on your time. Some days you probably wonder how you'll ever get everything done. One key to managing your time is developing a schedule. Here are some hints for making a schedule:

1. Make a list of all of your weekly commitments for the term. Include classes, work, regular meetings, extracurricular activities, etc. You may also find it helpful to list such things as laundry, regular workouts, grocery shopping, etc.

2. Next, estimate the time needed for each item on the list. Also make a note of how often you will need to do each item. Don't forget to include time estimates for the reading, studying, and homework you do outside of your classes. You may want to ask your instructor for help estimating the time needed.

3. In the exercise set that follows, you are asked to block out a typical week on the schedule grid given. Start with items with fixed time slots like classes and work.

4. Next, include the items on your list with flexible time slots. Think carefully about how best to schedule items such as study time.

5. Don't fill up every time slot on the schedule. Remember that you need to allow time for eating, sleeping, and relaxing! You should also allow a little extra time in case some items take longer than planned.

6. If you find that your weekly schedule is too full for you to handle, you may need to make some changes in your workload, classload, or other areas of your life. You may want to talk to your advisor, manager or supervisor at work, or someone in your college's academic counseling center for help with such decisions.

1.1 Exercise Set MyMathLab®

1. What is your instructor's name?

2. What are your instructor's office location and office hours?

3. What is the best way to contact your instructor?

4. Do you have the name and contact information of at least one other student in class?

5. Will your instructor allow you to use a calculator in this class?

6. Why is it important that you write step-by-step solutions to homework exercises and keep a hard copy of all work submitted?

7. Is there a tutoring service available on campus? If so, what are its hours? What services are available?

8. Have you attempted this course before? If so, write down ways that you might improve your chances of success during this next attempt.

9. List some steps that you can take if you begin having trouble understanding the material or completing an assignment. If you are completing your homework in MyMathLab® and MathXL®, list the resources you can use for help.

10. How many hours of studying does your instructor advise for each hour of instruction?

11. What does the ✎ icon in this text mean?

12. What does the △ icon in this text mean?

13. What does the ▶ icon in this text mean?

14. Search the minor columns in your text. What are Practice exercises?

15. When might be the best time to work a Practice exercise?

16. Where are the answers to Practice exercises?

17. What answers are contained in this text and where are they?

18. What are Study Skill Tips of the Day and where are they?

19. What and where are Integrated Reviews?

20. How many times is it suggested that you work through the homework exercises in MathXL® before the submission deadline?

21. How far in advance of the assigned due date is it suggested that homework be submitted online? Why?

22. Chapter Highlights are found at the end of each chapter. Find the Chapter 1 Highlights and explain how you might use it and how it might be helpful.

23. Chapter Reviews are found at the end of each chapter. Find the Chapter 1 Review and explain how you might use it and how it might be helpful.

24. Chapter Tests are found at the end of each chapter. Find the Chapter 1 Test and explain how you might use it and how it might be helpful when preparing for an exam on Chapter 1. Include how the Chapter Test Prep Videos may help. If you are working in MyMathLab® and MathXL®, how can you use previous homework assignments to study?

25. What is the Video Organizer? Explain the contents and how it might be used.

26. What is the Student Organizer? Explain the contents and how it might be used.

27. Read or reread objective **G** and fill out the schedule grid on the next page.

	Monday	Tuesday	Wednesday	Thursday	Friday	Saturday	Sunday
4:00 a.m.							
5:00 a.m.							
6:00 a.m.							
7:00 a.m.							
8:00 a.m.							
9:00 a.m.							
10:00 a.m.							
11:00 a.m.							
12:00 p.m.							
1:00 p.m.							
2:00 p.m.							
3:00 p.m.							
4:00 p.m.							
5:00 p.m.							
6:00 p.m.							
7:00 p.m.							
8:00 p.m.							
9:00 p.m.							
10:00 p.m.							
11:00 p.m.							
Midnight							
1:00 a.m.							
2:00 a.m.							
3:00 a.m.							

1.2   Place Value, Names for Numbers, and Reading Tables

Objectives

A Find the Place Value of a Digit in a Whole Number.

B Write a Whole Number in Words and in Standard Form.

C Write a Whole Number in Expanded Form.

D Read Tables.

The **digits** 0, 1, 2, 3, 4, 5, 6, 7, 8, and 9 can be used to write numbers. For example, the **whole numbers** are

0, 1, 2, 3, 4, 5, 6, 7, 8, 9, 10, 11, . . .

and the **natural numbers** are 1, 2, 3, 4, 5, 6, 7, 8, 9, 10, 11, . . .

The three dots (. . .) after the 11 mean that this list continues indefinitely. That is, there is no largest whole number. The smallest whole number is 0.

Objective A Finding the Place Value of a Digit in a Whole Number

The position of each digit in a number determines its **place value.** For example, the average distance (in miles) between the planet Mercury and the planet Earth can be represented by the whole number 48,337,000. On the next page, is a place-value chart for this whole number.

The two 3s in 48,337,000 represent different amounts because of their different placements. The place value of the 3 on the left is hundred-thousands. The place value of the 3 on the right is ten-thousands.

Examples Find the place value of the digit 3 in each whole number.

1. 396,418
↑
hundred-thousands

2. 93,192
↑
thousands

3. 534,275,866
↑
ten-millions

▶ **Work Practice 1–3**

Practice 1–3

Find the place value of the digit 8 in each whole number.
1. 38,760,005
2. 67,890
3. 481,922

Objective **B** Writing a Whole Number in Words and in Standard Form ▶

A whole number such as 1,083,664,500 is written in **standard form.** Notice that commas separate the digits into groups of three, starting from the right. Each group of three digits is called a **period.** The names of the first four periods are shown in red.

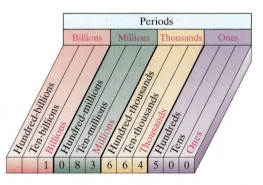

Writing a Whole Number in Words

To write a whole number in words, write the number in each period followed by the name of the period. (The ones period name is usually not written.) This same procedure can be used to read a whole number.

Fox example, we write 1,083,664,500 as

one **billion,**
eighty-three **million,**
six hundred sixty-four **thousand,**
five **hundred**

> **Helpful Hint** Notice the commas after the name of each period.

Answers
1. millions **2.** hundreds
3. ten-thousands

Helpful Hint

The name of the ones period is not used when reading and writing whole numbers. For example,

   9,265

is read as

   "nine **thousand**, two hundred sixty-five."

Practice 4-6

Write each whole number in words.

4. 54

5. 678

6. 93,205

| **Examples** | Write each whole number in words. |

4. 72   seventy-two

5. 546   five hundred forty-six

6. 27,034   twenty-seven thousand, thirty-four

◼ **Work Practice 4–6**

Helpful Hint

The word "and" is *not* used when reading and writing whole numbers. It is used only when reading and writing mixed numbers and some decimal values, as shown later in this text.

Practice 7

Write 679,430,105 in words.

| **Example 7** | Write 308,063,557 in words. |

Solution: 308,063,557 is written as

   three hundred eight **million**, sixty-three **thousand**, five hundred fifty-seven

◼ **Work Practice 7**

✓**Concept Check** True or false? When writing a check for $2600, the word name we write for the dollar amount of the check is "two thousand sixty." Explain your answer.

Practice 8-11

Write each whole number in standard form.

8. thirty-seven

9. two hundred twelve

10. eight thousand, two hundred seventy-four

11. five million, fifty-seven thousand, twenty-six

Writing a Whole Number in Standard Form

To write a whole number in standard form, write the number in each period, followed by a comma.

Answers

4. fifty-four

5. six hundred seventy-eight

6. ninety-three thousand, two hundred five

7. six hundred seventy-nine million, four hundred thirty thousand, one hundred five

8. 37   **9.** 212

10. 8,274 or 8274   **11.** 5,057,026

✓**Concept Check Answer**

false

| **Examples** | Write each whole number in standard form. |

8. forty-one   41   **9.** seven hundred eight   708

10. six thousand, four hundred ninety-three

   6,493   or   6493

11. three million, seven hundred forty-six thousand, five hundred twenty-two

   3,746,522

◼ **Work Practice 8–11**

Helpful Hint

A comma may or may not be inserted in a four-digit number. For example, both

    6,493   and   6493

are acceptable ways of writing six thousand, four hundred ninety-three.

Objective C Writing a Whole Number in Expanded Form

The place value of a digit can be used to write a number in expanded form. The **expanded form** of a number shows each digit of the number with its place value. For example, 5672 is written in expanded form as

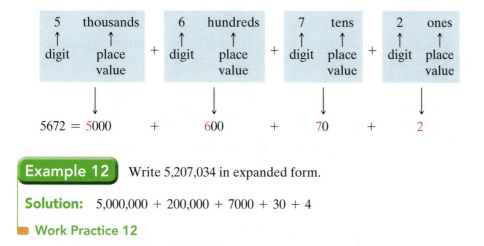

5	thousands		6	hundreds		7	tens		2	ones
↑	↑	+	↑	↑	+	↑	↑	+	↑	↑
digit	place value		digit	place value		digit	place value		digit	place value

$$5672 = 5000 \quad + \quad 600 \quad + \quad 70 \quad + \quad 2$$

Example 12   Write 5,207,034 in expanded form.

Solution: 5,000,000 + 200,000 + 7000 + 30 + 4

▸ **Work Practice 12**

We can visualize whole numbers by points on a line. The line below is called a **number line.** This number line has equally spaced marks for each whole number. The arrow to the right simply means that the whole numbers continue indefinitely. In other words, there is no largest whole number.

Number Line

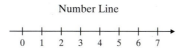

We will study number lines further in Section 1.4.

Objective D Reading Tables

Now that we know about place value and names for whole numbers, we introduce one way that whole number data may be presented. **Tables** are often used to organize and display facts that contain numbers. The following table shows the countries that won the most medals during the Summer Olympic Games in London in 2012. (Although the medals are truly won by athletes from the various countries, for simplicity we will state that countries have won the medals.)

Practice 12

Write 4,026,301 in expanded form.

Answer

12. 4,000,000 + 20,000 + 6000 + 300 + 1

2012 London Summer Olympics Medal Count

Country	Gold	Silver	Bronze	Total
United States	46	29	29	104
China	38	27	23	88
Russian Federation	24	26	32	82
Great Britain	29	17	19	65
Germany	11	19	14	44
Japan	7	14	17	38

Country	Gold	Silver	Bronze	Total
Australia	7	16	12	35
France	11	11	12	34
South Korea	13	8	7	28
Italy	8	9	11	28
Netherlands	6	6	8	20
Ukraine	6	5	9	20

(*Source:* International Olympic Committee)

For example, by reading from left to right along the row marked "Japan," we find that Japan won 7 gold, 14 silver, and 17 bronze medals during the 2012 Summer Olympic Games.

Practice 13

Use the 2012 Summer Games table to answer each question.

a. How many bronze medals did Great Britain win during the 2012 Summer Olympic Games?

b. Which countries won more than 80 medals?

Answers

13. **a.** 19   **b.** United States, China, Russia

Example 13   Use the 2012 Summer Games table to answer each question.

a. How many silver medals did the Russian Federation win during the 2012 Summer Olympic Games?

b. Which countries shown won fewer gold medals than Italy?

Solution:

a. Find "Russian Federation" in the left-hand column. Then read from left to right until the "silver" column is reached. We find that the Russian Federation won 26 silver medals.

b. Italy won 8 gold medals. Of the countries shown, Australia, the Netherlands, Ukraine, and Japan each won fewer than 8 gold medals.

■ **Work Practice 13**

Vocabulary, Readiness & Video Check

Use the choices below to fill in each blank.

standard form	period	whole
expanded form	place value	words

1. The numbers 0, 1, 2, 3, 4, 5, 6, 7, 8, 9, 10, 11, 12, … are called _____ numbers.
2. The number 1286 is written in _____.
3. The number "twenty-one" is written in _____.
4. The number 900 + 60 + 5 is written in _____.
5. In a whole number, each group of three digits is called a(n) _____.
6. The _____ of the digit 4 in the whole number 264 is ones.

Martin-Gay Interactive Videos Watch the section lecture video and answer the following questions.

Objective A 7. In ▣ Example 1, what is the place value of the digit 6? ▶

Objective B 8. Complete this statement based on ▣ Example 3. To read (or write) a number, read from _____ to _____. ▶

Objective C 9. In ▣ Example 5, what is the expanded form value of the digit 8? ▶

Objective D 10. Use the table given in ▣ Example 6 to determine which breed shown has the fewest American Kennel Club registrations. ▶

See Video 1.2 🔵

1.2 Exercise Set MyMathLab® ▶

Objective A *Determine the place value of the digit 5 in each whole number. See Examples 1 through 3.*

▶ **1.** 657

2. 905

▶ **3.** 5423

4. 6527

5. 43,526,000

6. 79,050,000

7. 5,408,092

8. 51,682,700

Objective B *Write each whole number in words. See Examples 4 through 7.*

9. 354

10. 316

11. 8279

12. 5445

▶ **13.** 26,990

14. 42,009

15. 2,388,000

16. 3,204,000

17. 24,350,185

18. 47,033,107

Write each number in the sentence in words. (Do not write the years in words.) See Examples 4 through 7.

19. In 2013, the population of Iceland was 321,800. (*Source: Statistics: Iceland*)

20. Between 1990 and 2010, Brazil lost 553,170 square kilometers of forest area.

21. The Burj Khalifa, in Dubai, United Arab Emirates, a hotel and office building, is currently the tallest in the world at a height of 2720 feet. (*Source:* Council on Tall Buildings and Urban Habitat)

22. In 2012, there were 124,681 patients in the United States waiting for an organ transplant. (*Source:* United Network for Organ Sharing)

23. Each day, UPS delivers an average of 16,300,000 packages and documents worldwide. (*Source:* UPS)

24. Each day, FedEx delivers an average of 3,300,000 packages and documents worldwide. (*Source:* FedEx)

25. The highest point in Colorado is Mount Elbert, at an elevation of 14,433 feet. (*Source:* U.S. Geological Survey)

26. The highest point in Oregon is Mount Hood, at an elevation of 11,239 feet. (*Source:* U.S. Geological Survey)

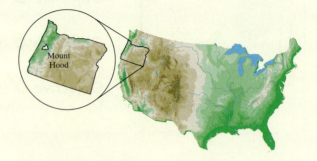

27. The average low price for a 2013 Honda Civic was $18,027. (*Source: U.S. News*)

28. The Goodyear blimp *Eagle* holds 202,700 cubic feet of helium. (*Source:* The Goodyear Tire & Rubber Company)

Write each whole number in standard form. See Examples 8 through 11.

29. Six thousand, five hundred eighty-seven

30. Four thousand, four hundred sixty-eight

31. Fifty-nine thousand, eight hundred

32. Seventy-three thousand, two

33. Thirteen million, six hundred one thousand, eleven

34. Sixteen million, four hundred five thousand, sixteen

35. Seven million, seventeen

36. Two million, twelve

37. Two hundred sixty thousand, nine hundred ninety-seven

38. Six hundred forty thousand, eight hundred eighty-one

Write the whole number in each sentence in standard form. See Examples 8 through 11.

39. The Mir Space Station orbits above Earth at an average altitude of three hundred ninety-five kilometers. (*Source:* Heavens Above)

40. The average distance between the surfaces of the Earth and the Moon is about two hundred thirty-four thousand miles.

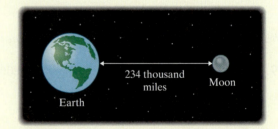

41. The world's tallest free-standing tower is the Tokyo Sky Tree in Tokyo, Japan. Its height is two thousand eighty feet. (*Source:* Council on Tall Buildings and Urban Habitat)

42. The world's second-tallest free-standing tower is the Canton Tower in Guanzhou, China. Its height is one thousand nine hundred sixty-nine feet. (*Source:* Council on Tall Buildings and Urban Habitat)

43. The film *The Avengers* grabbed hold of the world record for opening weekend income when it took in two hundred million, three hundred thousand dollars in 2012. (*Source: Guinness World Records*)

44. The film *Harry Potter and the Deathly Hallows—Part II* holds the record for second-highest opening weekend income; it took in one hundred sixty-nine million, two hundred thousand dollars in 2011. (*Source: Guinness World Records*)

45. Morten Anderson, who played in the National Football League in 1982–2007, holds the record for most career field goals at five hundred sixty-five. (*Source:* NFL)

46. Morten Anderson also holds the record for the most field goals attempted in a career at seven hundred nine. (*Source:* NFL)

Objective C *Write each whole number in expanded form. See Example 12.*

47. 209

48. 789

49. 3470

50. 6040

▶ **51.** 80,774

52. 20,215

53. 66,049

54. 99,032

55. 39,680,000

56. 47,703,029

Objective D *The table shows the beginning year of recent eruptions of major volcanoes in the Cascade Mountains. Use this table to answer Exercises 57 through 62. See Example 13.*

Recent Eruptions of Major Cascade Mountain Volcanoes (1750–present) *							
		Year of Eruption (beginning year)					
Volcano	**State Location**	**1750–1799**	**1800–1849**	**1850–1899**	**1900–1949**	**1950–1999**	**2000–present**
Mt. Baker	Washington	1792	1843	1870, 1880			
Glacier Peak	Washington	1750(?)					
Mt. Rainier	Washington		1841, 1843	1854			
Mt. St. Helens	Washington		1800			1980	
Mt. Hood	Oregon			1854, 1859, 1865			
Three Sisters	Oregon			1853(?)			
Medicine Lake	California				1910		
Mt. Shasta	California	1786		1855			
Cinder Cone*	California			1850			
Lassen Peak	California				1914		
Chaos Crags*	California			1854			

Other major volcanoes in the Cascades have had no eruptions from 1750 to present; *Source:* Harris
*Cinder Cone and Chaos Crags are located by Lassen Peak.

Baker
Glacier Park
Rainier
St. Helens
Adams
Hood
Jefferson
Three Sisters
Newberry
Crater Lake
Medicine Lake
Shasta
Lassen

Washington
Pacific Ocean
Oregon
California

57. Mount Shasta erupted in the 1700s. Locate, then write this eruption year in standard form.

58. Mount Baker erupted in the 1700s. Locate, then write this eruption year in standard form.

59. Which volcano in the table has had the most eruptions?

60. Which volcano(es) in the table has had two eruptions?

61. Which volcano in the table had the earliest eruption?

62. Which volcano in the table had the most recent eruption?

The table shows the top ten popular breeds of dogs in 2012 according to the American Kennel Club. The breeds are listed in the order of number of registrations. Use this table to answer Exercises 63 through 68. See Example 13.

Top Ten American Kennel Club Registrations in 2012		
Breed	**Average Dog Maximum Height (in inches)**	**Average Dog Maximum Weight (in pounds)**
Labrador retriever	25	75
German shepherd	26	95
Golden retriever	24	80
Beagle	15	30
Bulldog	26	90
Yorkshire terrier	9	7
Boxer	25	70
Poodle (standard, miniature, and toy)	standard: 26	standard: 70
Rottweiler	26	none given
Dachshund	9	25

(*Source:* American Kennel Club)

63. Which breed has a greater average weight, the Bulldog or the German shepherd?

64. Which breed has more dogs registered, Golden retriever or German shepherd?

65. Which breed is the most popular dog? Write the maximum weight for this breed in words.

66. Which of the listed breeds has the fewest registrations? Write the average weight for this breed in words.

67. What is the maximum weight of an average-size Boxer?

68. What is the maximum height of an average-size standard poodle?

Concept Extensions

69. Write the largest four-digit number that can be made from the digits 1, 9, 8, and 6 if each digit must be used once. _____ _____ _____ _____

70. Write the largest five-digit number that can be made using the digits 5, 3, and 7 if each digit must be used at least once. _____ _____ _____ _____ _____

Check to see whether each number written in standard form matches the number written in words. If not, correct the number in words. See the Concept Check in this section.

71.

```
                          60–8124/7233              1401
                          1000613331
                        DATE _____
PAY TO
THE ORDER OF _____ $ 105.00
One Hundred Fifty and 00/100 ~~~~~~ DOLLARS
FIRST STATE BANK
OF  FARTHINGTON
FARTHINGTON, IL 64422
MEMO _____
⑈621497260⑈  1000613331⑈  1401
```

72.

```
                          60–8124/7233              1402
                          1000613331
                        DATE _____
PAY TO
THE ORDER OF _____ $ 7030.00
Seven Thousand Thirty and 00/100 ~~~~~~ DOLLARS
FIRST STATE BANK
OF  FARTHINGTON
FARTHINGTON, IL 64422
MEMO _____
⑈621497260⑈  1000613331⑈  1402
```

73. If a number is given in words, describe the process used to write this number in standard form.

74. If a number is written in standard form, describe the process used to write this number in expanded form.

75. How large is a trillion? To get an idea, one trillion seconds is over 31 thousand years. Look up "trillion" in a dictionary and use the definition to write this number in standard form.

76. How large is a quadrillion? To get an idea, if one quadrillion dollars were divided among the population of the United States, we would each receive over 9 million dollars. Look up "quadrillion" (in the American system) and write this number in standard form.

1.3 Adding and Subtracting Whole Numbers, and Perimeter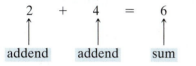

Objective A Adding Whole Numbers

An iPod is a hard drive–based portable audio player. In 2013, it was still the most popular digital music player in the United States.

Suppose that an electronics store received a shipment of two boxes of iPods one day and an additional four boxes of iPods the next day. The **total** shipment in the two days can be found by adding 2 and 4.

2 boxes of iPods + 4 boxes of iPods = 6 boxes of iPods

The **sum** (or total) is 6 boxes of iPods. Each of the numbers 2 and 4 is called an **addend,** and the process of finding the sum is called **addition.**

$$2 \quad + \quad 4 \quad = \quad 6$$

addend addend sum

To add whole numbers, we add the digits in the ones place, then the tens place, then the hundreds place, and so on. For example, let's add 2236 + 160.

```
  2236
+  160
  2396
```
sum of ones
sum of tens
sum of hundreds
sum of thousands

Line up numbers vertically so that the place values correspond. Then add digits in corresponding place values, starting with the ones place.

Example 1   Add: 46 + 713

Solution:
```
   46
+ 713
  759
```

Work Practice 1

Adding by Carrying

When the sum of digits in corresponding place values is more than 9, **carrying** is necessary. For example, to add 365 + 89, add the ones-place digits first.

Carrying
```
   1
  365
+  89
    4
```
5 ones + 9 ones = **14 ones** or **1 ten** + **4 ones**
Write the 4 ones in the ones place and carry the 1 ten to the tens place.

Next, add the tens-place digits.
```
  11
  365
+  89
   54
```
1 ten + 6 tens + 8 tens = **15 tens** or **1 hundred** + **5 tens**
Write the 5 tens in the tens place and carry the 1 hundred to the hundreds place.

Objectives

A Add Whole Numbers.

B Subtract Whole Numbers.

C Find the Perimeter of a Polygon.

D Solve Problems by Adding or Subtracting Whole Numbers.

Practice 1
Add: 4135 + 252

Answer

1. 4387

Next, add the hundreds-place digits.

$$
\begin{array}{r}
\overset{1\ 1}{3\ 6\ 5} \\
+\ \ 8\ 9 \\
\hline
4\ 5\ 4
\end{array}
$$
1 hundred + 3 hundreds = 4 hundreds
Write the 4 hundreds in the hundreds place.

Practice 2

Add: 47,364 + 135,898

Example 2 Add: 46,278 + 124,931

Solution:
$$
\begin{array}{r}
\overset{1\ 1\ 1}{46{,}278} \\
+\ 124{,}931 \\
\hline
171{,}209
\end{array}
$$

■ Work Practice 2

✓**Concept Check** What is wrong with the following computation?

$$
\begin{array}{r}
394 \\
+\ 283 \\
\hline
577
\end{array}
$$

Before we continue adding whole numbers, let's review some properties of addition that you may have already discovered. The first property that we will review is the **addition property of 0.** This property reminds us that the sum of 0 and any number is that same number.

Addition Property of 0

The sum of 0 and any number is that number. For example,

$7 + 0 = 7$

$0 + 7 = 7$

Next, notice that we can add any two whole numbers in any order and the sum is the same. For example,

$4 + 5 = 9$ and $5 + 4 = 9$

We call this special property of addition the **commutative property of addition.**

Commutative Property of Addition

Changing the **order** of two addends does not change their sum. For example,

$2 + 3 = 5$ and $3 + 2 = 5$

Another property that can help us when adding numbers is the **associative property of addition.** This property states that when adding numbers, the grouping of the numbers can be changed without changing the sum. We use parentheses to group numbers. They indicate which numbers to add first. For example, let's use two different groupings to find the sum of $2 + 1 + 5$.

$(2 + 1) + 5 = 3 + 5 = 8$

Also,

$2 + (1 + 5) = 2 + 6 = 8$

Both groupings give a sum of 8.

Answer

2. 183,262

✓**Concept Check Answer**

forgot to carry 1 hundred to the hundreds place

Copyright 2015 Pearson Education, Inc.

Associative Property of Addition

Changing the **grouping** of addends does not change their sum. For example,

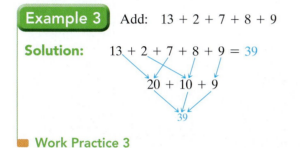

$$3 + (5 + 7) = 3 + 12 = 15 \quad \text{and} \quad (3 + 5) + 7 = 8 + 7 = 15$$

The commutative and associative properties tell us that we can add whole numbers using any order and grouping that we want.

When adding several numbers, it is often helpful to look for two or three numbers whose sum is 10, 20, and so on. Why? Adding multiples of 10 such as 10 and 20 is easier.

Example 3 Add: $13 + 2 + 7 + 8 + 9$

Solution: $13 + 2 + 7 + 8 + 9 = 39$

$$20 + 10 + 9$$

$$39$$

Practice 3
Add: $12 + 4 + 8 + 6 + 5$

■ **Work Practice 3**

Feel free to use the process of Example 3 anytime when adding.

Example 4 Add: $1647 + 246 + 32 + 85$

Solution:
$$
\begin{array}{r}
\overset{1\,2\,2}{1647} \\
246 \\
32 \\
+\quad 85 \\
\hline
2010
\end{array}
$$

Practice 4
Add: $6432 + 789 + 54 + 28$

■ **Work Practice 4**

Objective B Subtracting Whole Numbers ▶

If you have $5 and someone gives you $3, you have a total of $8, since $5 + 3 = 8$. Similarly, if you have $8 and then someone borrows $3, you have $5 left. **Subtraction** is finding the **difference** of two numbers.

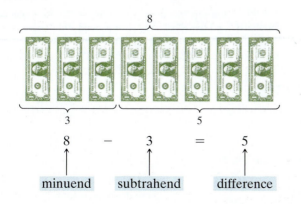

$$\underset{\text{minuend}}{8} \quad - \quad \underset{\text{subtrahend}}{3} \quad = \quad \underset{\text{difference}}{5}$$

In this example, 8 is the **minuend,** and 3 is the **subtrahend.** The **difference** between these two numbers, 8 and 3, is 5.

Notice that addition and subtraction are very closely related. In fact, subtraction is defined in terms of addition.

$$8 - 3 = 5 \text{ because } 5 + 3 = 8$$

This means that subtraction can be *checked* by addition, and we say that addition and subtraction are reverse operations.

Example 5   Subtract. Check each answer by adding.

a. $12 - 9$ **b.** $22 - 7$ **c.** $35 - 35$ **d.** $70 - 0$

Solution:

a. $12 - 9 = 3$ because $3 + 9 = 12$

b. $22 - 7 = 15$ because $15 + 7 = 22$

c. $35 - 35 = 0$ because $0 + 35 = 35$

d. $70 - 0 = 70$ because $70 + 0 = 70$

◼ **Work Practice 5**

Look again at Examples 5(c) and 5(d).

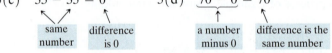

5(c) $35 - 35 = 0$ 5(d) $70 - 0 = 70$

same number / difference is 0   a number minus 0 / difference is the same number

These two examples illustrate the subtraction properties of 0.

> **Subtraction Properties of 0**
>
> The difference of any number and that same number is 0. For example,
>
> $$11 - 11 = 0$$
>
> The difference of any number and 0 is that same number. For example,
>
> $$45 - 0 = 45$$

To subtract whole numbers we subtract the digits in the ones place, then the tens place, then the hundreds place, and so on. When subtraction involves numbers of two or more digits, it is more convenient to subtract vertically. For example, to subtract $893 - 52$,

$$
\begin{array}{r}
8\,9\,3 \\
-\ 5\,2 \\
\hline
8\,4\,1
\end{array}
$$

⟵ minuend   Line up the numbers vertically so that the minuend is on top
⟵ subtrahend   and the place values correspond. Subtract in corresponding
⟵ difference   place values, starting with the ones place.

$3 - 2$
$9 - 5$
$8 - 0$

To check, add.

$$
\begin{array}{r}
\text{difference} \\
+\ \text{subtrahend} \\
\hline
\text{minuend}
\end{array}
\quad \text{or} \quad
\begin{array}{r}
841 \\
+\ 52 \\
\hline
893
\end{array}
$$

⟵ Since this is the original minuend, the problem checks.

Example 6 Subtract: $7826 - 505$. Check by adding.

Solution:

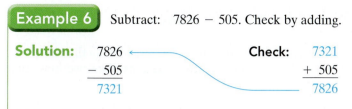

$$\begin{array}{r} 7826 \\ -\ 505 \\ \hline 7321 \end{array} \qquad \textbf{Check:} \begin{array}{r} 7321 \\ +\ 505 \\ \hline 7826 \end{array}$$

■ **Work Practice 6**

Subtracting by Borrowing

When subtracting vertically, if a digit in the second number (subtrahend) is larger than the corresponding digit in the first number (minuend), **borrowing** is necessary. For example, consider

$$\begin{array}{r} 8\,|1 \\ -\ 6\,|3 \end{array}$$

Since the 3 in the ones place of 63 is larger than the 1 in the ones place of 81, borrowing is necessary. We borrow 1 ten from the tens place and add it to the ones place.

Borrowing

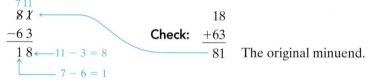

$$\begin{array}{ccc} 8 & - & 1 & = & 7 \\ \text{tens} & & \text{ten} & & \text{tens} \end{array} \rightarrow \begin{array}{c} {}^{7\ 11}\!\!\!\!\!\! \\ 8\!\!\!/\,1\!\!\!/ \\ -\ 6\ 3 \end{array} \leftarrow 1\ \text{ten} + 1\ \text{one} = 11\ \text{ones}$$

Now we subtract the ones-place digits and then the tens-place digits.

$$\begin{array}{r} {}^{7\ 11} \\ 8\!\!\!/\,1\!\!\!/ \\ -6\ 3 \\ \hline 1\ 8 \end{array} \begin{array}{l} \\ \leftarrow 11 - 3 = 8 \\ \\ \llcorner\quad 7 - 6 = 1 \end{array} \qquad \textbf{Check:} \begin{array}{r} 18 \\ +63 \\ \hline 81 \end{array} \quad \text{The original minuend.}$$

Example 7 Subtract: $543 - 29$. Check by adding.

Solution:

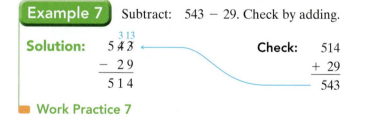

$$\begin{array}{r} {}^{3\ 13} \\ 5\,4\!\!\!/\,3\!\!\!/ \\ -\ 2\ 9 \\ \hline 5\ 1\ 4 \end{array} \qquad \textbf{Check:} \begin{array}{r} 514 \\ +\ 29 \\ \hline 543 \end{array}$$

■ **Work Practice 7**

Sometimes we may have to borrow from more than one place. For example, to subtract $7631 - 152$, we first borrow from the tens place.

$$\begin{array}{r} {}^{2\ 11} \\ 76\,3\!\!\!/\,1\!\!\!/ \\ -\ 1\ 5\ 2 \\ \hline 9 \end{array} \leftarrow 11 - 2 = 9$$

In the tens place, 5 is greater than 2, so we borrow again. This time we borrow from the hundreds place.

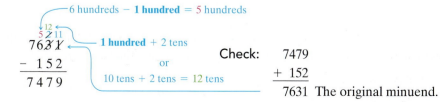

6 hundreds − **1 hundred** = 5 hundreds

$$\begin{array}{r} {}^{5\ \overset{12}{\cancel{2}}\ 11} \\ 76\,3\!\!\!/\,1\!\!\!/ \\ -\ 1\ 5\ 2 \\ \hline 7\ 4\ 7\ 9 \end{array}$$

1 hundred + 2 tens
or
10 tens + 2 tens = 12 tens

Check: $\begin{array}{r} 7479 \\ +\ 152 \\ \hline 7631 \end{array}$ The original minuend.

Practice 6

Subtract. Check by adding.

a. $9143 - 122$

b. $978 - 851$

Practice 7

Subtract. Check by adding.

a. $\begin{array}{r} 697 \\ -\ 49 \end{array}$

b. $\begin{array}{r} 326 \\ -\ 245 \end{array}$

c. $\begin{array}{r} 1234 \\ -\ 822 \end{array}$

Answers

6. a. 9021 **b.** 127

7. a. 648 **b.** 81 **c.** 412

Practice 8

Subtract. Check by adding.

a.   400
        − 164

b.   1000
        − 762

Copyright 2015 Pearson Education, Inc.

Example 8   Subtract: 900 − 174. Check by adding.

Solution: In the ones place, 4 is larger than 0, so we borrow from the tens place. But the tens place of 900 is 0, so to borrow from the tens place, we must first borrow from the hundreds place.

$$\begin{array}{r} \overset{8}{\cancel{9}}\ \overset{10}{\cancel{0}}\ 0 \\ -1\ 7\ 4 \\ \hline \end{array}$$

Now borrow from the tens place.

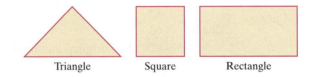

$$\begin{array}{r} \overset{8}{\cancel{9}}\ \overset{\overset{9}{10}}{\cancel{0}}\ \overset{10}{\cancel{0}} \\ -1\ 7\ 4 \\ \hline 7\ 2\ 6 \end{array}$$

Check: 726
          +174
           900

■ **Work Practice 8**

Objective C Finding the Perimeter of a Polygon ▶

In geometry, addition is used to find the perimeter of a polygon. A **polygon** can be described as a flat figure formed by line segments connected at their ends. (For more review, see Appendix A.1.) Geometric figures such as triangles, squares, and rectangles are called polygons.

Triangle   Square   Rectangle

The **perimeter** of a polygon is the *distance around* the polygon, shown in red above. This means that the perimeter of a polygon is the sum of the lengths of its sides.

Practice 9

Find the perimeter of the polygon shown. (A centimeter is a unit of length in the metric system.)

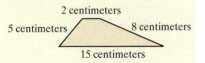

2 centimeters
5 centimeters   8 centimeters
      15 centimeters

△ **Example 9**   Find the perimeter of the polygon shown.

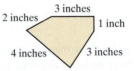

3 inches
2 inches   1 inch
4 inches   3 inches

Solution: To find the perimeter (distance around), we add the lengths of the sides.

2 in. + 3 in. + 1 in. + 3 in. + 4 in. = 13 in.

The perimeter is 13 inches.

■ **Work Practice 9**

Answers

8. a. 236   **b.** 238

9. 30 cm

To make the addition appear simpler, we will often not include units with the addends. If you do this, make sure units are included in the final answer.

Example 10   Calculating the Perimeter of a Building

The largest commercial building in the world un- under one roof is the flower auction building of the cooperative VBA in Aalsmeer, Netherlands. The floor plan is a rectangle that measures 776 meters by 639 meters. Find the perimeter of this building. (A meter is a unit of length in the metric system.) (*Source: The Handy Science Answer Book*, Visible Ink Press)

Solution: Recall that opposite sides of a rectangle have the same length. To find the perimeter of this building, we add the lengths of the sides. The sum of the lengths of its sides is

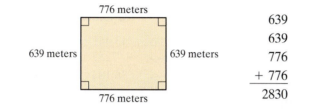

$$\begin{array}{r} 639 \\ 639 \\ 776 \\ + 776 \\ \hline 2830 \end{array}$$

The perimeter of the building is 2830 meters.

■ **Work Practice 10**

Objective D Solving Problems by Adding or Subtracting ▶

Often, real-life problems occur that can be solved by adding or subtracting. The first step in solving any word problem is to *understand* the problem by reading it carefully.

Descriptions in problems solved through addition or subtraction *may* include any of these key words or phrases:

Addition		
Key Words or Phrases	**Examples**	**Symbols**
added to	5 added to 7	7 + 5
plus	0 plus 78	0 + 78
increased by	12 increased by 6	12 + 6
more than	11 more than 25	25 + 11
total	the total of 8 and 1	8 + 1
sum	the sum of 4 and 133	4 + 133

Subtraction		
Key Words or Phrases	**Examples**	**Symbols**
subtract	subtract 5 from 8	8 − 5
difference	the difference of 10 and 2	10 − 2
less	17 less 3	17 − 3
less than	2 less than 20	20 − 2
take away	14 take away 9	14 − 9
decreased by	7 decreased by 5	7 − 5
subtracted from	9 subtracted from 12	12 − 9

✓**Concept Check** In each of the following problems, identify which number is the minuend and which number is the subtrahend.

a. What is the result when 6 is subtracted from 40?
b. What is the difference of 15 and 8?
c. Find a number that is 15 fewer than 23.

To solve a word problem that involves addition or subtraction, we first use the facts given to write an addition or subtraction statement. Then we write the corresponding solution of the real-life problem. It is sometimes helpful to write the statement in words (brief phrases) and then translate to numbers.

Practice 10

A park is in the shape of a triangle. Each of the park's three sides is 647 feet. Find the perimeter of the park.

Helpful Hint Be careful when solving applications that suggest subtraction. Although order *does not* matter when adding, order *does* matter when subtracting. For example, 20 − 15 and 15 − 20 do not simplify to the same number.

Answer
10. 1941 ft

✓**Concept Check Answers**
a. minuend: 40; subtrahend: 6
b. minuend: 15; subtrahend: 8
c. minuend: 23; subtrahend: 15

Practice 11

The radius of Uranus is 15,759 miles. The radius of Neptune is 458 miles less than the radius of Uranus. What is the radius of Neptune? (*Source:* National Space Science Data Center)

Example 11   Finding the Radius of a Planet

The radius of Jupiter is 43,441 miles. The radius of Saturn is 7257 miles less than the radius of Jupiter. Find the radius of Saturn. (*Source:* National Space Science Data Center)

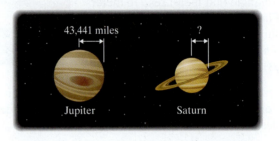

43,441 miles   ?

Jupiter   Saturn

Solution:

In Words		Translate to Numbers
radius of Jupiter	$\longrightarrow$	43,441
$-$ 7257	$\longrightarrow$	$-$ 7257
radius of Saturn	$\longrightarrow$	36,184

The radius of Saturn is 36,184 miles.

■ Work Practice 11

Helpful Hint   Since subtraction and addition are reverse operations, don't forget that a subtraction problem can be checked by adding.

Graphs can be used to visualize data. The graph shown next is called a **bar graph.** For this bar graph, the height of each bar is labeled above the bar. To check this height, follow the top of each bar to the vertical line to the left. For example, the first bar is labeled 214. Follow the top of that bar to the left until the vertical line is reached, a bit more than halfway between 200 and 225, or 214.

Practice 12

Use the graph in Example 12 to answer the following:

a. Which country shown has the fewest threatened amphibians?

b. Find the total number of threatened amphibians for Brazil, Peru, and Mexico.

Example 12   Reading a Bar Graph

As years pass, the number of endangered species per country increases. In the following graph, each bar represents a country and the height of each bar represents the number of endangered species identified in that country.

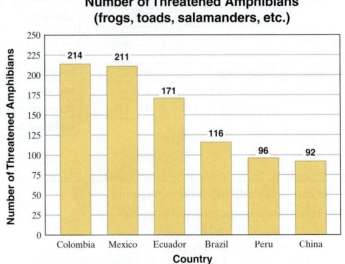

Number of Threatened Amphibians (frogs, toads, salamanders, etc.)

(*Source: The Top 10 of Everything,* 2013)

a. Which country shown has the greatest number of threatened amphibians?

b. Find the total number of threatened amphibians for Ecuador, China, and Colombia.

Answers

11. 15,301 miles

12. a. China   **b.** 423

Solution:

a. The country with the greatest number of threatened amphibians corresponds to the tallest bar, which is Colombia.

b. The key word here is "total." To find the total number of threatened amphibians for Ecuador, China, and Colombia, we add.

In Words		Translate to Numbers
Ecuador	⟶	171
China	⟶	92
Colombia	⟶	+ 214
		Total 477

The total number of threatened amphibians for Ecuador, China, and Colombia is 477.

■ **Work Practice 12**

 Calculator Explorations **Adding and Subtracting Numbers**

Adding Numbers

To add numbers on a calculator, find the keys marked ⊞ and ⊟ or ⊡ENTER⊡.

For example, to add 5 and 7 on a calculator, press the keys ⊡5⊡ ⊞ ⊡7⊡ then ⊟ or ⊡ENTER⊡.

The display will read ⊡ 12⊡.

Thus, 5 + 7 = 12.

To add 687, 981, and 49 on a calculator, press the keys ⊡687⊡ ⊞ ⊡981⊡ ⊞ ⊡49⊡ then ⊟ or ⊡ENTER⊡.

The display will read ⊡ 1717⊡.

Thus, 687 + 981 + 49 = 1717. (Although entering 687, for example, requires pressing more than one key, here numbers are grouped together for easier reading.)

Use a calculator to add.

1. 89 + 45

2. 76 + 97

3. 285 + 55

4. 8773 + 652

5.
    985
  1210
   562
+   77

6.
   465
 9888
  620
+ 1550

Subtracting Numbers

To subtract numbers on a calculator, find the keys marked ⊟ and ⊟ or ⊡ENTER⊡.

For example, to find 83 − 49 on a calculator, press the keys ⊡83⊡ ⊟ ⊡49⊡ then ⊟ or ⊡ENTER⊡.

The display will read ⊡ 34⊡. Thus, 83 − 49 = 34.

Use a calculator to subtract.

7. 865 − 95

8. 76 − 27

9. 147 − 38

10. 366 − 87

11. 9625 − 647

12. 10,711 − 8925

Vocabulary, Readiness & Video Check

Use the choices below to fill in each blank. Some choices may be used more than once and some may not be used at all.

0	order	addend	associative
sum	number	grouping	commutative
perimeter	minuend	subtrahend	difference

1. The sum of 0 and any number is the same _____ .

2. In $35 + 20 = 55$, the number 55 is called the _____ and 35 and 20 are each called a(n) _____ .

3. The difference of any number and that same number is _____ .

4. The difference of any number and 0 is the same _____ .

5. In $37 - 19 = 18$, the number 37 is the _____ , the 19 is the _____ , and the 18 is the _____ .

6. The distance around a polygon is called its _____ .

7. Since $7 + 10 = 10 + 7$, we say that changing the _____ in addition does not change the sum. This property is called the _____ property of addition.

8. Since $(3 + 1) + 20 = 3 + (1 + 20)$, we say that changing the _____ in addition does not change the sum. This property is called the _____ property of addition.

Martin-Gay Interactive Videos   Watch the section lecture video and answer the following questions.

See Video 1.3

Objective A   9. Complete this statement based on the lecture before Example 1. To add whole numbers, we line up _____ values and add from _____ to _____ .

Objective B   10. In Example 5, explain how we end up subtracting 7 from 12 in the ones place.

Objective C   11. In Example 7, the perimeter of what type of polygon is found? How many addends are in the resulting addition problem?

Objective D   12. Complete this statement based on Example 8. To find the sale price, subtract the _____ from the _____ price.

1.3 Exercise Set MyMathLab®

Objective A   *Add. See Examples 1 through 4.*

1. $14 + 22$

2. $27 + 31$

3. $\begin{array}{r} 62 \\ +230 \\ \hline \end{array}$

4. $\begin{array}{r} 37 \\ +542 \\ \hline \end{array}$

5. $\begin{array}{r} 12 \\ 13 \\ +24 \\ \hline \end{array}$

6. $\begin{array}{r} 23 \\ 45 \\ +30 \\ \hline \end{array}$

▶ 7. $\begin{array}{r} 5267 \\ +\ \ 132 \\ \hline \end{array}$

8. $\begin{array}{r} 236 \\ +6243 \\ \hline \end{array}$

9. $22{,}781 + 186{,}297$

10. $17{,}427 + 821{,}059$

11.
```
   8
   9
   2
   5
 + 1
```

12.
```
   3
   5
   8
   5
 + 7
```

13.
```
  81
  17
  23
  79
 +12
```

14.
```
  64
  28
  56
  25
 +32
```

15. $24 + 9006 + 489 + 2407$

16. $16 + 1056 + 748 + 7770$

17.
```
  6820
  4271
+ 5626
```

18.
```
  6789
  4321
+ 5555
```

19.
```
      49
     628
   5 762
 + 29,462
```

20.
```
      26
     582
   4 763
 + 62,511
```

21.
```
  121,742
   57,279
   26,586
+ 426,782
```

22.
```
  504,218
  321,920
   38,507
+ 594,687
```

Objective B *Subtract. Check by adding. See Examples 5 through 8.*

23.
```
  749
- 149
```

24.
```
  957
- 257
```

25.
```
  62
- 37
```

26.
```
  55
- 29
```

27.
```
  922
- 634
```

28.
```
  674
- 299
```

29.
```
  600
- 432
```

30.
```
  300
- 149
```

31.
```
  6283
-  560
```

32.
```
  5349
-  720
```

33.
```
  533
-  29
```

34.
```
  724
-  16
```

35. $1983 - 1904$

36. $1983 - 1914$

37. $50,000 - 17,289$

38. $40,000 - 23,582$

39. $7020 - 1979$

40. $6050 - 1878$

41. $51,111 - 19,898$

42. $62,222 - 39,898$

Objectives A B Mixed Practice *Add or subtract as indicated. See Examples 1 through 8.*

43.
```
  986
+  48
```

44.
```
  986
-  48
```

45. $76 - 67$

46. $80 + 93 + 17 + 9 + 2$

47.
```
  9000
-  482
```

48.
```
  10,000
-  1786
```

49.
```
  10,962
   4851
 + 7063
```

50.
```
  12,468
   3211
 + 1988
```

Objective C *Find the perimeter of each figure. See Examples 9 and 10.*

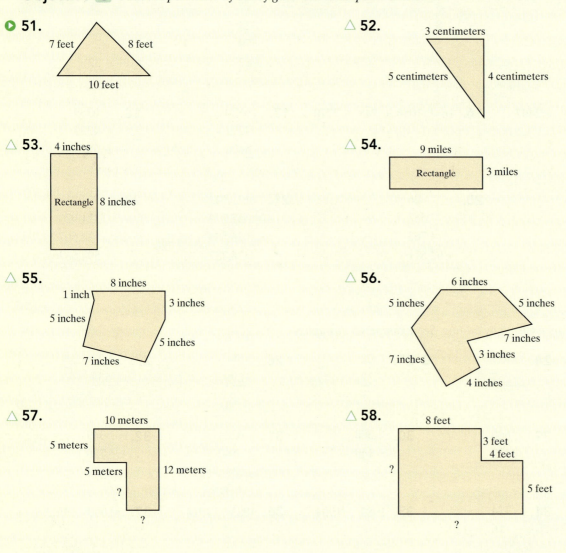

51.
7 feet   8 feet
10 feet

△ 52.
3 centimeters
5 centimeters   4 centimeters

△ 53.
4 inches
Rectangle | 8 inches

△ 54.
9 miles
Rectangle   3 miles

△ 55.
8 inches
1 inch   3 inches
5 inches
5 inches
7 inches

△ 56.
6 inches
5 inches   5 inches
7 inches
7 inches   3 inches
4 inches

△ 57.
10 meters
5 meters
5 meters   12 meters
?
?

△ 58.
8 feet
3 feet
4 feet
?
5 feet
?

Objectives A B D Mixed Practice–Translating *Solve. See Examples 9 through 12.*

59. Find the sum of 297 and 1796.

60. Find the sum of 802 and 6487.

61. Find the total of 76, 39, 8, 17, and 126.

62. Find the total of 89, 45, 2, 19, and 341.

63. Find the difference of 41 and 21.

64. Find the difference of 16 and 5.

65. What is 452 increased by 92?

66. What is 712 increased by 38?

67. Find 108 less 36.

68. Find 25 less 12.

69. Find 12 subtracted from 100.

70. Find 86 subtracted from 90.

Solve.

71. The population of Florida is projected to grow from 19,308 thousand in 2010 to 22,478 thousand in 2020. What is Florida's projected population increase over this time period?

72. The population of California is projected to grow from 39,136 thousand in 2010 to 44,126 thousand in 2020. What is California's projected population increase over this time period?

73. A new DVD player with remote control costs $295. A college student has $914 in her savings account. How much will she have left in her savings account after she buys the DVD player?

▶ 74. A stereo that regularly sells for $547 is discounted by $99 in a sale. What is the sale price?

A river basin is the geographic area drained by a river and its tributaries. The Mississippi River Basin is the third largest in the world and is divided into six sub-basins, whose areas are shown in the following bar graph. Use this graph for Exercises 75 through 78.

75. Find the total U.S. land area drained by the Upper Mississippi and Lower Mississippi sub-basins.

76. Find the total U.S. land area drained by the Ohio and Tennessee sub-basins.

77. How many more square miles of land are drained by the Missouri sub-basin than the Arkansas Red-White sub-basin?

78. How many more square miles of land are drained by the Upper Mississippi sub-basin than the Lower Mississippi sub-basin?

Mississippi River Basin

Area (in square miles)

Missouri	Arkansas Red-White	Upper Mississippi	Lower Mississippi	Ohio	Tennessee
530,000	247,000	189,000	75,000	164,000	40,000

Sub-Basins

△ **79.** A homeowner is installing a fence in his backyard. How many feet of fencing are needed to enclose the yard below?

70 feet   78 feet   90 feet   102 feet

△ **80.** A homeowner is considering installing gutters around her home. Find the perimeter of her rectangular home.

60 feet   45 feet

81. Professor Graham is reading a 503-page book. If she has just finished reading page 239, how many more pages must she read to finish the book?

82. When a couple began a trip, the odometer read 55,492. When the trip was over, the odometer read 59,320. How many miles did they drive on their trip?

83. In 2013, the country of New Zealand had 26,820,424 more sheep than people. If the human population of New Zealand in 2013 was 4,479,576, what was the sheep population? (*Source: Statistics: New Zealand*)

84. During one month in 2013, the two top-selling vehicles in the United States were the Ford F-Series and the Chevrolet Silverado, both trucks. There were 60,449 F-Series trucks and 42,080 Silverados sold that month. What was the total number of these trucks sold in that month? (*Source:* cars.com)

The decibel (dB) is a unit of measurement for sound. Every increase of 10 dB is a tenfold increase in sound intensity. The bar graph shows the decibel levels for some common sounds. Use this graph for Exercises 85 through 88.

85. What is the dB level for live rock music?

86. Which is the quietest of all the sounds shown in the graph?

87. How much louder is the sound of snoring than normal conversation?

88. What is the difference in sound intensity between live rock music and loud television?

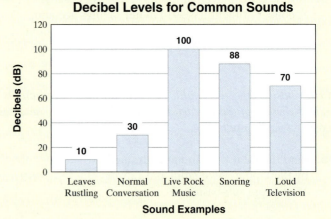

Decibel Levels for Common Sounds

89. In 2013, there were 2410 Gap Inc. (Gap, Banana Republic, Old Navy) stores located in the United States and 1034 located outside the United States. How many Gap Inc. stores were located worldwide? (*Source:* Gap, Inc.)

90. Automobile classes are defined by the amount of interior room. A subcompact car is defined as a car with a maximum interior space of 99 cubic feet. A midsize car is defined as a car with a maximum interior space of 119 cubic feet. What is the difference in volume between a midsize and a subcompact car?

91. The largest permanent Monopoly board is made of granite and is located in San Jose, California. It is in the shape of a square with side lengths of 31 ft. Find the perimeter of the square playing board.

92. The smallest commercially available jigsaw puzzle is a 1000-piece puzzle manufactured in Spain. It is in the shape of a rectangle with length of 18 inches and width of 12 inches. Find the perimeter of this rectangular-shaped puzzle.

The table shows the number of Target stores in ten states. Use this table to answer Exercises 93 through 98.

The Top States for Target Stores in 2012	
State	**Number of Stores**
Pennsylvania	63
California	257
Florida	123
Virginia	57
Illinois	89
New York	67
Michigan	59
Minnesota	75
Ohio	64
Texas	149

(*Source:* Target Corporation)

93. Which state has the most Target stores?

94. Which of the states listed in the table has the fewest Target stores?

95. What is the total number of Target stores located in the three states with the most Target stores?

96. How many Target stores are located in the ten states listed in the table?

97. Which pair of neighboring states have more Target stores combined, Pennsylvania and New York or Michigan and Ohio?

98. There are 775 Target stores located in the states not listed in the table. How many Target stores are in the United States?

99. The state of Delaware has 2029 miles of urban highways and 3865 miles of rural highways. Find the total highway mileage in Delaware. (*Source:* U.S. Federal Highway Administration)

100. The state of Rhode Island has 5193 miles of urban highways and 1222 miles of rural highways. Find the total highway mileage in Rhode Island. (*Source:* U.S. Federal Highway Administration)

Concept Extensions

For Exercises 101–104, identify which number is the minuend and which number is the subtrahend. See the second Concept Check in this section.

101.
$$\begin{array}{r} 48 \\ -\ 1 \\ \hline \end{array}$$

102.
$$\begin{array}{r} 2863 \\ -1904 \\ \hline \end{array}$$

103. Subtract 7 from 70.

104. Find 86 decreased by 25.

105. In your own words, explain the commutative property of addition.

106. In your own words, explain the associative property of addition.

Check each addition below. If it is incorrect, find the correct answer. See the first Concept Check in this section.

107.
$$\begin{array}{r} 566 \\ 932 \\ +\ 871 \\ \hline 2369 \end{array}$$

108.
$$\begin{array}{r} 773 \\ 659 \\ +\ 481 \\ \hline 1913 \end{array}$$

109.
$$\begin{array}{r} 14 \\ 173 \\ 86 \\ +\ 257 \\ \hline 520 \end{array}$$

110.
$$\begin{array}{r} 19 \\ 214 \\ 49 \\ +\ 651 \\ \hline 923 \end{array}$$

Identify each answer as correct or incorrect. Use addition to check. If the answer is incorrect, write the correct answer.

111.
$$\begin{array}{r} 741 \\ -\ 56 \\ \hline 675 \end{array}$$

112.
$$\begin{array}{r} 478 \\ -\ 89 \\ \hline 389 \end{array}$$

113.
$$\begin{array}{r} 1029 \\ -\ 888 \\ \hline 141 \end{array}$$

114.
$$\begin{array}{r} 7615 \\ -\ 547 \\ \hline 7168 \end{array}$$

Fill in the missing digits in each problem.

115.
$$\begin{array}{r} 526\_ \\ -\ 2\_85 \\ \hline 28\_4 \end{array}$$

116.
$$\begin{array}{r} 10,\_4\_ \\ -\ 8\_5\ 4 \\ \hline \_710 \end{array}$$

117. Is there a commutative property of subtraction? In other words, does order matter when subtracting? Why or why not?

118. Explain why the phrase "Subtract 7 from 10" translates to "10 − 7."

119. The local college library is having a Million Pages of Reading promotion. The freshmen have read a total of 289,462 pages; the sophomores have read a total of 369,477 pages; the juniors have read a total of 218,287 pages; and the seniors have read a total of 121,685 pages. Have they reached a goal of one million pages? If not, how many more pages need to be read?

Objectives

A Round Whole Numbers.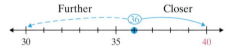

B Use Rounding to Estimate Sums and Differences.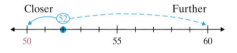

C Solve Problems by Estimating.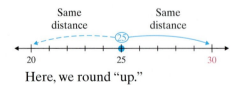

Colorado Population: 5,188,000 or about 5 million

Objective A Rounding Whole Numbers

Rounding a whole number means approximating it. A rounded whole number is often easier to use, understand, and remember than the precise whole number. For example, instead of trying to remember the Colorado state population as 5,188,000, it is much easier to remember it rounded to the nearest million: 5,000,000, or 5 million people. (*Source:* U.S. census)

Recall from Section 1.2 that the line below is called a number line. To **graph** a whole number on this number line, we darken the point representing the location of the whole number. For example, the number 4 is graphed below.

```
+---+---+---+---●---+---+---+--->
0   1   2   3   4   5   6   7
```

On the number line, the whole number 36 is closer to 40 than 30, so 36 rounded to the nearest ten is 40.

```
        Further          Closer
              ⌢ (36) ⌢
◄---+---+---+---+---+---●---+---+---+---+---►
   30          35                  40
```

The whole number 52 is closer to 50 than 60, so 52 rounded to the nearest ten is 50.

```
   Closer                  Further
    ⌢ (52) ⌢
◄---+---●---+---+---+---+---+---+---+---+---►
   50          55                  60
```

In trying to round 25 to the nearest ten, we see that 25 is halfway between 20 and 30. It is not closer to either number. In such a case, we round to the larger ten, that is, to 30.

```
      Same                   Same
    distance               distance
        ⌢    (25)    ⌢
◄---+---+---+---+---●---+---+---+---+---►
   20              25              30
```

Here, we round "up."

To round a whole number without using a number line, follow these steps:

Rounding a Whole Number to a Given Place Value

Step 1: Locate the digit to the right of the given place value.

Step 2: If this digit is 5 or greater, add 1 to the digit in the given place value and replace each digit to its right by 0.

Step 3: If this digit is less than 5, replace it and each digit to its right by 0.

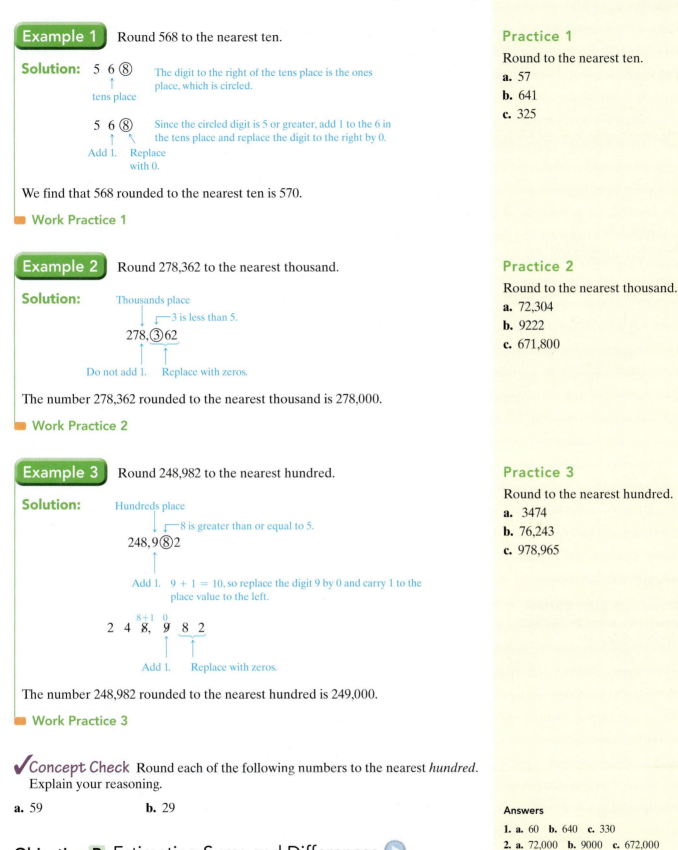

Example 1 Round 568 to the nearest ten.

Solution: 5 6 ⑧   The digit to the right of the tens place is the ones place, which is circled.
↑
tens place

5 6 ⑧   Since the circled digit is 5 or greater, add 1 to the 6 in the tens place and replace the digit to the right by 0.
↑ ↖
Add 1. Replace with 0.

We find that 568 rounded to the nearest ten is 570.

◼ **Work Practice 1**

Example 2 Round 278,362 to the nearest thousand.

Solution:
Thousands place
│ ┌─3 is less than 5.
278,③62
↑   ↑
Do not add 1.   Replace with zeros.

The number 278,362 rounded to the nearest thousand is 278,000.

◼ **Work Practice 2**

Example 3 Round 248,982 to the nearest hundred.

Solution:
Hundreds place
│ ┌─8 is greater than or equal to 5.
248,9⑧2
↑

Add 1.   9 + 1 = 10, so replace the digit 9 by 0 and carry 1 to the place value to the left.

8+1 0
2 4 8, 9̶ 8 2
↑   ↑
Add 1.   Replace with zeros.

The number 248,982 rounded to the nearest hundred is 249,000.

◼ **Work Practice 3**

✔**Concept Check** Round each of the following numbers to the nearest *hundred*. Explain your reasoning.

a. 59   **b.** 29

Objective B Estimating Sums and Differences ▶

By rounding addends, minuends, and subtrahends, we can **estimate** sums and differences. An estimated sum or difference is appropriate when the exact number is not necessary. Also, an estimated sum or difference can help us determine if we made a mistake in calculating an exact amount. To estimate the sum below, round each number to the nearest hundred and then add.

$$
\begin{array}{rll}
768 & \text{rounds to} & 800 \\
1952 & \text{rounds to} & 2000 \\
225 & \text{rounds to} & 200 \\
+\,149 & \text{rounds to} & +\,100 \\
\hline
& & 3100
\end{array}
$$

The estimated sum is 3100, which is close to the **exact** sum of 3094.

Practice 4

Round each number to the nearest ten to find an estimated sum.

$$
\begin{array}{r}
49 \\
25 \\
32 \\
51 \\
+\,98 \\
\hline
\end{array}
$$

Example 4 Round each number to the nearest hundred to find an estimated sum.

$$
\begin{array}{r}
294 \\
625 \\
1071 \\
+\,349 \\
\hline
\end{array}
$$

Solution:

Exact:		Estimate:
294	rounds to	300
625	rounds to	600
1071	rounds to	1100
+ 349	rounds to	+ 300
		2300

The estimated sum is 2300. (The exact sum is 2339.)

▶ **Work Practice 4**

Practice 5

Round each number to the nearest thousand to find an estimated difference.

$$
\begin{array}{r}
3785 \\
-\,2479 \\
\hline
\end{array}
$$

Example 5 Round each number to the nearest hundred to find an estimated difference.

$$
\begin{array}{r}
4725 \\
-\,2879 \\
\hline
\end{array}
$$

Solution:

Exact:		Estimate:
4725	rounds to	4700
−2879	rounds to	−2900
		1800

The estimated difference is 1800. (The exact difference is 1846.)

▶ **Work Practice 5**

Objective C Solving Problems by Estimating ▶

Making estimates is often the quickest way to solve real-life problems when solutions do not need to be exact.

Answers

4. 260
5. 2000

Example 6 Estimating Distances

A driver is trying to quickly estimate the distance from Temple, Texas, to Brenham, Texas. Round each distance given on the map to the nearest ten to estimate the total distance.

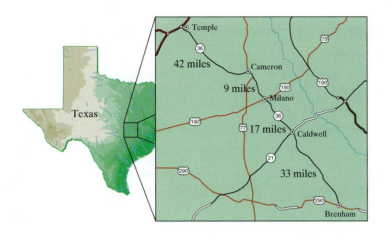

Practice 6

Tasha Kilbey is trying to estimate how far it is from Gove, Kansas, to Hays, Kansas. Round each given distance on the map to the nearest ten to estimate the total distance.

Solution:

Exact Distance:		Estimate:
42	rounds to	40
9	rounds to	10
17	rounds to	20
+33	rounds to	+30
		100

It is approximately 100 miles from Temple to Brenham. (The exact distance is 101 miles.)

■ Work Practice 6

Example 7 Estimating Data

In three recent years the numbers of tons of air cargo and mail that went through Hartsfield-Jackson Atlanta International Airport were 629,700, 685,550, and 737,655. Round each number to the nearest thousand to estimate the tons of mail that passed through this airport.

Solution:

Exact Tons of Cargo Mail:		Estimate:
629,700	rounds to	630,000
685,550	rounds to	686,000
+737,655	rounds to	+738,000
		2,054,000

The approximate tonnage of mail that moved through Atlanta's airport over this 3-year period was 2,054,000 tons. (The exact tonnage was 2,052,905 tons.)

■ Work Practice 7

Practice 7

In 2010, there were 15,427 reported cases of chicken pox, 2612 reported cases of mumps, and 27,550 reported cases of pertussis (whooping cough). Round each number to the nearest thousand to estimate the total number of cases reported for these preventable diseases. (*Source:* Centers for Disease Control and Prevention)

Answers

6. 80 mi

7. 46,000 total cases

Vocabulary, Readiness & Video Check

Use the choices below to fill in each blank.

60	rounding	exact
70	estimate	graph

1. To _____ a number on a number line, darken the point representing the location of the number.
2. Another word for approximating a whole number is _____.
3. The number 65 rounded to the nearest ten is _____, but the number 61 rounded to the nearest ten is _____.
4. A(n) _____ number of products is 1265, but a(n) _____ is 1000.

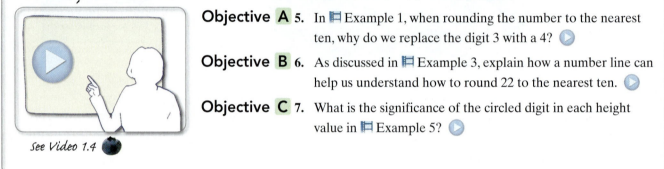

Martin-Gay Interactive Videos   *Watch the section lecture video and answer the following questions.*

Objective A 5. In ▦ Example 1, when rounding the number to the nearest ten, why do we replace the digit 3 with a 4? ▸

Objective B 6. As discussed in ▦ Example 3, explain how a number line can help us understand how to round 22 to the nearest ten. ▸

Objective C 7. What is the significance of the circled digit in each height value in ▦ Example 5? ▸

See Video 1.4

1.4 Exercise Set MyMathLab® ▸

Objective A *Round each whole number to the given place. See Examples 1 through 3.*

1. 423 to the nearest ten

2. 273 to the nearest ten

▸ 3. 635 to the nearest ten

4. 846 to the nearest ten

5. 2791 to the nearest hundred

6. 8494 to the nearest hundred

7. 495 to the nearest ten

8. 898 to the nearest ten

9. 21,094 to the nearest thousand

10. 82,198 to the nearest thousand

11. 33,762 to the nearest thousand

12. 42,682 to the nearest ten-thousand

13. 328,495 to the nearest hundred

14. 179,406 to the nearest hundred

▸ 15. 36,499 to the nearest thousand

16. 96,501 to the nearest thousand

17. 39,994 to the nearest ten

18. 99,995 to the nearest ten

19. 29,834,235 to the nearest ten-million

20. 39,523,698 to the nearest million

Complete the table by estimating the given number to the given place value.

		Tens	Hundreds	Thousands
21.	5281			
22.	7619			
23.	9444			
24.	7777			
25.	14,876			
26.	85,049			

Round each number to the indicated place.

27. The state of Texas contains 310,850 miles of urban and rural highways. Round this number to the nearest thousand. (*Source:* U.S. Federal Highway Administration)

28. The state of California contains 171,874 miles of urban and rural highways. Round this number to the nearest thousand. (*Source:* U.S. Federal Highway Administration)

29. It takes 60,149 days for Neptune to make a complete orbit around the Sun. Round this number to the nearest hundred. (*Source:* National Space Science Data Center)

30. Kareem Abdul-Jabbar holds the NBA record for points scored, a total of 38,387 over his NBA career. Round this number to the nearest thousand. (*Source:* National Basketball Association)

31. In 2013, the most valuable brand in the world was Apple, Inc. The estimated brand value at this time of Apple was $185,000,000,000. Round this to the nearest ten billion. (*Source:* Millward Brown)

32. According to the U.S. Population Clock, the population of the United States was 316,539,415 in August 2013. Round this population figure to the nearest million. (*Source:* U.S. Census population clock)

33. The average salary for a baseball player in 2012 was $3,213,479. Round this average salary to the nearest hundred-thousand. (*Source:* Major League Baseball Players Association)

34. The average salary for a football player in 2012 was $1,900,000. Round this average salary to the nearest million. (*Source:* Businessweek.com)

35. The United States currently has 331,600,000 cellular phone users, while India has 893,862,500 users. Round each of the user numbers to the nearest million. (*Source:* World Almanac, 2013)

36. U.S. farms produced 3,112,500,000 bushels of soybeans in 2011. Round the soybean production figure to the nearest ten-million. (*Source:* U.S. Department of Agriculture)

Objective **B** *Estimate the sum or difference by rounding each number to the nearest ten. See Examples 4 and 5.*

▶ **37.**
39
45
22
+ 17

38.
52
33
15
+ 29

39.
449
− 373

40.
555
− 235

Estimate the sum or difference by rounding each number to the nearest hundred. See Examples 4 and 5.

41.
1913
1886
+ 1925

42.
4050
3133
+ 1220

▶ **43.**
1774
− 1492

44.
1989
− 1870

45.
3995
2549
+ 4944

46.
799
1655
+ 271

Three of the given calculator answers below are incorrect. Find them by estimating each sum.

47. 463 + 219   600

48. 522 + 785   1307

49. 229 + 443 + 606   1278

50. 542 + 789 + 198   2139

51. 7806 + 5150   12,956

52. 5233 + 4988   9011

> **Helpful Hint**
> Estimation is useful to check for incorrect answers when using a calculator. For example, pressing a key too hard may result in a double digit, while pressing a key too softly may result in the digit not appearing in the display.

Objective **C** *Solve each problem by estimating. See Examples 6 and 7.*

53. An appliance store advertises three refrigerators on sale at $899, $1499, and $999. Round each cost to the nearest hundred to estimate the total cost.

54. Suppose you scored 89, 97, 100, 79, 75, and 82 on your biology tests. Round each score to the nearest ten to estimate your total score.

55. The distance from Kansas City to Boston is 1429 miles and from Kansas City to Chicago is 530 miles. Round each distance to the nearest hundred to estimate how much farther Boston is from Kansas City than Chicago is.

56. The Gonzales family took a trip and traveled 588, 689, 277, 143, 59, and 802 miles on six consecutive days. Round each distance to the nearest hundred to estimate the distance they traveled.

▶ **57.** The peak of Mt. McKinley, in Alaska, is 20,320 feet above sea level. The top of Mt. Rainier, in Washington, is 14,410 feet above sea level. Round each height to the nearest thousand to estimate the difference in elevation of these two peaks. (*Source:* U.S. Geological Survey)

58. A student is pricing new car stereo systems. One system sells for $1895 and another system sells for $1524. Round each price to the nearest hundred dollars to estimate the difference in price of these systems.

59. In 2012, the population of Springfield, Illinois, was 117,126, and the population of Champaign, Illinois, was 82,517. Round each population to the nearest ten-thousand to estimate how much larger Springfield was than Champaign. (*Source:* U.S. Census Bureau)

60. Round each distance given on the map to the nearest ten to estimate the total distance from North Platte, Nebraska, to Lincoln, Nebraska.

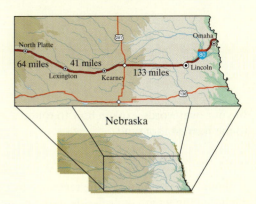

61. Head Start is a national program that provides developmental and social services for America's low-income preschool children ages three to five. Enrollment figures in Head Start programs showed an increase from 1,073,440 in 2010 to 1,128,030 in 2012. Round each number of children to the nearest thousand to estimate this increase. (*Source:* U.S. Department of Health and Human Services)

62. Enrollment figures at a local community college showed an increase from 49,713 credit hours in 2005 to 51,746 credit hours in 2006. Round each number to the nearest thousand to estimate the increase.

Mixed Practice (Sections 1.2 and 1.4) *The following table shows a few of the airports in the United States with the largest volumes of passengers. Complete this table. The first line is completed for you. (Source: 2011 World Annual Trafic Report)*

	City Location of Airport	Total Passengers in 2011 (in hundred-thousands of passengers)	Amount Written in Standard Form	Standard Form Rounded to the Nearest Million	Standard Form Rounded to the Nearest Ten-Million
	Atlanta, GA	924	92,400,000	92,000,000	90,000,000
63.	Chicago, IL	667			
64.	Los Angeles, CA	619			
65.	Dallas/Fort Worth, TX	578			
66.	Denver, CO	528			

Concept Extensions

67. Find one number that when rounded to the nearest hundred is 5700.

68. Find one number that when rounded to the nearest ten is 5700.

69. A number rounded to the nearest hundred is 8600.

    a. Determine the smallest possible number.

    b. Determine the largest possible number.

70. On August 23, 1989, it was estimated that 1,500,000 people joined hands in a human chain stretching 370 miles to protest the fiftieth anniversary of the pact that allowed what was then the Soviet Union to annex the Baltic nations in 1939. If the estimate of the number of people is to the nearest hundred-thousand, determine the largest possible number of people in the chain.

71. In your own words, explain how to round a number to the nearest thousand.

72. In your own words, explain how to round 9660 to the nearest thousand.

△ **73.** Estimate the perimeter of the rectangle by first rounding the length of each side to the nearest ten.

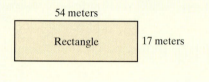

54 meters

Rectangle   17 meters

△ **74.** Estimate the perimeter of the triangle by first rounding the length of each side to the nearest hundred.

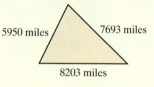

5950 miles   7693 miles

8203 miles

1.5   Multiplying Whole Numbers and Area

Objectives

A Use the Properties of Multiplication.

B Multiply Whole Numbers.

C Find the Area of a Rectangle.

D Solve Problems by Multiplying Whole Numbers.

Multiplication Shown as Repeated Addition Suppose that we wish to count the number of laptops provided in a computer class. The laptops are arranged in 5 rows, and each row has 6 laptops.

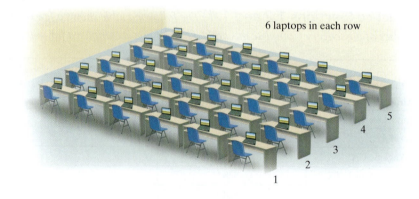

6 laptops in each row

Adding 5 sixes gives the total number of laptops. We can write this as $6 + 6 + 6 + 6 + 6 = 30$ laptops. When each addend is the same, we refer to this as **repeated addition.**

    Multiplication is repeated addition but with different notation.

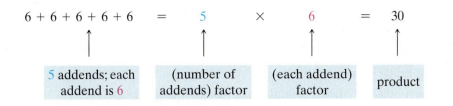

$$6 + 6 + 6 + 6 + 6 \quad = \quad 5 \quad \times \quad 6 \quad = \quad 30$$

| 5 addends; each addend is 6 | (number of addends) factor | (each addend) factor | product |

The $\times$ is called a **multiplication sign.** The numbers 5 and 6 are called **factors.** The number 30 is called the **product.** The notation 5×6 is read as "five times six." The symbols $\cdot$ and $(\ )$ can also be used to indicate multiplication.

$$5 \times 6 = 30, \quad 5 \cdot 6 = 30, \quad (5)(6) = 30, \quad \text{and} \quad 5(6) = 30$$

✔Concept Check

a. Rewrite $5 + 5 + 5 + 5 + 5 + 5 + 5$ using multiplication.

b. Rewrite 3×16 as repeated addition. Is there more than one way to do this? If so, show all ways.

Objective A Using the Properties of Multiplication ▶

As with addition, we memorize products of one-digit whole numbers and then use certain properties of multiplication to multiply larger numbers. (If necessary, review the multiplication of one-digit numbers.)

Notice that when any number is multiplied by 0, the result is always 0. This is called the **multiplication property of 0.**

Multiplication Property of 0

The product of 0 and any number is 0. For example,

$$5 \cdot 0 = 0 \quad \text{and} \quad 0 \cdot 8 = 0$$

Also notice that when any number is multiplied by 1, the result is always the original number. We call this result the **multiplication property of 1.**

Multiplication Property of 1

The product of 1 and any number is that same number. For example,

$$1 \cdot 9 = 9 \quad \text{and} \quad 6 \cdot 1 = 6$$

Example 1 Multiply.

a. 4×1 **b.** $0(3)$ **c.** $1 \cdot 64$ **d.** $(48)(0)$

Solution:

a. $4 \times 1 = 4$ **b.** $0(3) = 0$

c. $1 \cdot 64 = 64$ **d.** $(48)(0) = 0$

■ Work Practice 1

Like addition, multiplication is commutative and associative. Notice that when multiplying two numbers, the order of these numbers can be changed without changing the product. For example,

$$3 \cdot 5 = 15 \quad \text{and} \quad 5 \cdot 3 = 15$$

This property is called the **commutative property of multiplication.**

Commutative Property of Multiplication

Changing the **order** of two factors does not change their product. For example,

$$9 \cdot 2 = 18 \quad \text{and} \quad 2 \cdot 9 = 18$$

Another property that can help us when multiplying is the **associative property of multiplication.** This property states that when multiplying numbers, the grouping of the numbers can be changed without changing the product. For example,

$$(2 \cdot 3) \cdot 4 = 6 \cdot 4 = 24$$

Also,

$$2 \cdot (3 \cdot 4) = 2 \cdot 12 = 24$$

Both groupings give a product of 24.

Associative Property of Multiplication

Changing the **grouping** of factors does not change their product. From the previous work, we know that, for example,

$$(2 \cdot 3) \cdot 4 = 2 \cdot (3 \cdot 4)$$

With these properties, along with the **distributive property,** we can find the product of any whole numbers. The distributive property says that multiplication **distributes** over addition. For example, notice that $3(2 + 5)$ simplifies to the same number as $3 \cdot 2 + 3 \cdot 5$.

$$3(2 + 5) = 3(7) = 21$$

$$3 \cdot 2 + 3 \cdot 5 = 6 + 15 = 21$$

Since $3(2 + 5)$ and $3 \cdot 2 + 3 \cdot 5$ both simplify to 21, then

$$3(2 + 5) = 3 \cdot 2 + 3 \cdot 5$$

Notice in $3(2 + 5) = 3 \cdot 2 + 3 \cdot 5$ that each number inside the parentheses is multiplied by 3.

Distributive Property

Multiplication distributes over addition. For example,

$$2(3 + 4) = 2 \cdot 3 + 2 \cdot 4$$

Example 2 Rewrite each using the distributive property.

a. $5(6 + 5)$ **b.** $20(4 + 7)$ **c.** $2(7 + 9)$

Solution: Using the distributive property, we have

a. $5(6 + 5) = 5 \cdot 6 + 5 \cdot 5$
b. $20(4 + 7) = 20 \cdot 4 + 20 \cdot 7$
c. $2(7 + 9) = 2 \cdot 7 + 2 \cdot 9$

■ **Work Practice 2**

Objective B Multiplying Whole Numbers ▶

Let's use the distributive property to multiply 7(48). To do so, we begin by writing the expanded form of 48 (see Section 1.2) and then applying the distributive property.

$$
\begin{aligned}
7(48) &= 7(40 + 8) && \text{Write 48 in expanded form.}\\
&= 7 \cdot 40 + 7 \cdot 8 && \text{Apply the distributive property.}\\
&= 280 + 56 && \text{Multiply.}\\
&= 336 && \text{Add.}
\end{aligned}
$$

This is how we multiply whole numbers. When multiplying whole numbers, we will use the following notation.

First:

$$
\begin{array}{r}
\overset{5}{48}\\
\times\ 7\\
\hline
336
\end{array}
$$
$7 \cdot 8 = 56$ — Write 6 in the ones place and carry 5 to the tens place.

Next:

$$
\begin{array}{r}
\overset{5}{48}\\
\times\ 7\\
\hline
336
\end{array}
$$
$7 \cdot 4 + 5 = 28 + 5 = 33$

The product of 48 and 7 is 336.

Example 3 Multiply:

a.
$$
\begin{array}{r}
25\\
\times\ 8\\
\hline
\end{array}
$$

b.
$$
\begin{array}{r}
246\\
\times\ 5\\
\hline
\end{array}
$$

Solution:

a.
$$
\begin{array}{r}
\overset{4}{25}\\
\times\ 8\\
\hline
200
\end{array}
$$

b.
$$
\begin{array}{r}
\overset{23}{246}\\
\times\ 5\\
\hline
1230
\end{array}
$$

■ **Work Practice 3**

Practice 2
Rewrite each using the distributive property.
a. $6(4 + 5)$
b. $30(2 + 3)$
c. $7(2 + 8)$

Practice 3
Multiply.
a. $\begin{array}{r} 29 \\ \times\ 6 \\ \hline \end{array}$ **b.** $\begin{array}{r} 648 \\ \times\ 5 \\ \hline \end{array}$

Answers
2. **a.** $6(4 + 5) = 6 \cdot 4 + 6 \cdot 5$
   b. $30(2 + 3) = 30 \cdot 2 + 30 \cdot 3$
   c. $7(2 + 8) = 7 \cdot 2 + 7 \cdot 8$
3. **a.** 174 **b.** 3240

To multiply larger whole numbers, use the following similar notation. Multiply 89×52.

Step 1	Step 2	Step 3
$\overset{1}{89}$	$\overset{4}{89}$	89
$\times\ 52$	$\times\ 52$	$\times\ 52$
178 ← Multiply 89×2.	178	178
	$\times 4450$ ← Multiply 89×50.	4450
		4628 Add.

The numbers 178 and 4450 are called **partial products.** The sum of the partial products, 4628, is the product of 89 and 52.

Practice 4

Multiply.

$\quad\ 306$
$\times\ \ 81$

Example 4 Multiply: 236×86

Solution:

$$
\begin{array}{r}
236 \\
\times\ \ 86 \\
\hline
1416 \quad \leftarrow 6(236) \\
18880 \quad \leftarrow 80(236) \\
\hline
20{,}296 \quad \text{Add.}
\end{array}
$$

■ Work Practice 4

Practice 5

Multiply.

$\quad\ 726$
$\times 142$

Example 5 Multiply: 631×125

Solution:

$$
\begin{array}{r}
631 \\
\times\ \ 125 \\
\hline
3155 \quad \leftarrow 5(631) \\
12620 \quad \leftarrow 20(631) \\
63100 \quad \leftarrow 100(631) \\
\hline
78{,}875 \quad \text{Add.}
\end{array}
$$

■ Work Practice 5

✓**Concept Check** Find and explain the error in the following multiplication problem.

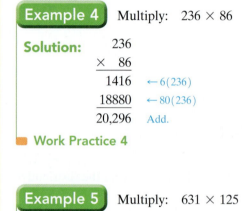

$$
\begin{array}{r}
102 \\
\times\ \ 33 \\
\hline
306 \\
306 \\
\hline
612
\end{array}
$$

Answers
4. 24,786   **5.** 103,092

✓**Concept Check Answer**

$$
\begin{array}{r}
102 \\
\times\ \ 33 \\
\hline
306 \\
3060 \\
\hline
3366
\end{array}
$$

Objective C Finding the Area of a Rectangle

A special application of multiplication is finding the **area** of a region. Area measures the amount of surface of a region. For example, we measure a plot of land or the living space of a home by its area. The figures on the next page show two examples of units of area measure. (A centimeter is a unit of length in the metric system.)

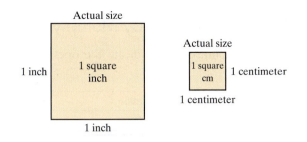

For example, to measure the area of a geometric figure such as the rectangle below, count the number of square units that cover the region.

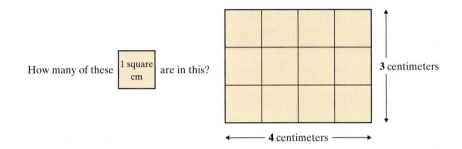

This rectangular region contains 12 square units, each 1 square centimeter. Thus, the area is 12 square centimeters. This total number of squares can be found by counting or by multiplying **4 · 3** (length · width).

$$\text{Area of a rectangle} = \text{length} \cdot \text{width}$$
$$= (4 \text{ centimeters})(3 \text{ centimeters})$$
$$= 12 \text{ square centimeters}$$

In this section, we find the areas of rectangles only. In later sections, we will find the areas of other geometric regions.

Helpful Hint

Notice that area is measured in **square** units while perimeter is measured in units.

Example 6   Finding the Area of a State

The state of Colorado is in the shape of a rectangle whose length is 380 miles and whose width is 280 miles. Find its area.

Solution: The area of a rectangle is the product of its length and its width.

$$\text{Area} = \text{length} \cdot \text{width}$$
$$= (380 \text{ miles})(280 \text{ miles})$$
$$= 106,400 \text{ square miles}$$

The area of Colorado is 106,400 square miles.

Work Practice 6

Practice 6

The state of Wyoming is in the shape of a rectangle whose length is 360 miles and whose width is 280 miles. Find its area.

Answer
6. 100,800 sq mi

Objective D Solving Problems by Multiplying

There are several words or phrases that indicate the operation of multiplication. Some of these are as follows:

Multiplication		
Key Words or Phrases	**Examples**	**Symbols**
multiply	multiply 5 by 7	$5 \cdot 7$
product	the product of 3 and 2	$3 \cdot 2$
times	10 times 13	$10 \cdot 13$

Many key words or phrases describing real-life problems that suggest addition might be better solved by multiplication instead. For example, to find the **total** cost of 8 shirts, each selling for $27, we can either add

$$27 + 27 + 27 + 27 + 27 + 27 + 27 + 27$$

or we can multiply 8(27).

Practice 7

A particular computer printer can print 16 pages per minute in color. How many pages can it print in 45 minutes?

Example 7 Finding DVD Space

A digital video disc (DVD) can hold about 4800 megabytes (MB) of information. How many megabytes can 12 DVDs hold?

Solution: Twelve DVDs will hold 12×4800 megabytes.

In Words		Translate to Numbers
megabytes per disc	→	4800
$\times$ DVDs	→	$\times$ 12
		9600
		48000
total megabytes		57,600

Twelve DVDs will hold 57,600 megabytes.

■ **Work Practice 7**

Practice 8

A professor of history purchased DVDs and CDs through a club. Each DVD was priced at $11 and each CD cost $9. He bought eight DVDs and five CDs. Find the total cost of the order.

Example 8 Budgeting Money

A woman and her friend plan to take their children to the Georgia Aquarium in Atlanta, the world's largest aquarium. The ticket price for each child is $22 and for each adult, $26. If five children and two adults plan to go, how much money is needed for admission? (*Source:* GeorgiaAquarium.org)

Solution: If the price of one child's ticket is $22, the price for 5 children is $5 \times 22 = \$110$. The price of one adult ticket is $26, so the price for two adults is $2 \times 26 = \$52$. The total cost is:

In Words		Translate to Numbers
cost for 5 children	→	110
+ cost for 2 adults	→	+ 52
total cost		162

The total cost is $162.

■ **Work Practice 8**

Answers

7. 720 pages **8.** $133

Example 9 Estimating Word Count

The average page of a book contains 259 words. Estimate, rounding each number to the nearest hundred, the total number of words contained on 212 pages.

Solution: The exact number of words is 259×212. Estimate this product by rounding each factor to the nearest hundred.

259 rounds to 300
$\times 212$ rounds to $\times 200$, $300 \times 200 = 60,000$

$3 \cdot 2 = 6$

There are approximately 60,000 words contained on 212 pages.

Work Practice 9

Practice 9

If an average page in a book contains 163 words, estimate, rounding each number to the nearest hundred, the total number of words contained on 391 pages.

Answer

9. 80,000 words

Calculator Explorations **Multiplying Numbers**

To multiply numbers on a calculator, find the keys marked $\times$ and $=$ or ENTER. For example, to find $31 \cdot 66$ on a calculator, press the keys 31 $\times$ 66 then $=$ or ENTER. The display will read 2046. Thus, $31 \cdot 66 = 2046$.

Use a calculator to multiply.

1. 72×48 **2.** 81×92

3. $163 \cdot 94$ **4.** $285 \cdot 144$

5. $983(277)$ **6.** $1562(843)$

Vocabulary, Readiness & Video Check

Use the choices below to fill in each blank.

area   grouping   commutative   1   product   length

factor   order   associative   0   distributive   number

1. The product of 0 and any number is _____.

2. The product of 1 and any number is the _____.

3. In $8 \cdot 12 = 96$, the 96 is called the _____ and 8 and 12 are each called a(n) _____.

4. Since $9 \cdot 10 = 10 \cdot 9$, we say that changing the _____ in multiplication does not change the product. This property is called the _____ property of multiplication.

5. Since $(3 \cdot 4) \cdot 6 = 3 \cdot (4 \cdot 6)$, we say that changing the _____ in multiplication does not change the product. This property is called the _____ property of multiplication.

6. _____ measures the amount of surface of a region.

7. Area of a rectangle $=$ _____ $\cdot$ width.

8. We know $9(10 + 8) = 9 \cdot 10 + 9 \cdot 8$ by the _____ property.

Martin-Gay Interactive Videos   Watch the section lecture video and answer the following questions.

Objective A 9. The expression in ▣ Example 3 is rewritten using what property? ▶

Objective B 10. During the multiplication process for ▣ Example 5, why is a single zero placed at the end of the second partial product? ▶

Objective C 11. Why are the units to the answer to ▣ Example 6 not just meters? What are the correct units? ▶

Objective D 12. In ▣ Example 7, why can "total" imply multiplication as well as addition? ▶

See Video 1.5

1.5 Exercise Set MyMathLab® ▶

Objective A *Multiply. See Example 1.*

▶ **1.** $1 \cdot 24$
2. $55 \cdot 1$
▶ **3.** $0 \cdot 19$
4. $27 \cdot 0$

5. $8 \cdot 0 \cdot 9$
6. $7 \cdot 6 \cdot 0$
7. $87 \cdot 1$
8. $1 \cdot 41$

Use the distributive property to rewrite each expression. See Example 2.

9. $6(3 + 8)$
10. $5(8 + 2)$
11. $4(3 + 9)$

12. $6(1 + 4)$
▶ **13.** $20(14 + 6)$
14. $12(12 + 3)$

Objective B *Multiply. See Example 3.*

15. $\begin{array}{r} 64 \\ \times\ 8 \\ \hline \end{array}$
16. $\begin{array}{r} 79 \\ \times\ 3 \\ \hline \end{array}$
17. $\begin{array}{r} 613 \\ \times\ 6 \\ \hline \end{array}$
18. $\begin{array}{r} 638 \\ \times\ 5 \\ \hline \end{array}$

▶ **19.** 277×6
20. 882×2
21. 1074×6
22. 9021×3

Objectives A B Mixed Practice *Multiply. See Examples 1 through 5.*

23. $\begin{array}{r} 89 \\ \times 13 \\ \hline \end{array}$
24. $\begin{array}{r} 91 \\ \times 72 \\ \hline \end{array}$
25. $\begin{array}{r} 421 \\ \times\ 58 \\ \hline \end{array}$
26. $\begin{array}{r} 526 \\ \times\ 23 \\ \hline \end{array}$
27. $\begin{array}{r} 306 \\ \times\ 81 \\ \hline \end{array}$
28. $\begin{array}{r} 708 \\ \times\ 21 \\ \hline \end{array}$

29. $(780)(20)$
30. $(720)(80)$
31. $(495)(13)(0)$
32. $(593)(47)(0)$
33. $(640)(1)(10)$

34. (240)(1)(20)   **35.** 1234 × 39   **36.** 1357 × 79   **37.** 609 × 234   **38.** 807 × 127

39. 8649   **40.** 1234   **41.**   589   **42.**   426   **43.** 1941   **44.** 1876
        × 274   × 567   ×110   ×110   ×2035   ×1407

Objective C Mixed Practice (*Section 1.3*) *Find the area and the perimeter of each rectangle. See Example 6.*

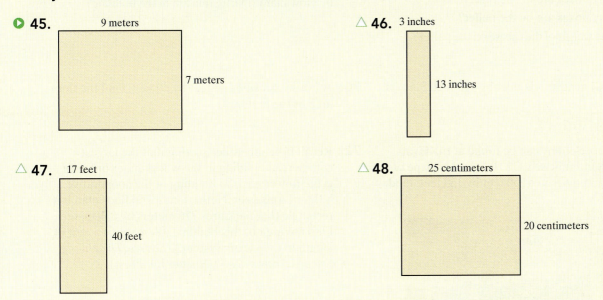

45. 9 meters / 7 meters

46. 3 inches / 13 inches

47. 17 feet / 40 feet

48. 25 centimeters / 20 centimeters

Objective D *Estimate the products by rounding each factor to the nearest hundred. See Example 9.*

49. 576 × 354   **50.** 982 × 650   **51.** 604 × 451   **52.** 111 × 999

Without actually calculating, mentally round, multiply, and choose the best estimate.

53. 38 × 42 =
  a. 16
  b. 160
  c. 1600
  d. 16,000

54. 2872 × 12 =
  a. 2872
  b. 28,720
  c. 287,200
  d. 2,872,000

55. 612 × 29 =
  a. 180
  b. 1800
  c. 18,000
  d. 180,000

56. 706 × 409 =
  a. 280
  b. 2800
  c. 28,000
  d. 280,000

Objectives C D Mixed Practice–Translating *Solve. See Examples 6 through 9.*

57. Multiply 80 by 11.   **58.** Multiply 70 by 12.   **59.** Find the product of 6 and 700.

60. Find the product of 9 and 900.   **61.** Find 2 times 2240.   **62.** Find 3 times 3310.

63. One tablespoon of olive oil contains 125 calories. How many calories are in 3 tablespoons of olive oil? (*Source: Home and Garden Bulletin No. 72, U.S. Department of Agriculture*).

64. One ounce of hulled sunflower seeds contains 14 grams of fat. How many grams of fat are in 8 ounces of hulled sunflower seeds? (*Source: Home and Garden Bulletin No. 72, U.S. Department of Agriculture*).

65. The textbook for a course in biology costs $94. There are 35 students in the class. Find the total cost of the biology books for the class.

66. The seats in a large lecture hall are arranged in 14 rows with 34 seats in each row. Find how many seats are in this room.

67. Cabot Creamery is packing a pallet of 20-lb boxes of cheddar cheese to send to a local restaurant. There are five layers of boxes on the pallet, and each layer is four boxes wide by five boxes deep.
   a. How many boxes are in one layer?
   b. How many boxes are on the pallet?
   c. What is the weight of the cheese on the pallet?

68. An apartment building has *three floors*. Each floor has five rows of apartments with four apartments in each row.
   a. How many apartments are on 1 floor?
   b. How many apartments are in the building?

△ **69.** A plot of land measures 80 feet by 110 feet. Find its area.

△ **70.** A house measures 45 feet by 60 feet. Find the floor area of the house.

△ **71.** The largest hotel lobby can be found at the Hyatt Regency in San Francisco, CA. It is in the shape of a rectangle that measures 350 feet by 160 feet. Find its area.

△ **72.** Recall from an earlier section that the largest commercial building in the world under one roof is the flower auction building of the cooperative VBA in Aalsmeer, Netherlands. The floor plan is a rectangle that measures 776 meters by 639 meters. Find the area of this building. (A meter is a unit of length in the metric system.) (*Source: The Handy Science Answer Book,* Visible Ink Press)

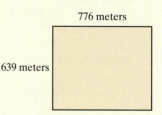

73. A pixel is a rectangular dot on a graphing calculator screen. If a graphing calculator screen contains 62 pixels in a row and 94 pixels in a column, find the total number of pixels on a screen.

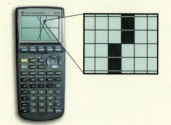

74. A certain compact disc (CD) can hold 700 megabytes (MB) of information. How many MBs can 17 discs hold?

75. A line of print on a computer contains 60 characters (letters, spaces, punctuation marks). Find how many characters there are in 35 lines.

76. An average cow eats 3 pounds of grain per day. Find how much grain a cow eats in a year. (Assume 365 days in 1 year.)

77. One ounce of Planters® Dry Roasted Peanuts has 160 calories. How many calories are in 8 ounces? (*Source:* RJR Nabisco, Inc.)

78. One ounce of Planters® Dry Roasted Peanuts has 13 grams of fat. How many grams of fat are in 16 ounces? (*Source:* RJR Nabisco, Inc.)

79. The Thespian club at a local community college is ordering T-shirts. T-shirts size S, M, or L cost $10 each and T-shirts size XL or XXL cost $12 each. Complete the table below and use it to find the total cost. (The first row is filled in for you.)

T-Shirt Size	Number of Shirts Ordered	Cost per Shirt	Cost per Size Ordered
S	4	$10	$40
M	6		
L	20		
XL	3		
XXL	3		

80. The student activities group at North Shore Community College is planning a trip to see the local minor league baseball team. Tickets cost $5 for students, $7 for non-students, and $2 for children under 12. Complete the following table and use it to find the total cost.

Person	Number of Persons	Cost per Person	Cost per Category
Student	24	$5	$120
Nonstudent	4		
Children under 12	5		

81. Celestial Seasonings of Boulder, Colorado, is a tea company that specializes in herbal teas. Their plant in Boulder has bagging machines capable of bagging over 1000 bags of tea per minute. If the plant runs 24 hours day, how many tea bags are produced in one day? (*Source:* Celestial Seasonings)

82. There were about 3 million "older" Americans (ages 65 and older) in 1900. By 2020, this number is projected to increase eighteen times. Find this projected number of "older" Americans in 2020. (*Source:* Administration on Aging, U.S. Census Bureau)

Mixed Practice (*Sections 1.3, 1.5*) *Perform each indicated operation.*

83.
$$\begin{array}{r} 128 \\ + 7 \\ \hline \end{array}$$

84.
$$\begin{array}{r} 126 \\ - 8 \\ \hline \end{array}$$

85.
$$\begin{array}{r} 134 \\ \times 16 \\ \hline \end{array}$$

86. $47 + 26 + 10 + 231 + 50$

87. Find the sum of 19 and 4.

88. Find the product of 19 and 4.

89. Find the difference of 19 and 4.

90. Find the total of 14 and 9.

Concept Extensions

Solve. See the first Concept Check in this section.

91. Rewrite $6 + 6 + 6 + 6 + 6$ using multiplication.

92. Rewrite $11 + 11 + 11 + 11 + 11 + 11$ using multiplication.

93. **a.** Rewrite $3 \cdot 5$ as repeated addition.
b. Explain why there is more than one way to do this.

94. **a.** Rewrite $4 \cdot 5$ as repeated addition.
b. Explain why there is more than one way to do this.

Find and explain the error in each multiplication problem. See the second Concept Check in this section.

95.
```
   203
×   14
   812
   203
  1015
```

96.
```
    31
×   50
   155
```

Fill in the missing digits in each problem.

97.
```
     4_
×   _3
    126
   3780
   3906
```

98.
```
    _7
×   6_
    171
   3420
   3591
```

99. Explain how to multiply two 2-digit numbers using partial products.

100. In your own words, explain the meaning of the area of a rectangle and how this area is measured.

101. A window washer in New York City is bidding for a contract to wash the windows of a 23-story building. To write a bid, the number of windows in the building is needed. If there are 7 windows in each row of windows on 2 sides of the building and 4 windows per row on the other 2 sides of the building, find the total number of windows.

102. During the NBA's 2012–2013 regular season, Carmelo Anthony of the New York Knicks scored 157 three-point field goals, 512 two-point field goals, and 425 free throws (worth one point each). How many points did Carmelo Anthony score during the 2012–2013 regular season? (*Source:* NBA)

1.6 Dividing Whole Numbers

Objectives

A Divide Whole Numbers. ▶

B Perform Long Division. ▶

C Solve Problems That Require Dividing by Whole Numbers. ▶

D Find the Average of a List of Numbers. ▶

Suppose three people pooled their money and bought a raffle ticket at a local fund-raiser. Their ticket was the winner and they won a $75 cash prize. They then divided the prize into three equal parts so that each person received $25.

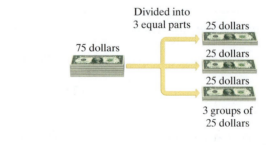

Objective A Dividing Whole Numbers ▶

The process of separating a quantity into equal parts is called **division.** The division above can be symbolized by several notations.

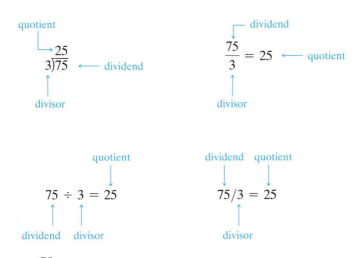

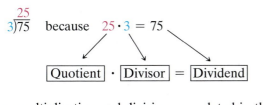

(In the notation $\dfrac{75}{3}$, the bar separating 75 and 3 is called a **fraction bar.**) Just as subtraction is the reverse of addition, division is the reverse of multiplication. This means that division can be checked by multiplication.

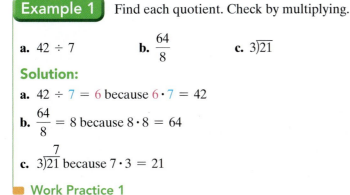

Since multiplication and division are related in this way, you can use your knowledge of multiplication facts to review quotients of one-digit divisors if necessary.

Example 1   Find each quotient. Check by multiplying.

a. $42 \div 7$ **b.** $\dfrac{64}{8}$ **c.** $3\overline{)21}$

Solution:

a. $42 \div 7 = 6$ because $6 \cdot 7 = 42$

b. $\dfrac{64}{8} = 8$ because $8 \cdot 8 = 64$

c. $3\overline{)21}^{7}$ because $7 \cdot 3 = 21$

🟧 **Work Practice 1**

Example 2   Find each quotient. Check by multiplying.

a. $1\overline{)7}$ **b.** $12 \div 1$ **c.** $\dfrac{6}{6}$ **d.** $9 \div 9$ **e.** $\dfrac{20}{1}$ **f.** $18\overline{)18}$

Solution:

a. $1\overline{)7}^{7}$ because $7 \cdot 1 = 7$ **b.** $12 \div 1 = 12$ because $12 \cdot 1 = 12$

c. $\dfrac{6}{6} = 1$ because $1 \cdot 6 = 6$ **d.** $9 \div 9 = 1$ because $1 \cdot 9 = 9$

e. $\dfrac{20}{1} = 20$ because $20 \cdot 1 = 20$ **f.** $18\overline{)18}^{1}$ because $1 \cdot 18 = 18$

🟧 **Work Practice 2**

Example 2 illustrates the important properties of division described next:

Division Properties of 1

The quotient of any number (except 0) and that same number is 1. For example,

$$8 \div 8 = 1 \qquad \frac{5}{5} = 1 \qquad 4\overline{)4}^{\,1}$$

The quotient of any number and 1 is that same number. For example,

$$9 \div 1 = 9 \qquad \frac{6}{1} = 6 \qquad 1\overline{)3}^{\,3} \qquad \frac{0}{1} = 0$$

Practice 3

Find each quotient. Check by multiplying.

a. $\dfrac{0}{7}$ **b.** $8\overline{)0}$

c. $7 \div 0$ **d.** $0 \div 14$

Example 3 Find each quotient. Check by multiplying.

a. $9\overline{)0}$ **b.** $0 \div 12$ **c.** $\dfrac{0}{5}$ **d.** $\dfrac{3}{0}$

Solution:

a. $9\overline{)0}^{\,0}$ because $0 \cdot 9 = 0$

b. $0 \div 12 = 0$ because $0 \cdot 12 = 0$

c. $\dfrac{0}{5} = 0$ because $0 \cdot 5 = 0$

d. If $\dfrac{3}{0} = $ a *number*, then the *number* times $0 = 3$. Recall from Section 1.5 that any number multiplied by 0 is 0 and not 3. We say, then, that $\dfrac{3}{0}$ is **undefined.**

■ **Work Practice 3**

Example 3 illustrates important division properties of 0.

Division Properties of 0

The quotient of 0 and any number (except 0) is 0. For example,

$$0 \div 9 = 0 \qquad \frac{0}{5} = 0 \qquad 14\overline{)0}^{\,0}$$

The quotient of any number and 0 is not a number. We say that

$$\frac{3}{0}, \quad 0\overline{)3}, \quad \text{and} \quad 3 \div 0$$

are **undefined.**

Objective B Performing Long Division ▶

When dividends are larger, the quotient can be found by a process called **long division.** For example, let's divide 2541 by 3.

$$\text{divisor} \longrightarrow 3\overline{)2541}$$
$$\underset{\uparrow}{\phantom{3\overline{)25}}}$$
$$\text{dividend}$$

We can't divide 3 into 2, so we try dividing 3 into the first two digits.

$$\begin{array}{r} 8 \\ 3\overline{)2541} \end{array}$$ $25 \div 3 = 8$ with 1 left, so our best estimate is 8. We place 8 over the 5 in 25.

Next, multiply 8 and 3 and subtract this product from 25. Make sure that this difference is less than the divisor.

```
     8
3)2541
 −24        8(3) = 24
   1        25 − 24 = 1, and 1 is less than the divisor 3.
```

Bring down the next digit and go through the process again.

```
    84       14 ÷ 3 = 4 with 2 left
3)2541
 −24↓
   14
  −12        4(3) = 12
    2        14 − 12 = 2
```

Once more, bring down the next digit and go through the process.

```
   847       21 ÷ 3 = 7
3)2541
 −24
   14
  −12
   21
  −21        7(3) = 21
    0        21 − 21 = 0
```

The quotient is 847. To check, see that 847 × 3 = 2541.

Example 4   Divide: 3705 ÷ 5. Check by multiplying.

Solution:

```
     7       37 ÷ 5 = 7 with 2 left. Place this estimate, 7, over the 7 in 37.
5)3705
 −35↓        7(5) = 35
   20        37 − 35 = 2, and 2 is less than the divisor 5.
              Bring down the 0.
    74       20 ÷ 5 = 4
5)3705
 −35
   20
  −20↓       4(5) = 20
   05        20 − 20 = 0. and 0 is less than the divisor 5.
              Bring down the 5.
   741       5 ÷ 5 = 1
5)3705
 −35
   20
  −20↓
    5
   −5        1(5) = 5
    0        5 − 5 = 0
```

Practice 4

Divide. Check by multiplying.
a. 4908 ÷ 6

b. 2212 ÷ 4

c. 753 ÷ 3

(Continued on next page)

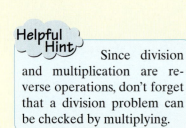

Helpful Hint   Since division and multiplication are reverse operations, don't forget that a division problem can be checked by multiplying.

Check:

$$\begin{array}{r} 741 \\ \times 5 \\ \hline 3705 \end{array}$$

🟧 **Work Practice 4**

Practice 5

Divide and check by multiplying.

a. $7\overline{)2128}$

b. $9\overline{)45,900}$

Example 5   Divide and check: $1872 \div 9$

Solution:

$$\begin{array}{r} 208 \\ 9\overline{)1872} \end{array}$$

$$
\begin{array}{rl}
-18\downarrow & \quad 2(9) = 18 \\
07 & \quad 18 - 18 = 0;\ \text{bring down the 7.} \\
-0\downarrow & \quad 0(9) = 0 \\
72 & \quad 7 - 0 = 7;\ \text{bring down the 2.} \\
-72 & \quad 8(9) = 72 \\
0 & \quad 72 - 72 = 0
\end{array}
$$

Check: $208 \cdot 9 = 1872$

🟧 **Work Practice 5**

Naturally, quotients don't always "come out even." Making 4 rows out of 26 chairs, for example, isn't possible if each row is supposed to have exactly the same number of chairs. Each of 4 rows can have 6 chairs, but 2 chairs are still left over.

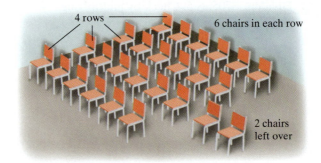

4 rows — 6 chairs in each row

2 chairs left over

We signify "leftovers" or **remainders** in this way:

$$\begin{array}{r} 6 \ \text{R}\,2 \\ 4\overline{)26} \end{array}$$

The **whole number part of the quotient** is 6; the **remainder part of the quotient** is 2. Checking by multiplying,

whole number part	·	divisor	+	remainder part	=	dividend
↓		↓		↓		↓
6	·	4	+	2		
		24	+	2	=	26

Answers

5. a. 304   **b.** 5100

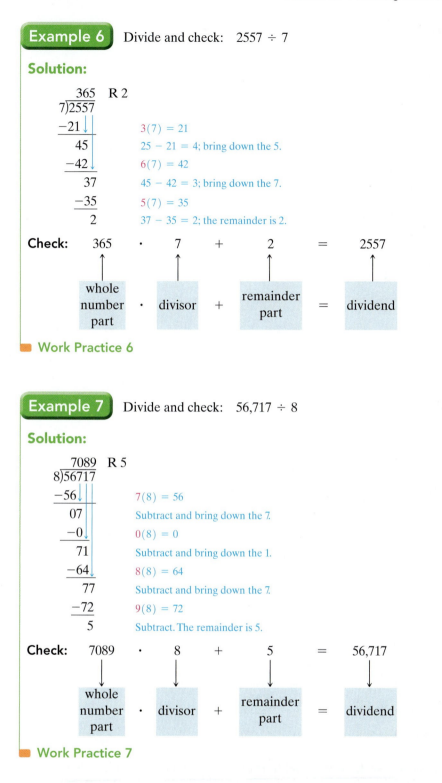

Example 6 Divide and check: 2557 ÷ 7

Solution:

$$
\begin{array}{r}
365 \quad \text{R } 2 \\
7\overline{)2557} \\
\end{array}
$$

$-21\downarrow$ $3(7) = 21$
 45          $25 - 21 = 4$; bring down the 5.
$-42\downarrow$ $6(7) = 42$
  37         $45 - 42 = 3$; bring down the 7.
 -35 $5(7) = 35$
   2        $37 - 35 = 2$; the remainder is 2.

Check: 365 · 7 + 2 = 2557

whole number part · divisor + remainder part = dividend

■ **Work Practice 6**

Example 7 Divide and check: 56,717 ÷ 8

Solution:

$$
\begin{array}{r}
7089 \quad \text{R } 5 \\
8\overline{)56717} \\
\end{array}
$$

$-56\downarrow$ $7(8) = 56$
 07   Subtract and bring down the 7.
$-0\downarrow$ $0(8) = 0$
 71   Subtract and bring down the 1.
$-64\downarrow$ $8(8) = 64$
 77   Subtract and bring down the 7.
-72 $9(8) = 72$
  5   Subtract. The remainder is 5.

Check: 7089 · 8 + 5 = 56,717

whole number part · divisor + remainder part = dividend

■ **Work Practice 7**

When the divisor has more than one digit, the same pattern applies. For example, let's find 1358 ÷ 23.

$$
\begin{array}{r}
5 \\
23\overline{)1358} \\
\end{array}
$$

$135 \div 23 = 5$ with 20 left over. Our estimate is 5.

$-115\downarrow$ $5(23) = 115$
 208        $135 - 115 = 20$. Bring down the 8.

(Continued on next page)

Practice 6
Divide and check.
a. $4\overline{)939}$
b. $5\overline{)3287}$

Practice 7
Divide and check.
a. $9\overline{)81,605}$
b. $4\overline{)23,310}$

Answers
6. a. 234 R 3 **b.** 657 R 2
7. a. 9067 R 2 **b.** 5827 R 2

Now we continue estimating.

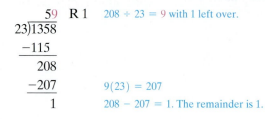

To check, see that $59 \cdot 23 + 1 = 1358$.

Practice 8

Divide: $8920 \div 17$

Example 8   Divide: $6819 \div 17$

Solution:

$$
\begin{array}{r}
401 \quad \text{R } 2 \\
17\overline{)6819}
\end{array}
$$

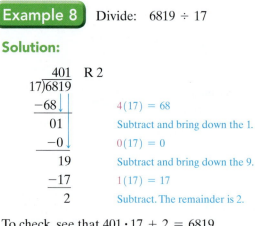

$4(17) = 68$

Subtract and bring down the 1.

$0(17) = 0$

Subtract and bring down the 9.

$1(17) = 17$

Subtract. The remainder is 2.

To check, see that $401 \cdot 17 + 2 = 6819$.

Work Practice 8

Practice 9

Divide: $33{,}282 \div 678$

Example 9   Divide: $51{,}600 \div 403$

Solution:

$$
\begin{array}{r}
128 \quad \text{R } 16 \\
403\overline{)51600} \\
-403 \quad \\
\hline
1130 \quad \\
-806 \quad \\
\hline
3240 \\
-3224 \\
\hline
16
\end{array}
$$

$1(403) = 403$

Subtract and bring down the 0.

$2(403) = 806$

Subtract and bring down the 0.

$8(403) = 3224$

Subtract. The remainder is 16.

To check, see that $128 \cdot 403 + 16 = 51{,}600$.

Work Practice 9

Division Shown as Repeated Subtraction   To further understand division, recall from Section 1.5 that addition and multiplication are related in the following manner:

$$\underbrace{3 + 3 + 3 + 3}_{4 \text{ addends; each addend is } 3} = 4 \times 3 = 12$$

In other words, multiplication is repeated addition. Likewise, division is repeated subtraction.

For example, let's find

$$35 \div 8$$

Answers

8. 524 R 12   **9.** 49 R 60

by repeated subtraction. Keep track of the number of times 8 is subtracted from 35. We are through when we can subtract no more because the difference is less than 8.

35 ÷ 8:   Repeated subtraction

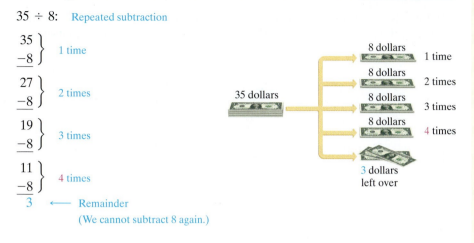

Thus, 35 ÷ 8 = 4 R 3.

To check, perform the same multiplication as usual, but finish by adding in the remainder.

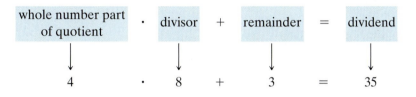

$$\text{whole number part of quotient} \cdot \text{divisor} + \text{remainder} = \text{dividend}$$

$$4 \cdot 8 + 3 = 35$$

Objective C Solving Problems by Dividing ▶

Below are some key words and phrases that may indicate the operation of division:

Division		
Key Words or Phrases	**Examples**	**Symbols**
divide	divide 10 by 5	$10 \div 5$ or $\dfrac{10}{5}$
quotient	the quotient of 64 and 4	$64 \div 4$ or $\dfrac{64}{4}$
divided by	9 divided by 3	$9 \div 3$ or $\dfrac{9}{3}$
divided or shared equally among	$100 divided equally among five people	$100 \div 5$ or $\dfrac{100}{5}$
per	100 miles per 2 hours	$\dfrac{100 \text{ miles}}{2 \text{ hours}}$

✓Concept Check Which of the following is the correct way to represent "the quotient of 60 and 12"? Or are both correct? Explain your answer.

a. $12 \div 60$

b. $60 \div 12$

Practice 10

Three students bought 171 blank CDs to share equally. How many CDs did each person get?

Example 10   Finding Shared Earnings

Three college freshmen share a paper route to earn money for expenses. The total in their fund after expenses is $2895. How much is each person's equal share?

Solution:

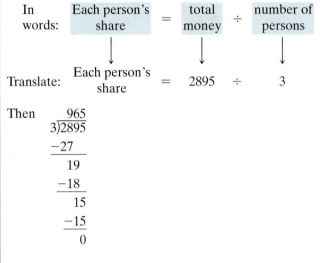

In words: Each person's share   =   total money   ÷   number of persons

Translate: Each person's share   =   2895   ÷   3

Then

$$
\begin{array}{r}
965 \\
3\overline{)2895} \\
-27 \\
\hline
19 \\
-18 \\
\hline
15 \\
-15 \\
\hline
0
\end{array}
$$

Each person's share is $965.

▶ **Work Practice 10**

Practice 11

Printers can be packed 12 to a box. If 532 printers are to be packed but only full boxes are shipped, how many full boxes will be shipped? How many printers are left over and not shipped?

Example 11   Dividing Number of Downloads

As part of a promotion, an executive receives 238 cards, each good for one free song download. If she wants to share them evenly with 19 friends, how many download cards will each friend receive? How many will be left over?

Solution:

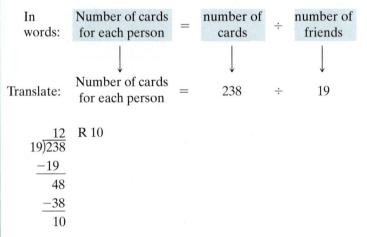

In words: Number of cards for each person   =   number of cards   ÷   number of friends

Translate: Number of cards for each person   =   238   ÷   19

$$
\begin{array}{r}
12 \;\; \text{R } 10 \\
19\overline{)238} \\
-19 \\
\hline
48 \\
-38 \\
\hline
10
\end{array}
$$

Each friend will receive 12 download cards. The cards cannot be divided equally among her friends since there is a nonzero remainder. There will be 10 download cards left over.

▶ **Work Practice 11**

Answers

10. 57 CDs

11. 44 full boxes; 4 printers left over

Objective D Finding Averages ▶

A special application of division (and addition) is finding the average of a list of numbers. The **average** of a list of numbers is the sum of the numbers divided by the *number* of numbers.

$$\text{average} = \frac{\text{sum of numbers}}{\textit{number} \text{ of numbers}}$$

Example 12 Averaging Scores

A mathematics instructor is checking a simple program she wrote for averaging the scores of her students. To do so, she averages a student's scores of 75, 96, 81, and 88 by hand. Find this average score.

Solution: To find the average score, we find the sum of the student's scores and divide by 4, the number of scores.

$$
\begin{array}{r}
75 \\
96 \\
81 \\
+88 \\
\hline
340 \quad \text{sum}
\end{array}
\qquad
\text{average} = \frac{340}{4} = 85
\qquad
\begin{array}{r}
85 \\
4\overline{)340} \\
-32 \\
\hline
20 \\
-20 \\
\hline
0
\end{array}
$$

The average score is 85.

■ **Work Practice 12**

Practice 12

To compute a safe time to wait for reactions to occur after allergy shots are administered, a lab technician is given a list of elapsed times between administered shots and reactions. Find the average of the times 4 minutes, 7 minutes, 35 minutes, 16 minutes, 9 minutes, 3 minutes, and 52 minutes.

Answer

12. 18 min

▦ **Calculator Explorations Dividing Numbers**

To divide numbers on a calculator, find the keys marked ÷ and = or ENTER . For example, to find $435 \div 5$ on a calculator, press the keys 435 ÷ 5 then = or ENTER . The display will read 87 . Thus, $435 \div 5 = 87$.

Use a calculator to divide.

1. $848 \div 16$ 2. $564 \div 12$

3. $95\overline{)5890}$ 4. $27\overline{)1053}$

5. $\dfrac{32{,}886}{126}$ 6. $\dfrac{143{,}088}{264}$

7. $0 \div 315$ 8. $315 \div 0$

Vocabulary, Readiness & Video Check

Use the choices below to fill in each blank. Some choices may be used more than once.

1	number	divisor	dividend
0	undefined	average	quotient

1. In $90 \div 2 = 45$, the answer 45 is called the _____ , 90 is called the _____ , and 2 is called the

   _____ .

2. The quotient of any number and 1 is the same _____ .

3. The quotient of any number (except 0) and the same number is _____ .

4. The quotient of 0 and any number (except 0) is _____ .

5. The quotient of any number and 0 is _____ .

6. The _____ of a list of numbers is the sum of the numbers divided by the _____ of numbers.

Martin-Gay Interactive Videos   *Watch the section lecture video and answer the following questions.*

Objective A **7.** Look at ⊞ Examples 6–8. What number can never be the divisor in division? ▶

Objective B **8.** In ⊞ Example 10, how many 102s are in 21? How does this result affect the quotient? ▶

  9. What calculation would you use to check the answer in ⊞ Example 10?

Objective C **10.** In ⊞ Example 11, what is the importance of knowing that the distance to each hole is the same? ▶

Objective D **11.** As shown in ⊞ Example 12, what two operations are used when finding an average? ▶

See Video 1.6

1.6 Exercise Set MyMathLab® ▶

Objective A *Find each quotient. See Examples 1 through 3.*

1. $54 \div 9$ **2.** $72 \div 9$ ▶ **3.** $36 \div 3$ **4.** $24 \div 3$ **5.** $0 \div 8$

6. $0 \div 4$ ▶ **7.** $31 \div 1$ **8.** $38 \div 1$ ▶ **9.** $\dfrac{18}{18}$ **10.** $\dfrac{49}{49}$

11. $\dfrac{24}{3}$ **12.** $\dfrac{45}{9}$ ▶ **13.** $26 \div 0$ **14.** $\dfrac{12}{0}$ **15.** $26 \div 26$

16. $6 \div 6$ ▶ **17.** $0 \div 14$ **18.** $7 \div 0$ **19.** $18 \div 2$ **20.** $18 \div 3$

Objectives A B Mixed Practice *Divide and then check by multiplying. See Examples 1 through 5.*

21. $3\overline{)87}$ **22.** $5\overline{)85}$ **23.** $3\overline{)222}$ **24.** $8\overline{)640}$ **25.** $3\overline{)1014}$ **26.** $4\overline{)2104}$

27. $\dfrac{30}{0}$ **28.** $\dfrac{0}{30}$ **29.** $63 \div 7$ **30.** $56 \div 8$ **31.** $150 \div 6$ **32.** $121 \div 11$

Divide and then check by multiplying. See Examples 6 and 7.

33. $7\overline{)479}$ **34.** $7\overline{)426}$ **35.** $6\overline{)1421}$ **36.** $3\overline{)1240}$

37. $305 \div 8$ **38.** $167 \div 3$ **39.** $2286 \div 7$ **40.** $3333 \div 4$

Divide and then check by multiplying. See Examples 8 and 9.

41. $55\overline{)715}$ **42.** $23\overline{)736}$ **43.** $23\overline{)1127}$ **44.** $42\overline{)2016}$ **45.** $97\overline{)9417}$

46. $44\overline{)1938}$ **47.** $3146 \div 15$ **48.** $7354 \div 12$ **49.** $6578 \div 13$ **50.** $5670 \div 14$

51. $9299 \div 46$ **52.** $2505 \div 64$ **53.** $\dfrac{12{,}744}{236}$ **54.** $\dfrac{5781}{123}$ **55.** $\dfrac{10{,}297}{103}$

56. $\dfrac{23{,}092}{240}$ **57.** $20{,}619 \div 102$ **58.** $40{,}853 \div 203$ **59.** $244{,}989 \div 423$ **60.** $164{,}592 \div 543$

Divide. See Examples 1 through 9.

61. $7\overline{)119}$ **62.** $8\overline{)104}$ **63.** $7\overline{)3580}$ **64.** $5\overline{)3017}$

65. $40\overline{)85{,}312}$ **66.** $50\overline{)85{,}747}$ **67.** $142\overline{)863{,}360}$ **68.** $214\overline{)650{,}560}$

Objective C Translating *Solve. See Examples 10 and 11.*

69. Find the quotient of 117 and 5.

70. Find the quotient of 94 and 7.

71. Find 200 divided by 35.

72. Find 116 divided by 32.

73. Find the quotient of 62 and 3.

74. Find the quotient of 78 and 5.

Solve.

75. Martin Thieme teaches American Sign Language classes for $65 per student for a 7-week session. He collects $2145 from the group of students. Find how many students are in the group.

76. Kathy Gomez teaches Spanish lessons for $85 per student for a 5-week session. From one group of students, she collects $4930. Find how many students are in the group.

77. The gravity of Jupiter is 318 times as strong as the gravity of Earth, so objects on Jupiter weigh 318 times as much as they weigh on Earth. If a person would weigh 52,470 pounds on Jupiter, find how much the person weighs on Earth.

78. Twenty-one people pooled their money and bought lottery tickets. One ticket won a prize of $5,292,000. Find how many dollars each person received.

79. An 18-hole golf course is 5580 yards long. If the distance to each hole is the same, find the distance between holes.

80. A truck hauls wheat to a storage granary. It carries a total of 5768 bushels of wheat in 14 trips. How much does the truck haul each trip if each trip it hauls the same amount?

81. There is a bridge over highway I-35 every three miles. The first bridge is at the beginning of a 265-mile stretch of highway. Find how many bridges there are over 265 miles of I-35.

82. The white stripes dividing the lanes on a highway are 25 feet long, and the spaces between them are 25 feet long. Let's call a "lane divider" a stripe followed by a space. Find how many whole "lane dividers" there are in 1 mile of highway. (A mile is 5280 feet.)

83. Ari Trainor is in the requisitions department of Central Electric Lighting Company. Light poles along a highway are placed 492 feet apart. The first light pole is at the beginning of a 1-mile strip. Find how many poles he should order for the 1-mile strip of highway. (A mile is 5280 feet.)

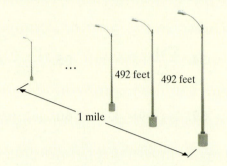

84. Professor Lopez has a piece of rope 185 feet long that she wants to cut into pieces for an experiment in her physics class. Each piece of rope is to be 8 feet long. Determine whether she has enough rope for her 22-student class. Determine the amount extra or the amount short.

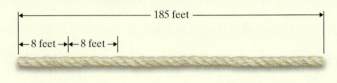

85. Broad Peak in Pakistan is the twelfth-tallest mountain in the world. Its elevation is 26,400 feet. A mile is 5280 feet. How many miles tall is Broad Peak? (*Source:* National Geographic Society)

86. Randy Moss of the New England Patriots led the NFL in touchdowns during the 2007 regular football season, scoring a total of 138 points from touchdowns. If a touchdown is worth 6 points, how many touchdowns did Moss make during the 2007 season? (*Source:* NFL)

87. Find how many yards are in 1 mile. (A mile is 5280 feet; a yard is 3 feet.)

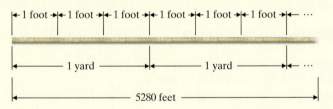

88. Find how many whole feet are in 1 rod. (A mile is 5280 feet; 1 mile is 320 rods.)

Objective D *Find the average of each list of numbers. See Example 12.*

89. 10, 24, 35, 22, 17, 12

90. 37, 26, 15, 29, 51, 22

91. 205, 972, 210, 161

92. 121, 200, 185, 176, 163

▷ 93. 86, 79, 81, 69, 80

94. 92, 96, 90, 85, 92, 79

The normal monthly temperatures in degrees Fahrenheit for Salt Lake City, Utah, are given in the graph. Use this graph to answer Exercises 95 and 96. (Source: National Climatic Data Center)

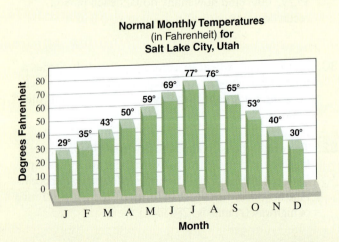

95. Find the average temperature for June, July, and August.

96. Find the average temperature for October, November, and December.

Mixed Practice (*Sections 1.3, 1.5, 1.6*) *Perform each indicated operation. Watch the operation symbol.*

97. 82 + 463 + 29 + 8704

98. 23 + 407 + 92 + 7011

99. 546
       × 28

100. 712
        × 54

101. 722
        − 43

102. 712
        − 54

103. $\dfrac{45}{0}$

104. $\dfrac{0}{23}$

105. 228 ÷ 24

106. 304 ÷ 31

Concept Extensions

Match each word phrase to the correct translation. (Not all letter choices will be used.) See the Concept Check in this section.

107. The quotient of 40 and 8

108. The quotient of 200 and 20

a. 20 ÷ 200   **b.** 200 ÷ 20

c. 40 ÷ 8   **d.** 8 ÷ 40

109. 200 divided by 20

110. 40 divided by 8

The following table shows the top five leading U.S. television advertisers during the first half of 2013 and the amount of money spent that half-year on advertising. Use this table to answer Exercises 111 and 112. (Source: Local Media Marketing Solutions)

Advertiser	Amount Spent on Television Advertising in 1st half of 2013
Ford Motor Company, dealers and corporate	$191,055,900
AT&T Inc.	$170,634,500
Comcast Corp	$157,780,500
Toyota, Dealers and corporate	$153,946,200
Chrysler-Cerberus	$122,784,200

111. Find the average amount of money spent on television ads for the half-year by the top two advertisers.

112. Find the average amount of money spent on television advertising by the top four advertisers.

In Example 12 in this section, we found that the average of 75, 96, 81, and 88 is 85. Use this information to answer Exercises 113 and 114.

113. If the number 75 is removed from the list of numbers, does the average increase or decrease? Explain why.

114. If the number 96 is removed from the list of numbers, does the average increase or decrease? Explain why.

115. Without computing it, tell whether the average of 126, 135, 198, and 113 is 86. Explain why it is possible or why it is not.

116. Without computing it, tell whether the average of 38, 27, 58, and 43 is 17. Explain why it is possible or why it is not.

117. If the area of a rectangle is 60 square feet and its width is 5 feet, what is its length?

118. If the area of a rectangle is 84 square inches and its length is 21 inches, what is its width?

119. Write down any two numbers whose quotient is 25.

120. Write down any two numbers whose quotient is 1.

121. Find 26 ÷ 5 using the process of repeated subtraction.

122. Find 86 ÷ 10 using the process of repeated subtraction.

Operations on Whole Numbers

Answers

Perform each indicated operation.

1. _____

2. _____

3. _____

4. _____

5. _____

6. _____

7. _____

8. _____

9. _____

10. _____

11. _____

12. _____

13. _____

14. _____

15. _____

16. _____

17. _____

18. _____

19. _____

20. _____

21. _____

22. _____

23. _____

24. _____

25. _____

26. _____

27. _____

28. _____

29. _____

30. _____

1.
$$\begin{array}{r} 42 \\ 63 \\ +\,89 \\ \hline \end{array}$$

2.
$$\begin{array}{r} 7006 \\ -\,451 \\ \hline \end{array}$$

3.
$$\begin{array}{r} 87 \\ \times\,52 \\ \hline \end{array}$$

4. $8\overline{)4496}$

5. $1 \cdot 67$

6. $\dfrac{36}{0}$

7. $16 \div 16$

8. $5 \div 1$

9. $0 \cdot 21$

10. $7 \cdot 0 \cdot 8$

11. $0 \div 7$

12. $12 \div 4$

13. $9 \cdot 7$

14. $45 \div 5$

15.
$$\begin{array}{r} 207 \\ -\,69 \\ \hline \end{array}$$

16.
$$\begin{array}{r} 207 \\ +\,69 \\ \hline \end{array}$$

17. $3718 - 2549$

18. $1861 + 7965$

19. $7\overline{)1278}$

20.
$$\begin{array}{r} 1259 \\ \times\,63 \\ \hline \end{array}$$

21. $7\overline{)7695}$

22. $9\overline{)1000}$

23. $32\overline{)21,240}$

24. $65\overline{)70,000}$

25. $4000 - 2963$

26. $10,000 - 101$

27.
$$\begin{array}{r} 303 \\ \times\,101 \\ \hline \end{array}$$

28. $(475)(100)$

29. Find the total of 62 and 9.

30. Find the product of 62 and 9.

31. Find the quotient of 62 and 9.

32. Find the difference of 62 and 9.

33. Subtract 17 from 200.

34. Find the difference of 432 and 201.

Complete the table by rounding the given number to the given place value.

		Tens	Hundrds	Thousands
35.	9735			
36.	1429			
37.	20,801			
38.	432,198			

Find the perimeter and area of each figure.

△ **39.**

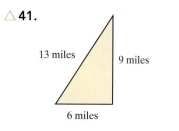

Square   6 feet

△ **40.**

14 inches

Rectangle   7 inches

Find the perimeter of each figure.

△ **41.**

13 miles   9 miles

6 miles

△ **42.**

3 meters

4 meters

3 meters

3 meters

Find the average of each list of numbers.

43. 19, 15, 25, 37, 24

44. 108, 131, 98, 159

45. The Mackinac Bridge is a suspension bridge that connects the lower and upper peninsulas of Michigan across the Straits of Mackinac. Its total length is 26,372 feet. The Lake Pontchartrain Bridge is a twin concrete trestle bridge in Slidell, Louisiana. Its total length is 28,547 feet. Which bridge is longer and by how much? (*Sources:* Mackinac Bridge Authority and Federal Highway Administration, Bridge Division)

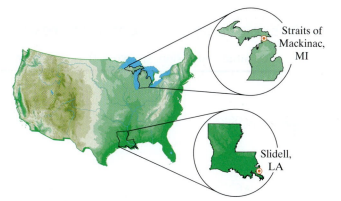

Straits of Mackinac, MI

Slidell, LA

46. In the United States, the average toy expenditure per child is $309 per year. On average, how much is spent on toys for a child by the time he or she reaches age 18? (*Source:* statista)

31. _____

32. _____

33. _____

34. _____

35. _____

36. _____

37. _____

38. _____

39. _____

40. _____

41. _____

42. _____

43. _____

44. _____

45. _____

46. _____

Objectives

A Write Repeated Factors Using Exponential Notation. ▶

B Evaluate Expressions Containing Exponents. ▶

C Use the Order of Operations. ▶

D Find the Area of a Square. ▶

Objective A Using Exponential Notation ▶

In the product $3 \cdot 3 \cdot 3 \cdot 3 \cdot 3$, notice that 3 is a factor several times. When this happens, we can use a shorthand notation, called an **exponent,** to write the repeated multiplication.

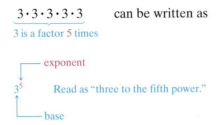

$\underbrace{3 \cdot 3 \cdot 3 \cdot 3 \cdot 3}_{\text{3 is a factor 5 times}}$ can be written as

3^5 Read as "three to the fifth power."

This is called **exponential notation.** The **exponent,** 5, indicates how many times the **base,** 3, is a factor.

The table below shows examples of reading exponential notation in words.

Expression	In Words
5^2	"five to the second power" or "five squared"
5^3	"five to the third power" or "five cubed"
5^4	"five to the fourth power"

Usually, an exponent of 1 is not written, so when no exponent appears, we assume that the exponent is 1. For example, $2 = 2^1$ and $7 = 7^1$.

Practice 1–4

Write using exponential notation.

1. $8 \cdot 8 \cdot 8 \cdot 8$

2. $3 \cdot 3 \cdot 3$

3. $10 \cdot 10 \cdot 10 \cdot 10 \cdot 10$

4. $5 \cdot 5 \cdot 4 \cdot 4 \cdot 4 \cdot 4 \cdot 4 \cdot 4$

Examples Write using exponential notation.

1. $7 \cdot 7 \cdot 7 = 7^3$

2. $3 \cdot 3 = 3^2$

3. $6 \cdot 6 \cdot 6 \cdot 6 \cdot 6 = 6^5$

4. $3 \cdot 3 \cdot 3 \cdot 3 \cdot 9 \cdot 9 \cdot 9 = 3^4 \cdot 9^3$

◼ **Work Practice 1–4**

Objective B Evaluating Exponential Expressions ▶

To **evaluate** an exponential expression, we write the expression as a product and then find the value of the product.

Practice 5–8

Evaluate.

5. 4^2 **6.** 7^3

7. 11^1 **8.** $2 \cdot 3^2$

Examples Evaluate.

5. $9^2 = 9 \cdot 9 = 81$

6. $6^1 = 6$

7. $3^4 = 3 \cdot 3 \cdot 3 \cdot 3 = 81$

8. $5 \cdot 6^2 = 5 \cdot 6 \cdot 6 = 180$

◼ **Work Practice 5–8**

Answers

1. 8^4 **2.** 3^3 **3.** 10^5 **4.** $5^2 \cdot 4^6$

5. 16 **6.** 343 **7.** 11 **8.** 18

Example 8 illustrates an important property: An exponent applies only to its base. The exponent 2, in $5 \cdot 6^2$, applies only to its base, 6.

Helpful Hint

An exponent applies only to its base. For example, $4 \cdot 2^3$ means $4 \cdot 2 \cdot 2 \cdot 2$.

Helpful Hint

Don't forget that 2^4, for example, is *not* $2 \cdot 4$. The expression 2^4 means repeated multiplication of the same factor.

$2^4 = 2 \cdot 2 \cdot 2 \cdot 2 = 16,$ whereas $2 \cdot 4 = 8$

✓**Concept Check** Which of the following statements is correct?

a. 3^5 is the same as $5 \cdot 5 \cdot 5$.
b. "Ten cubed" is the same as 10^2.
c. "Six to the fourth power" is the same as 6^4.
d. 12^2 is the same as $12 \cdot 2$.

Objective C Using the Order of Operations ▶

Suppose that you are in charge of taking inventory at a local cell phone store. An employee has given you the number of a certain cell phone in stock as the expression

$6 + 2 \cdot 30$

To calculate the value of this expression, do you add first or multiply first? If you add first, the answer is 240. If you multiply first, the answer is 66.

Mathematical symbols wouldn't be very useful if two values were possible for one expression. Thus, mathematicians have agreed that, given a choice, we multiply first.

$6 + 2 \cdot 30 = 6 + 60$ Multiply.
$ = 66$ Add.

This agreement is one of several **order of operations** agreements.

Order of Operations

1. Perform all operations within parentheses ( ), brackets [ ], or other grouping symbols such as fraction bars, starting with the innermost set.
2. Evaluate any expressions with exponents.
3. Multiply or divide in order from left to right.
4. Add or subtract in order from left to right.

✓**Concept Check Answer**

c

Below we practice using order of operations to simplify expressions.

Practice 9
Simplify: $9 \cdot 3 - 8 \div 4$

Example 9 Simplify: $2 \cdot 4 - 3 \div 3$

Solution: There are no parentheses and no exponents, so we start by multiplying and dividing, from left to right.

$$2 \cdot 4 - 3 \div 3 = 8 - 3 \div 3 \quad \text{Multiply.}$$
$$= 8 - 1 \quad \text{Divide.}$$
$$= 7 \quad \text{Subtract.}$$

■ Work Practice 9

Practice 10
Simplify: $48 \div 3 \cdot 2^2$

Example 10 Simplify: $4^2 \div 2 \cdot 4$

Solution: We start by evaluating 4^2.

$$4^2 \div 2 \cdot 4 = 16 \div 2 \cdot 4 \quad \text{Write } 4^2 \text{ as 16.}$$

Next we multiply or divide *in order* from left to right. Since division appears before multiplication from left to right, we divide first, then multiply.

$$16 \div 2 \cdot 4 = 8 \cdot 4 \quad \text{Divide.}$$
$$= 32 \quad \text{Multiply.}$$

■ Work Practice 10

Practice 11
Simplify: $(10 - 7)^4 + 2 \cdot 3^2$

Example 11 Simplify: $(8 - 6)^2 + 2^3 \cdot 3$

Solution:
$$(8 - 6)^2 + 2^3 \cdot 3 = 2^2 + 2^3 \cdot 3 \quad \text{Simplify inside parentheses.}$$
$$= 4 + 8 \cdot 3 \quad \text{Write } 2^2 \text{ as 4 and } 2^3 \text{ as 8.}$$
$$= 4 + 24 \quad \text{Multiply.}$$
$$= 28 \quad \text{Add.}$$

■ Work Practice 11

Practice 12
Simplify:
$36 \div [20 - (4 \cdot 2)] + 4^3 - 6$

Example 12 Simplify: $4^3 + [3^2 - (10 \div 2)] - 7 \cdot 3$

Solution: Here we begin with the innermost set of parentheses.

$$4^3 + [3^2 - (10 \div 2)] - 7 \cdot 3 = 4^3 + [3^2 - 5] - 7 \cdot 3 \quad \text{Simplify inside parentheses.}$$
$$= 4^3 + [9 - 5] - 7 \cdot 3 \quad \text{Write } 3^2 \text{ as 9.}$$
$$= 4^3 + 4 - 7 \cdot 3 \quad \text{Simplify inside brackets.}$$
$$= 64 + 4 - 7 \cdot 3 \quad \text{Write } 4^3 \text{ as 64.}$$
$$= 64 + 4 - 21 \quad \text{Multiply.}$$
$$= 47 \quad \text{Add and subtract from left to right.}$$

■ Work Practice 12

Example 13 Simplify: $\dfrac{7 - 2 \cdot 3 + 3^2}{5(2 - 1)}$

Solution: Here, the fraction bar is a grouping symbol. We simplify above and below the fraction bar separately.

$$\frac{7 - 2 \cdot 3 + 3^2}{5(2 - 1)} = \frac{7 - 2 \cdot 3 + 9}{5(1)} \quad \text{Evaluate } 3^2 \text{ and } (2 - 1).$$

$$= \frac{7 - 6 + 9}{5} \quad \text{Multiply } 2 \cdot 3 \text{ in the numerator and multiply 5 and 1 in the denominator.}$$

$$= \frac{10}{5} \quad \text{Add and subtract from left to right.}$$

$$= 2 \quad \text{Divide.}$$

■ **Work Practice 13**

Example 14 Simplify: $64 \div 8 \cdot 2 + 4$

Solution: $64 \div 8 \cdot 2 + 4 = \underline{8 \cdot 2} + 4$ Divide.
$\qquad\qquad\qquad\qquad = 16 + 4$ Multiply.
$\qquad\qquad\qquad\qquad = 20$ Add.

■ **Work Practice 14**

Objective D Finding the Area of a Square

Since a square is a special rectangle, we can find its area by finding the product of its length and its width.

$\qquad$ Area of a rectangle = length $\cdot$ width

By recalling that each side of a square has the same measurement, we can use the following procedure to find its area:

$\qquad$ Area of a square = length $\cdot$ width
$\qquad\qquad\qquad\qquad = $ side $\cdot$ side
$\qquad\qquad\qquad\qquad = ($ side $)^2$

Helpful Hint

Recall from Section 1.5 that area is measured in **square** units while perimeter is measured in units.

Example 15 Find the area of a square whose side measures 4 inches.

Solution: Area of a square $= ($ side $)^2$
$\qquad\qquad\qquad\qquad = ($ 4 inches $)^2$
$\qquad\qquad\qquad\qquad = 16$ square inches

The area of the square is 16 square inches.

■ **Work Practice 15**

Practice 13

Simplify: $\dfrac{25 + 8 \cdot 2 - 3^3}{2(3 - 2)}$

Practice 14

Simplify: $36 \div 6 \cdot 3 + 5$

Practice 15

Find the area of a square whose side measures 12 centimeters.

Answers
13. 7 **14.** 23 **15.** 144 sq cm

Calculator Explorations Exponents

To evaluate an exponential expression such as 4^7 on a calculator, find the keys marked $\boxed{y^x}$ or $\boxed{\wedge}$ and $\boxed{=}$ or $\boxed{\text{ENTER}}$. To evaluate 4^7, press the keys $\boxed{4}$ $\boxed{y^x}$ (or $\boxed{\wedge}$) $\boxed{7}$ then $\boxed{=}$ or $\boxed{\text{ENTER}}$. The display will read $\boxed{16384}$. Thus, $4^7 = 16{,}384$.

Use a calculator to evaluate.

1. 4^6 **2.** 5^6 **3.** 5^5

4. 7^6 **5.** 2^{11} **6.** 6^8

Order of Operations

To see whether your calculator has the order of operations built in, evaluate $5 + 2 \cdot 3$ by pressing the keys $\boxed{5}$ $\boxed{+}$ $\boxed{2}$ $\boxed{\times}$ $\boxed{3}$ then $\boxed{=}$ or $\boxed{\text{ENTER}}$. If the display reads $\boxed{11}$, your calculator does have the order of operations built in. This means that most of the time,

you can key in a problem exactly as it is written, and the calculator will perform operations in the proper order. When evaluating an expression containing parentheses, key in the parentheses. (If an expression contains brackets, key in parentheses.) For example, to evaluate $2[25 - (8 + 4)] - 11$, press the keys $\boxed{2}$ $\boxed{\times}$ $\boxed{(}$ $\boxed{25}$ $\boxed{-}$ $\boxed{(}$ $\boxed{8}$ $\boxed{+}$ $\boxed{4}$ $\boxed{)}$ $\boxed{)}$ $\boxed{-}$ $\boxed{11}$ then $\boxed{=}$ or $\boxed{\text{ENTER}}$.

The display will read $\boxed{15}$.

Use a calculator to evaluate.

7. $7^4 + 5^3$

8. $12^4 - 8^4$

9. $63 \cdot 75 - 43 \cdot 10$

10. $8 \cdot 22 + 7 \cdot 16$

11. $4(15 \div 3 + 2) - 10 \cdot 2$

12. $155 - 2(17 + 3) + 185$

Vocabulary, Readiness & Video Check

Use the choices below to fill in each blank.

addition   multiplication   exponent

subtraction   division   base

1. In $2^5 = 32$, the 2 is called the _____ and the 5 is called the _____ .

2. To simplify $8 + 2 \cdot 6$, which operation should be performed first? _____

3. To simplify $(8 + 2) \cdot 6$, which operation should be performed first? _____

4. To simplify $9(3 - 2) \div 3 + 6$, which operation should be performed first? _____

5. To simplify $8 \div 2 \cdot 6$, which operation should be performed first? _____

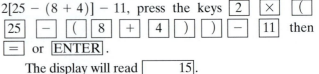

Martin-Gay Interactive Videos

See Video 1.7

Watch the section lecture video and answer the following questions.

Objective A   **6.** In the ▣ Example 1 expression, what is the 3 called and what is the 12 called? ▶

Objective B   **7.** As mentioned in ▣ Example 4, what "understood exponent" does any number we've worked with before have? ▶

Objective C   **8.** List the three operations needed to evaluate ▣ Example 7 in the order they should be performed. ▶

Objective D   **9.** As explained in the lecture before ▣ Example 10, why does the area of a square involve an exponent whereas the area of a rectangle usually does not? ▶

1.7 Exercise Set MyMathLab® ▶

Objective A *Write using exponential notation. See Examples 1 through 4.*

1. $4 \cdot 4 \cdot 4$ **2.** $5 \cdot 5 \cdot 5 \cdot 5$ **3.** $7 \cdot 7 \cdot 7 \cdot 7 \cdot 7 \cdot 7$ **4.** $6 \cdot 6 \cdot 6 \cdot 6 \cdot 6 \cdot 6 \cdot 6$

▶ **5.** $12 \cdot 12 \cdot 12$ **6.** $10 \cdot 10 \cdot 10$ ▶ **7.** $6 \cdot 6 \cdot 5 \cdot 5 \cdot 5$ **8.** $4 \cdot 4 \cdot 3 \cdot 3 \cdot 3$

9. $9 \cdot 8 \cdot 8$ **10.** $7 \cdot 4 \cdot 4 \cdot 4$ **11.** $3 \cdot 2 \cdot 2 \cdot 2 \cdot 2$ **12.** $4 \cdot 6 \cdot 6 \cdot 6 \cdot 6$

13. $3 \cdot 2 \cdot 2 \cdot 2 \cdot 2 \cdot 5 \cdot 5 \cdot 5 \cdot 5 \cdot 5$ **14.** $6 \cdot 6 \cdot 2 \cdot 9 \cdot 9 \cdot 9 \cdot 9$

Objective B *Evaluate. See Examples 5 through 8.*

15. 8^2 **16.** 6^2 ▶ **17.** 5^3 **18.** 6^3 **19.** 2^5 **20.** 3^5

21. 1^{10} **22.** 1^{12} ▶ **23.** 7^1 **24.** 8^1 **25.** 2^7 **26.** 5^4

27. 2^8 **28.** 3^3 **29.** 4^4 **30.** 4^3 **31.** 9^3 **32.** 8^3

33. 12^2 **34.** 11^2 ▶ **35.** 10^2 **36.** 10^3 **37.** 20^1 **38.** 14^1

39. 3^6 **40.** 4^5 **41.** $3 \cdot 2^6$ **42.** $5 \cdot 3^2$ **43.** $2 \cdot 3^4$ **44.** $2 \cdot 7^2$

Objective C *Simplify. See Examples 9 through 14.*

▶ **45.** $15 + 3 \cdot 2$ **46.** $24 + 6 \cdot 3$ ▶ **47.** $14 \div 7 \cdot 2 + 3$ **48.** $100 \div 10 \cdot 5 + 4$

49. $32 \div 4 - 3$ **50.** $42 \div 7 - 6$ **51.** $13 + \dfrac{24}{8}$ **52.** $32 + \dfrac{8}{2}$

53. $6 \cdot 5 + 8 \cdot 2$ **54.** $3 \cdot 4 + 9 \cdot 1$ **55.** $\dfrac{5 + 12 \div 4}{1^7}$ **56.** $\dfrac{6 + 9 \div 3}{3^2}$

57. $(7 + 5^2) \div 4 \cdot 2^3$ **58.** $6^2 \cdot (10 - 8)$ **59.** $5^2 \cdot (10 - 8) + 2^3 + 5^2$

60. $5^3 \div (10 + 15) + 9^2 + 3^3$ **61.** $\dfrac{18 + 6}{2^4 - 2^2}$ **62.** $\dfrac{40 + 8}{5^2 - 3^2}$

63. $(3 + 5) \cdot (9 - 3)$ **64.** $(9 - 7) \cdot (12 + 18)$ ▶ **65.** $\dfrac{7(9 - 6) + 3}{3^2 - 3}$

66. $\dfrac{5(12 - 7) - 4}{5^2 - 18}$

67. $8 \div 0 + 37$

68. $18 - 7 \div 0$

69. $2^4 \cdot 4 - (25 \div 5)$

70. $2^3 \cdot 3 - (100 \div 10)$

71. $3^4 - [35 - (12 - 6)]$

72. $[40 - (8 - 2)] - 2^5$

▶ **73.** $(7 \cdot 5) + [9 \div (3 \div 3)]$

74. $(18 \div 6) + [(3 + 5) \cdot 2]$

75. $8 \cdot [2^2 + (6 - 1) \cdot 2] - 50 \cdot 2$

76. $35 \div [3^2 + (9 - 7) - 2^2] + 10 \cdot 3$

77. $\dfrac{9^2 + 2^2 - 1^2}{8 \div 2 \cdot 3 \cdot 1 \div 3}$

78. $\dfrac{5^2 - 2^3 + 1^4}{10 \div 5 \cdot 4 \cdot 1 \div 4}$

79. $\dfrac{2 + 4^2}{5(20 - 16) - 3^2 - 5}$

80. $\dfrac{3 + 9^2}{3(10 - 6) - 2^2 - 1}$

81. $9 \div 3 + 5^2 \cdot 2 - 10$

82. $10 \div 2 + 3^3 \cdot 2 - 20$

83. $[13 \div (20 - 7) + 2^5] - (2 + 3)^2$

84. $[15 \div (11 - 6) + 2^2] + (5 - 1)^2$

85. $7^2 - \{18 - [40 \div (5 \cdot 1) + 2] + 5^2\}$

86. $29 - \{5 + 3[8 \cdot (10 - 8)] - 50\}$

△**Objective D** **Mixed Practice (Section 1.3)** *Find the area and perimeter of each square. See Example 15.*

▶ **87.**   7 meters

△**88.**   9 centimeters

△**89.**   23 miles

△**90.**   41 feet

Concept Extensions

Answer the following true or false. See the Concept Check in this section.

91. "Six to the fifth power" is the same as 6^5.

92. "Seven squared" is the same as 7^2.

93. 2^5 is the same as $5 \cdot 5$.

94. 4^9 is the same as $4 \cdot 9$.

Insert grouping symbols (parentheses) so that each given expression evaluates to the given number.

95. $2 + 3 \cdot 6 - 2$; evaluates to 28

96. $2 + 3 \cdot 6 - 2$; evaluates to 20

97. $24 \div 3 \cdot 2 + 2 \cdot 5$; evaluates to 14

98. $24 \div 3 \cdot 2 + 2 \cdot 5$; evaluates to 15

△ **99.** A building contractor is bidding on a contract to install gutters on seven homes in a retirement community, all in the shape shown. To estimate the cost of materials, she needs to know the total perimeter of all seven homes. Find the total perimeter.

100. The building contractor from Exercise 99 plans to charge $4 per foot for installing vinyl gutters. Find the total charge for the seven homes given the total perimeter answer to Exercise 99.

Simplify.

101. $(7 + 2^4)^5 - (3^5 - 2^4)^2$

102. $25^3 \cdot (45 - 7 \cdot 5) \cdot 5$

103. Write an expression that simplifies to 5. Use multiplication, division, addition, subtraction, and at least one set of parentheses. Explain the process you would use to simplify the expression.

104. Explain why $2 \cdot 3^2$ is not the same as $(2 \cdot 3)^2$.

1.8 **Introduction to Variables, Algebraic Expressions, and Equations** ▶

Objective A Evaluating Algebraic Expressions ▶

Perhaps the most important quality of mathematics is that it is a science of patterns. Communicating about patterns is often made easier by using a letter to represent all the numbers fitting a pattern. We call such a letter a **variable.** For example, in Section 1.3 we presented the addition property of 0, which states that the sum of 0 and any number is that number. We might write

$0 + 1 = 1$
$0 + 2 = 2$
$0 + 3 = 3$
$0 + 4 = 4$
$0 + 5 = 5$
$0 + 6 = 6$
$\vdots$

Objectives

A Evaluate Algebraic Expressions Given Replacement Values. ▶

B Identify Solutions of Equations. ▶

C Translate Phrases into Variable Expressions. ▶

continuing indefinitely. This is a pattern, and all whole numbers fit the pattern. We can communicate this pattern for all whole numbers by letting a letter, such as a, represent all whole numbers. We can then write

$$0 + a = a$$

Using variable notation is a primary goal of learning **algebra.** We now take some important first steps in beginning to use variable notation.

A combination of operations on letters (variables) and numbers is called an **algebraic expression** or simply an **expression.**

Algebraic Expressions

$$3 + x \qquad 5 \cdot y \qquad 2 \cdot z - 1 + x$$

If two variables or a number and a variable are next to each other, with no operation sign between them, the operation is multiplication. For example,

$$2x \quad \text{means} \quad 2 \cdot x$$

and

$$xy \text{ or } x(y) \quad \text{means} \quad x \cdot y$$

Also, the meaning of an exponent remains the same when the base is a variable. For example,

$$x^2 = \underbrace{x \cdot x}_{2 \text{ factors of } x} \qquad \text{and} \qquad y^5 = \underbrace{y \cdot y \cdot y \cdot y \cdot y}_{5 \text{ factors of } y}$$

Algebraic expressions such as $3x$ have different values depending on replacement values for x. For example, if x is 2, then $3x$ becomes

$$3x = 3 \cdot 2$$
$$= 6$$

If x is 7, then $3x$ becomes

$$3x = 3 \cdot 7$$
$$= 21$$

Replacing a variable in an expression by a number and then finding the value of the expression is called **evaluating the expression** for the variable. When finding the value of an expression, remember to follow the order of operations given in Section 1.7.

Practice 1

Evaluate $x - 2$ if x is 7.

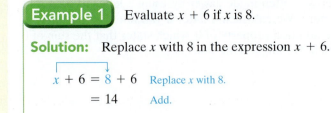

Example 1 Evaluate $x + 6$ if x is 8.

Solution: Replace x with 8 in the expression $x + 6$.

$$x + 6 = 8 + 6 \qquad \text{Replace } x \text{ with 8.}$$
$$= 14 \qquad \text{Add.}$$

Work Practice 1

When we write a statement such as "x is 5," we can use an equal sign $(=)$ to represent "is" so that

$$x \text{ is 5} \quad \text{can be written as} \quad x = 5.$$

Answer

1. 5

Example 2   Evaluate $2(x - y)$ for $x = 6$ and $y = 3$.

Solution: $2(x - y) = 2(6 - 3)$ Replace x with 6 and y with 3.

$\qquad\qquad\quad = 2(3)$ Subtract.

$\qquad\qquad\quad = 6$ Multiply.

■ **Work Practice 2**

Practice 2

Evaluate $y(x - 3)$ for $x = 8$ and $y = 4$.

Example 3   Evaluate $\dfrac{x - 5y}{y}$ for $x = 35$ and $y = 5$.

Solution: $\dfrac{x - 5y}{y} = \dfrac{35 - 5(5)}{5}$ Replace x with 35 and y with 5.

$\qquad\qquad = \dfrac{35 - 25}{5}$ Multiply.

$\qquad\qquad = \dfrac{10}{5}$ Subtract.

$\qquad\qquad = 2$ Divide.

■ **Work Practice 3**

Practice 3

Evaluate $\dfrac{y + 6}{x}$ for $x = 6$ and $y = 18$.

Example 4   Evaluate $x^2 + z - 3$ for $x = 5$ and $z = 4$.

Solution: $x^2 + z - 3 = 5^2 + 4 - 3$ Replace x with 5 and z with 4.

$\qquad\qquad\qquad = 25 + 4 - 3$ Evaluate 5^2.

$\qquad\qquad\qquad = 26$ Add and subtract from left to right.

■ **Work Practice 4**

Practice 4

Evaluate $25 - z^3 + x$ for $z = 2$ and $x = 1$.

Helpful Hint

If you are having difficulty replacing variables with numbers, first replace each variable with a set of parentheses, then insert the replacement number within the parentheses.

Example:

$x^2 + z - 3 = (\quad)^2 + (\quad) - 3$

$\qquad\qquad\quad = (5)^2 + (4) - 3$

$\qquad\qquad\quad = 25 + 4 - 3$

$\qquad\qquad\quad = 26$

✔**Concept Check**   What's wrong with the solution to the following problem?

Evaluate $3x + 2y$ for $x = 2$ and $y = 3$.

Solution: $3x + 2y = 3(3) + 2(2)$

$\qquad\qquad\quad = 9 + 4$

$\qquad\qquad\quad = 13$

Practice 5

Evaluate $\dfrac{5(F - 32)}{9}$ for $F = 41$.

Example 5   The expression $\dfrac{5(F - 32)}{9}$ can be used to write degrees Fahrenheit F as degrees Celsius C. Find the value of this expression for $F = 86$.

Solution:

$$\dfrac{5(F - 32)}{9} = \dfrac{5(86 - 32)}{9}$$

$$= \dfrac{5(54)}{9}$$

$$= \dfrac{270}{9}$$

$$= 30$$

Thus $86°F$ is the same temperature as $30°C$.

◼ Work Practice 5

Objective B   Identifying Solutions of Equations

In Objective **A** , we learned that a combination of operations on variables and numbers is called an algebraic expression or simply an expression. Frequently in this book, we have written statements like $7 + 4 = 11$ or area $=$ length $\cdot$ width. Each of these statements is called an **equation.** An equation is of the form

> **Helpful Hint**
> An equation contains "=," while an expression does not.

$$\underbrace{\textbf{expression}} = \textbf{expression}$$

An equation can be labeled as

$$\underbrace{x + 7}_{\text{left side}} \overset{\text{equal sign}}{=} \underset{\text{right side}}{10}$$

When an equation contains a variable, deciding which values of the variable make an equation a true statement is called **solving** an equation for the variable. A **solution** of an equation is a value for the variable that makes an equation a true statement. For example, 2 is a solution of the equation $x + 5 = 7$, since replacing x with 2 results in the *true* statement $2 + 5 = 7$. Similarly, 3 is not a solution of $x + 5 = 7$, since replacing x with 3 results in the *false* statement $3 + 5 = 7$.

Practice 6

Determine whether 8 is a solution of the equation $3(y - 6) = 6$.

Example 6   Determine whether 6 is a solution of the equation $4(x - 3) = 12$.

Solution: We replace x with 6 in the equation.

$$4(x - 3) = 12$$
$$\downarrow$$
$$4(6 - 3) \overset{?}{=} 12 \quad \text{Replace } x \text{ with 6.}$$
$$4(3) \overset{?}{=} 12$$
$$12 = 12 \quad \text{True}$$

Since $12 = 12$ is a true statement, 6 *is* a solution of the equation.

◼ Work Practice 6

Answers

5. 5   **6.** yes

A collection of numbers enclosed by braces is called a set. For example,

$$\{0, 1, 2, 3, \ldots\}$$

is the set of whole numbers that we are studying about in this chapter. The three dots after the number 3 in the set mean that this list of numbers continues in the same manner indefinitely.

The next example contains set notation.

Example 7   Determine which numbers in the set $\{26, 40, 20\}$ are solutions of the equation $2n - 30 = 10$.

Solution: Replace n with each number from the set to see if a true statement results.

Let n be 26.	Let n be 40.	Let n be 20.
$2n - 30 = 10$	$2n - 30 = 10$	$2n - 30 = 10$
$2 \cdot 26 - 30 \stackrel{?}{=} 10$	$2 \cdot 40 - 30 \stackrel{?}{=} 10$	$2 \cdot 20 - 30 \stackrel{?}{=} 10$
$52 - 30 \stackrel{?}{=} 10$	$80 - 30 \stackrel{?}{=} 10$	$40 - 30 \stackrel{?}{=} 10$
$22 = 10$ False	$50 = 10$ False	$10 = 10$ True ✓

Thus, 20 is a solution while 26 and 40 are not solutions.

◼ **Work Practice 7**

Objective C   Translating Phrases into Variable Expressions ▶

To aid us in solving problems later, we practice translating verbal phrases into algebraic expressions. Certain key words and phrases suggesting addition, subtraction, multiplication, or division are reviewed next.

Addition (+)	Subtraction (−)	Multiplication (·)	Division (÷)
sum	difference	product	quotient
plus	minus	times	divide
added to	subtract	multiply	shared equally among
more than	less than	multiply by	per
increased by	decreased by	of	divided by
total	less	double/triple	divided into

Example 8   Write as an algebraic expression. Use x to represent "a number."

a. 7 increased by a number
b. 15 decreased by a number
c. The product of 2 and a number
d. The quotient of a number and 5
e. 2 subtracted from a number

Solution:

a. In words: | 7 | increased by | a number |
     Translate: 7   + x

(Continued on next page)

Practice 7

Determine which numbers in the set $\{10, 6, 8\}$ are solutions of the equation $5n + 4 = 34$.

Practice 8

Write as an algebraic expression. Use x to represent "a number."

a. Twice a number
b. 8 increased by a number
c. 10 minus a number
d. 10 subtracted from a number
e. The quotient of 6 and a number

Answers

7. 6 is a solution.
8. a. $2x$ **b.** $8 + x$ **c.** $10 - x$
d. $x - 10$ **e.** $6 \div x$ or $\dfrac{6}{x}$

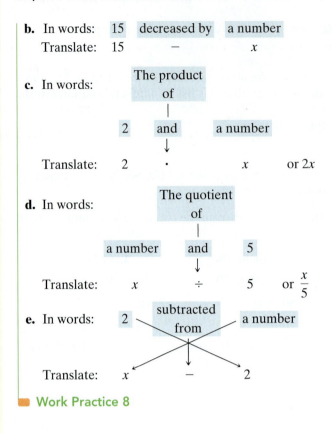

b. In words: 15   decreased by   a number

Translate: 15   − x

c. In words: The product of

2   and   a number

Translate: 2   · x or $2x$

d. In words: The quotient of

a number   and   5

Translate: x ÷   5   or $\dfrac{x}{5}$

e. In words: 2   subtracted from   a number

Translate: x −   2

■ **Work Practice 8**

Helpful Hint

Remember that order is important when subtracting. Study the order of numbers and variables below.

Phrase	Translation
a number *decreased by* 5	$x - 5$
a number *subtracted from* 5	$5 - x$

Vocabulary, Readiness & Video Check

Use the choices below to fill in each blank. You may use each choice more than once.

evaluating the expression   variable(s)   expression   equation   solution

1. A combination of operations on letters (variables) and numbers is a(n) _____.

2. A letter that represents a number is a(n) _____.

3. $3x - 2y$ is called a(n) _____ and the letters x and y are _____.

4. Replacing a variable in an expression by a number and then finding the value of the expression is called _____.

5. A statement of the form "expression = expression" is called a(n) _____.

6. A value for the variable that makes an equation a true statement is called a(n) _____.

Martin-Gay Interactive Videos   *Watch the section lecture video and answer the following questions.*

See Video 1.8

Objective A 7. Complete this statement based on the lecture before Example 1: When a letter and a variable are next to each other, the operation is an understood _____.

Objective B 8. In Example 5, why is a question mark written over the equal sign?

Objective C 9. In Example 6, what phrase translates to subtraction?

1.8 Exercise Set MyMathLab®

Objective A *Complete the table. The first row has been done for you. See Examples 1 through 5.*

	a	b	$a + b$	$a - b$	$a \cdot b$	$a \div b$
	45	9	54	36	405	5
1.	21	7				
2.	24	6				
3.	152	0				
4.	298	0				
5.	56	1				
6.	82	1				

Evaluate each following expression for $x = 2$, $y = 5$, and $z = 3$. See Examples 1 through 5.

7. $3 + 2z$

8. $7 + 3z$

9. $3xz - 5x$

10. $4yz + 2x$

11. $z - x + y$

12. $x + 5y - z$

13. $4x - z$

14. $2y + 5z$

15. $y^3 - 4x$

16. $y^3 - z$

17. $2xy^2 - 6$

18. $3yz^2 + 1$

19. $8 - (y - x)$

20. $3 + (2y - 4)$

21. $x^5 + (y - z)$

22. $x^4 - (y - z)$

23. $\dfrac{6xy}{z}$

24. $\dfrac{8yz}{15}$

25. $\dfrac{2y - 2}{x}$

26. $\dfrac{6 + 3x}{z}$

27. $\dfrac{x + 2y}{z}$

28. $\dfrac{2z + 6}{3}$

29. $\dfrac{5x}{y} - \dfrac{10}{y}$

30. $\dfrac{70}{2y} - \dfrac{15}{z}$

31. $2y^2 - 4y + 3$

32. $3x^2 + 2x - 5$

33. $(4y - 5z)^3$

34. $(4y + 3z)^2$

35. $(xy + 1)^2$

36. $(xz - 5)^4$

37. $2y(4z - x)$

38. $3x(y + z)$

39. $xy(5 + z - x)$ **40.** $xz(2y + x - z)$ **41.** $\dfrac{7x + 2y}{3x}$ **42.** $\dfrac{6z + 2y}{4}$

43. The expression $16t^2$ gives the distance in feet that an object falls after t seconds. Complete the table by evaluating $16t^2$ for each given value of t.

t	1	2	3	4
$16t^2$				

44. The expression $\dfrac{5(F - 32)}{9}$ gives the equivalent degrees Celsius for F degrees Fahrenheit. Complete the table by evaluating this expression for each given value of F.

F	50	59	68	77
$\dfrac{5(F - 32)}{9}$				

Objective B *Decide whether the given number is a solution of the given equation. See Example 6.*

45. Is 10 a solution of $n - 8 = 2$?

46. Is 9 a solution of $n - 2 = 7$?

47. Is 3 a solution of $24 = 80n$?

48. Is 50 a solution of $250 = 5n$?

49. Is 7 a solution of $3n - 5 = 10$?

50. Is 8 a solution of $11n + 3 = 91$?

51. Is 20 a solution of $2(n - 17) = 6$?

52. Is 0 a solution of $5(n + 9) = 40$?

53. Is 0 a solution of $5x + 3 = 4x + 13$?

54. Is 2 a solution of $3x - 6 = 5x - 10$?

▶ **55.** Is 8 a solution of $7f = 64 - f$?

56. Is 5 a solution of $8x - 30 = 2x$?

Determine which numbers in each set are solutions to the corresponding equations. See Example 7.

57. $n - 2 = 10$; $\{10, 12, 14\}$

58. $n + 3 = 16$; $\{9, 11, 13\}$

59. $5n = 30$; $\{6, 25, 30\}$

60. $3n = 45$; $\{15, 30, 45\}$

61. $6n + 2 = 26$; $\{0, 2, 4\}$

62. $4n - 14 = 6$; $\{0, 5, 10\}$

63. $3(n - 4) = 10$; $\{5, 7, 10\}$

64. $6(n + 2) = 23$; $\{1, 3, 5\}$

▶ **65.** $7x - 9 = 5x + 13$; $\{3, 7, 11\}$

66. $9x - 15 = 5x + 1$; $\{2, 4, 11\}$

Objective C **Translating** *Write each phrase as a variable expression. Use x to represent "a number." See Example 8.*

67. Eight more than a number

68. The sum of three and a number

69. The total of a number and eight

70. The difference of a number and five hundred

71. Twenty decreased by a number

72. A number less thirty

73. The product of 512 and a number

74. A number times twenty

75. The quotient of eight and a number

76. A number divided by 11

77. The sum of seventeen and a number added to the product of five and the number

78. The quotient of twenty and a number, decreased by three

79. The product of five and a number

80. The difference of twice a number, and four

81. A number subtracted from 11

82. Twelve subtracted from a number

83. A number less 5

84. The sum of a number and 7

85. 6 divided by a number

86. The product of a number and 7

87. Fifty decreased by eight times a number

88. Twenty decreased by twice a number

Concept Extensions

For Exercises 89 through 92, use a calculator to evaluate each expression for $x = 23$ and $y = 72$.

89. $x^4 - y^2$

90. $2(x + y)^2$

91. $x^2 + 5y - 112$

92. $16y - 20x + x^3$

93. If x is a whole number, which expression is the largest: $2x, 5x,$ or $\dfrac{x}{3}$?
Explain your answer.

94. If x is a whole number, which expression is the smallest: $2x, 5x,$ or $\dfrac{x}{3}$?
Explain your answer.

95. In Exercise 43, what do you notice about the value of $16t^2$ as t gets larger?

96. In Exercise 44, what do you notice about the value of $\dfrac{5(F - 32)}{9}$ as F gets larger?

Chapter 1 Group Activity

Investigating Endangered and Threatened Species

An **endangered** species is one that is thought to be in danger of becoming extinct throughout all or a major part of its habitat. A **threatened** species is one that may become endangered. The Division of Endangered Species at the U.S. Fish and Wildlife Service keeps close tabs on the state of threatened and endangered wildlife in the United States and around the world. The table below was compiled from 2013 data in the Division of Endangered Species' box score. The "Total Species" column gives the total number of endangered and threatened species for each group.

1. Round each number of *endangered animal species* to the nearest ten to estimate the Animal Total.

2. Round each number of *endangered plant species* to the nearest ten to estimate the Plant Total.

3. Add the exact numbers of endangered animal species to find the exact Animal Total and record it in the table in the Endangered Species column. Add the exact numbers of endangered plant species to find the Plant Total and record it in the table in the Endangered Species column. Then find the total number of endangered species (animals and plants combined) and record this number in the table as the Grand Total in the Endangered Species column.

4. Find the Animal Total, Plant Total, and Grand Total for the Total Species column. Record these values in the table.

5. Use the data in the table to complete the Threatened Species column.

6. Write a paragraph discussing the conclusions that can be drawn from the table.

Endangered and Threatened Species Worldwide				
	Group	Endangered Species	Threatened Species	Total Species
Animals	Mammals	325		361
	Birds	286		317
	Reptiles	84		126
	Amphibians	25		37
	Fishes	95		166
	Snails	34		47
	Clams	74		86
	Crustaceans	20		23
	Insects	61		71
	Arachnids	12		12
	Corals	0		2
	Animal Total			
Plants	Flowering Plants	571		820
	Conifers	2		5
	Ferns and others	26		30
	Lichens	2		2
	Plant Total			
	Grand Total			

Chapter 1   Vocabulary Check

Fill in each blank with one of the words or phrases listed below.

difference	factor	perimeter	dividend	minuend
place value	whole numbers	equation	divisor	variable
sum	set	addend	exponent	expression
solution	quotient	subtrahend	product	digits	area

1. The _____ are 0, 1, 2, 3, . . .

2. The _____ of a polygon is its distance around or the sum of the lengths of its sides.

3. The position of each digit in a number determines its _____.

4. A(n) _____ is a shorthand notation for repeated multiplication of the same factor.

5. To find the _____ of a rectangle, multiply length times width.

6. The _____ used to write numbers are 0, 1, 2, 3, 4, 5, 6, 7, 8, and 9.

7. A letter used to represent a number is called a(n) _____.

8. A(n) _____ can be written in the form "expression = expression."

9. A combination of operations on variables and numbers is called a(n) _____.

10. A(n) _____ of an equation is a value of the variable that makes the equation a true statement.

11. A collection of numbers (or objects) enclosed by braces is called a(n) _____.

Use the facts below for Exercises 12 through 21.

$$2 \cdot 3 = 6 \quad 4 + 17 = 21 \quad 20 - 9 = 11 \quad 5\overline{)35}\;{}^{7}$$

12. The 21 above is called the _____.

13. The 5 above is called the _____.

14. The 35 above is called the _____.

15. The 7 above is called the _____.

16. The 3 above is called a(n) _____.

17. The 6 above is called the _____.

18. The 20 above is called the _____.

19. The 9 above is called the _____.

20. The 11 above is called the _____.

21. The 4 above is called a(n) _____.

Helpful Hint

▶ Are you preparing for your test? Don't forget to take the Chapter 1 Test on page 95. Then check your answers at the back of the text and use the Chapter Test Prep Videos to see the fully worked-out solutions to any of the exercises you want to review.

1 Chapter Highlights

Definitions and Concepts	Examples
Section 1.2 Place Value, Names for Numbers, and Reading Tables	
The **whole numbers** are 0, 1, 2, 3, 4, 5, ... The position of each digit in a number determines its **place value.** A place-value chart is shown next with the names of the periods given.	Examples of whole numbers: 0, 14, 968, 5,268,619

Periods — Billions | Millions | Thousands | Ones

Hundred-billions, Ten-billions, Billions, Hundred-millions, Ten-millions, Millions, Hundred-thousands, Ten-thousands, Thousands, Hundreds, Tens, Ones

1 0 8 3 6 6 4 5 0 0

(continued)

Definitions and Concepts	Examples
Section 1.2 Place Value, Names for Numbers, and Reading Tables (*continued*)	

To write a whole number in words, write the number in each period followed by the name of the period. (The name of the ones period is not included.)	9,078,651,002 is written as nine billion, seventy-eight million, six hundred fifty-one thousand, two.
To write a whole number in standard form, write the number in each period, followed by a comma.	Four million, seven hundred six thousand, twenty-eight is written as 4,706,028.

Section 1.3 Adding and Subtracting Whole Numbers, and Perimeter	

To add whole numbers, add the digits in the ones place, then the tens place, then the hundreds place, and so on, carrying when necessary.

Find the sum:

$$\begin{array}{r} \overset{211}{2689} \leftarrow \text{addend} \\ 1735 \leftarrow \text{addend} \\ +\ \ 662 \leftarrow \text{addend} \\ \hline 5086 \leftarrow \text{sum} \end{array}$$

To subtract whole numbers, subtract the digits in the ones place, then the tens place, then the hundreds place, and so on, borrowing when necessary.

Subtract:

$$\begin{array}{r} \overset{8\ 15}{79\cancel{5}4} \leftarrow \text{minuend} \\ -\ 5673 \leftarrow \text{subtrahend} \\ \hline 2281 \leftarrow \text{difference} \end{array}$$

The **perimeter** of a polygon is its distance around or the sum of the lengths of its sides.

△ Find the perimeter of the polygon shown.

The perimeter is
5 feet + 3 feet + 9 feet + 2 feet = 19 feet.

Section 1.4 Rounding and Estimating	

Rounding Whole Numbers to a Given Place Value

Step 1: Locate the digit to the right of the given place value.

Step 2: If this digit is 5 or greater, add 1 to the digit in the given place value and replace each digit to its right with 0.

Step 3: If this digit is less than 5, replace it and each digit to its right with 0.

Round 15,721 to the nearest thousand.

15, ⑦ 21

Add 1 ⎯⎺⏋ Replace with zeros.

Since the circled digit is 5 or greater, add 1 to the given place value and replace digits to its right with zeros.

15,721 rounded to the nearest thousand is 16,000.

Definitions and Concepts	Examples

Section 1.5   Multiplying Whole Numbers and Area

To multiply 73 and 58, for example, multiply 73 and 8, then 73 and 50. The sum of these partial products is the product of 73 and 58. Use the notation to the right.

$$
\begin{array}{rl}
73 & \leftarrow \quad \text{factor} \\
\times\ 58 & \leftarrow \quad \text{factor} \\
\hline
584 & \leftarrow \quad 73 \times 8 \\
3650 & \leftarrow \quad 73 \times 50 \\
\hline
4234 & \leftarrow \quad \text{product}
\end{array}
$$

To find the **area** of a rectangle, multiply length times width.

△ Find the area of the rectangle shown.

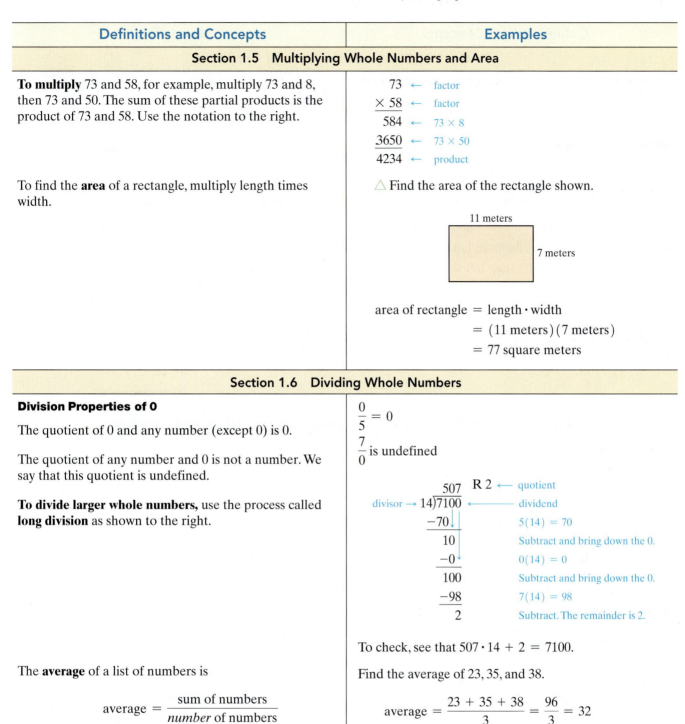

11 meters

7 meters

area of rectangle = length · width

= (11 meters)(7 meters)

= 77 square meters

Section 1.6   Dividing Whole Numbers

Division Properties of 0

The quotient of 0 and any number (except 0) is 0.

The quotient of any number and 0 is not a number. We say that this quotient is undefined.

To divide larger whole numbers, use the process called **long division** as shown to the right.

$$\frac{0}{5} = 0$$

$$\frac{7}{0} \text{ is undefined}$$

$$
\begin{array}{r}
507 \ \ \text{R } 2 \leftarrow \text{ quotient} \\
\text{divisor} \rightarrow 14{\overline{)7100}} \longleftarrow \text{ dividend} \\
-70 \qquad 5(14) = 70 \\
\hline
10 \qquad \text{Subtract and bring down the 0.} \\
-0 \qquad 0(14) = 0 \\
\hline
100 \qquad \text{Subtract and bring down the 0.} \\
-98 \qquad 7(14) = 98 \\
\hline
2 \qquad \text{Subtract. The remainder is 2.}
\end{array}
$$

To check, see that $507 \cdot 14 + 2 = 7100$.

The **average** of a list of numbers is

$$\text{average} = \frac{\text{sum of numbers}}{\textit{number} \text{ of numbers}}$$

Find the average of 23, 35, and 38.

$$\text{average} = \frac{23 + 35 + 38}{3} = \frac{96}{3} = 32$$

Definitions and Concepts	Examples

Section 1.7   Exponents and Order of Operations

An **exponent** is a shorthand notation for repeated multiplication of the same factor.

$$3^4 = \underbrace{3 \cdot 3 \cdot 3 \cdot 3}_{\text{4 factors of 3}} = 81$$

base

Order of Operations

1. Perform all operations within parentheses ( ), brackets [ ], or other grouping symbols such as fraction bars, starting with the innermost set.

2. Evaluate any expressions with exponents.

3. Multiply or divide in order from left to right.

4. Add or subtract in order from left to right.

Simplify: $\dfrac{5 + 3^2}{2(7 - 6)}$

Simplify above and below the fraction bar separately.

$$\frac{5 + 3^2}{2(7 - 6)} = \frac{5 + 9}{2(1)} \quad \begin{array}{l}\text{Evaluate } 3^2 \text{ above the fraction bar.}\\ \text{Subtract } 7 - 6 \text{ below the fraction bar.}\end{array}$$

$$= \frac{14}{2} \quad \begin{array}{l}\text{Add.}\\ \text{Multiply.}\end{array}$$

$$= 7 \quad \text{Divide.}$$

The **area of a square** is $(\text{side})^2$.

Find the area of a square with side length 9 inches.

$$\text{Area of the square} = (\text{side})^2$$
$$= (9 \text{ inches})^2$$
$$= 81 \text{ square inches}$$

Section 1.8   Introduction to Variables, Algebraic Expressions, and Equations

A letter used to represent a number is called a **variable.**

Variables:

$$x, \quad y, \quad z, \quad a, \quad b$$

A combination of operations on variables and numbers is called an **algebraic expression.**

Algebraic expressions:

$$3 + x, \quad 7y, \quad x^3 + y - 10$$

Replacing a variable in an expression by a number, and then finding the value of the expression, is called **evaluating the expression** for the variable.

Evaluate $2x + y$ for $x = 22$ and $y = 4$.

$$2x + y = 2 \cdot 22 + 4 \quad \text{Replace } x \text{ with 22 and } y \text{ with 4.}$$
$$= 44 + 4 \quad \text{Multiply.}$$
$$= 48 \quad \text{Add.}$$

A statement written in the form "expression = expression" is an **equation.**

Equations:

$$n - 8 = 12$$
$$2(20 - 7n) = 32$$
$$\text{Area} = \text{length} \cdot \text{width}$$

A **solution** of an equation is a value for the variable that makes the equation a true statement.

Determine whether 2 is a solution of the equation $4(x - 1) = 7$.

$$4(2 - 1) \stackrel{?}{=} 7 \quad \text{Replace } x \text{ with 2.}$$
$$4(1) \stackrel{?}{=} 7 \quad \text{Subtract.}$$
$$4 = 7 \quad \text{False}$$

No, 2 is not a solution.

(1.2) *Determine the place value of the digit 4 in each whole number.*

1. 7640

2. 46,200,120

Write each whole number in words.

3. 7640

4. 46,200,120

Write each whole number in expanded form.

5. 3158

6. 403,225,000

Write each whole number in standard form.

7. Eighty-one thousand, nine hundred

8. Six billion, three hundred four million

The following table shows the Internet use by world regions. Use this table to answer Exercises 9 through 12 and other exercises throughout this review. (Source: International Telecommunications Union and Internet World Stats)

Internet Use by World Regions (in millions)			
World Region	**2004**	**2008**	**2013**
Africa	21	51	140
Asia	296	579	1268
Europe	241	385	467
Middle East	29	42	141
North America	218	248	302
Latin America/ Caribbean	51	139	280
Oceania/ Australia	12	20	145

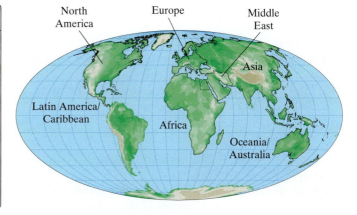

9. Find the number of Internet users in 2013 in Europe. Write your answer in standard form.

10. Find the number of Internet users in Oceania/ Australia in 2013. Write your answer in standard form.

11. Which world region had the smallest number of Internet users in 2008?

12. Which world region had the greatest number of Internet users in 2008?

(1.3) *Add or subtract as indicated.*

13. 18 + 49

14. 28 + 39

15. 462 − 397

16. 583 − 279

17. 428 + 21

18. 819 + 21

19. 4000 − 86

20. 8000 − 92

21. 91 + 3623 + 497

22. 82 + 1647 + 238

Translating *Solve.*

23. Find the sum of 74, 342, and 918.

24. Find the sum of 49, 529, and 308.

25. Subtract 7965 from 25,862.

26. Subtract 4349 from 39,007.

27. The distance from Washington, DC, to New York City is 205 miles. The distance from New York City to New Delhi, India, is 7318 miles. Find the total distance from Washington, DC, to New Delhi if traveling by air through New York City.

28. Susan Summerline earned salaries of $62,589, $65,340, and $69,770 during the years 2004, 2005, and 2006, respectively. Find her total earnings during those three years.

Find the perimeter of each figure.

△ **29.**

△ **30.**

11 kilometers   20 kilometers
          35 kilometers

Use the Internet Use by World Regions table for Exercises 31 and 32.

31. Find the increase in Internet users in Europe from 2008 to 2013.

32. Find the difference in the number of Internet users in 2013 between Oceania/Australia and the Middle East.

The following bar graph shows the monthly savings account balances for a freshman attending a local community college. Use this graph to answer Exercises 33 through 36.

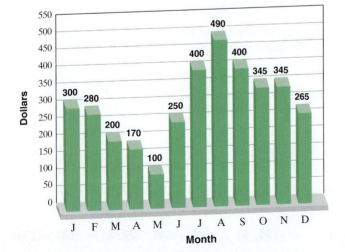

33. During what month was the balance the least?

34. During what month was the balance the greatest?

35. By how much did the balance decrease from February to April?

36. By how much did the balance increase from June to August?

(1.4) *Round to the given place.*

37. 43 to the nearest ten

38. 45 to the nearest ten

39. 876 to the nearest ten

40. 493 to the nearest hundred

41. 3829 to the nearest hundred

42. 57,534 to the nearest thousand

43. 39,583,819 to the nearest million

44. 768,542 to the nearest hundred-thousand

Estimate the sum or difference by rounding each number to the nearest hundred.

45. $3785 + 648 + 2866$

46. $5925 - 1787$

47. A group of students took a week-long driving trip and traveled 630, 192, 271, 56, 703, 454, and 329 miles on seven consecutive days. Round each distance to the nearest hundred to estimate the distance they traveled.

48. In 2012, the population of Europe was 820,918,446 while the population of Latin America/Caribbean was 593,688,638. Round each number to the nearest million, and estimate their difference in population in 2012. (*Source:* Internet World Stats)

(1.5) *Multiply.*

49.
$$\begin{array}{r} 276 \\ \times\ \ 8 \\ \hline \end{array}$$

50.
$$\begin{array}{r} 349 \\ \times\ \ 4 \\ \hline \end{array}$$

51.
$$\begin{array}{r} 57 \\ \times\ 40 \\ \hline \end{array}$$

52.
$$\begin{array}{r} 69 \\ \times\ 42 \\ \hline \end{array}$$

53. $20(7)(4)$

54. $25(9)(4)$

55. $26 \cdot 34 \cdot 0$

56. $62 \cdot 88 \cdot 0$

57.
$$\begin{array}{r} 586 \\ \times\ 29 \\ \hline \end{array}$$

58.
$$\begin{array}{r} 242 \\ \times\ 37 \\ \hline \end{array}$$

59.
$$\begin{array}{r} 642 \\ \times\ 177 \\ \hline \end{array}$$

60.
$$\begin{array}{r} 347 \\ \times\ 129 \\ \hline \end{array}$$

61.
$$\begin{array}{r} 1026 \\ \times\ 401 \\ \hline \end{array}$$

62.
$$\begin{array}{r} 2107 \\ \times\ 302 \\ \hline \end{array}$$

Translating *Solve.*

63. Find the product of 6 and 250.

64. Find the product of 6 and 820.

65. A golf pro orders shirts for the company sponsoring a local charity golfing event. Shirts size large cost $32 while shirts size extra large cost $38. If 15 large shirts and 11 extra-large shirts are ordered, find the cost.

66. The cost for a South Dakota resident to attend Black Hills State University full-time is $6112 per semester. Determine the cost for 20 students to attend full-time. (*Source:* Black Hills State University)

Find the area of each rectangle.

△ **67.** 13 miles

7 miles

△ **68.** 20 centimeters

25 centimeters

(1.6) *Divide and then check.*

69. $\dfrac{49}{7}$

70. $\dfrac{36}{9}$

71. $27 \div 5$

72. $18 \div 4$

73. $918 \div 0$

74. $0 \div 668$

75. $5\overline{)167}$

76. $8\overline{)159}$

77. $26\overline{)626}$

78. $19\overline{)680}$

79. $47\overline{)23{,}792}$

80. $53\overline{)48{,}111}$

81. $207\overline{)578{,}291}$

82. $306\overline{)615{,}732}$

Translating   *Solve.*

83. Find the quotient of 92 and 5.

84. Find the quotient of 86 and 4.

85. A box can hold 24 cans of corn. How many boxes can be filled with 648 cans of corn?

86. One mile is 1760 yards. Find how many miles there are in 22,880 yards.

87. Find the average of the numbers 76, 49, 32, and 47.

88. Find the average of the numbers 23, 85, 62, and 66.

(1.7) *Simplify.*

89. 8^2

90. 5^3

91. $5 \cdot 9^2$

92. $4 \cdot 10^2$

93. $18 \div 2 + 7$

94. $12 - 8 \div 4$

95. $\dfrac{5(6^2 - 3)}{3^2 + 2}$

96. $\dfrac{7(16 - 8)}{2^3}$

97. $48 \div 8 \cdot 2$

98. $27 \div 9 \cdot 3$

99. $2 + 3[1^5 + (20 - 17) \cdot 3] + 5 \cdot 2$

100. $21 - [2^4 - (7 - 5) - 10] + 8 \cdot 2$

101. $19 - 2(3^2 - 2^2)$

102. $16 - 2(4^2 - 3^2)$

103. $4 \cdot 5 - 2 \cdot 7$

104. $8 \cdot 7 - 3 \cdot 9$

105. $(6 - 4)^3 \cdot [10^2 \div (3 + 17)]$

106. $(7 - 5)^3 \cdot [9^2 \div (2 + 7)]$

107. $\dfrac{5 \cdot 7 - 3 \cdot 5}{2(11 - 3^2)}$

108. $\dfrac{4 \cdot 8 - 1 \cdot 11}{3(9 - 2^3)}$

Find the area of each square.

△ **109.** A square with side length of 7 meters.

△ **110.**

3 inches

(1.8) *Evaluate each expression for $x = 5$, $y = 0$, and $z = 2$.*

111. $\dfrac{2x}{z}$

112. $4x - 3$

113. $\dfrac{x + 7}{y}$

114. $\dfrac{y}{5x}$

115. $x^3 - 2z$

116. $\dfrac{7 + x}{3z}$

117. $(y + z)^2$

118. $\dfrac{100}{x} + \dfrac{y}{3}$

Translating *Translate each phrase into a variable expression. Use x to represent a number.*

119. Five subtracted from a number

120. Seven more than a number

121. Ten divided by a number

122. The product of 5 and a number

Decide whether the given number is a solution of the given equation.

123. Is 5 a solution of $n + 12 = 20 - 3$?

124. Is 23 a solution of $n - 8 = 10 + 6$?

125. Is 14 a solution of $30 = 3(n - 3)$?

126. Is 20 a solution of $5(n - 7) = 65$?

Determine which numbers in each set are solutions to the corresponding equations.

127. $7n = 77$; $\{6, 11, 20\}$

128. $n - 25 = 150$; $\{125, 145, 175\}$

129. $5(n + 4) = 90$; $\{14, 16, 26\}$

130. $3n - 8 = 28$; $\{3, 7, 15\}$

Mixed Review

Perform the indicated operations.

131. $485 - 68$

132. $729 - 47$

133. 732×3

134. 629×4

135. $374 + 29 + 698$

136. $593 + 52 + 766$

137. $13\overline{)5962}$

138. $18\overline{)4267}$

139. 1968×36

140. 5324×18

141. $2000 - 356$

142. $9000 - 519$

Round to the given place.

143. 842 to the nearest ten

144. 258,371 to the nearest hundred-thousand

Simplify.

145. $24 \div 4 \cdot 2$

146. $\dfrac{(15 + 3) \cdot (8 - 5)}{2^3 + 1}$

Solve.

147. Is 9 a solution of $5n - 6 = 40$?

148. Is 3 a solution of $2n - 6 = 5n - 15$?

149. A manufacturer of drinking glasses ships his delicate stock in special boxes that can hold 32 glasses. If 1714 glasses are manufactured, how many full boxes are filled? Are there any glasses left over?

150. A teacher orders 2 small whiteboards for $27 each and 8 boxes of dry erase pens for $4 each. What is her total bill before taxes?

Simplify.

Answers

1. Write 82,426 in words.

2. Write "four hundred two thousand, five hundred fifty" in standard form.

3. $59 + 82$ **4.** $600 - 487$ **5.** $\begin{array}{r} 496 \\ \times\ \ 30 \\ \hline \end{array}$

6. $52,896 \div 69$ **7.** $2^3 \cdot 5^2$ **8.** $98 \div 1$

9. $0 \div 49$ **10.** $62 \div 0$ **11.** $(2^4 - 5) \cdot 3$

12. $16 + 9 \div 3 \cdot 4 - 7$ **13.** $6^1 \cdot 2^3$

14. $2[(6 - 4)^2 + (22 - 19)^2] + 10$ **15.** $5698 \cdot 1000$

16. Find the average of 62, 79, 84, 90, and 95.   **17.** Round 52,369 to the nearest thousand.

Estimate each sum or difference by rounding each number to the nearest hundred.

18. $6289 + 5403 + 1957$ **19.** $4267 - 2738$

Solve.

20. Subtract 15 from 107.   **21.** Find the sum of 15 and 107.

22. Find the product of 15 and 107.   **23.** Find the quotient of 107 and 15.

1. _____

2. _____

3. _____

4. _____

5. _____

6. _____

7. _____

8. _____

9. _____

10. _____

11. _____

12. _____

13. _____

14. _____

15. _____

16. _____

17. _____

18. _____

19. _____

20. _____

21. _____

22. _____

23. _____

24. _____

25. _____

26. _____

27. _____

28. _____

29. _____

30. _____

31. _____

32. a. _____

b. _____

33. _____

34. _____

24. Twenty-nine cans of Sherwin-Williams paint cost $493. How much was each can?

25. Jo McElory is looking at two new refrigerators for her apartment. One costs $599 and the other costs $725. How much more expensive is the higher-priced one?

26. One tablespoon of white granulated sugar contains 45 calories. How many calories are in 8 tablespoons of white granulated sugar? (_Source: Home and Garden Bulletin No. 72_, U.S. Department of Agriculture)

27. A small business owner recently ordered 16 digital cameras that cost $430 each and 5 printers that cost $205 each. Find the total cost for these items.

Find the perimeter and the area of each figure.

△ **28.**

Square | 5 centimeters

△ **29.**

20 yards

Rectangle | 10 yards

30. Evaluate $5(x^3 - 2)$ for $x = 2$.

31. Evaluate $\dfrac{3x - 5}{2y}$ for $x = 7$ and $y = 8$.

32. Translate the following phrases into mathematical expressions. Use x to represent "a number."

a. The quotient of a number and 17

b. Twice a number, decreased by 20

33. Is 6 a solution of the equation $5n - 11 = 19$?

34. Determine which number in the set is a solution to the given equation.
$n + 20 = 4n - 10; \{0, 10, 20\}$

Integers and Introduction to Solving Equations

Director James Cameron made the deepest solo descent so far into the Mariana Trench in the Pacific Ocean. He reached a depth of 35,756 feet in the Deepsea Challenger, shown above. Next, may be Richard Branson?

The Krubera Cave now holds the title of deepest. In this cave, many new depth records have been set—each one deeper than the last. The latest record is 7188 feet but who knows how deeply this cave will be explored next?

Where do we explore next? Throughout this chapter, we present many applications having to do with water depths below sea level and land depths below the surface of Earth by way of mines and caves. Recently, there has been a surge of interest in exploring these depths. Although we have already reached the deepest-known part of our oceans—the Mariana Trench in the Pacific Ocean—cave exploration is a little more tricky. For example, with so many "branches" of a cave, we are never certain that it has been totally explored. New caves are being discovered and explored even as this is written. The deepest-known cave in the world, the Krubera, was not discovered until 2001 by Ukrainian cave explorers. See exercises throughout this chapter.

Thus far, we have studied whole numbers, but these numbers are not sufficient for representing many situations in real life. For example, to express 5 degrees below zero or $100 in debt, numbers less than 0 are needed. This chapter is devoted to integers, which include numbers less than 0, and operations on these numbers.

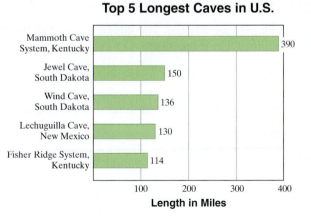

Top 5 Longest Caves in U.S.

Source: Top 10 of Everything, 2013

2.1 Introduction to Integers

Objectives

A Represent Real-Life Situations with Integers.

B Graph Integers on a Number Line.

C Compare Integers.

D Find the Absolute Value of a Number.

E Find the Opposite of a Number.

F Read Bar Graphs Containing Integers.

Objective A Representing Real-Life Situations

Thus far in this text, all numbers have been 0 or greater than 0. Numbers greater than 0 are called **positive numbers.** However, sometimes situations exist that cannot be represented by a number greater than 0. For example,

0° — 5 degrees below 0°

Sea level

20 feet below sea level

To represent these situations, we need numbers less than 0.

Extending the number line to the left of 0 allows us to picture **negative numbers,** which are numbers that are less than 0.

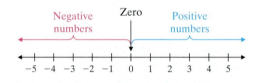

When a single + sign or no sign is in front of a number, the number is a positive number. When a single − sign is in front of a number, the number is a negative number. Together, we call positive numbers, negative numbers, and zero the **signed numbers.**

−5 indicates "negative five."

5 and +5 both indicate "positive five."

The number 0 is neither positive nor negative.

Some signed numbers are integers. The **integers** consist of the numbers labeled on the number line above. The integers are

..., −3, −2, −1, 0, 1, 2, 3, ...

Now we have numbers to represent the situations previously mentioned.

5 degrees below 0   −5°

20 feet below sea level   −20 feet

> **Helpful Hint**   Notice that 0 is neither positive nor negative.

> **Helpful Hint**
>
> A − sign, such as the one in −1, tells us that the number is to the left of 0 on the number line. −1 is read "negative one."
>
> A + sign or no sign tells us that a number lies to the right of 0 on the number line. For example, 3 and +3 both mean "positive three."

Example 1 Representing Depth with an Integer

The world's deepest cave is Krubera (or Voronja), in the country of Georgia, located by the Black Sea in Asia. It has been explored to a depth of 7188 feet below the surface of Earth. Represent this position using an integer. (*Source:* MessagetoEagle.com and Wikipedia)

Solution: If 0 represents the surface of Earth, then 7188 feet below the surface can be represented by −7188.

■ **Work Practice 1**

Objective B Graphing Integers ▶

Example 2 Graph 0, −3, 5, and −5 on the number line.

Solution:

<----●--+--●--+--+--+--+--+--+--+--●---->
      −5 −4 −3 −2 −1 0 1 2 3 4 5

■ **Work Practice 2**

Objective C Comparing Integers ▶

We can compare integers by using a number line. For any two numbers graphed on a number line, the number to the **right** is the **greater number** and the number to the **left** is the **smaller number.** Also, the symbols < and > are called **inequality symbols.**

The inequality symbol > means **"is greater than"** and
the inequality symbol < means **"is less than."**

For example, both −5 and −7 are graphed on the number line below.

<----+--●--+--●--+--+--+--+--+--+--+---->
    −8 −7 −6 −5 −4 −3 −2 −1 0 1 2

On the graph, −7 is **to the left of** −5, so −7 **is less than** −5, written as

  −7 < −5

We can also write

  −5 > −7

since −5 is **to the right of** −7, so −5 **is greater than** −7.

✔**Concept Check** Is there a largest positive number? Is there a smallest negative number? Explain.

Practice 1

a. The world's deepest bat colony spends each winter in a New York zinc mine at a depth of 3805 feet. Represent this position with an integer. (*Source: Guinness Book of World Records*)

b. The tamarack tree, a type of conifer, commonly grows at the edge of the arctic tundra and survives winter temperatures of 85 degrees below zero, Fahrenheit. Represent this temperature with an integer in degrees Fahrenheit.

Practice 2

Graph −4, −1, 2, and −2 on the number line.

<----+--+--+--+--+--+--+--+--+--+--+---->
    −5 −4 −3 −2 −1 0 1 2 3 4 5

Answers

1. a. −3805 **b.** −85°F

2. <----+--●--+--●--+--+--+--●--+--+---->
        −5 −4 −3 −2 −1 0 1 2 3 4 5

✔**Concept Check Answer**

no

Practice 3

Insert < or > between each pair of numbers to make a true statement.

a. 0 −5   **b.** −3 3

c. −7 −12

Example 3 Insert < or > between each pair of numbers to make a true statement.

a. −7 7   **b.** 0 −4   **c.** −9 −11

Solution:

a. −7 is to the left of 7 on a number line, so −7 < 7.

b. 0 is to the right of −4 on a number line, so 0 > −4.

c. −9 is to the right of −11 on a number line, so −9 > −11.

■ **Work Practice 3**

Helpful Hint

If you think of < and > as arrowheads, notice that in a true statement the arrow always points to the smaller number.

$$5 > -4 \qquad -3 < -1$$

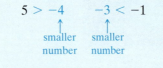

Objective D Finding the Absolute Value of a Number

The **absolute value** of a number is the number's distance from 0 on the number line. The symbol for absolute value is | |. For example, |3| is read as "the absolute value of 3."

$|3| = 3$ because 3 is 3 units from 0.

$|-3| = 3$ because −3 is 3 units from 0.

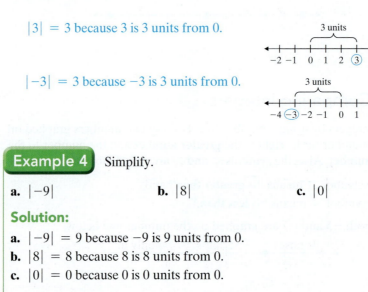

Practice 4

Simplify.

a. |−6|

b. |4|

c. |−12|

Example 4 Simplify.

a. |−9| **b.** |8| **c.** |0|

Solution:

a. |−9| = 9 because −9 is 9 units from 0.

b. |8| = 8 because 8 is 8 units from 0.

c. |0| = 0 because 0 is 0 units from 0.

■ **Work Practice 4**

Helpful Hint

Since the absolute value of a number is that number's *distance* from 0, the absolute value of a number is always 0 or positive. It is never negative.

$$|0| = 0 \qquad |-6| = 6$$

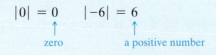

Answers

3. a. >   **b.** <   **c.** >

4. a. 6   **b.** 4   **c.** 12

Objective E Finding Opposites

Two numbers that are the same distance from 0 on the number line but are on opposite sides of 0 are called **opposites.**

4 and −4 are opposites.

4 units   4 units

$$\begin{array}{ccccccccccc} & & & & & & & & & & \\ -5 & -4 & -3 & -2 & -1 & 0 & 1 & 2 & 3 & 4 & 5 \end{array}$$

When two numbers are opposites, we say that each is the opposite of the other. Thus **4 is the opposite of −4** and **−4 is the opposite of 4.**

The phrase "the opposite of" is written in symbols as "−". For example,

The opposite of	5	is	−5
↓	↓	↓	↓
−	(5)	=	−5, or −(5) = −5

The opposite of	−3	is	3
↓	↓	↓	↓
−	(−3)	=	3 or

$$-(-3) = 3$$

In general, we have the following:

Opposites

If a is a number, then $-(-a) = a$.

Notice that because "the opposite of" is written as "−", to find the opposite of a number we place a "−" sign in front of the number.

Example 5 Find the opposite of each number.

a. 13　　　　　**b.** −2　　　　　**c.** 0

Solution:

a. The opposite of 13 is −13.

b. The opposite of −2 is −(−2) or 2.

c. The opposite of 0 is 0.

Helpful Hint Remember that 0 is neither positive nor negative.

■ **Work Practice 5**

✓**Concept Check** True or false? The number 0 is the only number that is its own opposite.

Example 6 Simplify.

a. −(−4)　　　　　**b.** −|−5|　　　　　**c.** −|6|

Solution:

a. −(−4) = 4　　The opposite of negative 4 is 4.

b. −|−5| = −5　　The opposite of the absolute value of −5 is the opposite of 5, or −5.

c. −|6| = −6　　The opposite of the absolute value of 6 is the opposite of 6, or −6.

■ **Work Practice 6**

Practice 5

Find the opposite of each number.

a. 14　　**b.** −9

Practice 6

Simplify.

a. −|−7|

b. −|4|

c. −(−12)

Answers

5. a. −14　**b.** 9

6. a. −7　**b.** −4　**c.** 12

✓**Concept Check Answer**

true

Practice 7

Evaluate $-|x|$ if $x = -6$.

> **Example 7** Evaluate $-|-x|$ if $x = -2$.
>
> **Solution:** Carefully replace x with -2; then simplify.
>
> $$-|-x| = -|-(-2)| \qquad \text{Replace } x \text{ with } -2.$$
>
> Then $-|-(-2)| = -|2| = -2$.
>
> ■ **Work Practice 7**

Objective F Reading Bar Graphs Containing Integers

The bar graph below shows the average daytime surface temperatures (in degrees Fahrenheit) of the eight planets, excluding the newly classified "dwarf planet," Pluto. Notice that a negative temperature is illustrated by a bar below the horizontal line representing 0°F, and a positive temperature is illustrated by a bar above the horizontal line representing 0°F.

Average Daytime Surface Temperatures of Planets*

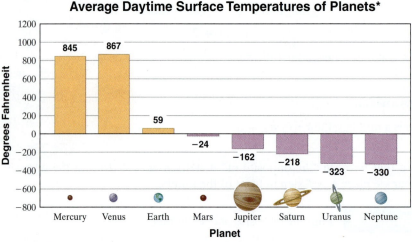

Source: *The World Almanac*, 2013
* For some planets, the temperature given is the temperature where the atmospheric pressure equals 1 Earth atmosphere.

Practice 8

Which planet has the highest average daytime surface temperature?

> **Example 8** Which planet has the lowest average daytime surface temperature?
>
> **Solution:** The planet with the lowest average daytime surface temperature is the one that corresponds to the bar that extends the farthest in the negative direction (downward). Neptune has the lowest average daytime surface temperature, $-330°F$.
>
> ■ **Work Practice 8**

Answers

7. -6 **8.** Venus

Vocabulary, Readiness & Video Check

Use the choices below to fill in each blank. Not all choices will be used.

opposites	absolute value	right	is less than
inequality symbols	negative	positive	left
signed	integers	is greater than	

1. The numbers ... $-3, -2, -1, 0, 1, 2, 3, ...$ are called _____.
2. Positive numbers, negative numbers, and zero together are called _____ numbers.
3. The symbols "<" and ">" are called _____.

4. Numbers greater than 0 are called _____ numbers while numbers less than 0 are called _____ numbers.

5. The sign "<" means _____ and ">" means _____ .

6. On a number line, the greater number is to the _____ of the lesser number.

7. A number's distance from 0 on a number line is the number's _____ .

8. The numbers −5 and 5 are called _____ .

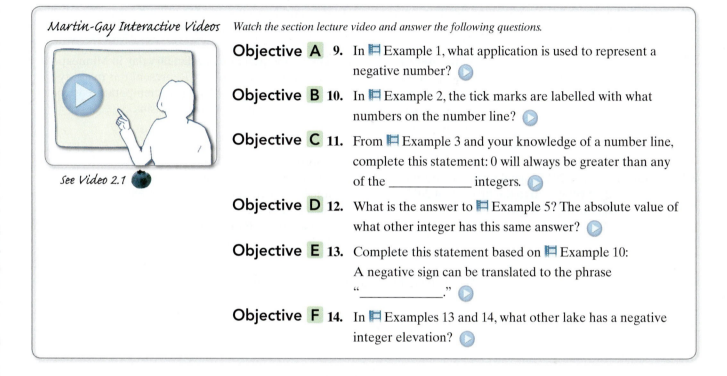

Martin-Gay Interactive Videos   Watch the section lecture video and answer the following questions.

See Video 2.1

Objective A   9. In ▤ Example 1, what application is used to represent a negative number? ▷

Objective B 10. In ▤ Example 2, the tick marks are labelled with what numbers on the number line? ▷

Objective C 11. From ▤ Example 3 and your knowledge of a number line, complete this statement: 0 will always be greater than any of the _____ integers. ▷

Objective D 12. What is the answer to ▤ Example 5? The absolute value of what other integer has this same answer? ▷

Objective E 13. Complete this statement based on ▤ Example 10: A negative sign can be translated to the phrase "_____ ." ▷

Objective F 14. In ▤ Examples 13 and 14, what other lake has a negative integer elevation? ▷

2.1 Exercise Set MyMathLab® ▷

Objective A *Represent each quantity by an integer. See Example 1.*

▷ **1.** A worker in a silver mine in Nevada works 1235 feet underground.

2. A scuba diver is swimming 25 feet below the surface of the water in the Gulf of Mexico.

3. The peak of Mount Elbert in Colorado is 14,433 feet above sea level. (*Source:* U.S. Geological Survey)

4. The lowest elevation in the United States is found at Death Valley, California, at an elevation of 282 feet below sea level. (*Source:* U.S. Geological Survey)

5. The record high temperature in Arkansas is 120 degrees above zero Fahrenheit. (*Source:* National Climatic Data Center)

6. The record high temperature in California is 134 degrees above zero Fahrenheit. (*Source:* National Climatic Data Center.)

7. The average depth of the Atlantic Ocean is 11,810 feet below its surface. (*Source: The World Almanac,* 2013)

8. The average depth of the Pacific Ocean is 14,040 feet below its surface. (*Source: The World Almanac,* 2013)

9. Sears had a loss of $3140 million for the fiscal year 2011. (*Source:* CNN Money)

10. Rite Aid had a loss of $555 million for the fiscal year 2011. (*Source:* CNN Money)

11. Two divers are exploring the wreck of the *Andrea Doria,* south of Nantucket Island, Massachusetts. Guillermo is 160 feet below the surface of the ocean and Luigi is 147 feet below the surface. Represent each quantity by an integer and determine who is deeper.

12. The temperature on one January day in Minneapolis was 10° below 0° Celsius. Represent this quantity by an integer and tell whether this temperature is cooler or warmer than 5° below 0° Celsius.

13. For the first half of 2013, digital track sales declined 2 percent when compared to the first half of 2012. (*Source:* Nielsen Sound Scan)

14. In a recent year, the number of CDs shipped to music retailers reflected a 23 percent decrease from the previous year. Write an integer to represent the percent decrease in CDs shipped. (*Source:* Recording Industry Association of America)

Objective B *Graph each integer in the list on the same number line. See Example 2.*

15. 0, 3, 4, 6

16. 7, 5, 2, 0

▶ **17.** 1, −1, 2, −2, −4

18. 3, −3, 5, −5, 6

19. 0, 1, 9, 14

20. 0, 3, 10, 11

21. 0, −2, −7, −5

22. 0, −7, 3, −6

Objective C *Insert < or > between each pair of integers to make a true statement. See Example 3.*

▶ **23.** 0   −7

24. −8   0

▶ **25.** −7   −5

26. −12   −10

27. −30   −35

28. −27   −29

29. −26   26

30. 13   −13

Objective D *Simplify. See Example 4.*

▶ **31.** $|5|$

32. $|7|$

▶ **33.** $|-8|$

34. $|-19|$

▶ **35.** $|0|$

36. $|100|$

37. $|-55|$

38. $|-10|$

Objective E Find the opposite of each integer. See Example 5.

39. 5

40. 8

41. −4

42. −6

43. 23

44. 123

45. −85

46. −13

Objectives D E Mixed Practice Simplify. See Example 6.

47. $|-7|$

48. $|-11|$

49. $-|20|$

50. $-|43|$

51. $-|-3|$

52. $-|-18|$

53. $-(-43)$

54. $-(-27)$

55. $|-15|$

56. $-(-14)$

57. $-(-33)$

58. $-|-29|$

Evaluate. See Example 7.

59. $|-x|$ if $x = -6$

60. $-|x|$ if $x = -8$

61. $-|-x|$ if $x = 2$

62. $-|-x|$ if $x = 10$

63. $|x|$ if $x = -32$

64. $|x|$ if $x = 32$

65. $-|x|$ if $x = 7$

66. $|-x|$ if $x = 1$

Insert $<$, $>$, *or* $=$ *between each pair of numbers to make a true statement. See Examples 3 through 6.*

67. -12 -6

68. -4 -17

69. $|-8|$ $|-11|$

70. $|-8|$ $|-4|$

71. $|-47|$ $-(-47)$

72. $-|17|$ $-(-17)$

73. $-|-12|$ $-(-12)$

74. $|-24|$ $-(-24)$

75. 0 -9

76. -45 0

77. $|0|$ $|-9|$

78. $|-45|$ $|0|$

79. $-|-2|$ $-|-10|$

80. $-|-8|$ $-|-4|$

81. $-(-12)$ $-(-18)$

82. -22 $-(-38)$

Objectives D E Mixed Practice Fill in the chart. See Examples 4 through 7.

	Number	Absolute Value of Number	Opposite of Number
83.	31		
85.			−28

	Number	Absolute Value of Number	Opposite of Number
84.	−13		
86.			90

Objective F The bar graph shows the elevations of selected lakes. Use this graph For Exercises 87 through 90. (*Source:* U.S. Geological Survey) See Example 8.

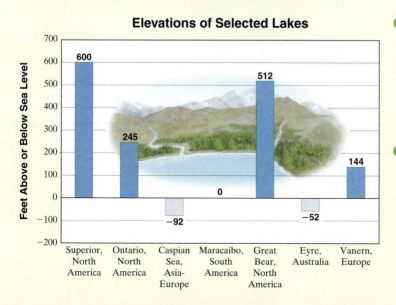

Elevations of Selected Lakes

87. Which lake shown has the lowest elevation?

88. Which lake has an elevation at sea level?

89. Which lake shown has the highest elevation?

90. Which lake shown has the second-lowest elevation?

The following bar graph represents the boiling temperature, the temperature at which a substance changes from liquid to gas at standard atmospheric pressure. Use this graph to answer Exercises 91 through 94.

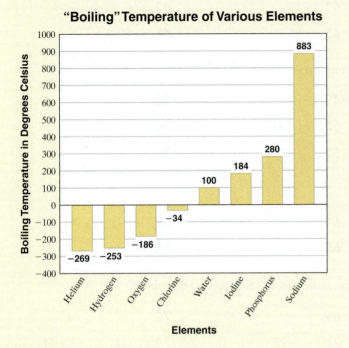

"Boiling" Temperature of Various Elements

91. Which element has a positive boiling temperature closest to that of water?

92. Which element has the lowest boiling temperature?

93. Which element has a boiling temperature closest to −200°C?

94. Which element has an average boiling temperature closest to +300°C?

Review

Add. See Section 1.3.

95. $0 + 13$

96. $9 + 0$

97. $15 + 20$

98. $20 + 15$

99. $47 + 236 + 77$

100. $362 + 37 + 90$

Concept Extensions

Write the given numbers in order from least to greatest.

101. $2^2, -|3|, -(-5), -|-8|$

102. $|10|, 2^3, -|-5|, -(-4)$

103. $|-1|, -|-6|, -(-6), -|1|$

104. $1^4, -(-3), -|7|, |-20|$

105. $-(-2), 5^2, -10, -|-9|, |-12|$

106. $3^3, -|-11|, -(-10), -4, -|2|$

Choose all numbers for x from each given list that make each statement true.

107. $|x| > 8$
   a. -9 **b.** -5 **c.** 8 **d.** -12

108. $|x| > 4$
   a. 0 **b.** -4 **c.** 5 **d.** -100

109. Evaluate: $-(-|-8|)$

110. Evaluate: $(-|-(-7)|)$

Answer true or false for Exercises 111 through 115.

111. If $a > b$, then a must be a positive number.

112. The absolute value of a number is *always* a positive number.

113. A positive number is always greater than a negative number.

114. Zero is always less than a positive number.

115. The number $-a$ is always a negative number. (*Hint:* Read "$-$" as "the opposite of.")

116. Given the number line is it true that $b < a$?

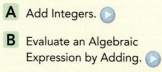

117. Write in your own words how to find the absolute value of a signed number.

118. Explain how to determine which of two signed numbers is larger.

For Exercises 119 and 120, see the first Concept Check in this section.

119. Is there a largest negative number? If so, what is it?

120. Is there a smallest positive number? If so, what is it?

2.2 Adding Integers

Objective A Adding Integers

Adding integers can be visualized using a number line. A positive number can be represented on the number line by an arrow of appropriate length pointing to the right, and a negative number by an arrow of appropriate length pointing to the left.

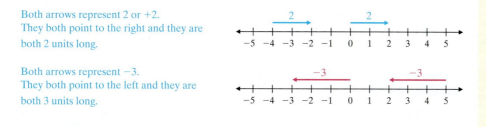

Both arrows represent 2 or +2. They both point to the right and they are both 2 units long.

Both arrows represent −3. They both point to the left and they are both 3 units long.

Objectives

A Add Integers.

B Evaluate an Algebraic Expression by Adding.

C Solve Problems by Adding Integers.

Example 1 Add using a number line: $5 + (-2)$

Solution: To add integers on a number line, such as $5 + (-2)$, we start at 0 on the number line and draw an arrow representing 5. From the tip of this arrow, we draw another arrow representing −2. The tip of the second arrow ends at their sum, 3.

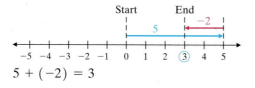

$5 + (-2) = 3$

 Work Practice 1

Practice 1

Add using a number line:
$5 + (-1)$

Answer

1.

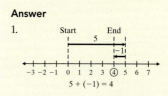

$5 + (-1) = 4$

Practice 2

Add using a number line:
$-6 + (-2)$

Practice 3

Add using a number line:
$-8 + 3$

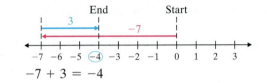

Example 2 Add using a number line: $-1 + (-4)$

Start at 0 and draw an arrow representing -1. From the tip of this arrow, we draw another arrow representing -4. The tip of the second arrow ends at their sum, -5.

Solution:

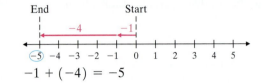

$-1 + (-4) = -5$

■ Work Practice 2

Example 3 Add using a number line: $-7 + 3$

Solution:

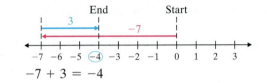

$-7 + 3 = -4$

■ Work Practice 3

Using a number line each time we add two numbers can be time consuming. Instead, we can notice patterns in the previous examples and write rules for adding signed numbers.

Rules for adding signed numbers depend on whether we are adding numbers with the same sign or different signs. When adding two numbers with the same sign, as in Example 2, notice that the sign of the sum is the same as the sign of the addends.

Adding Two Numbers with the Same Sign

Step 1: Add their absolute values.

Step 2: Use their common sign as the sign of the sum.

Practice 4

Add: $(-3) + (-19)$

Practice 5–6

Add.

5. $-12 + (-30)$ **6.** $9 + 4$

Answers

2.

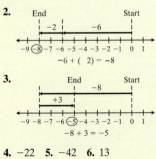

3.

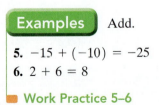

4. -22 **5.** -42 **6.** 13

Example 4 Add: $-2 + (-21)$

Solution:

Step 1: $|-2| = 2, |-21| = 21$, and $2 + 21 = 23$.

Step 2: Their common sign is negative, so the sum is negative:

$$-2 + (-21) = -23$$

■ Work Practice 4

Examples Add.

5. $-15 + (-10) = -25$
6. $2 + 6 = 8$

■ Work Practice 5–6

When adding two numbers with different signs, as in Examples 1 and 3, the sign of the result may be positive or negative, or the result may be 0.

Adding Two Numbers with Different Signs

Step 1: Find the larger absolute value minus the smaller absolute value.

Step 2: Use the sign of the number with the larger absolute value as the sign of the sum.

Example 7 Add: $-2 + 25$

Solution:

Step 1: $|-2| = 2, |25| = 25$, and $25 - 2 = 23$.

Step 2: 25 has the larger absolute value and its sign is an understood $+$:

$-2 + 25 = +23$ or 23

🟫 **Work Practice 7**

Practice 7

Add: $-1 + 26$

Example 8 Add: $3 + (-17)$

Solution:

Step 1: $|3| = 3, |-17| = 17$, and $17 - 3 = 14$.

Step 2: -17 has the larger absolute value and its sign is $-$:

$3 + (-17) = -14$

🟫 **Work Practice 8**

Practice 8

Add: $2 + (-18)$

Examples Add.

9. $-18 + 10 = -8$

10. $12 + (-8) = 4$

11. $0 + (-5) = -5$ The sum of 0 and any number is the number.

🟫 **Work Practice 9–11**

Practice 9–11

Add.

9. $-54 + 20$

10. $7 + (-2)$

11. $-3 + 0$

Recall that numbers such as 7 and -7 are called opposites. In general, the sum of a number and its opposite is always 0.

$$7 + (-7) = 0 \qquad -26 + 26 = 0 \qquad 1008 + (-1008) = 0$$

opposites   opposites   opposites

If a is a number, then

$-a$ is its opposite. Also,

$$\left. \begin{array}{l} a + (-a) = 0 \\ -a + a = 0 \end{array} \right\}$$ The sum of a number and its opposite is 0.

Examples Add.

12. $-21 + 21 = 0$

13. $36 + (-36) = 0$

🟫 **Work Practice 12–13**

Practice 12–13

Add.

12. $18 + (-18)$

23. $-64 + 64$

Answers

7. 25 8. -16 9. -34 10. 5

11. -3 12. 0 13. 0

✓ **Concept Check Answer**

$5 + (-22) = -17$

✓**Concept Check** What is wrong with the following calculation?

$5 + (-22) = 17$

In the following examples, we add three or more integers. Remember that by the associative and commutative properties for addition, we may add numbers in any order that we wish. In Examples 14 and 15, let's add the numbers from left to right.

Practice 14

Add: $6 + (-2) + (-15)$

Example 14 Add: $(-3) + 4 + (-11)$

Solution: $(-3) + 4 + (-11) = 1 + (-11)$
$$= -10$$

▪ Work Practice 14

Practice 15

Add: $5 + (-3) + 12 + (-14)$

Example 15 Add: $1 + (-10) + (-8) + 9$

Solution: $1 + (-10) + (-8) + 9 = -9 + (-8) + 9$
$$= -17 + 9$$
$$= -8$$

▪ Work Practice 15

A sum is the same if we add the numbers in any order. To see this, let's add the numbers in Example 15 by first adding the positive numbers together and the negative numbers together.

> **Helpful Hint** Don't forget that addition is commutative and associative. In other words, numbers may be added in any order.

$1 + (-10) + (-8) + 9 = 10 + (-18)$
$$= -8$$

Add the positive numbers: $1 + 9 = 10$.
Add the negative numbers: $(-10) + (-8) = -18$.
Add these results.

The sum is -8.

Objective B Evaluating Algebraic Expressions ▶

We can continue our work with algebraic expressions by evaluating expressions given integer replacement values.

Practice 16

Evaluate $x + 3y$ for $x = -6$ and $y = 2$.

Example 16 Evaluate $2x + y$ for $x = 3$ and $y = -5$.

Solution: Replace x with 3 and y with -5 in $2x + y$.

$2x + y = 2 \cdot 3 + (-5)$
$$= 6 + (-5)$$
$$= 1$$

▪ Work Practice 16

Practice 17

Evaluate $x + y$ for $x = -13$ and $y = -9$.

Example 17 Evaluate $x + y$ for $x = -2$ and $y = -10$.

Solution: $x + y = (-2) + (-10)$ Replace x with -2 and y with -10.
$$= -12$$

▪ Work Practice 17

Practice 18

If the temperature was $-7°$ Fahrenheit at 6 a.m., and it rose 4 degrees by 7 a.m. and then rose another 7 degrees in the hour from 7 a.m. to 8 a.m., what was the temperature at 8 a.m.?

Objective C Solving Problems by Adding Integers ▶

Next, we practice solving problems that require adding integers.

Example 18 Calculating Temperature

In Philadelphia, Pennsylvania, the record extreme high temperature is 104°F. Decrease this temperature by 111 degrees, and the result is the record extreme low temperature. Find this temperature. (*Source:* National Climatic Data Center)

Answers

14. -11 **15.** 0 **16.** 0 **17.** -22
18. 4°F

Solution:

In words:	extreme low temperature	=	extreme high temperature	+	decrease of of 111°
	↓		↓		↓
Translate:	extreme low temperature	=	104	+	(−111)

$$= -7$$

The record extreme low temperature in Philadelphia, Pennsylvania, is −7°F.

🟧 **Work Practice 18**

 Calculator Explorations Entering Negative Numbers

To enter a negative number on a calculator, find the key marked ⎡+/−⎤. (Some calculators have a key marked ⎡CHS⎤ and some calculators have a special key ⎡(−)⎤ for entering a negative sign.) To enter the number −2, for example, press the keys ⎡2⎤ ⎡+/−⎤. The display will read ⎡   −2⎤.

To find −32 + (−131), press the keys
⎡32⎤ ⎡+/−⎤ ⎡+⎤ ⎡131⎤ ⎡+/−⎤ ⎡=⎤ or
⎡(−)⎤ ⎡32⎤ ⎡+⎤ ⎡(−)⎤ ⎡131⎤ ⎡ENTER⎤
The display will read ⎡   −163⎤.
Thus −32 + (−131) = −163.

Use a calculator to perform each indicated operation.

1. −256 + 97

2. 811 + (−1058)

3. 6(15) + (−46)

4. −129 + 10(48)

5. −108,650 + (−786,205)

6. −196,662 + (−129,856)

Vocabulary, Readiness & Video Check

Use the choices below to fill in each blank. Not all choices will be used.

−a	a	0	commutative	associative

1. If n is a number, then $-n + n =$ _____.

2. Since $x + n = n + x$, we say that addition is _____.

3. If a is a number, then $-(-a) =$ _____.

4. Since $n + (x + a) = (n + x) + a$, we say that addition is _____.

Martin-Gay Interactive Videos   *Watch the section lecture video and answer the following questions.*

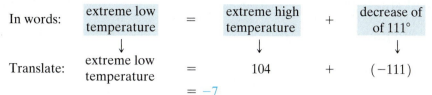

Objective A **5.** What is the sign of the sum in ⊞ Example 6 and why? ▶

Objective B **6.** What is the sign of the sum in ⊞ Example 8 and why? ▶

Objective C **7.** What does the answer to ⊞ Example 10, −231, mean in the context of the application? ▶

See Video 2.2 🔵

2.2 Exercise Set MyMathLab®

Objective A *Add using a number line. See Examples 1 through 3.*

1. $-1 + (-6)$

2. $-6 + (-5)$

3. $-4 + 7$

4. $10 + (-3)$

5. $-13 + 7$

6. $9 + (-4)$

Add. See Examples 4 through 13.

7. $46 + 21$

8. $15 + 42$

9. $-8 + (-2)$

10. $-5 + (-4)$

11. $-43 + 43$

12. $-62 + 62$

13. $6 + (-2)$

14. $8 + (-3)$

15. $-6 + 0$

16. $-8 + 0$

17. $3 + (-5)$

18. $5 + (-9)$

19. $-2 + (-7)$

20. $-6 + (-1)$

21. $-12 + (-12)$

22. $-23 + (-23)$

23. $-640 + (-200)$

24. $-400 + (-256)$

25. $12 + (-5)$

26. $24 + (-10)$

27. $-6 + 3$

28. $-8 + 4$

29. $-56 + 26$

30. $-89 + 37$

31. $-45 + 85$

32. $-32 + 62$

33. $124 + (-144)$

34. $325 + (-375)$

35. $-82 + (-43)$

36. $-56 + (-33)$

Add. See Examples 14 and 15.

37. $-4 + 2 + (-5)$

38. $-1 + 5 + (-8)$

39. $-52 + (-77) + (-117)$

40. $-103 + (-32) + (-27)$

41. $12 + (-4) + (-4) + 12$

42. $18 + (-9) + 5 + (-2)$

43. $(-10) + 14 + 25 + (-16)$

44. $34 + (-12) + (-11) + 213$

Objective A Mixed Practice *Add. See Examples 1 through 15.*

45. $-6 + (-15) + (-7)$

46. $-12 + (-3) + (-5)$

47. $-26 + 15$

48. $-35 + (-12)$

49. $5 + (-2) + 17$

50. $3 + (-23) + 6$

51. $-13 + (-21)$

52. $-100 + 70$

53. $3 + 14 + (-18)$

54. $(-45) + 22 + 20$

55. $-92 + 92$

56. $-87 + 0$

57. $-13 + 8 + (-10) + (-27)$

58. $-16 + 6 + (-14) + (-20)$

Objective B *Evaluate $x + y$ for the given replacement values. See Examples 16 and 17.*

▶ **59.** $x = -20$ and $y = -50$

60. $x = -1$ and $y = -29$

Evaluate $3x + y$ for the given replacement values. See Examples 16 and 17.

▶ **61.** $x = 2$ and $y = -3$

62. $x = 7$ and $y = -11$

63. $x = 3$ and $y = -30$

64. $x = 13$ and $y = -17$

Objective C Translating *Translate each phrase; then simplify. See Example 18.*

65. Find the sum of -6 and 25.

66. Find the sum of -30 and 15.

67. Find the sum of -31, -9, and 30.

68. Find the sum of -49, -2, and 40.

Solve. See Example 18.

▶ **69.** Suppose a deep-sea diver dives from the surface to 215 feet below the surface. He then dives down 16 more feet. Use positive and negative numbers to represent this situation. Then find the diver's present depth.

70. Suppose a diver dives from the surface to 248 meters below the surface and then swims up 8 meters, down 16 meters, down another 28 meters, and then up 32 meters. Use positive and negative numbers to represent this situation. Then find the diver's depth after these movements.

In golf, it is possible to have positive and negative scores. The following table shows the results of the eighteen-hole Round 2 for Jim Furyk and Jason Dufner at the 2013 PGA Championship in Rochester, New York. Use the table to answer Exercises 71 and 72.

Player/Hole	1	2	3	4	5	6	7	8	9	10	11	12	13	14	15	16	17	18
Furyk	−1	0	0	0	0	0	0	0	0	−1	0	0	0	0	0	−1	1	0
Dufner	0	−2	0	−1	−1	0	0	0	0	0	−1	0	−1	0	0	−1	0	0

(*Source:* Professional Golfers' Association)

71. Find the total score for each of the athletes in the round.

72. In golf, the lower score is the winner. Use the result of Exercise 71 to determine who won Round 2.

The following bar graph shows the yearly net income for Apple, Inc. Net income is one indication of a company's health. It measures revenue (money taken in) minus cost (money spent). Use this graph to answer Exercises 73 through 76. (Source: Apple, Inc.)

73. What was the net income (in dollars) for Apple, Inc. in 2012?

74. What was the net income (in dollars) for Apple, Inc. in 2001?

75. Find the total net income for the years 2011 and 2012.

76. Find the total net income for the years 2007, 2009, and 2011.

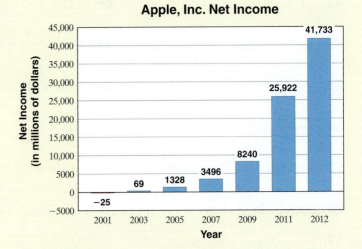

Apple, Inc. Net Income

77. The temperature at 4 p.m. on February 2 was −10° Celsius. By 11 p.m. the temperature had risen 12 degrees. Find the temperature at 11 p.m.

78. In some card games, it is possible to have both positive and negative scores. After four rounds of play, Michelle had scores of 14, −5, −8, and 7. What was her total score for the game?

A small business reports the following net incomes. Use this table to answer Exercises 79 and 80.

Year	Net Income (in dollars)
2009	−$10,412
2010	−$1786
2011	$15,395
2012	$31,418

79. Find the sum of the net incomes for 2010 and 2011.

80. Find the sum of the net incomes for all four years shown.

81. The all-time record low temperature for Texas is −23°F. Florida's all-time record low temperature is 21°F higher than Texas' record low. What is Florida's record low temperature? (*Source:* National Climatic Data Center)

82. The all-time record low temperature for California is −45°F. In Pennsylvania, the lowest temperature ever recorded is 3°F higher than California's all-time low temperature. What is the all-time record low temperature for Pennsylvania? (*Source:* National Climatic Data Center)

83. The deepest spot in the Atlantic Ocean is the Puerto Rico Trench, which has an elevation of 8605 meters below sea level. The bottom of the Atlantic's Cayman Trench has an elevation 1070 meters above the level of the Puerto Rico Trench. Use a negative number to represent the depth of the Cayman Trench. (*Source:* Defense Mapping Agency)

84. The deepest spot in the Pacific Ocean is the Mariana Trench, which has an elevation of 10,924 meters below sea level. The bottom of the Pacific's Aleutian Trench has an elevation 3245 meters higher than that of the Mariana Trench. Use a negative number to represent the depth of the Aleutian Trench. (*Source:* Defense Mapping Agency)

Review

Subtract. See Section 1.3.

85. $44 - 0$ **86.** $91 - 0$ **87.** $200 - 59$ **88.** $400 - 18$

Concept Extensions

89. Name 2 numbers whose sum is -17.

90. Name 2 numbers whose sum is -30.

Each calculation below is incorrect. Find the error and correct it. See the Concept Check in this section.

91. $7 + (-10) \overset{?}{=} 17$

92. $-4 + 14 \overset{?}{=} -18$

93. $-10 + (-12) \overset{?}{=} -120$

94. $-15 + (-17) \overset{?}{=} 32$

For Exercises 95 through 98, determine whether each statement is true or false.

95. The sum of two negative numbers is always a negative number.

96. The sum of two positive numbers is always a positive number.

97. The sum of a positive number and a negative number is always a negative number.

98. The sum of zero and a negative number is always a negative number.

99. In your own words, explain how to add two negative numbers.

100. In your own words, explain how to add a positive number and a negative number.

2.3 Subtracting Integers

In Section 2.1, we discussed the opposite of an integer.

    The opposite of 3 is -3.
    The opposite of -6 is 6.

In this section, we use opposites to subtract integers.

Objective A Subtracting Integers

To subtract integers, we will write the subtraction problem as an addition problem. To see how to do this, study the examples below.

$$10 - 4 = 6$$
$$10 + (-4) = 6$$

Since both expressions simplify to 6, this means that

$$10 - 4 = 10 + (-4) = 6$$

Also,

$$3 - 2 = 3 + (-2) = 1$$
$$15 - 1 = 15 + (-1) = 14$$

Objectives

A Subtract Integers.

B Add and Subtract Integers.

C Evaluate an Algebraic Expression by Subtracting.

D Solve Problems by Subtracting Integers.

Thus, to subtract two numbers, we add the first number to the opposite of the second number. (The opposite of a number is also known as its **additive inverse**.)

> ### Subtracting Two Numbers
> If a and b are numbers, then $a - b = a + (-b)$.

Practice 1–4

Subtract.

1. $13 - 4$

2. $-8 - 2$

3. $11 - (-15)$

4. $-9 - (-1)$

Examples Subtract.

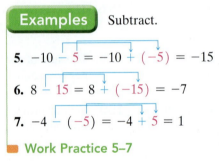

subtraction	=	first number	+	opposite of the second number		
1. $8 - 5$	=	8	+	(-5)	=	3
2. $-4 - 10$	=	-4	+	(-10)	=	-14
3. $6 - (-5)$	=	6	+	5	=	11
4. $-11 - (-7)$	=	-11	+	7	=	-4

🟧 **Work Practice 1–4**

Practice 5–7

Subtract.

5. $6 - 9$

6. $-14 - 5$

7. $-3 - (-4)$

Examples Subtract.

5. $-10 - 5 = -10 + (-5) = -15$

6. $8 - 15 = 8 + (-15) = -7$

7. $-4 - (-5) = -4 + 5 = 1$

🟧 **Work Practice 5–7**

> ### Helpful Hint
> To visualize subtraction, try the following:
>
> The difference between $5°F$ and $-2°F$ can be found by subtracting. That is,
>
> $5 - (-2) = 5 + 2 = 7$
>
> Can you visually see from the thermometer on the right that there are actually 7 degrees between $5°F$ and $-2°F$?

✔**Concept Check** What is wrong with the following calculation?

$-9 - (-5) = -14$

Practice 8

Subtract 6 from -15.

Answers

1. 9   **2.** -10 **3.** 26   **4.** -8

5. -3 **6.** -19 **7.** 1   **8.** -21

✔**Concept Check Answer**

$-9 - (-5) = -9 + 5 = -4$

Example 8 Subtract 7 from -3.

Solution: To subtract 7 *from* -3, we find

$-3 - 7 = -3 + (-7) = -10$

🟧 **Work Practice 8**

Objective B Adding and Subtracting Integers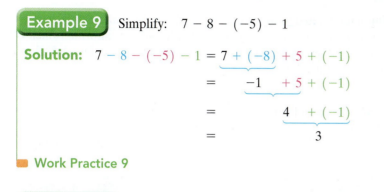

If a problem involves adding or subtracting more than two integers, we rewrite differences as sums and add. Recall that by associative and commutative properties, we may add numbers in any order. In Examples 9 and 10, we will add from left to right.

Example 9 Simplify: $7 - 8 - (-5) - 1$

Solution: $7 - 8 - (-5) - 1 = 7 + (-8) + 5 + (-1)$

$$= \quad -1 \quad + 5 + (-1)$$

$$= \quad\quad\quad 4 \quad + (-1)$$

$$= \quad\quad\quad\quad 3$$

■ Work Practice 9

Practice 9

Simplify: $-6 - 5 - 2 - (-3)$

Example 10 Simplify: $7 + (-12) - 3 - (-8)$

Solution: $7 + (-12) - 3 - (-8) = 7 + (-12) + (-3) + 8$

$$= \quad -5 \quad + (-3) + 8$$

$$= \quad\quad\quad -8 \quad\quad + 8$$

$$= \quad\quad\quad\quad 0$$

■ Work Practice 10

Practice 10

Simplify: $8 + (-2) - 9 - (-7)$

Objective C Evaluating Expressions

Now let's practice evaluating expressions when the replacement values are integers.

Example 11 Evaluate $x - y$ for $x = -3$ and $y = 9$.

Solution: Replace x with -3 and y with 9 in $x - y$.

$$\begin{array}{ccc} x & - & y \\ \downarrow & \downarrow & \downarrow \end{array}$$

$$= (-3) - \quad 9$$

$$= (-3) + (-9)$$

$$= -12$$

■ Work Practice 11

Practice 11

Evaluate $x - y$ for $x = -5$ and $y = 13$.

Example 12 Evaluate $2a - b$ for $a = 8$ and $b = -6$.

Solution: Watch your signs carefully!

$$\begin{array}{ccc} 2a & - & b \\ \downarrow & \downarrow & \downarrow \end{array}$$

$= 2 \cdot 8 - (-6)$ Replace a with 8 and b with -6.

$= 16 + 6$ Multiply.

$= 22$ Add.

Helpful Hint Watch carefully when replacing variables in the expression $2a - b$. Make sure that all symbols are inserted and accounted for.

■ Work Practice 12

Practice 12

Evaluate $3y - z$ for $y = 9$ and $z = -4$.

Answers

9. -10 **10.** 4   **11.** -18 **12.** 31

Objective D Solving Problems by Subtracting Integers ▶

Solving problems often requires subtraction of integers.

Practice 13

The highest point in Asia is the top of Mount Everest, at a height of 29,028 feet above sea level. The lowest point is the Dead Sea, which is 1312 feet below sea level. How much higher is Mount Everest than the Dead Sea? (*Source: National Geographic Society*)

Example 13   Finding a Change in Elevation

The highest point in the United States is the top of Mount McKinley, at a height of 20,320 feet above sea level. The lowest point is Death Valley, California, which is 282 feet below sea level. How much higher is Mount McKinley than Death Valley? (*Source:* U.S. Geological Survey)

Solution:

1. UNDERSTAND. Read and reread the problem. To find "how much higher," we subtract. Don't forget that since Death Valley is 282 feet *below* sea level, we represent its height by -282. Draw a diagram to help visualize the problem.

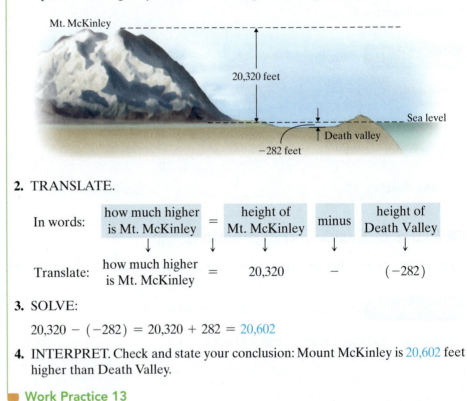

2. TRANSLATE.

In words:	how much higher is Mt. McKinley	=	height of Mt. McKinley	minus	height of Death Valley
	↓	↓	↓	↓	↓
Translate:	how much higher is Mt. McKinley	=	20,320	−	(−282)

3. SOLVE:

$$20{,}320 - (-282) = 20{,}320 + 282 = 20{,}602$$

4. INTERPRET. Check and state your conclusion: Mount McKinley is 20,602 feet higher than Death Valley.

Answer

13. 30,340 ft

■ **Work Practice 13**

Vocabulary, Readiness & Video Check

Multiple choice: Select the correct lettered response following each exercise.

1. It is true that $a - b =$ _____.
   a. $b - a$ **b.** $a + (-b)$ **c.** $a + b$

2. The opposite of n is _____.
   a. $-n$ **b.** $-(-n)$ **c.** n

3. To evaluate $x - y$ for $x = -10$ and $y = -14$, we replace x with -10 and y with -14 and evaluate _____.
   a. $10 - 14$ **b.** $-10 - 14$ **c.** $-14 - 10$ **d.** $-10 - (-14)$

4. The expression $-5 - 10$ equals _____.
   a. $5 - 10$ **b.** $5 + 10$ **c.** $-5 + (-10)$ **d.** $10 - 5$

Martin-Gay Interactive Videos Watch the section lecture video and answer the following questions.

Objective A 5. In the lecture before ▥ Example 1, what can the "opposite" of a number also be called? ▶

Objective B 6. In ▥ Example 7, how is the example rewritten in the first step of simplifying and why? ▶

Objective C 7. In ▥ Example 8, why do we multiply first? ▶

Objective D 8. What does the answer to ▥ Example 9, 265, mean in the context of the application? ▶

See Video 2.3

2.3 Exercise Set MyMathLab® ▶

Objective A *Subtract. See Examples 1 through 7.*

1. $-8 - (-8)$ **2.** $-6 - (-6)$ **3.** $19 - 16$ **4.** $15 - 12$

▶ **5.** $3 - 8$ **6.** $2 - 5$ **7.** $11 - (-11)$ **8.** $12 - (-12)$

9. $-4 - (-7)$ **10.** $-25 - (-25)$ **11.** $-16 - 4$ **12.** $-2 - 42$

13. $3 - 15$ **14.** $8 - 9$ **15.** $42 - 55$ **16.** $17 - 63$

17. $478 - (-30)$ **18.** $844 - (-20)$ ▶ **19.** $-4 - 10$ **20.** $-5 - 8$

▶ **21.** $-7 - (-3)$ **22.** $-12 - (-5)$ **23.** $17 - 29$ **24.** $16 - 45$

Translating *Translate each phrase; then simplify. See Example 8.*

▶ **25.** Subtract 17 from -25.   **26.** Subtract 10 from -22.   **27.** Find the difference of -22 and -3.

28. Find the difference of -8 and -13.   **29.** Subtract -12 from 2.   **30.** Subtract -50 from -50.

Mixed Practice (*Sections 2.2, 2.3*) *Add or subtract as indicated.*

31. $-37 + (-19)$ **32.** $-35 + (-11)$ **33.** $8 - 13$ **34.** $4 - 21$

35. $-56 - 89$ **36.** $-105 - 68$ **37.** $30 - 67$ **38.** $86 - 98$

Objective B *Simplify. See Examples 9 and 10.*

39. $8 - 3 - 2$ **40.** $8 - 4 - 1$ **41.** $13 - 5 - 7$

42. $30 - 18 - 12$ **43.** $-5 - 8 - (-12)$ **44.** $-10 - 6 - (-9)$

45. $-11 + (-6) - 14$ **46.** $-15 + (-8) - 4$ **47.** $18 - (-32) + (-6)$

48. $23 - (-17) + (-9)$ **49.** $-(-5) - 21 + (-16)$ **50.** $-(-9) - 14 + (-23)$

51. $-10 - (-12) + (-7) - 4$ **52.** $-6 - (-8) + (-12) - 7$ ◗ **53.** $-3 + 4 - (-23) - 10$

54. $5 + (-18) - (-21) - 2$

Objective **C** *Evaluate* $x - y$ *for the given replacement values. See Examples 11 and 12.*

55. $x = -4$ and $y = 7$ **56.** $x = -7$ and $y = 1$

57. $x = 8$ and $y = -23$ **58.** $x = 9$ and $y = -2$

Evaluate $2x - y$ *for the given replacement values. See Examples 11 and 12.*

◗ **59.** $x = 4$ and $y = -4$ **60.** $x = 8$ and $y = -10$

61. $x = 1$ and $y = -18$ **62.** $x = 14$ and $y = -12$

Objective **D** *Solve. See Example 13.*

The bar graph shows the monthly average temperatures in Fairbanks, Alaska. Notice that a negative temperatures is illustrated by a bar below the horizontal line representing 0°F. Use this graph to answer Exercises 63 through 66.

63. Find the difference in temperature between the months of March and February.

64. Find the difference in temperature between the months of November and December.

65. Find the difference in temperature between the two months with the lowest temperatures.

66. Find the difference in temperature between the month with the warmest temperature and the month with the coldest temperature.

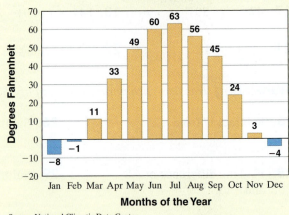

Monthly Average Temperatures in Fairbanks, AK

Source: National Climatic Data Center

Solve.

◗ **67.** The coldest temperature ever recorded on Earth was −129°F in Antarctica. The warmest temperature ever recorded was 134°F in Death Valley, California. How many degrees warmer is 134°F than −129°F? (*Source: The World Almanac,* 2013)

68. The coldest temperature ever recorded in the United States was −80°F in Alaska. The warmest temperature ever recorded was 134°F in California. How many degrees warmer is 134°F than −80°F? (*Source: The World Almanac,* 2013)

69. Adam Scott from Australia finished first in the 2013 PGA Master's Tournament at Augusta National Golf Club with a score of −9, or nine strokes under par. Tied for 46th place was Thomas Bjorn from Denmark, with a score of +5, or 5 strokes over par. What was the difference in scores between Scott and Bjorn?

70. A woman received a statement of her charge account at Old Navy. She spent $93 on purchases last month. She returned an $18 blouse because she didn't like the color. She also returned a $26 pajama set because it was damaged. What does she actually owe on her account?

71. The temperature on a February morning was −4° Celsius at 6 a.m. If the temperature drops 3 degrees by 7 a.m., rises 4 degrees between 7 a.m. and 8 a.m., and then drops 7 degrees between 8 a.m. and 9 a.m., find the temperature at 9 a.m.

72. Mauna Kea in Hawaii has an elevation of 13,796 feet above sea level. The Mid-America Trench in the Pacific Ocean has an elevation of 21,857 feet below sea level. Find the difference in elevation between those two points. (*Source:* National Geographic Society and Defense Mapping Agency)

Some places on Earth lie below sea level, which is the average level of the surface of the oceans. Use this diagram to answer Exercises 73 through 76. (Source: Fantastic Book of Comparisons, Russell Ash)

73. Find the difference in elevation between Death Valley and Qattâra Depression.

74. Find the difference in elevation between the Danakil and Turfan Depressions.

75. Find the difference in elevation between the two lowest elevations shown.

76. Find the difference in elevation between the highest elevation shown and the lowest elevation shown.

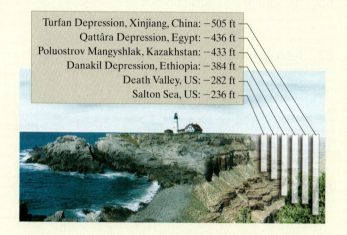

Turfan Depression, Xinjiang, China: −505 ft
Qattâra Depression, Egypt: −436 ft
Poluostrov Mangyshlak, Kazakhstan: −433 ft
Danakil Depression, Ethiopia: −384 ft
Death Valley, US: −282 ft
Salton Sea, US: −236 ft

The bar graph from Section 2.1 shows heights of selected lakes. For Exercises 77 through 80, find the difference in elevation for the lakes listed. (Source: U.S. Geological Survey)

77. Lake Superior and Lake Eyre

78. Great Bear Lake and Caspian Sea

79. Lake Maracaibo and Lake Vanern

80. Lake Eyre and Caspian Sea

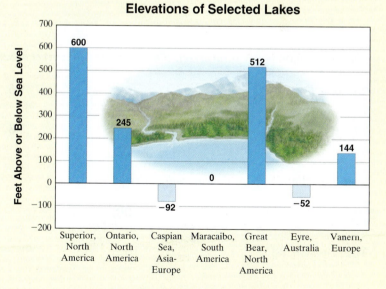

Elevations of Selected Lakes

Feet Above or Below Sea Level

600 — Superior, North America
245 — Ontario, North America
−92 — Caspian Sea, Asia-Europe
0 — Maracaibo, South America
512 — Great Bear, North America
−52 — Eyre, Australia
144 — Vanern, Europe

Solve.

81. The average daytime surface temperature of the hottest planet, Venus, is 867°F, while the average daytime surface temperature of the coldest planet, Neptune, is −330°F. Find the difference in temperatures.

82. The average daytime surface temperature of Mercury is 845°F, while the average daytime surface temperature of Jupiter is −162°F. Find the difference in temperatures.

83. The difference between a country's exports and imports is called the country's *trade balance*. In June 2013, the United States had $191 billion in exports and $225 billion in imports. What was the U.S. trade balance in June 2013? (*Source:* U.S. Department of Commerce)

84. In 2012, the United States exported 1165 million barrels of petroleum products and imported 3878 million barrels of petroleum products. What was the U.S. trade balance for petroleum products in 2012? (*Source:* U.S. Energy Information Administration)

Mixed Practice–Translating (Sections 2.2, 2.3) *Translate each phrase to an algebraic expression. Use "x" to represent "a number."*

85. The sum of −5 and a number.

86. The difference of −3 and a number.

87. Subtract a number from −20.

88. Add a number and −36.

Review

Multiply or divide as indicated. See Sections 1.5 and 1.6.

89. $\dfrac{100}{20}$

90. $\dfrac{96}{3}$

91. $\begin{array}{r} 23 \\ \times\ 46 \\ \hline \end{array}$

92. $\begin{array}{r} 51 \\ \times\ 89 \\ \hline \end{array}$

Concept Extensions

93. Name two numbers whose difference is −3.

94. Name two numbers whose difference is −10.

*Each calculation below is **incorrect**. Find the error and correct it. See the Concept Check in this section.*

95. $9 - (-7) \overset{?}{=} 2$

96. $-4 - 8 \overset{?}{=} 4$

97. $10 - 30 \overset{?}{=} 20$

98. $-3 - (-10) \overset{?}{=} -13$

Simplify. (Hint: Find the absolute values first.)

▶ **99.** $|-3| - |-7|$

100. $|-12| - |-5|$

101. $|-5| - |5|$

102. $|-8| - |8|$

103. $|-15| - |-29|$

104. $|-23| - |-42|$

For Exercises 105 and 106, determine whether each statement is true or false.

105. $|-8 - 3| = 8 - 3$

106. $|-2 - (-6)| = |-2| - |-6|$

107. In your own words, explain how to subtract one signed number from another.

108. A student explains to you that the first step to simplify $8 + 12 \cdot 5 - 100$ is to add 8 and 12. Is the student correct? Explain why or why not.

2.4 Multiplying and Dividing Integers

Multiplying and dividing integers is similar to multiplying and dividing whole numbers. One difference is that we need to determine whether the result is a positive number or a negative number.

Objective A Multiplying Integers

Consider the following pattern of products.

First factor decreases by 1 each time.

$$3 \cdot 2 = 6$$
$$2 \cdot 2 = 4$$
$$1 \cdot 2 = 2$$
$$0 \cdot 2 = 0$$

Product decreases by 2 each time.

This pattern can be continued, as follows.

$$-1 \cdot 2 = -2$$
$$-2 \cdot 2 = -4$$
$$-3 \cdot 2 = -6$$

This suggests that the product of a negative number and a positive number is a negative number.

What is the sign of the product of two negative numbers? To find out, we form another pattern of products. Again, we decrease the first factor by 1 each time, but this time the second factor is negative.

$$2 \cdot (-3) = -6$$
$$1 \cdot (-3) = -3$$
$$0 \cdot (-3) = 0$$

Product increases by 3 each time.

This pattern continues as:

$$-1 \cdot (-3) = 3$$
$$-2 \cdot (-3) = 6$$
$$-3 \cdot (-3) = 9$$

This suggests that the product of two negative numbers is a positive number. Thus we can determine the sign of a product when we know the signs of the factors.

Multiplying Numbers

The product of two numbers having the same sign is a positive number.

The product of two numbers having different signs is a negative number.

Product of Like Signs

$$(+)(+) = +$$
$$(-)(-) = +$$

Product of Different Signs

$$(-)(+) = -$$
$$(+)(-) = -$$

Examples Multiply.

1. $-7 \cdot 3 = -21$

2. $-3(-5) = 15$

3. $0 \cdot (-4) = 0$

4. $10(-8) = -80$

■ **Work Practice 1–4**

Practice 1–4

Multiply.

1. $-3 \cdot 8$ **2.** $-5(-2)$
3. $0 \cdot (-20)$ **4.** $10(-5)$

Answers

1. -24 **2.** 10 **3.** 0 **4.** -50

Recall that by the associative and commutative properties for multiplication, we may multiply numbers in any order that we wish. In Example 5, we multiply from left to right.

Practice 5–7

Multiply.

5. $8(-6)(-2)$

6. $(-9)(-2)(-1)$

7. $(-3)(-4)(-5)(-1)$

Examples   Multiply.

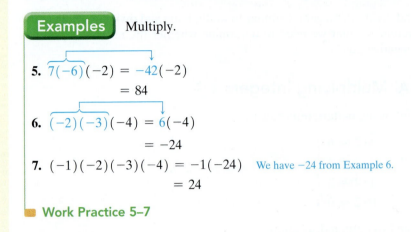

5. $7(-6)(-2) = -42(-2)$
   $= 84$

6. $(-2)(-3)(-4) = 6(-4)$
   $= -24$

7. $(-1)(-2)(-3)(-4) = -1(-24)$ We have -24 from Example 6.
   $= 24$

Work Practice 5–7

✓**Concept Check** What is the sign of the product of five negative numbers? Explain.

Recall from our study of exponents that $2^3 = 2 \cdot 2 \cdot 2 = 8$. We can now work with bases that are negative numbers. For example,

$$(-2)^3 = (-2)(-2)(-2) = -8$$

Practice 8

Evaluate $(-2)^4$.

Example 8 Evaluate: $(-5)^2$

Solution: Remember that $(-5)^2$ means 2 factors of -5.

$$(-5)^2 = (-5)(-5) = 25$$

Work Practice 8

Helpful Hint

Have you noticed a pattern when multiplying signed numbers?

If we let $(-)$ represent a negative number and $(+)$ represent a positive number, then

The product of an even number of negative numbers is a positive result.

$$(-)(-) = (+)$$
$$(-)(-)(-) = (-)$$
$$(-)(-)(-)(-) = (+)$$
$$(-)(-)(-)(-)(-) = (-)$$

The product of an odd number of negative numbers is a negative result.

Notice in Example 8 the parentheses around -5 in $(-5)^2$. With these parentheses, -5 is the base that is squared. Without parentheses, such as -5^2, only the 5 is squared. In other words, $-5^2 = -(5 \cdot 5) = -25$.

Practice 9

Evaluate: -8^2

Example 9 Evaluate: -7^2

Solution: Remember that without parentheses, only the 7 is squared.

$$-7^2 = -(7 \cdot 7) = -49$$

Work Practice 9

Answers

5. 96 **6.** -18 **7.** 60 **8.** 16 **9.** -64

✓**Concept Check Answer**

negative; answers may vary

Helpful Hint

Make sure you understand the difference between Examples 8 and 9.

$$\overbrace{(-5)^2}^{\text{parentheses, so }-5\text{ is squared}} = (-5)(-5) = 25$$

$$\overbrace{-7^2}^{\text{no parentheses, so only the 7 is squared}} = -(7\cdot7) = -49$$

Objective B Dividing Integers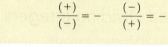

Division of integers is related to multiplication of integers. The sign rules for division can be discovered by writing a related multiplication problem. For example,

$$\frac{6}{2} = 3 \qquad \text{because } 3\cdot2 = 6$$

$$\frac{-6}{2} = -3 \qquad \text{because } -3\cdot2 = -6$$

$$\frac{6}{-2} = -3 \qquad \text{because } -3\cdot(-2) = 6$$

$$\frac{-6}{-2} = 3 \qquad \text{because } 3\cdot(-2) = -6$$

Helpful Hint Just as for whole numbers, division can be checked by multiplication.

Dividing Numbers

The quotient of two numbers having the same sign is a positive number.

The quotient of two numbers having different signs is a negative number.

Quotient of Like Signs

$$\frac{(+)}{(+)} = + \qquad \frac{(-)}{(-)} = +$$

Quotient of Different Signs

$$\frac{(+)}{(-)} = - \qquad \frac{(-)}{(+)} = -$$

Examples Divide.

10. $\dfrac{-12}{6} = -2$

11. $-20 \div (-4) = 5$

12. $\dfrac{48}{-3} = -16$

■ **Work Practice 10–12**

✔**Concept Check** What is wrong with the following calculation?

$$\frac{-36}{-9} = -4$$

Practice 10–12

Divide.

10. $\dfrac{42}{-7}$

11. $-16 \div (-2)$

12. $\dfrac{-80}{10}$

Answers

10. -6 **11.** 8 **12.** -8

✔**Concept Check Answer**

$$\frac{-36}{-9} = 4$$

Practice 13–14

Divide, if possible.

13. $\dfrac{-6}{0}$ **14.** $\dfrac{0}{-7}$

Examples   Divide, if possible.

13. $\dfrac{0}{-5} = 0$ because $0 \cdot -5 = 0$

14. $\dfrac{-7}{0}$ is undefined because there is no number that gives a product of -7 when multiplied by 0.

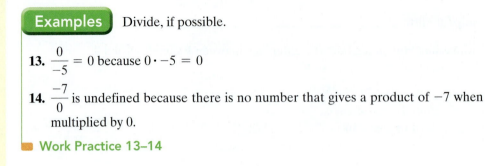 **Work Practice 13–14**

Objective C Evaluating Expressions

Next, we practice evaluating expressions given integer replacement values.

Practice 15

Evaluate xy for $x = 5$ and $y = -8$.

Example 15   Evaluate xy for $x = -2$ and $y = 7$.

Solution: Recall that xy means $x \cdot y$.

Replace x with -2 and y with 7.

$$xy = -2 \cdot 7$$
$$= -14$$

Work Practice 15

Practice 16

Evaluate $\dfrac{x}{y}$ for $x = -12$ and $y = -3$.

Example 16   Evaluate $\dfrac{x}{y}$ for $x = -24$ and $y = 6$.

Solution: $\dfrac{x}{y} = \dfrac{-24}{6}$ Replace x with -24 and y with 6.

$$= -4$$

Work Practice 16

Objective D Solving Problems by Multiplying and Dividing Integers

Many real-life problems involve multiplication and division of signed numbers.

Practice 17

A card player had a score of -13 for each of four games. Find the total score.

Example 17   Calculating a Total Golf Score

A professional golfer finished seven strokes under par (-7) for each of three days of a tournament. What was his total score for the tournament?

Solution:

1. UNDERSTAND. Read and reread the problem. Although the key word is "total," since this is repeated addition of the same number, we multiply.

2. TRANSLATE.

In words:	golfer's total score	=	number of days	·	score each day
	↓	↓	↓	↓	↓
Translate:	golfer's total	=	3	·	(-7)

3. SOLVE: $3 \cdot (-7) = -21$

4. INTERPRET. Check and state your conclusion: The golfer's total score was -21, or 21 strokes under par.

Work Practice 17

Answers

13. undefined   **14.** 0   **15.** -40
16. 4   **17.** -52

Vocabulary, Readiness & Video Check

Use the choices below to fill in each blank. Each choice may be used more than once.

negative   0
positive   undefined

1. The product of a negative number and a positive number is a(n) _____ number.
2. The product of two negative numbers is a(n) _____ number.
3. The quotient of two negative numbers is a(n) _____ number.
4. The quotient of a negative number and a positive number is a(n) _____ number.
5. The product of a negative number and zero is _____.
6. The quotient of 0 and a negative number is _____.
7. The quotient of a negative number and 0 is _____.

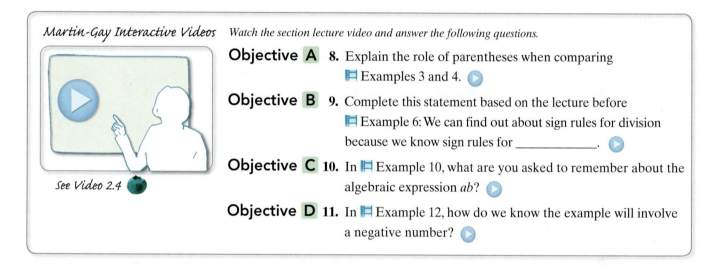

Martin-Gay Interactive Videos   Watch the section lecture video and answer the following questions.

Objective A 8. Explain the role of parentheses when comparing Examples 3 and 4.

Objective B 9. Complete this statement based on the lecture before Example 6: We can find out about sign rules for division because we know sign rules for _____.

Objective C 10. In Example 10, what are you asked to remember about the algebraic expression *ab*?

Objective D 11. In Example 12, how do we know the example will involve a negative number?

See Video 2.4

2.4 Exercise Set MyMathLab®

Objective A *Multiply. See Examples 1 through 4.*

1. $-6(-2)$ 2. $5(-3)$ 3. $-4(9)$ 4. $-7(-2)$

5. $9(-9)$ 6. $-9(7)$ 7. $0(-11)$ 8. $-6(0)$

Multiply. See Examples 5 through 7.

9. $6(-2)(-4)$ 10. $-2(3)(-7)$ 11. $-1(-3)(-4)$ 12. $-8(-3)(-3)$

13. $-4(4)(-5)$ 14. $2(-5)(-4)$ 15. $10(-5)(0)(-7)$ 16. $3(0)(-4)(-8)$

17. $-5(3)(-1)(-1)$ 18. $-2(-1)(3)(-2)$

Evaluate. See Examples 8 and 9.

19. -3^2

20. -2^4

21. $(-3)^3$

22. $(-1)^4$

23. -6^2

24. -4^3

25. $(-4)^3$

26. $(-3)^2$

Objective **B** *Find each quotient. See Examples 10 through 14.*

27. $-24 \div 3$

28. $90 \div (-9)$

29. $\dfrac{-30}{6}$

30. $\dfrac{56}{-8}$

31. $\dfrac{-77}{-11}$

32. $\dfrac{-32}{4}$

33. $\dfrac{0}{-21}$

34. $\dfrac{-13}{0}$

35. $\dfrac{-10}{0}$

36. $\dfrac{0}{-15}$

37. $\dfrac{56}{-4}$

38. $\dfrac{-24}{-12}$

Objectives **A** **B** **Mixed Practice** *Multiply or divide as indicated. See Examples 1 through 14.*

39. $-14(0)$

40. $0(-100)$

41. $-5(3)$

42. $-6 \cdot 2$

43. $-9 \cdot 7$

44. $-12(13)$

45. $-7(-6)$

46. $-9(-5)$

47. $-3(-4)(-2)$

48. $-7(-5)(-3)$

49. $(-7)^2$

50. $(-5)^2$

51. $-\dfrac{25}{5}$

52. $-\dfrac{30}{5}$

53. $-\dfrac{72}{8}$

54. $-\dfrac{49}{7}$

55. $-18 \div 3$

56. $-15 \div 3$

57. $4(-10)(-3)$

58. $6(-5)(-2)$

59. $-30(6)(-2)(-3)$

60. $-20 \cdot 5 \cdot (-5) \cdot (-3)$

61. $\dfrac{-25}{0}$

62. $\dfrac{0}{-14}$

63. $\dfrac{120}{-20}$

64. $\dfrac{63}{-9}$

65. $280 \div (-40)$

66. $480 \div (-8)$

67. $\dfrac{-12}{-4}$

68. $\dfrac{-36}{-3}$

69. -1^4

70. -2^3

71. $(-2)^5$

72. $(-11)^2$

73. $-2(3)(5)(-6)$

74. $-1(2)(7)(-3)$

75. $(-1)^{32}$

76. $(-1)^{33}$

77. $-2(-3)(-5)$

78. $-2(-2)(-3)(-2)$

79. $-48 \cdot 23$

80. $-56 \cdot 43$

81. $35 \cdot (-82)$

82. $70 \cdot (-23)$

Objective C *Evaluate ab for the given replacement values. See Example 15.*

83. $a = -8$ and $b = 7$ **84.** $a = 5$ and $b = -1$ ▶ **85.** $a = 9$ and $b = -2$

86. $a = -8$ and $b = 8$ **87.** $a = -7$ and $b = -5$ **88.** $a = -9$ and $b = -6$

Evaluate $\dfrac{x}{y}$ for the given replacement values. See Example 16.

▶ **89.** $x = 5$ and $y = -5$ **90.** $x = 9$ and $y = -3$ **91.** $x = -15$ and $y = 0$

92. $x = 0$ and $y = -5$ **93.** $x = -36$ and $y = -6$ **94.** $x = -10$ and $y = -10$

Evaluate xy and also $\dfrac{x}{y}$ for the given replacement values. See Examples 15 and 16.

95. $x = -8$ and $y = -2$ **96.** $x = 20$ and $y = -5$ **97.** $x = 0$ and $y = -8$ **98.** $x = -3$ and $y = 0$

Objective D **Translating** *Translate each phrase; then simplify. See Example 17.*

99. Find the quotient of -54 and 9.   **100.** Find the quotient of -63 and -3.

101. Find the product of -42 and -6.   **102.** Find the product of -49 and 5.

Translating *Translate each phrase to an expression. Use x to represent "a number." See Example 17.*

103. The product of -71 and a number   **104.** The quotient of -8 and a number

105. Subtract a number from -16.   **106.** The sum of a number and -12

107. -29 increased by a number   **108.** The difference of a number and -10

109. Divide a number by -33.   **110.** Multiply a number by -17.

Solve. See Example 17.

▶ **111.** A football team lost four yards on each of three consecutive plays. Represent the total loss as a product of signed numbers and find the total loss.

112. An investor lost $400 on each of seven consecutive days in the stock market. Represent his total loss as a product of signed numbers and find his total loss.

113. A deep-sea diver must move up or down in the water in short steps in order to keep from getting a physical condition called the "bends." Suppose a diver moves down from the surface in five steps of 20 feet each. Represent his total movement as a product of signed numbers and find the product.

114. A weather forecaster predicts that the temperature will drop five degrees each hour for the next six hours. Represent this drop as a product of signed numbers and find the total drop in temperature.

The graph shows melting points in degrees Celsius of selected elements. Use this graph to answer Exercises 115 through 118.

115. The melting point of nitrogen is 3 times the melting point of radon. Find the melting point of nitrogen.

116. The melting point of rubidium is −1 times the melting point of mercury. Find the melting point of rubidium.

117. The melting point of argon is −3 times the melting point of potassium. Find the melting point of argon.

118. The melting point of strontium is −11 times the melting point of radon. Find the melting point of strontium.

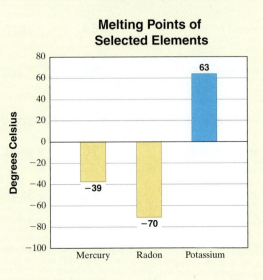

Melting Points of Selected Elements

Solve. See Example 17.

119. For the first quarter of 2013, Wal-Mart, Inc. posted a loss of $33 million in membership and other income. If this trend was consistent for each month of the quarter, how much would you expect this loss to have been for each month? (*Source*: Wal-Mart Stores, Inc.)

120. For the first quarter of 2013, Chrysler Group LLC, maker of Jeep vehicles, posted a loss of about 30,000 Jeep Liberty shipments because they had stopped producing the vehicle in 2012. If this trend was consistent for each month of the quarter, how much would you expect this loss to have been for each month? (*Source*: Chrysler Group, LLC)

121. In 2008, there were 33,319 analog (nondigital) U.S. movie screens. In 2012, this number of screens dropped to 6387. (*Source: Motion Picture Association: Worldwide Market Research*)

   a. Find the change in the number of U.S. analog movie screens from 2008 to 2012.

   b. Find the average change per year in the number of analog movie screens over this period.

122. In 1987, the California Condor was all but extinct in the wild, with about 30 condors in the world. The condors in the wild were captured by the U.S. Fish and Wildlife Service in an aggressive move to rebuild the population by breeding them in captivity and releasing the chicks in to the wild. The condor population increased to approximately 405 birds in 2012. (*Source: Arizona Game and Fish Department*)

   a. Find the change in the number of California Condors from 1987 to 2012.

   b. Find the average change per year in the California Condor population over the period in part **a.**

California Condor: Average life span, 60 years

Review

Perform each indicated operation. See Section 1.7.

123. $90 + 12^2 - 5^3$ $\qquad$ **124.** $3 \cdot (7 - 4) + 2 \cdot 5^2$ $\qquad$ **125.** $12 \div 4 - 2 + 7$ $\qquad$ **126.** $12 \div (4 - 2) + 7$

Concept Extensions

Mixed Practice (Sections 2.2, 2.3, 2.4) *Perform the indicated operations.*

127. $-57 \div 3$ $\qquad$ **128.** $-9(-11)$ $\qquad$ **129.** $-8 - 20$

130. $-4 + (-3) + 21$ $\qquad$ **131.** $-4 - 15 - (-11)$ $\qquad$ **132.** $-16 - (-2)$

Solve. For Exercises 133 and 134, see the first Concept Check in this section.

133. What is the sign of the product of seven negative numbers?

134. What is the sign of the product of ten negative numbers?

Without actually finding the product, write the list of numbers in Exercises 135 and 136 in order from least to greatest. For help, see a helpful hint box in this section.

135. $(-2)^{12}, (-2)^{17}, (-5)^{12}, (-5)^{17}$

136. $(-1)^{50}, (-1)^{55}, 0^{15}, (-7)^{20}, (-7)^{23}$

137. In your own words, explain how to divide two integers.

138. In your own words, explain how to multiply two integers.

Integers

Answers

1. _____

2. _____

3. _____

4. _____

5. _____

6. _____

7. _____

8. _____

9. _____

10. _____

11. _____

12. _____

13. _____

14. _____

15. _____

16. _____

17. _____

18. _____

19. _____

20. _____

21. _____

22. _____

1. The record low temperature in New Mexico is 50 degrees Fahrenheit below zero. The highest temperature in that state is 122 degrees above zero. Represent each quantity by an integer.

2. Graph the signed numbers on the given number line. $-4, 0, -1, 3$

New Mexico

Insert $<$ or $>$ between each pair of numbers to make a true statement.

3. 0 -10 **4.** -4 4   **5.** -15 -5 **6.** -2 -7

Simplify.

7. $|-3|$ **8.** $|-9|$ **9.** $-|-4|$ **10.** $-(-5)$

Find the opposite of each number.

11. 11   **12.** -3 **13.** 64   **14.** 0

Perform the indicated operation.

15. $-3 + 15$ **16.** $-9 + (-11)$ **17.** $-8(-6)(-1)$ **18.** $-18 \div 2$

19. $65 + (-55)$ **20.** $1000 - 1002$ **21.** $53 - (-53)$ **22.** $-2 - 1$

23. $\dfrac{0}{-47}$ **24.** $\dfrac{-36}{-9}$

25. $-17 - (-59)$ **26.** $-8 + (-6) + 20$

27. $\dfrac{-95}{-5}$ **28.** $-9(100)$

29. $-12 - 6 - (-6)$ **30.** $-4 + (-8) - 16 - (-9)$

31. $\dfrac{-105}{0}$ **32.** $7(-16)(0)(-3)$

Translating *Translate each phrase; then simplify.*

33. Subtract -8 from -12.   **34.** Find the sum of -17 and -27.

35. Find the product of -5 and -25.

36. Find the quotient of -100 and -5.

Translating *Translate each phrase to an expression. Use x to represent "a number."*

37. Divide a number by -17 **38.** The sum of -3 and a number

39. A number decreased by -18 **40.** The product of -7 and a number

Evaluate the expressions below for $x = -3$ and $y = 12$.

41. $x + y$ **42.** $x - y$

43. $2y - x$ **44.** $3y + x$

45. $5x$ **46.** $\dfrac{y}{x}$

23. _____

24. _____

25. _____

26. _____

27. _____

28. _____

29. _____

30. _____

31. _____

32. _____

33. _____

34. _____

35. _____

36. _____

37. _____

38. _____

39. _____

40. _____

41. _____

42. _____

43. _____

44. _____

45. _____

46. _____

Objectives

A Simplify Expressions by Using the Order of Operations.

B Evaluate an Algebraic Expression.

C Find the Average of a List of Numbers.

Objective A Simplifying Expressions

We first discussed the order of operations in Chapter 1. In this section, you are given an opportunity to practice using the order of operations when expressions contain signed numbers. The rules for the order of operations from Section 1.7 are repeated here.

Order of Operations

1. Perform all operations within parentheses ( ), brackets [ ], or other grouping symbols such as fraction bars, starting with the innermost set.
2. Evaluate any expressions with exponents.
3. Multiply or divide in order from left to right.
4. Add or subtract in order from left to right.

Before simplifying other expressions, make sure you are confident simplifying Examples 1 through 3.

Practice 1-3

Find the value of each expression.

1. $(-2)^4$
2. -2^4
3. $3 \cdot 6^2$

Examples Find the value of each expression.

1. $(-3)^2 = (-3)(-3) = 9$ The base of the exponent is -3.
2. $-3^2 = -(3)(3) = -9$ The base of the exponent is 3.
3. $2 \cdot 5^2 = 2 \cdot (5 \cdot 5) = 2 \cdot 25 = 50$ The base of the exponent is 5.

■ Work Practice 1–3

Helpful Hint

When simplifying expressions with exponents, remember that parentheses make an important difference.

$(-3)^2$ and -3^2 **do not** mean the same thing.

$(-3)^2$ means $(-3)(-3) = 9$.

-3^2 means the opposite of $3 \cdot 3$, or -9.

Only with parentheses around it is the -3 squared.

Practice 4

Simplify: $\dfrac{-25}{5(-1)}$

Example 4 Simplify: $\dfrac{-6(2)}{-3}$

Solution: First we multiply -6 and 2. Then we divide.

$$\frac{-6(2)}{-3} = \frac{-12}{-3}$$
$$= 4$$

■ Work Practice 4

Answers

1. 16   **2.** −16   **3.** 108   **4.** 5

134

Example 5 Simplify: $\dfrac{12 - 16}{-1 + 3}$

Solution: We simplify above and below the fraction bar separately. Then we divide.

$$\frac{12 - 16}{-1 + 3} = \frac{-4}{2}$$
$$= -2$$

■ Work Practice 5

Practice 5

Simplify: $\dfrac{-18 + 6}{-3 - 1}$

Example 6 Simplify: $60 + 30 + (-2)^3$

Solution: $60 + 30 + (-2)^3 = 60 + 30 + (-8)$ Write $(-2)^3$ as -8.
$$= 90 + (-8)$$ Add from left to right.
$$= 82$$

■ Work Practice 6

Practice 6

Simplify: $30 + 50 + (-4)^3$

Example 7 Simplify: $-4^2 + (-3)^2 - 1^3$

Solution:

$$-4^2 + (-3)^2 - 1^3 = -16 + 9 - 1$$ Simplify expressions with exponents.
$$= -7 - 1$$ Add or subtract from left to right.
$$= -8$$

■ Work Practice 7

Practice 7

Simplify: $-2^3 + (-4)^2 + 1^5$

Example 8 Simplify: $3(4 - 7) + (-2) - 5$

Solution:

$$3(4 - 7) + (-2) - 5 = 3(-3) + (-2) - 5$$ Simplify inside parentheses.
$$= -9 + (-2) - 5$$ Multiply.
$$= -11 - 5$$ Add or subtract from left to right.
$$= -16$$

■ Work Practice 8

Practice 8

Simplify:
$2(2 - 9) + (-12) - 3$

Example 9 Simplify: $(-3) \cdot |-5| - (-2) + 4^2$

Solution:

$$(-3) \cdot |-5| - (-2) + 4^2 = (-3) \cdot 5 - (-2) + 4^2$$ Write $|-5|$ as 5.
$$= (-3) \cdot 5 - (-2) + 16$$ Write 4^2 as 16.
$$= -15 - (-2) + 16$$ Multiply.
$$= -13 + 16$$ Add or subtract from left to right.
$$= 3$$

■ Work Practice 9

Practice 9

Simplify:
$(-5) \cdot |-8| + (-3) + 2^3$

Answers

5. 3 **6.** 16 **7.** 9 **8.** −29 **9.** −35

Practice 10

Simplify:
$-4[-6 + 5(-3 + 5)] - 7$

Example 10   Simplify: $-2[-3 + 2(-1 + 6)] - 5$

Solution: Here we begin with the innermost set of parentheses.

$$-2[-3 + 2(-1 + 6)] - 5 = -2[-3 + 2(5)] - 5 \quad \text{Write } -1 + 6 \text{ as 5.}$$
$$= -2[-3 + 10] - 5 \quad \text{Multiply.}$$
$$= -2(7) - 5 \quad \text{Add.}$$
$$= -14 - 5 \quad \text{Multiply.}$$
$$= -19 \quad \text{Subtract.}$$

Work Practice 10

✓**Concept Check** True or false? Explain your answer. The result of

$$-4(3 - 7) - 8(9 - 6)$$

is positive because there are four negative signs.

Objective B Evaluating Expressions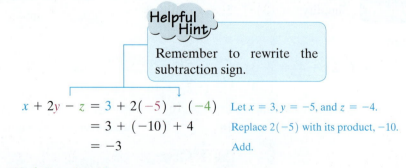

Now we practice evaluating expressions.

Practice 11

Evaluate x^2 and $-x^2$ for $x = -15$.

Example 11   Evaluate x^2 and $-x^2$ for $x = -11$.

Solution: $x^2 = (-11)^2 = (-11)(-11) = 121$

$-x^2 = -(-11)^2 = -(-11)(-11) = -121$

Work Practice 11

Practice 12

Evaluate $5y^2$ for $y = 4$ and $y = -4$.

Example 12   Evaluate $6z^2$ for $z = 2$ and $z = -2$.

Solution: $6z^2 = 6(2)^2 = 6(4) = 24$

$6z^2 = 6(-2)^2 = 6(4) = 24$

Work Practice 12

Practice 13

Evaluate $x^2 + y$ for $x = -6$ and $y = -3$.

Example 13   Evaluate $x + 2y - z$ for $x = 3$, $y = -5$, and $z = -4$.

Solution: Replace x with 3, y with -5, z with -4, and simplify.

> **Helpful Hint**
>
> Remember to rewrite the subtraction sign.

$$x + 2y - z = 3 + 2(-5) - (-4) \quad \text{Let } x = 3, y = -5, \text{ and } z = -4.$$
$$= 3 + (-10) + 4 \quad \text{Replace } 2(-5) \text{ with its product, } -10.$$
$$= -3 \quad \text{Add.}$$

Work Practice 13

Answers
10. -23 **11.** $225; -225$ **12.** $80; 80$
13. 33

✓**Concept Check Answer**
false; $-4(3 - 7) - 8(9 - 6) = -8$

Example 14 Evaluate $7 - x^2$ for $x = -4$.

Solution: Replace x with -4 and simplify carefully!

$$7 - x^2 = 7 - (-4)^2$$

$$= 7 - 16 \qquad (-4)^2 = (-4)(-4) = 16$$

$$= -9 \qquad \text{Subtract.}$$

■ **Work Practice 14**

Practice 14

Evaluate $4 - x^2$ for $x = -8$.

Objective C Finding Averages

Recall from Chapter 1 that the average of a list of numbers is

$$\text{average} = \frac{\text{sum of numbers}}{number \text{ of numbers}}$$

Example 15 The graph shows the monthly normal temperatures for Barrow, Alaska. Use this graph to find the average of the temperatures for the months January through April.

Practice 15

Find the average of the temperatures for the months October through April.

Monthly Normal Temperatures for Barrow, Alaska

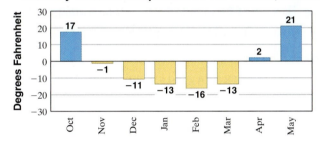

Solution: By reading the graph, we have

$$\text{average} = \frac{-13 + (-16) + (-13) + 2}{4} \qquad \text{There are 4 months from January through April.}$$

$$= \frac{-40}{4}$$

$$= -10$$

The average of the temperatures is $-10°F$.

■ **Work Practice 15**

Answers

14. -60 15. $-5°F$

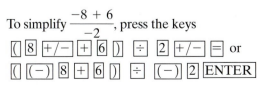 **Calculator Explorations** **Simplifying an Expression Containing a Fraction Bar**

Recall that even though most calculators follow the order of operations, parentheses must sometimes be inserted. For example, to simplify $\dfrac{-8 + 6}{-2}$ on a calculator, enter parentheses around the expression above the fraction bar so that it is simplified separately.

To simplify $\dfrac{-8 + 6}{-2}$, press the keys

$$[\,(\,]\,[\,8\,]\,[\,+/-\,]\,[\,+\,]\,[\,6\,]\,[\,)\,]\,[\,\div\,]\,[\,2\,]\,[\,+/-\,]\,[\,=\,] \text{ or}$$

$$[\,(\,]\,[\,(-)\,]\,[\,8\,]\,[\,+\,]\,[\,6\,]\,[\,)\,]\,[\,\div\,]\,[\,(-)\,]\,[\,2\,]\,[\,\text{ENTER}\,]$$

The display will read $\boxed{\quad 1}$.

Thus, $\dfrac{-8 + 6}{-2} = 1$.

Use a calculator to simplify.

1. $\dfrac{-120 - 360}{-10}$

2. $\dfrac{4750}{-2 + (-17)}$

3. $\dfrac{-316 + (-458)}{28 + (-25)}$

4. $\dfrac{-234 + 86}{-18 + 16}$

Vocabulary, Readiness & Video Check

Use the choices below to fill in each blank. Not all choices will be used.

average	subtraction	division	$-7 - 3(1)$
addition	multiplication	$-7 - 3(-1)$	

1. To simplify $-2 \div 2 \cdot (3)$, which operation should be performed first? _____

2. To simplify $-9 - 3 \cdot 4$, which operation should be performed first? _____

3. The _____ of a list of numbers is $\dfrac{\text{sum of numbers}}{\text{number of numbers}}$.

4. To simplify $5[-9 + (-3)] \div 4$, which operation should be performed first? _____

5. To simplify $-2 + 3(10 - 12) \cdot (-8)$, which operation should be performed first? _____

6. To evaluate $x - 3y$ for $x = -7$ and $y = -1$, replace x with -7 and y with -1 and evaluate _____.

Martin-Gay Interactive Videos Watch the section lecture video and answer the following questions.

See Video 2.5

Objective A 7. In ▣ Example 1, what two things about the fraction bar are we reminded of? ▶

Objective B 8. In ▣ Example 5, why is it important to place the replacement value for x within parentheses? ▶

Objective C 9. From the lecture before ▣ Example 6, explain why finding the average is a good example of an application for this section. ▶

2.5 Exercise Set MyMathLab®

Objective A *Simplify. See Examples 1 through 10.*

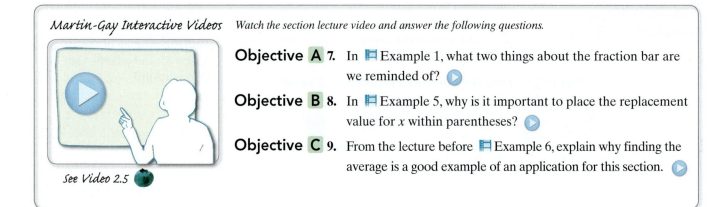

1. $(-5)^3$

2. -2^4

3. -4^3

4. $(-2)^4$

5. $8 \cdot 2^2$

6. $5 \cdot 2^3$

7. $8 - 12 - 4$

8. $10 - 23 - 12$

9. $7 + 3(-6)$

10. $-8 + 4(3)$

11. $5(-9) + 2$

12. $7(-6) + 3$

13. $-10 + 4 \div 2$

14. $-12 + 6 \div 3$

15. $6 + 7 \cdot 3 - 10$

16. $5 + 9 \cdot 4 - 20$

17. $\dfrac{16 - 13}{-3}$

18. $\dfrac{20 - 15}{-1}$

▶ 19. $\dfrac{24}{10 + (-4)}$

20. $\dfrac{88}{-8 - 3}$

21. $5(-3) - (-12)$

22. $7(-4) - (-6)$

23. $[8 + (-4)]^2$

24. $[9 + (-2)]^3$

25. $8 \cdot 6 - 3 \cdot 5 + (-20)$ **26.** $7 \cdot 6 - 6 \cdot 5 + (-10)$ ▶ **27.** $4 - (-3)^4$ **28.** $7 - (-5)^2$

29. $|7 + 3| \cdot 2^3$ **30.** $|-3 + 7| \cdot 7^2$ **31.** $7 \cdot 6^2 + 4$ **32.** $10 \cdot 5^3 + 7$

33. $7^2 - (4 - 2^3)$ **34.** $8^2 - (5 - 2)^4$ **35.** $|3 - 15| \div 3$ **36.** $|12 - 19| \div 7$

37. $-(-2)^6$ **38.** $-(-2)^3$ **39.** $(5 - 9)^2 \div (4 - 2)^2$ **40.** $(2 - 7)^2 \div (4 - 3)^4$

▶ **41.** $|8 - 24| \cdot (-2) \div (-2)$ **42.** $|3 - 15| \cdot (-4) \div (-16)$ **43.** $(-12 - 20) \div 16 - 25$

44. $(-20 - 5) \div 5 - 15$ **45.** $5(5 - 2) + (-5)^2 - 6$ **46.** $3 \cdot (8 - 3) + (-4) - 10$

47. $(2 - 7) \cdot (6 - 19)$ **48.** $(4 - 12) \cdot (8 - 17)$ **49.** $(-36 \div 6) - (4 \div 4)$

50. $(-4 \div 4) - (8 \div 8)$ **51.** $(10 - 4^2)^2$ **52.** $(11 - 3^2)^3$

53. $2(8 - 10)^2 - 5(1 - 6)^2$ **54.** $-3(4 - 8)^2 + 5(14 - 16)^3$ **55.** $3(-10) \div [5(-3) - 7(-2)]$

56. $12 - [7 - (3 - 6)] + (2 - 3)^3$ **57.** $\dfrac{(-7)(-3) - (4)(3)}{3[7 \div (3 - 10)]}$ **58.** $\dfrac{10(-1) - (-2)(-3)}{2[-8 \div (-2 - 2)]}$

▶ **59.** $-3[5 + 2(-4 + 9)] + 15$ **60.** $-2[6 + 4(2 - 8)] - 25$

Objective **B** *Evaluate each expression for* $x = -2, y = 4,$ *and* $z = -1$. *See Examples 11 through 14.*

61. $x + y + z$ **62.** $x - y - z$ **63.** $2x - 3y - 4z$ **64.** $5x - y + 4z$

▶ **65.** $x^2 - y$ **66.** $x^2 + z$ **67.** $\dfrac{5y}{z}$ **68.** $\dfrac{4x}{y}$

Evaluate each expression for $x = -3$ *and* $z = -4$. *See Examples 11 through 14.*

69. x^2 **70.** z^2 **71.** $-z^2$ **72.** $-x^2$

73. $2z^3$ **74.** $3x^2$ **75.** $10 - x^2$ **76.** $3 - z^2$

77. $2x^3 - z$ **78.** $3z^2 - x$

Objective C *Find the average of each list of numbers. See Example 15.*

▶ **79.** $-10, 8, -4, 2, 7, -5, -12$

80. $-18, -8, -1, -1, 0, 4$

81. $-17, -26, -20, -13$

82. $-40, -20, -10, -15, -5$

Scores in golf can be 0 (also called par), a positive integer (also called above par), or a negative integer (also called below par). The bar graph shows final scores of selected golfers from a 2013 tournament. Use this graph for Exercises 83 through 88. (Source: LPGA)

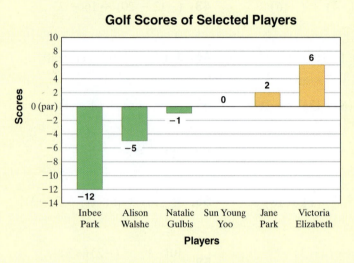

Golf Scores of Selected Players

83. Find the difference between the lowest score shown and the highest score shown.

84. Find the difference between the two lowest scores.

85. Find the average of the scores for Walshe, Gulbis, Yoo, and Jane Park. (*Hint:* Here, the average is the sum of the scores divided by the number of players.)

86. Find the average of the scores for Inbee Park, Walshe, Gulbis, and Elizabeth.

87. Can the average for the scores in Exercise **86** be greater than the highest score, 6? Explain why or why not.

88. Can the average of the scores in Exercise **86** be less than the lowest score, −12? Explain why or why not.

Review

Perform each indicated operation. See Sections 1.3, 1.5, and 1.6.

89. $45 \cdot 90$

90. $90 \div 45$

91. $90 - 45$

92. $45 + 90$

Find the perimeter of each figure. See Section 1.3.

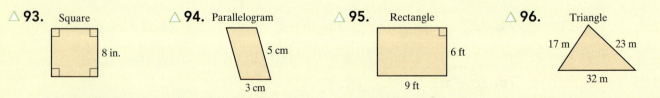

△ **93.** Square
8 in.

△ **94.** Parallelogram
5 cm
3 cm

△ **95.** Rectangle
6 ft
9 ft

△ **96.** Triangle
17 m   23 m
32 m

Concept Extensions

Insert parentheses where needed so that each expression evaluates to the given number.

97. $2 \cdot 7 - 5 \cdot 3$; evaluates to 12

98. $7 \cdot 3 - 4 \cdot 2$; evaluates to 34

99. $-6 \cdot 10 - 4$; evaluates to -36

100. $2 \cdot 8 \div 4 - 20$; evaluates to -36

101. Are parentheses necessary in the expression $3 + (4 \cdot 5)$? Explain your answer.

102. Are parentheses necessary in the expression $(3 + 4) \cdot 5$? Explain your answer.

103. Discuss the effect parentheses have in an exponential expression. For example, what is the difference between $(-6)^2$ and -6^2?

104. Discuss the effect parentheses have in an exponential expression. For example, what is the difference between $(2 \cdot 4)^2$ and $2 \cdot 4^2$?

Evaluate.

105. $(-12)^4$

106. $(-17)^6$

107. $x^3 - y^2$ for $x = 21$ and $y = -19$

108. $3x^2 + 2x - y$ for $x = -18$ and $y = 2868$

109. $(xy + z)^x$ for $x = 2, y = -5,$ and $z = 7$

110. $5(ab + 3)^b$ for $a = -2, b = 3$

2.6   Solving Equations: The Addition and Multiplication Properties

In this section, we introduce properties of equations and we use these properties to begin solving equations. Now that we know how to perform operations on integers, this is an excellent way to practice these operations.

First, let's recall the difference between an equation and an expression. From Section 1.8, a combination of operations on variables and numbers is an expression, and an equation is of the form "expression = expression."

Equations	Expressions
$3x - 1 = -17$	$3x - 1$
area $=$ length $\cdot$ width	$5(20 - 3) + 10$
$8 + 16 = 16 + 8$	y^3
$-9a + 11b = 14b + 3$	$-x^2 + y - 2$

Objectives

A   Identify Solutions of Equations. ▶

B   Use the Addition Property of Equality to Solve Equations. ▶

C   Use the Multiplication Property of Equality to Solve Equations. ▶

> **Helpful Hint**
>
> Simply stated, an equation contains "=" while an expression does not. Also, we *simplify* expressions and *solve* equations.

Objective A Identifying Solutions of Equations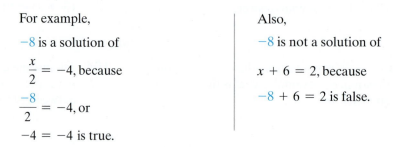

Let's practice identifying solutions of equations. Recall from Section 1.8 that a solution of an equation is a number that when substituted for a variable makes the equation a true statement.

For example,

-8 is a solution of

$\dfrac{x}{2} = -4$, because

$\dfrac{-8}{2} = -4$, or

$-4 = -4$ is true.

Also,

-8 is not a solution of

$x + 6 = 2$, because

$-8 + 6 = 2$ is false.

Let's practice determining whether a number is a solution of an equation. In this section, we will be performing operations on integers.

Practice 1

Determine whether -2 is a solution of the equation $-4x - 3 = 5$.

> **Example 1** Determine whether -1 is a solution of the equation $3y + 1 = 3$.
>
> **Solution:**
> $$3y + 1 = 3$$
> $$3(-1) + 1 \stackrel{?}{=} 3$$
> $$-3 + 1 \stackrel{?}{=} 3$$
> $$-2 = 3 \quad \text{False}$$
>
> Since $-2 = 3$ is false, -1 is *not* a solution of the equation.

◼ **Work Practice 1**

Now we know how to check whether a number is a solution. But, given an equation, how do we find its **solution**? In other words, how do we find a number that makes the equation true? How do we solve an equation?

Objective B Using the Addition Property to Solve Equations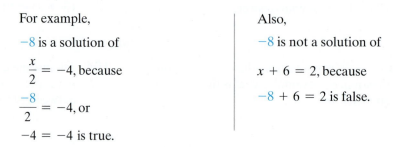

To solve an equation, we use properties of equality to write simpler equations, all equivalent to the original equation, until the final equation has the form

$x = $ **number**   or   **number** $= x$

Equivalent equations have the same solution, so the word "number" above represents the solution of the original equation. The first property of equality to help us write simpler, equivalent equations is the **addition property of equality.**

Answer

1. yes

Addition Property of Equality

Let a, b, and c represent numbers. Then

$a = b$	Also,	$a = b$
and $a + c = b + c$		and $a - c = b - c$
are equivalent equations.		are equivalent equations.

In other words, the same number may be added to or subtracted from both sides of an equation without changing the solution of the equation. (Recall from Section 2.3 that we defined subtraction as addition of the first number and the opposite of the second number. Because of this, the addition property of equality also allows us to subtract the same number from both sides.)

A good way to visualize a true equation is to picture a balanced scale. Since it is balanced, each side of the scale weighs the same amount. Similarly, in a true equation the expressions on each side have the same value. Picturing our balanced scale, if we add the same weight to each side, the scale remains balanced.

Example 2 Solve: $x - 2 = -1$ for x.

Solution: To solve the equation for x, we need to rewrite the equation in the form $x =$ number. In other words, our goal is to get x alone on one side of the equation. To do so, we add 2 to both sides of the equation.

$$x - 2 = -1$$
$$x - 2 + 2 = -1 + 2 \quad \text{Add 2 to both sides of the equation.}$$
$$x + 0 = 1 \quad \text{Replace } -2 + 2 \text{ with 0.}$$
$$x = 1 \quad \text{Simplify by replacing } x + 0 \text{ with } x.$$

Check: To check, we replace x with 1 in the *original* equation.

$$x - 2 = -1 \quad \text{Original equation}$$
$$1 - 2 \stackrel{?}{=} -1 \quad \text{Replace } x \text{ with 1.}$$
$$-1 = -1 \quad \text{True}$$

Since $-1 = -1$ is a true statement, 1 is the solution of the equation.

■ **Work Practice 2**

Practice 2

Solve the equation for y: $y - 6 = -2$

Helpful Hint

Note that it is always a good idea to check the solution in the *original* equation to see that it makes the equation a true statement.

Let's visualize how we used the addition property of equality to solve an equation. Picture the equation $x - 2 = 1$ as a balanced scale. The left side of the equation has the same value as the right side.

Answer

2. 4

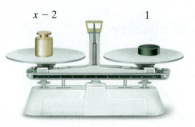

If the same weight is added to each side of a scale, the scale remains balanced. Likewise, if the same number is added to each side of an equation, the left side continues to have the same value as the right side.

Practice 3
Solve: $-2 = z + 8$

Example 3   Solve: $-8 = n + 1$

Solution: To get n alone on one side of the equation, we subtract 1 from both sides of the equation.

$$-8 = n + 1$$
$$-8 - 1 = n + 1 - 1 \qquad \text{Subtract 1 from both sides.}$$
$$-9 = n + 0 \qquad \text{Replace } 1 - 1 \text{ with 0.}$$
$$-9 = n \qquad \text{Simplify.}$$

Check:

$$-8 = n + 1$$
$$-8 \stackrel{?}{=} -9 + 1 \qquad \text{Replace } n \text{ with } -9.$$
$$-8 = -8 \qquad \text{True}$$

The solution is -9.

■ **Work Practice 3**

> **Helpful Hint**
> Remember that we can get the variable alone on either side of the equation. For example, the equations $-9 = n$ and $n = -9$ both have the solution of -9.

✓**Concept Check** What number should be added to or subtracted from both sides of the equation in order to solve the equation $-3 = y + 2$?

Practice 4
Solve: $x = -2 + 90 + (-100)$

Example 4   Solve: $x = -60 + 4 + 10$

Solution: Study this equation for a moment. Notice that our variable x is alone on the left side. Thus, we only need to add on the right side to find the value of x.

$$x = -60 + 4 + 10$$
$$x = -56 + 10 \qquad \text{Add } -60 \text{ and 4.}$$
$$x = -46 \qquad \text{Add } -56 \text{ and 10.}$$

Check to see that -46 is the solution.

■ **Work Practice 4**

Answers
3. -10 **4.** -12

✓**Concept Check Answer**
Subtract 2 from both sides.

Objective C Using the Multiplication Property to Solve Equations ▶

Although the addition property of equality is a powerful tool for helping us solve equations, it cannot help us solve all types of equations. For example, it cannot help us solve an equation such as $2x = 6$. To solve this equation, we use a second property of equality called the **multiplication property of equality.**

> ### Multiplication Property of Equality
>
> Let $a, b,$ and c represent numbers and let $c \neq 0$. Then
>
> $a = b$
> and $a \cdot c = b \cdot c$
> are equivalent equations.
>
> Also, $a = b$
> and $\dfrac{a}{c} = \dfrac{b}{c}$
> are equivalent equations.

In other words, both sides of an equation may be multiplied or divided by the same nonzero number without changing the solution of the equation. (We will see in Chapter 4 how the multiplication property allows us to divide both sides of an equation by the same nonzero number.)

To solve an equation like $2x = 6$ for x, notice that 2 is *multiplied* by x. To get x alone, we use the multiplication property of equality to *divide* both sides of the equation by 2, and simplify as follows:

$$2x = 6$$

$$\frac{2 \cdot x}{2} = \frac{6}{2} \quad \text{Divide both sides by 2.}$$

Then it can be shown that an expression such as $\dfrac{2 \cdot x}{2}$ is equivalent to $\dfrac{2}{2} \cdot x$, so

$$\frac{2 \cdot x}{2} = \frac{6}{2} \quad \text{can be written as} \quad \frac{2}{2} \cdot x = \frac{6}{2}$$

$$1 \cdot x = 3 \quad \text{or} \quad x = 3$$

Picturing again our balanced scale, if we multiply or divide the weight on each side by the same nonzero number, the scale (or equation) remains balanced.

$2x \qquad\qquad 6 \qquad\qquad\qquad \frac{2x}{2} \text{ or } x \qquad \frac{6}{2} \text{ or } 3$

Example 5 Solve: $-5x = 15$

Solution: To get x alone, divide both sides by -5.

$$-5x = 15 \qquad \text{Original equation}$$

$$\frac{-5x}{-5} = \frac{15}{-5} \qquad \text{Divide both sides by } -5.$$

$$\frac{-5}{-5} \cdot x = \frac{15}{-5}$$

$$1x = -3 \quad \text{or} \quad x = -3 \quad \text{Simplify.} \qquad (\textit{Continued on next page})$$

Practice 5

Solve: $3y = -18$

Answer
5. -6

Check: To check, replace x with -3 in the original equation.

$$-5x = 15 \quad \text{Original equation}$$
$$-5(-3) \stackrel{?}{=} 15 \quad \text{Let } x = -3.$$
$$15 = 15 \quad \text{True}$$

The solution is -3.

🟧 **Work Practice 5**

Practice 6

Solve: $-32 = 8x$

Example 6   Solve: $-27 = 3y$

Solution: To get y alone, divide both sides of the equation by 3.

$$-27 = 3y$$
$$\frac{-27}{3} = \frac{3y}{3} \quad \text{Divide both sides by 3.}$$
$$\frac{-27}{3} = \frac{3}{3} \cdot y$$
$$-9 = 1y \quad \text{or} \quad y = -9$$

Check to see that -9 is the solution.

🟧 **Work Practice 6**

Practice 7

Solve: $-3y = -27$

Example 7   Solve: $-12x = -36$

Solution: To get x alone, divide both sides of the equation by -12.

$$-12x = -36$$
$$\frac{-12x}{-12} = \frac{-36}{-12}$$
$$\frac{-12}{-12} \cdot x = \frac{-36}{-12}$$
$$x = 3$$

Check: To check, replace x with 3 in the original equation.

$$-12x = -36$$
$$-12(3) \stackrel{?}{=} -36 \quad \text{Let } x = 3.$$
$$-36 = -36 \quad \text{True}$$

Since $-36 = -36$ is a true statement, the solution is 3.

🟧 **Work Practice 7**

✔**Concept Check**   Which operation is appropriate for solving each of the following equations, addition or division?

a. $12 = x - 3$

b. $12 = 3x$

The multiplication property also allows us to solve equations like

$$\frac{x}{5} = 2$$

Answers

6. -4 **7.** 9

✔ **Concept Check Answers**

a. addition

b. division

Here, x is *divided* by 5. To get x alone, we use the multiplication property to *multiply* both sides by 5.

$$5 \cdot \frac{x}{5} = 5 \cdot 2 \quad \text{Multiply both sides by 5.}$$

Then it can be shown that

$$5 \cdot \frac{x}{5} = 5 \cdot 2 \text{ can be written as } \frac{5}{5} \cdot x = 5 \cdot 2$$

$$1 \cdot x = 10 \quad \text{or} \quad x = 10$$

Example 8   Solve: $\dfrac{x}{3} = -2$

Solution: To get x alone, multiply both sides by 3.

$$\frac{x}{3} = -2$$

$$3 \cdot \frac{x}{3} = 3 \cdot (-2) \qquad \text{Multiply both sides by 3.}$$

$$\frac{3}{3} \cdot x = 3 \cdot (-2)$$

$$1x = -6 \quad \text{or} \quad x = -6 \quad \text{Simplify.}$$

Check: Replace x with -6 in the original equation.

$$\frac{x}{3} = -2 \quad \text{Original equation}$$

$$\frac{-6}{3} \overset{?}{=} -2 \quad \text{Let } x = -6.$$

$$-2 = -2 \quad \text{True}$$

The solution is -6.

■ **Work Practice 8**

Practice 8

Solve: $\dfrac{x}{-4} = 7$

Answer
8. -28

Vocabulary, Readiness & Video Check

Use the choices below to fill in each blank. Some choices may be used more than once.

equation	multiplication	addition
expression	solution	equivalent

1. A combination of operations on variables and numbers is called a(n) _____.

2. A statement of the form "expression = expression" is called a(n) _____.

3. A(n) _____ contains an equals sign ($=$) while a(n) _____ does not.

4. A(n) _____ may be simplified and evaluated while a(n) _____ may be solved.

5. A(n) _____ of an equation is a number that when substituted for a variable makes the equation a true statement.

6. _____ equations have the same solution.

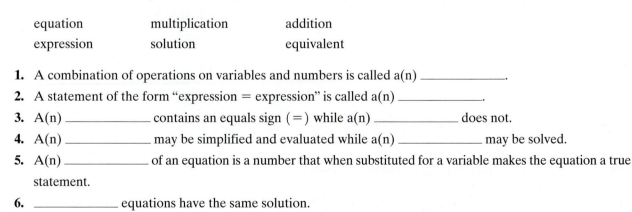

7. By the _____ property of equality, the same number may be added to or subtracted from both sides of an equation without changing the solution of the equation.

8. By the _____ property of equality, both sides of an equation may be multiplied or divided by the same non-zero number without changing the solution of the equation.

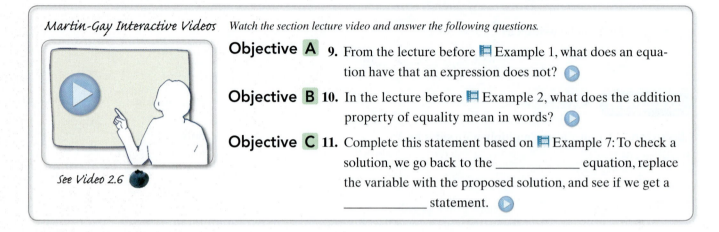

Martin-Gay Interactive Videos   Watch the section lecture video and answer the following questions.

Objective A 9. From the lecture before ▣ Example 1, what does an equation have that an expression does not? ▶

Objective B 10. In the lecture before ▣ Example 2, what does the addition property of equality mean in words? ▶

Objective C 11. Complete this statement based on ▣ Example 7: To check a solution, we go back to the _____ equation, replace the variable with the proposed solution, and see if we get a _____ statement. ▶

See Video 2.6

2.6 Exercise Set MyMathLab® ▶

Objective A *Determine whether the given number is a solution of the given equation. See Example 1.*

1. Is 6 a solution of $x - 8 = -2$?

2. Is 9 a solution of $y - 16 = -7$?

▶ **3.** Is -5 a solution of $x + 12 = 17$?

4. Is -7 a solution of $a + 23 = -16$?

5. Is -8 a solution of $-9f = 64 - f$?

6. Is -6 a solution of $-3k = 12 - k$?

7. Is 3 a solution of $5(c - 5) = -10$?

8. Is 1 a solution of $2(b - 3) = 10$?

Objective B *Solve. Check each solution. See Examples 2 through 4.*

▶ **9.** $a + 5 = 23$

10. $f + 4 = -6$

11. $d - 9 = -21$

12. $s - 7 = -15$

▶ **13.** $7 = y - 2$

14. $1 = y + 7$

▶ **15.** $-7 + 10 - 20 = x$

16. $-50 + 40 - 5 = z$

Objective C *Solve. Check each solution. See Examples 5 through 8.*

▶ **17.** $5x = 20$

18. $6y = 48$

▶ **19.** $-3z = 12$

20. $-2x = 26$

21. $\dfrac{n}{7} = -2$ **22.** $\dfrac{n}{11} = -5$ **23.** $2z = -34$ **24.** $7y = -21$

25. $-4y = 0$ **26.** $-9x = 0$ **27.** $-10x = -10$ **28.** $-31x = -31$

Objectives **B** **C** **Mixed Practice** *Solve. See Examples 2 through 8.*

29. $5x = -35$ **30.** $3y = -27$ **31.** $n - 5 = -55$ **32.** $n - 4 = -48$

33. $-15 = y + 10$ **34.** $-36 = y + 12$ **35.** $\dfrac{x}{-6} = -6$ **36.** $\dfrac{x}{-9} = -9$

37. $n = -10 + 31$ **38.** $z = -28 + 36$ **39.** $-12y = -144$ **40.** $-11x = -121$

▶ **41.** $\dfrac{n}{4} = -20$ **42.** $\dfrac{n}{5} = -20$ **43.** $-64 = 32y$ **44.** $-81 = 27x$

Review

Translate each phrase to an algebraic expression. Use x to represent "a number." See Section 1.8.

45. A number decreased by -2 **46.** A number increased by -5 **47.** The product of -6 and a number

48. The quotient of a number and -20 **49.** The sum of -15 and a number   **50.** -32 multiplied by a number

51. -8 divided by a number   **52.** Subtract a number from -18.

Concept Extensions

Solve.

53. $n - 42{,}860 = -1286$ **54.** $n + 961 = 120$

55. $-38x = 15{,}542$ **56.** $\dfrac{y}{-18} = 1098$

57. Explain the differences between an equation and an expression.

58. Explain the differences between the addition property of equality and the multiplication property of equality.

59. Write an equation that can be solved using the addition property of equality.

60. Write an equation that can be solved using the multiplication property of equality.

Chapter 2 Group Activity

Magic Squares

Sections 2.1–2.3

A magic square is a set of numbers arranged in a square table so that the sum of the numbers in each column, row, and diagonal is the same. For instance, in the magic square below, the sum of each column, row, and diagonal is 15. Notice that no number is used more than once in the magic square.

2	9	4
7	5	3
6	1	8

The properties of magic squares have been known for a very long time and once were thought to be good luck charms. The ancient Egyptians and Greeks understood their patterns. A magic square even made it into a famous work of art. The engraving titled *Melencolia I,* created by German artist Albrecht Dürer in 1514, features the following four-by-four magic square on the building behind the central figure.

16	3	2	13
5	10	11	8
9	6	7	12
4	15	14	1

Exercises

1. Verify that what is shown in the Dürer engraving is, in fact, a magic square. What is the common sum of the columns, rows, and diagonals?

2. Negative numbers can also be used in magic squares. Complete the following magic square:

	−1	
0		−4

3. Use the numbers $-16, -12, -8, -4, 0, 4, 8, 12,$ and 16 to form a magic square:

Chapter 2 Vocabulary Check

Fill in each blank with one of the words or phrases listed below.

inequality symbols	addition	solution	is less than	integers
expression	average	negative	absolute value	equation
positive	opposites	is greater than	multiplication	

1. Two numbers that are the same distance from 0 on the number line but are on opposite sides of 0 are called _____.

2. The _____ of a number is that number's distance from 0 on a number line.

3. The _____ are . . . , $-3, -2, -1, 0, 1, 2, 3, \ldots$.

4. The _____ numbers are numbers less than zero.

5. The _____ numbers are numbers greater than zero.

6. The symbols "$<$" and "$>$" are called _____.

7. A(n) _____ of an equation is a number that when substituted for a variable makes the equation a true statement.

8. The _____ of a list of numbers is $\dfrac{\text{sum of numbers}}{\text{number of numbers}}$.

9. A combination of operations on variables and numbers is called a(n) _____.

10. A statement of the form "expression = expression" is called a(n) _____.

11. The sign "$<$" means _____ and "$>$" means _____.

12. By the _____ property of equality, the same number may be added to or subtracted from both sides of an equation without changing the solution of the equation.

13. By the _____ property of equality, both sides of an equation may be multiplied or divided by the same nonzero number without changing the solution of the equation.

Helpful Hint

▶ Are you preparing for your test? Don't forget to take the Chapter 2 Test on page 158. Then check your answers at the back of the text and use the Chapter Test Prep Videos to see the fully worked-out solutions to any of the exercises you want to review.

2 Chapter Highlights

Definitions and Concepts	Examples
Section 2.1 Introduction to Integers	
Together, positive numbers, negative numbers, and 0 are called **signed numbers**.	$-432, -10, 0, 15$
The **integers** are . . . , $-3, -2, -1, 0, 1, 2, 3, \ldots$.	
The **absolute value** of a number is that number's distance from 0 on a number line. The symbol for absolute value is $\mid \ \mid$.	$\lvert -2 \rvert = 2$ 2 units $-3\ -2\ -1\ \ 0\ \ 1\ \ 2\ \ 3$ $\lvert 2 \rvert = 2$ 2 units $-3\ -2\ -1\ \ 0\ \ 1\ \ 2\ \ 3$

(continued)

Definitions and Concepts	Examples

Section 2.1 Introduction to Integers (continued)

Two numbers that are the same distance from 0 on the number line but are on opposite sides of 0 are called **opposites**.

5 and −5 are opposites.

If a is a number, then $-(-a) = a$.

$-(-11) = 11$. Do not confuse with $-|-3| = -3$

Section 2.2 Adding Integers

Adding Two Numbers with the Same Sign

Step 1: Add their absolute values.

Step 2: Use their common sign as the sign of the sum.

Add:

$$-3 + (-2) = -5$$
$$-7 + (-15) = -22$$

Adding Two Numbers with Different Signs

Step 1: Find the larger absolute value minus the smaller absolute value.

Step 2: Use the sign of the number with the larger absolute value as the sign of the sum.

$$-6 + 4 = -2$$
$$17 + (-12) = 5$$
$$-32 + (-2) + 14 = -34 + 14$$
$$= -20$$

Section 2.3 Subtracting Integers

Subtracting Two Numbers

If a and b are numbers, then $a - b = a + (-b)$.

Subtract:

$$-35 - 4 = -35 + (-4) = -39$$
$$3 - 8 = 3 + (-8) = -5$$
$$-10 - (-12) = -10 + 12 = 2$$
$$7 - 20 - 18 - (-3) = 7 + (-20) + (-18) + (+3)$$
$$= -13 + (-18) + 3$$
$$= -31 + 3$$
$$= -28$$

Section 2.4 Multiplying and Dividing Integers

Multiplying Numbers

The product of two numbers having the same sign is a positive number.
The product of two numbers having different signs is a negative number.

Multiply:

$$(-7)(-6) = 42$$
$$9(-4) = -36$$

Evaluate:

$$(-3)^2 = (-3)(-3) = 9$$

Dividing Numbers

The quotient of two numbers having the same sign is a positive number.
The quotient of two numbers having different signs is a negative number.

Divide:

$$-100 \div (-10) = 10$$
$$\frac{14}{-2} = -7, \quad \frac{0}{-3} = 0, \quad \frac{22}{0} \text{ is undefined.}$$

Definitions and Concepts	Examples
Section 2.5 Order of Operations	

Order of Operations

1. Perform all operations within parentheses ( ), brackets [ ], or other grouping symbols such as fraction bars, starting with the innermost set.
2. Evaluate any expressions with exponents.
3. Multiply or divide in order from left to right.
4. Add or subtract in order from left to right.

Simplify:

$$3 + 2 \cdot (-5) = 3 + (-10)$$
$$= -7$$

$$\frac{-2(5-7)}{-7 + |-3|} = \frac{-2(-2)}{-7 + 3}$$
$$= \frac{4}{-4}$$
$$= -1$$

| **Section 2.6 Solving Equations: The Addition and Multiplication Properties** ||

Addition Property of Equality

Let $a, b,$ and c represent numbers.

If $a = b$, then

$$a + c = b + c \quad \text{and} \quad a - c = b - c$$

In other words, the same number may be added to or subtracted from both sides of an equation without changing the solution of the equation.

Multiplication Property of Equality

Let $a, b,$ and c represent numbers and let $c \neq 0$.

If $a = b$, then

$$a \cdot c = b \cdot c \quad \text{and} \quad \frac{a}{c} = \frac{b}{c}$$

In other words, both sides of an equation may be multiplied or divided by the same nonzero number without changing the solution of the equation.

Solve:

$$x + 8 = 1$$
$$x + 8 - 8 = 1 - 8 \quad \text{Subtract 8 from both sides.}$$
$$x = -7 \quad \text{Simplify.}$$

The solution is -7.

Solve:

$$-6y = 30$$
$$\frac{-6y}{-6} = \frac{30}{-6} \quad \text{Divide both sides by } -6.$$
$$\frac{-6}{-6} \cdot y = \frac{30}{-6}$$
$$y = -5 \quad \text{Simplify.}$$

The solution is -5.

Chapter 2 Review

(2.1) *Represent each quantity by an integer.*

1. A gold miner is working 1572 feet down in a mine.

2. Mount Hood, in Oregon, has an elevation of 11,239 feet.

Graph each integer in the list on the same number line.

3. $-3, -5, 0, 7$

‹———+———+———+———+———+———+———+———+———+———+———+———+———+———+———›
$\quad$ -7 -6 -5 -4 -3 -2 -1 $\;\,0\;$ $\;1\;$ $\;2\;$ $\;3\;$ $\;4\;$ $\;5\;$ $\;6\;$ $\;7\;$

4. $-6, -1, 0, 5$

‹———+———+———+———+———+———+———+———+———+———+———+———+———+———+———›
$\quad$ -7 -6 -5 -4 -3 -2 -1 $\;\,0\;$ $\;1\;$ $\;2\;$ $\;3\;$ $\;4\;$ $\;5\;$ $\;6\;$ $\;7\;$

Simplify.

5. $|-11|$

6. $|0|$

7. $-|8|$

8. $-(-9)$

9. $-|-16|$

10. $-(-2)$

Insert $<$ *or* $>$ *between each pair of integers to make a true statement.*

11. $-18 \quad -20$

12. $-5 \quad 5$

13. $|-123| \quad -|-198|$

14. $|-12| \quad -|-16|$

Find the opposite of each integer.

15. -18

16. 42

Answer true or false for each statement.

17. If $a < b$, then a must be a negative number.

18. The absolute value of an integer is always 0 or a positive number.

19. A negative number is always less than a positive number.

20. If a is a negative number, then $-a$ is a positive number.

Evaluate.

21. $|y|$ if $y = -2$

22. $|-x|$ if $x = -3$

23. $-|-z|$ if $z = -5$

24. $-|-n|$ if $n = -10$

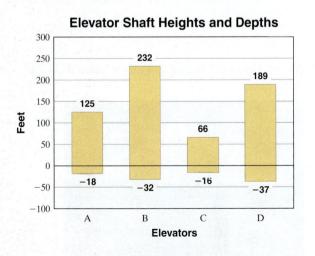

Elevator Shaft Heights and Depths

Elevator shafts in some buildings extend not only above ground, but in many cases below ground to accommodate basements, underground parking, etc. The bar graph shows four such elevators and their shaft distance above and below ground. Use the bar graph to answer Exercises 25 and 26.

25. Which elevator shaft extends the farthest below ground?

26. Which elevator shaft extends the highest above ground?

(2.2) *Add.*

27. $5 + (-3)$ **28.** $18 + (-4)$ **29.** $-12 + 16$ **30.** $-23 + 40$

31. $-8 + (-15)$ **32.** $-5 + (-17)$ **33.** $-24 + 3$ **34.** $-89 + 19$

35. $15 + (-15)$ **36.** $-24 + 24$ **37.** $-43 + (-108)$ **38.** $-100 + (-506)$

Solve.

39. The temperature at 5 a.m. on a day in January was $-15°$ Celsius. By 6 a.m. the temperature had fallen 5 degrees. Use a signed number to represent the temperature at 6 a.m.

40. A diver starts out at 127 feet below the surface and then swims downward another 23 feet. Use a signed number to represent the diver's current depth.

41. During the PGA 2008 Wyndham Championship tournament, the winner, Carl Pettersson, had scores of -6, -9, -4, and -2. What was his total score for the tournament? (*Source:* Professional Golfers' Association)

42. The Solheim Cup, a biannual pro women golfers' tournament between an American team and a European team, scores holes won. During the 2007 Solheim Cup, the winners, the American team, had a score of 16. The losing team, the Europeans, had a score 4 less than the Americans' score. What was the European team's score? (*Source:* Professional Golfers' Association)

(2.3) *Subtract.*

43. $12 - 4$ **44.** $-12 - 4$ **45.** $-7 - 17$ **46.** $7 - 17$

47. $7 - (-13)$ **48.** $-6 - (-14)$ **49.** $16 - 16$ **50.** $-16 - 16$

51. $-12 - (-12)$ **52.** $-5 - (-12)$

53. $-(-5) - 12 + (-3)$ **54.** $-8 + (-12) - 10 - (-3)$

Solve.

55. If the elevation of Lake Superior is 600 feet above sea level and the elevation of the Caspian Sea is 92 feet below sea level, find the difference of the elevations.

56. Josh Weidner has $142 in his checking account. He writes a check for $125, makes a deposit of $43, and then writes another check for $85. Represent the balance in his account by an integer.

57. Some roller coasters travel above and below ground. One such roller coaster is Tremors, located in Silverwood Theme Park, Athol, Idaho. If this coaster rises to a height of 85 feet above ground, then drops 99 feet, how many feet below ground are you at the end of the drop? (*Source:* ultimateroller-coaster.com)

58. Go to the bar graph for Review Exercises **25** and **26** and find the total length of the elevator shaft for Elevator C.

Answer true or false for each statement.

59. $|-5| - |-6| = 5 - 6$

60. $|-5 - (-6)| = 5 + 6$

(2.4) *Multiply.*

61. $-3(-7)$

62. $-6(3)$

63. $-4(16)$

64. $-5(-12)$

65. $(-5)^2$

66. $(-1)^5$

67. $12(-3)(0)$

68. $-1(6)(2)(-2)$

Divide.

69. $-15 \div 3$

70. $\dfrac{-24}{-8}$

71. $\dfrac{0}{-3}$

72. $\dfrac{-46}{0}$

73. $\dfrac{100}{-5}$

74. $\dfrac{-72}{8}$

75. $\dfrac{-38}{-1}$

76. $\dfrac{45}{-9}$

Solve.

77. A football team lost 5 yards on each of two consecutive plays. Represent the total loss by a product of integers, and find the product.

78. A horse race bettor lost $50 on each of four consecutive races. Represent the total loss by a product of integers, and find the product.

79. A person has a debt of $1024 and is ordered to pay it back in four equal payments. Represent the amount of each payment by a quotient of integers, and find the quotient.

80. Overnight, the temperature dropped 45 degrees Fahrenheit. If this took place over a time period of nine hours, represent the average temperature drop each hour by a quotient of integers. Then find the quotient.

(2.5) *Simplify.*

81. $(-7)^2$

82. -7^2

83. $5 - 8 + 3$

84. $-3 + 12 + (-7) - 10$

85. $-10 + 3 \cdot (-2)$

86. $5 - 10 \cdot (-3)$

87. $16 \div (-2) \cdot 4$

88. $-20 \div 5 \cdot 2$

89. $16 + (-3) \cdot 12 \div 4$

90. $-12 + 10 \div (-5)$

91. $4^3 - (8 - 3)^2$

92. $(-3)^3 - 90$

93. $\dfrac{(-4)(-3) - (-2)(-1)}{-10 + 5}$

94. $\dfrac{4(12 - 18)}{-10 \div (-2 - 3)}$

Find the average of each list of numbers.

95. $-18, 25, -30, 7, 0, -2$

96. $-45, -40, -30, -25$

Evaluate each expression for $x = -2$ and $y = 1$.

97. $2x - y$

98. $y^2 + x^2$

99. $\dfrac{3x}{6}$

100. $\dfrac{5y - x}{-y}$

(2.6) *For Exercises 101 and 102, answer "yes" or "no."*

101. Is -5 a solution of $2n - 6 = 16$?

102. Is -2 a solution of $2(c - 8) = -20$?

Solve.

103. $n - 7 = -20$

104. $-5 = n + 15$

105. $10x = -30$

106. $-8x = 72$

107. $-20 + 7 = y$

108. $x - 31 = -62$

109. $\dfrac{n}{-4} = -11$

110. $\dfrac{x}{-2} = 13$

111. $n + 12 = -7$

112. $n - 40 = -2$

113. $-36 = -6x$

114. $-40 = 8y$

Mixed Review

Perform the indicated operations.

115. $-6 + (-9)$

116. $-16 - 3$

117. $-4(-12)$

118. $\dfrac{84}{-4}$

119. $-76 - (-97)$

120. $-9 + 4$

Solve.

121. Wednesday's lowest temperature was $-18°C$. The cold weather continued and by Friday, it had dropped another $9°C$. What was the temperature on Friday?

122. The temperature at noon on a Monday in December was $-11°C$. By noon on Tuesday, it had warmed by $17°C$. What was the temperature at noon on Tuesday?

123. The top of a mountain has an altitude of 12,923 feet. The bottom of a valley is 195 feet below sea level. Find the difference between these two elevations.

124. Joe owed his mother \$32. He gave her \$23. Write his financial situation as a signed number.

Simplify.

125. $(3 - 7)^2 \div (6 - 4)^3$

126. $3(4 + 2) + (-6) - 3^2$

127. $2 - 4 \cdot 3 + 5$

128. $4 - 6 \cdot 5 + 1$

129. $\dfrac{-|-14| - 6}{7 + 2(-3)}$

130. $5(7 - 6)^3 - 4(2 - 3)^2 + 2^4$

Solve.

131. $n - 9 = -30$

132. $n + 18 = 1$

133. $-4x = -48$

134. $9x = -81$

135. $\dfrac{n}{-2} = 100$

136. $\dfrac{y}{-1} = -3$

Step-by-step test solutions are found on the Chapter Test Prep Videos. Where available: **MyMathLab®** or **You Tube**

Answers

Simplify each expression.

1. $-5 + 8$

2. $18 - 24$

3. $5 \cdot (-20)$

4. $-16 \div (-4)$

5. $-18 + (-12)$

6. $-7 - (-19)$

7. $-5 \cdot (-13)$

8. $\dfrac{-25}{-5}$

9. $|-25| + (-13)$

10. $14 - |-20|$

11. $|5| \cdot |-10|$

12. $\dfrac{|-10|}{-|-5|}$

13. $-8 + 9 \div (-3)$

14. $-7 + (-32) - 12 + 5$

15. $(-5)^3 - 24 \div (-3)$

16. $(5 - 9)^2 \cdot (8 - 2)^3$

17. $-(-7)^2 \div 7 \cdot (-4)$

18. $3 - (8 - 2)^3$

1. _____

2. _____

3. _____

4. _____

5. _____

6. _____

7. _____

8. _____

9. _____

10. _____

11. _____

12. _____

13. _____

14. _____

15. _____

16. _____

17. _____

18. _____

19. $\dfrac{4}{2} - \dfrac{8^2}{16}$

20. $\dfrac{-3(-2) + 12}{-1(-4 - 5)}$

21. $\dfrac{|25 - 30|^2}{2(-6) + 7}$

22. $5(-8) - [6 - (2 - 4)] + (12 - 16)^2$

Evaluate each expression for $x = 0$, $y = -3$, and $z = 2$.

23. $7x + 3y - 4z$

24. $10 - y^2$

25. $\dfrac{3z}{2y}$

26. Mary Dunstan, a diver, starts at sea level and then makes 4 successive descents of 22 feet. After the descents, what is her elevation?

27. Aaron Hawn has $129 in his checking account. He writes a check for $79, withdraws $40 from an ATM, and then deposits $35. Represent the new balance in his account by an integer.

28. Mt. Washington in New Hampshire has an elevation of 6288 feet above sea level. The Romanche Gap in the Atlantic Ocean has an elevation of 25,354 feet below sea level. Represent the difference in elevation between these two points by an integer.
(*Source:* National Geographic Society and Defense Mapping Agency)

29. Lake Baykal in Siberian Russia is the deepest lake in the world, with a maximum depth of 5315 feet. The elevation of the lake's surface is 1495 feet above sea level. What is the elevation (with respect to sea level) of the deepest point in the lake?
(*Source:* U.S. Geological Survey)

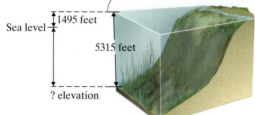

30. Find the average of $-12, -13, 0, 9$.

31. Translate the following phrases into mathematical expressions. Use x to represent "a number."

    a. The product of a number and 17

    b. A number subtracted from 20

Solve.

32. $-9n = -45$

33. $\dfrac{n}{-7} = 4$

34. $x - 16 = -36$

35. $-20 + 8 + 8 = x$

19. _____

20. _____

21. _____

22. _____

23. _____

24. _____

25. _____

26. _____

27. _____

28. _____

29. _____

30. _____

31. a. _____

    b. _____

32. _____

33. _____

34. _____

35. _____

Answers

1. _____
2. _____
3. _____
4. _____
5. _____
6. _____
7. a. _____
   b. _____
   c. _____
8. a. _____
   b. _____
   c. _____
9. _____
10. _____
11. _____
12. _____
13. _____
14. _____
15. _____
16. _____
17. _____
18. _____
19. a. _____
    b. _____
    c. _____
20. a. _____
    b. _____
    c. _____
21. _____
22. _____
23. a. _____
    b. _____
    c. _____
24. a. _____
    b. _____
    c. _____

Find the place value of the digit 3 in each whole number.

1. 396,418

2. 4308

3. 93,192

4. 693,298

5. 534,275,866

6. 267,301,818

7. Insert < or > to make a true statement.

  a. -7 7

  b. 0 -4

  c. -9 -11

8. Insert < or > to make a true statement

  a. 12 -4

  b. -13 -31

  c. -82 79

9. Add:
$13 + 2 + 7 + 8 + 9$

10. Add:
$11 + 3 + 9 + 16$

11. Subtract: $7826 - 505$
Check by adding.

12. Subtract: $3285 - 272$
Check by adding.

13. The radius of Jupiter is 43,441 miles. The radius of Saturn is 7257 miles less than the radius of Jupiter. Find the radius of Saturn. (*Source:* National Space Science Data Center)

14. C. J. Dufour wants to buy a digital camera. She has $762 in her savings account. If the camera costs $237, how much money will she have in her account after buying the camera?

15. Round 568 to the nearest ten.

16. Round 568 to the nearest hundred.

17. Round each number to the nearest hundred to find an estimated difference.

   4725
-2879

18. Round each number to the nearest thousand to find an estimated difference.

   8394
-2913

19. Rewrite each using the distributive property.

  a. $5(6 + 5)$

  b. $20(4 + 7)$

  c. $2(7 + 9)$

20. Rewrite each using the distributive property.

  a. $5(2 + 12)$

  b. $9(3 + 6)$

  c. $4(8 + 1)$

21. Multiply: 631×125

22. Multiply: 299×104

23. Find each quotient. Check by multiplying.

  a. $42 \div 7$

  b. $\dfrac{64}{8}$

  c. $3\overline{)21}$

24. Find each quotient. Check by multiplying.

  a. $\dfrac{35}{5}$

  b. $64 \div 8$

  c. $4\overline{)48}$

25. Divide: $3705 \div 5$. Check by multiplying.

26. Divide: $3648 \div 8$. Check by multiplying.

27. As part of a promotion, an executive receives 238 cards, each good for one free song download. If she wants to share them evenly with 19 friends, how many download cards will each friend receive? How many will be left over?

28. Mrs. Mallory's first-grade class is going to the zoo. She pays a total of $324 for 36 admission tickets. How much does each ticket cost?

Evaluate.

29. 9^2

30. 5^3

31. 6^1

32. 4^1

33. $5 \cdot 6^2$

34. $2^3 \cdot 7$

35. Simplify: $\dfrac{7 - 2 \cdot 3 + 3^2}{5(2 - 1)}$

36. Simplify: $\dfrac{6^2 + 4 \cdot 4 + 2^3}{37 - 5^2}$

37. Evaluate $x + 6$ if x is 8.

38. Evaluate $5 + x$ if x is 9.

39. Simplify:
   a. $|-9|$
   b. $|8|$
   c. $|0|$

40. Simplify:
   a. $|4|$
   b. $|-7|$

41. Add: $-2 + 25$

42. Add: $8 + (-3)$

43. Evaluate $2a - b$ for $a = 8$ and $b = -6$.

44. Evaluate $x - y$ for $x = -2$ and $y = -7$.

45. Multiply: $-7 \cdot 3$

46. Multiply: $5(-2)$

47. Multiply: $0 \cdot (-4)$

48. Multiply: $-6 \cdot 9$

49. Simplify: $3(4 - 7) + (-2) - 5$

50. Simplify: $4 - 8(7 - 3) - (-1)$

25. _____
26. _____
27. _____
28. _____
29. _____
30. _____
31. _____
32. _____
33. _____
34. _____
35. _____
36. _____
37. _____
38. _____
39. a. _____
    b. _____
    c. _____
40. a. _____
    b. _____
41. _____
42. _____
43. _____
44. _____
45. _____
46. _____
47. _____
48. _____
49. _____
50. _____

In this chapter, we continue making the transition from arithmetic to algebra. Recall that in algebra, letters (called variables) represent unknown quantities. Using variables is a very powerful method for solving problems that cannot be solved with arithmetic alone. This chapter introduces operations on algebraic expressions, and we continue solving variable equations.

3 Solving Equations and Problem Solving

Coca-Cola recently celebrated the 125th anniversary of its namesake brand. The illumination in the photo was on display for a month in Atlanta, Georgia. At this writing, the illumination holds the world record for largest single-building illumination. Also, Coca-Cola was recently named the Most Valuable Global Brand, as shown in the graph below. What is a global brand? There are many definitions, but a brand is the identity of a product or a service. The identity can be projected through its name, a symbol, a slogan, or something similar. A global brand, then, is a product whose brand is easily recognized throughout the world. Read the list of company names in the graph below and see how many of these brand names you know.

In Section 3.4, Exercises 31 and 32, we will solve applications that give us the unknown bar heights in the bar graph.

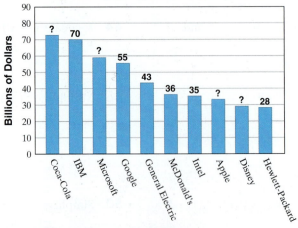

Top Ten Most Valuable Global Brands

Source: Interbrand, *Top 10 of Everything,* 2013

Simplifying Algebraic Expressions

Recall from Section 1.8 that a combination of numbers, letters (variables), and operation symbols is called an **algebraic expression** or simply an **expression.** Examples of expressions are below.

Algebraic Expressions

$4 \cdot x$, $n + 7$, and $3y - 5 - x$

Recall that if two variables or a number and a variable are next to each other, with no operation sign between them, the indicated operation is multiplication. For example,

$3y$ means $3 \cdot y$

and

xy or $x(y)$ means $x \cdot y$

Also, the meaning of an exponent remains the same when the base is a variable. For example,

$$y^2 = \underbrace{y \cdot y}_{2 \text{ factors of } y} \quad \text{and} \quad x^4 = \underbrace{x \cdot x \cdot x \cdot x}_{4 \text{ factors of } x}$$

Just as we can add, subtract, multiply, and divide numbers, we can add, subtract, multiply, and divide algebraic expressions. In previous sections we evaluated algebraic expressions like $x + 3$, $4x$, and $x + 2y$ for particular values of the variables. In this section, we explore working with variable expressions without evaluating them. We begin with a definition of a term.

Objective A Combining Like Terms ▶

The addends of an algebraic expression are called the **terms** of the expression.

$$\underset{\text{2 terms}}{x + 3}$$

$$\underset{\text{3 terms}}{3y^2 + (-6y) + 4}$$

A term that is only a number has a special name. It is called a **constant term,** or simply a **constant.** A term that contains a variable is called a **variable term.**

x	$+$	3	and	$3y^2 + (-6y)$	$+$	4
↑		↑		↑		↑
variable term		constant term		variable terms		constant term

The number factor of a variable term is called the **numerical coefficient.** A numerical coefficient of 1 is usually not written.

$5x$	x or $1x$	$3y^2$	$-6y$
↑	↑	↑	↑
Numerical coefficient is 5.	Understood numerical coefficient is 1.	Numerical coefficient is 3.	Numerical coefficient is -6.

> **Helpful Hint**
> Recall that $1 \cdot$ any number = that number. This means that
> $$1 \cdot x = x \quad \text{or that} \quad 1x = x$$
> Thus x can always be replaced by $1x$ or $1 \cdot x$.

Terms with the same variable factors, except that they may have different numerical coefficients, are called **like terms.**

Like Terms	Unlike Terms
$3x, -4x$	$5x, x^2$
$-6y, 2y, y$	$7x, 7y$

✓**Concept Check** True or false? The terms $-7xz^2$ and $3z^2x$ are like terms. Explain.

A sum or difference of like terms can be simplified using the **distributive property.** Recall from Section 1.5 that the distributive property says that multiplication distributes over addition (and subtraction). Using variables, we can write the distributive property as follows:

$$(a + b)c = ac + bc.$$

If we write the right side of the equation first, then the left side, we have the following:

Distributive Property

If a, b, and c are numbers, then

$$ac + bc = (a + b)c$$

Also,

$$ac - bc = (a - b)c$$

The distributive property guarantees that, no matter what number x is, $7x + 2x$ (for example) has the same value as $(7 + 2)x$, or $9x$. We then have that

$$7x + 2x = (7 + 2)x = 9x$$

This is an example of **combining like terms.** An algebraic expression is **simplified** when all like terms have been combined.

Practice 1

Simplify each expression by combining like terms.

a. $8m - 14m$

b. $6a + a$

c. $-y^2 + 3y^2 + 7$

Answers

1. **a.** $-6m$ **b.** $7a$ **c.** $2y^2 + 7$

✓**Concept Check Answer**
true

Example 1 Simplify each expression by combining like terms.

a. $4x + 6x$ **b.** $y - 5y$ **c.** $3x^2 + 5x^2 - 2$

Solution: Add or subtract like terms.

a. $4x + 6x = (4 + 6)x$
$= 10x$

Understood 1

b. $y - 5y = 1y - 5y$
$= (1 - 5)y$
$= -4y$

c. $3x^2 + 5x^2 - 2 = (3 + 5)x^2 - 2$
$= 8x^2 - 2$

Work Practice 1

Helpful Hint

In this section, we are simplifying expressions. Try not to confuse the two processes below.

Expression: $5y - 8y$
Simplify the expression:

$$5y - 8y = (5 - 8)y$$
$$= -3y$$

Equation: $8n = -40$
Solve the equation:

$$8n = -40$$
$$\frac{8n}{8} = \frac{-40}{8} \quad \text{Divide both sides by 8.}$$
$$n = -5 \quad \text{The solution is } -5.$$

The commutative and associative properties of addition and multiplication can also help us simplify expressions. We presented these properties in Sections 1.3 and 1.5 and state them again using variables.

Properties of Addition and Multiplication

If a, b, and c are numbers, then

$a + b = b + a$ Commutative property of addition

$a \cdot b = b \cdot a$ Commutative property of multiplication

That is, the **order** of adding or multiplying two numbers can be changed without changing their sum or product.

$(a + b) + c = a + (b + c)$ Associative property of addition

$(a \cdot b) \cdot c = a \cdot (b \cdot c)$ Associative property of multiplication

That is, the **grouping** of numbers in addition or multiplication can be changed without changing their sum or product.

Helpful Hint

- Examples of these properties are

$$2 + 3 = 3 + 2 \qquad \text{Commutative property of addition}$$
$$7 \cdot 9 = 9 \cdot 7 \qquad \text{Commutative property of multiplication}$$
$$(1 + 8) + 10 = 1 + (8 + 10) \qquad \text{Associative property of addition}$$
$$(4 \cdot 2) \cdot 3 = 4 \cdot (2 \cdot 3) \qquad \text{Associative property of multiplication}$$

- These properties are not true for subtraction or division.

Example 2   Simplify: $2y - 6 + 4y + 8$

Solution: We begin by writing subtraction as the opposite of addition.

$$
\begin{aligned}
2y - 6 + 4y + 8 &= 2y + (-6) + 4y + 8 & \text{Apply the commutative property} \\
&= 2y + 4y + (-6) + 8 & \text{of addition.} \\
&= (2 + 4)y + (-6) + 8 & \text{Apply the distributive property.} \\
&= 6y + 2 & \text{Simplify.}
\end{aligned}
$$

Work Practice 2

Practice 2

Simplify: $6z + 5 + z - 4$

Answer

2. $7z + 1$

Practice 3–5

Simplify each expression by combining like terms.

3. $6y + 12y - 6$

4. $7y - 5 + y + 8$

5. $-7y + 2 - 2y - 9x + 12 - x$

Examples Simplify each expression by combining like terms.

3. $6x + 2x - 5 = 8x - 5$

4. $4x + 2 - 5x + 3 = 4x + 2 + (-5x) + 3$
$$= 4x + (-5x) + 2 + 3$$
$$= -1x + 5 \quad \text{or} \quad -x + 5$$

5. $2x - 5 + 3y + 4x - 10y + 11$
$$= 2x + (-5) + 3y + 4x + (-10y) + 11$$
$$= 2x + 4x + 3y + (-10y) + (-5) + 11$$
$$= 6x - 7y + 6$$

Work Practice 3–5

As we practice combining like terms, keep in mind that some of the steps may be performed mentally.

Objective B Multiplying Expressions ▶

We can also use properties of numbers to multiply expressions such as $3(2x)$. By the associative property of multiplication, we can write the product $3(2x)$ as $(3 \cdot 2)x$, which simplifies to $6x$.

Practice 6–7

Multiply.

6. $6(4a)$

7. $-8(9x)$

Examples Multiply.

6. $5(3y) = (5 \cdot 3)y$ Apply the associative property of multiplication.
$$= 15y \quad \text{Multiply.}$$

7. $-2(4x) = (-2 \cdot 4)x$ Apply the associative property of multiplication.
$$= -8x \quad \text{Multiply.}$$

Work Practice 6–7

We can use the distributive property to combine like terms, which we have done, and also to multiply expressions such as $2(3 + x)$. By the distributive property, we have

$$2(3 + x) = 2 \cdot 3 + 2 \cdot x \quad \text{Apply the distributive property.}$$
$$= 6 + 2x \quad \text{Multiply.}$$

Practice 8

Use the distributive property to multiply: $8(y + 2)$

Example 8 Use the distributive property to multiply: $5(x + 4)$

Solution: By the distributive property,

$$5(x + 4) = 5 \cdot x + 5 \cdot 4 \quad \text{Apply the distributive property.}$$
$$= 5x + 20 \quad \text{Multiply.}$$

Work Practice 8

Answers

3. $18y - 6$ **4.** $8y + 3$

5. $-9y - 10x + 14$ **6.** $24a$

7. $-72x$ **8.** $8y + 16$

✔**Concept Check Answer**

did not distribute the 8;
$8(a - b) = 8a - 8b$

✔**Concept Check** What's wrong with the following?

$$8(a - b) = 8a - b$$

Example 9 Multiply: $-3(5a + 2)$

Solution: By the distributive property,

$$-3(5a + 2) = -3(5a) + (-3)(2) \quad \text{Apply the distributive property.}$$
$$= (-3 \cdot 5)a + (-6) \quad \text{Use the associative property and multiply.}$$
$$= -15a - 6 \quad \text{Multiply.}$$

■ **Work Practice 9**

Practice 9
Multiply: $3(7a - 5)$

Example 10 Multiply: $8(x - 4)$

Solution:

$$8(x - 4) = 8 \cdot x - 8 \cdot 4$$
$$= 8x - 32$$

■ **Work Practice 10**

Practice 10
Multiply: $6(5 - y)$

Objective C Simplifying Expressions ▶

Next we will **simplify** expressions by first using the distributive property to multiply and then **combining** any like terms.

Example 11 Simplify: $2(3 + 7x) - 15$

Solution: First we use the distributive property to remove parentheses.

$$2(3 + 7x) - 15 = 2(3) + 2(7x) - 15 \quad \text{Apply the distributive property.}$$
$$= 6 + 14x - 15 \quad \text{Multiply.}$$
$$= 14x + (-9) \quad \text{or} \quad 14x - 9 \quad \text{Combine like terms.}$$

■ **Work Practice 11**

Practice 11
Simplify: $5(2y - 3) - 8$

> **Helpful Hint**
> 2 is *not* distributed to the -15 since it is not within the parentheses.

Example 12 Simplify: $-2(x - 5) + 4(2x + 2)$

Solution: First we use the distributive property to remove parentheses.

$$-2(x - 5) + 4(2x + 2) = -2(x) - (-2)(5) + 4(2x) + 4(2) \quad \text{Apply the distributive property.}$$
$$= -2x + 10 + 8x + 8 \quad \text{Multiply.}$$
$$= 6x + 18 \quad \text{Combine like terms.}$$

■ **Work Practice 12**

Practice 12
Simplify:
$-7(x - 1) + 5(2x + 3)$

Example 13 Simplify: $-(x + 4) + 5x + 16$

Solution: The expression $-(x + 4)$ means $-1(x + 4)$.

$$-(x + 4) + 5x + 16 = -1(x + 4) + 5x + 16$$
$$= -1 \cdot x + (-1)(4) + 5x + 16 \quad \text{Apply the distributive property.}$$
$$= -x + (-4) + 5x + 16 \quad \text{Multiply.}$$
$$= 4x + 12 \quad \text{Combine like terms.}$$

■ **Work Practice 13**

Practice 13
Simplify: $-(y + 1) + 3y - 12$

Answers
9. $21a - 15$ **10.** $30 - 6y$
11. $10y - 23$ **12.** $3x + 22$
13. $2y - 13$

Objective **D** Finding Perimeter and Area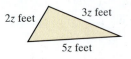

Example 14 Find the perimeter of the triangle.

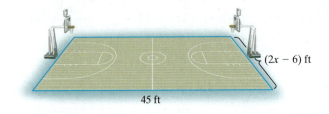

2z feet   3z feet

5z feet

Solution: Recall that the perimeter of a figure is the distance around the figure. To find the perimeter, then, we find the sum of the lengths of the sides. We use the letter P to represent perimeter.

$$P = 2z + 3z + 5z$$
$$= 10z$$

> **Helpful Hint**
> Don't forget to insert proper units.

The perimeter is $10z$ feet.

■ **Work Practice 14**

Practice 15

Find the area of the rectangular garden.

$(12y + 9)$ yards

3 yards

Example 15 Finding the Area of a Basketball Court

Find the area of this YMCA basketball court.

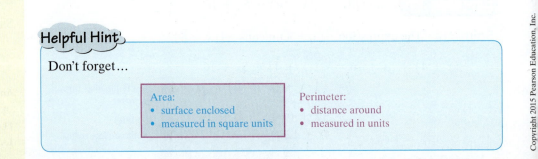

$(2x - 6)$ ft

45 ft

Solution: Recall how to find the area of a rectangle. **Area = Length · Width**, or if A represents area, l represents length, and w represents width, we have $A = l \cdot w$.

$$A = l \cdot w$$

$$= 45(2x - 6) \quad \text{Let length} = 45 \text{ and width} = (2x - 6).$$
$$= 90x - 270 \quad \text{Multiply.}$$

The area is $(90x - 270)$ *square* feet.

■ **Work Practice 15**

> **Helpful Hint**
> Don't forget…
>
> | Area: | Perimeter: |
> | --- | --- |
> | • surface enclosed | • distance around |
> | • measured in square units | • measured in units |

Answers

14. $8x$ cm   **15.** $(36y + 27)$ sq yd

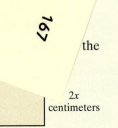

167

the

$2x$ centimeters

Vocabulary, Readiness & Video Check

Use the choices below to fill in each blank. Some choices may be used more than once.

numerical coefficient	combine like terms	like	term	variable	associative
constant	expression	unlike	distributive	commutative	

1. $14y^2 + 2x - 23$ is called a(n) _____ while $14y^2$, $2x$, and -23 are each called a(n) _____.

2. To multiply $3(-7x + 1)$, we use the _____ property.

3. To simplify an expression like $y + 7y$, we _____.

4. By the _____ properties, the *order* of adding or multiplying two numbers can be changed without changing their sum or product.

5. The term $5x$ is called a(n) _____ term while the term 7 is called a(n) _____ term.

6. The term z has an understood _____ of 1.

7. By the _____ properties, the *grouping* of adding or multiplying numbers can be changed without changing their sum or product.

8. The terms $-x$ and $5x$ are _____ terms and the terms $5x$ and $5y$ are _____ terms.

9. For the term $-3x^2 y$, -3 is called the _____.

Martin-Gay Interactive Videos   Watch the section lecture video and answer the following questions.

Objective **A** 10. In Example 2, why can't the expression be simplified after the first step? ▶

Objective **B** 11. In Example 5, what property is used to multiply? ▶

Objective **C** 12. In Example 6, why is the 20 not multiplied by the 2? ▶

Objective **D** 13. In Example 8, what operation is used to find P? What operation is used to find A? What do P and A stand for? ▶

See Video 3.1

3.1 Exercise Set MyMathLab® ▶

Objective **A** *Simplify each expression by combining like terms. See Examples 1 through 5.*

▶ **1.** $3x + 5x$ **2.** $8y + 3y$ **3.** $2n - 3n$ **4.** $7z - 10z$

5. $4c + c - 7c$ **6.** $5b - 8b - b$ **7.** $4x - 6x + x - 5x$ **8.** $8y + y - 2y - 8y$

▶ **9.** $3a + 2a + 7a - 5$ **10.** $5b - 4b + b - 15$

Objective **B** *Multiply. See Examples 6 and 7.*

11. $6(7x)$ **12.** $4(4x)$ ▶ **13.** $-3(11y)$ **14.** $-3(21z)$

15. $12(6a)$ **16.** $13(5b)$

Multiply. See Examples 8 through 10.

17. $2(y + 3)$ **18.** $3(x + 1)$ ◗ **19.** $3(a - 6)$ **20.** $4(y - 6)$

21. $-4(3x + 7)$ **22.** $-8(8y + 10)$

Objective C *Simplify each expression. First use the distributive property to multiply and remove parentheses. See Examples 11 through 13.*

23. $2(x + 4) - 7$ **24.** $5(6 - y) - 2$ **25.** $8 + 5(3c - 1)$ **26.** $10 + 4(6d - 2)$

27. $-4(6n - 5) + 3n$ **28.** $-3(5 - 2b) - 4b$ **29.** $3 + 6(w + 2) + w$ **30.** $8z + 5(6 + z) + 20$

31. $2(3x + 1) + 5(x - 2)$ **32.** $3(5x - 2) + 2(3x + 1)$ **33.** $-(2y - 6) + 10$ **34.** $-(5x - 1) - 10$

Objectives A B C **Mixed Practice** *Simplify each expression. See Examples 1 through 13.*

35. $18y - 20y$ **36.** $x + 12x$ **37.** $z - 8z$ **38.** $-12x + 8x$

39. $9d - 3c - d$ **40.** $8r + s - 7s$ **41.** $2y - 6 + 4y - 8$ **42.** $a + 4 - 7a - 5$

43. $5q + p - 6q - p$ **44.** $m - 8n + m + 8n$ ◗ **45.** $2(x + 1) + 20$ **46.** $5(x - 1) + 18$

47. $5(x - 7) - 8x$ **48.** $3(x + 2) - 11x$ **49.** $-5(z + 3) + 2z$

50. $-8(1 + v) + 6v$ **51.** $8 - x + 4x - 2 - 9x$ **52.** $5y - 4 + 9y - y + 15$

◗ **53.** $-7(x + 5) + 5(2x + 1)$ **54.** $-2(x + 4) + 8(3x - 1)$ **55.** $3r - 5r + 8 + r$

◗ **56.** $6x - 4 + 2x - x + 3$ **57.** $-3(n - 1) - 4n$ **58.** $5(c + 2) + 7c$

59. $4(z - 3) + 5z - 2$ **60.** $8(m + 3) - 20 + m$ **61.** $6(2x - 1) - 12x$

62. $5(2a + 3) - 10a$ **63.** $-(4x - 5) + 5$ **64.** $-(7y - 2) + 6$

65. $-(4x - 10) + 2(3x + 5)$ **66.** $-(12b - 10) + 5(3b - 2)$ **67.** $3a + 4(a + 3)$

68. $b + 2(b - 5)$ **69.** $5y - 2(y - 1) + 3$ **70.** $3x - 4(x + 2) + 1$

Objective **D** *Find the perimeter of each figure. See Example 14.*

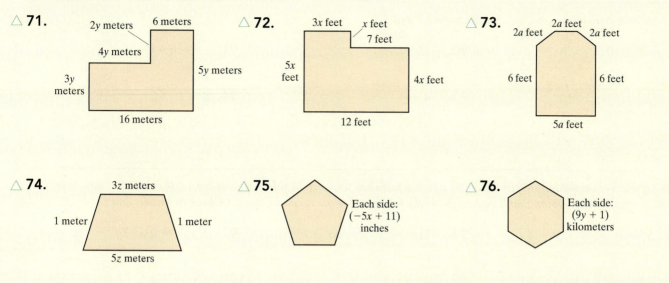

△ **71.** 2y meters 6 meters 4y meters 5y meters 3y meters 16 meters

△ **72.** 3x feet x feet 7 feet 5x feet 4x feet 12 feet

△ **73.** 2a feet 2a feet 2a feet 2a feet 6 feet 6 feet 5a feet

△ **74.** 3z meters 1 meter 1 meter 5z meters

△ **75.** Each side: $(-5x + 11)$ inches

△ **76.** Each side: $(9y + 1)$ kilometers

Find the area of each rectangle. See Example 15.

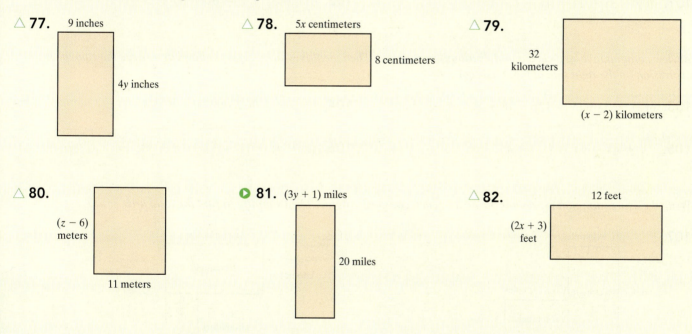

△ **77.** 9 inches 4y inches

△ **78.** 5x centimeters 8 centimeters

△ **79.** 32 kilometers $(x - 2)$ kilometers

△ **80.** $(z - 6)$ meters 11 meters

▶ **81.** $(3y + 1)$ miles 20 miles

△ **82.** 12 feet $(2x + 3)$ feet

Objectives **A** **B** **C** **D** **Mixed Practice** *Solve. See Examples 1 through 15.*

83. Find the area of a regulation NCAA basketball court that is 94 feet long and 50 feet wide.

84. Find the area of a rectangular movie screen that is 50 feet long and 40 feet high.

85. A decorator wishes to put a wallpaper border around a rectangular room that measures 14 feet by 18 feet. Find the room's perimeter.

86. How much fencing will a rancher need for a rectangular cattle lot that measures 80 feet by 120 feet?

87. Find the perimeter of a triangular garden that measures 5 feet by x feet by $(2x + 1)$ feet.

88. Find the perimeter of a triangular picture frame that measures x inches by x inches by $(x - 14)$ inches.

Review

Perform each indicated operation. See Sections 2.2 and 2.3.

89. $-13 + 10$

90. $-15 + 23$

91. $-4 - (-12)$

92. $-7 - (-4)$

93. $-4 + 4$

94. $8 + (-8)$

Concept Extensions

If the expression on the left side of the equal sign is equivalent to the right side, write "correct." If not, write "incorrect" and then write an expression that is equivalent to the left side. See the second Concept Check in this section.

95. $5(3x - 2) \stackrel{?}{=} 15x - 2$

96. $-2(4x - 1) \stackrel{?}{=} -8x - 2$

97. $2(xy) \stackrel{?}{=} 2x \cdot 2y$

98. $-8(ab) \stackrel{?}{=} -8a \cdot (-8b)$

99. $7x - (x + 2) \stackrel{?}{=} 7x - x - 2$

100. $12y - (3y - 1) \stackrel{?}{=} 12y - 3y + 1$

101. $4(y - 3) + 11 \stackrel{?}{=} 4y - 7 + 11$

102. $6(x + 5) + 2 \stackrel{?}{=} 6x + 30 + 12$

Review commutative, associative, and distributive properties. Then identify which property allows us to write the equivalent expression on the right side of the equal sign.

103. $6(2x - 3) + 5 = 12x - 18 + 5$

104. $9 + 7x + (-2) = 7x + 9 + (-2)$

105. $-7 + (4 + y) = (-7 + 4) + y$

106. $(x + y) + 11 = 11 + (x + y)$

Write the expression that represents the area of each composite figure. Then simplify to find the total area.

△ **107.**   △ **108.**

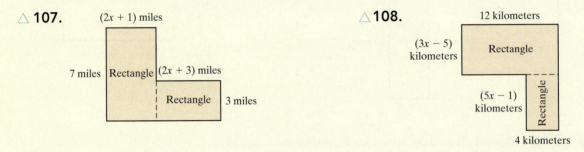

Simplify.

▦ **109.** $9684q - 686 - 4860q + 12{,}960$

▦ **110.** $76(268x + 592) - 2960$

✎ **111.** If x is a whole number, which expression is greater: $2x$ or $5x$? Explain your answer.

✎ **112.** If x is a whole number, which expression is greater: $-2x$ or $-5x$? Explain your answer.

✎ **113.** Explain what makes two terms "like terms."

✎ **114.** Explain how to combine like terms.

3.2 Solving Equations: Review of the Addition and Multiplication Properties ▶

Objective A Using the Addition Property or the Multiplication Property ▶

In this section, we continue solving equations using the properties first introduced in Section 2.6.

First, let's recall the difference between an **expression** and an **equation.** Remember—an equation contains an equal sign and an expression does not.

	Equations	Expressions	
equal signs	$7x = 6x + 4$	$7x - 6x + 4$	no equal signs
	$3(3y - 5) = 10y$	$y - 1 + 11y - 21$	

Thus far in this text, we have

Solved some equations (Section 2.6)

and

Simplified some expressions (Section 3.1)

As we will see in this section, the ability to simplify expressions will help us as we solve more equations.

The addition and multiplication properties are reviewed below.

Addition Property of Equality

Let a, b, and c represent numbers. Then

$$a = b$$
and $a + c = b + c$

are equivalent equations.

Also, $a = b$
and $a - c = b - c$

are equivalent equations.

Multiplication Property of Equality

Let a, b, and c represent numbers and let $c \neq 0$. Then

$$a = b$$
and $a \cdot c = b \cdot c$

are equivalent equations.

Also, $a = b$
and $\dfrac{a}{c} = \dfrac{b}{c}$

are equivalent equations.

In other words, the same number may be added to or subtracted from *both* sides of an equation without changing the solution of the equation. Also, *both* sides of an equation may be multiplied or divided by the same nonzero number without changing the solution of the equation.

Many equations in this section will contain expressions that can be simplified. If one or both sides of an equation can be simplified, do that first.

Example 1 Solve: $y - 5 = -2 - 6$

Solution: First we simplify the right side of the equation.

$y - 5 = -2 - 6$

$y - 5 = -8$ Combine like terms.

(Continued on next page)

Objectives

A Use the Addition Property or the Multiplication Property to Solve Equations. ▶

B Use Both Properties to Solve Equations. ▶

C Translate Word Phrases to Mathematical Expressions. ▶

Practice 1

Solve: $x + 6 = 1 - 3$

Answer
1. -8

Next we get y alone by using the addition property of equality. We add 5 to both sides of the equation.

$$y - 5 + 5 = -8 + 5 \quad \text{Add 5 to both sides.}$$
$$y = -3 \quad \text{Simplify.}$$

Check: To see that -3 is the solution, replace y with -3 in the original equation.

$$y - 5 = -2 - 6$$
$$-3 - 5 \overset{?}{=} -2 - 6 \quad \text{Replace } y \text{ with } -3.$$
$$-8 = -8 \quad \text{True}$$

Since $-8 = -8$ is true, the solution is -3.

🔶 **Work Practice 1**

Practice 2

Solve: $10 = 2m - 4m$

Example 2   Solve: $3y - 7y = 12$

Solution: First, simplify the left side of the equation by combining like terms.

$$3y - 7y = 12$$
$$-4y = 12 \quad \text{Combine like terms.}$$

Next, we get y alone by using the multiplication property of equality and dividing both sides by -4.

$$\frac{-4y}{-4} = \frac{12}{-4} \quad \text{Divide both sides by } -4.$$
$$y = -3 \quad \text{Simplify.}$$

Check: Replace y with -3 in the original equation.

$$3y - 7y = 12$$
$$3(-3) - 7(-3) \overset{?}{=} 12$$
$$-9 + 21 \overset{?}{=} 12$$
$$12 = 12 \quad \text{True}$$

The solution is -3.

🔶 **Work Practice 2**

✓**Concept Check**   What's wrong with the following solution?

$$4x - 6x = 10$$
$$2x = 10$$
$$\frac{2x}{2} = \frac{10}{2}$$
$$x = 5$$

Practice 3

Solve: $-8 + 6 = \dfrac{a}{3}$

Example 3   Solve: $\dfrac{z}{-4} = 11 - 5$

Solution: Simplify the right side of the equation first.

$$\frac{z}{-4} = 11 - 5$$
$$\frac{z}{-4} = 6$$

Answers

2. -5 3. -6

✓**Concept Check answer**

On the left side of the equation, $4x - 6x$ simplifies to $-2x$ *not* $2x$.

Next, to get z alone, multiply both sides by -4.

$$-4 \cdot \frac{z}{-4} = -4 \cdot 6 \qquad \text{Multiply both sides by } -4.$$

$$\frac{-4}{-4} \cdot z = -4 \cdot 6$$

$$1z = -24 \quad \text{or} \quad z = -24$$

Check to see that -24 is the solution.

■ **Work Practice 3**

Example 4 Solve: $5x + 2 - 4x = 7 - 19$

Solution: First we simplify each side of the equation separately.

$$5x + 2 - 4x = 7 - 19$$

$$\underbrace{5x - 4x} + 2 = \underbrace{7 - 19}$$

$$1x + 2 = -12 \qquad \text{Combine like terms.}$$

To get x alone on the left side, we subtract 2 from both sides.

$$1x + 2 - 2 = -12 - 2 \qquad \text{Subtract 2 from both sides.}$$

$$1x = -14 \quad \text{or} \quad x = -14 \quad \text{Simplify.}$$

Check to see that -14 is the solution.

■ **Work Practice 4**

Practice 4
Solve:
$-6y - 1 + 7y = 17 + 2$

Example 5 Solve: $8x - 9x = 12 - 17$

Solution: First combine like terms on each side of the equation.

$$8x - 9x = 12 - 17$$

$$-x = -5$$

Recall that $-x$ means $-1x$ and divide both sides by -1.

$$\frac{-1x}{-1} = \frac{-5}{-1} \qquad \text{Divide both sides by } -1.$$

$$x = 5 \qquad \text{Simplify.}$$

Check to see that the solution is 5.

■ **Work Practice 5**

Practice 5
Solve: $-4 - 10 = 4y - 5y$

Example 6 Solve: $3(3x - 5) = 10x$

Solution: First we multiply on the left side to remove the parentheses.

$$3(3x - 5) = 10x$$

$$3 \cdot 3x - 3 \cdot 5 = 10x \qquad \text{Use the distributive property.}$$

$$9x - 15 = 10x$$

Now we subtract $9x$ from both sides.

$$9x - 15 - 9x = 10x - 9x \qquad \text{Subtract } 9x \text{ from both sides.}$$

$$-15 = 1x \quad \text{or} \quad x = -15 \quad \text{Simplify.}$$

■ **Work Practice 6**

Practice 6
Solve: $13x = 4(3x - 1)$

Answers
4. 20   **5.** 14   **6.** -4

Objective B Using Both Properties to Solve Equations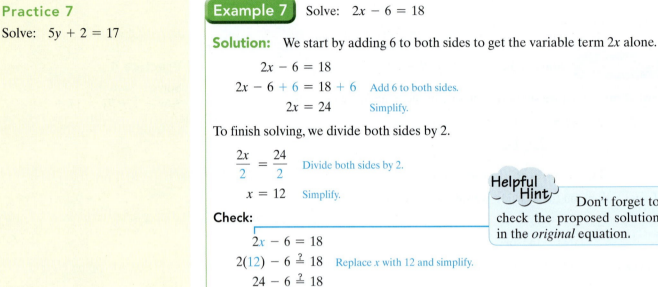

We now solve equations in one variable using more than one property of equality. To solve an equation such as $2x - 6 = 18$, we first get the variable term $2x$ alone on one side of the equation.

Practice 7

Solve: $5y + 2 = 17$

Example 7   Solve: $2x - 6 = 18$

Solution: We start by adding 6 to both sides to get the variable term $2x$ alone.

$$2x - 6 = 18$$
$$2x - 6 + 6 = 18 + 6 \quad \text{Add 6 to both sides.}$$
$$2x = 24 \quad \text{Simplify.}$$

To finish solving, we divide both sides by 2.

$$\frac{2x}{2} = \frac{24}{2} \quad \text{Divide both sides by 2.}$$
$$x = 12 \quad \text{Simplify.}$$

Check:

$$2x - 6 = 18$$
$$2(12) - 6 \stackrel{?}{=} 18 \quad \text{Replace } x \text{ with 12 and simplify.}$$
$$24 - 6 \stackrel{?}{=} 18$$
$$18 = 18 \quad \text{True}$$

Helpful Hint Don't forget to check the proposed solution in the *original* equation.

The solution is 12.

■ **Work Practice 7**

Don't forget, if one or both sides of an equation can be simplified, do that first.

Practice 8

Solve:
$-4(x + 2) - 60 = 2 - 10$

Example 8   Solve: $2 - 6 = -5(x + 4) - 39$

Solution: First, simplify each side of the equation.

$$2 - 6 = -5(x + 4) - 39$$
$$2 - 6 = -5x - 20 - 39 \quad \text{Use the distributive property.}$$
$$-4 = -5x - 59 \quad \text{Combine like terms on each side.}$$
$$-4 + 59 = -5x - 59 + 59 \quad \text{Add 59 to both sides to get the variable term alone.}$$
$$55 = -5x \quad \text{Simplify.}$$
$$\frac{55}{-5} = \frac{-5x}{-5} \quad \text{Divide both sides by } -5.$$
$$-11 = x \quad \text{or} \quad x = -11 \quad \text{Simplify.}$$

Check to see that -11 is the solution.

■ **Work Practice 8**

Answers

7. 3   **8.** −15

Objective C Translating Word Phrases into Expressions ▶

Section 3.4 in this chapter contains a formal introduction to problem solving. To prepare for this section, let's once again review writing phrases as algebraic expressions using the following key words and phrases as a guide:

Addition	Subtraction	Multiplication	Division
sum	difference	product	quotient
plus	minus	times	divided
added to	subtracted from	multiply	shared equally among
more than	less than	twice	per
increased by	decreased by	of	divided by
total	less	twice/double/triple	divided into

Example 9 Write each phrase as an algebraic expression. Use x to represent "a number."

a. a number increased by -5 **b.** the product of -7 and a number
c. a number less 20 **d.** the quotient of -18 and a number
e. a number subtracted from -2

Solution:

a. In words: a number | increased by | -5

Translate: x $+$ (-5) or $x - 5$

b. In words: the product of -7 and a number

Translate: $-7 \cdot x$ or $-7x$

c. In words: a number less 20

Translate: $x - 20$

d. In words: the quotient of -18 and a number

Translate: $-18 \div x$ or $\dfrac{-18}{x}$ or $-\dfrac{18}{x}$

e. In words: a number subtracted from -2

Translate: $-2 - x$

■ **Work Practice 9**

Practice 9

Write each phrase as an algebraic expression. Use x to represent "a number."

a. the sum of -3 and a number
b. -5 decreased by a number
c. three times a number
d. a number subtracted from 83
e. the quotient of a number and -4

Helpful Hint

As we reviewed in Chapter 1, don't forget that order is important when subtracting. Notice the translation order of numbers and variables below.

Phrase	Translation
a number less 9	$x - 9$
a number subtracted from 9	$9 - x$

Practice 10

Translate each phrase into an algebraic expression. Let x be the unknown number.

a. The product of 5 and a number, decreased by 25

b. Twice the sum of a number and 3

c. The quotient of 39 and twice a number

Example 10

Write each phrase as an algebraic expression. Let x be the unknown number.

a. Twice a number, increased by -9

b. Three times the difference of a number and 11

c. The quotient of 5 times a number and 17

Solution:

a. In words:

twice a number	increased by	-9
↓	↓	↓
$2x$	$+$	(-9) or $2x - 9$

Translate:

b. In words:

three times ⟶ the difference of

a number · and · 11
↓ · ↓ · ↓

Translate: 3 $(x$ $-$ $11)$

c. In words:

the quotient of

5 times a number · and · 17
↓ · ↓ · ↓

Translate: $5x$ $\div$ 17   or $\dfrac{5x}{17}$

Answers

10. **a.** $5x - 25$ **b.** $2(x + 3)$

c. $39 \div (2x)$ or $\dfrac{39}{2x}$

■ **Work Practice 10**

Vocabulary, Readiness & Video Check

Use the choices below to fill in each blank.

equation	multiplication	equivalent	expression
solving	addition	simplifying	

1. The equations $-3x = 51$ and $\dfrac{-3x}{-3} = \dfrac{51}{-3}$ are called _____ equations.

2. The difference between an equation and an expression is that a(n) _____ contains an equal sign, while a(n) _____ does not.

3. The process of writing $-3x + 10x$ as $7x$ is called _____ the expression.

4. For the equation $-5x - 1 = -21$, the process of finding that 4 is the solution is called _____ the equation.

5. By the _____ property of equality, $x = -2$ and $x + 7 = -2 + 7$ are equivalent equations.

6. By the _____ property of equality, $y = 8$ and $3 \cdot y = 3 \cdot 8$ are equivalent equations.

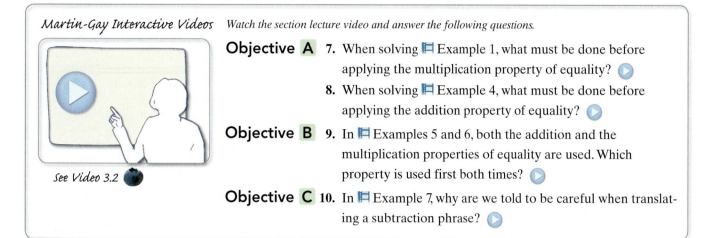

Martin-Gay Interactive Videos   Watch the section lecture video and answer the following questions.

Objective A 7. When solving ▣ Example 1, what must be done before applying the multiplication property of equality? ▶

8. When solving ▣ Example 4, what must be done before applying the addition property of equality? ▶

Objective B 9. In ▣ Examples 5 and 6, both the addition and the multiplication properties of equality are used. Which property is used first both times? ▶

Objective C 10. In ▣ Example 7, why are we told to be careful when translating a subtraction phrase? ▶

See Video 3.2

3.2 Exercise Set MyMathLab® ▶

Objective A *Solve each equation. First combine any like terms on each side of the equation. See Examples 1 through 5.*

1. $x - 3 = -1 + 4$

2. $x + 7 = 2 + 3$

▶ 3. $-7 + 10 = m - 5$

4. $1 - 8 = n + 2$

5. $2w - 12w = 40$

6. $10y - y = 45$

7. $24 = t + 3t$

8. $100 = 15y + 5y$

▶ 9. $2z = 12 - 14$

10. $-3x = 11 - 2$

11. $4 - 10 = \dfrac{z}{-3}$

12. $20 - 22 = \dfrac{z}{-4}$

13. $-3x - 3x = 50 - 2$

14. $5y - 9y = -14 + (-14)$

15. $\dfrac{x}{5} = -26 + 16$

16. $\dfrac{y}{3} = 32 - 52$

17. $7x + 7 - 6x = 10$

18. $-3 + 5x - 4x = 13$

19. $-8 - 9 = 3x + 5 - 2x$

20. $-7 + 10 = 4x - 6 - 3x$

Solve. First multiply to remove parentheses. See Example 6.

▶ 21. $2(5x - 3) = 11x$

22. $6(3x + 1) = 19x$

23. $3y = 2(y + 12)$

24. $17x = 4(4x - 6)$

25. $21y = 5(4y - 6)$

26. $28z = 9(3z - 2)$

27. $-3(-4 - 2z) = 7z$

28. $-2(-1 - 3y) = 7y$

Objective B *Solve each equation. See Examples 7 and 8.*

29. $2x - 8 = 0$

30. $3y - 12 = 0$

▶ 31. $7y + 3 = 24$

32. $5m + 1 = 46$

33. $-7 = 2x - 1$

34. $-11 = 3t - 2$

35. $6(6 - 4y) = 12y$

36. $4(3y - 5) = 14y$

37. $11(x - 6) = -4 - 7$

38. $5(x - 6) = -2 - 8$

39. $-3(x + 1) - 10 = 12 + 8$

40. $-2(x + 5) - 2 = -8 - 4$ **41.** $y - 20 = 6y$ **42.** $x - 63 = 10x$

43. $22 - 42 = 4(x - 1) - 4$ **44.** $35 - (-3) = 3(x - 2) + 17$

Objectives A B Mixed Practice *Solve each equation. See Examples 1 through 8.*

45. $-2 - 3 = -4 + x$ **46.** $7 - (-10) = x - 5$ **47.** $y + 1 = -3 + 4$

48. $y - 8 = -5 - 1$ ▶ **49.** $3w - 12w = -27$ **50.** $y - 6y = 20$

51. $-4x = 20 - (-4)$ **52.** $6x = 5 - 35$ **53.** $18 - 11 = \dfrac{x}{-5}$

54. $9 - 14 = \dfrac{x}{-12}$ **55.** $9x - 12 = 78$ **56.** $8x - 8 = 32$

57. $10 = 7t - 12t$ **58.** $-30 = t + 9t$ **59.** $5 - 5 = 3x + 2x$

60. $-42 + 20 = -2x + 13x$ **61.** $50y = 7(7y + 4)$ **62.** $65y = 8(8y - 9)$

63. $8x = 2(6x + 10)$ **64.** $10x = 6(2x - 3)$ **65.** $7x + 14 - 6x = -4 - 10$

66. $-10x + 11x + 5 = 9 - 5$ **67.** $\dfrac{x}{-4} = -1 - (-8)$ **68.** $\dfrac{y}{-6} = 6 - (-1)$

69. $23x + 8 - 25x = 7 - 9$ **70.** $8x - 4 - 6x = 12 - 22$ ▶ **71.** $-3(x + 9) - 41 = 4 - 60$

72. $-4(x + 7) - 30 = 3 - 37$

Objective C Translating *Write each phrase as a variable expression. Use x to represent "a number." See Examples 9 and 10.*

73. The sum of -7 and a number

74. Negative eight plus a number

▶ **75.** Eleven subtracted from a number

76. A number subtracted from twelve

77. The product of -13 and a number

78. Twice a number

79. A number divided by -12

80. The quotient of negative six and a number

▶ **81.** The product of -11 and a number, increased by 5

82. Negative four times a number, increased by 18

83. Negative ten decreased by 7 times a number

84. Twice a number, decreased by thirty

85. Seven added to the product of 4 and a number

86. The product of 7 and a number, added to 100

87. Twice a number, decreased by 17

88. The difference of -9 times a number, and 1

89. The product of -6 and the sum of a number and 15

90. Twice the sum of a number and -5

91. The quotient of 45 and the product of a number and -5

92. The quotient of ten times a number, and -4

93. The quotient of seventeen and a number, increased by -15

94. The quotient of -20 and a number, decreased by three

Review

This horizontal bar graph shows the top ten states for traveler spending in a recent year. Use this graph to answer Exercises 95 through 98. See Sections 1.2 and 1.3.

95. For what state do travelers spend the most money?

96. For the states shown, which states have traveler amounts less than $20 billion?

97. What is the combined spending for the neighboring states of Florida and Georgia?

98. What is the combined spending for the two largest states, Texas and California?

Top Ten U.S. States by Traveler Spending

State	Spending
California	$96
Florida	$67
New York	$52
Texas	$50
Illinois	$29
Nevada	$27
Georgia	$21
Pennsylvania	$21
Virginia	$19
New Jersey	$19

Domestic and International Traveler Spending Within State (in billions of dollars)

Source: U.S. Travel Association, 2010

Concept Extensions

99. In your own words, explain the addition property of equality.

100. Write an equation that can be solved using the addition property of equality.

101. Are the equations below equivalent? Why or why not?

$$x + 7 = 4 + (-9)$$
$$x + 7 \stackrel{?}{=} 5$$

102. Are the equations below equivalent? Why or why not?

$$3x - 6x = 12$$
$$3x \stackrel{?}{=} 12$$

103. Why does the multiplication property of equality not allow us to divide both sides of an equation by zero?

104. Is the equation $-x = 6$ solved for the variable? Explain why or why not.

Solve.

105. $\dfrac{y}{72} = -86 - (-1029)$

106. $\dfrac{x}{-13} = 4^6 - 5^7$

107. $\dfrac{x}{-2} = 5^2 - |-10| - (-9)$

108. $\dfrac{y}{10} = (-8)^2 - |20| + (-2)^2$

109. $|-13| + 3^2 = 100y - |-20| - 99y$

110. $4(x - 11) + |90| - |-86| + 2^5 = 5x$

Expressions and Equations

For the table below, identify each as an expression or an equation.

Expression or Equation	
1. $7x - 5y + 14$	
2. $7x = 35 + 14$	
3. $3(x - 2) = 5(x + 1) - 17$	
4. $-9(2x + 1) - 4(x - 2) + 14$	

Fill in each blank with "simplify" or "solve."

5. To _____ an expression, we combine any like terms.

6. To _____ an equation, we use the properties of equality to find any value of the variable that makes the equation a true statement.

Simplify each expression by combining like terms.

7. $7x + x$

8. $6y - 10y$

9. $2a + 5a - 9a - 2$

10. $6a - 12 - a - 14$

Multiply and simplify if possible.

11. $-2(4x + 7)$

12. $-3(2x - 10)$

13. $5(y + 2) - 20$

14. $12x + 3(x - 6) - 13$

△ **15.** Find the area of the rectangle.

Rectangle — 3 meters — $(4x - 2)$ meters

△ **16.** Find the perimeter of the triangle.

x feet, $(x + 2)$ feet, 7 feet

Solve and check.

17. $12 = 11x - 14x$

18. $8y + 7y = -45$

19. $x - 12 = -45 + 23$

20. $6 - (-5) = x + 5$

Solve and check.

21. $\dfrac{x}{3} = -14 + 9$

22. $\dfrac{z}{4} = -23 - 7$

23. $-6 + 2 = 4x + 1 - 3x$

24. $5 - 8 = 5x + 10 - 4x$

25. $6(3x - 4) = 19x$

26. $25x = 6(4x - 9)$

27. $-36x - 10 + 37x = -12 - (-14)$

28. $-8 + (-14) = -80y + 20 + 81y$

29. $3x - 16 = -10$

30. $4x - 21 = -13$

31. $-8z - 2z = 26 - (-4)$

32. $-12 + (-13) = 5x - 10x$

33. $-4(x + 8) - 11 = 3 - 26$

34. $-6(x - 2) + 10 = -4 - 10$

Translating *Write each phrase as an algebraic expression. Use x to represent "a number."*

35. The difference of a number and 10

36. The sum of -20 and a number

37. The product of 10 and a number

38. The quotient of 10 and a number

39. Five added to the product of -2 and a number

40. The product of -4 and the difference of a number and 1

21. _____

22. _____

23. _____

24. _____

25. _____

26. _____

27. _____

28. _____

29. _____

30. _____

31. _____

32. _____

33. _____

34. _____

35. _____

36. _____

37. _____

38. _____

39. _____

40. _____

Solving Linear Equations in One Variable ▶

Objectives

A Solve Linear Equations Using the Addition and Multiplication Properties. ▶

B Solve Linear Equations Containing Parentheses. ▶

C Write Numerical Sentences as Equations. ▶

In this chapter, the equations we are solving are called **linear equations in one variable** or **first-degree equations in one variable.** For example, an equation such as $5x - 2 = 6x$ is a linear equation in one variable. It is called linear or first degree because the exponent on each x is 1 and there is no variable below a fraction bar. It is an equation in one variable because it contains one variable, x.

Let's continue solving linear equations in one variable.

Objective **A** Solving Equations Using the Addition and Multiplication Properties ▶

If an equation contains variable terms on both sides, we use the addition property of equality to get all the variable terms on one side and all the constants or numbers on the other side.

Practice 1

Solve: $7x + 12 = 3x - 4$

Example 1 Solve: $3a - 6 = a + 4$

Solution: Although it makes no difference which side we choose, let's move variable terms to the left side and constants to the right side.

$$3a - 6 = a + 4$$
$$3a - 6 + 6 = a + 4 + 6 \qquad \text{Add 6 to both sides.}$$
$$3a = a + 10 \qquad \text{Simplify.}$$
$$3a - a = a + 10 - a \qquad \text{Subtract } a \text{ from both sides.}$$
$$2a = 10 \qquad \text{Simplify.}$$
$$\frac{2a}{2} = \frac{10}{2} \qquad \text{Divide both sides by 2.}$$
$$a = 5 \qquad \text{Simplify.}$$

Check:
$$3a - 6 = a + 4 \qquad \text{Original equation}$$
$$3 \cdot 5 - 6 \stackrel{?}{=} 5 + 4 \qquad \text{Replace } a \text{ with 5.}$$
$$15 - 6 \stackrel{?}{=} 9 \qquad \text{Simplify.}$$
$$9 = 9 \qquad \text{True}$$

The solution is 5.

▪ **Work Practice 1**

Helpful Hint

Make sure you understand which property to use to solve an equation.

Addition	Understood multiplication
$x + 2 = 10$	$2x = 10$
To undo addition of 2, we subtract 2 from both sides.	To undo multiplication of 2, we divide both sides by 2.
$x + 2 - 2 = 10 - 2$ Use addition property of equality.	$\frac{2x}{2} = \frac{10}{2}$ Use multiplication property of equality.
$x = 8$	$x = 5$
Check: $x + 2 = 10$	Check: $2x = 10$
$8 + 2 \stackrel{?}{=} 10$	$2 \cdot 5 \stackrel{?}{=} 10$
$10 = 10$ True	$10 = 10$ True

Answer

1. -4

Example 2   Solve: $17 - 7x + 3 = -3x + 21 - 3x$

Solution: First, simplify both sides of the equation.

$$17 - 7x + 3 = -3x + 21 - 3x$$
$$20 - 7x = -6x + 21 \qquad \text{Simplify.}$$

Next, move variable terms to one side of the equation and constants, or numbers, to the other side. To begin, let's add $6x$ to both sides.

$$20 - 7x + 6x = -6x + 21 + 6x \qquad \text{Add } 6x \text{ to each side.}$$
$$20 - x = 21 \qquad \text{Simplify.}$$
$$20 - x - 20 = 21 - 20 \qquad \text{Subtract 20 from both sides.}$$
$$-1x = 1 \qquad \text{Simplify. Recall that } -x \text{ means } -1x.$$
$$\frac{-1x}{-1} = \frac{1}{-1} \qquad \text{Divide both sides by } -1.$$
$$x = -1 \qquad \text{Simplify.}$$

Check:
$$17 - 7x + 3 = -3x + 21 - 3x$$
$$17 - 7(-1) + 3 \stackrel{?}{=} -3(-1) + 21 - 3(-1)$$
$$17 + 7 + 3 \stackrel{?}{=} 3 + 21 + 3$$
$$27 = 27 \qquad \text{True}$$

The solution is -1.

■ **Work Practice 2**

Objective B Solving Equations Containing Parentheses ▶

Recall from the previous section that if an equation contains parentheses, we will first use the distributive property to remove them.

Example 3   Solve: $7(x - 2) = 9x - 6$

Solution: First we apply the distributive property.

$$7(x - 2) = 9x - 6$$
$$7x - 14 = 9x - 6 \qquad \text{Apply the distributive property.}$$

Next we move variable terms to one side of the equation and constants to the other side.

$$7x - 14 - 9x = 9x - 6 - 9x \qquad \text{Subtract } 9x \text{ from both sides.}$$
$$-2x - 14 = -6 \qquad \text{Simplify.}$$
$$-2x - 14 + 14 = -6 + 14 \qquad \text{Add 14 to both sides.}$$
$$-2x = 8 \qquad \text{Simplify.}$$
$$\frac{-2x}{-2} = \frac{8}{-2} \qquad \text{Divide both sides by } -2.$$
$$x = -4 \qquad \text{Simplify.}$$

Check to see that -4 is the solution.

■ **Work Practice 3**

✓**Concept Check** In Example 3, the solution is -4. To check this solution, what equation should we use?

Practice 2

Solve:
$$40 - 5y + 5 = -2y - 10 - 4y$$

Practice 3

Solve: $6(a - 5) = 4a + 4$

Answers
2. -55 **3.** 17

✓**Concept Check Answer**
$7(x - 2) = 9x - 6$

You may want to use the following steps to solve equations.

> ## Steps for Solving an Equation
>
> **Step 1:** If parentheses are present, use the distributive property.
>
> **Step 2:** Combine any like terms on each side of the equation.
>
> **Step 3:** Use the addition property of equality to rewrite the equation so that variable terms are on one side of the equation and constant terms are on the other side.
>
> **Step 4:** Use the multiplication property of equality to divide both sides by the numerical coefficient of the variable to solve.
>
> **Step 5:** Check the solution in the *original equation*.

Practice 4

Solve: $4(x + 3) + 1 = 13$

Example 4   Solve: $3(2x - 6) + 6 = 0$

Solution: $3(2x - 6) + 6 = 0$

Step 1: $6x - 18 + 6 = 0$ Apply the distributive property.

Step 2: $6x - 12 = 0$ Combine like terms on the left side of the equation.

Step 3: $6x - 12 + 12 = 0 + 12$ Add 12 to both sides.

$6x = 12$ Simplify.

Step 4: $\dfrac{6x}{6} = \dfrac{12}{6}$ Divide both sides by 6.

$x = 2$ Simplify.

Check:

Step 5: $3(2x - 6) + 6 = 0$

$3(2 \cdot 2 - 6) + 6 \stackrel{?}{=} 0$

$3(4 - 6) + 6 \stackrel{?}{=} 0$

$3(-2) + 6 \stackrel{?}{=} 0$

$-6 + 6 \stackrel{?}{=} 0$

$0 = 0$ True

The solution is 2.

■ **Work Practice 4**

Objective C Writing Numerical Sentences as Equations

Next, we practice translating sentences into equations. Below are key words and phrases that translate to an equal sign. (*Note:* For a review of key words and phrases that translate to addition, subtraction, multiplication, and division, see Sections 1.8 and 3.2.)

Key Words or Phrases	Examples	Symbols
equals	3 equals 2 plus 1	$3 = 2 + 1$
gives	the quotient of 10 and -5 gives -2	$\dfrac{10}{-5} = -2$
is/was	17 minus 12 is 5	$17 - 12 = 5$
yields	11 plus 2 yields 13	$11 + 2 = 13$
amounts to	twice -15 amounts to -30	$2(-15) = -30$
is equal to	-24 is equal to 2 times -12	$-24 = 2(-12)$

Answer

4. 0

Example 5 Translate each sentence into an equation.

a. The product of 7 and 6 is 42.
b. Twice the sum of 3 and 5 is equal to 16.
c. The quotient of -45 and 5 yields -9.

Solution:

a. In words: the product of 7 and 6   is   42

Translate: $7 \cdot 6$ $=$ 42

b. In words: twice   the sum of 3 and 5   is equal to   16

Translate: 2 $(3 + 5)$ $=$ 16

c. In words: the quotient of -45 and 5   yields -9

Translate: $\dfrac{-45}{5}$ $=$ -9

■ **Work Practice 5**

Practice 5

Translate each sentence into an equation.
a. The difference of 110 and 80 is 30.
b. The product of 3 and the sum of -9 and 11 amounts to 6.
c. The quotient of 24 and -6 yields -4.

Answers

5. a. $110 - 80 = 30$
b. $3(-9 + 11) = 6$ **c.** $\dfrac{24}{-6} = -4$

🖩 Calculator Explorations   Checking Equations

A calculator can be used to check possible solutions of equations. To do this, replace the variable by the possible solution and evaluate each side of the equation separately. For example, to see whether 7 is a solution of the equation $52x = 15x + 259$, replace x with 7 and use your calculator to evaluate each side separately.

Equation: $52x = 15x + 259$
   $52 \cdot 7 \overset{?}{=} 15 \cdot 7 + 259$ *Replace x with 7.*

Evaluate left side: $\boxed{52}$ $\boxed{\times}$ $\boxed{7}$ then $\boxed{=}$ or $\boxed{\text{ENTER}}$.
Display: $\boxed{364}$.
Evaluate right side: $\boxed{15}$ $\boxed{\times}$ $\boxed{7}$ $\boxed{+}$ $\boxed{259}$ then $\boxed{=}$ or $\boxed{\text{ENTER}}$.
Display: $\boxed{364}$.

Since the left side equals the right side, 7 is a solution of the equation $52x = 15x + 259$.

Use a calculator to determine whether the numbers given are solutions of each equation.

1. $76(x - 25) = -988;$ 12
2. $-47x + 862 = -783;$ 35
3. $x + 562 = 3x + 900;$ -170
4. $55(x + 10) = 75x + 910;$ -18
5. $29x - 1034 = 61x - 362;$ -21
6. $-38x + 205 = 25x + 120;$ 25

Vocabulary, Readiness & Video Check

Use the choices below to fill in each blank. Some choices may be used more than once.

addition   multiplication   combine like terms
$5(2x + 6) - 1 = 39$ $3x - 9 + x - 16$ distributive

1. An example of an expression is _____ while an example of an equation is _____ .
2. To solve $\dfrac{x}{-7} = -10$, we use the _____ property of equality.
3. To solve $x - 7 = -10$, we use the _____ property of equality.

Use the order of the Steps for Solving an Equation in this section to answer Exercises 4 through 6.

4. To solve $9x - 6x = 10 + 6$, first _____.

5. To solve $5(x - 1) = 25$, first use the _____ property.

6. To solve $4x + 3 = 19$, first use the _____ property of equality.

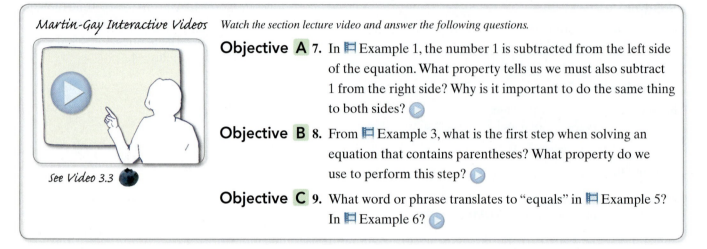

Martin-Gay Interactive Videos   *Watch the section lecture video and answer the following questions.*

Objective A 7. In ⊟ Example 1, the number 1 is subtracted from the left side of the equation. What property tells us we must also subtract 1 from the right side? Why is it important to do the same thing to both sides? ▶

Objective B 8. From ⊟ Example 3, what is the first step when solving an equation that contains parentheses? What property do we use to perform this step? ▶

See Video 3.3

Objective C 9. What word or phrase translates to "equals" in ⊟ Example 5? In ⊟ Example 6? ▶

3.3 Exercise Set MyMathLab® ▶

Objective A *Solve each equation. See Examples 1 and 2.*

1. $3x - 7 = 4x + 5$

2. $7x - 1 = 8x + 4$

▶ **3.** $10x + 15 = 6x + 3$

4. $5x - 3 = 2x - 18$

5. $19 - 3x = 14 + 2x$

6. $4 - 7m = -3m + 4$

7. $-14x - 20 = -12x + 70$

8. $57y + 140 = 54y - 100$

9. $x + 20 + 2x = -10 - 2x - 15$

10. $2x + 10 + 3x = -12 - x - 20$

11. $40 + 4y - 16 = 13y - 12 - 3y$

12. $19x - 2 - 7x = 31 + 6x - 15$

Objective B *Solve each equation. See Examples 3 and 4.*

13. $35 - 17 = 3(x - 2)$

14. $22 - 42 = 4(x - 1)$

▶ **15.** $3(x - 1) - 12 = 0$

16. $2(x + 5) + 8 = 0$

17. $2(y - 3) = y - 6$

18. $3(z + 2) = 5z + 6$

19. $-2(y + 4) = 2$

20. $-1(y + 3) = 10$

21. $2t - 1 = 3(t + 7)$

22. $-4 + 3c = 4(c + 2)$

23. $3(5c + 1) - 12 = 13c + 3$

24. $4(3t + 4) - 20 = 3 + 5t$

Mixed Practice *(Sections 2.6, 3.2, 3.3)* *Solve each equation.*

25. $-4x = 44$

26. $-3x = 51$

27. $x + 9 = 2$

28. $y - 6 = -11$

29. $8 - b = 13$

30. $7 - z = 15$

31. $-20 - (-50) = \dfrac{x}{9}$

32. $-2 - 10 = \dfrac{z}{10}$

33. $3r + 4 = 19$

34. $7y + 3 = 38$

▶ 35. $-7c + 1 = -20$

36. $-2b + 5 = -7$

37. $8y - 13y = -20 - 25$

38. $4x - 11x = -14 - 14$

39. $6(7x - 1) = 43x$

40. $5(3y - 2) = 16y$

41. $-4 + 12 = 16x - 3 - 15x$

42. $-9 + 20 = 19x - 4 - 18x$

43. $-10(x + 3) + 28 = -16 - 16$

44. $-9(x + 2) + 25 = -19 - 19$

45. $4x + 3 = 2x + 11$

46. $6y - 8 = 3y + 7$

47. $-2y - 10 = 5y + 18$

48. $7n + 5 = 12n - 10$

49. $-8n + 1 = -6n - 5$

50. $10w + 8 = w - 10$

51. $9 - 3x = 14 + 2x$

52. $4 - 7m = -3m$

53. $9a + 29 + 7 = 0$

54. $10 + 4v + 6 = 0$

55. $7(y - 2) = 4y - 29$

56. $2(z - 2) = 5z + 17$

57. $12 + 5t = 6(t + 2)$

58. $4 + 3c = 2(c + 2)$

▶ 59. $3(5c - 1) - 2 = 13c + 3$

60. $4(2t + 5) - 21 = 7t - 6$

61. $10 + 5(z - 2) = 4z + 1$

62. $14 + 4(w - 5) = 6 - 2w$

63. $7(6 + w) = 6(2 + w)$

64. $6(5 + c) = 5(c - 4)$

Objective C Translating *Write each sentence as an equation. See Example 5.*

65. The sum of -42 and 16 is -26.

66. The difference of -30 and 10 equals -40.

▶ 67. The product of -5 and -29 gives 145.

68. The quotient of -16 and 2 yields -8.

▶ 69. Three times the difference of -14 and 2 amounts to -48.

70. Negative 2 times the sum of 3 and 12 is -30.

71. The quotient of 100 and twice 50 is equal to 1.

72. Seventeen subtracted from -12 equals -29.

Review

The following bar graph shows the number of U.S. federal individual income tax returns that are filed electronically during the years shown. Electronically filed returns include Telefile and online returns. Use this graph to answer Exercises 73 through 76. Write number answers in standard form. See Sections 1.2 and 1.3.

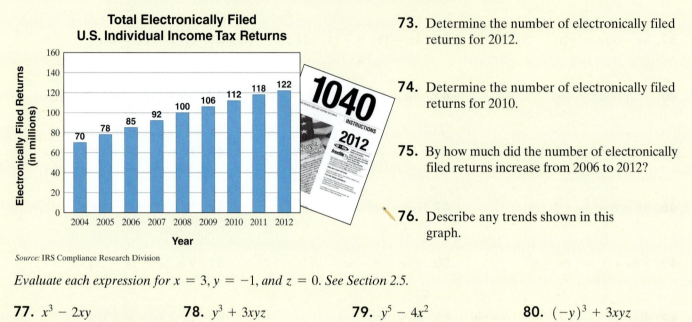

Source: IRS Compliance Research Division

73. Determine the number of electronically filed returns for 2012.

74. Determine the number of electronically filed returns for 2010.

75. By how much did the number of electronically filed returns increase from 2006 to 2012?

76. Describe any trends shown in this graph.

Evaluate each expression for $x = 3$, $y = -1$, and $z = 0$. See Section 2.5.

77. $x^3 - 2xy$

78. $y^3 + 3xyz$

79. $y^5 - 4x^2$

80. $(-y)^3 + 3xyz$

Concept Extensions

Using the Steps for Solving an Equation, choose the next operation for solving the given equation.

81. $2x - 5 = -7$
    a. Add 7 to both sides.
    b. Add 5 to both sides.
    c. Divide both sides by 2.

82. $3x + 2x = -x - 4$
    a. Add 4 to both sides.
    b. Subtract $2x$ from both sides.
    c. Add $3x$ and $2x$.

83. $-3x = -12$
    a. Divide both sides by -3.
    b. Add 12 to both sides.
    c. Add $3x$ to both sides.

84. $9 - 5x = 15$
    a. Divide both sides by -5.
    b. Subtract 15 from both sides.
    c. Subtract 9 from both sides.

A classmate shows you steps for solving an equation. The solution does not check, but the classmate is unable to find the error. For each set of steps, check the solution, find the error, and correct it.

85.
$$2(3x - 5) = 5x - 7$$
$$6x - 5 = 5x - 7$$
$$6x - 5 + 5 = 5x - 7 + 5$$
$$6x = 5x - 2$$
$$6x - 5x = 5x - 2 - 5x$$
$$x = -2$$

86.
$$37x + 1 = 9(4x - 7)$$
$$37x + 1 = 36x - 7$$
$$37x + 1 - 1 = 36x - 7 - 1$$
$$37x = 36x - 8$$
$$37x - 36x = 36x - 8 - 36x$$
$$x = -8$$

Solve.

87. $(-8)^2 + 3x = 5x + 4^3$

88. $3^2 \cdot x = (-9)^3$

89. $2^3(x + 4) = 3^2(x + 4)$

90. $x + 45^2 = 54^2$

91. A classmate tries to solve $3x = 39$ by subtracting 3 from both sides of the equation. Will this step solve the equation for x? Why or why not?

92. A classmate tries to solve $2 + x = 20$ by dividing both sides by 2. Will this step solve the equation for x? Why or why not?

Linear Equations in One Variable and Problem Solving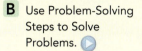

Objective A Writing Sentences as Equations

Now that we have practiced solving equations for a variable, we can extend considerably our problem-solving skills. We begin by writing sentences as equations using the following key words and phrases as a guide:

Addition	Subtraction	Multiplication	Division	Equal Sign
sum	difference	product	quotient	equals
plus	minus	times	divide	gives
added to	subtracted from	multiply	shared equally among	is/was
more than	less than	twice	per	yields
increased by	decreased by	of	divided by	amounts to
total	less	double	divided into	is equal to

Objectives

A Write Sentences as Equations.

B Use Problem-Solving Steps to Solve Problems.

Example 1 Write each sentence as an equation. Use x to represent "a number."

a. Twenty increased by a number is 5.

b. Twice a number equals -10.

c. A number minus 11 amounts to 168.

d. Three times the sum of a number and 5 is -30.

e. The quotient of twice a number and 8 is equal to 2.

Solution:

a. In words: twenty   increased by   a number   is   5

  Translate: 20 $+$ x $=$ 5

b. In words: twice a number   equals -10

  Translate: $2x$ $=$ -10

c. In words: a number   minus   11   amounts to   168

  Translate: x $-$ 11 $=$ 168

d. In words: three times   the sum of a number and 5   is -30

  Translate: 3 $(x + 5)$ $=$ -30

(*Continued on next page*)

Practice 1

Write each sentence as an equation. Use x to represent "a number."

a. Four times a number is 20.

b. The sum of a number and -5 yields 32.

c. Fifteen subtracted from a number amounts to -23.

d. Five times the difference of a number and 7 is equal to -8.

e. The quotient of triple a number and 5 gives 1.

Answers

1. a. $4x = 20$ **b.** $x + (-5) = 32$
c. $x - 15 = -23$
d. $5(x - 7) = -8$
e. $\dfrac{3x}{5} = 1$

e. In words:

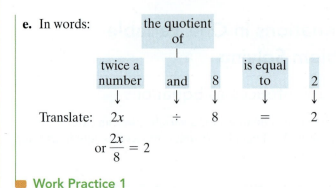

or $\dfrac{2x}{8} = 2$

■ **Work Practice 1**

Objective B Using Problem-Solving Steps to Solve Problems ▶

Our main purpose for studying arithmetic and algebra is to solve problems. In previous sections, we have prepared for problem solving by writing phrases as algebraic expressions and sentences as equations. We now draw upon this experience as we solve problems. The following problem-solving steps will be used throughout this text.

Problem-Solving Steps

1. **UNDERSTAND** the problem. During this step, become comfortable with the problem. Some ways of doing this are as follows:
   - Read and reread the problem.
   - Construct a drawing.
   - Propose a solution and check. Pay careful attention to how you check your proposed solution. This will help when writing an equation to model the problem.
   - Choose a variable to represent the unknown. Use this variable to represent any other unknowns.
2. **TRANSLATE** the problem into an equation.
3. **SOLVE** the equation.
4. **INTERPRET** the results: *Check* the proposed solution in the stated problem and *state* your conclusion.

The first problem that we solve consists of finding an unknown number.

Example 2 Finding an Unknown Number

Twice a number plus 3 is the same as the number minus 6. Find the unknown number.

Solution:

1. **UNDERSTAND** the problem. To do so, we read and reread the problem.

   Let's propose a solution to help us understand. Suppose the unknown number is 5. Twice this number plus 3 is $2 \cdot 5 + 3$ or 13. Is this the same as the number minus 6, or $5 - 6$, or -1? Since 13 is not the same as -1, we know that 5 is not the solution. However, remember that the purpose of proposing a solution is not to guess correctly, but to better understand the problem.

   Now let's choose a variable to represent the unknown. Let's let

   x = unknown number

Practice 2

Translate "the sum of a number and 2 equals 6 added to three times the number" into an equation and solve.

Answer

2. -2

2. TRANSLATE the problem into an equation.

In words:	twice a number	plus 3	is the same as	the number minus 6
	↓	↓	↓	↓
Translate:	$2x$	$+ 3$	$=$	$x - 6$

3. SOLVE the equation. To solve the equation, we first subtract x from both sides.

$$2x + 3 = x - 6$$
$$2x + 3 - x = x - 6 - x$$
$$x + 3 = -6 \qquad \text{Simplify.}$$
$$x + 3 - 3 = -6 - 3 \qquad \text{Subtract 3 from both sides.}$$
$$x = -9 \qquad \text{Simplify.}$$

4. INTERPRET the results. First, *check* the proposed solution in the stated problem. Twice "-9" is -18 and $-18 + 3$ is -15. This is equal to the number minus 6, or "-9" $- 6$, or -15. Then *state* your conclusion: The unknown number is -9.

■ **Work Practice 2**

✓ **Concept Check** Suppose you have solved an equation involving perimeter to find the length of a rectangular table. Explain why you would want to recheck your math if you obtain the result of -5.

Example 3   Determining Distances

The distance by road from Chicago, Illinois, to Los Angeles, California, is 1091 miles *more* than the distance from Chicago to Boston, Massachusetts. If the total of these two distances is 3017 miles, find the distance from Chicago to Boston. (*Source: World Almanac*)

Solution:

1. UNDERSTAND the problem. We read and reread the problem.

Let's propose and check a solution to help us better understand the problem. Suppose the distance from Chicago to Boston is 600 miles. Since the distance from Chicago to Los Angeles is 1091 miles *more*, then this distance is $600 + 1091 = 1691$ miles. With these numbers, the total of the distances is $600 + 1691 = 2291$ miles. This is less than the given total of 3017 miles, so we are incorrect. But not only do we have a better understanding of this exercise, we also know that the distance from Boston to Chicago is greater than 600 miles since this proposed solution led to a total too small. Now let's choose a variable to represent an unknown. Then we'll use this variable to represent any other unknown quantities. Let

$$x = \text{distance from Chicago to Boston}$$

Then

$$x + 1091 = \text{distance from Chicago to Los Angeles}$$

Since that distance is 1091 miles more.

2. TRANSLATE the problem into an equation.

In words:	Chicago to Boston distance	+	Chicago to Los Angeles distance	=	total miles
	↓		↓		↓
Translate:	x	+	$x + 1091$	=	3017

(Continued on next page)

Practice 3

The distance by road from Cincinnati, Ohio, to Denver, Colorado, is 71 miles *less* than the distance from Denver to San Francisco, California. If the total of these two distances is 2399 miles, find the distance from Denver to San Francisco.

Answer
3. 1235 miles

✓ **Concept Check Answer**
Length cannot be negative.

3. SOLVE the equation:

$$x + x + 1091 = 3017$$
$$2x + 1091 = 3017 \qquad \text{Combine like terms.}$$
$$2x + 1091 - 1091 = 3017 - 1091 \qquad \text{Subtract 1091 from both sides.}$$
$$2x = 1926 \qquad \text{Simplify.}$$
$$\frac{2x}{2} = \frac{1926}{2} \qquad \text{Divide both sides by 2.}$$
$$x = 963 \qquad \text{Simplify.}$$

4. INTERPRET the results. First *check* the proposed solution in the stated problem. Since x represents the distance from Chicago to Boston, this is 963 miles. The distance from Chicago to Los Angeles is $x + 1091 = 963 + 1091 = 2054$ miles. To check, notice that the total number of miles is $963 + 2054 = 3017$ miles, the given total of miles. Also, 2054 is 1091 more miles than 963, so the solution checks. Then, *state* your conclusion: The distance from Chicago to Boston is 963 miles.

■ **Work Practice 3**

Practice 4

A woman's $57,000 estate is to be divided so that her husband receives twice as much as her son. How much will each receive?

Example 4   Calculating Separate Costs

A salesperson at an electronics store sold a computer system and software for $2100, receiving four times as much money for the computer system as for the software. Find the price of each.

Solution:

1. UNDERSTAND the problem. We read and reread the problem. Then we choose a variable to represent an unknown. We use this variable to represent any other unknown quantities. We let

$$x = \text{the software price}$$

Then

$$4x = \text{the computer system price}$$

2. TRANSLATE the problem into an equation.

In words:	software price	and	computer price	is	2100
	↓	↓	↓	↓	↓
Translate:	x	$+$	$4x$	$=$	2100

3. SOLVE the equation:

$$x + 4x = 2100$$
$$5x = 2100 \qquad \text{Combine like terms.}$$
$$\frac{5x}{5} = \frac{2100}{5} \qquad \text{Divide both sides by 5.}$$
$$x = 420 \qquad \text{Simplify.}$$

4. INTERPRET the results. *Check* the proposed solution in the stated problem. The software sold for $420. The computer system sold for $4x = 4(\$420) = \$1680$. Since $\$420 + \$1680 = \$2100$, the total price, and $1680 is four times $420, the solution checks. *State* your conclusion: The software sold for $420, and the computer system sold for $1680.

■ **Work Practice 4**

Answer
4. husband: $38,000; son: $19,000

Vocabulary, Readiness & Video Check

Martin-Gay Interactive Videos Watch the section lecture video and answer the following questions.

See Video 3.4

Objective A 1. In ⊟ Example 2, why does the left side of the equation translate to $-20 - x$ and not $x - (-20)$? ▶

Objective B 2. Why are parentheses used in the translation of the left side of the equation in ⊟ Example 4? ▶

3. In ⊟ Example 5, the solution to the equation is $x = 37$. Why is this not the solution to the application? ▶

3.4 Exercise Set MyMathLab®

Objective A Translating *Write each sentence as an equation. Use x to represent "a number." See Example 1.*

1. A number added to -5 is -7.

2. Five subtracted from a number equals 10.

3. Three times a number yields 27.

4. The quotient of 8 and a number is -2.

5. A number subtracted from -20 amounts to 104.

6. Two added to twice a number gives -14.

7. Twice a number gives 108.

8. Five times a number is equal to -75.

9. The product of 5 and the sum of -3 and a number is -20.

10. Twice the sum of -17 and a number is -14.

Objective B Translating *Translate each sentence into an equation. Then solve the equation. See Example 2.*

11. Three times a number, added to 9, is 33. Find the number.

12. Twice a number, subtracted from 60, is 20. Find the number.

13. The sum of 3, 4, and a number amounts to 16. Find the number.

14. The sum of 7, 9, and a number is 40. Find the number.

15. The difference of a number and 3 is equal to the quotient of 10 and 5. Find the number.

16. Eight decreased by a number equals the quotient of 15 and 5. Find the number.

17. Thirty less a number is equal to the product of 3 and the sum of the number and 6. Find the number.

18. The product of a number and 3 is twice the sum of that number and 5. Find the number.

19. 40 subtracted from five times a number is 8 more than the number. Find the number.

20. Five times the sum of a number and 2 is 11 less than the number times 8. Find the number.

21. Three times the difference of some number and 5 amounts to the quotient of 108 and 12. Find the number.

22. Seven times the difference of some number and 1 gives the quotient of 70 and 10. Find the number.

23. The product of 4 and a number is the same as 30 less twice that same number. Find the number.

24. Twice a number equals 25 less triple that same number. Find the number.

Solve. For Exercises 25 and 26, the solutions have been started for you. See Examples 3 and 4.

25. Florida has 28 fewer electoral votes for president than California. If the total number of electoral votes for these two states is 82, find the number for each state. (*Source: The World Almanac* 2013)

Start the solution:

1. UNDERSTAND the problem. Reread it as many times as needed. Let's let

   x = number of electoral votes for California

   Then

   $x - 28$ = number of electoral votes for Florida

2. TRANSLATE into an equation. (Fill in the blanks below.)

$$\boxed{\text{votes for California}} + \boxed{\text{votes for Florida}} = 82$$
$$\downarrow \qquad\qquad \downarrow$$
$$\underline{\hspace{1.5cm}} + \underline{\hspace{1.5cm}} = 82$$

Now, you finish with

3. SOLVE the equation.
4. INTERPRET the results.

26. Texas has twice the number of electoral votes for president as Michigan. If the total number of electoral votes for these two states is 51, find the number for each state. (*Source: The World Almanac* 2013)

Start the solution:

1. UNDERSTAND the problem. Reread it as many times as needed. Let's let

   x = number of electoral votes for Michigan

   Then

   $2x$ = number of electoral votes for Texas

2. TRANSLATE into an equation. (Fill in the blanks below.)

$$\boxed{\text{votes for Michigan}} + \boxed{\text{votes for Texas}} = 51$$
$$\downarrow \qquad\qquad \downarrow$$
$$\underline{\hspace{1.5cm}} + \underline{\hspace{1.5cm}} = 51$$

Now, you finish with

3. SOLVE the equation.
4. INTERPRET the results.

27. A falcon, when diving, can travel five times as fast as a pheasant's top speed. If the total speed for these two birds is 222 miles per hour, find the fastest speed of the falcon and the fastest speed of the pheasant.
(*Source: Fantastic Book of Comparisons*)

28. Norway has had three times as many rulers as Liechtenstein. If the total number of rulers for both countries is 56, find the number of rulers for Norway and the number for Liechtenstein.

Norway

Liechtenstein

29. The U.S. Sunday newspaper with the greatest circulation is *The New York Times,* followed by *The Los Angeles Times*. If the Sunday circulation for *The New York Times* is 384 thousand more than the circulation for *The Los Angeles Times*, and their combined circulation is 2494 thousand, find the circulation for each newspaper. (*Source: The Top Ten of Everything*, 2013)

30. The average life expectancy for an elephant is 24 years longer than the life expectancy for a chimpanzee. If the total of these life expectancies is 130 years, find the life expectancy of each.

31. Coca-Cola was recently named the Most Valuable Global Brand. It has a recognition value worth $43 billion more than Disney's. If the total value of these two global brands is $101 billion, find the value of each of the brands. (See the Chapter 3 Opener.)

32. The global brand Microsoft has a recognition value worth $26 billion more than Apple's. If the total value of these two global brands is $92 billion, find the value of each of the brands. (See the Chapter 3 Opener.)

33. An Xbox 360 game system and several games are sold for $440. The cost of the Xbox 360 is 3 times as much as the cost of the games. Find the cost of the Xbox 360 and the cost of the games.

34. The two top-selling PC games are *Call of Duty: Black Ops 2* and *Madden NFL 13*. A price for *Call of Duty: Black Ops 2* is $13 more than a price for *Madden NFL 13*. If the total of these two prices is $47, find the price of each game.(*Source:* Internet Research)

35. By air, the distance from New York City to London is 2001 miles *less* than the distance from Los Angeles to Tokyo. If the total of these two distances is 8939 miles, find the distance from Los Angeles to Tokyo.

36. By air, the distance from Melbourne, Australia, to Cairo, Egypt, is 2338 miles *more* than the distance from Madrid, Spain, to Bangkok, Thailand. If the total of these distances is 15,012 miles, find the distance from Madrid to Bangkok.

37. The two NCAA stadiums with the largest capacities are Beaver Stadium (Penn State) and Michigan Stadium (Univ. of Michigan). Beaver Stadium has a capacity of 1081 more than Michigan Stadium. If the combined capacity for the two stadiums is 213,483, find the capacity for each stadium. (*Source:* National Collegiate Athletic Association)

38. A National Hot Rod Association (NHRA) top fuel dragster has a top speed of 95 mph faster than an Indy Racing League car. If the top speed for these two cars combined is 565 mph, find the top speed of each car. (*Source: USA Today*)

39. In 2020, China is projected to be the country with the greatest number of visiting tourists. This number is twice the number of tourists projected for Spain. If the total number of tourists for these two countries is projected to be 210 million, find the number projected for each. (*Source: The State of the World Atlas* by Dan Smith)

40. California contains the largest state population of native Americans. This population is three times the native American population of Washington state. If the total of these two populations is 412 thousand, find the native American population in each of these two states. (*Source:* U.S. Census Bureau)

41. In Germany, about twice as many cars are manufactured per day than in the United States. If the total number of these cars manufactured per day is 24,258, find the number manufactured in the United States and the number manufactured in Germany. (Based on data from the International Organization of Motor Vehicle Manufacturers)

42. A Toyota Camry is traveling twice as fast as a Dodge truck. If their combined speed is 105 miles per hour, find the speed of the car and find the speed of the truck.

43. A biker sold his used mountain bike and accessories for $270. If he received five times as much money for the bike as he did for the accessories, find how much money he received for the bike.

44. A tractor and a plow attachment are worth $1200. The tractor is worth seven times as much money as the plow. Find the value of the tractor and the value of the plow.

45. During the 2013 Women's NCAA Division I basketball championship game, the Connecticut Huskies scored 33 points more than the Louisville Cardinals. Together, both teams scored a total of 153 points. How many points did the 2013 Champion Connecticut Huskies score during this game? (*Source:* National Collegiate Athletic Association)

46. During the 2013 Men's NCAA Division I basketball championship game, the Michigan Wolverines scored 6 points fewer than the Louisville Cardinals. Together, both teams scored 158 points. How many points did the 2013 Champion Louisville Cardinals score during the game? (*Source:* National Collegiate Athletic Association)

47. The USA is the country with the most personal computers in use, followed by China. If the USA has 115,500 thousand more computers than China and the total number of computers for both countries is 505,780 thousand, find the number of computers for each country. (*Source:* Computer Industry Almanac)

48. The total number of personal computers in use for Italy and Russia is 98,240 thousand. If Russia has 8800 thousand more computers than Italy, find the number of computers for each country. (*Source:* Computer Industry Almanac)

Review

Round each number to the given place value. See Section 1.4.

49. 586 to the nearest ten

50. 82 to the nearest ten

51. 1026 to the nearest hundred

52. 52,333 to the nearest thousand

53. 2986 to the nearest thousand

54. 101,552 to the nearest hundred

Concept Extensions

55. Solve Example 3 again, but this time let x be the distance from Chicago to Los Angeles. Did you get the same results? Explain why or why not.

56. Solve Exercise 25 again, but this time let x be the number of electoral votes for Florida. Did you get the same results? Explain why or why not.

In real estate, a house's selling price P is found by adding the real estate agent's commission C to the amount A that the seller of the house receives: $P = A + C$.

57. A house sold for $230,000. The owner's real estate agent received a commission of $13,800. How much did the seller receive? (*Hint:* Substitute the known values into the equation, then solve the equation for the remaining unknown.)

58. A homeowner plans to use a real estate agent to sell his house. He hopes to sell the house for $165,000 and keep $156,750 of that. If everything goes as he has planned, how much will his real estate agent receive as a commission?

In retailing, the retail price P of an item can be computed using the equation $P = C + M$, where C is the wholesale cost of the item and M is the amount of markup.

59. The retail price of a computer system is $999 after a markup of $450. What is the wholesale cost of the computer system? (*Hint:* Substitute the known values into the equation, then solve the equation for the remaining unknown.)

60. Slidell Feed and Seed sells a bag of cat food for $12. If the store paid $7 for the cat food, what is the markup on the cat food?

Chapter 3 Group Activity

Modeling Equation Solving with Addition and Subtraction

Sections 3.1–3.4

We can use positive counters ● and negative counters ● to help us model the equation-solving process. We also need to use an object that represents a variable. We use small slips of paper with the variable name written on them.

Recall that taking a ● and ● together creates a neutral or zero pair. After a neutral pair has been formed, it can be removed from or added to an equation model without changing the overall value. We also need to remember that we can add or remove the same number of positive or negative counters from both sides of an equation without changing the overall value.

We can represent the equation $x + 5 = 2$ as follows:

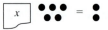

To get the variable by itself, we must remove 5 black counters from both sides of the model. Because there are only 2 counters on the right side, we must add 5 negative counters to both sides of the model. Then we can remove neutral pairs: 5 from the left side and 2 from the right side.

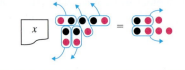

We are left with the following model, which represents the solution, $x = -3$.

Similarly, we can represent the equation $x - 4 = -6$ as follows:

To get the variable by itself, we must remove 4 red counters from both sides of the model.

We are left with the following model, which represents the solution, $x = -2$.

Use the counter model to solve each equation.

1. $x - 3 = -7$ **2.** $x - 1 = -9$

3. $x + 2 = 8$ **4.** $x + 4 = 5$

5. $x + 8 = 3$ **6.** $x - 5 = -1$

7. $x - 2 = 1$ **8.** $x - 5 = 10$

9. $x + 3 = -7$ **10.** $x + 8 = -2$

Chapter 3 Vocabulary Check

Fill in each blank with one of the words or phrases listed below.

variable	addition	constant	algebraic expression	equation
terms	simplified	multiplication	evaluating the expression	solution
like	combined	numerical coefficient	distributive	

1. An algebraic expression is _____ when all like terms have been _____.

2. Terms that are exactly the same, except that they may have different numerical coefficients, are called _____ terms.

3. A letter used to represent a number is called a(n) _____.

4. A combination of operations on variables and numbers is called a(n) _____.

5. The addends of an algebraic expression are called the _____ of the expression.

6. The number factor of a variable term is called the _____.

7. Replacing a variable in an expression by a number and then finding the value of the expression is called _____ for the variable.

8. A term that is a number only is called a(n) _____.

9. A(n) _____ is of the form expression = expression.

10. A(n) _____ of an equation is a value for the variable that makes the equation a true statement.

11. To multiply $-3(2x + 1)$, we use the _____ property.

12. By the _____ property of equality, we may multiply or divide both sides of an equation by any nonzero number without changing the solution of the equation.

13. By the _____ property of equality, the same number may be added to or subtracted from both sides of an equation without changing the solution of the equation.

> **Helpful Hint**
> ▶ Are you preparing for your test? Don't forget to take the Chapter 3 Test on page 207. Then check your answers at the back of the text and use the Chapter Test Prep Videos to see the fully worked-out solutions to any of the exercises you want to review.

3 Chapter Highlights

Definitions and Concepts	Examples

Section 3.1   Simplifying Algebraic Expressions

The addends of an algebraic expression are called the **terms** of the expression.

$$5x^2 + (-4x) + (-2)$$

↑   ↑   ↑ ——— 3 terms

The number factor of a variable term is called the **numerical coefficient.**

Term	Numerical Coefficient
$7x$	7
$-6y$	-6
x or $1x$	1

Terms that are exactly the same, except that they may have different numerical coefficients, are called **like terms.**

An algebraic expression is **simplified** when all like terms have been **combined.**

$$5x + 11x = (5 + 11)x = 16x$$

↑   ↑
like terms
↓   ↓
$$y - 6y = (1 - 6)y = -5y$$

Use the **distributive property** to multiply an algebraic expression by a term.

Simplify: $-4(x + 2) + 3(5x - 7)$
$$= -4 \cdot x + (-4) \cdot 2 + 3 \cdot 5x - 3 \cdot 7$$
$$= -4x + (-8) + 15x + (-21)$$
$$= 11x + (-29) \quad \text{or} \quad 11x - 29$$

Section 3.2   Solving Equations: Review of the Addition and Multiplication Properties

Addition Property of Equality

Let a, b, and c represent numbers. Then

$a = b$	Also, $a = b$
and $a + c = b + c$	and $a - c = b - c$
are equivalent equations.	are equivalent equations.

In other words, the same number may be added to or subtracted from both sides of an equation without changing the solution of the equation.

Solve for x:

$$x + 8 = 2 + (-1)$$
$$x + 8 = 1 \qquad \text{Combine like terms.}$$
$$x + 8 - 8 = 1 - 8 \qquad \text{Subtract 8 from both sides.}$$
$$x = -7 \qquad \text{Simplify.}$$

The solution is -7.

(continued)

Definitions and Concepts	Examples

Section 3.2   Solving Equations: Review of the Addition and Multiplication Properties   (*continued*)

Multiplication Property of Equality

Let a, b, and c represent numbers and let $c \neq 0$. Then

$a = b$	Also, $a = b$
and $a \cdot c = b \cdot c$	and $\dfrac{a}{c} = \dfrac{b}{c}$
are equivalent equations.	are equivalent equations.

In other words, both sides of an equation may be multiplied or divided by the same nonzero number without changing the solution of the equation.

Solve: $y - 7y = 30$

$$-6y = 30 \qquad \text{Combine like terms.}$$

$$\frac{-6y}{-6} = \frac{30}{-6} \qquad \text{Divide both sides by } -6.$$

$$y = -5 \qquad \text{Simplify.}$$

The solution is -5.

Section 3.3   Solving Linear Equations in One Variable

Steps for Solving an Equation

Step 1:   If parentheses are present, use the distributive property.

Step 2:   Combine any like terms on each side of the equation.

Step 3:   Use the addition property of equality to rewrite the equation so that variable terms are on one side of the equation and constant terms are on the other side.

Step 4:   Use the multiplication property of equality to divide both sides by the numerical coefficient of the variable to solve.

Step 5:   Check the solution in the *original equation*.

Solve for x: $5(3x - 1) + 15 = -5$

Step 1: $15x - 5 + 15 = -5$ Apply the distributive property.

Step 2: $15x + 10 = -5$ Combine like terms.

Step 3: $15x + 10 - 10 = -5 - 10$ Subtract 10 from both sides.

$$15x = -15$$

Step 4: $\dfrac{15x}{15} = \dfrac{-15}{15}$ Divide both sides by 15.

$$x = -1$$

Step 5:   Check to see that -1 is the solution.

Section 3.4   Linear Equations in One Variable and Problem Solving

Problem-Solving Steps

1. UNDERSTAND the problem. Some ways of doing this are

Read and reread the problem.

Construct a drawing.

Choose a variable to represent an unknown in the problem.

The incubation period for a golden eagle is three times the incubation period for a hummingbird. If the total of their incubation periods is 60 days, find the incubation period for each bird. (*Source: Wildlife Fact File,* International Masters Publishers)

1. UNDERSTAND the problem. Then choose a variable to represent an unknown. Let

x = incubation period of a hummingbird

Then

$3x$ = incubation period of a golden eagle

2. TRANSLATE the problem into an equation.

2. TRANSLATE.

incubation of hummingbird	+	incubation of golden eagle	is	60
↓		↓	↓	↓
x	+	$3x$	=	60

Definitions and Concepts	Examples
Section 3.4 Linear Equations in One Variable and Problem Solving (continued)	
3. SOLVE the equation.	**3.** SOLVE: $$x + 3x = 60$$ $$4x = 60$$ $$\frac{4x}{4} = \frac{60}{4}$$ $$x = 15$$
4. INTERPRET the results. *Check* the proposed solution in the stated problem and *state* your conclusion.	**4.** INTERPRET the solution in the stated problem. The incubation period for a hummingbird is 15 days. The incubation period for a golden eagle is $$3x = 3 \cdot 15 = 45 \text{ days.}$$ Since 15 days + 45 days = 60 days and 45 is 3(15), the solution checks. *State* your conclusion: The incubation period for a hummingbird is 15 days. The incubation period for a golden eagle is 45 days.

Chapter 3   Review

(3.1) *Simplify each expression by combining like terms.*

1. $3y + 7y - 15$

2. $2y - 10 - 8y$

3. $8a + a - 7 - 15a$

4. $y + 3 - 9y - 1$

Multiply.

5. $2(x + 5)$

6. $-3(y + 8)$

Simplify.

7. $7x + 3(x - 4) + x$

8. $-(3m + 2) - m - 10$

9. $3(5a - 2) - 20a + 10$

10. $6y + 3 + 2(3y - 6)$

11. $6y - 7 + 11y - y + 2$

12. $10 - x + 5x - 12 - 3x$

Find the perimeter of each figure.

△**13.**

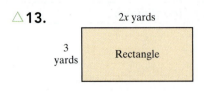

2x yards

3 yards

Rectangle

△**14.**

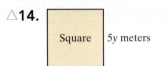

Square

5y meters

Find the area of each figure.

△**15.**

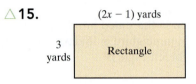

(2x − 1) yards

3 yards

Rectangle

△**16.**

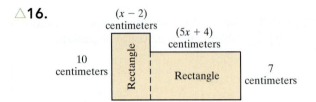

(x − 2) centimeters

(5x + 4) centimeters

10 centimeters

Rectangle

Rectangle

7 centimeters

(3.2) *Solve each equation.*

17. $z - 5 = -7$

18. $3x + 10 = 4x$

19. $3y = -21$

20. $-3a = -15$

21. $\dfrac{x}{-6} = 2$

22. $\dfrac{y}{-15} = -3$

23. $n + 18 = 10 - (-2)$

24. $c - 5 = -13 + 7$

25. $7x + 5 - 6x = -20$

26. $17x = 2(8x - 4)$

27. $5x + 7 = -3$

28. $-14 = 9y + 4$

29. $\dfrac{z}{4} = -8 - (-6)$

30. $-1 + (-8) = \dfrac{x}{5}$

31. $6y - 7y = 100 - 105$

32. $19x - 16x = 45 - 60$

33. $9(2x - 7) = 19x$

34. $-5(3x + 3) = -14x$

35. $3x - 4 = 11$

36. $6y + 1 = 73$

37. $2(x + 4) - 10 = -2(7)$

38. $-3(x - 6) + 13 = 20 - 1$

Translating *Translate each phrase into an algebraic expression. Let x represent "a number."*

39. The product of -5 and a number

40. Three subtracted from a number

41. The sum of -5 and a number

42. The quotient of -2 and a number

43. The product of -5 and a number, decreased by 50

44. Eleven added to twice a number

45. The quotient of 70 and the sum of a number and 6

46. Twice the difference of a number and 13

(3.3) *Solve each equation.*

47. $2x + 5 = 7x - 100$

48. $-6x - 4 = x + 66$

49. $2x + 7 = 6x - 1$

50. $5x - 18 = -4x$

51. $5(n - 3) = 7 + 3n$

52. $7(2 + x) = 4x - 1$

53. $6x + 3 - (-x) = -20 + 5x - 7$

54. $x - 25 + 2x = -5 + 2x - 10$

55. $3(x - 4) = 5x - 8$

56. $4(x - 3) = -2x - 48$

57. $6(2n - 1) + 18 = 0$

58. $7(3y - 2) - 7 = 0$

59. $95x - 14 = 20x - 10 + 10x - 4$

60. $32z + 11 - 28z = 50 + 2z - (-1)$

Translating *Write each sentence as an equation.*

61. The difference of 20 and -8 is 28.

62. Nineteen subtracted from -2 amounts to -21.

63. The quotient of -75 and the sum of 5 and 20 is equal to -3.

64. Five times the sum of 2 and -6 yields -20.

(3.4) Translating *Write each sentence as an equation using x as the variable.*

65. Twice a number minus 8 is 40.

66. The product of a number and 6 is equal to the sum of the number and 20.

67. Twelve subtracted from the quotient of a number and 2 is 10.

68. The difference of a number and 3 is the quotient of 8 and 4.

Solve.

69. Five times a number subtracted from 40 is the same as three times the number. Find the number.

70. The product of a number and 3 is twice the difference of that number and 8. Find the number.

71. In an election, the incumbent received 14,000 votes of the 18,500 votes cast. Of the remaining votes, the Democratic candidate received 272 more than the Independent candidate. Find how many votes the Democratic candidate received.

72. Rajiv Puri has twice as many movies on DVD as he has on Blu-ray Disc. Find the number of DVDs if he has a total of 126 movie recordings.

Mixed Review

Simplify.

73. $9x - 20x$

74. $-5(7x)$

75. $12x + 5(2x - 3) - 4$

76. $-7(x + 6) - 2(x - 5)$

Solve.

77. $c - 5 = -13 + 7$

78. $7x + 5 - 6x = -20$

79. $-7x + 3x = -50 - 2$

80. $-x + 8x = -38 - 4$

81. $9x + 12 - 8x = -6 + (-4)$

82. $-17x + 14 + 20x - 2x = 5 - (-3)$

83. $5(2x - 3) = 11x$

84. $\dfrac{y}{-3} = -1 - 5$

85. $12y - 10 = -70$

86. $4n - 8 = 2n + 14$

87. $-6(x - 3) = x + 4$

88. $9(3x - 4) + 63 = 0$

89. $-5z + 3z - 7 = 8z - 1 - 6$

90. $4x - 3 + 6x = 5x - 3 - 30$

91. Three times a number added to twelve is 27. Find the number.

92. Twice the sum of a number and four is ten. Find the number.

93. Out of the 50 states, Hawaii has the least number of roadway miles, followed by Delaware. If Delaware has 1931 more roadway miles than Hawaii and the total number of roadway miles for both states is 10,673, find the number of roadway miles for each state. (*Source:* U.S. Federal Highway Administration)

94. North and South Dakota both have over 80,000 roadway miles. North Dakota has 4489 more miles and the total number of roadway miles for both states is 169,197. Find the number of roadway miles for North Dakota and for South Dakota. (*Source:* U.S. Federal Highway Administration)

1. Simplify $7x - 5 - 12x + 10$ by combining like terms.

2. Multiply: $-2(3y + 7)$

Answers

3. Simplify: $-(3z + 2) - 5z - 18$

△ **4.** Write an expression that represents the perimeter of the equilateral triangle (a triangle with three sides of equal length). Simplify the expression.

$(5x + 5)$ inches

△ **5.** Write an expression that represents the area of the rectangle. Simplify the expression.

4 meters

Rectangle | $(3x - 1)$ meters

Solve each equation.

6. $12 = y - 3y$

7. $\dfrac{x}{2} = -5 - (-2)$

8. $5x + 12 - 4x - 14 = 22$

9. $-4x + 7 = 15$

10. $2(x - 6) = 0$

11. $-4(x - 11) - 34 = 10 - 12$

12. $5x - 2 = x - 10$

13. $4(5x + 3) = 2(7x + 6)$

14. $6 + 2(3n - 1) = 28$

Translate the following phrases into mathematical expressions. If needed, use x to represent "a number."

15. The sum of -23 and a number

16. Three times a number, subtracted from -2

1. _____

2. _____

3. _____

4. _____

5. _____

6. _____

7. _____

8. _____

9. _____

10. _____

11. _____

12. _____

13. _____

14. _____

15. _____

16. _____

17. _____

18. _____

19. _____

20. _____

21. _____

Translate each sentence into an equation. If needed, use x to represent "a number."

17. The sum of twice 5 and −15 is −5.

18. Six added to three times a number equals −30.

Solve.

19. The difference of three times a number and five times the same number is 4. Find the number.

20. In a championship basketball game, Paula made twice as many free throws as Maria. If the total number of free throws made by both women was 12, find how many free throws Paula made.

21. In a 10-kilometer race, there are 112 more men entered than women. Find the number of female runners if the total number of runners in the race is 600.

1. Write 308,063,557 in words.

2. Write 276,004 in words.

△ **3.** Find the perimeter of the polygon shown.

3 inches
2 inches
1 inch
4 inches
3 inches

△ **4.** Find the perimeter of the rectangle shown.

6 inches
3 inches

5. Subtract: $900 - 174$. Check by adding.

6. Subtract: $17,801 - 8216$. Check by adding.

7. Round 248,982 to the nearest hundred.

8. Round 844,497 to the nearest thousand.

9. Multiply: 25×8

10. Multiply: 395×74

11. Divide and check: $1872 \div 9$

12. Divide and check: $3956 \div 46$

13. Simplify: $2 \cdot 4 - 3 \div 3$

14. Simplify: $8 \cdot 4 + 9 \div 3$

15. Evaluate $x^2 + z - 3$ for $x = 5$ and $z = 4$.

16. Evaluate $2a^2 + 5 - c$ for $a = 2$ and $c = 3$.

17. Determine which numbers in the set $\{26, 40, 20\}$ are solutions of the equation $2n - 30 = 10$.

18. Insert $<$ or $>$ to make a true statement.

   a. $-14 \qquad 0$

   b. $-(-7) \qquad -8$

19. Add using a number line: $5 + (-2)$

20. Add using a number line: $-3 + (-4)$

Answers

1. _____

2. _____

3. _____

4. _____

5. _____

6. _____

7. _____

8. _____

9. _____

10. _____

11. _____

12. _____

13. _____

14. _____

15. _____

16. _____

17. _____

18. a. _____

   b. _____

19. _____

20. _____

21. _____

22. _____

23. _____

24. _____

25. _____

26. _____

27. _____

28. _____

29. _____

30. _____

31. _____

32. _____

33. _____

34. _____

35. _____

36. _____

37. _____

38. _____

39. _____

40. _____

41. _____

42. _____

43. _____

44. _____

45. _____

46. _____

47. _____

48. _____

49. _____

50. _____

Add.

21. $-15 + (-10)$

22. $3 + (-7)$

23. $-2 + 25$

24. $21 + 15 + (-19)$

Subtract.

25. $-4 - 10$

26. $-2 - 3$

27. $6 - (-5)$

28. $19 - (-10)$

29. $-11 - (-7)$

30. $-16 - (-13)$

Divide.

31. $\dfrac{-12}{6}$

32. $\dfrac{-30}{-5}$

33. $-20 \div (-4)$

34. $26 \div (-2)$

35. $\dfrac{48}{-3}$

36. $\dfrac{-120}{12}$

Find the value of each expression.

37. $(-3)^2$

38. -2^5

39. -3^2

40. $(-5)^2$

41. Simplify: $2y - 6 + 4y + 8$

42. Simplify: $6x + 2 - 3x + 7$

43. Determine whether -1 is a solution of the equation $3y + 1 = 3$.

44. Determine whether 2 is a solution of $5x - 3 = 7$.

45. Solve: $-12x = -36$

46. Solve: $-3y = 15$

47. Solve: $2x - 6 = 18$

48. Solve: $3a + 5 = -1$

49. A salesperson at an electronics store sold a computer system and software for $2100, receiving four times as much money for the computer system as for the software. Find the price of each.

50. Rose Daunis is thinking of a number. Two times the number plus four is the same amount as three times the number minus seven. Find Rose's number.

Fractions and Mixed Numbers

The following graph is called a circle graph or a pie chart. Each sector (shaped like a piece of pie) shows the fraction of entering college freshmen who choose to major in each discipline shown. Can you find your current choice of major in this graph? In Section 4.2, Exercises 93–96, we show this same circle graph, but in 3-D design. We simplify some of the fractions in it and also study sector size versus fraction value.

Fractions are numbers and, like whole numbers and integers, they can be added, subtracted, multiplied, and divided. Fractions are very useful and appear frequently in everyday language, in common phrases such as "half an hour," "quarter of a pound," and "third of a cup." This chapter reviews the concepts of fractions and mixed numbers and demonstrates how to add, subtract, multiply, and divide these numbers.

College Freshman Majors

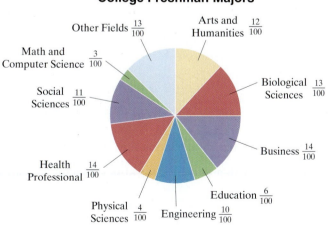

Other Fields $\frac{13}{100}$

Arts and Humanities $\frac{12}{100}$

Math and Computer Science $\frac{3}{100}$

Social Sciences $\frac{11}{100}$

Biological Sciences $\frac{13}{100}$

Business $\frac{14}{100}$

Health Professional $\frac{14}{100}$

Education $\frac{6}{100}$

Physical Sciences $\frac{4}{100}$

Engineering $\frac{10}{100}$

Source: The Higher Education Research Institute

Introduction to Fractions and Mixed Numbers

Objectives

A Identify the Numerator and the Denominator of a Fraction.

B Write a Fraction to Represent Parts of Figures or Real-Life Data.

C Graph Fractions on a Number Line.

D Review Division Properties of 0 and 1.

E Write Mixed Numbers as Improper Fractions.

F Write Improper Fractions as Mixed Numbers or Whole Numbers.

Objective A Identifying Numerators and Denominators

Whole numbers are used to count whole things or units, such as cars, horses, dollars, and people. To refer to a part of a whole, fractions can be used. Here are some examples of **fractions.** Study these examples for a moment.

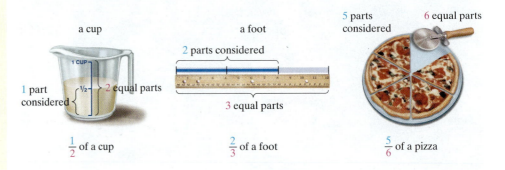

$\frac{1}{2}$ of a cup $\qquad$ $\frac{2}{3}$ of a foot $\qquad$ $\frac{5}{6}$ of a pizza

In a fraction, the top number is called the **numerator** and the bottom number is called the **denominator.** The bar between the numbers is called the **fraction bar.**

Name	Fraction	Meaning
numerator $\longrightarrow$	5	$\longleftarrow$ number of parts being considered
denominator $\longrightarrow$	6	$\longleftarrow$ number of equal parts in the whole

Examples Identify the numerator and the denominator of each fraction.

1. $\frac{3}{7}$ $\leftarrow$ numerator
   $\phantom{\frac{3}{7}}$ $\leftarrow$ denominator

2. $\frac{13}{5x}$ $\leftarrow$ numerator
   $\phantom{\frac{13}{5x}}$ $\leftarrow$ denominator

■ **Work Practice 1–2**

Practice 1–2

Identify the numerator and the denominator of each fraction.

1. $\frac{11}{2}$

2. $\frac{10y}{17}$

 Helpful Hint

$\frac{3}{7}$ $\leftarrow$ Remember that the bar in a fraction means division. Since division by 0 is undefined, a fraction with a denominator of 0 is undefined. For example, $\frac{3}{0}$ is undefined.

Objective B Writing Fractions to Represent Parts of Figures or Real-Life Data

One way to become familiar with the concept of fractions is to visualize fractions with shaded figures. We can then write a fraction to represent the shaded area of the figure (or diagram).

Answers

1. numerator = 11, denominator = 2
2. numerator = 10y, denominator = 17

Examples 　Write a fraction to represent the shaded part of each figure.

3. In this figure, 2 of the 5 equal parts are shaded. Thus, the fraction is $\frac{2}{5}$.

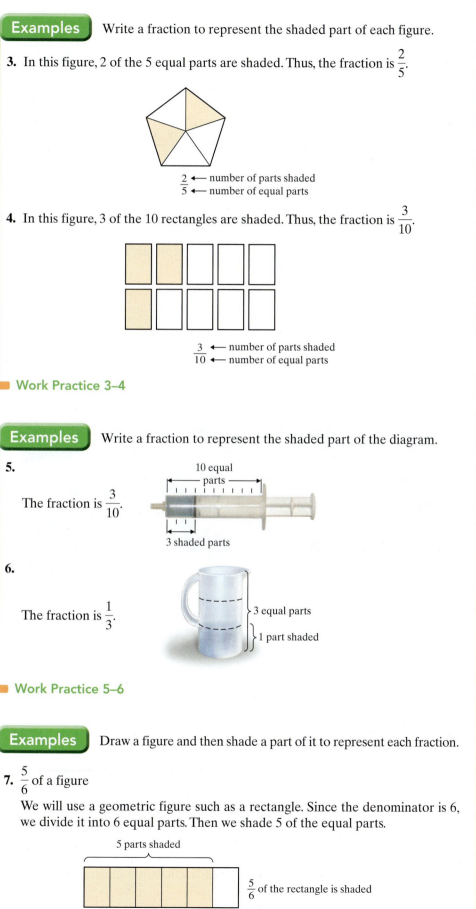

$$\frac{2}{5} \begin{array}{l} \leftarrow \text{number of parts shaded} \\ \leftarrow \text{number of equal parts} \end{array}$$

4. In this figure, 3 of the 10 rectangles are shaded. Thus, the fraction is $\frac{3}{10}$.

$$\frac{3}{10} \begin{array}{l} \leftarrow \text{number of parts shaded} \\ \leftarrow \text{number of equal parts} \end{array}$$

■ **Work Practice 3–4**

Examples 　Write a fraction to represent the shaded part of the diagram.

5.

The fraction is $\frac{3}{10}$.

10 equal parts

3 shaded parts

6.

The fraction is $\frac{1}{3}$.

3 equal parts

1 part shaded

■ **Work Practice 5–6**

Examples 　Draw a figure and then shade a part of it to represent each fraction.

7. $\frac{5}{6}$ of a figure

We will use a geometric figure such as a rectangle. Since the denominator is 6, we divide it into 6 equal parts. Then we shade 5 of the equal parts.

5 parts shaded

$\frac{5}{6}$ of the rectangle is shaded

6 equal parts

(Continued on next page)

(Continued on next page)

Practice 3–4

Write a fraction to represent the shaded part of each figure.

3.

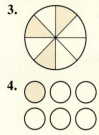

4.

○○○
○○○

Practice 5–6

Write a fraction to represent the part of the whole shown.

5. 　Just consider this part of the syringe

6. 　|←— Whole part —→|

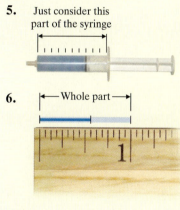

Practice 7

Draw and shade a part of a figure to represent the fraction.

7. $\frac{2}{3}$ of a figure

Answers

3. $\frac{3}{8}$ 　**4.** $\frac{1}{6}$ 　**5.** $\frac{7}{10}$ 　**6.** $\frac{9}{16}$

7. answers may vary; for example,

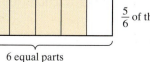

Practice 8

Draw and shade a part of a figure to represent the fraction.

8. $\frac{7}{11}$ of a figure

8. $\frac{3}{8}$ of a figure

If you'd like, our figure can consist of 8 triangles of the same size. We will shade 3 of the triangles.

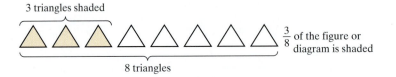

3 triangles shaded

8 triangles

$\frac{3}{8}$ of the figure or diagram is shaded

■ **Work Practice 7–8**

✓ **Concept Check** If [diagram of 6 shaded squares] represents $\frac{6}{7}$ of a whole diagram, sketch the whole diagram.

Practice 9

Of the eight planets in our solar system, five are farther from the Sun than Earth is. What fraction of the planets are farther from the Sun than Earth is?

Example 9 Writing Fractions from Real-Life Data

Of the eight planets in our solar system (Pluto is now a dwarf planet), three are closer to the Sun than Mars. What fraction of the planets are closer to the Sun than Mars?

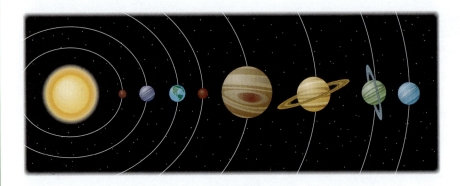

Solution: The fraction of planets closer to the Sun than Mars is:

$\frac{3}{8}$ ← number of planets closer
    ← number of planets in our solar system

Thus, $\frac{3}{8}$ of the planets in our solar system are closer to the Sun than Mars.

■ **Work Practice 9**

Answers

8. answers may vary; for example,

9. $\frac{5}{8}$

✓ **Concept Check Answer**

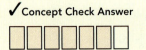

The definitions and statements below apply to positive fractions.

A **proper fraction** is a fraction whose numerator is less than its denominator. Proper fractions are less than 1. For example, the shaded portion of the triangle is represented by $\frac{2}{3}$.

An **improper fraction** is a fraction whose numerator is greater than or equal to its denominator. Improper fractions are greater than or equal to 1.

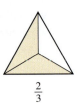

$\frac{2}{3}$

The shaded part of the group of circles below is $\frac{9}{4}$. The shaded part of the rectangle is $\frac{6}{6}$. Recall that $\frac{6}{6}$ simplifies to 1 and notice that the entire rectangle (1 whole figure) is shaded below.

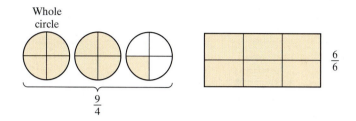

A **mixed number** contains a whole number and a fraction. Mixed numbers are greater than 1. Above, we wrote the shaded part of the group of circles below as the improper fraction $\frac{9}{4}$. Now let's write the shaded part as a mixed number. The shaded part of the group of circles' area is $2\frac{1}{4}$. Read this as "two and one-fourth."

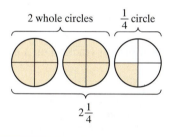

Note: The mixed number $2\frac{1}{4}$, diagrammed above, represents $2 + \frac{1}{4}$.

   The mixed number $-3\frac{1}{5}$ represents $-\left(3 + \frac{1}{5}\right)$ or $-3 - \frac{1}{5}$. We review this later in this chapter.

Examples Represent the shaded part of each figure group as both an improper fraction and a mixed number.

10.
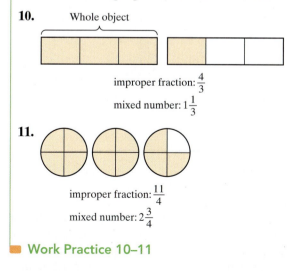

improper fraction: $\frac{4}{3}$

mixed number: $1\frac{1}{3}$

11.

improper fraction: $\frac{11}{4}$

mixed number: $2\frac{3}{4}$

■ **Work Practice 10–11**

✓**Concept Check** If you were to round $2\frac{3}{4}$, shown in Example 11 above, to the nearest whole number, would you choose 2 or 3? Why?

Practice 10–11

Represent the shaded part of each figure group as both an improper fraction and a mixed number.

10.
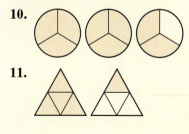

11.

Answers

10. $\frac{8}{3}, 2\frac{2}{3}$ **11.** $\frac{5}{4}, 1\frac{1}{4}$

✓**Concept Check Answer**

3; answers may vary

Objective C Graphing Fractions on a Number Line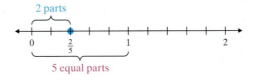

Another way to visualize fractions is to graph them on a number line. To do this, think of 1 unit on the number line as a whole. To graph $\frac{2}{5}$, for example, divide the distance from 0 to 1 into **5 equal parts**. Then start at 0 and count **2 parts** to the right.

Notice that the graph of $\frac{2}{5}$ lies between 0 and 1. This means

$$0 < \frac{2}{5}\left(\text{or } \frac{2}{5} > 0\right) \text{ and also } \frac{2}{5} < 1$$

Practice 12

Graph each proper fraction on a number line.

a. $\frac{5}{7}$ **b.** $\frac{2}{3}$ **c.** $\frac{4}{6}$

Example 12   Graph each proper fraction on a number line.

a. $\frac{3}{4}$ **b.** $\frac{1}{2}$ **c.** $\frac{3}{6}$

Solution:

a. To graph $\frac{3}{4}$, divide the distance from 0 to 1 into 4 parts. Then start at 0 and count over 3 parts.

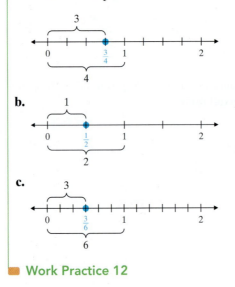

Work Practice 12

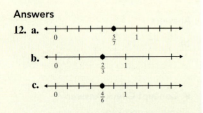
The statements below apply to positive fractions.

The fractions in Example 12 are all proper fractions. Notice that the value of each is less than 1. This is always true for proper fractions since the numerator of a proper fraction is less than the denominator.

On the next page, we graph improper fractions. Notice that improper fractions are greater than or equal to 1. This is always true since the numerator of an improper fraction is greater than or equal to the denominator.

Example 13 Graph each improper fraction on a number line.

a. $\dfrac{7}{6}$ b. $\dfrac{9}{5}$ c. $\dfrac{6}{6}$ d. $\dfrac{3}{1}$

Solution:

a.
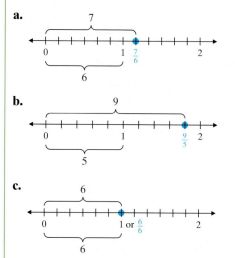

b.

c.

d. Each 1-unit distance has 1 equal part. Count over 3 parts.

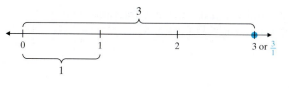

 **Work Practice 13**

Note: We will graph mixed numbers at the end of this chapter.

Objective D Reviewing Division Properties of 0 and 1

Before we continue further, don't forget from Section 1.6 that a fraction bar indicates division. Let's review some division properties of 1 and 0.

$$\dfrac{9}{9} = 1 \text{ because } 1 \cdot 9 = 9 \qquad \dfrac{-11}{1} = -11 \text{ because } -11 \cdot 1 = -11$$

$$\dfrac{0}{6} = 0 \text{ because } 0 \cdot 6 = 0 \qquad \dfrac{6}{0} \text{ is } \textit{undefined} \text{ because there is no number that when multiplied by 0 gives 6.}$$

In general, we can say the following.

Let n be any integer except 0.

$$\dfrac{n}{n} = 1 \qquad \dfrac{0}{n} = 0$$

$$\dfrac{n}{1} = n \qquad \dfrac{n}{0} \text{ is undefined.}$$

Practice 13

Graph each improper fraction on a number line.

a. $\dfrac{8}{3}$ b. $\dfrac{5}{4}$ c. $\dfrac{7}{7}$

Answers

13. a.

b.

c.

Practice 14–19

Simplify.

14. $\dfrac{9}{9}$ **15.** $\dfrac{-6}{-6}$

16. $\dfrac{0}{-1}$ **17.** $\dfrac{4}{1}$

18. $\dfrac{-13}{0}$ **19.** $\dfrac{-13}{1}$

Examples   Simplify.

14. $\dfrac{5}{5} = 1$ **15.** $\dfrac{-2}{-2} = 1$ **16.** $\dfrac{0}{-5} = 0$

17. $\dfrac{-5}{1} = -5$ **18.** $\dfrac{41}{1} = 41$ **19.** $\dfrac{19}{0}$ is undefined

■ **Work Practice 14–19**

Notice from Example 17 that we can have negative fractions. In fact,

$$\dfrac{-5}{1} = -5, \qquad \dfrac{5}{-1} = -5, \qquad \text{and} \qquad -\dfrac{5}{1} = -5$$

Because all of the fractions equal -5, we have

$$\dfrac{-5}{1} = \dfrac{5}{-1} = -\dfrac{5}{1}$$

This means that the negative sign in a fraction can be written in the numerator, in the denominator, or in front of the fraction. Remember this as we work with negative fractions.

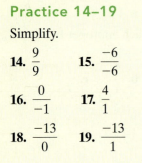

Helpful Hint   Remember, for example, that

$$-\dfrac{2}{3} = \dfrac{-2}{3} = \dfrac{2}{-3}$$

Objective E Writing Mixed Numbers as Improper Fractions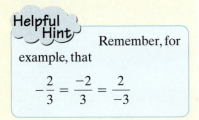

Earlier in this section, mixed numbers and improper fractions were both used to represent the shaded part of figure groups. For example,

$1\dfrac{2}{3}$ or $\dfrac{5}{3}$ Thus, $1\dfrac{2}{3} = \dfrac{5}{3}$.

The following steps may be used to write a mixed number as an improper fraction:

Writing a Mixed Number as an Improper Fraction

To write a mixed number as an improper fraction:

Step 1: Multiply the denominator of the fraction by the whole number.

Step 2: Add the numerator of the fraction to the product from Step 1.

Step 3: Write the sum from Step 2 as the numerator of the improper fraction over the original denominator.

Practice 20

Write each as an improper fraction.

a. $5\dfrac{2}{7}$ **b.** $6\dfrac{2}{3}$ **c.** $10\dfrac{9}{10}$ **d.** $4\dfrac{1}{5}$

For example,

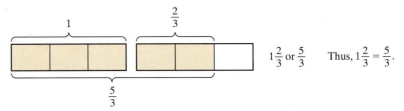

$$1\dfrac{2}{3} = \dfrac{3 \cdot 1 + 2}{3} = \dfrac{3 + 2}{3} = \dfrac{5}{3} \quad \text{or} \quad 1\dfrac{2}{3} = \dfrac{5}{3}, \text{ as stated above.}$$

Answers

14. 1 **15.** 1 **16.** 0 **17.** 4
18. undefined **19.** −13

20. a. $\dfrac{37}{7}$ **b.** $\dfrac{20}{3}$ **c.** $\dfrac{109}{10}$ **d.** $\dfrac{21}{5}$

Example 20   Write each as an improper fraction.

a. $4\dfrac{2}{9} = \dfrac{9 \cdot 4 + 2}{9} = \dfrac{36 + 2}{9} = \dfrac{38}{9}$ **b.** $1\dfrac{8}{11} = \dfrac{11 \cdot 1 + 8}{11} = \dfrac{11 + 8}{11} = \dfrac{19}{11}$

■ **Work Practice 20**

Objective F Writing Improper Fractions as Mixed Numbers or Whole Numbers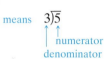

Just as there are times when an improper fraction is preferred, sometimes a mixed or a whole number better suits a situation. To write improper fractions as mixed or whole numbers, we use division. Recall once again from Section 1.6 that the fraction bar means division. This means that the fraction

$$\frac{5}{3} \quad \text{numerator} \atop \text{denominator}$$

means $3\overline{)5}$
$\qquad\qquad\qquad\uparrow\quad\uparrow$
$\qquad\qquad\qquad$ numerator
$\qquad\qquad$ denominator

> **Writing an Improper Fraction as a Mixed Number or a Whole Number**
>
> To write an improper fraction as a mixed number or a whole number:
>
> **Step 1:** Divide the denominator into the numerator.
>
> **Step 2:** The whole number part of the mixed number is the quotient. The fraction part of the mixed number is the remainder over the original denominator.
>
> $$\text{quotient}\frac{\text{remainder}}{\text{original denominator}}$$

For example,

Step 1
$$\frac{5}{3}: \quad 3\overline{)5} \atop \underline{-3} \atop 2$$
$\qquad\qquad$ quotient

Step 2
$$\frac{5}{3} = 1\frac{2}{3} \quad \leftarrow \text{remainder} \atop \leftarrow \text{original denominator}$$

Example 21 Write each as a mixed number or a whole number.

a. $\dfrac{30}{7}$ **b.** $\dfrac{16}{15}$ **c.** $\dfrac{84}{6}$

Solution:

a. $\dfrac{30}{7}: \quad 7\overline{)30} \atop \underline{-28} \atop 2$ $\qquad \dfrac{30}{7} = 4\dfrac{2}{7}$

b. $\dfrac{16}{15}: \quad 15\overline{)16} \atop \underline{-15} \atop 1$ $\qquad \dfrac{16}{15} = 1\dfrac{1}{15}$

c. $\dfrac{84}{6}: \quad 6\overline{)84} \atop \underline{-6} \atop 24 \atop \underline{-24} \atop 0$ $\qquad \dfrac{84}{6} = 14$ Since the remainder is 0, the result is the whole number 14.

> **Helpful Hint** When the remainder is 0, the improper fraction is a whole number. For example, $\dfrac{92}{4} = 23$.
>
> $$23 \atop 4\overline{)92} \atop \underline{-8} \atop 12 \atop \underline{-12} \atop 0$$

■ **Work Practice 21**

Practice 21

Write each as a mixed number or a whole number.

a. $\dfrac{9}{5}$ **b.** $\dfrac{23}{9}$ **c.** $\dfrac{48}{4}$

d. $\dfrac{62}{13}$ **e.** $\dfrac{51}{7}$ **f.** $\dfrac{21}{20}$

Answers

21. a. $1\dfrac{4}{5}$ **b.** $2\dfrac{5}{9}$ **c.** 12 **d.** $4\dfrac{10}{13}$

e. $7\dfrac{2}{7}$ **f.** $1\dfrac{1}{20}$

Vocabulary, Readiness & Video Check

Use the choices below to fill in each blank.

improper	fraction	proper	is undefined	mixed number	= 0
≥ 1	denominator	= 1	< 1	numerator	

1. The number $\frac{17}{31}$ is called a(n) _____ . The number 31 is called its _____ and 17 is called its _____ .

2. If we simplify each fraction, $\frac{-9}{-9}$ _____ , $\frac{0}{-4}$ _____ , and we say $\frac{-4}{0}$ _____ .

3. The fraction $\frac{8}{3}$ is called a(n) _____ fraction, the fraction $\frac{3}{8}$ is called a(n) _____ fraction, and $10\frac{3}{8}$ is called a(n) _____ .

4. The value of an improper fraction is always _____ , and the value of a proper fraction is always _____ .

Martin-Gay Interactive Videos   *Watch the section lecture video and answer the following questions.*

See Video 4.1

Objective A   5. Complete this statement based on ⊞ Example 1: When the numerator is greater than or _____ to the denominator, you have a(n) _____ fraction. ▶

Objective B   6. In ⊞ Example 4, there are two shapes in the diagram, so why do the representative fractions have a denominator 3? ▶

Objective C   7. From ⊞ Examples 6 and 7, when graphing a positive fraction on a number line, how does the denominator help? What does the denominator tell you? ▶

Objective D   8. From ⊞ Example 10, what can you conclude about any fraction where the numerator and denominator are the same nonzero number? ▶

Objective E   9. Complete this statement based on the lecture before ⊞ Example 13: The operation of _____ is understood in a mixed number notation; for example, $1\frac{1}{3}$ means 1 _____ $\frac{1}{3}$. ▶

Objective F   10. From the lecture before ⊞ Example 16, what operation is used to write an improper fraction as a mixed number? ▶

4.1   Exercise Set MyMathLab® ▶

Objectives A B *Identify the numerator and the denominator of each fraction and identify each fraction as proper or improper. See Examples 1, 2, 10, and 11.*

▶ 1. $\frac{1}{2}$

2. $\frac{1}{4}$

▶ 3. $\frac{10}{3}$

4. $\frac{53}{21}$

5. $\frac{15}{15}$

6. $\frac{26}{26}$

Objective B *Write a proper or improper fraction to represent the shaded part of each diagram. If an improper fraction is appropriate, write the shaded part of the diagram as (**a**) an improper fraction and (**b**) a mixed number. (Note to students: In case you know how to simplify fractions, none of the fractions in this section are simplified.) See Examples 3 through 6 and 10 and 11.*

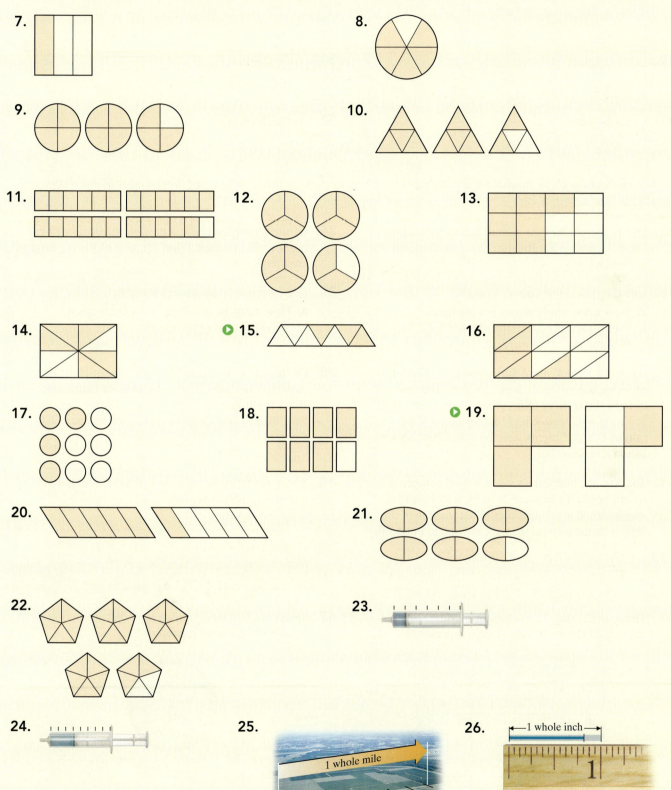

7.

8.

9.

10.

11.

12.

13.

14.

15.

16.

17.

18.

19.

20.

21.

22.

23.

24.

25.

1 whole mile

26.

1 whole inch

Objective **B** *Draw and shade a part of a diagram to represent each fraction. See Examples 7 and 8.*

27. $\frac{1}{5}$ of a diagram

28. $\frac{1}{16}$ of a diagram

29. $\frac{6}{7}$ of a diagram

30. $\frac{7}{9}$ of a diagram

31. $\frac{4}{4}$ of a diagram

32. $\frac{6}{6}$ of a diagram

Write each fraction. (Note to students: In case you know how to simplify fractions, none of the fractions in this section are simplified.) See Example 9.

33. Of the 131 students at a small private school, 42 are freshmen. What fraction of the students are freshmen?

34. Of the 63 employees at a new biomedical engineering firm, 22 are men. What fraction of the employees are men?

35. Use Exercise 33 to answer **a** and **b**.
   a. How many students are *not* freshmen?
   b. What fraction of the students are *not* freshmen?

36. Use Exercise 34 to answer **a** and **b**.
   a. How many of the employees are women?
   b. What fraction of the employees are women?

37. As of 2013, the United States has had 44 different presidents. A total of seven U.S. presidents were born in the state of Ohio, second only to the state of Virginia in producing U.S. presidents. What fraction of U.S. presidents were born in Ohio? (*Source: World Almanac*, 2013)

38. Of the eight planets in our solar system, four have days that are longer than the 24-hour Earth day. What fraction of the planets have longer days than Earth has? (*Source:* National Space Science Data Center)

39. The Atlantic hurricane season of 2005 rewrote the record books. There were 28 tropical storms, 15 of which turned into hurricanes. What fraction of the 2005 Atlantic tropical storms escalated to hurricanes?

40. There are 12 inches in a foot. What fractional part of a foot do 5 inches represent?

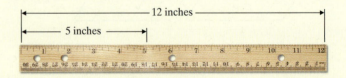

41. There are 31 days in the month of March. What fraction of the month do 11 days represent?

42. There are 60 minutes in an hour. What fraction of an hour do 37 minutes represent?

Mon.	Tue.	Wed.	Thu.	Fri.	Sat.	Sun.
					1	2
3	4	5	6	7	8	9
10	11	12	13	14	15	16
17	18	19	20	21	22	23
24	25	26	27	28	29	30
31						

43. In a prealgebra class containing 31 students, there are 18 freshmen, 10 sophomores, and 3 juniors. What fraction of the class is sophomores?

44. In a sports team with 20 children, there are 9 boys and 11 girls. What fraction of the team is boys?

45. Thirty-three out of the fifty total states in the United States contain federal Indian reservations.
   a. What fraction of the states contain federal Indian reservations?
   b. How many states do not contain federal Indian reservations?
   c. What fraction of the states do not contain federal Indian reservations? (*Source:* Tiller Research, Inc., Albuquerque, NM)

46. Consumer fireworks are legal in 44 out of the 50 total states in the United States.
   a. In what fraction of the states are consumer fireworks legal?
   b. In how many states are consumer fireworks illegal?
   c. In what fraction of the states are consumer fireworks illegal? (*Source:* United States Fireworks Safety Council)

47. A bag contains 50 red or blue marbles. If 21 marbles are blue, answer each question.
   a. What fraction of the marbles are blue?
   b. How many marbles are red?
   c. What fraction of the marbles are red?

48. An art dealer is taking inventory. His shop contains a total of 37 pieces, which are all sculptures, watercolor paintings, or oil paintings. If there are 15 watercolor paintings and 17 oil paintings, answer each question.
   a. What fraction of the inventory is watercolor paintings?
   b. What fraction of the inventory is oil paintings?
   c. How many sculptures are there?
   d. What fraction of the inventory is sculptures?

Objective **C** *Graph each fraction on a number line. See Examples 12 and 13.*

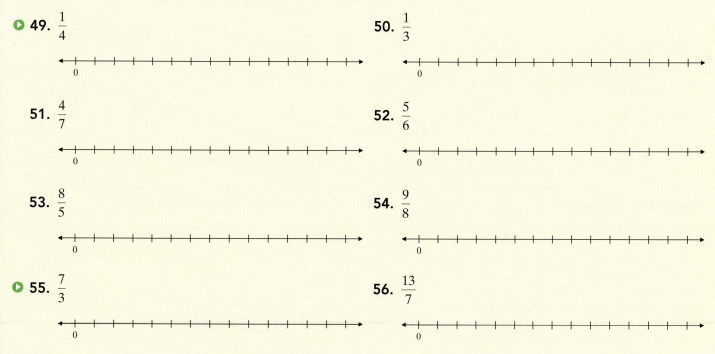

49. $\dfrac{1}{4}$

50. $\dfrac{1}{3}$

51. $\dfrac{4}{7}$

52. $\dfrac{5}{6}$

53. $\dfrac{8}{5}$

54. $\dfrac{9}{8}$

55. $\dfrac{7}{3}$

56. $\dfrac{13}{7}$

Objective **D** *Simplify by dividing. See Examples 14 through 19.*

57. $\dfrac{12}{12}$

58. $\dfrac{-3}{-3}$

59. $\dfrac{-5}{1}$

60. $\dfrac{-20}{1}$

61. $\dfrac{0}{-2}$ 　　**62.** $\dfrac{0}{-8}$ 　　**63.** $\dfrac{-8}{-8}$ 　　**64.** $\dfrac{-14}{-14}$

65. $\dfrac{-9}{0}$ 　　**66.** $\dfrac{-7}{0}$ 　　**67.** $\dfrac{3}{1}$ 　　**68.** $\dfrac{5}{5}$

Objective E *Write each mixed number as an improper fraction. See Example 20.*

69. $2\dfrac{1}{3}$ 　　**70.** $1\dfrac{13}{17}$ 　　**71.** $3\dfrac{3}{5}$ 　　**72.** $2\dfrac{5}{9}$

73. $6\dfrac{5}{8}$ 　　**74.** $7\dfrac{3}{8}$ 　　**75.** $11\dfrac{6}{7}$ 　　**76.** $12\dfrac{2}{5}$

77. $9\dfrac{7}{20}$ 　　**78.** $10\dfrac{14}{27}$ 　　**79.** $166\dfrac{2}{3}$ 　　**80.** $114\dfrac{2}{7}$

Objective F *Write each improper fraction as a mixed number or a whole number. See Example 21.*

81. $\dfrac{17}{5}$ 　　**82.** $\dfrac{13}{7}$ 　　**83.** $\dfrac{37}{8}$ 　　**84.** $\dfrac{64}{9}$

85. $\dfrac{47}{15}$ 　　**86.** $\dfrac{65}{12}$ 　　**87.** $\dfrac{225}{15}$ 　　**88.** $\dfrac{196}{14}$

89. $\dfrac{182}{175}$ 　　**90.** $\dfrac{149}{143}$ 　　**91.** $\dfrac{737}{112}$ 　　**92.** $\dfrac{901}{123}$

Review

Simplify. See Section 1.7.

93. 3^2 　　**94.** 4^3 　　**95.** 5^3 　　**96.** 3^4

Concept Extensions

Write each fraction in two other equivalent ways by inserting the negative sign in different places.

97. $-\dfrac{11}{2} = \quad = $ 　　**98.** $-\dfrac{21}{4} = \quad = $

99. $\dfrac{-13}{15} = \quad = $ 　　**100.** $\dfrac{45}{-57} = \quad = $

101. In your own words, explain why $\dfrac{0}{10} = 0$ and $\dfrac{10}{0}$ is undefined.

102. In your own words, explain why $\dfrac{0}{-3} = 0$ and $\dfrac{-3}{0}$ is undefined.

Solve. See the Concept Checks in this section.

103. If ◯◯◯◯ represents $\frac{4}{9}$ of a whole diagram, sketch the whole diagram.

104. If △ △ represents $\frac{1}{3}$ of a whole diagram, sketch the whole diagram.

105. Round the mixed number $7\frac{1}{8}$ to the nearest whole number.

106. Round the mixed number $5\frac{11}{12}$ to the nearest whole number.

107. The Gap Corporation owns stores with six different brand names, with the three most popular shown on the bar graph. What fraction of the total stores shown on the graph are named "Banana Republic"?

108. The Public Broadcasting Service (PBS) provides programming to the noncommercial public TV stations of the United States. The bar graph shows a breakdown of the public television licensees by type. Each licensee operates one or more PBS member TV stations. What fraction of the public television licensees are universities or colleges? (*Source:* The Public Broadcasting Service)

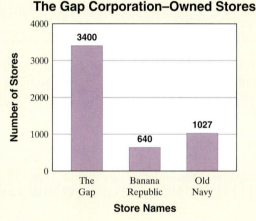

The Gap Corporation–Owned Stores

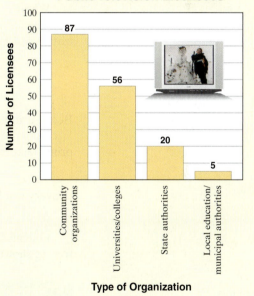

Public Television Licensees

109. Habitat for Humanity is a nonprofit organization that helps provide affordable housing to families in need. Habitat for Humanity does its work of building and renovating houses through 1500 local affiliates in the United States and 80 international affiliates. What fraction of the total Habitat for Humanity affiliates are located in the United States? (*Hint:* First find the total number of affiliates.) (*Source:* Habitat for Humanity International)

110. The United States Marine Corps (USMC) has five principal training centers in California, three in North Carolina, two in South Carolina, one in Arizona, one in Hawaii, and one in Virginia. What fraction of the total USMC principal training centers are located in California? (*Hint:* First find the total number of USMC training centers.) (*Source:* U.S. Department of Defense)

Objectives

A Write a Number as a Product of Prime Numbers.

B Write a Fraction in Simplest Form.

C Determine Whether Two Fractions Are Equivalent.

D Solve Problems by Writing Fractions in Simplest Form.

Objective A Writing a Number as a Product of Prime Numbers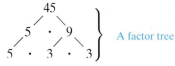

Recall from Section 1.5 that since $12 = 2 \cdot 2 \cdot 3$, the numbers 2 and 3 are called *factors* of 12. A **factor** is any number that divides a number evenly (with a remainder of 0).

To perform operations on fractions, it is necessary to be able to factor a number. Remember that factoring a number means writing a number as a product. We first practice writing a number as a product of prime numbers.

> A **prime number** is a natural number greater than 1 whose only factors are 1 and itself. The first few prime numbers are 2, 3, 5, 7, 11, 13, 17, 19, 23, 29, … .
> A **composite number** is a natural number greater than 1 that is not prime.

Helpful Hint

The natural number 1 is neither prime nor composite.

When a composite number is written as a product of prime numbers, this product is called the **prime factorization** of the number. For example, the prime factorization of 12 is $2 \cdot 2 \cdot 3$ because

$$12 = \underbrace{2 \cdot 2 \cdot 3}$$
This product is 12 and each number is a prime number.

Because multiplication is commutative, the order of the factors is not important. We can write the factorization $2 \cdot 2 \cdot 3$ as $2 \cdot 3 \cdot 2$ or $3 \cdot 2 \cdot 2$. Any of these is called the prime factorization of 12.

> Every whole number greater than 1 has exactly one prime factorization.

One method for finding the prime factorization of a number is by using a factor tree, as shown in the next example.

Practice 1

Use a factor tree to find the prime factorization of each number.

a. 30 **b.** 56 **c.** 72

Example 1 Write the prime factorization of 45.

Solution: We can begin by writing 45 as the product of two numbers, say, 5 and 9.

```
    45
   /  \
  5  ·  9
```

The number 5 is prime but 9 is not, so we write 9 as $3 \cdot 3$.

```
        45
       /  \
      5  ·  9      }  A factor tree
     / \   / \
    5 · 3 · 3
```

Each factor is now a prime number, so the prime factorization of 45 is $3 \cdot 3 \cdot 5$ or $3^2 \cdot 5$.

■ **Work Practice 1**

Answers

1. a. $2 \cdot 3 \cdot 5$ **b.** $2^3 \cdot 7$ **c.** $2^3 \cdot 3^2$

✓**Concept Check** True or false? Two different numbers can have exactly the same prime factorization. Explain your answer.

Example 2 Write the prime factorization of 80.

Solution: Write 80 as a product of two numbers. Continue this process until all factors are prime.

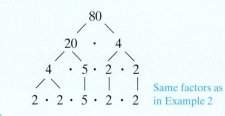

All factors are now prime, so the prime factorization of 80 is

$$2 \cdot 2 \cdot 2 \cdot 2 \cdot 5 \quad \text{or} \quad 2^4 \cdot 5.$$

■ **Work Practice 2**

Helpful Hint

It makes no difference which factors you start with. The prime factorization of a number will be the same.

$$
\begin{array}{c}
80 \\
\diagup\ \diagdown \\
20 \quad\cdot\quad 4 \\
\diagup\diagdown\ \quad \diagup\diagdown \\
4\ \cdot\ 5\ \cdot\ 2\ \cdot\ 2 \\
\diagup\diagdown\ \ |\quad |\quad | \\
2\ \cdot\ 2\ \cdot\ 5\ \cdot\ 2\ \cdot\ 2
\end{array}
$$

Same factors as in Example 2

There are a few quick **divisibility tests** to determine whether a number is divisible by the primes 2, 3, or 5. (A number is divisible by 2, for example, if 2 divides it evenly so that the remainder is 0.)

Divisibility Tests

A whole number is divisible by:

- 2 if the last digit is 0, 2, 4, 6, or 8.
   ↓
   13**2** is divisible by 2 since the last digit is a 2.
- 3 if the sum of the digits is divisible by 3.

   144 is divisible by 3 since $1 + 4 + 4 = 9$ is divisible by 3.
- 5 if the last digit is 0 or 5.
   ↓
   111**5** is divisible by 5 since the last digit is a 5.

Practice 2

Write the prime factorization of 60.

Answer
2. $2^2 \cdot 3 \cdot 5$

✓**Concept Check Answer**
false; answers may vary

> **Helpful Hint**
>
> Here are a few other divisibility tests you may want to use. A whole number is divisible by:
>
> - 4 if its last two digits are divisible by 4.
>
>   1712 is divisible by 4.
>
> - 6 if it's divisible by 2 and 3.
>
>   9858 is divisible by 6.
>
> - 9 if the sum of its digits is divisible by 9.
>
>   5238 is divisible by 9 since $5 + 2 + 3 + 8 = 18$ is divisible by 9.

When finding the prime factorization of larger numbers, you may want to use the procedure shown in Example 3.

Practice 3

Write the prime factorization of 297.

Example 3   Write the prime factorization of 252.

Solution: For this method, we divide prime numbers into the given number. Since the ones digit of 252 is 2, we know that 252 is divisible by 2.

$$\begin{array}{r} 126 \\ 2\overline{)252} \end{array}$$

126 is divisible by 2 also.

$$\begin{array}{r} 63 \\ 2\overline{)126} \\ 2\overline{)252} \end{array}$$

63 is not divisible by 2 but is divisible by 3. Divide 63 by 3 and continue in this same manner until the quotient is a prime number.

$$\begin{array}{r} 7 \\ 3\overline{)\ 21} \\ 3\overline{)\ 63} \\ 2\overline{)126} \\ 2\overline{)252} \end{array}$$

> **Helpful Hint**   The order of choosing prime numbers does not matter. For consistency, we use the order $2, 3, 5, 7, \ldots$.

The prime factorization of 252 is $2 \cdot 2 \cdot 3 \cdot 3 \cdot 7$ or $2^2 \cdot 3^2 \cdot 7$.

■ **Work Practice 3**

In this text, we will write the factorization of a number from the smallest factor to the largest factor.

✓ **Concept Check**   True or false? The prime factorization of 117 is $9 \cdot 13$. Explain your reasoning.

Objective B   Writing Fractions in Simplest Form ▶

Fractions that represent the same portion of a whole or the same point on a number line are called **equivalent fractions.** Study the table on the next page to see two ways to visualize equivalent fractions.

Answer

3. $3^3 \cdot 11$

✓ **Concept Check Answer**

False; 9 is not prime.

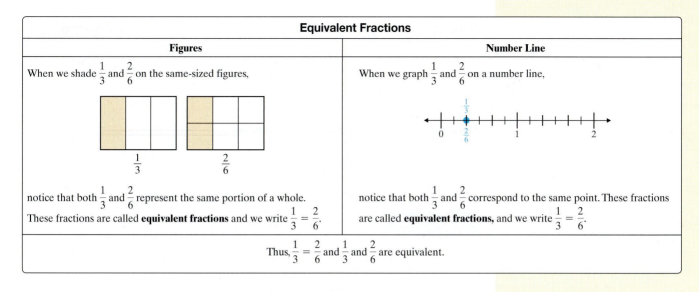

Equivalent Fractions	
Figures	**Number Line**
When we shade $\frac{1}{3}$ and $\frac{2}{6}$ on the same-sized figures, $\frac{1}{3}$ $\frac{2}{6}$ notice that both $\frac{1}{3}$ and $\frac{2}{6}$ represent the same portion of a whole. These fractions are called **equivalent fractions** and we write $\frac{1}{3} = \frac{2}{6}$.	When we graph $\frac{1}{3}$ and $\frac{2}{6}$ on a number line, notice that both $\frac{1}{3}$ and $\frac{2}{6}$ correspond to the same point. These fractions are called **equivalent fractions,** and we write $\frac{1}{3} = \frac{2}{6}$.
Thus, $\frac{1}{3} = \frac{2}{6}$ and $\frac{1}{3}$ and $\frac{2}{6}$ are equivalent.	

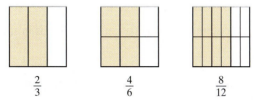

$$\frac{2}{3} \qquad \frac{4}{6} \qquad \frac{8}{12}$$

For example, $\frac{2}{3}, \frac{4}{6}$, and $\frac{8}{12}$ all represent the same shaded portion of the rectangle's area, so they are equivalent fractions. To show that these fractions are equivalent, we place an equal sign between them. In other words,

$$\frac{2}{3} = \frac{4}{6} = \frac{8}{12}$$

There are many equivalent forms of a fraction. A special equivalent form of a fraction is called **simplest form.**

Simplest Form of a Fraction

A fraction is written in **simplest form** or **lowest terms** when the numerator and the denominator have no common factors other than 1.

For example, the fraction $\frac{2}{3}$ *is* in simplest form because 2 and 3 have no common factor other than 1. The fraction $\frac{4}{6}$ *is not* in simplest form because 4 and 6 both have a factor of 2. That is, 2 is a common factor of 4 and 6. The process of writing a fraction in simplest form is called **simplifying** the fraction.

To simplify $\frac{4}{6}$ and write it as $\frac{2}{3}$, let's first study a few properties. Recall from Section 4.1 that any nonzero whole number n divided by itself is 1.

Any nonzero number n divided by itself is 1.

$$\frac{5}{5} = 1, \quad \frac{17}{17} = 1, \quad \frac{24}{24} = 1, \text{ or, in general, } \frac{n}{n} = 1$$

Copyright 2015 Pearson Education, Inc.

Also, in general, if $\frac{a}{b}$ and $\frac{c}{d}$ are fractions (with b and d not 0), the following is true.

$$\frac{a \cdot c}{b \cdot d} = \frac{a}{b} \cdot \frac{c}{d}^*$$

These properties allow us to do the following:

$$\frac{4}{6} = \frac{2 \cdot 2}{2 \cdot 3} = \frac{2}{2} \cdot \frac{2}{3} = 1 \cdot \frac{2}{3} = \frac{2}{3}$$

This is 1.

When 1 is multiplied by a number, the result is the same number.

Helpful Hint

These two properties together are called the Fundamental Property of Fractions

$$\frac{a \cdot c}{b \cdot c} = \frac{a}{b} \cdot \frac{c}{c} = \frac{a}{b}$$

Practice 4
Write in simplest form: $\frac{30}{45}$

Example 4 Write in simplest form: $\frac{12}{20}$

Solution: Notice that 12 and 20 have a common factor of 4.

$$\frac{12}{20} = \frac{4 \cdot 3}{4 \cdot 5} = \frac{4}{4} \cdot \frac{3}{5} = 1 \cdot \frac{3}{5} = \frac{3}{5}$$

Since 3 and 5 have no common factors (other than 1), $\frac{3}{5}$ is in simplest form.

■ **Work Practice 4**

If you have trouble finding common factors, write the prime factorization of the numerator and the denominator.

Practice 5
Write in simplest form: $\frac{39x}{51}$

Example 5 Write in simplest form: $\frac{42x}{66}$

Solution: Let's write the prime factorizations of 42 and 66. Remember that $42x$ means $42 \cdot x$.

$$\frac{42x}{66} = \frac{2 \cdot 3 \cdot 7 \cdot x}{2 \cdot 3 \cdot 11} = \frac{2}{2} \cdot \frac{3}{3} \cdot \frac{7x}{11} = 1 \cdot 1 \cdot \frac{7x}{11} = \frac{7x}{11}$$

■ **Work Practice 5**

In the example above, you may have saved time by noticing that 42 and 66 have a common factor of 6.

$$\frac{42x}{66} = \frac{6 \cdot 7x}{6 \cdot 11} = \frac{6}{6} \cdot \frac{7x}{11} = 1 \cdot \frac{7x}{11} = \frac{7x}{11}$$

Helpful Hint

Writing the prime factorizations of the numerator and the denominator is helpful in finding any common factors.

The method for simplifying negative fractions is the same as for positive fractions.

Practice 6
Write in simplest form: $-\frac{9}{50}$

Example 6 Write in simplest form: $-\frac{10}{27}$

Solution:

$$-\frac{10}{27} = -\frac{2 \cdot 5}{3 \cdot 3 \cdot 3}$$ Prime factorizations of 10 and 27

Since 10 and 27 have no common factors, $-\frac{10}{27}$ is already in simplest form.

■ **Work Practice 6**

Note: We will study this concept further in the next section.

Answers
4. $\frac{2}{3}$ 5. $\frac{13x}{17}$ 6. $-\frac{9}{50}$

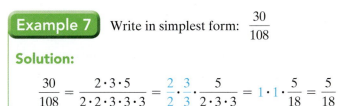

Example 7 Write in simplest form: $\dfrac{30}{108}$

Solution:

$$\dfrac{30}{108} = \dfrac{2 \cdot 3 \cdot 5}{2 \cdot 2 \cdot 3 \cdot 3 \cdot 3} = \dfrac{2}{2} \cdot \dfrac{3}{3} \cdot \dfrac{5}{2 \cdot 3 \cdot 3} = 1 \cdot 1 \cdot \dfrac{5}{18} = \dfrac{5}{18}$$

■ **Work Practice 7**

Practice 7
Write in simplest form: $\dfrac{49}{112}$

We can use a shortcut procedure with common factors when simplifying.

$$\dfrac{4}{6} = \dfrac{\overset{1}{\cancel{2}} \cdot 2}{\underset{1}{\cancel{2}} \cdot 3} = \dfrac{1 \cdot 2}{1 \cdot 3} = \dfrac{2}{3} \qquad \text{Divide out the common factor of 2 in the numerator and denominator.}$$

This procedure is possible because dividing out a common factor in the numerator and denominator is the same as removing a factor of 1 in the product.

Writing a Fraction in Simplest Form

To write a fraction in simplest form, write the prime factorization of the numerator and the denominator and then divide both by all common factors.

Example 8 Write in simplest form: $-\dfrac{72}{26}$

Solution:

$$-\dfrac{72}{26} = -\dfrac{\overset{1}{\cancel{2}} \cdot 2 \cdot 2 \cdot 3 \cdot 3}{\underset{1}{\cancel{2}} \cdot 13} = -\dfrac{1 \cdot 2 \cdot 2 \cdot 3 \cdot 3}{1 \cdot 13} = -\dfrac{36}{13}$$

■ **Work Practice 8**

Practice 8
Write in simplest form: $-\dfrac{64}{20}$

✓**Concept Check** Which is the correct way to simplify the fraction $\dfrac{15}{25}$? Or are both correct? Explain.

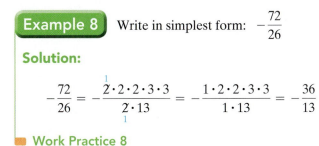

a. $\dfrac{15}{25} = \dfrac{3 \cdot \overset{1}{\cancel{5}}}{5 \cdot \underset{1}{\cancel{5}}} = \dfrac{3}{5}$ **b.** $\dfrac{1\cancel{5}}{2\cancel{5}} = \dfrac{11}{21}$

In this chapter, we will simplify and perform operations on fractions containing variables. When the denominator of a fraction contains a variable, such as $\dfrac{6x^2}{60x^3}$, we will assume that the variable does not represent 0. Recall that the denominator of a fraction cannot be 0.

Example 9 Write in simplest form: $\dfrac{6x^2}{60x^3}$

Solution: Notice that 6 and 60 have a common factor of 6. Let's also use the definition of an exponent to factor x^2 and x^3.

$$\dfrac{6x^2}{60x^3} = \dfrac{\overset{1}{\cancel{6}} \cdot \overset{1}{\cancel{x}} \cdot \overset{1}{\cancel{x}}}{\underset{1}{\cancel{6}} \cdot 10 \cdot \underset{1}{\cancel{x}} \cdot \underset{1}{\cancel{x}} \cdot x} = \dfrac{1 \cdot 1 \cdot 1}{1 \cdot 10 \cdot 1 \cdot 1 \cdot x} = \dfrac{1}{10x}$$

■ **Work Practice 9**

Practice 9
Write in simplest form: $\dfrac{7a^3}{56a^2}$

Answers

7. $\dfrac{7}{16}$ 8. $-\dfrac{16}{5}$ 9. $\dfrac{a}{8}$

Concept Check Answers
a. correct **b.** incorrect

Helpful Hint

Be careful when all factors of the numerator or denominator are divided out. In Example 9, the numerator was $1 \cdot 1 \cdot 1 = 1$, so the final result was $\dfrac{1}{10x}$.

Objective C Determining Whether Two Fractions Are Equivalent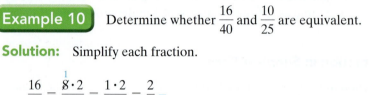

Recall from Objective B that two fractions are equivalent if they represent the same part of a whole. One way to determine whether two fractions are equivalent is to see whether they simplify to the same fraction.

Practice 10

Determine whether $\dfrac{7}{9}$ and $\dfrac{21}{27}$ are equivalent.

Example 10 Determine whether $\dfrac{16}{40}$ and $\dfrac{10}{25}$ are equivalent.

Solution: Simplify each fraction.

$$\frac{16}{40} = \frac{\overset{1}{\cancel{8}} \cdot 2}{\underset{1}{\cancel{8}} \cdot 5} = \frac{1 \cdot 2}{1 \cdot 5} = \frac{2}{5}$$

$$\frac{10}{25} = \frac{2 \cdot \overset{1}{\cancel{5}}}{5 \cdot \underset{1}{\cancel{5}}} = \frac{2 \cdot 1}{5 \cdot 1} = \frac{2}{5}$$

Since these fractions are the same, $\dfrac{16}{40} = \dfrac{10}{25}$.

■ **Work Practice 10**

There is a shortcut method you may use to check or test whether two fractions are equivalent. In the example above, we learned that the fractions are equivalent, or

$$\frac{16}{40} = \frac{10}{25}$$

In this example above, we call $25 \cdot 16$ and $40 \cdot 10$ **cross products** because they are the products one obtains by multiplying diagonally across the equal sign, as shown below.

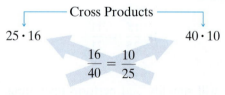

Cross Products

$$25 \cdot 16 \qquad \frac{16}{40} = \frac{10}{25} \qquad 40 \cdot 10$$

Notice that these cross products are equal:

$$25 \cdot 16 = 400, \quad 40 \cdot 10 = 400$$

In general, this is true for equivalent fractions.

Equality of Fractions

$$8 \cdot 6 \qquad \frac{6}{24} \overset{?}{=} \frac{2}{8} \qquad 24 \cdot 2$$

Since the cross products ($8 \cdot 6 = 48$ and $24 \cdot 2 = 48$) are equal, the fractions are equal.

Note: If the cross products are not equal, the fractions are not equal.

Answer

10. equivalent

Example 11 Determine whether $\frac{8}{11}$ and $\frac{19}{26}$ are equivalent.

Solution: Let's check cross products.

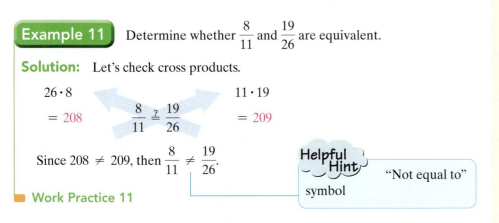

$26 \cdot 8$ $11 \cdot 19$

$= 208$ $\frac{8}{11} \stackrel{?}{=} \frac{19}{26}$ $= 209$

Since $208 \neq 209$, then $\frac{8}{11} \neq \frac{19}{26}$.

Helpful Hint

"Not equal to" symbol

■ Work Practice 11

Practice 11

Determine whether $\frac{4}{13}$ and $\frac{5}{18}$ are equivalent.

Objective D Solving Problems by Writing Fractions in Simplest Form

Many real-life problems can be solved by writing fractions. To make the answers clearer, these fractions should be written in simplest form.

Example 12 Calculating Fraction of Parks in Wyoming

There are currently 58 national parks in the United States. Two of these parks are located in the state of Wyoming. What fraction of the United States' national parks can be found in Wyoming? Write the fraction in simplest form. (*Source: World Almanac*, 2013)

Solution: First we determine the fraction of parks found in Wyoming state.

$\frac{2}{58}$ ← national parks in Wyoming
      ← total national parks

Next we simplify the fraction.

$$\frac{2}{58} = \frac{\overset{1}{\cancel{2}}}{\underset{1}{\cancel{2}} \cdot 29} = \frac{1}{1 \cdot 29} = \frac{1}{29}$$

Thus, $\frac{1}{29}$ of the United States' national parks are in Wyoming state.

■ Work Practice 12

Practice 12

There are eight national parks in Alaska. See Example 12 and determine what fraction of the United States' national parks are located in Alaska. Write the fraction in simplest form.

Answers

11. not equivalent   **12.** $\frac{4}{29}$

Calculator Explorations Simplifying Fractions

Scientific Calculator

Many calculators have a fraction key, such as $\boxed{a^b/c}$, that allows you to simplify a fraction on the calculator. For example, to simplify $\dfrac{324}{612}$, enter

$$\boxed{324} \quad \boxed{a^b/c} \quad \boxed{612} \quad \boxed{=}$$

The display will read

$$\boxed{\quad 9 \mid 17 \quad}$$

which represents $\dfrac{9}{17}$, the original fraction simplified.

> **Helpful Hint**
>
> The Calculator Explorations boxes in this chapter provide only an introduction to fraction keys on calculators. Any time you use a calculator, there are both advantages and limitations to its use. Never rely solely on your calculator. It is very important that you understand how to perform all operations on fractions by hand in order to progress through later topics. For further information, talk to your instructor.

Graphing Calculator

Graphing calculators also allow you to simplify fractions. The fraction option on a graphing calculator may be found under the $\boxed{\text{MATH}}$ menu.

To simplify $\dfrac{324}{612}$, enter

$$\boxed{324} \quad \boxed{\div} \quad \boxed{612} \quad \boxed{\text{MATH}} \quad \boxed{\text{ENTER}} \quad \boxed{\text{ENTER}}$$

The display will read

$$\boxed{324/612 \blacktriangleright \text{Frac } 9/17}$$

Use your calculator to simplify each fraction.

1. $\dfrac{128}{224}$ **2.** $\dfrac{231}{396}$ **3.** $\dfrac{340}{459}$

4. $\dfrac{999}{1350}$ **5.** $\dfrac{432}{810}$ **6.** $\dfrac{225}{315}$

7. $\dfrac{54}{243}$ **8.** $\dfrac{455}{689}$

Vocabulary, Readiness & Video Check

Use the choices below to fill in each blank.

cross products   equivalent   composite

simplest form   prime factorization   prime

1. The number 40 equals $2 \cdot 2 \cdot 2 \cdot 5$. Since each factor is prime, we call $2 \cdot 2 \cdot 2 \cdot 5$ the _____ of 40.

2. A natural number, other than 1, that is not prime is called a(n) _____ number.

3. A natural number that has exactly two different factors, 1 and itself, is called a(n) _____ number.

4. In $\dfrac{11}{48}$, since 11 and 48 have no common factors other than 1, $\dfrac{11}{48}$ is in _____.

5. Fractions that represent the same portion of a whole are called _____ fractions.

6. In the statement $\dfrac{5}{12} = \dfrac{15}{36}$, $5 \cdot 36$ and $12 \cdot 15$ are called _____.

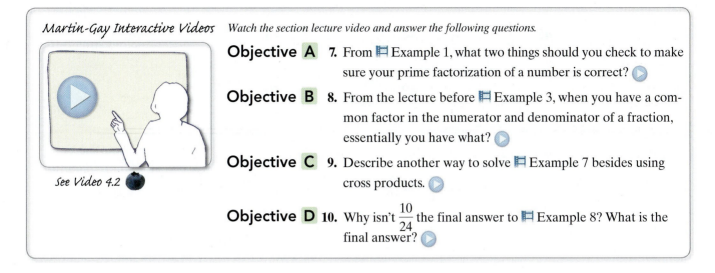

Martin-Gay Interactive Videos   *Watch the section lecture video and answer the following questions.*

Objective A 7. From ▣ Example 1, what two things should you check to make sure your prime factorization of a number is correct? ▶

Objective B 8. From the lecture before ▣ Example 3, when you have a common factor in the numerator and denominator of a fraction, essentially you have what? ▶

Objective C 9. Describe another way to solve ▣ Example 7 besides using cross products. ▶

Objective D 10. Why isn't $\frac{10}{24}$ the final answer to ▣ Example 8? What is the final answer? ▶

See Video 4.2

4.2 Exercise Set MyMathLab® ▶

Objective A *Write the prime factorization of each number. See Examples 1 through 3.*

1. 20

2. 12

▶ **3.** 48

4. 75

5. 81

6. 64

7. 162

8. 128

9. 110

10. 130

11. 85

12. 93

▶ **13.** 240

14. 836

15. 828

16. 504

Objective B *Write each fraction in simplest form. See Examples 4 through 9.*

17. $\frac{3}{12}$

18. $\frac{5}{30}$

19. $\frac{4x}{42}$

20. $\frac{9y}{48}$

21. $\frac{14}{16}$

22. $\frac{22}{34}$

23. $\frac{20}{30}$

24. $\frac{70}{80}$

25. $\frac{35a}{50a}$

26. $\frac{25z}{55z}$

▶ **27.** $-\frac{63}{81}$

28. $-\frac{21}{49}$

▶ **29.** $\frac{30x^2}{36x}$

30. $\frac{45b}{80b^2}$

31. $\frac{27}{64}$

32. $\frac{32}{63}$

33. $\frac{25xy}{40y}$

34. $\frac{36y}{42yz}$

35. $-\frac{40}{64}$

36. $-\frac{28}{60}$

37. $\frac{36x^3y^2}{24xy}$

38. $\frac{60a^2b}{36ab^3}$

39. $\frac{90}{120}$

40. $\frac{60}{150}$

▶ **41.** $\frac{40xy}{64xyz}$

42. $\dfrac{28abc}{60ac}$ **43.** $\dfrac{66}{308}$ **44.** $\dfrac{65}{234}$ **45.** $-\dfrac{55}{85y}$ **46.** $-\dfrac{78}{90x}$

47. $\dfrac{189z}{216z}$ **48.** $\dfrac{144y}{162y}$ **49.** $\dfrac{224a^3b^4c^2}{16ab^4c^2}$ **50.** $\dfrac{270x^4y^3z^3}{15x^3y^3z^3}$

Objective C *Determine whether each pair of fractions is equivalent. See Examples 10 and 11.*

51. $\dfrac{2}{6}$ and $\dfrac{4}{12}$ **52.** $\dfrac{3}{6}$ and $\dfrac{5}{10}$ ▶ **53.** $\dfrac{7}{11}$ and $\dfrac{5}{8}$

54. $\dfrac{2}{5}$ and $\dfrac{4}{11}$ **55.** $\dfrac{10}{15}$ and $\dfrac{6}{9}$ **56.** $\dfrac{4}{10}$ and $\dfrac{6}{15}$

▶ **57.** $\dfrac{3}{9}$ and $\dfrac{6}{18}$ **58.** $\dfrac{2}{8}$ and $\dfrac{7}{28}$ **59.** $\dfrac{10}{13}$ and $\dfrac{13}{15}$

60. $\dfrac{16}{20}$ and $\dfrac{9}{16}$ **61.** $\dfrac{8}{18}$ and $\dfrac{12}{24}$ **62.** $\dfrac{6}{21}$ and $\dfrac{14}{35}$

Objective D *Solve. Write each fraction in simplest form. See Example 12.*

63. A work shift for an employee at Starbucks consists of 8 hours. What fraction of the employee's work shift is represented by 2 hours?

64. Two thousand baseball caps were sold one year at the U.S. Open Golf Tournament. What fractional part of this total do 200 caps represent?

65. There are 5280 feet in a mile. What fraction of a mile is represented by 2640 feet?

66. There are 100 centimeters in 1 meter. What fraction of a meter is 20 centimeters?

67. Sixteen out of the total fifty states in the United States have Ritz-Carlton hotels. (*Source:* Ritz-Carlton Hotel Company, LLC)
   a. What fraction of states can claim at least one Ritz-Carlton hotel?
   b. How many states do not have a Ritz-Carlton hotel?
   c. Write the fraction of states without a Ritz-Carlton hotel.

68. There are 75 national monuments in the United States. Ten of these monuments are located in New Mexico. (*Source: World Almanac,* 2013)
   a. What fraction of the national monuments in the United States can be found in New Mexico?
   b. How many of the national monuments in the United States are found outside New Mexico?
   c. Write the fraction of national monuments found in states other than New Mexico.

▶ **69.** The outer wall of the Pentagon is 24 inches thick. Ten inches is concrete, 8 inches is brick, and 6 inches is limestone. What fraction of the wall is concrete?

70. There are 35 students in a biology class. If 10 students made an A on the first test, what fraction of the students made an A?

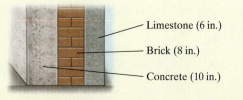

Limestone (6 in.)

Brick (8 in.)

Concrete (10 in.)

71. As Internet usage grows in the United States, more and more state governments are placing services on-line. Forty-two out of the total fifty states have Web sites that allow residents to file their federal and state income tax electronically at the same time.
  a. How many states do not have this type of Web site?
  b. What fraction of states do not have this type of Web site?(*Source:* MeF Federal/State Program)

72. Katy Biagini just bought a brand-new 2013 Toyota Camry hybrid for $24,200. Her old car was traded in for $12,000.
  a. How much of her purchase price was not covered by her trade-in?
  b. What fraction of the purchase price was not covered by the trade-in?

73. As of this writing, a total of 464 individuals from the United States are or have been astronauts. Of these, 26 were born in Texas. What fraction of U.S. astronauts were born in Texas? (*Source:* Spacefacts Web site)

74. Worldwide, Hallmark employs nearly 12,000 full-time employees. About 3200 employees work at the Hallmark headquarters in Kansas City, Missouri. What fraction of Hallmark employees work in Kansas City? (*Source:* Hallmark Cards, Inc.)

Review

Evaluate each expression using the given replacement numbers. See Section 2.5.

75. $\dfrac{x^3}{9}$ when $x = -3$

76. $\dfrac{y^3}{5}$ when $y = -5$

77. $2y$ when $y = -7$

78. $-5a$ when $a = -4$

Concept Extensions

79. In your own words, define equivalent fractions.

80. Given a fraction, say $\dfrac{15}{40}$, how many fractions are there that are equivalent to it, but in simplest form or lowest terms? Explain your answer.

Write each fraction in simplest form.

81. $\dfrac{3975}{6625}$

82. $\dfrac{9506}{12,222}$

There are generally considered to be eight basic blood types. The table shows the number of people with the various blood types in a typical group of 100 blood donors. Use the table to answer Exercises 83 through 86. Write each answer in simplest form.

Distribution of Blood Types in Blood Donors	
Blood Type	**Number of People**
O Rh-positive	37
O Rh-negative	7
A Rh-positive	36
A Rh-negative	6
B Rh-positive	9
B Rh-negative	1
AB Rh-positive	3
AB Rh-negative	1
(*Source:* American Red Cross Biomedical Services)	

83. What fraction of blood donors have blood type A Rh-positive?

84. What fraction of blood donors have an O blood type?

85. What fraction of blood donors have an AB blood type?

86. What fraction of blood donors have a B blood type?

Find the prime factorization of each number.

87. 34,020

88. 131,625

89. In your own words, define a prime number.

90. The number 2 is a prime number. All other even natural numbers are composite numbers. Explain why.

91. Two students have different prime factorizations for the same number. Is this possible? Explain.

92. Two students work to prime factor 120. One student starts by writing 120 as 12×10. The other student writes 120 as 24×5. Finish each prime factorization. Are they the same? Why or why not?

The following graph is called a circle graph or pie chart. Each sector (shaped like a piece of pie) shows the fraction of entering college freshmen who choose to major in each discipline shown. The whole circle represents the entire class of college freshmen. Use this graph to answer Exercises 93 through 96. Write each fraction answer in simplest form.

College Freshmen Majors

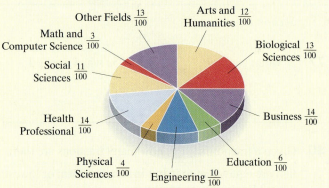

Other Fields $\frac{13}{100}$

Arts and Humanities $\frac{12}{100}$

Math and Computer Science $\frac{3}{100}$

Social Sciences $\frac{11}{100}$

Biological Sciences $\frac{13}{100}$

Health Professional $\frac{14}{100}$

Business $\frac{14}{100}$

Physical Sciences $\frac{4}{100}$

Engineering $\frac{10}{100}$

Education $\frac{6}{100}$

Source: The Higher Education Research Institute

93. What fraction of entering college freshmen plan to major in education?

94. What fraction of entering college freshmen plan to major in engineering?

95. Why is the Business sector the same size as the Health Professional sector?

96. Why is the Physical Sciences sector smaller than the Business sector?

Use this circle graph to answer Exercises 97 through 100. Write each fraction answer in simplest form.

Areas Maintained by the National Park Service

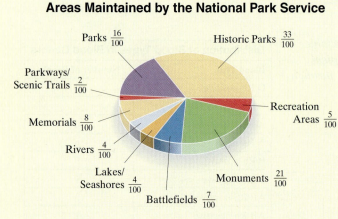

Parks $\frac{16}{100}$

Historic Parks $\frac{33}{100}$

Parkways/ Scenic Trails $\frac{2}{100}$

Memorials $\frac{8}{100}$

Recreation Areas $\frac{5}{100}$

Rivers $\frac{4}{100}$

Lakes/ Seashores $\frac{4}{100}$

Monuments $\frac{21}{100}$

Battlefields $\frac{7}{100}$

Source: National Park Service

97. What fraction of National Park Service areas are National Memorials?

98. What fraction of National Park Service areas are National Parks?

99. Why is the National Battlefields sector smaller than the National Monuments sector?

100. Why is the National Lakes/National Seashores sector the same size as the National Rivers sector?

Use the following numbers for Exercises 101 through 104.

8691   786   1235   2235   85   105   22   222   900   1470

101. List the numbers divisible by both 2 and 3.

102. List the numbers that are divisible by both 3 and 5.

103. The answers to Exercise 101 are also divisible by what number? Tell why.

104. The answers to Exercise 102 are also divisible by what number? Tell why.

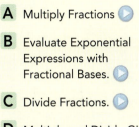

Objective A Multiplying Fractions

Let's use a diagram to discover how fractions are multiplied. For example, to multiply $\frac{1}{2}$ and $\frac{3}{4}$, we find $\frac{1}{2}$ of $\frac{3}{4}$. To do this, we begin with a diagram showing $\frac{3}{4}$ of a rectangle's area shaded.

$\frac{3}{4}$ of the rectangle's area is shaded.

To find $\frac{1}{2}$ of $\frac{3}{4}$, we heavily shade $\frac{1}{2}$ of the part that is already shaded.

By counting smaller rectangles, we see that $\frac{3}{8}$ of the larger rectangle is now heavily shaded, so that

$$\frac{1}{2} \text{ of } \frac{3}{4} \text{ is } \frac{3}{8}, \text{ or } \frac{1}{2} \cdot \frac{3}{4} = \frac{3}{8}$$ Notice that $\frac{1}{2} \cdot \frac{3}{4} = \frac{1 \cdot 3}{2 \cdot 4} = \frac{3}{8}$.

Multiplying Fractions

To multiply two fractions, multiply the numerators and multiply the denominators. If a, b, c, and d represent numbers, and b and d are not 0, we have

$$\frac{a}{b} \cdot \frac{c}{d} = \frac{a \cdot c}{b \cdot d}$$

Examples Multiply.

1. $\frac{2}{3} \cdot \frac{5}{11} = \frac{2 \cdot 5}{3 \cdot 11} = \frac{10}{33}$ Multiply numerators.
Multiply denominators.

This fraction is in simplest form since 10 and 33 have no common factors other than 1.

2. $\frac{1}{4} \cdot \frac{1}{2} = \frac{1 \cdot 1}{4 \cdot 2} = \frac{1}{8}$ This fraction is in simplest form.

■ Work Practice 1–2

Practice 1–2

Multiply.

1. $\frac{3}{7} \cdot \frac{5}{11}$ **2.** $\frac{1}{3} \cdot \frac{1}{9}$

Example 3 Multiply and simplify: $\frac{6}{7} \cdot \frac{14}{27}$

Solution:

$$\frac{6}{7} \cdot \frac{14}{27} = \frac{6 \cdot 14}{7 \cdot 27}$$

We can simplify by finding the prime factorizations and using our shortcut procedure of dividing out common factors in the numerator and denominator.

$$\frac{6 \cdot 14}{7 \cdot 27} = \frac{2 \cdot \overset{1}{\cancel{3}} \cdot 2 \cdot \overset{1}{\cancel{7}}}{\underset{1}{\cancel{7}} \cdot \underset{1}{\cancel{3}} \cdot 3 \cdot 3} = \frac{2 \cdot 2}{3 \cdot 3} = \frac{4}{9}$$

■ Work Practice 3

Practice 3

Multiply and simplify: $\frac{6}{77} \cdot \frac{7}{8}$

Answers

1. $\frac{15}{77}$ **2.** $\frac{1}{27}$ **3.** $\frac{3}{44}$

Copyright 2015 Pearson Education, Inc.

Helpful Hint

Remember that the shortcut procedure in Example 3 is the same as removing factors of 1 in the product.

$$\frac{6 \cdot 14}{7 \cdot 27} = \frac{2 \cdot 3 \cdot 2 \cdot 7}{7 \cdot 3 \cdot 3 \cdot 3} = \frac{7}{7} \cdot \frac{3}{3} \cdot \frac{2 \cdot 2}{3 \cdot 3} = 1 \cdot 1 \cdot \frac{4}{9} = \frac{4}{9}$$

Practice 4

Multiply and simplify: $\dfrac{4}{27} \cdot \dfrac{3}{8}$

Example 4 Multiply and simplify: $\dfrac{23}{32} \cdot \dfrac{4}{7}$

Solution: Notice that 4 and 32 have a common factor of 4.

$$\frac{23}{32} \cdot \frac{4}{7} = \frac{23 \cdot 4}{32 \cdot 7} = \frac{23 \cdot \overset{1}{\cancel{4}}}{\cancel{4} \cdot 8 \cdot 7} = \frac{23}{8 \cdot 7} = \frac{23}{56}$$

◼ **Work Practice 4**

Helpful Hint Don't forget that we may identify common factors that are not prime numbers.

After multiplying two fractions, always check to see whether the product can be simplified.

Practice 5

Multiply.

$$\frac{1}{2} \cdot \left(-\frac{11}{28}\right)$$

Example 5 Multiply: $-\dfrac{1}{4} \cdot \dfrac{1}{2}$

Solution: Recall that the product of a negative number and a positive number is a negative number.

$$-\frac{1}{4} \cdot \frac{1}{2} = -\frac{1 \cdot 1}{4 \cdot 2} = -\frac{1}{8}$$

◼ **Work Practice 5**

Practice 6–7

Multiply.

6. $\left(-\dfrac{4}{11}\right)\left(-\dfrac{33}{16}\right)$

7. $\dfrac{1}{6} \cdot \dfrac{3}{10} \cdot \dfrac{25}{16}$

Examples Multiply.

6. $\left(-\dfrac{6}{13}\right)\left(-\dfrac{26}{30}\right) = \dfrac{6 \cdot 26}{13 \cdot 30} = \dfrac{\overset{1}{\cancel{6}} \cdot \overset{1}{\cancel{13}} \cdot 2}{\cancel{13} \cdot \cancel{6} \cdot 5} = \dfrac{2}{5}$ The product of two negative numbers is a positive number.

7. $\dfrac{1}{3} \cdot \dfrac{2}{5} \cdot \dfrac{9}{16} = \dfrac{1 \cdot 2 \cdot 9}{3 \cdot 5 \cdot 16} = \dfrac{1 \cdot \overset{1}{\cancel{2}} \cdot \overset{1}{\cancel{3}} \cdot 3}{\cancel{3} \cdot 5 \cdot \cancel{2} \cdot 8} = \dfrac{3}{40}$

◼ **Work Practice 6–7**

We multiply fractions in the same way if variables are involved.

Practice 8

Multiply: $\dfrac{2}{3} \cdot \dfrac{3y}{2}$

Example 8 Multiply: $\dfrac{3x}{4} \cdot \dfrac{8}{5x}$

Solution: Notice that 8 and 4 have a common factor of 4.

$$\frac{3x}{4} \cdot \frac{8}{5x} = \frac{3 \cdot \overset{1}{\cancel{x}} \cdot \overset{1}{\cancel{4}} \cdot 2}{\cancel{4} \cdot 5 \cdot \cancel{x}} = \frac{3 \cdot 1 \cdot 1 \cdot 2}{1 \cdot 5 \cdot 1} = \frac{6}{5}$$

◼ **Work Practice 8**

Helpful Hint

Recall that when the denominator of a fraction contains a variable, such as $\dfrac{8}{5x}$, we assume that the variable does not represent 0.

Answers

4. $\dfrac{1}{18}$ **5.** $-\dfrac{11}{56}$ **6.** $\dfrac{3}{4}$ **7.** $\dfrac{5}{64}$ **8.** y

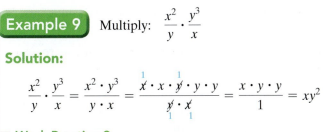

Example 9 Multiply: $\dfrac{x^2}{y} \cdot \dfrac{y^3}{x}$

Solution:

$$\frac{x^2}{y} \cdot \frac{y^3}{x} = \frac{x^2 \cdot y^3}{y \cdot x} = \frac{\overset{1}{\cancel{x}} \cdot x \cdot \overset{1}{\cancel{y}} \cdot y \cdot y}{\underset{1}{\cancel{y}} \cdot \underset{1}{\cancel{x}}} = \frac{x \cdot y \cdot y}{1} = xy^2$$

◼ **Work Practice 9**

Practice 9

Multiply: $\dfrac{a^3}{b^2} \cdot \dfrac{b}{a^2}$

Objective B Evaluating Expressions with Fractional Bases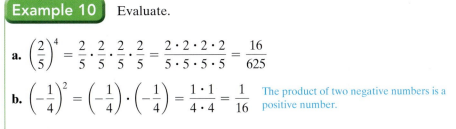

The base of an exponential expression can also be a fraction.

$$\left(\frac{1}{3}\right)^4 = \underbrace{\frac{1}{3} \cdot \frac{1}{3} \cdot \frac{1}{3} \cdot \frac{1}{3}}_{\frac{1}{3} \text{ is a factor 4 times.}} = \frac{1 \cdot 1 \cdot 1 \cdot 1}{3 \cdot 3 \cdot 3 \cdot 3} = \frac{1}{81}$$

Example 10 Evaluate.

a. $\left(\dfrac{2}{5}\right)^4 = \dfrac{2}{5} \cdot \dfrac{2}{5} \cdot \dfrac{2}{5} \cdot \dfrac{2}{5} = \dfrac{2 \cdot 2 \cdot 2 \cdot 2}{5 \cdot 5 \cdot 5 \cdot 5} = \dfrac{16}{625}$

b. $\left(-\dfrac{1}{4}\right)^2 = \left(-\dfrac{1}{4}\right) \cdot \left(-\dfrac{1}{4}\right) = \dfrac{1 \cdot 1}{4 \cdot 4} = \dfrac{1}{16}$ The product of two negative numbers is a positive number.

◼ **Work Practice 10**

Practice 10

Evaluate.

a. $\left(\dfrac{3}{4}\right)^3$ **b.** $\left(-\dfrac{4}{5}\right)^2$

Objective C Dividing Fractions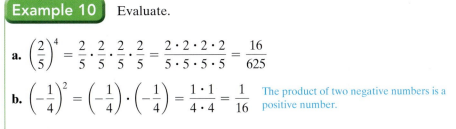

Before we can divide fractions, we need to know how to find the **reciprocal** of a fraction.

> **Reciprocal of a Fraction**
>
> Two numbers are **reciprocals** of each other if their product is 1. The reciprocal of the fraction $\dfrac{a}{b}$ is $\dfrac{b}{a}$ because $\dfrac{a}{b} \cdot \dfrac{b}{a} = \dfrac{a \cdot b}{b \cdot a} = 1$.

Helpful Hint

Every number has a reciprocal except 0. The number 0 has no reciprocal because there is no number such that $0 \cdot a = 1$.

For example,

The reciprocal of $\dfrac{2}{5}$ is $\dfrac{5}{2}$ because $\dfrac{2}{5} \cdot \dfrac{5}{2} = \dfrac{10}{10} = 1$.

The reciprocal of 5 is $\dfrac{1}{5}$ because $5 \cdot \dfrac{1}{5} = \dfrac{5}{1} \cdot \dfrac{1}{5} = \dfrac{5}{5} = 1$.

The reciprocal of $-\dfrac{7}{11}$ is $-\dfrac{11}{7}$ because $-\dfrac{7}{11} \cdot -\dfrac{11}{7} = \dfrac{77}{77} = 1$.

Answers

9. $\dfrac{a}{b}$ 10. a. $\dfrac{27}{64}$ b. $\dfrac{16}{25}$

Division of fractions has the same meaning as division of whole numbers. For example,

$10 \div 5$ means: How many 5s are there in 10?

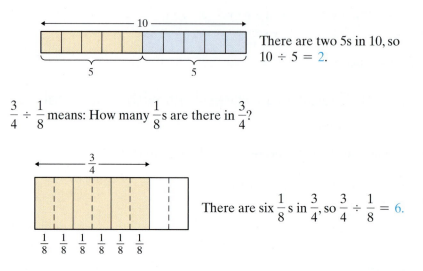

There are two 5s in 10, so $10 \div 5 = 2$.

$\dfrac{3}{4} \div \dfrac{1}{8}$ means: How many $\dfrac{1}{8}$s are there in $\dfrac{3}{4}$?

There are six $\dfrac{1}{8}$s in $\dfrac{3}{4}$, so $\dfrac{3}{4} \div \dfrac{1}{8} = 6$.

We use reciprocals to divide fractions.

Dividing Fractions

To divide two fractions, multiply the first fraction by the reciprocal of the second fraction.

   If $a, b, c,$ and d represent numbers, and $b, c,$ and d are not 0, then

$$\frac{a}{b} \div \frac{c}{d} = \frac{a}{b} \cdot \underset{\uparrow}{\frac{d}{c}} = \frac{a \cdot d}{b \cdot c}$$

   reciprocal

For example,

multiply by reciprocal

$$\frac{3}{4} \div \frac{1}{8} = \frac{3}{4} \cdot \frac{8}{1} = \frac{3 \cdot 8}{4 \cdot 1} = \frac{3 \cdot 2 \cdot \overset{1}{\cancel{4}}}{\underset{1}{\cancel{4}} \cdot 1} = \frac{6}{1} \text{ or } 6$$

After dividing fractions, *always* check to see whether the result can be simplified.

Practice 11–12

Divide and simplify.

11. $\dfrac{8}{7} \div \dfrac{2}{9}$ **12.** $\dfrac{4}{9} \div \dfrac{1}{2}$

Answers

11. $\dfrac{36}{7}$ **12.** $\dfrac{8}{9}$

> **Examples**   Divide and simplify.
>
> **11.** $\dfrac{5}{16} \div \dfrac{3}{4} = \dfrac{5}{16} \cdot \dfrac{4}{3} = \dfrac{5 \cdot 4}{16 \cdot 3} = \dfrac{5 \cdot \overset{1}{\cancel{4}}}{\underset{1}{\cancel{4}} \cdot 4 \cdot 3} = \dfrac{5}{12}$
>
> **12.** $\dfrac{2}{5} \div \dfrac{1}{2} = \dfrac{2}{5} \cdot \dfrac{2}{1} = \dfrac{2 \cdot 2}{5 \cdot 1} = \dfrac{4}{5}$
>
> ■ **Work Practice 11–12**

Helpful Hint

When dividing by a fraction, do not look for common factors to divide out until you rewrite the division as multiplication.

Do not try to divide out these two 2s.

$$\frac{1}{2} \div \frac{2}{3} = \frac{1}{2} \cdot \frac{3}{2} = \frac{3}{4}$$

Example 13 Divide: $-\dfrac{7}{12} \div -\dfrac{5}{6}$

Solution: Recall that the quotient (or product) of two negative numbers is a positive number.

$$-\frac{7}{12} \div -\frac{5}{6} = -\frac{7}{12} \cdot -\frac{6}{5} = \frac{7 \cdot \overset{1}{\cancel{6}}}{2 \cdot \cancel{6} \cdot 5} = \frac{7}{10}$$

■ **Work Practice 13**

Practice 13

Divide: $-\dfrac{10}{4} \div \dfrac{2}{9}$

Example 14 Divide: $\dfrac{2x}{3} \div 3x^2$

Solution:

$$\frac{2x}{3} \div 3x^2 = \frac{2x}{3} \div \frac{3x^2}{1} = \frac{2x}{3} \cdot \frac{1}{3x^2} = \frac{2 \cdot \overset{1}{\cancel{x}} \cdot 1}{3 \cdot 3 \cdot \cancel{x} \cdot x} = \frac{2}{9x}$$

■ **Work Practice 14**

Practice 14

Divide: $\dfrac{3y}{4} \div 5y^3$

Example 15 Simplify: $\left(\dfrac{4}{7} \cdot \dfrac{3}{8}\right) \div -\dfrac{3}{4}$

Solution: Remember to perform the operations inside the ( ) first.

$$\left(\frac{4}{7} \cdot \frac{3}{8}\right) \div -\frac{3}{4} = \left(\frac{\overset{1}{\cancel{4}} \cdot 3}{7 \cdot 2 \cdot \cancel{4}}\right) \div -\frac{3}{4} = \frac{3}{14} \div -\frac{3}{4}$$

Now divide.

$$\frac{3}{14} \div -\frac{3}{4} = \frac{3}{14} \cdot -\frac{4}{3} = -\frac{\overset{1}{\cancel{3}} \cdot \overset{1}{\cancel{2}} \cdot 2}{\underset{1}{\cancel{2}} \cdot 7 \cdot \underset{1}{\cancel{3}}} = -\frac{2}{7}$$

■ **Work Practice 15**

Practice 15

Simplify: $\left(-\dfrac{2}{3} \cdot \dfrac{9}{14}\right) \div \dfrac{7}{15}$

Answers

13. $-\dfrac{45}{4}$ **14.** $\dfrac{3}{20y^2}$ **15.** $-\dfrac{45}{49}$

✓**Concept Check** Which is the correct way to divide $\dfrac{3}{5}$ by $\dfrac{5}{12}$? Explain.

a. $\dfrac{3}{5} \div \dfrac{5}{12} = \dfrac{5}{3} \cdot \dfrac{5}{12}$ 　　 **b.** $\dfrac{3}{5} \div \dfrac{5}{12} = \dfrac{3}{5} \cdot \dfrac{12}{5}$

✓**Concept Check Answers**

a. incorrect **b.** correct

Objective D Multiplying and Dividing with Fractional Replacement Values

Recall the difference between an expression and an equation. For example, xy and $x \div y$ are expressions. They contain no equal signs. In Example 16, we practice *simplifying* expressions given fractional replacement values.

Practice 16

If $x = -\dfrac{3}{4}$ and $y = \dfrac{9}{2}$, evaluate

(a) xy and **(b)** $x \div y$.

Example 16 If $x = \dfrac{7}{8}$ and $y = -\dfrac{1}{3}$, evaluate **(a)** xy and **(b)** $x \div y$.

Solution: Replace x with $\dfrac{7}{8}$ and y with $-\dfrac{1}{3}$.

a. $xy = \dfrac{7}{8} \cdot -\dfrac{1}{3}$

$= -\dfrac{7 \cdot 1}{8 \cdot 3}$

$= -\dfrac{7}{24}$

b. $x \div y = \dfrac{7}{8} \div -\dfrac{1}{3}$

$= \dfrac{7}{8} \cdot -\dfrac{3}{1}$

$= -\dfrac{7 \cdot 3}{8 \cdot 1}$

$= -\dfrac{21}{8}$

Work Practice 16

Practice 17

Is $-\dfrac{9}{8}$ a solution of the equation $2x = -\dfrac{9}{4}$?

Example 17 Is $-\dfrac{2}{3}$ a solution of the equation $-\dfrac{1}{2}x = \dfrac{1}{3}$?

Solution: To check whether a number is a solution of an equation, recall that we replace the variable with the given number and see if a true statement results.

$-\dfrac{1}{2} \cdot x = \dfrac{1}{3}$ Recall that $-\dfrac{1}{2}x$ means $-\dfrac{1}{2} \cdot x$.

$-\dfrac{1}{2} \cdot -\dfrac{2}{3} \overset{?}{=} \dfrac{1}{3}$ Replace x with $-\dfrac{2}{3}$.

$\dfrac{1 \cdot \overset{1}{2}}{\underset{1}{2} \cdot 3} \overset{?}{=} \dfrac{1}{3}$ The product of two negative numbers is a positive number.

$\dfrac{1}{3} = \dfrac{1}{3}$ True

Since we have a true statement, $-\dfrac{2}{3}$ is a solution.

Work Practice 17

Helpful Hint 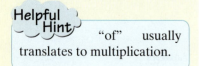 "of" usually translates to multiplication.

Objective E Solving Problems by Multiplying Fractions

To solve real-life problems that involve multiplying fractions, we use our four problem-solving steps from Chapter 3. In Example 18, a new key word that implies multiplication is used. That key word is "**of**."

Answers

16. a. $-\dfrac{27}{8}$ **b.** $-\dfrac{1}{6}$ **17.** yes

Example 18 Finding the Number of Roller Coasters in an Amusement Park

Cedar Point is an amusement park located in Sandusky, Ohio. Its collection of 72 rides is the largest in the world. Of the rides, $\frac{2}{9}$ are roller coasters. How many roller coasters are in Cedar Point's collection of rides? (*Source:* Wikipedia)

Practice 18

Hershey Park is an amusement park in Hershey, Pennsylvania. Of its 66 rides, $\frac{1}{6}$ of these are roller coasters. How many roller coasters are in Hershey Park?

Solution:

1. UNDERSTAND the problem. To do so, read and reread the problem. We are told that $\frac{2}{9}$ of Cedar Point's rides are roller coasters. The word "of" here means multiplication.

2. TRANSLATE.

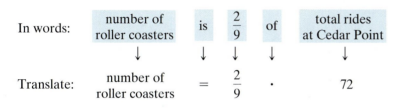

In words:

number of roller coasters	is	$\frac{2}{9}$	of	total rides at Cedar Point
↓	↓	↓	↓	↓

Translate:

$$\text{number of roller coasters} = \frac{2}{9} \cdot 72$$

3. SOLVE: Before we solve, let's estimate a reasonable answer. The fraction $\frac{2}{9}$ is less than $\frac{1}{4}$ (draw a diagram, if needed), and $\frac{1}{4}$ of 72 rides is 18 rides, so the number of roller coasters should be less than 18.

$$\frac{2}{9} \cdot 72 = \frac{2}{9} \cdot \frac{72}{1} = \frac{2 \cdot 72}{9 \cdot 1} = \frac{2 \cdot \overset{1}{\cancel{9}} \cdot 8}{\underset{1}{\cancel{9}} \cdot 1} = \frac{16}{1} \quad \text{or} \quad 16$$

4. INTERPRET. *Check* your work. From our estimate, our answer is reasonable. *State* your conclusion: The number of roller coasters at Cedar Point is 16.

■ **Work Practice 18**

Helpful Hint

To help visualize a fractional part of a whole number, look at the diagram below.

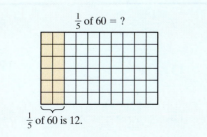

$\frac{1}{5}$ of 60 is 12.

Answer

18. 11 roller coasters

Vocabulary, Readiness & Video Check

Use the choices below to fill in each blank. Not all choices will be used.

multiplication $\dfrac{a \cdot d}{b \cdot c}$ $\dfrac{a \cdot c}{b \cdot d}$ $\dfrac{2 \cdot 2 \cdot 2}{7}$ $\dfrac{2}{7} \cdot \dfrac{2}{7} \cdot \dfrac{2}{7}$

division 0 reciprocals

1. To multiply two fractions, we write $\dfrac{a}{b} \cdot \dfrac{c}{d} =$ _____ .

2. Two numbers are _____ of each other if their product is 1.

3. The expression $\dfrac{2^3}{7} =$ _____ while $\left(\dfrac{2}{7}\right)^3 =$ _____ .

4. Every number has a reciprocal except _____ .

5. To divide two fractions, we write $\dfrac{a}{b} \div \dfrac{c}{d} =$ _____ .

6. The word "of" indicates _____ .

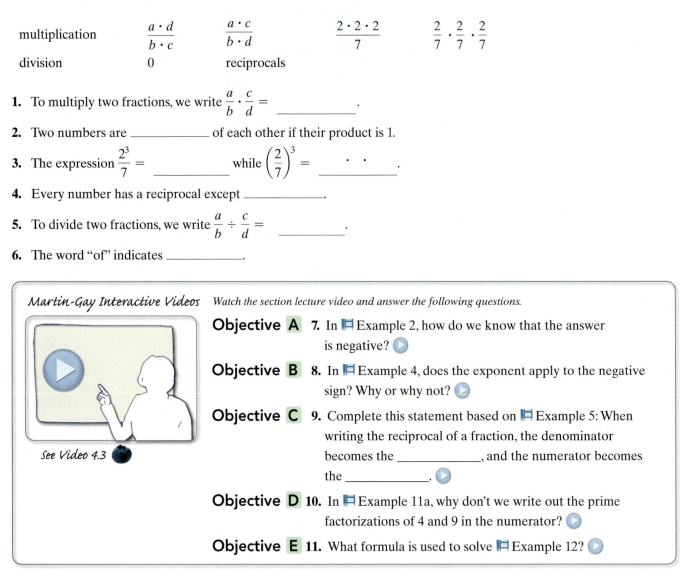

Martin-Gay Interactive Videos   *Watch the section lecture video and answer the following questions.*

See Video 4.3

Objective A   7. In ▣ Example 2, how do we know that the answer is negative? ▶

Objective B   8. In ▣ Example 4, does the exponent apply to the negative sign? Why or why not? ▶

Objective C   9. Complete this statement based on ▣ Example 5: When writing the reciprocal of a fraction, the denominator becomes the _____ , and the numerator becomes the _____ . ▶

Objective D   10. In ▣ Example 11a, why don't we write out the prime factorizations of 4 and 9 in the numerator? ▶

Objective E   11. What formula is used to solve ▣ Example 12? ▶

4.3   Exercise Set   MyMathLab® ▶

Objective A *Multiply. Write the product in simplest form. See Examples 1 through 9.*

1. $\dfrac{6}{11} \cdot \dfrac{3}{7}$

2. $\dfrac{5}{9} \cdot \dfrac{7}{4}$

▶ **3.** $-\dfrac{2}{7} \cdot \dfrac{5}{8}$

4. $\dfrac{4}{15} \cdot -\dfrac{1}{20}$

5. $-\dfrac{1}{2} \cdot -\dfrac{2}{15}$

6. $-\dfrac{3}{11} \cdot -\dfrac{11}{12}$

7. $\dfrac{18x}{20} \cdot \dfrac{36}{99}$

8. $\dfrac{5}{32} \cdot \dfrac{64y}{100}$

9. $3a^2 \cdot \dfrac{1}{4}$

10. $-\dfrac{2}{3} \cdot 6y^3$

11. $\dfrac{x^3}{y^3} \cdot \dfrac{y^2}{x}$

12. $\dfrac{a}{b^3} \cdot \dfrac{b}{a^3}$

13. $0 \cdot \dfrac{8}{9}$

14. $\dfrac{11}{12} \cdot 0$

15. $-\dfrac{17y}{20} \cdot \dfrac{4}{5y}$

16. $-\dfrac{13x}{20} \cdot \dfrac{5}{6x}$

17. $\dfrac{11}{20} \cdot \dfrac{1}{7} \cdot \dfrac{5}{22}$

18. $\dfrac{27}{32} \cdot \dfrac{10}{13} \cdot \dfrac{16}{30}$

Objective **B** *Evaluate. See Example 10.*

19. $\left(\dfrac{1}{5}\right)^3$

20. $\left(\dfrac{8}{9}\right)^2$

▶ **21.** $\left(-\dfrac{2}{3}\right)^2$

22. $\left(-\dfrac{1}{2}\right)^4$

23. $\left(-\dfrac{2}{3}\right)^3 \cdot \dfrac{1}{2}$

24. $\left(-\dfrac{3}{4}\right)^3 \cdot \dfrac{1}{3}$

Objective **C** *Divide. Write all quotients in simplest form. See Examples 11 through 14.*

25. $\dfrac{2}{3} \div \dfrac{5}{6}$

26. $\dfrac{5}{8} \div \dfrac{3}{4}$

27. $-\dfrac{6}{15} \div \dfrac{12}{5}$

28. $-\dfrac{4}{15} \div -\dfrac{8}{3}$

29. $-\dfrac{8}{9} \div \dfrac{x}{2}$

30. $\dfrac{10}{11} \div -\dfrac{4}{5x}$

▶ **31.** $\dfrac{11y}{20} \div \dfrac{3}{11}$

32. $\dfrac{9z}{20} \div \dfrac{2}{9}$

33. $-\dfrac{2}{3} \div 4$

34. $-\dfrac{5}{6} \div 10$

35. $\dfrac{1}{5x} \div \dfrac{5}{x^2}$

36. $\dfrac{3}{y^2} \div \dfrac{9}{y^3}$

Objectives **A** **B** **C** **Mixed Practice** *Perform each indicated operation. See Examples 1 through 15.*

▶ **37.** $\dfrac{2}{3} \cdot \dfrac{5}{9}$

38. $\dfrac{8}{15} \cdot \dfrac{5}{32}$

39. $\dfrac{3x}{7} \div \dfrac{5}{6x}$

40. $\dfrac{2}{5y} \div \dfrac{5y}{11}$

41. $\dfrac{16}{27y} \div \dfrac{8}{15y}$

42. $\dfrac{12y}{21} \div \dfrac{4y}{7}$

43. $-\dfrac{5}{28} \cdot \dfrac{35}{25}$

44. $\dfrac{24}{45} \cdot -\dfrac{5}{8}$

45. $\left(-\dfrac{3}{4}\right)^2$

46. $\left(-\dfrac{1}{2}\right)^5$

▶ **47.** $\dfrac{x^2}{y} \cdot \dfrac{y^3}{x}$

48. $\dfrac{b}{a^2} \cdot \dfrac{a^3}{b^3}$

49. $7 \div \dfrac{2}{11}$

50. $-100 \div \dfrac{1}{2}$

51. $-3x \div \dfrac{x^2}{12}$

52. $-7x^2 \div \dfrac{14x}{3}$

53. $\left(\dfrac{2}{7} \div \dfrac{7}{2}\right) \cdot \dfrac{3}{4}$

54. $\dfrac{1}{2} \cdot \left(\dfrac{5}{6} \div \dfrac{1}{12}\right)$

55. $-\dfrac{19}{63y} \cdot 9y^2$

56. $16a^2 \cdot -\dfrac{31}{24a}$

57. $-\dfrac{2}{3} \cdot -\dfrac{6}{11}$

58. $-\dfrac{1}{5} \cdot -\dfrac{6}{7}$

▶ **59.** $\dfrac{4}{8} \div \dfrac{3}{16}$

60. $\dfrac{9}{2} \div \dfrac{16}{15}$

61. $\dfrac{21x^2}{10y} \div \dfrac{14x}{25y}$

62. $\dfrac{17y^2}{24x} \div \dfrac{13y}{18x}$

63. $\left(1 \div \dfrac{3}{4}\right) \cdot \dfrac{2}{3}$

64. $\left(33 \div \dfrac{2}{11}\right) \cdot \dfrac{5}{9}$

65. $\dfrac{a^3}{2} \div 30a^3$

66. $15c^3 \div \dfrac{3c^2}{5}$

67. $\dfrac{ab^2}{c} \cdot \dfrac{c}{ab}$

68. $\dfrac{ac}{b} \cdot \dfrac{b^3}{a^2c}$

▶ **69.** $\left(\dfrac{1}{2} \cdot \dfrac{2}{3}\right) \div \dfrac{5}{6}$

70. $\left(\dfrac{3}{4} \cdot \dfrac{8}{9}\right) \div \dfrac{2}{5}$

71. $-\dfrac{4}{7} \div \left(\dfrac{4}{5} \cdot \dfrac{3}{7}\right)$

72. $\dfrac{5}{8} \div \left(\dfrac{4}{7} \cdot -\dfrac{5}{16}\right)$

Objective **D** *Given the following replacement values, evaluate (**a**) xy and (**b**) x ÷ y. See Example 16.*

73. $x = \dfrac{2}{5}$ and $y = \dfrac{5}{6}$ **74.** $x = \dfrac{8}{9}$ and $y = \dfrac{1}{4}$ ▶ **75.** $x = -\dfrac{4}{5}$ and $y = \dfrac{9}{11}$ **76.** $x = \dfrac{7}{6}$ and $y = -\dfrac{1}{2}$

Determine whether the given replacement values are solutions of the given equations. See Example 17.

77. Is $-\dfrac{5}{18}$ a solution to $3x = -\dfrac{5}{6}$?

78. Is $\dfrac{9}{11}$ a solution to $\dfrac{2}{3}y = \dfrac{6}{11}$?

79. Is $\dfrac{2}{5}$ a solution to $-\dfrac{1}{2}z = \dfrac{1}{10}$?

80. Is $\dfrac{3}{5}$ a solution to $5x = \dfrac{1}{3}$?

Objective **E** **Translating** *Solve. Write each answer in simplest form. For Exercises 81 through 84, recall that "of" translates to multiplication. See Example 18.*

81. Find $\dfrac{1}{4}$ of 200.   **82.** Find $\dfrac{1}{5}$ of 200.   **83.** Find $\dfrac{5}{6}$ of 24.   **84.** Find $\dfrac{5}{8}$ of 24.

Solve. For Exercises 85 and 86, the solutions have been started for you. See Example 18.

85. In the United States, $\dfrac{7}{50}$ of college freshmen major in business. A community college in Pennsylvania has a freshman enrollment of approximately 800 students. How many of these freshmen might we project are majoring in business?

Start the solution:

1. UNDERSTAND the problem. Reread it as many times as needed.
2. TRANSLATE into an equation. (Fill in the blank below.)

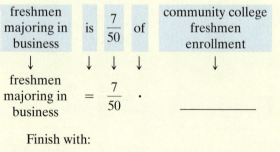

Finish with:
3. SOLVE
4. INTERPRET

86. A patient was told that, at most, $\dfrac{1}{5}$ of his calories should come from fat. If his diet consists of 3000 calories a day, find the maximum number of calories that can come from fat.

Start the solution:

1. UNDERSTAND the problem. Reread it as many times as needed.
2. TRANSLATE into an equation. (Fill in the blank below.)

patient's fat calories	is	$\dfrac{1}{5}$	of	his daily calories
↓	↓	↓	↓	↓
patient's fat calories	=	$\dfrac{1}{5}$	·	_____

Finish with:
3. SOLVE
4. INTERPRET

87. In 2012, there were approximately 225 million moviegoers in the United States and Canada. Of these, about $\dfrac{12}{25}$ viewed at least one 3-D movie. Find the approximate number of people who viewed at least one 3-D movie. (*Source:* Motion Picture Association of America)

88. In a recent year, movie theater owners received a total of $7660 million in movie admission tickets. About $\dfrac{7}{10}$ of this amount was for R-rated movies. Find the amount of money received from R-rated movies. (*Source:* Motion Picture Association of America)

89. The Oregon National Historic Trail is 2170 miles long. It begins in Independence, Missouri, and ends in Oregon City, Oregon. Manfred Coulon has hiked $\frac{2}{5}$ of the trail before. How many miles has he hiked?

(*Source:* National Park Service)

90. Each turn of a screw sinks it $\frac{3}{16}$ of an inch deeper into a piece of wood. Find how deep the screw is after 8 turns.

91. The radius of a circle is one-half of its diameter, as shown. If the diameter of a circle is $\frac{3}{8}$ of an inch, what is its radius?

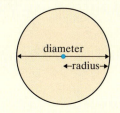

92. The diameter of a circle is twice its radius, as shown in the Exercise 91 illustration. If the radius of a circle is $\frac{7}{20}$ of a foot, what is its diameter?

93. A special on a cruise to the Bahamas is advertised to be $\frac{2}{3}$ of the regular price. If the regular price is $2757, what is the sale price?

94. A family recently sold their house for $102,000, but $\frac{3}{50}$ of this amount goes to the real estate companies that helped them sell their house. How much money does the family pay to the real estate companies?

95. The state of Mississippi houses $\frac{1}{184}$ of the total U.S. libraries. If there are about 9200 libraries in the United States, how many libraries are in Mississippi?

96. There have been about 410 contestants on the reality television show *Survivor* over 27 seasons. Some of these contestants have appeared in multiple seasons. If the number of repeat contestants is $\frac{6}{41}$ of the total number of participants in the first 27 seasons, how many contestants have participated more than once? (*Source:* Survivor.com)

Find the area of each rectangle. Recall that area = length · width.

97.

98.

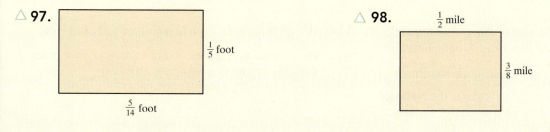

*Recall from Section 4.2 that the following graph is called a **circle graph** or **pie chart**. Each sector (shaped like a piece of pie) shows the fractional part of a car's total mileage that falls into a particular category. The whole circle represents a car's total mileage.*

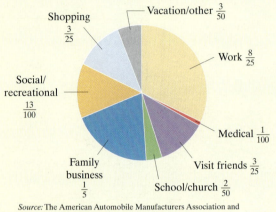

Shopping $\frac{3}{25}$

Vacation/other $\frac{3}{50}$

Work $\frac{8}{25}$

Social/ recreational $\frac{13}{100}$

Medical $\frac{1}{100}$

Family business $\frac{1}{5}$

Visit friends $\frac{3}{25}$

School/church $\frac{2}{50}$

Source: The American Automobile Manufacturers Association and The National Automobile Dealers Association

In one year, a family drove 12,000 miles in the family car. Use the circle graph to determine how many of these miles might be expected to fall in the categories shown in Exercises 99 through 102.

99. Work

100. Shopping

101. Family business

102. Medical

Review

Perform each indicated operation. See Section 1.3.

103.
$$\begin{array}{r} 27 \\ 76 \\ + 98 \\ \hline \end{array}$$

104.
$$\begin{array}{r} 811 \\ 42 \\ + 69 \\ \hline \end{array}$$

105.
$$\begin{array}{r} 968 \\ - 772 \\ \hline \end{array}$$

106.
$$\begin{array}{r} 882 \\ - 773 \\ \hline \end{array}$$

Concept Extensions

107. In your own words, describe how to divide fractions.

108. In your own words, explain how to multiply fractions.

Simplify.

109. $\frac{42}{25} \cdot \frac{125}{36} \div \frac{7}{6}$

110. $\left(\frac{8}{13} \cdot \frac{39}{16} \cdot \frac{8}{9}\right)^2 \div \frac{1}{2}$

111. Approximately $\frac{1}{8}$ of the U.S. population lives in the state of California. If the U.S. population is approximately 313,914,000, find the approximate population of California. (*Source:* U.S. Census Bureau)

112. In 2012, there were approximately 11,430 commercial radio stations broadcasting in the United States. Of these, approximately $\frac{9}{51}$ were country stations. How many radio stations were country stations in 2012? (Round to the nearest whole.) (*Source:* Federal Communications Commission)

113. The National Park Service is charged with maintaining 27,000 historic structures. Monuments and statues make up $\frac{63}{200}$ of these historic structures. How many monuments and statues is the National Park Service charged with maintaining? (*Source:* National Park Service)

114. If $\frac{3}{4}$ of 36 students on a first bus are girls and $\frac{2}{3}$ of the 30 students on a second bus are *boys*, how many students on the two buses are girls?

4.4 Adding and Subtracting Like Fractions, Least Common Denominator, and Equivalent Fractions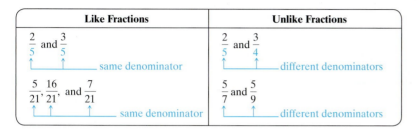

Fractions with the same denominator are called **like fractions.** Fractions that have different denominators are called **unlike fractions.**

Like Fractions	Unlike Fractions
$\dfrac{2}{5}$ and $\dfrac{3}{5}$ same denominator	$\dfrac{2}{5}$ and $\dfrac{3}{4}$ different denominators
$\dfrac{5}{21}, \dfrac{16}{21},$ and $\dfrac{7}{21}$ same denominator	$\dfrac{5}{7}$ and $\dfrac{5}{9}$ different denominators

Objectives

A Add or Subtract Like Fractions.

B Add or Subtract Given Fractional Replacement Values.

C Solve Problems by Adding or Subtracting Like Fractions.

D Find the Least Common Denominator of a List of Fractions.

E Write Equivalent Fractions.

Objective A Adding or Subtracting Like Fractions

To see how we add like fractions (fractions with the same denominator), study one or both illustrations below.

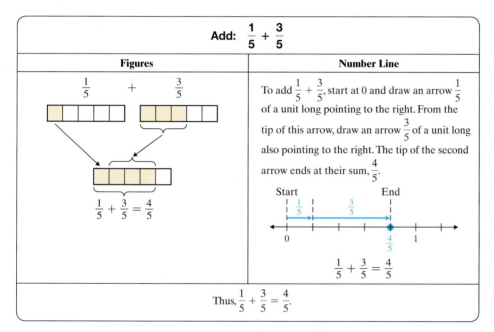

Add: $\dfrac{1}{5} + \dfrac{3}{5}$

Figures

$\dfrac{1}{5}$ $+$ $\dfrac{3}{5}$

$\dfrac{1}{5} + \dfrac{3}{5} = \dfrac{4}{5}$

Number Line

To add $\dfrac{1}{5} + \dfrac{3}{5}$, start at 0 and draw an arrow $\dfrac{1}{5}$ of a unit long pointing to the right. From the tip of this arrow, draw an arrow $\dfrac{3}{5}$ of a unit long also pointing to the right. The tip of the second arrow ends at their sum, $\dfrac{4}{5}$.

$\dfrac{1}{5} + \dfrac{3}{5} = \dfrac{4}{5}$

Thus, $\dfrac{1}{5} + \dfrac{3}{5} = \dfrac{4}{5}.$

Notice that the numerator of the sum is the sum of the numerators. Also, the denominator of the sum is the **common denominator.** This is how we add fractions. Similar illustrations can be shown for subtracting fractions.

> **Adding or Subtracting Like Fractions (Fractions with the Same Denominator)**
>
> If $a, b,$ and c are numbers and b is not 0, then
>
> $$\frac{a}{b} + \frac{c}{b} = \frac{a+c}{b} \qquad \text{and also} \qquad \frac{a}{b} - \frac{c}{b} = \frac{a-c}{b}$$

In other words, to add or subtract fractions with the same denominator, add or subtract their numerators and write the sum or difference over the **common denominator.**

For example,

$$\frac{1}{4} + \frac{2}{4} = \frac{1 + 2}{4} = \frac{3}{4} \qquad \text{Add the numerators.} \\ \text{Keep the denominator.}$$

$$\frac{4}{5} - \frac{2}{5} = \frac{4 - 2}{5} = \frac{2}{5} \qquad \text{Subtract the numerators.} \\ \text{Keep the denominator.}$$

Helpful Hint

As usual, don't forget to write all answers in simplest form.

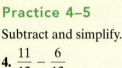

Practice 1–3

Add and simplify.

1. $\dfrac{6}{13} + \dfrac{2}{13}$

2. $\dfrac{5}{8x} + \dfrac{1}{8x}$

3. $\dfrac{20}{11} + \dfrac{6}{11} + \dfrac{7}{11}$

Examples   Add and simplify.

1. $\dfrac{2}{7} + \dfrac{3}{7} = \dfrac{2 + 3}{7} = \dfrac{5}{7}$ ← Add the numerators.
  ← Keep the common denominator.

2. $\dfrac{3}{16x} + \dfrac{7}{16x} = \dfrac{3 + 7}{16x} = \dfrac{10}{16x} = \dfrac{\overset{1}{\cancel{2}} \cdot 5}{\underset{1}{\cancel{2}} \cdot 8 \cdot x} = \dfrac{5}{8x}$

3. $\dfrac{7}{8} + \dfrac{6}{8} + \dfrac{3}{8} = \dfrac{7 + 6 + 3}{8} = \dfrac{16}{8}$ or 2

🔲 **Work Practice 1–3**

Concept Check Find and correct the error in the following:

$$\frac{1}{5} + \frac{1}{5} = \frac{2}{10}$$

Practice 4–5

Subtract and simplify.

4. $\dfrac{11}{12} - \dfrac{6}{12}$

5. $\dfrac{7}{15} - \dfrac{2}{15}$

Examples   Subtract and simplify.

4. $\dfrac{8}{9} - \dfrac{1}{9} = \dfrac{8 - 1}{9} = \dfrac{7}{9}$ ← Subtract the numerators.
  ← Keep the common denominator.

5. $\dfrac{7}{8} - \dfrac{5}{8} = \dfrac{7 - 5}{8} = \dfrac{2}{8} = \dfrac{\overset{1}{\cancel{2}}}{\underset{1}{2} \cdot 4} = \dfrac{1}{4}$

🔲 **Work Practice 4–5**

From our earlier work, we know that

$$\frac{-12}{6} = \frac{12}{-6} = -\frac{12}{6} \quad \text{since these all simplify to } -2.$$

In general, the following is true:

$$\frac{-a}{b} = \frac{a}{-b} = -\frac{a}{b} \quad \text{as long as } b \text{ is not 0.}$$

Practice 6

Add: $-\dfrac{8}{17} + \dfrac{4}{17}$

Example 6   Add: $-\dfrac{11}{8} + \dfrac{6}{8}$

Solution: $-\dfrac{11}{8} + \dfrac{6}{8} = \dfrac{-11 + 6}{8}$

$$= \frac{-5}{8} \text{ or } -\frac{5}{8}$$

🔲 **Work Practice 6**

Answers

1. $\dfrac{8}{13}$ **2.** $\dfrac{3}{4x}$ **3.** 3 **4.** $\dfrac{5}{12}$ **5.** $\dfrac{1}{3}$

6. $-\dfrac{4}{17}$

✔ **Concept Check Answer**

We don't add denominators together;

correct solution: $\dfrac{1}{5} + \dfrac{1}{5} = \dfrac{2}{5}$.

Example 7 Subtract: $\dfrac{3x}{4} - \dfrac{7}{4}$

Solution: $\dfrac{3x}{4} - \dfrac{7}{4} = \dfrac{3x - 7}{4}$

Recall from Section 3.1 that the terms in the numerator are unlike terms and cannot be combined.

■ Work Practice 7

Example 8 Subtract: $\dfrac{3}{7} - \dfrac{6}{7} - \dfrac{3}{7}$

Solution: $\dfrac{3}{7} - \dfrac{6}{7} - \dfrac{3}{7} = \dfrac{3 - 6 - 3}{7} = \dfrac{-6}{7}$ or $-\dfrac{6}{7}$

■ Work Practice 8

Helpful Hint

Recall that $\dfrac{-6}{7} = -\dfrac{6}{7}$ $\left(\text{Also, } \dfrac{6}{-7} = -\dfrac{6}{7}, \text{ if needed.}\right)$

Objective B Adding or Subtracting Given Fractional Replacement Values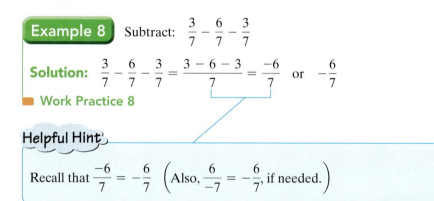

Example 9 Evaluate $y - x$ if $x = -\dfrac{3}{10}$ and $y = -\dfrac{8}{10}$.

Solution: Be very careful when replacing x and y with replacement values.

$y - x = -\dfrac{8}{10} - \left(-\dfrac{3}{10}\right)$ Replace x with $-\dfrac{3}{10}$ and y with $-\dfrac{8}{10}$.

$= \dfrac{-8 - (-3)}{10}$

$= \dfrac{-5}{10} = \dfrac{-1 \cdot 5}{2 \cdot 5} = \dfrac{-1}{2}$ or $-\dfrac{1}{2}$

■ Work Practice 9

✓**Concept Check** Fill in each blank with the best choice given.

expression   equation   simplified   solved

A(n) _____ contains an equal sign and may be _____ for the variable.

A(n) _____ does not contain an equal sign but may be _____.

Objective C Solving Problems by Adding or Subtracting Like Fractions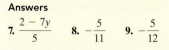

Many real-life problems involve finding the perimeters of square or rectangular-shaped figures such as pastures, swimming pools, and so on. We can use our knowledge of adding fractions to find perimeters.

Practice 7

Subtract: $\dfrac{2}{5} - \dfrac{7y}{5}$

Practice 8

Subtract: $\dfrac{4}{11} - \dfrac{6}{11} - \dfrac{3}{11}$

Practice 9

Evaluate $x + y$ if $x = -\dfrac{10}{12}$ and $y = \dfrac{5}{12}$.

Answers

7. $\dfrac{2 - 7y}{5}$ 8. $-\dfrac{5}{11}$ 9. $-\dfrac{5}{12}$

✓**Concept Check Answer**

equation; solved; expression; simplified

Practice 10

Find the perimeter of the square.

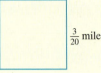

$\frac{3}{20}$ mile

△ **Example 10** Find the perimeter of the rectangle.

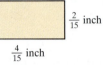

$\frac{2}{15}$ inch

$\frac{4}{15}$ inch

Solution: Recall that perimeter means distance around and that opposite sides of a rectangle are the same length.

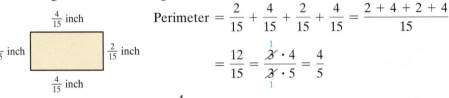

$\frac{4}{15}$ inch

$\frac{2}{15}$ inch $\frac{2}{15}$ inch

$\frac{4}{15}$ inch

$$\text{Perimeter} = \frac{2}{15} + \frac{4}{15} + \frac{2}{15} + \frac{4}{15} = \frac{2+4+2+4}{15}$$

$$= \frac{12}{15} = \frac{\cancel{3} \cdot 4}{\cancel{3} \cdot 5} = \frac{4}{5}$$

The perimeter of the rectangle is $\frac{4}{5}$ inch.

■ **Work Practice 10**

We can combine our skills in adding and subtracting fractions with our four problem-solving steps from Section 3.4 to solve many kinds of real-life problems.

Practice 11

A jogger ran $\frac{13}{4}$ miles on Monday and $\frac{11}{4}$ miles on Wednesday. How much farther did he run on Monday than on Wednesday?

Example 11 Calculating Distance

The distance from home to the World Gym is $\frac{7}{8}$ of a mile and from home to the post office is $\frac{3}{8}$ of a mile. How much farther is it from home to the World Gym than from home to the post office?

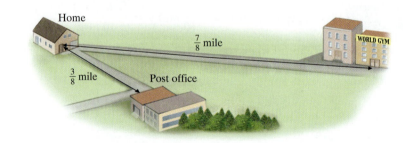

Home

$\frac{7}{8}$ mile

WORLD GYM

$\frac{3}{8}$ mile

Post office

Solution:

1. UNDERSTAND. Read and reread the problem. The phrase "How much farther" tells us to subtract distances.
2. TRANSLATE.

In words:	distance farther	is	home to World Gym distance	minus	home to post office distance
	↓	↓	↓	↓	↓
Translate:	distance farther	=	$\frac{7}{8}$	−	$\frac{3}{8}$

3. SOLVE: $\dfrac{7}{8} - \dfrac{3}{8} = \dfrac{7-3}{8} = \dfrac{4}{8} = \dfrac{\cancel{4}}{2 \cdot \cancel{4}} = \dfrac{1}{2}$

4. INTERPRET. *Check* your work. *State* your conclusion: The distance from home to the World Gym is $\frac{1}{2}$ mile farther than from home to the post office.

■ **Work Practice 11**

Answers

10. $\frac{3}{5}$ mi   **11.** $\frac{1}{2}$ mi

Objective D Finding the Least Common Denominator ▶

In the next section, we will add and subtract fractions that have different, or unlike, denominators. To do so, we first write them as equivalent fractions with a common denominator.

Although any common denominator can be used to add or subtract unlike fractions, we will use the **least common denominator (LCD).** The LCD of a list of fractions is the same as the **least common multiple (LCM)** of the denominators. Why do we use this number as the common denominator? Since the LCD is the *smallest* of all common denominators, operations are usually less tedious with this number.

> The **least common denominator (LCD)** of a list of fractions is the smallest positive number divisible by all the denominators in the list. (The least common denominator is also the **least common multiple (LCM) of the denominators.**)

For example, the LCD of $\frac{1}{4}$ and $\frac{3}{10}$ is 20 because 20 is the smallest positive number divisible by both 4 and 10.

Finding the LCD: Method 1

One way to find the LCD is to see whether the larger denominator is divisible by the smaller denominator. If so, the larger number is the LCD. If not, then check consecutive multiples of the larger denominator until the LCD is found.

> **Method 1: Finding the LCD of a List of Fractions Using Multiples of the Largest Number**
>
> **Step 1:** Write the multiples of the largest denominator (starting with the number itself) until a multiple common to all denominators in the list is found.
>
> **Step 2:** The multiple found in Step 1 is the LCD.

Example 12 Find the LCD of $\frac{3}{7}$ and $\frac{5}{14}$.

Solution: We write the multiples of 14 until we find one that is also a multiple of 7.

$14 \cdot 1 = 14$ A multiple of 7

The LCD is 14.

■ Work Practice 12

Example 13 Find the LCD of $\frac{11}{12}$ and $\frac{7}{20}$.

Solution: We write the multiples of 20 until we find one that is also a multiple of 12.

$20 \cdot 1 = 20$ Not a multiple of 12
$20 \cdot 2 = 40$ Not a multiple of 12
$20 \cdot 3 = 60$ A multiple of 12

The LCD is 60.

■ Work Practice 13

Practice 12

Find the LCD of $\frac{7}{8}$ and $\frac{11}{16}$.

Practice 13

Find the LCD of $\frac{23}{25}$ and $\frac{1}{30}$.

Answers

12. 16   **13.** 150

Method 1 for finding multiples works fine for smaller numbers, but may get tedious for larger numbers. For this reason, let's study a second method, which uses prime factorization.

Finding the LCD: Method 2

For example, to find the LCD of $\frac{11}{12}$ and $\frac{7}{20}$, such as in Example 13, let's look at the prime factorization of each denominator.

$$12 = 2 \cdot 2 \cdot 3$$
$$20 = 2 \cdot 2 \cdot 5$$

Recall that the LCD must be a multiple of both 12 and 20. Thus, to build the LCD, we will circle the greatest number of factors for each different prime number. The LCD is the product of the circled factors.

Prime Number Factors

$$12 = \boxed{2 \cdot 2} \quad \boxed{3}$$
$$20 = 2 \cdot 2 \qquad \boxed{5}$$

> Circle either pair of 2s, but not both.

$$\text{LCD} = 2 \cdot 2 \cdot 3 \cdot 5 = 60$$

The number 60 is the smallest number that both 12 and 20 divide into evenly. This method is summarized below:

Method 2: Finding the LCD of a List of Denominators Using Prime Factorization

Step 1: Write the prime factorization of each denominator.

Step 2: For each different prime factor in Step 1, circle the *greatest* number of times that factor occurs in any one factorization.

Step 3: The LCD is the product of the circled factors.

Example 14 Find the LCD of $-\frac{23}{72}$ and $\frac{17}{60}$.

Solution: First we write the prime factorization of each denominator.

$$72 = 2 \cdot 2 \cdot 2 \cdot 3 \cdot 3$$
$$60 = 2 \cdot 2 \cdot 3 \cdot 5$$

For the prime factors shown, we circle the greatest number of factors found in either factorization.

$$72 = \boxed{2 \cdot 2 \cdot 2} \cdot \boxed{3 \cdot 3}$$
$$60 = 2 \cdot 2 \cdot 3 \cdot \boxed{5}$$

The LCD is the product of the circled factors.

$$\text{LCD} = 2 \cdot 2 \cdot 2 \cdot 3 \cdot 3 \cdot 5 = 360$$

The LCD is 360.

■ **Work Practice 14**

Practice 14

Find the LCD of $-\frac{3}{40}$ and $\frac{11}{108}$.

Helpful Hint

If you prefer working with exponents, circle the factor with the greatest exponent.

Example 14:

$$72 = \boxed{2^3} \cdot \boxed{3^2}$$
$$60 = 2^2 \cdot 3 \cdot \boxed{5}$$
$$\text{LCD} = 2^3 \cdot 3^2 \cdot 5 = 360$$

Answer

14. 1080

Helpful Hint

If the number of factors of a prime number is equal, circle either one, but not both. For example,

$12 = \boxed{2 \cdot 2} \cdot \boxed{3}$

$15 = 3 \cdot \boxed{5}$ — Circle either 3 but not both.

The LCD is $2 \cdot 2 \cdot 3 \cdot 5 = 60$.

Example 15   Find the LCD of $\dfrac{1}{15}, \dfrac{5}{18}$, and $\dfrac{53}{54}$.

Solution: $15 = 3 \cdot \boxed{5}$

$18 = \boxed{2} \cdot 3 \cdot 3$

$54 = 2 \cdot \boxed{3 \cdot 3 \cdot 3}$

The LCD is $2 \cdot 3 \cdot 3 \cdot 3 \cdot 5$ or 270.

■ **Work Practice 15**

Practice 15

Find the LCD of $\dfrac{7}{20}, \dfrac{1}{24}$, and $\dfrac{13}{45}$.

Example 16   Find the LCD of $\dfrac{3}{5}, \dfrac{2}{x}$, and $\dfrac{7}{x^3}$.

Solution: $5 = \boxed{5}$

$x = x$

$x^3 = \boxed{x \cdot x \cdot x}$

$\text{LCD} = 5 \cdot x \cdot x \cdot x = 5x^3$

■ **Work Practice 16**

Practice 16

Find the LCD of $\dfrac{7}{y}$ and $\dfrac{6}{11}$.

✓**Concept Check** True or false? The LCD of the fractions $\dfrac{1}{6}$ and $\dfrac{1}{8}$ is 48.

Objective E Writing Equivalent Fractions ▶

To add or subtract unlike fractions in the next section, we first write equivalent fractions with the LCD as the denominator.

To write $\dfrac{1}{3}$ as an equivalent fraction with a denominator of 6, we multiply by 1 in the form of $\dfrac{2}{2}$. Why? Because $3 \cdot 2 = 6$, so the new denominator will become 6, as shown below.

$$\frac{1}{3} = \frac{1}{3} \cdot 1 = \frac{1}{3} \cdot \frac{2}{2} = \frac{1 \cdot 2}{3 \cdot 2} = \frac{2}{6}$$

$$\frac{2}{2} = 1$$

So $\dfrac{1}{3} = \dfrac{2}{6}$.

Helpful Hint Recall from the Helpful Hint on p. 230, that this is also called the Fundamental Property of Fractions.

$$\frac{a}{b} = \frac{a}{b} \cdot \frac{c}{c} = \frac{a \cdot c}{b \cdot c}$$

To write an equivalent fraction,

$$\frac{a}{b} = \frac{a}{b} \cdot \frac{c}{c} = \frac{a \cdot c}{b \cdot c}$$

where a, b, and c are nonzero numbers.

Answers
15. 360   **16.** $11y$

✓**Concept Check Answer**
false; it is 24

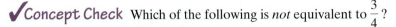

✓**Concept Check** Which of the following is *not* equivalent to $\dfrac{3}{4}$?

a. $\dfrac{6}{8}$ **b.** $\dfrac{18}{24}$ **c.** $\dfrac{9}{14}$ **d.** $\dfrac{30}{40}$

Practice 17

Write $\dfrac{7}{8}$ as an equivalent fraction with a denominator of 56.

$$\dfrac{7}{8} = \dfrac{}{56}$$

Example 17 Write $\dfrac{3}{4}$ as an equivalent fraction with a denominator of 20.

$$\dfrac{3}{4} = \dfrac{}{20}$$

Solution: In the denominators, since $4 \cdot 5 = 20$, we will multiply by 1 in the form of $\dfrac{5}{5}$.

$$\dfrac{3}{4} = \dfrac{3}{4} \cdot \dfrac{5}{5} = \dfrac{3 \cdot 5}{4 \cdot 5} = \dfrac{15}{20}$$

Thus, $\dfrac{3}{4} = \dfrac{15}{20}$.

🟧 **Work Practice 17**

Helpful Hint

To check Example 17, write $\dfrac{15}{20}$ in simplest form.

$$\dfrac{15}{20} = \dfrac{3 \cdot \overset{1}{\cancel{5}}}{4 \cdot \underset{1}{\cancel{5}}} = \dfrac{3}{4}\text{, the original fraction.}$$

If the original fraction is in lowest terms, we can check our work by writing the new, equivalent fraction in simplest form. This form should be the original fraction.

✓**Concept Check** True or false? When the fraction $\dfrac{2}{9}$ is rewritten as an equivalent fraction with 27 as the denominator, the result is $\dfrac{2}{27}$.

Practice 18

Write an equivalent fraction with the given denominator.

$$\dfrac{1}{4} = \dfrac{}{20}$$

Answers

17. $\dfrac{49}{56}$ **18.** $\dfrac{5}{20}$

✓**Concept Check Answers**

c

false; the correct result would be $\dfrac{6}{27}$

Example 18 Write an equivalent fraction with the given denominator.

$$\dfrac{2}{5} = \dfrac{}{15}$$

Solution: Since $5 \cdot 3 = 15$, we multiply by 1 in the form of $\dfrac{3}{3}$.

$$\dfrac{2}{5} = \dfrac{2}{5} \cdot \dfrac{3}{3} = \dfrac{2 \cdot 3}{5 \cdot 3} = \dfrac{6}{15}$$

Then $\dfrac{2}{5}$ is equivalent to $\dfrac{6}{15}$. They both represent the same part of a whole.

🟧 **Work Practice 18**

Example 19 Write an equivalent fraction with the given denominator.

$$\frac{9x}{11} = \frac{}{44}$$

Solution: Since $11 \cdot 4 = 44$, we multiply by 1 in the form of $\frac{4}{4}$.

$$\frac{9x}{11} = \frac{9x}{11} \cdot \frac{4}{4} = \frac{9x \cdot 4}{11 \cdot 4} = \frac{36x}{44}$$

Then $\frac{9x}{11}$ is equivalent to $\frac{36x}{44}$.

■ **Work Practice 19**

Example 20 Write an equivalent fraction with the given denominator.

$$3 = \frac{}{7}$$

Solution: Recall that $3 = \frac{3}{1}$. Since $1 \cdot 7 = 7$, multiply by 1 in the form $\frac{7}{7}$.

$$\frac{3}{1} = \frac{3}{1} \cdot \frac{7}{7} = \frac{3 \cdot 7}{1 \cdot 7} = \frac{21}{7}$$

■ **Work Practice 20**

Don't forget that when the denominator of a fraction contains a variable, such as $\frac{8}{3x}$, we will assume that the variable does not represent 0. Recall that the denominator of a fraction cannot be 0.

Example 21 Write an equivalent fraction with the given denominator.

$$\frac{8}{3x} = \frac{}{24x}$$

Solution: Since $3x \cdot 8 = 24x$, multiply by 1 in the form $\frac{8}{8}$.

$$\frac{8}{3x} = \frac{8}{3x} \cdot \frac{8}{8} = \frac{8 \cdot 8}{3x \cdot 8} = \frac{64}{24x}$$

■ **Work Practice 21**

✔**Concept Check** What is the first step in writing $\frac{3}{10}$ as an equivalent fraction whose denominator is 100?

Practice 19

Write an equivalent fraction with the given denominator.

$$\frac{3x}{7} = \frac{}{42}$$

Practice 20

Write an equivalent fraction with the given denominator.

$$4 = \frac{}{6}$$

Practice 21

Write an equivalent fraction with the given denominator.

$$\frac{9}{4x} = \frac{}{36x}$$

Answers

19. $\frac{18x}{42}$ 20. $\frac{24}{6}$ 21. $\frac{81}{36x}$

✔**Concept Check Answer**
answers may vary

Vocabulary, Readiness & Video Check

Use the choices below to fill in each blank. Not all choices will be used.

least common denominator (LCD)　　like　　$-\dfrac{a}{b}$　$\dfrac{a-c}{b}$　$\dfrac{a+c}{b}$　$-\dfrac{a}{-b}$

perimeter　　　　　　　　　　　　unlike

equivalent

1. The fractions $\dfrac{9}{11}$ and $\dfrac{13}{11}$ are called _____ fractions while $\dfrac{3}{4}$ and $\dfrac{1}{3}$ are called _____ fractions.

2. $\dfrac{a}{b} + \dfrac{c}{b} = $ _____ and $\dfrac{a}{b} - \dfrac{c}{b} = $ _____ .

3. As long as b is not 0, $\dfrac{-a}{b} = \dfrac{a}{-b} = $ _____ .

4. The distance around a figure is called its _____ .

5. The smallest positive number divisible by all the denominators of a list of fractions is called the

_____ .

6. Fractions that represent the same portion of a whole are called _____ fractions.

Martin-Gay Interactive Videos　　*Watch the section lecture video and answer the following questions.*

Objective A　**7.** Complete this statement based on the lecture before ◫ Example 1: To add like fractions, we add the _____ and keep the same _____ ▶.

Objective B　**8.** In ◫ Example 6, why are we told to be careful when substituting the replacement value for y? ▶

Objective C　**9.** What is the perimeter equation used to solve ◫ Example 7? What is the final answer? ▶

Objective D　**10.** In ◫ Example 8, the LCD is found to be 45. What does this mean in terms of the specific fractions in the problem? ▶

Objective E　**11.** From ◫ Example 10, why can we multiply a fraction by a form of 1 to get an equivalent fraction? ▶

See Video 4.4

4.4 Exercise Set MyMathLab® ▶

Objective A *Add and simplify. See Examples 1 through 3, and 6.*

▶ **1.** $\dfrac{5}{11} + \dfrac{2}{11}$　　　　**2.** $\dfrac{9}{17} + \dfrac{2}{17}$　　　　**3.** $\dfrac{2}{9} + \dfrac{4}{9}$　　　　**4.** $\dfrac{3}{10} + \dfrac{2}{10}$

▶ **5.** $-\dfrac{6}{20} + \dfrac{1}{20}$　　**6.** $-\dfrac{3}{8} + \dfrac{1}{8}$　　**7.** $-\dfrac{3}{14} + \left(-\dfrac{4}{14}\right)$　　**8.** $-\dfrac{5}{24} + \left(-\dfrac{7}{24}\right)$

9. $\dfrac{2}{9x} + \dfrac{4}{9x}$　　**10.** $\dfrac{3}{10y} + \dfrac{2}{10y}$　　**11.** $-\dfrac{7x}{18} + \dfrac{3x}{18} + \dfrac{2x}{18}$　　**12.** $-\dfrac{7z}{15} + \dfrac{3z}{15} + \dfrac{1z}{15}$

Subtract and simplify. See Examples 4, 5, 7, and 8.

13. $\dfrac{10}{11} - \dfrac{4}{11}$

14. $\dfrac{9}{13} - \dfrac{5}{13}$

15. $\dfrac{7}{8} - \dfrac{1}{8}$

16. $\dfrac{5}{6} - \dfrac{1}{6}$

17. $\dfrac{1}{y} - \dfrac{4}{y}$

18. $\dfrac{4}{z} - \dfrac{7}{z}$

19. $-\dfrac{27}{33} - \left(-\dfrac{8}{33}\right)$

20. $-\dfrac{37}{45} - \left(-\dfrac{18}{45}\right)$

21. $\dfrac{20}{21} - \dfrac{10}{21} - \dfrac{17}{21}$

22. $\dfrac{27}{28} - \dfrac{5}{28} - \dfrac{28}{28}$

23. $\dfrac{7a}{4} - \dfrac{3}{4}$

24. $\dfrac{18b}{5} - \dfrac{3}{5}$

Mixed Practice *Perform the indicated operation. See Examples 1 through 8.*

25. $-\dfrac{9}{100} + \dfrac{99}{100}$

26. $-\dfrac{15}{200} + \dfrac{85}{200}$

27. $-\dfrac{13x}{28} - \dfrac{13x}{28}$

28. $-\dfrac{15}{26y} - \dfrac{15}{26y}$

29. $\dfrac{9x}{15} + \dfrac{1}{15}$

30. $\dfrac{2x}{15} + \dfrac{7}{15}$

31. $\dfrac{7x}{16} - \dfrac{15x}{16}$

32. $\dfrac{3}{16z} - \dfrac{15}{16z}$

33. $\dfrac{9}{12} - \dfrac{7}{12} - \dfrac{10}{12}$

34. $\dfrac{1}{8} - \dfrac{15}{8} + \dfrac{2}{8}$

35. $\dfrac{x}{4} + \dfrac{3x}{4} - \dfrac{2x}{4} + \dfrac{x}{4}$

36. $\dfrac{9y}{8} + \dfrac{2y}{8} + \dfrac{5y}{8} - \dfrac{4y}{8}$

Objective B *Evaluate each expression for the given replacement values. See Example 9.*

37. $x + y; x = \dfrac{3}{4}, y = \dfrac{2}{4}$

38. $x - y; x = \dfrac{7}{8}, y = \dfrac{9}{8}$

39. $x - y; x = -\dfrac{1}{5}, y = -\dfrac{3}{5}$

40. $x + y; x = -\dfrac{1}{6}, y = \dfrac{5}{6}$

Objective C *Find the perimeter of each figure. (Hint: Recall that perimeter means distance around.) See Example 10.*

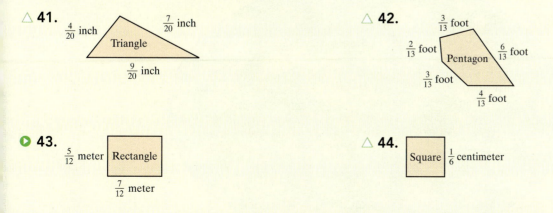

△ **41.**

$\frac{4}{20}$ inch $\frac{7}{20}$ inch

Triangle

$\frac{9}{20}$ inch

△ **42.**

$\frac{3}{13}$ foot

$\frac{2}{13}$ foot   Pentagon $\frac{6}{13}$ foot

$\frac{3}{13}$ foot

$\frac{4}{13}$ foot

43.

$\frac{5}{12}$ meter | Rectangle

$\frac{7}{12}$ meter

△ **44.**

Square | $\frac{1}{6}$ centimeter

Solve. For Exercises 45 and 46, the solutions have been started for you. Write each answer in simplest form. See Example 11.

45. A railroad inspector must inspect $\frac{19}{20}$ of a mile of railroad track. If she has already inspected $\frac{5}{20}$ of a mile, how much more does she need to inspect?

Start the solution:
1. UNDERSTAND the problem. Reread it as many times as needed.
2. TRANSLATE into an equation. (Fill in the blanks.)

distance left to inspect	is	distance needed to inspect	minus	distance already inspected
↓	↓	↓	↓	↓

$$\text{distance left to inspect} = \underline{\qquad} - \underline{\qquad}$$

Finish with:
3. SOLVE. and 4. INTERPRET.

46. Scott Davis has run $\frac{11}{8}$ miles already and plans to complete $\frac{16}{8}$ miles. To do this, how much farther must he run?

Start the solution:
1. UNDERSTAND the problem. Reread it as many times as needed.
2. TRANSLATE into an equation. (Fill in the blanks.)

distance left to run	is	planned to run	minus	distance already run
↓	↓	↓	↓	↓

$$\text{distance left to run} = \underline{\qquad} - \underline{\qquad}$$

Finish with:
3. SOLVE. and 4. INTERPRET.

47. As of 2013, the fraction of states in the United States with maximum interstate highway speed limits up to and including 70 mph was $\frac{33}{50}$. The fraction of states with 70 mph speed limits was $\frac{20}{50}$. What fraction of states had speed limits that were less than 70 mph? (*Source:* Insurance Institute for Highway Safety)

48. When people take aspirin, $\frac{31}{50}$ of the time it is used to treat some type of pain. Approximately $\frac{7}{50}$ of all aspirin use is for treating headaches. What fraction of aspirin use is for treating pain other than headaches? (*Source:* Bayer Market Research)

The map of the world below shows the fraction of the world's surface land area taken up by each continent. In other words, the continent of Africa, for example, makes up $\frac{20}{100}$ of the land in the world. Use this map to solve Exercises 49 through 52. Write answers in simplest form.

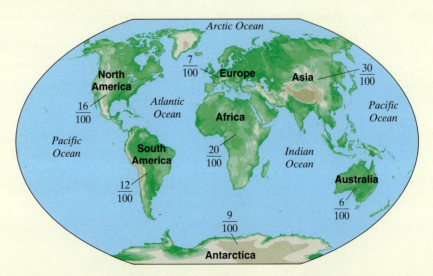

Copyright 2015 Pearson Education, Inc.

49. Find the fractional part of the world's land area within the continents of North America and South America.

50. Find the fractional part of the world's land area within the continents of Asia and Africa.

51. How much greater is the fractional part of the continent of Antarctica than the fractional part of the continent of Europe?

52. How much greater is the fractional part of the continent of Asia than the continent of Australia?

Objective D *Find the LCD of each list of fractions. See Examples 12 through 16.*

53. $\dfrac{2}{9}, \dfrac{6}{15}$

54. $\dfrac{7}{12}, \dfrac{3}{20}$

55. $-\dfrac{1}{36}, \dfrac{1}{24}$

56. $-\dfrac{1}{15}, \dfrac{1}{90}$

57. $\dfrac{2}{25}, \dfrac{3}{15}, \dfrac{5}{6}$

58. $\dfrac{3}{4}, \dfrac{1}{14}, \dfrac{13}{20}$

59. $-\dfrac{7}{24}, -\dfrac{5}{x}$

60. $-\dfrac{11}{y}, -\dfrac{13}{70}$

61. $\dfrac{23}{18}, \dfrac{1}{21}$

62. $\dfrac{45}{24}, \dfrac{2}{45}$

63. $\dfrac{4}{3}, \dfrac{8}{21}, \dfrac{3}{56}$

64. $\dfrac{12}{11}, \dfrac{20}{33}, \dfrac{12}{121}$

Objective E *Write each fraction as an equivalent fraction with the given denominator. See Examples 17 through 21.*

65. $\dfrac{2}{3} = \dfrac{}{21}$

66. $\dfrac{5}{6} = \dfrac{}{24}$

67. $\dfrac{4}{7} = \dfrac{}{35}$

68. $\dfrac{3}{5} = \dfrac{}{100}$

69. $\dfrac{1}{2} = \dfrac{}{50}$

70. $\dfrac{1}{5} = \dfrac{}{50}$

71. $\dfrac{14x}{17} = \dfrac{}{68}$

72. $\dfrac{19z}{21} = \dfrac{}{126}$

73. $\dfrac{2y}{3} = \dfrac{}{12}$

74. $\dfrac{3x}{2} = \dfrac{}{12}$

75. $\dfrac{5}{9} = \dfrac{}{36a}$

76. $\dfrac{7}{6} = \dfrac{}{36a}$

The table on the next page shows the fraction of goods sold online by type of goods in a particular year. Use this table to answer Exercises 77 through 80.

77. Complete the table by writing each fraction as an equivalent fraction with a denominator of 100.

78. Which of these types of goods has the largest fraction sold online?

79. Which of these types of goods has the smallest fraction sold online?

80. Which of the types of goods has **more than** $\frac{3}{5}$ of the goods sold online? (*Hint:* Write $\frac{3}{5}$ as an equivalent fraction with a denominator of 100.)

Type of Goods	Fraction of All Goods That Are Sold Online	Equivalent Fraction with a Denominator of 100
books and magazines	$\frac{27}{50}$	
clothing and accessories	$\frac{1}{2}$	
computer hardware	$\frac{23}{50}$	
computer software	$\frac{1}{2}$	
drugs, health and beauty aids	$\frac{3}{20}$	
electronics and appliances	$\frac{13}{20}$	
food, beer, and wine	$\frac{9}{20}$	
home furnishings	$\frac{13}{25}$	
music and videos	$\frac{3}{5}$	
office equipment and supplies	$\frac{61}{100}$	
sporting goods	$\frac{12}{25}$	
toys, hobbies, and games	$\frac{1}{2}$	

(*Source:* Fedstats.gov)

Review

Simplify. See Section 1.7.

81. 3^2

82. 4^3

83. 5^3

84. 3^4

85. 7^2

86. 5^4

87. $2^3 \cdot 3$

88. $4^2 \cdot 5$

Concept Extensions

Find and correct the error. See the first Concept Check in this section.

89. $\frac{2}{7} + \frac{9}{7} \, \cancel{=} \, \frac{11}{14}$

90. $\frac{3}{4} - \frac{1}{4} \, \cancel{=} \, \frac{2}{8} \, \cancel{=} \, \frac{1}{4}$

Solve.

91. In your own words, explain how to add like fractions.

92. In your own words, explain how to subtract like fractions.

93. Use the map of the world for Exercises 49 through 52 and find the sum of all the continents' fractions. Explain your answer.

94. Mike Cannon jogged $\frac{3}{8}$ of a mile from home and then rested. Then he continued jogging farther from home for another $\frac{3}{8}$ of a mile until he discovered his watch had fallen off. He walked back along the same path for $\frac{4}{8}$ of a mile until he found his watch. Find how far he was from his home.

Write each fraction as an equivalent fraction with the indicated denominator.

95. $\frac{37x}{165} = \frac{}{3630}$

96. $\frac{108}{215y} = \frac{}{4085y}$

97. In your own words, explain how to find the LCD of two fractions.

98. In your own words, explain how to write a fraction as an equivalent fraction with a given denominator.

Solve. See the fourth and fifth Concept Checks in this section.

99. Which of the following are equivalent to $\frac{2}{3}$?

a. $\frac{10}{15}$ b. $\frac{40}{60}$

c. $\frac{16}{20}$ d. $\frac{200}{300}$

100. True or false? When the fraction $\frac{7}{12}$ is rewritten with a denominator of 48, the result is $\frac{11}{48}$. If false, give the correct fraction.

4.5 Adding and Subtracting Unlike Fractions ▶

Objective A Adding and Subtracting Unlike Fractions ▶

In this section we add and subtract fractions with unlike denominators. To add or subtract these unlike fractions, we first write the fractions as equivalent fractions with a common denominator and then add or subtract the like fractions. Recall from the previous section that the common denominator we use is called the **least common denominator (LCD).**

To begin, let's add the unlike fractions $\frac{3}{4} + \frac{1}{6}$.

The LCD of these fractions is 12. So we write each fraction as an equivalent fraction with a denominator of 12, then add as usual. This addition process is shown next and also illustrated by figures.

Objectives

A Add or Subtract Unlike Fractions. ▶

B Write Fractions in Order. ▶

C Evaluate Expressions Given Fractional Replacement Values. ▶

D Solve Problems by Adding or Subtracting Unlike Fractions. ▶

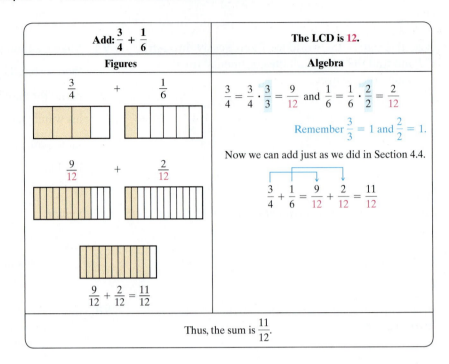

Add: $\dfrac{3}{4} + \dfrac{1}{6}$ The LCD is **12**.

Figures	Algebra

$\dfrac{3}{4} = \dfrac{3}{4} \cdot \dfrac{3}{3} = \dfrac{9}{12}$ and $\dfrac{1}{6} = \dfrac{1}{6} \cdot \dfrac{2}{2} = \dfrac{2}{12}$

Remember $\dfrac{3}{3} = 1$ and $\dfrac{2}{2} = 1$.

Now we can add just as we did in Section 4.4.

$$\dfrac{3}{4} + \dfrac{1}{6} = \dfrac{9}{12} + \dfrac{2}{12} = \dfrac{11}{12}$$

$\dfrac{9}{12} + \dfrac{2}{12} = \dfrac{11}{12}$

Thus, the sum is $\dfrac{11}{12}$.

Adding or Subtracting Unlike Fractions

Step 1: Find the least common denominator (LCD) of the fractions.

Step 2: Write each fraction as an equivalent fraction whose denominator is the LCD.

Step 3: Add or subtract the like fractions.

Step 4: Write the sum or difference in simplest form.

Practice 1

Add: $\dfrac{2}{7} + \dfrac{8}{21}$

Example 1   Add: $\dfrac{2}{5} + \dfrac{4}{15}$

Solution:

Step 1: The LCD of the fractions is 15. In later examples, we shall simply say, for example, that the LCD of the denominators 5 and 15 is 15.

Step 2: $\dfrac{2}{5} = \dfrac{2}{5} \cdot \dfrac{3}{3} = \dfrac{6}{15}$, $\dfrac{4}{15} = \dfrac{4}{15}$ ← This fraction already has a denominator of 15.

   └ Multiply by 1 in the form $\dfrac{3}{3}$.

Step 3: $\dfrac{2}{5} + \dfrac{4}{15} = \dfrac{6}{15} + \dfrac{4}{15} = \dfrac{10}{15}$

Step 4: Write in simplest form.

$$\dfrac{10}{15} = \dfrac{2 \cdot \cancel{5}}{3 \cdot \cancel{5}} = \dfrac{2}{3}$$

■ **Work Practice 1**

Answer

1. $\dfrac{2}{3}$

When the fractions contain variables, we add and subtract the same way.

 Add: $\dfrac{2x}{15} + \dfrac{3x}{10}$

Solution:

Step 1: The LCD of the denominators 15 and 10 is 30.

Step 2: $\dfrac{2x}{15} = \dfrac{2x}{15} \cdot \dfrac{2}{2} = \dfrac{4x}{30}$ $\dfrac{3x}{10} = \dfrac{3x}{10} \cdot \dfrac{3}{3} = \dfrac{9x}{30}$

Step 3: $\dfrac{2x}{15} + \dfrac{3x}{10} = \dfrac{4x}{30} + \dfrac{9x}{30} = \dfrac{13x}{30}$

Step 4: $\dfrac{13x}{30}$ is in simplest form.

◼ **Work Practice 2**

Practice 2

Add: $\dfrac{5y}{6} + \dfrac{2y}{9}$

Example 3 Add: $-\dfrac{1}{6} + \dfrac{1}{2}$

Solution: The LCD of the denominators 6 and 2 is 6.

$$-\dfrac{1}{6} + \dfrac{1}{2} = \dfrac{-1}{6} + \dfrac{1 \cdot 3}{2 \cdot 3}$$

$$= \dfrac{-1}{6} + \dfrac{3}{6}$$

$$= \dfrac{2}{6}$$

Helpful Hint

Recall that

$$-\dfrac{1}{6} = \dfrac{-1}{6} = \dfrac{1}{-6}$$

Next, simplify $\dfrac{2}{6}$.

$$\dfrac{2}{6} = \dfrac{\overset{1}{\cancel{2}}}{\underset{1}{2} \cdot 3} = \dfrac{1}{3}$$

◼ **Work Practice 3**

Practice 3

Add: $-\dfrac{1}{5} + \dfrac{9}{20}$

✓Concept Check Find and correct the error in the following:

$$\dfrac{2}{9} + \dfrac{4}{11} \;\cancel{=}\; \dfrac{6}{20} = \dfrac{3}{10}$$

Example 4 Subtract: $\dfrac{2}{3} - \dfrac{10}{11}$

Solution:

Step 1: The LCD of the denominators 3 and 11 is 33.

Step 2: $\dfrac{2}{3} = \dfrac{2}{3} \cdot \dfrac{11}{11} = \dfrac{22}{33}$ $\dfrac{10}{11} = \dfrac{10}{11} \cdot \dfrac{3}{3} = \dfrac{30}{33}$

Step 3: $\dfrac{2}{3} - \dfrac{10}{11} = \dfrac{22}{33} - \dfrac{30}{33} = \dfrac{-8}{33}$ or $-\dfrac{8}{33}$

Step 4: $-\dfrac{8}{33}$ is in simplest form.

◼ **Work Practice 4**

Practice 4

Subtract: $\dfrac{5}{7} - \dfrac{9}{10}$

Answers

2. $\dfrac{19y}{18}$ **3.** $\dfrac{1}{4}$ **4.** $-\dfrac{13}{70}$

✓Concept Check Answer

When adding fractions, we don't add the denominators. Correct solution:

$$\dfrac{2}{9} + \dfrac{4}{11} = \dfrac{22}{99} + \dfrac{36}{99} = \dfrac{58}{99}$$

Practice 5

Find: $\dfrac{5}{8} - \dfrac{1}{3} - \dfrac{1}{12}$

Example 5   Find: $-\dfrac{3}{4} - \dfrac{1}{14} + \dfrac{6}{7}$

Solution: The LCD of 4, 14, and 7 is 28.

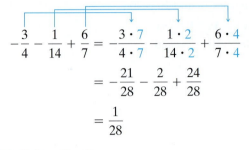

$$-\frac{3}{4} - \frac{1}{14} + \frac{6}{7} = -\frac{3 \cdot 7}{4 \cdot 7} - \frac{1 \cdot 2}{14 \cdot 2} + \frac{6 \cdot 4}{7 \cdot 4}$$

$$= -\frac{21}{28} - \frac{2}{28} + \frac{24}{28}$$

$$= \frac{1}{28}$$

■ **Work Practice 5**

✓**Concept Check**   Find and correct the error in the following:

$$\frac{7}{12} \;\cancel{-}\; \frac{3}{4} = \frac{4}{8} \;\cancel{=}\; \frac{1}{2}$$

Practice 6

Subtract: $5 - \dfrac{y}{4}$

Example 6   Subtract: $2 - \dfrac{x}{3}$

Solution: Recall that $2 = \dfrac{2}{1}$. The LCD of the denominators 1 and 3 is 3.

$$\frac{2}{1} - \frac{x}{3} = \frac{2 \cdot 3}{1 \cdot 3} - \frac{x}{3}$$

$$= \frac{6}{3} - \frac{x}{3}$$

$$= \frac{6 - x}{3}$$

> **Helpful Hint**
>
> The expression $\dfrac{6-x}{3}$ from Example 6 *does not simplify* to $2 - x$. The number 3 must be a factor of both terms in the numerator (not just 6) in order to simplify.

The numerator $6 - x$ cannot be simplified further since 6 and $-x$ are unlike terms.

■ **Work Practice 6**

Practice 7

Insert $<$ or $>$ to form a true sentence.

$$\frac{5}{8} \qquad \frac{11}{20}$$

Answers

5. $\dfrac{5}{24}$ **6.** $\dfrac{20 - y}{4}$ **7.** $>$

✓**Concept Check Answers**

When adding fractions, we don't add the denominators. Correct solutions:

$$\frac{7}{12} - \frac{3}{4} = \frac{7}{12} - \frac{9}{12} = -\frac{2}{12} = -\frac{1}{6}$$

Objective B   Writing Fractions in Order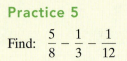

One important application of the least common denominator is to use the LCD to help order or compare fractions.

Example 7   Insert $<$ or $>$ to form a true sentence.

$$\frac{3}{4} \qquad \frac{9}{11}$$

Solution: The LCD for these fractions is 44. Let's write each fraction as an equivalent fraction with a denominator of 44.

$$\frac{3}{4} = \frac{3 \cdot 11}{4 \cdot 11} = \frac{33}{44} \qquad\qquad \frac{9}{11} = \frac{9 \cdot 4}{11 \cdot 4} = \frac{36}{44}$$

Since $33 < 36$, then

$$\frac{33}{44} < \frac{36}{44} \text{ or}$$

$$\frac{3}{4} < \frac{9}{11}$$

■ Work Practice 7

Example 8 Insert $<$ or $>$ to form a true sentence.

$$-\frac{2}{7} \qquad -\frac{1}{3}$$

Solution: The LCD is 21.

$$-\frac{2}{7} = -\frac{2 \cdot 3}{7 \cdot 3} = -\frac{6}{21} \text{ or } \frac{-6}{21} \qquad -\frac{1}{3} = -\frac{1 \cdot 7}{3 \cdot 7} = -\frac{7}{21} \text{ or } \frac{-7}{21}$$

Since $-6 > -7$, then

$$-\frac{6}{21} > -\frac{7}{21} \text{ or}$$

$$-\frac{2}{7} > -\frac{1}{3}$$

■ Work Practice 8

Objective C Evaluating Expressions Given Fractional Replacement Values

Example 9 Evaluate $x - y$ if $x = \frac{7}{18}$ and $y = \frac{2}{9}$.

Solution: Replace x with $\frac{7}{18}$ and y with $\frac{2}{9}$ in the expression $x - y$.

$$x - y = \frac{7}{18} - \frac{2}{9}$$

The LCD of the denominators 18 and 9 is 18. Then

$$\frac{7}{18} - \frac{2}{9} = \frac{7}{18} - \frac{2 \cdot 2}{9 \cdot 2}$$

$$= \frac{7}{18} - \frac{4}{18}$$

$$= \frac{3}{18} = \frac{1}{6} \qquad \text{Simplified}$$

■ Work Practice 9

Objective D Solving Problems by Adding or Subtracting Unlike Fractions

Very often, real-world problems involve adding or subtracting unlike fractions.

Practice 8
Insert $<$ or $>$ to form a true sentence.

$$-\frac{17}{20} \qquad -\frac{4}{5}$$

Practice 9
Evaluate $x - y$ if $x = \frac{5}{11}$ and $y = \frac{4}{9}$.

Answers

8. $<$ **9.** $\frac{1}{99}$

Practice 10

To repair her sidewalk, a homeowner must pour cement in three different locations. She needs $\frac{3}{5}$ of a cubic yard, $\frac{3}{10}$ of a cubic yard, and $\frac{1}{15}$ of a cubic yard for these locations. Find the total amount of cement the homeowner needs.

Copyright 2015 Pearson Education, Inc.

Example 10   Finding Total Weight

A freight truck has $\frac{1}{4}$ of a ton of computers, $\frac{1}{3}$ of a ton of televisions, and $\frac{3}{8}$ of a ton of small appliances. Find the total weight of its load.

Solution:

1. UNDERSTAND. Read and reread the problem. The phrase "total weight" tells us to add.
2. TRANSLATE.

In words:	total weight	is	weight of computers	plus	weight of televisions	plus	weight of appliances
	↓	↓	↓	↓	↓	↓	↓
Translate:	total weight	$=$	$\frac{1}{4}$	$+$	$\frac{1}{3}$	$+$	$\frac{3}{8}$

3. SOLVE: The LCD is 24.

$$\frac{1}{4} + \frac{1}{3} + \frac{3}{8} = \frac{1}{4} \cdot \frac{6}{6} + \frac{1}{3} \cdot \frac{8}{8} + \frac{3}{8} \cdot \frac{3}{3}$$

$$= \frac{6}{24} + \frac{8}{24} + \frac{9}{24}$$

$$= \frac{23}{24}$$

4. INTERPRET. *Check* the solution. *State* your conclusion: The total weight of the truck's load is $\frac{23}{24}$ ton.

■ **Work Practice 10**

Practice 11

Find the difference in length of two boards if one board is $\frac{3}{4}$ of a foot long and the other is $\frac{2}{3}$ of a foot long.

Answers

10. $\frac{29}{30}$ cu yd   **11.** $\frac{1}{12}$ ft

Example 11   Calculating Flight Time

A flight from Tucson to Phoenix, Arizona, requires $\frac{5}{12}$ of an hour. If the plane has been flying $\frac{1}{4}$ of an hour, find how much time remains before landing.

Solution:

1. UNDERSTAND. Read and reread the problem. The phrase "how much time remains" tells us to subtract.

2. TRANSLATE.

In words:	time remaining	is	flight time from Tucson to Phoenix	minus	flight time already passed
	↓	↓	↓	↓	↓
Translate:	time remaining	$=$	$\dfrac{5}{12}$	$-$	$\dfrac{1}{4}$

3. SOLVE: The LCD is 12.

$$\frac{5}{12} - \frac{1}{4} = \frac{5}{12} - \frac{1}{4} \cdot \frac{3}{3}$$

$$= \frac{5}{12} - \frac{3}{12}$$

$$= \frac{2}{12} = \frac{\overset{1}{\cancel{2}}}{\underset{1}{\cancel{2}} \cdot 6} = \frac{1}{6}$$

4. INTERPRET. *Check* the solution. *State* your conclusion: The remaining flight time is $\dfrac{1}{6}$ of an hour.

■ **Work Practice 11**

 Calculator Explorations **Performing Operations on Fractions**

Scientific Calculator

Many calculators have a fraction key, such as $\boxed{a^b/_c}$, that allows you to enter fractions and perform operations on fractions, and gives the result as a fraction. If your calculator has a fraction key, use it to calculate

$$\frac{3}{5} + \frac{4}{7}$$

Enter the keystrokes

$\boxed{3}\ \boxed{a^b/_c}\ \boxed{5}\ \boxed{+}\ \boxed{4}\ \boxed{a^b/_c}\ \boxed{7}\ \boxed{=}$

The display should read $\boxed{1\_6 \mid 35}$

which represents the mixed number $1\dfrac{6}{35}$. Let's write the result as a fraction. To convert from mixed number notation to fractional notation, press

$\boxed{2^{nd}}\ \boxed{d/c}$

The display now reads $\boxed{41 \mid 35}$

which represents $\dfrac{41}{35}$, the sum in fractional notation.

Graphing Calculator

Graphing calculators also allow you to perform operations on fractions and will give exact fractional results. The fraction option on a graphing calculator may be found under the $\boxed{\text{MATH}}$ menu. To perform the addition in the first column, try the keystrokes

$\boxed{3}\ \boxed{\div}\ \boxed{5}\ \boxed{+}\ \boxed{4}\ \boxed{\div}\ \boxed{7}\ \boxed{\text{MATH}}\ \boxed{\text{ENTER}}$
$\boxed{\text{ENTER}}$

The display should read

$\boxed{3/5 + 4/7 \blacktriangleright \text{Frac } 41/35}$

Use a calculator to add the following fractions. Give each sum as a fraction.

1. $\dfrac{1}{16} + \dfrac{2}{5}$ **2.** $\dfrac{3}{20} + \dfrac{2}{25}$ **3.** $\dfrac{4}{9} + \dfrac{7}{8}$

4. $\dfrac{9}{11} + \dfrac{5}{12}$ **5.** $\dfrac{10}{17} + \dfrac{12}{19}$ **6.** $\dfrac{14}{31} + \dfrac{15}{21}$

Vocabulary, Readiness & Video Check

Use the choices below to fill in each blank. Any numerical answers are not listed. Not all choices may be used.

expression	least common denominator	$>$
equation	equivalent	$<$

1. To add or subtract unlike fractions, we first write the fractions as _____ fractions with a common denominator. The common denominator we use is called the _____.

2. The LCD for $\dfrac{1}{6}$ and $\dfrac{5}{8}$ is _____.

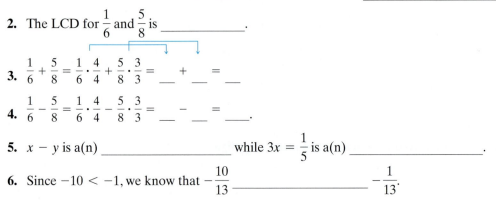

3. $\dfrac{1}{6} + \dfrac{5}{8} = \dfrac{1}{6} \cdot \dfrac{4}{4} + \dfrac{5}{8} \cdot \dfrac{3}{3} = \underline{\quad} + \underline{\quad} = \underline{\quad}$

4. $\dfrac{1}{6} - \dfrac{5}{8} = \dfrac{1}{6} \cdot \dfrac{4}{4} - \dfrac{5}{8} \cdot \dfrac{3}{3} = \underline{\quad} - \underline{\quad} = \underline{\quad}$.

5. $x - y$ is a(n) _____ while $3x = \dfrac{1}{5}$ is a(n) _____.

6. Since $-10 < -1$, we know that $-\dfrac{10}{13}$ _____ $-\dfrac{1}{13}$.

Martin-Gay Interactive Videos   *Watch the section lecture video and answer the following questions.*

Objective A   **7.** In ▤ Example 3, why can't we add the two terms in the numerator? ▶

Objective B   **8.** In ▤ Example 5, when comparing two fractions, how does writing each fraction with the same denominator help? ▶

Objective C   **9.** In ▤ Example 6, if we had chosen to simplify the first fraction before adding, what would our addition problem have become and what would our LCD have been? ▶

Objective D   **10.** What are the two forms of the answer to ▤ Example 7? ▶

See Video 4.5

4.5 Exercise Set MyMathLab® ▶

Objective A *Add or subtract as indicated. See Examples 1 through 6.*

1. $\dfrac{2}{3} + \dfrac{1}{6}$

2. $\dfrac{5}{6} + \dfrac{1}{12}$

3. $\dfrac{1}{2} - \dfrac{1}{3}$

4. $\dfrac{2}{3} - \dfrac{1}{4}$

5. $-\dfrac{2}{11} + \dfrac{2}{33}$

6. $-\dfrac{5}{9} + \dfrac{1}{3}$

7. $\dfrac{3}{14} - \dfrac{3}{7}$

8. $\dfrac{2}{15} - \dfrac{2}{5}$

9. $\dfrac{11x}{35} + \dfrac{2x}{7}$

10. $\dfrac{2y}{5} + \dfrac{3y}{25}$

11. $2 - \dfrac{y}{12}$

12. $5 - \dfrac{y}{20}$

13. $\dfrac{5}{12} - \dfrac{1}{9}$

14. $\dfrac{7}{12} - \dfrac{5}{18}$

15. $-7 + \dfrac{5}{7}$

16. $-10 + \dfrac{7}{10}$

17. $\dfrac{5a}{11} + \dfrac{4a}{9}$

18. $\dfrac{7x}{18} + \dfrac{2x}{9}$

19. $\dfrac{2y}{3} - \dfrac{1}{6}$

20. $\dfrac{5z}{6} - \dfrac{1}{12}$

21. $\dfrac{1}{2} + \dfrac{3}{x}$

22. $\dfrac{2}{5} + \dfrac{3}{x}$

23. $-\dfrac{2}{11} - \dfrac{2}{33}$

24. $-\dfrac{5}{9} - \dfrac{1}{3}$

25. $\dfrac{9}{14} - \dfrac{3}{7}$

26. $\dfrac{4}{5} - \dfrac{2}{15}$

27. $\dfrac{11y}{35} - \dfrac{2}{7}$

28. $\dfrac{2b}{5} - \dfrac{3}{25}$

29. $\dfrac{1}{9} - \dfrac{5}{12}$

30. $\dfrac{5}{18} - \dfrac{7}{12}$

31. $\dfrac{7}{15} - \dfrac{5}{12}$

32. $\dfrac{5}{8} - \dfrac{3}{20}$

33. $\dfrac{5}{7} - \dfrac{1}{8}$

34. $\dfrac{10}{13} - \dfrac{7}{10}$

35. $\dfrac{7}{8} + \dfrac{3}{16}$

36. $\dfrac{7}{18} + \dfrac{2}{9}$

37. $\dfrac{3}{9} - \dfrac{5}{9}$

38. $\dfrac{1}{13} - \dfrac{4}{13}$

39. $-\dfrac{2}{5} + \dfrac{1}{3} - \dfrac{3}{10}$

40. $-\dfrac{1}{3} - \dfrac{1}{4} + \dfrac{2}{5}$

41. $\dfrac{5}{11} + \dfrac{y}{3}$

42. $\dfrac{5z}{13} + \dfrac{3}{26}$

43. $-\dfrac{5}{6} - \dfrac{3}{7}$

44. $-\dfrac{1}{2} - \dfrac{3}{29}$

45. $\dfrac{x}{2} + \dfrac{x}{4} + \dfrac{2x}{16}$

46. $\dfrac{z}{4} + \dfrac{z}{8} + \dfrac{2z}{16}$

47. $\dfrac{7}{9} - \dfrac{1}{6}$

48. $\dfrac{9}{16} - \dfrac{3}{8}$

49. $\dfrac{2a}{3} + \dfrac{6a}{13}$

50. $\dfrac{3y}{4} + \dfrac{y}{7}$

51. $\dfrac{7}{30} - \dfrac{5}{12}$

52. $\dfrac{7}{30} - \dfrac{3}{20}$

53. $\dfrac{5}{9} + \dfrac{1}{y}$

54. $\dfrac{1}{12} - \dfrac{5}{x}$

55. $\dfrac{6}{5} - \dfrac{3}{4} + \dfrac{1}{2}$

56. $\dfrac{6}{5} + \dfrac{3}{4} - \dfrac{1}{2}$

57. $\dfrac{4}{5} + \dfrac{4}{9}$

58. $\dfrac{11}{12} - \dfrac{7}{24}$

59. $\dfrac{5}{9x} + \dfrac{1}{8}$

60. $\dfrac{3}{8} + \dfrac{5}{12x}$

61. $-\dfrac{9}{12} + \dfrac{17}{24} - \dfrac{1}{6}$

62. $-\dfrac{5}{14} + \dfrac{3}{7} - \dfrac{1}{2}$

63. $\dfrac{3x}{8} + \dfrac{2x}{7} - \dfrac{5}{14}$

64. $\dfrac{9x}{10} - \dfrac{1}{2} + \dfrac{x}{5}$

Objective **B** *Insert $<$ or $>$ to form a true sentence. See Examples 7 and 8.*

65. $\dfrac{2}{7} \quad \dfrac{3}{10}$

66. $\dfrac{5}{9} \quad \dfrac{6}{11}$

67. $-\dfrac{5}{6} \quad -\dfrac{13}{15}$

68. $-\dfrac{7}{8} \quad -\dfrac{5}{6}$

69. $-\dfrac{3}{4} \quad -\dfrac{11}{14}$

70. $-\dfrac{2}{9} \quad -\dfrac{3}{13}$

Objective **C** *Evaluate each expression if $x = \dfrac{1}{3}$ and $y = \dfrac{3}{4}$. See Example 9.*

71. $x + y$

72. $x - y$

73. xy

74. $x \div y$

75. $2y + x$

76. $2x + y$

Objective D *Find the perimeter of each geometric figure. (Hint: Recall that perimeter means distance around.)*

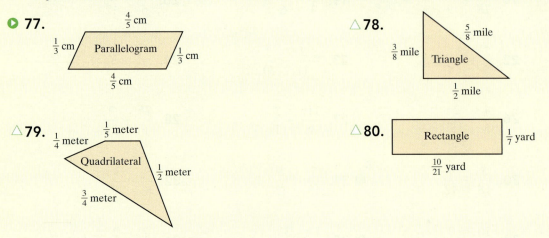

77.
$\frac{4}{5}$ cm
$\frac{1}{3}$ cm Parallelogram $\frac{1}{3}$ cm
$\frac{4}{5}$ cm

78.
$\frac{5}{8}$ mile
$\frac{3}{8}$ mile Triangle
$\frac{1}{2}$ mile

79.
$\frac{1}{5}$ meter
$\frac{1}{4}$ meter Quadrilateral $\frac{1}{2}$ meter
$\frac{3}{4}$ meter

80.
Rectangle $\frac{1}{7}$ yard
$\frac{10}{21}$ yard

Translating *Translate each phrase into an algebraic expression. Use "x" to represent "a number." See Examples 10 and 11.*

81. The sum of a number and $\frac{1}{2}$

82. A number increased by $-\frac{2}{5}$

83. A number subtracted from $-\frac{3}{8}$

84. The difference of a number and $\frac{7}{20}$

Solve. For Exercises 85 and 86, the solutions have been started for you. See Examples 10 and 11.

85. The slowest mammal is the three-toed sloth from South America. The sloth has an average ground speed of $\frac{1}{10}$ mph. In trees, it can accelerate to $\frac{17}{100}$ mph. How much faster can a sloth travel in trees? (*Source: The Guinness Book of World Records*)

Start the solution:

1. UNDERSTAND the problem. Reread it as many times as needed.
2. TRANSLATE into an equation. (Fill in the blanks.)

how much faster sloth travels in trees	is	sloth speed in trees	minus	sloth speed on ground
↓	↓	↓	↓	↓

how much faster sloth travels in trees = _____ − _____

Finish with:

3. SOLVE. and
4. INTERPRET.

86. Killer bees have been known to chase people for up to $\frac{1}{4}$ of a mile, while domestic European honeybees will normally chase a person for no more than 100 feet, or $\frac{5}{264}$ of a mile. How much farther will a killer bee chase a person than a domestic honeybee? (*Source:* Coachella Valley Mosquito & Vector Control District)

Start the solution:

1. UNDERSTAND the problem. Reread it as many times as needed.
2. TRANSLATE into an equation. (Fill in the blanks.)

how much farther killer bee will chase than honeybee	is	distance killer bee chases	minus	distance honeybee chases
↓	↓	↓	↓	↓

how much farther killer bee will chase than honeybee = _____ − _____

Finish with

3. SOLVE. and
4. INTERPRET.

Summary on Fractions and Operations on Fractions

Answers

Use a fraction to represent the shaded area of each figure. If the fraction is improper, also write the fraction as a mixed number.

1.

2.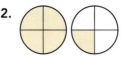

Solve.

3. In a survey, 73 people out of 85 get fewer than 8 hours of sleep each night. What fraction of people in the survey get fewer than 8 hours of sleep?

4. Sketch a diagram to represent $\dfrac{9}{13}$.

Simplify.

5. $\dfrac{11}{-11}$ **6.** $\dfrac{17}{1}$ **7.** $\dfrac{0}{-3}$ **8.** $\dfrac{7}{0}$

Write the prime factorization of each composite number. Write any repeated factors using exponents.

9. 65 **10.** 70 **11.** 315 **12.** 441

Write each fraction in simplest form.

13. $\dfrac{2}{14}$ **14.** $\dfrac{24}{20}$ **15.** $-\dfrac{56}{60}$ **16.** $-\dfrac{72}{80}$

17. $\dfrac{54x}{135}$ **18.** $\dfrac{90}{240y}$ **19.** $\dfrac{165z^3}{210z}$ **20.** $\dfrac{245ab}{385a^2b^3}$

Determine whether each pair of fractions is equivalent.

21. $\dfrac{7}{8}$ and $\dfrac{9}{10}$ **22.** $\dfrac{10}{12}$ and $\dfrac{15}{18}$

23. Of the 50 states, 2 states are not adjacent to any other states.

    a. What fraction of the states are not adjacent to other states?

    b. How many states are adjacent to other states?

    c. What fraction of the states are adjacent to other states?

24. There are approximately 90,000 digital movie screens in the world. Of these, about 36,000 are in the U.S./Canada. (*Source:* Motion Picture Association of America)

    a. What fraction of digital movie screens are in the U.S./Canada?

    b. How many digital movie screens are not in the U.S./Canada?

    c. What fraction of digital movie screens are not in the U.S./Canada?

1. _____

2. _____

3. _____

4. _____

5. _____

6. _____

7. _____

8. _____

9. _____

10. _____

11. _____

12. _____

13. _____

14. _____

15. _____

16. _____

17. _____

18. _____

19. _____

20. _____

21. _____

22. _____

23. a.   b.   c.

24. a.   b.   c.

25. _____ 26. _____

27. _____ 28. _____

29. _____ 30. _____

31. _____ 32. _____

33. _____ 34. _____

35. _____ 36. _____

37. _____

38. _____

39. _____

40. _____

41. _____

42. _____

43. _____

44. _____

45. _____

46. _____

47. _____

48. _____

49. _____

50. _____

51. _____

52. _____

53. _____

Find the LCM of each list of numbers.

25. $5, 6$ **26.** $2, 14$ **27.** $6, 18, 30$

Write each fraction as an equivalent fraction with the indicated denominator.

28. $\dfrac{7}{9} = \dfrac{}{36}$ **29.** $\dfrac{11}{15} = \dfrac{}{75}$ **30.** $\dfrac{5}{6} = \dfrac{}{48}$

The following summary will help you with the following review of operations on fractions.

Operations on Fractions

Let $a, b, c,$ and d be integers.

Addition: $\dfrac{a}{b} + \dfrac{c}{b} = \dfrac{a+c}{b}$

$b \neq 0$ ↑   ↑
common denominator

Subtraction: $\dfrac{a}{b} - \dfrac{c}{b} = \dfrac{a-c}{b}$

$(b \neq 0)$ ↑   ↑
common denominator

Multiplication: $\dfrac{a}{b} \cdot \dfrac{c}{d} = \dfrac{a \cdot c}{b \cdot d}$

$(b \neq 0, d \neq 0)$

Division: $\dfrac{a}{b} \div \dfrac{c}{d} = \dfrac{a}{b} \cdot \dfrac{d}{c} = \dfrac{a \cdot d}{b \cdot c}$

$(b \neq 0, d \neq 0, c \neq 0)$

Perform each indicated operation.

31. $\dfrac{1}{5} + \dfrac{3}{5}$ **32.** $\dfrac{1}{5} - \dfrac{3}{5}$ **33.** $\dfrac{1}{5} \cdot \dfrac{3}{5}$ **34.** $\dfrac{1}{5} \div \dfrac{3}{5}$

35. $\dfrac{2}{3} \div \dfrac{5}{6}$ **36.** $\dfrac{2a}{3} \cdot \dfrac{5}{6a}$ **37.** $\dfrac{2}{3y} - \dfrac{5}{6y}$ **38.** $\dfrac{2x}{3} + \dfrac{5x}{6}$

39. $-\dfrac{1}{7} \cdot -\dfrac{7}{18}$ **40.** $-\dfrac{4}{9} \cdot -\dfrac{3}{7}$ **41.** $-\dfrac{7z}{8} \div 6z^2$ **42.** $-\dfrac{9}{10} \div 5$

43. $\dfrac{7}{8} + \dfrac{1}{20}$ **44.** $\dfrac{5}{12} - \dfrac{1}{9}$ **45.** $\dfrac{2}{9} + \dfrac{1}{18} + \dfrac{1}{3}$ **46.** $\dfrac{3y}{10} + \dfrac{y}{5} + \dfrac{6}{25}$

Translating *Translate each to an expression. Use x to represent "a number."*

47. $\dfrac{2}{3}$ of a number

48. The quotient of a number and $-\dfrac{1}{5}$

49. A number subtracted from $-\dfrac{8}{9}$

50. $\dfrac{6}{11}$ increased by a number

Solve.

51. Find $\dfrac{2}{3}$ of 1530.

52. A contractor is using 18 acres of his land to sell $\dfrac{3}{4}$-acre lots. How many lots can he sell?

53. Suppose that the cross-section of a piece of pipe looks like the diagram shown. What is the inner diameter?

$\frac{7}{8}$ foot

$\frac{1}{16}$ foot $\frac{1}{16}$ foot

inner diameter

4.6 Complex Fractions and Review of Order of Operations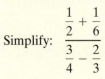

Objective A Simplifying Complex Fractions

Thus far, we have studied operations on fractions. We now practice simplifying fractions whose numerators or denominators themselves contain fractions. These fractions are called **complex fractions.**

Objectives

A Simplify Complex Fractions.

B Review the Order of Operations.

C Evaluate Expressions Given Replacement Values.

> **Complex Fraction**
>
> A fraction whose numerator or denominator or both numerator and denominator contain fractions is called a **complex fraction.**

Examples of complex fractions are

$$\frac{\frac{x}{4}}{\frac{3}{2}} \qquad \frac{\frac{1}{2}+\frac{3}{8}}{\frac{3}{4}-\frac{1}{6}} \qquad \frac{\frac{y}{5}-2}{\frac{3}{10}} \qquad \frac{-4z}{\frac{3}{5}}$$

Method 1 for Simplifying Complex Fractions

Two methods are presented to simplify complex fractions. The first method makes use of the fact that a fraction bar means division.

Example 1 Simplify: $\dfrac{\frac{x}{4}}{\frac{3}{2}}$

Solution: Since a fraction bar means division, the complex fraction $\dfrac{\frac{x}{4}}{\frac{3}{2}}$ can be written as $\dfrac{x}{4} \div \dfrac{3}{2}$. Then divide as usual to simplify.

$$\frac{x}{4} \div \frac{3}{2} = \frac{x}{4} \cdot \frac{2}{3} \qquad \text{Multiply by the reciprocal.}$$

$$= \frac{x \cdot \overset{1}{\cancel{2}}}{\underset{1}{2 \cdot 2 \cdot 3}}$$

$$= \frac{x}{6}$$

■ Work Practice 1

Practice 1

Simplify: $\dfrac{\frac{7y}{10}}{\frac{1}{5}}$

Example 2 Simplify: $\dfrac{\frac{1}{2}+\frac{3}{8}}{\frac{3}{4}-\frac{1}{6}}$

Solution: Recall the order of operations. Since the fraction bar is considered a grouping symbol, we simplify the numerator and the denominator of the complex fraction separately. Then we divide.

$$\frac{\frac{1}{2}+\frac{3}{8}}{\frac{3}{4}-\frac{1}{6}} = \frac{\frac{1 \cdot 4}{2 \cdot 4}+\frac{3}{8}}{\frac{3 \cdot 3}{4 \cdot 3}-\frac{1 \cdot 2}{6 \cdot 2}} = \frac{\frac{4}{8}+\frac{3}{8}}{\frac{9}{12}-\frac{2}{12}} = \frac{\frac{7}{8}}{\frac{7}{12}}$$

(Continued on next page)

Practice 2

Simplify: $\dfrac{\frac{1}{2}+\frac{1}{6}}{\frac{3}{4}-\frac{2}{3}}$

Answers

1. $\dfrac{7y}{2}$ **2.** $\dfrac{8}{1}$ or 8

Thus,

$$\frac{\dfrac{1}{2} + \dfrac{3}{8}}{\dfrac{3}{4} - \dfrac{1}{6}} = \frac{\dfrac{7}{8}}{\dfrac{7}{12}}$$

$$= \frac{7}{8} \div \frac{7}{12} \qquad \text{Rewrite the quotient using the} \div \text{ sign.}$$

$$= \frac{7}{8} \cdot \frac{12}{7} \qquad \text{Multiply by the reciprocal.}$$

$$= \frac{\overset{1}{7} \cdot 3 \cdot \overset{1}{4}}{2 \cdot \underset{1}{4} \cdot \underset{1}{7}} \qquad \text{Multiply.}$$

$$= \frac{3}{2} \qquad \text{Simplify.}$$

■ **Work Practice 2**

Method 2 for Simplifying Complex Fractions

The second method for simplifying complex fractions is to multiply the numerator and the denominator of the complex fraction by the LCD of all the fractions in its numerator and its denominator. This has the effect of leaving sums and differences of integers in the numerator and the denominator, as we shall see in the example below.

Let's use this second method to simplify the complex fraction in Example 2 again.

Practice 3

Use Method 2 to simplify:

$$\frac{\dfrac{1}{2} + \dfrac{1}{6}}{\dfrac{3}{4} - \dfrac{2}{3}}$$

Example 3   Simplify: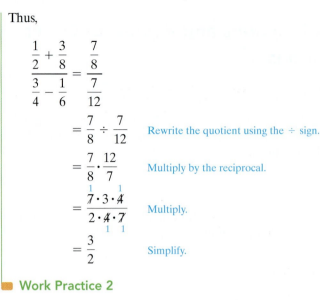

Solution: The complex fraction contains fractions with denominators 2, 8, 4, and 6. The LCD is 24. By the fundamental property of fractions (See the Helpful Hints on p. 230 and p. 257), we can multiply the numerator and the denominator of the complex fraction by 24. Notice below that by the distributive property, this means that we multiply each term in the numerator and denominator by 24.

$$\frac{\dfrac{1}{2} + \dfrac{3}{8}}{\dfrac{3}{4} - \dfrac{1}{6}} = \frac{24\left(\dfrac{1}{2} + \dfrac{3}{8}\right)}{24\left(\dfrac{3}{4} - \dfrac{1}{6}\right)}$$

$$= \frac{\left(\overset{12}{24} \cdot \dfrac{1}{\underset{1}{2}}\right) + \left(\overset{3}{24} \cdot \dfrac{3}{\underset{1}{8}}\right)}{\left(\overset{6}{24} \cdot \dfrac{3}{\underset{1}{4}}\right) - \left(\overset{4}{24} \cdot \dfrac{1}{\underset{1}{6}}\right)} \qquad \begin{array}{l}\text{Apply the distributive property. Then divide out} \\ \text{common factors to aid in multiplying.}\end{array}$$

$$= \frac{12 + 9}{18 - 4} \qquad \text{Multiply.}$$

$$= \frac{21}{14}$$

$$= \frac{\overset{1}{7} \cdot 3}{\underset{1}{7} \cdot 2} = \frac{3}{2} \qquad \text{Simplify.}$$

■ **Work Practice 3**

Answer

3. $\dfrac{8}{1}$ or 8

The simplified result is the same, of course, no matter which method is used.

 **Example 4** Simplify: $\dfrac{\dfrac{y}{5} - 2}{\dfrac{3}{10}}$

Solution: Use the second method and multiply the numerator and the denominator of the complex fraction by the LCD of all fractions. Recall that $2 = \dfrac{2}{1}$. The LCD of the denominators 5, 1, and 10 is 10.

$$\dfrac{\dfrac{y}{5} - \dfrac{2}{1}}{\dfrac{3}{10}} = \dfrac{10\left(\dfrac{y}{5} - \dfrac{2}{1}\right)}{10\left(\dfrac{3}{10}\right)}$$ Multiply the numerator and denominator by 10.

$$= \dfrac{\left(\overset{2}{\cancel{10}} \cdot \dfrac{y}{\cancel{5}}\right) - \left(10 \cdot \dfrac{2}{1}\right)}{\underset{1}{\cancel{10}} \cdot \dfrac{3}{\cancel{10}}}$$ Apply the distributive property. Then divide out common factors to aid in multiplying.

$$= \dfrac{2y - 20}{3}$$ Multiply.

■ **Work Practice 4**

Objective B Reviewing the Order of Operations ▶

At this time, it is probably a good idea to review the order of operations on expressions containing fractions. Before we do so, let's review how we perform operations on fractions.

	Review of Operations on Fractions	
Operation	**Procedure**	**Example**
Multiply	Multiply the numerators and multiply the denominators.	$\dfrac{5}{9} \cdot \dfrac{1}{2} = \dfrac{5 \cdot 1}{9 \cdot 2} = \dfrac{5}{18}$
Divide	Multiply the first fraction by the reciprocal of the second fraction.	$\dfrac{2}{3} \div \dfrac{11}{13} = \dfrac{2}{3} \cdot \dfrac{13}{11} = \dfrac{2 \cdot 13}{3 \cdot 11} = \dfrac{26}{33}$
Add or Subtract	1. Write each fraction as an equivalent fraction whose denominator is the LCD. 2. Add or subtract numerators and write the result over the common denominator.	$\dfrac{3}{4} + \dfrac{1}{8} = \dfrac{3}{4} \cdot \dfrac{2}{2} + \dfrac{1}{8} = \dfrac{6}{8} + \dfrac{1}{8} = \dfrac{7}{8}$

Now let's review order of operations.

Order of Operations

1. Perform all operations within parentheses ( ), brackets [ ], or other grouping symbols such as fraction bars, starting with the innermost set.
2. Evaluate any expressions with exponents.
3. Multiply or divide in order from left to right.
4. Add or subtract in order from left to right.

Example 5 Simplify: $\left(\dfrac{4}{5}\right)^2 - 1$

Solution: According to the order of operations, first evaluate $\left(\dfrac{4}{5}\right)^2$.

$$\left(\dfrac{4}{5}\right)^2 - 1 = \dfrac{16}{25} - 1$$ Write $\left(\dfrac{4}{5}\right)^2$ as $\dfrac{16}{25}$.

(Continued on next page)

(Continued on next page)

Practice 4

Simplify: $\dfrac{\dfrac{3}{4}}{\dfrac{x}{5} - 1}$

Helpful Hint Don't forget to multiply the numerator and the denominator of the complex fraction by the same number—the LCD.

Practice 5

Simplify: $\left(\dfrac{2}{3}\right)^3 - 2$

Answers

4. $\dfrac{15}{4x - 20}$ 5. $-\dfrac{46}{27}$

Next, combine the fractions. The LCD of 25 and 1 is 25.

$$\frac{16}{25} - 1 = \frac{16}{25} - \frac{25}{25} \qquad \text{Write 1 as } \frac{25}{25}.$$

$$= \frac{-9}{25} \text{ or } -\frac{9}{25} \qquad \text{Subtract.}$$

■ **Work Practice 5**

Practice 6

Simplify: $\left(-\dfrac{1}{2} + \dfrac{1}{5}\right)\left(\dfrac{7}{8} + \dfrac{1}{8}\right)$

Example 6 Simplify: $\left(\dfrac{1}{4} + \dfrac{2}{3}\right)\left(\dfrac{11}{12} + \dfrac{1}{4}\right)$

Solution: First perform operations inside parentheses. Then multiply.

$$\left(\frac{1}{4} + \frac{2}{3}\right)\left(\frac{11}{12} + \frac{1}{4}\right) = \left(\frac{1\cdot 3}{4\cdot 3} + \frac{2\cdot 4}{3\cdot 4}\right)\left(\frac{11}{12} + \frac{1\cdot 3}{4\cdot 3}\right) \quad \text{Each LCD is 12.}$$

$$= \left(\frac{3}{12} + \frac{8}{12}\right)\left(\frac{11}{12} + \frac{3}{12}\right)$$

$$= \left(\frac{11}{12}\right)\left(\frac{14}{12}\right) \qquad \text{Add.}$$

$$= \frac{11 \cdot \overset{1}{2} \cdot 7}{\underset{1}{2} \cdot 6 \cdot 12} \qquad \text{Multiply.}$$

$$= \frac{77}{72} \qquad \text{Simplify.}$$

■ **Work Practice 6**

Helpful Hint If you find it difficult replacing a variable with a number, try the following. First, replace the variable with a set of parentheses, then place the replacement number between the parentheses.

If $x = \dfrac{4}{5}$, find $2x + x^2$.

$2x + x^2 = 2(\ ) + (\ )^2$

$\qquad = 2\left(\dfrac{4}{5}\right) + \left(\dfrac{4}{5}\right)^2 \cdots$

then continue simplifying.

✓**Concept Check** What should be done first to simplify the expression $\dfrac{1}{5} \cdot \dfrac{5}{2} - \left(\dfrac{2}{3} + \dfrac{4}{5}\right)^2$?

Objective C Evaluating Algebraic Expressions ▶

Practice 7

Evaluate $-\dfrac{3}{5} - xy$ if $x = \dfrac{3}{10}$ and $y = \dfrac{2}{3}$.

Example 7 Evaluate $2x + y^2$ if $x = -\dfrac{1}{2}$ and $y = \dfrac{1}{3}$.

Solution: Replace x and y with the given values and simplify.

$$2x + y^2 = 2\left(-\frac{1}{2}\right) + \left(\frac{1}{3}\right)^2 \quad \text{Replace } x \text{ with } -\frac{1}{2} \text{ and } y \text{ with } \frac{1}{3}.$$

$$= 2\left(-\frac{1}{2}\right) + \frac{1}{9} \qquad \text{Write } \left(\frac{1}{3}\right)^2 \text{ as } \frac{1}{9}.$$

$$= -1 + \frac{1}{9} \qquad \text{Multiply.}$$

$$= -\frac{9}{9} + \frac{1}{9} \qquad \text{The LCD is 9.}$$

$$= -\frac{8}{9} \qquad \text{Add.}$$

■ **Work Practice 7**

Answers

6. $-\dfrac{3}{10}$ 7. $-\dfrac{4}{5}$

✓**Concept Check Answer**

Add inside parentheses.

Vocabulary, Readiness & Video Check

Use the choices below to fill in each blank.

addition   multiplication   evaluate the exponential expression

subtraction   division   complex

1. A fraction whose numerator or denominator or both numerator and denominator contain fractions is called a(n)
   _____ fraction.

2. To simplify $-\dfrac{1}{2} + \dfrac{2}{3} \cdot \dfrac{7}{8}$, which operation do we perform first? _____

3. To simplify $-\dfrac{1}{2} \div \dfrac{2}{3} \cdot \dfrac{7}{8}$, which operation do we perform first? _____

4. To simplify $\dfrac{7}{8} \cdot \left(\dfrac{1}{2} - \dfrac{2}{3} \right)$, which operation do we perform first? _____

5. To simplify $\dfrac{1}{3} \div \dfrac{1}{4} \cdot \left(\dfrac{9}{11} + \dfrac{3}{8} \right)^3$, which operation do we perform first? _____

6. To simplify $9 - \left(-\dfrac{3}{4} \right)^2$, which operation do we perform first? _____

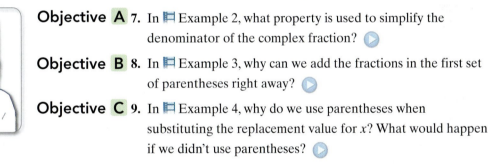

Martin-Gay Interactive Videos   Watch the section lecture video and answer the following questions.

Objective A   7. In ⊟ Example 2, what property is used to simplify the denominator of the complex fraction? ▶

Objective B   8. In ⊟ Example 3, why can we add the fractions in the first set of parentheses right away? ▶

Objective C   9. In ⊟ Example 4, why do we use parentheses when substituting the replacement value for *x*? What would happen if we didn't use parentheses? ▶

See Video 4.6

4.6   Exercise Set   MyMathLab® ▶

Objective A   *Simplify each complex fraction. See Examples 1 through 4.*

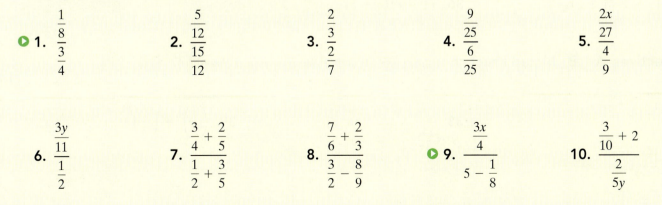

▶ 1. $\dfrac{\frac{1}{8}}{\frac{3}{4}}$
     2. $\dfrac{\frac{5}{12}}{\frac{15}{12}}$
     3. $\dfrac{\frac{2}{3}}{\frac{2}{7}}$
     4. $\dfrac{\frac{9}{25}}{\frac{6}{25}}$
     5. $\dfrac{\frac{2x}{27}}{\frac{4}{9}}$

6. $\dfrac{\frac{3y}{11}}{\frac{1}{2}}$
     7. $\dfrac{\frac{3}{4} + \frac{2}{5}}{\frac{1}{2} + \frac{3}{5}}$
     8. $\dfrac{\frac{7}{6} + \frac{2}{3}}{\frac{3}{2} - \frac{8}{9}}$
     ▶ 9. $\dfrac{\frac{3x}{4}}{5 - \frac{1}{8}}$
     10. $\dfrac{\frac{3}{10} + 2}{\frac{2}{5y}}$

Objective B *Use the order of operations to simplify each expression. See Examples 5 and 6.*

11. $\dfrac{1}{5} + \dfrac{1}{3} \cdot \dfrac{1}{4}$

12. $\dfrac{1}{2} + \dfrac{1}{6} \cdot \dfrac{1}{3}$

13. $\dfrac{5}{6} \div \dfrac{1}{3} \cdot \dfrac{1}{4}$

14. $\dfrac{7}{8} \div \dfrac{1}{4} \cdot \dfrac{1}{7}$

15. $2^2 - \left(\dfrac{1}{3}\right)^2$

16. $3^2 - \left(\dfrac{1}{2}\right)^2$

▶ 17. $\left(\dfrac{2}{9} + \dfrac{4}{9}\right)\left(\dfrac{1}{3} - \dfrac{9}{10}\right)$

18. $\left(\dfrac{1}{5} - \dfrac{1}{10}\right)\left(\dfrac{1}{5} + \dfrac{1}{10}\right)$

19. $\left(\dfrac{7}{8} - \dfrac{1}{2}\right) \div \dfrac{3}{11}$

20. $\left(-\dfrac{2}{3} - \dfrac{7}{3}\right) \div \dfrac{4}{9}$

21. $2 \cdot \left(\dfrac{1}{4} + \dfrac{1}{5}\right) + 2$

22. $\dfrac{2}{5} \cdot \left(5 - \dfrac{1}{2}\right) - 1$

23. $\left(\dfrac{3}{4}\right)^2 \div \left(\dfrac{3}{4} - \dfrac{1}{12}\right)$

24. $\left(\dfrac{8}{9}\right)^2 \div \left(2 - \dfrac{2}{3}\right)$

25. $\left(\dfrac{2}{5} - \dfrac{3}{10}\right)^2$

26. $\left(\dfrac{3}{2} - \dfrac{4}{3}\right)^3$

27. $\left(\dfrac{3}{4} + \dfrac{1}{8}\right)^2 - \left(\dfrac{1}{2} + \dfrac{1}{8}\right)$

28. $\left(\dfrac{1}{6} + \dfrac{1}{3}\right)^3 + \left(\dfrac{2}{5} \cdot \dfrac{3}{4}\right)^2$

Objective C *Evaluate each expression if $x = -\dfrac{1}{3}$, $y = \dfrac{2}{5}$, and $z = \dfrac{5}{6}$. See Example 7.*

29. $5y - z$

30. $2z - x$

31. $\dfrac{x}{z}$

32. $\dfrac{y + x}{z}$

▶ 33. $x^2 - yz$

34. $x^2 - z^2$

35. $(1 + x)(1 + z)$

36. $(1 - x)(1 - z)$

Objectives A B Mixed Practice *Simplify the following. See Examples 1 through 6.*

37. $\dfrac{\dfrac{5a}{24}}{\dfrac{1}{12}}$

38. $\dfrac{\dfrac{7}{10}}{\dfrac{14z}{25}}$

39. $\left(\dfrac{3}{2}\right)^3 + \left(\dfrac{1}{2}\right)^3$

40. $\left(\dfrac{5}{21} \div \dfrac{1}{2}\right) + \left(\dfrac{1}{7} \cdot \dfrac{1}{3}\right)$

41. $\left(-\dfrac{1}{2}\right)^2 + \dfrac{1}{5}$

42. $\left(-\dfrac{3}{4}\right)^2 + \dfrac{3}{8}$

43. $\dfrac{2 + \dfrac{1}{6}}{1 - \dfrac{4}{3}}$

44. $\dfrac{3 - \dfrac{1}{2}}{4 + \dfrac{1}{5}}$

45. $\left(1 - \dfrac{2}{5}\right)^2$

46. $\left(-\dfrac{1}{2}\right)^2 - \left(\dfrac{3}{4}\right)^2$

47. $\left(\dfrac{3}{4} - 1\right)\left(\dfrac{1}{8} + \dfrac{1}{2}\right)$

48. $\left(\dfrac{1}{10} + \dfrac{3}{20}\right)\left(\dfrac{1}{5} - 1\right)$

49. $\left(-\dfrac{2}{9} - \dfrac{7}{9}\right)^4$

50. $\left(\dfrac{5}{9} - \dfrac{2}{3}\right)^2$

51. $\dfrac{\dfrac{1}{3} - \dfrac{5}{6}}{\dfrac{3}{4} + \dfrac{1}{2}}$

52. $\dfrac{\dfrac{7}{10} + \dfrac{1}{2}}{\dfrac{4}{5} + \dfrac{3}{4}}$

53. $\left(\dfrac{3}{4} \div \dfrac{6}{5}\right) - \left(\dfrac{3}{4} \cdot \dfrac{6}{5}\right)$

54. $\left(\dfrac{1}{2} \cdot \dfrac{2}{7}\right) - \left(\dfrac{1}{2} \div \dfrac{2}{7}\right)$

55. $\dfrac{\dfrac{x}{3} + 2}{5 + \dfrac{1}{3}}$

56. $\dfrac{1 - \dfrac{x}{4}}{2 + \dfrac{3}{8}}$

Review

Perform each indicated operation. If the result is an improper fraction, also write the improper fraction as a mixed number. See Sections 4.1 and 4.5.

57. $3 + \dfrac{1}{2}$

58. $2 + \dfrac{2}{3}$

59. $9 - \dfrac{5}{6}$

60. $4 - \dfrac{1}{5}$

Concept Extensions

61. Calculate $\dfrac{2^3}{3}$ and $\left(\dfrac{2}{3}\right)^3$. Do both of these expressions simplify to the same number? Explain why or why not.

62. Calculate $\left(\dfrac{1}{2}\right)^2 \cdot \left(\dfrac{3}{4}\right)^2$ and $\left(\dfrac{1}{2} \cdot \dfrac{3}{4}\right)^2$. Do both of these expressions simplify to the same number? Explain why or why not.

Recall that to find the average of two numbers, we find their sum and divide by 2. For example, the average of $\dfrac{1}{2}$ and $\dfrac{3}{4}$ is $\dfrac{\dfrac{1}{2} + \dfrac{3}{4}}{2}$. Find the average of each pair of numbers.

63. $\dfrac{1}{2}, \dfrac{3}{4}$

64. $\dfrac{3}{5}, \dfrac{9}{10}$

65. $\dfrac{1}{4}, \dfrac{2}{14}$

66. $\dfrac{5}{6}, \dfrac{7}{9}$

67. Two positive numbers, a and b, are graphed below. Where should the graph of their average lie?

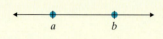

a b

68. Study Exercise 67. Without calculating, can $\frac{1}{3}$ be the average of $\frac{1}{2}$ and $\frac{8}{9}$? Explain why or why not.

Answer true or false for each statement.

69. It is possible for the average of two numbers to be greater than both numbers.

70. It is possible for the average of two numbers to be less than both numbers.

71. The sum of two negative fractions is always a negative number.

72. The sum of a negative fraction and a positive fraction is always a positive number.

73. It is possible for the sum of two fractions to be a whole number.

74. It is possible for the difference of two fractions to be a whole number.

75. What operation should be performed first to simplify

$$\frac{1}{5} \cdot \frac{5}{2} - \left(\frac{2}{3} + \frac{4}{5}\right)^2 ?$$

Explain your answer.

76. A student is to evaluate $x - y$ when $x = \frac{1}{5}$ and $y = -\frac{1}{7}$. This student is asking you if he should evaluate $\frac{1}{5} - \frac{1}{7}$. What do you tell this student and why?

Each expression contains one addition, one subtraction, one multiplication, and one division. Write the operations in the order that they should be performed. Do not actually simplify. See the Concept Check in this section.

77. $[9 + 3(4 - 2)] \div \dfrac{10}{21}$

78. $[30 - 4(3 + 2)] \div \dfrac{5}{2}$

79. $\dfrac{1}{3} \div \left(\dfrac{2}{3}\right)\left(\dfrac{4}{5}\right) - \dfrac{1}{4} + \dfrac{1}{2}$

80. $\left(\dfrac{5}{6} - \dfrac{1}{3}\right) \cdot \dfrac{1}{3} + \dfrac{1}{2} \div \dfrac{9}{8}$

Evaluate each expression if $x = \dfrac{3}{4}$ and $y = -\dfrac{4}{7}$.

81. $\dfrac{2 + x}{y}$

82. $4x + y$

83. $x^2 + 7y$

84. $\dfrac{\frac{9}{14}}{x + y}$

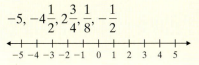

Objective A Graphing Fractions and Mixed Numbers

Let's review graphing fractions and practice graphing mixed numbers on a number line. This will help us visualize rounding and estimating operations with mixed numbers.

> Recall that $5\frac{2}{3}$ means $5 + \frac{2}{3}$ and
>
> $-4\frac{1}{6}$ means $-\left(4 + \frac{1}{6}\right)$ or $-4 - \frac{1}{6}$ or $-4 + \left(-\frac{1}{6}\right)$

Example 1 Graph the numbers on a number line:

$$\frac{1}{2}, -\frac{3}{4}, 2\frac{2}{3}, -3, -3\frac{1}{8}$$

Solution: Remember that $2\frac{2}{3}$ means $2 + \frac{2}{3}$.

Also, $-3\frac{1}{8}$ means $-3 - \frac{1}{8}$, so $-3\frac{1}{8}$ lies to the left of -3.

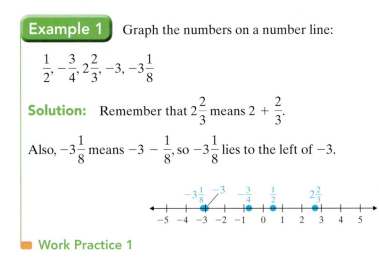

Work Practice 1

✓**Concept Check** Which of the following is/are equivalent to 9?

a. $7\frac{6}{3}$ **b.** $8\frac{4}{4}$ **c.** $8\frac{9}{9}$ **d.** $\frac{18}{2}$ **e.** all of these

Objective B Multiplying or Dividing with Mixed Numbers or Whole Numbers

When multiplying or dividing a fraction and a mixed or a whole number, remember that mixed and whole numbers can be written as improper fractions.

> **Multiplying or Dividing Fractions and Mixed Numbers or Whole Numbers**
>
> To multiply or divide with mixed numbers or whole numbers, first write any mixed or whole numbers as improper fractions and then multiply or divide as usual.

(*Note:* If an exercise contains a mixed number, we will write the answer as a mixed number, if possible.)

Objectives

A Graph Positive and Negative Fractions and Mixed Numbers

B Multiply or Divide Mixed or Whole Numbers.

C Add or Subtract Mixed Numbers.

D Solve Problems Containing Mixed Numbers.

E Perform Operations on Negative Mixed Numbers.

Practice 1

Graph the numbers on a number line.

$$-5, -4\frac{1}{2}, 2\frac{3}{4}, \frac{1}{8}, -\frac{1}{2}$$

Answer

1.

✓**Concept Check Answer**
e

Practice 2

Multiply and simplify: $1\dfrac{2}{3} \cdot \dfrac{11}{15}$

Example 2   Multiply: $3\dfrac{1}{3} \cdot \dfrac{7}{8}$

Solution: Recall from Section 4.1 that the mixed number $3\dfrac{1}{3}$ can be written as the fraction $\dfrac{10}{3}$. Then

$$3\frac{1}{3} \cdot \frac{7}{8} = \frac{10}{3} \cdot \frac{7}{8} = \frac{\overset{1}{\cancel{2}} \cdot 5 \cdot 7}{3 \cdot \underset{1}{\cancel{2}} \cdot 4} = \frac{35}{12} \quad \text{or} \quad 2\frac{11}{12}$$

■ **Work Practice 2**

Don't forget that a whole number can be written as a fraction by writing the whole number over 1. For example,

$$20 = \frac{20}{1} \quad \text{and} \quad 7 = \frac{7}{1}$$

Practice 3

Multiply: $\dfrac{5}{6} \cdot 18$

Example 3   Multiply: $\dfrac{3}{4} \cdot 20$

Solution: $\dfrac{3}{4} \cdot 20 = \dfrac{3}{4} \cdot \dfrac{20}{1} = \dfrac{3 \cdot 20}{4 \cdot 1} = \dfrac{3 \cdot \overset{1}{\cancel{4}} \cdot 5}{\underset{1}{\cancel{4}} \cdot 1} = \dfrac{15}{1} \quad \text{or} \quad 15$

■ **Work Practice 3**

When both numbers to be multiplied are mixed or whole numbers, it is a good idea to estimate the product to see if your answer is reasonable. To do this, we first practice rounding mixed numbers to the nearest whole. If the fraction part of the mixed number is $\dfrac{1}{2}$ or greater, we round the whole number part up. If the fraction part of the mixed number is less than $\dfrac{1}{2}$, then we do not round the whole number part up. Study the table below for examples.

Mixed Number	Rounding
$5\dfrac{1}{4}$ $\dfrac{1}{4}$ is less than $\dfrac{1}{2}$ $\dfrac{1}{4}$ $\dfrac{1}{2}$	Thus, $5\dfrac{1}{4}$ rounds to 5.
$3\dfrac{9}{16}$ ← 9 is greater than 8. → Half of 16 is 8.	Thus, $3\dfrac{9}{16}$ rounds to 4.
$1\dfrac{3}{7}$ ← 3 is less than $3\frac{1}{2}$. → Half of 7 is $3\frac{1}{2}$.	Thus, $1\dfrac{3}{7}$ rounds to 1.

Practice 4

Multiply. Check by estimating.

$3\dfrac{1}{5} \cdot 2\dfrac{3}{4}$

Example 4   Multiply $1\dfrac{2}{3} \cdot 2\dfrac{1}{4}$. Check by estimating.

Solution: $1\dfrac{2}{3} \cdot 2\dfrac{1}{4} = \dfrac{5}{3} \cdot \dfrac{9}{4} = \dfrac{5 \cdot 9}{3 \cdot 4} = \dfrac{5 \cdot \overset{1}{\cancel{3}} \cdot 3}{\cancel{3} \cdot 4} = \dfrac{15}{4} \text{ or } 3\dfrac{3}{4}$ Exact

Let's check by estimating.

$1\dfrac{2}{3}$ rounds to 2, $2\dfrac{1}{4}$ rounds to 2, and $2 \cdot 2 = 4$ Estimate

The estimate is close to the exact value, so our answer is reasonable.

Answers

2. $1\dfrac{2}{9}$ **3.** 15   **4.** $8\dfrac{4}{5}$

■ **Work Practice 4**

Example 5 Multiply: $7 \cdot 2\frac{11}{14}$. Check by estimating.

Solution: $7 \cdot 2\frac{11}{14} = \frac{7}{1} \cdot \frac{39}{14} = \frac{7 \cdot 39}{1 \cdot 14} = \frac{\overset{1}{7} \cdot 39}{1 \cdot 2 \cdot \underset{1}{7}} = \frac{39}{2}$ or $19\frac{1}{2}$ Exact

To estimate,

$2\frac{11}{14}$ rounds to 3 and $7 \cdot 3 = 21$.   Estimate

The estimate is close to the exact value, so our answer is reasonable.

▪ **Work Practice 5**

✓**Concept Check** Find the error.

$$2\frac{1}{4} \cdot \frac{1}{2} = 2\frac{1 \cdot 1}{4 \cdot 2} = 2\frac{1}{8}$$

Examples Divide.

6. $\frac{3}{4} \div 5 = \frac{3}{4} \div \frac{5}{1} = \frac{3}{4} \cdot \frac{1}{5} = \frac{3 \cdot 1}{4 \cdot 5} = \frac{3}{20}$

7. $\frac{11}{18} \div 2\frac{5}{6} = \frac{11}{18} \div \frac{17}{6} = \frac{11}{18} \cdot \frac{6}{17} = \frac{11 \cdot 6}{18 \cdot 17} = \frac{11 \cdot \overset{1}{6}}{\underset{1}{6} \cdot 3 \cdot 17} = \frac{11}{51}$

8. $5\frac{2}{3} \div 2\frac{5}{9} = \frac{17}{3} \div \frac{23}{9} = \frac{17}{3} \cdot \frac{9}{23} = \frac{17 \cdot 9}{3 \cdot 23} = \frac{17 \cdot \overset{1}{\cancel{3}} \cdot 3}{\underset{1}{\cancel{3}} \cdot 23} = \frac{51}{23}$ or $2\frac{5}{23}$

▪ **Work Practice 6–8**

Objective C Adding or Subtracting Mixed Numbers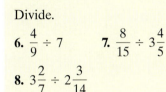

We can add or subtract mixed numbers, too, by first writing each mixed number as an improper fraction. But it is often easier to add or subtract the whole-number parts and add or subtract the proper-fraction parts vertically.

Adding or Subtracting Mixed Numbers

To add or subtract mixed numbers, add or subtract the fraction parts and then add or subtract the whole number parts.

Example 9 Add: $2\frac{1}{3} + 5\frac{3}{8}$. Check by estimating.

Solution: The LCD of the denominators 3 and 8 is 24.

$$
\begin{array}{rcl}
2\frac{1 \cdot 8}{3 \cdot 8} &=& 2\frac{8}{24} \\
+ 5\frac{3 \cdot 3}{8 \cdot 3} &=& + 5\frac{9}{24} \\
\hline
&=& 7\frac{17}{24}
\end{array}
$$
← Add the fractions.
← Add the whole numbers.

To check by estimating, we round as usual. The fraction $2\frac{1}{3}$ rounds to 2, $5\frac{3}{8}$ rounds to 5, and $2 + 5 = 7$, our estimate.

Our exact answer is close to 7, so our answer is reasonable.

▪ **Work Practice 9**

Helpful Hint

When adding or subtracting mixed numbers and whole numbers, it is a good idea to estimate to see if your answer is reasonable.

For the rest of this section, we leave most of the checking by estimating to you.

Practice 10

Add: $3\dfrac{5}{14} + 2\dfrac{6}{7}$

Example 10   Add: $3\dfrac{4}{5} + 1\dfrac{4}{15}$

Solution: The LCD of the denominators 5 and 15 is 15.

$$
\begin{array}{rcl}
3\dfrac{4}{5} & = & 3\dfrac{12}{15} \\[2mm]
+1\dfrac{4}{15} & = & +1\dfrac{4}{15} \\[2mm]
\hline
& & 4\dfrac{16}{15}
\end{array}
$$

Add the fractions; then add the whole numbers.

Notice that the fraction part is improper.

Since $\dfrac{16}{15}$ is $1\dfrac{1}{15}$ we can write the sum as

$$4\dfrac{16}{15} + 4 + 1\dfrac{1}{15} = 5\dfrac{1}{15}$$

■ **Work Practice 10**

✔**Concept Check** Explain how you could estimate the following sum:

$$5\dfrac{1}{9} + 14\dfrac{10}{11}.$$

Practice 11

Add: $12 + 3\dfrac{6}{7} + 2\dfrac{1}{5}$

Example 11   Add: $2\dfrac{4}{5} + 5 + 1\dfrac{1}{2}$

Solution: The LCD of the denominators 5 and 2 is 10.

$$
\begin{array}{rcl}
2\dfrac{4}{5} & = & 2\dfrac{8}{10} \\[2mm]
5 & = & 5 \\[2mm]
+1\dfrac{1}{2} & = & +1\dfrac{5}{10} \\[2mm]
\hline
& & 8\dfrac{13}{10} = 8 + 1\dfrac{3}{10} = 9\dfrac{3}{10}
\end{array}
$$

■ **Work Practice 11**

Practice 12

Subtract: $32\dfrac{7}{9} - 16\dfrac{5}{18}$

Example 12   Subtract: $8\dfrac{3}{7} - 5\dfrac{2}{21}$. Check by estimating.

Solution: The LCD of the denominators 7 and 21 is 21.

$$
\begin{array}{rcl}
8\dfrac{3}{7} & = & 8\dfrac{9}{21} \quad \leftarrow \text{The LCD of 7 and 21 is 21.} \\[2mm]
-5\dfrac{2}{21} & = & -5\dfrac{2}{21} \\[2mm]
\hline
& & 3\dfrac{7}{21} \quad \leftarrow \text{Subtract the fractions.}
\end{array}
$$

↑ Subtract the whole numbers.

Answers

10. $6\dfrac{3}{14}$ **11.** $18\dfrac{2}{35}$ **12.** $16\dfrac{1}{2}$

✔**Concept Check Answer**

Round each mixed number to the nearest whole number and add. $5\dfrac{1}{9}$ rounds to 5 and $14\dfrac{10}{11}$ rounds to 15, and the estimated sum is $5 + 15 = 20$.

Then $3\frac{7}{21}$ simplifies to $3\frac{1}{3}$. The difference is $3\frac{1}{3}$.

To check, $8\frac{3}{7}$ rounds to 8, $5\frac{2}{21}$ rounds to 5, and $8 - 5 = 3$, our estimate.

Our exact answer is close to 3, so our answer is reasonable.

■ **Work Practice 12**

When subtracting mixed numbers, borrowing may be needed, as shown in the next example.

Example 13   Subtract: $7\frac{3}{14} - 3\frac{6}{7}$

Solution: The LCD of the denominators 7 and 14 is 14.

$$7\frac{3}{14} = \quad 7\frac{3}{14} \qquad \text{Notice that we cannot subtract } \frac{12}{14} \text{ from } \frac{3}{14}, \text{ so we borrow}$$
$$-3\frac{6}{7} = -3\frac{12}{14} \qquad \text{from the whole number 7.}$$

borrow 1 from 7

$$7\frac{3}{14} = 6 + 1\frac{3}{14} = 6 + \frac{17}{14} \text{ or } 6\frac{17}{14}$$

Now subtract.

$$7\frac{3}{14} = \quad 7\frac{3}{14} = \quad 6\frac{17}{14}$$
$$-3\frac{6}{7} = -3\frac{12}{14} = -3\frac{12}{14}$$
$$\overline{\qquad\qquad\qquad\qquad\qquad\quad 3\frac{5}{14}} \leftarrow \text{Subtract the fractions.}$$
$$\uparrow$$
$$\text{Subtract the whole numbers.}$$

■ **Work Practice 13**

✓**Concept Check** In the subtraction problem $5\frac{1}{4} - 3\frac{3}{4}$, $5\frac{1}{4}$ must be rewritten because $\frac{3}{4}$ cannot be subtracted from $\frac{1}{4}$. Why is it incorrect to rewrite $5\frac{1}{4}$ as $5\frac{5}{4}$?

Example 14   Subtract: $14 - 8\frac{3}{7}$

Solution: $14 = \quad 13\frac{7}{7} \qquad$ Borrow 1 from 14 and write it as $\frac{7}{7}$.

$$-8\frac{3}{7} = -8\frac{3}{7}$$
$$\overline{\qquad\qquad\qquad\quad 5\frac{4}{7}} \leftarrow \text{Subtract the fractions.}$$
$$\uparrow$$
$$\text{Subtract the whole numbers.}$$

■ **Work Practice 14**

Practice 13

Subtract: $9\frac{7}{15} - 4\frac{3}{5}$

Practice 14

Subtract: $25 - 10\frac{2}{9}$

Answers

13. $4\frac{13}{15}$ **14.** $14\frac{7}{9}$

✓**Concept Check Answer**

Rewrite $5\frac{1}{4}$ as $4\frac{5}{4}$ by borrowing from the 5.

Objective D Solving Problems Containing Mixed Numbers

Now that we know how to perform operations on mixed numbers, we can solve real-life problems.

Practice 15

The measurement around the trunk of a tree just below shoulder height is called its girth. The largest known American beech tree in the United States has a girth of $23\frac{1}{4}$ feet. The largest known sugar maple tree in the United States has a girth of $19\frac{5}{12}$ feet. How much larger is the girth of the largest known American beech tree than the girth of the largest known sugar maple tree? (*Source: American Forests*)

Girth

Example 15   Finding Legal Lobster Size

Lobster fishermen must measure the upper body shells of the lobsters they catch. Lobsters that are too small are thrown back into the ocean. Each state has its own size standard for lobsters to help control the breeding stock. Massachusetts divides its waters into four Lobster Conservation Management Areas, with a different minimum lobster size permitted in each area. In area three, the legal lobster size increased from $3\frac{13}{32}$ inches to $3\frac{1}{2}$ inches. How much of an increase was this? (*Source: Massachusetts Division of Marine Fisheries*)

Solution:

1. UNDERSTAND. Read and reread the problem carefully. The word "increase" found in the problem might make you think that we add to solve the problem. But the phrase "how much of an increase" tells us to subtract to find the increase.

2. TRANSLATE.

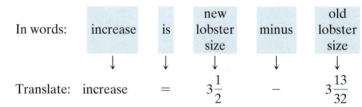

In words:	increase	is	new lobster size	minus	old lobster size
	↓	↓	↓	↓	↓
Translate:	increase	=	$3\frac{1}{2}$	−	$3\frac{13}{32}$

3. SOLVE. Before we solve, let's estimate by rounding to the nearest wholes. The fraction $3\frac{1}{2}$ can be rounded up to 4, $3\frac{13}{32}$ rounds to 3, and $4 - 3 = 1$. The increase is not 1, but will be smaller since we rounded $3\frac{1}{2}$ up and rounded $3\frac{13}{32}$ down.

$$3\frac{1}{2} = 3\frac{16}{32}$$
$$-3\frac{13}{32} = -3\frac{13}{32}$$
$$\frac{3}{32}$$

4. INTERPRET. *Check* your work. Our estimate tells us that the exact increase of $\frac{3}{32}$ is reasonable. *State* your conclusion: The increase in lobster size was $\frac{3}{32}$ of an inch.

Answer

15. $3\frac{5}{6}$ ft

■ Work Practice 15

Example 16 Calculating Manufacturing Materials Needed

In a manufacturing process, a metal-cutting machine cuts strips $1\frac{3}{5}$ inches long from a piece of metal stock. How many such strips can be cut from a 48-inch piece of stock?

Solution:

1. UNDERSTAND the problem. To do so, read and reread the problem. Then draw a diagram:

   We want to know how many $1\frac{3}{5}$s there are in 48.

2. TRANSLATE.

In words:	number of strips	is	48	divided by	$1\frac{3}{5}$
	↓	↓	↓	↓	↓
Translate:	number of strips	=	48	÷	$1\frac{3}{5}$

3. SOLVE: Let's estimate a reasonable answer. The mixed number $1\frac{3}{5}$ rounds to 2 and $48 \div 2 = 24$.

$$48 \div 1\frac{3}{5} = 48 \div \frac{8}{5} = \frac{48}{1} \cdot \frac{5}{8} = \frac{48 \cdot 5}{1 \cdot 8} = \frac{\overset{1}{\cancel{8}} \cdot 6 \cdot 5}{1 \cdot \underset{1}{\cancel{8}}} = \frac{30}{1} \quad \text{or} \quad 30$$

4. INTERPRET. *Check* your work. Since the exact answer of 30 is close to our estimate of 24, our answer is reasonable. *State* your conclusion: Thirty strips can be cut from the 48-inch piece of stock.

■ **Work Practice 16**

Objective E Operating on Negative Mixed Numbers

To perform operations on negative mixed numbers, let's first practice writing these numbers as negative fractions and negative fractions as negative mixed numbers.

To understand negative mixed numbers, we simply need to know that, for example,

$$-3\frac{2}{5} \text{ means } -\left(3\frac{2}{5}\right)$$

Thus, to write a negative mixed number as a fraction, we do the following.

$$-3\frac{2}{5} = -\left(3\frac{2}{5}\right) = -\left(\frac{5 \cdot 3 + 2}{5}\right) = -\left(\frac{17}{5}\right) \quad \text{or} \quad -\frac{17}{5}$$

Examples Write each as a fraction.

17. $-1\frac{7}{8} = -\frac{8 \cdot 1 + 7}{8} = -\frac{15}{8}$ Write $1\frac{7}{8}$ as an improper fraction and keep the negative sign.

18. $-23\frac{1}{2} = -\frac{2 \cdot 23 + 1}{2} = -\frac{47}{2}$ Write $23\frac{1}{2}$ as an improper fraction and keep the negative sign.

■ **Work Practice 17–18**

Practice 16

A designer of women's clothing designs a woman's dress that requires $3\frac{1}{7}$ yards of material. How many dresses can be made from a 44-yard bolt of material?

Practice 17–18

Write each as a fraction.

17. $-9\frac{3}{7}$ 18. $-5\frac{10}{11}$

Answers

16. 14 dresses   17. $-\frac{66}{7}$

18. $-\frac{65}{11}$

To write a negative fraction as a negative mixed number, we use a similar procedure. We simply disregard the negative sign, convert the improper fraction to a mixed number, and then reinsert the negative sign.

Practice 19–20

Write each as a mixed number.

19. $-\dfrac{37}{8}$ **20.** $-\dfrac{46}{5}$

Examples Write each as a mixed number.

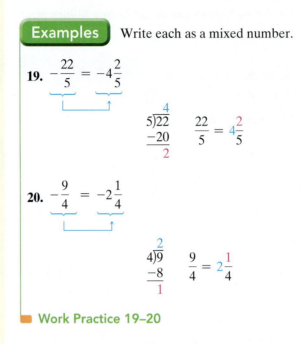

19. $-\dfrac{22}{5} = -4\dfrac{2}{5}$

$$5\overline{)22} \qquad \dfrac{22}{5} = 4\dfrac{2}{5}$$
$$\dfrac{-20}{2}$$

20. $-\dfrac{9}{4} = -2\dfrac{1}{4}$

$$4\overline{)9} \qquad \dfrac{9}{4} = 2\dfrac{1}{4}$$
$$\dfrac{-8}{1}$$

■ **Work Practice 19–20**

We multiply or divide with negative mixed numbers the same way that we multiply or divide with positive mixed numbers. We first write each mixed number as a fraction.

Practice 21–22

21. $2\dfrac{3}{4} \cdot \left(-3\dfrac{3}{5}\right)$

22. $-4\dfrac{2}{7} \div 1\dfrac{1}{4}$

Examples Perform the indicated operations.

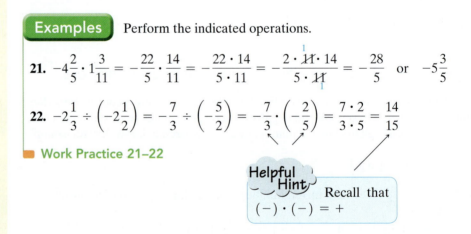

21. $-4\dfrac{2}{5} \cdot 1\dfrac{3}{11} = -\dfrac{22}{5} \cdot \dfrac{14}{11} = -\dfrac{22 \cdot 14}{5 \cdot 11} = -\dfrac{2 \cdot \cancel{11} \cdot 14}{5 \cdot \cancel{11}} = -\dfrac{28}{5}$ or $-5\dfrac{3}{5}$

22. $-2\dfrac{1}{3} \div \left(-2\dfrac{1}{2}\right) = -\dfrac{7}{3} \div \left(-\dfrac{5}{2}\right) = -\dfrac{7}{3} \cdot \left(-\dfrac{2}{5}\right) = \dfrac{7 \cdot 2}{3 \cdot 5} = \dfrac{14}{15}$

■ **Work Practice 21–22**

Helpful Hint Recall that $(-) \cdot (-) = +$

To add or subtract with negative mixed numbers, we must be very careful! Problems arise because recall that

$$-3\dfrac{2}{5} \text{ means } -\left(3\dfrac{2}{5}\right)$$

This means that

$$-3\dfrac{2}{5} = -\left(3\dfrac{2}{5}\right) = -\left(3 + \dfrac{2}{5}\right) = -3 - \dfrac{2}{5}$$ This can sometimes be easily overlooked.

To avoid problems, we will add or subtract negative mixed numbers by rewriting as addition and recalling how to add signed numbers.

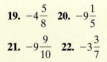

Answers

19. $-4\dfrac{5}{8}$ **20.** $-9\dfrac{1}{5}$

21. $-9\dfrac{9}{10}$ **22.** $-3\dfrac{3}{7}$

Example 23   Add: $6\dfrac{3}{5} + \left(-9\dfrac{7}{10}\right)$

Solution: Here we are adding two numbers with different signs. Recall that we then subtract the absolute values and keep the sign of the larger absolute value.

Since $-9\dfrac{7}{10}$ has the larger absolute value, the answer is negative.

First, subtract absolute values:

$$
\begin{array}{r}
9\dfrac{7}{10} = 9\dfrac{7}{10} \\[2mm]
-\,6\dfrac{3\cdot 2}{5\cdot 2} = -6\dfrac{6}{10} \\[2mm]
\hline
3\dfrac{1}{10}
\end{array}
$$

Thus,

$$6\dfrac{3}{5} + \left(-9\dfrac{7}{10}\right) = -3\dfrac{1}{10}$$ The result is negative since $-9\dfrac{7}{10}$ has the larger absolute value.

■ **Work Practice 23**

Example 24   Subtract: $-11\dfrac{5}{6} - 20\dfrac{4}{9}$

Solution: Let's write as an equivalent addition: $-11\dfrac{5}{6} + \left(-20\dfrac{4}{9}\right)$. Here, we are adding two numbers with like signs. Recall that we add their absolute values and keep the common negative sign.

First, add absolute values:

$$
\begin{array}{r}
11\dfrac{5\cdot 3}{6\cdot 3} = 11\dfrac{15}{18} \\[2mm]
+\,20\dfrac{4\cdot 2}{9\cdot 2} = +20\dfrac{8}{18} \\[2mm]
\hline
31\dfrac{23}{18} \text{ or } 32\dfrac{5}{18}
\end{array}
$$

Since $\dfrac{23}{18} = 1\dfrac{5}{18}$

Thus,

$$-11\dfrac{5}{6} - 20\dfrac{4}{9} = -32\dfrac{5}{18}$$

Keep the common sign.

■ **Work Practice 24**

Practice 23

Add: $6\dfrac{2}{3} + \left(-12\dfrac{3}{4}\right)$

Practice 24

Subtract: $-9\dfrac{2}{7} - 30\dfrac{11}{14}$

Answers

23. $-6\dfrac{1}{12}$ **24.** $-40\dfrac{1}{14}$

Calculator Explorations Converting Between Mixed Number and Fraction Notation

If your calculator has a fraction key, such as $\boxed{a^b/c}$, you can use it to convert between mixed number notation and fraction notation.

To write $13\frac{7}{16}$ as an improper fraction, press

$\boxed{13}$ $\boxed{a^b/c}$ $\boxed{7}$ $\boxed{a^b/c}$ $\boxed{16}$ $\boxed{\text{2nd}}$ $\boxed{d/c}$

The display will read

$\boxed{215\,|\,16}$

which represents $\frac{215}{16}$. Thus $13\frac{7}{16} = \frac{215}{16}$.

To convert $\frac{190}{13}$ to a mixed number, press

$\boxed{190}$ $\boxed{a^b/c}$ $\boxed{13}$ $\boxed{=}$

The display will read

$\boxed{14\_8/13}$

which represents $14\frac{8}{13}$. Thus $\frac{190}{13} = 14\frac{8}{13}$.

Write each mixed number as a fraction and each fraction as a mixed number.

1. $25\frac{5}{11}$ 2. $67\frac{14}{15}$ 3. $107\frac{31}{35}$

4. $186\frac{17}{21}$ 5. $\frac{365}{14}$ 6. $\frac{290}{13}$

7. $\frac{2769}{30}$ 8. $\frac{3941}{17}$

Vocabulary, Readiness & Video Check

Use the choices below to fill in each blank.

round   fraction   whole number
improper   mixed number

1. The number $5\frac{3}{4}$ is called a(n) _____.

2. For $5\frac{3}{4}$, the 5 is called the _____ part and $\frac{3}{4}$ is called the _____ part.

3. To estimate operations on mixed numbers, we _____ mixed numbers to the nearest whole number.

4. The mixed number $2\frac{5}{8}$ written as a(n) _____ fraction is $\frac{21}{8}$.

Martin-Gay Interactive Videos

See Video 4.7

Watch the section lecture video and answer the following questions.

Objective A 5. In Example 1, why is the unit distance between -4 and -3 on the number line split into 5 equal parts?

Objective B 6. Why do we need to know how to multiply fractions to solve Example 2?

Objective C 7. In Example 4, why is the first form of the answer not an appropriate form?

Objective D 8. Why do we need to know how to subtract fractions to solve Example 5?

Objective E 9. In Example 6, how is it determined whether the answer is positive or negative?

4.7 Exercise Set MyMathLab® ▶

Objective A *Graph each list of numbers on the given number line. See Example 1.*

1. $-2, -2\frac{2}{3}, 0, \frac{7}{8}, -\frac{1}{3}$

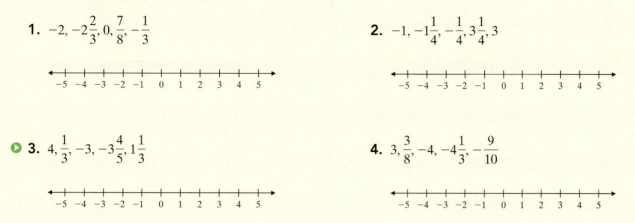

2. $-1, -1\frac{1}{4}, -\frac{1}{4}, 3\frac{1}{4}, 3$

▶ 3. $4, \frac{1}{3}, -3, -3\frac{4}{5}, 1\frac{1}{3}$

4. $3, \frac{3}{8}, -4, -4\frac{1}{3}, -\frac{9}{10}$

Objective B *Choose the best estimate for each product or quotient. See Examples 4 and 5.*

5. $2\frac{11}{12} \cdot 1\frac{1}{4}$

   a. 2   **b.** 3   **c.** 1   **d.** 12

6. $5\frac{1}{6} \cdot 3\frac{5}{7}$

   a. 9   **b.** 15   **c.** 8   **d.** 20

7. $12\frac{2}{11} \div 3\frac{9}{10}$

   a. 3   **b.** 4   **c.** 36   **d.** 9

8. $20\frac{3}{14} \div 4\frac{8}{11}$

   a. 5   **b.** 80   **c.** 4   **d.** 16

Multiply or divide. For Exercises 13 through 16, find an exact answer and an estimated answer. See Examples 2 through 8.

9. $2\frac{2}{3} \cdot \frac{1}{7}$

10. $\frac{5}{9} \cdot 4\frac{1}{5}$

▶ 11. $7 \div 1\frac{3}{5}$

12. $9 \div 1\frac{2}{3}$

13. $2\frac{1}{5} \cdot 3\frac{1}{2}$

Exact:

Estimate:

14. $2\frac{1}{4} \cdot 7\frac{1}{8}$

Exact:

Estimate:

15. $3\frac{4}{5} \cdot 6\frac{2}{7}$

Exact:

Estimate:

16. $5\frac{5}{6} \cdot 7\frac{3}{5}$

Exact:

Estimate:

17. $5 \cdot 2\frac{1}{2}$

18. $6 \cdot 3\frac{1}{3}$

▶ 19. $3\frac{2}{3} \cdot 1\frac{1}{2}$

20. $2\frac{4}{5} \cdot 2\frac{5}{8}$

21. $2\frac{2}{3} \div \frac{1}{7}$

22. $\frac{5}{9} \div 4\frac{1}{5}$

Objective C *Choose the best estimate for each sum or difference. See Examples 9 and 12.*

23. $3\frac{7}{8} + 2\frac{1}{5}$

    a. 6 **b.** 5 **c.** 1 **d.** 2

24. $3\frac{7}{8} - 2\frac{1}{5}$

    a. 6 **b.** 5 **c.** 1 **d.** 2

25. $8\frac{1}{3} + 1\frac{1}{2}$

    a. 4 **b.** 10 **c.** 6 **d.** 16

26. $8\frac{1}{3} - 1\frac{1}{2}$

    a. 4 **b.** 10 **c.** 6 **d.** 16

Add. For Exercises 27 through 30, find an exact sum and an estimated sum. See Examples 9 through 11.

27. $4\frac{7}{12}$
$+\,2\frac{1}{12}$

Exact:

Estimate:

28. $7\frac{4}{11}$
$+\,3\frac{2}{11}$

Exact:

Estimate:

29. $10\frac{3}{14}$
$+\,\;3\frac{4}{7}$

Exact:

Estimate:

30. $12\frac{5}{12}$
$+\,\;4\frac{1}{6}$

Exact:

Estimate:

31. $9\frac{1}{5}$
$+\,8\frac{2}{25}$

32. $6\frac{2}{13}$
$+\,8\frac{7}{26}$

33. $12\frac{3}{14}$
10
$+\,25\frac{5}{12}$

34. $8\frac{2}{9}$
32
$+\,9\frac{10}{21}$

▶ **35.** $15\frac{4}{7}$
$+\,9\frac{11}{14}$

36. $23\frac{3}{5}$
$+\,8\frac{8}{15}$

37. $3\frac{5}{8}$
$2\frac{1}{6}$
$+\,7\frac{3}{4}$

38. $4\frac{1}{3}$
$9\frac{2}{5}$
$+\,3\frac{1}{6}$

Subtract. For Exercises 39 through 42, find an exact difference and an estimated difference. See Examples 12 through 14.

39. $4\frac{7}{10}$
$-\,2\frac{1}{10}$

Exact:

Estimate:

40. $7\frac{4}{9}$
$-\,3\frac{2}{9}$

Exact:

Estimate:

41. $10\frac{13}{14}$
$-\,\;3\frac{4}{7}$

Exact:

Estimate:

42. $12\frac{5}{12}$
$-\,\;4\frac{1}{6}$

Exact:

Estimate:

43. $9\frac{1}{5}$
$-\ 8\frac{6}{25}$

44. $5\frac{2}{13}$
$-\ 4\frac{7}{26}$

45. 6
$-\ 2\frac{4}{9}$

46. 8
$-\ 1\frac{7}{10}$

47. $63\frac{1}{6}$
$-\ 47\frac{5}{12}$

48. $86\frac{2}{15}$
$-\ 27\frac{3}{10}$

Objectives **B** **C** **Mixed Practice** *Perform each indicated operation. See Examples 2 through 14.*

49. $2\frac{3}{4}$
$+\ 1\frac{1}{4}$

50. $5\frac{5}{8}$
$+\ 2\frac{3}{8}$

51. $15\frac{4}{7}$
$-\ 9\frac{11}{14}$

52. $23\frac{8}{15}$
$-\ 8\frac{3}{5}$

53. $3\frac{1}{9} \cdot 2$

54. $4\frac{1}{2} \cdot 3$

55. $1\frac{2}{3} \div 2\frac{1}{5}$

56. $5\frac{1}{5} \div 3\frac{1}{4}$

57. $22\frac{4}{9} + 13\frac{5}{18}$

58. $15\frac{3}{25} + 5\frac{2}{5}$

59. $5\frac{2}{3} - 3\frac{1}{6}$

60. $5\frac{3}{8} - 2\frac{3}{16}$

61. $15\frac{1}{5}$
$20\frac{3}{10}$
$+\ 37\frac{2}{15}$

62. $3\frac{7}{16}$
$6\frac{1}{2}$
$+\ 9\frac{3}{8}$

63. $6\frac{4}{7} - 5\frac{11}{14}$

64. $47\frac{5}{12} - 23\frac{19}{24}$

65. $4\frac{2}{7} \cdot 1\frac{3}{10}$

66. $6\frac{2}{3} \cdot 2\frac{3}{4}$

67. $6\frac{2}{11}$
3
$+\ 4\frac{10}{33}$

68. $7\frac{3}{7}$
15
$+\ 20\frac{1}{2}$

Objective D Translating *Translate each phrase into an algebraic expression. Use x to represent "a number." See Examples 15 and 16.*

69. $-5\frac{2}{7}$ decreased by a number

70. The sum of $8\frac{3}{4}$ and a number

71. Multiply $1\frac{9}{10}$ by a number.

72. Divide a number by $-6\frac{1}{11}$.

Solve. For Exercises 73 and 74, the solutions have been started for you. Write each answer in simplest form. See Examples 15 and 16.

73. A heart attack patient in rehabilitation walked on a treadmill $12\frac{3}{4}$ miles over 4 days. How many miles is this per day on average?

Start the solution:
1. UNDERSTAND the problem. Reread it as many times as needed.
2. TRANSLATE into an equation. (Fill in the blanks.)

miles per day	is	total miles	divided by	number of days
↓	↓	↓	↓	↓
miles per day	=	_____	÷	_____

Finish with:
3. SOLVE and 4. INTERPRET

74. A local restaurant is selling hamburgers from a booth on Memorial Day. A total of $27\frac{3}{4}$ pounds of hamburger have been ordered. How many quarter-pound hamburgers can this make?

Start the solution:
1. UNDERSTAND the problem. Reread it as many times as needed.
2. TRANSLATE into an equation. (Fill in the blanks.)

how many quarter-pound hamburgers	is	total pounds of hamburger	divided by	a quarter pound
↓	↓	↓	↓	↓
how many quarter-pound hamburgers	=	_____	÷	_____

Finish with:
3. SOLVE and 4. INTERPRET

75. The Gauge Act of 1846 set the standard gauge for U.S. railroads at $56\frac{1}{2}$ inches. (See figure.) If the standard gauge in Spain is $65\frac{9}{10}$ inches, how much wider is Spain's standard gauge than the U.S. standard gauge? (*Source:* San Diego Railroad Museum)

76. The standard railroad track gauge (see figure) in Spain is $65\frac{9}{10}$ inches, while in neighboring Portugal it is $65\frac{11}{20}$ inches. Which gauge is wider and by how much? (*Source:* San Diego Railroad Museum)

Track gauge (U.S. $56\frac{1}{2}$ inches)

$\frac{5}{8}$ inch

Point of measurement of gauge

77. If Tucson's average rainfall is $11\frac{1}{4}$ inches and Yuma's is $3\frac{3}{5}$ inches, how much more rain, on the average, does Tucson get than Yuma?

78. A pair of crutches needs adjustment. One crutch is 43 inches and the other is $41\frac{5}{8}$ inches. Find how much the shorter crutch should be lengthened to make both crutches the same length.

For Exercises 79 and 80, find the area of each figure.

79.

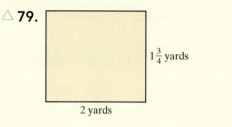

$1\frac{3}{4}$ yards

2 yards

80.

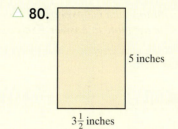

5 inches

$3\frac{1}{2}$ inches

81. A model for a proposed computer chip measures $\frac{3}{4}$ inch by $1\frac{1}{4}$ inches. Find its area.

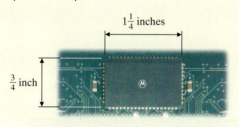

$1\frac{1}{4}$ inches

$\frac{3}{4}$ inch

82. The Saltalamachios are planning to build a deck that measures $4\frac{1}{2}$ yards by $6\frac{1}{3}$ yards. Find the area of their proposed deck.

$4\frac{1}{2}$ yards

$6\frac{1}{3}$ yards

For Exercises 83 and 84, find the perimeter of each figure.

83.

$5\frac{1}{3}$ meters

3 meters

5 meters

$7\frac{7}{8}$ meters

84.

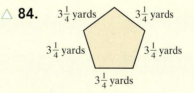

$3\frac{1}{4}$ yards $3\frac{1}{4}$ yards

$3\frac{1}{4}$ yards $3\frac{1}{4}$ yards

$3\frac{1}{4}$ yards

85. A homeowner has $15\frac{2}{3}$ feet of plastic pipe. She cuts off a $2\frac{1}{2}$-foot length and then a $3\frac{1}{4}$-foot length. If she now needs a 10-foot piece of pipe, will the remaining piece do? If not, by how much will the piece be short?

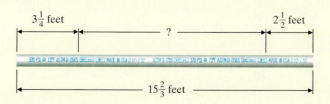

$3\frac{1}{4}$ feet $2\frac{1}{2}$ feet

?

$15\frac{2}{3}$ feet

86. A trim carpenter cuts a board $3\frac{3}{8}$ feet long from one 6 feet long. How long is the remaining piece?

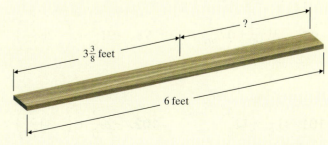

?

$3\frac{3}{8}$ feet

6 feet

△ **87.** The area of the rectangle below is 12 square meters. △ **88.** The perimeter of the square below is $23\frac{1}{2}$ feet. Find the length of each side.

If its width is $2\frac{4}{7}$ meters, find its length.

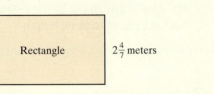

Rectangle $2\frac{4}{7}$ meters

Square

The following table lists three upcoming total eclipses of the Sun that will be visible in North America. The duration of each eclipse is listed in the table. Use the table to answer Exercises 89 through 92.

Total Solar Eclipses Visible from North America	
Date of Eclipse	**Duration (in minutes)**
August 21, 2017	$2\frac{2}{3}$
April 8, 2024	$4\frac{7}{15}$
March 30, 2033	$2\frac{37}{60}$

(*Source:* NASA/Goddard Space Flight Center)

89. What is the total duration for the three eclipses?

90. What is the total duration for the two eclipses occurring in odd-numbered years?

91. How much longer will the April 8, 2024, eclipse be than the August 21, 2017, eclipse?

92. How much longer will the April 8, 2024, eclipse be than the March 30, 2033, eclipse?

Objective E *Perform the indicated operations. See Examples 17 through 24.*

93. $-4\frac{2}{5} \cdot 2\frac{3}{10}$

94. $-3\frac{5}{6} \div \left(-3\frac{2}{3}\right)$

95. $-5\frac{1}{8} - 19\frac{3}{4}$

96. $17\frac{5}{9} + \left(-14\frac{2}{3}\right)$

▸ **97.** $-31\frac{2}{15} + 17\frac{3}{20}$

98. $-31\frac{7}{8} - \left(-26\frac{5}{12}\right)$

99. $-1\frac{5}{7} \cdot \left(-2\frac{1}{2}\right)$

100. $1\frac{3}{4} \div \left(-3\frac{1}{2}\right)$

101. $11\frac{7}{8} - 13\frac{5}{6}$

102. $-20\frac{2}{5} + \left(-30\frac{3}{10}\right)$

103. $-7\frac{3}{10} \div (-100)$

104. $-4\frac{1}{4} \div 2\frac{3}{8}$

Review

Multiply. See Section 4.3.

105. $\frac{1}{3}(3x)$

106. $\frac{1}{5}(5y)$

107. $\frac{2}{3}\left(\frac{3}{2}a\right)$

108. $-\frac{9}{10}\left(-\frac{10}{9}m\right)$

Concept Extensions

Solve. See the first Concept Check in this section.

109. Which of the following are equivalent to 10?

    a. $9\frac{5}{5}$ **b.** $9\frac{100}{100}$ **c.** $6\frac{44}{11}$ **d.** $8\frac{13}{13}$

110. Which of the following are equivalent to $7\frac{3}{4}$?

    a. $6\frac{7}{4}$ **b.** $5\frac{11}{4}$ **c.** $7\frac{12}{16}$ **d.** all of them

Solve. See the second Concept Check in this section.

111. A student asked you to check her work below. Is it correct? If not, where is the error?

$$20\frac{2}{3} \div 10\frac{1}{2} \stackrel{?}{=} 2\frac{1}{3}$$

112. A student asked you to check his work below. Is it correct? If not, where is the error?

$$3\frac{2}{3} \cdot 1\frac{1}{7} \stackrel{?}{=} 3\frac{2}{21}$$

113. In your own words, describe how to divide mixed numbers.

114. In your own words, explain how to multiply
    a. fractions
    b. mixed numbers

Solve. See the third Concept Check in this section.

115. In your own words, explain how to round a mixed number to the nearest whole number.

116. Use rounding to estimate the best sum for
$$11\frac{19}{20} + 9\frac{1}{10}.$$
    a. 2   **b.** 3   **c.** 20   **d.** 21

Solve.

117. Explain in your own words why $9\frac{13}{9}$ is equal to $10\frac{4}{9}$.

118. In your own words, explain
    a. when to borrow when subtracting mixed numbers, and
    b. how to borrow when subtracting mixed numbers.

4.8 **Solving Equations Containing Fractions**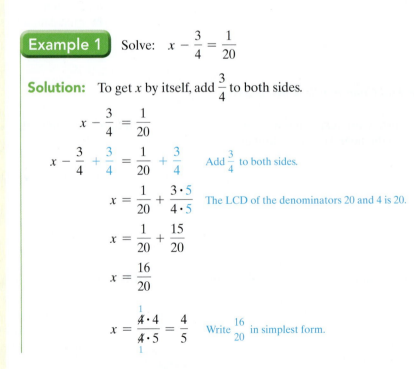

Objectives

A Solve Equations Containing Fractions.

B Solve Equations by Multiplying by the LCD.

C Review Adding and Subtracting Fractions.

Objective A Solving Equations Containing Fractions

In Chapter 3, we solved linear equations in one variable. In this section, we practice this skill by solving linear equations containing fractions. To help us solve these equations, let's review the properties of equality.

Addition Property of Equality

Let a, b, and c represent numbers. Then

$a = b$	Also, $\quad a = b$
and $\quad a + c = b + c$	and $\quad a - c = b - c$
are equivalent equations.	are equivalent equations.

Multiplication Property of Equality

Let a, b, and c represent numbers and let $c \neq 0$. Then

$a = b$	Also, $\quad a = b$
and $\quad a \cdot c = b \cdot c$	and $\quad \dfrac{a}{c} = \dfrac{b}{c}$
are equivalent equations.	are equivalent equations.

In other words, the same number may be added to or subtracted from both sides of an equation without changing the solution of the equation. Also, both sides of an equation may be multiplied or divided by the same nonzero number without changing the solution.

Also, don't forget that to solve an equation in x, our goal is to use properties of equality to write simpler equations, all equivalent to the original equation, until the final equation has the form

$$x = \textbf{number} \quad \text{or} \quad \textbf{number} = x$$

Practice 1

Solve: $y - \dfrac{2}{3} = \dfrac{5}{12}$

Example 1 Solve: $x - \dfrac{3}{4} = \dfrac{1}{20}$

Solution: To get x by itself, add $\dfrac{3}{4}$ to both sides.

$$x - \frac{3}{4} = \frac{1}{20}$$

$$x - \frac{3}{4} + \frac{3}{4} = \frac{1}{20} + \frac{3}{4} \qquad \text{Add } \frac{3}{4} \text{ to both sides.}$$

$$x = \frac{1}{20} + \frac{3 \cdot 5}{4 \cdot 5} \qquad \text{The LCD of the denominators 20 and 4 is 20.}$$

$$x = \frac{1}{20} + \frac{15}{20}$$

$$x = \frac{16}{20}$$

$$x = \frac{\overset{1}{\cancel{4}} \cdot 4}{\underset{1}{\cancel{4}} \cdot 5} = \frac{4}{5} \qquad \text{Write } \frac{16}{20} \text{ in simplest form.}$$

Answer

1. $\dfrac{13}{12}$

Check: To check, replace x with $\frac{4}{5}$ in the original equation.

$$x - \frac{3}{4} = \frac{1}{20}$$

$$\frac{4}{5} - \frac{3}{4} \stackrel{?}{=} \frac{1}{20} \qquad \text{Replace } x \text{ with } \frac{4}{5}.$$

$$\frac{4 \cdot 4}{5 \cdot 4} - \frac{3 \cdot 5}{4 \cdot 5} \stackrel{?}{=} \frac{1}{20} \qquad \text{The LCD of 5 and 4 is 20.}$$

$$\frac{16}{20} - \frac{15}{20} \stackrel{?}{=} \frac{1}{20}$$

$$\frac{1}{20} = \frac{1}{20} \qquad \text{True}$$

Thus $\frac{4}{5}$ is the solution of $x - \frac{3}{4} = \frac{1}{20}$.

■ **Work Practice 1**

Example 2 Solve: $\frac{1}{3}x = 7$

Solution: Recall that isolating x means that we want the coefficient of x to be 1. To do so, we use the multiplication property of equality and multiply both sides of the equation by the reciprocal of $\frac{1}{3}$, or 3. Since $\frac{1}{3} \cdot 3 = 1$, we will have isolated x.

$$\frac{1}{3}x = 7$$

$$3 \cdot \frac{1}{3}x = 3 \cdot 7 \qquad \text{Multiply both sides by 3.}$$

$$1 \cdot x = 21 \text{ or } x = 21 \qquad \text{Simplify.}$$

Check: To check, replace x with 21 in the original equation.

$$\frac{1}{3}x = 7 \qquad \text{Original equation}$$

$$\frac{1}{3} \cdot 21 \stackrel{?}{=} 7 \qquad \text{Replace } x \text{ with 21.}$$

$$7 = 7 \qquad \text{True}$$

Since $7 = 7$ is a true statement, 21 is the solution of $\frac{1}{3}x = 7$.

■ **Work Practice 2**

Example 3 Solve: $\frac{3}{5}a = 9$

Solution: Multiply both sides by $\frac{5}{3}$, the reciprocal of $\frac{3}{5}$, so that the coefficient of a is 1.

$$\frac{3}{5}a = 9$$

$$\frac{5}{3} \cdot \frac{3}{5}a = \frac{5}{3} \cdot 9 \qquad \text{Multiply both sides by } \frac{5}{3}.$$

$$1a = \frac{5 \cdot 9}{3} \qquad \text{Multiply.}$$

$$a = 15 \qquad \text{Simplify.}$$

(Continued on next page)

Practice 2

Solve: $\frac{1}{5}y = 2$

Practice 3

Solve: $\frac{5}{7}b = 25$

Answers

2. 10 $\qquad$ **3.** 35

4. $\frac{1}{4}$

5.

Check: To check, replace a with 15 in the original equation.

$$\frac{3}{5}a = 9$$

$$\frac{3}{5} \cdot 15 \stackrel{?}{=} 9 \qquad \text{Replace } a \text{ with 15.}$$

$$\frac{3 \cdot 15}{5} \stackrel{?}{=} 9 \qquad \text{Multiply.}$$

$$9 = 9 \qquad \text{True}$$

Since $9 = 9$ is true, 15 is the solution of $\frac{3}{5}a = 9$.

🟧 **Work Practice 3**

Practice 4

Solve: $-\dfrac{7}{10}x = \dfrac{2}{5}$

Example 4   Solve: $\dfrac{3}{4}x = -\dfrac{1}{8}$

Solution: Multiply both sides of the equation by $\frac{4}{3}$, the reciprocal of $\frac{3}{4}$.

$$\frac{3}{4}x = -\frac{1}{8}$$

$$\frac{4}{3} \cdot \frac{3}{4}x = \frac{4}{3} \cdot -\frac{1}{8} \qquad \text{Multiply both sides by } \frac{4}{3}.$$

$$1x = -\frac{4 \cdot 1}{3 \cdot 8} \qquad \text{Multiply.}$$

$$x = -\frac{1}{6} \qquad \text{Simplify.}$$

Check: To check, replace x with $-\dfrac{1}{6}$ in the original equation.

$$\frac{3}{4}x = -\frac{1}{8} \qquad \text{Original equation}$$

$$\frac{3}{4} \cdot -\frac{1}{6} \stackrel{?}{=} -\frac{1}{8} \qquad \text{Replace } x \text{ with } -\frac{1}{6}.$$

$$-\frac{1}{8} = -\frac{1}{8} \qquad \text{True}$$

Since we arrived at a true statement, $-\dfrac{1}{6}$ is the solution of $\dfrac{3}{4}x = -\dfrac{1}{8}$.

🟧 **Work Practice 4**

Practice 5

Solve: $5x = -\dfrac{3}{4}$

Example 5   Solve: $3y = -\dfrac{2}{11}$

Solution: We can either divide both sides by 3 or multiply both sides by the reciprocal of 3, which is $\dfrac{1}{3}$.

$$3y = -\frac{2}{11}$$

$$\frac{1}{3} \cdot 3y = \frac{1}{3} \cdot -\frac{2}{11} \qquad \text{Multiply both sides by } \frac{1}{3}.$$

$$1y = -\frac{1 \cdot 2}{3 \cdot 11} \qquad \text{Multiply.}$$

$$y = -\frac{2}{33} \qquad \text{Simplify.}$$

Check to see that the solution is $-\dfrac{2}{33}$.

🟧 **Work Practice 5**

Objective B Solving Equations by Multiplying by the LCD ▶

Solving equations with fractions can be tedious. If an equation contains fractions, it is often helpful to first multiply both sides of the equation by the LCD of the fractions. This has the effect of eliminating the fractions in the equation, as shown in the next example.

Let's solve the equation in Example 4 again. This time, we will multiply both sides by the LCD.

Example 6 Solve: $\dfrac{3}{4}x = -\dfrac{1}{8}$ (Example 4 solved an alternate way.)

Solution: First, multiply both sides of the equation by the LCD of the fractions $\dfrac{3}{4}$ and $-\dfrac{1}{8}$. The LCD of the denominators is 8.

$$\frac{3}{4}x = -\frac{1}{8}$$

$$8 \cdot \frac{3}{4}x = 8 \cdot -\frac{1}{8} \qquad \text{Multiply both sides by 8.}$$

$$\frac{8 \cdot 3}{1 \cdot 4}x = -\frac{\overset{1}{8} \cdot 1}{1 \cdot \underset{1}{8}} \qquad \text{Multiply the fractions.}$$

$$\frac{2 \cdot \overset{1}{\cancel{4}} \cdot 3}{1 \cdot \underset{1}{\cancel{4}}}x = -\frac{1 \cdot 1}{1 \cdot 1} \qquad \text{Simplify.}$$

$$6x = -1$$

$$\frac{6x}{6} = \frac{-1}{6} \qquad \text{Divide both sides by 6.}$$

$$x = \frac{-1}{6} \text{ or } -\frac{1}{6} \qquad \text{Simplify.}$$

As seen in Example 4, the solution is $-\dfrac{1}{6}$.

■ **Work Practice 6**

Practice 6

Solve: $\dfrac{11}{15}x = -\dfrac{3}{5}$

Example 7 Solve: $\dfrac{x}{6} + 1 = \dfrac{4}{3}$

Solution:

Solve by multiplying by the LCD:	Solve with fractions:
The LCD of the denominators 6 and 3 is 6.	$\dfrac{x}{6} + 1 = \dfrac{4}{3}$
$\dfrac{x}{6} + 1 = \dfrac{4}{3}$	$\dfrac{x}{6} + 1 - 1 = \dfrac{4}{3} - 1$ Subtract 1 from both sides.
$6\left(\dfrac{x}{6} + 1\right) = 6\left(\dfrac{4}{3}\right)$ Multiply both sides by 6.	$\dfrac{x}{6} = \dfrac{4}{3} - \dfrac{3}{3}$ Write 1 as $\dfrac{3}{3}$.
$\overset{1}{\cancel{6}}\left(\dfrac{x}{\underset{1}{\cancel{6}}}\right) + 6(1) = \overset{2}{\cancel{6}}\left(\dfrac{4}{\underset{1}{\cancel{3}}}\right)$ Apply the distributive property.	$\dfrac{x}{6} = \dfrac{1}{3}$ Subtract.
$x + 6 = 8$ Simplify.	$6 \cdot \dfrac{x}{6} = 6 \cdot \dfrac{1}{3}$ Multiply both sides by 6.
$x + 6 + (-6) = 8 + (-6)$ Add −6 to both sides.	$\overset{1}{\cancel{6}} \cdot \dfrac{1}{\underset{1}{\cancel{6}}} \cdot x = \dfrac{\overset{2}{\cancel{6}} \cdot 1}{1 \cdot \underset{1}{\cancel{3}}}$
$x = 2$ Simplify.	
	$1x = 2 \quad \text{or} \quad x = 2$

Practice 7

Solve: $\dfrac{y}{8} + \dfrac{3}{4} = 2$

Answers

6. $-\dfrac{9}{11}$ **7.** 10

(Continued on next page)

Check: To check, replace x with 2 in the original equation.

$$\frac{x}{6} + 1 = \frac{4}{3} \qquad \text{Original equation}$$

$$\frac{2}{6} + 1 \stackrel{?}{=} \frac{4}{3} \qquad \text{Replace } x \text{ with 2.}$$

$$\frac{1}{3} + \frac{3}{3} \stackrel{?}{=} \frac{4}{3} \qquad \text{Simplify } \frac{2}{6}. \text{ The LCD of 3 and 1 is 3.}$$

$$\frac{4}{3} = \frac{4}{3} \qquad \text{True}$$

Since we arrived at a true statement, 2 is the solution of $\dfrac{x}{6} + 1 = \dfrac{4}{3}$.

■ Work Practice 7

Let's review the steps for solving equations in x. An extra step is now included to handle equations containing fractions.

> ### Solving an Equation in x
>
> **Step 1:** If fractions are present, multiply both sides of the equation by the LCD of the fractions.
>
> **Step 2:** If parentheses are present, use the distributive property.
>
> **Step 3:** Combine any like terms on each side of the equation.
>
> **Step 4:** Use the addition property of equality to rewrite the equation so that variable terms are on one side of the equation and constant terms are on the other side.
>
> **Step 5:** Divide both sides of the equation by the numerical coefficient of x to solve.
>
> **Step 6:** Check the answer in the **original equation.**

Practice 8

Solve: $\dfrac{x}{5} - x = \dfrac{1}{5}$

Helpful Hint   Don't forget to multiply *both* sides of the equation by the LCD.

Example 8   Solve: $\dfrac{z}{5} - \dfrac{z}{3} = 6$

Solution:

$$\frac{z}{5} - \frac{z}{3} = 6$$

$$15\left(\frac{z}{5} - \frac{z}{3}\right) = 15(6) \qquad \text{Multiply both sides by the LCD, 15.}$$

$$\overset{3}{\cancel{15}}\left(\frac{z}{\cancel{5}}\right) - \overset{5}{\cancel{15}}\left(\frac{z}{\cancel{3}}\right) = 15(6) \qquad \text{Apply the distributive property.}$$

$$3z - 5z = 90 \qquad \text{Simplify.}$$

$$-2z = 90 \qquad \text{Combine like terms.}$$

$$\frac{-2z}{-2} = \frac{90}{-2} \qquad \text{Divide both sides by } -2, \text{ the coefficient of } z.$$

$$z = -45 \qquad \text{Simplify.}$$

To check, replace z with -45 in the **original equation** to see that a true statement results.

■ Work Practice 8

Answer

8. $-\dfrac{1}{4}$

Example 9   Solve: $\dfrac{x}{2} = \dfrac{x}{3} + \dfrac{1}{2}$

Solution: First multiply both sides by the LCD, 6.

$$\dfrac{x}{2} = \dfrac{x}{3} + \dfrac{1}{2}$$

$$6\left(\dfrac{x}{2}\right) = 6\left(\dfrac{x}{3} + \dfrac{1}{2}\right) \qquad \text{Multiply both sides by the LCD, 6.}$$

$$\overset{3}{\cancel{6}}\left(\dfrac{x}{2}\right) = \overset{2}{\cancel{6}}\left(\dfrac{x}{3}\right) + \overset{3}{\cancel{6}}\left(\dfrac{1}{2}\right) \qquad \text{Apply the distributive property.}$$

$$3x = 2x + 3 \qquad \text{Simplify.}$$

$$3x - 2x = 2x + 3 - 2x \qquad \text{Subtract } 2x \text{ from both sides.}$$

$$x = 3 \qquad \text{Simplify.}$$

To check, replace x with 3 in the original equation to see that a true statement results.

■ **Work Practice 9**

Objective C Review of Adding and Subtracting Fractions ▶

Make sure you understand the difference between **solving an equation** containing fractions and **adding or subtracting two fractions.** To solve an equation containing fractions, we use the multiplication property of equality and multiply both sides by the LCD of the fractions, thus eliminating the fractions. This method does not apply to adding or subtracting fractions. The multiplication property of equality applies only to equations. To add or subtract unlike fractions, we write each fraction as an equivalent fraction using the LCD of the fractions as the denominator. See the next example for a review.

Example 10   Add: $\dfrac{x}{3} + \dfrac{2}{5}$

Solution: This is an expression, not an equation. Here, we are adding two unlike fractions. To add unlike fractions, we need to find the LCD. The LCD of the denominators 3 and 5 is 15. Write each fraction as an equivalent fraction with a denominator of 15.

$$\dfrac{x}{3} + \dfrac{2}{5} = \dfrac{x}{3} \cdot \dfrac{5}{5} + \dfrac{2}{5} \cdot \dfrac{3}{3} = \dfrac{x \cdot 5}{3 \cdot 5} + \dfrac{2 \cdot 3}{5 \cdot 3}$$

$$= \dfrac{5x}{15} + \dfrac{6}{15}$$

$$= \dfrac{5x + 6}{15}$$

■ **Work Practice 10**

✓**Concept Check** Which of the following are equations and which are expressions?

a. $\dfrac{1}{2} + 3x = 5$ **b.** $\dfrac{2}{3}x - \dfrac{x}{5}$

c. $\dfrac{x}{12} + \dfrac{5x}{24}$ **d.** $\dfrac{x}{5} = \dfrac{1}{10}$

Practice 9

Solve: $\dfrac{y}{2} = \dfrac{y}{5} + \dfrac{3}{2}$

Practice 10

Subtract: $\dfrac{9}{10} - \dfrac{y}{3}$

Answers

9. 5   **10.** $\dfrac{27 - 10y}{30}$

✓**Concept Check Answers**

equations: a, d; expressions: b, c

Vocabulary, Readiness & Video Check

Fill in the blank with the least common denominator (LCD). Do not solve these equations.

1. Equation: $\dfrac{2}{3} + x = \dfrac{5}{6}$; LCD = ____

2. Equation: $\dfrac{x}{21} - 1 = \dfrac{1}{7}$; LCD = ____

3. Equation: $\dfrac{y}{5} + \dfrac{1}{3} = 2$; LCD = ____

4. Equation: $\dfrac{-2n}{11} + \dfrac{1}{2} = 5$; LCD = ____

Martin-Gay Interactive Videos   *Watch the section lecture video and answer the following questions.*

Objective **A** **5.** In ▣ Example 1, what property is used to get *x* by itself on one side of the equation? ▷

     6. Explain how a reciprocal is used to solve ▣ Example 2. ▷

Objective **B** **7.** Why are both sides of the equation multiplied by 12 in ▣ Example 3? What effect does this have on the fractions in the equation? ▷

Objective **C** **8.** In ▣ Example 5, why can't we multiply by the LCD of all fractions? ▷

See Video 4.8 🫐

4.8   **Exercise Set** MyMathLab® ▷

Objective **A**   *Solve each equation. Check your proposed solution. See Example 1.*

1. $x + \dfrac{1}{3} = -\dfrac{1}{3}$

2. $x + \dfrac{1}{9} = -\dfrac{7}{9}$

3. $y - \dfrac{3}{13} = -\dfrac{2}{13}$

4. $z - \dfrac{5}{14} = \dfrac{4}{14}$

5. $3x - \dfrac{1}{5} - 2x = \dfrac{1}{5} + \dfrac{2}{5}$

6. $5x + \dfrac{1}{11} - 4x = \dfrac{2}{11} - \dfrac{5}{11}$

▷ 7. $x - \dfrac{1}{12} = \dfrac{5}{6}$

8. $y - \dfrac{8}{9} = \dfrac{1}{3}$

9. $\dfrac{2}{5} + y = -\dfrac{3}{10}$

10. $\dfrac{1}{2} + a = -\dfrac{3}{8}$

11. $7z + \dfrac{1}{16} - 6z - \dfrac{3}{4}$

12. $9x - \dfrac{2}{7} - 8x = \dfrac{11}{14}$

13. $-\dfrac{2}{9} = x - \dfrac{5}{6}$

14. $-\dfrac{1}{4} = y - \dfrac{7}{10}$

Solve each equation. See Examples 2 through 5.

15. $7x = 2$

16. $-5x = 4$

17. $\frac{1}{4}x = 3$

18. $\frac{1}{3}x = 6$

19. $\frac{2}{9}y = -6$

20. $\frac{4}{7}x = -8$

21. $-\frac{4}{9}z = -\frac{3}{2}$

22. $-\frac{11}{10}x = -\frac{2}{7}$

23. $7a = \frac{1}{3}$

24. $2z = -\frac{5}{12}$

25. $-3x = -\frac{6}{11}$

26. $-4z = -\frac{12}{25}$

Objective B *Solve each equation. See Examples 6 through 9.*

27. $\frac{5}{9}x = -\frac{3}{18}$

28. $\frac{3}{5}y = -\frac{7}{20}$

29. $\frac{x}{3} + 2 = \frac{7}{3}$

30. $\frac{x}{5} - 1 = \frac{7}{5}$

31. $\frac{x}{5} - x = -8$

32. $\frac{x}{3} - x = -6$

33. $\frac{1}{2} - \frac{3}{5} = \frac{x}{10}$

34. $\frac{2}{3} - \frac{1}{4} = \frac{x}{12}$

35. $\frac{x}{3} = \frac{x}{5} - 2$

36. $\frac{a}{2} = \frac{a}{7} + \frac{5}{2}$

Objective C *Add or subtract as indicated. See Example 10.*

37. $\frac{x}{7} - \frac{4}{3}$

38. $-\frac{5}{9} + \frac{y}{8}$

39. $\frac{y}{2} + 5$

40. $2 + \frac{7x}{3}$

41. $\frac{3x}{10} + \frac{x}{6}$

42. $\frac{9x}{8} - \frac{5x}{6}$

Objectives A B C **Mixed Practice** *Solve. If no equation is given, perform the indicated operation. See Examples 1 through 10.*

43. $\frac{3}{8}x = \frac{1}{2}$

44. $\frac{2}{5}y = \frac{3}{10}$

45. $\frac{2}{3} - \frac{x}{5} = \frac{4}{15}$

46. $\frac{4}{5} + \frac{x}{4} = \frac{21}{20}$

47. $\frac{9}{14}z = \frac{27}{20}$

48. $\frac{5}{16}a = \frac{5}{6}$

49. $-3m - 5m = \frac{4}{7}$

50. $30n - 34n = \frac{3}{20}$

51. $\dfrac{x}{4} + 1 = \dfrac{1}{4}$

52. $\dfrac{y}{7} - 2 = \dfrac{1}{7}$

53. $\dfrac{5}{9} - \dfrac{2}{3}$

54. $\dfrac{8}{11} - \dfrac{1}{2}$

55. $\dfrac{1}{5}y = 10$

56. $\dfrac{1}{4}x = -2$

57. $\dfrac{5}{7}y = -\dfrac{15}{49}$

58. $-\dfrac{3}{4}x = \dfrac{9}{2}$

59. $\dfrac{x}{2} - x = -2$

60. $\dfrac{y}{3} = -4 + y$

61. $-\dfrac{5}{8}y = \dfrac{3}{16} - \dfrac{9}{16}$

62. $-\dfrac{7}{9}x = -\dfrac{5}{18} - \dfrac{4}{18}$

63. $17x - 25x = \dfrac{1}{3}$

64. $27x - 30x = \dfrac{4}{9}$

65. $\dfrac{7}{6}x = \dfrac{1}{4} - \dfrac{2}{3}$

66. $\dfrac{5}{4}y = \dfrac{1}{2} - \dfrac{7}{10}$

67. $\dfrac{b}{4} = \dfrac{b}{12} + \dfrac{2}{3}$

68. $\dfrac{a}{6} = \dfrac{a}{3} + \dfrac{1}{2}$

69. $\dfrac{x}{3} + 2 = \dfrac{x}{2} + 8$

70. $\dfrac{y}{5} - 2 = \dfrac{y}{3} - 4$

Review

Round each number to the given place value. See Section 1.4.

71. 57,236 to the nearest hundred

72. 576 to the nearest hundred

73. 327 to the nearest ten

74. 2333 to the nearest ten

Concept Extensions

75. Explain why the method for eliminating fractions in an **equation** does not apply to simplifying **expressions** containing fractions.

76. Think about which exercise (part **a** or part **b**) may be completed by multiplying by 6. Now complete each exercise.

    a. Solve: $\dfrac{x}{6} - \dfrac{5}{3} = 2$

    b. Subtract: $\dfrac{x}{6} - \dfrac{5}{3}$

Solve.

77. $\dfrac{14}{11} + \dfrac{3x}{8} = \dfrac{x}{2}$

78. $\dfrac{19}{53} = \dfrac{353x}{1431} + \dfrac{23}{27}$

79. Find the area and the perimeter of the rectangle. Remember to attach proper units.

$\frac{1}{4}$ inch

$\frac{3}{4}$ inch

80. The area of the rectangle is $\dfrac{5}{12}$ square inch. Find its length, x.

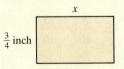

x

$\frac{3}{4}$ inch

Chapter 4 Group Activity

Lobster Classification

Sections 4.1, 4.7, 4.8

This activity may be completed by working in groups or individually.

Lobsters are normally classified by weight. Use the weight classification table to answer the questions in this activity.

Classification of Lobsters	
Class	**Weight (in pounds)**
Chicken	1 to $1\frac{1}{8}$
Quarter	$1\frac{1}{4}$
Half	$1\frac{1}{2}$ to $1\frac{3}{4}$
Select	$1\frac{3}{4}$ to $2\frac{1}{2}$
Large select	$2\frac{1}{2}$ to $3\frac{1}{2}$
Jumbo	Over $3\frac{1}{2}$

(*Source:* The Maine Lobster Promotion Council)

1. A lobster fisher has kept four lobsters from a lobster trap. Classify each lobster if they have the following weights:

   a. $1\frac{7}{8}$ pounds

   b. $1\frac{9}{16}$ pounds

   c. $2\frac{3}{4}$ pounds

   d. $2\frac{3}{8}$ pounds

2. A recipe requires 5 pounds of lobster. Using the minimum weight for each class, decide whether a chicken, half, and select lobster will be enough for the recipe, and explain your reasoning. If not, suggest a better choice of lobsters to meet the recipe requirements.

3. A lobster market customer has selected two chickens, a select, and a large select. What is the most that these four lobsters could weigh? What is the least that these four lobsters could weigh?

4. A lobster market customer wishes to buy three quarters. If lobsters sell for $7 per pound, how much will the customer owe for her purchase?

5. Why do you think there is no classification for lobsters weighing under 1 pound?

Chapter 4 Vocabulary Check

Fill in each blank with one of the words or phrases listed below.

mixed number	complex fraction	like	numerator	prime factorization
composite number	equivalent	cross products	least common denominator	denominator
prime number	improper fraction	simplest form	undefined	0
reciprocals	proper fraction			

1. Two numbers are _____ of each other if their product is 1.

2. A(n) _____ is a natural number greater than 1 that is not prime.

3. Fractions that represent the same portion of a whole are called _____ fractions.

4. A(n) _____ is a fraction whose numerator is greater than or equal to its denominator.

5. A(n) _____ is a natural number greater than 1 whose only factors are 1 and itself.

6. A fraction is in _____ when the numerator and the denominator have no factors in common other than 1.

7. A(n) _____ is one whose numerator is less than its denominator.

8. A(n) _____ contains a whole number part and a fraction part.

9. In the fraction $\frac{7}{9}$, the 7 is called the _____ and the 9 is called the _____ .

10. The _____ of a number is the factorization in which all the factors are prime numbers.

11. The fraction $\dfrac{3}{0}$ is _____.

12. The fraction $\dfrac{0}{5}$ = _____.

13. Fractions that have the same denominator are called _____ fractions.

14. The LCM of the denominators in a list of fractions is called the _____.

15. A fraction whose numerator or denominator or both numerator and denominator contain fractions is called a(n) _____.

16. In $\dfrac{a}{b} = \dfrac{c}{d}$, $a \cdot d$ and $b \cdot c$ are called _____.

Helpful Hint

◯ Are you preparing for your test? Don't forget to take the Chapter 4 Test on page 318. Then check your answers at the back of the text and use the Chapter Test Prep Videos to see the fully worked-out solutions to any of the exercises you want to review.

4 Chapter Highlights

Definitions and Concepts	Examples
Section 4.1 Introduction to Fractions and Mixed Numbers	

A **fraction** is of the form

$$\dfrac{\text{numerator}}{\text{denominator}} \quad \begin{array}{l} \leftarrow \text{number of parts being considered} \\ \leftarrow \text{number of equal parts in the whole} \end{array}$$

Write a fraction to represent the shaded part of the figure.

$$\dfrac{3}{8} \quad \begin{array}{l} \leftarrow \text{number of parts shaded} \\ \leftarrow \text{number of equal parts} \end{array}$$

A fraction is called a **proper fraction** if its numerator is less than its denominator.

A fraction is called an **improper fraction** if its numerator is greater than or equal to its denominator.

A **mixed number** contains a whole number and a fraction.

Proper Fractions: $\dfrac{1}{3}, \dfrac{2}{5}, \dfrac{7}{8}, \dfrac{100}{101}$

Improper Fractions: $\dfrac{5}{4}, \dfrac{2}{2}, \dfrac{9}{7}, \dfrac{101}{100}$

Mixed Numbers: $1\dfrac{1}{2}, 5\dfrac{7}{8}, 25\dfrac{9}{10}$

To Write a Mixed Number As an Improper Fraction

1. Multiply the denominator of the fraction by the whole number.

2. Add the numerator of the fraction to the product from Step 1.

3. Write the sum from Step 2 as the numerator of the improper fraction over the original denominator.

$$5\dfrac{2}{7} = \dfrac{7 \cdot 5 + 2}{7} = \dfrac{35 + 2}{7} = \dfrac{37}{7}$$

To Write an Improper Fraction As a Mixed Number or a Whole Number

1. Divide the denominator into the numerator.

2. The whole number part of the mixed number is the quotient. The fraction is the remainder over the original denominator.

$$\text{quotient}\dfrac{\text{remainder}}{\text{original denominator}}$$

$$\dfrac{17}{3} = 5\dfrac{2}{3}$$

$$\begin{array}{r} 5 \\ 3\overline{)17} \\ -15 \\ \hline 2 \end{array}$$

Definitions and Concepts	Examples

Section 4.2 Factors and Simplest Form

A **prime number** is a natural number that has exactly two different factors, 1 and itself.

$2, 3, 5, 7, 11, 13, 17, \ldots$

A **composite number** is any natural number other than 1 that is not prime.

$4, 6, 8, 9, 10, 12, 14, 15, 16, \ldots$

The **prime factorization** of a number is the factorization in which all the factors are prime numbers.

Write the prime factorization of 60.
$$60 = 6 \cdot 10$$
$$= 2 \cdot 3 \cdot 2 \cdot 5 \quad \text{or} \quad 2^2 \cdot 3 \cdot 5$$

Fractions that represent the same portion of a whole are called **equivalent fractions.**

$$\frac{3}{4} \quad = \quad \frac{12}{16}$$

A fraction is in **simplest form** or **lowest terms** when the numerator and the denominator have no common factors other than 1.

The fraction $\frac{2}{3}$ is in simplest form.

To write a fraction in simplest form, write the prime factorizations of the numerator and the denominator and then divide both by all common factors.

Write in simplest form: $\dfrac{30}{36}$

$$\frac{30}{36} = \frac{2 \cdot 3 \cdot 5}{2 \cdot 2 \cdot 3 \cdot 3} = \frac{2}{2} \cdot \frac{3}{3} \cdot \frac{5}{2 \cdot 3} = 1 \cdot 1 \cdot \frac{5}{6} = \frac{5}{6}$$

$$\text{or} \quad \frac{30}{36} = \frac{\overset{1}{\cancel{2}} \cdot \overset{1}{\cancel{3}} \cdot 5}{2 \cdot 2 \cdot \underset{1}{\cancel{3}} \cdot \underset{1}{\cancel{3}}} = \frac{5}{6}$$

Two fractions are equivalent if

Determine whether $\dfrac{7}{8}$ and $\dfrac{21}{24}$ are equivalent.

Method 1. They simplify to the same fraction.

$\dfrac{7}{8}$ is in simplest form.

Method 2. Their cross products are equal.

$$\frac{21}{24} = \frac{\overset{1}{\cancel{3}} \cdot 7}{\underset{1}{\cancel{3}} \cdot 8} = \frac{1 \cdot 7}{1 \cdot 8} = \frac{7}{8}$$

$$\begin{array}{cc} 24 \cdot 7 & 8 \cdot 21 \\ = 168 & \dfrac{7}{8} \diagup\!\!\!\diagdown \dfrac{21}{24} \quad = 168 \end{array}$$

Since both simplify to $\dfrac{7}{8}$, then $\dfrac{7}{8} = \dfrac{21}{24}$.

Since $168 = 168$, $\dfrac{7}{8} = \dfrac{21}{24}$.

Section 4.3 Multiplying and Dividing Fractions

To multiply two fractions, multiply the numerators and multiply the denominators.

Multiply.

$$\frac{2x}{3} \cdot \frac{5}{7} = \frac{2x \cdot 5}{3 \cdot 7} = \frac{10x}{21}$$

$$\frac{3}{4} \cdot \frac{1}{6} = \frac{3 \cdot 1}{4 \cdot 6} = \frac{\overset{1}{\cancel{3}} \cdot 1}{4 \cdot \underset{1}{\cancel{3}} \cdot 2} = \frac{1}{8}$$

To find the **reciprocal** of a fraction, interchange its numerator and denominator.

The reciprocal of $\dfrac{3}{5}$ is $\dfrac{5}{3}$.

To divide two fractions, multiply the first fraction by the reciprocal of the second fraction.

Divide.

$$-\frac{3}{10} \div \frac{7}{9} = -\frac{3}{10} \cdot \frac{9}{7} = -\frac{3 \cdot 9}{10 \cdot 7} = -\frac{27}{70}$$

Definitions and Concepts	Examples

Section 4.4   Adding and Subtracting Like Fractions, Least Common Denominator, and Equivalent Fractions

Fractions that have the same denominator are called **like fractions**.

$$-\frac{1}{3} \text{ and } \frac{2}{3}; \frac{5x}{7} \text{ and } \frac{6}{7}$$

To add or subtract like fractions, combine the numerators and place the sum or difference over the common denominator.

$$\frac{2}{7} + \frac{3}{7} = \frac{5}{7} \quad \leftarrow \text{Add the numerators.}$$
$$\phantom{\frac{2}{7} + \frac{3}{7} = \frac{5}{7}} \quad \leftarrow \text{Keep the common denominator.}$$
$$\frac{7}{8} - \frac{4}{8} = \frac{3}{8} \quad \leftarrow \text{Subtract the numerators.}$$
$$\phantom{\frac{7}{8} - \frac{4}{8} = \frac{3}{8}} \quad \leftarrow \text{Keep the common denominator.}$$

The **least common denominator (LCD)** of a list of fractions is the smallest positive number divisible by all the denominators in the list.

The LCD of $\frac{1}{2}$ and $\frac{5}{6}$ is 6 because 6 is the smallest positive number that is divisible by both 2 and 6.

Method 1 for finding the LCD of a list of fractions using multiples

Find the LCD of $\frac{1}{4}$ and $\frac{5}{6}$ using Method 1.

Step 1: Write the multiples of the largest denominator (starting with the number itself) until a multiple common to all denominators in the list is found.

$$6 \cdot 1 = 6 \quad \text{Not a multiple of 4}$$
$$6 \cdot 2 = 12 \quad \text{A multiple of 4}$$

Step 2: The multiple found in Step 1 is the LCD.

The LCD is 12.

Method 2 for finding the LCD of a list of a fractions using prime factorization

Find the LCD of $\frac{5}{6}$ and $\frac{11}{20}$ using Method 2.

Step 1: Write the prime factorization of each denominator.

$$6 = 2 \cdot \textcircled{3}$$
$$20 = \textcircled{2 \cdot 2} \cdot \textcircled{5}$$

Step 2: For each different prime factor in Step 1, circle the greatest number of times that factor occurs in any one factorization.

Step 3: The LCD is the product of the circled factors.

The LCD is

$$2 \cdot 2 \cdot 3 \cdot 5 = 60$$

Equivalent fractions represent the same portion of a whole.

Write an equivalent fraction with the indicated denominator.

$$\frac{2}{8} = \frac{}{16}$$
$$\frac{2 \cdot 2}{8 \cdot 2} = \frac{4}{16}$$

Section 4.5   Adding and Subtracting Unlike Fractions

To add or subtract fractions with unlike denominators

Add: $\frac{3}{20} + \frac{2}{5}$

Step 1: Find the LCD.

Step 1: The LCD of the denominators 20 and 5 is 20.

Step 2: Write each fraction as an equivalent fraction whose denominator is the LCD.

Step 2: $\frac{3}{20} = \frac{3}{20}; \frac{2}{5} = \frac{2}{5} \cdot \frac{4}{4} = \frac{8}{20}$

Step 3: Add or subtract the like fractions.

Step 3: $\frac{3}{20} + \frac{2}{5} = \frac{3}{20} + \frac{8}{20} = \frac{11}{20}$

Step 4: Write the sum or difference in simplest form.

Step 4: $\frac{11}{20}$ is in simplest form.

Copyright 2015 Pearson Education, Inc.

Definitions and Concepts	Examples

Section 4.6 Complex Fractions and Review of Order of Operations

A fraction whose numerator or denominator or both contain fractions is called a **complex fraction.**

Complex Fractions:

$$\frac{\dfrac{11}{4}}{\dfrac{7}{10}}, \quad \frac{\dfrac{y}{6} - 11}{\dfrac{4}{3}}$$

One method for simplifying complex fractions is to multiply the numerator and the denominator of the complex fraction by the LCD of all fractions in its numerator and its denominator.

$$\frac{\dfrac{y}{6} - 11}{\dfrac{4}{3}} = \frac{6\left(\dfrac{y}{6} - 11\right)}{6\left(\dfrac{4}{3}\right)} = \frac{6\left(\dfrac{y}{6}\right) - 6(11)}{6\left(\dfrac{4}{3}\right)}$$

$$= \frac{y - 66}{8}$$

Section 4.7 Operations on Mixed Numbers

To multiply with mixed numbers or whole numbers, first write any mixed or whole numbers as improper fractions and then multiply as usual.

$$2\frac{1}{3} \cdot \frac{1}{9} = \frac{7}{3} \cdot \frac{1}{9} = \frac{7 \cdot 1}{3 \cdot 9} = \frac{7}{27}$$

To divide with mixed numbers or whole numbers, first write any mixed or whole numbers as fractions and then divide as usual.

$$2\frac{5}{8} \div 3\frac{7}{16} = \frac{21}{8} \div \frac{55}{16} = \frac{21}{8} \cdot \frac{16}{55} = \frac{21 \cdot 16}{8 \cdot 55}$$

$$= \frac{21 \cdot 2 \cdot \overset{1}{8}}{\underset{1}{8} \cdot 55} = \frac{42}{55}$$

To add or subtract with mixed numbers, add or subtract the fractions and then add or subtract the whole numbers.

Add: $2\dfrac{1}{2} + 5\dfrac{7}{8}$

$$2\frac{1}{2} = 2\frac{4}{8}$$

$$+ 5\frac{7}{8} = + 5\frac{7}{8}$$

$$\overline{\qquad\qquad 7\frac{11}{8}} = 7 + 1\frac{3}{8} = 8\frac{3}{8}$$

Section 4.8 Solving Equations Containing Fractions

To Solve an Equation in x

Step 1: If fractions are present, multiply both sides of the equation by the LCD of the fractions.

Step 2: If parentheses are present, use the distributive property.

Step 3: Combine any like terms on each side of the equation.

Step 4: Use the addition property of equality to rewrite the equation so that variable terms are on one side of the equation and constant terms are on the other side.

Step 5: Divide both sides by the numerical coefficient of x to solve.

Step 6: Check the answer in the *original equation.*

Solve: $\dfrac{x}{15} + 2 = \dfrac{7}{3}$

$$15\left(\frac{x}{15} + 2\right) = 15\left(\frac{7}{3}\right) \quad \text{Multiply by the LCD 15.}$$

$$15\left(\frac{x}{15}\right) + 15 \cdot 2 = 15\left(\frac{7}{3}\right)$$

$$x + 30 = 35$$

$$x + 30 - 30 = 35 - 30$$

$$x = 5$$

Check to see that 5 is the solution.

(4.1) *Write a fraction to represent the shaded area. If the fraction is improper, write the shaded area as a mixed number also. Do not simplify these answers.*

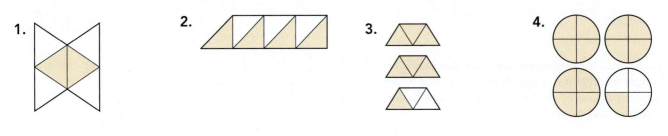

1.   2.   3.   4.

Solve.

5. A basketball player made 11 free throws out of 12 tries during a game. What fraction of free throws did the player make?

6. A new car lot contains 23 blue cars out of a total of 131 cars.
    a. How many cars on the lot are not blue?
    b. What fraction of cars on the lot are not blue?

Simplify by dividing.

7. $-\dfrac{3}{3}$

8. $\dfrac{-20}{-20}$

9. $\dfrac{0}{-1}$

10. $\dfrac{4}{0}$

Graph each fraction on a number line.

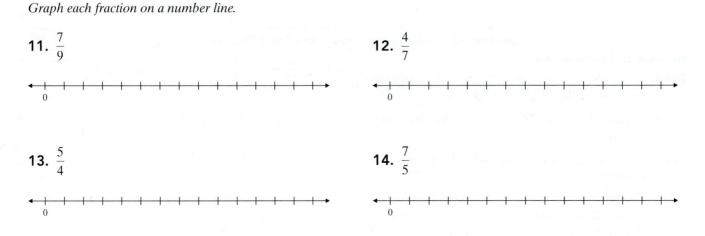

11. $\dfrac{7}{9}$

12. $\dfrac{4}{7}$

13. $\dfrac{5}{4}$

14. $\dfrac{7}{5}$

Write each improper fraction as a mixed number or a whole number.

15. $\dfrac{15}{4}$

16. $\dfrac{39}{13}$

Write each mixed number as an improper fraction.

17. $2\frac{1}{5}$

18. $3\frac{8}{9}$

(4.2) *Write each fraction in simplest form.*

19. $\frac{12}{28}$

20. $\frac{15}{27}$

21. $-\frac{25x}{75x^2}$

22. $-\frac{36y^3}{72y}$

23. $\frac{29ab}{32abc}$

24. $\frac{18xyz}{23xy}$

25. $\frac{45x^2y}{27xy^3}$

26. $\frac{42ab^2c}{30abc^3}$

27. There are 12 inches in a foot. What fractional part of a foot does 8 inches represent?

28. Six out of 15 cars are white. What fraction of the cars are *not* white?

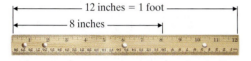

Determine whether each two fractions are equivalent.

29. $\frac{10}{34}$ and $\frac{4}{14}$

30. $\frac{30}{50}$ and $\frac{9}{15}$

(4.3) *Multiply.*

31. $\frac{3}{5} \cdot \frac{1}{2}$

32. $-\frac{6}{7} \cdot \frac{5}{12}$

33. $-\frac{24x}{5} \cdot -\frac{15}{8x^3}$

34. $\frac{27y^3}{21} \cdot \frac{7}{18y^2}$

35. $\left(-\frac{1}{3}\right)^3$

36. $\left(-\frac{5}{12}\right)^2$

Divide.

37. $-\frac{3}{4} \div \frac{3}{8}$

38. $\frac{21a}{4} \div \frac{7a}{5}$

39. $-\frac{9}{2} \div -\frac{1}{3}$

40. $-\frac{5}{3} \div 2y$

41. Evaluate $x \div y$ if $x = \frac{9}{7}$ and $y = \frac{3}{4}$.

42. Evaluate ab if $a = -7$ and $b = \frac{9}{10}$.

Find the area of each figure.

△ **43.**

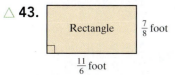

△ **44.**

(4.4) *Add or subtract as indicated.*

45. $\dfrac{7}{11} + \dfrac{3}{11}$

46. $\dfrac{4}{9} + \dfrac{2}{9}$

47. $\dfrac{1}{12} - \dfrac{5}{12}$

48. $\dfrac{11x}{15} + \dfrac{x}{15}$

49. $\dfrac{4y}{21} - \dfrac{3}{21}$

50. $\dfrac{4}{15} - \dfrac{3}{15} - \dfrac{2}{15}$

Find the LCD of each list of fractions.

51. $\dfrac{2}{3}, \dfrac{5}{x}$

52. $\dfrac{3}{4}, \dfrac{3}{8}, \dfrac{7}{12}$

Write each fraction as an equivalent fraction with the given denominator.

53. $\dfrac{2}{3} = \dfrac{?}{30}$

54. $\dfrac{5}{8} = \dfrac{?}{56}$

55. $\dfrac{7a}{6} = \dfrac{?}{42}$

56. $\dfrac{9b}{4} = \dfrac{?}{20}$

57. $\dfrac{4}{5x} = \dfrac{?}{50x}$

58. $\dfrac{5}{9y} = \dfrac{?}{18y}$

Solve.

59. One evening Mark Alorenzo did $\dfrac{3}{8}$ of his homework before supper, another $\dfrac{2}{8}$ of it while his children did their homework, and $\dfrac{1}{8}$ after his children went to bed. What part of his homework did he do that evening?

△ **60.** The Simpsons will be fencing in their land, which is in the shape of a rectangle. In order to do this, they need to find its perimeter. Find the perimeter of their land.

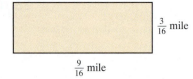

$\dfrac{3}{16}$ mile

$\dfrac{9}{16}$ mile

(4.5) *Add or subtract as indicated.*

61. $\dfrac{7}{18} + \dfrac{2}{9}$

62. $\dfrac{4}{13} - \dfrac{1}{26}$

63. $-\dfrac{1}{3} + \dfrac{1}{4}$

64. $-\dfrac{2}{3} + \dfrac{1}{4}$

65. $\dfrac{5x}{11} + \dfrac{2}{55}$

66. $\dfrac{4}{15} + \dfrac{b}{5}$

67. $\dfrac{5y}{12} - \dfrac{2y}{9}$

68. $\dfrac{7x}{18} + \dfrac{2x}{9}$

69. $\dfrac{4}{9} + \dfrac{5}{y}$

70. $-\dfrac{9}{14} - \dfrac{3}{7}$

71. $\dfrac{4}{25} + \dfrac{23}{75} + \dfrac{7}{50}$

72. $\dfrac{2}{3} - \dfrac{2}{9} - \dfrac{1}{6}$

Find the perimeter of each figure.

△ **73.**

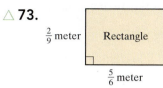

$\frac{2}{9}$ meter   Rectangle

$\frac{5}{6}$ meter

△ **74.** $\frac{1}{5}$ foot $\frac{3}{5}$ foot

$\frac{7}{10}$ foot

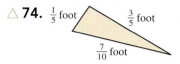

75. In a group of 100 blood donors, typically $\frac{9}{25}$ have type A Rh-positive blood and $\frac{3}{50}$ have type A Rh-negative blood. What fraction have type A blood?

76. Find the difference in length of two scarves if one scarf is $\frac{5}{12}$ of a yard long and the other is $\frac{2}{3}$ of a yard long.

$\frac{2}{3}$ of a yard

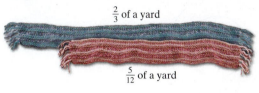

$\frac{5}{12}$ of a yard

(4.6) *Simplify each complex fraction.*

77. $\dfrac{\dfrac{2x}{5}}{\dfrac{7}{10}}$

78. $\dfrac{\dfrac{3y}{7}}{\dfrac{11}{7}}$

79. $\dfrac{\dfrac{2}{5}-\dfrac{1}{2}}{\dfrac{3}{4}-\dfrac{7}{10}}$

80. $\dfrac{\dfrac{5}{6}-\dfrac{1}{4}}{\dfrac{-1}{12y}}$

Evaluate each expression if $x = \frac{1}{2}$, $y = -\frac{2}{3}$, and $z = \frac{4}{5}$.

81. $\dfrac{x}{y+z}$

82. $\dfrac{x+y}{z}$

Evaluate each expression. Use the order of operations to simplify.

83. $\dfrac{5}{13} \div \dfrac{1}{2} \cdot \dfrac{4}{5}$

84. $\dfrac{2}{27} - \left(\dfrac{1}{3}\right)^2$

85. $\dfrac{9}{10} \cdot \dfrac{1}{3} - \dfrac{2}{5} \cdot \dfrac{1}{11}$

86. $-\dfrac{2}{7} \cdot \left(\dfrac{1}{5} + \dfrac{3}{10}\right)$

(4.7) *Perform operations as indicated. Simplify your answers. Estimate where noted.*

87. $7\dfrac{3}{8}$
         $9\dfrac{5}{6}$
      $+\ 3\dfrac{1}{12}$

88. $8\dfrac{1}{5}$
       $-\ 5\dfrac{3}{11}$

Exact:

Estimate:

89. $1\dfrac{5}{8} \cdot 3\dfrac{1}{5}$

Exact:

Estimate:

90. $6\dfrac{3}{4} \div 1\dfrac{2}{7}$

91. A truck traveled 341 miles on $15\frac{1}{2}$ gallons of gas. How many miles might we expect the truck to travel on 1 gallon of gas?

92. There are $7\frac{1}{3}$ grams of fat in each ounce of hamburger. How many grams of fat are in a 5-ounce hamburger patty?

Find the unknown measurements.

△**93.**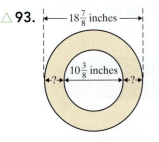

△**94.**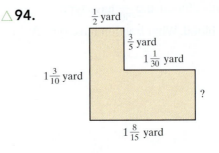

Perform the indicated operations.

95. $-12\frac{1}{7} + \left(-15\frac{3}{14}\right)$

96. $23\frac{7}{8} - 24\frac{7}{10}$

97. $-3\frac{1}{5} \div \left(-2\frac{7}{10}\right)$

98. $-2\frac{1}{4} \cdot 1\frac{3}{4}$

(4.8) *Solve each equation.*

99. $a - \frac{2}{3} = \frac{1}{6}$

100. $9x + \frac{1}{5} - 8x = -\frac{7}{10}$

101. $-\frac{3}{5}x = 6$

102. $\frac{2}{9}y = -\frac{4}{3}$

103. $\frac{x}{7} - 3 = -\frac{6}{7}$

104. $\frac{y}{5} + 2 = \frac{11}{5}$

105. $\frac{1}{6} + \frac{x}{4} = \frac{17}{12}$

106. $\frac{x}{5} - \frac{5}{4} = \frac{x}{2} - \frac{1}{20}$

Mixed Review

Perform the indicated operations. Write each answer in simplest form. Estimate where noted.

107. $\frac{6}{15} \cdot \frac{5}{8}$

108. $\frac{5x^2}{y} \div \frac{10x^3}{y^3}$

109. $\frac{3}{10} - \frac{1}{10}$

110. $\frac{7}{8x} \cdot -\frac{2}{3}$

111. $\frac{2x}{3} + \frac{x}{4}$

112. $-\frac{5}{11} + \frac{2}{55}$

113. $-1\frac{3}{5} \div \frac{1}{4}$

114. $2\frac{7}{8}$
$+ 9\frac{1}{2}$ Exact:

         Estimate:

115. $12\frac{1}{7}$
         Exact:
$- 9\frac{3}{5}$
         Estimate:

116. Simplify: $\dfrac{2 + \dfrac{3}{4}}{1 - \dfrac{1}{8}}$

117. Evaluate: $-\frac{3}{8} \cdot \left(\frac{2}{3} - \frac{4}{9}\right)$

Solve.

118. $11x - \dfrac{2}{7} - 10x = -\dfrac{13}{14}$

119. $-\dfrac{3}{5}x = \dfrac{4}{15}$

120. $\dfrac{x}{12} + \dfrac{5}{6} = -\dfrac{3}{4}$

121. A ribbon $5\dfrac{1}{2}$ yards long is cut from a reel of ribbon with 50 yards on it. Find the length of the piece remaining on the reel.

△ **122.** A slab of natural granite is purchased and a rectangle with length $7\dfrac{4}{11}$ feet and width $5\dfrac{1}{2}$ feet is cut from it. Find the area of the rectangle.

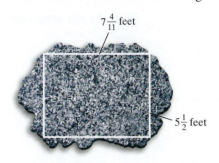

$7\dfrac{4}{11}$ feet

$5\dfrac{1}{2}$ feet

Answers

Write a fraction to represent the shaded area.

1.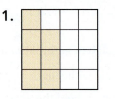

Write the mixed number as an improper fraction.

2. $7\dfrac{2}{3}$

Write the improper fraction as a mixed number.

3. $\dfrac{75}{4}$

Write each fraction in simplest form.

4. $\dfrac{24}{210}$

5. $-\dfrac{42x}{70}$

Determine whether these fractions are equivalent.

6. $\dfrac{5}{7}$ and $\dfrac{8}{11}$

7. $\dfrac{6}{27}$ and $\dfrac{14}{63}$

Find the prime factorization of each number.

8. 84

9. 495

Perform each indicated operation and write the answers in simplest form.

10. $\dfrac{4}{4} \div \dfrac{3}{4}$

11. $-\dfrac{4}{3} \cdot \dfrac{4}{4}$

12. $\dfrac{7x}{9} + \dfrac{x}{9}$

13. $\dfrac{1}{7} - \dfrac{3}{x}$

14. $\dfrac{xy^3}{z} \cdot \dfrac{z}{xy}$

15. $-\dfrac{2}{3} \cdot -\dfrac{8}{15}$

16. $\dfrac{9a}{10} + \dfrac{2}{5}$

17. $-\dfrac{8}{15y} - \dfrac{2}{15y}$

18. $\dfrac{3a}{8} \cdot \dfrac{16}{6a^3}$

19. $\dfrac{11}{12} - \dfrac{3}{8} + \dfrac{5}{24}$

20. $3\dfrac{7}{8}$
$7\dfrac{2}{5}$
$+\,2\dfrac{3}{4}$

21. 19
$-2\dfrac{3}{11}$

Answers

1. _____

2. _____

3. _____

4. _____

5. _____

6. _____

7. _____

8. _____

9. _____

10. _____

11. _____

12. _____

13. _____

14. _____

15. _____

16. _____

17. _____

18. _____

19. _____

20. _____

21. _____

22. $-\dfrac{16}{3} \div -\dfrac{3}{12}$ **23.** $3\dfrac{1}{3} \cdot 6\dfrac{3}{4}$ **24.** $-\dfrac{2}{7} \cdot \left(6 - \dfrac{1}{6}\right)$ **25.** $\dfrac{1}{2} \div \dfrac{2}{3} \cdot \dfrac{3}{4}$

26. $\left(-\dfrac{3}{4}\right)^2 \div \left(\dfrac{2}{3} + \dfrac{5}{6}\right)$ **27.** Find the average of $\dfrac{5}{6}, \dfrac{4}{3}$, and $\dfrac{7}{12}$.

Simplify each complex fraction.

28. $\dfrac{\dfrac{5x}{7}}{\dfrac{20x^2}{21}}$ **29.** $\dfrac{5 + \dfrac{3}{7}}{2 - \dfrac{1}{2}}$

Solve.

30. $-\dfrac{3}{8}x = \dfrac{3}{4}$ **31.** $\dfrac{x}{5} + x = -\dfrac{24}{5}$ **32.** $\dfrac{2}{3} + \dfrac{x}{4} = \dfrac{5}{12} + \dfrac{x}{2}$

Evaluate each expression for the given replacement values.

33. $-5x; x = -\dfrac{1}{2}$ **34.** $x \div y; x = \dfrac{1}{2}, y = 3\dfrac{7}{8}$

Solve.

35. A carpenter cuts a piece $2\dfrac{3}{4}$ feet long from a cedar plank that is $6\dfrac{1}{2}$ feet long. How long is the remaining piece?

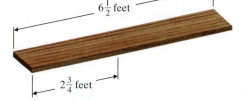

$6\dfrac{1}{2}$ feet

$2\dfrac{3}{4}$ feet

The circle graph below shows us how the average consumer spends money. For example, $\dfrac{7}{50}$ of spending goes for food. Use this information for Exercises 36 through 38.

Consumer Spending

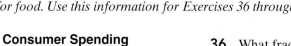

Other $\frac{2}{25}$
Education $\frac{1}{50}$
Insurance and pension $\frac{1}{10}$
Entertainment $\frac{1}{25}$
Health care $\frac{3}{50}$
Transportation $\frac{1}{5}$
Clothing $\frac{1}{25}$
Housing $\frac{8}{25}$
Food $\frac{7}{50}$

Source: U.S. Bureau of Labor Statistics; based on survey

36. What fraction of spending goes for housing and food combined?

37. What fraction of spending goes for education, transportation, and clothing?

38. Suppose your family spent $47,000 on the items in the graph. How much might we expect was spent on health care?

Find the perimeter and area of the figure.

△ **39.**

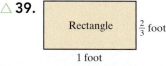

Rectangle $\frac{2}{3}$ foot

1 foot

40. During a 258-mile trip, a car used $10\dfrac{3}{4}$ gallons of gas. How many miles would we expect the car to travel on 1 gallon of gas?

22. _____

23. _____

24. _____

25. _____

26. _____

27. _____

28. _____

29. _____

30. _____

31. _____

32. _____

33. _____

34. _____

35. _____

36. _____

37. _____

38. _____

39. _____

40. _____

Answers

1. _____

2. _____

3. _____

4. _____

5. _____

6. _____

7. _____

8. _____

9. _____

10. _____

11. _____

12. _____

13. _____

14. _____

15. _____

16. _____

17. _____

18. _____

19. _____

20. _____

21. _____

22. _____

Write each number in words.

1. 546

2. 115

3. 27,034

4. 6573

5. Add: $46 + 713$

6. Add: $587 + 44$

7. Subtract: $543 - 29$. Check by adding.

8. Subtract: $995 - 62$. Check by adding.

9. Round 278,362 to the nearest thousand.

10. Round 1436 to the nearest ten.

11. A digital video disc (DVD) can hold about 4800 megabytes (MB) of information. How many megabytes can 12 DVDs hold?

12. On a trip across the country, Daniel Daunis travels 435 miles per day. How many total miles does he travel in 3 days?

13. Divide and check: $56,717 \div 8$

14. Divide and check: $4558 \div 12$

Write using exponential notation.

15. $7 \cdot 7 \cdot 7$

16. $7 \cdot 7$

17. $3 \cdot 3 \cdot 3 \cdot 3 \cdot 9 \cdot 9 \cdot 9$

18. $9 \cdot 9 \cdot 9 \cdot 9 \cdot 5 \cdot 5$

19. Evaluate $2(x - y)$ for $x = 6$ and $y = 3$.

20. Evaluate $8a + 3(b - 5)$ for $a = 5$ and $b = 9$.

21. The world's deepest cave is Krubera (or Voronja), in the country of Georgia, located by the Black Sea in Asia. It has been explored to a depth of 7188 feet below the surface of Earth. Represent this position using an integer. (*Source:* messagetoeagle.com and Wikipedia)

22. The temperature on a cold day in Minneapolis, MN, was 21°F below zero. Represent this temperature using an integer.

23. Add using a number line: $-7 + 3$

24. Add using a number line: $-3 + 8$

25. Simplify: $7 - 8 - (-5) - 1$

26. Simplify: $6 + (-8) - (-9) + 3$

27. Evaluate: $(-5)^2$

28. Evaluate: -2^4

29. Simplify: $3(4 - 7) + (-2) - 5$

30. Simplify: $(20 - 5^2)^2$

31. Simplify: $2y - 6 + 4y + 8$

32. Simplify: $5x - 1 + x + 10$

Solve.

33. $5x + 2 - 4x = 7 - 19$

34. $9y + 1 - 8y = 3 - 20$

35. $17 - 7x + 3 = -3x + 21 - 3x$

36. $9x - 2 = 7x - 24$

37. Write a fraction to represent the shaded part of the figure.

38. Write the prime factorization of 156.

39. Write each as an improper fraction.

   a. $4\frac{2}{9}$ **b.** $1\frac{8}{11}$

40. Write $\frac{39}{5}$ as a mixed number.

41. Write in simplest form: $\frac{42x}{66}$

42. Write in simplest form: $\frac{70}{105y}$

43. Multiply: $3\frac{1}{3} \cdot \frac{7}{8}$

44. Multiply: $\frac{2}{3} \cdot 4$

45. Divide and simplify: $\frac{5}{16} \div \frac{3}{4}$

46. Divide: $1\frac{1}{10} \div 5\frac{3}{5}$

23. _____

24. _____

25. _____

26. _____

27. _____

28. _____

29. _____

30. _____

31. _____

32. _____

33. _____

34. _____

35. _____

36. _____

37. _____

38. _____

39. a. _____ b. _____

40. _____

41. _____

42. _____

43. _____

44. _____

45. _____

46. _____

Decimal numbers represent parts of a whole, just like fractions. For example, one penny is 0.01 or $\frac{1}{100}$ of a dollar. In this chapter, we learn to perform arithmetic operations on decimals and to analyze the relationship between fractions and decimals. We also learn how decimals are used in the real world.

5 Decimals

The graph below shows the age group distribution for average daily texting. (Check your age group and see if the data are accurate based on your own experiences.) While we have practiced calculating averages before in this text, an average certainly does not usually simplify to a whole number. While fractions are useful, decimals are also an important system of numbers that can be used to show values between whole numbers. Data are easy to round when they are in the form of a decimal. In Section 5.2, Exercises 83 and 84, we continue to study text messaging and how it is increasing each year.

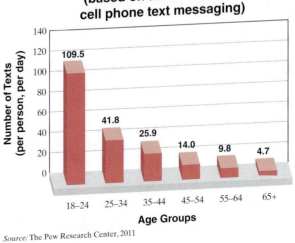

Average Number of Texts Sent/Received per Day (based on adults who use cell phone text messaging)

Source: The Pew Research Center, 2011

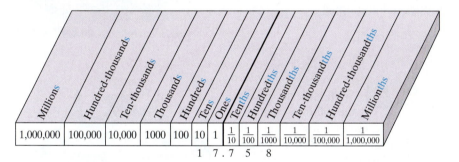
Objective A Decimal Notation and Writing Decimals in Words

Like fractional notation, decimal notation is used to denote a part of a whole. Numbers written in decimal notation are called **decimal numbers,** or simply **decimals.** The decimal 17.758 has three parts.

$$\underbrace{1\ 7}_{\substack{\text{Whole} \\ \text{number} \\ \text{part}}}\ .\ \underbrace{7\ 5\ 8}_{\substack{\text{Decimal} \\ \text{part}}}$$

Decimal point

In Section 1.2, we introduced place value for whole numbers. Place names and place values for the whole number part of a decimal number are exactly the same. Place names and place values for the decimal part are shown below.

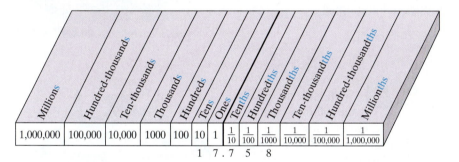

Millions	Hundred-thousands	Ten-thousands	Thousands	Hundreds	Tens	Ones	Tenths	Hundredths	Thousandths	Ten-thousandths	Hundred-thousandths	Millionths
1,000,000	100,000	10,000	1000	100	10	1	$\frac{1}{10}$	$\frac{1}{100}$	$\frac{1}{1000}$	$\frac{1}{10,000}$	$\frac{1}{100,000}$	$\frac{1}{1,000,000}$

1   7 . 7   5   8

Notice that the value of each place is $\frac{1}{10}$ of the value of the place to its left. For example,

$$\underset{\text{ones}}{1} \cdot \underset{\text{tenths}}{\frac{1}{10}} = \underset{\text{tenths}}{\frac{1}{10}} \quad \text{and} \quad \underset{\text{tenths}}{\frac{1}{10}} \cdot \frac{1}{10} = \underset{\text{hundredths}}{\frac{1}{100}}$$

The decimal number 17.758 means

1 ten	+	7 ones	+	7 tenths	+	5 hundredths	+	8 thousandths

or $1 \cdot 10$ + $7 \cdot 1$ + $7 \cdot \dfrac{1}{10}$ + $5 \cdot \dfrac{1}{100}$ + $8 \cdot \dfrac{1}{1000}$

or 10 + 7 + $\dfrac{7}{10}$ + $\dfrac{5}{100}$ + $\dfrac{8}{1000}$

Writing (or Reading) a Decimal in Words

Step 1: Write the whole number part in words.

Step 2: Write "and" for the decimal point.

Step 3: Write the decimal part in words as though it were a whole number, followed by the place value of the last digit.

Objectives

A Know the Meaning of Place Value for a Decimal Number and Write Decimals in Words.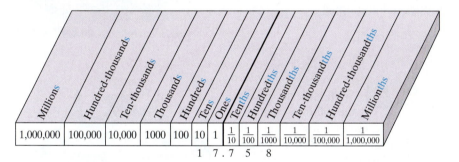

B Write Decimals in Standard Form.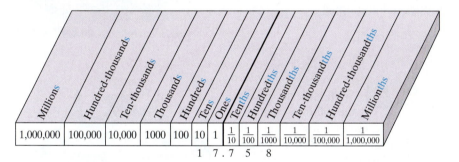

C Write Decimals as Fractions.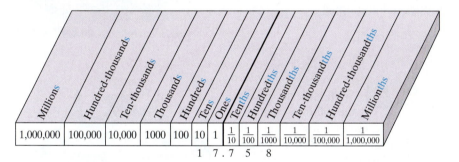

D Compare Decimals.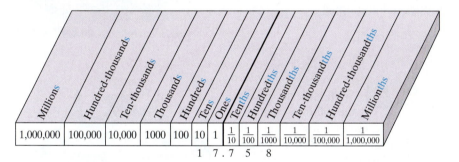

E Round Decimals to Given Place Values.

Helpful Hint   Notice that place values to the left of the decimal point end in "s." Place values to the right of the decimal point end in "ths."

Practice 1

Write each decimal in words.
a. 0.06
b. −200.073
c. 0.0829

Example 1 Write each decimal in words.

a. 0.7   b. −50.82   c. 21.093

Solution:

a. seven tenths
b. negative fifty and eighty-two hundredths
c. twenty-one and ninety-three thousandths

■ **Work Practice 1**

Practice 2

Write the decimal 87.31 in words.

Example 2 Write the decimal in the following sentence in words: The Golden Jubilee Diamond is a 545.67-carat cut diamond. (*Source: The Guinness Book of Records*)

Solution: five hundred forty-five and sixty-seven hundredths

■ **Work Practice 2**

Practice 3

Write the decimal 52.1085 in words.

Example 3 Write the decimal in the following sentence in words: The oldest known fragments of the Earth's crust are Zircon crystals; they were discovered in Australia and are thought to be 4.276 billion years old. (*Source: The Guinness Book of Records*)

Solution: four and two hundred seventy-six thousandths

■ **Work Practice 3**

Suppose that you are paying for a purchase of $368.42 at Circuit City by writing a check. Checks are usually written using the following format.

Answers

1. a. six hundredths b. negative two hundred and seventy-three thousandths c. eight hundred twenty-nine ten-thousandths

2. eighty-seven and thirty-one hundredths

3. fifty-two and one thousand eighty-five ten-thousandths

Example 4 Fill in the check to Camelot Music to pay for your purchase of $92.98.

Solution:

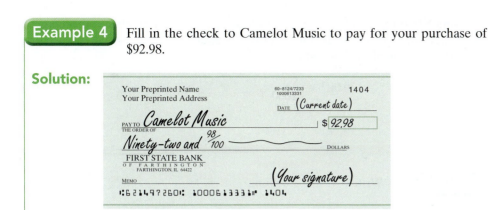

Work Practice 4

Practice 4

Fill in the check to CLECO (Central Louisiana Electric Company) to pay for your monthly electric bill of $207.40.

Objective B Writing Decimals in Standard Form

A decimal written in words can be written in standard form by reversing the procedure in Objective A.

Examples Write each decimal in standard form.

5. Forty-eight and twenty-six hundredths is

48.26

— hundredths place

6. Six and ninety-five thousandths is

6.095

— thousandths place

Work Practice 5–6

Practice 5–6

Write each decimal in standard form.

5. Five hundred and ninety-six hundredths

6. Thirty-nine and forty-two thousandths

Helpful Hint

When converting a decimal from words to decimal notation, make sure the last digit is in the correct place by inserting 0s if necessary. For example,

Two and thirty-eight thousandths is 2.038

thousandths place

Objective C Writing Decimals as Fractions

Once you master reading and writing decimals, writing a decimal as a fraction follows naturally.

Decimal	In Words	Fraction
0.7	seven tenths	$\dfrac{7}{10}$
0.51	fifty-one hundredths	$\dfrac{51}{100}$
0.009	nine thousandths	$\dfrac{9}{1000}$
0.05	five hundredths	$\dfrac{5}{100} = \dfrac{1}{20}$

Answers

4.

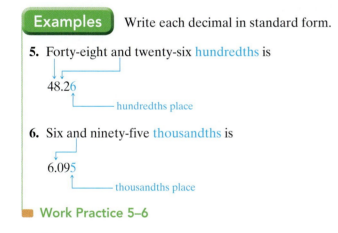

5. 500.96

6. 39.042

Notice that the number of decimal places in a decimal number is the same as the number of zeros in the denominator of the equivalent fraction. We can use this fact to write decimals as fractions.

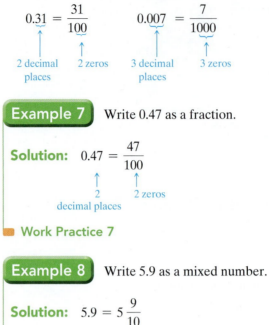

$$0.31 = \frac{31}{100} \qquad 0.007 = \frac{7}{1000}$$

2 decimal places   2 zeros   3 decimal places   3 zeros

Practice 7

Write 0.051 as a fraction.

Example 7 Write 0.47 as a fraction.

Solution: $0.47 = \frac{47}{100}$

2 decimal places   2 zeros

■ **Work Practice 7**

Practice 8

Write 29.97 as a mixed number.

Example 8 Write 5.9 as a mixed number.

Solution: $5.9 = 5\frac{9}{10}$

1 decimal place   1 zero

■ **Work Practice 8**

Practice 9–11

Write each decimal as a fraction or mixed number. Write your answer in simplest form.

9. 0.12

10. 64.8

11. −209.986

Examples Write each decimal as a fraction or a mixed number. Write your answer in simplest form.

9. $0.125 = \frac{125}{1000} = \frac{\overset{1}{\cancel{125}}}{8 \cdot \cancel{125}} = \frac{1}{8}$

10. $43.5 = 43\frac{5}{10} = 43\frac{\overset{1}{\cancel{5}}}{2 \cdot \cancel{5}} = 43\frac{1}{2 \cdot 1} = 43\frac{1}{2}$

11. $-105.083 = -105\frac{83}{1000}$

■ **Work Practice 9–11**

Later in the chapter, we write fractions as decimals. If you study Examples 7–11, you already know how to write fractions with denominators of 10, 100, 1000, and so on, as decimals.

Objective D Comparing Decimals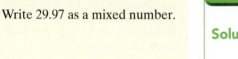

One way to compare positive decimals is by comparing digits in corresponding places. To see why this works, let's compare 0.5 or $\frac{5}{10}$ and 0.8 or $\frac{8}{10}$. We know

$$\frac{5}{10} < \frac{8}{10} \text{ since } 5 < 8, \text{ so}$$

$$0.5 < 0.8 \text{ since } 5 < 8$$

This leads to the following.

Comparing Two Positive Decimals

Compare digits in the same places from left to right. When two digits are not equal, the number with the larger digit is the larger decimal. If necessary, insert 0s after the last digit to the right of the decimal point to continue comparing.

Compare hundredths place digits

28.2**5**3   28.2**6**3
          ↑   ↑
          5   <   6
so 28.253   <   28.263

Helpful Hint

For any decimal, writing 0s after the last digit to the right of the decimal point does not change the value of the number.

$7.6 = 7.60 = 7.600$, and so on

When a whole number is written as a decimal, the decimal point is placed to the right of the ones digit.

$25 = 25.0 = 25.00$, and so on

Example 12 Insert $<$, $>$, or $=$ to form a true statement.

0.378   0.368

Solution:

0.3 78   0.3 68   The tenths places are the same.

0.3 7 8   0.3 6 8   The hundredths places are different.

Since $7 > 6$, then $0.378 > 0.368$.

■ Work Practice 12

Practice 12

Insert $<$, $>$, or $=$ to form a true statement.

26.208   26.28

Example 13 Insert $<$, $>$, or $=$ to form a true statement.

0.052   0.236

Solution: 0. 0 52 $<$ 0. 2 36   0 is smaller than 2 in the tenths place.
          ↑   ↑

■ Work Practice 13

Practice 13

Insert $<$, $>$, or $=$ to form a true statement.

0.12   0.026

We can also use a number line to compare decimals. This is especially helpful when comparing negative decimals. Remember, the number whose graph is to the left is smaller, and the number whose graph is to the right is larger.

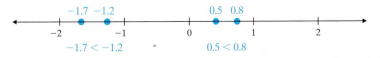

$-1.7 < -1.2$ $0.5 < 0.8$

Answers

12. $<$ **13.** $>$

Helpful Hint

If you have trouble comparing two negative decimals, try the following: Compare their absolute values. Then to correctly compare the negative decimals, reverse the direction of the inequality symbol.

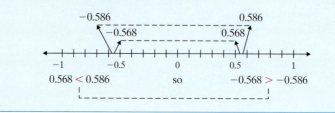

$$0.568 < 0.586 \quad \text{so} \quad -0.568 > -0.586$$

Practice 14

Insert $<$, $>$, or $=$ to form a true statement.

$-0.039 \quad -0.0309$

Example 14   Insert $<$, $>$, or $=$ to form a true statement.

$-0.0101 \quad -0.00109$

Solution: Since $0.0101 > 0.00109$, then $-0.0101 \; < \; -0.00109$.

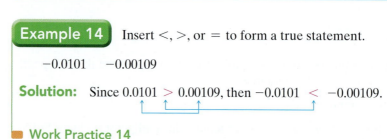

■ **Work Practice 14**

Objective E Rounding Decimals

We **round the decimal part** of a decimal number in nearly the same way as we round whole numbers. The only difference is that we drop digits to the right of the rounding place, instead of replacing these digits with 0s. For example,

36.954   rounded to the nearest hundredth is   36.95.

Rounding Decimals to a Place Value to the Right of the Decimal Point

Step 1: Locate the digit to the right of the given place value.

Step 2: If this digit is 5 or greater, add 1 to the digit in the given place value and drop all digits to its right. If this digit is less than 5, drop all digits to the right of the given place.

Practice 15

Round 482.7817 to the nearest thousandth.

Example 15   Round 736.2359 to the nearest tenth.

Solution:

Step 1: We locate the digit to the right of the tenths place.

tenths place

736.2 **3** 59

digit to the right

Step 2: Since the digit to the right is less than 5, we drop it and all digits to its right.

Thus, 736.2359 rounded to the nearest tenth is 736.2.

■ **Work Practice 15**

The same steps for rounding can be used when the decimal is negative.

Answers

14. $<$ **15.** 482.782

Example 16 Round −0.027 to the nearest hundredth.

Solution:

Step 1: Locate the digit to the right of the hundredths place.

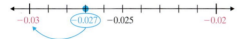

$$- \ 0.02\,\boxed{7}$$

Step 2: Since the digit to the right is 5 or greater, we add 1 to the hundredths digit and drop all digits to its right.

Thus, −0.027 is −0.03 rounded to the nearest hundredth.

■ Work Practice 16

The following number line illustrates the rounding of negative decimals.

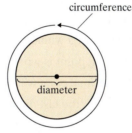

In Section 5.3, we will introduce a formula for the distance around a circle. The distance around a circle is given the special name **circumference.**

The symbol π is the Greek letter pi, pronounced "pie." We use π to denote the following constant:

$$\pi = \frac{\text{circumference of a circle}}{\text{diameter of a circle}}$$

The value π is an **irrational number.** This means if we try to write it as a decimal, it neither ends nor repeats in a pattern.

Example 17 $\pi \approx 3.14159265$. Round π to the nearest hundredth.

Solution:

hundredths place ⟶ | | ⟵ 1 is less than 5.

3.14159265

↑ ⟶ Delete these digits.

Thus, 3.14159265 rounded to the nearest hundredth is 3.14. In other words, $\pi \approx 3.14$.

■ Work Practice 17

Rounding often occurs with money amounts. Since there are 100 cents in a dollar, each cent is $\frac{1}{100}$ of a dollar. This means that if we want to round to the nearest cent, we round to the nearest hundredth of a dollar.

✓ **Concept Check** 1756.0894 rounded to the nearest *ten* is

a. 1756.1 **b.** 1760.0894 **c.** 1760 **d.** 1750

Practice 16

Round −0.032 to the nearest hundredth.

Practice 17

$\pi \approx 3.14159265$. Round π to the nearest ten-thousandth.

Answers

16. −0.03 **17.** $\pi \approx 3.1416$

✓ **Concept Check Answer**

c

Practice 18

Water bills in Mexia are always rounded to the nearest dollar. Round a water bill of $24.62 to the nearest dollar.

Answer

18. $25

Example 18   Determining State Taxable Income

A high school teacher's taxable income is $41,567.72. The tax tables in the teacher's state use amounts rounded to the nearest dollar. Round the teacher's income to the nearest whole dollar.

Solution: Rounding to the nearest whole dollar means rounding to the ones place.

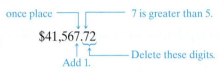

once place ──┐   ┌── 7 is greater than 5.

$41,567.72

Add 1. ↑ ↑── Delete these digits.

Thus, the teacher's income rounded to the nearest dollar is $41,568.

■ **Work Practice 18**

Vocabulary, Readiness & Video Check

Use the choices below to fill in each blank.

| words | decimals | tenths | after |
| tens | circumference | and | standard form |

1. The number "twenty and eight hundredths" is written in _____ and "20.08" is written in _____.
2. Another name for the distance around a circle is its _____.
3. Like fractions, _____ are used to denote part of a whole.
4. When writing a decimal number in words, the decimal point is written as _____.
5. The place value _____ is to the right of the decimal point while _____ is to the left of the decimal point.
6. The decimal point in a whole number is _____ the last digit.

Martin-Gay Interactive Videos

See Video 5.1 ●

Watch the section lecture video and answer the following questions.

Objective A   7. In ▦ Example 1, how is the decimal point written? ▶

Objective B   8. Why is 9.8 not the correct answer to ▦ Example 3? What is the correct answer? ▶

Objective C   9. From ▦ Example 5, why does reading a decimal number correctly help you write it as an equivalent fraction? ▶

Objective D 10. In ▦ Example 7, we compare place value by place value in which direction? ▶

Objective E 11. ▦ Example 8 is being rounded to the nearest tenth, so why is the digit 7, which is not in the tenths place, looked at? ▶

5.1   **Exercise Set** MyMathLab® ▶

Objective A *Write each decimal number in words. See Examples 1 through 3.*

1. 5.62

2. 9.57

3. 16.23

4. 47.65

5. −0.205

6. −0.495

▶ **7.** 167.009

8. 233.056

9. 3000.04

10. 5000.02

11. 105.6

12. 410.3

13. The Akashi Kaikyo Bridge, between Kobe and Awaji-Shima, Japan, is approximately 2.43 miles long.

14. The English Channel Tunnel is a 31.04 miles long undersea rail tunnel connecting England and France. (*Source: Railway Directory & Year Book*)

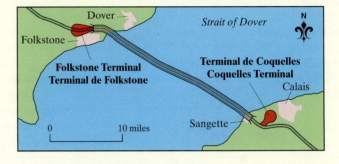

Fill in each check for the described purchase. See Example 4.

15. Your monthly car loan of $321.42 to R. W. Financial.

Your Preprinted Name	60–8124/7233	1407
Your Preprinted Address	1000613331	
	DATE	
PAY TO THE ORDER OF		$
		DOLLARS
FIRST STATE BANK OF FARTHINGTON FARTHINGTON, IL 64422		
MEMO		
⑆621497260⑆ 1000613331⑈ 1407		

16. Your part of the monthly apartment rent, which is $213.70. You pay this to Amanda Dupre.

Your Preprinted Name	60–8124/7233	1408
Your Preprinted Address	1000613331	
	DATE	
PAY TO THE ORDER OF		$
		DOLLARS
FIRST STATE BANK OF FARTHINGTON FARTHINGTON, IL 64422		
MEMO		
⑆621497260⑆ 1000613331⑈ 1408		

17. Your bill of $91.68 to Verizon wireless.

Your Preprinted Name	60–8124/7233	1409
Your Preprinted Address	1000613331	
	DATE	
PAY TO THE ORDER OF		$
		DOLLARS
FIRST STATE BANK OF FARTHINGTON FARTHINGTON, IL 64422		
MEMO		
⑆621497260⑆ 1000613331⑈ 1409		

18. Your grocery bill of $387.49 at Kroger.

Your Preprinted Name	60–8124/7233	1410
Your Preprinted Address	1000613331	
	DATE	
PAY TO THE ORDER OF		$
		DOLLARS
FIRST STATE BANK OF FARTHINGTON FARTHINGTON, IL 64422		
MEMO		
⑆621497260⑆ 1000613331⑈ 1410		

Objective B *Write each decimal number in standard form. See Examples 5 and 6.*

19. Two and eight tenths

20. Five and one tenth

▶ **21.** Nine and eight hundredths

22. Twelve and six hundredths

23. Negative seven hundred five and six hundred twenty-five thousandths

24. Negative eight hundred four and three hundred ninety-nine thousandths

▶ **25.** Forty-six ten-thousandths

26. Eighty-three ten-thousandths

Objective C *Write each decimal as a fraction or a mixed number. Write your answer in simplest form. See Examples 7 through 11.*

27. 0.7

28. 0.9

▶ **29.** 0.27

30. 0.39

31. 0.4

32. 0.8

33. 5.4

34. 6.8

35. −0.058

36. −0.024

▶ **37.** 7.008

38. 9.005

39. 15.802

40. 11.406

41. 0.3005

42. 0.2006

Objectives A B C Mixed Practice *Fill in the chart. The first row is completed for you. See Examples 1 through 11.*

	Decimal Number in Standard Form	In Words	Fraction
	0.37	thirty-seven hundredths	$\frac{37}{100}$
43.		eight tenths	
44.		five tenths	
45.	0.077		
46.	0.019		

Objective D *Insert <, >, or = between each pair of numbers to form a true statement. See Examples 12 through 14.*

47. 0.15   0.16

48. 0.12   0.15

49. −0.57   −0.54

50. −0.59   −0.52

51. 0.098   0.1

52. 0.0756   0.2

53. 0.54900   0.549

54. 0.98400   0.984

▶ **55.** 167.908   167.980

56. 519.3405   519.3054

57. −1.062   −1.07

58. −18.1   −18.01

59. −7.052   7.0052

60. 0.01   −0.1

61. −0.023   −0.024

62. −0.562   −0.652

Objective E *Round each decimal to the given place value. See Examples 15 through 18.*

▶ **63.** 0.57, nearest tenth

64. 0.64, nearest tenth

65. 98,207.23, nearest ten

66. 68,934.543, nearest ten

67. −0.234, nearest hundredth

68. −0.892, nearest hundredth

▶ **69.** 0.5942, nearest thousandth

70. 63.4523, nearest thousandth

Recall that the number π, written as a decimal, neither ends nor repeats in a pattern. Given that π ≈ 3.14159265, round π to the given place values below. (We study π further in Section 5.3.) See Example 17.

71. tenth   **72.** ones   **73.** thousandth   **74.** hundred-thousandth

Round each monetary amount to the nearest cent or dollar as indicated. See Example 18.

75. $26.95, to the nearest dollar

76. $14,769.52, to the nearest dollar

77. $0.1992, to the nearest cent

78. $0.7633, to the nearest cent

Round each number to the given place value. See Example 18.

79. At the time of this writing, the Apple MacBook Air is the thinnest Mac in production. At its thickest point, it measures 0.68 in. Round this number to the nearest tenth. (*Source:* Apple, Inc.)

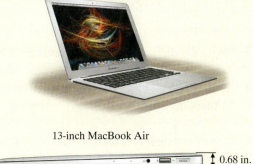

13-inch MacBook Air

0.68 in.

80. A large tropical cockroach of the family Dictyoptera is the fastest-moving insect. This insect was clocked at a speed of 3.36 miles per hour. Round this number to the nearest tenth. (*Source:* University of California, Berkeley)

81. Missy Franklin of the United States won the gold medal for the 200 m backstroke in the 2012 London Summer Olympics with a record time of 2.0677 minutes. Round this time to the nearest hundredth of a minute.

82. The population density of the state of Utah is 34.745 people per square mile. Round this population density to the nearest tenth. (*Source:* U.S. Census Bureau)

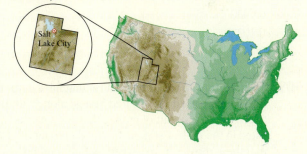

Salt Lake City

83. A used biology textbook is priced at $67.89. Round this price to the nearest dollar.

84. A used office desk is advertised at $19.95 by Drawley's Office Furniture. Round this price to the nearest dollar.

85. Venus makes a complete orbit around the Sun every 224.695 days. Round this figure to the nearest whole day. (*Source:* National Space Science Data Center)

86. The length of a day on Mars, a full rotation about its axis, is 24.6229 hours. Round this figure to the nearest thousandth. (*Source:* National Space Science Data Center)

Review

Perform each indicated operation. See Section 1.3.

87. 3452 + 2314   **88.** 8945 + 4536   **89.** 82 − 47   **90.** 4002 − 3897

Concept Extensions

Solve. See the Concept Check in this section.

91. 2849.1738 rounded to the nearest hundred is

    a. 2849.17   **b.** 2800   **c.** 2850   **d.** 2849.174

92. 146.059 rounded to the nearest ten is

    a. 146.0   **b.** 146.1   **c.** 140   **d.** 150

93. 2849.1738 rounded to the nearest hundredth is

    a. 2849.17   **b.** 2800   **c.** 2850   **d.** 2849.18

94. 146.059 rounded to the nearest tenth is

    a. 146.0   **b.** 146.1   **c.** 140   **d.** 150

95. In your own words, describe how to write a decimal as a fraction or a mixed number.

96. Explain how to identify the value of the 9 in the decimal 486.3297.

97. Write $7\dfrac{12}{100}$ as a decimal.

98. Write $17\dfrac{268}{1000}$ as a decimal.

99. Write 0.00026849577 as a fraction.

100. Write 0.00026849576 in words.

101. Write a 5-digit number that rounds to 1.7.

102. Write a 4-digit number that rounds to 26.3.

103. Write a decimal number that is greater than 8 but less than 9.

104. Write a decimal number that is greater than 48.1, but less than 48.2.

105. Which number(s) rounds to 0.26?

    0.26559   0.26499   0.25786   0.25186

106. Which number(s) rounds to 0.06?

    0.0612   0.066   0.0586   0.0506

Write these numbers from smallest to largest.

107. 0.9

0.1038

0.10299

0.1037

108. 0.01

0.0839

0.09

0.1

109. The all-time top six movies (those that have earned the most money in the United States) along with the approximate amount of money they have earned are listed in the table. Estimate the total amount of money that these movies have earned by first rounding each earning to the nearest hundred-million. (*Source:* The Internet Movie Database)

Top All-Time American Movies	
Movie	**Gross Domestic Earnings**
Avatar (2009)	$760.5 million
Titanic (1997)	$658.7 million
The Avengers (2012)	$623.3 million
The Dark Knight (2008)	$533.3 million
Star Wars: The Phantom Menace (1999)	$474.5 million
Star Wars (1977)	$460.9 million

110. In 2012, there were 1392.2 million singles downloaded at an average price of $1.20 each. Find an estimate of the total revenue from downloaded singles by answering parts **a–c**. (*Source:* Recording Industry Association of America)

    a. Round 1392.2 million to the nearest ten million.

    b. Multiply the rounded value in part **a** by 12.

    c. Move the decimal point in the product from part **b** one place to the left. This number is the total revenue in million dollars.

5.2 Adding and Subtracting Decimals

Objective A Adding or Subtracting Decimals

Adding or subtracting decimals is similar to adding or subtracting whole numbers. We add or subtract digits in corresponding place values from right to left, carrying or borrowing if necessary. To make sure that digits in corresponding place values are added or subtracted, we line up the decimal points vertically.

Objectives

A Add or Subtract Decimals.

B Estimate when Adding or Subtracting Decimals.

C Evaluate Expressions with Decimal Replacement Values.

D Simplify Expressions Containing Decimals.

E Solve Problems That Involve Adding or Subtracting Decimals.

> **Adding or Subtracting Decimals**
>
> **Step 1:** Write the decimals so that the decimal points line up vertically.
>
> **Step 2:** Add or subtract as with whole numbers.
>
> **Step 3:** Place the decimal point in the sum or difference so that it lines up vertically with the decimal points in the problem.

In this section, we will insert zeros in decimal numbers so that place-value digits line up neatly. This is shown in Example 1.

Example 1 Add: $23.85 + 1.604$

Solution: First we line up the decimal points vertically.

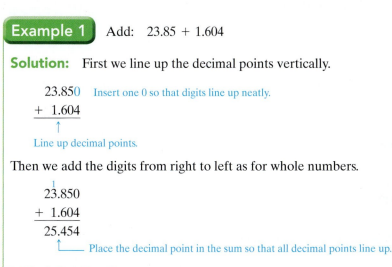

```
   23.850    Insert one 0 so that digits line up neatly.
+   1.604
    ↑
Line up decimal points.
```

Then we add the digits from right to left as for whole numbers.

```
     1
   23.850
+   1.604
   25.454
       └── Place the decimal point in the sum so that all decimal points line up.
```

■ **Work Practice 1**

Practice 1

Add.
a. $19.52 + 5.371$
b. $40.08 + 17.612$
c. $0.125 + 422.8$

> ☁ **Helpful Hint**
>
> Recall that 0s may be placed after the last digit to the right of the decimal point without changing the value of the decimal. This may be used to help line up place values when adding decimals.
>
> ```
>     3.2     becomes      3.200    Insert two 0s.
>    15.567               15.567
> +   0.11              +  0.110    Insert one 0.
>                         18.877    Add.
> ```

Answers
1. a. 24.891 **b.** 57.692 **c.** 422.925

341

Practice 2

Add.

a. $34.567 + 129.43 + 2.8903$

b. $11.21 + 46.013 + 362.526$

Example 2 Add: $763.7651 + 22.001 + 43.89$

Solution: First we line up the decimal points.

$$
\begin{array}{r}
\overset{1\ 1\ 1}{763.7651} \\
22.0010 \quad \text{Insert one 0.}\\
+\ \ 43.8900 \quad \text{Insert two 0s.}\\
\hline
829.6561 \quad \text{Add.}
\end{array}
$$

■ **Work Practice 2**

> **Helpful Hint**
>
> Don't forget that the decimal point in a whole number is positioned after the last digit.

Practice 3

Add: $19 + 26.072$

Example 3 Add: $45 + 2.06$

Solution:

$$
\begin{array}{r}
45.00 \quad \text{Insert a decimal point and two 0s.}\\
+\ \ 2.06 \quad \text{Line up decimal points.}\\
\hline
47.06 \quad \text{Add.}
\end{array}
$$

■ **Work Practice 3**

✓**Concept Check** What is wrong with the following calculation of the sum of 7.03, 2.008, 19.16, and 3.1415?

$$
\begin{array}{r}
7.03 \\
2.008 \\
19.16 \\
+\ 3.1415 \\
\hline
3.6042
\end{array}
$$

Practice 4

Add: $7.12 + (-9.92)$

Example 4 Add: $3.62 + (-4.78)$

Solution: Recall from Chapter 2 that to add two numbers with different signs, we find the difference of the larger absolute value and the smaller absolute value. The sign of the answer is the same as the sign of the number with the larger absolute value.

$$
\begin{array}{r}
4.78 \\
-\ 3.62 \\
\hline
1.16 \quad \text{Subtract the absolute values.}
\end{array}
$$

Thus, $3.62 + (-4.78) = -1.16$

The sign of the number with the larger absolute value; -4.78 has the larger absolute value.

■ **Work Practice 4**

Subtracting decimals is similar to subtracting whole numbers. We line up digits and subtract from right to left, borrowing when needed.

Answers

2. a. 166.8873 **b.** 419.749

3. 45.072 **4.** −2.8

✓**Concept Check Answer**

The decimal places are not lined up properly.

Example 5 Subtract: 3.5 − 0.068. Check your answer.

Solution:
$$
\begin{array}{r}
\overset{\scriptstyle 9}{}\\
4\,\overset{\scriptstyle 10}{\cancel{5}}\,10\\
3\,.\,\cancel{5}\,\cancel{0}\,\cancel{0}\\
-\,0\,.\,0\,6\,8\\
\hline
3\,.\,4\,3\,2
\end{array}
$$

Insert two 0s.
Line up decimal points.
Subtract.

Check: Recall that we can check a subtraction problem by adding.

$$
\begin{array}{r}
3.432 \\
+\,0.068 \\
\hline
3.500
\end{array}
$$

Difference
Subtrahend
Minuend

■ **Work Practice 5**

Practice 5
Subtract. Check your answers.
a. 6.7 − 3.92
b. 9.72 − 4.068

Example 6 Subtract: 85 − 17.31. Check your answer.

Solution:
$$
\begin{array}{r}
\overset{14\;\;9}{7\,\cancel{4}\,\cancel{5}}\,10\\
8\,\cancel{5}\,.\,\cancel{0}\,\cancel{0}\\
-\,1\,7\,.\,3\,1\\
\hline
6\,7\,.\,6\,9
\end{array}
$$

Check:
$$
\begin{array}{r}
67.69 \\
+\,17.31 \\
\hline
85.00
\end{array}
$$

Difference
Subtrahend
Minuend

■ **Work Practice 6**

Practice 6
Subtract. Check your answers.
a. 73 − 29.31
b. 210 − 68.22

Example 7 Subtract 3 from 6.98.

Solution:
$$
\begin{array}{r}
6.98 \\
-\,3.00 \\
\hline
3.98
\end{array}
$$

Insert two 0s.

Check:
$$
\begin{array}{r}
3.98 \\
+\,3.00 \\
\hline
6.98
\end{array}
$$

Difference
Subtrahend
Minuend

■ **Work Practice 7**

Practice 7
Subtract 19 from 25.91

Example 8 Subtract: −5.8 − 1.7

Solution: Recall from Chapter 2 that to subtract 1.7, we add the opposite of 1.7, or −1.7. Thus

$$-5.8 - 1.7 = -5.8 + (-1.7)$$ To subtract, add the opposite of 1.7, which is −1.7.

Add the absolute values.

$$= -7.5.$$

Use the common negative sign.

■ **Work Practice 8**

Practice 8
Subtract: −5.4 − 9.6

Example 9 Subtract: −2.56 − (−4.01)

Solution: $$-2.56 - (-4.01) = -2.56 + 4.01$$ To subtract, add the opposite of −4.01, which is 4.01.

Subtract the absolute values.

$$= 1.45$$

The answer is positive since 4.01 has the larger absolute value.

■ **Work Practice 9**

Practice 9
Subtract: −1.05 − (−7.23)

Answers
5. a. 2.78 **b.** 5.652
6. a. 43.69 **b.** 141.78
7. 6.91 **8.** −15 **9.** 6.18

Objective B Estimating when Adding or Subtracting Decimals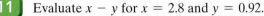

To help avoid errors, we can also estimate to see if our answer is reasonable when adding or subtracting decimals. Although only one estimate is needed per operation, we show two for variety.

Practice 10

Add or subtract as indicated. Then estimate to see if the answer is reasonable by rounding the given numbers and adding or subtracting the rounded numbers.

a. $58.1 + 326.97$

b. $16.08 - 0.925$

Example 10 Add or subtract as indicated. Then estimate to see if the answer is reasonable by rounding the given numbers and adding or subtracting the rounded numbers.

a. $27.6 + 519.25$

Exact		Estimate 1		Estimate 2
$\overset{1}{27.60}$ rounds to		30		30
$+\,519.25$ rounds to		$+\,500$ or		$+\,520$
546.85		530		550

Since the exact answer is close to either estimate, it is reasonable. (In the first estimate, each number is rounded to the place value of the leftmost digit. In the second estimate, each number is rounded to the nearest ten.)

b. $11.01 - 0.862$

Exact		Estimate 1		Estimate 2
$\overset{0\ \ \overset{9}{\cancel{10}}\ \cancel{0}\ \overset{10}{\cancel{1}}}{1\cancel{1}.\cancel{0}\cancel{1}\cancel{0}}$ rounds to		10		11
$-\ 0\ .8\ 6\ 2$ rounds to		$-\ 1$ or		$-\ 1$
$10\ .1\ 4\ 8$		9		10

In the first estimate, we rounded the first number to the nearest ten and the second number to the nearest one. In the second estimate, we rounded both numbers to the nearest one. Both estimates show us that our answer is reasonable.

■ **Work Practice 10**

 Helpful Hint Remember that estimates are used for our convenience to quickly check the reasonableness of an answer.

✓**Concept Check** Why shouldn't the sum $21.98 + 42.36$ be estimated as $30 + 50 = 80$?

Objective C Using Decimals as Replacement Values ▶

Let's review evaluating expressions with given replacement values. This time the replacement values are decimals.

Practice 11

Evaluate $y - z$ for $y = 11.6$ and $z = 10.8$.

Example 11 Evaluate $x - y$ for $x = 2.8$ and $y = 0.92$.

Solution: Replace x with 2.8 and y with 0.92 and simplify.

$$x - y = 2.8 - 0.92$$
$$= 1.88$$

$$\begin{array}{r} 2.80 \\ -0.92 \\ \hline 1.88 \end{array}$$

Answers

10. a. Exact: 385.07; an Estimate: 390
b. Exact: 15.155; an Estimate: 15   **11.** 0.8

■ **Work Practice 11**

✓ **Concept Check Answer**

Each number is rounded incorrectly. The estimate is too high.

Example 12   Is 2.3 a solution of the equation $6.3 = x + 4$?

Solution: Replace x with 2.3 in the equation $6.3 = x + 4$ to see if the result is a true statement.

$$6.3 = x + 4$$
$$6.3 \stackrel{?}{=} 2.3 + 4 \quad \text{\textcolor{blue}{Replace x with 2.3.}}$$
$$6.3 = 6.3 \quad \text{\textcolor{blue}{True}}$$

Since $6.3 = 6.3$ is a true statement, 2.3 is a solution of $6.3 = x + 4$.

■ **Work Practice 12**

Practice 12
Is 12.1 a solution of the equation $y - 4.3 = 7.8$?

Objective D Simplifying Expressions Containing Decimals ▶

Example 13   Simplify by combining like terms:

$$11.1x - 6.3 + 8.9x - 4.6$$

Solution:

$$11.1x - 6.3 + 8.9x - 4.6 = 11.1x + 8.9x + (-6.3) + (-4.6)$$
$$= 20x + (-10.9)$$
$$= 20x - 10.9$$

■ **Work Practice 13**

Practice 13
Simplify by combining like terms:
$$-4.3y + 7.8 - 20.1y + 14.6$$

Objective E Solving Problems by Adding or Subtracting Decimals ▶

Decimals are very common in real-life problems.

Example 14   Calculating the Cost of Owning an Automobile

Find the total monthly cost of owning and operating a certain automobile given the expenses shown.

Monthly car payment:	$256.63
Monthly insurance cost:	$47.52
Average gasoline bill per month:	$195.33

Solution:
1. **UNDERSTAND.** Read and reread the problem. The phrase "total monthly cost" tells us to add.
2. **TRANSLATE.**

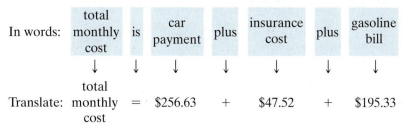

In words:	total monthly cost	is	car payment	plus	insurance cost	plus	gasoline bill
	↓	↓	↓	↓	↓	↓	↓
Translate:	total monthly cost	=	$256.63	+	$47.52	+	$195.33

(Continued on next page)

Practice 14
Find the total monthly cost of owning and operating a certain automobile given the expenses shown.

Monthly car payment:	$563.52
Monthly insurance cost:	$52.68
Average gasoline bill per month:	$127.50

Answers
12. yes   **13.** $-24.4y + 22.4$
14. $743.70

3. SOLVE: Let's also estimate by rounding each number to the nearest ten.

$$
\begin{array}{r}
\overset{1\ 1\ 1}{256.63} \quad \text{rounds to} \quad 260 \\
47.52 \quad \text{rounds to} \quad 50 \\
+\ 195.33 \quad \text{rounds to} \quad 200 \\
\hline
499.48 \quad \text{Exact} \qquad 510 \quad \text{Estimate}
\end{array}
$$

4. INTERPRET. *Check* your work. Since our estimate is close to our exact answer, our answer is reasonable. *State* your conclusion: The total monthly cost is $499.48.

🔲 **Work Practice 14**

The next bar graph has horizontal bars. To visualize the value represented by a bar, see how far it extends to the right. The value of each bar is labeled, and we will study bar graphs further in a later chapter.

Practice 15

Use the bar graph in Example 15. How much greater is the average height in the Netherlands than the average height in Czechoslovakia?

> **Example 15** Comparing Average Heights

The bar graph shows the current average heights for adults in various countries. How much greater is the average height in Denmark than the average height in the United States?

Average Adult Height

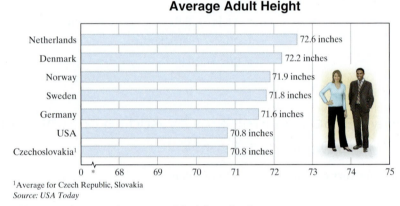

¹Average for Czech Republic, Slovakia
Source: USA Today

* The 〰 means that some numbers are purposefully missing on the axis.

Solution:

1. UNDERSTAND. Read and reread the problem. Since we want to know "how much greater," we subtract.

2. TRANSLATE.

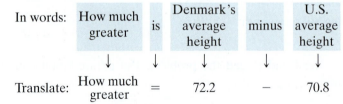

In words:	How much greater	is	Denmark's average height	minus	U.S. average height
	↓	↓	↓	↓	↓
Translate:	How much greater	=	72.2	−	70.8

3. SOLVE: We estimate by rounding each number to the nearest whole.

$$
\begin{array}{r}
\overset{1\ \ 12}{7\cancel{2}.\cancel{2}} \quad \text{rounds to} \quad 72 \\
-\ 7\ 0\ .\ 8 \quad \text{rounds to} \quad -71 \\
\hline
1\ .\ 4 \quad \text{Exact} \qquad 1 \quad \text{Estimate}
\end{array}
$$

4. INTERPRET. *Check* your work. Since our estimate is close to our exact answer, 1.4 inches is reasonable. *State* your conclusion: The average height in Denmark is 1.4 inches greater than the average U.S. height.

🔲 **Work Practice 15**

Answer

15. 1.8 in.

Calculator Explorations Decimals

Entering Decimal Numbers

To enter a decimal number, find the key marked $\boxed{\cdot}$. To enter the number 2.56, for example, press the keys $\boxed{2}\,\boxed{\cdot}\,\boxed{56}$.

The display will read $\boxed{\quad 2.56\quad}$.

Operations on Decimal Numbers

Operations on decimal numbers are performed in the same way as operations on whole or signed numbers. For example, to find $8.625 - 4.29$, press the keys $\boxed{8.625}\;\boxed{-}\;\boxed{4.29}$ then $\boxed{=}$ or $\boxed{\text{ENTER}}$.

The display will read $\boxed{\quad 4.335\quad}$. (Although entering 8.625, for example, requires pressing more than one key, we group numbers together here for easier reading.)

Use a calculator to perform each indicated operation.

1. $315.782 + 12.96$ **2.** $29.68 + 85.902$

3. $6.249 - 1.0076$ **4.** $5.238 - 0.682$

5.
$$
\begin{array}{r}
12.555 \\
224.987 \\
5.2 \\
+\,622.65 \\
\end{array}
$$

6.
$$
\begin{array}{r}
47.006 \\
0.17 \\
313.259 \\
+\,139.088 \\
\end{array}
$$

Vocabulary, Readiness & Video Check

Use the choices below to fill in each blank. Not all choices will be used.

minuend	vertically	like	true
difference	subtrahend	last	false

1. The decimal point in a whole number is positioned after the _____ digit.
2. In $89.2 - 14.9 = 74.3$, the number 74.3 is called the _____, 89.2 is the _____, and 14.9 is the _____.
3. To simplify an expression, we combine any _____ terms.
4. To add or subtract decimals, we line up the decimal points _____.
5. True or false: If we replace x with 11.2 and y with -8.6 in the expression $x - y$, we have $11.2 - 8.6$. _____
6. True or false: If we replace x with -9.8 and y with -3.7 in the expression $x + y$, we have $-9.8 + 3.7$. _____

Martin-Gay Interactive Videos Watch the section lecture video and answer the following questions.

See Video 5.2

Objective A **7.** From Examples 1–3, why do you think we line up decimal points?

Objective B **8.** In Example 4, estimating is used to check whether the answer to the subtraction problem is reasonable, but what is the best way to fully check?

Objective C **9.** In Example 5, why is the actual subtraction performed to the side?

Objective D **10.** How many sets of like terms are there in Example 6?

Objective E **11.** In Example 7, to calculate the amount of border material needed, we are actually calculating the _____ of the triangle.

5.2 Exercise Set MyMathLab® ▶

Objectives A B Mixed Practice *Add. See Examples 1 through 4 and 10. For those exercises marked, also estimate to see if the answer is reasonable.*

1. 5.6 + 2.1

2. 3.6 + 4.1

3. 8.2 + 2.15

4. 5.17 + 3.7

▶ **5.** 24.6 + 2.39 + 0.0678

6. 32.4 + 1.58 + 0.0934

7. −2.6 + (−5.97)

8. −18.2 + (−10.8)

9. 18.56 + (−8.23)

10. 4.38 + (−6.05)

11.
$$234.89$$
$$+\ 230.67$$
Exact: Estimate:

12.
$$734.89$$
$$+\ 640.56$$
Exact: Estimate:

13.
$$100.009$$
$$6.08$$
$$+\ \ \ \ 9.034$$
Exact: Estimate:

14.
$$200.89$$
$$7.49$$
$$+\ \ 62.83$$
Exact: Estimate:

15. Find the sum of 39, 3.006, and 8.403

16. Find the sum of 65, 5.0903, and 6.9003

Subtract and check. See Examples 5 through 10. For those exercises marked, also estimate to see if the answer is reasonable.

17. 12.6 − 8.2

18. 8.9 − 3.1

19. 18 − 2.7

20. 28 − 3.3

▶ **21.**
$$654.9$$
$$-\ \ 56.67$$

22.
$$863.23$$
$$-\ \ 39.453$$

23. 5.9 − 4.07
Exact:
Estimate:

24. 6.4 − 3.04
Exact:
Estimate:

▶ **25.**
$$1000$$
$$-\ \ 123.4$$
Exact:

Estimate:

26.
$$2000$$
$$-\ \ 327.47$$
Exact:

Estimate:

27. 200 − 5.6

28. 800 − 8.9

▶ **29.** −1.12 − 5.2

30. −8.63 − 5.6

31. 5.21 − 11.36

32. 8.53 − 17.84

33. −2.6 − (−5.7)

34. −9.4 − (−10.4)

35. 3 − 0.0012

36. 7 − 0.097

37. Subtract 6.7 from 23.

38. Subtract 9.2 from 45.

Objective A *Perform the indicated operation. See Examples 1 through 9.*

39. $0.9 + 2.2$ **40.** $0.7 + 3.4$ **41.** $-6.06 + 0.44$ **42.** $-5.05 + 0.88$

43. $500.21 - 136.85$ **44.** $600.47 - 254.68$ **45.** $50.2 - 600$ **46.** $40.3 - 700$

47. Subtract 61.9 from 923.5.   **48.** Subtract 45.8 from 845.9.   **49.** Add 100.009 and 6.08 and 9.034.

50. Add 200.89 and 7.49 and 62.83.   **51.** $-0.003 + 0.091$ **52.** $-0.004 + 0.085$

53. $-102.4 - 78.04$ **54.** $-36.2 - 10.02$ **55.** $-2.9 - (-1.8)$ **56.** $-6.5 - (-3.3)$

Objective C *Evaluate each expression for $x = 3.6$, $y = 5$, and $z = 0.21$. See Example 11.*

57. $x + z$ **58.** $y + x$ ▶ **59.** $x - z$

60. $y - z$ **61.** $y - x + z$ **62.** $x + y + z$

Determine whether the given values are solutions to the given equations. See Example 12.

63. Is 7 a solution to $x + 2.7 = 9.3$?   **64.** Is 3.7 a solution to $x + 5.9 = 8.6$?

65. Is -11.4 a solution to $27.4 + y = 16$?   **66.** Is -22.9 a solution to $45.9 + z = 23$?

67. Is 1 a solution to $2.3 + x = 5.3 - x$?   **68.** Is 0.9 a solution to $1.9 - x = x + 0.1$?

Objective D *Simplify by combining like terms. See Example 13.*

▶ **69.** $30.7x + 17.6 - 23.8x - 10.7$ **70.** $14.2z + 11.9 - 9.6z - 15.2$

71. $-8.61 + 4.23y - 2.36 - 0.76y$ **72.** $-8.96x - 2.31 - 4.08x + 9.68$

Objective E *Solve. For Exercises 73 and 74, the solutions have been started for you. See Examples 14 and 15.*

73. Ann-Margaret Tober bought a book for $32.48. If she paid with two $20 bills, what was her change?

74. Phillip Guillot bought a car part for $18.26. If he paid with two $10 bills, what was his change?

Start the solution:

1. UNDERSTAND the problem. Reread it as many times as needed.
2. TRANSLATE into an equation. (Fill in the blank.)

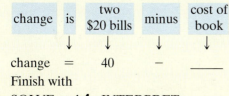

change	is	two $20 bills	minus	cost of book
↓	↓	↓	↓	↓

change = 40 − _____

Finish with
3. SOLVE and 4. INTERPRET

Start the solution:

1. UNDERSTAND the problem. Reread it as many times as needed.
2. TRANSLATE into an equation. (Fill in the blank.)

change	is	two $10 bills	minus	cost of car part
↓	↓	↓	↓	↓

change = 20 − _____

Finish with
3. SOLVE and 4. INTERPRET

75. Microsoft stock opened the day at $35.17 per share, and the closing price the same day was $34.75. By how much did the price of each share change?

76. A pair of eyeglasses costs a total of $347.89. The frames of the glasses are $97.23. How much do the lenses of the eyeglasses cost?

△ **77.** Find the perimeter.

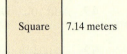

Square   7.14 meters

△ **78.** Find the perimeter.

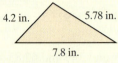

4.2 in.   5.78 in.   7.8 in.

79. The Apple iPhone 5 was released in 2012. It measures 4.87 inches by 2.31 inches. Find the perimeter of this phone. (*Source:* Apple.com)

80. The Google Nexus 4, released in 2012, is the newest Google phone (at this writing). It measures 5.27 inches by 2.7 inches. Find the perimeter of the phone. (*Source:* Google.com)

81. The average wind speed at the weather station on Mt. Washington in New Hampshire is 35.2 miles per hour. The highest speed ever recorded at the station is 231.0 miles per hour. How much faster is the highest speed than the average wind speed? (*Source:* National Climatic Data Center)

82. The average annual rainfall in Omaha, Nebraska, is 30.08 inches. The average annual rainfall in New Orleans, Louisiana, is 64.16 inches. On average, how much more rain does New Orleans receive annually than Omaha? (*Source:* National Climatic Data Center)

This bar graph shows the predicted increase in the total number of text messages per person per day in the United States. Use this graph for Exercises 83 and 84. (Source: Pew Research Center.)

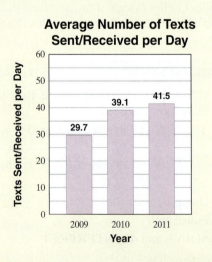

Average Number of Texts Sent/Received per Day

Texts Sent/Received per Day

29.7 (2009)   39.1 (2010)   41.5 (2011)

Year

83. Find the increase in the number of texts sent or received per day from 2009 to 2011.

84. Find the increase in the number of texts sent or received per day from 2010 to 2011.

85. As of this writing, the top three U.S. movies that made the most money through movie ticket sales are *Avatar* (2009), $760.5 million; *Titanic* (1997), $658.7 million; and *The Avengers* (2012), $623.3 million. What was the total amount of ticket sales for these three movies? (*Source:* MovieWeb)

86. In 2010, the average credit card late fee was $35. In 2011, the average credit card late fee had decreased by about $11.85. Find the average credit card late fee in 2011. (*Source:* Consumer Financial Protection Bureau)

87. The snowiest city in the United States is Valdez, AK, which receives an average of 110.5 more inches of snow than the second snowiest city. The second snowiest city in the United States is Crested Butte, CO. Crested Butte receives an average of 215.8 inches annually. How much snow does Valdez receive on average each year?(*Source:* The Weather Channel)

88. The driest place in the world is the Atacama Desert in Chile, which receives an average of only 0.004 inch of rain per year. Yuma, Arizona, is the driest city in the United States. Yuma receives an average of 3.006 more inches of rain each year than the Atacama Desert. What is the average annual rainfall in Yuma? (*Source:* National Climatic Data Center)

▶ 89. A landscape architect is planning a border for a flower garden shaped like a triangle. The sides of the garden measure 12.4 feet, 29.34 feet, and 25.7 feet. Find the amount of border material needed.

△**90.** A contractor purchased enough railing to completely enclose the newly built deck shown below. Find the amount of railing purchased.

29.34 feet

12.4 feet   25.7 feet

15.7 feet

10.6 feet

The table shows the average speeds for the Daytona 500 winners for the years shown. Use this table to answer Exercises 91 and 92. (*Source:* Daytona International Speedway)

Daytona 500 Winners		
Year	**Winner**	**Average Speed**
1978	Bobby Allison	159.73
1988	Bobby Allison	137.531
1998	Dale Earnhardt	172.712
2008	Ryan Newman	152.672
2013	Jimmie Johnson	159.250

91. How much slower was the average Daytona 500 winning speed in 2013 than in 1998?

92. How much faster was Bobby Allison's average Daytona 500 winning speed in 1978 than his average Daytona 500 winning speed in 1988?

The bar graph shows the top five chocolate-consuming nations in the world. Use this table to answer Exercises 93 through 97.

The World's Top Chocolate-Consuming Countries

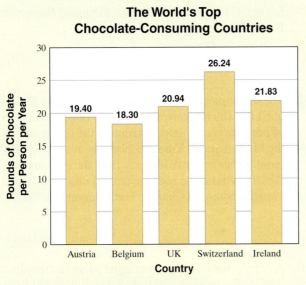

Source: Confectionary News and Leatherhead Food Research

93. Which country in the table has the greatest chocolate consumption per person?

94. Which country in the table has the least chocolate consumption per person?

95. How much more is the greatest chocolate consumption than the least chocolate consumption shown in the table?

96. How much more chocolate does the average Irishman or woman consume per year than the average Austrian?

97. Make a new chart listing the countries and their corresponding chocolate consumptions in order from greatest to least.

Review

Multiply. See Sections 1.5 and 4.3.

98. $23 \cdot 2$

99. $46 \cdot 3$

100. $39 \cdot 3$

101. $\left(\dfrac{2}{3}\right)^2$

102. $\left(\dfrac{1}{5}\right)^3$

Concept Extensions

A friend asks you to check his calculations for Exercises 103 and 104. Are they correct? If not, explain your friend's errors and correct the calculations. See the first Concept Check in this section.

103.
$$
\begin{array}{r}
\overset{1}{9}.2 \\
\overset{1}{8}.6\,3 \\
+\ 4.0\,0\,5 \\
\hline
4.9\,6\,0
\end{array}
$$

104.
$$
\begin{array}{r}
\overset{8\ 9\ 9\ 9}{9\,0\,0.0} \\
-\ \ \ 96.4 \\
\hline
8\,03.5
\end{array}
$$

Find the unknown length in each figure.

△**105.**

2.3 inches   ?   2.3 inches

10.68 inches

△**106.**

| ←5.26→ | ←7.82→ | ←?→ |
| meters | meters | meters |

17.67 meters

Let's review the values of these common U.S. coins in order to answer the following exercises.

Penny   Nickel   Dime   Quarter

$0.01 $0.05 $0.10 $0.25

For Exercises 107 and 108, write the value of each group of coins. To do so, it is usually easiest to start with the coin(s) of greatest value and end with the coin(s) of least value.

107.

108.

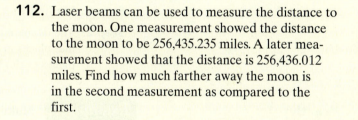

109. Name the different ways that coins can have a value of $0.17 given that you may use no more than 10 coins.

110. Name the different ways that coin(s) can have a value of $0.25 given that there are no pennies.

111. Why shouldn't the sum

$$82.95 + 51.26$$

be estimated as $90 + 60 = 150$?
See the second Concept Check in this section.

112. Laser beams can be used to measure the distance to the moon. One measurement showed the distance to the moon to be 256,435.235 miles. A later measurement showed that the distance is 256,436.012 miles. Find how much farther away the moon is in the second measurement as compared to the first.

113. Explain how adding or subtracting decimals is similar to adding or subtracting whole numbers.

114. Can the sum of two negative decimals ever be a positive decimal? Why or why not?

Combine like terms and simplify.

115. $-8.689 + 4.286x - 14.295 - 12.966x + 30.861x$

116. $14.271 - 8.968x + 1.333 - 201.815x + 101.239x$

5.3 Multiplying Decimals and Circumference of a Circle

Objectives

A Multiply Decimals.

B Estimate when Multiplying Decimals.

C Multiply Decimals by Powers of 10.

D Evaluate Expressions with Decimal Replacement Values.

E Find the Circumference of Circles.

F Solve Problems by Multiplying Decimals.

Objective A Multiplying Decimals

Multiplying decimals is similar to multiplying whole numbers. The only difference is that we place a decimal point in the product. To discover where a decimal point is placed in a product, let's multiply 0.6×0.03. We first write each decimal as an equivalent fraction and then multiply.

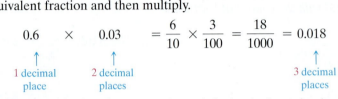

$$0.6 \quad \times \quad 0.03 \quad = \frac{6}{10} \times \frac{3}{100} = \frac{18}{1000} = 0.018$$

1 decimal place   2 decimal places   3 decimal places

Notice that $1 + 2 = 3$, the number of decimal places in the product. Now let's multiply 0.03×0.002.

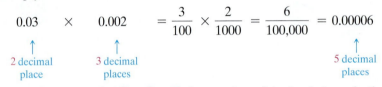

$$0.03 \quad \times \quad 0.002 \quad = \frac{3}{100} \times \frac{2}{1000} = \frac{6}{100,000} = 0.00006$$

2 decimal place   3 decimal places   5 decimal places

Again, we see that $2 + 3 = 5$, the number of decimal places in the product.

Instead of writing decimals as fractions each time we want to multiply, we notice a pattern from these examples and state a rule that we can use:

Multiplying Decimals

Step 1: Multiply the decimals as though they are whole numbers.

Step 2: The decimal point in the product is placed so that the number of decimal places in the product is equal to the *sum* of the number of decimal places in the factors.

Practice 1

Multiply: 34.8×0.62

Example 1 Multiply: 23.6×0.78

Solution:

$$
\begin{array}{r}
23.6 \\
\times\ 0.78 \\
\hline
1888 \\
16520 \\
\hline
18.408
\end{array}
$$

23.6   1 decimal place
$\times$ 0.78   2 decimal places

Since $1 + 2 = 3$, insert the decimal point in the product so that there are 3 decimal places.

■ **Work Practice 1**

Practice 2

Multiply: 0.0641×27

Example 2 Multiply: 0.0531×16

Solution:

$$
\begin{array}{r}
0.0531 \\
\times\ \ \ \ 16 \\
\hline
3186 \\
5310 \\
\hline
0.8496
\end{array}
$$

0.0531   4 decimal places
$\times$ 16   0 decimal places

4 decimal places $(4 + 0 = 4)$

■ **Work Practice 2**

Answers
1. 21.576   **2.** 1.7307

Copyright 2015 Pearson Education, Inc.

✓**Concept Check** True or false? The number of decimal places in the product of 0.261 and 0.78 is 6. Explain.

Example 3 Multiply: $(-2.6)(0.8)$

Solution: Recall that the product of a negative number and a positive number is a negative number.

$$(-2.6)(0.8) = -2.08$$

■ **Work Practice 3**

Practice 3

Multiply: $(7.3)(-0.9)$

Objective B Estimating when Multiplying Decimals ▶

Just as for addition and subtraction, we can estimate when multiplying decimals to check the reasonableness of our answer.

Example 4 Multiply: 28.06×1.95. Then estimate to see whether the answer is reasonable by rounding each factor, then multiplying the rounded numbers.

Solution:

Exact	Estimate 1	or	Estimate 2
28.06	28		30
$\times$ 1.95	$\times$ 2 Rounded to ones		$\times$ 2 Rounded to one nonzero digit
14030	56		60
252540			
280600			
54.7170			

The answer 54.7170 or 54.717 is reasonable.

■ **Work Practice 4**

Practice 4

Multiply: 30.26×2.89. Then estimate to see whether the answer is reasonable.

As shown in Example 4, estimated results will vary depending on what estimates are used. Notice that estimating results is a good way to see whether the decimal point has been correctly placed.

Objective C Multiplying Decimals by Powers of 10 ▶

There are some patterns that occur when we multiply a number by a power of 10 such as 10, 100, 1000, 10,000, and so on.

$23.6951 \times 10 = 236.951$ Move the decimal point *1 place* to the *right*.
    ↑
    1 zero

$23.6951 \times 100 = 2369.51$ Move the decimal point *2 places* to the *right*.
    ↑
    2 zeros

$23.6951 \times 100,000 = 2,369,510.$ Move the decimal point *5 places* to the *right* (insert a 0).
    ↑
    5 zeros

Notice that we move the decimal point the same number of places as there are zeros in the power of 10.

Answers
3. -6.57 **4.** Exact: 87.4514; Estimate: $30 \cdot 3 = 90$

✓**Concept Check Answer**

false: 3 decimal places and 2 decimal places means 5 decimal places in the product

Multiplying Decimals by Powers of 10 Such as 10, 100, 1000, 10,000

Move the decimal point to the *right* the same number of places as there are *zeros* in the power of 10.

Practice 5–7

Multiply.

5. 46.8×10

6. 203.004×100

7. $(-2.33)(1000)$

Examples   Multiply.

5. $7.68 \times 10 = 76.8$ 7.68

6. $23.702 \times 100 = 2370.2$ 23.702

7. $(-76.3)(1000) = -76,300$ 76.300

■ **Work Practice 5–7**

There are also powers of 10 that are less than 1. The decimals 0.1, 0.01, 0.001, 0.0001, and so on, are examples of powers of 10 less than 1. Notice the pattern when we multiply by these powers of 10:

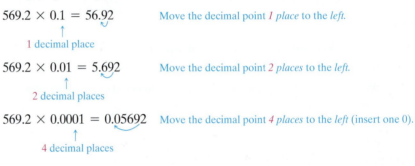

$569.2 \times 0.1 = 56.92$ Move the decimal point *1 place* to the *left.*

     1 decimal place

$569.2 \times 0.01 = 5.692$ Move the decimal point *2 places* to the *left.*

     2 decimal places

$569.2 \times 0.0001 = 0.05692$ Move the decimal point *4 places* to the *left* (insert one 0).

     4 decimal places

Multiplying Decimals by Powers of 10 Such as 0.1, 0.01, 0.001, 0.0001

Move the decimal point to the *left* the same number of places as there are *decimal places* in the power of 10.

Practice 8–10

Multiply.

8. 6.94×0.1

9. 3.9×0.01

10. $(-7682)(-0.001)$

Examples   Multiply.

8. $42.1 \times 0.1 = 4.21$ 42.1

9. $76,805 \times 0.01 = 768.05$ 76,805.

10. $(-9.2)(-0.001) = 0.0092$ 0009.2

■ **Work Practice 8–10**

Many times we see large numbers written, for example, in the form 297.9 million rather than in the longer standard notation. The next example shows us how to interpret these numbers.

Answers

5. 468   **6.** 20,300.4   **7.** −2330

8. 0.694   **9.** 0.039   **10.** 7.682

Example 11 In 2050, the population of the United States is projected to be 420.3 million. Write this number in standard notation. (*Source:* U.S. Census Bureau)

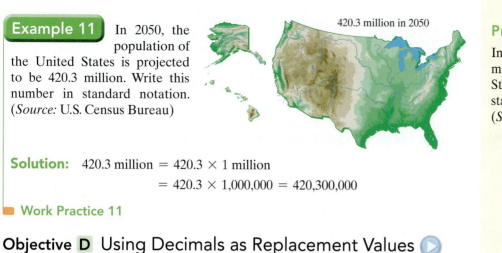

420.3 million in 2050

Practice 11

In 2012, there were 58.9 million married couples in the United States. Write this number in standard notation. (*Source:* U.S. Census Bureau)

Solution: 420.3 million = 420.3 × 1 million

= 420.3 × 1,000,000 = 420,300,000

▪ **Work Practice 11**

Objective D Using Decimals as Replacement Values ▶

Now let's practice working with variables.

Example 12 Evaluate xy for $x = 2.3$ and $y = 0.44$.

Practice 12

Evaluate $7y$ for $y = -0.028$.

Solution: Recall that xy means $x \cdot y$.

$xy = (2.3)(0.44)$

$$\begin{array}{r} 2.3 \\ \times\, 0.44 \\ \hline 92 \\ 920 \\ \hline \end{array}$$

$= 1.012 \longleftarrow \quad 1.012$

▪ **Work Practice 12**

Example 13 Is -9 a solution of the equation $3.7y = -3.33$?

Practice 13

Is -5.5 a solution of the equation $-6x = 33$?

Solution: Replace y with -9 in the equation $3.7y = -3.33$ to see if a true equation results.

$3.7y = -3.33$

$3.7(-9) \overset{?}{=} -3.33$ Replace y with -9.

$-33.3 = -3.33$ False

Since $-33.3 = -3.33$ is a false statement, -9 is **not** a solution of $3.7y = -3.33$.

▪ **Work Practice 13**

Objective E Finding the Circumference of a Circle ▶

Recall from Section 1.3 that the distance around a polygon is called its perimeter. The distance around a circle is given the special name **circumference,** and this distance depends on the radius or the diameter of the circle.

Circumference of a Circle

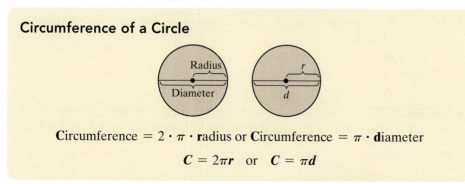

Circumference = 2 · π · radius or Circumference = π · diameter

$$C = 2\pi r \quad \text{or} \quad C = \pi d$$

In Section 5.1, we learned about the symbol π as the Greek letter pi, pronounced "pie." It is a constant between 3 and 4. A decimal approximation for π is 3.14. Also, a fraction approximation for π is $\frac{22}{7}$.

Practice 14

Find the circumference of a circle whose radius is 11 meters. Then use the approximation 3.14 for π to approximate this circumference.

Example 14   Circumference of a Circle

Find the circumference of a circle whose radius is 5 inches. Then use the approximation 3.14 for π to approximate the circumference.

Solution: Let $r = 5$ in the formula $C = 2\pi r$.

$$C = 2\pi r$$
$$= 2\pi \cdot 5$$
$$= 10\pi$$

5 inches

Next, replace π with the approximation 3.14.

$$C = 10\pi$$
(is approximately) $\longrightarrow \approx 10(3.14)$
$$\approx 31.4$$

The **exact** circumference or distance around the circle is 10π inches, which is **approximately** 31.4 inches.

🟧 Work Practice 14

Objective F Solving Problems by Multiplying Decimals ▶

The solutions to many real-life problems are found by multiplying decimals. We continue using our four problem-solving steps to solve such problems.

Practice 15

A biology major is fertilizing her personal garden. She uses 5.6 ounces of fertilizer per square yard. The garden measures 60.5 square yards. How much fertilizer does she need?

Example 15   Finding the Total Cost of Materials for a Job

A college student is hired to paint a billboard with paint costing $2.49 per quart. If the job requires 3 quarts of paint, what is the total cost of the paint?

Solution:

1. UNDERSTAND. Read and reread the problem. The phrase "total cost" might make us think addition, but since this problem requires repeated addition, let's multiply.

2. TRANSLATE.

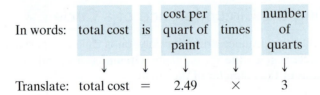

In words:	total cost	is	cost per quart of paint	times	number of quarts
	↓	↓	↓	↓	↓
Translate:	total cost	=	2.49	×	3

3. SOLVE. We can estimate to check our calculations. The number 2.49 rounds to 2 and $2 \times 3 = 6$.

$$\begin{array}{r} 2.49 \\ \times 3 \\ \hline 7.47 \end{array}$$

4. INTERPRET.   *Check* your work. Since 7.47 is close to our estimate of 6, our answer is reasonable. *State* your conclusion: The total cost of the paint is $7.47.

🟧 Work Practice 15

Answers
14. 22π m $\approx$ 69.08 m   15. 338.8 oz

Vocabulary, Readiness & Video Check

Use the choices below to fill in each blank.

circumference	left	sum	zeros
decimal places	right	product	factor

1. When multiplying decimals, the number of decimal places in the product is equal to the _____ of the number of decimal places in the factors.

2. In $8.6 \times 5 = 43$, the number 43 is called the _____, while 8.6 and 5 are each called a _____.

3. When multiplying a decimal number by powers of 10 such as 10, 100, 1000, and so on, we move the decimal point in the number to the _____ the same number of places as there are _____ in the power of 10.

4. When multiplying a decimal number by powers of 10 such as 0.1, 0.01, and so on, we move the decimal point in the number to the _____ the same number of places as there are _____ in the power of 10.

5. The distance around a circle is called its _____.

Martin-Gay Interactive Videos   Watch the section lecture video and answer the following questions.

See Video 5.3

Objective A 6. From the lecture before ▣ Example 1, what's the main difference between multiplying whole numbers and multiplying decimal numbers? ▶

Objective B 7. From ▣ Example 3, what does estimating especially help us with? ▶

Objective C 8. Why don't we do any actual multiplying in ▣ Example 5? ▶

Objective D 9. In ▣ Example 8, once all replacement values are inserted in the variable expression, what is the resulting expression to evaluate? ▶

Objective E 10. Why is 31.4 cm not the exact answer to ▣ Example 9? ▶

Objective F 11. In ▣ Example 10, why is 24.8 not the complete answer? What is the complete answer? ▶

5.3 Exercise Set MyMathLab® ▶

Objectives A B Mixed Practice *Multiply. See Examples 1 through 4. For those exercises marked, also estimate to see if the answer is reasonable.*

1. 0.17×8

2. 0.23×9

3. $\begin{array}{r} 1.2 \\ \times\, 0.5 \\ \hline \end{array}$

4. $\begin{array}{r} 6.8 \\ \times\, 0.3 \\ \hline \end{array}$

▶ 5. $(-2.3)(7.65)$

6. $(4.7)(-9.02)$

7. $(-5.73)(-9.6)$

8. $(-7.84)(-3.5)$

▶ 9. 6.8×4.2

Exact:

Estimate:

10. 8.3×2.7

Exact:

Estimate:

11. $\begin{array}{r} 0.347 \\ \times\quad 0.3 \\ \hline \end{array}$

12. $\begin{array}{r} 0.864 \\ \times\quad 0.4 \\ \hline \end{array}$

13. 1.0047
× 8.2

Exact: ____   Estimate: ____

14. 2.0005
× 5.5

Exact: ____   Estimate: ____

15. 490.2
×0.023

16. 300.9
×0.032

Objective C *Multiply. See Examples 5 through 10.*

▶ **17.** 6.5 × 10

18. 7.2 × 100

▶ **19.** 8.3 × 0.1

20. 23.4 × 0.1

21. (−7.093)(1000)

22. (−1.123)(1000)

23. 0.7 × 100

24. 0.5 × 100

▶ **25.** (−9.83)(−0.01)

26. (−4.72)(−0.01)

27. 25.23 × 0.001

28. 36.41 × 0.001

Objectives A B C Mixed Practice *Multiply. See Examples 1 through 10.*

29. 0.123 × 0.4

30. 0.216 × 0.3

▶ **31.** (147.9)(100)

32. (345.2)(100)

33. 8.6 × 0.15

34. 0.42 × 5.7

35. (937.62)(−0.01)

36. (−0.001)(562.01)

37. 562.3 × 0.001

38. 993.5 × 0.001

39. 6.32
× 5.7

40. 9.21
× 3.8

Write each number in standard notation. See Example 11.

41. The cost of the Hubble Space Telescope at launch was $1.5 billion. (*Source:* NASA)

42. About 56.7 million American households own at least one dog. (*Source:* American Pet Products Manufacturers Association)

43. The Blue Streak is the oldest roller coaster at Cedar Point, an amusement park in Sandusky, Ohio. Since 1964, it has given more than 49.8 million rides. (*Source:* Cedar Fair, L.P.)

44. In 2013, the restaurant industry had projected sales of $660.5 billion. (*Source:* National Restaurant Association)

Objective D *Evaluate each expression for x = 3, y = −0.2, and z = 5.7. See Example 12.*

45. xy

46. yz

▶ **47.** $xz - y$

48. $-5y + z$

Determine whether the given value is a solution of each given equation. See Example 13.

49. Is 14.2 a solution of $0.6x = 4.92$?

50. Is 1414 a solution of $100z = 14.14$?

51. Is −4 a solution of $3.5y = -14$?

52. Is −3.6 a solution of $0.7x = -2.52$?

Objective E *Find the circumference of each circle. Then use the approximation 3.14 for π and approximate each circumference. See Example 14.*

▶ **53.**

△ **54.**

△ **55.**

△ **56.**

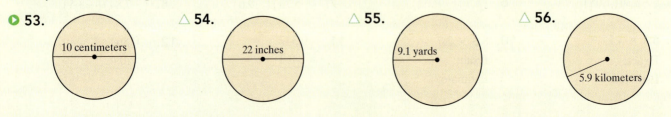

10 centimeters   22 inches   9.1 yards   5.9 kilometers

Objectives E F Mixed Practice *Solve. For Exercises 57 and 58, the solutions have been started for you. For circumference applications find the exact circumference and then use 3.14 for π to approximate the circumference. See Examples 14 and 15.*

57. An electrician for Central Power and Light worked 40 hours last week. Calculate his pay before taxes for last week if his hourly wage is $17.88.

Start the solution:

1. UNDERSTAND the problem. Reread it as many times as needed.
2. TRANSLATE into an equation. (Fill in the blanks.)

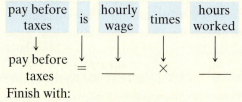

Finish with:

3. SOLVE and 4. INTERPRET.

58. An assembly line worker worked 20 hours last week. Her hourly rate is $19.52 per hour. Calculate her pay before taxes.

Start the solution:

1. UNDERSTAND the problem. Reread it as many times as needed.
2. TRANSLATE into an equation. (Fill in the blanks.)

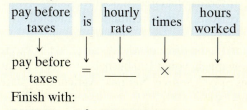

Finish with:

3. SOLVE and 4. INTERPRET.

▷ 59. A 1-ounce serving of cream cheese contains 6.2 grams of saturated fat. How much saturated fat is in 4 ounces of cream cheese? (*Source: Home and Garden Bulletin No. 72; U.S. Department of Agriculture*)

60. A 3.5-ounce serving of lobster meat contains 0.1 gram of saturated fat. How much saturated fat do 3 servings of lobster meat contain? (*Source: The National Institutes of Health*)

△ 61. Recall that the face of the Apple iPhone 5 (see Section 5.2) measures 4.87 inches by 2.3 inches rounded. Find the approximate area of the face of the Apple iPhone 5.

△ 62. Recall that the face of the Google Nexus 4 (see Section 5.2) measures 5.27 inches by 2.7 inches. Find the area of the face of the Google Nexus 4.

△ 63. In 1893, the first ride called a Ferris wheel was constructed by Washington Gale Ferris. Its diameter was 250 feet. Find its circumference. Give an exact answer and an approximation using 3.14 for π. (*Source: The Handy Science Answer Book*, Visible Ink Press, 1994)

△ 64. The radius of Earth is approximately 3950 miles. Find the distance around Earth at the equator. Give an exact answer and an approximation using 3.14 for π. (*Hint:* Find the circumference of a circle with radius 3950 miles.)

△ 65. The London Eye, built for the millennium celebration in London, resembles a gigantic Ferris wheel with a diameter of 135 meters. If Adam Hawn rides the Eye for one revolution, find how far he travels. Give an exact answer and an approximation using 3.14 for π. (*Source:* Londoneye.com)

△ 66. The world's longest suspension bridge is the Akashi Kaikyo Bridge in Japan. This bridge has two circular caissons, which are underwater foundations. If the diameter of a caisson is 80 meters, find its circumference. Give an exact answer and an approximation using 3.14 for π. (*Source: Scientific American;* How Things Work Today)

67. A meter is a unit of length in the metric system that is approximately equal to 39.37 inches. Sophia Wagner is 1.65 meters tall. Find her approximate height in inches.

68. The doorway to a room is 2.15 meters tall. Approximate this height in inches. (*Hint:* See Exercise **67.**)

△ **69. a.** Approximate the circumference of each circle.

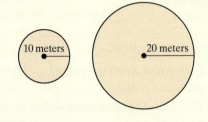

　　b. If radius of a circle is doubled, is its corresponding circumference doubled?

△ **70. a.** Approximate the circumference of each circle.

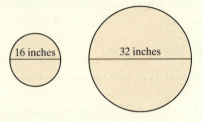

　　b. If the diameter of a circle is doubled, is its corresponding circumference doubled?

71. In 2011, the price of wheat was $7.30 per bushel. How much would 100 bushels of wheat cost at this price?(*Source:* National Agricultural Statistics Service)

72. In 2011, the price of soybeans was $11.70 per bushel. How much would a company pay for 10,000 bushels of soybeans?(*Source:* National Agricultural Statistics Service)

The table shows currency exchange rates for various countries on August 26, 2013. To find the amount of foreign currency equivalent to an amount of U.S. dollars, multiply the U.S. dollar amount by the exchange rate listed in the table. Use this table to answer Exercises 73 through 76. If needed, round answers to the nearest hundredth.

Foreign Currency Exchange Rates	
Country	**Exchange Rate**
Canadian dollar	1.04920
European Union euro	0.74697
New Zealand dollar	1.28084
Chinese yuan	6.15231
Japanese yen	98.68
Swiss franc	0.92090

73. How many Canadian dollars are equivalent to $750 U.S.?

74. Suppose you wish to exchange 300 American dollars for Chinese yuan. How much money, in Chinese yuan, would you receive?

75. The Scarpulla family is traveling to New Zealand. How many New Zealand dollars can they "buy" with 800 U.S. dollars?

76. A French tourist to the United States spent $130 for souvenirs at the *Head of the Charles Regatta* in Boston. How much money did he spend in euros? Round to the nearest hundredth.

Review

Divide. See Sections 1.6 and 4.3.

77. $2916 \div 6$

78. $2920 \div 365$

79. $-\dfrac{24}{7} \div \dfrac{8}{21}$

80. $\dfrac{162}{25} \div -\dfrac{9}{75}$

Concept Extensions

Mixed Practice (Sections 5.2, 5.3) *Perform the indicated operations.*

81. $3.6 + 0.04$

82. $7.2 + 0.14 + 98.6$

83. $3.6 - 0.04$

84. $100 - 48.6$

85. -0.221×0.5

86. -3.6×0.04

87. Find how far radio waves travel in 20.6 seconds. (Radio waves travel at a speed of 186,000 miles per second.)

88. If it takes radio waves approximately 8.3 minutes to travel from the sun to the earth, find approximately how far it is from the sun to the earth. (*Hint:* See Exercise **87.**)

89. In your own words, explain how to find the number of decimal places in a product of decimal numbers.

90. In your own words, explain how to multiply by a power of 10.

91. Write down two decimal numbers whose product will contain 5 decimal places. Without multiplying, explain how you know your answer is correct.

5.4 Dividing Decimals

Objective A Dividing Decimals

Dividing decimal numbers is similar to dividing whole numbers. The only difference is that we place a decimal point in the quotient. If the divisor is a whole number, we place the decimal point in the quotient directly above the decimal point in the dividend, and then divide as with whole numbers. Recall that division can be checked by multiplication.

Objectives

A Divide Decimals.

B Estimate when Dividing Decimals.

C Divide Decimals by Powers of 10.

D Evaluate Expressions with Decimal Replacement Values.

E Solve Problems by Dividing Decimals.

Example 1 Divide: $270.2 \div 7$. Check your answer.

Solution: We divide as usual. The decimal point in the quotient is directly above the decimal point in the dividend.

```
                 ┌─Write the decimal point
            38.6  ← quotient
divisor → 7)270.2  ← dividend
          −21↓|
            60|
          −56↓
            4 2
          −4 2
             0
```

Check:
```
      6 4
     38.6
   ×    7
    270.2
```

The quotient is 38.6.

■ **Work Practice 1**

Practice 1

Divide: $370.4 \div 8$. Check your answer.

Example 2 Divide: $32\overline{)8.32}$

Solution: We divide as usual. The decimal point in the quotient is directly above the decimal point in the dividend.

```
               0.26  ← quotient
dividend → 32)8.32  ← divisor
             −64
             192
            −192
               0
```

Check:
```
       0.26  quotient
     ×  32   divisor
        52
      7 80
      8.32  dividend
```

■ **Work Practice 2**

Practice 2

Divide: $48\overline{)34.08}$. Check your answer.

Answers
1. 46.3 **2.** 0.71

Chapter 5 | Decimals

Sometimes to continue dividing we need to insert zeros after the last digit in the dividend.

Copyright 2015 Pearson Education, Inc.

Practice 3

Divide and check.
a. $-15.89 \div 14$
b. $-2.808 \div (-104)$

 Example 3   Divide: $-5.98 \div 115$

Solution: Recall that a negative number divided by a positive number gives a negative quotient.

$$
\begin{array}{r}
0.052 \\
115\overline{)5.980} \quad \leftarrow \text{Insert one 0.}\\
\underline{-575} \\
230 \\
\underline{-230} \\
0
\end{array}
$$

Thus $-5.98 \div 115 = -0.052$.

■ **Work Practice 3**

If the divisor is not a whole number, before we divide we need to move the decimal point to the right until the divisor is a whole number.

$$
1.5\overline{)64.85}
$$

divisor ⌐   └ dividend

To understand how this works, let's rewrite

$$
1.5\overline{)64.85} \quad \text{as} \quad \frac{64.85}{1.5}
$$

and then multiply by 1 in the form of $\dfrac{10}{10}$. We use the form $\dfrac{10}{10}$ so that the denominator (divisor) becomes a whole number.

$$
\frac{64.85}{1.5} = \frac{64.85}{1.5} \cdot 1 = \frac{64.85}{1.5} \cdot \frac{10}{10} = \frac{64.85 \cdot 10}{1.5 \cdot 10} = \frac{648.5}{15},
$$

which can be written as $15.\overline{)648.5}$. Notice that

$$
1.5\overline{)64.85} \text{ is equivalent to } 15.\overline{)648.5}
$$

The decimal points in the dividend and the divisor were both moved one place to the right, and the divisor is now a whole number. This procedure is summarized next:

Dividing by a Decimal

Step 1: Move the decimal point in the divisor to the right until the divisor is a whole number.

Step 2: Move the decimal point in the dividend to the right the *same number of places* as the decimal point was moved in Step 1.

Step 3: Divide. Place the decimal point in the quotient directly over the moved decimal point in the dividend.

Answers
3. **a.** -1.135 **b.** 0.027

Example 4 Divide: $10.764 \div 2.3$

Solution: We move the decimal points in the divisor and the dividend one place to the right so that the divisor is a whole number.

$2.3\overline{)10.764}$ becomes

$$
\begin{array}{r}
4.68 \\
23\overline{)107.64} \\
-92 \\
\hline
15\,6 \\
-13\,8 \\
\hline
1\,84 \\
-1\,84 \\
\hline
0
\end{array}
$$

■ **Work Practice 4**

Practice 4
Divide: $166.88 \div 5.6$

Example 5 Divide: $5.264 \div 0.32$

Solution:

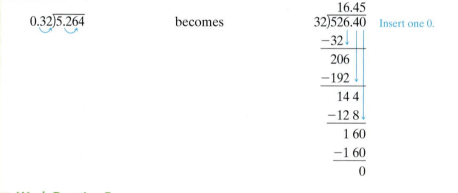

$0.32\overline{)5.264}$ becomes

$$
\begin{array}{r}
16.45 \\
32\overline{)526.40} \quad \text{Insert one 0.} \\
-32 \\
\hline
206 \\
-192 \\
\hline
14\,4 \\
-12\,8 \\
\hline
1\,60 \\
-1\,60 \\
\hline
0
\end{array}
$$

■ **Work Practice 5**

Practice 5
Divide: $1.976 \div 0.16$

✓**Concept Check** Is it always true that the number of decimal places in a quotient equals the sum of the decimal places in the dividend and divisor?

Example 6 Divide: $17.5 \div 0.48$. Round the quotient to the nearest hundredth.

Solution: First we move the decimal points in the divisor and the dividend two places. Then we divide and round the quotient to the nearest hundredth.

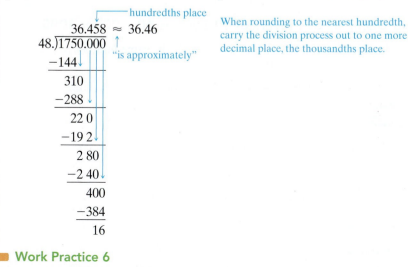

hundredths place

$$36.458 \approx 36.46$$

$$
\begin{array}{r}
48\overline{)1750.000} \\
-144 \\
\hline
310 \\
-288 \\
\hline
22\,0 \\
-19\,2 \\
\hline
2\,80 \\
-2\,40 \\
\hline
400 \\
-384 \\
\hline
16
\end{array}
$$

"is approximately"

When rounding to the nearest hundredth, carry the division process out to one more decimal place, the thousandths place.

■ **Work Practice 6**

Practice 6
Divide: $23.4 \div 0.57$. Round the quotient to the nearest hundredth.

Answers
4. 29.8 **5.** 12.35 **6.** 41.05

✓**Concept Check Answer**
no

✓**Concept Check** If a quotient is to be rounded to the nearest thousandth, to what place should the division be carried out? (Assume that the division carries out to your answer.)

Objective B Estimating when Dividing Decimals

Just as for addition, subtraction, and multiplication of decimals, we can estimate when dividing decimals to check the reasonableness of our answer.

Practice 7

Divide: $713.7 \div 91.5$. Then estimate to see whether the proposed answer is reasonable.

> **Example 7** Divide: $272.356 \div 28.4$. Then estimate to see whether the proposed result is reasonable.
>
> **Solution:**
>
> | **Exact:** | **Estimate 1** | | **Estimate 2** |
> |---|---|---|---|
>
> $$\begin{array}{r} 9.59 \\ 284.\overline{)2723.56} \\ -2556 \\ \hline 167\,5 \\ -142\,0 \\ \hline 25\,56 \\ -25\,56 \\ \hline 0 \end{array}$$
>
> Estimate 1: $\dfrac{9}{30\overline{)270}}$ or   Estimate 2: $\dfrac{10}{30\overline{)300}}$
>
> The estimate is 9 or 10, so 9.59 is reasonable.

🟧 **Work Practice 7**

Objective C Dividing Decimals by Powers of 10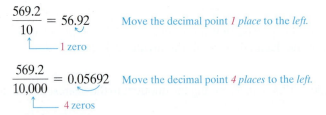

As with multiplication, there are patterns that occur when we divide decimals by powers of 10 such as 10, 100, 1000, and so on.

$$\frac{569.2}{10} = 56.92 \qquad \text{Move the decimal point } 1 \text{ place to the } \textit{left.}$$

└──── 1 zero

$$\frac{569.2}{10,000} = 0.05692 \qquad \text{Move the decimal point } 4 \text{ places to the } \textit{left.}$$

└──── 4 zeros

This pattern suggests the following rule:

> **Dividing Decimals by Powers of 10 Such as 10, 100, or 1000**
>
> Move the decimal point of the dividend to the *left* the same number of places as there are *zeros* in the power of 10.

Practice 8–9

Divide.

8. $\dfrac{362.1}{1000}$ **9.** $-\dfrac{0.49}{10}$

Answers

7. 7.8   **8.** 0.3621   **9.** −0.049

✓**Concept Check Answer**

ten-thousandths place

> **Examples** Divide.
>
> **8.** $\dfrac{786.1}{1000} = 0.7861$ Move the decimal point *3 places* to the *left.*
>
> └──── 3 zeros
>
> **9.** $\dfrac{0.12}{10} = 0.012$ Move the decimal point *1 place* to the *left.*
>
> └──── 1 zero

🟧 **Work Practice 8–9**

Objective D Using Decimals as Replacement Values ▶

Example 10 Evaluate $x \div y$ for $x = 2.5$ and $y = 0.05$.

Solution: Replace x with 2.5 and y with 0.05.

$$x \div y = 2.5 \div 0.05 \qquad 0.05\overline{)2.5} \text{ becomes } 5\overline{)250} \quad \genfrac{}{}{0pt}{}{50}{}$$
$$= 50$$

■ **Work Practice 10**

Practice 10

Evaluate $x \div y$ for $x = 0.035$ and $y = 0.02$.

Example 11 Is 720 a solution of the equation $\dfrac{y}{100} = 7.2$?

Solution: Replace y with 720 to see if a true statement results.

$$\frac{y}{100} = 7.2 \quad \text{Original equation}$$

$$\frac{720}{100} \stackrel{?}{=} 7.2 \quad \text{Replace } y \text{ with 720.}$$

$$7.2 = 7.2 \quad \text{True}$$

Since $7.2 = 7.2$ is a true statement, 720 is a solution of the equation.

■ **Work Practice 11**

Practice 11

Is 39 a solution of the equation

$$\frac{x}{100} = 3.9?$$

Objective E Solving Problems by Dividing Decimals ▶

Many real-life problems involve dividing decimals.

△ **Example 12** Calculating Materials Needed for a Job

A gallon of paint covers a 250-square-foot area. If Betty Adkins wishes to paint a wall that measures 1450 square feet, how many gallons of paint does she need? If she can buy only gallon containers of paint, how many gallon containers does she need?

Solution:

1. UNDERSTAND. Read and reread the problem. We need to know how many 250s are in 1450, so we divide.

2. TRANSLATE.

In words:

number of gallons	is	square feet	divided by	square feet per gallon
↓	↓	↓	↓	↓

Translate:

$$\text{number of gallons} = 1450 \div 250$$

3. SOLVE. Let's see if our answer is reasonable by estimating. The dividend 1450 rounds to 1500 and the divisor 250 rounds to 300. Then $1500 \div 300 = 5$.

$$\begin{array}{r} 5.8 \\ 250\overline{)1450.0} \\ -\underline{1250} \\ 200\ 0 \\ -\underline{200\ 0} \\ 0 \end{array}$$

Practice 12

A bag of fertilizer covers 1250 square feet of lawn. Tim Parker's lawn measures 14,800 square feet. How many bags of fertilizer does he need? If he can buy only whole bags of fertilizer, how many whole bags does he need?

(Continued on next page)

4. **INTERPRET.** *Check* your work. Since our estimate is close to our answer of 5, our answer is reasonable. *State* your conclusion: Betty needs 5.8 gallons of paint. If she can buy only gallon containers of paint, she needs 6 gallon containers of paint to complete the job.

■ **Work Practice 12**

Calculator Explorations Estimation

Calculator errors can easily be made by pressing an incorrect key or by not pressing a correct key hard enough. Estimation is a valuable tool that can be used to check calculator results.

Example   Use estimation to determine whether the calculator result is reasonable or not. (For example, a result that is not reasonable can occur if proper keys are not pressed.)

Divide: 82.064 ÷ 23
Calculator display: [35.68   ]

Solution: Round each number to the nearest 10. Since 80 ÷ 20 = 4, the calculator display 35.68 is not reasonable.

Use estimation to determine whether each result is reasonable or not.

1. 102.62 × 41.8; Result: 428.9516

2. 174.835 ÷ 47.9; Result: 3.65

3. 1025.68 − 125.42; Result: 900.26

4. 562.781 + 2.96; Result: 858.781

Vocabulary, Readiness & Video Check

Use the choices below to fill in each blank. Some choices may be used more than once, and some not used at all.

dividend	divisor	quotient	true
zeros	left	right	false

1. In 6.5 ÷ 5 = 1.3, the number 1.3 is called the _____ , 5 is the _____ , and 6.5 is the _____ .

2. To check a division exercise, we can perform the following multiplication: quotient · _____ = _____ .

3. To divide a decimal number by a power of 10 such as 10, 100, 1000, and so on, we move the decimal point in the number to the _____ the same number of places as there are _____ in the power of 10.

4. True or false: If we replace x with -12.6 and y with 0.3 in the expression $y \div x$, we have $0.3 \div (-12.6)$. _____

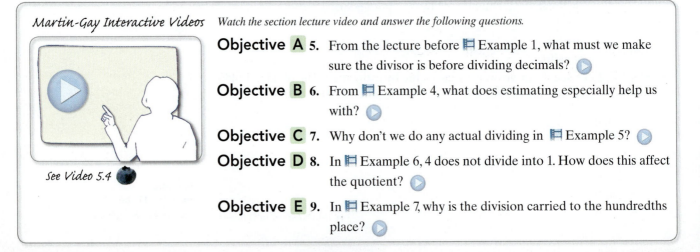

Martin-Gay Interactive Videos   Watch the section lecture video and answer the following questions.

Objective A 5. From the lecture before ▣ Example 1, what must we make sure the divisor is before dividing decimals? ▶

Objective B 6. From ▣ Example 4, what does estimating especially help us with? ▶

Objective C 7. Why don't we do any actual dividing in ▣ Example 5? ▶

Objective D 8. In ▣ Example 6, 4 does not divide into 1. How does this affect the quotient? ▶

Objective E 9. In ▣ Example 7, why is the division carried to the hundredths place? ▶

See Video 5.4

5.4 Exercise Set MyMathLab®

Objectives A B Mixed Practice *Divide. See Examples 1 through 5 and 7. For those exercises marked, also estimate to see if the answer is reasonable.*

1. $6\overline{)27.6}$

2. $4\overline{)23.6}$

3. $5\overline{)0.47}$

4. $6\overline{)0.51}$

5. $0.06\overline{)18}$

6. $0.04\overline{)20}$

7. $0.42\overline{)3.066}$

8. $0.36\overline{)1.764}$

9. $5.5\overline{)36.3}$
Exact:
Estimate:

10. $2.2\overline{)21.78}$
Exact:
Estimate:

11. $7.434 \div 18$

12. $8.304 \div 16$

13. $36 \div (-0.06)$

14. $36 \div (-0.04)$

15. Divide -4.2 by -0.6.

16. Divide -3.6 by -0.9.

17. $0.27\overline{)1.296}$

18. $0.34\overline{)2.176}$

19. $0.02\overline{)42}$

20. $0.03\overline{)24}$

21. $4.756 \div 0.82$

22. $3.312 \div 0.92$

23. $-36.3 \div (-6.6)$

24. $-21.78 \div (-9.9)$

25. $7.2\overline{)70.56}$
Exact:
Estimate:

26. $6.3\overline{)54.18}$
Exact:
Estimate:

27. $5.4\overline{)51.84}$

28. $7.7\overline{)33.88}$

29. $\dfrac{1.215}{0.027}$

30. $\dfrac{3.213}{0.051}$

31. $0.25\overline{)13.648}$

32. $0.75\overline{)49.866}$

33. $3.78\overline{)0.02079}$

34. $2.96\overline{)0.01332}$

Divide. Round the quotients as indicated. See Example 6.

35. Divide: $0.549 \div 0.023$. Round the quotient to the nearest hundredth.

36. Divide: $0.0453 \div 0.98$. Round the quotient to the nearest thousandth.

37. Divide: $68.39 \div 0.6$. Round the quotient to the nearest tenth.

38. Divide: $98.83 \div 3.5$. Round the quotient to the nearest tenth.

Objective C *Divide. See Examples 8 and 9.*

39. $\dfrac{83.397}{100}$

40. $\dfrac{64.423}{100}$

41. $\dfrac{26.87}{10}$

42. $\dfrac{13.49}{10}$

43. $12.9 \div (-1000)$

44. $13.49 \div (-10,000)$

Objectives A C Mixed Practice *Divide. See Examples 1 through 5, 8, and 9.*

45. $7\overline{)88.2}$

46. $9\overline{)130.5}$

47. $\dfrac{13.1}{10}$

48. $\dfrac{17.7}{10}$

49. $\dfrac{456.25}{10,000}$

50. $\dfrac{986.11}{10,000}$

51. $1.239 \div 3$

52. $0.54 \div 12$

53. Divide 4.8 by −0.6.   **54.** Divide 4.9 by −0.7.   **55.** −1.224 ÷ 0.17   **56.** −1.344 ÷ 0.42

57. Divide 42 by 0.03.   **58.** Divide 27 by 0.03.   **59.** Divide −18 by −0.6.   **60.** Divide 20 by 0.4.

61. Divide 87 by −0.0015.   **62.** Divide 35 by −0.0007.   **63.** −1.104 ÷ 1.6   **64.** −2.156 ÷ 0.98

65. −2.4 ÷ (−100)   **66.** −86.79 ÷ (−1000)   **67.** $\dfrac{4.615}{0.071}$ **68.** $\dfrac{23.8}{0.035}$

Objective D *Evaluate each expression for* $x = 5.65$, $y = -0.8$, *and* $z = 4.52$. *See Example 10.*

69. $z \div y$ **70.** $z \div x$ **71.** $x \div y$ **72.** $y \div 2$

Determine whether the given values are solutions of the given equations. See Example 11.

73. $\dfrac{x}{4} = 3.04$; $x = 12.16$ **74.** $\dfrac{y}{8} = 0.89$; $y = 7.12$ **75.** $\dfrac{z}{100} = 0.8$; $z = 8$ **76.** $\dfrac{x}{10} = 0.23$; $x = 23$

Objective E *Solve. For Exercises 77 and 78, the solutions have been started for you. See Example 12.*

△ **77.** A new homeowner is painting the walls of a room. The walls have a total area of 546 square feet. A quart of paint covers 52 square feet. If the paint is sold in whole quarts only, how many quarts are needed?

Start the solution:

1. UNDERSTAND the problem. Reread it as many times as needed.
2. TRANSLATE into an equation. (Fill in the blanks.)

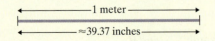

number of quarts	is	square feet	divided by	square feet per quart
↓	↓	↓	↓	↓

number of quarts = ____ ÷ ____

3. SOLVE. Don't forget to round up your quotient.
4. INTERPRET.

78. A shipping box can hold 36 books. If 486 books must be shipped, how many boxes are needed?

Start the solution:

1. UNDERSTAND the problem. Reread it as many times as needed.
2. TRANSLATE into an equation. (Fill in the blanks.)

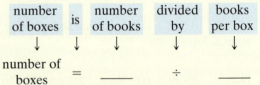

number of boxes	is	number of books	divided by	books per box
↓	↓	↓	↓	↓

number of boxes = ____ ÷ ____

3. SOLVE. Don't forget to round up your quotient.
4. INTERPRET.

79. There are approximately 39.37 inches in 1 meter. How many meters, to the nearest tenth of a meter, are there in 200 inches?

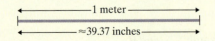

80. There are 2.54 centimeters in 1 inch. How many inches are there in 50 centimeters? Round to the nearest tenth.

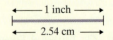

81. In the United States, an average child will wear down 730 crayons by his or her tenth birthday. Find the number of boxes of 64 crayons this is equivalent to. Round to the nearest tenth. (*Source:* Binney & Smith Inc.)

82. In 2011, American farmers received an average of $68.20 per hundred pounds of turkey. What was the average price per pound for turkeys? Round to the nearest cent. (*Source:* National Agricultural Statistics Service)

A child is to receive a dose of 0.5 teaspoon of cough medicine every 4 hours. If the bottle contains 4 fluid ounces, answer Exercises 83 through 86.

83. A fluid ounce equals 6 teaspoons. How many teaspoons are in 4 fluid ounces?

84. The bottle of medicine contains how many doses for the child? (*Hint:* See Exercise **83**.)

85. If the child takes a dose every four hours, how many days will the medicine last?

86. If the child takes a dose every six hours, how many days will the medicine last?

87. Americans ages 16–19 drive, on average, 7624 miles per year. About how many miles each week is that? Round to the nearest tenth. (*Note:* There are 52 weeks in a year.) (*Source:* Federal Highway Administration)

88. Drake Saucier was interested in the gas mileage on his "new" used car. He filled the tank, drove 423.8 miles, and filled the tank again. When he refilled the tank, it took 19.35 gallons of gas. Calculate the miles per gallon for Drake's car. Round to the nearest tenth.

89. The book *Harry Potter and the Deathly Hallows* was released to the public on July 21, 2007. Booksellers in the United States sold approximately 8292 thousand copies in the first 24 hours after release. If the same number of books was sold each hour, calculate the number of books sold each hour in the United States for that first day.

90. The leading money winner in men's professional golf in 2012 was Rory McIlroy. He earned approximately $8,048,000. Suppose he had earned this working 40 hours each week for a year. Determine his hourly wage to the nearest cent. (*Note:* There are 52 weeks in a year.) (*Source:* Professional Golfers' Association)

Review

Perform the indicated operation. See Sections 4.3 and 4.5.

91. $\dfrac{3}{5} \cdot \dfrac{7}{10}$

92. $\dfrac{3}{5} \div \dfrac{7}{10}$

93. $\dfrac{3}{5} - \dfrac{7}{10}$

94. $-\dfrac{3}{4} - \dfrac{1}{14}$

Concept Extensions

Mixed Practice (*Sections 5.2, 5.3, 5.4*) *Perform the indicated operation.*

95. $1.278 \div 0.3$

96. 1.278×0.3

97. $1.278 + 0.3$

98. $1.278 - 0.3$

99. $(-8.6)(3.1)$

100. $7.2 + 0.05 + 49.1$

101. $\begin{array}{r} 1000 \\ -95.71 \\ \hline \end{array}$

102. $\dfrac{87.2}{-10{,}000}$

Choose the best estimate.

103. 8.62×41.7
 a. 36
 b. 32
 c. 360
 d. 3.6

104. $1.437 + 20.69$
 a. 34
 b. 22
 c. 3.4
 d. 2.2

105. $78.6 \div 97$
 a. 7.86
 b. 0.786
 c. 786
 d. 7860

106. $302.729 - 28.697$
 a. 270
 b. 20
 c. 27
 d. 300

Recall from Section 1.6 that the average of a list of numbers is their total divided by how many numbers there are in the list. Use this procedure to find the average of the test scores listed in Exercises 107 and 108. If necessary, round to the nearest tenth.

107. 86, 78, 91, 87

108. 56, 75, 80

△ **109.** The area of a rectangle is 38.7 square feet. If its width is 4.5 feet, find its length.

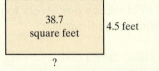

△ **110.** The perimeter of a square is 180.8 centimeters. Find the length of a side.

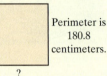

111. When dividing decimals, describe the process you use to place the decimal point in the quotient.

112. In your own words, describe how to quickly divide a number by a power of 10 such as 10, 100, 1000, etc.

To convert wind speeds in miles per hour to knots, divide by 1.15. Use this information and the Saffir-Simpson Hurricane Intensity chart below to answer Exercises 113 and 114. Round to the nearest tenth.

Saffir-Simpson Hurricane Intensity Scale				
Category	**Wind Speed**	**Barometric Pressure [inches of mercury (Hg)]**	**Storm Surge**	**Damage Potential**
1 (Weak)	75–95 mph	≥28.94 in.	4–5 ft	Minimal damage to vegetation
2 (Moderate)	96–110 mph	28.50–28.93 in.	6–8 ft	Moderate damage to houses
3 (Strong)	111–130 mph	27.91–28.49 in.	9–12 ft	Extensive damage to small buildings
4 (Very Strong)	131–155 mph	27.17–27.90 in.	13–18 ft	Extreme structural damage
5 (Devastating)	>155 mph	<27.17 in.	>18 ft	Catastrophic building failures possible

113. The chart gives wind speeds in miles per hour. What is the range of wind speeds for a Category 1 hurricane in knots?

114. What is the range of wind speeds for a Category 4 hurricane in knots?

△ **115.** A rancher is building a horse corral that's shaped like a rectangle with a width of 24.3 meters. He plans to make a four-wire fence; that is, he will string four wires around the corral. If he plans to use all of his 412.8 meters of wire, find the length of the corral he can construct.

116. A college student signed up for a new credit card that guarantees her no interest charges on transferred balances for a year. She transferred over a $2523.86 balance from her old credit card. Her minimum payment is $185.35 per month. If she pays only the minimum, will she pay off her balance before interest charges start again?

Integrated Review

Operations on Decimals

Perform the indicated operation.

Answers

1. $1.6 + 0.97$ **2.** $3.2 + 0.85$ **3.** $9.8 - 0.9$ **4.** $10.2 - 6.7$

1. _____

2. _____

3. _____

4. _____

5. $\begin{array}{r} 0.8 \\ \times\, 0.2 \\ \hline \end{array}$ **6.** $\begin{array}{r} 0.6 \\ \times\, 0.4 \\ \hline \end{array}$ **7.** $8\overline{)2.16}$ **8.** $6\overline{)3.12}$

5. _____

6. _____

7. _____

8. _____

9. _____

9. $(9.6)(-0.5)$ **10.** $(-8.7)(-0.7)$ **11.** $\begin{array}{r} 123.6 \\ -\;\; 48.04 \\ \hline \end{array}$ **12.** $\begin{array}{r} 325.2 \\ -\;\; 36.08 \\ \hline \end{array}$

10. _____

11. _____

12. _____

13. _____

13. $-25 + 0.026$ **14.** $0.125 + (-44)$ **15.** $29.24 \div (-3.4)$ **16.** $-10.26 \div (-1.9)$

14. _____

15. _____

16. _____

17. _____

17. -2.8×100 **18.** 1.6×1000 **19.** $\begin{array}{r} 96.21 \\ 7.028 \\ +\; 121.7 \\ \hline \end{array}$ **20.** $\begin{array}{r} 0.268 \\ 1.93 \\ +\; 142.881 \\ \hline \end{array}$

18. _____

19. _____

20. _____

21. _____

22. _____

23. _____

21. $-25.76 \div (-46)$ **22.** $-27.09 \div 43$ **23.** $\begin{array}{r} 12.004 \\ \times\;\;\; 2.3 \\ \hline \end{array}$ **24.** $\begin{array}{r} 28.006 \\ \times\;\;\; 5.2 \\ \hline \end{array}$

24. _____

25. _____

26. _____

27. _____

28. _____

29. _____

30. _____

31. _____

32. _____

33. _____

34. _____

35. _____

36. _____

37. _____

38. _____

39. _____

25. Subtract 4.6 from 10. **26.** Subtract 18 from 0.26. **27.** $-268.19 - 146.25$

28. $-860.18 - 434.85$ **29.** $\dfrac{2.958}{-0.087}$ **30.** $\dfrac{-1.708}{0.061}$

31. $160 - 43.19$ **32.** $120 - 101.21$ **33.** 15.62×10

34. $15.62 \div 10$ **35.** $15.62 + 10$ **36.** $15.62 - 10$

37. Find the distance in miles between Garden City, Kansas, and Wichita, Kansas. Next, estimate the distance by rounding each given distance to the nearest ten.

38. It costs $7.29 to send a 5-pound package locally via parcel post at a U.S. Post Office. To send the same package Priority Mail costs $8.10. How much more does it cost to send a package as Priority Mail? (*Source:* United States Postal Service)

39. In 2011, sales of Blu-ray Discs were $2 billion, but DVD sales dropped to $6.8 billion. Find the total spent to buy Blu-ray Discs or DVDs in 2011. Write the total in billions of dollars and also in standard notation. (*Source: USA Today*)

5.5 Fractions, Decimals, and Order of Operations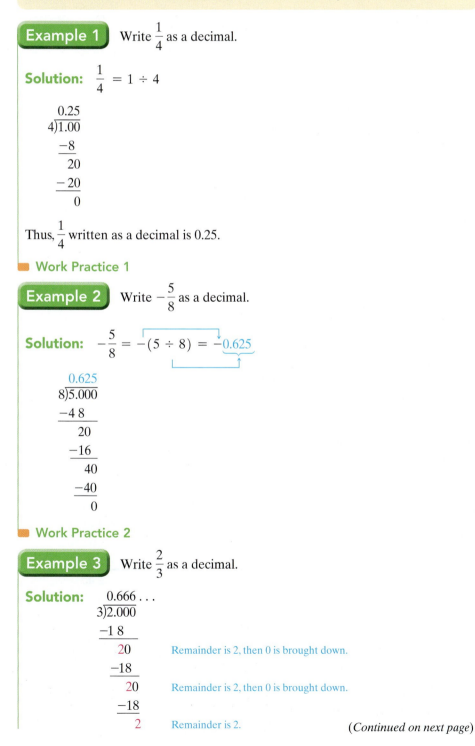

Objective A Writing Fractions as Decimals

To write a fraction as a decimal, we interpret the fraction bar to mean division and find the quotient.

> **Writing Fractions as Decimals**
>
> To write a fraction as a decimal, divide the numerator by the denominator.

Example 1 Write $\frac{1}{4}$ as a decimal.

Solution: $\frac{1}{4} = 1 \div 4$

$$
\begin{array}{r}
0.25 \\
4\overline{)1.00} \\
-8 \\
\hline
20 \\
-20 \\
\hline
0
\end{array}
$$

Thus, $\frac{1}{4}$ written as a decimal is 0.25.

■ **Work Practice 1**

Example 2 Write $-\frac{5}{8}$ as a decimal.

Solution: $-\frac{5}{8} = -(5 \div 8) = -0.625$

$$
\begin{array}{r}
0.625 \\
8\overline{)5.000} \\
-4\,8 \\
\hline
20 \\
-16 \\
\hline
40 \\
-40 \\
\hline
0
\end{array}
$$

■ **Work Practice 2**

Example 3 Write $\frac{2}{3}$ as a decimal.

Solution:
$$
\begin{array}{r}
0.666\ldots \\
3\overline{)2.000}
\end{array}
$$

$-1\,8$ Remainder is 2, then 0 is brought down.

20

-18 Remainder is 2, then 0 is brought down.

20

-18

2 Remainder is 2.

(Continued on next page)

Objectives

A Write Fractions as Decimals.

B Compare Decimals and Fractions.

C Simplify Expressions Containing Decimals and Fractions Using Order of Operations.

D Solve Area Problems Containing Fractions and Decimals.

E Evaluate Expressions Given Decimal Replacement Values.

Practice 1

a. Write $\frac{2}{5}$ as a decimal.

b. Write $\frac{9}{40}$ as a decimal.

Practice 2

Write $-\frac{3}{8}$ as a decimal.

Practice 3

a. Write $\frac{5}{6}$ as a decimal.

b. Write $\frac{2}{9}$ as a decimal.

Answers

1. a. 0.4   **b.** 0.225   **2.** −0.375

3. a. $0.8\overline{3}$ **b.** $0.\overline{2}$

Notice that the digit 2 keeps occurring as the remainder. This will continue so that the digit 6 will keep repeating in the quotient. We place a bar over the digit 6 to indicate that it repeats.

$$\frac{2}{3} = 0.666\ldots = 0.\overline{6}$$

We can also write a decimal approximation for $\frac{2}{3}$. For example, $\frac{2}{3}$ rounded to the nearest hundredth is 0.67. This can be written as $\frac{2}{3} \approx 0.67$.

■ Work Practice 3

Practice 4

Write $\frac{28}{13}$ as a decimal. Round to the nearest thousandth.

Example 4 Write $\frac{22}{7}$ as a decimal. (Recall that the fraction $\frac{22}{7}$ is an approximation for π.) Round to the nearest hundredth.

Solution:

$$\begin{array}{r} 3.142 \approx 3.14 \\ 7\overline{)22.000} \\ -21 \\ \hline 10 \\ -7 \\ \hline 30 \\ -28 \\ \hline 20 \\ -14 \\ \hline 6 \end{array}$$ Carry the division out to the thousandths place.

The fraction $\frac{22}{7}$ in decimal form is approximately 3.14.

■ Work Practice 4

Practice 5

Write $3\frac{5}{16}$ as a decimal.

Example 5 Write $2\frac{3}{16}$ as a decimal.

Solution:

Option 1. Write the fractional part only as a decimal.

$$\frac{3}{16} \longrightarrow \begin{array}{r} 0.1875 \\ 16\overline{)3.0000} \\ -1\,6 \\ \hline 1\,40 \\ -1\,28 \\ \hline 120 \\ -112 \\ \hline 80 \\ -80 \\ \hline 0 \end{array}$$

Thus $2\frac{3}{16} = 2.1875$

Option 2. Write $2\frac{3}{16}$ as an improper fraction, and divide.

$$2\frac{3}{16} = \frac{35}{16} \longrightarrow \begin{array}{r} 2.1875 \\ 16\overline{)35.0000} \\ -32 \\ \hline 3\,0 \\ -1\,6 \\ \hline 1\,40 \\ -1\,28 \\ \hline 120 \\ -112 \\ \hline 80 \\ -80 \\ \hline 0 \end{array}$$

■ Work Practice 5

Answers

4. 2.154 **5.** 3.3125

Some fractions may be written as decimals using our knowledge of decimals. From Section 5.1, we know that if the denominator of a fraction is 10, 100, 1000, or so on, we can immediately write the fraction as a decimal. For example,

$$\frac{4}{10} = 0.4, \quad \frac{12}{100} = 0.12, \text{ and so on}$$

Example 6 Write $\frac{4}{5}$ as a decimal.

Solution: Let's write $\frac{4}{5}$ as an equivalent fraction with a denominator of 10.

$$\frac{4}{5} = \frac{4}{5} \cdot \frac{2}{2} = \frac{8}{10} = 0.8$$

■ **Work Practice 6**

Practice 6

Write $\frac{3}{5}$ as a decimal.

Example 7 Write $\frac{1}{25}$ as a decimal.

Solution: $\frac{1}{25} = \frac{1}{25} \cdot \frac{4}{4} = \frac{4}{100} = 0.04$

■ **Work Practice 7**

Practice 7

Write $\frac{3}{50}$ as a decimal.

✓**Concept Check** Suppose you are writing the fraction $\frac{9}{16}$ as a decimal. How do you know you have made a mistake if your answer is 1.735?

Objective B Comparing Decimals and Fractions ▶

Now we can compare decimals and fractions by writing fractions as equivalent decimals.

Example 8 Insert $<$, $>$, or $=$ to form a true statement.

$$\frac{1}{8} \qquad 0.12$$

Solution: First we write $\frac{1}{8}$ as an equivalent decimal. Then we compare decimal places.

```
  0.125
8)1.000
 -8
  ──
  20
 -16
  ──
  40
 -40
  ──
   0
```

Original numbers	$\frac{1}{8}$	0.12
Decimals	0.125	0.120
Compare	0.125 > 0.12	

Thus, $\frac{1}{8} > 0.12$

■ **Work Practice 8**

Practice 8

Insert $<$, $>$, or $=$ to form a true statement.

$$\frac{1}{5} \qquad 0.25$$

Practice 9

Insert <, >, or = to form a true statement.

a. $\dfrac{1}{2}$ 0.54 **b.** $0.\overline{5}$ $\dfrac{5}{9}$

c. $\dfrac{5}{7}$ 0.72

Copyright 2015 Pearson Education, Inc.

Answers

9. **a.** < **b.** = **c.** <

10. **a.** $0.302, \dfrac{1}{3}, \dfrac{3}{8}$ **b.** $1\dfrac{1}{4}, 1.26, 1\dfrac{2}{5}$

c. $0.4, 0.41, \dfrac{3}{7}$

Example 9 Insert <, >, or = to form a true statement.

$$0.\overline{7} \quad \dfrac{7}{9}$$

Solution: We write $\dfrac{7}{9}$ as a decimal and then compare.

$$
\begin{array}{r}
0.77\ldots = 0.\overline{7} \\
9\overline{)7.00} \\
-6.3 \\
\hline
70 \\
-63 \\
\hline
7
\end{array}
$$

Original numbers	$0.\overline{7}$	$\dfrac{7}{9}$
Decimals	$0.\overline{7}$	$0.\overline{7}$
Compare	$0.\overline{7} = 0.\overline{7}$	

Thus, $\quad 0.\overline{7} = \dfrac{7}{9}$

■ **Work Practice 9**

Example 10 Write the numbers in order from smallest to largest.

$$\dfrac{9}{20}, \dfrac{4}{9}, 0.456$$

Solution:

	$\dfrac{9}{20}$	$\dfrac{4}{9}$	0.456
Original numbers			
Decimals	0.450	0.444 ...	0.456
Compare in order	2nd	1st	3rd

Written in order, we have

1st 2nd 3rd
↓ ↓ ↓

$$\dfrac{4}{9}, \dfrac{9}{20}, 0.456$$

■ **Work Practice 10**

Objective C Simplifying Expressions with Decimals and Fractions

In the remaining examples, we will review the order of operations by simplifying expressions that contain decimals.

Order of Operations

1. Perform all operations within parentheses ( ), brackets [ ], or other grouping symbols such as fraction bars.
2. Evaluate any expressions with exponents.
3. Multiply or divide in order from left to right.
4. Add or subtract in order from left to right.

Practice 10

Write the numbers in order from smallest to largest.

a. $\dfrac{1}{3}, 0.302, \dfrac{3}{8}$

b. $1.26, 1\dfrac{1}{4}, 1\dfrac{2}{5}$

c. $0.4, 0.41, \dfrac{3}{7}$

Example 11 Simplify: $723.6 \div 1000 \times 10$

Solution: Multiply or divide in order from left to right.

$$723.6 \div 1000 \times 10 = 0.7236 \times 10 \quad \text{Divide.}$$
$$= 7.236 \quad \text{Multiply.}$$

■ Work Practice 11

Practice 11
Simplify: $897.8 \div 100 \times 10$

Example 12 Simplify: $-0.5(8.6 - 1.2)$

Solution: According to the order of operations, we simplify inside the parentheses first.

$$-0.5(8.6 - 1.2) = -0.5(7.4) \quad \text{Subtract.}$$
$$= -3.7 \quad \text{Multiply.}$$

■ Work Practice 12

Practice 12
Simplify: $-8.69(3.2 - 1.8)$

Example 13 Simplify: $(-1.3)^2 + 2.4$

Solution: Recall the meaning of an exponent.

$$(-1.3)^2 + 2.4 = (-1.3)(-1.3) + 2.4 \quad \text{Use the definition of an exponent.}$$
$$= 1.69 + 2.4 \quad \text{Multiply. The product of two negative numbers is a positive number.}$$
$$= 4.09 \quad \text{Add.}$$

■ Work Practice 13

Practice 13
Simplify: $(-0.7)^2 + 2.1$

Example 14 Simplify: $\dfrac{5.68 + (0.9)^2 \div 100}{0.2}$

Solution: First we simplify the numerator of the fraction. Then we divide.

$$\frac{5.68 + (0.9)^2 \div 100}{0.2} = \frac{5.68 + 0.81 \div 100}{0.2} \quad \text{Simplify } (0.9)^2.$$
$$= \frac{5.68 + 0.0081}{0.2} \quad \text{Divide.}$$
$$= \frac{5.6881}{0.2} \quad \text{Add.}$$
$$= 28.4405 \quad \text{Divide.}$$

■ Work Practice 14

Practice 14
Simplify: $\dfrac{20.06 - (1.2)^2 \div 10}{0.02}$

Objective D Solving Area Problems Containing Fractions and Decimals ▶

Sometimes real-life problems contain both fractions and decimals. In the next example, we review the area of a triangle, and when values are substituted, the result may be an expression containing both fractions and decimals.

Answers
11. 89.78 **12.** −12.166
13. 2.59 **14.** 995.8

Practice 15

Find the area of the triangle.

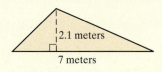

2.1 meters

7 meters

 Example 15   The area of a triangle is Area $= \frac{1}{2} \cdot$ base $\cdot$ height. Find the area of the triangle shown.

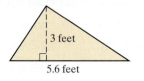

3 feet

5.6 feet

Solution:

$$\text{Area} = \frac{1}{2} \cdot \text{base} \cdot \text{height}$$

$$= \frac{1}{2} \cdot 5.6 \cdot 3$$

$$= 0.5 \cdot 5.6 \cdot 3 \qquad \text{Write } \frac{1}{2} \text{ as the decimal 0.5.}$$

$$= 8.4$$

The area of the triangle is 8.4 square feet.

■ **Work Practice 15**

Objective E Using Decimals as Replacement Values ▶

Practice 16

Evaluate $1.7y - 2$ for $y = 2.3$.

Example 16   Evaluate $-2x + 5$ for $x = 3.8$.

Solution: Replace x with 3.8 in the expression $-2x + 5$ and simplify.

$$-2x + 5 = -2(3.8) + 5 \quad \text{Replace } x \text{ with 3.8.}$$
$$= -7.6 + 5 \qquad \text{Multiply.}$$
$$= -2.6 \qquad \text{Add.}$$

■ **Work Practice 16**

Answers

15. 7.35 sq m   **16.** 1.91

Vocabulary, Readiness & Video Check

Answer each exercise "true" or "false."

1. The number $0.\overline{5}$ means 0.555. _____

2. To write $\frac{9}{19}$ as a decimal, perform the division $19\overline{)9}$. _____

3. $(-1.2)^2$ means $(-1.2)(-1.2)$ or -1.44. _____

4. To simplify $8.6(4.8 - 9.6)$, we first subtract. _____

Martin-Gay Interactive Videos *Watch the section lecture video and answer the following questions.*

Objective A 5. In ▦ Example 2, why is the bar placed over just the 6? ▶

Objective B 6. In ▦ Example 3, why do we write the fraction as a decimal rather than the decimal as a fraction? ▶

Objective C 7. In ▦ Example 4, besides meaning division, what other purpose does the fraction bar serve? ▶

Objective D 8. What formula is used to solve ▦ Example 5? What is the final answer? ▶

Objective E 9. In ▦ Example 6, once all replacement values are put into the variable expression, what is the resulting expression to evaluate? ▶

See Video 5.5

5.5 Exercise Set MyMathLab® ▶

Objective A *Write each number as a decimal. See Examples 1 through 7.*

1. $\dfrac{1}{5}$

2. $\dfrac{1}{20}$

3. $\dfrac{17}{25}$

4. $\dfrac{13}{25}$

▶ 5. $\dfrac{3}{4}$

6. $\dfrac{1}{8}$

7. $-\dfrac{2}{25}$

8. $-\dfrac{3}{25}$

9. $\dfrac{9}{4}$

10. $\dfrac{8}{5}$

▶ 11. $\dfrac{11}{12}$

12. $\dfrac{5}{12}$

13. $\dfrac{17}{40}$

14. $\dfrac{19}{25}$

15. $\dfrac{18}{40}$

16. $\dfrac{31}{40}$

17. $-\dfrac{1}{3}$

18. $-\dfrac{7}{9}$

19. $\dfrac{7}{16}$

20. $\dfrac{9}{16}$

21. $\dfrac{7}{11}$

22. $\dfrac{9}{11}$

23. $5\dfrac{17}{20}$

24. $4\dfrac{7}{8}$

25. $\dfrac{78}{125}$

26. $\dfrac{159}{375}$

Round each number as indicated. See Example 4.

27. Round your decimal answer to Exercise **17** to the nearest hundredth.

28. Round your decimal answer to Exercise **18** to the nearest hundredth.

29. Round your decimal answer to Exercise **19** to the nearest hundredth.

30. Round your decimal answer to Exercise **20** to the nearest hundredth.

31. Round your decimal answer to Exercise **21** to the nearest tenth.

32. Round your decimal answer to Exercise **22** to the nearest tenth.

Write each fraction as a decimal. If necessary, round to the nearest hundredth. See Examples 1 through 7.

33. Of the U.S. mountains that are over 14,000 feet in elevation, $\dfrac{56}{91}$ are located in Colorado. (*Source:* U.S. Geological Survey)

34. About $\dfrac{9}{20}$ of all U.S. citizens have type O blood. (*Source:* American Red Cross Biomedical Services)

35. About $\frac{43}{50}$ of Americans are Internet users. (*Source:* Digitalcenter)

36. About $\frac{14}{25}$ of Americans use the Internet through a wireless device. (*Source:* Digitalcenter)

37. When first launched, the Hubble Space Telescope's primary mirror was out of shape on the edges by $\frac{1}{50}$ of a human hair. This very small defect made it difficult to focus faint objects being viewed. Because the HST was in low Earth orbit, it was serviced by a shuttle and the defect was corrected.

38. The two mirrors currently in use in the Hubble Space Telescope were ground so that they do not deviate from a perfect curve by more than $\frac{1}{800,000}$ of an inch. Do not round this number.

Objective B *Insert $<$, $>$, or $=$ to form a true statement. See Examples 8 and 9.*

39. 0.562   0.569

40. 0.983   0.988

41. 0.215 $\frac{43}{200}$

42. $\frac{29}{40}$ 0.725

43. -0.0932 -0.0923

44. -0.00563 -0.00536

45. $0.\overline{6}$ $\frac{5}{6}$

46. $0.\overline{1}$ $\frac{2}{17}$

47. $\frac{51}{91}$ $0.56\overline{4}$

48. $0.58\overline{3}$ $\frac{6}{11}$

49. $\frac{4}{7}$ 0.14

50. $\frac{5}{9}$ 0.557

▶ **51.** 1.38 $\frac{18}{13}$

52. 0.372 $\frac{22}{59}$

53. 7.123 $\frac{456}{64}$

54. 12.713 $\frac{89}{7}$

Write the numbers in order from smallest to largest. See Example 10.

55. 0.34, 0.35, 0.32

56. 0.47, 0.42, 0.40

57. 0.49, 0.491, 0.498

58. 0.72, 0.727, 0.728

59. $5.23, \frac{42}{8}, 5.34$

60. $7.56, \frac{67}{9}, 7.562$

61. $\frac{5}{8}, 0.612, 0.649$

62. $\frac{5}{6}, 0.821, 0.849$

Objective C *Simplify each expression. See Examples 11 through 14.*

63. $(0.3)^2 + 0.5$

64. $(-2.5)(3) - 4.7$

65. $\dfrac{1 + 0.8}{-0.6}$

66. $(-0.05)^2 + 3.13$

67. $(-2.3)^2(0.3 + 0.7)$

68. $(8.2)(100) - (8.2)(10)$

69. $(5.6 - 2.3)(2.4 + 0.4)$

70. $\dfrac{0.222 - 2.13}{12}$

71. $\dfrac{(4.5)^2}{100}$

72. $0.9(5.6 - 6.5)$

▶ **73.** $\dfrac{7 + 0.74}{-6}$

74. $(1.5)^2 + 0.5$

Find the value of each expression. Give the result as a decimal. See Examples 11 through 14.

75. $\frac{1}{5} - 2(7.8)$

76. $\frac{3}{4} - (9.6)(5)$

77. $\frac{1}{4}(-9.6 - 5.2)$

78. $\frac{3}{8}(4.7 - 5.9)$

Objective D *Find the area of each triangle or rectangle. See Example 15.*

△ **79.**

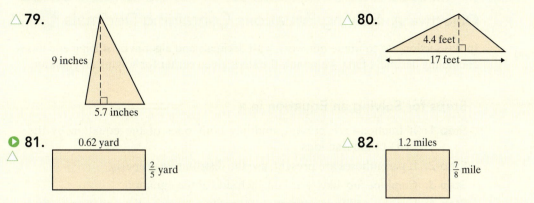

9 inches

5.7 inches

△ **80.**

4.4 feet

17 feet

⊙ **81.**

0.62 yard

$\frac{2}{5}$ yard

△ **82.**

1.2 miles

$\frac{7}{8}$ mile

Objective E *Evaluate each expression for $x = 6$, $y = 0.3$, and $z = -2.4$. See Example 16.*

83. z^2

84. y^2

85. $x - y$

86. $x - z$

⊙ **87.** $4y - z$

88. $\frac{x}{y} + 2z$

Review

Simplify. See Sections 4.3 and 4.5.

89. $\frac{9}{10} + \frac{16}{25}$

90. $\frac{4}{11} - \frac{19}{22}$

91. $\left(\frac{2}{5}\right)\left(\frac{5}{2}\right)^2$

92. $\left(\frac{2}{3}\right)^2\left(\frac{3}{2}\right)^3$

Concept Extensions

Without calculating, describe each number as < 1, $= 1$, or > 1. See the Concept Check in this section.

93. 1.0

94. 1.0000

95. 1.00001

96. $\frac{101}{99}$

97. $\frac{99}{100}$

98. $\frac{99}{99}$

In 2012, there were 11,434 commercial radio stations in the United States. The most popular formats are listed in the table along with their counts. Use this graph to answer Exercises 99 through 102.

99. Write as a decimal the fraction of radio stations with a classic hits music format. Round to the nearest thousandth.

100. Write as a decimal the fraction of radio stations with a Spanish format. Round to the nearest hundredth.

101. Estimate, by rounding each number in the table to the nearest hundred, the total number of stations with the top six formats in 2012.

102. Use your estimate from Exercise **101** to write the fraction of radio stations accounted for by the top six formats as a decimal. Round to the nearest hundredth.

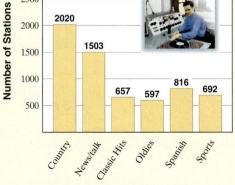

Top Commercial Radio Station Formats in 2012

Format (Total stations: 11,434)

103. Describe two ways to write fractions as decimals.

104. Describe two ways to write mixed numbers as decimals.

Objective

A Solve Equations Containing Decimals.

Objective A Solving Equations Containing Decimals

In this section, we continue our work with decimals and algebra by solving equations containing decimals. First, we review the steps given earlier for solving an equation.

> **Steps for Solving an Equation in x**
>
> **Step 1:** If fractions are present, multiply both sides of the equation by the LCD of the fractions.
>
> **Step 2:** If parentheses are present, use the distributive property.
>
> **Step 3:** Combine any like terms on each side of the equation.
>
> **Step 4:** Use the addition property of equality to rewrite the equation so that variable terms are on one side of the equation and constant terms are on the other side.
>
> **Step 5:** Divide both sides by the numerical coefficient of x to solve.
>
> **Step 6:** Check your answer in the **original equation**.

Practice 1

Solve: $z + 0.9 = 1.3$

Example 1 Solve: $x - 1.5 = 8$

Solution: Steps 1 through 3 are not needed for this equation, so we begin with Step 4. To get x alone on one side of the equation, add 1.5 to both sides.

$$x - 1.5 = 8 \qquad \text{Original equation}$$
$$x - 1.5 + 1.5 = 8 + 1.5 \qquad \text{Add 1.5 to both sides.}$$
$$x = 9.5 \qquad \text{Simplify.}$$

Check: To check, replace x with 9.5 in the *original equation*.

$$x - 1.5 = 8 \qquad \text{Original equation}$$
$$9.5 - 1.5 \stackrel{?}{=} 8 \qquad \text{Replace } x \text{ with 9.5.}$$
$$8 = 8 \qquad \text{True}$$

Since $8 = 8$ is a true statement, 9.5 is a solution of the equation.

■ **Work Practice 1**

Practice 2

Solve: $0.17x = -0.34$

Example 2 Solve: $-2y = 6.7$

Solution: Steps 1 through 4 are not needed for this equation, so we begin with Step 5. To solve for y, divide both sides by the coefficient of y, which is -2.

$$-2y = 6.7 \qquad \text{Original equation}$$
$$\frac{-2y}{-2} = \frac{6.7}{-2} \qquad \text{Divide both sides by } -2.$$
$$y = -3.35 \qquad \text{Simplify.}$$

Check: To check, replace y with -3.35 in the original equation.

$$-2y = 6.7 \qquad \text{Original equation}$$
$$-2(-3.35) \stackrel{?}{=} 6.7 \qquad \text{Replace } y \text{ with } -3.35.$$
$$6.7 = 6.7 \qquad \text{True}$$

Thus -3.35 is a solution of the equation $-2y = 6.7$.

■ **Work Practice 2**

Answers

1. 0.4 **2.** −2

384

Example 3   Solve: $1.2x + 5.8 = 8.2$

Solution: We begin with Step 4 and get the variable term alone by subtracting 5.8 from both sides.

$$1.2x + 5.8 = 8.2$$
$$1.2x + 5.8 - 5.8 = 8.2 - 5.8 \quad \text{Subtract 5.8 from both sides.}$$
$$1.2x = 2.4 \qquad\qquad \text{Simplify.}$$
$$\frac{1.2x}{1.2} = \frac{2.4}{1.2} \qquad\quad \text{Divide both sides by 1.2.}$$
$$x = 2 \qquad\qquad\quad \text{Simplify.}$$

To check, replace x with 2 in the original equation. The solution is 2.

■ **Work Practice 3**

Practice 3

Solve: $2.9 = 1.7 + 0.3x$

Example 4   Solve: $7x + 3.2 = 4x - 1.6$

Solution: We start with Step 4 to get variable terms on one side and numerical terms on the other.

$$7x + 3.2 = 4x - 1.6$$
$$7x + 3.2 - 3.2 = 4x - 1.6 - 3.2 \quad \text{Subtract 3.2 from both sides.}$$
$$7x = 4x - 4.8 \qquad\qquad \text{Simplify.}$$
$$7x - 4x = 4x - 4.8 - 4x \qquad \text{Subtract } 4x \text{ from both sides.}$$
$$3x = -4.8 \qquad\qquad\quad \text{Simplify.}$$
$$\frac{3x}{3} = -\frac{4.8}{3} \qquad\qquad \text{Divide both sides by 3.}$$
$$x = -1.6 \qquad\qquad\quad \text{Simplify.}$$

Check to see that -1.6 is the solution.

■ **Work Practice 4**

Practice 4

Solve: $8x + 4.2 = 10x + 11.6$

Example 5   Solve: $5(x - 0.36) = -x + 2.4$

Solution: First use the distributive property to distribute the factor 5.

$$5(x - 0.36) = -x + 2.4 \quad \text{Original equation}$$
$$5x - 1.8 = -x + 2.4 \quad \text{Apply the distributive property.}$$

Next, get the variable terms alone on the left side of the equation by adding 1.8 to both sides of the equation and then adding x to both sides of the equation.

$$5x - 1.8 + 1.8 = -x + 2.4 + 1.8 \quad \text{Add 1.8 to both sides.}$$
$$5x = -x + 4.2 \qquad\qquad \text{Simplify.}$$
$$5x + x = -x + 4.2 + x \qquad \text{Add } x \text{ to both sides.}$$
$$6x = 4.2 \qquad\qquad\quad \text{Simplify.}$$
$$\frac{6x}{6} = \frac{4.2}{6} \qquad\qquad \text{Divide both sides by 6.}$$
$$x = 0.7 \qquad\qquad\quad \text{Simplify.}$$

To verify that 0.7 is the solution, replace x with 0.7 in the original equation.

■ **Work Practice 5**

Practice 5

Solve: $6.3 - 5x = 3(x + 2.9)$

Answers

3. 4   **4.** -3.7 **5.** -0.3

Instead of solving equations with decimals, sometimes it may be easier to first rewrite the equation so that it contains integers only. Recall that multiplying a decimal by a power of 10 such as 10, 100, or 1000 has the effect of moving the decimal point to the right. We can use the multiplication property of equality to multiply both sides of the equation by an appropriate power of 10. The resulting equivalent equation will contain integers only.

Practice 6

Solve: $0.2y + 2.6 = 4$

> **Example 6** Solve: $0.5y + 2.3 = 1.65$
>
> **Solution:** Multiply both sides of the equation by 100. This will move the decimal point in each term two places to the right.
>
> $$0.5y + 2.3 = 1.65 \qquad \text{Original equation}$$
> $$100(0.5y + 2.3) = 100(1.65) \qquad \text{Multiply both sides by 100.}$$
> $$100(0.5y) + 100(2.3) = 100(1.65) \qquad \text{Apply the distributive property.}$$
> $$50y + 230 = 165 \qquad \text{Simplify.}$$
>
> Now the equation contains integers only. Continue solving by subtracting 230 from both sides.
>
> $$50y + 230 = 165$$
> $$50y + 230 - 230 = 165 - 230 \qquad \text{Subtract 230 from both sides.}$$
> $$50y = -65 \qquad \text{Simplify.}$$
> $$\frac{50y}{50} = \frac{-65}{50} \qquad \text{Divide both sides by 50.}$$
> $$y = -1.3 \qquad \text{Simplify.}$$
>
> Check to see that -1.3 is the solution by replacing y with -1.3 in the original equation.
>
> ■ **Work Practice 6**

Answer

6. 7

✓**Concept Check Answer**

Multiply by 1000.

✓**Concept Check** By what number would you multiply both sides of the following equation to make calculations easier? Explain your choice.

$$1.7x + 3.655 = -14.2$$

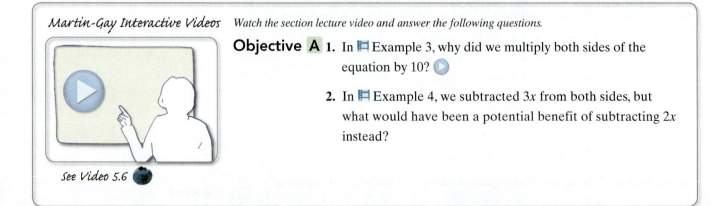

Martin-Gay Interactive Videos   *Watch the section lecture video and answer the following questions.*

Objective A **1.** In ▤ Example 3, why did we multiply both sides of the equation by 10? ▶

2. In ▤ Example 4, we subtracted $3x$ from both sides, but what would have been a potential benefit of subtracting $2x$ instead?

See Video 5.6 ●

5.6 Exercise Set MyMathLab® ▶

Objective **A** *Solve each equation. See Examples 1 and 2.*

1. $x + 1.2 = 7.1$ **2.** $y - 0.5 = 9$ **3.** $-5y = 2.15$ **4.** $-0.4x = 50$

5. $6.2 = y - 4$ **6.** $9.7 = x + 11.6$ **7.** $3.1x = -13.95$ **8.** $3y = -25.8$

Solve each equation. See Examples 3 through 5.

9. $-3.5x + 2.8 = -11.2$ **10.** $7.1 - 0.2x = 6.1$ ▶ **11.** $6x + 8.65 = 3x + 10$

12. $7x - 9.64 = 5x + 2.32$ **13.** $2(x - 1.3) = 5.8$ **14.** $5(x + 2.3) = 19.5$

Solve each equation by first multiplying both sides by an appropriate power of 10 so that the equation contains integers only. See Example 6.

15. $0.4x + 0.7 = -0.9$ **16.** $0.7x + 0.1 = 1.5$ **17.** $7x - 10.8 = x$

18. $3y = 7y + 24.4$ ▶ **19.** $2.1x + 5 - 1.6x = 10$ **20.** $1.5x + 2 - 1.2x = 12.2$

Solve. See Examples 1 through 6.

21. $y - 3.6 = 4$ **22.** $x + 5.7 = 8.4$ **23.** $-0.02x = -1.2$

24. $-9y = -0.162$ **25.** $6.5 = 10x + 7.2$ **26.** $2x - 4.2 = 8.6$

27. $2.7x - 25 = 1.2x + 5$ **28.** $9y - 6.9 = 6y - 11.1$ **29.** $200x - 0.67 = 100x + 0.81$

30. $2.3 + 500x = 600x - 0.2$ ▶ **31.** $3(x + 2.71) = 2x$ **32.** $7(x + 8.6) = 6x$

33. $8x - 5 = 10x - 8$ **34.** $24y - 10 = 20y - 17$ ▶ **35.** $1.2 + 0.3x = 0.9$

36. $1.5 = 0.4x + 0.5$ **37.** $-0.9x + 2.65 = -0.5x + 5.45$ **38.** $-50x + 0.81 = -40x - 0.48$

39. $4x + 7.6 = 2(3x - 3.2)$ **40.** $4(2x - 1.6) = 5x - 6.4$ **41.** $0.7x + 13.8 = x - 2.16$

42. $y - 5 = 0.3y + 4.1$

Review

Simplify each expression by combining like terms. See Section 3.1.

43. $2x - 7 + x - 9$

44. $x + 14 - 5x - 17$

Perform the indicated operation. See Sections 4.3, 4.5, and 4.7.

45. $\dfrac{6x}{5} \cdot \dfrac{1}{2x^2}$

46. $5\dfrac{1}{3} \div 9\dfrac{1}{6}$

47. $\dfrac{x}{3} + \dfrac{2x}{7}$

48. $50 - 14\dfrac{9}{13}$

Concept Extensions

Mixed Practice (Sections 5.2 and 5.6) *This section of exercises contains equations and expressions. If the exercise contains an equation, solve it for the variable. If the exercise contains an expression, simplify it by combining any like terms.*

49. $b + 4.6 = 8.3$

50. $y - 15.9 = -3.8$

51. $2x - 0.6 + 4x - 0.01$

52. $-x - 4.1 - x - 4.02$

53. $5y - 1.2 - 7y + 8$

54. $9a - 5.6 - 3a + 6$

55. $2.8 = z - 6.3$

56. $9.7 = x + 4.3$

57. $4.7x + 8.3 = -5.8$

58. $2.8x + 3.4 = -13.4$

59. $7.76 + 8z - 12z + 8.91$

60. $9.21 + x - 4x + 11.33$

61. $5(x - 3.14) = 4x$

62. $6(x + 1.43) = 5x$

63. $2.6y + 8.3 = 4.6y - 3.4$

64. $8.4z - 2.6 = 5.4z + 10.3$

65. $9.6z - 3.2 - 11.7z - 6.9$

66. $-3.2x + 12.6 - 8.9x - 15.2$

67. Explain in your own words the property of equality that allows us to multiply both sides of an equation by a power of 10.

68. By what number would you multiply both sides of $8x - 7.6 = 4.23$ to make calculations easier? Explain your choice.

69. Construct an equation whose solution is 1.4.

70. Construct an equation whose solution is -8.6.

Solve.

71. $-5.25x = -40.33575$

72. $7.68y = -114.98496$

73. $1.95y + 6.834 = 7.65y - 19.8591$

74. $6.11x + 4.683 = 7.51x + 18.235$

5.7 Decimal Applications: Mean, Median, and Mode

Objective A Finding the Mean

Sometimes we want to summarize data by displaying them in a graph, but sometimes it is also desirable to be able to describe a set of data, or a set of numbers, by a single "middle" number. Three such **measures of central tendency** are the **mean,** the **median,** and the **mode.**

The most common measure of central tendency is the mean (sometimes called the "arithmetic mean" or the "average"). Recall that we first introduced finding the average of a list of numbers in Section 1.6.

Objectives

A Find the Mean of a List of Numbers.

B Find the Median of a List of Numbers.

C Find the Mode of a List of Numbers.

> The **mean (average)** of a set of numbered items is the sum of the items divided by the number of items.
>
> $$\text{mean} = \frac{\text{sum of items}}{\text{number of items}}$$

Example 1 Finding the Mean Time in an Experiment

Seven students in a psychology class conducted an experiment on mazes. Each student was given a pencil and asked to successfully complete the same maze. The timed results are below:

Student	Ann	Thanh	Carlos	Jesse	Melinda	Ramzi	Dayni
Time (Seconds)	13.2	11.8	10.7	16.2	15.9	13.8	18.5

a. Who completed the maze in the shortest time? Who completed the maze in the longest time?

b. Find the mean time.

c. How many students took longer than the mean time? How many students took shorter than the mean time?

Solution:

a. Carlos completed the maze in 10.7 seconds, the shortest time. Dayni completed the maze in 18.5 seconds, the longest time.

b. To find the mean (or average), we find the sum of the items and divide by 7, the number of items.

$$\text{mean} = \frac{13.2 + 11.8 + 10.7 + 16.2 + 15.9 + 13.8 + 18.5}{7}$$

$$= \frac{100.1}{7} = 14.3$$

c. Three students, Jesse, Melinda, and Dayni, had times longer than the mean time. Four students, Ann, Thanh, Carlos, and Ramzi, had times shorter than the mean time.

■ Work Practice 1

✓ **Concept Check** Estimate the mean of the following set of data:

5, 10, 10, 10, 10, 15

Often in college, the calculation of a **grade point average** (GPA) is a **weighted mean** and is calculated as shown in Example 2.

Practice 1

Find the mean of the following test scores: 87, 75, 96, 91, and 78.

Answer

1. 85.4

✓ **Concept Check Answer**

10

389

Practice 2

Find the grade point average if the following grades were earned in one semester.

Grade	Credit Hours
A	2
B	4
C	5
D	2
A	2

Example 2 Calculating Grade Point Average (GPA)

The following grades were earned by a student during one semester. Find the student's grade point average.

Course	Grade	Credit Hours
College mathematics	A	3
Biology	B	3
English	A	3
PE	C	1
Social studies	D	2

Solution: To calculate the grade point average, we need to know the point values for the different possible grades. The point values of grades commonly used in colleges and universities are given below:

A: 4, B: 3, C: 2, D: 1, F: 0

Now, to find the grade point average, we multiply the number of credit hours for each course by the point value of each grade. The grade point average is the sum of these products divided by the sum of the credit hours.

Course	Grade	Point Value of Grade	Credit Hours	Point Value of Credit Hours
College mathematics	A	4	3	12
Biology	B	3	3	9
English	A	4	3	12
PE	C	2	1	2
Social studies	D	1	2	2
		Totals:	12	37

$$\text{grade point average} = \frac{37}{12} \approx 3.08 \text{ rounded to two decimal places}$$

The student earned a grade point average of 3.08.

■ **Work Practice 2**

Objective B Finding the Median ▶

You may have noticed that a very low number or a very high number can affect the mean of a list of numbers. Because of this, you may sometimes want to use another measure of central tendency. A second measure of central tendency is called the **median.** The median of a list of numbers is not affected by a low or high number in the list.

> The **median** of a set of numbers in numerical order is the middle number. If the number of items is odd, the median is the middle number. If the number of items is even, the median is the mean of the two middle numbers.

Practice 3

Find the median of the list of numbers: 5, 11, 14, 23, 24, 35, 38, 41, 43

Example 3 Find the median of the following list of numbers:

25, 54, 56, 57, 60, 71, 98

Solution: Because this list is in numerical order, the median is the middle number, 57.

■ **Work Practice 3**

Answers

2. 2.67 **3.** 24

Example 4   Find the median of the following list of scores: 67, 91, 75, 86, 55, 91

Solution: First we list the scores in numerical order and then we find the middle number.

55, 67, 75, 86, 91, 91

Since there is an even number of scores, there are two middle numbers, 75 and 86. The median is the mean of the two middle numbers.

$$\text{median} = \frac{75 + 86}{2} = 80.5$$

The median is 80.5.

Helpful Hint   Don't forget to write the numbers in order from smallest to largest before finding the median.

▪ **Work Practice 4**

Objective C Finding the Mode ▶

The last common measure of central tendency is called the **mode.**

The **mode** of a set of numbers is the number that occurs most often. (It is possible for a set of numbers to have more than one mode or to have no mode.)

Example 5   Find the mode of the list of numbers:

11, 14, 14, 16, 31, 56, 65, 77, 77, 78, 79

Solution: There are two numbers that occur the most often. They are 14 and 77. This list of numbers has two modes, 14 and 77.

▪ **Work Practice 5**

Example 6   Find the median and the mode of the following set of numbers. These numbers were high temperatures for 14 consecutive days in a city in Montana.

76, 80, 85, 86, 89, 87, 82, 77, 76, 79, 82, 89, 89, 92

Solution: First we write the numbers in numerical order.

76, 76, 77, 79, 80, 82, 82, 85, 86, 87, 89, 89, 89, 92

Since there is an even number of items, the median is the mean of the two middle numbers, 82 and 85.

$$\text{median} = \frac{82 + 85}{2} = 83.5$$

The mode is 89, since 89 occurs most often.

▪ **Work Practice 6**

✓**Concept Check**   True or false? Every set of numbers *must* have a mean, median, and mode. Explain your answer.

Practice 4

Find the median of the list of scores:
36, 91, 78, 65, 95, 95, 88, 71

Practice 5

Find the mode of the list of numbers:
14, 10, 10, 13, 15, 15, 15, 17, 18, 18, 20

Practice 6

Find the median and the mode of the list of numbers:
26, 31, 15, 15, 26, 30, 16, 18, 15, 35

Answers
4. 83   **5.** 15   **6.** median: 22; mode: 15

✓**Concept Check Answer**
false; a set of numbers may have no mode

> **Helpful Hint**
>
> Don't forget that it is possible for a list of numbers to have no mode. For example, the list
>
> 2, 4, 5, 6, 8, 9
>
> has no mode. There is no number or numbers that occur more often than the others.

Vocabulary, Readiness & Video Check

Use the choices below to fill in each blank. Some choices may be used more than once.

mean	mode	grade point average
median	average	

1. Another word for "mean" is _____.

2. The number that occurs most often in a set of numbers is called the _____.

3. The _____ of a set of number items is $\dfrac{\text{sum of items}}{\text{number of items}}$.

4. The _____ of a set of numbers is the middle number. If the number of numbers is even, it is the _____ of the two middle numbers.

5. An example of weighted mean is a calculation of _____.

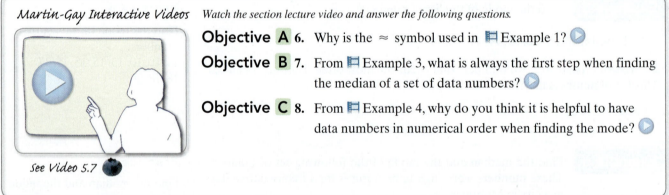

Martin-Gay Interactive Videos   *Watch the section lecture video and answer the following questions.*

Objective A 6. Why is the ≈ symbol used in ▯ Example 1? ▶

Objective B 7. From ▯ Example 3, what is always the first step when finding the median of a set of data numbers? ▶

Objective C 8. From ▯ Example 4, why do you think it is helpful to have data numbers in numerical order when finding the mode? ▶

See Video 5.7 🫐

5.7 Exercise Set MyMathLab® ▶

Objectives A B C Mixed Practice *For each set of numbers, find the mean, median, and mode. If necessary, round the mean to one decimal place. See Examples 1 and 3 through 6.*

1. 15, 23, 24, 18, 25

2. 45, 36, 28, 46, 52

▶ 3. 7.6, 8.2, 8.2, 9.6, 5.7, 9.1

4. 4.9, 7.1, 6.8, 6.8, 5.3, 4.9

5. 0.5, 0.2, 0.2, 0.6, 0.3, 1.3, 0.8, 0.1, 0.5

6. 0.6, 0.6, 0.8, 0.4, 0.5, 0.3, 0.7, 0.8, 0.1

7. 231, 543, 601, 293, 588, 109, 334, 268

8. 451, 356, 478, 776, 892, 500, 467, 780

The ten tallest buildings in the world, completed as of the start of 2012, are listed in the following table. Use this table to answer Exercises 9 through 14. If necessary, round results to one decimal place. See Examples 1 and 3 through 6.

9. Find the mean height of the five tallest buildings.

Building	Height (in feet)
Burj Khalifa, Dubai	2717
Makkah Royal Clock Tower Hotel, Saudi Arabia	1972
Taipei 101	1667
Shanghai World Financial Center	1614
International Commerce Centre, Hong Kong	1588
Petronas Tower 1, Kuala Lumpur	1483
Petronas Tower 2, Kuala Lumpur	1483
Zifeng Tower, China	1476
Willis Tower, Chicago	1451
KK 100 Development, China	1449

(*Source:* Council on Tall Buildings and Urban Habitat)

10. Find the median height of the five tallest buildings.

11. Find the median height of the eight tallest buildings.

12. Find the mean height of the eight tallest buildings.

13. Given the building heights, explain how you know, without calculating, that the answer to Exercise **10** is greater than the answer to Exercise **11**.

14. Given the building heights, explain how you know, without calculating, that the answer to Exercise **12** is less than the answer to Exercise **9**.

For Exercises 15 through 18, the grades are given for a student for a particular semester. Find the grade point average. If necessary, round the grade point average to the nearest hundredth. See Example 2.

15.

Grade	Credit Hours
B	3
C	3
A	4
C	4

16.

Grade	Credit Hours
D	1
F	1
C	4
B	5

17.

Grade	Credit Hours
A	3
A	3
A	4
B	3
C	1

18.

Grade	Credit Hours
B	2
B	2
C	3
A	3
B	3

During an experiment, the following times (in seconds) were recorded:

7.8, 6.9, 7.5, 4.7, 6.9, 7.0.

19. Find the mean. Round to the nearest tenth.

20. Find the median.

21. Find the mode.

In a mathematics class, the following test scores were recorded for a student: 93, 85, 89, 79, 88, 91.

22. Find the mean. Round to the nearest hundredth.

23. Find the median.

24. Find the mode.

The following pulse rates were recorded for a group of 15 students:

78, 80, 66, 68, 71, 64, 82, 71, 70, 65, 70, 75, 77, 86, 72.

25. Find the mean.

26. Find the median.

27. Find the mode.

28. How many pulse rates were higher than the mean?

29. How many pulse rates were lower than the mean?

Review

Write each fraction in simplest form. See Section 4.2.

30. $\dfrac{12}{20}$

31. $\dfrac{6}{18}$

32. $\dfrac{4x}{36}$

33. $\dfrac{18}{30y}$

34. $\dfrac{35a^3}{100a^2}$

35. $\dfrac{55y^2}{75y^2}$

Concept Extensions

Find the missing numbers in each set of numbers.

36. 16, 18, _____, _____, _____. The mode is 21. The median is 20.

37. _____, _____, _____, 40, _____. The mode is 35. The median is 37. The mean is 38.

38. Write a list of numbers for which you feel the median would be a better measure of central tendency than the mean.

39. Without making any computations, decide whether the median of the following list of numbers will be a whole number. Explain your reasoning.

36, 77, 29, 58, 43

Chapter 5 Group Activity

Maintaining a Checking Account

(Sections 5.1, 5.2, 5.3, 5.4)

This activity may be completed by working in groups or individually.

A checking account is a convenient way of handling money and paying bills. To open a checking account, the bank or savings and loan association requires a customer to make a deposit. Then the customer receives a checkbook that contains checks, deposit slips, and a register for recording checks written and deposits made. It is important to record all payments and deposits that affect the account. It is also important to keep the checkbook balance current by subtracting checks written and adding deposits made.

About once a month, checking customers receive a statement from the bank listing all activity that the account has had in the last month. The statement lists a beginning balance, all checks and deposits, any service charges made against the account, and an ending balance. Because it may take several days for checks that a customer has written to clear the banking system, the check register may list checks that do not appear on the monthly bank statement. These checks are called **outstanding checks.** Deposits that are recorded in the check register but do not appear on the statement are called **deposits in transit.** Because of these differences, it is important to balance, or reconcile, the checkbook against the monthly statement. The steps for doing so are listed below.

Balancing or Reconciling a Checkbook

Step 1: Place a check mark in the checkbook register next to each check and deposit listed on the monthly bank statement. Any entries in the register without a check mark are outstanding checks or deposits in transit.

Step 2: Find the ending checkbook register balance and add to it any outstanding checks and any interest paid on the account.

Step 3: From the total in Step 2, subtract any deposits in transit and any service charges.

Step 4: Compare the amount found in Step 3 with the ending balance listed on the bank statement. If they are the same, the checkbook balances with the bank statement. Be sure to update the check register with service charges and interest.

Step 5: If the checkbook does not balance, recheck the balancing process. Next, make sure that the running checkbook register balance was calculated correctly. Finally, compare the checkbook register with the statement to make sure that each check was recorded for the correct amount.

For the checkbook register and monthly bank statement given:

a. *update the checkbook register*   **b.** *list the outstanding checks and their total, and deposits in transit*

c. *balance the checkbook—be sure to update the register with any interest or service fees*

Checkbook Register

#	Date	Description	Payment	✓	Deposit	Balance 425.86
114	4/1	Market Basket	30.27			
115	4/3	May's Texaco	8.50			
	4/4	Cash at ATM	50.00			
116	4/6	UNO Bookstore	121.38			
	4/7	Deposit			100.00	
117	4/9	MasterCard	84.16			
118	4/10	Redbox	6.12			
119	4/12	Kroger	18.72			
120	4/14	Parking sticker	18.50			
	4/15	Direct deposit			294.36	
121	4/20	Rent	395.00			
122	4/25	Student fees	20.00			
	4/28	Deposit			75.00	

First National Bank Monthly Statement 4/30

BEGINNING BALANCE:		425.86
Date	Number	Amount
CHECKS AND ATM WITHDRAWALS		
4/3	114	30.27
4/4	ATM	50.00
4/11	117	84.16
4/13	115	8.50
4/15	119	18.72
4/22	121	395.00
DEPOSITS		
4/7		100.00
4/15	Direct deposit	294.36
SERVICE CHARGES		
Low balance fee		7.50
INTEREST		
Credited 4/30		1.15
ENDING BALANCE:		227.22

Chapter 5 Vocabulary Check

Fill in each blank with one of the choices listed below. Some choices may be used more than once and some not at all.

vertically	decimal	and	right triangle
standard form	mean	median	circumference
sum	denominator	numerator	mode

1. Like fractional notation, _____ notation is used to denote a part of a whole.

2. To write fractions as decimals, divide the _____ by the _____.

3. To add or subtract decimals, write the decimals so that the decimal points line up _____.

4. When writing decimals in words, write "_____" for the decimal point.

5. When multiplying decimals, the decimal point in the product is placed so that the number of decimal places in the product is equal to the _____ of the number of decimal places in the factors.

6. The _____ of a set of numbers is the number that occurs most often.

7. The distance around a circle is called the _____.

8. The _____ of a set of numbers in numerical order is the middle number. If there is an even number of numbers, the median is the _____ of the two middle numbers.

9. The _____ of a list of items with number values is $\dfrac{\text{sum of items}}{\text{number of items}}$.

10. When 2 million is written as 2,000,000, we say it is written in _____.

Helpful Hint

▶ Are you preparing for your test? Don't forget to take the Chapter 5 Test on page 405. Then check your answers at the back of the text and use the Chapter Test Prep Videos to see the fully worked-out solutions to any of the exercises you want to review.

5 Chapter Highlights

Definitions and Concepts	Examples
Section 5.1 Introduction to Decimals	

Place-Value Chart

hundreds	tens	ones	decimal point	tenths	hundredths	thousandths	ten-thousandths	hundred-thousandths
		4	.	2	6	5		
100	10	1		$\dfrac{1}{10}$	$\dfrac{1}{100}$	$\dfrac{1}{1000}$	$\dfrac{1}{10,000}$	$\dfrac{1}{100,000}$

4.265 means

$$4 \cdot 1 + 2 \cdot \frac{1}{10} + 6 \cdot \frac{1}{100} + 5 \cdot \frac{1}{1000}$$

or

$$4 + \frac{2}{10} + \frac{6}{100} + \frac{5}{1000}$$

Definitions and Concepts	Examples

Section 5.1   Introduction to Decimals   *(continued)*

Writing (or Reading) a Decimal in Words

Step 1: Write the whole number part in words.

Step 2: Write "and" for the decimal point.

Step 3: Write the decimal part in words as though it were a whole number, followed by the place value of the last digit.

Write 3.08 in words.

Three and eight hundredths

A decimal written in words can be written in standard form by reversing the above procedure.

Write "negative four and twenty-one thousandths" in standard form.

$$-4.021$$

To Round a Decimal to a Place Value to the Right of the Decimal Point

Step 1: Locate the digit to the right of the given place value.

Step 2: If this digit is 5 or greater, add 1 to the digit in the given place value and drop all digits to its right. If this digit is less than 5, drop all digits to the right of the given place value.

Round 86.1256 to the nearest hundredth.

Step 1: 86.12 5 6
— hundredths place
— digit to the right

Step 2: Since the digit to the right is 5 or greater, we add 1 to the digit in the hundredths place and drop all digits to its right.

86.1256 rounded to the nearest hundredth is 86.13.

Section 5.2   Adding and Subtracting Decimals

To Add or Subtract Decimals

Step 1: Write the decimals so that the decimal points line up vertically.

Step 2: Add or subtract as with whole numbers.

Step 3: Place the decimal point in the sum or difference so that it lines up vertically with the decimal points in the problem.

Add: $4.6 + 0.28$ Subtract: $2.8 - 1.04$

$$
\begin{array}{r}
4.60 \\
+\ 0.28 \\
\hline
4.88
\end{array}
\qquad
\begin{array}{r}
2.\overset{7\ 10}{8\ 0} \\
-\ 1.04 \\
\hline
1.76
\end{array}
$$

Section 5.3   Multiplying Decimals and Circumference of a Circle

To Multiply Decimals

Step 1: Multiply the decimals as though they are whole numbers.

Step 2: The decimal point in the product is placed so that the number of decimal places in the product is equal to the *sum* of the number of decimal places in the factors.

Multiply: 1.48×5.9

$$
\begin{array}{r}
1.48 \quad \leftarrow 2 \text{ decimal places} \\
\times\ 5.9 \quad \leftarrow 1 \text{ decimal place} \\
\hline
1332 \\
7400 \\
\hline
8.732 \quad \leftarrow 3 \text{ decimal places}
\end{array}
$$

(continued)

Definitions and Concepts	Examples

Section 5.3   Multiplying Decimals and Circumference of a Circle   (continued)

The **circumference** of a circle is the distance around the circle.

$$C = \pi d \text{ or } C = 2\pi r$$

where $\pi \approx 3.14$ or $\pi \approx \dfrac{22}{7}$.

or

Find the exact circumference of a circle with radius 5 miles and an approximation by using 3.14 for π.

$$
\begin{aligned}
C &= 2\pi r \\
&= 2\pi(5) \\
&= 10\pi \\
&\approx 10(3.14) \\
&\approx 31.4
\end{aligned}
$$

The circumference is exactly 10π miles and approximately 31.4 miles.

Section 5.4   Dividing Decimals

To Divide Decimals

Step 1: If the divisor is not a whole number, move the decimal point in the divisor to the right until the divisor is a whole number.

Step 2: Move the decimal point in the dividend to the right the *same number of places* as the decimal point was moved in Step 1.

Step 3: Divide. The decimal point in the quotient is directly over the moved decimal point in the dividend.

Divide: $1.118 \div 2.6$

```
         0.43
  2.6)1.118
      -104
        78
       -78
         0
```

Section 5.5   Fractions, Decimals, and Order of Operations

To **write fractions as decimals,** divide the numerator by the denominator.

Write $\dfrac{3}{8}$ as a decimal.

```
        0.375
   8)3.000
     -24
       60
      -56
       40
      -40
        0
```

Order of Operations

1. Perform all operations within parentheses ( ), brackets [ ], or other grouping symbols such as fraction bars.
2. Evaluate any expressions with exponents.
3. Multiply or divide in order from left to right.
4. Add or subtract in order from left to right.

Simplify.

$$
\begin{aligned}
-1.9(12.8 - 4.1) &= -1.9(8.7) \quad \text{Subtract.} \\
&= -16.53 \quad \text{Multiply.}
\end{aligned}
$$

Definitions and Concepts	Examples

Section 5.6   Solving Equations Containing Decimals

Steps for Solving an Equation in x

Step 1: If fractions are present, multiply both sides of the equation by the LCD of the fractions.

Step 2: If parentheses are present, use the distributive property.

Step 3: Combine any like terms on each side of the equation.

Step 4: Use the addition property of equality to rewrite the equation so that variable terms are on one side of the equation and constant terms are on the other side.

Step 5: Divide both sides by the numerical coefficient of x to solve.

Step 6: Check your answer in the *original equation*.

Solve:

$$3(x + 2.6) = 10.92$$
$$3x + 7.8 = 10.92 \quad \text{Apply the distributive property.}$$
$$3x + 7.8 - 7.8 = 10.92 - 7.8 \quad \text{Subtract 7.8 from both sides.}$$
$$3x = 3.12 \quad \text{Simplify.}$$
$$\frac{3x}{3} = \frac{3.12}{3} \quad \text{Divide both sides by 3.}$$
$$x = 1.04 \quad \text{Simplify.}$$

Check 1.04 in the original equation.

Section 5.7   Decimal Applications: Mean, Median, and Mode

The **mean** (or **average**) of a set of number items is

$$\text{mean} = \frac{\text{sum of items}}{\text{number of items}}$$

The **median** of a set of numbers in numerical order is the middle number. If the number of items is even, the median is the mean of the two middle numbers.

The **mode** of a set of numbers is the number that occurs most often. (A set of numbers may have no mode or more than one mode.)

Find the mean, median, and mode of the following set of numbers: 33, 35, 35, 43, 68, 68

$$\text{mean} = \frac{33 + 35 + 35 + 43 + 68 + 68}{6} = 47$$

The median is the mean of the two middle numbers, 35 and 43

$$\text{median} = \frac{35 + 43}{2} = 39$$

There are two modes because there are two numbers that occur twice:

$$35 \text{ and } 68$$

Chapter 5 Review

(5.1) *Determine the place value of the number 4 in each decimal.*

1. 23.45

2. 0.000345

Write each decimal in words.

3. −23.45

4. 0.00345

5. 109.23

6. 200.000032

Write each decimal in standard form.

7. Eight and six hundredths

8. Negative five hundred three and one hundred two thousandths

9. Sixteen thousand twenty-five and fourteen ten-thousandths

10. Fourteen and eleven thousandths

Write each decimal as a fraction or a mixed number.

11. 0.16

12. -12.023

Write each fraction or mixed number as a decimal.

13. $\dfrac{231}{100,000}$

14. $25\dfrac{1}{4}$

Insert $<$, $>$, or $=$ between each pair of numbers to make a true statement.

15. 0.49   0.43

16. 0.973   0.9730

17. -38.0027 -38.00056

18. -0.230505 -0.23505

Round each decimal to the given place value.

19. 0.623, nearest tenth

20. 0.9384, nearest hundredth

21. -42.895, nearest hundredth

22. 16.34925, nearest thousandth

Write each number in standard notation.

23. Saturn is a distance of about 887 million miles from the Sun.

24. The tail of a comet can be over 600 thousand miles long.

(5.2) *Add.*

25. $8.6 + 9.5$

26. $3.9 + 1.2$

27. $-6.4 + (-0.88)$

28. $-19.02 + 6.98$

29. $200.49 + 16.82 + 103.002$

30. $0.00236 + 100.45 + 48.29$

Subtract.

31. $4.9 - 3.2$

32. $5.23 - 2.74$

33. $-892.1 - 432.4$

34. $0.064 - 10.2$

35. $100 - 34.98$

36. $200 - 0.00198$

Solve.

37. Find the total distance between Grove City and Jerome.

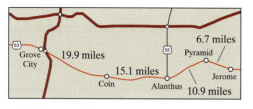

38. Evaluate $x - y$ for $x = 1.2$ and $y = 6.9$.

△ **39.** Find the perimeter.

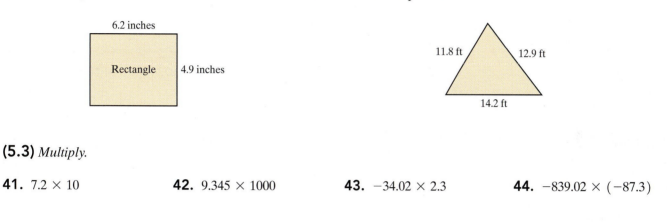

6.2 inches

Rectangle   4.9 inches

△ **40.** Find the perimeter.

11.8 ft   12.9 ft

14.2 ft

(5.3) *Multiply.*

41. 7.2×10

42. 9.345×1000

43. -34.02×2.3

44. $-839.02 \times (-87.3)$

Find the exact circumference of each circle. Then use the approximation 3.14 for π and approximate the circumference.

△ **45.**

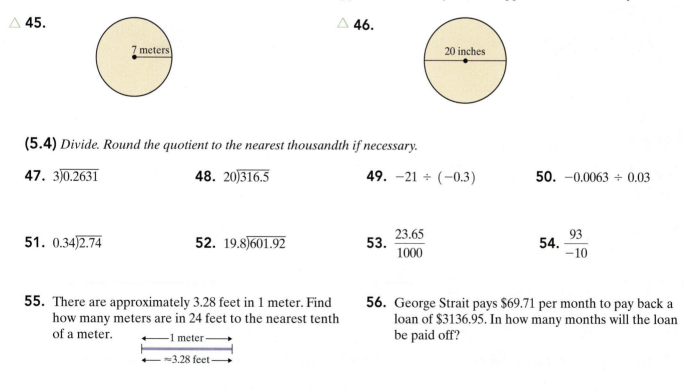

7 meters

△ **46.**

20 inches

(5.4) *Divide. Round the quotient to the nearest thousandth if necessary.*

47. $3\overline{)0.2631}$

48. $20\overline{)316.5}$

49. $-21 \div (-0.3)$

50. $-0.0063 \div 0.03$

51. $0.34\overline{)2.74}$

52. $19.8\overline{)601.92}$

53. $\dfrac{23.65}{1000}$

54. $\dfrac{93}{-10}$

55. There are approximately 3.28 feet in 1 meter. Find how many meters are in 24 feet to the nearest tenth of a meter.

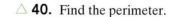

←—1 meter—→

←— ≈3.28 feet —→

56. George Strait pays $69.71 per month to pay back a loan of $3136.95. In how many months will the loan be paid off?

(5.5) *Write each fraction or mixed number as a decimal. Round to the nearest thousandth if necessary.*

57. $\dfrac{4}{5}$

58. $-\dfrac{12}{13}$

59. $2\dfrac{1}{3}$

60. $\dfrac{13}{60}$

Insert <, >, or = to make a true statement.

61. 0.392   0.39200

62. −0.0231   −0.0221

63. $\dfrac{4}{7}$ 0.625

64. 0.293 $\dfrac{5}{17}$

Write the numbers in order from smallest to largest.

65. 0.837, 0.839, 0.832

66. 0.685, 0.626, $\dfrac{5}{8}$

67. $\dfrac{3}{7}$, 0.42, 0.43

68. $\dfrac{18}{11}$, 1.63, $\dfrac{19}{12}$

Simplify each expression.

69. $-7.6 \times 1.9 + 2.5$

70. $(-2.3)^2 - 1.4$

71. $0.0726 \div 10 \times 1000$

72. $0.9(6.5 - 5.6)$

73. $\dfrac{(1.5)^2 + 0.5}{0.05}$

74. $\dfrac{7 + 0.74}{-0.06}$

Find each area.

△ **75.**

△ **76.**

(5.6) *Solve.*

77. $x + 3.9 = 4.2$

78. $70 = y - 22.81$

79. $2x = 17.2$

80. $-1.1y = 88$

81. $3x - 0.78 = 1.2 + 2x$

82. $-x + 0.6 - 2x = -4x - 0.9$

83. $-1.3x - 9.4 = -0.4x + 8.6$

84. $3(x - 1.1) = 5x - 5.3$

(5.7) *Find the mean, median, and any mode(s) for each list of numbers. If necessary, round to the nearest tenth.*

85. 13, 23, 33, 14, 6

86. 45, 86, 21, 60, 86, 64, 45

87. 14,000, 20,000, 12,000, 20,000, 36,000, 45,000

88. 560, 620, 123, 400, 410, 300, 400, 780, 430, 450

For Exercises 89 and 90, the grades are given for a student for a particular semester. Find each grade point average. If necessary, round the grade point average to the nearest hundredth.

89.

Grade	Credit Hours
A	3
A	3
C	2
B	3
C	1

90.

Grade	Credit Hours
B	3
B	4
C	2
D	2
B	3

Mixed Review

91. Write 200.0032 in words.

92. Write negative sixteen and nine hundredths in standard form.

93. Write 0.0847 as a fraction or a mixed number.

94. Write the numbers $\frac{6}{7}, \frac{8}{9}, 0.75$ in order from smallest to largest.

Write each fraction as a decimal. Round to the nearest thousandth, if necessary.

95. $-\frac{7}{100}$

96. $\frac{9}{80}$ (Do not round.)

97. $\frac{8935}{175}$

Insert $<$, $>$, or $=$ to make a true statement.

98. -402.000032 ___ -402.00032

99. $\frac{6}{11}$ ___ 0.55

Round each decimal to the given place value.

100. 86.905, nearest hundredth

101. 3.11526, nearest thousandth

Round each money amount to the nearest dollar.

102. $123.46

103. $3645.52

Add or subtract as indicated.

104. $3.2 - 4.9$

105. $9.12 - 3.86$

106. $-102.06 + 89.3$

107. $-4.021 + (-10.83) + (-0.056)$

Multiply or divide as indicated. Round to the nearest thousandth, if necessary.

108.
$$\begin{array}{r} 2.54 \\ \times\ 3.2 \\ \hline \end{array}$$

109. $(-3.45)(2.1)$

110. $0.005\overline{)24.5}$

111. $2.3\overline{)54.98}$

Solve.

△ **112.** Tomaso is going to fertilize his lawn, a rectangle that measures 77.3 feet by 115.9 feet. Approximate the area of the lawn by rounding each measurement to the nearest ten feet.

77.3 feet

115.9 feet

113. Estimate the cost of the items to see whether the groceries can be purchased with a $10 bill.

$3.79

$2.49

3 cans for $1.99

Simplify each expression.

114. $\dfrac{(3.2)^2}{100}$

115. $(2.6 + 1.4)(4.5 - 3.6)$

Find the mean, median, and any mode(s) for each list of numbers. If needed, round answers to the nearest hundredth.

116. $73, 82, 95, 68, 54$

117. $952, 327, 566, 814, 327, 729$

Write each decimal as indicated.

Answers

1. 45.092, in words

2. Three thousand and fifty-nine thousandths, in standard form

Perform each indicated operation. Round the result to the nearest thousandth if necessary.

3. $2.893 + 4.21 + 10.492$

4. $-47.92 - 3.28$

5. $9.83 - 30.25$

6. 10.2×4.01

7. $-0.00843 \div (-0.23)$

Round each decimal to the indicated place value.

8. 34.8923, nearest tenth

9. 0.8623, nearest thousandth

Insert $<$, $>$, or $=$ between each pair of numbers to form a true statement.

10. 25.0909   25.9090

11. $\dfrac{4}{9}$ 0.445

Write each decimal as a fraction or a mixed number.

12. 0.345

13. -24.73

Write each fraction as a decimal. If necessary, round to the nearest thousandth.

14. $-\dfrac{13}{26}$

15. $\dfrac{16}{17}$

Simplify.

16. $(-0.6)^2 + 1.57$

17. $\dfrac{0.23 + 1.63}{-0.3}$

18. $2.4x - 3.6 - 1.9x - 9.8$

1. _____

2. _____

3. _____

4. _____

5. _____

6. _____

7. _____

8. _____

9. _____

10. _____

11. _____

12. _____

13. _____

14. _____

15. _____

16. _____

17. _____

18. _____

Solve.

19. $0.2x + 1.3 = 0.7$ **20.** $2(x + 5.7) = 6x - 3.4$

Find the mean, median, and mode of each list of numbers.

21. $26, 32, 42, 43, 49$ **22.** $8, 10, 16, 16, 14, 12, 12, 13$

Find the grade point average. If necessary, round to the nearest hundredth.

23.

Grade	Credit Hours
A	3
B	3
C	3
B	4
A	1

Solve.

24. At its farthest, Pluto is 4583 million miles from the Sun. Write this number using standard notation.

△ **25.** Find the area.

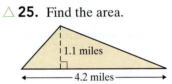

1.1 miles

4.2 miles

△ **26.** Find the exact circumference of the circle. Then use the approximation 3.14 for π and approximate the circumference.

9 miles

27. Vivian Thomas is going to put insecticide on her lawn to control grubworms. The lawn is a rectangle that measures 123.8 feet by 80 feet. The amount of insecticide required is 0.02 ounce per square foot.

a. Find the area of her lawn.

b. Find how much insecticide Vivian needs to purchase.

28. Find the total distance from Bayette to Center City.

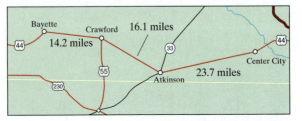

Bayette   Crawford   16.1 miles

44   14.2 miles   33   44

55   Center City

Atkinson   23.7 miles

230

Sidebar answer lines:

19. _____
20. _____
21. _____
22. _____
23. _____
24. _____
25. _____
26. _____
27. a. _____
    b. _____
28. _____

Write each number in words.

1. 72

2. 107

3. 546

4. 5026

5. Add: 46 + 713

6. Find the perimeter.

3 in.   7 in.

9 in.

7. Subtract: 543 − 29. Then check by adding.

8. Divide: 3268 ÷ 27

9. Round 278,362 to the nearest thousand.

10. Write the prime factorization of 30.

11. Multiply: 236 × 86

12. Multiply: 236 × 86 × 0

13. Find each quotient and then check the answer by multiplying.

   a. $1\overline{)7}$

   b. 12 ÷ 1

   c. $\dfrac{6}{6}$

   d. 9 ÷ 9

   e. $\dfrac{20}{1}$

   f. $18\overline{)18}$

14. Find the average of 25, 17, 19, and 39.

15. Simplify: $2 \cdot 4 - 3 \div 3$

16. Simplify: $77 \div 11 \cdot 7$

Evaluate.

17. 9^2

18. 5^3

Answers

1. _____

2. _____

3. _____

4. _____

5. _____

6. _____

7. _____

8. _____

9. _____

10. _____

11. _____

12. _____

13. a. _____

  b. _____

  c. _____

  d. _____

  e. _____

  f. _____

14. _____

15. _____

16. _____

17. _____

18. _____

19. 3^4

20. 10^3

21. Evaluate $\dfrac{x - 5y}{y}$ for $x = 35$ and $y = 5$.

22. Evaluate $\dfrac{2a + 4}{c}$ for $a = 7$ and $c = 3$.

23. Find the opposite of each number.

    a. 13   **b.** -2 **c.** 0

24. Find the opposite of each number.

    a. -7 **b.** 4   **c.** -1

25. Add: $-2 + (-21)$

26. Add: $-7 + (-15)$

Find the value of each expression.

27. $5 \cdot 6^2$

28. $4 \cdot 2^3$

29. -7^2

30. $(-2)^5$

31. $(-5)^2$

32. -3^2

Represent the shaded part as an improper fraction and a mixed number.

33.

34.

35.

36.

37. Write the prime factorization of 252.

38. Find the difference of 87 and 25.

39. Write $-\dfrac{72}{26}$ in simplest form.

40. Write $9\dfrac{7}{8}$ as an improper fraction.

41. Determine whether $\dfrac{16}{40}$ and $\dfrac{10}{25}$ are equivalent.

42. Insert $<$ or $>$ to form a true statement. $\dfrac{4}{7}$ $\qquad$ $\dfrac{5}{9}$

Multiply.

43. $\dfrac{2}{3} \cdot \dfrac{5}{11}$

44. $2\dfrac{5}{8} \cdot \dfrac{4}{7}$

45. $\dfrac{1}{4} \cdot \dfrac{1}{2}$

46. $7 \cdot 5\dfrac{2}{7}$

Solve.

47. $\dfrac{z}{-4} = 11 - 5$

48. $6x - 12 - 5x = -20$

49. Add: $763.7651 + 22.001 + 43.89$

50. Add: $89.27 + 14.361 + 127.2318$

51. Multiply: 23.6×0.78

52. Multiply: 43.8×0.645

39. _____

40. _____

41. _____

42. _____

43. _____

44. _____

45. _____

46. _____

47. _____

48. _____

49. _____

50. _____

51. _____

52. _____

Ratio, Proportion, and Triangle Applications

Check Your Progress

Having studied fractions in Chapter 4, we are ready to explore the useful notations of ratio and proportion. *Ratio* is another name for *quotient* and is usually written in fraction form. A proportion is an equation with two equal ratios.

s it "disk" or "disc"? Actually, either spelling can be used unless there is a trademark involved. Disk (or disc) storage is, then, a general category that includes various storage mechanisms where data are recorded and stored by various methods to a surface layer of a rotating disk(s). Disk storage has certainly changed or evolved over the years, and below we show just a few examples.

In Section 6.1, Exercises 25–28, we find the ratios of diameters of certain disk storage mechanisms.

Disk or Disc?

8" floppy disk

$5\frac{1}{4}$" floppy disk

$3\frac{1}{2}$" floppy disk

12 cm audio Compact Disc (CD)

8 cm mini CD

12 cm Digital Versatile Disc (DVD)

8 cm mini DVD

12 cm Blu-ray Disc (BD)

8 cm mini BD

12" vinyl record

10" vinyl record

7" vinyl record

20 cm LaserDisc

30 cm LaserDisc

12 cm LaserDisc

6.1 Ratios and Rates

Objective A Writing Ratios as Fractions

A **ratio** is the quotient of two quantities. A ratio, in fact, is no different from a fraction, except that a ratio is sometimes written using notation other than fractional notation. For example, the ratio of 1 to 2 can be written as

$$1 \text{ to } 2 \quad \text{or} \quad \frac{1}{2} \quad \text{or} \quad 1:2$$

fractional notation   colon notation

These ratios are all read as, "the ratio of 1 to 2."

✓**Concept Check** How should each ratio be read aloud?

a. $\dfrac{8}{5}$ **b.** $\dfrac{5}{8}$

In this section, we write ratios using fractional notation. If the fraction happens to be an improper fraction, do not write the fraction as a mixed number. Why? The mixed number form is not a ratio or quotient of two quantities.

Writing a Ratio as a Fraction

The order of the quantities is important when writing ratios. To write a ratio as a fraction, write the *first number* of the ratio as the *numerator* of the fraction and the *second number* as the *denominator*.

Helpful Hint

The ratio of 6 to 11 is $\dfrac{6}{11}$, *not* $\dfrac{11}{6}$.

Example 1 Write the ratio of 21 to 29 using fractional notation.

Solution: The ratio is $\dfrac{21}{29}$.

■ Work Practice 1

To simplify a ratio, we just write the fraction in simplest form. Common factors as well as common units can be divided out.

Example 2 Write the ratio of \$15 to \$10 as a fraction in simplest form.

Solution:

$$\frac{\$15}{\$10} = \frac{15}{10} = \frac{3 \cdot \cancel{5}^{1}}{2 \cdot \cancel{5}_{1}} = \frac{3}{2}$$

■ Work Practice 2

Objectives

A Write Ratios as Fractions.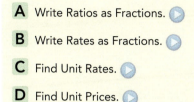

B Write Rates as Fractions.

C Find Unit Rates.

D Find Unit Prices.

Practice 1

Write the ratio of 19 to 30 using fractional notation.

Practice 2

Write the ratio of \$16 to \$12 as a fraction in simplest form.

Answers

1. $\dfrac{19}{30}$ **2.** $\dfrac{4}{3}$

✓**Concept Check Answers**

a. "the ratio of eight to five"
b. "the ratio of five to eight"

411

Helpful Hint

In the previous example, although $\frac{3}{2} = 1\frac{1}{2}$, a ratio is a quotient of *two* quantities. For that reason, ratios are not written as mixed numbers.

If a ratio contains decimal numbers or mixed numbers, we simplify by writing the ratio as a ratio of whole numbers.

Practice 3

Write the ratio of 1.68 to 4.8 as a fraction in simplest form.

Example 3 Write the ratio of 2.5 to 3.15 as a fraction in simplest form.

Solution: The ratio in fraction form is

$$\frac{2.5}{3.15}$$

Now let's clear the ratio of decimals.

$$\frac{2.5}{3.15} = \frac{2.5}{3.15} \cdot 1 = \frac{2.5}{3.15} \cdot \frac{100}{100} = \frac{2.5 \cdot 100}{3.15 \cdot 100} = \frac{250}{315} = \frac{50}{63} \quad \text{Simplest form}$$

■ **Work Practice 3**

Practice 4

Write the ratio of $2\frac{2}{3}$ to $1\frac{13}{15}$ as a fraction in simplest form.

Example 4 Write the ratio of $2\frac{5}{8}$ to $8\frac{3}{4}$ as a fraction in simplest form.

Solution: The ratio in fraction form is $\dfrac{2\frac{5}{8}}{8\frac{3}{4}}$.

To simplify, remember that the fraction bar means division.

$$\frac{2\frac{5}{8}}{8\frac{3}{4}} = 2\frac{5}{8} \div 8\frac{3}{4} = \frac{21}{8} \div \frac{35}{4} = \frac{21}{8} \cdot \frac{4}{35} = \frac{3 \cdot \overset{1}{\cancel{7}} \cdot \overset{1}{\cancel{4}}}{2 \cdot \underset{1}{\cancel{4}} \cdot 5 \cdot \underset{1}{\cancel{7}}} = \frac{3}{10} \quad \text{Simplest form}$$

■ **Work Practice 4**

Practice 5

Use the circle graph for Example 5 to write the ratio of work miles to total miles as a fraction in simplest form.

Example 5 Writing a Ratio from a Circle Graph

The circle graph at the right shows the part of a car's total mileage that falls into a particular category. Write the ratio of family business miles to total miles as a fraction in simplest form.

Solution:

$$\frac{\text{family business miles}}{\text{total miles}} = \frac{3000 \text{ miles}}{15{,}000 \text{ miles}}$$

$$= \frac{3000}{15{,}000}$$

$$= \frac{\overset{1}{\cancel{3000}}}{5 \cdot \underset{1}{\cancel{3000}}}$$

$$= \frac{1}{5}$$

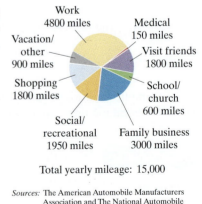

Sources: The American Automobile Manufacturers Association and The National Automobile Dealers Association.

Total yearly mileage: 15,000

■ **Work Practice 5**

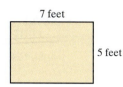 **Example 6**   Given the rectangle shown:

a. Find the ratio of its width (shorter side) to its length (longer side).
b. Find the ratio of its length to its perimeter.

7 feet

5 feet

Solution:

a. The ratio of its width to its length is

$$\frac{\text{width}}{\text{length}} = \frac{5 \text{ feet}}{7 \text{ feet}} = \frac{5}{7}$$

b. Recall that the perimeter of a rectangle is the distance around the rectangle: $7 + 5 + 7 + 5 = 24$ feet. The ratio of its length to its perimeter is

$$\frac{\text{length}}{\text{perimeter}} = \frac{7 \text{ feet}}{24 \text{ feet}} = \frac{7}{24}$$

◼ **Work Practice 6**

Objective B Writing Rates as Fractions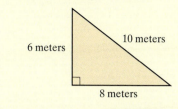

A special type of ratio is a rate. **Rates** are used to compare *different* kinds of quantities. For example, suppose that a recreational runner can run 3 miles in 33 minutes. If we write this rate as a fraction, we have

$$\frac{3 \text{ miles}}{33 \text{ minutes}} = \frac{1 \text{ mile}}{11 \text{ minutes}} \quad \text{In simplest form}$$

Helpful Hint

When comparing quantities with different units, write the units as part of the comparison. Units do not divide out unless they are the same.

Same Units: $\frac{3 \text{ inches}}{12 \text{ inches}} = \frac{1}{4}$ Units are the same and divide out.

Different Units: $\frac{2 \text{ miles}}{20 \text{ minutes}} = \frac{1 \text{ mile}}{10 \text{ minutes}}$ Units are still written.

Examples   Write each rate as a fraction in simplest form.

7. $2160 for 12 weeks is $\dfrac{2160 \text{ dollars}}{12 \text{ weeks}} = \dfrac{180 \text{ dollars}}{1 \text{ week}}$

8. 360 miles on 16 gallons of gasoline is $\dfrac{360 \text{ miles}}{16 \text{ gallons}} = \dfrac{45 \text{ miles}}{2 \text{ gallons}}$

◼ **Work Practice 7–8**

✓**Concept Check**   True or false? $\dfrac{16 \text{ gallons}}{4 \text{ gallons}}$ is a rate. Explain.

Practice 6

Given the triangle shown:

6 meters   10 meters

8 meters

a. Find the ratio of the length of the shortest side to the length of the longest side.

b. Find the ratio of the length of the longest side to the perimeter of the triangle.

Practice 7–8

Write each rate as a fraction in simplest form.

7. $1350 for 6 weeks

8. 295 miles on 15 gallons of gasoline

Answers

6. a. $\dfrac{3}{5}$ b. $\dfrac{5}{12}$ 7. $\dfrac{\$225}{1 \text{ wk}}$ 8. $\dfrac{59 \text{ mi}}{3 \text{ gal}}$

✓**Concept Check Answer**

false; a rate compares different kinds of quantities

Objective C Finding Unit Rates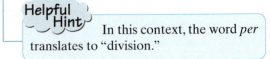

A **unit rate** is a rate with a denominator of 1. A familiar example of a unit rate is 55 mph, read as "55 **miles per hour.**" This means 55 miles per 1 hour or

$$\frac{55 \text{ miles}}{1 \text{ hour}}$$ Denominator of 1

> **Helpful Hint**
> In this context, the word *per* translates to "division."

Writing a Rate as a Unit Rate

To write a rate as a unit rate, divide the numerator of the rate by the denominator.

Practice 9

Write as a unit rate: 3200 feet every 8 seconds

Example 9 Write as a unit rate: $31,500 every 7 months

Solution:

$$\frac{31,500 \text{ dollars}}{7 \text{ months}} \qquad \begin{array}{r} 4,500 \\ 7\overline{)31,500} \end{array}$$

The unit rate is

$$\frac{4500 \text{ dollars}}{1 \text{ month}} \text{ or } 4500 \text{ dollars/month}$$ Read as, "4500 dollars per month."

■ **Work Practice 9**

Practice 10

Write as a unit rate: 78 bushels of fruit from 12 trees

Example 10 Write as a unit rate: 337.5 miles every 15 gallons of gas

Solution:

$$\frac{337.5 \text{ miles}}{15 \text{ gallons}} \qquad \begin{array}{r} 22.5 \\ 15\overline{)337.5} \end{array}$$

The unit rate is

$$\frac{22.5 \text{ miles}}{1 \text{ gallon}} \text{ or } 22.5 \text{ miles/gallon}$$ Read as, "22.5 miles per gallon."

■ **Work Practice 10**

Objective D Finding Unit Prices

Rates are used extensively in sports, business, medicine, and science. One of the most common uses of rates is in consumer economics. When a unit rate is "money per item," it is also called a **unit price.**

$$\text{unit price} = \frac{\text{price}}{\text{number of units}}$$

Answers

9. $\dfrac{400 \text{ ft}}{1 \text{ sec}}$ or 400 ft/sec

10. $\dfrac{6.5 \text{ bushels}}{1 \text{ tree}}$ or 6.5 bushels/tree

Example 11   Finding Unit Price

A store charges $3.36 for a 16-ounce jar of picante sauce. What is the unit price in dollars per ounce?

Solution:

$$\text{unit price} = \frac{\$3.36}{16 \text{ ounces}} = \frac{\$0.21}{1 \text{ ounce}} \quad \text{or} \quad \$0.21 \text{ per ounce}$$

■ **Work Practice 11**

Practice 11

An automobile rental agency charges $170 for 5 days for a certain model car. What is the unit price in dollars per day?

Example 12   Finding the Best Buy

Approximate each unit price to decide which is the better buy: 4 bars of soap for $0.99 or 5 bars of soap for $1.19.

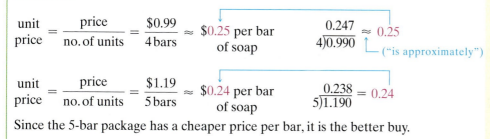

Solution:

$$\frac{\text{unit}}{\text{price}} = \frac{\text{price}}{\text{no. of units}} = \frac{\$0.99}{4 \text{ bars}} \approx \$0.25 \text{ per bar of soap} \qquad \frac{0.247}{4)0.990} \approx 0.25$$

("is approximately")

$$\frac{\text{unit}}{\text{price}} = \frac{\text{price}}{\text{no. of units}} = \frac{\$1.19}{5 \text{ bars}} \approx \$0.24 \text{ per bar of soap} \qquad \frac{0.238}{5)1.190} = 0.24$$

Since the 5-bar package has a cheaper price per bar, it is the better buy.

■ **Work Practice 12**

Practice 12

Approximate each unit price to decide which is the better buy for a bag of nacho chips: 11 ounces for $3.99 or 16 ounces for $5.99.

Answers

11. $34 per day   **12.** 11-oz bag

Vocabulary, Readiness & Video Check

Use the choices below to fill in each blank. Not all choices will be used.

rate	different	denominator	unit
numerator	unit price	true	
division	ratio	false	

1. A rate with a denominator of 1 is called a(n) _____ rate.

2. When a rate is written as money per item, a unit rate is called a(n) _____.

3. The word *per* translates to "_____."

4. Rates are used to compare _____ types of quantities.

5. To write a rate as a unit rate, divide the _____ of the rate by the _____.

6. The quotient of two quantities is called a(n) _____.

7. Answer true or false: The ratio $\frac{7}{5}$ means the same as the ratio $\frac{5}{7}$. _____

Martin-Gay Interactive Videos   *Watch the section lecture video and answer the following questions.*

See Video 6.1 🫐

Objective A 8. How is the ratio in ▯ Example 2 rewritten as an equivalent ratio containing no decimals? ▶

Objective B 9. Why can't we divide out the units in ▯ Example 5 as we did in ▯ Example 4? ▶

Objective C 10. Why did we divide the first quantity of the rate in ▯ Example 8 by the second quantity? ▶

Objective D 11. From ▯ Example 9, unit prices can be especially helpful when? ▶

6.1 Exercise Set MyMathLab® ▶

Objective A *Write each ratio as a ratio of whole numbers using fractional notation. Write the fraction in simplest form. See Examples 1 through 6.*

▶ **1.** 16 to 24

2. 25 to 150

▶ **3.** 7.7 to 10

4. 8.1 to 10

5. 4.63 to 8.21

6. 9.61 to 7.62

7. 6 ounces to 16 ounces

8. 35 meters to 100 meters

9. \$32 to \$100

10. \$46 to \$102

▶ **11.** 24 days to 14 days

12. 120 miles to 80 miles

▶ **13.** $3\frac{1}{2}$ to $12\frac{1}{4}$

14. $3\frac{1}{3}$ to $4\frac{1}{6}$

15. $7\frac{3}{5}$ hours to $1\frac{9}{10}$ hours

16. $25\frac{1}{2}$ days to $2\frac{5}{6}$ days

Find the ratio described in each exercise as a fraction in simplest form. See Examples 5 and 6.

17.

Average Weight of Mature Whales	
Blue Whale	**Fin Whale**
145 tons	50 tons

Use the table to find the ratio of the weight of an average mature Fin Whale to the weight of an average mature Blue Whale.

18.

Countries with Small Land Areas	
Tuvalu	**San Marino**
10 sq mi	24 sq mi

(*Source: World Almanac*)

Use the table to find the ratio of the land area of Tuvalu to the land area of San Marino.

△ **19.** Find the ratio of the width of a regulation size basketball court to its perimeter.

△ **20.** Find the ratio of the width to the perimeter shown of the swimming pool.

50 feet (width)
94 feet (length)

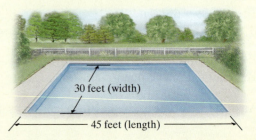

30 feet (width)
45 feet (length)

At the Hidalgo County School Board meeting one night, there were 125 women and 100 men present.

21. Find the ratio of women to men.

22. Find the ratio of men to the total number of people present.

Blood contains three types of cells: red blood cells, white blood cells, and platelets. For approximately every 600 red blood cells in healthy humans, there are 40 platelets and 1 white blood cell. Use this information for Exercises 23 and 24. (Source: American Red Cross Biomedical Services)

23. Write the ratio of red blood cells to platelet cells.

24. Write the ratio of white blood cells to red blood cells.

Exercises 25–28 have to do with disk storage. See the Chapter Opener for more information on disk storage.

25. A standard DVD has a diameter of 12 centimeters while a miniDVD has a diameter of 8 centimeters. Write the ratio of the miniDVD diameter to the standard DVD diameter.

26. LaserDiscs have a diameter of 12 centimeters and 20 centimeters. Write the ratio of the smaller diameter to the larger diameter.

27. Hard drives come in many diameters. Find the ratio of a $2\frac{1}{2}''$ diameter to a $5\frac{1}{4}''$ diameter hard drive.

28. Floppy disks come in many diameters. Find the ratio of a $3\frac{1}{2}''$ diameter to a $5\frac{1}{4}''$ diameter floppy disk.

△ **29.** Find the ratio of the longest side to the perimeter of the right-triangular-shaped billboard.

△ **30.** Find the ratio of the base to the perimeter of the triangular mainsail.

31. Citizens of Mexico consume the most Coca-Cola beverages per capita of any country in the world. Mexicans drink an average of the equivalent of 573 8-oz beverages per year each. In contrast, the average American consumes about the equivalent of 423 8-oz beverages per year. Find the ratio of the average amount of Coca-Cola beverages drunk by Mexicans to the average amount of Coca-Cola beverages drunk by Americans. (*Source:* Coca-Cola Company)

32. A large order of McDonald's french fries has 500 calories. Of this total, 220 calories are from fat. Find the ratio of the calories from fat to total calories in a large order of McDonald's french fries. (*Source:* McDonald's Corporation)

Objective B   *Write each rate as a fraction in simplest form. See Examples 7 and 8.*

▶ 33. 5 shrubs every 15 feet

34. 14 lab tables for 28 students

35. 15 returns for 100 sales

36. 8 phone lines for 36 employees

▶ 37. 6 laser printers for 28 computers

38. 4 inches of rain in 18 hours

39. 18 gallons of pesticide for 4 acres of crops

40. 150 graduate students for 8 advisors

Objective C   *Write each rate as a unit rate. See Examples 9 and 10.*

41. 330 calories in a 3-ounce serving

42. 275 miles in 11 hours

▶ 43. 375 riders in 5 subway cars

44. 18 signs in 6 blocks

45. A hummingbird moves its wings at a rate of 5400 wingbeats a minute. Write this rate in wingbeats per second.

46. A bat moves its wings at a rate of 1200 wingbeats a minute. Write this rate in wingbeats per second.

▶ 47. A $1,000,000 lottery winning paid over 20 years

48. 400,000 library books for 8000 students

49. The state of Delaware has 631,500 registered voters for two senators. (*Source:* Delaware.gov)

50. The 2020 projected population of Louisiana is approximately 4,588,800 residents for 64 parishes. (*Note:* Louisiana is the only U.S. state with parishes instead of counties.) (*Source:* Louisiana.gov)

51. 12,000 good assembly line products to 40 defective products

52. 5,000,000 lottery tickets for 4 lottery winners

53. The combined salary for the 20 highest-paid players of the 2012 World Series Champion San Francisco Giants was approximately $118,494,000. (*Source:* ESPN)

54. The top-grossing concert tour was the 2009–2011 U2 360° tour, which grossed over $735,900 thousand for 110 shows worldwide. (*Source:* Pollstar)

55. Charlie Catlett can assemble 250 computer boards in an 8-hour shift while Suellen Catlett can assemble 402 computer boards in a 12-hour shift.
 a. Find the unit rate of Charlie.
 b. Find the unit rate of Suellen.
 c. Who can assemble computer boards faster, Charlie or Suellen?

56. Jerry Stein laid 713 bricks in 46 minutes while his associate, Bobby Burns, laid 396 bricks in 30 minutes.
 a. Find the unit rate of Jerry.
 b. Find the unit rate of Bobby.
 c. Who is the faster bricklayer?

For Exercises 57 and 58, round the rates to the nearest tenth.

57. One student drove 400 miles in his car on 14.5 gallons of gasoline. His sister drove 270 miles in her truck on 9.25 gallons of gasoline.
 a. Find the unit rate of the car.
 b. Find the unit rate of the truck.
 c. Which vehicle gets better gas mileage?

58. Charlotte Leal is a grocery scanner who can scan an average of 100 items in 3.5 minutes while her cousin Leo can scan 148 items in 5.5 minutes.
 a. Find the unit rate of Charlotte.
 b. Find the unit rate of Leo.
 c. Who is the faster scanner?

Objective **D** *Find each unit price. See Example 11.*

59. $57.50 for 5 compact discs

60. $0.87 for 3 apples

61. $1.19 for 7 bananas

62. $73.50 for 6 lawn chairs

Find each unit price and decide which is the better buy. Round to three decimal places. Assume that we are comparing different sizes of the same brand. See Example 12.

63. Crackers:
 $3.29 for 8 ounces
 $4.79 for 12 ounces

64. Pickles:
 $2.79 for 32 ounces
 $1.49 for 18 ounces

65. Frozen orange juice:
 $1.89 for 16 ounces
 $0.69 for 6 ounces

66. Eggs:
 $1.56 for a dozen
 $3.69 for a flat $\left(2\frac{1}{2}\text{dozen} \right)$

▶ **67.** Soy sauce:
 12 ounces for $2.29
 8 ounces for $1.49

68. Shampoo:
 20 ounces for $1.89
 32 ounces for $3.19

69. Napkins:
 100 for $0.59
 180 for $0.93

70. Crackers:
 20 ounces for $2.39
 8 ounces for $0.99

Review

See Section 5.4.

71. $9\overline{)20.7}$

72. $7\overline{)60.2}$

73. $3.7\overline{)0.555}$

74. $4.6\overline{)1.15}$

Concept Extensions

75. Is the ratio $\dfrac{11}{15}$ the same as the ratio $\dfrac{15}{11}$? Explain your answer.

76. Explain why the ratio $\dfrac{5}{1}$ is incorrect for Exercise **20.**

Decide whether each value is a ratio written as a fraction in simplest form. If not, write it as a fraction in simplest form.

77. $\dfrac{6 \text{ inches}}{15 \text{ inches}}$

78. $4\dfrac{1}{2}$

79. A panty hose manufacturing machine will be repaired if the ratio of defective panty hose to good panty hose is at least 1 to 20. A quality control engineer found 10 defective panty hose in a batch of 200. Determine whether the machine should be repaired.

80. A grocer will refuse a shipment of tomatoes if the ratio of bruised tomatoes to the total batch is at least 1 to 10. A sample is found to contain 3 bruised tomatoes and 33 good tomatoes. Determine whether the shipment should be refused.

Fill in the table to calculate miles per gallon.

	Beginning Odometer Reading	Ending Odometer Reading	Miles Driven	Gallons of Gas Used	Miles per Gallon (round to the nearest tenth)
81.	29,286	29,543		13.4	
82.	16,543	16,895		15.8	
83.	79,895	80,242		16.1	
84.	31,623	32,056		11.9	

Find each unit rate.

85. The longest stairway is the service stairway for the Niesenbahn Cable railway near Spiez, Switzerland. It has 11,674 steps and rises to a height of 7759 feet. Find the unit rate of steps per foot rounded to the nearest tenth of a step. (*Source: Guinness World Records*)

86. In the United States, the total number of students enrolled in public schools is 49,373,000. There are 98,817 public schools. Write a unit rate in students per school. Round to the nearest whole. (*Source: National Center for Education Statistics*)

87. In 2013, 22 states had mandatory helmet laws. (*Source:* Insurance Institute for Highway Safety)

   a. Find the ratio of states with mandatory helmet laws to total U.S. states.

   b. Find the ratio of states with mandatory helmet laws to states without mandatory helmet laws.

   c. Are your ratios for parts **a** and **b** the same? Explain why or why not.

88. Suppose that the amount of a product decreases, say from an 80-ounce container to a 70-ounce container, but the price of the container remains the same. Does the unit price increase or decrease? Explain why.

89. In your own words, define the phrase unit rate.

90. In your own words, define the phrase unit price.

91. Should the rate $\dfrac{3 \text{ lights}}{2 \text{ feet}}$ be written as $\dfrac{3}{2}$? Explain why or why not.

92. Find an item in the grocery store and calculate its unit price.

6.2 Proportions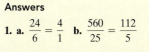

Objective A Writing Proportions

A **proportion** is a statement that two ratios or rates are equal.

> **Proportion**
>
> A proportion states that two ratios are equal. If $\frac{a}{b}$ and $\frac{c}{d}$ are two ratios, then
>
> $$\frac{a}{b} = \frac{c}{d}$$
>
> is a proportion.

For example,

$$\frac{5}{6} = \frac{10}{12}$$

is a proportion. We can read this as, "5 is to 6 as 10 is to 12."

Example 1 Write each sentence as a proportion.

a. 12 diamonds is to 15 rubies as 4 diamonds is to 5 rubies.
b. 5 hits is to 9 at bats as 20 hits is to 36 at bats.

Solution:

a. diamonds → $\dfrac{12}{15} = \dfrac{4}{5}$ ← diamonds
rubies → ← rubies

b. hits → $\dfrac{5}{9} = \dfrac{20}{36}$ ← hits
at bats → ← at bats

■ **Work Practice 1**

Helpful Hint

Notice in the above examples of proportions that the numerators contain the same units and the denominators contain the same units. In this text, proportions will be written so that this is the case.

Objective B Determining Whether Proportions Are True

Like other mathematical statements, a proportion may be either true or false. A proportion is true if its ratios are equal. Since ratios are fractions, one way to determine whether a proportion is true is to write both fractions in simplest form and compare them.

Another way is to compare cross products as we did in Section 4.2.

> **Using Cross Products to Determine Whether Proportions Are True or False**
>
>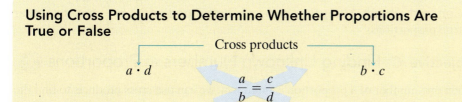

If cross products are *equal*, the proportion is *true*. If cross products are *not equal*, the proportion is *false*.

Objectives

A Write Sentences as Proportions.

B Determine Whether Proportions Are True.

C Find an Unknown Number in a Proportion.

Practice 1

Write each sentence as a proportion.
a. 24 cups is to 6 cups as 4 cups is to 1 cup.
b. 560 students is to 25 instructors as 112 students is to 5 instructors.

Answers

1. **a.** $\dfrac{24}{6} = \dfrac{4}{1}$ **b.** $\dfrac{560}{25} = \dfrac{112}{5}$

Practice 2

Is $\dfrac{4}{8} = \dfrac{10}{20}$ a true proportion?

Example 2 Is $\dfrac{2}{3} = \dfrac{4}{6}$ a true proportion?

Solution:

Cross products

$2 \cdot 6$ $\qquad\qquad\qquad\qquad$ $3 \cdot 4$

$$\dfrac{2}{3} = \dfrac{4}{6}$$

$2 \cdot 6 \overset{?}{=} 3 \cdot 4$ Are cross products equal?

$12 = 12$ Equal, so proportion is true.

Since the cross products are equal, the proportion is true.

■ **Work Practice 2**

Practice 3

Is $\dfrac{4.2}{6} = \dfrac{4.8}{8}$ a true proportion?

Example 3 Is $\dfrac{4.1}{7} = \dfrac{2.9}{5}$ a true proportion?

Solution:

Cross products

$4.1 \cdot 5$ $\qquad\qquad\qquad\qquad$ $7 \cdot 2.9$

$$\dfrac{4.1}{7} = \dfrac{2.9}{5}$$

$4.1 \cdot 5 \overset{?}{=} 7 \cdot 2.9$ Are cross products equal?

$20.5 \neq 20.3$ Not equal, so proportion is false.

Since the cross products are not equal, $\dfrac{4.1}{7} \neq \dfrac{2.9}{5}$. The proportion is false.

■ **Work Practice 3**

Practice 4

Is $\dfrac{3\frac{3}{10}}{1\frac{5}{6}} = \dfrac{4\frac{1}{5}}{2\frac{1}{3}}$ a true proportion?

Example 4 Is $\dfrac{1\frac{1}{6}}{10\frac{1}{2}} = \dfrac{\frac{1}{2}}{4\frac{1}{2}}$ a true proportion?

Solution:

$$\dfrac{1\frac{1}{6}}{10\frac{1}{2}} = \dfrac{\frac{1}{2}}{4\frac{1}{2}}$$

$1\frac{1}{6} \cdot 4\frac{1}{2} \overset{?}{=} 10\frac{1}{2} \cdot \frac{1}{2}$ Are cross products equal?

$\dfrac{7}{6} \cdot \dfrac{9}{2} \overset{?}{=} \dfrac{21}{2} \cdot \dfrac{1}{2}$ Write mixed numbers as improper fractions.

$\dfrac{21}{4} = \dfrac{21}{4}$ Equal, so proportion is true.

Since the cross products are equal, the proportion is true.

■ **Work Practice 4**

✔ **Concept Check** Using the numbers in the proportion $\dfrac{21}{27} = \dfrac{7}{9}$, write two other true proportions.

Objective C Finding Unknown Numbers in Proportions ▶

When one number of a proportion is unknown, we can use cross products to find the unknown number. For example, to find the unknown number x in the proportion $\dfrac{2}{3} = \dfrac{x}{30}$, we use cross products.

Answers

2. yes **3.** no **4.** yes

✔**Concept Check Answer**

possible answers: $\dfrac{27}{21} = \dfrac{9}{7}$ and $\dfrac{9}{27} = \dfrac{7}{21}$

Example 5 Solve $\dfrac{2}{3} = \dfrac{x}{30}$ for x.

Solution: If the cross products are equal, then the proportion is true. We begin, then, by setting cross products equal to each other.

$$\dfrac{2}{3} = \dfrac{x}{30}$$

$2 \cdot 30 = 3 \cdot x$ Set cross products equal.

$60 = 3x$ Multiply.

Recall that to find x, we divide both sides of the equation by 3.

$\dfrac{60}{3} = \dfrac{3x}{3}$ Divide both sides by 3.

$20 = x$ Simplify.

Check: To check, we replace x with 20 in the original proportion to see if the result is a true statement.

$\dfrac{2}{3} = \dfrac{x}{30}$ Original proportion

$\dfrac{2}{3} \overset{?}{=} \dfrac{20}{30}$ Replace x with 20.

$\dfrac{2}{3} = \dfrac{2}{3}$ True

Since $\dfrac{2}{3} = \dfrac{2}{3}$ is a true statement, 20 is the solution.

■ **Work Practice 5**

Example 6 Solve $\dfrac{51}{34} = \dfrac{-3}{x}$ for x.

Solution:

$$\dfrac{51}{34} = \dfrac{-3}{x}$$

$51 \cdot x = 34 \cdot -3$ Set cross products equal.

$51x = -102$ Multiply.

$\dfrac{51x}{51} = \dfrac{-102}{51}$ Divide both sides by 51.

$x = -2$ Simplify.

Check: $\dfrac{51}{34} = \dfrac{-3}{x}$ Original proportion

$\dfrac{51}{34} \overset{?}{=} \dfrac{-3}{-2}$ Replace x with -2.

$\dfrac{51}{34} \overset{?}{=} \dfrac{-3 \cdot -17}{-2 \cdot -17}$

$\dfrac{51}{34} = \dfrac{51}{34}$ True

■ **Work Practice 6**

Practice 5

Solve $\dfrac{2}{5} = \dfrac{x}{25}$ for x.

Practice 6

Solve $\dfrac{-15}{2} = \dfrac{60}{x}$ for x.

Answers

5. 10   **6.** -8

Practice 7

Solve for z:

$$\frac{\frac{7}{8}}{z} = \frac{\frac{2}{3}}{\frac{4}{7}}$$

Practice 8

Solve for y: $\dfrac{y}{9} = \dfrac{0.6}{1.2}$

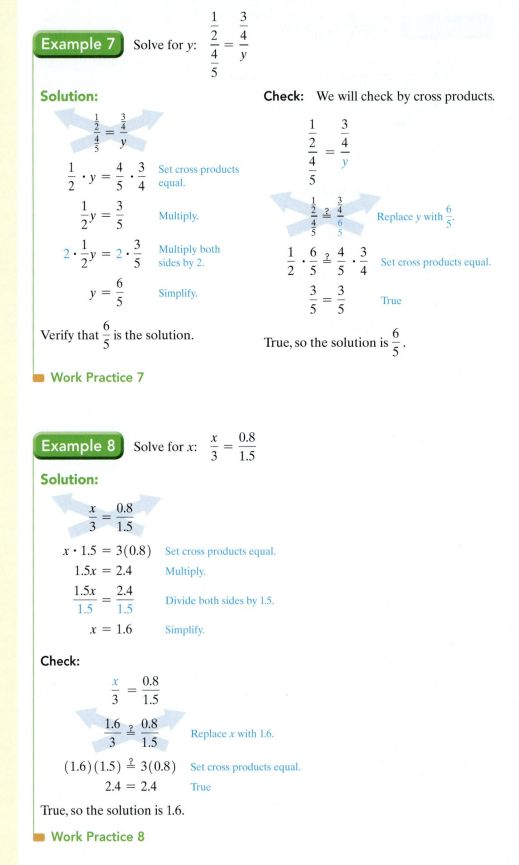

Example 7 Solve for y: $\dfrac{\frac{1}{2}}{\frac{4}{5}} = \dfrac{\frac{3}{4}}{y}$

Solution:

$$\frac{\frac{1}{2}}{\frac{4}{5}} = \frac{\frac{3}{4}}{y}$$

$\dfrac{1}{2} \cdot y = \dfrac{4}{5} \cdot \dfrac{3}{4}$ Set cross products equal.

$\dfrac{1}{2}y = \dfrac{3}{5}$ Multiply.

$2 \cdot \dfrac{1}{2}y = 2 \cdot \dfrac{3}{5}$ Multiply both sides by 2.

$y = \dfrac{6}{5}$ Simplify.

Verify that $\dfrac{6}{5}$ is the solution.

Check: We will check by cross products.

$$\frac{\frac{1}{2}}{\frac{4}{5}} = \frac{\frac{3}{4}}{y}$$

$\dfrac{\frac{1}{2}}{\frac{4}{5}} \stackrel{?}{=} \dfrac{\frac{3}{4}}{\frac{6}{5}}$ Replace y with $\dfrac{6}{5}$.

$\dfrac{1}{2} \cdot \dfrac{6}{5} \stackrel{?}{=} \dfrac{4}{5} \cdot \dfrac{3}{4}$ Set cross products equal.

$\dfrac{3}{5} = \dfrac{3}{5}$ True

True, so the solution is $\dfrac{6}{5}$.

■ **Work Practice 7**

Example 8 Solve for x: $\dfrac{x}{3} = \dfrac{0.8}{1.5}$

Solution:

$$\frac{x}{3} = \frac{0.8}{1.5}$$

$x \cdot 1.5 = 3(0.8)$ Set cross products equal.

$1.5x = 2.4$ Multiply.

$\dfrac{1.5x}{1.5} = \dfrac{2.4}{1.5}$ Divide both sides by 1.5.

$x = 1.6$ Simplify.

Check:

$$\frac{x}{3} = \frac{0.8}{1.5}$$

$\dfrac{1.6}{3} \stackrel{?}{=} \dfrac{0.8}{1.5}$ Replace x with 1.6.

$(1.6)(1.5) \stackrel{?}{=} 3(0.8)$ Set cross products equal.

$2.4 = 2.4$ True

True, so the solution is 1.6.

■ **Work Practice 8**

Answers

7. $\dfrac{3}{4}$ 8. 4.5

Example 9   Solve for y: $\dfrac{14}{y} = \dfrac{12}{16}$

Solution:

$$\dfrac{14}{y} = \dfrac{12}{16}$$

$14 \cdot 16 = y \cdot 12$ Set cross products equal.

$224 = 12y$ Multiply.

$\dfrac{224}{12} = \dfrac{12y}{12}$ Divide both sides by 12.

$\dfrac{56}{3} = y$ Simplify.

Check to see that the solution is $\dfrac{56}{3}$.

🔲 **Work Practice 9**

Helpful Hint

In Example 9, the fraction $\dfrac{12}{16}$ may be simplified to $\dfrac{3}{4}$ before solving the equation. The solution will remain the same.

✔**Concept Check** True or false: The first step in solving the proportion $\dfrac{4}{z} = \dfrac{12}{15}$ yields the equation $4z = 180$. If false, give the correct cross product equation.

Example 10   Solve for x: $\dfrac{1.6}{1.1} = \dfrac{x}{0.3}$. Round the solution to the nearest hundredth.

Solution:

$$\dfrac{1.6}{1.1} = \dfrac{x}{0.3}$$

$(1.6)(0.3) = 1.1 \cdot x$ Set cross products equal.

$0.48 = 1.1x$ Multiply.

$\dfrac{0.48}{1.1} = \dfrac{1.1x}{1.1}$ Divide both sides by 1.1.

$0.44 \approx x$ Round to the nearest hundredth.

🔲 **Work Practice 10**

Vocabulary, Readiness & Video Check

Use the words and phrases below to fill in each blank.

ratio   cross products   true

false   proportion

1. $\dfrac{4.2}{8.4} = \dfrac{1}{2}$ is called a _____ while $\dfrac{7}{8}$ is called a _____.

2. In $\dfrac{a}{b} = \dfrac{c}{d}$, $a \cdot d$ and $b \cdot c$ are called _____.

3. In a proportion, if cross products are equal, the proportion is _____.

4. In a proportion, if cross products are not equal, the proportion is _____.

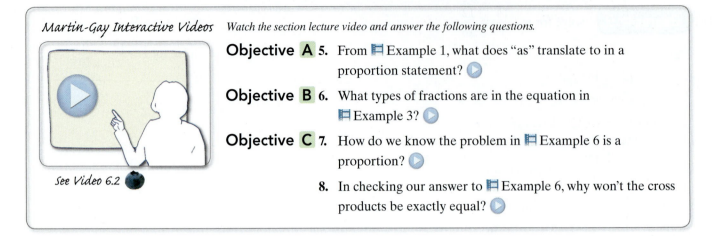

Martin-Gay Interactive Videos   *Watch the section lecture video and answer the following questions.*

See Video 6.2

Objective A **5.** From ▣ Example 1, what does "as" translate to in a proportion statement? ▶

Objective B **6.** What types of fractions are in the equation in ▣ Example 3? ▶

Objective C **7.** How do we know the problem in ▣ Example 6 is a proportion? ▶

8. In checking our answer to ▣ Example 6, why won't the cross products be exactly equal? ▶

6.2 Exercise Set MyMathLab® ▶

Objective A **Translating** *Write each sentence as a proportion. See Example 1.*

1. 10 diamonds is to 6 opals as 5 diamonds is to 3 opals.

2. 1 raisin is to 5 cornflakes as 8 raisins is to 40 cornflakes.

▶ **3.** 20 students is to 5 microscopes as 4 students is to 1 microscope.

4. 4 hit songs is to 16 releases as 1 hit song is to 4 releases.

5. 6 eagles is to 58 sparrows as 3 eagles is to 29 sparrows.

6. 12 errors is to 8 pages as 1.5 errors is to 1 page.

7. $2\frac{1}{4}$ cups of flour is to 24 cookies as $6\frac{3}{4}$ cups of flour is to 72 cookies.

8. $1\frac{1}{2}$ cups milk is to 10 bagels as $\frac{3}{4}$ cup milk is to 5 bagels.

9. 22 vanilla wafers is to 1 cup of cookie crumbs as 55 vanilla wafers is to 2.5 cups of cookie crumbs. (*Source:* Based on data from *Family Circle* magazine)

10. 1 cup of instant rice is to 1.5 cups cooked rice as 1.5 cups of instant rice is to 2.25 cups of cooked rice. (*Source:* Based on data from *Family Circle* magazine)

Objective B *Determine whether each proportion is true or false. See Examples 2 through 4.*

11. $\frac{15}{9} = \frac{5}{3}$

12. $\frac{8}{6} = \frac{20}{15}$

▶ **13.** $\frac{5}{8} = \frac{4}{7}$

14. $\frac{7}{3} = \frac{9}{5}$

15. $\frac{9}{36} = \frac{2}{8}$

16. $\frac{8}{24} = \frac{3}{9}$

17. $\frac{5}{8} = \frac{625}{1000}$

18. $\frac{30}{50} = \frac{600}{1000}$

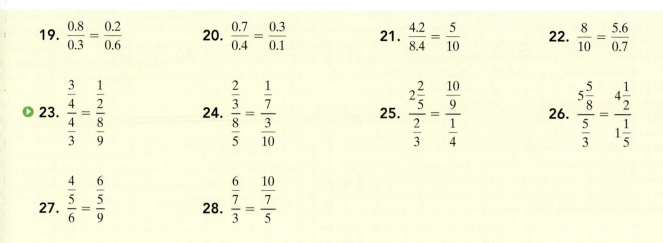

19. $\dfrac{0.8}{0.3} = \dfrac{0.2}{0.6}$

20. $\dfrac{0.7}{0.4} = \dfrac{0.3}{0.1}$

21. $\dfrac{4.2}{8.4} = \dfrac{5}{10}$

22. $\dfrac{8}{10} = \dfrac{5.6}{0.7}$

▶ 23. $\dfrac{\frac{3}{4}}{\frac{4}{3}} = \dfrac{\frac{1}{2}}{\frac{8}{9}}$

24. $\dfrac{\frac{2}{3}}{\frac{8}{5}} = \dfrac{\frac{1}{7}}{\frac{3}{10}}$

25. $\dfrac{2\frac{2}{5}}{\frac{2}{3}} = \dfrac{\frac{10}{9}}{\frac{1}{4}}$

26. $\dfrac{5\frac{5}{8}}{\frac{5}{3}} = \dfrac{4\frac{1}{2}}{1\frac{1}{5}}$

27. $\dfrac{\frac{4}{5}}{\frac{5}{6}} = \dfrac{\frac{6}{5}}{9}$

28. $\dfrac{\frac{6}{7}}{\frac{3}{3}} = \dfrac{\frac{10}{7}}{5}$

Objectives A B **Mixed Practice–Translating** *Write each sentence as a proportion. Then determine whether the proportion is a true proportion. See Examples 1 through 4.*

29. Ten is to fifteen as four is to six.

30. Six is to eight as nine is to twelve.

31. Eleven is to four as five is to two.

32. Five is to three as seven is to five.

33. Fifteen hundredths is to three as thirty-five hundredths is to seven.

34. One and eight tenths is to two as four and five tenths is to five.

35. Two thirds is to one fifth as two fifths is to one ninth.

36. Ten elevenths is to three fourths as one fourth is to one half.

Objective C *Solve each proportion for the given variable. Round the solution where indicated. See Examples 5 through 10.*

37. $\dfrac{x}{5} = \dfrac{6}{10}$

38. $\dfrac{x}{3} = \dfrac{12}{9}$

▶ 39. $\dfrac{-18}{54} = \dfrac{3}{n}$

40. $\dfrac{25}{100} = \dfrac{-7}{n}$

41. $\dfrac{30}{10} = \dfrac{15}{y}$

42. $\dfrac{16}{20} = \dfrac{z}{35}$

43. $\dfrac{8}{15} = \dfrac{z}{6}$

44. $\dfrac{12}{10} = \dfrac{z}{16}$

45. $\dfrac{24}{x} = \dfrac{60}{96}$

46. $\dfrac{26}{x} = \dfrac{28}{49}$

47. $\dfrac{-3.5}{12.5} = \dfrac{-7}{n}$

48. $\dfrac{-0.2}{0.7} = \dfrac{-8}{n}$

▶ 49. $\dfrac{n}{0.6} = \dfrac{0.05}{12}$

50. $\dfrac{7.8}{13} = \dfrac{n}{2.6}$

51. $\dfrac{8}{\frac{1}{3}} = \dfrac{24}{n}$

52. $\dfrac{12}{\frac{3}{4}} = \dfrac{48}{n}$

53. $\dfrac{\frac{1}{3}}{\frac{3}{8}} = \dfrac{\frac{2}{5}}{n}$

54. $\dfrac{\frac{7}{9}}{\frac{8}{27}} = \dfrac{\frac{1}{4}}{n}$

55. $\dfrac{12}{n} = \dfrac{\frac{2}{3}}{\frac{6}{9}}$

56. $\dfrac{24}{n} = \dfrac{\frac{8}{15}}{\frac{5}{9}}$

57. $\dfrac{n}{1\frac{1}{5}} = \dfrac{4\frac{1}{6}}{6\frac{2}{3}}$

58. $\dfrac{n}{3\frac{1}{8}} = \dfrac{7\frac{3}{5}}{2\frac{3}{8}}$

59. $\dfrac{25}{n} = \dfrac{3}{\frac{7}{30}}$

60. $\dfrac{9}{n} = \dfrac{5}{\frac{11}{15}}$

61. $\dfrac{3.2}{0.3} = \dfrac{x}{1.4}$
Round to the nearest tenth.

62. $\dfrac{1.8}{z} = \dfrac{2.5}{8.4}$
Round to the nearest tenth.

63. $\dfrac{z}{5.2} = \dfrac{0.08}{6}$
Round to the nearest hundredth.

64. $\dfrac{4.25}{6.03} = \dfrac{5}{y}$
Round to the nearest hundredth.

▶ **65.** $\dfrac{7}{18} = \dfrac{x}{5}$
Round to the nearest tenth.

66. $\dfrac{17}{x} = \dfrac{9}{4}$
Round to the nearest thousandth.

67. $\dfrac{43}{17} = \dfrac{8}{z}$
Round to the nearest thousandth.

68. $\dfrac{x}{12} = \dfrac{18}{7}$
Round to the nearest hundredth.

Review

Insert < or > to form a true statement. See Sections 4.7 and 5.1.

69. 8.01   8.1

70. 7.26   7.026

71. $2\frac{1}{2}$ $2\frac{1}{3}$

72. $9\frac{1}{5}$ $9\frac{1}{4}$

Simplify each fraction. See Section 4.2.

73. $\dfrac{75}{125}$

74. $\dfrac{11y}{99y}$

75. $\dfrac{12x}{42}$

76. $\dfrac{28y^2}{42y^3}$

Concept Extensions

Use the numbers in each proportion to write two other true proportions. See the first Concept Check in this section.

77. $\dfrac{9}{15} = \dfrac{3}{5}$

78. $\dfrac{1}{4} = \dfrac{5}{20}$

79. $\dfrac{6}{18} = \dfrac{1}{3}$

80. $\dfrac{2}{7} = \dfrac{4}{14}$

81. If the proportion $\dfrac{a}{b} = \dfrac{c}{d}$ is a true proportion, write two other true proportions using the same letters.

82. Write a true proportion.

83. Explain the difference between a ratio and a proportion.

84. Explain how to find the unknown number in a proportion such as $\dfrac{n}{18} = \dfrac{12}{8}$.

For each proportion, solve for the variable.

85. $\dfrac{x}{7} = \dfrac{0}{8}$

86. $\dfrac{0}{2} = \dfrac{y}{3.5}$

87. $\dfrac{z}{1150} = \dfrac{588}{483}$

88. $\dfrac{585}{x} = \dfrac{117}{474}$

89. $\dfrac{222}{1515} = \dfrac{37}{y}$

90. $\dfrac{1425}{1062} = \dfrac{z}{177}$

Integrated Review

Ratio, Rate, and Proportion

Answers

Write each ratio as a ratio of whole numbers using fractional notation. Write the fraction in simplest form.

1. 27 to 30

2. 18 to 50

3. 9.4 to 10

4. 3.2 to 9.2

5. 8.65 to 6.95

6. 3.6 to 4.2

7. $\frac{7}{2}$ to 13

8. $1\frac{2}{3}$ to $2\frac{3}{4}$

9. 16 inches to 24 inches

10. 5 hours to 40 hours

Find the ratio described in each problem.

11. Find the ratio of the width (shorter side) to the length (longer side) of the sign below.

12. The circle graph below shows the ratings of films released for the first four months of 2013. Use this graph to answer the questions.

  a. How many films were rated R?

  b. Find the ratio of PG-13 films to total films.

2013 Films Released Through April

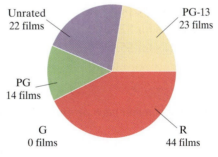

Source: Internet search

Write each rate as a fraction in simplest form.

13. 4 professors for every 20 graduate assistants

14. 6 lights every 20 feet

15. 100 U.S. Senators for 50 states

16. 5 teachers for every 140 students

17. 21 inches every 7 seconds

18. $40 every 5 hours

1. _____

2. _____

3. _____

4. _____

5. _____

6. _____

7. _____

8. _____

9. _____

10. _____

11. _____

12. a. _____

  b. _____

13. _____

14. _____

15. _____

16. _____

17. _____

18. _____

19. _____

20. _____

21. _____

22. _____

23. _____

24. _____

25. _____

26. _____

27. _____

28. _____

29. _____

30. _____

31. _____

32. _____

33. _____

34. _____

35. _____

36. _____

37. _____

38. _____

19. 76 households with computers for every 100 households

20. 538 electoral votes for 50 states

Write each rate as a unit rate.

21. 560 feet in 4 seconds

22. 195 miles in 3 hours

23. 63 employees per 3 fax lines

24. 85 phone calls for 5 teenagers

25. 156 miles per 6 gallons

26. 112 teachers for 7 computers

27. 8125 books for 1250 college students

28. 2310 pounds for 14 adults

Write each unit price and decide which is the better buy. Round to 3 decimal places.

29. Cat food:
   8 pounds for $2.16
   18 pounds for $4.99

30. Paper plates:
   100 for $1.98
   500 for $8.99

31. Microwave popcorn:
   3 packs for $2.39
   8 packs for $5.99

32. AA batteries:
   4 for $4.69
   10 for $14.89

Determine whether each proportion is true.

33. $\dfrac{7}{4} = \dfrac{5}{3}$

34. $\dfrac{8.2}{2} = \dfrac{16.4}{4}$

Solve each proportion for the given variable.

35. $\dfrac{5}{3} = \dfrac{40}{x}$

36. $\dfrac{y}{10} = \dfrac{13}{4}$

37. $\dfrac{6}{11} = \dfrac{z}{5}$

38. $\dfrac{21}{x} = \dfrac{\frac{7}{2}}{3}$

6.3 Proportions and Problem Solving

Objective A Solving Problems by Writing Proportions

Writing proportions is a powerful tool for solving problems in almost every field, including business, chemistry, biology, health sciences, and engineering, as well as in daily life. Given a specified ratio (or rate) of two quantities, a proportion can be used to determine an unknown quantity.

In this section, we use the same problem-solving steps that we have used earlier in this text.

Objective

A Solve Problems by Writing Proportions.

Example 1 Determining Distances from a Map

On a chamber of commerce map of Abita Springs, 5 miles corresponds to 2 inches. How many miles correspond to 7 inches?

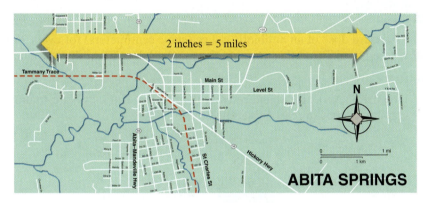

2 inches = 5 miles

ABITA SPRINGS

Practice 1

On an architect's blueprint, 1 inch corresponds to 4 feet. How long is a wall represented by a $4\frac{1}{4}$-inch line on the blueprint?

Solution:

1. UNDERSTAND. Read and reread the problem. You may want to draw a diagram.

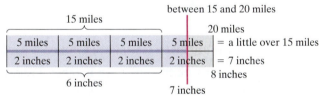

From the diagram we can see that a reasonable solution should be between 15 and 20 miles.

2. TRANSLATE. We will let x be our unknown number. Since 5 miles corresponds to 2 inches as x miles corresponds to 7 inches, we have the proportion

$$\text{miles} \rightarrow \quad \frac{5}{2} = \frac{x}{7} \quad \leftarrow \text{miles}$$
$$\text{inches} \rightarrow \qquad\qquad \leftarrow \text{inches}$$

3. SOLVE: In earlier sections, we estimated to obtain a reasonable answer. Notice we did this in Step 1 above.

$$\frac{5}{2} = \frac{x}{7}$$

$$5 \cdot 7 = 2 \cdot x \qquad \text{Set the cross products equal to each other.}$$

$$35 = 2x \qquad \text{Multiply.}$$

$$\frac{35}{2} = \frac{2x}{2} \qquad \text{Divide both sides by 2.}$$

$$17\frac{1}{2} = x \text{ or } x = 17.5 \qquad \text{Simplify.}$$

(Continued on next page)

Answer
1. 17 ft

431

4. INTERPRET. *Check* your work. This result is reasonable since it is between 15 and 20 miles. *State* your conclusion: 7 inches corresponds to 17.5 miles.

■ **Work Practice 1**

Helpful Hint

We can also solve Example 1 by writing the proportion

$$\frac{2 \text{ inches}}{5 \text{ miles}} = \frac{7 \text{ inches}}{x \text{ miles}}$$

Although other proportions may be used to solve Example 1, we will solve by writing proportions so that the numerators have the same unit measures and the denominators have the same unit measures.

Practice 2

An auto mechanic recommends that 5 ounces of isopropyl alcohol be mixed with a tankful of gas (16 gallons) to increase the octane of the gasoline for better engine performance. At this rate, how many gallons of gas can be treated with an 8-ounce bottle of alcohol?

Example 2   Finding Medicine Dosage

The standard dose of an antibiotic is 4 cc (cubic centimeters) for every 25 pounds (lb) of body weight. At this rate, find the standard dose for a 140-lb woman.

Solution:

1. UNDERSTAND. Read and reread the problem. You may want to draw a diagram to estimate a reasonable solution.

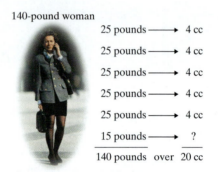

140-pound woman

25 pounds ⟶	4 cc
25 pounds ⟶	4 cc
25 pounds ⟶	4 cc
25 pounds ⟶	4 cc
25 pounds ⟶	4 cc
15 pounds ⟶	?
140 pounds over	20 cc

From the diagram, we can see that a reasonable solution is a little over 20 cc.

2. TRANSLATE. We will let x be the unknown number. From the problem, we know that 4 cc is to 25 pounds as x cc is to 140 pounds, or

$$\begin{array}{c} \text{cubic centimeters} \rightarrow \\ \text{pounds} \rightarrow \end{array} \frac{4}{25} = \frac{x}{140} \begin{array}{c} \leftarrow \text{cubic centimeters} \\ \leftarrow \text{pounds} \end{array}$$

3. SOLVE:

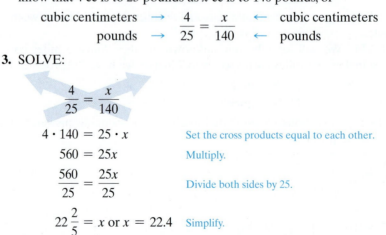

$$\frac{4}{25} = \frac{x}{140}$$

$4 \cdot 140 = 25 \cdot x$ Set the cross products equal to each other.

$560 = 25x$ Multiply.

$\dfrac{560}{25} = \dfrac{25x}{25}$ Divide both sides by 25.

$22\dfrac{2}{5} = x \text{ or } x = 22.4$ Simplify.

4. INTERPRET. *Check* your work. This result is reasonable since it is a little over 20 cc. *State* your conclusion: The standard dose for a 140-lb woman is 22.4 cc.

■ **Work Practice 2**

Answer

2. $25\dfrac{3}{5}$ or 25.6 gal

△ **Example 3** Calculating Supplies Needed to Fertilize a Lawn

A 50-pound bag of fertilizer covers 2400 square feet of lawn. How many bags of fertilizer are needed to cover a town square containing 15,360 square feet of lawn? Round the answer up to the nearest whole bag.

Solution:

1. UNDERSTAND. Read and reread the problem. Draw a picture.

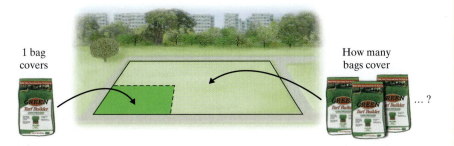

1 bag covers

How many bags cover

... ?

Since one bag covers 2400 square feet, let's see how many 2400s there are in 15,360. We will estimate. The number 15,360 rounded to the nearest thousand is 15,000 and 2400 rounded to the nearest thousand is 2000. Then

$$15,000 \div 2000 = 7\frac{1}{2} \text{ or } 7.5.$$

2. TRANSLATE. We'll let x be the unknown number. From the problem, we know that 1 bag is to 2400 square feet as x bags is to 15,360 square feet.

$$
\begin{array}{rcl}
\text{bags} \rightarrow & \dfrac{1}{2400} = \dfrac{x}{15,360} & \leftarrow \text{bags} \\
\text{square feet} \rightarrow & & \leftarrow \text{square feet}
\end{array}
$$

3. SOLVE:

$$\frac{1}{2400} = \frac{x}{15,360}$$

$1 \cdot 15,360 = 2400 \cdot x$ Set the cross products equal to each other.

$15,360 = 2400 \cdot x$ Multiply.

$\dfrac{15,360}{2400} = \dfrac{2400x}{2400}$ Divide both sides by 2400.

$6.4 = x$ Simplify.

4. INTERPRET. *Check* that replacing x with 6.4 makes the proportion true. Is the answer reasonable? Yes, since it's close to $7\frac{1}{2}$ or 7.5. Since we must buy whole bags of fertilizer, 7 bags are needed. *State* your conclusion: To cover 15,360 square feet of lawn, 7 bags are needed.

■ **Work Practice 3**

✓ **Concept Check** You are told that 12 ounces of ground coffee will brew enough coffee to serve 20 people. How could you estimate how much ground coffee will be needed to serve 95 people?

Practice 3

If a gallon of paint covers 450 square feet, how many gallons are needed to paint a retaining wall that is 270 feet long and 11 feet high? Round the answer up to the nearest whole gallon.

Answer

3. 7 gal

✓ **Concept Check Answer**

Find how much will be needed for 100 people (20×5) by multiplying 12 ounces by 5, which is 60 ounces.

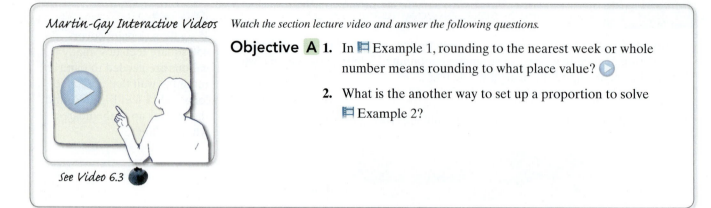

Martin-Gay Interactive Videos   *Watch the section lecture video and answer the following questions.*

Objective A   **1.** In Example 1, rounding to the nearest week or whole number means rounding to what place value?

     2. What is the another way to set up a proportion to solve Example 2?

See Video 6.3

6.3 Exercise Set MyMathLab®

Objective A *Solve. For Exercises 1 and 2, the solutions have been started for you. See Examples 1 through 3. An NBA basketball player averages 45 baskets for every 100 attempts.*

1. If he attempted 800 field goals, how many field goals did he make?

Start the solution:

   1. UNDERSTAND the problem. Reread it as many times as needed. Let's let

     x = how many field goals he made

   2. TRANSLATE into an equation.

$$\text{baskets (field goals)} \rightarrow \frac{45}{100} = \frac{x}{800} \leftarrow \text{baskets (field goals)} \atop \leftarrow \text{attempts}$$

attempts $\rightarrow$

   3. SOLVE the equation. Set cross products equal to each other and solve.

$$\frac{45}{100} \diagdown\!\!\!\!\diagup \frac{x}{800}$$

Finish by SOLVING and **4.** INTERPRET.

2. If he made 225 baskets, how many did he attempt?

Start the solution:

   1. UNDERSTAND the problem. Reread it as many times as needed. Let's let

     x = how many baskets attempted

   2. TRANSLATE into an equation.

$$\text{baskets} \rightarrow \frac{45}{100} = \frac{225}{x} \leftarrow \text{baskets} \atop \leftarrow \text{attempts}$$

attempts $\rightarrow$

   3. SOLVE the equation. Set cross products equal to each other and solve.

$$\frac{45}{100} \diagdown\!\!\!\!\diagup \frac{225}{x}$$

Finish by SOLVING and **4.** INTERPRET.

It takes a word processor 30 minutes to word process and spell check 4 pages.

3. Find how long it takes her to word process and spell check 22 pages.

4. Find how many pages she can word process and spell check in 4.5 hours.

University Law School accepts 2 out of every 7 applicants.

5. If the school accepted 180 students, find how many applications they received.

6. If the school accepted 150 students, find how many applications they received.

On an architect's blueprint, 1 inch corresponds to 8 feet.

7. Find the length of a wall represented by a line $2\frac{7}{8}$ inches long on the blueprint.

8. Find the length of a wall represented by a line $5\frac{1}{4}$ inches long on the blueprint.

A human-factors expert recommends that there be at least 9 square feet of floor space in a college classroom for every student in the class.

△**9.** Find the minimum floor space that 30 students require.

△**10.** Due to a lack of space, a university converts a 21-by-15-foot conference room into a classroom. Find the maximum number of students the room can accommodate.

A Honda Civic Hybrid averages 627 miles on a 12.3-gallon tank of gas.

11. Manuel Lopez is planning a 1250-mile vacation trip in his Honda Civic Hybrid. Find how many gallons of gas he can expect to burn. Round to the nearest gallon.

12. Ramona Hatch has enough money to put 6.9 gallons of gas in her Honda Civic Hybrid. She is planning on driving home from college for the weekend. If her home is 290 miles away, should she make it home before she runs out of gas?

The scale on an Italian map states that 1 centimeter corresponds to 30 kilometers.

13. Find how far apart Milan and Rome are if their corresponding points on the map are 15 centimeters apart.

14. On the map, a small Italian village is located 0.4 centimeter from the Mediterranean Sea. Find the actual distance.

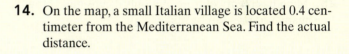

A bag of Scott fertilizer covers 3000 square feet of lawn.

△**15.** Find how many bags of fertilizer should be purchased to cover a rectangular lawn 260 feet by 180 feet.

△**16.** Find how many bags of fertilizer should be purchased to cover a square lawn measuring 160 feet on each side.

A Cubs baseball player gets 3 hits every 8 times at bat.

17. If this Cubs player comes up to bat 40 times in the World Series, find how many hits he would be expected to get.

18. At this rate, if he got 12 hits, find how many times he batted.

A survey reveals that 2 out of 3 people prefer Coke to Pepsi.

19. In a room of 40 people, how many people are likely to prefer Coke? Round the answer to the nearest person.

20. In a college class of 36 students, find how many students are likely to prefer Pepsi.

A self-tanning lotion advertises that a 3-oz bottle will provide four applications.

21. Jen Haddad found a great deal on a 14-oz bottle of the self-tanning lotion she had been using. Based on the advertising claims, how many applications of the self-tanner should Jen expect? Round down to the nearest whole number.

22. The Community College thespians need fake tans for a play they are doing. If the play has a cast of 35, how many ounces of self-tanning lotion should the cast purchase? Round up to the next whole number of ounces.

The school's computer lab goes through 5 reams of printer paper every 3 weeks.

▶ 23. Find out how long a case of printer paper is likely to last (a case of paper holds 8 reams of paper). Round to the nearest week.

24. How many cases of printer paper should be purchased to last the entire semester of 15 weeks? Round up to the next case.

A recipe for pancakes calls for 2 cups flour and $1\frac{1}{2}$ cups milk to make a serving for four people.

25. Ming has plenty of flour, but only 4 cups milk. How many servings can he make?

26. The swim team has a weekly breakfast after early practice. How much flour will it take to make pancakes for 18 swimmers?

27. In the Seattle Space Needle, the elevators whisk you to the revolving restaurant at a speed of 800 feet in 60 seconds. If the revolving restaurant is 500 feet up, how long does it take you to reach the restaurant by elevator? (*Source:* Seattle Space Needle)

28. A 16-oz grande Tazo Black Iced Tea at Starbucks has 80 calories. How many calories are there in a 24-oz venti Tazo Black Iced Tea? (*Source:* Starbucks Coffee Company)

29. Mosquitos are annoying insects. To eliminate mosquito larvae, a certain granular substance can be applied to standing water in a ratio of 1 tsp per 25 sq ft of standing water.
   a. At this rate, find how many teaspoons of granules must be used for 450 square feet.
   b. If 3 tsp = 1 tbsp, how many tablespoons of granules must be used?

30. Another type of mosquito control is liquid, where 3 oz of pesticide is mixed with 100 oz of water. This mixture is sprayed on roadsides to control mosquito breeding grounds hidden by tall grass.
   a. If one mixture of water with this pesticide can treat 150 feet of roadway, how many ounces of pesticide are needed to treat one mile? (*Hint:* 1 mile = 5280 feet)
   b. If 8 liquid ounces equals one cup, write your answer to part **a** in cups. Round to the nearest cup.

31. The daily supply of oxygen for one person is provided by 625 square feet of lawn. A total of 3750 square feet of lawn would provide the daily supply of oxygen for how many people? (*Source:* Professional Lawn Care Association of America)

32. In 2012, approximately $20 billion of the $50 billion Americans spent on their pets was spent on pet food. Petsmart had $6,758,237 in net sales that year. How much of Petsmart's net sales would you expect to have been spent on pet food? (*Source:* American Pet Products Manufacturers Association and Petsmart)

33. A student would like to estimate the height of the Statue of Liberty in New York City's harbor. The length of the Statue of Liberty's right arm is 42 feet. The student's right arm is 2 feet long and her height is $5\frac{1}{3}$ feet. Use this information to estimate the height of the Statue of Liberty. How close is your estimate to the statue's actual height of 111 feet, 1 inch from heel to top of head? (*Source:* National Park Service)

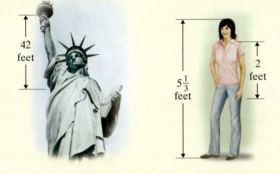

34. The length of the Statue of Liberty's index finger is 8 feet while the height to the top of the head is about 111 feet. Suppose your measurements are proportionally the same as this statue's and your height is 5 feet.

a. Use this information to find the proposed length of your index finger. Give an exact measurement and then a decimal rounded to the nearest hundredth.

b. Measure your index finger and write it as a decimal in feet rounded to the nearest hundredth. How close is the length of your index finger to the answer to part **a**? Explain why.

35. There are 72 milligrams of cholesterol in a 3.5-ounce serving of lobster. How much cholesterol is in 5 ounces of lobster? Round to the nearest tenth of a milligram. (*Source:* The National Institutes of Health)

36. There are 76 milligrams of cholesterol in a 3-ounce serving of skinless chicken. How much cholesterol is in 8 ounces of chicken? (*Source:* USDA)

37. Trump World Tower in New York City is 881 feet tall and contains 72 stories. The Empire State Building contains 102 stories. If the Empire State Building has the same number of feet per floor as the Trump World Tower, approximate its height rounded to the nearest foot. (*Source:* Skyscrapers.com)

38. Two out of every 5 men blame their poor eating habits on too much fast food. In a room of 40 men, how many would you expect to blame their not eating well on fast food? (*Source:* Healthy Choice Mixed Grills survey)

39. Medication is prescribed in 7 out of every 10 hospital emergency room visits that involve an injury. If a large urban hospital had 620 emergency room visits involving an injury in the past month, how many of these visits would you expect to have included a prescription for medication? (*Source:* National Center for Health Statistics)

40. Currently in the American population of people aged 65 years old and older, there are approximately 130 women for every 100 men. In a nursing home with 280 male residents over the age of 65, how many female residents over the age of 65 would be expected? (*Source:* U.S. Bureau of the Census)

41. One out of three American adults got his or her first job in the restaurant industry. In an office of 84 workers, how many of these people would you expect to have gotten their first job in the restaurant industry? (*Source:* National Restaurant Association)

42. One pound of firmly packed brown sugar yields $2\frac{1}{4}$ cups. How many pounds of brown sugar will be required in a recipe that calls for 6 cups of firmly packed brown sugar? (*Source:* Based on data from *Family Circle* magazine)

When making homemade ice cream in a hand-cranked freezer, the tub containing the ice cream mix is surrounded by a brine (water/salt) solution. To freeze the ice cream mix rapidly so that smooth and creamy ice cream results, the brine solution should combine crushed ice and rock salt in a ratio of 5 to 1. Use this for Exercises 43 and 44. (Source: White Mountain Freezers, The Rival Company)

43. A small ice cream freezer requires 12 cups of crushed ice. How much rock salt should be mixed with the ice to create the necessary brine solution?

44. A large ice cream freezer requires $18\frac{3}{4}$ cups of crushed ice. How much rock salt will be needed to create the necessary brine solution?

45. The gas/oil ratio for a certain chainsaw is 50 to 1.
  a. How much oil (in gallons) should be mixed with 5 gallons of gasoline?
  b. If 1 gallon equals 128 fluid ounces, write the answer to part **a** in fluid ounces. Round to the nearest whole ounce.

46. The gas/oil ratio for a certain tractor mower is 20 to 1.
  a. How much oil (in gallons) should be mixed with 10 gallons of gas?
  b. If 1 gallon equals 4 quarts, write the answer to part **a** in quarts.

47. The adult daily dosage for a certain medicine is 150 mg (milligrams) of medicine for every 20 pounds of body weight.
  a. At this rate, find the daily dose for a man who weighs 275 pounds.
  b. If the man is to receive 500 mg of this medicine every 8 hours, is he receiving the proper dosage?

48. The adult daily dosage for a certain medicine is 80 mg (milligrams) for every 25 pounds of body weight.
  a. At this rate, find the daily dose for a woman who weighs 190 pounds.
  b. If she is to receive this medicine every 6 hours, find the amount to be given every 6 hours.

Review

Find the prime factorization of each number. See Section 4.2.

49. 200   **50.** 300   **51.** 32   **52.** 81

Concept Extensions

As we have seen earlier, proportions are often used in medicine dosage calculations. The exercises below have to do with liquid drug preparations, where the weight of the drug is contained in a volume of solution. The description of mg and ml below will help. We will study metric units further in Chapter 9.

mg means milligrams (A paper clip weighs about a gram. A milligram is about the weight of $\frac{1}{1000}$ of a paper clip.)

ml means milliliter (A liter is about a quart. A milliliter is about the amount of liquid in $\frac{1}{1000}$ of a quart.)

One way to solve the applications below is to set up the proportion $\frac{mg}{ml} = \frac{mg}{ml}$.

A solution strength of 15 mg of medicine in 1 ml of solution is available.

53. If a patient needs 12 mg of medicine, how many ml do you administer?

54. If a patient needs 33 mg of medicine, how many ml do you administer?

A solution strength of 8 mg of medicine in 1 ml of solution is available.

55. If a patient needs 10 mg of medicine, how many ml do you administer?

56. If a patient needs 6 mg of medicine, how many ml do you administer?

Estimate the following. See the Concept Check in this section.

57. It takes 1.5 cups of milk to make 11 muffins. Estimate the amount of milk needed to make 8 dozen muffins. Explain your calculation.

58. A favorite chocolate chip cookie recipe calls for $2\frac{1}{2}$ cups of flour to make 2 dozen cookies. Estimate the amount of flour needed to make 50 cookies. Explain your calculation.

A board such as the one pictured below will balance if the following proportion is true:

$$\frac{\text{first weight}}{\text{second distance}} = \frac{\text{second weight}}{\text{first distance}}$$

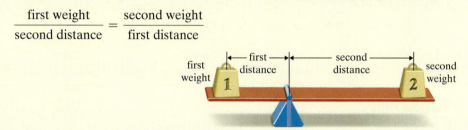

Use this proportion to solve Exercises 59 and 60.

59. Find the distance *n* that will allow the board to balance.

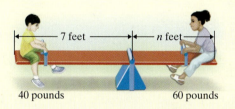

60. Find the length *n* needed to lift the weight below.

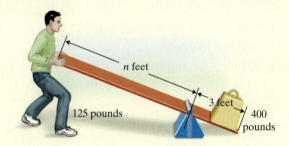

✎ **61.** Describe a situation in which writing a proportion might solve a problem related to driving a car.

△ **6.4** **Square Roots and the Pythagorean Theorem** ▶

Now that we know how to write ratios and solve proportions, in Section 6.5 we use proportions to help us find unknown sides of similar triangles. In this section, we prepare for work on triangles by studying right triangles and their applications.

First, let's practice finding square roots.

Objectives

A Find the Square Root of a Number. ▶

B Approximate Square Roots. ▶

C Use the Pythagorean Theorem. ▶

Objective **A** Finding Square Roots ▶

The **square** of a number is the number times itself. For example,

The square of 5 is 25 because 5^2 or $5 \cdot 5 = 25$.
The square of -5 is also 25 because $(-5)^2$ or $(-5)(-5) = 25$.

The reverse process of squaring is finding a **square root.** For example,

A square root of 25 is 5 because $5 \cdot 5$ or $5^2 = 25$.
A square root of 25 is also -5 because $(-5)(-5)$ or $(-5)^2 = 25$.

Every positive number has two square roots. We see on the previous page that the square roots of 25 are 5 and -5.

We use the symbol $\sqrt{}$, called a **radical sign,** to indicate the positive square root of a nonnegative number. For example,

$\sqrt{25} = 5$ because $5^2 = 25$ and 5 is positive.
$\sqrt{9} = 3$ because $3^2 = 9$ and 3 is positive.

Square Root of a Number

The square root, $\sqrt{}$, of a positive number a is the positive number b whose square is a. In symbols,

$$\sqrt{a} = b, \quad \text{if } b^2 = a$$

Also, $\sqrt{0} = 0$.

Helpful Hint

Remember that the radical sign $\sqrt{}$ is used to indicate the **positive** (or principal) **square root** of a nonnegative number.

Practice 1–6

Find each square root.

1. $\sqrt{100}$ 2. $\sqrt{64}$
3. $\sqrt{169}$ 4. $\sqrt{0}$
5. $\sqrt{\dfrac{1}{4}}$ 6. $\sqrt{\dfrac{9}{16}}$

Examples   Find each square root.

1. $\sqrt{49} = 7$ because $7^2 = 49$.
2. $\sqrt{36} = 6$ because $6^2 = 36$.
3. $\sqrt{1} = 1$ because $1^2 = 1$.
4. $\sqrt{81} = 9$ because $9^2 = 81$.
5. $\sqrt{\dfrac{1}{36}} = \dfrac{1}{6}$ because $\left(\dfrac{1}{6}\right)^2$ or $\dfrac{1}{6} \cdot \dfrac{1}{6} = \dfrac{1}{36}$.
6. $\sqrt{\dfrac{4}{25}} = \dfrac{2}{5}$ because $\left(\dfrac{2}{5}\right)^2$ or $\dfrac{2}{5} \cdot \dfrac{2}{5} = \dfrac{4}{25}$.

▪ **Work Practice 1–6**

Objective B Approximating Square Roots ▶

Thus far, we have found square roots of perfect squares. Numbers like $\dfrac{1}{4}$, 36, $\dfrac{4}{25}$, and 1 are called **perfect squares** because their square root is a whole number or a fraction. A square root such as $\sqrt{5}$ cannot be written as a whole number or a fraction since 5 is not a perfect square.

Although $\sqrt{5}$ cannot be written as a whole number or a fraction, it can be approximated by estimating, by using a table (as in the appendix), or by using a calculator.

Practice 7

Use Appendix A.4 or a calculator to approximate each square root to the nearest thousandth.

a. $\sqrt{10}$ b. $\sqrt{62}$

Example 7   Use Appendix A.4 or a calculator to approximate each square root to the nearest thousandth.

a. $\sqrt{43} \approx 6.557$ is approximately
b. $\sqrt{80} \approx 8.944$

▪ **Work Practice 7**

Answers

1. 10   2. 8   3. 13   4. 0   5. $\dfrac{1}{2}$ 6. $\dfrac{3}{4}$
7. a. 3.162   b. 7.874

Helpful Hint

$\sqrt{80}$, on the previous page, is *approximately* 8.944. This means that if we multiply 8.944 by 8.944, the product is *close* to 80.

$$8.944 \times 8.944 \approx 79.995$$

It is possible to approximate a square root to the nearest whole number without the use of a calculator or table. To do so, study the number line below and look for patterns.

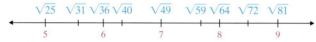

Above the number line, notice that as the numbers under the radical signs increase, their value, and thus their placement on the number line, increase also.

Example 8 Without a calculator or table:

a. Determine which two whole numbers $\sqrt{78}$ is between.

b. Use part **a** to approximate $\sqrt{78}$ to the nearest whole.

Practice 8

Without a calculator or table, approximate $\sqrt{62}$ to the nearest whole.

Solution:

a. Review perfect squares and recall that $\sqrt{64} = 8$ and $\sqrt{81} = 9$. Since 78 is between 64 and 81, $\sqrt{78}$ is between $\sqrt{64}$ (or 8) and $\sqrt{81}$ (or 9).

Thus, $\sqrt{78}$ is between 8 and 9.

b. Since $\sqrt{78}$ is closer to $\sqrt{81}$ (or 9) than $\sqrt{64}$ (or 8), then (as our number line shows) $\sqrt{78}$ approximate to the nearest whole is 9.

■ Work Practice 8

Objective C Using the Pythagorean Theorem ▶

One important application of square roots has to do with right triangles. Recall that a **right triangle** is a triangle in which one of the angles is a right angle, or measures 90° (degrees). The **hypotenuse** of a right triangle is the side opposite the right angle. The **legs** of a right triangle are the other two sides. These are shown in the following figure. The right angle in the triangle is indicated by the small square drawn in that angle.

The following theorem is true for all right triangles.

Pythagorean Theorem

If a and b are the lengths of the legs of a right triangle and c is the length of the hypotenuse, then

$$a^2 + b^2 = c^2$$

Leg a

Hypotenuse

Leg b

In other words, $(\text{leg})^2 + (\text{other leg})^2 = (\text{hypotenuse})^2$.

Answer

8. 8

Practice 9

Find the length of the hypotenuse of the given right triangle.

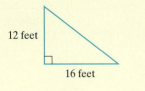

12 feet

16 feet

Copyright 2015 Pearson Education, Inc.

Example 9   Find the length of the hypotenuse of the given right triangle.

Solution: Let $a = 6$ and $b = 8$. According to the Pythagorean theorem,

$$a^2 + b^2 = c^2$$
$$6^2 + 8^2 = c^2 \quad \text{Let } a = 6 \text{ and } b = 8.$$
$$36 + 64 = c^2 \quad \text{Evaluate } 6^2 \text{ and } 8^2.$$
$$100 = c^2 \quad \text{Add.}$$

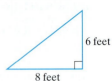

6 feet

8 feet

In the equation $c^2 = 100$, the solutions of c are the square roots of 100. Since $10 \cdot 10 = 100$ and $(-10)(-10) = 100$, both 10 and -10 are square roots of 100. Since c represents a length, we are only interested in the positive square root of c^2.

$$c = \sqrt{100}$$
$$= 10$$

The hypotenuse is 10 feet long.

■ **Work Practice 9**

Practice 10

Approximate the length of the hypotenuse of the given right triangle. Round to the nearest whole unit.

7 kilometers

9 kilometers

Example 10   Approximate the length of the hypotenuse of the given right triangle. Round the length to the nearest whole unit.

Solution: Let $a = 17$ and $b = 10$.

$$a^2 + b^2 = c^2$$
$$17^2 + 10^2 = c^2$$
$$289 + 100 = c^2$$
$$389 = c^2$$
$$\sqrt{389} = c \text{ or } c \approx 20 \quad \text{From Appendix A.4 or a calculator}$$

10 meters

17 meters

The hypotenuse is exactly $\sqrt{389}$ meters, which is approximately 20 meters.

■ **Work Practice 10**

Practice 11

Find the length of the leg in the given right triangle. Give the exact length and a two-decimal-place approximation.

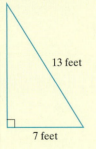

13 feet

7 feet

Example 11   Find the length of the leg in the given right triangle. Give the exact length and a two-decimal-place approximation.

5 inches

7 inches

Solution: Notice that the hypotenuse measures 7 inches and that the length of one leg measures 5 inches. Thus, let $c = 7$ and a or b be 5. We will let $a = 5$.

$$a^2 + b^2 = c^2$$
$$5^2 + b^2 = 7^2 \quad \text{Let } a = 5 \text{ and } c = 7.$$
$$25 + b^2 = 49 \quad \text{Evaluate } 5^2 \text{ and } 7^2.$$
$$b^2 = 24 \quad \text{Subtract 25 from both sides.}$$
$$b = \sqrt{24} \approx 4.90$$

The length of the leg is exactly $\sqrt{24}$ inches and approximately 4.90 inches.

■ **Work Practice 11**

Answers

9. 20 feet   **10.** 11 kilometers
11. $\sqrt{120}$ feet ≈ 10.95 feet

✓**Concept Check Answer**

a

✓**Concept Check** The following lists are the lengths of the sides of two triangles. Which set forms a right triangle?

a. 8, 15, 17   **b.** 24, 30, 40

△ **Example 12** Finding the Dimensions of a Park

An inner-city park is in the shape of a square that measures 300 feet on a side. A sidewalk is to be constructed along the diagonal of the park. Find the length of the sidewalk rounded to the nearest whole foot.

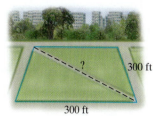

300 ft

300 ft

Solution: The diagonal is the hypotenuse of a right triangle, which we label c.

$$a^2 + b^2 = c^2$$
$$300^2 + 300^2 = c^2 \qquad \text{Let } a = 300 \text{ and } b = 300.$$
$$90{,}000 + 90{,}000 = c^2 \qquad \text{Evaluate } (300)^2.$$
$$180{,}000 = c^2 \qquad \text{Add.}$$
$$\sqrt{180{,}000} = c \text{ or } c \approx 424$$

The length of the sidewalk is approximately 424 feet.

🔲 **Work Practice 12**

Practice 12

A football field is a rectangle measuring 100 yards by 53 yards. Draw a diagram and find the length of the diagonal of a football field to the nearest yard.

Answer

12. 113 yards

🖩 **Calculator Explorations Finding and Approximating Square Roots**

To simplify or approximate square roots using a calculator, locate the key marked $\boxed{\sqrt{}}$.

To simplify $\sqrt{64}$, for example, press the keys

$\boxed{64}$ $\boxed{\sqrt{}}$ or $\boxed{\sqrt{}}$ $\boxed{64}$

The display should read $\boxed{\qquad 8}$. Then

$\sqrt{64} = 8$

To *approximate* $\sqrt{10}$, press the keys

$\boxed{10}$ $\boxed{\sqrt{}}$ or $\boxed{\sqrt{}}$ $\boxed{10}$

The display should read $\boxed{3.16227766}$. This is an *approximation* for $\sqrt{10}$. A three-decimal-place approximation is

$\sqrt{10} \approx 3.162$

Is this answer reasonable? Since 10 is between the perfect squares 9 and 16, $\sqrt{10}$ is between $\sqrt{9} = 3$ and $\sqrt{16} = 4$. Our answer is reasonable since 3.162 is between 3 and 4.

Simplify.

1. $\sqrt{1024}$
2. $\sqrt{676}$

Approximate each square root. Round each answer to the nearest thousandth.

3. $\sqrt{15}$
4. $\sqrt{19}$
5. $\sqrt{97}$
6. $\sqrt{56}$

Vocabulary, Readiness & Video Check

Use the choices below to fill in each blank. Some choices will be used more than once.

squaring	Pythagorean theorem	radical	-10	leg
hypotenuse	perfect squares	10	c^2	b^2

1. The square roots of 100 are _____ and _____ because $10 \cdot 10 = 100$ and $(-10)(-10) = 100$.

2. $\sqrt{100} =$ _____ because $10 \cdot 10 = 100$ and 10 is positive.

3. The _____ sign is used to denote the positive square root of a nonnegative number.

4. The reverse process of _____ a number is finding a square root of a number.

5. The numbers 9, 1, and $\frac{1}{25}$ are called _____.

6. Label the parts of the right triangle.

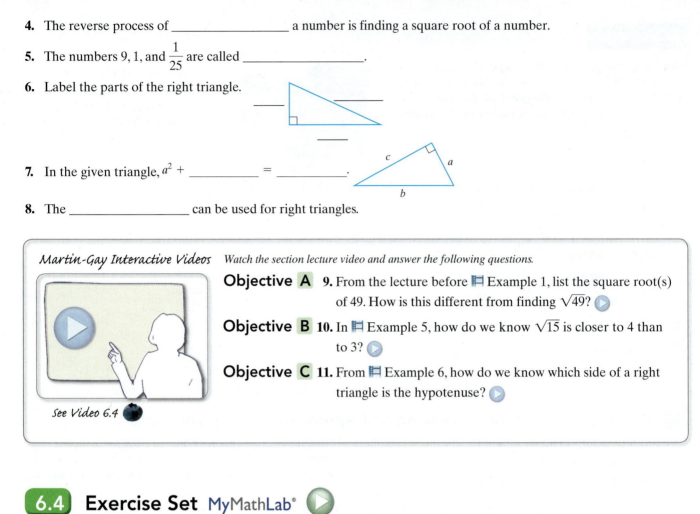

7. In the given triangle, $a^2 +$ _____ = _____.

8. The _____ can be used for right triangles.

Martin-Gay Interactive Videos   Watch the section lecture video and answer the following questions.

Objective A **9.** From the lecture before ⊟ Example 1, list the square root(s) of 49. How is this different from finding $\sqrt{49}$? ▶

Objective B **10.** In ⊟ Example 5, how do we know $\sqrt{15}$ is closer to 4 than to 3? ▶

Objective C **11.** From ⊟ Example 6, how do we know which side of a right triangle is the hypotenuse? ▶

See Video 6.4

6.4 **Exercise Set** MyMathLab® ▶

Objective A *Find each square root. See Examples 1 through 6.*

▶ **1.** $\sqrt{4}$ **2.** $\sqrt{9}$ ▶ **3.** $\sqrt{121}$ **4.** $\sqrt{144}$

▶ **5.** $\sqrt{\dfrac{1}{81}}$ **6.** $\sqrt{\dfrac{1}{64}}$ **7.** $\sqrt{\dfrac{16}{64}}$ **8.** $\sqrt{\dfrac{36}{81}}$

Objective B *Use Appendix A.4 or a calculator to approximate each square root. Round the square root to the nearest thousandth. See Example 7.*

9. $\sqrt{3}$ **10.** $\sqrt{5}$ ▶ **11.** $\sqrt{15}$ **12.** $\sqrt{17}$

13. $\sqrt{31}$ **14.** $\sqrt{85}$ **15.** $\sqrt{26}$ **16.** $\sqrt{35}$

Determine what two whole numbers each square root is between without using a calculator or table. Then use a calculator or Appendix A.4 to check. See Example 8.

▶ **17.** $\sqrt{38}$ **18.** $\sqrt{27}$ **19.** $\sqrt{101}$ **20.** $\sqrt{85}$

Objectives A B Mixed Practice *Find each square root. If necessary, round the square root to the nearest thousandth. See Examples 1 through 8.*

21. $\sqrt{256}$ **22.** $\sqrt{625}$ **23.** $\sqrt{92}$ **24.** $\sqrt{18}$

25. $\sqrt{\dfrac{49}{144}}$ **26.** $\sqrt{\dfrac{121}{169}}$ **27.** $\sqrt{71}$ **28.** $\sqrt{62}$

Objective C *Find the unknown length in each right triangle. If necessary, approximate the length to the nearest thousandth. See Examples 9 through 12.*

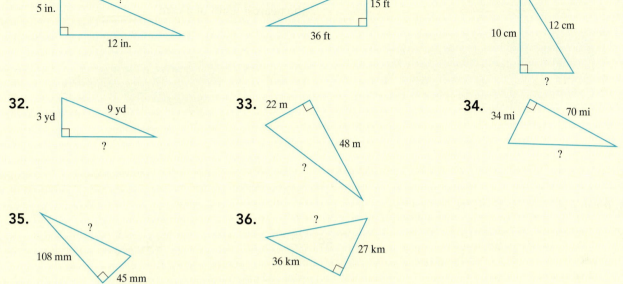

29. 5 in. ? 12 in.

30. ? 15 ft 36 ft

31. 10 cm 12 cm ?

32. 3 yd 9 yd ?

33. 22 m 48 m ?

34. 34 mi 70 mi ?

35. ? 108 mm 45 mm

36. ? 27 km 36 km

Sketch each right triangle and find the length of the side not given. If necessary, approximate the length to the nearest thousandth. (Each length is in units.) See Examples 9 through 12.

37. leg = 3, leg = 4

38. leg = 9, leg = 12

39. leg = 5, hypotenuse = 13

40. leg = 6, hypotenuse = 10

41. leg = 10, leg = 14

42. leg = 2, leg = 16

43. leg = 35, leg = 28

44. leg = 30, leg = 15

45. leg = 30, leg = 30

46. leg = 21, leg = 21

47. hypotenuse = 2, leg = 1

48. hypotenuse = 9, leg = 8

49. leg = 7.5, leg = 4

50. leg = 12, leg = 22.5

Solve. See Example 12.

51. A standard city block is a square with each side measuring 100 yards. Find the length of the diagonal of a city block to the nearest hundredth yard.

52. A section of land is a square with each side measuring 1 mile. Find the length of the diagonal of the section of land to the nearest thousandth mile.

53. Find the height of the tree. Round the height to one decimal place.

? 32 feet 20 feet

54. Find the height of the antenna. Round the height to one decimal place.

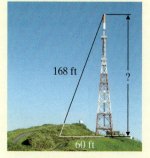

168 ft ? 60 ft

55. The playing field for football is a rectangle that is 300 feet long by 160 feet wide. Find the length of a straight-line run that started at one corner and went diagonally to end at the opposite corner. Round to the nearest foot, if necessary.

56. A soccer field is in the shape of a rectangle and its dimensions depend on the age of the players. The dimensions of the soccer field below are the minimum dimensions for international play. Find the length of the diagonal of this rectangle. Round the answer to the nearest tenth of a yard.

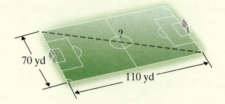

Review

Write each fraction in simplest form. See Section 4.2.

57. $\dfrac{10}{12}$

58. $\dfrac{10}{15}$

59. $\dfrac{2x}{60}$

60. $\dfrac{35}{75y}$

Perform the indicated operations. See Sections 4.3 and 4.4.

61. $\dfrac{9}{13y} + \dfrac{12}{13y}$

62. $\dfrac{3x}{9} - \dfrac{5}{9}$

63. $\dfrac{9}{8} \cdot \dfrac{x}{8}$

64. $\dfrac{7x}{11} \div \dfrac{8x}{11}$

Concept Extensions

*Use the results of Exercises **17–20** and approximate each square root to the nearest whole without using a calculator or table. Then use a calculator or Appendix A.4 to check. See Example 8.*

65. $\sqrt{38}$

66. $\sqrt{27}$

67. $\sqrt{101}$

68. $\sqrt{85}$

69. Without using a calculator, explain how you know that $\sqrt{105}$ is *not* approximately 9.875.

70. Without using a calculator, explain how you know that $\sqrt{27}$ is *not* approximately 3.296.

Does the set form the lengths of the sides of a right triangle? See the Concept Check in this section.

71. 25, 60, 65

72. 20, 45, 50

73. Find the exact length of x. Then give a two-decimal-place approximation.

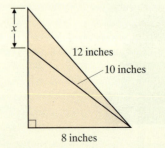

6.5 Congruent and Similar Triangles

Objective A Deciding Whether Two Triangles Are Congruent

Two triangles are **congruent** when they have the same shape and the same size. In congruent triangles, the measures of corresponding angles are equal and the lengths of corresponding sides are equal. The following triangles are congruent:

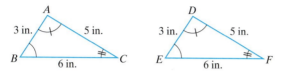

Since these triangles are congruent, the measures of corresponding angles are equal.

Angles with equal measure: ∠A and ∠D, ∠B and ∠E, ∠C and ∠F. Also, the lengths of corresponding sides are equal.

Equal corresponding sides: $\overline{AB}$ and $\overline{DE}$, $\overline{BC}$ and $\overline{EF}$, $\overline{CA}$ and $\overline{FD}$

Any one of the following may be used to determine whether two triangles are congruent:

Objectives

A Decide Whether Two Triangles Are Congruent.

B Find the Ratio of Corresponding Sides in Similar Triangles.

C Find Unknown Lengths of Sides in Similar Triangles.

Congruent Triangles

Angle-Side-Angle (ASA)

If the measures of two angles of a triangle equal the measures of two angles of another triangle, and the lengths of the sides between each pair of angles are equal, the triangles are congruent.

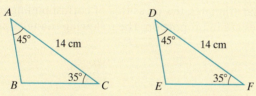

For example, these two triangles are congruent by Angle-Side-Angle.

Side-Side-Side (SSS)

If the lengths of the three sides of a triangle equal the lengths of the corresponding sides of another triangle, the triangles are congruent.

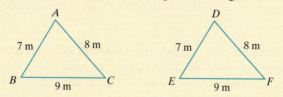

For example, these two triangles are congruent by Side-Side-Side.

Side-Angle-Side (SAS)

If the lengths of two sides of a triangle equal the lengths of corresponding sides of another triangle, and the measures of the angles between each pair of sides are equal, the triangles are congruent.

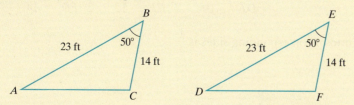

For example, these two triangles are congruent by Side-Angle-Side.

447

Practice 1

a. Determine whether triangle *MNO* is congruent to triangle *RQS*.

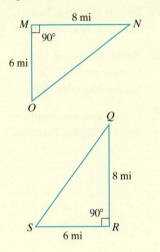

b. Determine whether triangle *GHI* is congruent to triangle *JKL*.

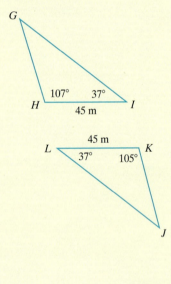

Example 1   Determine whether triangle *ABC* is congruent to triangle *DEF*.

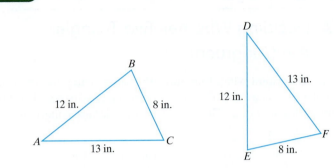

Solution: Since the lengths of all three sides of triangle *ABC* equal the lengths of all three sides of triangle *DEF*, the triangles are congruent.

▶ **Work Practice 1**

In Example 1, notice that as soon as we know that the two triangles are congruent, we know that all three corresponding angles are congruent.

Objective B Finding the Ratios of Corresponding Sides in Similar Triangles ▶

Two triangles are **similar** when they have the same shape but not necessarily the same size. In similar triangles, the measures of corresponding angles are equal and corresponding sides are in proportion. The following triangles are similar:

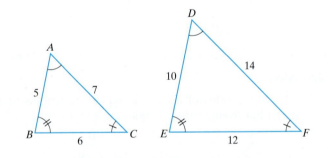

Since these triangles are similar, the measures of corresponding angles are equal. (Note: The triangles above are not drawn to scale.)

Angles with equal measure: ∠*A* and ∠*D*, ∠*B* and ∠*E*, ∠*C* and ∠*F*. Also, the lengths of corresponding sides are in proportion.

Sides in proportion: $\dfrac{AB}{DE} = \dfrac{BC}{EF} = \dfrac{CA}{FD}$ or, in this particular case,

$$\frac{AB}{DE} = \frac{5}{10} = \frac{1}{2}, \frac{BC}{EF} = \frac{6}{12} = \frac{1}{2}, \frac{CA}{FD} = \frac{7}{14} = \frac{1}{2}$$

The ratio of corresponding sides is $\dfrac{1}{2}$.

Answers
1. a. congruent **b.** not congruent

Example 2 Find the ratio of corresponding sides for the similar triangles ABC and DEF.

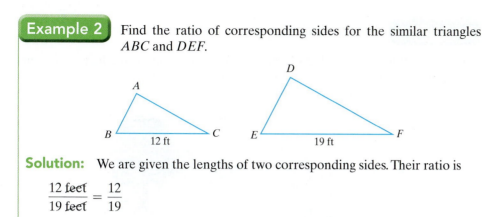

Solution: We are given the lengths of two corresponding sides. Their ratio is

$$\frac{12 \text{ feet}}{19 \text{ feet}} = \frac{12}{19}$$

■ **Work Practice 2**

Objective C Finding Unknown Lengths of Sides in Similar Triangles ▶

Because the ratios of lengths of corresponding sides are equal, we can use proportions to find unknown lengths in similar triangles.

Example 3 Given that the triangles are similar, find the missing length y.

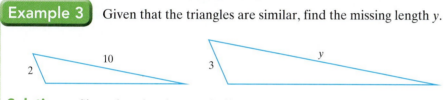

Solution: Since the triangles are similar, corresponding sides are in proportion. Thus, the ratio of 2 to 3 is the same as the ratio of 10 to y, or

$$\frac{2}{3} = \frac{10}{y}$$

To find the unknown length y, we set cross products equal.

$$\frac{2}{3} = \frac{10}{y}$$

$2 \cdot y = 3 \cdot 10$ Set cross products equal.

$2y = 30$ Multiply.

$\dfrac{2y}{2} = \dfrac{30}{2}$ Divide both sides by 2.

$y = 15$ Simplify.

The missing length is 15 units.

■ **Work Practice 3**

✓**Concept Check** The following two triangles are similar. Which vertices of the first triangle appear to correspond to which vertices of the second triangle?

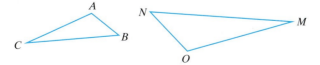

Many applications involve a diagram containing similar triangles. Surveyors, astronomers, and many other professionals continually use similar triangles in their work.

Practice 2
Find the ratio of corresponding sides for the similar triangles QRS and XYZ.

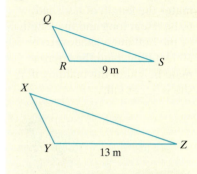

Practice 3
Given that the triangles are similar, find the missing length x.

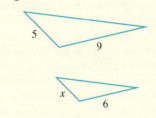

Answers

2. $\dfrac{9}{13}$ **3.** $x = \dfrac{10}{3}$ or $3\dfrac{1}{3}$ units

✓**Concept Check Answer**

A corresponds to O; B corresponds to N; C corresponds to M

Practice 4

Tammy Shultz, a firefighter, needs to estimate the height of a burning building. She estimates the length of her shadow to be 8 feet long and the length of the building's shadow to be 60 feet long. Find the approximate height of the building if she is 5 feet tall.

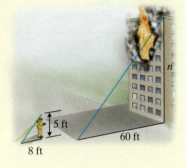

Example 4   Finding the Height of a Tree

Mel Wagstaff is a 6-foot-tall park ranger who needs to know the height of a particular tree. He measures the shadow of the tree to be 69 feet long when his own shadow is 9 feet long. Find the height of the tree.

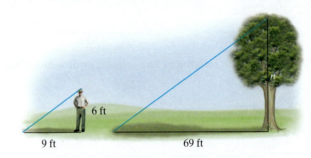

9 ft   69 ft

Solution:

1. **UNDERSTAND.** Read and reread the problem. Notice that the triangle formed by the sun's rays, Mel, and his shadow is similar to the triangle formed by the sun's rays, the tree, and its shadow.

2. **TRANSLATE.** Write a proportion from the similar triangles formed.

$$\text{Mel's height} \rightarrow \frac{6}{n} = \frac{9}{69} \leftarrow \text{length of Mel's shadow} \\ \text{height of tree} \rightarrow \qquad\qquad \leftarrow \text{length of tree's shadow}$$

or $\dfrac{6}{n} = \dfrac{3}{23}$ Simplify $\dfrac{9}{69}$ (ratio in lowest terms).

3. **SOLVE** for n:

$$\frac{6}{n} = \frac{3}{23}$$

$6 \cdot 23 = n \cdot 3$ Set cross products equal.

$138 = 3n$ Multiply.

$\dfrac{138}{3} = \dfrac{3n}{3}$ Divide both sides by 3.

$46 = n$

4. **INTERPRET.** *Check* to see that replacing n with 46 in the proportion makes the proportion true. *State* your conclusion: The height of the tree is 46 feet.

■ **Work Practice 4**

Answer

4. approximately 37.5 ft

Vocabulary, Readiness & Video Check

Answer each question true or false.

1. Two triangles that have the same shape but not necessarily the same size are congruent. _____

2. Two triangles are congruent if they have the same shape and size. _____

3. Congruent triangles are also similar. _____

4. Similar triangles are also congruent. _____

5. For the two similar triangles, the ratio of corresponding sides is $\dfrac{5}{6}$. _____

15 in.

18 in.

5 in.

6 in.

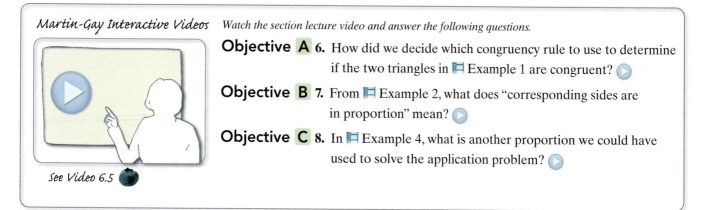

Martin-Gay Interactive Videos   Watch the section lecture video and answer the following questions.

See Video 6.5

Objective A   **6.** How did we decide which congruency rule to use to determine if the two triangles in ▯ Example 1 are congruent? ▶

Objective B   **7.** From ▯ Example 2, what does "corresponding sides are in proportion" mean? ▶

Objective C   **8.** In ▯ Example 4, what is another proportion we could have used to solve the application problem? ▶

6.5   Exercise Set   MyMathLab® ▶

Objective A   *Determine whether each pair of triangles is congruent. If congruent, state the reason why, such as SSS, SAS, or ASA. See Example 1.*

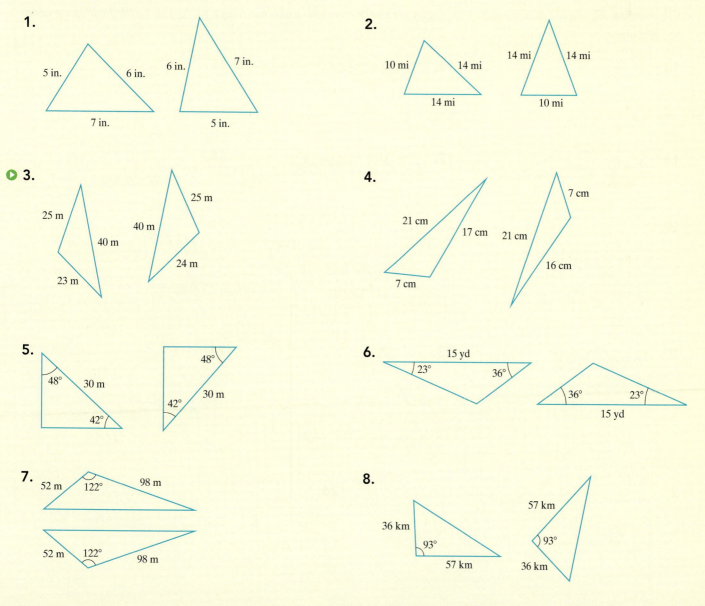

1.

5 in.   6 in.   7 in.

6 in.   7 in.   5 in.

2.

10 mi   14 mi   14 mi

14 mi   14 mi   10 mi

3.

25 m   25 m   40 m   40 m   24 m   23 m

4.

21 cm   17 cm   21 cm   7 cm   16 cm   7 cm

5.

48°   30 m   42°   48°   42°   30 m

6.

15 yd   23°   36°   36°   23°   15 yd

7.

52 m   122°   98 m   52 m   122°   98 m

8.

36 km   93°   57 km   57 km   93°   36 km

Objective **B** *Find each ratio of the corresponding sides of the given similar triangles. See Example 2.*

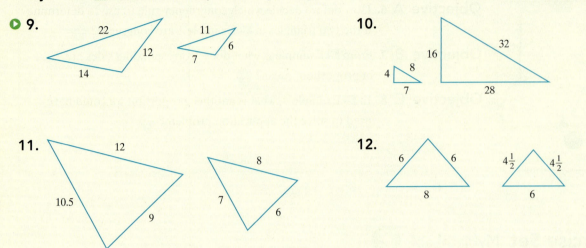

9.

10.

11.

12.

Objective **C** *Given that the pairs of triangles are similar, find the unknown length of the side labeled with a variable. See Example 3.*

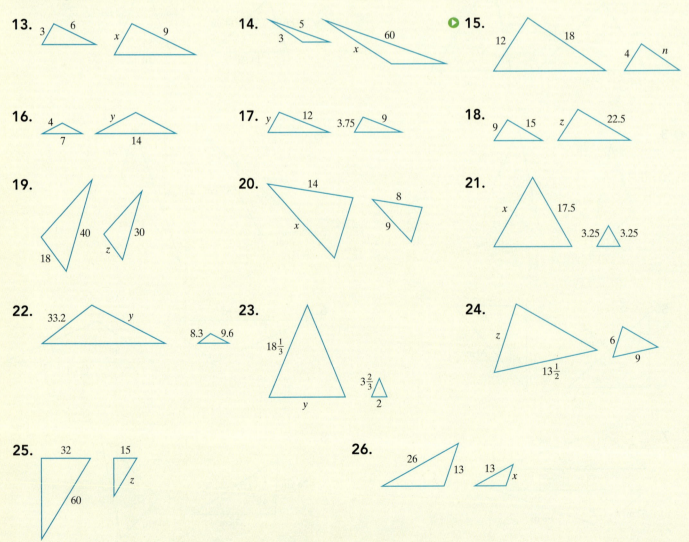

13.

14.

15.

16.

17.

18.

19.

20.

21.

22.

23.

24.

25.

26.

27.

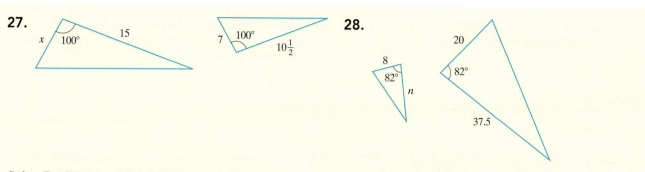

28.

Solve. *For Exercises 29 and 30, the solutions have been started for you. See Example 4.*

29. Given the following diagram, approximate the height of the observation deck in the Seattle Space Needle in Seattle, Washington. (*Source:* Seattle Space Needle)

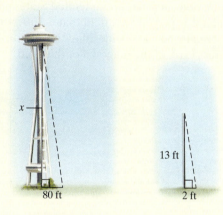

13 ft

80 ft   2 ft

Start the solution:

1. UNDERSTAND the problem. Reread it as many times as needed.

2. TRANSLATE into a proportion using the similar triangles formed. (Fill in the blanks.)

height of
observation deck → $\dfrac{x}{13} = \dfrac{}{}$ ← length of Space Needle shadow
height of pole → ← length of pole shadow

3. SOLVE by setting cross products equal.
4. INTERPRET.

30. Fountain Hills, Arizona, boasts the tallest fountain in the world. The fountain sits in a 28-acre lake and shoots up a column of water every hour. Based on the diagram below, what is the approximate height of the fountain?

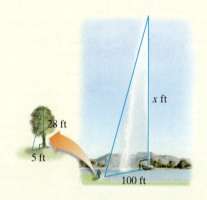

x ft

28 ft

5 ft

100 ft

Start the solution:

1. UNDERSTAND the problem. Reread it as many times as needed.

2. TRANSLATE into a proportion using the similar triangles formed. (Fill in the blanks.)

height of tree → $\dfrac{28}{x} = \dfrac{}{}$ ← length of tree shadow
height of fountain → ← length of fountain shadow

3. SOLVE by setting cross products equal.
4. INTERPRET.

31. Given the following diagram, approximate the height of the Bank One Tower in Oklahoma City, Oklahoma. (*Source: The World Almanac*)

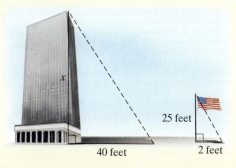

25 feet

40 feet   2 feet

32. The tallest tree currently growing is Hyperion, a redwood located in the Redwood National Park in California. Given the following diagram, approximate its height. (*Source: Guinness World Records*) (*Note:* The tree's current recorded height is 379.1 ft.)

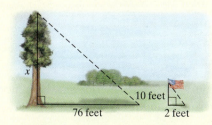

x

10 feet

76 feet   2 feet

33. If a 30-foot tree casts an 18-foot shadow, find the length of the shadow cast by a 24-foot tree.

34. If a 24-foot flagpole casts a 32-foot shadow, find the length of the shadow cast by a 44-foot antenna. Round to the nearest tenth.

Review

Solve. See Section 6.3.

35. For the health of his fish, Pete's Sea World uses the standard that a 20-gallon tank should house only 19 neon tetras. Find the number of neon tetras that Pete would place into a 55-gallon tank.

36. A local package express deliveryman is traveling the city expressway at 45 mph when he is forced to slow down due to traffic ahead. His truck slows at the rate of 3 mph every 5 seconds. Find his speed 8 seconds after braking.

Solve. See Section 6.4.

37. Launch Umbilical Tower 1 is the name of the gantry used for the *Apollo* launch that took Neil Armstrong and Buzz Aldrin to the moon. Find the height of the gantry to the nearest whole foot.

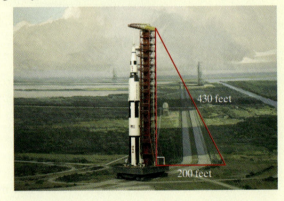

430 feet

200 feet

38. Arena polo, popular in the United States and England, is played on a field that is 100 yards long and usually 50 yards wide. Find the length, to the nearest yard, of the diagonal of this field.

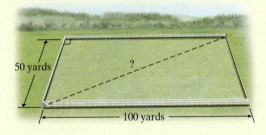

50 yards

?

100 yards

Perform the indicated operation. See Sections 5.2 through 5.4.

39. $3.6 + 0.41$ **40.** $0.41 - 3.6$ **41.** $(0.41)(-3)$ **42.** $-0.48 \div 3$

Concept Extensions

43. The print area on a particular page measures 7 inches by 9 inches. A printing shop is to copy the page and reduce the print area so that its length is 5 inches. What will its width be? Will the print now fit on a 3-by-5-inch index card?

44. The art sample for a banner measures $\frac{1}{3}$ foot in width by $1\frac{1}{2}$ feet in length. If the completed banner is to have a length of 9 feet, find its width.

Given that the pairs of triangles are similar, find the length of the side labeled n. Round your results to 1 decimal place.

45.

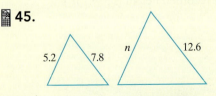

5.2   7.8 n 12.6

46.

11.6 n 20.8   58.7

47. In your own words, describe any differences in similar triangles and congruent triangles.

48. Describe a situation where similar triangles would be useful for a contractor building a house.

49. A triangular park is planned and waiting to be approved by the city zoning commission. A drawing of the park shows sides of length 5 inches, $7\frac{1}{2}$ inches, and $10\frac{5}{8}$ inches. If the scale on the drawing is $\frac{1}{4}$ in. = 10 ft, find the actual proposed dimensions of the park.

50. John and Robyn Costello draw a triangular deck on their house plans. Robyn measures sides of the deck drawing on the plans to be 3 inches, $4\frac{1}{2}$ inches, and 6 inches. If the scale on the drawing is $\frac{1}{4}$ in. = 1 foot, find the lengths of the sides of the deck they want built.

Chapter 6 Group Activity

Investigating Scale Drawings

Sections 6.1, 6.2, and 6.3

Materials:

- ruler
- tape measure
- grid paper (optional)

This activity may be completed by working in groups or individually.

Scale drawings are used by architects, engineers, interior designers, ship builders, and others. In a scale drawing, each unit measurement on the drawing represents a fixed length on the object being drawn. For instance, in an architect's scale drawing, 1 inch on the drawing may represent 10 feet on a building. The scale describes the relationship between the measurements. If the measurements have the same units, the scale can be expressed as a ratio. In this case, the ratio would be 1 : 120, representing 1 inch to 120 inches (or 10 feet).

Use a ruler and the scale drawing of a college building below to answer the following questions.

1. How wide is each of the front doors of the college building?

2. How long is the front of the college building?

3. How tall is the front of the college building?

Now you will draw your own scale floor plan. First choose a room to draw—it can be your math classroom, your living room, your dormitory room, or any room that can be easily measured. Start by using a tape measure to measure the distances around the base of the walls in the room you are drawing.

4. Choose a scale for your floor plan.

5. Convert each measurement in the room you are drawing to the corresponding lengths needed for the scale drawing.

6. Complete your floor plan (you may find it helpful to use grid paper). Mark the locations of doors and windows on your floor plan. Be sure to indicate on the drawing the scale used in your floor plan.

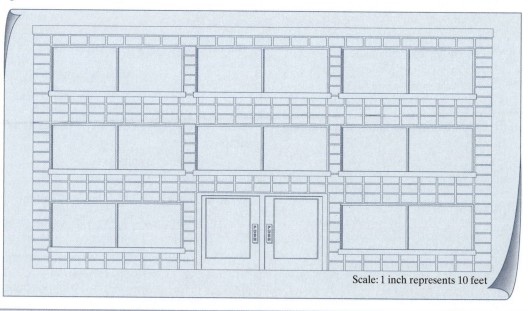

Scale: 1 inch represents 10 feet

Chapter 6 Vocabulary Check

Fill in each blank with one of the words or phrases listed below. Some choices may be used more than once.

not equal	equal	cross products	rate	right	unit rate	congruent
ratio	unit price	proportion	hypotenuse	Pythagorean	similar	leg

1. A(n) _____ is the quotient of two numbers. It can be written as a fraction, using a colon, or using the word *to*.

2. $\dfrac{x}{2} = \dfrac{7}{16}$ is an example of a(n) _____.

3. A(n) _____ is a rate with a denominator of 1.

4. A(n) _____ is a "money per item" unit rate.

5. A(n) _____ is used to compare different kinds of quantities.

6. In the proportion $\dfrac{x}{2} = \dfrac{7}{16}$, $x \cdot 16$ and $2 \cdot 7$ are called _____.

7. If cross products are _____, the proportion is true.

8. If cross products are _____, the proportion is false.

9. _____ triangles have the same shape and the same size.

10. _____ triangles have exactly the same shape but not necessarily the same size.

11–13. Label the sides of the right triangle.   **11.** _____   **13.** _____

                 12. _____

14. A triangle with one right angle is called a(n) _____ triangle.

15. In the right triangle $\begin{array}{c} a \quad c \\ \boxed{} \diagdown \\ b \end{array}$, $a^2 + b^2 = c^2$ is called the _____ theorem.

> **Helpful Hint**
> ▶ Are you preparing for your test? Don't forget to take the Chapter 6 Test on page 463. Then check your answers at the back of the text and use the Chapter Test Prep Videos to see the fully worked-out solutions to any of the exercises you want to review.

6   Chapter Highlights

Definitions and Concepts	Examples
Section 6.1 Ratios and Rates	
A **ratio** is the quotient of two quantities.	The ratio of 3 to 4 can be written as $$\dfrac{3}{4} \qquad \text{or} \qquad 3:4$$ ↑ fraction notation ↑ colon notation
Rates are used to compare different kinds of quantities.	Write the rate 12 spikes every 8 inches as a fraction in simplest form. $$\dfrac{12 \text{ spikes}}{8 \text{ inches}} = \dfrac{3 \text{ spikes}}{2 \text{ inches}}$$
A **unit rate** is a rate with a denominator of 1.	Write as a unit rate: 117 miles on 5 gallons of gas $$\dfrac{117 \text{ miles}}{5 \text{ gallons}} = \dfrac{23.4 \text{ miles}}{1 \text{ gallon}} \quad \text{or 23.4 miles per gallon}$$ or 23.4 miles/gallon
A **unit price** is a "money per item" unit rate.	Write as a unit price: $5.88 for 42 ounces of detergent $$\dfrac{\$5.88}{42 \text{ ounces}} = \dfrac{\$0.14}{1 \text{ ounce}} = \$0.14 \text{ per ounce}$$

Definitions and Concepts	Examples

Section 6.2   Proportions

A **proportion** is a statement that two ratios or rates are equal.

Using Cross Products to Determine Whether Proportions Are True or False

Cross products

$a \cdot d$ $b \cdot c$

$$\frac{a}{b} = \frac{c}{d}$$

If cross products are equal, the proportion is true.
If $ad = bc$, then the proportion is true.
If cross products are not equal, the proportion is false.
If $ad \neq bc$, then the proportion is false.

To find an unknown value x in a proportion, we set the cross products equal to each other and solve the resulting equation.

$\dfrac{1}{2} = \dfrac{4}{8}$ is a proportion.

Is $\dfrac{6}{10} = \dfrac{9}{15}$ a true proportion?

Cross products

$6 \cdot 15$ $10 \cdot 9$

$$\frac{6}{10} = \frac{9}{15}$$

$6 \cdot 15 \overset{?}{=} 10 \cdot 9$ Are cross products equal?

$90 = 90$

Since cross products are equal, the proportion is a true proportion.

Find x: $\dfrac{x}{7} = \dfrac{5}{8}$

$$\frac{x}{7} = \frac{5}{8}$$

$x \cdot 8 = 7 \cdot 5$ Set the cross products equal to each other.

$8x = 35$ Multiply.

$\dfrac{8x}{8} = \dfrac{35}{8}$ Divide both sides by 8.

$x = 4\dfrac{3}{8}$

Section 6.3   Proportions and Problem Solving

Given a specified ratio (or rate) of two quantities, a proportion can be used to determine an unknown quantity.

On a map, 50 miles corresponds to 3 inches. How many miles correspond to 10 inches?

1. **UNDERSTAND.** Read and reread the problem.

2. **TRANSLATE.** We let x represent the unknown number. We are given that 50 miles is to 3 inches as x miles is to 10 inches.

miles $\rightarrow$ $\dfrac{50}{3} = \dfrac{x}{10}$ $\leftarrow$ miles
inches $\rightarrow$ $\phantom{\dfrac{50}{3}}$ $\phantom{\dfrac{x}{10}}$ $\leftarrow$ inches

3. **SOLVE:**

$$\frac{50}{3} = \frac{x}{10}$$

$50 \cdot 10 = 3 \cdot x$ Set the cross products equal to each other.

$500 = 3x$ Multiply.

$\dfrac{500}{3} = \dfrac{3x}{3}$ Divide both sides by 3.

$x = 166\dfrac{2}{3}$

4. **INTERPRET.** *Check* your work. *State* your conclusion:

On the map, $166\dfrac{2}{3}$ miles corresponds to 10 inches.

Definitions and Concepts	Examples

Section 6.4   Square Roots and the Pythagorean Theorem

Square Root of a Number

The square root of a positive number a is the positive number b whose square is a. In symbols,

$$\sqrt{a} = b, \quad \text{if} \quad b^2 = a$$

Also, $\sqrt{0} = 0$.

$$\sqrt{9} = 3, \qquad \sqrt{100} = 10, \qquad \sqrt{1} = 1$$

Pythagorean Theorem

If a and b are the lengths of the legs of a right triangle and c is the length of the hypotenuse, then

$$a^2 + b^2 = c^2$$

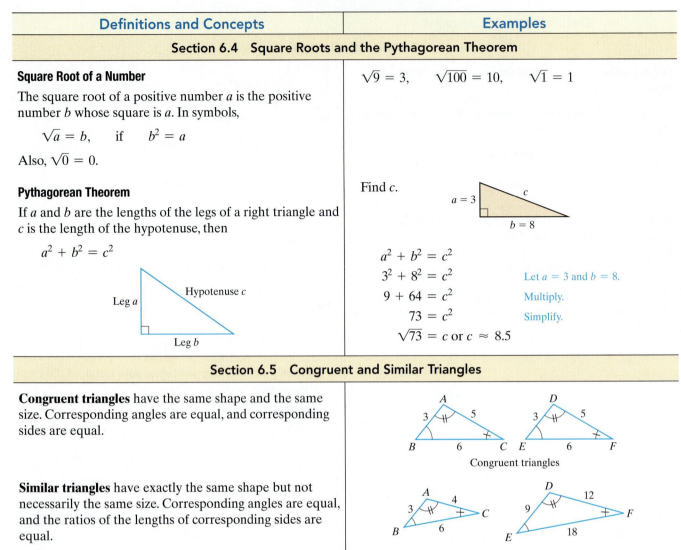

Find c.

$$a^2 + b^2 = c^2$$
$$3^2 + 8^2 = c^2 \qquad \text{Let } a = 3 \text{ and } b = 8.$$
$$9 + 64 = c^2 \qquad \text{Multiply.}$$
$$73 = c^2 \qquad \text{Simplify.}$$
$$\sqrt{73} = c \text{ or } c \approx 8.5$$

Section 6.5   Congruent and Similar Triangles

Congruent triangles have the same shape and the same size. Corresponding angles are equal, and corresponding sides are equal.

Congruent triangles

Similar triangles have exactly the same shape but not necessarily the same size. Corresponding angles are equal, and the ratios of the lengths of corresponding sides are equal.

Similar triangles

$$\frac{AB}{DE} = \frac{3}{9} = \frac{1}{3}, \frac{BC}{EF} = \frac{6}{18} = \frac{1}{3},$$

$$\frac{CA}{FD} = \frac{4}{12} = \frac{1}{3}$$

(6.1) *Write each ratio as a fraction in simplest form.*

1. 23 to 37

2. $121 to $143

3. 4.25 yards to 8.75 yards

4. $2\frac{1}{4}$ to $4\frac{3}{8}$

Use the garden shown for Exercises 5 and 6. Write each ratio as a fraction in simplest form.

4.5 meters   2 meters

5. Find the ratio of the garden's length (longer side) to the garden's width (shorter side).

△ **6.** Find the ratio of the width to the perimeter of the rectangular garden.

Write each rate as a fraction in simplest form.

7. 6000 people to 2400 pets

8. 15 pages printed in 6 minutes

Write each rate as a unit rate.

9. 468 miles in 9 hours

10. 180 feet in 12 seconds

11. $6.96 for 4 diskettes

12. 104 bushels of fruit from 8 trees

Find each unit price and decide which is the better buy. Round to 3 decimal places. Assume that we are comparing different sizes of the same brand.

13. Taco sauce: 8 ounces for $0.99 or 12 ounces for $1.69

14. Peanut butter: 18 ounces for $1.49 or 28 ounces for $2.39

(6.2) Translating *Write each sentence as a proportion.*

15. 24 uniforms is to 8 players as 3 uniforms is to 1 player.

16. 12 tires is to 3 cars as 4 tires is to 1 car.

Determine whether each proportion is true.

17. $\dfrac{19}{8} = \dfrac{14}{6}$

18. $\dfrac{3.75}{3} = \dfrac{7.5}{6}$

Find the unknown number x in each proportion.

19. $\dfrac{x}{3} = \dfrac{30}{18}$

20. $\dfrac{x}{9} = \dfrac{7}{3}$

21. $\dfrac{-8}{5} = \dfrac{9}{x}$

22. $\dfrac{-27}{\frac{9}{4}} = \dfrac{x}{-5}$

23. $\dfrac{0.4}{x} = \dfrac{2}{4.7}$

24. $\dfrac{x}{4\frac{1}{2}} = \dfrac{2\frac{1}{10}}{8\frac{2}{5}}$

25. $\dfrac{x}{0.4} = \dfrac{4.7}{3}$

Round to the nearest hundredth.

26. $\dfrac{0.07}{0.3} = \dfrac{7.2}{n}$

Round to the nearest tenth.

(6.3) *Solve.*

The ratio of a quarterback's completed passes to attempted passes is 3 to 7.

27. If he attempted 32 passes, find how many passes he completed. Round to the nearest whole pass.

28. If he completed 15 passes, find how many passes he attempted.

One bag of pesticide covers 4000 square feet of garden.

△ **29.** Find how many bags of pesticide should be purchased to cover a rectangular garden that is 180 feet by 175 feet.

△ **30.** Find how many bags of pesticide should be purchased to cover a square garden that is 250 feet on each side.

On a road map of Texas, 0.75 inch represents 80 miles.

31. Find the distance from Houston to Corpus Christi if the distance on the map is about 2 inches.

32. The distance from El Paso to Dallas is 1025 miles. Find the distance between these cities on the map. Round to the nearest tenth of an inch.

(6.4) *Find each square root. If necessary, round the square root to the nearest thousandth.*

33. $\sqrt{64}$

34. $\sqrt{144}$

35. $\sqrt{12}$

36. $\sqrt{15}$

37. $\sqrt{0}$

38. $\sqrt{1}$

39. $\sqrt{50}$

40. $\sqrt{65}$

41. $\sqrt{\dfrac{4}{25}}$

42. $\sqrt{\dfrac{1}{100}}$

Find the unknown length in each given right triangle. If necessary, round to the nearest tenth.

△ **43.** leg = 12, leg = 5

△ **44.** leg = 20, leg = 21

△ **45.** leg = 9, hypotenuse = 14

△ **46.** leg = 66, hypotenuse = 86

△**47.** Find the length, to the nearest hundredth, of the diagonal of a square that has a side of length 20 centimeters.

△ **48.** Find the height of the building rounded to the nearest tenth.

126 ft

← 90 ft →

(6.5) *Determine whether each pair of triangles is congruent. If congruent, state the reason why, such as SSS, SAS, or ASA.*

49.

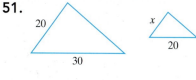

60°   1.7 cm
1.7 cm   60°
85°
85°

50.

12°   14°
154°

12°   14°
154°

Given that the pairs of triangles are similar, find the unknown length x.

51.

20

x

20

30

52.

x   24

5.8   8

Solve.

53. A housepainter needs to estimate the height of a condominium. He estimates the length of his shadow to be 7 feet long and the length of the building's shadow to be 42 feet long. Find the approximate height of the building if the housepainter is $5\frac{1}{2}$ feet tall.

54. A design company is making a triangular sail for a model sailboat. The model sail is to be the same shape as a life-size sailboat's sail. Use the following diagram to find the unknown lengths x and y.

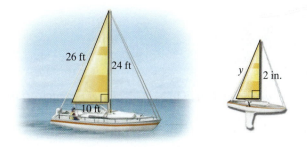

26 ft   24 ft
10 ft

y   2 in.
x

Mixed Review

Write each ratio as a fraction in simplest form.

55. 15 to 25

56. 3 pints to 81 pints

Write each rate as a fraction in simplest form.

57. 2 teachers for 18 students

58. 6 nurses for 24 patients

Write each rate as a unit rate.

59. 136 miles in 4 hours

60. 12 gallons of milk from 6 cows

For Exercises 61 and 62, write the ratio or rate in simplest form.

61. During the 2012 Summer Olympics in London, a total of 962 medals were awarded. Athletes from the Netherlands won a total of 20 medals. Find the ratio of medals won by the Netherlands to the total medals awarded. (*Source:* International Olympic Committee)

62. Paul Crake of Canberra, Australia, holds the record for the Empire State Building Run Up in New York. He ran up 1576 steps in about 9.5 minutes. Find the number of steps per minute. Round to the nearest step. (*Source: Guinness Book of World Records*)

Find each unit price and decide which is the better buy. Round to 3 decimal places. Assume that we are comparing different sizes of the same brand.

63. Cold medicine:
$4.94 for 4 oz.
$9.98 for 8 oz.

64. Juice:
12 oz for $0.65
64 oz for $2.98

Translating *Write each sentence as a proportion.*

65. 2 cups of cookie dough is to 30 cookies as 4 cups of cookie dough is to 60 cookies.

66. 5 nickels is to 3 dollars as 20 nickels is to 12 dollars.

Find the unknown number x in each proportion.

67. $\dfrac{3}{x} = \dfrac{15}{8}$

68. $\dfrac{5}{4} = \dfrac{x}{20}$

69. $\dfrac{x}{3} = \dfrac{7.5}{6}$

70. $\dfrac{\frac{1}{3}}{25} = \dfrac{x}{30}$

Find each square root. If necessary, approximate and round to the nearest thousandth.

71. $\sqrt{36}$

72. $\sqrt{\dfrac{16}{81}}$

73. $\sqrt{105}$

74. $\sqrt{32}$

Find the unknown length in each given right triangle. If necessary, round to the nearest tenth.

75. leg = 66, leg = 56

76. leg = 12, hypotenuse = 24

Given that the pairs of triangles are similar, find the unknown length n.

77.

78.

Write each ratio or rate as a fraction in simplest form.

1. 4500 trees to 6500 trees

2. 9 inches of rain in 30 days

3. 8.6 to 10

4. $5\frac{7}{8}$ to $9\frac{3}{4}$

5. The world's largest yacht, the Azzam, measures 590 feet. A Boeing 787-8 Dreamliner measures 186 feet long. Find the ratio of the Azzam to the length of a 787-8. (*Source:* CNN)

186 ft

590 ft

Find each unit rate.

6. 650 kilometers in 8 hours

7. 140 students for 5 teachers

8. The Sojourner is a 6-wheeled vehicle that was used in the exploration of Mars, but was remotely controlled from Earth. Each wheel was designed to move independently, allowing the capability of traversing various obstacles and the ability to be turned around in place. This little vehicle was capable of traveling about 960 inches each 60 minutes. (*Source:* NASA)

Find each unit price and decide which is the better buy. Round to three decimal places.

9. Steak sauce:
8 ounces for $1.19
12 ounces for $1.89

10. Jelly:
16 ounces for $1.49
24 ounces for $2.39

Determine whether the proportion is true.

11. $\dfrac{28}{16} = \dfrac{14}{8}$

12. $\dfrac{3.6}{2.2} = \dfrac{1.9}{1.2}$

Answers

1. _____

2. _____

3. _____

4. _____

5. _____

6. _____

7. _____

8. _____

9. _____

10. _____

11. _____

12. _____

13. _____

14. _____

15. _____

16. _____

17. _____

18. _____

19. _____

20. _____

21. _____

22. _____

23. _____

24. _____

25. _____

Solve each proportion for the given variable.

13. $\dfrac{n}{3} = \dfrac{15}{9}$

14. $\dfrac{8}{x} = \dfrac{11}{6}$

15. $\dfrac{4}{\frac{3}{7}} = \dfrac{y}{\frac{1}{4}}$

16. $\dfrac{1.5}{5} = \dfrac{2.4}{n}$

Solve.

17. On an architect's drawing, 2 inches corresponds to 9 feet. Find the length of a home represented by a line that is 11 inches long.

18. If a car can be driven 80 miles in 3 hours, how long will it take to travel 100 miles?

19. The standard dose of medicine for a dog is 10 grams for every 15 pounds of body weight. What is the standard dose for a dog that weighs 80 pounds?

Find each square root and simplify. Round to the nearest thousandth if necessary.

20. $\sqrt{49}$

21. $\sqrt{157}$

22. $\sqrt{\dfrac{64}{100}}$

Solve.

△**23.** Approximate, to the nearest hundredth of a centimeter, the unknown length of the side of a right triangle with legs of 4 centimeters each.

△**24.** Given that the following triangles are similar, find the unknown length n.

△**25.** A surveyor needs to estimate the height of a tower. She estimates the length of her shadow to be 4 feet long and the length of the tower's shadow to be 48 feet long. Find the height of the tower if she is $5\dfrac{3}{4}$ feet tall.

Answers

1. Subtract. Check each answer by adding.
   a. $12 - 9$
   b. $22 - 7$
   c. $35 - 35$
   d. $70 - 0$

2. Multiply
   a. $20 \cdot 0$
   b. $20 \cdot 1$
   c. $0 \cdot 20$
   d. $1 \cdot 20$

3. Round 248,982 to the nearest hundred.

4. Round 248,982 to the nearest thousand.

Perform the indicated operations.

5. a. $\begin{array}{r} 25 \\ \times\ 8 \\ \hline \end{array}$ b. $\begin{array}{r} 246 \\ \times\ \ 5 \\ \hline \end{array}$

6. $10{,}468 \div 28$

7. $1 + (-10) + (-8) + 9$

8. $-12(7)$

9. Write the prime factorization of 80.

10. Find $3^2 - 1^2$.

11. Write $\dfrac{12}{20}$ in simplest form.

12. Find $9^2 \cdot 3$.

Multiply.

13. $\left(-\dfrac{6}{13}\right)\left(-\dfrac{26}{30}\right)$

14. $3\dfrac{3}{8} \cdot 4\dfrac{5}{9}$

Perform the indicated operation and simplify.

15. $\dfrac{7}{8} + \dfrac{6}{8} + \dfrac{3}{8}$

16. $\dfrac{7}{10} - \dfrac{3}{10} + \dfrac{4}{10}$

17. Find the LCD of $\dfrac{3}{7}$ and $\dfrac{5}{14}$.

18. Add: $\dfrac{17}{25} + \dfrac{3}{10}$

19. Write an equivalent fraction with the indicated denominator. $\dfrac{3}{4} = \dfrac{}{20}$

20. Determine whether these fractions are equivalent.
    $\dfrac{10}{55}, \dfrac{6}{33}$

Answers
1. a.
b.
c.
d.
2. a.
b.
c.
d.
3.
4.
5. a.
b.
6.
7.
8.
9.
10.
11.
12.
13.
14.
15.
16.
17.
18.
19.
20.

465

21. _____

22. _____

23. _____

24. _____

25. _____

26. _____

27. _____

28. _____

29. _____

30. _____

31. _____

32. _____

33. _____

34. _____

35. _____

36. _____

37. _____

38. _____

39. _____

40. _____

41. _____

42. _____

43. _____

44. _____

45. _____

46. _____

47. _____

48. _____

49. _____

50. _____

21. Subtract: $\dfrac{2}{3} - \dfrac{10}{11}$

22. Subtract: $17\dfrac{5}{24} - 9\dfrac{5}{9}$

23. A flight from Tucson to Phoenix, Arizona, requires $\dfrac{5}{12}$ of an hour. If the plane has been flying $\dfrac{1}{4}$ of an hour, find how much time remains before landing.

24. Simplify: $80 \div 8 \cdot 2 + 7$

25. Add: $2\dfrac{1}{3} + 5\dfrac{3}{8}$

26. Find the average of $\dfrac{3}{5}, \dfrac{4}{9}$, and $\dfrac{11}{15}$.

27. Insert $<$ or $>$ to form a true statement.

$$\dfrac{3}{4} \quad \dfrac{9}{11}$$

Solve.

28. $5y - 8y = 24$

29. $y - 5 = -2 - 6$

30. $3y - 6 = 7y - 6$

31. $3a - 6 = a + 4$

32. $4(y + 1) - 3 = 21$

33. $3(2x - 6) + 6 = 0$

34. Write "seventy-five thousandths" in standard form.

35. Round 736.2359 to the nearest tenth.

36. Round 736.2359 to the nearest thousandth.

37. Add: $23.85 + 1.604$

38. Subtract: $700 - 18.76$

39. Multiply: 0.0531×16

40. Write $\dfrac{3}{8}$ as a decimal.

41. Divide: $-5.98 \div 115$

42. Write 7.9 as an improper fraction.

43. Simplify: $-0.5(8.6 - 1.2)$

44. Find the unknown number n.

$$\dfrac{n}{4} = \dfrac{12}{16}$$

45. Write the numbers in order from smallest to largest.

$$\dfrac{9}{20}, \dfrac{4}{9}, 0.456$$

46. Write the rate as a unit rate.
700 meters in 5 seconds

Write each ratio as a fraction in simplest form.

47. The ratio of $15 to $10

48. The ratio of 7 to 21

49. The ratio of 2.5 to 3.15

50. The ratio of 900 to 9000

Percent

Watching movies at a cinema is still a popular pastime in the United States, with 68% (68 percent) of us attending at least once in the year 2012. The bar graph below tells us much, especially about the growth in the number of digital cinema screens and the decline in the number of analog screens. In this chapter, we calculate the percent of different types of screens in use. Notice that by studying these bars, we can estimate whether our answers are reasonable.

In Section 7.4, Exercises 7, 8, 71, and 72 have to do with the numbers of the different types of cinema screens.

This chapter is devoted to percent, a concept used virtually every day in ordinary and business life. Understanding percent and using it efficiently depend on understanding ratios, because a percent is a ratio whose denominator is 100. We present techniques to write percents as fractions and as decimals. We then solve problems relating to interest rates, sales tax, discounts, and other real-life situations by writing percent equations.

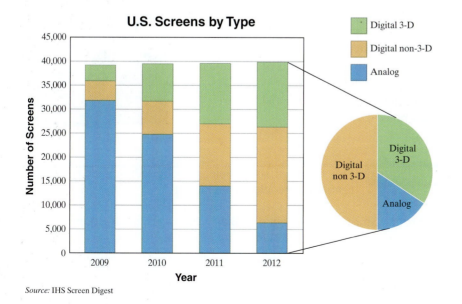

Source: IHS Screen Digest

Objectives

A Understand Percent.

B Write Percents as Decimals or Fractions.

C Write Decimals or Fractions as Percents.

D Solve Applications with Percents, Decimals, and Fractions.

Objective A Understanding Percent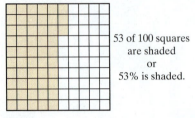

The word **percent** comes from the Latin phrase *per centum*, which means **"per 100."** For example, 53% (53 percent) means 53 per 100. In the square below, 53 of the 100 squares are shaded. Thus, 53% of the figure is shaded.

53 of 100 squares are shaded
or
53% is shaded.

Since 53% means 53 per 100, 53% is the ratio of 53 to 100, or $\frac{53}{100}$.

$$53\% = \frac{53}{100}$$

Also,

$$7\% = \frac{7}{100} \quad \text{7 parts per 100 parts}$$

$$73\% = \frac{73}{100} \quad \text{73 parts per 100 parts}$$

$$109\% = \frac{109}{100} \quad \text{109 parts per 100 parts}$$

> **Percent**
>
> **Percent** means **per one hundred.** The "%" symbol is used to denote percent.

Percent is used in a variety of everyday situations. For example,

- 84.1% of the U.S. population uses the Internet.
- The store is having a 20%-off sale.
- For the past two years, the enrollment in community colleges increased 15%.
- The South is the home of 49% of all frequent paintball participants.
- 50% of the cinema screens in the U.S. are digital non-3-D.

Practice 1

Of 100 students in a club, 27 are freshmen. What percent of the students are freshmen?

Example 1 In a survey of 100 people, 17 people drive blue cars. What percent of the people drive blue cars?

Solution: Since 17 people out of 100 drive blue cars, the fraction is $\frac{17}{100}$. Then

$$\frac{17}{100} = 17\%$$

■ **Work Practice 1**

Answer

1. 27%

Example 2 45 out of every 100 young college graduates return home to live (at least temporarily). What percent of college graduates is this? (*Source:* Independent Insurance Agents of America)

Solution:

$$\frac{45}{100} = 45\%$$

🔶 **Work Practice 2**

Practice 2

31 out of 100 college instructors are in their forties. What percent of these instructors are in their forties?

Objective B Writing Percents as Decimals or Fractions ▶

Since percent means "per hundred," we have that

$$1\% = \frac{1}{100} = 0.01$$

In other words, the percent symbol means "per hundred" or, equivalently, "$\frac{1}{100}$" or "0.01." Thus

Write 87% as a fraction: $87\% = 87 \times \frac{1}{100} = \frac{87}{100}$

or

Write 87% as a decimal: $87\% = 87 \times (0.01) = 0.87$

> Results are the same.

Of course, we know that the end results are the same; that is,

$$\frac{87}{100} = 0.87$$

The above gives us two options for converting percents. We can replace the percent symbol, %, by $\frac{1}{100}$ or by 0.01 and then multiply.

> For consistency, when we
> - convert from a percent to a *decimal,* we will drop the % symbol and multiply by 0.01.
> - convert from a percent to a *fraction,* we will drop the % symbol and multiply by $\frac{1}{100}$.

Let's practice writing percents as decimals, then writing percents as fractions.

> ### Writing a Percent as a Decimal
>
> Replace the percent symbol with its decimal equivalent, 0.01; then multiply.
>
> $43\% = 43(0.01) = 0.43$

Helpful Hint

If it helps, think of writing a percent as a decimal by

Percent → | Remove the % symbol and move the decimal point 2 places to the left. | → Decimal

Answer

2. 31%

Practice 3–7

Write each percent as a decimal.

3. 49%

4. 3.1%

5. 175%

6. 0.46%

7. 600%

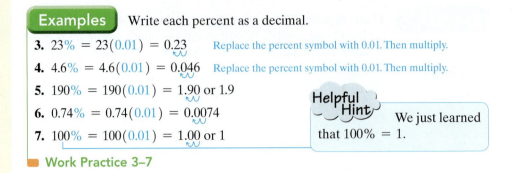

Examples   Write each percent as a decimal.

3. $23\% = 23(0.01) = 0.23$ Replace the percent symbol with 0.01. Then multiply.

4. $4.6\% = 4.6(0.01) = 0.046$ Replace the percent symbol with 0.01. Then multiply.

5. $190\% = 190(0.01) = 1.90$ or 1.9

6. $0.74\% = 0.74(0.01) = 0.0074$

7. $100\% = 100(0.01) = 1.00$ or 1

Helpful Hint   We just learned that $100\% = 1$.

■ **Work Practice 3–7**

✓**Concept Check**   Why is it incorrect to write the percent 0.033% as 3.3 in decimal form?

Now let's write percents as fractions.

> ### Writing a Percent as a Fraction
>
> Replace the percent symbol with its fraction equivalent, $\frac{1}{100}$; then multiply. Don't forget to simplify the fraction if possible.
>
> $$43\% = 43 \cdot \frac{1}{100} = \frac{43}{100}$$

Practice 8–12

Write each percent as a fraction or mixed number in simplest form.

8. 50%

9. 2.3%

10. 150%

11. $66\frac{2}{3}\%$

12. 12%

Examples   Write each percent as a fraction or mixed number in simplest form.

8. $40\% = 40 \cdot \frac{1}{100} = \frac{40}{100} = \frac{2 \cdot \overset{1}{\cancel{20}}}{5 \cdot \cancel{20}} = \frac{2}{5}$

9. $1.9\% = 1.9 \cdot \frac{1}{100} = \frac{1.9}{100}$. We don't want the numerator of the fraction to contain a decimal, so we multiply by 1 in the form of $\frac{10}{10}$.

$$= \frac{1.9}{100} \cdot \frac{10}{10} = \frac{1.9 \cdot 10}{100 \cdot 10} = \frac{19}{1000}$$

10. $125\% = 125 \cdot \frac{1}{100} = \frac{125}{100} = \frac{5 \cdot \overset{1}{\cancel{25}}}{4 \cdot \cancel{25}} = \frac{5}{4}$ or $1\frac{1}{4}$

11. $33\frac{1}{3}\% = 33\frac{1}{3} \cdot \frac{1}{100} = \frac{100}{3} \cdot \frac{1}{100} = \frac{\overset{1}{\cancel{100}} \cdot 1}{3 \cdot \cancel{100}} = \frac{1}{3}$

    └→Write as an improper fraction.

Helpful Hint   Just as in Example 7, we confirm that $100\% = 1$.

12. $100\% = 100 \cdot \frac{1}{100} = \frac{100}{100} = 1$

■ **Work Practice 8–12**

Answers

3. 0.49   **4.** 0.031   **5.** 1.75   **6.** 0.0046

7. 6   **8.** $\frac{1}{2}$ **9.** $\frac{23}{1000}$ **10.** $\frac{3}{2}$ or $1\frac{1}{2}$

11. $\frac{2}{3}$ **12.** $\frac{3}{25}$

✓**Concept Check**

To write a percent as a decimal, the decimal point should be moved two places to the left, not to the right. So the correct answer is 0.00033.

Objective C   Writing Decimals or Fractions as Percents ▶

To write a decimal or fraction as a percent, we use the result of Examples 7 and 12. In these examples, we found that $1 = 100\%$.

Write 0.38 as a percent: $0.38 = 0.38(1) = 0.38(100\%) = 38.\%$

Write $\frac{1}{5}$ as a percent: $\frac{1}{5} = \frac{1}{5}(1) = \frac{1}{5} \cdot 100\% = \frac{100}{5}\% = 20\%$

First, let's practice writing decimals as percents.

Writing a Decimal as a Percent

Multiply by 1 in the form of 100%.

$0.27 = 0.27(100\%) = 27.\%$

Helpful Hint

If it helps, think of writing a decimal as a percent by reversing the steps in the Helpful Hint on page 469.

Percent ← | Move the decimal point 2 places to the right and attach a % symbol. | ← Decimal

---

Examples Write each decimal as a percent.

13. $0.65 = 0.65(100\%) = 65.\%$ or 65% Multiply by 100%.

14. $1.25 = 1.25(100\%) = 125.\%$ or 125%

15. $0.012 = 0.012(100\%) = 001.2\%$ or 1.2% **Helpful Hint**

16. $0.6 = 0.6(100\%) = 060.\%$ or 60% A zero was inserted as a placeholder.

■ **Work Practice 13–16**

✔ **Concept Check** Why is it incorrect to write the decimal 0.0345 as 34.5% in percent form?

Now let's write fractions as percents.

Writing a Fraction as a Percent

Multiply by 1 in the form of 100%.

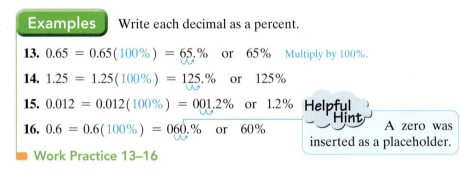

$$\frac{1}{8} = \frac{1}{8} \cdot 100\% = \frac{1}{8} \cdot \frac{100}{1}\% = \frac{100}{8}\% = 12\frac{1}{2}\% \quad \text{or} \quad 12.5\%$$

Helpful Hint

From Examples 7 and 12, we know that

$100\% = 1$

Recall that when we multiply a number by 1, we are not changing the value of that number. This means that when we multiply a number by 100%, we are not changing its value but rather writing the number as an equivalent percent.

Practice 13–16

Write each decimal as a percent.

13. 0.14   **14.** 1.75

15. 0.057   **16.** 0.5

Answers

13. 14%   **14.** 175%   **15.** 5.7%
16. 50%

✔ **Concept Check Answer**

To change a decimal to a percent, multiply by 100%, or move the decimal point *only* two places to the right. So the correct answer is 3.45%.

Practice 17–19

Write each fraction or mixed number as a percent.

17. $\dfrac{3}{25}$ **18.** $\dfrac{9}{40}$ **19.** $5\dfrac{1}{2}$

Examples   Write each fraction or mixed number as a percent.

17. $\dfrac{7}{20} = \dfrac{7}{20} \cdot 100\% = \dfrac{7}{20} \cdot \dfrac{100}{1}\% = \dfrac{700}{20}\% = 35\%$

18. $\dfrac{2}{3} = \dfrac{2}{3} \cdot 100\% = \dfrac{2}{3} \cdot \dfrac{100}{1}\% = \dfrac{200}{3}\% = 66\dfrac{2}{3}\%$

19. $2\dfrac{1}{4} = \dfrac{9}{4} \cdot 100\% = \dfrac{9}{4} \cdot \dfrac{100}{1}\% = \dfrac{900}{4}\% = 225\%$

> **Helpful Hint**
> $\dfrac{200}{3} = 66.\overline{6}.$
> Thus, another way to write $\dfrac{200}{3}\%$ is $66.\overline{6}\%$.

■ **Work Practice 17–19**

✔ **Concept Check** Which digit in the percent 76.4582% represents

a. A tenth percent?   **b.** A thousandth percent?

c. A hundredth percent?   **d.** A ten percent?

Example 20   Write $\dfrac{1}{12}$ as a percent. Round to the nearest hundredth percent.

Solution:

$\dfrac{1}{12} = \dfrac{1}{12} \cdot 100\% = \dfrac{1}{12} \cdot \dfrac{100\%}{1} = \dfrac{100}{12}\% \underset{\text{“approximately”}}{\approx} 8.33\%$

$$
\begin{array}{r}
8.333 \approx 8.33 \\
12\overline{)100.000} \\
-\,96 \\
\hline
4\,0 \\
-\,3\,6 \\
\hline
40 \\
-\,36 \\
\hline
40 \\
-\,36 \\
\hline
4
\end{array}
$$

Thus, $\dfrac{1}{12}$ is approximately 8.33%.

■ **Work Practice 20**

Practice 20

Write $\dfrac{3}{17}$ as a percent. Round to the nearest hundredth percent.

Objective D Solving Applications with Percents, Decimals, and Fractions

Let's summarize what we have learned so far about percents, decimals, and fractions:

> **Summary of Converting Percents, Decimals, and Fractions**
>
> - *To write a percent as a decimal,* replace the % symbol with its decimal equivalent, 0.01; then multiply.
> - *To write a percent as a fraction,* replace the % symbol with its fraction equivalent, $\dfrac{1}{100}$; then multiply.
> - *To write a decimal or fraction as a percent,* multiply by 100%.

Answers

17. 12%   **18.** $22\dfrac{1}{2}\%$ **19.** 550%

20. 17.65%

✔ **Concept Check Answers**

a. 4   **b.** 8   **c.** 5   **d.** 7

If we let x represent a number, below we summarize using symbols.

Write a percent as a decimal:	Write a percent as a fraction:	Write a number as a percent:
$x\% = x(0.01)$	$x\% = x \cdot \dfrac{1}{100}$	$x = x \cdot 100\%$

Example 21 In the last ten years, automobile thefts in the continental United States have decreased 39.8%. Write this percent as a decimal and as a fraction. (*Source:* The American Automobile Manufacturers Association)

Solution:

As a decimal: $39.8\% = 39.8(0.01) = 0.398$

As a fraction: $39.8\% = 39.8 \cdot \dfrac{1}{100} = \dfrac{39.8}{100} = \dfrac{39.8}{100} \cdot \dfrac{10}{10} = \dfrac{398}{1000} = \dfrac{\overset{1}{\cancel{2}} \cdot 199}{\underset{1}{\cancel{2}} \cdot 500} = \dfrac{199}{500}.$

Thus, 39.8% written as a decimal is 0.398, and written as a fraction is $\dfrac{199}{500}$.

■ **Work Practice 21**

Example 22 An advertisement for a stereo system reads "$\frac{1}{4}$ off." What percent off is this?

Solution: Write $\dfrac{1}{4}$ as a percent.

$$\dfrac{1}{4} = \dfrac{1}{4} \cdot 100\% = \dfrac{1}{4} \cdot \dfrac{100\%}{1} = \dfrac{100}{4}\% = 25\%$$

Thus, "$\dfrac{1}{4}$ off" is the same as "25% off."

■ **Work Practice 22**

Note: It is helpful to know a few basic percent conversions. Appendix A.2 contains a handy reference of percent, decimal, and fraction equivalencies.

Also, Appendix A.3 shows how to find common percents of a number.

Practice 21

A family decides to spend no more than 27.5% of its monthly income on rent. Write 27.5% as a decimal and as a fraction.

Practice 22

Provincetown's budget for waste disposal increased by $1\frac{3}{4}$ times over the budget from last year. What percent increase is this?

Answers

21. $0.275, \dfrac{11}{40}$ **22.** 175%

Vocabulary, Readiness & Video Check

Use the choices below to fill in each blank. Some choices may be used more than once.

$\dfrac{1}{100}$ 0.01   100%   percent

1. _____ means "per hundred."

2. _____ = 1.

3. The % symbol is read as _____.

4. To write a decimal or a fraction as a percent, multiply by 1 in the form of _____.

5. To write a percent as a *decimal*, drop the % symbol and multiply by _____.

6. To write a percent as a *fraction*, drop the % symbol and multiply by _____.

Martin-Gay Interactive Videos   Watch the section lecture video and answer the following questions.

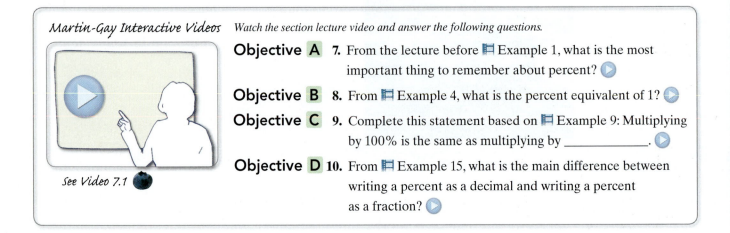

See Video 7.1

Objective A 7. From the lecture before ▣ Example 1, what is the most important thing to remember about percent? ▶

Objective B 8. From ▣ Example 4, what is the percent equivalent of 1? ▶

Objective C 9. Complete this statement based on ▣ Example 9: Multiplying by 100% is the same as multiplying by _____. ▶

Objective D 10. From ▣ Example 15, what is the main difference between writing a percent as a decimal and writing a percent as a fraction? ▶

7.1 **Exercise Set** MyMathLab® ▶

Objective A *Solve. See Examples 1 and 2.*

▶ 1. In a survey of 100 college students, 96 use the Internet. What percent use the Internet?

2. A basketball player makes 81 out of 100 attempted free throws. What percent of free throws are made?

One hundred adults were asked to name their favorite sport, and the results are shown in the circle graph.

3. What sport was preferred by most adults? What percent preferred this sport?

4. What sport was preferred by the least number of adults? What percent preferred this sport?

5. What percent of adults preferred football or soccer?

6. What percent of adults preferred basketball or baseball?

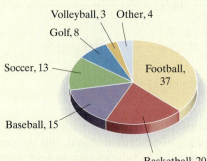

Volleyball, 3   Other, 4
Golf, 8
Soccer, 13
Football, 37
Baseball, 15
Basketball, 20

Objective B *Write each percent as a decimal. See Examples 3 through 7.*

▶ 7. 41% **8.** 62% **▶ 9.** 6% **10.** 3%

▶ 11. 100% **12.** 136% **13.** 73.6% **14.** 45.7%

▶ 15. 2.8% **16.** 1.4% **17.** 0.6% **18.** 0.9%

19. 300% **20.** 500% **21.** 32.58% **22.** 72.18%

Write each percent as a fraction or mixed number in simplest form. See Examples 8 through 12.

▶ 23. 8% **24.** 22% **25.** 4% **26.** 2%

27. 4.5% **28.** 7.5% **▶ 29.** 175% **30.** 275%

31. 6.25% **32.** 8.75% **▶ 33.** $10\frac{1}{3}$% **34.** $7\frac{3}{4}$%

35. $22\frac{3}{8}$% **36.** $21\frac{7}{8}$%

Objective C *Write each decimal as a percent. See Examples 13 through 16.*

▶ 37. 0.22 **38.** 0.44 **▶ 39.** 0.006 **40.** 0.008

41. 5.3 **42.** 2.7 **43.** 0.056 **44.** 0.019

45. 0.2228 **46.** 0.1115 **▶ 47.** 3.00 **48.** 9.00

49. 0.7 **50.** 0.8

Write each fraction or mixed number as a percent. See Examples 17 through 19.

51. $\frac{7}{10}$ **52.** $\frac{9}{10}$ **▶ 53.** $\frac{4}{5}$ **54.** $\frac{2}{5}$

55. $\frac{34}{50}$ **56.** $\frac{41}{50}$ **57.** $\frac{3}{8}$ **58.** $\frac{5}{16}$

▶ 59. $\frac{1}{3}$ **60.** $\frac{5}{6}$ **61.** $4\frac{1}{2}$ **62.** $6\frac{1}{5}$

63. $1\frac{9}{10}$ **64.** $2\frac{7}{10}$

Write each fraction as a percent. Round to the nearest hundredth percent. See Example 20.

65. $\frac{9}{11}$ **66.** $\frac{11}{12}$ **▶ 67.** $\frac{4}{15}$ **68.** $\frac{10}{11}$

Objectives A B C Mixed Practice *Complete each table. See Examples 1 through 20.*

69.

Percent	Decimal	Fraction
60%		
	0.235	
		$\frac{4}{5}$
$33\frac{1}{3}\%$		
		$\frac{7}{8}$
7.5%		

70.

Percent	Decimal	Fraction
	0.525	
		$\frac{3}{4}$
$66\frac{1}{2}\%$		
		$\frac{5}{6}$
100%		
		$\frac{7}{50}$

71.

Percent	Decimal	Fraction
200%		
	2.8	
705%		
		$4\frac{27}{50}$

72.

Percent	Decimal	Fraction
800%		
	3.2	
608%		
		$9\frac{13}{50}$

Objective D *Write each percent as a decimal and a fraction. See Examples 21 and 22.*

▶ 73. People take aspirin for a variety of reasons. The most common use of aspirin is to prevent heart disease, accounting for 38% of all aspirin use. (*Source:* Bayer Market Research)

74. Japan exports 80.5% of all motorcycles manufactured there. (*Source:* Japan Automobile Manufacturers Association)

75. In the United States recently, 35.8% of households had no landlines, just cell phones. (*Source:* CTIA—The Wireless Association)

76. From 2000 to 2011, use of smoked tobacco products in the United States decreased by 27.5%. (*Source:* Centers for Disease Control and Prevention)

77. Approximately 91% of all eighth-grade students in public schools recently reported that they use a computer at home. (*Source:* National Center for Education Statistics)

78. Approximately 15.7% of the American population is not covered by health insurance. (*Source:* U.S. Census Current Population Study)

In Exercises 79 through 82, write the percent from the circle graph as a decimal and a fraction.

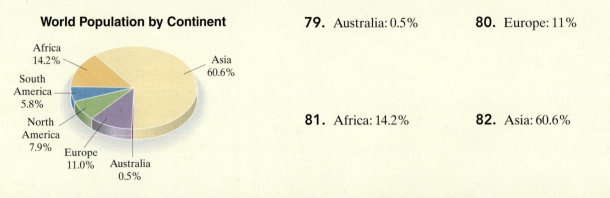

World Population by Continent

Africa 14.2%
South America 5.8%
North America 7.9%
Europe 11.0%
Australia 0.5%
Asia 60.6%

79. Australia: 0.5%

80. Europe: 11%

81. Africa: 14.2%

82. Asia: 60.6%

Solve. See Examples 21 and 22.

83. In a particular year, 0.781 of all electricity produced in France was nuclear generated. Write this decimal as a percent.

84. The United States' share of the total world motor vehicle production is 0.123. Write this decimal as a percent. (*Source:* OICA)

85. The mirrors on the Hubble Space Telescope are able to lock onto a target without deviating more than $\frac{7}{1000}$ of an arc-second. Write this fraction as a percent. (*Source:* NASA)

86. In a particular year, $\frac{1}{4}$ of light trucks sold in the United States were white. Write this fraction as a percent.

87. In 2012, the U.S. Postal Service handled 40% of the world's card and letter mail volume. Write this percent as a decimal. (*Source:* U.S. Postal Service)

88. In 2013, 20% of viewers bought a high-definition television to watch the Super Bowl. Write this percent as a fraction. (*Source:* Nielsen Media)

Review

Perform the indicated operations. See Sections 4.6 and 4.7.

89. $\frac{3}{4} - \frac{1}{2} \cdot \frac{8}{9}$

90. $\left(\frac{2}{11} + \frac{5}{11}\right)\left(\frac{2}{11} - \frac{5}{11}\right)$

91. $6\frac{2}{3} - 4\frac{5}{6}$

92. $6\frac{2}{3} \div 4\frac{5}{6}$

Concept Extensions

Solve. See the Concept Checks in this section.

93. Given the percent 52.8647%, round as indicated.
 a. Round to the nearest tenth percent.
 b. Round to the nearest hundredth percent.

94. Given the percent 0.5269%, round as indicated.
 a. Round to the nearest tenth percent.
 b. Round to the nearest hundredth percent.

95. Which of the following are correct?
 a. 6.5% = 0.65
 b. 7.8% = 0.078
 c. 120% = 0.12
 d. 0.35% = 0.0035

96. Which of the following are correct?
 a. 0.231 = 23.1%
 b. 5.12 = 0.0512%
 c. 3.2 = 320%
 d. 0.0175 = 0.175%

Recall that 1 = 100%. This means that 1 whole is 100%. Use this for Exercises 97 and 98. (Source: Some Body, by Dr. Pete Rowen)

97. The four blood types are A, B, O, and AB. (Each blood type can also be further classified as Rh-positive or Rh-negative depending upon whether your blood contains protein or not.) Given the percent blood types for people in the United States below, calculate the percent of the U.S. population with AB blood type.

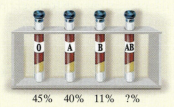

45%   40%   11%   ?%

98. The top four components of bone are below. Find the missing percent.
 1. Minerals—45%
 2. Living tissue—30%
 3. Water—20%
 4. Other—?

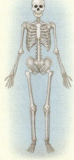

What percent of the figure is shaded?

99.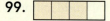

100.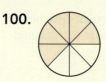

Fill in the blanks.

101. A fraction written as a percent is greater than 100% when the numerator is _____ than the denominator. (greater/less)

102. A decimal written as a percent is less than 100% when the decimal is _____ than 1. (greater/less)

Write each fraction as a decimal and then write each decimal as a percent. Round the decimal to three decimal places (nearest thousandth) and the percent to the nearest tenth percent.

103. $\dfrac{21}{79}$

104. $\dfrac{56}{102}$

The bar graph shows the predicted fastest-growing occupations by percent that require an associate degree or more education. Use this graph for Exercises 105 through 108.

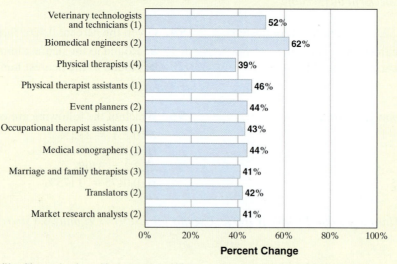

Fastest-Growing Occupations 2010–2020 (projected)

(*Note:* (1) = associate degree; (2) = bachelor degree; (3) = master degree; (4) = doctoral or professional degree)

Source: Bureau of Labor Statistics

105. What occupation is predicted to be the fastest growing?

106. What occupation is predicted to be the second fastest growing?

107. Write the percent change for physical therapists as a decimal.

108. Write the percent change for event planners as a decimal.

109. In your own words, explain how to write a percent as a decimal.

110. In your own words, explain how to write a decimal as a percent.

7.2 Solving Percent Problems with Equations

Sections 7.2 and 7.3 introduce two methods for solving percent problems. It may not be necessary for you to study both sections. You may want to check with your instructor for further advice.

To solve percent problems in this section, we will translate the problems into mathematical statements, or equations.

Objective A Writing Percent Problems as Equations

Recognizing key words in a percent problem is helpful in writing the problem as an equation. Three key words in the statement of a percent problem and their meanings are as follows:

of means **multiplication** (·)

is means **equal** (=)

what (or some equivalent) means **the unknown number**

In our examples, we will let the letter x stand for the unknown number.

Objectives

A Write Percent Problems as Equations.

B Solve Percent Problems.

Example 1 Translate to an equation:

5 is what percent of 20?

Solution: 5   is   what percent of   20?
    ↓   ↓   ↓   ↓   ↓
    5   = x ·   20

■ Work Practice 1

Practice 1

Translate: 8 is what percent of 48?

Helpful Hint

Remember that an equation is simply a mathematical statement that contains an equal sign (=).

$$5 = 20x$$
  ↑
equal sign

Example 2 Translate to an equation:

1.2 is 30% of what number?

Solution: 1.2   is   30%   of   what number?
     ↓   ↓   ↓   ↓   ↓
     1.2 =   30%   · x

■ Work Practice 2

Practice 2

Translate: 2.6 is 40% of what number?

Example 3 Translate to an equation:

What number is 25% of 0.008?

Solution: What number   is   25%   of   0.008?
       ↓   ↓   ↓   ↓   ↓
       x =   25%   ·   0.008

■ Work Practice 3

Practice 3

Translate: What number is 90% of 0.045?

Answers

1. $8 = x \cdot 48$ **2.** $2.6 = 40\% \cdot x$
3. $x = 90\% \cdot 0.045$

479

Practice 4–6

Translate each question to an equation.

4. 56% of 180 is what number?

5. 12% of what number is 21?

6. What percent of 95 is 76?

Examples Translate each question to an equation.

4. 38%   of   200   is   what number?

   38%   ·   200   = x

5. 40%   of   what number   is   80?

   40%   · x =   80

6. What percent   of   85   is   34?

   x ·   85   =   34

■ **Work Practice 4–6**

✓**Concept Check** In the equation $2x = 10$, what step is taken to solve the equation?

Objective B Solving Percent Problems

You may have noticed by now that each percent problem has contained three numbers—in our examples, two are known and one is unknown. Each of these numbers is given a special name.

15%   of   60   is   9

15% percent   ·   60 base   =   9 amount

We call this equation the **percent equation.**

> ### Percent Equation
>
> percent · base = amount

Once a percent problem has been written as a percent equation, we can use the equation to find the unknown number, whether it is the percent, the base, or the amount.

Practice 7

What number is 25% of 90?

Example 7 Solving Percent Equations for the Amount

What number   is   35%   of   60?

Solution:

$$x = 35\% \cdot 60 \quad \text{Translate to an equation.}$$
$$x = 0.35 \cdot 60 \quad \text{Write 35\% as 0.35.}$$
$$x = 21 \quad \text{Multiply:} \quad \begin{array}{r} 60 \\ \times\, 0.35 \\ \hline 300 \\ 1800 \\ \hline 21.00 \end{array}$$

Then 21 is 35% of 60. Is this reasonable? To see, round 35% to 40%. Then 40% of 60 or 0.40(60) is 24. Our result is reasonable since 21 is close to 24.

■ **Work Practice 7**

Helpful Hint

When solving a percent equation, write the percent as a decimal or fraction.

Example 8   85% of 300 is what number?

Solution: $85\% \cdot 300 = x$ Translate to an equation.

$0.85 \cdot 300 = x$ Write 85% as 0.85.

$255 = x$ Multiply: $0.85 \cdot 300 = 255$.

Then 85% of 300 is 255. Is this result reasonable? To see, round 85% to 90%. Then 90% of 300 or $0.90(300) = 270$, which is close to 255.

■ **Work Practice 8**

Practice 8

95% of 400 is what number?

Example 9   Solving Percent Equations for the Base

12% of what number is 0.6?

Solution: $12\% \cdot x = 0.6$ Translate to an equation.

$0.12 \cdot x = 0.6$ Write 12% as 0.12.

$$\frac{0.12 \cdot x}{0.12} = \frac{0.6}{0.12}$$ Divide both sides by 0.12.

$x = 5$

$$\begin{array}{r} 5. \\ 0.12\overline{)0.60} \\ \underline{60} \\ 0 \end{array}$$

Then 12% of 5 is 0.6. Is this reasonable? To see, round 12% to 10%. Then 10% of 5 or $0.10(5) = 0.5$, which is close to 0.6.

■ **Work Practice 9**

Practice 9

15% of what number is 2.4?

Example 10   13 is $6\frac{1}{2}\%$ of what number?

Solution: $13 = 6\frac{1}{2}\% \cdot x$ Translate to an equation.

$13 = 0.065 \cdot x$ $6\frac{1}{2}\% = 6.5\% = 0.065$

$$\frac{13}{0.065} = \frac{0.065 \cdot x}{0.065}$$ Divide both sides by 0.065.

$$\begin{array}{r} 200. \\ 0.065\overline{)13.000} \\ \underline{130} \\ 0 \end{array}$$

$200 = x$

Then 13 is $6\frac{1}{2}\%$ of 200. Check to see if this result is reasonable.

■ **Work Practice 10**

Practice 10

18 is $4\frac{1}{2}\%$ of what number?

Answers

8. 380 **9.** 16 **10.** 400

Practice 11
What percent of 90 is 27?

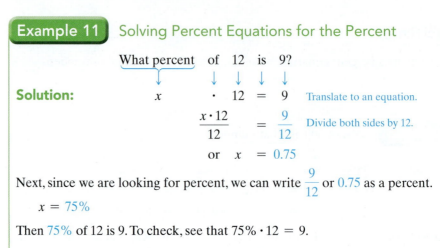

Example 11 Solving Percent Equations for the Percent

What percent of 12 is 9?

Solution:

$$x \cdot 12 = 9 \quad \text{Translate to an equation.}$$

$$\frac{x \cdot 12}{12} = \frac{9}{12} \quad \text{Divide both sides by 12.}$$

$$\text{or} \quad x = 0.75$$

Next, since we are looking for percent, we can write $\frac{9}{12}$ or 0.75 as a percent.

$$x = 75\%$$

Then 75% of 12 is 9. To check, see that $75\% \cdot 12 = 9$.

■ **Work Practice 11**

Helpful Hint

If your unknown in the percent equation is a percent, don't forget to convert your answer to a percent.

Practice 12
63 is what percent of 45?

Example 12 78 is what percent of 65?

Solution:

$$78 = x \cdot 65 \quad \text{Translate to an equation.}$$

$$\frac{78}{65} = \frac{x \cdot 65}{65} \quad \text{Divide both sides by 65.}$$

$$1.2 = x$$

$$120\% = x \quad \text{Write 1.2 as a percent.}$$

Then 78 is 120% of 65. Check this result.

■ **Work Practice 12**

✓**Concept Check** Consider these problems.

1. 75% of 50 =
   a. 50   **b.** a number greater than 50   **c.** a number less than 50
2. 40% of a number is 10. Is the number
   a. 10?   **b.** less than 10?   **c.** greater than 10?
3. 800 is 120% of what number? Is the number
   a. 800?   **b.** less than 800?   **c.** greater than 800?

Helpful Hint

Use the following to see if your answers are reasonable.

$$(100\%) \text{ of a number} = \text{the number}$$

$$\left(\begin{array}{c}\text{a percent}\\\text{greater than}\\100\%\end{array}\right) \text{of a number} = \begin{array}{c}\text{a number greater}\\\text{than the original number}\end{array}$$

$$\left(\begin{array}{c}\text{a percent}\\\text{less than }100\%\end{array}\right) \text{of a number} = \begin{array}{c}\text{a number less}\\\text{than the original number}\end{array}$$

Answers
11. 30%   **12.** 140%

✓**Concept Check Answers**
1. c   **2.** c   **3.** b

Vocabulary, Readiness & Video Check

Use the choices below to fill in each blank.

percent	amount	of	less
base	the number	is	greater

1. The word _____ translates to "=".
2. The word _____ usually translates to "multiplication."
3. In the statement "10% of 90 is 9," the number 9 is called the _____ , 90 is called the _____ , and 10 is called the _____ .
4. 100% of a number = _____ .
5. Any "percent greater than 100%" of "a number" = "a number _____ than the original number."
6. Any "percent less than 100%" of "a number" = "a number _____ than the original number."

Martin-Gay Interactive Videos   *Watch the section lecture video and answer the following questions.*

Objective A 7. What are the three translations we need to remember from the lecture before ▥ Example 1? ▶

Objective B 8. What is different about the translated equation in ▥ Example 5? ▶

See Video 7.2

7.2 Exercise Set MyMathLab® ▶

Objective A Translating *Translate each to an equation. Do not solve. See Examples 1 through 6.*

1. 18% of 81 is what number?

2. 36% of 72 is what number?

3. 20% of what number is 105?

4. 40% of what number is 6?

5. 0.6 is 40% of what number?

6. 0.7 is 20% of what number?

7. What percent of 80 is 3.8?

8. 9.2 is what percent of 92?

9. What number is 9% of 43?

10. What number is 25% of 55?

11. What percent of 250 is 150?

12. What percent of 375 is 300?

Objective B   *Solve. See Examples 7 and 8.*

▶ **13.** 10% of 35 is what number?

14. 25% of 68 is what number?

15. What number is 14% of 205?

16. What number is 18% of 425?

Solve. See Examples 9 and 10.

▶ **17.** 1.2 is 12% of what number?

18. 0.22 is 44% of what number?

19. $8\frac{1}{2}$% of what number is 51?

20. $4\frac{1}{2}$% of what number is 45?

Solve. See Examples 11 and 12.

21. What percent of 80 is 88?

22. What percent of 40 is 60?

23. 17 is what percent of 50?

24. 48 is what percent of 50?

Objectives A B Mixed Practice   *Solve. See Examples 1 through 12.*

25. 0.1 is 10% of what number?

26. 0.5 is 5% of what number?

27. 150% of 430 is what number?

28. 300% of 56 is what number?

29. 82.5 is $16\frac{1}{2}$% of what number?

30. 7.2 is $6\frac{1}{4}$% of what number?

▶ **31.** 2.58 is what percent of 50?

32. 2.64 is what percent of 25?

33. What number is 42% of 60?

34. What number is 36% of 80?

35. What percent of 184 is 64.4?

36. What percent of 120 is 76.8?

37. 120% of what number is 42?

38. 160% of what number is 40?

39. 2.4% of 26 is what number?

40. 4.8% of 32 is what number?

41. What percent of 600 is 3?

42. What percent of 500 is 2?

43. 6.67 is 4.6% of what number?

44. 9.75 is 7.5% of what number?

45. 1575 is what percent of 2500?

46. 2520 is what percent of 3500?

47. 2 is what percent of 50?

48. 2 is what percent of 40?

Review

Find the value of x in each proportion. See Section 6.2.

49. $\dfrac{27}{x} = \dfrac{9}{10}$

50. $\dfrac{35}{x} = \dfrac{7}{5}$

51. $\dfrac{x}{5} = \dfrac{8}{11}$

52. $\dfrac{x}{3} = \dfrac{6}{13}$

Write each sentence as a proportion. See Section 6.2.

53. 17 is to 12 as x is to 20.   **54.** 20 is to 25 as x is to 10.   **55.** 8 is to 9 as 14 is to x.   **56.** 5 is to 6 as 15 is to x.

Concept Extensions

For each equation, determine the next step taken to find the value of n. See the first Concept Check in this section.

57. $5 \cdot n = 32$

    a. $n = 5 \cdot 32$ **b.** $n = \dfrac{5}{32}$ **c.** $n = \dfrac{32}{5}$ **d.** none of these

58. $68 = 8 \cdot n$

    a. $n = 8 \cdot 68$ **b.** $n = \dfrac{68}{8}$ **c.** $n = \dfrac{8}{68}$ **d.** none of these

59. $0.06 = n \cdot 7$

    a. $n = 0.06 \cdot 7$ **b.** $n = \dfrac{0.06}{7}$ **c.** $n = \dfrac{7}{0.06}$ **d.** none of these

60. $n = 0.7 \cdot 12$

    a. $n = 8.4$ **b.** $n = \dfrac{12}{0.7}$ **c.** $n = \dfrac{0.7}{12}$ **d.** none of these

61. Write a word statement for the equation $20\% \cdot x = 18.6$. Use the phrase "what number" for "x."

62. Write a word statement for the equation $x = 33\frac{1}{3}\% \cdot 24$. Use the phrase "what number" for "x."

For each exercise, determine whether the percent, x, is (a) 100%, (b) greater than 100%, or (c) less than 100%. See the second Concept Check in this section.

63. $x\%$ of 20 is 30   **64.** $x\%$ of 98 is 98   **65.** $x\%$ of 120 is 85   **66.** $x\%$ of 35 is 50

For each exercise, determine whether the number, y, is (a) equal to 45, (b) greater than 45, or (c) less than 45.

67. 55% of 45 is y **68.** 230% of 45 is y **69.** 100% of 45 is y

70. 30% of y is 45   **71.** 100% of y is 45   **72.** 180% of y is 45

73. In your own words, explain how to solve a percent equation.

74. Write a percent problem that uses the percent 50%.

Solve.

75. 1.5% of 45,775 is what number?

76. What percent of 75,528 is 27,945.36?

77. 22,113 is 180% of what number?

7.3 Solving Percent Problems with Proportions

Objectives

A Write Percent Problems as Proportions. ▶

B Solve Percent Problems. ▶

There is more than one method that can be used to solve percent problems. (See the note at the beginning of Section 7.2.) In the last section, we used the percent equation. In this section, we will use proportions.

Objective A Writing Percent Problems as Proportions ▶

To understand the proportion method, recall that 70% means the ratio of 70 to 100, or $\dfrac{70}{100}$.

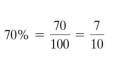

$$70\% = \frac{70}{100} = \frac{7}{10}$$

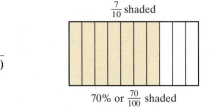

$\frac{7}{10}$ shaded

70% or $\frac{70}{100}$ shaded

Since the ratio $\dfrac{70}{100}$ is equal to the ratio $\dfrac{7}{10}$, we have the proportion

$$\frac{7}{10} = \frac{70}{100}$$

We call this proportion the **percent proportion.** In general, we can name the parts of this proportion as follows:

Percent Proportion

$$\frac{\text{amount}}{\text{base}} = \frac{\text{percent}}{100} \quad \leftarrow \text{always 100}$$

or

$$\text{amount} \rightarrow \quad \frac{a}{b} = \frac{p}{100} \quad \leftarrow \text{percent}$$
$$\text{base} \rightarrow$$

When we translate percent problems to proportions, the **percent**, p, can be identified by looking for the symbol % or the word *percent*. The **base**, b, usually follows the word *of*. The **amount**, a, is the part compared to the whole.

Helpful Hint

This table may be useful when identifying the parts of a proportion.

Part of Proportion	How It's Identified
Percent	% or *percent*
Base	Appears after *of*
Amount	Part compared to whole

Example 1 Translate to a proportion.

12% of what number is 47?

Solution:

| percent | base
It appears after the word *of*. | amount
It is the part compared to the whole. |

amount → $\dfrac{47}{b} = \dfrac{12}{100}$ ← percent
base →

■ **Work Practice 1**

Example 2 Translate to a proportion.

101 is what number of 200?

Solution:

| amount
It is the part compared to the whole. | percent | base
It appears after the word *of*. |

amount → $\dfrac{101}{200} = \dfrac{p}{100}$ ← percent
base →

■ **Work Practice 2**

Example 3 Translate to a proportion.

What number is 90% of 45?

Solution:

| amount
It is the part compared to the whole. | percent | base
It appears after the word *of*. |

amount → $\dfrac{a}{45} = \dfrac{90}{100}$ ← percent
base →

■ **Work Practice 3**

Example 4 Translate to a proportion.

238 is 40% of what number?

Solution: amount   percent   base

$\dfrac{238}{b} = \dfrac{40}{100}$

■ **Work Practice 4**

Practice 1

Translate to a proportion.
27% of what number is 54?

Practice 2

Translate to a proportion.
30 is what percent of 90?

Practice 3

Translate to a proportion.
What number is 25% of 116?

Practice 4

Translate to a proportion.
680 is 65% of what number?

Answers

1. $\dfrac{54}{b} = \dfrac{27}{100}$ **2.** $\dfrac{30}{90} = \dfrac{p}{100}$

3. $\dfrac{a}{116} = \dfrac{25}{100}$ **4.** $\dfrac{680}{b} = \dfrac{65}{100}$

Practice 5

Translate to a proportion.
What percent of 40 is 75?

Example 5 Translate to a proportion.

$$\underline{\text{What percent}} \quad \text{of} \quad 30 \quad \text{is} \quad 75?$$

Solution: percent   base   amount

$$\frac{75}{30} = \frac{p}{100}$$

■ **Work Practice 5**

Practice 6

Translate to a proportion.
46% of 80 is what number?

Example 6 Translate to a proportion.

$$45\% \quad \text{of} \quad 105 \quad \text{is} \quad \underline{\text{what number?}}$$

Solution: percent   base   amount

$$\frac{a}{105} = \frac{45}{100}$$

■ **Work Practice 6**

✓**Concept Check** Consider the statement "78 is what percent of 350?" Which part of the percent proportion is unknown?

a. the amount

b. the base

c. the percent

Consider another statement: "14 is 10% of some number."
Which part of the percent proportion is unknown?

a. the amount

b. the base

c. the percent

Objective B Solving Percent Problems ▶

The proportions that we have written in this section contain three values that can change: the percent, the base, and the amount. If any two of these values are known, we can find the third (the unknown) value. To do this, we write a percent proportion and find the unknown value as we did in Section 6.2.

Practice 7

What number is 8% of 120?

Example 7 Solving Percent Proportions for the Amount

$$\underline{\text{What number}} \quad \text{is} \quad 30\% \quad \text{of} \quad 9?$$

Solution: amount   percent   base

$$\frac{a}{9} = \frac{30}{100}$$

To solve, we set cross products equal to each other.

$$\frac{a}{9} = \frac{30}{100}$$

$a \cdot 100 = 9 \cdot 30$ Set cross products equal.

$100a = 270$ Multiply.

$$\frac{100a}{100} = \frac{270}{100}$$ Divide both sides by 100, the coefficient of a.

$a = 2.7$ Simplify.

Thus, 2.7 is 30% of 9.

■ **Work Practice 7**

Helpful Hint The proportion in Example 7 contains the ratio $\frac{30}{100}$. A ratio in a proportion may be simplified before solving the proportion. The unknown number in both

$\frac{a}{9} = \frac{30}{100}$ and $\frac{a}{9} = \frac{3}{10}$ is 2.7.

Example 8 Solving Percent Problems for the Base

150% of what number is 30?

Solution: percent   base   amount

$$\frac{30}{b} = \frac{150}{100}$$ Write the proportion.

$$\frac{30}{b} = \frac{3}{2}$$ Simplify $\frac{150}{100}$ and write as $\frac{3}{2}$.

$30 \cdot 2 = b \cdot 3$ Set cross products equal.

$60 = 3b$ Multiply.

$$\frac{60}{3} = \frac{3b}{3}$$ Divide both sides by 3.

$20 = b$ Simplify.

Thus, 150% of 20 is 30.

■ **Work Practice 8**

Practice 8

65% of what number is 52?

✓**Concept Check** When solving a percent problem by using a proportion, describe how you can check the result.

Example 9

20.8 is 40% of what number?

Solution: amount percent   base

$$\frac{20.8}{b} = \frac{40}{100}$$ or $\frac{20.8}{b} = \frac{2}{5}$ Write the proportion and simplify $\frac{40}{100}$.

$20.8 \cdot 5 = b \cdot 2$ Set cross products equal.

$104 = 2b$ Multiply.

$$\frac{104}{2} = \frac{2b}{2}$$ Divide both sides by 2.

$52 = b$ Simplify.

So, 20.8 is 40% of 52.

■ **Work Practice 9**

Practice 9

15.4 is 5% of what number?

Answers

8. 80   **9.** 308

✓**Concept Check Answer**

by putting the result into the proportion and checking that the proportion is true

Practice 10
What percent of 40 is 8?

Example 10 Solving Percent Problems for the Percent

$$\underbrace{\text{What percent of }}_{} \quad \underbrace{50}_{} \text{ is } \quad \underbrace{8}_{}?$$

Solution: percent   base   amount

$$\frac{8}{50} = \frac{p}{100} \quad \text{or} \quad \frac{4}{25} = \frac{p}{100} \qquad \text{Write the proportion and simplify } \frac{8}{50}.$$

$$4 \cdot 100 = 25 \cdot p \qquad \text{Set cross products equal.}$$

$$400 = 25p \qquad \text{Multiply.}$$

$$\frac{400}{25} = \frac{25p}{25} \qquad \text{Divide both sides by 25.}$$

$$16 = p \qquad \text{Simplify.}$$

So, 16% of 50 is 8.

■ **Work Practice 10**

> **Helpful Hint** Recall from our percent proportion that this number already is a percent. Just keep the number as is and attach a % symbol.

Practice 11
414 is what percent of 180?

Example 11

$$\underbrace{504}_{} \text{ is } \underbrace{\text{what percent}}_{} \text{ of } \underbrace{360}_{}?$$

Solution: amount   percent   base

$$\frac{504}{360} = \frac{p}{100}$$

Let's choose not to simplify the ratio $\frac{504}{360}$.

$$504 \cdot 100 = 360 \cdot p \qquad \text{Set cross products equal.}$$

$$50{,}400 = 360p \qquad \text{Multiply.}$$

$$\frac{50{,}400}{360} = \frac{360p}{360} \qquad \text{Divide both sides by 360.}$$

$$140 = p \qquad \text{Simplify.}$$

Notice that by choosing not to simplify $\frac{504}{360}$, we had larger numbers in our equation. Either way, we find that 504 is 140% of 360.

■ **Work Practice 11**

> **Helpful Hint**
>
> Use the following to see whether your answers to the above examples and practice problems are reasonable.
>
> 100% of a number = the number
>
> $\left(\begin{array}{c}\text{a percent}\\\text{greater than}\\100\%\end{array}\right)$ of a number = $\begin{array}{c}\text{a number larger}\\\text{than the original number}\end{array}$
>
> $\left(\begin{array}{c}\text{a percent}\\\text{less than }100\%\end{array}\right)$ of a number = $\begin{array}{c}\text{a number less}\\\text{than the original number}\end{array}$

Answers
10. 20% **11.** 230%

Vocabulary, Readiness & Video Check

Use the choices below to fill in each blank. These choices will be used more than once.

amount   base   percent

1. When translating the statement "20% of 15 is 3" to a proportion, the number 3 is called the _____,
    15 is the _____, and 20 is the _____.

2. In the question "50% of what number is 28?", which part of the percent proportion is unknown? _____

3. In the question "What number is 25% of 200?", which part of the percent proportion is unknown? _____

4. In the question "38 is what percent of 380?", which part of the percent proportion is unknown? _____

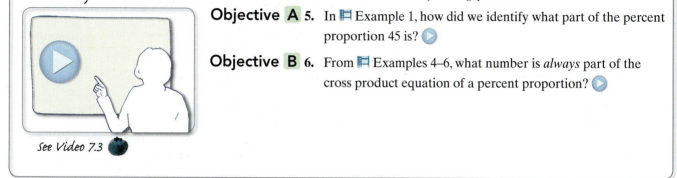

Martin-Gay Interactive Videos   *Watch the section lecture video and answer the following questions.*

Objective A 5. In ▯ Example 1, how did we identify what part of the percent
                     proportion 45 is? ▶

Objective B 6. From ▯ Examples 4–6, what number is *always* part of the
                     cross product equation of a percent proportion? ▶

See Video 7.3

7.3 Exercise Set MyMathLab® ▶

Objective A Translating *Translate each to a proportion. Do not solve. See Examples 1 through 6.*

▶ **1.** 98% of 45 is what number?

2. 92% of 30 is what number?

3. What number is 4% of 150?

4. What number is 7% of 175?

5. 14.3 is 26% of what number?

6. 1.2 is 47% of what number?

7. 35% of what number is 84?

8. 85% of what number is 520?

▶ **9.** What percent of 400 is 70?

10. What percent of 900 is 216?

11. 8.2 is what percent of 82?

12. 9.6 is what percent of 96?

Objective B *Solve. See Example 7.*

13. 40% of 65 is what number?

14. 25% of 84 is what number?

15. What number is 18% of 105?

16. What number is 60% of 29?

Solve. See Examples 8 and 9.

17. 15% of what number is 90?

18. 55% of what number is 55?

19. 7.8 is 78% of what number?

20. 1.1 is 44% of what number?

Solve. See Examples 10 and 11.

21. What percent of 35 is 42?

22. What percent of 98 is 147?

23. 14 is what percent of 50?

24. 24 is what percent of 50?

Objectives A B Mixed Practice *Solve. See Examples 1 through 11.*

25. 3.7 is 10% of what number?

26. 7.4 is 5% of what number?

27. 2.4% of 70 is what number?

28. 2.5% of 90 is what number?

29. 160 is 16% of what number?

30. 30 is 6% of what number?

31. 394.8 is what percent of 188?

32. 550.4 is what percent of 172?

33. What number is 89% of 62?

34. What number is 53% of 130?

35. What percent of 6 is 2.7?

36. What percent of 5 is 1.6?

37. 140% of what number is 105?

38. 170% of what number is 221?

39. 1.8% of 48 is what number?

40. 7.8% of 24 is what number?

41. What percent of 800 is 4?

42. What percent of 500 is 3?

43. 3.5 is 2.5% of what number?

44. 9.18 is 6.8% of what number?

45. 20% of 48 is what number?

46. 75% of 14 is what number?

47. 2486 is what percent of 2200?

48. 9310 is what percent of 3800?

Review

Add or subtract the fractions. See Sections 4.4, 4.5, and 4.7.

49. $-\dfrac{11}{16} + \left(-\dfrac{3}{16}\right)$

50. $\dfrac{7}{12} - \dfrac{5}{8}$

51. $3\dfrac{1}{2} - \dfrac{11}{30}$

52. $2\dfrac{2}{3} + 4\dfrac{1}{2}$

Add or subtract the decimals. See Section 5.2.

53. $\begin{array}{r} 0.41 \\ +\,0.29 \\ \hline \end{array}$

54. $\begin{array}{r} 10.78 \\ 4.3 \\ +\ 0.21 \\ \hline \end{array}$

55. $\begin{array}{r} 2.38 \\ -\,0.19 \\ \hline \end{array}$

56. $\begin{array}{r} 16.37 \\ -\ 2.61 \\ \hline \end{array}$

Concept Extensions

57. Write a word statement for the proportion $\dfrac{x}{28} = \dfrac{25}{100}$. Use the phrase "what number" for "x."

58. Write a percent statement that translates to $\dfrac{16}{80} = \dfrac{20}{100}$.

Suppose you have finished solving four percent problems using proportions that you set up correctly. Check each answer to see if each makes the proportion a true proportion. If any proportion is not true, solve it to find the correct solution. See the Concept Checks in this section.

59. $\dfrac{a}{64} = \dfrac{25}{100}$
Is the amount equal to 17?

60. $\dfrac{520}{b} = \dfrac{65}{100}$
Is the base equal to 800?

61. $\dfrac{p}{100} = \dfrac{13}{52}$
Is the percent equal to 25 (25%)?

62. $\dfrac{36}{12} = \dfrac{p}{100}$
Is the percent equal to 50 (50%)?

63. In your own words, describe how to identify the percent, the base, and the amount in a percent problem.

64. In your own words, explain how to use a proportion to solve a percent problem.

Solve. Round to the nearest tenth, if necessary.

65. What number is 22.3% of 53,862?

66. What percent of 110,736 is 88,542?

67. 8652 is 119% of what number?

Percent and Percent Problems

Answers

1. _____

2. _____

3. _____

4. _____

5. _____

6. _____

7. _____

8. _____

9. _____

10. _____

11. _____

12. _____

13. _____

14. _____

15. _____

16. _____

17. _____

18. _____

19. _____

20. _____

21. _____

22. _____

23. _____

24. _____

Write each number as a percent.

1. 0.94

2. 0.17

3. $\dfrac{3}{8}$

4. $\dfrac{7}{2}$

5. 4.7

6. 8

7. $\dfrac{9}{20}$

8. $\dfrac{53}{50}$

9. $6\dfrac{3}{4}$

10. $3\dfrac{1}{4}$

11. 0.02

12. 0.06

Write each percent as a decimal.

13. 71%

14. 31%

15. 3%

16. 4%

17. 224%

18. 700%

19. 2.9%

20. 6.6%

Write each percent as a decimal and as a fraction or mixed number in simplest form. (If necessary when writing as a decimal, round to the nearest thousandth.)

21. 7%

22. 5%

23. 6.8%

24. 11.25%

25. 74%   **26.** 45%   **27.** $16\frac{1}{3}\%$ **28.** $12\frac{2}{3}\%$

Solve each percent problem.

29. 15% of 90 is what number?   **30.** 78 is 78% of what number?

31. 297.5 is 85% of what number?   **32.** 78 is what percent of 65?

33. 23.8 is what percent of 85?   **34.** 38% of 200 is what number?

35. What number is 40% of 85?   **36.** What percent of 99 is 128.7?

37. What percent of 250 is 115?   **38.** What number is 45% of 84?

39. 42% of what number is 63?   **40.** 95% of what number is 58.9?

25. _____

26. _____

27. _____

28. _____

29. _____

30. _____

31. _____

32. _____

33. _____

34. _____

35. _____

36. _____

37. _____

38. _____

39. _____

40. _____

Objectives

A Solve Applications Involving Percent. ▶

B Find Percent Increase and Percent Decrease. ▶

Objective A Solving Applications Involving Percent ▶

Percent is used in a variety of everyday situations. The next examples show just a few ways that percent occurs in real-life settings. (Each of these examples shows two ways of solving these problems. If you studied Section 7.2 only, see *Method 1*. If you studied Section 7.3 only, see *Method 2*.)

The next example has to do with the Appalachian Trail, a hiking trail conceived by a forester in 1921 and diagrammed to the right. (*Note:* The trail mileage changes from year to year as maintenance groups reroute the trail as needed.)

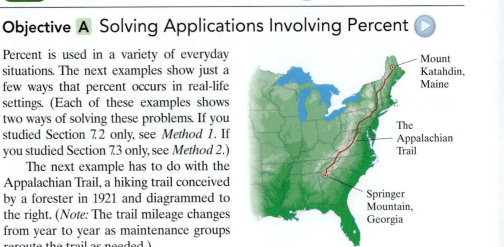

Mount Katahdin, Maine

The Appalachian Trail

Springer Mountain, Georgia

Practice 1

If the total mileage of the Appalachian Trail is 2174, use the circle graph to determine the number of miles in the state of Virginia.

Appalachian Trail Mileage by State Percent

Georgia 4%
Maine 13% North Carolina 4%
Tennessee 14%
New Hampshire 7%
Vermont 7%
Virginia 25%
Massachusetts 4%
Connecticut 2%
New York 4%
New Jersey 3% West Virginia 0.2%
Pennsylvania 11% Maryland 2%

Total miles: 2174
(*Due to rounding, these percents have a sum greater than 100%.)
Source: purebound.com

Example 1 The circle graph in the margin shows the Appalachian Trail mileage by state. If the total mileage of the trail is 2174, use the circle graph to determine the number of miles in the state of New York. Round to the nearest whole mile.

Solution: *Method 1.* First, we state the problem in words.

In words: What number is 4% of 2174?

Translate: x = 4% · 2174

To solve for x, we find 4% · 2174.

$x = 0.04 \cdot 2174$ Write 4% as a decimal.

$x = 86.96$ Multiply.

$x \approx 87$ Round to the nearest whole.

Rounded to the nearest whole mile, we have that approximately 87 miles of the Appalachian Trail is in New York state.

Method 2. State the problem in words; then translate.

In words: What number is 4% of 2174?

amount percent base

Translate: amount → $\dfrac{a}{2174} = \dfrac{4}{100}$ ← percent
  base →

Next, we solve for a.

$a \cdot 100 = 2174 \cdot 4$ Set cross products equal.

$100a = 8696$ Multiply.

$\dfrac{100a}{100} = \dfrac{8696}{100}$ Divide both sides by 100.

$a = 86.96$ Simplify.

$a \approx 87$ Round to the nearest whole.

Rounded to the nearest whole mile, we have that approximately 87 miles of the Appalachian Trail is in New York state.

Work Practice 1

Answer

1. 543.5 mi

Example 2 Finding Percent of Nursing School Applications Accepted

There continues to be a shortage of nursing school facilities. Recently, of the 256,000 applications to bachelor degree nursing schools, 101,000 of these were accepted. What percent of these applications were accepted? Round to the nearest percent. (*Source:* Bureau of Labor Statistics)

Solution: *Method 1.* First, we state the problem in words.

In words:　　101　　is　　what percent　　of　　256?

Translate:　　101　　=　　x　　·　　256

　　　　or　　101　　=　　256x

Next, solve for x.

$$\frac{101}{256} = \frac{256x}{256}$$　　Divide both sides by 256.

$0.39 \approx x$　　Divide and round to the nearest hundredth.

$39\% \approx x$　　Write as a percent.

About 39% of nursing school applications were accepted.

Method 2.

In words:　　101　　is　　what percent　　of　　256?

　　　　　　amount　　　　percent　　　　base

Translate:　amount → $\dfrac{101}{256} = \dfrac{p}{100}$ ← percent
　　　　　base →

Next, solve for p.

$101 \cdot 100 = 256 \cdot p$　　Set cross products equal.

$10{,}100 = 256p$　　Multiply.

$$\frac{10{,}100}{256} = \frac{256p}{256}$$　　Divide both sides by 256.

$39 \approx p$

About 39% of nursing school applications were accepted.

■ **Work Practice 2**

Example 3 Finding the Base Number of Absences

Mr. Percy, the principal at Slidell High School, counted 31 freshmen absent during a particular day. If this is 4% of the total number of freshmen, how many freshmen are there at Slidell High School?

Solution: *Method 1.* First we state the problem in words; then we translate.

In words:　　31　　is　　4%　　of　　what number?

Translate:　　31　　=　　4%　　·　　x

Next, we solve for x.

$31 = 0.04 \cdot x$　　Write 4% as a decimal.

$$\frac{31}{0.04} = \frac{0.04x}{0.04}$$　　Divide both sides by 0.04.

$775 = x$　　Simplify.

There are 775 freshmen at Slidell High School.

(*Continued on next page*)

Practice 2

From 2010 to 2020, it is projected that the number of employed nurses will grow by 710,000. If the number of nurses employed in 2010 was 2,740,000, find the percent increase in nurses employed from 2010 to 2020. Round to the nearest percent. (*Source:* Bureau of Labor Statistics)

Practice 3

The freshmen class of 864 students is 32% of all students at Euclid University. How many students go to Euclid University?

Answers

2. 26%　　**3.** 2700

Method 2.   First we state the problem in words; then we translate.

In words: 31   is   4%   of   what number?
                ↓   ↓   ↓
              amount   percent   base

Translate: amount → $\dfrac{31}{b} = \dfrac{4}{100}$ ← percent
             base →

Next, we solve for b.

$31 \cdot 100 = b \cdot 4$ Set cross products equal.

$3100 = 4b$ Multiply.

$\dfrac{3100}{4} = \dfrac{4b}{4}$ Divide both sides by 4.

$775 = b$ Simplify.

There are 775 freshmen at Slidell High School.

■ **Work Practice 3**

Example 4   Finding the Base Increase in Licensed Drivers

From 2000 to 2011, the number of licensed drivers on the road in the United States increased by 11.6%. In 2000, there were about 190 million licensed drivers on the road.

a. Find the increase in licensed drivers from 2000 to 2011.

b. Find the number of licensed drivers on the road in 2011. (*Source:* Federal Highway Administration)

Solution: *Method 1.*   First we find the increase in licensed drivers.

In words: What number   is   11.6%   of   190?
                 ↓   ↓   ↓   ↓   ↓
Translate: x =   11.6%   ·   190

Next, we solve for x.

$x = 0.116 \cdot 190$ Write 11.6% as a decimal.

$x = 22.04$ Multiply.

a. The increase in licensed drivers was 22.04 million.

b. This means that the number of licensed drivers in 2011 was

Number of   Number of   Increase
licensed drivers = licensed drivers + in number of
in 2011   in 2000   licensed drivers

$= 190 \text{ million} + 22.04 \text{ million}$

$= 212.04 \text{ million}$

Method 2. First we find the increase in licensed drivers.

In words: What number is 11.6% of 190?

amount   percent   base

Translate: $\begin{array}{c} \text{amount} \rightarrow \\ \text{base} \rightarrow \end{array} \dfrac{a}{190} = \dfrac{11.6}{100} \leftarrow \text{percent}$

Next, we solve for a.

$a \cdot 100 = 190 \cdot 11.6$ Set cross products equal.

$100a = 2204$ Multiply.

$\dfrac{100a}{100} = \dfrac{2204}{100}$ Divide both sides by 100.

$a = 22.04$ Simplify.

a. The increase in licensed drivers was 22.04 million.

b. This means that the number of licensed drivers in 2011 was

$$\begin{array}{c} \text{Number of} \\ \text{licensed drivers} \\ \text{in 2011} \end{array} = \begin{array}{c} \text{Number of} \\ \text{licensed drivers} \\ \text{in 2000} \end{array} + \begin{array}{c} \text{Increase} \\ \text{in number of} \\ \text{licensed drivers} \end{array}$$

$$= 190 \text{ million} + 22.04 \text{ million}$$

$$= 212.04 \text{ million}$$

■ **Work Practice 4**

Objective B Finding Percent Increase and Percent Decrease ▶

We often use percents to show how much an amount has increased or decreased.

Suppose that the population of a town is 10,000 people and then it increases by 2000 people. The **percent of increase** is

$$\begin{array}{c} \text{amount of increase} \rightarrow \\ \text{original amount} \rightarrow \end{array} \dfrac{2000}{10,000} = 0.2 = 20\%$$

In general, we have the following.

Percent of Increase

$$\text{percent of increase} = \dfrac{\text{amount of increase}}{\text{original amount}}$$

Then write the quotient as a percent.

Example 5   Finding Percent Increase

The number of applications for a mathematics scholarship at one university increased from 34 to 45 in one year. What is the percent increase? Round to the nearest whole percent.

Solution: First we find the amount of increase by subtracting the original number of applicants from the new number of applicants.

amount of increase $= 45 - 34 = 11$

The amount of increase is 11 applicants. To find the percent of increase,

(Continued on next page)

Practice 5

The number of people attending the local play, *Peter Pan*, increased from 285 on Friday to 333 on Saturday. Find the percent increase in attendance. Round to the nearest tenth percent.

Answer

5. 16.8%

Copyright 2015 Pearson Education, Inc.

$$\text{percent of increase} = \frac{\text{amount of increase}}{\text{original amount}} = \frac{11}{34} \approx 0.32 = 32\%$$

The number of applications increased by about 32%.

■ Work Practice 5

✓**Concept Check** A student is calculating the percent increase in enrollment from 180 students one year to 200 students the next year. Explain what is wrong with the following calculations:

$$\begin{aligned}\text{Amount of increase} &= 200 - 180 = 20 \\ \text{Percent of increase} &= \frac{20}{200} = 0.1 = 10\%\end{aligned}$$

Suppose that your income was $300 a week and then it decreased by $30. The **percent of decrease** is

$$\begin{aligned}\text{amount of decrease} &\rightarrow \\ \text{original amount} &\rightarrow\end{aligned} \frac{\$30}{\$300} = 0.1 = 10\%$$

Percent of Decrease

$$\text{percent of decrease} = \frac{\text{amount of decrease}}{\text{original amount}}$$

Then write the quotient as a percent.

Helpful Hint

Make sure that this number is the original number and not the new number.

Practice 6

A town with a population of 20,200 in 2003 decreased to 18,483 in 2013. What was the percent decrease?

Answer

6. 8.5%

✓**Concept Check Answers**

To find the percent of increase, you have to divide the amount of increase (20) by the original amount (180); 10% decrease.

Example 6   Finding Percent Decrease

In response to a decrease in sales, a company with 1500 employees reduces the number of employees to 1230. What is the percent decrease?

Solution: First we find the amount of decrease by subtracting 1230 from 1500.

$$\text{amount of decrease} = 1500 - 1230 = 270$$

The amount of decrease is 270. To find the percent of decrease,

$$\text{percent of decrease} = \frac{\text{amount of decrease}}{\text{original amount}} = \frac{270}{1500} = 0.18 = 18\%$$

The number of employees decreased by 18%.

■ Work Practice 6

✓**Concept Check** An ice cream stand sold 6000 ice cream cones last summer. This year the same stand sold 5400 cones. Was there a 10% increase, a 10% decrease, or neither? Explain.

Martin-Gay Interactive Videos   *Watch the section lecture video and answer the following questions.*

Objective A 1. How do we interpret the answer 175,000 in ▣ Example 1? ▸

Objective B 2. In ▣ Example 3, what does the improper fraction tell us? ▸

See Video 7.4 🔵

7.4 Exercise Set MyMathLab®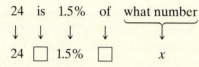

Objective A *Solve. For Exercises 1 and 2, the solutions have been started for you. See Examples 1 through 4.*
If necessary, round percents to the nearest tenth and all other answers to the nearest whole.

1. An inspector found 24 defective bolts during an inspection. If this is 1.5% of the total number of bolts inspected, how many bolts were inspected?

Start the solution:

1. UNDERSTAND the problem. Reread it as many times as needed.

Go to *Method* 1 or *Method* 2.

Method 1.

2. TRANSLATE into an equation. (Fill in the boxes.)

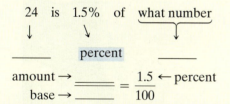

24   is   1.5%   of   what number

↓   ↓   ↓   ↓   ↓

24   ☐   1.5%   ☐ x

3. SOLVE for x. (See Example 3, Method 1, for help.)

4. INTERPRET. The total number of bolts inspected was _____.

Method 2.

2. TRANSLATE into a proportion. (Fill in the blanks with "amount" or "base.")

24   is   1.5%   of   what number

↓   ↘   ↓

_____   percent   _____

$\dfrac{\text{amount} \rightarrow \underline{\quad\quad}}{\text{base} \rightarrow \underline{\quad\quad}} = \dfrac{1.5}{100} \leftarrow \text{percent}$

3. SOLVE the proportion. (See Example 3, Method 2, for help.)

4. INTERPRET. The total number of bolts inspected was _____.

2. A day care worker found 28 children absent one day during an epidemic of chicken pox. If this was 35% of the total number of children attending the day care center, how many children attend this day care center?

Start the solution:

1. UNDERSTAND the problem. Reread it as many times as needed.

Go to *Method* 1 or *Method* 2.

Method 1.

2. TRANSLATE into an equation. (Fill in the boxes.)

28   is   35%   of   what number

↓   ↓   ↓   ↓   ↓

28   ☐   35%   ☐ x

3. SOLVE for x. (See Example 3, Method 1, for help.)

4. INTERPRET. The total number of children attending the day care center is _____.

Method 2.

2. TRANSLATE into a proportion. (Fill in the blanks with "amount" or "base.")

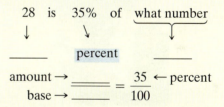

28   is   35%   of   what number

↓   ↘   ↓

_____   percent   _____

$\dfrac{\text{amount} \rightarrow \underline{\quad\quad}}{\text{base} \rightarrow \underline{\quad\quad}} = \dfrac{35}{100} \leftarrow \text{percent}$

3. SOLVE the proportion. (See Example 3, Method 2, for help.)

4. INTERPRET. The total number of children attending the day care center is _____.

3. One model of Total Gym® provides weight resistance through adjustments of incline. The minimum weight resistance is 4% of the weight of the person using the Total Gym. Find the minimum weight resistance possible for a 220-pound man. (*Source:* Total Gym)

4. The maximum weight resistance for one model of Total Gym is 60% of the weight of the person using it. Find the maximum weight resistance possible for a 220-pound man. (See Exercise **3** if needed.)

5. A student's cost for last semester at her community college was $2700. She spent $378 of that on books. What percent of last semester's college costs was spent on books?

6. Pierre Sampeau belongs to his local food cooperative, where he receives a percentage of what he spends each year as a dividend. He spent $3850 last year at the food cooperative store and received a dividend of $154. What percent of his total spending at the food cooperative did he receive as a dividend?

In 2012, there were about 40,000 cinema screens in the United States. Use this information for Exercises 7 and 8. See the Chapter Opener.

7. If about 34% of the total screens in the United States were digital 3-D screens, find the number of digital 3-D screens.

8. If about 16% of the total screens in the United States were analog screens, find the number of analog screens.

9. Approximately 160,650 of America's 945,000 restaurants are pizza restaurants. Determine the percent of restaurants in America that are pizza restaurants. (*Source:* Pizza Marketplace, National Restaurant Association)

10. Of the 97,100 veterinarians in private practice in the United States, approximately 52,434 are female. Determine the percent of female veterinarians in private practice in the United States. (*Source:* American Veterinary Medical Association)

11. A furniture company currently produces 6200 chairs per month. If production decreases by 8%, find the decrease and the new number of chairs produced each month.

12. The enrollment at a local college decreased by 5% over last year's enrollment of 7640. Find the decrease in enrollment and the current enrollment.

13. From 2010 to 2020, the number of people employed as physician assistants in the United States is expected to increase by 30%. The number of people employed as physician assistants in 2010 was 83,600. Find the predicted number of physician assistants in 2020. (*Source:* Bureau of Labor Statistics)

14. From 2007 to 2012, the number of households owning turtles increased by 19.3%. The number of households owning turtles in 2007 was 1,106,000. Find the number of households owning turtles in 2012. (*Source:* American Veterinary Medical Association)

Two states, Michigan and Rhode Island, decreased in population from 2010 to 2012. Their locations are shown on the U.S. map below. (Source: U.S. Dept. of Commerce)

15. In 2010, the population of Rhode Island was approximately 1,053,000. If the population decrease was 0.3%, find the population of Rhode Island in 2012.

16. In 2010, the population in Michigan was approximately 9,938,000. If the population decrease was about 0.6%, find the population of Michigan in 2012.

A popular extreme sport is snowboarding. Ski trails are marked with difficulty levels of easy ●, intermediate ■, difficult ◆, expert ◆◆, and other variations. Use this information for Exercises 17 and 18. Round each percent to the nearest whole.

17. At Keystone ski area in Colorado, approximately 41 of the 135 total ski runs are rated intermediate. What percent of the runs are intermediate?

18. At Telluride ski area in Colorado, about 28 of the 115 total ski runs are rated easy. What percent of the runs are easy?

For each food described, find the percent of total calories from fat. If necessary, round to the nearest tenth percent. See Example 2.

19. Ranch dressing serving size of 2 tablespoons

	Calories
Total	40
From fat	20

20. Unsweetened cocoa powder serving size of 1 tablespoon

	Calories
Total	20
From fat	5

21.

Nutrition Facts

Serving Size 1 pouch (20g)
Servings Per Container 6

Amount Per Serving

Calories	80
Calories from fat	10

	% Daily Value*
Total Fat 1g	**2%**
Sodium 45mg	**2%**
Total Carbohydrate 17g	**6%**
Sugars 9g	
Protein 0g	

Vitamin C	25%

Not a significant source of saturated fat, cholesterol, dietary fiber, vitamin A, calcium and iron.

*Percent Daily Values are based on a 2,000 calorie diet.

Artificial Fruit Snacks

22.

Nutrition Facts

Serving Size $\frac{1}{4}$ cup (33g)
Servings Per Container About 9

Amount Per Serving

Calories 190 **Calories from Fat** 130

	% Daily Value
Total Fat 16g	**24%**
Saturated Fat 3g	**16%**
Cholesterol 0mg	**0%**
Sodium 135mg	**6%**
Total Carbohydrate 9g	**3%**
Dietary Fiber 1g	**5%**
Sugars 2g	
Protein 5g	

Vitamin A 0% • Vitamin C 0%
Calcium 0% • Iron 8%

Peanut Mixture

23.

Nutrition Facts

Serving Size 18 crackers (29g)
Servings Per Container About 9

Amount Per Serving

Calories 120 **Calories from Fat** 35

	% Daily Value*
Total Fat 4g	**6%**
Saturated Fat 0.5g	**3%**
Polyunsaturated Fat 0g	
Monounsaturated Fat 1.5g	
Cholesterol 0mg	**0%**
Sodium 220mg	**9%**
Total Carbohydrate 21g	**7%**
Dietary Fiber 2g	**7%**
Sugars 3g	
Protein 2g	

Vitamin A 0% • Vitamin C 0%
Calcium 2% • Iron 4%
Phosphorus 10%

Snack Crackers

24.

Nutrition Facts

Serving Size 28 crackers (31g)
Servings Per Container About 6

Amount Per Serving

Calories 130 **Calories from Fat** 35

	% Daily Value*
Total Fat 4g	**6%**
Saturated Fat 2g	**10%**
Polyunsaturated Fat 1g	
Monounsaturated Fat 1g	
Cholesterol 0mg	**0%**
Sodium 470mg	**20%**
Total Carbohydrate 23g	**8%**
Dietary Fiber 1g	**4%**
Sugars 4g	
Protein 2g	

Vitamin A 0% • Vitamin C 0%
Calcium 0% • Iron 2%

Snack Crackers

Solve. If necessary, round money amounts to the nearest cent and all other amounts to the nearest tenth. See Examples 1 through 4.

25. A family paid $26,250 as a down payment for a home. If this represents 15% of the price of the home, find the price of the home.

26. A banker learned that $842.40 is withheld from his monthly check for taxes and insurance. If this represents 18% of his total pay, find the total pay.

27. An owner of a repair service company estimates that for every 40 hours a repairperson is on the job, he can bill for only 78% of the hours. The remaining hours, the repairperson is idle or driving to or from a job. Determine the number of hours per 40-hour week the owner can bill for a repairperson.

28. A manufacturer of electronic components expects 1.04% of its products to be defective. Determine the number of defective components expected in a batch of 28,350 components. Round to the nearest whole component.

29. A car manufacturer announced that next year, the price of a certain model of car will increase by 4.5%. This year the price is $19,286. Find the increase in price and the new price.

30. A union contract calls for a 6.5% salary increase for all employees. Determine the increase and the new salary that a worker currently making $58,500 under this contract can expect.

A popular extreme sport is artificial wall climbing. The photo shown is an artificial climbing wall. Exercises 31–32 are about the Footsloggers Climbing Tower in Boone, North Carolina.

31. A climber is resting at a height of 21 feet while on the Footsloggers Climbing Tower. If this is 60% of the tower's total height, find the height of the tower.

32. A group plans to climb the Footsloggers Climbing Tower at the group rate, once they save enough money. Thus far, $126 has been saved. If this is 70% of the total amount needed for the group, find the total price.

33. Tuition for an Ohio resident at the Columbus campus of Ohio State University was $8679 in 2008. The tuition increased by 15.3% during the period from 2008 to 2013. Find the increase and the tuition for the 2013–2014 school year. Round the increase to the nearest whole dollar. (*Source:* Ohio State University)

34. The population of Americans aged 65 and older was 40 million in 2010. That population is projected to increase by 80.5% by 2030. Find the increase and the projected 2030 population. (*Source:* Bureau of the Census)

35. From 2013–2014 to 2020–2021, the number of associate degrees awarded is projected to increase by 17.4%. If the number of associate degrees awarded in 2013–2014 was 943,000, find the increase and the projected number of associate degrees awarded in the 2020–2021 school year. (*Source:* National Center for Education Statistics)

36. From 2013–2014 to 2020–2021, the number of bachelor degrees awarded is projected to increase by 7.6%. If the number of bachelor degrees awarded in 2013–2014 was 1,836,000, find the increase and the projected number of bachelor degrees awarded in the 2020–2021 school year. (*Source:* National Center for Education Statistics)

Objective B   *Find the amount of increase and the percent increase. See Example 5.*

	Original Amount	New Amount	Amount of Increase	Percent Increase
37.	50	80		
38.	8	12		
39.	65	117		
40.	68	170		

Find the amount of decrease and the percent decrease. See Example 6.

	Original Amount	New Amount	Amount of Decrease	Percent Decrease
41.	8	6		
42.	25	20		
43.	160	40		
44.	200	162		

Solve. Round percents to the nearest tenth, if necessary. See Examples 5 and 6.

45. There are 150 calories in a cup of whole milk and only 84 in a cup of skim milk. In switching to skim milk, find the percent decrease in number of calories per cup.

46. In reaction to a slow economy, the number of employees at a soup company decreased from 530 to 477. What was the percent decrease in the number of employees?

47. Before taking a typing course, Geoffry Landers could type 32 words per minute. By the end of the course, he was able to type 76 words per minute. Find the percent increase.

48. The number of cable TV systems recently decreased from 10,845 to 10,700. Find the percent decrease.

49. The number of cell sites in the United States was 178,025 in 2005. By 2012, the number of cell sites had increased to 301,779. What was the percent increase? (*Source:* CTIA—The Wireless Association)

50. The population of Japan is expected to decrease from 127,799 thousand in 2011 to 97,076 thousand in 2050. Find the percent decrease. (*Source:* International Programs Center, Bureau of the Census, U.S. Dept. of Commerce)

Japan

Tokyo

51. In 2010, there were 3725 thousand elementary and secondary teachers employed in the United States. This number is expected to increase to 4205 thousand teachers in 2018. What is the percent increase? (*Source:* National Center for Education Statistics)

52. In 2010, approximately 493 thousand correctional officers were employed in the United States. By 2020, this number is expected to increase to 518 thousand correctional officers. What is the percent increase? (*Source:* Bureau of Labor Statistics)

53. In a recent 10-year period, the number of indoor cinema sites in the United States decreased from 5813 to 5331. What is this percent decrease? (*Source:* National Association of Theater Owners)

54. As the largest health care occupation, registered nurses held about 2.7 million jobs in 2010. The number of registered nurses is expected to be 7.1 million by 2020. What is the percent increase? (*Source:* Bureau of Labor Statistics)

Semiconductors are the foundation for solid-state electronics, including digital televisions (DTV). Businesses measure the overall decline in the production of televisions by studying DTV semiconductor revenue. (Examples of semiconductor use in DTVs include the main board for DTVs, power, LED backlighting, flat-panel screens, and integrated circuits used in remote controls, just to name a few.) Use this graph to answer Exercises 55 and 56.

Worldwide DTV Semiconductor Revenue

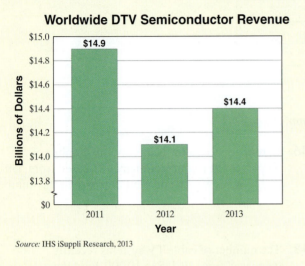

Source: IHS iSuppli Research, 2013

55. Find the percent decrease in DTV semiconductor revenue from 2011 to 2012.

56. Find the percent decrease in DTV semiconductor revenue from 2011 to 2013.

▶ 57. In 1940, the average size of a farm in the United States was 174 acres. In a recent year, the average size of a farm in the United States had increased to 420 acres. What is this percent increase? (*Source: National Agricultural Statistics Service*)

58. In 1994, there were 784 deaths from boating accidents in the United States. By 2012, the number of deaths from boating accidents had decreased to 651. What was the percent decrease? (*Source:* U.S. Coast Guard)

In 1999, Napster, a free online file-sharing service, debuted. iTunes, which debuted in 2003, is given credit for getting people to start paying for digital music. This 4-year gap between the starts of these two companies was only the beginning of the period of the decline in music revenue.

59. In 1999, total revenue from U.S. music sales and licensing was $14.6 billion. It was forecasted that this number would continue to drop until it reached $5.5 billion in 2014. Find this percent decrease in music revenue.

60. By comparing prices, a particular music album downloads from a low of $2.99 to a high of $7.99. Find the percent increase from $2.99 to $7.99.

Review

Perform each indicated operation. See Sections 4.3 through 4.5, 4.7, and 5.2 through 5.4.

61. 0.12×38

62. $29.4 \div 0.7$

63. $9.20 + 1.98$

64. $78 - 19.46$

65. $-\dfrac{3}{8} + \dfrac{5}{12}$

66. $\left(-\dfrac{3}{8}\right)\left(-\dfrac{5}{12}\right)$

67. $2\dfrac{4}{5} \div 3\dfrac{9}{10}$

68. $2\dfrac{4}{5} - 3\dfrac{9}{10}$

Concept Extensions

69. If a number is increased by 100%, how does the increased number compare with the original number? Explain your answer.

70. In your own words, explain what is wrong with the following statement: "Last year we had 80 students attend. This year we have a 50% increase or a total of 160 students attend."

71. Check the Chapter Opener graph. Use the last bar in the bar graph, the 2012 bar. Are your answers for Exercises **7** and **8** reasonable? Explain why or why not.

72. Check the Chapter Opener graph. Use the circle graph found there. If the 40,000 screens represent 100% of this circle graph, do the percents given and your answers in Exercises **7** and **8** seem reasonable? Explain why or why not.

Explain what errors were made by each student when solving percent of increase or decrease problems and then correct the errors. See the Concept Checks in this section.
"The population of a certain rural town was 150 in 1990, 180 in 2000, and 150 in 2010."

73. Find the percent of increase in population from 1990 to 2000.

Miranda's solution: Percent of increase $= \dfrac{30}{180} = 0.1\overline{6} \approx 16.7\%$

74. Find the percent of decrease in population from 2000 to 2010.

Jeremy's solution: Percent of decrease $= \dfrac{30}{150} = 0.20 = 20\%$

75. The percent of increase from 1990 to 2000 is the same as the percent decrease from 2000 to 2010. True or false?

Chris's answer: True because they had the same amount of increase as the amount of decrease.

Objectives

A Calculate Sales Tax and Total Price. ▶

B Calculate Commissions. ▶

C Calculate Discount and Sale Price. ▶

Objective A Calculating Sales Tax and Total Price ▶

Percents are frequently used in the retail trade. For example, most states charge a tax on certain items when purchased. This tax is called a **sales tax,** and retail stores collect it for the state. Sales tax is almost always stated as a percent of the purchase price.

A 9% sales tax rate on a purchase of a $10 calculator gives a sales tax of

$$\text{sales tax} = 9\% \text{ of } \$10 = 0.09 \cdot \$10.00 = \$0.90$$

The total price to the customer would be

purchase price   plus   sales tax

$$\$10.00 \quad + \quad \$0.90 = \$10.90$$

This example suggests the following equations:

> **Sales Tax and Total Price**
>
> $$\text{sales tax} = \text{tax rate} \cdot \text{purchase price}$$
> $$\text{total price} = \text{purchase price} + \text{sales tax}$$

In this section we round dollar amounts to the nearest cent.

Practice 1

If the sales tax rate is 8.5%, what is the sales tax and the total amount due on a $59.90 Goodgrip tire? (Round the sales tax to the nearest cent.)

Example 1   Finding Sales Tax and Purchase Price

Find the sales tax and the total price on the purchase of an $85.50 atlas in a city where the sales tax rate is 7.5%.

Solution: The purchase price is $85.50 and the tax rate is 7.5%.

sales tax = tax rate · purchase price

$$\text{sales tax} = 7.5\% \cdot \$85.50$$
$$= 0.075 \cdot \$85.5 \quad \text{Write 7.5\% as a decimal.}$$
$$\approx \$6.41 \quad \text{Rounded to the nearest cent}$$

Thus, the sales tax is $6.41. Next find the total price.

total price = purchase price + sales tax

$$\text{total price} = \$85.50 + \$6.41$$
$$= \$91.91$$

The sales tax on $85.50 is $6.41, and the total price is $91.91.

▮ **Work Practice 1**

Answer

1. tax: $5.09; total: $64.99

✓**Concept Check** The purchase price of a textbook is $50 and sales tax is 10%. If you are told by the cashier that the total price is $75, how can you tell that a mistake has been made?

Example 2 Finding a Sales Tax Rate

The sales tax on a $310 Sony flat-screen digital 32-inch television is $26.35. Find the sales tax rate.

Solution: Let r represent the unknown sales tax rate. Then

sales tax = tax rate · purchase price

$$\$26.35 = r \cdot \$310$$

$$\frac{26.35}{310} = \frac{r \cdot 310}{310} \qquad \text{Divide both sides by 310.}$$

$$0.085 = r \qquad \text{Simplify.}$$

$$8.5\% = r \qquad \text{Write 0.085 as a percent.}$$

The sales tax rate is 8.5%.

■ **Work Practice 2**

Practice 2

The sales tax on an $18,500 automobile is $1665. Find the sales tax rate.

Objective B Calculating Commissions

A **wage** is payment for performing work. Hourly wage, commissions, and salary are some of the ways wages can be paid. Many people who work in sales are paid a commission. An employee who is paid a **commission** is paid a percent of his or her total sales.

> **Commission**
>
> commission = commission rate · sales

Example 3 Finding the Amount of Commission

Sherry Souter, a real estate broker for Wealth Investments, sold a house for $214,000 last week. If her commission is 1.5% of the selling price of the home, find the amount of her commission.

Solution:

commission = commission rate · sales

commission =   1.5%   · $214,000

=   0.015   · $214,000   Write 1.5% as 0.015.

= $3210   Multiply.

Her commission on the house is $3210.

■ **Work Practice 3**

Practice 3

A sales representative for Office Product Copiers sold $47,632 worth of copy equipment and supplies last month. What is his commission for the month if he is paid a commission of 6.6% of his total sales for the month?

Answers

2. 9%   **3.** $3143.71

✓**Concept Check Answer**

Since $10\% = \frac{1}{10}$, the sales tax is $\frac{\$50}{10} = \$5$. The total price should have been $55.

Practice 4

A salesperson earns $645 for selling $4300 worth of appliances. Find the commission rate.

Example 4   Finding a Commission Rate

A salesperson earned $1560 for selling $13,000 worth of electronics equipment. Find the commission rate.

Solution: Let r stand for the unknown commission rate. Then

$$\text{commission} = \text{commission rate} \cdot \text{sales}$$

$$\$1560 = r \cdot \$13,000$$

$$\frac{1560}{13,000} = r \quad \text{Divide 1560 by 13,000, the number multiplied by } r.$$

$$0.12 = r \quad \text{Simplify.}$$

$$12\% = r \quad \text{Write 0.12 as a percent.}$$

The commission rate is 12%.

■ **Work Practice 4**

Objective C Calculating Discount and Sale Price

Suppose that an item that normally sells for $40 is on sale for 25% off. This means that the **original price** of $40 is reduced, or **discounted,** by 25% of $40, or $10. The **discount rate** is 25%, the **amount of discount** is $10, and the **sale price** is $40 − $10, or $30. Study the diagram below to visualize these terms.

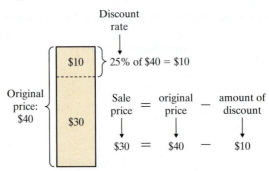

To calculate discounts and sale prices, we can use the following equations:

Discount and Sale Price

amount of discount = discount rate · original price

sale price = original price − amount of discount

Practice 5

A discontinued washer and dryer combo is advertised on sale for 35% off the regular price of $700. Find the amount of discount and the sale price.

Example 5   Finding a Discount and a Sale Price

An electric rice cooker that normally sells for $65 is on sale for 25% off. What is the amount of discount and what is the sale price?

Solution: First we find the amount of discount, or simply the discount.

$$\text{amount of discount} = \text{discount rate} \cdot \text{original price}$$

$$\text{amount of discount} = 25\% \cdot \$65$$

$$= 0.25 \cdot \$65 \quad \text{Write 25\% as 0.25.}$$

$$= \$16.25 \quad \text{Multiply.}$$

Answers

4. 15%   **5.** $245; $455

The discount is $16.25. Next, find the sale price.

sale price = original price − discount

sale price = $65 − $16.25

= $48.75   Subtract.

The sale price is $48.75.

◼ **Work Practice 5**

Vocabulary, Readiness & Video Check

Use the choices below to fill in each blank.

amount of discount	sale price	sales tax
commission	total price	

1. _____ = tax rate · purchase price

2. _____ = purchase price + sales tax

3. _____ = commission rate · sales

4. _____ = discount rate · original price

5. _____ = original price − amount of discount

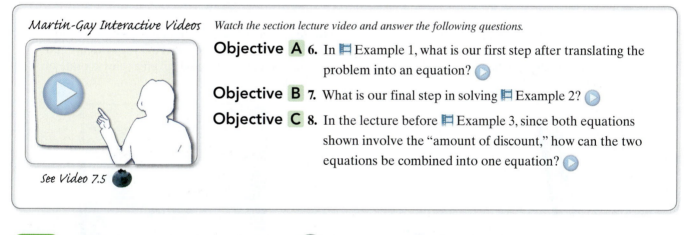

Martin-Gay Interactive Videos   Watch the section lecture video and answer the following questions.

Objective A 6. In ⊞ Example 1, what is our first step after translating the problem into an equation? ▶

Objective B 7. What is our final step in solving ⊞ Example 2? ▶

Objective C 8. In the lecture before ⊞ Example 3, since both equations shown involve the "amount of discount," how can the two equations be combined into one equation? ▶

See Video 7.5 🔵

7.5 **Exercise Set** MyMathLab® ▶

Objective A *Solve. See Examples 1 and 2.*

1. What is the sales tax on a jacket priced at $150 if the sales tax rate is 5%?

2. If the sales tax rate is 6%, find the sales tax on a microwave oven priced at $188.

3. The purchase price of a camcorder is $799. What is the total price if the sales tax rate is 7.5%?

4. A stereo system has a purchase price of $426. What is the total price if the sales tax rate is 8%?

5. A new large-screen television has a purchase price of $4790. If the sales tax on this purchase is $335.30, find the sales tax rate.

6. The sales tax on the purchase of a $6800 used car is $374. Find the sales tax rate.

7. The sales tax on a table saw is $10.20.

   a. What is the purchase price of the table saw (before tax) if the sales tax rate is 8.5%? (*Hint:* Use the sales tax equation and insert the replacement values.)

   b. Find the total price of the table saw.

8. The sales tax on a one-half-carat diamond ring is $76.

   a. Find the purchase price of the ring (before tax) if the sales tax rate is 9.5%. (See the hint for Exercise **7a.**)

   b. Find the total price of the ring.

9. A gold and diamond bracelet sells for $1800. Find the sales tax and the total price if the sales tax rate is 6.5%.

10. The purchase price of a personal computer is $1890. If the sales tax rate is 8%, what is the sales tax and the total price?

▶ 11. The sales tax on the purchase of a futon is $24.25. If the tax rate is 5%, find the purchase price of the futon.

12. The sales tax on the purchase of a TV-DVD combination is $32.85. If the tax rate is 9%, find the purchase price of the TV-DVD.

13. The sales tax is $98.70 on a stereo sound system purchase of $1645. Find the sales tax rate.

14. The sales tax is $103.50 on a necklace purchase of $1150. Find the sales tax rate.

15. A cell phone costs $210, a battery recharger costs $15, and batteries cost $5. What is the sales tax and total price for purchasing these items if the sales tax rate is 7%?

16. Ms. Warner bought a blouse for $35, a skirt for $55, and a blazer for $95. Find the sales tax and the total price she paid, given a sales tax rate of 6.5%.

Objective B   *Solve. See Examples 3 and 4.*

17. A sales representative for a large furniture warehouse is paid a commission rate of 4%. Find her commission if she sold $1,329,401 worth of furniture last year.

18. Rosie Davis-Smith is a beauty consultant for a home cosmetic business. She is paid a commission rate of 12.8%. Find her commission if she sold $1638 in cosmetics last month.

▶ 19. A salesperson earned a commission of $1380.40 for selling $9860 worth of paper products. Find the commission rate.

20. A salesperson earned a commission of $3575 for selling $32,500 worth of books to various bookstores. Find the commission rate.

21. How much commission will Jack Pruet make on the sale of a $325,900 house if he receives 1.5% of the selling price?

22. Frankie Lopez sold $9638 of jewelry this week. Find her commission for the week if she receives a commission rate of 5.6%.

23. A real estate agent earned a commission of $5565 for selling a house. If his rate is 3%, find the selling price of the house. (*Hint:* Use the commission equation and insert the replacement values.)

24. A salesperson earned $1750 for selling fertilizer. If her commission rate is 7%, find the selling price of the fertilizer. (See the hint for Exercise **23.**)

Objective **C** *Find the amount of discount and the sale price. See Example 5.*

	Original Price	Discount Rate	Amount of Discount	Sale Price
25.	$89	10%		
26.	$74	20%		
27.	$196.50	50%		
28.	$110.60	40%		
29.	$410	35%		
30.	$370	25%		
31.	$21,700	15%		
32.	$17,800	12%		

▶ **33.** A $300 fax machine is on sale for 15% off. Find the amount of discount and the sale price.

34. A $4295 designer dress is on sale for 30% off. Find the amount of discount and the sale price.

Objectives **A** **B** **Mixed Practice** *Complete each table.*

	Purchase Price	Tax Rate	Sales Tax	Total Price
35.	$305	9%		
36.	$243	8%		
37.	$56	5.5%		
38.	$65	8.4%		

	Sale	Commission Rate	Commission
39.	$235,800	3%	
40.	$195,450	5%	
41.	$17,900		$1432
42.	$25,600		$2304

Review

Multiply. See Sections 4.3, 5.3, and 5.5.

43. $2000 \cdot \dfrac{3}{10} \cdot 2$

44. $500 \cdot \dfrac{2}{25} \cdot 3$

45. $400 \cdot \dfrac{3}{100} \cdot 11$

46. $1000 \cdot \dfrac{1}{20} \cdot 5$

47. $600 \cdot 0.04 \cdot \dfrac{2}{3}$

48. $6000 \cdot 0.06 \cdot \dfrac{3}{4}$

Concept Extensions

Solve. See the Concept Check in this section.

49. Your purchase price is $68 and the sales tax rate is 9.5%. Round each amount and use the rounded amounts to estimate the total price. Choose the best estimate.
a. $105   **b.** $58   **c.** $93   **d.** $77

50. Your purchase price is $200 and the tax rate is 10%. Choose the best estimate of the total price.
a. $190   **b.** $210   **c.** $220   **d.** $300

Tipping *One very useful application of percent is mentally calculating a tip. Recall that to find 10% of a number, simply move the decimal point one place to the left. To find 20% of a number, just double 10% of the number. To find 15% of a number, find 10% and then add to that number half of the 10% amount. Mentally fill in the chart below. To do so, start by rounding the bill amount to the nearest dollar.*

Tipping Chart

	Bill Amount	10%	15%	20%
51.	$40.21			
52.	$15.89			
53.	$72.17			
54.	$9.33			

55. Suppose that the original price of a shirt is $50. Which is better, a 60% discount or a discount of 30% followed by a discount of 35% of the reduced price? Explain your answer.

56. Which is better, a 30% discount followed by an additional 25% off or a 20% discount followed by an additional 40% off? To see, suppose an item costs $100 and calculate each discounted price. Explain your answer.

57. A diamond necklace sells for $24,966. If the tax rate is 7.5%, find the total price.

58. A house recently sold for $562,560. The commission rate on the sale is 5.5%. If the real estate agent is to receive 60% of the commission, find the amount received by the agent.

7.6   Percent and Problem Solving: Interest

Objectives

A Calculate Simple Interest.

B Calculate Compound Interest.

Objective A Calculating Simple Interest

Interest is money charged for using other people's money. When you borrow money, you pay interest. When you loan or invest money, you earn interest. The money borrowed, loaned, or invested is called the **principal amount,** or simply **principal.** Interest is normally stated in terms of a percent of the principal for a given period of time. The **interest rate** is the percent used in computing the interest. Unless stated otherwise, *the rate is understood to be per year.* When the interest is computed on the original principal, it is called **simple interest.** Simple interest is calculated using the following equation:

Simple Interest

Simple Interest = Principal · Rate · Time

$$I = P \cdot R \cdot T$$

where the rate is understood to be per year and time is in years.

Example 1 Finding Simple Interest

Find the simple interest after 2 years on $500 at an interest rate of 12%.

Solution: In this example, $P = \$500$, $R = 12\%$, and $T = 2$ years. Replace the variables with values in the formula $I = PRT$.

$$I = P \cdot R \cdot T$$
$$I = \$500 \cdot 12\% \cdot 2 \quad \text{Let } P = \$500, R = 12\%, \text{ and } T = 2.$$
$$= \$500 \cdot (0.12) \cdot 2 \quad \text{Write 12\% as a decimal.}$$
$$= \$120 \quad \text{Multiply.}$$

The simple interest is $120.

◼ Work Practice 1

If time is not given in years, we need to convert the given time to years.

Example 2 Finding Simple Interest

A recent college graduate borrowed $2400 at 10% simple interest for 8 months to buy a used Toyota Corolla. Find the simple interest he paid.

Solution: Since there are 12 months in a year, we first find what part of a year 8 months is.

$$8 \text{ months} = \frac{8}{12} \text{ year} = \frac{2}{3} \text{ year}$$

Now we find the simple interest.

$$I = P \cdot R \cdot T$$
$$= \$2400 \cdot (0.10) \cdot \frac{2}{3} \quad \text{Let } P = \$2400, R = 10\% \text{ or } 0.10, \text{ and } T = \frac{2}{3}.$$
$$= \$160$$

The interest on his loan is $160.

◼ Work Practice 2

✓**Concept Check** Suppose in Example 2 you had obtained an answer of $16,000. How would you know that you had made a mistake in this problem?

When money is borrowed, the borrower pays the original amount borrowed, or the principal, as well as the interest. When money is invested, the investor receives the original amount invested, or the principal, as well as the interest. In either case, the **total amount** is the sum of the principal and the interest.

Finding the Total Amount of a Loan or Investment

total amount (paid or received) = principal + interest

Practice 1

Find the simple interest after 5 years on $875 at an interest rate of 7%.

Practice 2

A student borrowed $1500 for 9 months on her credit card at a simple interest rate of 20%. How much interest did she pay?

Answers
1. $306.25 **2.** $225

✓**Concept Check Answer**
$16,000 is too much interest.

Practice 3

If $2100 is borrowed at a simple interest rate of 13% for 6 months, find the total amount paid.

Example 3 Finding the Total Amount of an Investment

An accountant invested $2000 at a simple interest rate of 10% for 2 years. What total amount of money will she have from her investment in 2 years?

Solution: First we find her interest.

$$I = P \cdot R \cdot T$$
$$= \$2000 \cdot (0.10) \cdot 2 \quad \text{Let } P = \$2000, R = 10\% \text{ or } 0.10, \text{ and } T = 2.$$
$$= \$400$$

The interest is $400.

Next, we add the interest to the principal.

total amount = principal + interest
↓   ↓   ↓
total amount = $2000 + $400
= $2400

After 2 years, she will have a total amount of $2400.

■ Work Practice 3

✓**Concept Check** Which investment would earn more interest: an amount of money invested at 8% interest for 2 years, or the same amount of money invested at 8% for 3 years? Explain.

Objective B Calculating Compound Interest ▶

Recall that simple interest depends on the original principal only. Another type of interest is compound interest. **Compound interest** is computed not only on the principal, but also on the interest already earned in previous compounding periods. Compound interest is used more often than simple interest.

Let's see how compound interest differs from simple interest. Suppose that $2000 is invested at 7% interest **compounded annually** for 3 years. This means that interest is added to the principal at the end of each year and that next year's interest is computed on this new amount. In this section, we round dollar amounts to the nearest cent.

	Amount at Beginning of Year	Principal	·	Rate	·	Time	= Interest	Amount at End of Year
1st year	$2000	$2000	·	0.07	·	1	= $140	$2000 + 140 = $2140
2nd year	$2140	$2140	·	0.07	·	1	= $149.80	$2140 + 149.80 = $2289.80
3rd year	$2289.80	$2289.80	·	0.07	·	1	= $160.29	$2289.80 + 160.29 = $2450.09

The compound interest earned can be found by

total amount − original principal = compound interest
↓   ↓   ↓
$2450.09   − $2000   = $450.09

The simple interest earned would have been

principal · rate · time = interest
↓   ↓   ↓   ↓
$2000 · 0.07 · 3   = $420

Since compound interest earns "interest on interest," compound interest earns more than simple interest.

Computing compound interest using the method above can be tedious. We can use a calculator and the compound interest formula on the next page to compute compound interest more quickly.

Answer

3. $2236.50

✓**Concept Check Answers**

8% for 3 years. Since the interest rate is the same, the longer you keep the money invested, the more interest you earn.

Compound Interest Formula

The total amount A in an account is given by

$$A = P\left(1 + \frac{r}{n}\right)^{n \cdot t}$$

where P is the principal, r is the interest rate written as a decimal, t is the length of time in years, and n is the number of times compounded per year.

Example 4 $1800 is invested at 2% interest compounded annually. Find the total amount after 3 years.

Solution: "Compounded annually" means 1 time a year, so

$n = 1$. Also, $P = \$1800$, $r = 2\% = 0.02$, and $t = 3$ years.

$$A = P\left(1 + \frac{r}{n}\right)^{n \cdot t}$$

$$= 1800\left(1 + \frac{0.02}{1}\right)^{1 \cdot 3}$$

$$= 1800(1.02)^3$$

$$\approx 1910.17 \qquad \text{Round to 2 decimal places.}$$

Helpful Hint Remember order of operations. **First** evaluate $(1.02)^3$, then multiply by 1800.

The total amount at the end of 3 years is $1910.17.

Work Practice 4

Practice 4

$3000 is invested at 4% interest compounded annually. Find the total amount after 6 years.

Example 5 Finding Total Amount Received from an Investment

$4000 is invested at 5.3% compounded quarterly for 10 years. Find the total amount at the end of 10 years.

Solution: "Compounded quarterly" means 4 times a year, so

$n = 4$. Also, $P = \$4000$, $r = 5.3\% = 0.053$, and $t = 10$ years.

$$A = P\left(1 + \frac{r}{n}\right)^{n \cdot t}$$

$$= 4000\left(1 + \frac{0.053}{4}\right)^{4 \cdot 10}$$

$$= 4000(1.01325)^{40}$$

$$\approx 6772.12$$

The total amount after 10 years is $6772.12.

Work Practice 5

Practice 5

$5500 is invested at $6\frac{1}{4}\%$ compounded *daily* for 5 years. Find the total amount at the end of 5 years. (Use 1 year = 365 days.)

Answers

4. $3795.96 **5.** $7517.41

Calculator Explorations Compound Interest Formula

For a review of using your calculator to evaluate compound interest, see this box.

Let's review the calculator keys pressed to evaluate the expression in Example 5,

$$4000\left(1 + \frac{0.053}{4}\right)^{40}$$

To evaluate, press the keys

[4000] [×] [(] [(] [1] [+] [0.053] [÷] [4] [)] [y^x] or [∧] [40] then [=] or [ENTER]. The display will read [6772.117549]. Rounded to 2 decimal places, this is 6772.12.

Find the compound interest.

1. $600, 5 years, 9%, compounded quarterly

2. $10,000, 15 years, 4%, compounded daily

3. $1200, 20 years, 11%, compounded annually

4. $5800, 1 year, 7%, compounded semiannually

5. $500, 4 years, 6%, compounded quarterly

6. $2500, 19 years, 5%, compounded daily

Vocabulary, Readiness & Video Check

Use the choices below to fill in each blank. Choices may be used more than once.

| total amount | simple | principal amount | compound |

1. To calculate _____ interest, use $I = P \cdot R \cdot T$.

2. To calculate _____ interest, use $A = P\left(1 + \dfrac{r}{n}\right)^{n \cdot t}$.

3. _____ interest is computed not only on the original principal, but also on interest already earned in previous compounding periods.

4. When interest is computed on the original principal only, it is called _____ interest.

5. _____ (paid or received) = principal + interest.

6. The _____ is the money borrowed, loaned, or invested.

Martin-Gay Interactive Videos   Watch the section lecture video and answer the following questions.

Objective A **7.** Complete this statement based on the lecture before ⊞ Example 1: Simple interest is charged on the _____ only. ▶

Objective B **8.** In ⊞ Example 2, how often is the interest compounded and what number does this translate to in the formula? ▶

See Video 7.6

7.6 Exercise Set MyMathLab® ▶

Objective A *Find the simple interest. See Examples 1 and 2.*

	Principal	Rate	Time
1.	$200	8%	2 years
3.	$160	11.5%	4 years
5.	$5000	10%	$1\frac{1}{2}$ years
7.	$375	18%	6 months
9.	$2500	16%	21 months

	Principal	Rate	Time
2.	$800	9%	3 years
4.	$950	12.5%	5 years
6.	$1500	14%	$2\frac{1}{4}$ years
8.	$775	15%	8 months
10.	$1000	10%	18 months

Solve. See Examples 1 through 3.

▶ **11.** A company borrows $162,500 for 5 years at a simple interest rate of 12.5%. Find the interest paid on the loan and the total amount paid back.

12. $265,000 is borrowed to buy a house. If the simple interest rate on the 30-year loan is 8.25%, find the interest paid on the loan and the total amount paid back.

13. A money market fund advertises a simple interest rate of 9%. Find the total amount received on an investment of $5000 for 15 months.

14. The Real Service Company takes out a 270-day (9-month) short-term, simple interest loan of $4500 to finance the purchase of some new equipment. If the interest rate is 14%, find the total amount that the company pays back.

15. Marsha borrows $8500 and agrees to pay it back in 4 years. If the simple interest rate is 17%, find the total amount she pays back.

16. An 18-year-old is given a high school graduation gift of $2000. If this money is invested at 8% simple interest for 5 years, find the total amount.

Objective **B** *Find the total amount in each compound interest account. See Examples 4 and 5.*

17. $6150 is compounded semiannually at a rate of 14% for 15 years.

18. $2060 is compounded annually at a rate of 15% for 10 years.

19. $1560 is compounded daily at a rate of 8% for 5 years.

20. $1450 is compounded quarterly at a rate of 10% for 15 years.

21. $10,000 is compounded semiannually at a rate of 9% for 20 years.

22. $3500 is compounded daily at a rate of 8% for 10 years.

23. $2675 is compounded annually at a rate of 9% for 1 year.

24. $6375 is compounded semiannually at a rate of 10% for 1 year.

25. $2000 is compounded annually at a rate of 8% for 5 years.

26. $2000 is compounded semiannually at a rate of 8% for 5 years.

27. $2000 is compounded quarterly at a rate of 8% for 5 years.

28. $2000 is compounded daily at a rate of 8% for 5 years.

Review

Find the perimeter of each figure. See Section 1.3.

△ **29.**
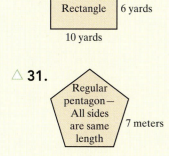

△ **30.**
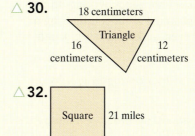

△ **31.**

Regular pentagon— All sides are same length   7 meters

△ **32.**

Square   21 miles

Perform the indicated operations. See Section 4.3 through 4.6.

33. $\dfrac{x}{4} + \dfrac{x}{5}$

34. $-\dfrac{x}{4} \div \left(-\dfrac{x}{5}\right)$

35. $\left(\dfrac{2}{3}\right)\left(-\dfrac{1}{3}\right) - \left(\dfrac{9}{10}\right)\left(\dfrac{2}{5}\right)$

36. $\dfrac{3}{11} \div \dfrac{9}{22} \cdot \dfrac{1}{3}$

Concept Extensions

37. Review Exercises **25** through **28**. As the number of compoundings increases per year, how is the total amount affected?

38. Explain how to find the amount of interest in a compounded account.

39. Compare the following accounts: Account 1: $1000 is invested for 10 years at a simple interest rate of 6%. Account 2: $1000 is compounded semiannually at a rate of 6% for 10 years. Discuss how the interest is computed for each account. Determine which account earns more interest. Why?

Chapter 7 Group Activity

Fastest-Growing Occupations

According to U.S. Bureau of Labor Statistics projections, the careers listed below are the top ten fastest-growing jobs ranked by expected percent increase through the year 2020. (Source: Bureau of Labor Statistics)

Occupation	Employment in 2010	Percent Change	Expected Employment in 2020
Personal care aides	861,000	70.5%	
Home health aides	1,017,700	69.4%	
Biomedical engineers	15,700	61.7%	
Helpers—brickmasons, blockmasons, stonemasons, and tile and marble setters	29,400	60.1%	
Helpers—carpenters	46,500	55.7%	
Veterinary technologists and technicians	80,200	52.0%	
Reinforcing iron and rebar workers	19,100	48.6%	
Physical therapist assistants	67,400	45.7%	
Helpers—pipelayers, plumbers, pipefitters, and steamfitters	57,900	45.4%	
Meeting, convention, and event planners	71,600	43.7%	

What do most of these fast-growing occupations have in common? They require knowledge of math! For some careers, such as home health aides, veterinary technicians, and biomedical engineers, the ways math is used on the job may be obvious. For other occupations, the use of math may not be quite as apparent. However, tasks common to many jobs—filling in a time sheet, writing up an expense or mileage report, planning a budget, figuring a bill, ordering supplies, and even making a work schedule—all require math.

This activity may be completed by working in groups or individually.

1. List the top five occupations by order of employment figures for 2010.

2. Using the 2010 employment figures and the percent increase from 2010 to 2020, find the expected 2020 employment figures for each occupation listed in the table. Round to the nearest hundred.

3. List the top five occupations by order of employment figures for 2020. Did the order change at all from 2010? Explain.

Chapter 7 Vocabulary Check

Fill in each blank with one of the words or phrases listed below. Some words may be used more than once.

percent	sales tax	is		0.01	$\dfrac{1}{100}$		amount of discount	percent of decrease	total price
base	of		amount	100%	compound interest	percent of increase	sale price		commission

1. In a mathematical statement, _____ usually means "multiplication."

2. In a mathematical statement, _____ means "equal."

3. _____ means "per hundred."

4. _____ is computed not only on the principal, but also on interest already earned in previous compounding periods.

5. In the percent proportion, $\dfrac{\rule{2cm}{0.4pt}}{\rule{2cm}{0.4pt}} = \dfrac{\text{percent}}{100}$.

6. To write a decimal or fraction as a percent, multiply by _____.

7. The decimal equivalent of the % symbol is _____.

8. The fraction equivalent of the % symbol is _____.

9. The percent equation is _____ · percent = _____.

10. _____ $= \dfrac{\text{amount of decrease}}{\text{original amount}}$.

11. _____ $= \dfrac{\text{amount of increase}}{\text{original amount}}$.

12. _____ = tax rate · purchase price.

13. _____ = purchase price + sales tax.

14. _____ = commission rate · sales.

15. _____ = discount rate · original price.

16. _____ = original price − amount of discount.

> **Helpful Hint** ▶ Are you preparing for your test? Don't forget to take the Chapter 7 Test on page 527. Then check your answers at the back of the text and use the Chapter Test Prep Videos to see the fully worked-out solutions to any of the exercises you want to review.

7 Chapter Highlights

Definitions and Concepts	Examples
Section 7.1 Percents, Decimals, and Fractions	
Percent means "per hundred." The % symbol denotes percent.	$51\% = \dfrac{51}{100}$ 51 per 100 $7\% = \dfrac{7}{100}$ 7 per 100
To write a percent as a decimal, replace the % symbol with its decimal equivalent, 0.01, and multiply. **To write a decimal as a percent,** multiply by 100%.	$32\% = 32(0.01) = 0.32$ $0.08 = 0.08(100\%) = 08.\% = 8\%$
To write a percent as a fraction, replace the % symbol with its fraction equivalent, $\dfrac{1}{100}$, and multiply. **To write a fraction as a percent,** multiply by 100%.	$25\% = 25 \cdot \dfrac{1}{100} = \dfrac{25}{100} = \dfrac{25}{4 \cdot 25} = \dfrac{1}{4}$ $\dfrac{1}{6} = \dfrac{1}{6} \cdot 100\% = \dfrac{1}{6} \cdot \dfrac{100}{1}\% = \dfrac{100}{6}\% = 16\dfrac{2}{3}\%$

(Continued)

Definitions and Concepts	Examples

Section 7.2   Solving Percent Problems with Equations

Three key words in the statement of a percent problem are

of, which means multiplication ($\cdot$)

is, which means equal ($=$)

what (or some equivalent word or phrase), which stands for the unknown (x)

Solve:

6	is	12%	of	what number?
↓	↓	↓	↓	↓
6	$=$	12%	$\cdot$	x

$6 = 0.12 \cdot x$ Write 12% as a decimal.

$\dfrac{6}{0.12} = \dfrac{0.12 \cdot x}{0.12}$ Divide both sides by 0.12.

$50 = x$

Thus, 6 is 12% of 50.

Section 7.3   Solving Percent Problems with Proportions

Percent Proportion

$$\frac{\text{amount}}{\text{base}} = \frac{\text{percent}}{100} \quad \leftarrow \text{always 100}$$

or

$$\text{amount} \rightarrow \frac{a}{b} = \frac{p}{100} \leftarrow \text{percent}$$
$$\text{base} \rightarrow$$

Solve:

20.4 is what percent of 85?

amount   percent   base

$$\text{amount} \rightarrow \frac{20.4}{85} = \frac{p}{100} \leftarrow \text{percent}$$
$$\text{base} \rightarrow$$

$20.4 \cdot 100 = 85 \cdot p$ Set cross products equal.

$2040 = 85 \cdot p$ Multiply.

$\dfrac{2040}{85} = \dfrac{85 \cdot p}{85}$ Divide both sides by 85.

$24 = p$ Simplify.

Thus, 20.4 is 24% of 85.

Section 7.4   Applications of Percent

Percent Increase

$$\text{percent increase} = \frac{\text{amount of increase}}{\text{original amount}}$$

Percent Decrease

$$\text{percent decrease} = \frac{\text{amount of decrease}}{\text{original amount}}$$

A town with a population of 16,480 decreased to 13,870 over a 12-year period. Find the percent decrease. Round to the nearest whole percent.

$$\text{amount of decrease} = 16,480 - 13,870$$
$$= 2610$$

$$\text{percent decrease} = \frac{\text{amount of decrease}}{\text{original amount}}$$

$$= \frac{2610}{16,480}$$

$$\approx 0.16$$

$$= 16\%$$

The town's population decreased by about 16%.

Definitions and Concepts	Examples

Section 7.5   Percent and Problem Solving: Sales Tax, Commission, and Discount

Sales Tax

    sales tax = sales tax rate · purchase price

  total price = purchase price + sales tax

Find the sales tax and the total price of a purchase of $42.00 if the sales tax rate is 9%.

$$\text{sales tax} = \text{sales tax rate} \cdot \text{purchase price}$$
$$\downarrow \qquad\qquad \downarrow \qquad\qquad \downarrow$$
$$\text{sales tax} = \qquad 9\% \qquad \cdot \qquad \$42$$
$$= 0.09 \cdot \$42$$
$$= \$3.78$$

The total price is

$$\text{total price} = \text{purchase price} + \text{sales tax}$$
$$\downarrow \qquad\qquad \downarrow \qquad\qquad \downarrow$$
$$\text{total price} = \qquad \$42.00 \qquad + \qquad \$3.78$$
$$= \$45.78$$

The total price is $45.78.

Commission

  commission = commission rate · sales

A salesperson earns a commission of 3%. Find the commission from sales of $12,500 worth of appliances.

$$\text{commission} = \text{commission rate} \cdot \text{sales}$$
$$\downarrow \qquad\qquad \downarrow \qquad\qquad \downarrow$$
$$\text{commission} = \qquad 3\% \qquad \cdot \$12{,}500$$
$$= 0.03 \cdot 12{,}500$$
$$= \$375$$

The commission is $375.

Discount and Sale Price

  amount of discount = discount rate · original price

  sale price = original price − amount of discount

A suit is priced at $320 and is on sale today for 25% off. What is the sale price?

$$\begin{array}{c}\text{amount of}\\\text{discount}\end{array} = \text{discount rate} \cdot \text{original price}$$
$$\downarrow \qquad\qquad \downarrow \qquad\qquad \downarrow$$
$$\begin{array}{c}\text{amount of}\\\text{discount}\end{array} = \qquad 25\% \qquad \cdot \qquad \$320$$
$$= 0.25 \cdot 320$$
$$= \$80$$

$$\text{sale price} = \text{original price} - \begin{array}{c}\text{amount of}\\\text{discount}\end{array}$$
$$\downarrow \qquad\qquad \downarrow \qquad\qquad \downarrow$$
$$\text{sale price} = \qquad \$320 \qquad - \qquad \$80$$
$$= \$240$$

The sale price is $240.

(Continued)

Definitions and Concepts	Examples

Section 7.6   Percent and Problem Solving: Interest

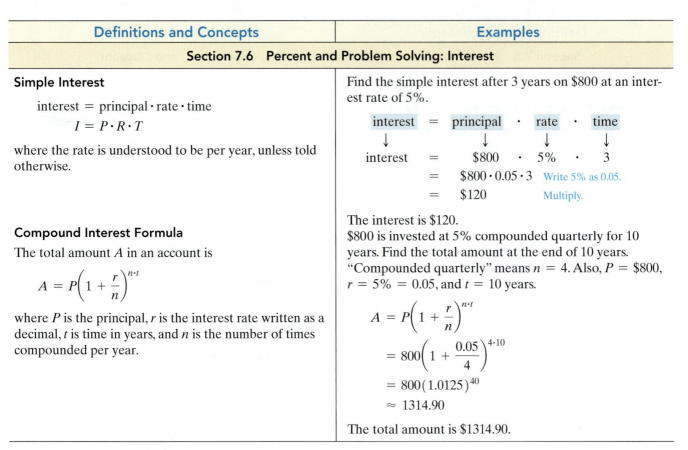

Simple Interest

$$\text{interest} = \text{principal} \cdot \text{rate} \cdot \text{time}$$
$$I = P \cdot R \cdot T$$

where the rate is understood to be per year, unless told otherwise.

Find the simple interest after 3 years on $800 at an interest rate of 5%.

interest	=	principal	·	rate	·	time
↓		↓		↓		↓
interest	=	$800	·	5%	·	3
	=	$800 · 0.05 · 3	Write 5% as 0.05.			
	=	$120	Multiply.			

The interest is $120.

Compound Interest Formula

The total amount A in an account is

$$A = P\left(1 + \frac{r}{n}\right)^{n \cdot t}$$

where P is the principal, r is the interest rate written as a decimal, t is time in years, and n is the number of times compounded per year.

$800 is invested at 5% compounded quarterly for 10 years. Find the total amount at the end of 10 years. "Compounded quarterly" means $n = 4$. Also, $P = \$800$, $r = 5\% = 0.05$, and $t = 10$ years.

$$A = P\left(1 + \frac{r}{n}\right)^{n \cdot t}$$
$$= 800\left(1 + \frac{0.05}{4}\right)^{4 \cdot 10}$$
$$= 800(1.0125)^{40}$$
$$\approx 1314.90$$

The total amount is $1314.90.

Chapter 7   Review

(7.1) *Solve.*

1. In a survey of 100 adults, 37 preferred pepperoni on their pizzas. What percent preferred pepperoni?

2. A basketball player made 77 out 100 attempted free throws. What percent of free throws was made?

Write each percent as a decimal.

3. 26%
4. 75%
5. 3.5%
6. 1.5%

7. 275%
8. 400%
9. 47.85%
10. 85.34%

Write each decimal as a percent.

11. 1.6
12. 0.055
13. 0.076
14. 0.085

15. 0.71
16. 0.65
17. 6
18. 9

Write each percent as a fraction or mixed number in simplest form.

19. 7%
20. 15%
21. 25%
22. 8.5%

23. 10.2%
24. $16\frac{2}{3}\%$
25. $33\frac{1}{3}\%$
26. 110%

Write each fraction or mixed number as a percent.

27. $\dfrac{2}{5}$ **28.** $\dfrac{7}{10}$ **29.** $\dfrac{7}{12}$ **30.** $1\dfrac{2}{3}$

31. $1\dfrac{1}{4}$ **32.** $\dfrac{3}{5}$ **33.** $\dfrac{1}{16}$ **34.** $\dfrac{5}{8}$

(7.2) Translating *Translate each to an equation and solve.*

35. 1250 is 1.25% of what number?

36. What number is $33\dfrac{1}{3}$% of 24,000?

37. 124.2 is what percent of 540?

38. 22.9 is 20% of what number?

39. What number is 17% of 640?

40. 693 is what percent of 462?

(7.3) Translating *Translate each to a proportion and solve.*

41. 104.5 is 25% of what number?

42. 16.5 is 5.5% of what number?

43. What number is 30% of 532?

44. 63 is what percent of 35?

45. 93.5 is what percent of 85?

46. What number is 33% of 500?

(7.4) *Solve.*

47. In a survey of 2000 people, it was found that 1320 have a microwave oven. Find the percent of people who own microwaves.

48. Of the 12,360 freshmen entering County College, 2000 are enrolled in prealgebra. Find the percent of entering freshmen who are enrolled in prealgebra. Round to the nearest whole percent.

49. The number of violent crimes in a city decreased from 675 to 534. Find the percent decrease. Round to the nearest tenth percent.

50. The current charge for dumping waste in a local landfill is $16 per cubic foot. To cover new environmental costs, the charge will increase to $33 per cubic foot. Find the percent increase.

51. A local union negotiated a new contract that increases the hourly pay 15% over last year's pay. The old hourly rate was $11.50. Find the new hourly rate rounded to the nearest cent.

52. This year the fund drive for a charity collected $215,000. Next year, a 4% decrease is expected. Find how much is expected to be collected in next year's drive.

(7.5) *Solve.*

53. If the sales tax rate is 9.5%, what is the total amount charged for a $250 coat?

54. Find the sales tax paid on a $25.50 purchase if the sales tax rate is 8.5%. Round to the nearest cent.

55. Russ James is a sales representative for a chemical company and is paid a commission rate of 5% on all sales. Find his commission if he sold $100,000 worth of chemicals last month.

56. Carol Sell is a sales clerk in a clothing store. She receives a commission of 7.5% on all sales. Find her commission for the week if her sales for the week were $4005. Round to the nearest cent.

57. A $3000 mink coat is on sale for 30% off. Find the discount and the sale price.

58. A $90 calculator is on sale for 10% off. Find the discount and the sale price.

(7.6) *Solve.*

59. Find the simple interest due on $4000 loaned for 4 months with a 12% interest rate.

60. Find the simple interest due on $6500 loaned for 3 months with a 20% interest rate.

61. Find the total amount in an account if $5500 is compounded annually at 12% for 15 years.

62. Find the total amount in an account if $6000 is compounded semiannually at 11% for 10 years.

63. Find the total amount in an account if $100 is compounded quarterly at 12% for 5 years.

64. Find the total amount in an account if $1000 is compounded quarterly at 18% for 20 years.

Mixed Review

Write each percent as a decimal.

65. 3.8%

66. 124.5%

Write each decimal as a percent.

67. 0.54

68. 95.2

Write each percent as a fraction or mixed number in simplest form.

69. 47%

70. 5.6%

Write each fraction or mixed number as a percent.

71. $\dfrac{1}{8}$

72. $\dfrac{6}{5}$

Translating *Translate each into an equation and solve.*

73. 43 is 16% of what number?

74. 27.5 is what percent of 25?

75. What number is 36% of 1968?

76. 67 is what percent of 50?

Translate each into a proportion and solve.

77. 75 is what percent of 25?

78. What number is 16% of 240?

79. 28 is 5% of what number?

80. 52 is what percent of 16?

Solve.

81. The total number of cans in a soft drink machine is 300. If 78 soft drinks have been sold, find the percent of soft drink cans that have been sold.

82. A home valued at $96,950 last year has lost 7% of its value this year. Find the loss in value.

83. A dinette set sells for $568.00. If the sales tax rate is 8.75%, find the total price of the dinette set.

84. The original price of a video game is $23.00. It is on sale for 15% off. What is the amount of the discount?

85. A candy salesman makes a commission of $1.60 from each case of candy he sells. If a case of candy costs $12.80, what is his rate of commission?

86. Find the total amount due on a 6-month loan of $1400 at a simple interest rate of 13%.

87. Find the total amount due on a loan of $5500 for 9 years at 12.5% simple interest.

Answers

Write each percent as a decimal.

1. 85%

2. 500%

3. 0.6%

Write each decimal as a percent.

4. 0.056

5. 6.1

6. 0.35

Write each percent as a fraction or a mixed number in simplest form.

7. 120%

8. 38.5%

9. 0.2%

Write each fraction or mixed number as a percent.

10. $\dfrac{11}{20}$

11. $\dfrac{3}{8}$

12. $1\dfrac{3}{4}$

13. Bottled water accounts for $\dfrac{1}{5}$ of the total noncarbonated beverage category in convenience stores. Write $\dfrac{1}{5}$ as a percent. (*Source:* Grocery Manufacturers of America)

14. In small firms in the United States, 64% of full-time employees receive medical insurance benefits. Write 64% as a fraction. (*Source:* U.S. Bureau of Labor Statistics)

Solve.

15. What number is 42% of 80?

16. 0.6% of what number is 7.5?

17. 567 is what percent of 756?

1. _____

2. _____

3. _____

4. _____

5. _____

6. _____

7. _____

8. _____

9. _____

10. _____

11. _____

12. _____

13. _____

14. _____

15. _____

16. _____

17. _____

527

18. _____

19. _____

20. _____

21. _____

22. _____

23. _____

24. _____

25. _____

26. _____

27. _____

28. _____

Solve. If necessary, round percents to the nearest tenth, dollar amounts to the nearest cent, and all other numbers to the nearest whole.

18. An alloy is 12% copper. How much copper is contained in 320 pounds of this alloy?

19. A farmer in Nebraska estimates that 20% of his potential crop, or $11,350 worth, has been lost to a hard freeze. Find the total value of his potential crop.

20. If the local sales tax rate is 8.25%, find the total amount charged for a stereo system priced at $354.

21. A town's population increased from 25,200 to 26,460. Find the percent increase.

22. A $120 framed picture is on sale for 15% off. Find the discount and the sale price.

23. Randy Nguyen is paid a commission rate of 4% on all sales. Find Randy's commission if his sales were $9875.

24. A sales tax of $13.77 is added to an item's price of $152.99. Find the sales tax rate. Round to the nearest whole percent.

25. Find the simple interest earned on $2000 saved for $3\frac{1}{2}$ years at an interest rate of 9.25%.

26. $1365 is compounded annually at 8%. Find the total amount in the account after 5 years.

27. A couple borrowed $400 from a bank at 13.5% simple interest for 6 months for car repairs. Find the total amount due the bank at the end of the 6-month period.

28. In a recent 5-year period, the number of major crimes reported in New York City decreased from 149,488 to 142,760. Find the percent decrease. (_Source:_ New York State Division of Criminal Justice Services)

1. Multiply: 236×86

2. Multiply: 409×76

3. Subtract 7 from -3.

4. Subtract -2 from 8.

5. Solve: $x - 2 = -1$

6. Solve: $x + 4 = 3$

7. Solve: $3(2x - 6) + 6 = 0$

8. Solve: $5(x - 2) = 3x$

9. Write an equivalent fraction with the given denominator.

$$3 = \frac{}{7}$$

10. Write an equivalent fraction with the given denominator.

$$8 = \frac{}{5}$$

11. Simplify: $-\dfrac{10}{27}$

12. Simplify: $\dfrac{10y}{32}$

13. Divide: $-\dfrac{7}{12} \div -\dfrac{5}{6}$

14. Divide: $-\dfrac{2}{5} \div \dfrac{7}{10}$

15. Evaluate $y - x$ if $x = -\dfrac{3}{10}$ and $y = -\dfrac{8}{10}$.

16. Evaluate $2x + 3y$ if $x = \dfrac{2}{5}$ and $y = -\dfrac{1}{5}$.

17. Find: $-\dfrac{3}{4} - \dfrac{1}{14} + \dfrac{6}{7}$

18. Find: $\dfrac{2}{9} + \dfrac{7}{15} - \dfrac{1}{3}$

19. Simplify: $\dfrac{\dfrac{1}{2} + \dfrac{3}{8}}{\dfrac{3}{4} - \dfrac{1}{6}}$

20. Simplify: $\dfrac{\dfrac{2}{3} + \dfrac{1}{6}}{\dfrac{3}{4} - \dfrac{3}{5}}$

21. Solve: $\dfrac{x}{2} = \dfrac{x}{3} + \dfrac{1}{2}$

22. Solve: $\dfrac{x}{2} + \dfrac{1}{5} = 3 - \dfrac{x}{5}$

Answers

1. _____

2. _____

3. _____

4. _____

5. _____

6. _____

7. _____

8. _____

9. _____

10. _____

11. _____

12. _____

13. _____

14. _____

15. _____

16. _____

17. _____

18. _____

19. _____

20. _____

21. _____

22. _____

529

23. a. _____

b. _____

24. a. _____

b. _____

25. _____

26. _____

27. _____

28. _____

29. _____

30. _____

31. _____

32. _____

33. _____

34. _____

35. _____

36. _____

37. _____

38. _____

39. _____

40. _____

41. _____

42. _____

43. _____

44. _____

45. _____

46. _____

47. _____

48. _____

49. _____

50. _____

23. Write each mixed number as an improper fraction.

   a. $4\frac{2}{9}$

   b. $1\frac{8}{11}$

24. Write each mixed number as an improper fraction.

   a. $3\frac{2}{5}$

   b. $6\frac{2}{7}$

Write each decimal as a fraction or mixed number. Write your answer in simplest form.

25. 0.125

26. 0.85

27. −105.083

28. 17.015

29. Subtract: $85 - 17.31$. Check your answer.

30. Subtract: $38 - 10.06$. Check your answer.

Multiply.

31. 7.68×10

32. 12.483×100

33. $(-76.3)(1000)$

34. -853.75×10

35. Evaluate $x \div y$ for $x = 2.5$ and $y = 0.05$.

36. Is 470 a solution of the equation

$$\frac{x}{100} = 4.75?$$

37. Find the median of the scores: 67, 91, 75, 86, 55, 91

38. Find the mean of 36, 40, 86, and 30.

39. Write the ratio of 2.5 to 3.15 as a fraction in simplest form.

40. Write the ratio of 5.8 to 7.6 as a fraction in simplest form.

41. A store charges $3.36 for a 16-ounce jar of picante sauce. What is the unit price in dollars per ounce?

42. Flooring tiles cost $90 for a box with 40 tiles. Each tile is 1 square foot. Find the unit price in dollars per square foot.

43. Is $\dfrac{4.1}{7} = \dfrac{2.9}{5}$ a true proportion?

44. Is $\dfrac{6.3}{9} = \dfrac{3.5}{5}$ a true proportion?

45. On a chamber of commerce map of Abita Springs, 5 miles corresponds to 2 inches. How many miles correspond to 7 inches?

46. A student doing math homework can complete 7 problems in about 6 minutes. At this rate, how many problems can be completed in 30 minutes?

Write each percent as a fraction or mixed number in simplest form.

47. 1.9%

48. 2.3%

49. $33\frac{1}{3}\%$

50. 108%

Graphing and Introduction to Statistics

Hybrid Electric Vehicle (HEV)

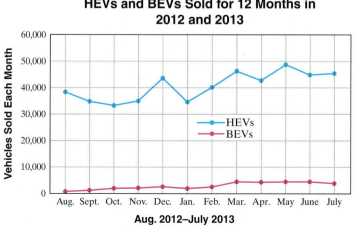

Battery Electric Vehicle (BEV)

The graph below is a double line graph of hybrid electric vehicles (HEVs) and battery electric vehicles (BEVs) sold for 12 months during 2012–2013 in the United States. A graph like this helps us visually see trends and make predictions. In Section 8.4, Exercises 55–60, we use this graph for such applications.

We often need to make decisions based on known statistics or the probability of an event occurring. For example, we decide whether or not to bring an umbrella to work based on the probability of rain. We interpret numerical data, often presented in graph form, from newspaper and magazine articles and in television news reports. We can predict which football team will win based on the trend in its previous wins and losses. This chapter reviews presenting data in a usable form on a graph and the basic ideas of probability.

HEVs and BEVs Sold for 12 Months in 2012 and 2013

Source: Electric Drive Transportation Association

8.1 Reading Pictographs, Bar Graphs, Histograms, and Line Graphs

Objectives

A Read Pictographs.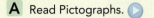

B Read and Construct Bar Graphs.

C Read and Construct Histograms.

D Read Line Graphs.

Often data are presented visually in a graph. In this section, we practice reading several kinds of graphs including pictographs, bar graphs, and line graphs.

Objective A Reading Pictographs

A **pictograph** such as the one below is a graph in which pictures or symbols are used. This type of graph contains a key that explains the meaning of the symbol used. An advantage of using a pictograph to display information is that comparisons can easily be made. A disadvantage of using a pictograph is that it is often hard to tell what fractional part of a symbol is shown. For example, in the pictograph below, Arabic shows a part of a symbol, but it's hard to read with any accuracy what fractional part of a symbol is shown.

Practice 1

Use the pictograph shown in Example 1 to answer the following questions:

a. Approximate the number of people who primarily speak Spanish.

b. Approximate how many more people primarily speak Spanish than Arabic.

Example 1 Calculating Languages Spoken

The following pictograph shows the top eight most-spoken (primary) languages. Use this pictograph to answer the questions.

Top 8 Most-Spoken (Primary) Languages

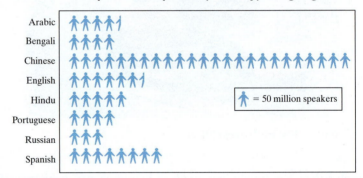

Source: www.ethnologue.com

a. Approximate the number of people who primarily speak Russian.

b. Approximate how many more people primarily speak English than Russian.

Solution:

a. Russian corresponds to 3 symbols, and each symbol represents 50 million speakers. This means that the number of people who primarily speak Russian is approximately $3 \cdot (50 \text{ million})$ or 150 million people.

b. English shows $3\frac{1}{2}$ more symbols than Russian. This means that $3\frac{1}{2} \cdot (50 \text{ million})$ or 175 million more people primarily speak English than Russian.

Work Practice 1

Answers

1. a. 400 million people

b. 175 million people

Objective B Reading and Constructing Bar Graphs ▶

Another way to visually present data is with a **bar graph.** Bar graphs can appear with vertical bars or horizontal bars. Although we have studied bar graphs in previous sections, we now practice reading the height or length of the bars contained in a bar graph. An advantage to using bar graphs is that a scale is usually included for greater accuracy. Care must be taken when reading bar graphs, as well as other types of graphs—they may be misleading, as shown later in this section.

Example 2   Finding the Number of Endangered Species

The following bar graph shows the number of endangered species in the United States in 2013. Use this graph to answer the questions.

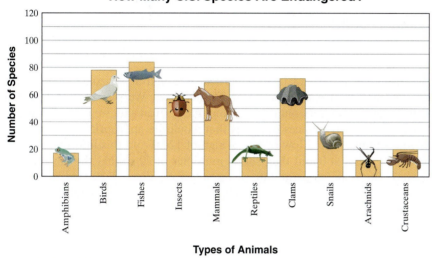

How Many U.S. Species Are Endangered?

Source: U.S. Fish and Wildlife Service

a. Approximate the number of endangered species that are clams.

b. Which category has the most endangered species?

Solution:

a. To approximate the number of endangered species that are clams, we go to the top of the bar that represents clams. From the top of this bar, we move horizontally to the left until the scale is reached. We read the height of the bar on the scale as approximately 72. There are approximately 72 clam species that are endangered, as shown.

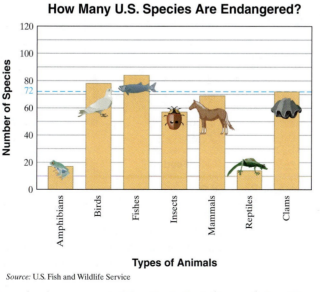

How Many U.S. Species Are Endangered?

Source: U.S. Fish and Wildlife Service

b. The most endangered species is represented by the tallest (longest) bar. The tallest bar corresponds to fishes.

■ **Work Practice 2**

Practice 2

Use the bar graph in Example 2 to answer the following questions:

a. Approximate the number of endangered species that are birds.

b. Which category shows the fewest endangered species?

Answers

2. a. 78   **b.** arachnids

Practice 3

Draw a vertical bar graph using the information in the table about electoral votes for selected states.

Total Electoral Votes by Selected States	
State	**Electoral Votes**
Texas	34
California	55
Florida	27
Nebraska	5
Indiana	11
Georgia	15

(*Source: World Almanac* 2013)

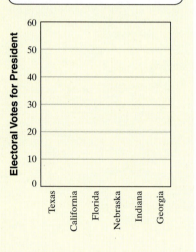

Next, we practice constructing a bar graph.

Example 3 Draw a vertical bar graph using the information in the table below, which gives the caffeine content of selected foods.

Average Caffeine Content of Selected Foods			
Food	**Milligrams**	**Food**	**Milligrams**
Brewed coffee (percolator, 8 ounces)	124	Instant coffee (8 ounces)	104
Brewed decaffeinated coffee (8 ounces)	3	Brewed tea (U.S. brands, 8 ounces)	64
Coca-Cola Classic (8 ounces)	31	Mr. Pibb (8 ounces)	27
Dark chocolate (semisweet, $1\frac{1}{2}$ ounces)	30	Milk chocolate (8 ounces)	9

(*Sources:* International Food Information Council and the Coca-Cola Company)

Solution: We draw and label a vertical line and a horizontal line as shown below on the left. These lines are also called axes. We place the different food categories along the horizontal axis. Along the vertical axis, we place a scale.

There are many choices of scales that would be appropriate. Notice that the milligrams range from a low of 3 to a high of 124. From this information, we use a scale that starts at 0 and then shows multiples of 20 so that the scale is not too cluttered. The scale stops at 140, the smallest multiple of 20 that will allow all milligrams to be graphed. It may also be helpful to draw horizontal lines along the scale markings to help draw the vertical bars at the correct height. The finished bar graph is shown below on the right.

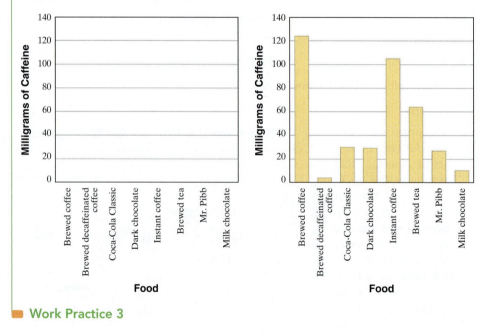

Work Practice 3

As mentioned previously, graphs can be misleading. Both graphs on the next page show the same information, but with different scales. Special care should be taken when forming conclusions from the appearance of a graph.

Answer

3.

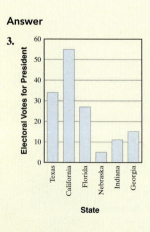

Notice the ⌇ symbol on each vertical scale on the graphs below. This symbol alerts us that numbers are missing from that scale

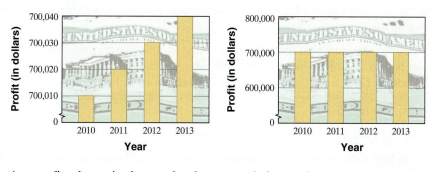

Are profits shown in the graphs above greatly increasing, or are they remaining about the same?

Objective C Reading and Constructing Histograms ▶

Suppose that the test scores of 36 students are summarized in the table below:

Student Scores	Frequency (Number of Students)
40–49	1
50–59	3
60–69	2
70–79	10
80–89	12
90–99	8

The results in the table can be displayed in a histogram. A **histogram** is a special bar graph. The width of each bar represents a range of numbers called a **class interval.** The height of each bar corresponds to how many times a number in the class interval occurs and is called the **class frequency.** The bars in a histogram lie side by side with no space between them.

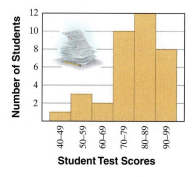

Example 4   Reading a Histogram on Student Test Scores

Use the preceding histogram to determine how many students scored 50–59 on the test.

Solution: We find the bar representing 50–59. The height of this bar is 3, which means 3 students scored 50–59 on the test.

■ **Work Practice 4**

Practice 4

Use the histogram above Example 4 to determine how many students scored 80–89 on the test.

Answer

4. 12

Practice 5

Use the histogram above Example 4 to determine how many students scored less than 80 on the test.

Practice 6

Complete the frequency distribution table for the data below. Each number represents a credit card owner's unpaid balance for one month.

0	53	89	125
265	161	37	76
62	201	136	42

Class Intervals (Credit Card Balances)	Tally	Class Frequency (Number of Months)
$0–$49	_____	_____
$50–$99	_____	_____
$100–$149	_____	_____
$150–$199	_____	_____
$200–$249	_____	_____
$250–$299	_____	_____

Practice 7

Construct a histogram from the frequency distribution table for Practice 6.

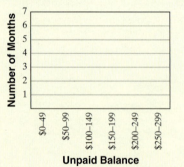

Example 5 | Reading a Histogram on Student Test Scores

Use the histogram above Example 4 to determine how many students scored 80 or above on the test.

Solution: We see that two different bars fit this description. There are 12 students who scored 80–89 and 8 students who scored 90–99. The sum of these two categories is 12 + 8 or 20 students. Thus, 20 students scored 80 or above on the test.

■ **Work Practice 5**

Now we will look at a way to construct histograms.

The daily high temperatures for 1 month in New Orleans, Louisiana, are recorded in the following list:

85°	90°	95°	89°	88°	94°
87°	90°	95°	92°	95°	94°
82°	92°	96°	91°	94°	92°
89°	89°	90°	93°	95°	91°
88°	90°	88°	86°	93°	89°

The data in this list have not been organized and can be hard to interpret. One way to organize the data is to place them in a **frequency distribution table.** We will do this in Example 6.

Example 6 | Completing a Frequency Distribution on Temperature

Complete the frequency distribution table for the preceding temperature data.

Solution: Go through the data and place a tally mark in the second column of the table next to the class interval. Then count the tally marks and write each total in the third column of the table.

Class Intervals (Temperatures)	Tally	Class Frequency (Number of Days)
82°–84°	I	1
85°–87°	III	3
88°–90°	T+++ T+++ I	11
91°–93°	T+++ II	7
94°–96°	T+++ III	8

■ **Work Practice 6**

Example 7 | Constructing a Histogram

Construct a histogram from the frequency distribution table in Example 6.

Solution:

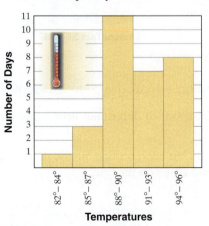

■ **Work Practice 7**

Answers

5. 16

6.

Tally	Class Frequency (Number Months)	Tally	Class Frequency (Number Months)
III	3	I	1
IIII	4	I	1
II	2	I	1

7.

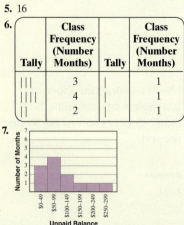

✓**Concept Check** Which of the following sets of data is better suited to representation by a histogram? Explain.

Set 1		Set 2	
Grade on Final	# of Students	Section Number	Avg. Grade on Final
51–60	12	150	78
61–70	18	151	83
71–80	29	152	87
81–90	23	153	73
91–100	25		

Objective D Reading Line Graphs ▶

Another common way to display information with a graph is by using a **line graph.** An advantage of a line graph is that it can be used to visualize relationships between two quantities. A line graph can also be very useful in showing changes over time.

Example 8 Reading Temperatures from a Line Graph

The following line graph shows the average daily temperature for each month in Omaha, Nebraska. Use this graph to answer the questions below.

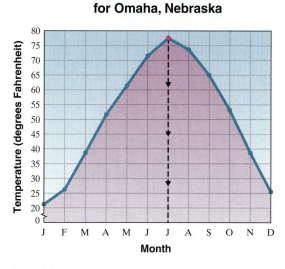

Average Daily Temperature for Omaha, Nebraska

Source: National Climatic Data Center

a. During what month is the average daily temperature the highest?

b. During what month, from July through December, is the average daily temperature 65°F?

c. During what months is the average daily temperature less than 30°F?

Solution:

a. The month with the highest temperature corresponds to the highest point. This is the red point shown on the graph above. We follow this highest point downward to the horizontal month scale and see that this point corresponds to July.

(Continued on next page)

Practice 8

Use the temperature graph in Example 8 to answer the following questions:

a. During what month is the average daily temperature the lowest?

b. During what month is the average daily temperature 25°F?

c. During what months is the average daily temperature greater than 70°F?

Answers

8. a. January **b.** December
c. June, July, and August

✓**Concept Check Answer**

Set 1; the grades are arranged in ranges of scores.

b. The months July through December correspond to the right side of the graph. We find the 65°F mark on the vertical temperature scale and move to the right until a point on the right side of the graph is reached. From that point, we move downward to the horizontal month scale and read the corresponding month. During the month of September, the average daily temperature is 65°F.

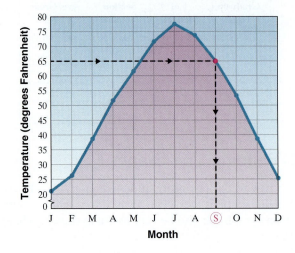

Source: National Climatic Data Center

c. To see what months the temperature is less than 30°F, we find what months correspond to points that fall below the 30°F mark on the vertical scale. These months are January, February, and December.

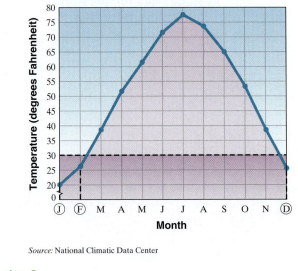

Source: National Climatic Data Center

■ **Work Practice 8**

Vocabulary, Readiness & Video Check

Fill in each blank with one of the choices below.

pictograph	bar	class frequency
histogram	line	class interval

1. A _____ graph presents data using vertical or horizontal bars.

2. A _____ is a graph in which pictures or symbols are used to visually present data.

3. A _____ graph displays information with a line that connects data points.

4. A _____ is a special bar graph in which the width of each bar represents a _____ and the height of each bar represents the _____.

Martin-Gay Interactive Videos   Watch the section lecture video and answer the following questions.

Objective A 5. From the pictograph in ▣ Example 1, how would you approximate the number of wildfires for any given year? ▷

Objective B 6. What is one advantage of displaying data in a bar graph? ▷

Objective C 7. Complete this statement based on the lecture before ▣ Example 6: A histogram is a special kind of _____. ▷

See Video 8.1 ◉

Objective D 8. From the line graph in ▣ Examples 10–13, what year averaged the greatest number of goals per game average and what was this average? ▷

8.1 Exercise Set MyMathLab® ▷

Objective A *The following pictograph shows the number of acres devoted to wheat production in selected states. Use this graph to answer Exercises 1 through 8. See Example 1. (Source: U.S. Department of Agriculture)*

1. Which state plants the greatest quantity of acreage in wheat?

2. Which of the states shown plant the least amount of wheat acreage?

3. Approximate the number of acres of wheat planted in Oklahoma.

4. Approximate the number of acres of wheat planted in Kansas.

5. Which state plants about 6,000,000 acres of wheat?

6. Which state plants about 2,000,000 acres of wheat?

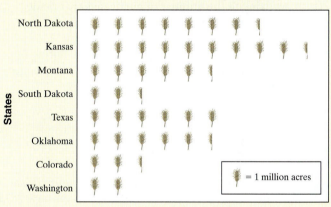

Annual Wheat Acreage in Selected Top States

7. Which two states together plant about the same acreage of wheat as North Dakota?

8. Which two states together plant about the same acreage of wheat as Kansas?

The following pictograph shows the average number of wildfires in the United States between 2006 and 2012. Use this graph to answer Exercises 9 through 16. See Example 1. (Source: National Interagency Fire Center)

▷ **9.** Approximate the number of wildfires in 2008.

10. Approximately how many wildfires were there in 2012?

▷ **11.** Which year, of the years shown, had the most wildfires?

12. In what years were the number of wildfires greater than 72,000?

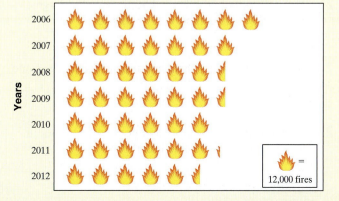

Wildfires in the United States

13. What was the amount of decrease in wildfires from 2006 to 2012?

14. What was the amount of decrease in wildfires from 2010 to 2012?

15. What was the average annual number of wildfires from 2006 to 2008? (*Hint:* How do you calculate the average?)

16. Give a possible explanation for the overall decrease in the number of wildfires since 2006.

Objective **B** *The National Weather Service has exacting definitions for hurricanes; they are tropical storms with winds in excess of 74 mph. The following bar graph shows the number of hurricanes, by month, that have made landfall on the mainland United States between 1851 and 2012. Use this graph to answer Exercises 17 through 22. See Example 2. (Source: National Weather Service: National Hurricane Center)*

17. In which month did the most hurricanes make landfall in the United States?

18. In which month did the fewest hurricanes make landfall in the United States?

19. Approximate the number of hurricanes that made landfall in the United States during the month of August.

20. Approximate the number of hurricanes that made landfall in the United States in September.

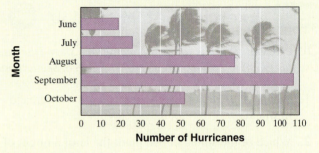

Hurricanes Making Landfall in the United States, by Month, 1851–2012

21. In 2008 alone, two hurricanes made landfall during the month of August. What fraction of all the 77 hurricanes that made landfall during August is this?

22. In 2007, only one hurricane made landfall on the United States during the entire season, in the month of September. If there have been 107 hurricanes to make landfall in the month of September since 1851, approximately what percent of these arrived in 2007?

The following horizontal bar graph shows the approximate 2012 population of the world's largest cities (including their suburbs). Use this graph to answer Exercises 23 through 28. See Example 2. (Source: CityPopulation)

23. Name the city with the largest population, and estimate its population.

24. Name the cities whose population is between 19 million and 22 million.

25. Name the city in the United States with the largest population, and estimate its population.

26. Name the two cities that have approximately the same population.

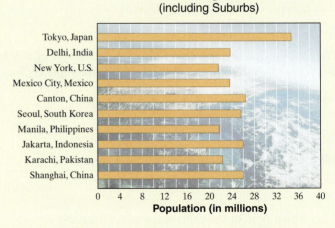

World's Largest Cities
(including Suburbs)

27. How much larger (in terms of population) is Seoul, South Korea, than Delhi, India?

28. How much larger (in terms of population) is, Shanghai, China, than Mexico City, Mexico?

Use the information given to draw a vertical bar graph. Clearly label the bars. See Example 3.

29.

Fiber Content of Selected Foods

Food	Grams of Total Fiber
Kidney beans $\left(\frac{1}{2}\,c\right)$	4.5
Oatmeal, cooked $\left(\frac{3}{4}\,c\right)$	3.0
Peanut butter, chunky (2 tbsp)	1.5
Popcorn (1 c)	1.0
Potato, baked with skin (1 med)	4.0
Whole wheat bread (1 slice)	2.5

(*Sources:* American Dietetic Association and National Center for Nutrition and Dietetics)

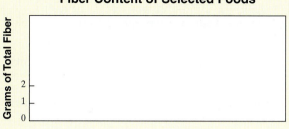

Fiber Content of Selected Foods

30.

U.S. Annual Food Sales

Year	Sales in Billions of Dollars
2009	1086
2010	1139
2011	1274
2012	1357

(*Source:* U.S. Department of Agriculture)

U.S. Annual Food Sales

31.

Best-Selling Albums of All Time (U.S. Sales)

Album	Estimated Sales (in millions)
Pink Floyd: *The Wall* (1979)	23
Michael Jackson: *Thriller* (1982)	29
Billy Joel: *Greatest Hits Volumes I - II* (1985)	23
Eagles: *Their Greatest Hits* (1976)	29
Led Zeppelin: *Led Zeppelin IV* (1971)	23

(*Source:* Recording Industry Association of America)

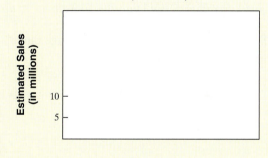

Best-Selling Albums of All Time
(U.S. sales)

32.

Selected Worldwide Commercial Space Launches

Location or Name	Total Commercial Space Launches 1990–2012
United States	156
Europe	153
Russia	141
China	23
Sea Launch[*]	39

[*]Sea Launch is an international venture involving 4 countries that uses its own launch facility outside national borders.

Source: Bureau of Transportation Statistics

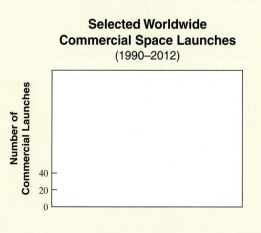

Selected Worldwide Commercial Space Launches
(1990–2012)

Objective C *The following histogram shows the number of miles that each adult, from a survey of 100 adults, drives per week. Use this histogram to answer Exercises 33 through 42. See Examples 4 and 5.*

33. How many adults drive 100–149 miles per week?

34. How many adults drive 200–249 miles per week?

▶ **35.** How many adults drive fewer than 150 miles per week?

36. How many adults drive 200 miles or more per week?

37. How many adults drive 100–199 miles per week?

38. How many adults drive 150–249 miles per week?

▶ **39.** How many more adults drive 250–299 miles per week than 200–249 miles per week?

40. How many more adults drive 0–49 miles per week than 50–99 miles per week?

41. What is the ratio of adults who drive 150–199 miles per week to the total number of adults surveyed?

42. What is the ratio of adults who drive 50–99 miles per week to the total number of adults surveyed?

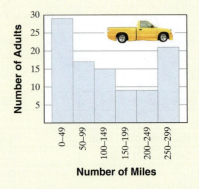

The following histogram shows the ages of householders for the year 2010. Use this histogram to answer Exercises 43 through 50. For Exercises 45 through 50, estimate to the nearest whole million. See Examples 4 and 5.

43. The most householders were in what age range?

44. The least number of householders were in what age range?

45. How many householders were 55–64 years old?

46. How many householders were 35–44 years old?

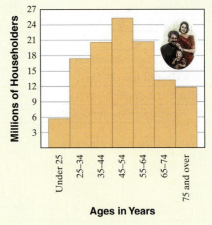

Source: U.S. Bureau of the Census, *Current Population Reports*

47. How many householders were 44 years old or younger?

48. How many householders were 55 years old or older?

49. How many more householders were 45–54 years old than 55–64 years old?

50. How many more householders were 45–54 years old than 75 and over?

The following list shows the golf scores for an amateur golfer. Use this list to complete the frequency distribution table to the right. See Example 6.

78	84	91	93	97
97	95	85	95	96
101	89	92	89	100

Class Intervals (Scores)	Tally	Class Frequency (Number of Games)
51. 70–79		
52. 80–89		
53. 90–99		
54. 100–109		

Twenty-five people in a survey were asked to give their current checking account balances. Use the balances shown in the following list to complete the frequency distribution table to the right. See Example 6.

$53	$105	$162	$443	$109
$468	$47	$259	$316	$228
$207	$357	$15	$301	$75
$86	$77	$512	$219	$100
$192	$288	$352	$166	$292

Class Intervals (Account Balances)	Tally	Class Frequency (Number of People)
55. $0–$99		
56. $100–$199		
57. $200–$299		
58. $300–$399		
59. $400–$499		
60. $500–$599		

61. Use the frequency distribution table from Exercises **51** through **54** to construct a histogram. See Example 7.

Golf Scores

62. Use the frequency distribution table from Exercises **55** through **60** to construct a histogram. See Example 7.

Account Balances

Objective D *Beach Soccer World Cup is now held every two years. The following line graph shows the World Cup goals per game average for beach soccer during the years shown. Use this graph to answer Exercises 63 through 70. See Example 8.*

63. Find the average number of goals per game in 2011.

64. Find the average number of goals per game in 2009.

65. During what year shown was the average number of goals per game the highest?

66. During what year shown was the average number of goals per game the lowest?

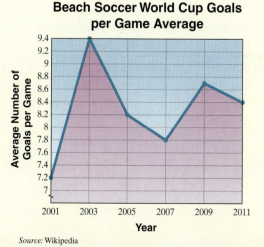

Beach Soccer World Cup Goals per Game Average

Source: Wikipedia

67. From 2007 to 2009, did the average number of goals per game increase or decrease?

68. From 2009 to 2011, did the average number of goals per game increase or decrease?

69. During what year(s) shown were the average goals per game less than 8?

70. During what year(s) shown were the average goals per game greater than 8?

Review

Find each percent. See Sections 7.2 and 7.3.

71. 30% of 12

72. 45% of 120

73. 10% of 62

74. 95% of 50

Write each fraction as a percent. See Section 7.1.

75. $\frac{1}{4}$

76. $\frac{2}{5}$

77. $\frac{17}{50}$

78. $\frac{9}{10}$

Concept Extensions

The following double line graph shows temperature highs and lows for a week. Use this graph to answer Exercises 79 through 84.

79. What was the high temperature reading on Thursday?

80. What was the low temperature reading on Thursday?

81. What day was the temperature the lowest? What was this low temperature?

82. What day of the week was the temperature the highest? What was this high temperature?

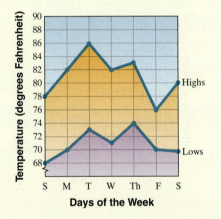

83. On what day of the week was the difference between the high temperature and the low temperature the greatest? What was this difference in temperature?

84. On what day of the week was the difference between the high temperature and the low temperature the least? What was this difference in temperature?

85. True or false? With a bar graph, the width of the bar is just as important as the height of the bar. Explain your answer.

86. Kansas plants about 17% of the wheat acreage in the United States. About how many acres of wheat are planted in the United States, according to the pictograph for Exercises **1** through **8**? Round to the nearest million acre.

8.2 Reading Circle Graphs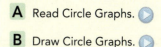

Objective A Reading Circle Graphs

Objectives

A Read Circle Graphs.

B Draw Circle Graphs.

In Exercise Set 7.1, the following **circle graph** was shown. This particular graph shows the favorite sport for 100 adults.

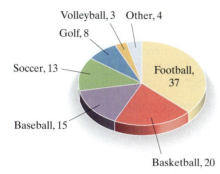

Volleyball, 3   Other, 4
Golf, 8
Soccer, 13
Football, 37
Baseball, 15
Basketball, 20

Each sector of the graph (shaped like a piece of pie) shows a category and the relative size of the category. In other words, the most popular sport is football, and it is represented by the largest sector.

Example 1 Find the ratio of adults preferring basketball to total adults. Write the ratio as a fraction in simplest form.

Solution: The ratio is

$$\frac{\text{people preferring basketball}}{\text{total adults}} = \frac{20}{100} = \frac{1}{5}$$

■ Work Practice 1

Practice 1

Find the ratio of adults preferring golf to total adults. Write the ratio as a fraction in simplest form.

A circle graph is often used to show percents in different categories, with the whole circle representing 100%.

Example 2 Using a Circle Graph

The following graph shows the percent of visitors to the United States in a recent year by various regions. Using the circle graph shown, determine the percent of visitors who came to the United States from Mexico or Canada.

Solution: To find this percent, we add the percents corresponding to Mexico and Canada. The percent of visitors to the United States that came from Mexico or Canada is

$$34\% + 21\% = 55\%$$

■ Work Practice 2

Visitors to U.S. by Region

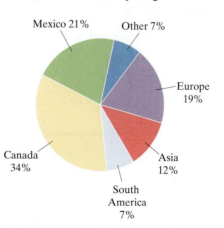

Mexico 21%   Other 7%
Europe 19%
Canada 34%
Asia 12%
South America 7%

Source: Office of Travel and Tourism Industries, 2012

Practice 2

Using the circle graph shown in Example 2, determine the percent of visitors to the United States that came from Europe, Asia, or South America.

Answers

1. $\frac{2}{25}$ 2. 38%

Helpful Hint

Since a circle graph represents a whole, the percents should add to 100% or 1. Notice this is true for Example 2.

Practice 3

Use the information in Example 3 and the circle graph from Example 2 to predict the number of tourists from Mexico in 2017.

Example 3 Finding Percent of Population

The U.S. Department of Commerce forecasts 81 million international visitors to the United States in 2017. Use the circle graph from Example 2 and predict the number of tourists that might be from Europe.

Solution: We use the percent equation.

$$\text{amount} = \boxed{\text{percent}} \cdot \boxed{\text{base}}$$
$$\text{amount} = 0.19 \cdot 81{,}000{,}000$$
$$= 0.19(81{,}000{,}000)$$
$$= 15{,}390{,}000$$

Thus, 15,390,000 tourists might come from Europe in 2017.

■ **Work Practice 3**

✓**Concept Check** Can the following data be represented by a circle graph? Why or why not?

Responses to the Question, "In Which Activities Are You Involved?"	
Intramural sports	60%
On-campus job	42%
Fraternity/sorority	27%
Academic clubs	21%
Music programs	14%

Objective B Drawing Circle Graphs ▶

To draw a circle graph, we use the fact that a whole circle contains 360° (degrees).

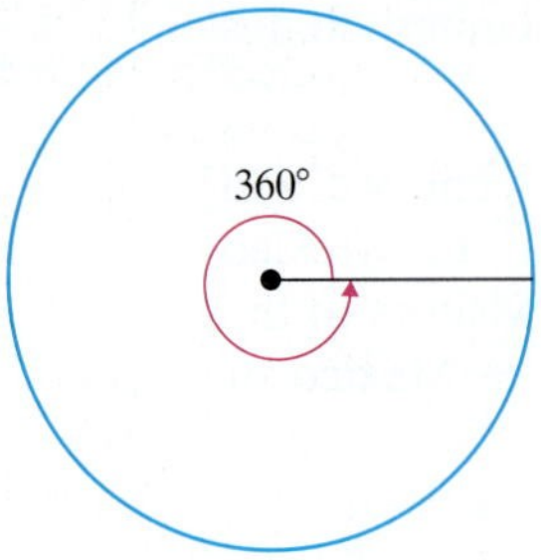

Answer
3. 17,010,000 tourists from Mexico

✓**Concept Check Answer**
no; the percents add up to more than 100%

Example 4   Drawing a Circle Graph for U.S. Armed Forces Personnel

The following table shows the percent of U.S. armed forces personnel that were in each branch of service in 2008. (*Source:* U.S. Department of Defense)

Branch of Service	Percent
Army	40
Navy	23
Marine Corps	15
Air Force	22
(Note: The Coast Guard is now under the Department of Homeland Security.)	

Draw a circle graph showing this data.

Solution: First we find the number of degrees in each sector representing each branch of service. Remember that the whole circle contains 360°. (We will round degrees to the nearest whole.)

Sector	Degrees in Each Sector
Army	40% × 360° = 0.40 × 360° = 144°
Navy	23% × 360° = 0.23 × 360° = 82.8° ≈ 83°
Marine Corps	15% × 360° = 0.15 × 360° = 54°
Air Force	22% × 360° = 0.22 × 360° = 79.2° ≈ 79°

Helpful Hint

Check your calculations by finding the sum of the degrees.

144° + 83° + 54° + 79° = 360°

The sum should be 360°. (It may vary only slightly because of rounding.)

Next we draw a circle and mark its center. Then we draw a line from the center of the circle to the circle itself.

To construct the sectors, we will use a **protractor.** A protractor measures the number of degrees in an angle. We place the hole in the protractor over the center of the circle. Then we adjust the protractor so that 0° on the protractor is aligned with the line that we drew.

It makes no difference which sector we draw first. To construct the "Army" sector, we find 144° on the protractor and mark our circle. Then we remove the protractor and use this mark to draw a second line from the center to the circle itself.

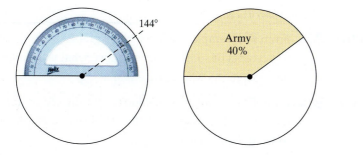

(Continued on next page)

Practice 4

Use the data shown to draw a circle graph.

Freshmen	30%
Sophomores	27%
Juniors	25%
Seniors	18%

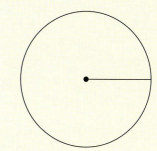

Answer

4.

To construct the "Navy" sector, we follow the same procedure as above, except that we line up 0° with the second line we drew and mark the protractor at 83°.

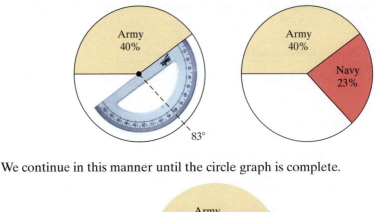

We continue in this manner until the circle graph is complete.

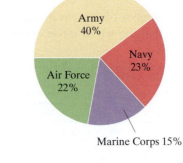

Work Practice 4

✔**Concept Check** True or false? The larger a sector in a circle graph, the larger the percent of the total it represents. Explain your answer.

Vocabulary, Readiness & Video Check

Use the choices below to fill in each blank.

sector   circle   100   360

1. In a _____ graph, each section (shaped like a piece of pie) shows a category and the relative size of the category.

2. A circle graph contains pie-shaped sections, each called a _____.

3. The number of degrees in a whole circle is _____.

4. If a circle graph has percent labels, the percents should add up to _____.

Martin-Gay Interactive Videos

See Video 8.2

Watch the section lecture video and answer the following questions.

Objective A 5. From ▣ Example 3, when a circle graph shows different parts or percents of some whole category, what is the sum of the percents in the whole circle graph? ▶

Objective B 6. From ▣ Example 6, when looking at the sector degree measures of a circle graph, the whole circle graph corresponds to what degree measure? ▶

8.2 Exercise Set MyMathLab®

Objective A *The following circle graph is a result of surveying 700 college students. They were asked where they live while attending college. Use this graph to answer Exercises 1 through 6. Write all ratios as fractions in simplest form. See Example 1.*

1. Where do most of these college students live?

2. Besides the category "Other arrangements," where do the fewest of these college students live?

3. Find the ratio of students living in campus housing to total students.

4. Find the ratio of students living in off-campus rentals to total students.

5. Find the ratio of students living in campus housing to students living in a parent or guardian's home.

6. Find the ratio of students living in off-campus rentals to students living in a parent or guardian's home.

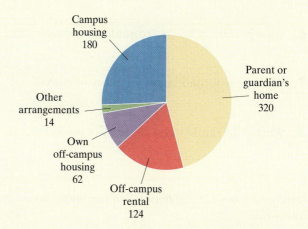

The following circle graph shows the percent of the land area of the continents of Earth. Use this graph for Exercises 7 through 14. See Example 2.

7. Which continent is the largest?

8. Which continent is the smallest?

9. What percent of the land on Earth is accounted for by Asia and Europe together?

10. What percent of the land on Earth is accounted for by North and South America?

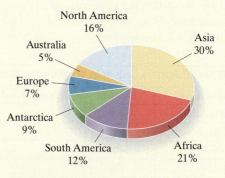

Source: National Geographic Society

The total amount of land from the continents is approximately 57,000,000 square miles. Use the graph to find the area of the continents given in Exercises 11 through 14. See Example 3.

11. Asia   12. South America   13. Australia   14. Europe

The following circle graph shows the percent of the types of books available at Midway Memorial Library. Use this graph for Exercises 15 through 24. See Example 2.

15. What percent of books are classified as some type of fiction?

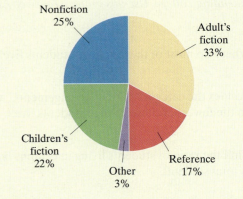

16. What percent of books are nonfiction or reference?

17. What is the second-largest category of books?

18. What is the third-largest category of books?

If this library has 125,600 books, find how many books are in each category given in Exercises 19 through 24. See Example 3.

19. Nonfiction

20. Reference

21. Children's fiction

22. Adult's fiction

23. Reference or other

24. Nonfiction or other

Objective B *Fill in the table. Round to the nearest degree. Then draw a circle graph to represent the information given in each table. (Remember: The total of "Degrees in Sector" column should equal 360° or very close to 360° because of rounding.) See Example 4.*

25.

Types of Apples Grown in Washington State		
Type of Apple	**Percent**	**Degrees in Sector**
Red Delicious	37%	
Golden Delicious	13%	
Fuji	14%	
Gala	15%	
Granny Smith	12%	
Other varieties	6%	
Braeburn	3%	
(*Source:* U.S. Apple Association)		

26.

Color Distribution of M&M's Milk Chocolate		
Color	**Percent**	**Degrees in Sector**
Blue	24%	
Orange	20%	
Green	16%	
Yellow	14%	
Red	13%	
Brown	13%	
(*Source:* M&M Mars)		

27.

Distribution of Large Dams by Continent		
Continent	Percent	Degrees in Sector
Europe	19%	
North America	32%	
South America	3%	
Asia	39%	
Africa	5%	
Australia	2%	

(*Source:* International Commission on Large Dams)

28.

2010 Hybrid Sales by Make of Car		
Company	Percent	Degrees in Sector
Toyota	69%	
Ford	13%	
Honda	12%	
GM	3%	
Nissan	2%	
Other	1%	

(*Source:* U.S. Department of Energy, 2010 Data)

Review

Write the prime factorization of each number. See Section 4.2.

29. 20

30. 25

31. 40

32. 16

33. 85

34. 105

Concept Extensions

The following circle graph shows the relative sizes of the great oceans.

35. Without calculating, determine which ocean is the largest. How can you answer this question by looking at the circle graph?

36. Without calculating, determine which ocean is the smallest. How can you answer this question by looking at the circle graph?

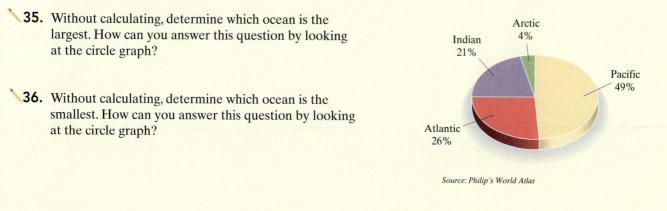

Source: Philip's World Atlas

These oceans together make up 264,489,800 square kilometers of Earth's surface. Find the square kilometers for each ocean.

37. Pacific Ocean

38. Atlantic Ocean

39. Indian Ocean

40. Arctic Ocean

The following circle graph summarizes the results of online spending in America. Let's use these results to make predictions about the online spending behavior of a community of 2800 Internet users age 18 and over. Use this graph for Exercises 41 through 46. Round to the nearest whole. (Note: Because of rounding, these percents do not have a sum of 100%.)

41. How many of the survey respondents said that they spend $0 online each month?

42. How many of the survey repondents said that they spend $1–$100 online each month?

43. How many of the survey respondents said that they spend $0 to $100 online each month?

44. How many of the survey respondents said that they spend $1 to $1000 online each month?

Online Spending per Month

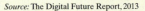

$101–$1000
14%

> $1000
0.2%

$0
24%

$1–$100
62%

Source: The Digital Future Report, 2013

45. Find the ratio of *number* of respondents who spend $0 online to *number* of respondents who spend $1–$100 online. Write the ratio as a fraction. Simplify the fraction if possible.

46. Find the ratio of *percent* of respondents who spend $101–$1000 online to *percent* of those who spend $1–$100. Write the ratio as a fraction with integers in the numerator and denominator. Simplify the fraction if possible.

See the Concept Checks in this section.

47. Can the data below be represented by a circle graph? Why or why not?

Responses to the Question, "What Classes Are You Taking?"	
Math	80%
English	72%
History	37%
Biology	21%
Chemistry	14%

48. True or false? The smaller a sector in a circle graph, the smaller the percent of the total it represents. Explain why.

Objective A Plotting Points

In Section 8.1, we saw how bar and line graphs can be used to show relationships between items listed on the horizontal and vertical axes.

The table and line graph below both show the U.S. spending on DVD home entertainment for the years shown.

U.S. Spending on DVD Home Entertainment	
Year	**Money Spent (in billions)**
2000	$3.9
2001	$7.0
2002	$11.0
2003	$17.4
2004	$22.0
2005	$23.3
2006	$24.1
2007	$23.5
2008	$22.4
2009	$16.4
2010	$18.4
2011	$18.0

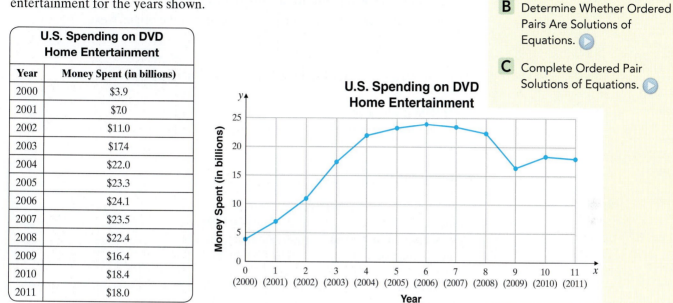

Both the table and the line graph show the relationship between two quantities (year and money), called **paired data.** These data are called paired data because each year corresponds to, or is paired with, a specific amount of money. In other words, the year 2010 is paired with $18.4 billion. Notice that the dots, or points, on the line graph visually show the same pairing. For the year 2010, the height of the dot is 18.4, representing $18.4 billion.

We can call the "year" axis the horizontal axis and the "money" axis the vertical axis.

In general, we can use this same horizontal and vertical axis idea to describe the location of points in a plane. A **plane** is a flat surface that extends indefinitely. Surfaces like a plane (except that they don't extend indefinitely) are a classroom floor, a blackboard or whiteboard, and this sheet of paper.

The system that we use to describe the locations of points in a plane is called the **rectangular coordinate system.** It consists of two number lines, one horizontal and one vertical, intersecting at the point 0 on each number line. This point of intersection is called the **origin.** We call the horizontal number line the *x*-axis and the vertical number line the *y*-axis. Notice that the axes divide the plane into four regions, called **quadrants.** They are numbered as shown.

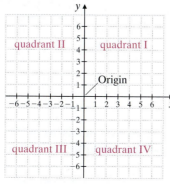

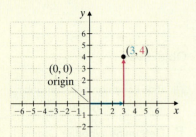

Every point in the rectangular coordinate system corresponds to an **ordered pair of numbers,** such as $(3, 4)$. The first number, 3, of an ordered pair is associated with the x-axis and is called the **x-coordinate** or **x-value.** The second number, 4, is associated with the y-axis and is called the **y-coordinate** or **y-value.** To find the **single point** on the rectangular coordinate system corresponding to the ordered pair $(3, 4)$, start at the origin. Move 3 units in the positive direction along the x-axis. From there, move 4 units in the positive direction parallel to the y-axis as shown to the left. This process of locating a point on the rectangular coordinate system is called **plotting the point.** Since the origin is located at 0 on the x-axis and 0 on the y-axis, the origin corresponds to the ordered pair $(0, 0)$.

In general, to plot the ordered pair (x, y), start at the origin. Next,

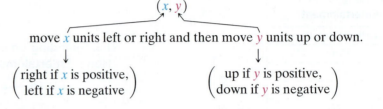

(x, y)

move x units left or right and then move y units up or down.

$$\begin{pmatrix} \text{right if } x \text{ is positive,} \\ \text{left if } x \text{ is negative} \end{pmatrix} \qquad \begin{pmatrix} \text{up if } y \text{ is positive,} \\ \text{down if } y \text{ is negative} \end{pmatrix}$$

Helpful Hint

Since the first number, or x-coordinate, of an ordered pair is associated with the x-axis, it tells how many units to move left or right. Similarly, the second number, or y-coordinate, tells how many units to move up or down.

To plot $(-1, 5)$, start at the origin. Move 1 unit left (because the x-coordinate is negative) and then 5 units up and draw a dot at that point. This dot is the graph of the point that corresponds to the ordered pair $(-1, 5)$.

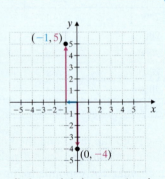

To plot $(0, -4)$, start at the origin, move 0 units (left or right), then 4 units down, and then draw a dot.

Below are some more plotted points with their corresponding ordered pairs. Another name for the rectangular coordinate system is the Cartesian coordinate system. This name comes from French scientist, philosopher, and mathematician René Descartes (1596–1650), who introduced the system.

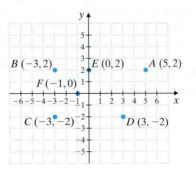

Helpful Hint

Remember that **each point** in the rectangular coordinate system corresponds to exactly **one ordered pair** and that **each ordered pair** corresponds to exactly **one point.**

Example 1 Plot each point corresponding to the ordered pairs on the same set of axes.

$$(5, 4), (-2, 3), (-1, -2), (6, -3), (0, 2), (-4, 0), \left(3, \frac{1}{2}\right)$$

Solution:

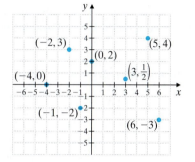

Work Practice 1

Example 2 Find the ordered pair corresponding to each point plotted below.

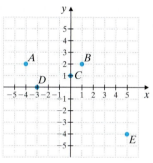

Solution:

Point A has coordinates $(-4, 2)$.

Point B has coordinates $(1, 2)$.

Point C has coordinates $(0, 1)$.

Point D has coordinates $(-3, 0)$.

Point E has coordinates $(5, -4)$.

Work Practice 2

Have you noticed a pattern from the preceding examples? **If an ordered pair has a y-coordinate of 0, its graph lies on the x-axis. If an ordered pair has an x-coordinate of 0, its graph lies on the y-axis.**

Helpful Hint

Order is the key word in "ordered pair." The first value always corresponds to the x-value and the second value always corresponds to the y-value. For example, $(-2, 4)$ does not describe the same point as $(4, -2)$.

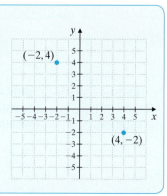

Practice 1

Plot each point corresponding to the ordered pairs on the same set of axes. $(2, 3), (-4, 2),$ $(-6, -5), (2, -2), (5, 0),$ $(0, -3), \left(-\frac{1}{2}, 4\right)$

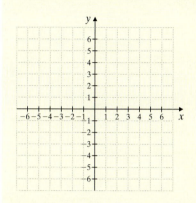

Practice 2

Find the ordered pair corresponding to each point plotted on the rectangular coordinate system.

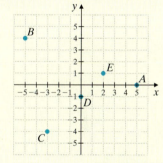

Answers

1.

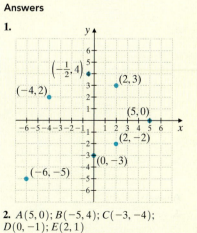

2. $A(5, 0); B(-5, 4); C(-3, -4);$ $D(0, -1); E(2, 1)$

Objective B Determining Whether an Ordered Pair Is a Solution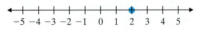

Let's see how we can use ordered pairs of numbers to record solutions of equations containing two variables.

Recall that an equation with one variable, such as $3 + x = 5$, has one solution—in this case, 2 because 2 is the only value that can be substituted for x so that the resulting equation $3 + 2 = 5$ is a true statement. We can graph this solution on a number line.

$$\xleftarrow{\hspace{0.3em}} \overset{\bullet}{\underset{-5\ -4\ -3\ -2\ -1\ \ 0\ \ 1\ \ 2\ \ 3\ \ 4\ \ 5}{\rule{0pt}{0pt}}} \xrightarrow{\hspace{0.3em}}$$

An equation with two variables such as $x + y = 7$ has many solutions. Each solution is a pair of numbers, one for each variable, that makes the equation a true statement. For example, $x = 4$ and $y = 3$ is a solution of the equation $x + y = 7$ because, when x is replaced with 4 and y is replaced with 3, $x + y = 7$ becomes $4 + 3 = 7$, which is a true statement. We can write the solution $x = 4$ and $y = 3$ as the ordered pair $(4, 3)$. Study the chart below for more solutions.

Replacement Values	$x + y = 7$	True or False	Ordered Pair Solution
$x = 4, y = 3$	$4 + 3 = 7$	True	$(4, 3)$
$x = 0, y = 7$	$0 + 7 = 7$	True	$(0, 7)$
$x = 7, y = 1$	$7 + 1 = 7$	False	No
$x = 11, y = -4$	$11 + (-4) = 7$	True	$(11, -4)$

In general, we say that an ordered pair of numbers is a solution of an equation if the equation is a true statement when the variables are replaced by the coordinates of the ordered pair.

Let's practice deciding whether an ordered pair of numbers is a solution of an equation in two variables.

Practice 3

Is $(0, -4)$ a solution of the equation $x + 3y = -12$?

Example 3 Is $(-1, 6)$ a solution of the equation $2x + y = 4$?

Solution: Replace x with -1 and y with 6 in the equation $2x + y = 4$ to see if the result is a true statement.

$$
\begin{array}{ll}
2x + y = 4 & \text{Original equation} \\
2(-1) + 6 \overset{?}{=} 4 & \text{Replace } x \text{ with } -1 \text{ and } y \text{ with 6.} \\
-2 + 6 \overset{?}{=} 4 & \text{Multiply.} \\
4 = 4 & \text{True}
\end{array}
$$

Since $4 = 4$ is true, $(-1, 6)$ is a solution of the equation $2x + y = 4$.

■ Work Practice 3

Unlike equations with one variable, equations with two variables, such as $x + y = 7$, have so many solutions we cannot simply list them. Instead, to "see" all the solutions of a two-variable equation, we draw a graph of its solutions.

Answer

3. yes

Example 4 Each ordered pair listed is a solution of the equation $x + y = 7$. Plot them on the same set of axes.

a. $(3, 4)$ **b.** $(0, 7)$ **c.** $(-1, 8)$

d. $(7, 0)$ **e.** $(5, 2)$ **f.** $(4, 3)$

Solution:

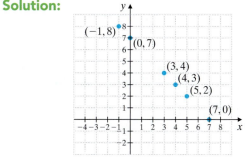

■ **Work Practice 4**

Notice that the points in Example 4 all seem to lie on the same line. We will discuss this more in the next section.

An equation such as $x + y = 7$ is called a **linear** or **first-degree equation in two variables.** It is called a **first-degree equation** because the exponent on x and y is an understood 1. The equation is called an equation **in two variables** because it contains two different variables, x and y.

Objective C Completing Ordered Pair Solutions

In the next section, we will graph linear equations in two variables. Before that, we need to practice finding ordered pair solutions of these equations. If one coordinate of an ordered pair solution is known, the other coordinate can be determined. To find the unknown coordinate, replace the appropriate variable with the known coordinate in the equation. Doing so results in an equation with one variable that we can solve.

Example 5 Complete each ordered pair solution of the equation $y = 2x$.

a. $(3, \ )$ **b.** $(\ , 0)$ **c.** $(-2, \ )$

Solution:

a. In the ordered pair $(3, \ )$, the x-value is 3. To find the corresponding y-value, let $x = 3$ in the equation $y = 2x$ and calculate the value of y.

$$y = 2x \qquad \text{Original equation}$$
$$y = 2(3) \qquad \text{Replace } x \text{ with 3.}$$
$$y = 6 \qquad \text{Multiply.}$$

Check: To check, replace x with 3 and y with 6 in the original equation to see that a true statement results.

$$y = 2x \qquad \text{Original equation}$$
$$6 \overset{?}{=} 2(3) \qquad \text{Let } x = 3 \text{ and } y = 6.$$
$$6 = 6 \qquad \text{True}$$

The ordered pair solution is $(3, 6)$.

(Continued on next page)

Practice 4

Each ordered pair listed is a solution of the equation $x - y = 5$. Plot them on the same set of axes.

a. $(6, 1)$ **b.** $(5, 0)$
c. $(0, -5)$ **d.** $(7, 2)$
e. $(-1, -6)$ **f.** $(2, -3)$

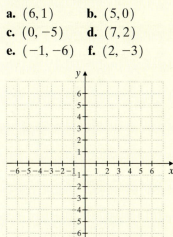

Practice 5

Complete each ordered pair solution of the equation $y = -5x$.

a. $(5, \ )$ **b.** $(\ , 0)$
c. $(-3, \ )$

Answers

4.

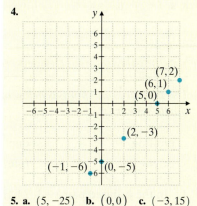

5. a. $(5, -25)$ **b.** $(0, 0)$ **c.** $(-3, 15)$

b. In the ordered pair $(\ \ ,0)$, the y-value is 0. To find the corresponding x-value, replace y with 0 in the equation and solve for x.

$y = 2x$ Original equation   **Check:** $y = 2x$ Original equation

$0 = 2x$ Replace y with 0. $0 \overset{?}{=} 2 \cdot 0$ Let $x = 0$ and $y = 0$.

$\dfrac{0}{2} = \dfrac{2x}{2}$ Divide both sides by 2. $0 = 0$ True

$0 = x$ Simplify.

The ordered pair solution is $(0,0)$, the origin.

c. In the ordered pair $(-2,\ \ )$, the x-value is -2. To find the corresponding y-value, replace x with -2 in the equation and calculate the value for y.

$y = 2x$ Original equation   **Check:** $y = 2x$ Original equation

$y = 2(-2)$ Replace x with -2. $-4 \overset{?}{=} 2(-2)$ Let $x = -2$ and $y = -4$.

$y = -4$ Multiply. $-4 = -4$ True

The ordered pair solution is $(-2, -4)$.

■ **Work Practice 5**

Practice 6

Complete each ordered pair solution of the equation $y = 5x + 2$.

a. $(0,\ \ )$

b. $(\ \ ,-3)$

Example 6   Complete each ordered pair solution of the equation $y = 3x - 4$.

a. $(2,\ \ )$ **b.** $(\ \ , -16)$

Solution:

a. Replace x with 2 in the equation and calculate the value for y.

$y = 3x - 4$ Original equation

$y = 3 \cdot 2 - 4$ Replace x with 2.

$y = 2$ Simplify.

The ordered pair solution is $(2,2)$.

b. Replace y with -16 in the equation and solve for x.

$y = 3x - 4$ Original equation

$-16 = 3x - 4$ Replace y with -16.

$-16 + 4 = 3x - 4 + 4$ Add 4 to both sides.

$-12 = 3x$ Simplify.

$\dfrac{-12}{3} = \dfrac{3x}{3}$ Divide both sides by 3.

$-4 = x$ Simplify.

The ordered pair solution is $(-4, -16)$.

■ **Work Practice 6**

Answers

6. a. $(0,2)$ **b.** $(-1,-3)$

Vocabulary, Readiness & Video Check

Use the words and phrases below to fill in each blank. The questions and statements all have to do with the rectangular coordinate system.

plotting	x	one	$(0,0)$	ordered pair of numbers
origin	y	four	plane	

1. In the ordered pair $(-1, 2)$, the _____-value is -1 and the _____-value is 2.

2. In the rectangular coordinate system, the point of intersection of the x-axis and the y-axis is called the _____.

3. The axes divide the plane into _____ regions, called quadrants.

4. Every ordered pair of numbers, such as $\left(\frac{1}{2}, -3\right)$, corresponds to how many points in the plane? _____

5. The process of locating a point on the rectangular coordinate system is called _____ the point.

6. The origin corresponds to the ordered pair _____.

7. A(n) _____ is a flat surface that extends indefinitely in all directions.

8. Every point in the rectangular coordinate system corresponds to a(n) _____.

Martin-Gay Interactive Videos   *Watch the section lecture video and answer the following questions.*

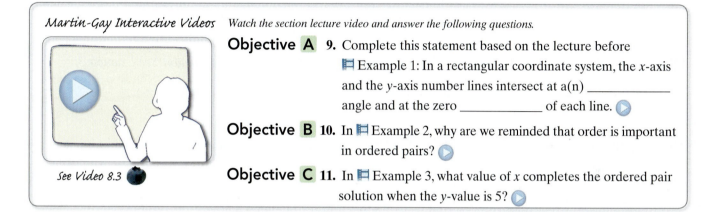

See Video 8.3

Objective A **9.** Complete this statement based on the lecture before Example 1: In a rectangular coordinate system, the *x*-axis and the *y*-axis number lines intersect at a(n) _____ angle and at the zero _____ of each line.

Objective B **10.** In Example 2, why are we reminded that order is important in ordered pairs?

Objective C **11.** In Example 3, what value of *x* completes the ordered pair solution when the *y*-value is 5?

8.3 **Exercise Set** MyMathLab®

Objective A *Use the rectangular coordinate system below each exercise to plot the points corresponding to the ordered pairs. See Example 1.*

1. $(1, 3), (-2, 4), (0, 2), (-5, 0), (-3, -3), (5, -5)$

2. $(5, 2), (3, -4), (-1, -1), (0, -1), (4, 0), (-2, 4)$

3. $\left(2\frac{1}{2}, 3\right), (0, -3), (-4, -6), \left(-1, 5\frac{1}{2}\right), (1, 0), (3, -5)$

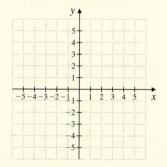

4. $\left(5, \frac{1}{2}\right), \left(-3\frac{1}{2}, 0\right), (-1, 4), (4, -1), (0, 2), (-5, -5)$

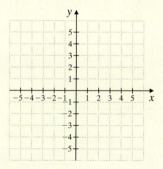

Find the x- and y-coordinates of each labeled point. See Example 2.

5.

6.

Objective B *Determine whether each ordered pair is a solution of the given linear equation. See Example 3.*

7. $(0, 0); y = -20x$

8. $(1, 14); x = 14y$

9. $(1, 2); x - y = 3$

10. $(-1, 9); x + y = 8$

▶ **11.** $(-2, -3); y = 2x + 1$

12. $(1, 1); y = -x$

13. $(6, -2); x = -3y$

14. $(-9, 1); x = -9y$

15. $(5, 0); 3y + 2x = 10$

16. $(1, 1); -5y + 4x = -1$

17. $(3, 1); x - 5y = -1$

18. $(0, 2); x - 7y = -15$

Use the rectangular coordinate system below each exercise to plot the three ordered pair solutions of the given equation. See Example 4.

19. $2x + y = 5; (1, 3), (0, 5), (3, -1)$

20. $3x + y = 5; (1, 2), (0, 5), (2, -1)$

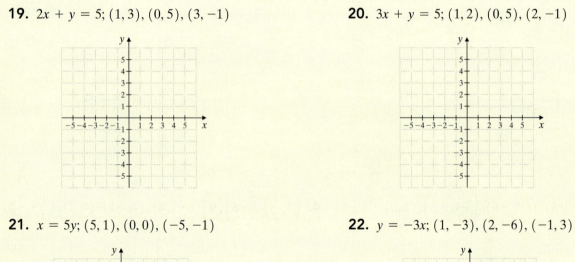

21. $x = 5y; (5, 1), (0, 0), (-5, -1)$

22. $y = -3x; (1, -3), (2, -6), (-1, 3)$

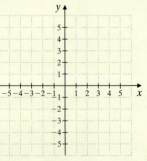

23. $y = -2x + 3;\ (4, -5),\ (2, -1),\ (0, 3)$

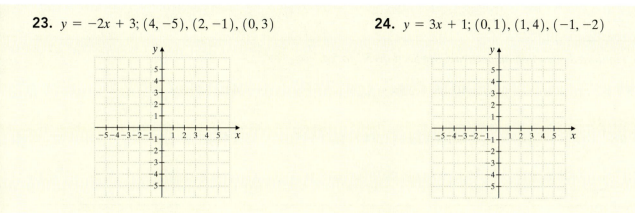

24. $y = 3x + 1;\ (0, 1),\ (1, 4),\ (-1, -2)$

Objective **C** *Complete each ordered pair solution of the given equations. See Examples 5 and 6.*

25. $y = -9x;\ (1,\ \ ),\ (0,\ \ ),\ (\ \ , -18)$

26. $x = -7y;\ (\ \ , 2),\ (14,\ \ ),\ (\ \ , -1)$

27. $x - y = 14;\ (2,\ \ ),\ (\ \ , -8),\ (0,\ \ )$

28. $x - y = 8;\ (0,\ \ ),\ (\ \ , 0),\ (5,\ \ )$

▶ **29.** $x + y = -2;\ (-2,\ \ ),\ (1,\ \ ),\ (\ \ , 5)$

30. $x + y = -3;\ (\ \ , 0),\ (0,\ \ ),\ (4,\ \ )$

31. $x = y - 4;\ (\ \ , -12),\ (\ \ , 3),\ (100,\ \ )$

32. $y = x + 5;\ (1,\ \ ),\ (\ \ , 7),\ (3,\ \ )$

▶ **33.** $y = 3x - 5;\ (1,\ \ ),\ (2,\ \ ),\ (\ \ , 4)$

34. $x = -12y + 5;\ (\ \ , -1),\ (\ \ , 1),\ (41,\ \ )$

35. $x = -y;\ (\ \ , 0),\ (3,\ \ ),\ (\ \ , -9)$

36. $y = -x;\ (\ \ , 0),\ (2,\ \ ),\ (\ \ , 2)$

37. $x + 2y = -8;\ (4,\ \ ),\ (\ \ , -3),\ (0,\ \ )$

38. $-5x + y = 1;\ (0,\ \ ),\ (-1,\ \ ),\ (2,\ \ )$

Review

Perform each indicated operation on decimals. See Sections 5.2 to 5.4.

39. $5.6 - 3.9$

40. $5 + 2.54 + 8.7$

41. $5.6 \cdot 3.9$

42. $0.56 \div 0.8$

43. $(0.236)(-100)$

44. $44.72 \div 100$

Concept Extensions

Recall that the axes divide the plane into four quadrants as shown. For Exercises 45 through 52, if a and b are both positive numbers, determine whether each statement is true or false.

45. (a, b) lies in quadrant I.

46. $(-a, -b)$ lies in quadrant IV.

47. $(0, b)$ lies on the *y*-axis.

48. $(a, 0)$ lies on the *x*-axis.

49. $(0, -b)$ lies on the *x*-axis.

50. $(-a, 0)$ lies on the *y*-axis.

51. $(-a, b)$ lies in quadrant III.

52. $(a, -b)$ lies in quadrant IV.

53. Is the ordered pair $(4, -3)$ plotted to the left or right of the *y*-axis? Explain.

54. Is the ordered pair $(6, -2)$ plotted above or below the *x*-axis? Explain.

Plot the points $A(4, 3)$, $B(-2, 3)$, $C(-2, -1)$, and $D(4, -1)$ on a rectangular coordinate system.

55. Draw a line segment connecting A and B, B and C, C and D, and D and A. Name the figure drawn.

56. Find the length and the width of the figure.

57. Find the perimeter of the figure.

58. Find the area of the figure.

Integrated Review

Reading Graphs

The following pictograph shows the six occupations with the largest estimated numerical increase (not salary increase) in employment in the United States between 2010 and 2020. Use this graph to answer Exercises 1 through 4.

**Jobs with Projected Highest
Numerical Increase: 2010–2020**

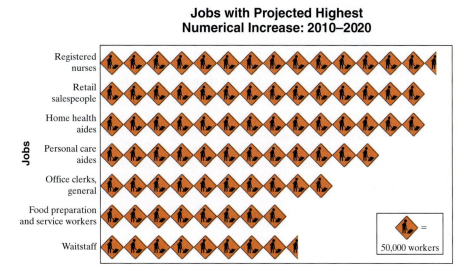

Registered nurses
Retail salespeople
Home health aides
Personal care aides
Office clerks, general
Food preparation and service workers
Waitstaff

Jobs

= 50,000 workers

Source: Bureau of Labor Statistics

1. Approximate the increase in the number of retail salespeople from 2010 to 2020.

2. Approximate the increase in the number of registered nurses from 2010 to 2020.

3. Which occupation is expected to show the greatest increase in numbers of employees between the years shown?

4. Which of the listed occupations is expected to show the smallest increase in numbers of employees between the years shown?

The following bar graph shows the tallest U.S. dams. Use this graph to answer Exercises 5 through 8.

Highest U.S. Dams

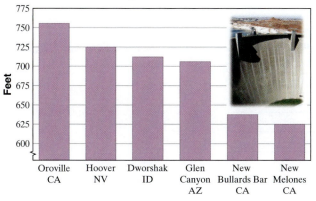

775
750
725
700
675
650
625
600

Feet

Oroville CA
Hoover NV
Dworshak ID
Glen Canyon AZ
New Bullards Bar CA
New Melones CA

Source: Committee on Register of Dams

5. Name the U.S. dam with the greatest height and estimate its height.

6. Name the U.S. dam whose height is between 625 and 650 feet and estimate its height.

7. Estimate how much taller the Hoover Dam is than the Glen Canyon Dam.

8. How many U.S. dams have heights over 700 feet?

9. _____

10. _____

11. _____

12. _____

13. _____

14. _____

15. _____

16. _____

17. _____

18. _____

19. _____

20. _____

21. _____

The following line graph shows the daily high temperatures for one week in Annapolis, Maryland. Use this graph to answer Exercises 9 through 12.

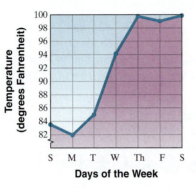

Days of the Week

9. Name the day(s) of the week with the highest temperature and give that high temperature.

10. Name the day(s) of the week with the lowest temperature and give that low temperature.

11. On what days of the week was the temperature less than 90° Fahrenheit?

12. On what days of the week was the temperature greater than 90° Fahrenheit?

The following circle graph shows the types of milk beverage consumed in the United States. Use this graph for Exercises 13 through 16. If a store in Kerrville, Texas, sells 200 quart containers of milk per week, estimate how many quart containers are sold in each category below.

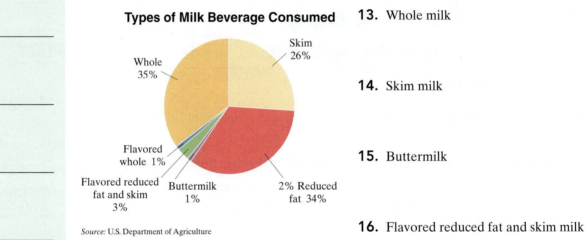

Types of Milk Beverage Consumed

Skim 26%

Whole 35%

Flavored whole 1%

Flavored reduced fat and skim 3%

Buttermilk 1%

2% Reduced fat 34%

Source: U.S. Department of Agriculture

13. Whole milk

14. Skim milk

15. Buttermilk

16. Flavored reduced fat and skim milk

The following list shows weekly quiz scores for a student in prealgebra. Use this list to complete the frequency distribution table below.

50	80	80	90	88	97
67	89	71	78	93	99
75	53	93	83	86	

	Class Intervals (Scores)	Tally	Class Frequency (Number of Quizzes)
17.	50–59		
18.	60–69		
19.	70–79		
20.	80–89		
21.	90–99		

22. Use the frequency distribution table from Exercises **17** through **21** to construct a histogram.

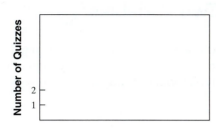

Quiz Scores

22. _____

23. Plot the points corresponding to the ordered pairs on the same set of axes. $(0, 2), (-1, 4), (2, 1), (-3, 0),$ $(-3, -5), (4, -1)$

24. Determine whether the ordered pair $(1, 3)$ is a solution of the linear equation $x = 3y$.

23. _____

24. _____

25. Determine whether the ordered pair $(-2, -4)$ is a solution of the linear equation $x + y = -6$.

26. Complete each ordered pair solution of the equation $x - y = 6$. $(0, \ ),$ $(\ , 0), (2, \ )$

25. _____

26. _____

Graphing Linear Equations in Two Variables

Objective

A Graph Linear Equations by Plotting Points.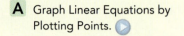

Now that we know how to plot points in a rectangular coordinate system and how to find ordered pair solutions, we are ready to graph linear equations in two variables. First, we give a formal definition.

> **Linear Equation in Two Variables**
>
> A *linear equation in two variables* is an equation that can be written in the form
>
> $ax + by = c$ Examples: $\begin{cases} 2x + 3y = 6 \\ \quad\ 2x = 5 \\ \qquad\ y = 3 \end{cases}$
>
> where a, b, and c are numbers, and a and b are not both 0.

In the last section, we discovered that a linear equation in two variables has many solutions. For the linear equation $x + y = 7$, we listed, for example, the solutions $(3, 4)$, $(0, 7)$, $(-1, 8)$, $(7, 0)$, $(5, 2)$, and $(4, 3)$. Are these all the solutions? No. There are infinitely many solutions of the equation $x + y = 7$ since there are infinitely many pairs of numbers whose sum is 7. Every linear equation in two variables has infinitely many ordered pair solutions. Since it is impossible to list every solution, we graph the solutions instead.

Objective **A** Graphing Linear Equations in Two Variables by Plotting Points ▶

Fortunately, the pattern described by the solutions of a linear equation makes "seeing" the solutions possible by graphing. This is so because **all the solutions of a linear equation in two variables correspond to points on a single straight line.** If we plot just a few of these points and draw the straight line connecting them, we have a complete graph of all the solutions.

To graph the equation $x + y = 7$, then, we plot a few ordered pair solutions, say $(3, 4)$, $(0, 7)$, and $(-1, 8)$.

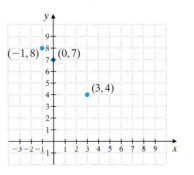

Now we connect these three points by drawing a straight line through them. The arrows at both ends of the line indicate that the line goes on forever in both directions.

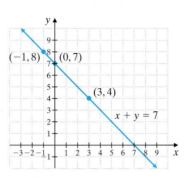

Every point on this line corresponds to an ordered pair solution of the equation $x + y = 7$. Also, every ordered pair solution of the equation $x + y = 7$ corresponds to a point on this line. In other words, this line is the graph of the equation $x + y = 7$. Although a line can be drawn using just two points, we will graph a third solution to check our work.

Example 1 Graph the equation $y = 3x$ by plotting the following points, which satisfy the equation, and drawing a line through the points.

$$(1, 3), (0, 0), (-1, -3)$$

Solution: Plot the points and draw a line through them. The line is the graph of the linear equation $y = 3x$.

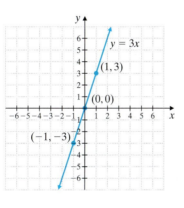

 Work Practice 1

Graphing Linear Equations and Finding Ordered Pair Solutions

- To graph a linear equation in two variables, find three ordered pair solutions, graph the solutions, and draw a line through the plotted points.
- To find an ordered pair solution of an equation, choose either an x-value or a y-value of the ordered pair and complete the ordered pair as we did in the previous section.

Practice 1

Graph the equation $x - y = 1$ by plotting the following points, which satisfy the equation, and drawing a line through the points.

$$(3, 2), (0, -1), (1, 0)$$

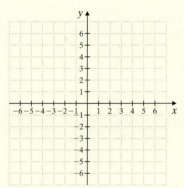

Answer

1.

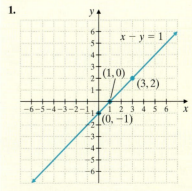

Practice 2

Graph: $y - x = 4$

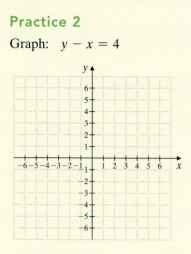

Practice 3

Graph: $y = -3x + 2$

Copyright 2015 Pearson Education, Inc.

Answers

2.

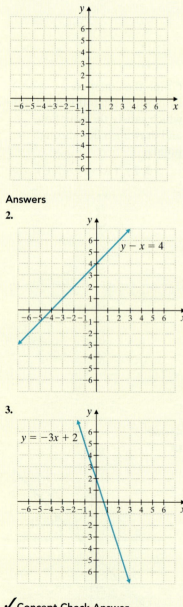

$y - x = 4$

3.

$y = -3x + 2$

✓ **Concept Check Answer**

no

Example 2   Graph: $x - y = 6$

Solution: Find any three ordered pair solutions. For each solution, we choose a value for x or y, and replace x or y by its chosen value in the equation. Then we solve the resulting equation for the other variable.

For example, let $x = 3$.

$x - y = 6$	Original equation
$3 - y = 6$	Let $x = 3$.
$-y = 6 - 3$	Subtract 3 from both sides.
$-y = 3$	Simplify.
$\dfrac{-y}{-1} = \dfrac{3}{-1}$	Divide both sides by -1.
$y = -3$	Simplify.

The ordered pair is (3, -3).

Also, let $y = 0$.

$x - y = 6$	Original equation
$x - 0 = 6$	Let $y = 0$.
$x = 6$	Simplify.

The ordered pair is (6, 0).

If we let $x = 7$, and solve the resulting equation for y, we will see that $y = 1$. **A third ordered pair solution is, then, (7, 1).**

Plot the three ordered pair solutions on a rectangular coordinate system and then draw a line through the three points. This line is the graph of $x - y = 6$. (For organization purposes, the ordered pairs are entered in the table below.)

x	y
3	-3
6	0
7	1

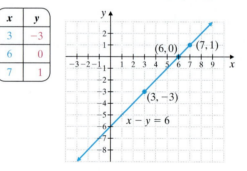

■ **Work Practice 2**

Helpful Hint

All three points should fall on the same straight line. If they do not, check your ordered pair solutions for a mistake, since the graph of every linear equation is a line.

✓ **Concept Check**   In Example 2, is the point $(9, -3)$ on the graph of $x - y = 6$?

Example 3   Graph: $y = 5x + 1$

Solution: For this equation, we will choose three x-values and find the corresponding y-values. We choose $x = 0$, $x = 1$, and $x = 2$. The first table on the next page shows our chosen x-values.

If $x = 0$, then
$y = 5x + 1$ becomes
$y = 5 \cdot 0 + 1$ or
$y = 1$

If $x = 1$, then
$y = 5x + 1$ becomes
$y = 5 \cdot 1 + 1$ or
$y = 6$

If $x = 2$, then
$y = 5x + 1$ becomes
$y = 5 \cdot 2 + 1$ or
$y = 11$

Now we complete the table of values, plot the ordered pair solutions, and graph the equation $y = 5x + 1$.

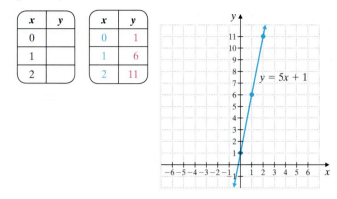

■ **Work Practice 3**

Next, we will graph a few special linear equations.

Example 4   Graph: $y = 4$

Solution: The equation $y = 4$ can be written as $0x + y = 4$. When the equation is written in this form, notice that no matter what value we choose for x, y is always 4.

$$0 \cdot x + y = 4$$
$$0 \cdot (\text{any number}) + y = 4$$
$$0 + y = 4$$
$$y = 4$$

Fill in a table listing ordered pair solutions of $y = 4$. Choose any three x-values. The y-values are always 4. Plot the ordered pair solutions and graph $y = 4$.

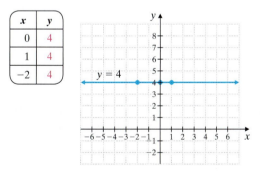

The graph is a horizontal line that crosses the y-axis at 4.

■ **Work Practice 4**

Practice 4

Graph: $y = -4$

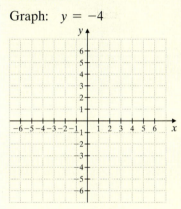

Answer

4.

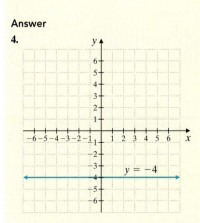

Horizontal Lines

If a is a number, then the graph of $y = a$ is a *horizontal line* that crosses the y-axis at a. For example, the graph of $y = 2$ is a horizontal line that crosses the y-axis at 2.

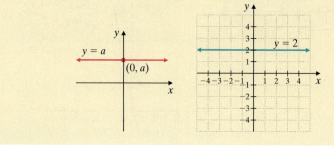

Practice 5

Graph: $x = 5$

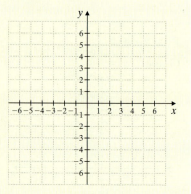

Example 5 Graph: $x = -2$

Solution: The equation $x = -2$ can be written as $x + 0y = -2$. No matter what y-value we choose, x is always -2. Fill in a table listing ordered pair solutions of $x = -2$. Choose any three y-values. The x-values are always -2. Plot the ordered pair solutions and graph $x = -2$.

x	y
-2	3
-2	0
-2	-2

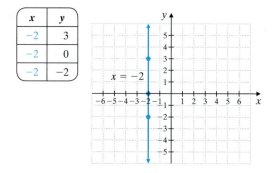

The graph is a vertical line that crosses the x-axis at -2.

■ **Work Practice 5**

Vertical Lines

If a is a number, then the graph of $x = a$ is a *vertical line* that crosses the x-axis at a. For example, the graph of $x = -3$ is a vertical line that crosses the x-axis at -3.

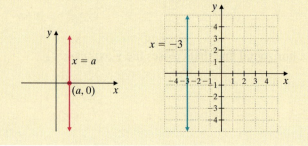

Answer

5.

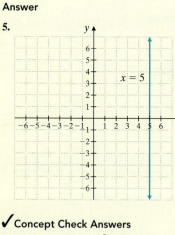

✓ **Concept Check** Determine whether the following equations represent vertical lines, horizontal lines, or neither.

a. $y = -6$ **b.** $x + y = 7$ **c.** $x = \dfrac{1}{2}$ **d.** $x = \dfrac{1}{2}y$

Vocabulary, Readiness & Video Check

Use the choices below to fill in each blank.

horizontal   line
vertical   linear

1. A _____ equation in two variables can be written in the form $ax + by = c$.

2. The graph of the equation $x = 5$ is a _____ line.

3. The graph of the equation $y = -2$ is a _____ line.

4. The graph of the equation $3x - 2y = 6$ is a _____ that is not vertical or horizontal.

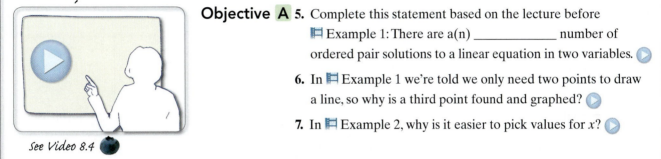

Martin-Gay Interactive Videos   *Watch the section lecture video and answer the following question.*

Objective A **5.** Complete this statement based on the lecture before
◫ Example 1: There are a(n) _____ number of
ordered pair solutions to a linear equation in two variables. ▶

6. In ◫ Example 1 we're told we only need two points to draw
a line, so why is a third point found and graphed? ▶

7. In ◫ Example 2, why is it easier to pick values for x? ▶

See Video 8.4

8.4 Exercise Set MyMathLab® ▶

Objective A *Graph each equation. See Examples 1 through 3.*

1. $x + y = 4$

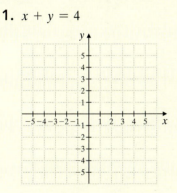

2. $x + y = 2$

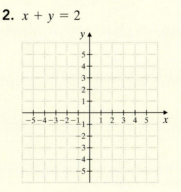

3. $x - y = -6$

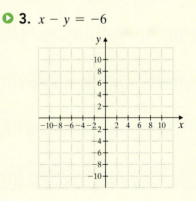

4. $y - x = 6$

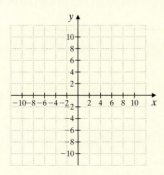

5. $y = 4x$

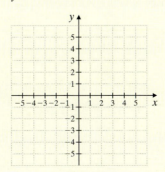

6. $x = 2y$

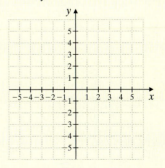

7. $y = 2x - 1$

8. $y = 2x + 5$

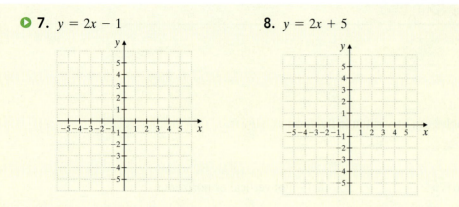

Graph each equation. See Examples 4 and 5.

9. $x = -3$

10. $y = 1$

11. $y = -3$

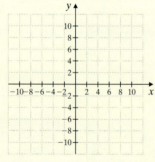

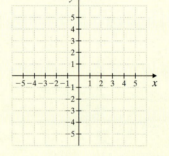

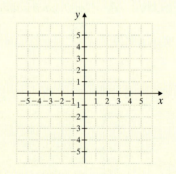

12. $x = -7$

13. $x = 0$

14. $y = 0$

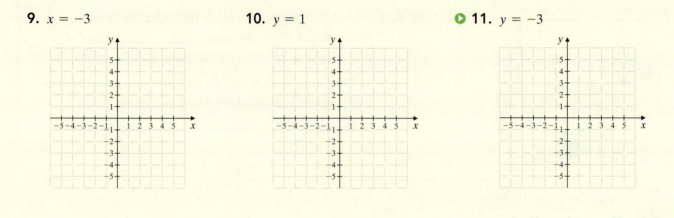

Mixed Practice *Graph each equation. See Examples 1 through 5.*

15. $y = -2x$

16. $x = y$

17. $y = -2$

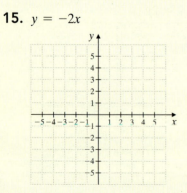
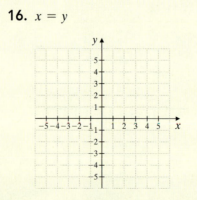
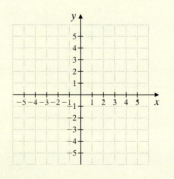

18. $x = 1$

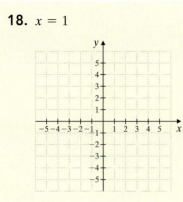

19. $x + 2y = 12$

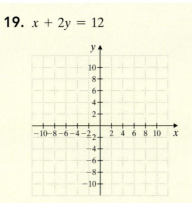

20. $3x - y = 3$

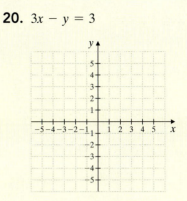

21. $x = 6$

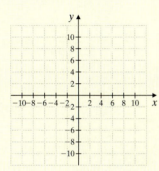

22. $y = 7$

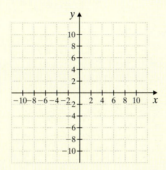

23. $y = x - 3$

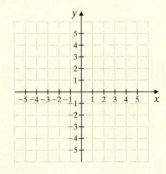

24. $y = x - 1$

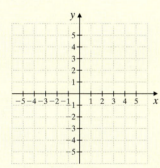

25. $x = y - 4$

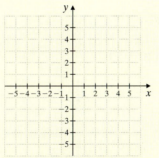

26. $x = 5 - y$

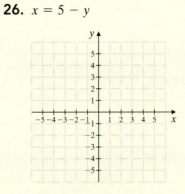

27. $x + 3 = 0$

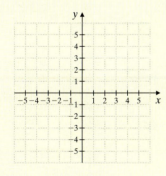

28. $y + 4 = 0$

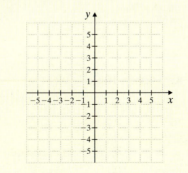

29. $y = -\dfrac{1}{4}x$

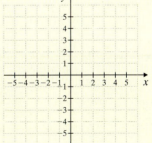

30. $y = \dfrac{1}{4}x$

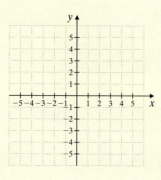

31. $y = \dfrac{1}{3}x$

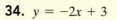

32. $y = -\dfrac{1}{3}x$

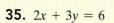

33. $y = 4x + 2$

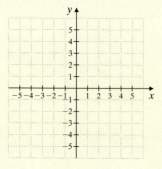

34. $y = -2x + 3$

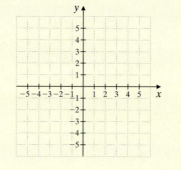

35. $2x + 3y = 6$

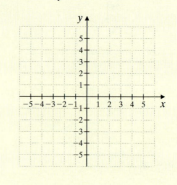

36. $5x - 2y = 10$

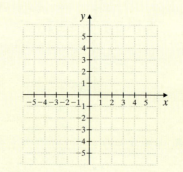

37. $x = -3.5$

38. $y = 5.5$

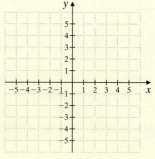

39. $3x - 4y = 24$

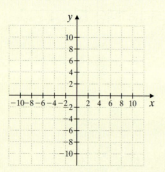

40. $4x + 2y = 16$

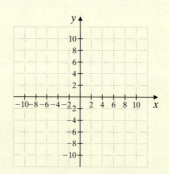

41. $y = \dfrac{1}{2}$

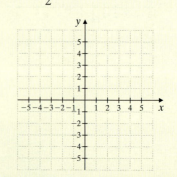

42. $x = -\dfrac{1}{2}$ **43.** $y = \dfrac{1}{3}x - 2$ **44.** $x = \dfrac{1}{3}y - 2$

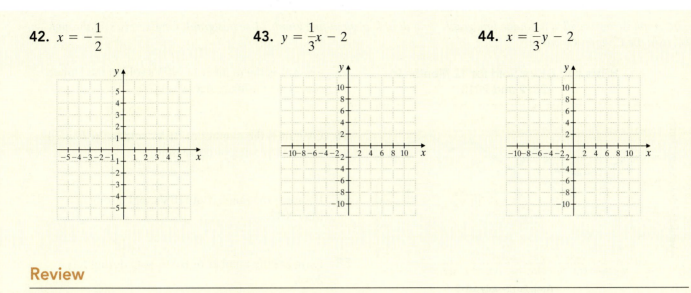

Review

Perform the indicated operations. Give the result as a fraction in simplest form. See Sections 4.3 through 4.5, and 5.5.

45. $(0.5)\left(-\dfrac{1}{8}\right)$ **46.** $\dfrac{3}{4} - \dfrac{19}{20}$ **47.** $\dfrac{3}{4} \div \left(-\dfrac{19}{20}\right)$

48. $0.75 + \dfrac{3}{10}$ **49.** $\dfrac{2}{11} - \dfrac{x}{11}$ **50.** $\dfrac{2}{11}\left(-\dfrac{x}{11}\right)$

Concept Extensions

51. Fill in the table and use it to graph $y = |x|$ on the coordinate system.

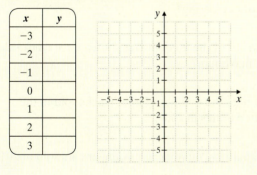

x	y
−3	
−2	
−1	
0	
1	
2	
3	

52. Graph: $15x - 18y = 270$

To do so, complete the table and find at least 2 additional ordered pair solutions.

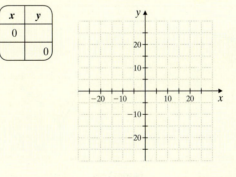

x	y
0	
	0

53. In your own words, explain how to graph a linear equation in two variables.

54. Suppose that a classmate tries to graph a linear equation in two variables and plots three ordered pair solutions. If the three points do not all lie on the same line, what should the student do next?

The graph below is called a double line graph (or a double broken-line graph.) Use this graph for Exercises 55 through 60. (See the Chapter Opener.)

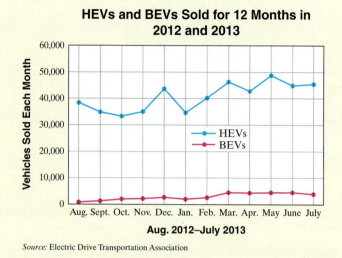

HEVs and BEVs Sold for 12 Months in 2012 and 2013

Aug. 2012–July 2013

Source: Electric Drive Transportation Association

55. Overall, is the number of HEVs sold in the United States increasing or decreasing?

56. Overall, is the number of BEVs sold in the United States increasing or decreasing?

57. Estimate the number of HEVs sold during May 2013.

58. Estimate the number of BEVs sold during July 2013.

59. Place a straight edge along the points for HEVs sold in Aug. 2012 and June 2013. Draw a line using these two points and use this line to estimate the number of HEVs sold in Oct. 2013.

60. Place a straight edge along the points for BEVs sold in Aug. 2012 and June 2013. Draw a line using these two points and use this line to estimate the number of BEVs sold in Oct. 2013.

Use this double line graph to answer Exercises 61 through 66.

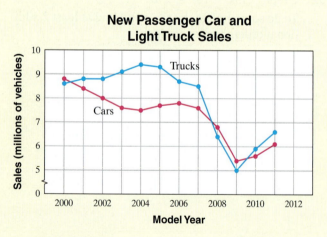

New Passenger Car and Light Truck Sales

Model Year

Source: Wards Auto Group

61. From 2000 to 2004, were passenger car sales increasing or decreasing?

62. In which of the model years shown were truck sales at a maximum?

63. Estimate the number of light trucks sold in the United States in 2009.

64. Estimate the number of passenger cars sold in 2011.

65. Describe any trends you see in this graph.

66. In which three years was the number of passenger cars sold greater than the number of light trucks sold?

Counting and Introduction to Probability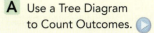

Objective A Using a Tree Diagram

In our daily conversations, we often talk about the likelihood or **probability** of a given result occurring. For example:

The *chance* of thundershowers is 70 percent.

What are the *odds* that the New Orleans Saints will go to the Super Bowl?

What is the *probability* that you will finish cleaning your room today?

Each of these chance happenings—thundershowers, the New Orleans Saints playing in the Super Bowl, and finishing cleaning your room today—is called an **experiment.** The possible results of an experiment are called **outcomes.** For example, flipping a coin is an experiment, and the possible outcomes are heads (H) or tails (T).

One way to picture the outcomes of an experiment is to draw a **tree diagram.** Each outcome is shown on a separate branch. For example, the outcomes of flipping a coin are

Heads   Tails

 Example 1 Draw a tree diagram for tossing a coin twice. Then use the diagram to find the number of possible outcomes.

Solution:

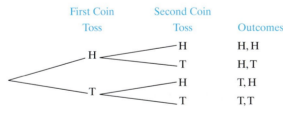

There are 4 possible outcomes when tossing a coin twice.

◼ **Work Practice 1**

Example 2 Draw a tree diagram for an experiment consisting of rolling a die and then tossing a coin. Then use the diagram to find the number of possible outcomes.

Die

(Continued on next page)

Objectives

A Use a Tree Diagram to Count Outcomes.

B Find the Probability of an Event.

Practice 1

Draw a tree diagram for tossing a coin three times. Then use the diagram to find the number of possible outcomes.

Practice 2

Draw a tree diagram for an experiment consisting of tossing a coin and then rolling a die. Then use the diagram to find the number of possible outcomes. (Answer appears on the next page.)

Answer

1.

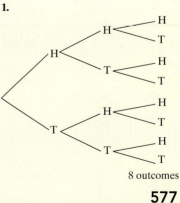

8 outcomes

577

Solution: Recall that a die has six sides and that each side represents a number, 1 through 6.

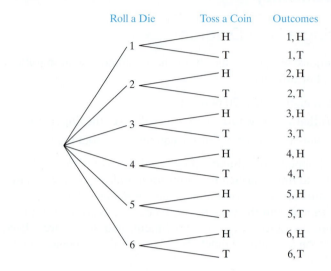

There are 12 possible outcomes for rolling a die and then tossing a coin.

▶ **Work Practice 2**

Any number of outcomes considered together is called an **event.** For example, when tossing a coin twice, H, H is an event. The event is tossing heads first and tossing heads second. Another event would be tossing tails first and then heads (T, H), and so on.

Objective B Finding the Probability of an Event ▶

As we mentioned earlier, the **probability of an event is a measure of the chance or likelihood of it occurring.** For example, if a coin is tossed, what is the probability that heads occurs? Since one of two equally likely possible outcomes is heads, the probability is $\frac{1}{2}$.

The Probability of an Event

$$\text{probability of an event} = \frac{\text{number of ways that the event can occur}}{\text{number of possible outcomes}}$$

Note from the definition of probability that the probability of an event is always between 0 and 1, inclusive (i.e., including 0 and 1). A probability of 0 means that an event won't occur, and a probability of 1 means that an event is certain to occur.

Example 3 If a coin is tossed twice, find the probability of tossing heads on the first toss and then heads again on the second toss (H, H).

Solution: 1 way the event can occur
↓

H, T   H, H   T, H   T, T

4 possible outcomes

Practice 3

If a coin is tossed three times, find the probability of tossing tails, then heads, then tails (T, H, T).

Answers

2.

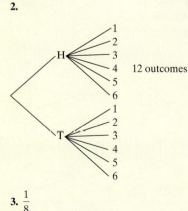

12 outcomes

3. $\frac{1}{8}$

$$\text{probability} = \frac{1}{4} \quad \begin{matrix} \text{Number of ways the event can occur} \\ \text{Number of possible outcomes} \end{matrix}$$

The probability of tossing heads and then heads is $\frac{1}{4}$.

■ **Work Practice 3**

Example 4   If a die is rolled one time, find the probability of rolling a 3 or a 4.

Solution: Recall that there are 6 possible outcomes when rolling a die.

2 ways that the event can occur

possible outcomes: 1,   2,   3,   4,   5,   6

6 possible outcomes

$$\text{probability of a 3 or a 4} = \frac{2}{6} \quad \begin{matrix} \text{Number of ways the event can occur} \\ \text{Number of possible outcomes} \end{matrix}$$

$$= \frac{1}{3} \quad \text{Simplest form}$$

■ **Work Practice 4**

✔**Concept Check**   Suppose you have calculated a probability of $\frac{11}{9}$. How do you know that you have made an error in your calculation?

Example 5   Find the probability of choosing a red marble from a box containing 1 red, 1 yellow, and 2 blue marbles.

Solution: 1 way that event can occur

red   yellow   blue   blue

4 possible outcomes

$$\text{probability} = \frac{1}{4}$$

■ **Work Practice 5**

Practice 4

If a die is rolled one time, find the probability of rolling a 2 or a 5.

Practice 5

Use the diagram and information in Example 5 and find the probability of choosing a blue marble from the box.

Answers

4. $\frac{1}{3}$ **5.** $\frac{1}{2}$

✔**Concept Check Answer**

The number of ways an event can occur can't be larger than the number of possible outcomes.

Vocabulary, Readiness & Video Check

Use the choices below to fill in each blank. Choices may be used more than once.

    0   probability   tree diagram

    1   outcome

1. A possible result of an experiment is called a(n) _____.

2. A(n) _____ shows each outcome of an experiment as a separate branch.

3. The _____ of an event is a measure of the likelihood of it occurring.

4. _____ is calculated by the number of ways that an event can occur divided by the number of possible outcomes.

5. A probability of _____ means that an event won't occur.

6. A probability of _____ means that an event is certain to occur.

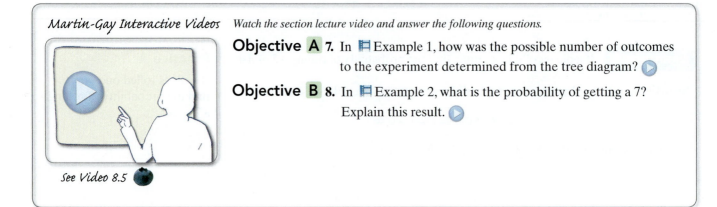

Martin-Gay Interactive Videos   *Watch the section lecture video and answer the following questions.*

Objective A 7. In ⊞ Example 1, how was the possible number of outcomes to the experiment determined from the tree diagram? ▶

Objective B 8. In ⊞ Example 2, what is the probability of getting a 7? Explain this result. ▶

See Video 8.5 🫐

8.5 Exercise Set MyMathLab® ▶

Objective A *Draw a tree diagram for each experiment. Then use the diagram to find the number of possible outcomes. See Examples 1 and 2.*

▶ **1.** Choosing a letter in the word MATH, then a number (1, 2, or 3)

2. Choosing a number (1 or 2) and then a vowel (a, e, i, o, u)

Spinner A   Spinner B

3. Spinning Spinner A once

4. Spinning Spinner B once

5. Spinning Spinner B twice

6. Spinning Spinner A twice

7. Spinning Spinner A and then Spinner B

8. Spinning Spinner B and then Spinner A

9. Tossing a coin and then spinning Spinner B

10. Spinning Spinner A and then tossing a coin

Objective **B** *If a single die is tossed once, find the probability of each event. See Examples 3 through 5.*

11. A 5

12. A 9

13. A 1 or a 6

14. A 2 or a 3

15. An even number

16. An odd number

17. A number greater than 2

18. A number less than 6

Suppose the spinner shown is spun once. Find the probability of each event. See Examples 3 through 5.

19. The result of the spin is 2.

20. The result of the spin is 3.

21. The result of the spin is 1, 2, or 3.

22. The result of the spin is not 3.

23. The result of the spin is an odd number.

24. The result of the spin is an even number.

If a single choice is made from the bag of marbles shown, find the probability of each event. See Examples 3 through 5.

25. A red marble is chosen.

26. A blue marble is chosen.

27. A yellow marble is chosen.

28. A green marble is chosen.

29. A green or red marble is chosen.

30. A blue or yellow marble is chosen.

A new drug is being tested that is supposed to lower blood pressure. This drug was given to 200 people and the results are shown below.

Lower Blood Pressure	Higher Blood Pressure	Blood Pressure Not Changed
152	38	10

31. If a person is testing this drug, what is the probability that his or her blood pressure will be higher?

32. If a person is testing this drug, what is the probability that his or her blood pressure will be lower?

33. If a person is testing this drug, what is the probability that his or her blood pressure will not change?

34. What is the sum of the answers to Exercises **31**, **32**, and **33**? In your own words, explain why.

Review

Perform each indicated operation. See Sections 4.3 and 4.5.

35. $\dfrac{1}{2} + \dfrac{1}{3}$ **36.** $\dfrac{7}{10} - \dfrac{2}{5}$ **37.** $\dfrac{1}{2} \cdot \dfrac{1}{3}$ **38.** $\dfrac{7}{10} \div \dfrac{2}{5}$ **39.** $5 \div \dfrac{3}{4}$ **40.** $\dfrac{3}{5} \cdot 10$

Concept Extensions

Recall that a deck of cards contains 52 cards. These cards consist of four suits (hearts, spades, clubs, and diamonds) of each of the following: 2, 3, 4, 5, 6, 7, 8, 9, 10, jack, queen, king, and ace. If a card is chosen from a deck of cards, find the probability of each event.

41. The king of hearts

42. The 10 of spades

43. A king

44. A 10

45. A heart

46. A club

47. A card in black ink

48. A queen or ace

Two dice are tossed. Find the probability of each sum of the dice. (Hint: Draw a tree diagram of the possibilities of two tosses of a die, and then find the sum of the numbers on each branch.)

49. A sum of 6

50. A sum of 10

51. A sum of 13

52. A sum of 2

Solve. See the Concept Check in this section.

53. In your own words, explain why the probability of an event cannot be greater than 1.

54. In your own words, explain when the probability of an event is 0.

Chapter 8 Group Activity

Scatter Diagrams

Sections 8.1–8.3

In Section 8.1, we learned about presenting data visually in a graph, such as a pictograph, bar graph, or line graph. Now we learn about another type of graph, based on ordered pairs, that can be used to present data.

Data that can be represented as an ordered pair are called paired data. Many types of data collected from the real world are paired data. For instance, the total amount of rainfall a location receives each year can be written as an ordered pair of the form (year, total rainfall in inches) and is paired data. The graph of paired data as points in the rectangular coordinate system is called a **scatter diagram,** or scatter plot. Scatter diagrams can be used to look for patterns and trends in paired data.

For example, the data shown in the table are paired data that can be written as a set of ordered pairs of the form (year, number of restaurants in thousands), such as (2010, 33) and (2012, 37). A scatter diagram of the paired data is shown below. Notice that the horizontal axis is labeled "Year" to describe the x-coordinates in the ordered pairs. The vertical axis is labeled "Restaurants (in thousands)" to describe the y-coordinates in the ordered pairs.

Number of Subway Restaurants	
Year	Number of Restaurants (in thousands)
2009	31
2010	33
2011	36
2012	37
2013	39

(*Source*: Subway)

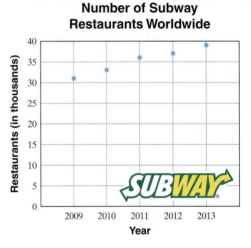

Number of Subway Restaurants Worldwide

The table to the right gives the number of Wal-Mart Supercenters each year. Use the table to answer the following questions.

1. Write this paired data as a set of ordered pairs of the form (year, number in hundreds).

2. Create a scatter diagram of the paired data.

Number of Wal-Mart Supercenters	
Year	Number of Supercenters (in hundreds)
2009	26
2010	27
2011	29
2012	30
2013	32

(*Source:* Wal-Mart Corporation)

3. What trend in the paired data does the scatter diagram show?

4. Find or collect your own paired data and present them in a scatter diagram. What does the graph show?

Chapter 8 Vocabulary Check

Fill in each blank with one of the words or phrases listed below.

outcomes	quadrants	line	*y*	class frequency	probability
pictograph	plotting	origin	experiment	linear	class interval
histogram	bar	*x*	circle	tree diagram	

1. A(n) _____ graph presents data using vertical or horizontal bars.

2. The possible results of an experiment are the _____.

3. A(n) _____ is a graph in which pictures or symbols are used to visually present data.

4. A(n) _____ graph displays information with a line that connects data points.

5. In the ordered pair (*a, b*) the _____-value is *a* and the _____-value is *b*.

6. A(n) _____ is one way to picture and count outcomes.

7. A(n) _____ is an activity being considered, such as tossing a coin or rolling a die.

8. In a(n) _____ graph, each section (shaped like a piece of pie) shows a category and the relative size of the category.

9. The _____ of an event is $\dfrac{\text{number of ways that the event can occur}}{\text{number of possible outcomes}}$.

10. A(n) _____ is a special bar graph in which the width of each bar represents a(n) _____ and the height of each bar represents the _____.

11. The point of intersection of the *x*-axis and the *y*-axis in the rectangular coordinate system is called the _____.

12. The axes divide the plane into four regions called _____.

13. The process of locating a point on the rectangular coordinate system is called _____ the point.

14. A(n) _____ equation in two variables can be written in the form $ax + by = c$.

> **Helpful Hint**
> ● Are you preparing for your test? Don't forget to take the Chapter 8 Test on page 595. Then check your answers at the back of the text and use the Chapter Test Prep Videos to see the fully worked-out solutions to any of the exercises you want to review.

8 Chapter Highlights

Definitions and Concepts	Examples
Section 8.1 Reading Pictographs, Bar Graphs, Histograms, and Line Graphs	

A **pictograph** is a graph in which pictures or symbols are used to visually present data.

A **line graph** displays information with a line that connects data points.

A **bar graph** presents data using vertical or horizontal bars.

The bar graph on the right shows the number of acres of corn harvested in 2012 for leading states.

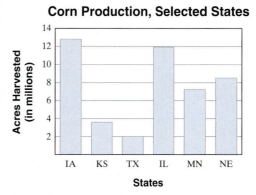

Corn Production, Selected States

Source: U.S. Department of Agriculture

(continued)

Definitions and Concepts	Examples

Section 8.1   Reading Pictographs, Bar Graphs, Histograms, and Line Graphs   (continued)

	1. Approximately how many acres of corn were harvested in Iowa? 12,800,000 acres **2.** About how many more acres of corn were harvested in Illinois than Nebraska? 12 million $-$ 8.5 million 3.5 million or 3,500,000 acres
A **histogram** is a special bar graph in which the width of each bar represents a **class interval** and the height of each bar represents the **class frequency.** The histogram on the right shows student quiz scores.	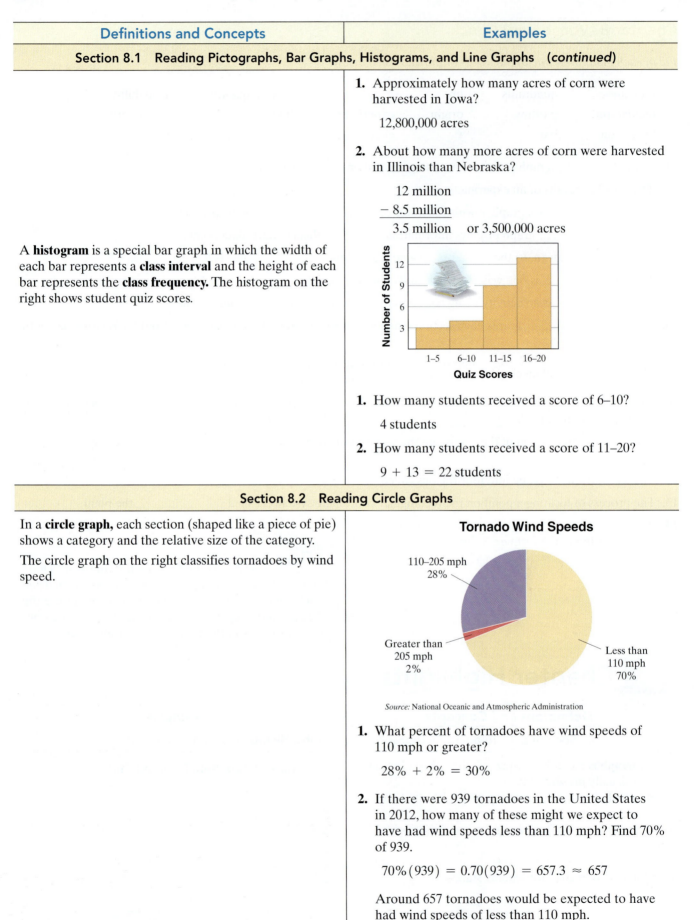 **1.** How many students received a score of 6–10? 4 students **2.** How many students received a score of 11–20? $9 + 13 = 22$ students

Section 8.2   Reading Circle Graphs

In a **circle graph,** each section (shaped like a piece of pie) shows a category and the relative size of the category. The circle graph on the right classifies tornadoes by wind speed.	**Tornado Wind Speeds** 110–205 mph 28% Greater than 205 mph 2% Less than 110 mph 70% *Source:* National Oceanic and Atmospheric Administration **1.** What percent of tornadoes have wind speeds of 110 mph or greater? $28\% + 2\% = 30\%$ **2.** If there were 939 tornadoes in the United States in 2012, how many of these might we expect to have had wind speeds less than 110 mph? Find 70% of 939. $70\%(939) = 0.70(939) = 657.3 \approx 657$ Around 657 tornadoes would be expected to have had wind speeds of less than 110 mph.

Definitions and Concepts	Examples

Section 8.3   The Rectangular Coordinate System and Paired Data

The **rectangular coordinate system** consists of two number lines intersecting at the point 0 on each number line. The horizontal number line is called the **x-axis** and the vertical number line is called the **y-axis.**

Every point in the rectangular coordinate system corresponds to an **ordered pair of numbers** such as

$$(2, -2)$$

   x-coordinate   y-coordinate

   or x-value   or y-value

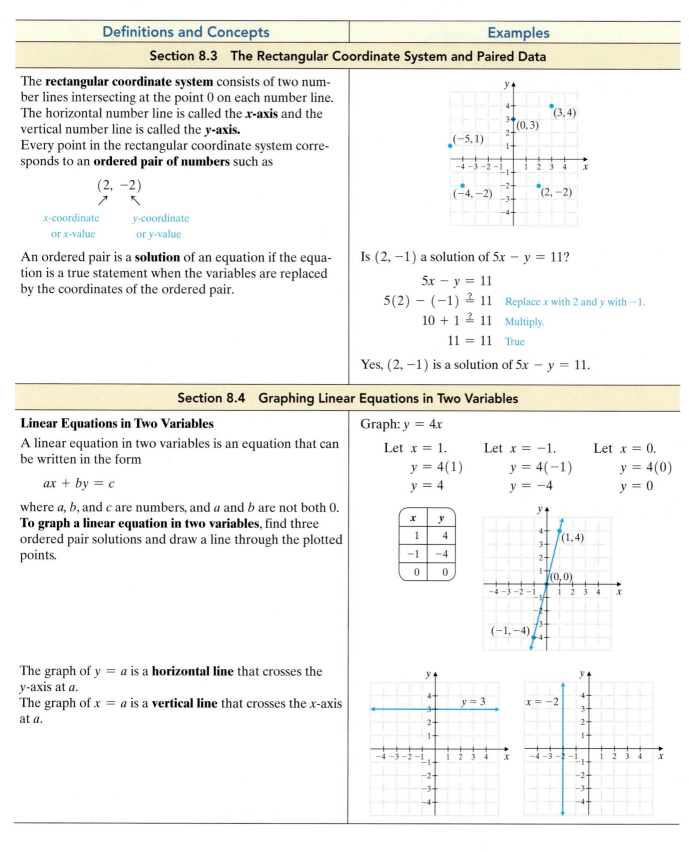

An ordered pair is a **solution** of an equation if the equation is a true statement when the variables are replaced by the coordinates of the ordered pair.

Is $(2, -1)$ a solution of $5x - y = 11$?

$$5x - y = 11$$
$$5(2) - (-1) \overset{?}{=} 11 \quad \text{Replace } x \text{ with 2 and } y \text{ with } -1.$$
$$10 + 1 \overset{?}{=} 11 \quad \text{Multiply.}$$
$$11 = 11 \quad \text{True}$$

Yes, $(2, -1)$ is a solution of $5x - y = 11$.

Section 8.4   Graphing Linear Equations in Two Variables

Linear Equations in Two Variables

A linear equation in two variables is an equation that can be written in the form

$$ax + by = c$$

where a, b, and c are numbers, and a and b are not both 0.
To graph a linear equation in two variables, find three ordered pair solutions and draw a line through the plotted points.

Graph: $y = 4x$

Let $x = 1$.   Let $x = -1$.   Let $x = 0$.
$$y = 4(1) \qquad y = 4(-1) \qquad y = 4(0)$$
$$y = 4 \qquad\quad y = -4 \qquad\quad y = 0$$

x	y
1	4
-1	-4
0	0

The graph of $y = a$ is a **horizontal line** that crosses the y-axis at a.
The graph of $x = a$ is a **vertical line** that crosses the x-axis at a.

Definitions and Concepts	Examples
Section 8.5 Counting and Introduction to Probability	

An **experiment** is an activity being considered, such as tossing a coin or rolling a die. The possible results of an experiment are the **outcomes**. A **tree diagram** is one way to picture and count outcomes.

Draw a tree diagram for tossing a coin and then choosing a number from 1 to 4.

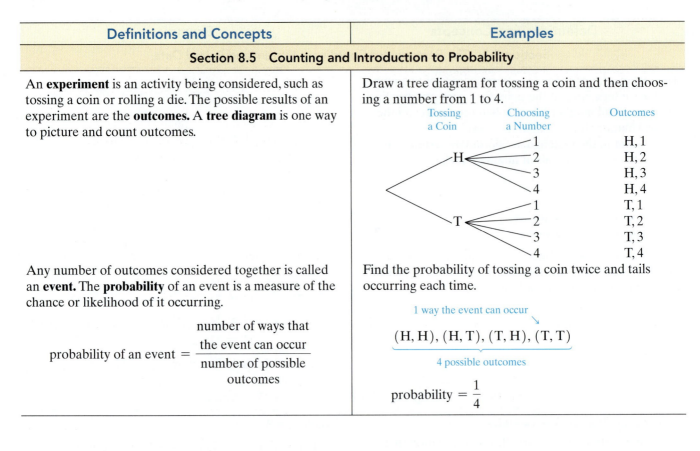

Any number of outcomes considered together is called an **event**. The **probability** of an event is a measure of the chance or likelihood of it occurring.

$$\text{probability of an event} = \frac{\text{number of ways that the event can occur}}{\text{number of possible outcomes}}$$

Find the probability of tossing a coin twice and tails occurring each time.

1 way the event can occur

$$(H, H), (H, T), (T, H), (T, T)$$

4 possible outcomes

$$\text{probability} = \frac{1}{4}$$

Chapter 8   Review

(8.1) *The following pictograph shows the number of new homes constructed from July 2012 to July 2013, by region. Use this graph to answer Exercises 1 through 6.*

**New Home Construction
July 2012–July 2013**

Source: U.S. Census Bureau

1. How many new homes were constructed in the West during the given year?

2. How many new homes were constructed in the South during the given year?

3. Which region had the most new homes constructed?

4. Which region had the fewest new homes constructed?

5. Which region(s) had 2,000,000 or more new homes constructed?

6. Which region(s) had fewer than 2,000,000 new homes constructed?

The following bar graph shows the percent of persons age 25 or over who completed four or more years of college. Use this graph to answer Exercises 7 through 10.

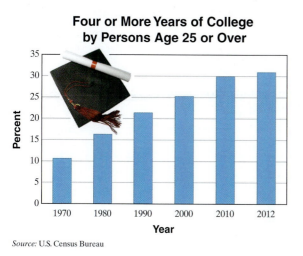

Four or More Years of College by Persons Age 25 or Over

Year

Source: U.S. Census Bureau

7. Approximate the percent of persons who had completed four or more years of college by 2010.

8. What year shown had the greatest percent of persons completing four or more years of college?

9. What years shown had 20% or more of persons completing four or more years of college?

10. Describe any patterns you notice in this graph.

The following line graph shows the total number of Olympic medals awarded during the Summer Olympics since 1992. Use this graph to answer Exercises 11 through 16.

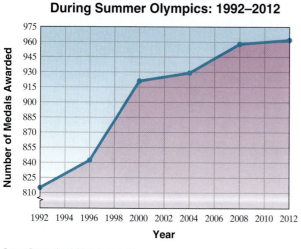

Number of Medals Awarded During Summer Olympics: 1992–2012

Year

Source: International Olympic Committee

11. Approximate the number of medals awarded during the Summer Olympics of 2012.

12. Approximate the number of medals awarded during the Summer Olympics of 2000.

13. Approximate the number of medals awarded during the Summer Olympics of 2004.

14. Approximate the number of medals awarded during the Summer Olympics of 1992.

15. How many more medals were awarded at the Summer Olympics of 1996 than at the Summer Olympics of 1992?

16. How many more medals were awarded at the Summer Olympics of 2012 than at the Summer Olympics of 1992?

The following histogram shows the hours worked per week by the employees of Southern Star Furniture. Use this histogram to answer Exercises 17 through 20.

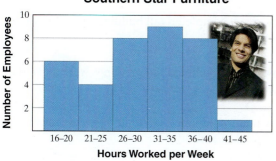

Southern Star Furniture

Hours Worked per Week

17. How many employees work 41–45 hours per week?

18. How many employees work 21–25 hours per week?

19. How many employees work 30 hours or less per week?

20. How many employees work 36 hours or more per week?

Following is a list of monthly record high temperatures for New Orleans, Louisiana. Use this list to complete the frequency distribution table below.

83	96	101	92
85	100	92	102
89	101	87	84

	Class Intervals (Temperatures)	Tally	Class Frequency (Number of Months)
21.	80°-89°		
22.	90°-99°		
23.	100°-109°		

24. Use the table from Exercises **21–23** to draw a histogram.

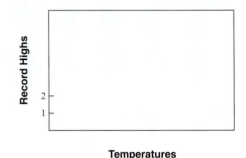

(8.2) *The following circle graph shows a family's $4000 monthly budget. Use this graph to answer Exercises 25 through 30. Write all ratios as fractions in simplest form.*

25. What is the largest budget item?

26. What is the smallest budget item?

27. How much money is budgeted for the mortgage payment and utilities?

28. How much money is budgeted for savings and contributions?

29. Find the ratio of the mortage payment to the total monthly budget.

30. Find the ratio of food to the total monthly budget.

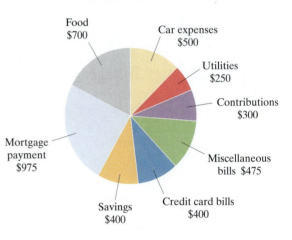

In 2012, there were approximately 62 buildings 200 meters or taller completed in the world. The following circle graph shows the percent of these buildings by area. Use this graph to answer Exercises 31 through 34. Round each answer to the nearest whole.

Percent of Tall Buildings Completed in 2012 200 Meters or Taller by Continent

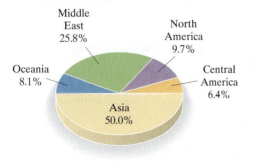

Source: Council on Tall Buildings and Urban Habitats

31. How many completed tall buildings were located in Asia?

32. How many completed tall buildings were located in North America?

33. How many completed tall buildings were located in Oceania?

34. How many completed tall buildings were located in the Middle East?

(8.3) *Complete and graph the ordered pair solutions of each given equation.*

35. $x = -6y$; $(0, \quad)$, $(\quad, -1)$, $(-6, \quad)$

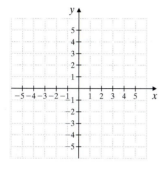

36. $y = 3x - 2$; $(0, \quad)$, $(-1, \quad)$, $(2, \quad)$

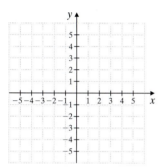

37. $x + y = -4$; $(-1, \quad)$, $(\quad, 0)$, $(-5, \quad)$

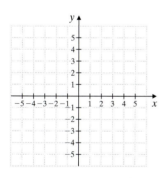

38. $x - y = 3$; $(4, \quad)$, $(0, \quad)$, $(\quad, 3)$

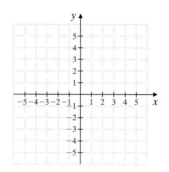

39. $y = 3x$; $(1, \quad)$, $(-2, \quad)$, $(\quad, 0)$

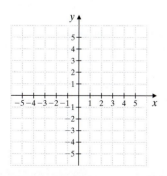

40. $x = y + 6$; $(1, \quad)$, $(6, \quad)$, $(\quad, -4)$

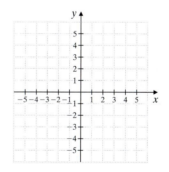

(8.4) *Graph each linear equation.*

41. $x = -5$

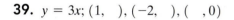

42. $y = \dfrac{3}{2}$

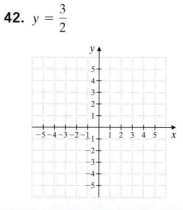

43. $x + y = 11$

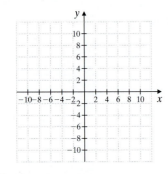

44. $x - y = 11$

45. $y = 4x - 2$

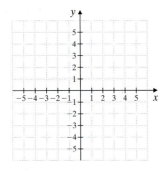

46. $y = 5x$

47. $x = -2y$

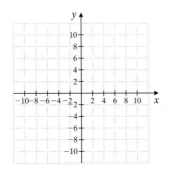

48. $x + y = -1$

49. $2x - 3y = 12$

50. $x = \dfrac{1}{2}y$

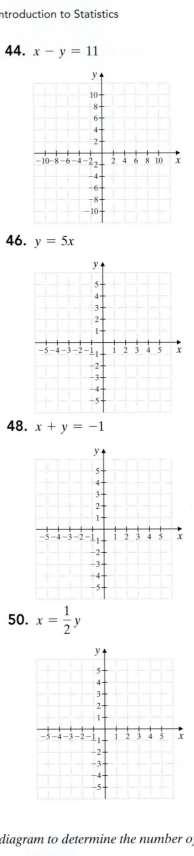

(8.5) *Draw a tree diagram for each experiment. Then use the diagram to determine the number of outcomes.*

Spinner 1   Spinner 2

51. Tossing a coin and then spinning Spinner 1

52. Spinning Spinner 2 and then tossing a coin

53. Spinning Spinner 1 twice

54. Spinning Spinner 2 twice

55. Spinning Spinner 1 and then Spinner 2

Find the probability of each event.

56. Rolling a 4 on a die

57. Rolling a 3 on a die

58. Spinning a 4 on Spinner 1

59. Spinning a 3 on Spinner 1

60. Spinning either a 1, 3, or 5 on Spinner 1

61. Spinning either a 2 or a 4 on Spinner 1

Die

Mixed Review

Given a bag containing 2 red marbles, 2 blue marbles, 3 yellow marbles, and 1 green marble, find the following:

62. The probability of choosing a blue marble from the bag

63. The probability of choosing a yellow marble from the bag

64. The probability of choosing a red marble from the bag

65. The probability of choosing a green marble from the bag

Graph each equation.

66. $x = -4$

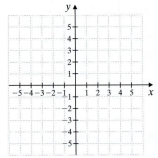

67. $y = 3$

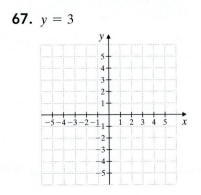

68. $x - 2y = 8$

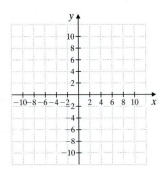

69. $2x + y = 6$

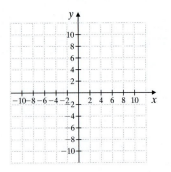

70. $x = y + 3$

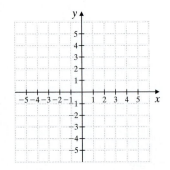

71. $x + y = -4$

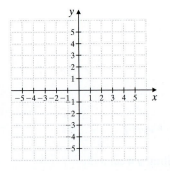

72. $y = \dfrac{3}{4}x$

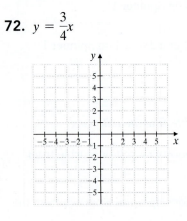

73. $y = -\dfrac{3}{4}x$

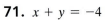
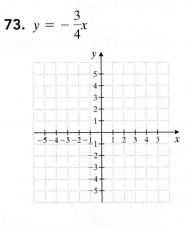

The following pictograph shows the money collected each week from a wrapping paper fundraiser. Use this graph to answer Exercises 1 through 3.

Weekly Wrapping Paper Sales

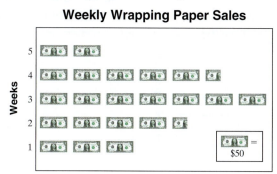

1. How much money was collected during the second week?

2. During which week was the most money collected? How much money was collected during that week?

3. What was the total amount of money collected for the fundraiser?

The bar graph shows the normal monthly precipitation in centimeters for Chicago, Illinois. Use this graph to answer Exercises 4 through 6.

Chicago Precipitation

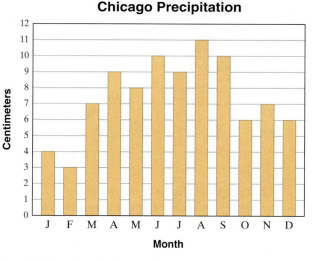

Source: U.S. National Oceanic and Atmospheric Administration, *Climatography of the United States*, No. 81

4. During which month(s) does Chicago normally have more than 9 centimeters of precipitation?

5. During which month does Chicago normally have the least amount of precipitation? How much precipitation occurs during that month?

Answers

1. _____

2. _____

3. _____

4. _____

5. _____

6. _____

6. During which month(s) does 7 centimeters of precipitation normally occur?

7. _____

7. Use the information in the table to draw a bar graph. Clearly label each bar.

Most Common Blood Types	
Blood Type	**% of Population with This Blood Type**
O+	38%
A+	34%
B+	9%
O−	7%
A−	6%
AB+	3%
B−	2%
AB−	1%

Most Common Blood Types by Percent in the Population

Percent of Population

10

5

Blood Type

8. _____

The following line graph shows the annual inflation rate in the United States for the years 2003–2013. Use this graph to answer Exercises 8 through 10.

9. _____

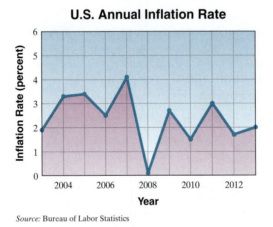

U.S. Annual Inflation Rate

Inflation Rate (percent)

Year

Source: Bureau of Labor Statistics

10. _____

8. Approximate the annual inflation rate in 2012.

9. During which of the years shown was the inflation rate greater than 3%?

10. During which sets of years was the inflation rate decreasing?

11. _____

The result of a survey of 200 people is shown in the following circle graph. Each person was asked to tell his or her favorite type of music. Use this graph to answer Exercises 11 and 12.

12. _____

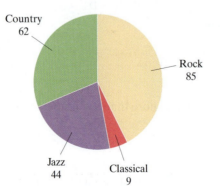

Country
62

Rock
85

Jazz
44

Classical
9

11. Find the ratio of those who prefer rock music to the total number surveyed.

12. Find the ratio of those who prefer country music to those who prefer jazz.

13. _____

The following circle graph shows the projected age distribution of the population of the United States in 2015. There are projected to be 326 million people in the United States in 2015. Use the graph to find how many people are expected to be in the age groups given.

U.S. Population in 2015 by Age Groups

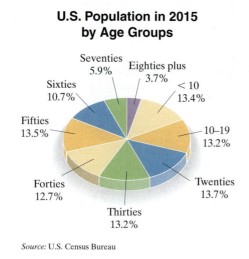

Source: U.S. Census Bureau

14. _____

13. Twenties (Round to nearest whole million.)

14. Eighties plus (Round to nearest whole million.)

A professor measures the heights of the students in her class. The results are shown in the following histogram. Use this histogram to answer Exercises 15 and 16.

15. _____

Student Heights

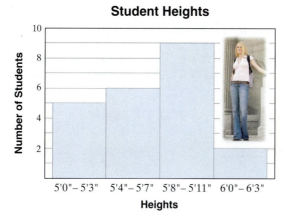

15. How many students are 5'8"–5'11" tall?

16. How many students are 5'7" or shorter?

16. _____

17. The history test scores of 25 students are shown below. Use these scores to complete the frequency distribution table.

70	86	81	65	92
43	72	85	69	97
82	51	75	50	68
88	83	85	77	99
77	63	59	84	90

Class Intervals (Scores)	Tally	Class Frequency (Number of Students)
40–49		
50–59		
60–69		
70–79		
80–89		
90–99		

17. _____

18. _____

18. Use the results of Exercise **17** to draw a histogram.

19. _____

20. _____

Find the coordinates of each point in the graph below.

19. *A*   **20.** *B*   **21.** *C*   **22.** *D*

21. _____

22. _____

Complete and graph the ordered pair solutions of each given equation.

23. $x = -6y;\ (0,\ \ ),\ (\ \ ,1),\ (12,\ \ )$ **24.** $y = 7x - 4;\ (2,\ \ ),\ (-1,\ \ ),\ (0,\ \ )$

23. _____

24. _____

Graph each linear equation.

25. $y + x = -4$ **26.** $y = -4$

25. _____

26. _____

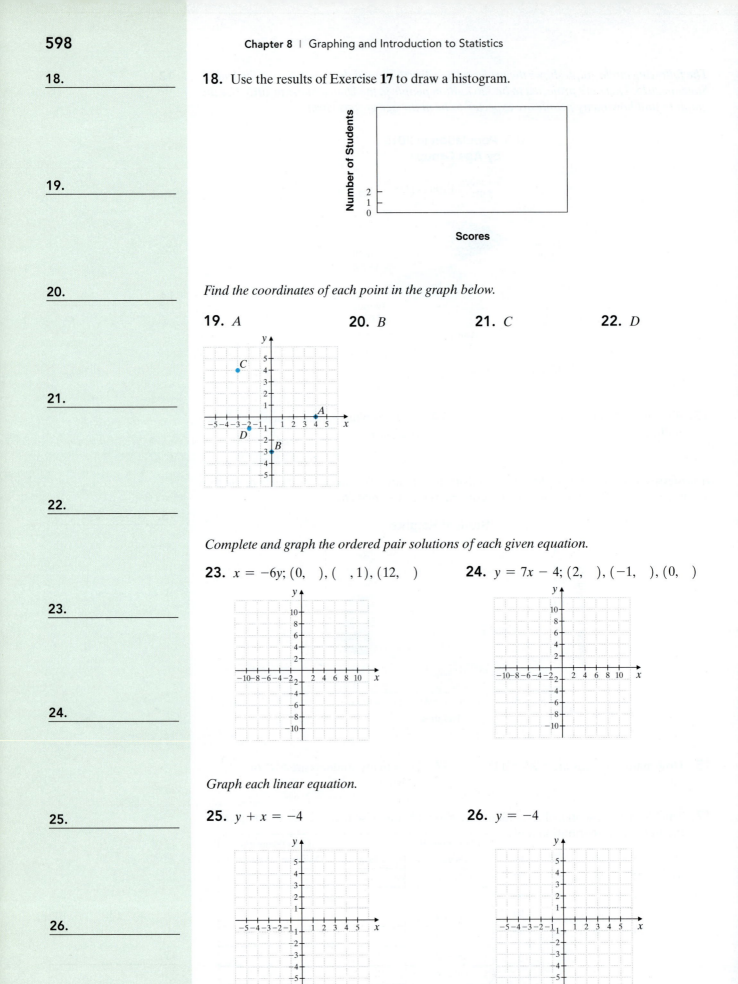

27. $y = 3x - 5$

28. $x = 5$

29. $y = -\dfrac{1}{2}x$

30. $3x - 2y = 12$

31. Draw a tree diagram for the experiment of spinning the spinner twice. Then use the diagram to determine the number of outcomes.

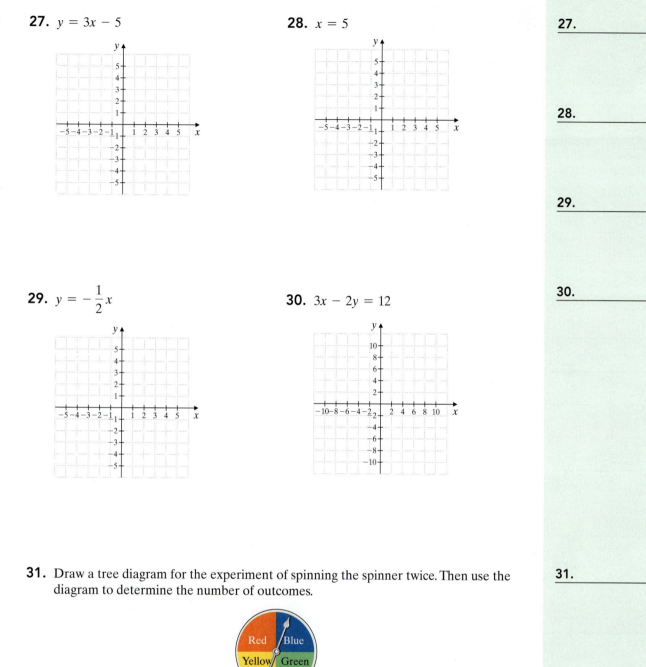

32. Draw a tree diagram for the experiment of tossing a coin twice. Then use the diagram to determine the number of outcomes.

Suppose that the numbers 1 through 10 are each written on same-size sheets of paper and placed in a bag. You then select one sheet of paper from the bag.

33. What is the probability of choosing a 6 from the bag?

34. What is the probability of choosing a 3 or a 4 from the bag?

27. _____

28. _____

29. _____

30. _____

31. _____

32. _____

33. _____

34. _____

Answers

1. _____

2. _____

3. _____

4. _____

5. _____

6. _____

7. _____

8. _____

9. _____

10. _____

11. _____

12. _____

13. _____

14. _____

15. _____

16. _____

17. _____

18. _____

19. _____

20. _____

21. _____

22. _____

23. _____

24. _____

25. _____

26. _____

1. Simplify: $4^3 + [3^2 - (10 \div 2)] - 7 \cdot 3$

2. $7^2 + [5^3 - (6 \div 3)] + 4 \cdot 2$

3. Evaluate $x - y$ for $x = -3$ and $y = 9$.

4. Evaluate $x - y$ for $x = 7$ and $y = -2$.

5. Solve: $3y - 7y = 12$

6. Solve: $2x - 6x = 24$

7. Solve: $\dfrac{x}{6} + 1 = \dfrac{4}{3}$

8. Solve: $\dfrac{7}{2} + \dfrac{a}{4} = 1$

9. Add: $2\dfrac{1}{3} + 5\dfrac{3}{8}$

10. Add: $3\dfrac{2}{5} + 4\dfrac{3}{4}$

11. Write 5.9 as a mixed number.

12. Write 2.8 as a mixed number.

13. Subtract: $3.5 - 0.068$. Check your answer.

14. Subtract: $7.4 - 0.073$. Check your answer.

15. Multiply: 0.0531×16

16. Multiply: 0.147×0.2

17. Divide: $-5.98 \div 115$

18. Divide: $27.88 \div 205$

19. Simplify: $(-1.3)^2 + 2.4$

20. Simplify: $(-2.7)^2$

21. Write $\dfrac{1}{4}$ as a decimal.

22. Write $\dfrac{3}{8}$ as a decimal.

23. Solve: $5(x - 0.36) = -x + 2.4$

24. Solve: $4(0.35 - x) = x - 7$

25. Use the appendix or a calculator to approximate $\sqrt{80}$ to the nearest thousandth.

26. Use the appendix or a calculator to approximate $\sqrt{60}$ to the nearest thousandth.

27. Write the ratio of 21 to 29 using fractional notation.

28. Write the ratio of 7 to 15 using fractional notation.

29. Write as a unit rate: 337.5 miles every 15 gallons of gas

30. Write as a unit price: $1.59 for 3 ounces

31. Solve $\dfrac{51}{34} = \dfrac{-3}{x}$ for x.

32. Solve $\dfrac{8}{5} = \dfrac{x}{10}$ for x.

△ **33.** Find the ratio of corresponding sides for the similar triangles ABC and DEF.

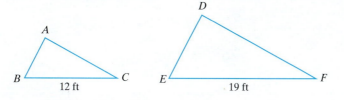

△ **34.** Find the ratio of corresponding sides for the similar triangles GHJ and KLM.

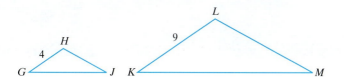

Write each percent as a decimal.

35. 4.6%

36. 32%

37. 0.74%

38. 2.7%

39. What number is 35% of 60?

40. What number is 40% of 36?

41. 20.8 is 40% of what number?

42. 9.5 is 25% of what number?

43. The sales tax on a $310 Sony flat-screen digital 32-inch television is $26.35. Find the sales tax rate.

44. The sales tax on a $2.00 yo-yo is $0.13. Find the sales tax rate.

45. $4000 is invested at 5.3% compounded quarterly for 10 years. Find the total amount at the end of 10 years.

46. Linda Bonnett borrows $1600 for 1 year. If the interest is $128.60, find the monthly payment. (*Hint:* Find the total amount due and divide by the number of months in a year.)

47. Find the median of the list of numbers: 25, 54, 56, 57, 60, 71, 98

48. Find the median of the list of numbers: 43, 46, 47, 50, 52, 83

49. If a die is rolled one time, find the probability of rolling a 3 or a 4.

50. If a die is rolled one time, find the probability of rolling an even number.

27. _____

28. _____

29. _____

30. _____

31. _____

32. _____

33. _____

34. _____

35. _____

36. _____

37. _____

38. _____

39. _____

40. _____

41. _____

42. _____

43. _____

44. _____

45. _____

46. _____

47. _____

48. _____

49. _____

50. _____

9 Geometry and Measurement

The word *geometry* is formed from the Greek words *geo*, meaning "Earth," and *metron*, meaning "measure." Geometry literally means to measure the Earth. In this chapter we learn about various geometric figures and their properties, such as perimeter, area, and volume. Knowledge of geometry can help us solve practical problems in all types of real-life situations.

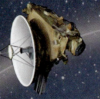

2007-2014
For most of the eight year trek from Jupiter to Pluto, the craft spun slowly in a state of "hibernation," signaling once a week. But for about 50 days each year, it was awakened to conduct an intensive set of calibration and science observations.

New Horizons spacecraft

July 2015
During the flyby of Pluto, scientists expect a frenzied 24 hours of data gathering. At its closest, New Horizons will pass within 6000 miles of the frozen dwarf.

2017-2020
With NASA approval, the spacecraft will be directed toward one or more Kuiper Belt objects beyond Pluto.

New Horizons is NASA's robotic spacecraft mission, New Horizons, shown above. The spacecraft is about the size and shape of a grand piano with a satellite dish attached. It was launched in January 2006 and is now heading toward Pluto. How do we receive the images of Pluto on Earth when this spacecraft is so far into deep space and how long before we receive the images?

The Deep Space Network (DSN) is a worldwide network of large antennas and communication facilities. When a mission is in deep space, fewer sites are needed for sending and receiving transmissions; thus, the DSN has only three sites—one in California (Goldstone), one in Spain (Madrid), and one in Australia (Canberra). The diagram below shows an overview of Earth from the vantage point of the North Pole and the location of these three sites.

We study some geometry of the DSN in Section 9.1, Exercises 65 and 66, and Section 9.7, Exercises 23–26.

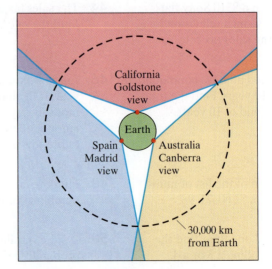

△ 9.1 Lines and Angles ▷

Objective A Identifying Lines, Line Segments, Rays, and Angles ▷

Let's begin with a review of two important concepts—space and plane.

Space extends in all directions indefinitely. Examples of objects in space are houses, grains of salt, bushes, your *Prealgebra* textbook, and you.

A **plane** is a flat surface that extends indefinitely. Surfaces like a plane are a classroom floor or a blackboard or whiteboard.

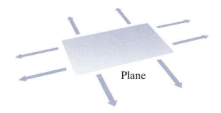
Plane

The most basic concept of geometry is the idea of a point in space. A **point** has no length, no width, and no height, but it does have location. We represent a point by a dot, and we usually label points with capital letters.

P

Point *P*

A **line** is a set of points extending indefinitely in two directions. A line has no width or height, but it does have length. We can name a line by any two of its points or by a single lowercase letter. A **line segment** is a piece of a line with two endpoints.*

Line *AB*, $\overleftrightarrow{AB}$, or line *l*＊   Line segment *AB* or $\overline{AB}$

A **ray** is a part of a line with one endpoint. A ray extends indefinitely in one direction. An **angle** is made up of two rays that share the same endpoint. The common endpoint is called the **vertex.**

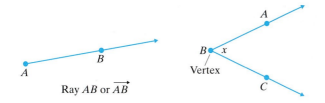

Ray *AB* or $\overrightarrow{AB}$

The angle in the figure above can be named

$\angle ABC$ $\angle CBA$ $\angle B$ or $\angle x$
        ↑   ↑
   The vertex is the
   middle point.

Rays *BA* and *BC* are **sides** of the angle.

*Although line *l* is also line *BA* or $\overleftrightarrow{BA}$, we will use only one order of points to name a line or line segment.

Naming an Angle
When there is no confusion as to what angle is being named, you may use the vertex alone.

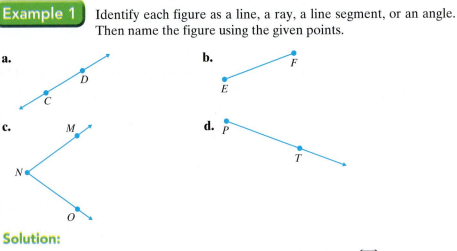

Name of ∠B is all right.
There is no confusion. ∠B means ∠1.

Name of ∠B is *not* all right.
There is confusion. Does ∠B mean ∠1, ∠2, ∠3, or ∠4?

Practice 1

Identify each figure as a line, a ray, a line segment, or an angle. Then name the figure using the given points.

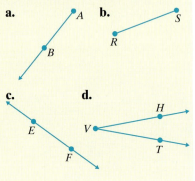

a.   **b.**

c.   **d.**

Example 1   Identify each figure as a line, a ray, a line segment, or an angle. Then name the figure using the given points.

a.   **b.**

c.   **d.**

Solution:

Figure (a) extends indefinitely in two directions. It is line *CD* or $\overleftrightarrow{CD}$.

Figure (b) has two endpoints. It is line segment *EF* or $\overline{EF}$.

Figure (c) has two rays with a common endpoint. It is ∠*MNO*, ∠*ONM*, or ∠*N*.

Figure (d) is part of a line with one endpoint. It is ray *PT* or $\overrightarrow{PT}$.

Work Practice 1

Practice 2

Use the figure in Example 2 to list other ways to name ∠z.

Example 2   List other ways to name ∠y.

Solution: Two other ways to name ∠y are ∠*QTR* and ∠*RTQ*. We may *not* use the vertex alone to name this angle because three different angles have *T* as their vertex.

Work Practice 2

Objective B Classifying Angles as Acute, Right, Obtuse, or Straight ▶

An angle can be measured in **degrees.** The symbol for degrees is a small, raised circle, °. There are 360° in a full revolution, or a full circle.

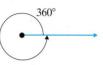
360°

$\frac{1}{2}$ of a revolution measures $\frac{1}{2}(360°) = 180°$. An angle that measures 180° is called a **straight angle.**

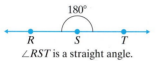
180°
∠RST is a straight angle.

$\frac{1}{4}$ of a revolution measures $\frac{1}{4}(360°) = 90°$. An angle that measures 90° is called a **right angle.** The symbol ∟ is used to denote a right angle.

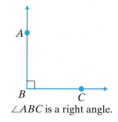
∠ABC is a right angle.

An angle whose measure is between 0° and 90° is called an **acute angle.**

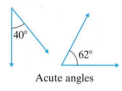
40°
62°
Acute angles

An angle whose measure is between 90° and 180° is called an **obtuse angle.**

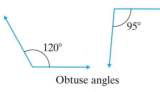
95°
120°
Obtuse angles

Example 3 Classify each angle as acute, right, obtuse, or straight.

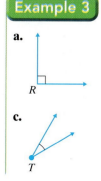
a.
R

b.
S

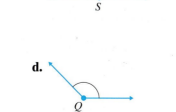
c.
T

d.
Q

(Continued on next page)

(Continued on next page)

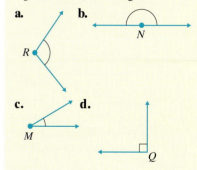

Practice 3

Classify each angle as acute, right, obtuse, or straight.

a.
R

b.
N

c.
M

d.
Q

Answers

3. **a.** obtuse **b.** straight **c.** acute
d. right

Solution:

a. $\angle R$ is a right angle, denoted by ⌐. It measures 90°.

b. $\angle S$ is a straight angle. It measures 180°.

c. $\angle T$ is an acute angle. It measures between 0° and 90°.

d. $\angle Q$ is an obtuse angle. It measures between 90° and 180°.

■ Work Practice 3

Let's look at $\angle B$ below, whose measure is 62°.

There is a shorthand notation for writing the measure of this angle. To write "The measure of $\angle B$ is 62°, we can write,

$$m\angle B = 62°.$$

By the way, note that $\angle B$ is an acute angle because $m\angle B$ is between 0° and 90°.

Objective C Identifying Complementary and Supplementary Angles

Two angles that have a sum of 90° are called **complementary angles.** We say that each angle is the **complement** of the other.

$\angle R$ and $\angle S$ are complementary angles because

$$m\angle R + m\angle S = 60° + 30° = 90°$$

Complementary angles
60° + 30° = 90°

Two angles that have a sum of 180° are called **supplementary angles.** We say that each angle is the **supplement** of the other.

$\angle M$ and $\angle N$ are supplementary angles because

$$m\angle M + m\angle N = 125° + 55° = 180°$$

Supplementary angles
125° + 55° = 180°

Practice 4

Find the complement of a 29° angle.

Example 4 Find the complement of a 48° angle.

Solution: Two angles that have a sum of 90° are complementary. This means that the complement of an angle that measures 48° is an angle that measures $90° - 48° = 42°$.

■ Work Practice 4

Answer

4. 61°

Example 5 Find the supplement of a 107° angle.

Solution: Two angles that have a sum of 180° are supplementary. This means that the *supplement* of an angle that measures 107° is an angle that measures $180° - 107° = 73°$.

■ **Work Practice 5**

✓**Concept Check** True or false? The supplement of a 48° angle is 42°. Explain.

Objective D Finding Measures of Angles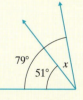

Measures of angles can be added or subtracted to find measures of related angles.

Example 6 Find the measure of $\angle x$. Then classify $\angle x$ as an acute, obtuse, or right angle.

Solution:
$$
\begin{aligned}
m\angle x &= m\angle QTS - m\angle RTS \\
&= 87° - 52° \\
&= 35°
\end{aligned}
$$

Thus, the measure of $\angle x$ ($m\angle x$) is 35°.
    Since $\angle x$ measures between 0° and 90°, it is an acute angle.

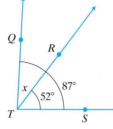

■ **Work Practice 6**

Two lines in a plane can be either parallel or intersecting. **Parallel lines** never meet. **Intersecting lines** meet at a point. The symbol ∥ is used to indicate "is parallel to." For example, in the figure, $p \parallel q$.

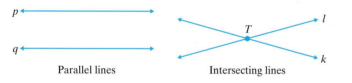

Parallel lines   Intersecting lines

Some intersecting lines are perpendicular. Two lines are **perpendicular** if they form right angles when they intersect. The symbol ⊥ is used to denote "is perpendicular to." For example, in the figure below, $n \perp m$.

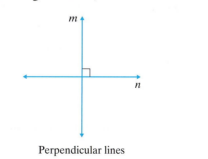

Perpendicular lines

When two lines intersect, four angles are formed. Two angles that are opposite each other are called **vertical angles.** Vertical angles have the same measure.
    Two angles that share a common side are called **adjacent angles.** Adjacent angles formed by intersecting lines are supplementary. That is, they have a sum of 180°.

Vertical angles:
$\angle a$ and $\angle c$
$\angle d$ and $\angle b$

Adjacent angles:
$\angle a$ and $\angle b$
$\angle b$ and $\angle c$
$\angle c$ and $\angle d$
$\angle d$ and $\angle a$

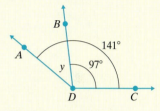

Here are a few real-life examples of the lines we just discussed.

Parallel lines   Vertical angles   Perpendicular lines

Practice 7

Find the measure of $\angle a$, $\angle b$, and $\angle c$.

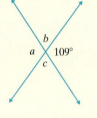

Example 7   Find the measure of $\angle x$, $\angle y$, and $\angle z$ if the measure of $\angle t$ is 42°.

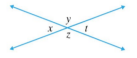

Solution: Since $\angle t$ and $\angle x$ are vertical angles, they have the same measure, so $\angle x$ measures 42°.

Since $\angle t$ and $\angle y$ are adjacent angles, their measures have a sum of 180°. So $\angle y$ measures $180° - 42° = 138°$.

Since $\angle y$ and $\angle z$ are vertical angles, they have the same measure. So $\angle z$ measures 138°.

■ **Work Practice 7**

A line that intersects two or more lines at different points is called a **transversal.** Line l is a transversal that intersects lines m and n. The eight angles formed have special names. Some of these names are:

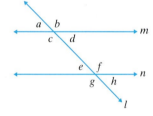

Corresponding angles: $\angle a$ and $\angle e$, $\angle c$ and $\angle g$, $\angle b$ and $\angle f$, $\angle d$ and $\angle h$

Alternate interior angles: $\angle c$ and $\angle f$, $\angle d$ and $\angle e$

When two lines cut by a transversal are *parallel,* the following statement is true:

Practice 8

Given that $m \parallel n$ and that the measure of $\angle w = 45°$, find the measures of all the angles shown.

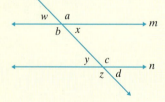

Parallel Lines Cut by a Transversal

If two parallel lines are cut by a transversal, then the measures of **corresponding angles are equal** and the measures of the **alternate interior angles are equal.**

Example 8   Given that $m \parallel n$ and that the measure of $\angle w$ is 100°, find the measures of $\angle x$, $\angle y$, and $\angle z$.

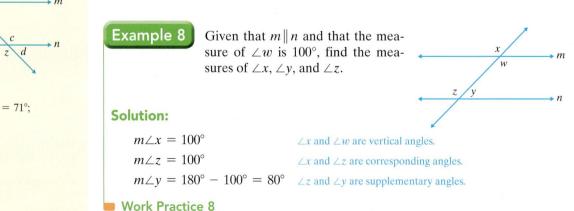

Answers

7. $m\angle a = 109°$; $m\angle b = 71°$;
   $m\angle c = 71°$

8. $m\angle x = 45°$;
   $m\angle y = 45°$;
   $m\angle z = 135°$;
   $m\angle a = 135°$;
   $m\angle b = 135°$;
   $m\angle c = 135°$;
   $m\angle d = 45°$

Solution:

$m\angle x = 100°$ $\angle x$ and $\angle w$ are vertical angles.

$m\angle z = 100°$ $\angle x$ and $\angle z$ are corresponding angles.

$m\angle y = 180° - 100° = 80°$ $\angle z$ and $\angle y$ are supplementary angles.

■ **Work Practice 8**

Vocabulary, Readiness & Video Check

Use the choices below to fill in each blank.

acute	straight	degrees	adjacent	parallel	intersecting
obtuse	space	plane	point	vertical	vertex
right	angle	ray	line	perpendicular	transversal

1. A(n) _____ is a flat surface that extends indefinitely.

2. A(n) _____ has no length, no width, and no height.

3. _____ extends in all directions indefinitely.

4. A(n) _____ is a set of points extending indefinitely in two directions.

5. A(n) _____ is part of a line with one endpoint.

6. A(n) _____ is made up of two rays that share a common endpoint. The common endpoint is called the _____ .

7. A(n) _____ angle measures 180°.

8. A(n) _____ angle measures 90°.

9. A(n) _____ angle measures between 0° and 90°.

10. A(n) _____ angle measures between 90° and 180°.

11. _____ lines never meet and _____ lines meet at a point.

12. Two intersecting lines are _____ if they form right angles when they intersect.

13. An angle can be measured in _____ .

14. A line that intersects two or more lines at different points is called a(n) _____ .

15. When two lines intersect, four angles are formed, called _____ angles.

16. Two angles that share a common side are called _____ angles.

Martin-Gay Interactive Videos   *Watch the section lecture video and answer the following questions.*

See Video 9.1

Objective A **17.** In the lecture after ▣ Example 2, what are the four ways we can name the angle shown? ▶

Objective B **18.** In the lecture before ▣ Example 3, what type of angle forms a line? What is its measure? ▶

Objective C **19.** What calculation is used to find the answer to ▣ Example 6? ▶

Objective D **20.** In the lecture before ▣ Example 7, two lines in a plane that aren't parallel must what? ▶

9.1 Exercise Set MyMathLab® ▶

Objective A *Identify each figure as a line, a ray, a line segment, or an angle. Then name the figure using the given points. See Examples 1 and 2.*

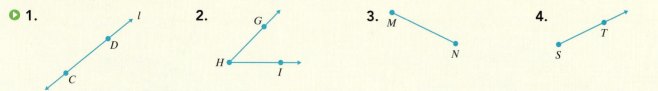

▶ **1.** *l* / *D* / *C*   **2.** *G* / *H* / *I*   **3.** *M* / *N*   **4.** *T* / *S*

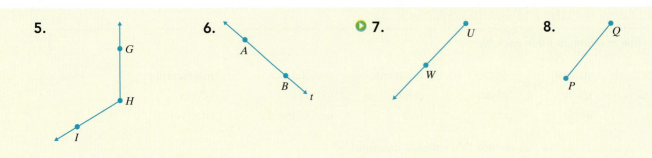

5.   **6.**   **7.**   **8.**

List two other ways to name each angle. See Example 2.

9. ∠x

10. ∠w

11. ∠z

12. ∠y

Objective B *Classify each angle as acute, right, obtuse, or straight. See Example 3.*

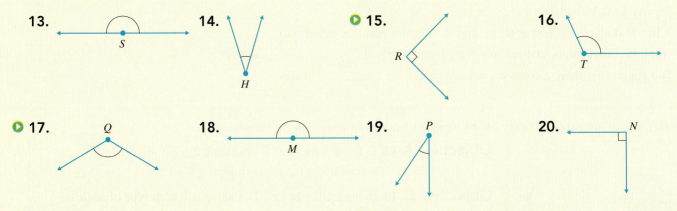

13.   **14.**   **15.**   **16.**

17.   **18.**   **19.**   **20.**

Objective C *Find each complementary or supplementary angle as indicated. See Examples 4 and 5.*

21. Find the complement of a 23° angle.

22. Find the complement of a 77° angle.

23. Find the supplement of a 17° angle.

24. Find the supplement of a 77° angle.

25. Find the complement of a 58° angle.

26. Find the complement of a 22° angle.

27. Find the supplement of a 150° angle.

28. Find the supplement of a 130° angle.

29. Identify the pairs of complementary angles.

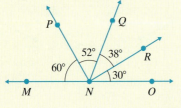

30. Identify the pairs of complementary angles.

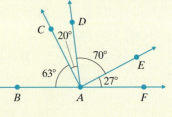

31. Identify the pairs of supplementary angles.

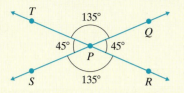

32. Identify the pairs of supplementary angles.

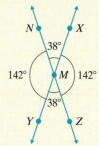

Objective D *Find the measure of ∠x in each figure. See Example 6.*

33.

34.

35.

36.

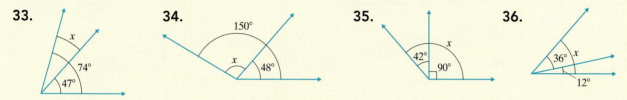

Find the measures of angles x, y, and z in each figure. See Examples 7 and 8.

37.

38.

39.

40.

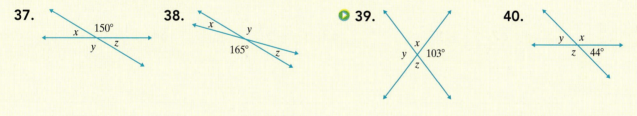

41. *m ∥ n*

42. *m ∥ n*

43. *m ∥ n*

44. *m ∥ n*

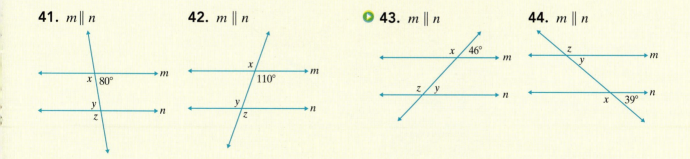

Objectives A D Mixed Practice *Find two other ways of naming each angle. See Example 2.*

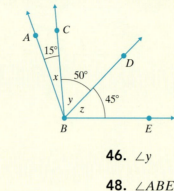

45. ∠x

46. ∠y

47. ∠z

48. ∠ABE (just name one other way)

Find the measure of each angle in the figure above. See Example 6.

49. ∠ABC

50. ∠EBD

51. ∠CBD

52. ∠CBA

53. ∠DBA

54. ∠EBC

55. ∠CBE

56. ∠ABE

Review

Perform each indicated operation. See Sections 4.3, 4.5, and 4.7.

57. $\dfrac{7}{8} + \dfrac{1}{4}$

58. $\dfrac{7}{8} - \dfrac{1}{4}$

59. $\dfrac{7}{8} \cdot \dfrac{1}{4}$

60. $\dfrac{7}{8} \div \dfrac{1}{4}$

61. $3\dfrac{1}{3} - 2\dfrac{1}{2}$

62. $3\dfrac{1}{3} + 2\dfrac{1}{2}$

63. $3\dfrac{1}{3} \div 2\dfrac{1}{2}$

64. $3\dfrac{1}{3} \cdot 2\dfrac{1}{2}$

Concept Extensions

Use this North Pole overhead view of the three sites of the Deep Space Network to answer Exercises 65 and 66. (See the Chapter Opener.)

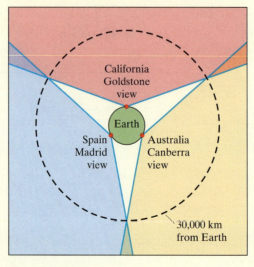

65. How many degrees are there around the Earth at the equator?

66. If the three sites of the Deep Space Network (red dots shown) are about the same number of degrees apart, how many degrees apart are they?

67. The angle between the two walls of the Vietnam Veterans Memorial in Washington, D.C., is 125.2°. Find the supplement of this angle. (*Source:* National Park Service)

68. The faces of Khafre's Pyramid at Giza, Egypt, are inclined at an angle of 53.13°. Find the complement of this angle. (*Source:* PBS *NOVA* Online)

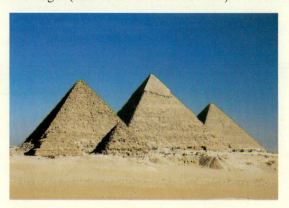

Answer true or false for Exercises 69 through 72. See the Concept Check in this section. If false, explain why.

69. The complement of a 100° angle is an 80° angle.

70. It is possible to find the complement of a 120° angle.

71. It is possible to find the supplement of a 120° angle.

72. The supplement of a 5° angle is a 175° angle.

73. If lines m and n are parallel, find the measures of angles a through e.

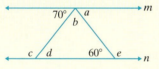

74. Below is a rectangle. List which segments, if extended, would be parallel lines.

75. Can two supplementary angles both be acute? Explain why or why not.

76. In your own words, describe how to find the complement and the supplement of a given angle.

77. Find two complementary angles with the same measure.

78. Is the figure below possible? Why or why not?

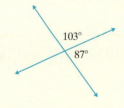

Objectives

A Use Formulas to Find Perimeters.

B Use Formulas to Find Circumferences.

Practice 1

a. Find the perimeter of the rectangle.

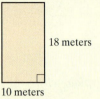

18 meters

10 meters

b. Find the perimeter of the rectangular lot shown below:

50 feet

125 feet

Objective A Using Formulas to Find Perimeters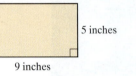

Recall from Section 1.3 that the perimeter of a polygon is the distance around the polygon. This means that the perimeter of a polygon is the sum of the lengths of its sides.

Example 1 Find the perimeter of the rectangle below.

5 inches

9 inches

Solution:

$$\text{perimeter} = 9 \text{ inches} + 9 \text{ inches} + 5 \text{ inches} + 5 \text{ inches}$$
$$= 28 \text{ inches}$$

■ **Work Practice 1**

Notice that the perimeter of the rectangle in Example 1 can be written as $2 \cdot (9 \text{ inches}) + 2 \cdot (5 \text{ inches})$.

↑ length   ↑ width

In general, we can say that the perimeter of a rectangle is always

$$2 \cdot \text{length} + 2 \cdot \text{width}$$

As we have just seen, the perimeters of some special figures such as rectangles form patterns. These patterns are given as **formulas.** The formula for the perimeter of a rectangle is shown next:

Perimeter of a Rectangle

Perimeter $= 2 \cdot \textbf{length} + 2 \cdot \textbf{width}$

In symbols, this can be written as

$$P = 2l + 2w$$

length

width   width

length

Practice 2

Find the perimeter of a rectangle with a length of 32 centimeters and a width of 15 centimeters.

Example 2 Find the perimeter of a rectangle with a length of 11 inches and a width of 3 inches.

11 in.

3 in.

Solution: We use the formula for perimeter and replace the letters by their known lengths.

$$P = 2l + 2w$$
$$= 2 \cdot 11 \text{ in.} + 2 \cdot 3 \text{ in.} \quad \text{Replace } l \text{ with 11 in. and } w \text{ with 3 in.}$$
$$= 22 \text{ in.} + 6 \text{ in.}$$
$$= 28 \text{ in.}$$

The perimeter is 28 inches.

■ **Work Practice 2**

Answers

1. a. 56 m **b.** 350 ft **2.** 94 cm

Recall that a square is a special rectangle with all four sides the same length. The formula for the perimeter of a square is shown next:

Perimeter of a Square

$$\textbf{Perimeter} = \textbf{side} + \textbf{side} + \textbf{side} + \textbf{side}$$
$$= 4 \cdot \textbf{side}$$

In symbols,

$$P = 4s$$

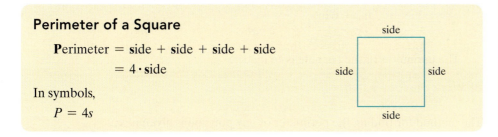

Example 3 Finding the Perimeter of a Field

How much fencing is needed to enclose a square field 50 yards on a side?

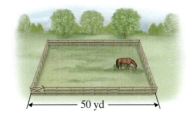

Solution: To find the amount of fencing needed, we find the distance around, or perimeter. The formula for the perimeter of a square is $P = 4s$. We use this formula and replace s by 50 yards.

$$P = 4s$$
$$= 4 \cdot 50 \text{ yd}$$
$$= 200 \text{ yd}$$

The amount of fencing needed is 200 yards.

■ **Work Practice 3**

The formula for the perimeter of a triangle with sides of lengths a, b, and c is given next:

Perimeter of a Triangle

$$\textbf{Perimeter} = \textbf{side } \textbf{a} + \textbf{side } \textbf{b} + \textbf{side } \textbf{c}$$

In symbols,

$$P = a + b + c$$

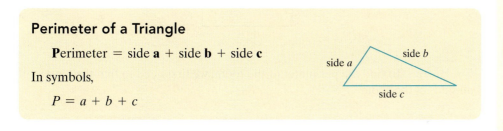

Example 4 Find the perimeter of a triangle if the sides are 3 inches, 7 inches, and 6 inches.

7 in.

6 in.

3 in.

(Continued on next page)

Practice 3

Find the perimeter of a square tabletop if each side is 4 feet long.

Practice 4

Find the perimeter of a triangle if the sides are 6 centimeters, 10 centimeters, and 8 centimeters in length.

Answers

3. 16 ft   **4.** 24 cm

Solution: The formula for the perimeter is $P = a + b + c$, where a, b, and c are the lengths of the sides. Thus,

$$P = a + b + c$$
$$= 3 \text{ in.} + 7 \text{ in.} + 6 \text{ in.}$$
$$= 16 \text{ in.}$$

The perimeter of the triangle is 16 inches.

■ **Work Practice 4**

The method for finding the perimeter of any polygon is given next:

Perimeter of a Polygon

The perimeter of a polygon is the sum of the lengths of its sides.

Practice 5

Find the perimeter of the trapezoid shown.

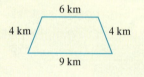

Example 5   Find the perimeter of the trapezoid shown below:

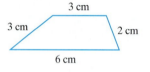

Solution: To find the perimeter, we find the sum of the lengths of its sides.

perimeter $= 3 \text{ cm} + 2 \text{ cm} + 6 \text{ cm} + 3 \text{ cm} = 14 \text{ cm}$

The perimeter is 14 centimeters.

■ **Work Practice 5**

Practice 6

Find the perimeter of the room shown.

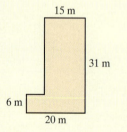

Example 6   Finding the Perimeter of a Room

Find the perimeter of the room shown below:

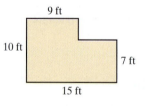

Solution: To find the perimeter of the room, we first need to find the lengths of all sides of the room.

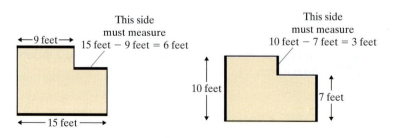

Now that we know the measures of all sides of the room, we can add the measures to find the perimeter.

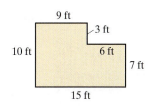

9 ft
3 ft
10 ft
6 ft
7 ft
15 ft

perimeter = 10 ft + 9 ft + 3 ft + 6 ft + 7 ft + 15 ft
= 50 ft

The perimeter of the room is 50 feet.

■ Work Practice 6

Example 7 Calculating the Cost of Wallpaper Border

A rectangular room measures 10 feet by 12 feet. Find the cost to hang a wallpaper border on the walls close to the ceiling if the cost of the wallpaper border is $1.09 per foot.

Solution: First we find the perimeter of the room.

$P = 2l + 2w$
$= 2 \cdot 12 \text{ ft} + 2 \cdot 10 \text{ ft}$ Replace *l* with 12 feet and *w* with 10 feet.
$= 24 \text{ ft} + 20 \text{ ft}$
$= 44 \text{ ft}$

The cost of the wallpaper is

cost = $1.09 \cdot 44$ ft = 47.96

The cost of the wallpaper is $47.96.

■ Work Practice 7

Practice 7

A rectangular lot measures 60 feet by 120 feet. Find the cost to install fencing around the lot if the cost of fencing is $1.90 per foot.

Objective B Using Formulas to Find Circumferences ▶

Recall from Section 5.3 that the distance around a circle is called the **circumference.** This distance depends on the radius or the diameter of the circle.

The formulas for circumference are shown next:

Circumference of a Circle

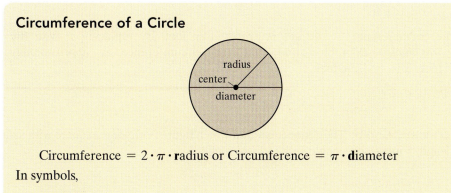

radius
center
diameter

Circumference = $2 \cdot \pi \cdot$ **radius** or Circumference = $\pi \cdot$ **diameter**

In symbols,

$C = 2\pi r$ or $C = \pi d$

where $\pi \approx 3.14$ or $\pi \approx \dfrac{22}{7}$.

Answer
7. $684

To better understand circumference and π (pi), try the following experiment. Take any can and measure its circumference and its diameter.

The can in the figure above has a circumference of 23.5 centimeters and a diameter of 7.5 centimeters. Now divide the circumference by the diameter.

$$\frac{\text{circumference}}{\text{diameter}} = \frac{23.5 \text{ cm}}{7.5 \text{ cm}} \approx 3.13$$

Try this with other sizes of cylinders and circles—you should always get a number close to 3.1. The exact ratio of circumference to diameter is π. (Recall that $\pi \approx 3.14$ or $\approx \frac{22}{7}$.)

Practice 8

An irrigation device waters a circular region with a diameter of 20 yards. Find the exact circumference of the watered region, then use $\pi \approx 3.14$ to give an approximation.

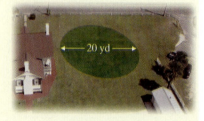

20 yd

Example 8 Finding Circumference of a Circular Spa

A homeowner plans to install a border of new tiling around the circumference of her circular spa. If her spa has a diameter of 14 feet, find its exact circumference. Then use the approximation 3.14 for π to approximate the circumference.

14 feet

Solution: Because we are given the diameter, we use the formula $C = \pi d$.

$$C = \pi d$$
$$= \pi \cdot 14 \text{ ft} \quad \text{\textcolor{blue}{Replace } d \text{ with 14 feet.}}$$
$$= 14\pi \text{ ft}$$

The circumference of the spa is *exactly* 14π feet. By replacing π with the *approximation* 3.14, we find that the circumference is *approximately* 14 feet $\cdot$ 3.14 = 43.96 feet.

■ **Work Practice 8**

✓**Concept Check** The distance around which figure is greater: a square with side length 5 inches or a circle with radius 3 inches?

Answer

8. exactly 20π yd ≈ 62.8 yd

✓**Concept Check Answer**
a square with side length 5 in.

Vocabulary, Readiness & Video Check

Use the choices below to fill in each blank.

circumference	radius	π	$\frac{22}{7}$
diameter	perimeter	3.14	

1. The _____ of a polygon is the sum of the lengths of its sides.

2. The distance around a circle is called the _____.

3. The exact ratio of circumference to diameter is _____.

4. The diameter of a circle is double its _____.

5. Both _____ and _____ are approximations for π.

6. The radius of a circle is half its _____.

Martin-Gay Interactive Videos

Watch the section lecture video and answer the following questions.

Objective A 7. In ▣ Example 1, how can the perimeter be found if we forget the formula? ▸

Objective B 8. From the lecture before ▣ Example 6, circumference is a special name for what? ▸

See Video 9.2

9.2 Exercise Set MyMathLab® ▸

Objective A *Find the perimeter of each figure. (See Appendix A.1 for any unknown geometric figures.) See Examples 1 through 6.*

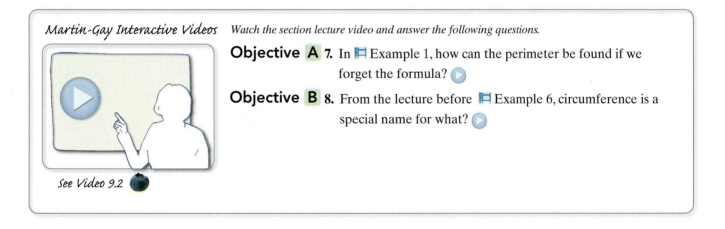

1. Rectangle, 15 ft, 17 ft

2. Rectangle, 14 m, 5 m

3. Parallelogram, 25 cm, 35 cm

4. Parallelogram, 3 yd, 2 yd

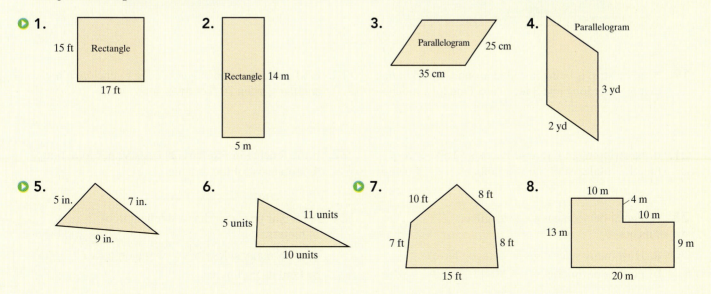

5. 5 in., 7 in., 9 in.

6. 5 units, 11 units, 10 units

7. 10 ft, 8 ft, 7 ft, 8 ft, 15 ft

8. 10 m, 4 m, 10 m, 13 m, 9 m, 20 m

Find the perimeter of each regular polygon. (The sides of a regular polygon have the same length.)

9.
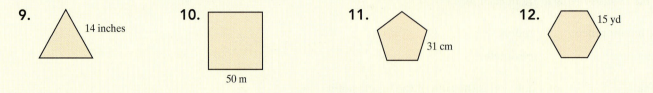

14 inches

10.

50 m

11.

31 cm

12.

15 yd

Solve. See Examples 1 through 7.

13. A polygon has sides of length 5 feet, 3 feet, 2 feet, 7 feet, and 4 feet. Find its perimeter.

14. A triangle has sides of length 8 inches, 12 inches, and 10 inches. Find its perimeter.

15. A line-marking machine lays down lime powder to mark both foul lines on a baseball field. If each foul line for this field measures 312 feet, how many feet of lime powder will be deposited?

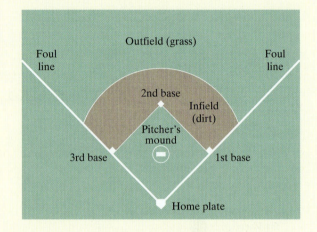

16. A baseball diamond has 4 sides, with each side length 90 feet. If a baseball player hits a home run, how far does the player run (home plate, around the bases, then back to home plate)?

17. If a football field is 53 yards wide and 120 yards long, what is the perimeter?

18. A stop sign has eight equal sides of length 12 inches. Find its perimeter.

19. A metal strip is being installed around a workbench that is 8 feet long and 3 feet wide. Find how much stripping is needed for this project.

20. Find how much fencing is needed to enclose a rectangular garden 70 feet by 21 feet.

21. If the stripping in Exercise **19** costs $2.50 per foot, find the total cost of the stripping.

22. If the fencing in Exercise **20** costs $2 per foot, find the total cost of the fencing.

23. A regular octagon has a side length of 9 inches.

   a. How many sides does an octagon have?

   b. Find its perimeter.

24. A regular pentagon has a side length of 14 meters.

   a. How many sides does a pentagon have?

   b. Find its perimeter.

25. Find the perimeter of the top of a square compact disc case if the length of one side is 7 inches.

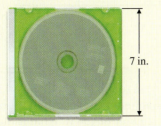

7 in.

26. Find the perimeter of a square ceramic tile with a side of length 3 inches.

3 in.

27. A rectangular room measures 10 feet by 11 feet. Find the cost of installing a strip of wallpaper around the room if the wallpaper costs $0.86 per foot.

28. A rectangular house measures 85 feet by 70 feet. Find the cost of installing gutters around the house if the cost is $2.36 per foot.

Find the perimeter of each figure. See Example 6.

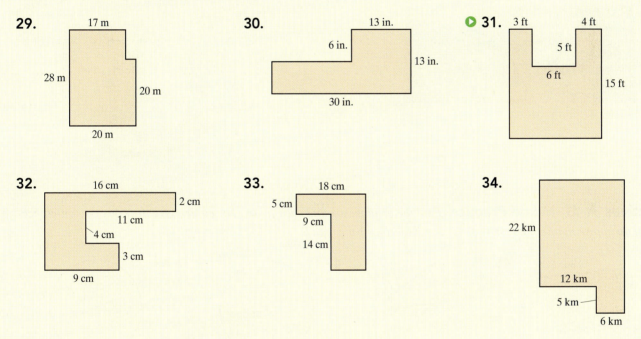

29.
17 m
28 m
20 m
20 m

30.
13 in.
6 in.
13 in.
30 in.

31.
3 ft   4 ft
5 ft
6 ft
15 ft

32.
16 cm
2 cm
11 cm
4 cm
3 cm
9 cm

33.
18 cm
5 cm
9 cm
14 cm

34.
22 km
12 km
5 km
6 km

Objective B *Find the circumference of each circle. Give the exact circumference and then an approximation. Use* $\pi \approx 3.14$. *See Example 8.*

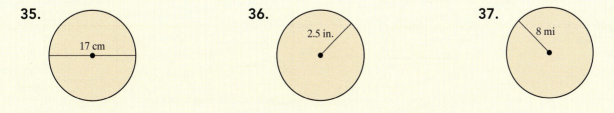

35.
17 cm

36.
2.5 in.

37.
8 mi

38.

50 ft

39.

26 m

40.

10 yd

41. Wyley Robinson just bought a trampoline for his children to use. The trampoline has a diameter of 15 feet. If Wyley wishes to buy netting to go around the outside of the trampoline, how many feet of netting does he need?

42. The largest round barn in the world is located at the Marshfield Fairgrounds in Wisconsin. The barn has a diameter of 150 ft. What is the circumference of the barn? (*Source: The Milwaukee Journal Sentinel*)

43. Meteor Crater, near Winslow, Arizona, is 4000 feet in diameter. Approximate the distance around the crater. Use 3.14 for π. (*Source: The Handy Science Answer Book*)

44. The *Pearl of Lao-tze* has a diameter of $5\frac{1}{2}$ inches. Approximate the distance around the pearl. Use $\frac{22}{7}$ for π. (*Source: The Guinness World Records*)

$5\frac{1}{2}$ in.

Objectives A B Mixed Practice *Find the distance around each figure. For circles, give the exact circumference and then an approximation. Use $\pi \approx 3.14$. See Examples 1 through 8.*

45.

9 mi
4.7 mi
6 mi
11 mi

46.

4.5 yd
7 yd
9 yd

47.

14 cm

48.

11 m

49.

Regular
Pentagon
8 mm

50.

Regular
Parallelogram
19 km

51.

7 ft
8 ft
22 ft
20 ft

52.

44 mi
40 mi
9 mi

Review

Simplify. See Section 1.7.

53. $5 + 6 \cdot 3$

54. $25 - 3 \cdot 7$

55. $(20 - 16) \div 4$

56. $6 \cdot (8 + 2)$

57. $72 \div (2 \cdot 6)$

58. $(72 \div 2) \cdot 6$

59. $(18 + 8) - (12 + 4)$

60. $4^1 \cdot (2^3 - 8)$

Concept Extensions

There are a number of factors that determine the dimensions of a rectangular soccer field. Use the table below to answer Exercises 61 and 62.

Soccer Field Width and Length		
Age	**Width Min–Max**	**Length Min–Max**
Under 6/7:	15–20 yards	25–30 yards
Under 8:	20–25 yards	30–40 yards
Under 9:	30–35 yards	40–50 yards
Under 10:	40–50 yards	60–70 yards
Under 11:	40–50 yards	70–80 yards
Under 12:	40–55 yards	100–105 yards
Under 13:	50–60 yards	100–110 yards
International:	70–80 yards	110–120 yards

61. a. Find the minimum length and width of a soccer field for 8-year-old children. (Carefully consider the age.)

b. Find the perimeter of this field.

62. a. Find the maximum length and width of a soccer field for 12-year-old children.

b. Find the perimeter of this field.

Solve. See the Concept Check in this section. Choose the figure that has the greater distance around.

63. a. A square with side length 3 inches

b. A circle with diameter 4 inches

64. a. A circle with diameter 7 inches

b. A square with side length 7 inches

65. a. Find the circumference of each circle. Approximate the circumference by using 3.14 for π.

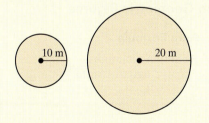

b. If the radius of a circle is doubled, is its corresponding circumference doubled?

66. a. Find the circumference of each circle. Approximate the circumference by using 3.14 for π.

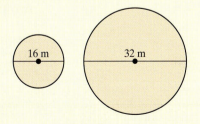

b. If the diameter of a circle is doubled, is its corresponding circumference doubled?

67. In your own words, explain how to find the perimeter of any polygon.

68. In your own words, explain how perimeter and circumference are the same and how they are different.

Find the perimeter. Round your results to the nearest tenth.

69.

6 meters

6 meters

70.

6 meters

6 meters

71.

ROYALS

5 m

22 m

72.

5 feet

7 feet

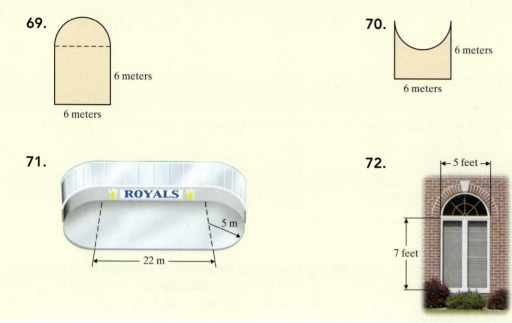

9.3   Area, Volume, and Surface Area

Objectives

A Find the Area of Plane Regions.

B Find the Volume and Surface Area of Solids.

Objective A Finding Area of Plane Regions

Recall that area measures the number of square units that cover the surface of a plane region; that is, a region that lies in a plane. Thus far, we know how to find the areas of a rectangle and a square. These formulas, as well as formulas for finding the areas of other common geometric figures, are given next.

Area Formulas of Common Geometric Figures

Geometric Figure	**Area Formula**
RECTANGLE width length	Area of a rectangle: **A**rea = **length · width** $A = lw$
SQUARE side side	Area of a square: **A**rea = **side · side** $A = s \cdot s = s^2$

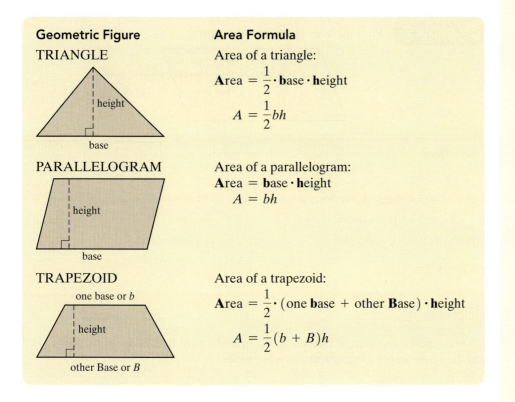

Geometric Figure	Area Formula

TRIANGLE

Area of a triangle:

$$\text{Area} = \frac{1}{2} \cdot \text{base} \cdot \text{height}$$

$$A = \frac{1}{2}bh$$

PARALLELOGRAM

Area of a parallelogram:

$$\text{Area} = \text{base} \cdot \text{height}$$

$$A = bh$$

TRAPEZOID

Area of a trapezoid:

$$\text{Area} = \frac{1}{2} \cdot (\text{one base} + \text{other Base}) \cdot \text{height}$$

$$A = \frac{1}{2}(b + B)h$$

Use these formulas for the following examples.

Helpful Hint

Area is always measured in square units.

Example 1 Find the area of the triangle.

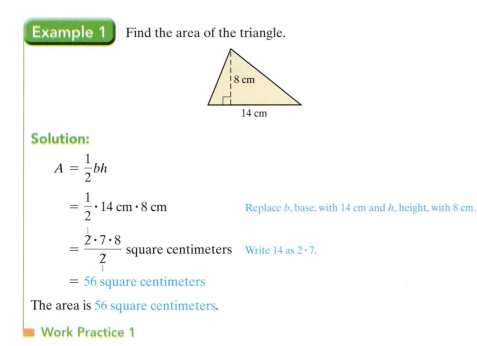

8 cm

14 cm

Solution:

$$A = \frac{1}{2}bh$$

$$= \frac{1}{2} \cdot 14 \text{ cm} \cdot 8 \text{ cm} \qquad \text{Replace } b, \text{ base, with 14 cm and } h, \text{ height, with 8 cm.}$$

$$= \frac{\overset{1}{\cancel{2}} \cdot 7 \cdot 8}{\underset{1}{\cancel{2}}} \text{ square centimeters} \qquad \text{Write 14 as } 2 \cdot 7.$$

$$= 56 \text{ square centimeters}$$

The area is 56 square centimeters.

■ **Work Practice 1**

Practice 1

Find the area of the triangle.

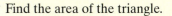
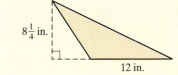

$8\frac{1}{4}$ in.

12 in.

Answer

1. $49\frac{1}{2}$ sq in.

Practice 2

Find the area of the trapezoid.

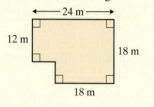

5 yd

6.1 yd

11 yd

Example 2 Find the area of the parallelogram.

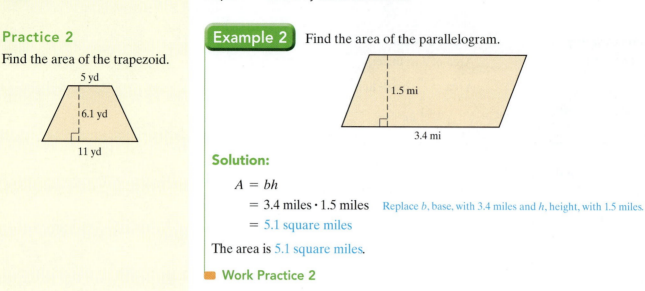

1.5 mi

3.4 mi

Solution:

$A = bh$

$= 3.4 \text{ miles} \cdot 1.5 \text{ miles}$ Replace b, base, with 3.4 miles and h, height, with 1.5 miles.

$= 5.1 \text{ square miles}$

The area is 5.1 square miles.

◼ **Work Practice 2**

Helpful Hint

When finding the area of figures, check to make sure that all measurements are in the same units before calculations are made.

Practice 3

Find the area of the figure.

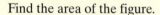

24 m

12 m

18 m

18 m

Example 3 Find the area of the figure.

4 ft

8 ft

5 ft

12 ft

Solution: Split the figure into two rectangles. To find the area of the figure, we find the sum of the areas of the two rectangles.

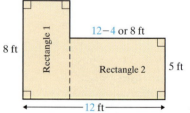

4 ft

8 ft

Rectangle 1

12−4 or 8 ft

Rectangle 2

5 ft

12 ft

area of Rectangle 1 $= lw$

$= 8 \text{ feet} \cdot 4 \text{ feet}$

$= 32 \text{ square feet}$

Notice that the length of Rectangle 2 is 12 feet − 4 feet or 8 feet.

area of Rectangle 2 $= lw$

$= 8 \text{ feet} \cdot 5 \text{ feet}$

$= 40 \text{ square feet}$

area of the figure $=$ area of Rectangle 1 $+$ area of Rectangle 2

$= 32 \text{ square feet} + 40 \text{ square feet}$

$= 72 \text{ square feet}$

◼ **Work Practice 3**

Answers

2. 48.8 sq yd   **3.** 396 sq m

Helpful Hint

The figure in Example 3 could also be split into two rectangles as shown.

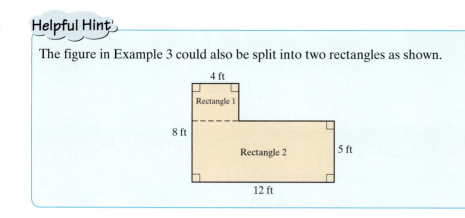

To better understand the formula for area of a circle, try the following. Cut a circle into many pieces, as shown.

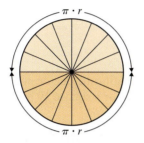

The circumference of a circle is $2\pi r$. This means that the circumference of half a circle is half of $2\pi r$, or πr.

Then unfold the two halves of the circle and place them together, as shown.

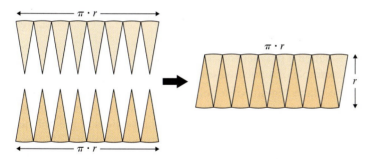

The figure on the right is almost a parallelogram with a base of πr and a height of r. The area is

$$A = \boxed{\text{base}} \cdot \boxed{\text{height}}$$

$$= (\pi r) \cdot r$$

$$= \pi r^2$$

This is the formula for the area of a circle.

Area Formula of a Circle

Circle

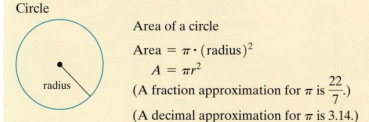

Area of a circle

Area $= \pi \cdot (\text{radius})^2$

$A = \pi r^2$

(A fraction approximation for π is $\dfrac{22}{7}$.)

(A decimal approximation for π is 3.14.)

Practice 4

Find the area of the given circle. Find the exact area and an approximation. Use 3.14 as an approximation for π.

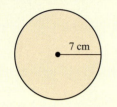

7 cm

 Example 4 Find the area of a circle with a radius of 3 feet. Find the exact area and an approximation. Use 3.14 as an approximation for π.

3 ft

Solution: We let $r = 3$ feet and use the formula

$$A = \pi r^2$$
$$= \pi \cdot (3 \text{ feet})^2 \quad \text{Replace } r \text{ with 3 feet.}$$
$$= 9 \cdot \pi \text{ square feet} \quad \text{Replace } (3 \text{ feet})^2 \text{ with 9 sq ft.}$$

To approximate this area, we substitute 3.14 for π.

$$9 \cdot \pi \text{ square feet} \approx 9 \cdot 3.14 \text{ square feet}$$
$$= 28.26 \text{ square feet}$$

The *exact* area of the circle is 9π square feet, which is *approximately* 28.26 square feet.

■ **Work Practice 4**

✓**Concept Check** Use estimation to decide which figure would have a larger area: a circle of diameter 10 in. or a square 10 in. long on each side.

Objective B Finding Volume and Surface Area of Solids ▶

A **convex solid** is a set of points, *S*, not all in one plane, such that for any two points *A* and *B* in *S*, all points between *A* and *B* are also in *S*. In this section, we will find the volume and surface area of special types of solids called polyhedrons. A solid formed by the intersection of a finite number of planes is called a **polyhedron.** The box to the right is an example of a polyhedron.

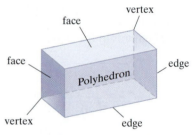

vertex
face
face
edge
Polyhedron
vertex
edge

Each of the plane regions of a polyhedron is called a **face** of the polyhedron. If the intersection of two faces is a line segment, this line segment is an **edge** of the polyhedron. The intersections of the edges are the **vertices** of the polyhedron.

Volume is a measure of the space of a region. The volume of a box or can, for example, is the amount of space inside. Volume can be used to describe the amount of juice in a pitcher or the amount of concrete needed to pour a foundation for a house.

The volume of a solid is the number of **cubic units** in the solid. A cubic centimeter and a cubic inch are illustrated.

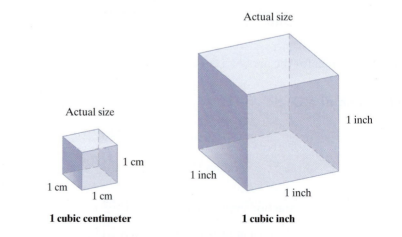

Actual size

Actual size

1 cm
1 cm
1 cm

1 inch
1 inch
1 inch
1 inch

1 cubic centimeter

1 cubic inch

Answer

4. 49π sq cm ≈ 153.86 sq cm

✓ **Concept Check Answer**

A square 10 in. long on each side would have a larger area.

The **surface area** of a polyhedron is the sum of the areas of the faces of the polyhedron. For example, each face of the cube to the left on the previous page has an area of 1 square centimeter. Since there are 6 faces of the cube, the sum of the areas of the faces is 6 square centimeters. Surface area can be used to describe the amount of material needed to cover a solid. Surface area is measured in square units.

Formulas for finding the volumes, V, and surface areas, SA, of some common solids are given next. (Note: Spheres, circular cylinders, and cones are not polyhedrons, but they are solids and we will calculate surface areas and volumes of these solids.)

Volume and Surface Area Formulas of Common Solids	
Solid	**Formulas**
RECTANGULAR SOLID height width length	$V = lwh$ $SA = 2lh + 2wh + 2lw$ where h = height, w = width, l = length
CUBE side side side	$V = s^3$ $SA = 6s^2$ where s = side
SPHERE radius	$V = \dfrac{4}{3}\pi r^3$ $SA = 4\pi r^2$ where r = radius
CIRCULAR CYLINDER height radius	$V = \pi r^2 h$ $SA = 2\pi rh + 2\pi r^2$ where h = height, r = radius
CONE height radius	$V = \dfrac{1}{3}\pi r^2 h$ $SA = \pi r\sqrt{r^2 + h^2} + \pi r^2$ where h = height, r = radius
SQUARE-BASED PYRAMID slant height height side	$V = \dfrac{1}{3}s^2 h$ $SA = B + \dfrac{1}{2}pl$ where B = area of base, p = perimeter of base, h = height, s = side, l = slant height

Practice 5

Find the volume and surface area of a rectangular box that is 7 feet long, 3 feet wide, and 4 feet high.

Example 5 Find the volume and surface area of a rectangular box that is 12 inches long, 6 inches wide, and 3 inches high.

3 in.

6 in.   12 in.

Solution: Let $h = 3$ in., $l = 12$ in., and $w = 6$ in.

$$V = lwh$$

$$V = 12 \text{ inches} \cdot 6 \text{ inches} \cdot 3 \text{ inches} = 216 \text{ cubic inches}$$

The volume of the rectangular box is 216 cubic inches.

$$SA = 2lh + 2wh + 2lw$$
$$= 2(12 \text{ in.})(3 \text{ in.}) + 2(6 \text{ in.})(3 \text{ in.}) + 2(12 \text{ in.})(6 \text{ in.})$$
$$= 72 \text{ sq in.} + 36 \text{ sq in.} + 144 \text{ sq in.}$$
$$= 252 \text{ sq in.}$$

The surface area of the rectangular box is 252 square inches.

■ **Work Practice 5**

✔ **Concept Check** Juan is calculating the volume of the following rectangular solid. Find the error in his calculation.

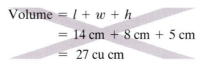

Volume $= l + w + h$
$$= 14 \text{ cm} + 8 \text{ cm} + 5 \text{ cm}$$
$$= 27 \text{ cu cm}$$

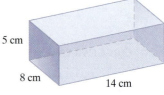

5 cm

8 cm

14 cm

Practice 6

Find the volume and surface area of a ball of radius $\frac{1}{2}$ centimeter. Give the exact volume and surface area. Then use $\frac{22}{7}$ for π and approximate the values.

Example 6 Find the volume and surface area of a ball of radius 2 inches. Give the exact volume and surface area. Then use the approximation $\frac{22}{7}$ for π.

2 in.

Solution:

$$V = \frac{4}{3}\pi r^3 \qquad \text{Formula for volume of a sphere}$$

$$V = \frac{4}{3} \cdot \pi (2 \text{ in.})^3 \qquad \text{Let } r = 2 \text{ inches.}$$

$$= \frac{32}{3}\pi \text{ cu in.} \qquad \text{Exact volume}$$

$$\approx \frac{32}{3} \cdot \frac{22}{7} \text{ cu in.} \qquad \text{Approximate } \pi \text{ with } \frac{22}{7}.$$

$$= \frac{704}{21} \text{ or } 33\frac{11}{21} \text{ cu in.} \qquad \text{Approximate volume}$$

Answers

5. $V = 84$ cu ft; $SA = 122$ sq ft

6. $V = \frac{1}{6}\pi$ cu cm $\approx \frac{11}{21}$ cu cm;

$SA = \pi$ sq cm $\approx 3\frac{1}{7}$ sq cm

✔ **Concept Check Answer**

Volume $= lwh$
$$= 14 \text{ cm} \cdot 8 \text{ cm} \cdot 5 \text{ cm}$$
$$= 560 \text{ cu cm}$$

The volume of the sphere is exactly $\frac{32}{3}\pi$ cubic inches or approximately $33\frac{11}{21}$ cubic inches.

$$SA = 4\pi r^2 \qquad \text{Formula for surface area}$$
$$SA = 4 \cdot \pi (2\,\text{in.})^2 \qquad \text{Let } r = 2 \text{ inches.}$$
$$= 16\pi \text{ sq in.} \qquad \text{Exact surface area}$$
$$\approx 16 \cdot \frac{22}{7} \text{ sq in.} \qquad \text{Approximate } \pi \text{ with } \frac{22}{7}.$$
$$= \frac{352}{7} \text{ or } 50\frac{2}{7} \text{ sq in.} \qquad \text{Approximate surface area}$$

The surface area of the sphere is exactly 16π square inches or approximately $50\frac{2}{7}$ square inches.

■ **Work Practice 6**

Example 7   Find the volume of a can that has a $3\frac{1}{2}$-inch radius and a height of 6 inches. Give an exact volume and an approximate volume. Use $\frac{22}{7}$ for π.

$3\frac{1}{2}$ in.

6 in.

Solution: Using the formula for a circular cylinder, we have

$$V = \pi \cdot r^2 \cdot h \qquad 3\frac{1}{2} = \frac{7}{2}$$
$$= \pi \cdot \left(\frac{7}{2}\,\text{in.}\right)^2 \cdot 6\,\text{in.}$$
$$= \pi \cdot \frac{49}{4}\,\text{sq in.} \cdot 6\,\text{in.}$$
$$= \frac{\pi \cdot 49 \cdot \overset{1}{\cancel{2}} \cdot 3}{\underset{1}{\cancel{2}} \cdot 2}\,\text{cu in.}$$
$$= 73\frac{1}{2}\pi \text{ cu in. or } 73.5\pi \text{ cu in.}$$

This is the exact volume. To approximate the volume, use the approximation $\frac{22}{7}$ for π.

$$V = 73\frac{1}{2}\pi \text{ or } \frac{147}{2} \cdot \frac{22}{7}\,\text{cu in.} \qquad \text{Replace } \pi \text{ with } \frac{22}{7}.$$
$$= \frac{21 \cdot \overset{1}{\cancel{7}} \cdot \overset{1}{\cancel{2}} \cdot 11}{\underset{1}{\cancel{2}} \cdot \underset{1}{\cancel{7}}}\,\text{cu in.}$$
$$= 231 \text{ cu in.}$$

The volume is approximately 231 cubic inches.

■ **Work Practice 7**

Practice 7

Find the volume of a cylinder of radius 5 inches and height 9 inches. Give an exact answer and an approximate answer. Use 3.14 for π.

Answer

7. 225π cu in. ≈ 706.5 cu in.

Practice 8

Find the volume of a square-based pyramid that has a 3-meter side and a height of 5.1 meters.

5.1 m

3 m

Answer

8. 15.3 cu m

Example 8   Find the volume of a cone that has a height of 14 centimeters and a radius of 3 centimeters. Give an exact answer and an approximate answer. Use 3.14 for π.

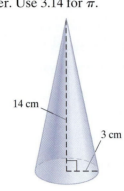

14 cm

3 cm

Solution: Using the formula for volume of a cone, we have

$$V = \frac{1}{3} \cdot \pi \cdot r^2 \cdot h$$

$$= \frac{1}{3} \cdot \pi \cdot (3 \text{ cm})^2 \cdot 14 \text{ cm} \quad \text{Replace } r \text{ with 3 cm and } h \text{ with 14 cm.}$$

$$= 42\pi \text{ cu cm}$$

Thus, 42π cubic centimeters is the exact volume. To approximate the volume, use the approximation 3.14 for π.

$$V \approx 42 \cdot 3.14 \text{ cu cm} \quad \text{Replace } \pi \text{ with 3.14.}$$

$$= 131.88 \text{ cu cm}$$

The volume is approximately 131.88 cubic centimeters.

■ **Work Practice 8**

Vocabulary, Readiness & Video Check

Use the choices below to fill in each blank. Some choices may be used more than once.

area   surface area   cubic

volume   square

1. The _____ of a polyhedron is the sum of the areas of its faces.

2. The measure of the amount of space inside a solid is its _____.

3. _____ measures the amount of surface enclosed by a region.

4. Volume is measured in _____ units.

5. Area is measured in _____ units.

6. Surface area is measured in _____ units.

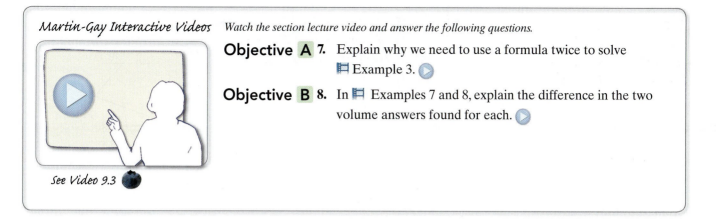

Martin-Gay Interactive Videos   *Watch the section lecture video and answer the following questions.*

Objective A 7. Explain why we need to use a formula twice to solve
  Example 3. ◉

Objective B 8. In Examples 7 and 8, explain the difference in the two
  volume answers found for each. ◉

See Video 9.3 🫐

9.3 Exercise Set MyMathLab® ◉

Objective A *Find the area of each geometric figure. If the figure is a circle, give an exact area and then use the given approximation for π to approximate the area. See Examples 1 through 4.*

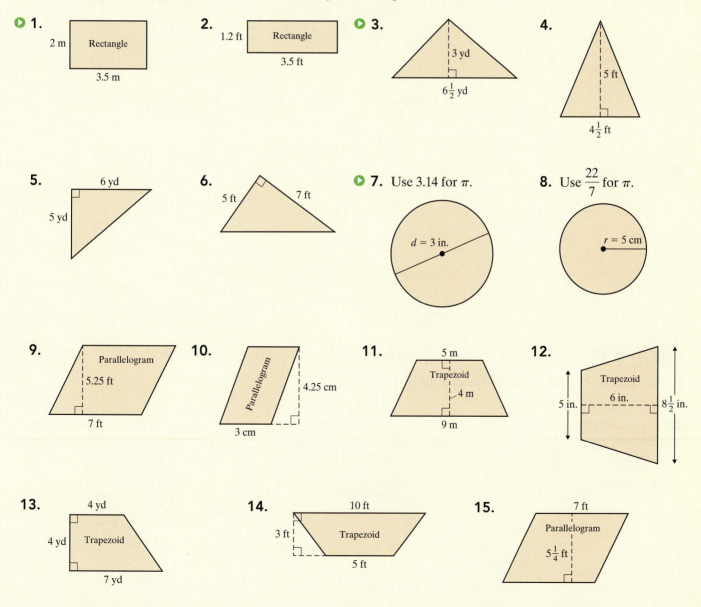

◉ **1.**

2 m | Rectangle
3.5 m

2.

1.2 ft | Rectangle
3.5 ft

◉ **3.**

3 yd
$6\frac{1}{2}$ yd

4.

5 ft
$4\frac{1}{2}$ ft

5.

6 yd
5 yd

6.

5 ft   7 ft

◉ **7.** Use 3.14 for π.

d = 3 in.

8. Use $\frac{22}{7}$ for π.

r = 5 cm

9.

Parallelogram
5.25 ft
7 ft

10.

Parallelogram
4.25 cm
3 cm

11.

5 m
Trapezoid
4 m
9 m

12.

Trapezoid
6 in.
5 in. $8\frac{1}{2}$ in.

13.

4 yd
4 yd | Trapezoid
7 yd

14.

10 ft
3 ft | Trapezoid
5 ft

15.

7 ft
Parallelogram
$5\frac{1}{4}$ ft

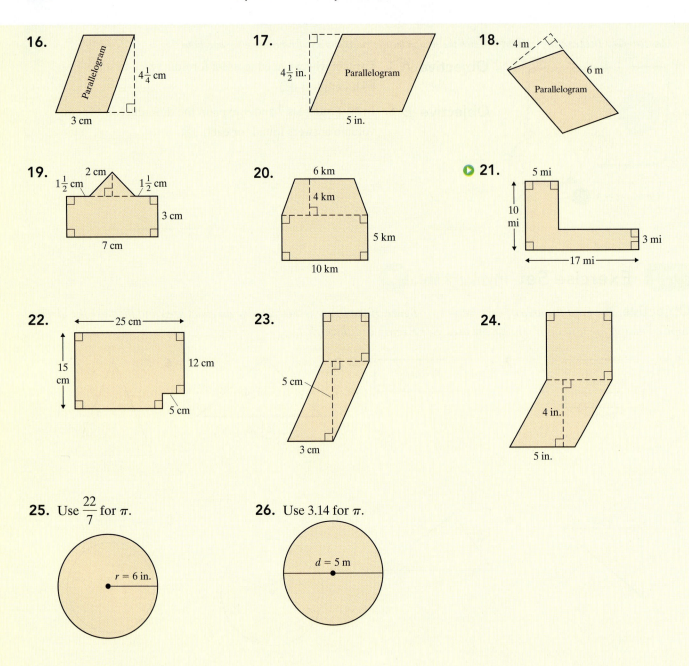

16. Parallelogram, $4\frac{1}{4}$ cm, 3 cm

17. $4\frac{1}{2}$ in., Parallelogram, 5 in.

18. 4 m, 6 m, Parallelogram

19. 2 cm, $1\frac{1}{2}$ cm, $1\frac{1}{2}$ cm, 3 cm, 7 cm

20. 6 km, 4 km, 5 km, 10 km

21. 5 mi, 10 mi, 3 mi, 17 mi

22. 25 cm, 15 cm, 12 cm, 5 cm

23. 5 cm, 3 cm

24. 4 in., 5 in.

25. Use $\frac{22}{7}$ for π. $r = 6$ in.

26. Use 3.14 for π. $d = 5$ m

Objective B *Find the volume and surface area of each solid. See Examples 5 through 8. For formulas containing π, give an exact answer and then approximate using $\frac{22}{7}$ for π.*

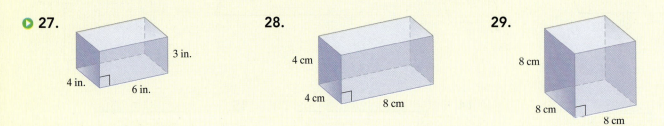

27. 3 in., 4 in., 6 in.

28. 4 cm, 4 cm, 8 cm

29. 8 cm, 8 cm, 8 cm

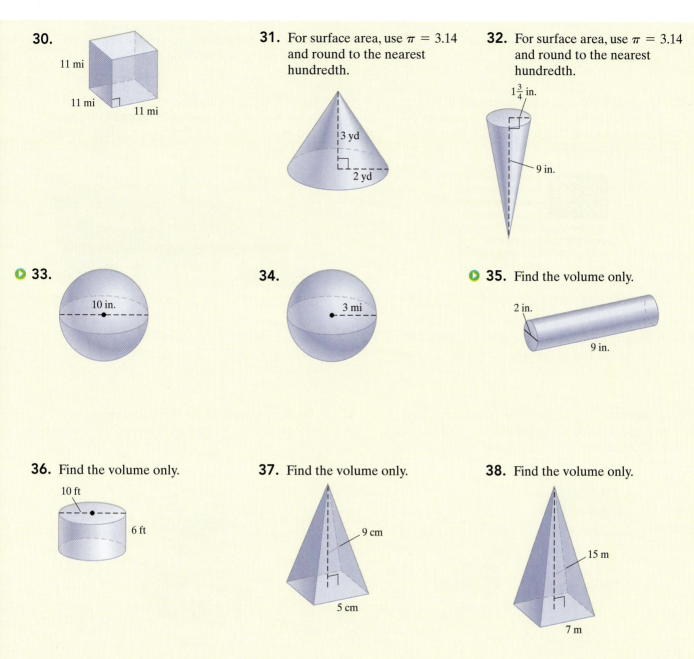

30.

11 mi
11 mi
11 mi

31. For surface area, use $\pi = 3.14$ and round to the nearest hundredth.

3 yd
2 yd

32. For surface area, use $\pi = 3.14$ and round to the nearest hundredth.

$1\frac{3}{4}$ in.
9 in.

33.

10 in.

34.

3 mi

35. Find the volume only.

2 in.
9 in.

36. Find the volume only.

10 ft
6 ft

37. Find the volume only.

9 cm
5 cm

38. Find the volume only.

15 m
7 m

Objectives A B Mixed Practice *Solve. See Examples 1 through 8.*

39. Find the volume of a cube with edges of $1\frac{1}{3}$ inches.

$1\frac{1}{3}$ inches

40. A water storage tank is in the shape of a cone with the pointed end down. If the radius is 14 ft and the depth of the tank is 15 ft, approximate the volume of the tank in cubic feet. Use $\frac{22}{7}$ for π.

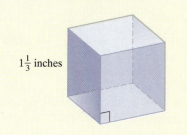

14 ft
15 ft

41. Find the volume and surface area of a rectangular box 2 ft by 1.4 ft by 3 ft.

42. Find the volume and surface area of a box in the shape of a cube that is 5 ft on each side.

43. The largest American flag measures 505 feet by 225 feet. It's the U.S. "Super flag" owned by "Ski" Demski of Long Beach, California. Find its area. (*Source: Guinness World Records*)

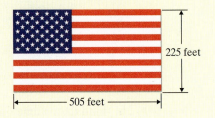

44. The largest indoor illuminated sign is a billboard at Dubai Airport. It measures 28 meters by 6.2 meters. Find its area. (*Source: The Guinness Book of World Records*)

▶ **45.** A drapery panel measures 6 ft by 7 ft. Find how many square feet of material are needed for *four* panels.

46. A page in this book measures 27.6 cm by 21.5 cm. Find its area.

47. A paperweight is in the shape of a square-based pyramid 20 centimeters tall. If an edge of the base is 12 centimeters, find the volume of the paperweight.

48. A birdbath is made in the shape of a hemisphere (half-sphere). If its radius is 10 inches, approximate its volume. Use $\frac{22}{7}$ for π.

49. Find how many square feet of land are in the following plot:

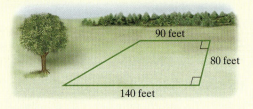

50. For Gerald Gomez to determine how much grass seed he needs to buy, he must know the size of his yard. Use the drawing to determine how many square feet are in his yard.

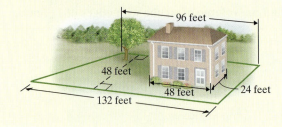

51. Find the exact volume and surface area of a sphere with a radius of 7 inches.

52. A tank is in the shape of a cylinder 8 feet tall and 3 feet in radius. Find the exact volume and surface area of the tank.

53. The outlined part of the roof shown is in the shape of a trapezoid and needs to be shingled. The number of shingles to buy depends on the area.

   a. Use the dimensions given to find the area of the outlined part of the roof to the nearest whole square foot.

   b. Shingles are packaged in a unit called a "square." If a "square" covers 100 square feet, how many whole squares need to be purchased to shingle this part of the roof?

54. The entire side of the building shaded in the drawing is to be bricked. The number of bricks to buy depends on the area.

   a. Find the area.

   b. If the side area of each brick (including mortar room) is $\frac{1}{6}$ square ft, find the number of bricks that are needed to brick the end of the building.

55. Find the exact volume of a waffle ice cream cone with a 3-in. diameter and a height of 7 inches.

56. A snow globe has a diameter of 6 inches. Find its exact volume. Then approximate its volume using 3.14 for π.

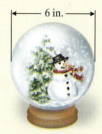

57. Paul Revere's Pizza in the USA will bake and deliver a round pizza with a 4-foot diameter. This pizza is called the "Ultimate Party Pizza" and its current price is $99.99. Find the exact area of the top of the pizza and an approximation. Use 3.14 as an approximation for π.

58. The face of a circular watch has a diameter of 2 centimeters. What is its area? Find the exact area and an approximation. Use 3.14 as an approximation for π.

59. Zorbing is an extreme sport invented by two New Zealanders who joke that they were looking for a way to walk on water. A Zorb is a large sphere inside a second sphere with the space between the spheres pumped full of air. There is a tunnel-like opening so a person can crawl into the inner sphere. You are strapped in and sent down a Zorbing hill. A standard Zorb is approximately 3 m in diameter. Find the exact volume of a Zorb, and approximate the volume using 3.14 for π.

60. Mount Fuji, in Japan, is considered the most beautiful composite volcano in the world. The mountain is in the shape of a cone whose height is about 3.5 kilometers and whose base radius is about 3 kilometers. Approximate the volume of Mt. Fuji in cubic kilometers. Use $\frac{22}{7}$ for π.

61. A $10\frac{1}{2}$-foot by 16-foot concrete wall is to be built using concrete blocks. Find the area of the wall.

62. The floor of Terry's attic is 24 feet by 35 feet. Find how many square feet of insulation are needed to cover the attic floor.

63. Find the volume of a pyramid with a square base 5 inches on a side and a height of $1\frac{3}{10}$ inches.

64. Approximate to the nearest hundredth the volume of a sphere with a radius of 2 centimeters. Use 3.14 for π.

The Space Cube is supposed to be the world's smallest computer, with dimensions of 2 inches by 2 inches by 2.2 inches.

65. Find the volume of the Space Cube.

66. Find the surface area of the Space Cube.

Review

Evaluate. See Section 1.7.

67. 5^2

68. 7^2

69. 3^2

70. 20^2

71. $1^2 + 2^2$

72. $5^2 + 3^2$

73. $4^2 + 2^2$

74. $1^2 + 6^2$

Concept Extensions

Given the following situations, tell whether you are more likely to be concerned with area or perimeter.

75. ordering fencing to fence a yard

76. ordering grass seed to plant in a yard

77. buying carpet to install in a room

78. buying gutters to install on a house

79. ordering paint to paint a wall

80. ordering baseboards to install in a room

81. buying a wallpaper border to go on the walls around a room

82. buying fertilizer for your yard

Solve.

83. A pizza restaurant recently advertised two specials. The first special was a 12-inch pizza for $10. The second special was two 8-inch pizzas for $9. Determine the better buy. (*Hint:* First compare the areas of the two specials and then find a price per square inch for both specials.)

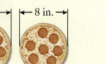
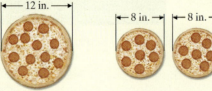

84. Find the approximate area of the state of Utah.

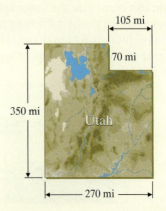

85. The Hayden Planetarium, at the Museum of Natural History in New York City, boasts a dome that has a diameter of 20 m. The dome is a hemisphere, or half a sphere. What is the volume enclosed by the dome at the Hayden Planetarium? Use 3.14 for π and round to the nearest hundredth. (*Source:* Hayden Planetarium)

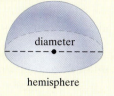

hemisphere

86. The Adler Museum in Chicago has a planetarium, its StarRider Theater, that has a diameter of 55 feet. Find the surface area of its hemispheric (half a sphere) dome. Use 3.14 for π. (*Source:* The Adler Museum)

87. Can you compute the volume of a rectangle? Why or why not?

88. In your own words, explain why perimeter is measured in units and area is measured in square units. (*Hint:* See Section 1.5 for an introduction to the meaning of area.)

89. Find the area of the shaded region. Use the approximation 3.14 for π.

90. The largest pumpkin pie on record was made in New Breman, Ohio, by the New Breman Giant Pumpkin Growers, in September 2010. The pie had a diameter of 240 inches. Find the exact area of the top of the pie and an approximation. Use π ≈ 3.14. (*Source:* World Record Academy)

Find the area of each figure. If needed, use π ≈ 3.14 and round results to the nearest tenth.

91. Find the skating area.

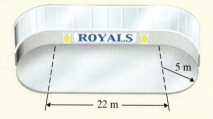

92.

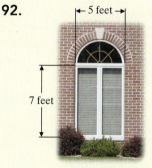

93. Do two rectangles with the same perimeter have the same area? To see, find the perimeter and the area of each rectangle.

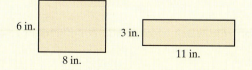

94. Do two rectangular solids with the same volume have the same surface area? To see, find the volume and surface area of each rectangular solid.

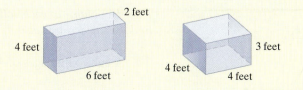

Geometry Concepts

Answers

1. _____

2. _____

3. _____

4. _____

5. _____

6. _____

7. _____

8. _____

9. _____

10. _____

11. _____

12. _____

13. _____

14. _____

15. _____

16. _____

1. Find the supplement and the complement of a 27° angle.

Find the measures of angles x, y, and z in each figure in Exercises 2 and 3.

2.
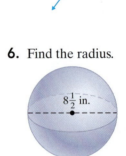
105°

3. $m \parallel n$

x / 52°   m

y
z   n

4. Find the measure of ∠x. (*Hint:* The sum of the angle measures of a triangle is 180°.)

x

38°

5. Find the diameter.

2.3 in.

6. Find the radius.

$8\frac{1}{2}$ in.

For Exercises 7 through 11, find the perimeter (or circumference) and area of each figure. For the circle, give an exact circumference and area. Then use $\pi \approx 3.14$ to approximate each. Don't forget to attach correct units.

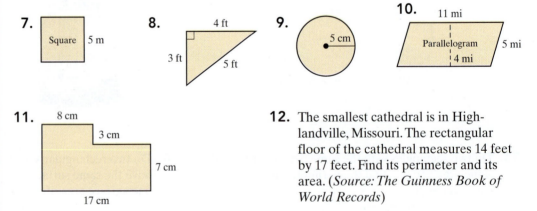

7.
Square 5 m

8.
4 ft
3 ft
5 ft

9.
5 cm

10.
11 mi
Parallelogram 5 mi
4 mi

11.
8 cm
3 cm
7 cm
17 cm

12. The smallest cathedral is in High-landville, Missouri. The rectangular floor of the cathedral measures 14 feet by 17 feet. Find its perimeter and its area. (*Source: The Guinness Book of World Records*)

Find the volume of each solid. Don't forget to attach correct units. For Exercises 13 and 14, find the surface area also.

13. A cube with edges of 4 inches each

14. A rectangular box 2 feet by 3 feet by 5.1 feet

15. A pyramid with a square base 10 centimeters on a side and a height of 12 centimeters

16. A sphere with a diameter of 3 miles. Give the exact volume and then use $\pi \approx \frac{22}{7}$ to approximate.

9.4 Linear Measurement

Objective A Defining and Converting U.S. System Units of Length

In the United States, two systems of measurement are commonly used. They are the **United States (U.S.), or English, measurement system** and the **metric system.** The U.S. measurement system is familiar to most Americans. Units such as feet, miles, ounces, and gallons are used. However, the metric system is also commonly used in fields such as medicine, sports, international marketing, and certain physical sciences. We are accustomed to buying 2-liter bottles of soft drinks, watching televised coverage of the 100-meter dash at the Olympic Games, or taking a 200-milligram dose of pain reliever.

The U.S. system of measurement uses the **inch, foot, yard,** and **mile** to measure **length.** The following is a summary of equivalencies between units of length:

> ### U.S. Units of Length
>
> 12 inches (in.) = 1 foot (ft)
> 3 feet = 1 yard (yd)
> 36 inches = 1 yard
> 5280 feet = 1 mile (mi)

To convert from one unit of length to another, we will use **unit fractions.** We define a unit fraction to be a fraction that is equivalent to 1. Examples of unit fractions are as follows:

> ### Unit Fractions
>
> $\dfrac{12 \text{ in.}}{1 \text{ ft}} = 1$ or $\dfrac{1 \text{ ft}}{12 \text{ in.}} = 1$ (since 12 in. = 1 ft)
>
> $\dfrac{3 \text{ ft}}{1 \text{ yd}} = 1$ or $\dfrac{1 \text{ yd}}{3 \text{ ft}} = 1$ (since 3 ft = 1 yd)
>
> $\dfrac{5280 \text{ ft}}{1 \text{ mi}} = 1$ or $\dfrac{1 \text{ mi}}{5280 \text{ ft}} = 1$ (since 5280 ft = 1 mi)

Remember that multiplying a number by 1 does not change the value of the number.

Example 1 Convert 8 feet to inches.

Solution: We multiply 8 feet by a unit fraction that uses the equality 12 inches = 1 foot. The unit fraction should be in the form $\dfrac{\text{units to convert to}}{\text{original units}}$ or, in this case, $\dfrac{12 \text{ inches}}{1 \text{ foot}}$. We do this so that like units will divide out to 1, as shown.

$8 \text{ ft} = \dfrac{8 \text{ ft}}{1} \cdot 1$ Multiply by 1 in the form of $\dfrac{12 \text{ in.}}{1 \text{ ft}}$.

$= \dfrac{8 \text{ ft}}{1} \cdot \dfrac{12 \text{ in.}}{1 \text{ ft}}$

$= 8 \cdot 12 \text{ in.}$

$= 96 \text{ in.}$ Multiply.   *(Continued on next page)*

Objectives

A Define U.S. Units of Length and Convert from One Unit to Another.

B Use Mixed U.S. Units of Length.

C Perform Arithmetic Operations on U.S. Units of Length.

D Define Metric Units of Length and Convert from One Unit to Another.

E Perform Arithmetic Operations on Metric Units of Length.

Practice 1

Convert 6 feet to inches.

Answer
1. 72 in.

641

Thus, 8 ft = 96 in., as shown in the diagram:

8 feet = 96 inches

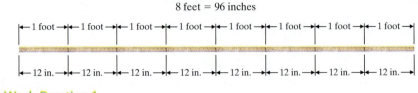

Work Practice 1

Practice 2

Convert 8 yards to feet.

Example 2   Convert 7 feet to yards.

Solution: We multiply by a unit fraction that compares 1 yard to 3 feet.

$$7 \text{ ft} = \frac{7 \text{ ft}}{1} \cdot 1$$

$$= \frac{7 \text{ ft}}{1} \cdot \frac{1 \text{ yd}}{3 \text{ ft}} \quad \leftarrow \text{Units to convert to}$$
$$\qquad\qquad\qquad \leftarrow \text{Original units}$$

$$= \frac{7}{3} \text{ yd}$$

$$= 2\frac{1}{3} \text{ yd} \qquad \text{Divide.}$$

> **Helpful Hint**   When converting from one unit to another, select a unit fraction with the properties below:
>
> $$\frac{\text{units you are converting to}}{\text{original units}}$$
>
> By using this unit fraction, the original units will divide out, as wanted.

Thus, $7 \text{ ft} = 2\frac{1}{3} \text{ yd}$, as shown in the diagram.

$$7 \text{ feet} = 2\frac{1}{3} \text{ yards}$$

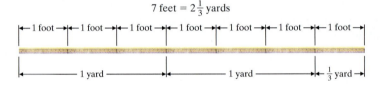

Work Practice 2

Practice 3

Suppose the pelican's bill in the photo measures 18 inches. Convert 18 inches to feet, using decimals.

Example 3   Finding the Length of a Pelican's Bill

The Australian pelican has the longest bill, measuring from 13 to 18.5 inches long. The pelican in the photo has a 15-inch bill. Convert 15 inches to feet, using decimals in your final answer.

Solution:

$$15 \text{ in.} = \frac{15 \text{ in.}}{1} \cdot \frac{1 \text{ ft}}{12 \text{ in.}} \quad \begin{array}{l} \leftarrow \text{Units to convert to} \\ \leftarrow \text{Original units} \end{array}$$

$$= \frac{15}{12} \text{ ft}$$

$$= \frac{5}{4} \text{ ft} \qquad \text{Simplify } \frac{15}{12}.$$

$$= 1.25 \text{ ft} \qquad \text{Divide.}$$

Thus, 15 in. = 1.25 ft, as shown in the diagram.

15 inches = 1.25 ft

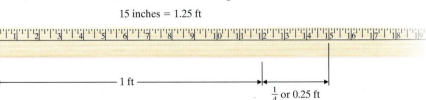

Answers

2. 24 ft   **3** 1.5 ft

Work Practice 3

Objective B Using Mixed U.S. System Units of Length ▶

Sometimes it is more meaningful to express a measurement of length with mixed units such as 1 ft and 5 in. We usually condense this and write 1 ft 5 in.

In Example 2, we found that 7 feet is the same as $2\frac{1}{3}$ yards. The measurement can also be written as a mixture of yards and feet. That is,

7 ft = ____ yd ____ ft

Because 3 ft = 1 yd, we divide 3 into 7 to see how many whole yards are in 7 feet. The quotient is the number of yards, and the remainder is the number of feet.

$$\begin{array}{r} 2 \text{ yd } 1 \text{ ft} \\ 3\overline{)7} \quad \uparrow \\ \underline{-6} \quad | \\ 1 \underline{\quad} \end{array}$$

Thus, 7 ft = 2 yd 1 ft, as seen in the diagram:

Example 4 Convert: 134 in. = _____ ft _____ in.

Solution: Because 12 in. = 1 ft, we divide 12 into 134. The quotient is the number of feet. The remainder is the number of inches. To see why we divide 12 into 134, notice that

$$134 \text{ in.} = \frac{134 \text{ in.}}{1} \cdot \frac{1 \text{ ft}}{12 \text{ in.}} = \frac{134}{12} \text{ ft}$$

$$\begin{array}{r} 11 \text{ ft } 2 \text{ in.} \\ 12\overline{)134} \quad \uparrow \\ \underline{-12} \quad | \\ 14 \quad | \\ \underline{-12} \quad | \\ 2 \underline{\quad} \end{array}$$

Thus, 134 in. = 11 ft 2 in.

■ **Work Practice 4**

Example 5 Convert 3 feet 7 inches to inches.

Solution: First, we convert 3 feet to inches. Then we add 7 inches.

$$3 \text{ ft} = \frac{3 \text{ ft}}{1} \cdot \frac{12 \text{ in.}}{1 \text{ ft}} = 36 \text{ in.}$$

Then

3 ft 7 in. = 36 in. + 7 in. = 43 in.

■ **Work Practice 5**

Practice 4

Convert: 68 in. = _____ ft _____ in.

Practice 5

Convert 5 yards 2 feet to feet.

Answers

4. 5 ft 8 in. **5.** 17 ft

Objective C Performing Operations on U.S. System Units of Length ▶

Finding sums or differences of measurements often involves converting units, as shown in the next example. Just remember that, as usual, only like units can be added or subtracted.

Practice 6

Add 4 ft 8 in. to 8 ft 11 in.

Example 6   Add 3 ft 2 in. and 5 ft 11 in.

Solution: To add, we line up the similar units.

$$
\begin{array}{r}
3 \text{ ft} \;\; 2 \text{ in.} \\
+\, 5 \text{ ft } 11 \text{ in.} \\
\hline
8 \text{ ft } 13 \text{ in.}
\end{array}
$$

Since 13 inches is the same as 1 ft 1 in., we have

$$8 \text{ ft } 13 \text{ in.} = 8 \text{ ft} + 1 \text{ ft } 1 \text{ in.}$$
$$= 9 \text{ ft } 1 \text{ in.}$$

■ **Work Practice 6**

✓**Concept Check** How could you estimate the following sum?

$$
\begin{array}{r}
7 \text{ yd} \;\; 4 \text{ in.} \\
+\, 3 \text{ yd } 27 \text{ in.}
\end{array}
$$

Practice 7

Multiply 4 ft 7 in. by 4.

Example 7   Multiply 8 ft 9 in. by 3.

Solution: By the distributive property, we multiply 8 ft by 3 and 9 in. by 3.

$$
\begin{array}{r}
8 \text{ ft} \;\; 9 \text{ in.} \\
\times \qquad\;\; 3 \\
\hline
24 \text{ ft } 27 \text{ in.}
\end{array}
$$

Since 27 in. is the same as 2 ft 3 in., we simplify the product as

$$24 \text{ ft } 27 \text{ in.} = 24 \text{ ft} + 2 \text{ ft } 3 \text{ in.}$$
$$= 26 \text{ ft } 3 \text{ in.}$$

■ **Work Practice 7**

We divide in a similar manner as above.

Practice 8

A carpenter cuts 1 ft 9 in. from a board of length 5 ft 8 in. Find the remaining length of the board.

Example 8   Finding the Length of a Piece of Rope

A rope of length 6 yd 1 ft has 2 yd 2 ft cut from one end. Find the length of the remaining rope.

Solution: Subtract 2 yd 2 ft from 6 yd 1 ft.

$$
\begin{array}{rcl}
\text{beginning length} & \rightarrow & 6 \text{ yd } 1 \text{ ft} \\
-\quad \text{amount cut} & \rightarrow & -2 \text{ yd } 2 \text{ ft} \\
\hline
\text{remaining length} & &
\end{array}
$$

We cannot subtract 2 ft from 1 ft, so we borrow 1 yd from the 6 yd. One yard is converted to 3 ft and combined with the 1 ft already there.

Answers

6. 13 ft 7 in.   7. 18 ft 4 in.
8. 3 ft 11 in.

✓**Concept Check Answer**

round each to the nearest yard:
7 yd + 4 yd = 11 yd

Borrow 1 yd = 3 ft

5 yd + (1 yd)(3 ft)

$$
\begin{array}{r}
6 \text{ yd } 1 \text{ ft} = 5 \text{ yd } 4 \text{ ft} \\
-2 \text{ yd } 2 \text{ ft} = -2 \text{ yd } 2 \text{ ft} \\
\hline
3 \text{ yd } 2 \text{ ft}
\end{array}
$$

The remaining rope is 3 yd 2 ft long.

■ Work Practice 8

Objective D Defining and Converting Metric System Units of Length ▶

The basic unit of length in the metric system is the **meter.** A meter is slightly longer than a yard. It is approximately 39.37 inches long. Recall that a yard is 36 inches long.

1 yard = 36 inches

1 meter ≈ 39.37 inches

All units of length in the metric system are based on the meter. The following is a summary of the prefixes used in the metric system. Also shown are equivalencies between units of length. Like the decimal system, the metric system uses powers of 10 to define units.

Metric Units of Length
1 **kilo**meter (km) = 1000 meters (m)
1 **hecto**meter (hm) = 100 m
1 **deka**meter (dam) = 10 m
1 meter (m) = 1 m
1 **deci**meter (dm) = 1/10 m or 0.1 m
1 **centi**meter (cm) = 1/100 m or 0.01 m
1 **milli**meter (mm) = 1/1000 m or 0.001 m

The figure below will help you with decimeters, centimeters, and millimeters.

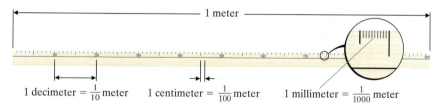

1 decimeter = $\frac{1}{10}$ meter　　1 centimeter = $\frac{1}{100}$ meter　　1 millimeter = $\frac{1}{1000}$ meter

Helpful Hint

Study the figure above for other equivalencies between metric units of length.

10 decimeters = 1 meter　　　10 millimeters = 1 centimeter
100 centimeters = 1 meter　　　10 centimeters = 1 decimeter
1000 millimeters = 1 meter

These same prefixes are used in the metric system for mass and capacity. The most commonly used measurements of length in the metric system are the **meter, millimeter, centimeter,** and **kilometer.**

✓**Concept Check** Is this statement reasonable? "The screen of a home television set has a 30-meter diagonal." Why or why not?

Being comfortable with the metric units of length means gaining a "feeling" for metric lengths, just as you have a "feeling" for the lengths of an inch, a foot, and a mile. To help you accomplish this, study the following examples:

- A millimeter is about the thickness of a large paper clip.

- A centimeter is about the width of a large paper clip.

- A meter is slightly longer than a yard.

- A kilometer is about two-thirds of a mile.

- The width of this book is approximately 21.5 centimeters.

- The distance between New York City and Philadelphia is about 160 kilometers.

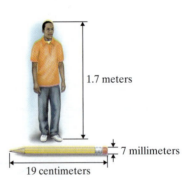

1.7 meters

19 centimeters

7 millimeters

$2\frac{1}{2}$ centimeters is about 1 inch.

2.54 cm

1 inch

As with the U.S. system of measurement, unit fractions may be used to convert from one unit of length to another. For example, let's convert 1200 meters to kilometers. To do so, we will multiply by 1 in the form of the unit fraction

$$\frac{1 \text{ km}}{1000 \text{ m}} \quad \begin{array}{l}\leftarrow \text{Units to convert to}\\ \leftarrow \text{Original units}\end{array}$$

Unit fraction

$$1200 \text{ m} = \frac{1200 \text{ m}}{1} \cdot 1 = \frac{1200 \text{ m}}{1} \cdot \frac{1 \text{ km}}{1000 \text{ m}} = \frac{1200 \text{ km}}{1000} = 1.2 \text{ km}$$

The metric system does, however, have a distinct advantage over the U.S. system of measurement: the ease of converting from one unit of length to another. Since all units of length are powers of 10 of the meter, converting from one unit of length to another is as simple as moving the decimal point. Listing units of length in

✓**Concept Check Answer**

no; answers may vary

order from largest to smallest helps to keep track of how many places to move the decimal point when converting.

Let's again convert 1200 meters to kilometers. This time, to convert from meters to kilometers, we move along the chart shown, 3 units to the left, from meters to kilometers. This means that we move the decimal point 3 places to the left.

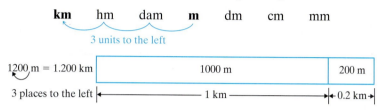

Thus, 1200 m = 1.2 km, as shown in the diagram.

Example 9 Convert 2.3 m to centimeters.

Solution: First we will convert by using a unit fraction.

$$2.3 \text{ m} = \frac{2.3 \ \cancel{m}}{1} \cdot \overbrace{\frac{100 \text{ cm}}{1 \ \cancel{m}}}^{\text{Unit fraction}} = 230 \text{ cm}$$

Now we will convert by listing the units of length in order from left to right and moving from meters to centimeters.

km   hm   dam   m   dm   cm   mm

2 units to the right

2.30 m = 230. cm

2 places to the right

With either method, we get 230 cm.

■ **Work Practice 9**

Example 10 Convert 450,000 mm to meters.

Solution: We list the units of length in order from left to right and move from millimeters to meters.

km   hm   dam   m   dm   cm   mm

3 units to the left

Thus, move the decimal point 3 places to the left.

450,000 mm = 450.000 m or 450 m

■ **Work Practice 10**

✓**Concept Check** What is wrong with the following conversion of 150 cm to meters?

150.00 cm = 15,000 m

Objective E Performing Operations on Metric System Units of Length ▶

To add, subtract, multiply, or divide with metric measurements of length, we write all numbers using the same unit of length and then add, subtract, multiply, or divide as with decimals.

Practice 9

Convert 2.5 m to millimeters.

Practice 10

Convert 3500 m to kilometers.

Practice 11

Subtract 640 m from 2.1 km.

Example 11   Subtract 430 m from 1.3 km.

Solution: First we convert both measurements to kilometers or both to meters.

$$430\text{ m} = 0.43\text{ km} \qquad \text{or} \qquad 1.3\text{ km} = 1300\text{ m}$$

$$\begin{array}{r} 1.30\text{ km} \\ -0.43\text{ km} \\ \hline 0.87\text{ km} \end{array} \qquad\qquad \begin{array}{r} 1300\text{ m} \\ -430\text{ m} \\ \hline 870\text{ m} \end{array}$$

The difference is 0.87 km or 870 m.

■ **Work Practice 11**

Practice 12

Multiply 18.3 hm by 5.

Example 12   Multiply 5.7 mm by 4.

Solution: Here we simply multiply the two numbers. Note that the unit of measurement remains the same.

$$\begin{array}{r} 5.7\text{ mm} \\ \times\quad 4 \\ \hline 22.8\text{ mm} \end{array}$$

■ **Work Practice 12**

Practice 13

Doris Blackwell is knitting a scarf that is currently 0.8 meter long. If she knits an additional 45 centimeters, how long will the scarf be?

Answers

11. 1.46 km or 1460 m   **12.** 91.5 hm
13. 125 cm or 1.25 m

Example 13   Finding a Person's Height

Fritz Martinson was 1.2 meters tall on his last birthday. Since then, he has grown 14 centimeters. Find his current height in meters.

Solution:

$$\begin{array}{lll} \text{original height} & \rightarrow & 1.20\text{ m} \\ +\ \text{height grown} & \rightarrow & +\ 0.14\text{ m} \quad (\text{Since }14\text{ cm} = 0.14\text{ m}) \\ \hline \text{current height} & & 1.34\text{ m} \end{array}$$

Fritz is now 1.34 meters tall.

■ **Work Practice 13**

Vocabulary, Readiness & Video Check

Use the choices below to fill in each blank. Some choices may be used more than once.

inches   yard   unit fraction

feet   meter

1. The basic unit of length in the metric system is the _____.

2. The expression $\dfrac{1\text{ foot}}{12\text{ inches}}$ is an example of a(n) _____.

3. A meter is slightly longer than a(n) _____.

4. One foot equals 12 _____.

5. One yard equals 3 _____.

6. One yard equals 36 _____.

7. One mile equals 5280 _____.

Martin-Gay Interactive Videos   Watch the section lecture video and answer the following questions.

See Video 9.4

Objective A   **8.** In Example 3, what units are used in the denominator of the unit fraction and why was this decided?

Objective B   **9.** In Example 4, how is a mixed unit similar to a mixed number? Use examples in your answer.

Objective C   **10.** In Example 5, why is the sum of the addition problem not the final answer? Reference the sum in your answer.

Objective D   **11.** In the lecture before Example 6, why is it easier to convert metric units than U.S. units?

Objective E   **12.** What two answers did we get for Example 8? Explain why both answers are correct.

9.4 Exercise Set MyMathLab®

Objective A *Convert each measurement as indicated. See Examples 1 through 3.*

1. 60 in. to feet

2. 84 in. to feet

3. 12 yd to feet

4. 18 yd to feet

5. 42,240 ft to miles

6. 36,960 ft to miles

7. $8\frac{1}{2}$ ft to inches

8. $12\frac{1}{2}$ ft to inches

9. 10 ft to yards

10. 25 ft to yards

11. 6.4 mi to feet

12. 3.8 mi to feet

13. 162 in. to yd (Write answer as a decimal.)

14. 7216 yd to mi (Write answer as a decimal.)

15. 3 in. to ft (Write answer as a decimal.)

16. 129 in. to ft (Write answer as a decimal.)

Objective B *Convert each measurement as indicated. See Examples 4 and 5.*

17. 40 ft = _____ yd _____ ft

18. 100 ft = _____ yd _____ ft

19. 85 in. = _____ ft _____ in.

20. 59 in. = _____ ft _____ in.

21. 10,000 ft = _____ mi _____ ft

22. 25,000 ft = _____ mi _____ ft

23. 5 ft 2 in. = _____ in.

24. 4 ft 11 in. = _____ in.

25. 8 yd 2 ft = _____ ft

26. 4 yd 1 ft = _____ ft

27. 2 yd 1 ft = _____ in.

28. 1 yd 2 ft = _____ in.

Objective C *Perform each indicated operation. Simplify the result if possible. See Examples 6 through 8.*

29. 3 ft 10 in. + 7 ft 4 in.

30. 12 ft 7 in. + 9 ft 11 in.

▶ **31.** 12 yd 2 ft + 9 yd 2 ft

32. 16 yd 2 ft + 8 yd 2 ft

33. 22 ft 8 in. − 16 ft 3 in.

34. 15 ft 5 in. − 8 ft 2 in.

35. 18 ft 3 in. − 10 ft 9 in.

36. 14 ft 8 in. − 3 ft 11 in.

37. 28 ft 8 in. ÷ 2

38. 34 ft 6 in. ÷ 2

39. 16 yd 2 ft × 5

40. 15 yd 1 ft × 8

Objective D *Convert as indicated. See Examples 9 and 10.*

41. 60 m to centimeters

42. 46 m to centimeters

43. 40 mm to centimeters

44. 14 mm to centimeters

45. 500 m to kilometers

46. 400 m to kilometers

47. 1700 mm to meters

48. 6400 mm to meters

▶ **49.** 1500 cm to meters

50. 6400 cm to meters

51. 0.42 km to centimeters

52. 0.95 km to centimeters

53. 7 km to meters

54. 5 km to meters

55. 8.3 cm to millimeters

56. 4.6 cm to millimeters

57. 20.1 mm to decimeters

58. 140.2 mm to decimeters

▶ **59.** 0.04 m to millimeters

60. 0.2 m to millimeters

Objective E *Perform each indicated operation. Remember to insert units when writing your answers. See Examples 11 through 13.*

61. 8.6 m + 0.34 m

62. 14.1 cm + 3.96 cm

63. 2.9 m + 40 mm

64. 30 cm + 8.9 m

▶ **65.** 24.8 mm − 1.19 cm

66. 45.3 m − 2.16 dam

67. 15 km − 2360 m

68. 14 cm − 15 mm

69. 18.3 m × 3

70. 14.1 m × 4

71. 6.2 km ÷ 4

72. 9.6 m ÷ 5

Objectives A C D E Mixed Practice *Solve. Remember to insert units when writing your answers. For Exercises 73 through 82, complete the charts. See Examples 1 through 13.*

		Yards	Feet	Inches
73.	Chrysler Building in New York City		1046	
74.	4-story building			792
75.	Python length		35	
76.	Ostrich height			108

		Meters	Millimeters	Kilometers	Centimeters
77.	Length of elephant	5			
78.	Height of grizzly bear	3			
79.	Tennis ball diameter				6.5
80.	Golf ball diameter				4.6
81.	Distance from London to Paris			342	
82.	Distance from Houston to Dallas			396	

83. The National Zoo maintains a small patch of bamboo, which it grows as a food supply for its pandas. Two weeks ago, the bamboo was 6 ft 10 in. tall. Since then, the bamboo has grown 3 ft 8 in. How tall is the bamboo now?

84. While exploring in the Marianas Trench, a submarine probe was lowered to a point 1 mile 1400 feet below the ocean's surface. Later it was lowered an additional 1 mile 4000 feet below this point. How far was the probe below the surface of the Pacific?

85. At its deepest point, the Grand Canyon of the Colorado River in Arizona is about 6000 ft. The Grand Canyon of the Yellowstone River, which is in Yellowstone National Park in Wyoming, is at most 900 feet deep. How much deeper is the Grand Canyon of the Colorado River than the Grand Canyon of the Yellowstone River? (*Source:* National Park Service)

86. The Grand Canyon of the Gunnison River, in Colorado, is often called the Black Canyon of the Gunnison because it is so steep that light rarely penetrates the depth of the canyon. The Black Canyon of the Gunnison is only 1150 ft wide at its narrowest point. At its narrowest, the Grand Canyon of the Yellowstone is $\frac{1}{2}$ mile wide. Find the difference in width between the Grand Canyon of the Yellowstone and the Black Canyon of the Gunnison. (*Note:* Notice that the dimensions are different.) (*Source:* National Park Service)

87. The tallest man in the world is recorded as Robert Pershing Wadlow of Alton, Illinois. Born in 1918, he measured 8 ft 11 in. at his tallest. The shortest man in the world is Chandra Bahadur Dangi of Nepal, who measures 21.5 in. How many times taller than Chandra is Robert? Round to one decimal place. (*Source: Guinness World Records*)

88. A 3.4-m rope is attached to a 5.8-m rope. However, when the ropes are tied, 8 cm of length is lost to form the knot. What is the length of the tied ropes?

89. The ice on a pond is 5.33 cm thick. For safe skating, the owner of the pond insists that it be 80 mm thick. How much thicker must the ice be before skating is allowed?

90. The sediment on the bottom of the Towamencin Creek is normally 14 cm thick, but a recent flood washed away 22 mm of sediment. How thick is it now?

91. The Amana Corporation stacks up its microwave ovens in a distribution warehouse. Each stack is 1 ft 9 in. wide. How far from the wall would 9 of these stacks extend?

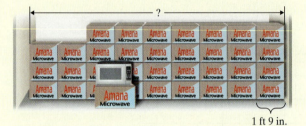

1 ft 9 in.

92. The highway commission is installing concrete sound barriers along a highway. Each barrier is 1 yd 2 ft long. Find the total length of 25 barriers placed end to end.

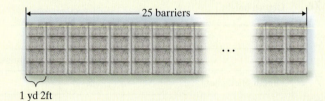

1 yd 2ft

93. A logging firm needs to cut a 67-m-long redwood log into 20 equal pieces before loading it onto a truck for shipment. How long will each piece be?

94. A 112.5-foot-tall dead pinewood tree is removed by starting at the top and cutting off 9-foot-long sections. How many whole sections are removed?

95. The longest truck in the world is operated by Gould Transport in Australia, and is the 182-ft Road Train. How many *yards* long are 2 of these trucks? (*Source: Guinness World Records*)

96. The large Coca-Cola sign in the Tokyo Ginza shopping district is in the shape of a rectangle whose length is 31 yards and whose width is 49 feet. Find the area of the sign in square feet. (*Source:* Coca-Cola Company) (*Hint:* Recall that the area of a rectangle is the product length times width.)

Review

Write each decimal as a fraction and each fraction as a decimal. See Sections 5.1 and 5.5.

97. 0.21

98. 0.86

99. $\dfrac{13}{100}$

100. $\frac{47}{100}$ **101.** $\frac{1}{4}$ **102.** $\frac{3}{20}$

Concept Extensions

Determine whether the measurement in each statement is reasonable.

103. The width of a twin-size bed is 20 meters.

104. A window measures 1 meter by 0.5 meter.

105. A drinking glass is made of glass 2 millimeters thick.

106. A paper clip is 4 kilometers long.

107. The distance across the Colorado River is 50 kilometers.

108. A model's hair is 30 centimeters long.

Estimate each sum or difference. See the first Concept Check in this section.

109.
$$\begin{array}{r} 5 \text{ yd } 2 \text{ in.} \\ + 7 \text{ yd } 30 \text{ in.} \\ \hline \end{array}$$

110.
$$\begin{array}{r} 45 \text{ ft } 1 \text{ in.} \\ - 10 \text{ ft } 11 \text{ in.} \\ \hline \end{array}$$

111. Using a unit other than the foot, write a length that is equivalent to 4 feet. (*Hint:* There are many possibilities.)

112. Using a unit other than the meter, write a length that is equivalent to 7 meters. (*Hint:* There are many possibilities.)

113. To convert from meters to centimeters, the decimal point is moved two places to the right. Explain how this relates to the fact that the prefix *centi* means $\frac{1}{100}$.

114. Explain why conversions in the metric system are easier to make than conversions in the U.S. system of measurement.

115. An advertisement sign outside Fenway Park in Boston measures 18.3 m by 18.3 m. What is the area of this sign?

Copyright 2015 Pearson Education, Inc.

Objectives

A Define U.S. Units of Weight and Convert from One Unit to Another.

B Perform Arithmetic Operations on U.S. Units of Weight.

C Define Metric Units of Mass and Convert from One Unit to Another.

D Perform Arithmetic Operations on Metric Units of Mass.

Objective A Defining and Converting U.S. System Units of Weight

Whenever we talk about how heavy an object is, we are concerned with the object's **weight.** We discuss weight when we refer to a 12-ounce box of Rice Krispies, a 15-pound tabby cat, or a barge hauling 24 tons of garbage.

12 ounces

15 pounds

24 tons of garbage

The most common units of weight in the U.S. measurement system are the **ounce,** the **pound,** and the **ton.** The following is a summary of equivalencies between units of weight:

U.S. Units of Weight	Unit Fractions
16 ounces (oz) = 1 pound (lb)	$\dfrac{16 \text{ oz}}{1 \text{ lb}} = \dfrac{1 \text{ lb}}{16 \text{ oz}} = 1$
2000 pounds = 1 ton	$\dfrac{2000 \text{ lb}}{1 \text{ ton}} = \dfrac{1 \text{ ton}}{2000 \text{ lb}} = 1$

✔**Concept Check** If you were describing the weight of a fully loaded semi-trailer, which type of unit would you use: ounce, pound, or ton? Why?

Unit fractions that equal 1 are used to convert between units of weight in the U.S. system. When converting using unit fractions, recall that the numerator of a unit fraction should contain the units we are converting to and the denominator should contain the original units.

Practice 1

Convert 6500 pounds to tons.

Answer

1. $3\dfrac{1}{4}$ tons

✔**Concept Check Answer**
ton

Example 1 Convert 9000 pounds to tons.

Solution: We multiply 9000 lb by a unit fraction that uses the equality 2000 pounds = 1 ton.

Remember, the unit fraction should be $\dfrac{\text{units to convert to}}{\text{original units}}$ or $\dfrac{1 \text{ ton}}{2000 \text{ lb}}$.

$$9000 \text{ lb} = \frac{9000 \text{ lb}}{1} \cdot 1 = \frac{9000 \text{ l\!b}}{1} \cdot \frac{1 \text{ ton}}{2000 \text{ l\!b}} = \frac{9000 \text{ tons}}{2000} = \frac{9}{2} \text{ tons or } 4\frac{1}{2} \text{ tons}$$

2000 lb   2000 lb   2000 lb   2000 lb   1000 lb

9000 lb = $4\frac{1}{2}$ tons

1 ton   1 ton   1 ton   1 ton $\frac{1}{2}$ ton

■ **Work Practice 1**

Example 2   Convert 3 pounds to ounces.

Solution: We multiply by the unit fraction $\frac{16 \text{ oz}}{1 \text{ lb}}$ to convert from pounds to ounces.

$$3 \text{ lb} = \frac{3 \text{ lb}}{1} \cdot 1 = \frac{3 \text{ lb}}{1} \cdot \frac{16 \text{ oz}}{1 \text{ lb}} = 3 \cdot 16 \text{ oz} = 48 \text{ oz}$$

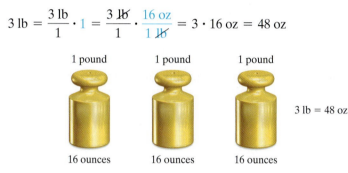

1 pound   1 pound   1 pound

3 lb = 48 oz

16 ounces   16 ounces   16 ounces

■ **Work Practice 2**

As with length, it is sometimes useful to simplify a measurement of weight by writing it in terms of mixed units.

Example 3   Convert: 33 ounces = _____ lb _____ oz

Solution: Because 16 oz = 1 lb, divide 16 into 33 to see how many pounds are in 33 ounces. The quotient is the number of pounds, and the remainder is the number of ounces. To see why we divide 16 into 33, notice that

$$33 \text{ oz} = 33 \text{ oz} \cdot \frac{1 \text{ lb}}{16 \text{ oz}} = \frac{33}{16} \text{ lb}$$

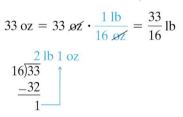

Thus, 33 ounces is the same as 2 lb 1 oz.

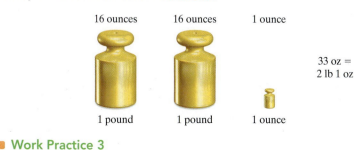

16 ounces   16 ounces   1 ounce

33 oz = 2 lb 1 oz

1 pound   1 pound   1 ounce

■ **Work Practice 3**

Practice 2

Convert 72 ounces to pounds.

Practice 3

Convert:
47 ounces = _____ lb _____ oz

Answers

2. $4\frac{1}{2}$ lb   **3.** 2 lb 15 oz

Objective B Performing Operations on U.S. System Units of Weight ▶

Performing arithmetic operations on units of weight works the same way as performing arithmetic operations on units of length.

Practice 4

Subtract 5 tons 1200 lb from 8 tons 100 lb.

Example 4   Subtract 3 tons 1350 lb from 8 tons 1000 lb.

Solution: To subtract, we line up similar units.

$$
\begin{array}{r}
8 \text{ tons } 1000 \text{ lb} \\
- \ 3 \text{ tons } 1350 \text{ lb} \\
\hline
\end{array}
$$

Since we cannot subtract 1350 lb from 1000 lb, we borrow 1 ton from the 8 tons. To do so, we write 1 ton as 2000 lb and combine it with the 1000 lb.

7 tons + 1 ton 2000 lb

$$
\begin{array}{rclcr}
8 \text{ tons } 1000 \text{ lb} & = & 7 \text{ tons } 3000 \text{ lb} \\
- \ 3 \text{ tons } 1350 \text{ lb} & = & - \ 3 \text{ tons } 1350 \text{ lb} \\
\hline
& & 4 \text{ tons } 1650 \text{ lb}
\end{array}
$$

To check, see that the sum of 4 tons 1650 lb and 3 tons 1350 lb is 8 tons 1000 lb.

▪ **Work Practice 4**

Practice 5

Divide 5 lb 8 oz by 4.

Example 5   Divide 9 lb 6 oz by 2.

Solution: We divide each of the units by 2.

$$
\begin{array}{r}
4 \text{ lb} \quad 11 \text{ oz} \\
2 \overline{) \ 9 \text{ lb} \quad \ 6 \text{ oz}} \\
-8 \qquad\qquad \\
\hline
1 \text{ lb} = 16 \text{ oz} \\
\hline
22 \text{ oz} \quad \text{Divide 2 into 22 oz to get 11 oz.}
\end{array}
$$

To check, multiply 4 pounds 11 ounces by 2. The result is 9 pounds 6 ounces.

▪ **Work Practice 5**

Practice 6

A 5-lb 14-oz batch of cookies is packed into a 6-oz container before it is mailed. Find the total weight.

Example 6   Finding the Weight of a Child

Bryan weighed 8 lb 8 oz at birth. By the time he was 1 year old, he had gained 11 lb 14 oz. Find his weight at age 1 year.

Solution:

$$
\begin{array}{rcl}
\text{birth weight} & \rightarrow & 8 \text{ lb } \ 8 \text{ oz} \\
+ \text{ weight gained} & \rightarrow & + \ 11 \text{ lb } 14 \text{ oz} \\
\hline
\text{total weight} & \rightarrow & 19 \text{ lb } 22 \text{ oz}
\end{array}
$$

Since 22 oz equals 1 lb 6 oz,

$$
19 \text{ lb } 22 \text{ oz} = 19 \text{ lb} + 1 \text{ lb } 6 \text{ oz}
$$
$$
= 20 \text{ lb } 6 \text{ oz}
$$

Bryan weighed 20 lb 6 oz on his first birthday.

▪ **Work Practice 6**

Objective C Defining and Converting Metric System Units of Mass

In scientific and technical areas, a careful distinction is made between **weight** and **mass. Weight** is really a measure of the pull of gravity. The farther from Earth an object gets, the less it weighs. However, **mass** is a measure of the amount of substance in the object and does not change. Astronauts orbiting Earth weigh much less than they weigh on Earth, but they have the same mass in orbit as they do on Earth. Here on Earth, weight and mass are the same, so either term may be used.

The basic unit of mass in the metric system is the **gram.** It is defined as the mass of water contained in a cube 1 centimeter (cm) on each side.

Actual size

1 cm

1 cm

1 cm

1 cubic centimeter

The following examples may help you get a feeling for metric masses:

- A tablet contains 200 milligrams of ibuprofen.

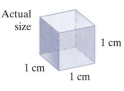

- A large paper clip weighs approximately 1 gram.

- A box of crackers weighs 453 grams.

A kilogram is slightly over 2 pounds. An adult woman may weigh 60 kilograms.

The prefixes for units of mass in the metric system are the same as for units of length, as shown in the following table:

Metric Units of Mass
1 **kilo**gram (kg) = 1000 grams (g)
1 **hecto**gram (hg) = 100 g
1 **deka**gram (dag) = 10 g
1 gram (g) = 1 g
1 **deci**gram (dg) = 1/10 g or 0.1 g
1 **centi**gram (cg) = 1/100 g or 0.01 g
1 **milli**gram (mg) = 1/1000 g or 0.001 g

✓**Concept Check** True or false? A decigram is larger than a dekagram. Explain.

The **milligram,** the **gram,** and the **kilogram** are the three most commonly used units of mass in the metric system.

As with lengths, all units of mass are powers of 10 of the gram, so converting from one unit of mass to another only involves moving the decimal point. To convert from one unit of mass to another in the metric system, list the units of mass in order from largest to smallest.

✓**Concept Check Answer**
false

Let's convert 4300 milligrams to grams. To convert from milligrams to grams, we move along the list 3 units to the left.

kg   hg   dag   **g**   dg   cg   **mg**

3 units to the left

This means that we move the decimal point 3 places to the left to convert from milligrams to grams.

$$4300 \text{ mg} = 4.3 \text{ g}$$

Don't forget, the same conversion can be done with unit fractions.

$$4300 \text{ mg} = \frac{4300 \text{ mg}}{1} \cdot 1 = \frac{4300 \text{ mg}}{1} \cdot \frac{0.001 \text{ g}}{1 \text{ mg}}$$

$$= 4300 \cdot 0.001 \text{ g}$$

$$= 4.3 \text{ g} \quad \text{To multiply by 0.001, move the decimal point 3 places to the left.}$$

To see that this is reasonable, study the diagram:

1000 mg   1000 mg   1000 mg   1000 mg   300 mg

1 g   1 g   1 g   1 g   0.3 g

4300 mg = 4.3 g

Thus, 4300 mg = 4.3 g

Practice 7

Convert 3.41 g to milligrams.

Example 7   Convert 3.2 kg to grams.

Solution: First we convert by using a unit fraction.

Unit fraction

$$3.2 \text{ kg} = 3.2 \text{ kg} \cdot 1 = 3.2 \text{ kg} \cdot \frac{1000 \text{ g}}{1 \text{ kg}} = 3200 \text{ g}$$

Now let's list the units of mass in order from left to right and move from kilograms to grams.

kg   hg   dag   g   dg   cg   mg

3 units to the right

$$3.200 \text{ kg} = 3200. \text{ g}$$

3 places to the right

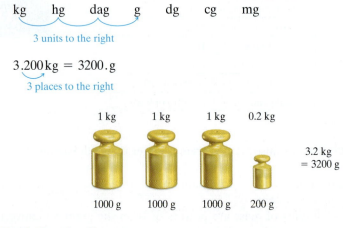

1 kg   1 kg   1 kg   0.2 kg

3.2 kg = 3200 g

1000 g   1000 g   1000 g   200 g

■ **Work Practice 7**

Answer

7. 3410 mg

Example 8 Convert 2.35 cg to grams.

Solution: We list the units of mass in a chart and move from centigrams to grams.

kg   hg   dag   g   dg   cg   mg

2 units to the left

02.35 cg = 0.0235 g

2 places to the left

■ **Work Practice 8**

Objective D Performing Operations on Metric System Units of Mass

Arithmetic operations can be performed with metric units of mass just as we performed operations with metric units of length. We convert each number to the same unit of mass and add, subtract, multiply, or divide as with decimals.

Example 9 Subtract 5.4 dg from 1.6 g.

Solution: We convert both numbers to decigrams or to grams before subtracting.

5.4 dg = 0.54 g   or   1.6 g = 16 dg

1.60 g   16.0 dg
−0.54 g   −5.4 dg
1.06 g   10.6 g

The difference is 1.06 g or 10.6 dg.

■ **Work Practice 9**

Example 10 Calculating Allowable Weight in an Elevator

An elevator has a weight limit of 1400 kg. A sign posted in the elevator indicates that the maximum capacity of the elevator is 17 persons. What is the average allowable weight for each passenger, rounded to the nearest kilogram?

Solution: To solve, notice that the total weight of 1400 kilograms ÷ 17 = average weight.

$$
\begin{array}{r}
82.3 \text{ kg} \approx 82 \text{ kg} \\
17\overline{)1400.0 \text{ kg}} \\
-136 \\
\overline{40} \\
-34 \\
\overline{60} \\
-51 \\
\overline{9}
\end{array}
$$

PASSENGER
ELEVATOR
17 PERSONS
1400 kg

11A

Each passenger can weigh an average of 82 kg. (Recall that a kilogram is slightly over 2 pounds, so 82 kilograms is over 164 pounds.)

■ **Work Practice 10**

Practice 8

Convert 56.2 cg to grams.

Practice 9

Subtract 3.1 dg from 2.5 g.

Practice 10

Twenty-four bags of cement weigh a total of 550 kg. Find the average weight of 1 bag, rounded to the nearest kilogram.

Answers
8. 0.562 g   **9.** 2.19 g or 21.9 dg
10. 23 kg

Object	Grams	Kilograms	Milligrams	Centigrams
71. Capsule of amoxicillin (antibiotic)			500	
72. Tablet of Topamax (epilepsy and migraine uses)			25	
73. A six-year-old boy		21		
74. A golf ball	45			

75. A can of 7-Up weighs 336 grams. Find the weight in kilograms of 24 cans.

76. Guy Green normally weighs 73 kg, but he lost 2800 grams after being sick with the flu. Find Guy's new weight.

77. Sudafed is a decongestant that comes in two strengths. Regular strength contains 60 mg of medication. Extra strength contains 0.09 g of medication. How much extra medication is in the extra-strength tablet?

78. A small can of Planters sunflower seeds weighs 177 g. If each can contains 6 servings, find the weight of one serving.

79. Doris Johnson has two open containers of rice. If she combines 1 lb 10 oz from one container with 3 lb 14 oz from the other container, how much total rice does she have?

80. Dru Mizel maintains the records of the amount of coal delivered to his department in the steel mill. In January, 3 tons 1500 lb were delivered. In February, 2 tons 1200 lb were delivered. Find the total amount delivered in these two months.

81. Carla Hamtini was amazed when she grew a 28-lb 10-oz zucchini in her garden, but later she learned that the heaviest zucchini ever grown weighed 64 lb 8 oz in Llanharry, Wales, by B. Lavery in 1990. How far below the record weight was Carla's zucchini? (*Source: Guinness World Records*)

82. The heaviest baby born in good health weighed an incredible 22 lb 8 oz. He was born in Italy in September 1955. How much heavier is this than a 7-lb 12-oz baby? (*Source: Guinness World Records*)

83. The smallest baby born in good health weighed only 8.6 ounces, less than a can of soda. She was born in Chicago in December 2004. How much lighter was she than an average baby, who weighs about 7 lb 8 ounces?

84. A large bottle of Hire's Root Beer weighs 1900 grams. If a carton contains 6 large bottles of root beer, find the weight in kilograms of 5 cartons.

Example 8 Convert 2.35 cg to grams.

Solution: We list the units of mass in a chart and move from centigrams to grams.

kg　　hg　　dag　　g　　dg　　cg　　mg

2 units to the left

$02.35 \, \text{cg} = 0.0235 \, \text{g}$

2 places to the left

◼ Work Practice 8

Objective D Performing Operations on Metric System Units of Mass ▶

Arithmetic operations can be performed with metric units of mass just as we performed operations with metric units of length. We convert each number to the same unit of mass and add, subtract, multiply, or divide as with decimals.

Example 9 Subtract 5.4 dg from 1.6 g.

Solution: We convert both numbers to decigrams or to grams before subtracting.

$5.4 \, \text{dg} = 0.54 \, \text{g}$　　or　　$1.6 \, \text{g} = 16 \, \text{dg}$

$$\begin{array}{r} 1.60 \, \text{g} \\ -0.54 \, \text{g} \\ \hline 1.06 \, \text{g} \end{array} \qquad \begin{array}{r} 16.0 \, \text{dg} \\ -5.4 \, \text{dg} \\ \hline 10.6 \, \text{g} \end{array}$$

The difference is 1.06 g or 10.6 dg.

◼ Work Practice 9

Example 10 Calculating Allowable Weight in an Elevator

An elevator has a weight limit of 1400 kg. A sign posted in the elevator indicates that the maximum capacity of the elevator is 17 persons. What is the average allowable weight for each passenger, rounded to the nearest kilogram?

Solution: To solve, notice that the total weight of 1400 kilograms ÷ 17 = average weight.

$$\begin{array}{r} 82.3 \, \text{kg} \approx 82 \, \text{kg} \\ 17\overline{)1400.0 \, \text{kg}} \\ \underline{-136} \\ 40 \\ \underline{-34} \\ 60 \\ \underline{-51} \\ 9 \end{array}$$

PASSENGER
ELEVATOR
17 PERSONS
1400 kg

11A

Each passenger can weigh an average of 82 kg. (Recall that a kilogram is slightly over 2 pounds, so 82 kilograms is over 164 pounds.)

◼ Work Practice 10

Vocabulary, Readiness & Video Check

Use the choices below to fill in each blank.

    mass   weight   gram

1. _____ is a measure of the amount of substance in an object. This measure does not change.

2. _____ is the measure of the pull of gravity.

3. The basic unit of mass in the metric system is the _____.

Fill in these blanks with the correct number. Choices for these blanks are not shown in the list of terms above.

4. One pound equals _____ ounces.

5. One ton equals _____ pounds.

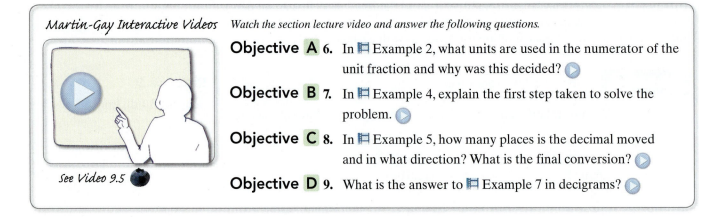

Martin-Gay Interactive Videos

See Video 9.5

Watch the section lecture video and answer the following questions.

Objective A 6. In ▣ Example 2, what units are used in the numerator of the unit fraction and why was this decided? ▶

Objective B 7. In ▣ Example 4, explain the first step taken to solve the problem. ▶

Objective C 8. In ▣ Example 5, how many places is the decimal moved and in what direction? What is the final conversion? ▶

Objective D 9. What is the answer to ▣ Example 7 in decigrams? ▶

9.5   Exercise Set   MyMathLab® ▶

Objective A *Convert as indicated. See Examples 1 through 3.*

1. 2 pounds to ounces

2. 5 pounds to ounces

3. 5 tons to pounds

4. 7 tons to pounds

5. 18,000 pounds to tons

6. 28,000 pounds to tons

▶ **7.** 60 ounces to pounds

8. 90 ounces to pounds

9. 3500 pounds to tons

10. 11,000 pounds to tons

11. 12.75 pounds to ounces

12. 9.5 pounds to ounces

▶ **13.** 4.9 tons to pounds

14. 8.3 tons to pounds

15. $4\frac{3}{4}$ pounds to ounces

16. $9\frac{1}{8}$ pounds to ounces

17. 2950 pounds to the nearest tenth of a ton

18. 51 ounces to the nearest tenth of a pound

19. $\frac{4}{5}$ oz to pounds

20. $\frac{1}{4}$ oz to pounds

21. $5\frac{3}{4}$ lb to ounces

22. $2\frac{1}{4}$ lb to ounces

23. 10 lb 1 oz to ounces

24. 7 lb 6 oz to ounces

25. 89 oz = ____ lb ____ oz

26. 100 oz = ____ lb ____ oz

Objective B *Perform each indicated operation. See Examples 4 through 6.*

27. 34 lb 12 oz + 18 lb 14 oz

28. 6 lb 10 oz + 10 lb 8 oz

29. 3 tons 1820 lb + 4 tons 930 lb

30. 1 ton 1140 lb + 5 tons 1200 lb

31. 5 tons 1050 lb − 2 tons 875 lb

32. 4 tons 850 lb − 1 ton 260 lb

33. 12 lb 4 oz − 3 lb 9 oz

34. 45 lb 6 oz − 26 lb 10 oz

35. 5 lb 3 oz × 6

36. 2 lb 5 oz × 5

37. 6 tons 1500 lb ÷ 5

38. 5 tons 400 lb ÷ 4

Objective C *Convert as indicated. See Examples 7 and 8.*

39. 500 g to kilograms

40. 820 g to kilograms

41. 4 g to milligrams

42. 9 g to milligrams

43. 25 kg to grams

44. 18 kg to grams

45. 48 mg to grams

46. 112 mg to grams

47. 6.3 g to kilograms

48. 4.9 g to kilograms

49. 15.14 g to milligrams

50. 16.23 g to milligrams

51. 6.25 kg to grams

52. 3.16 kg to grams

53. 35 hg to centigrams

54. 4.26 cg to dekagrams

Objective D *Perform each indicated operation. Remember to insert units when writing your answers. See Examples 9 and 10.*

55. 3.8 mg + 9.7 mg

56. 41.6 g + 9.8 g

57. 205 mg + 5.61 g

58. 2.1 g + 153 mg

59. 9 g − 7150 mg

60. 6.13 g − 418 mg

61. 1.61 kg − 250 g

62. 4 kg − 2410 g

63. 5.2 kg × 2.6

64. 4.8 kg × 9.3

65. 17 kg ÷ 8

66. 8.25 g ÷ 6

Objectives A B C D **Mixed Practice** *Solve. Remember to insert units when writing your answers. For Exercises 67 through 74, complete the chart. See Examples 1 through 10.*

	Object	Tons	Pounds	Ounces
67.	Statue of Liberty—weight of copper sheeting	100		
68.	Statue of Liberty—weight of steel	125		
69.	A 12-inch cube of osmium (heaviest metal)		1345	
70.	A 12-inch cube of lithium (lightest metal)		32	

	Object	Grams	Kilograms	Milligrams	Centigrams
71.	Capsule of amoxicillin (antibiotic)			500	
72.	Tablet of Topamax (epilepsy and migraine uses)			25	
73.	A six-year-old boy		21		
74.	A golf ball	45			

75. A can of 7-Up weighs 336 grams. Find the weight in kilograms of 24 cans.

76. Guy Green normally weighs 73 kg, but he lost 2800 grams after being sick with the flu. Find Guy's new weight.

77. Sudafed is a decongestant that comes in two strengths. Regular strength contains 60 mg of medication. Extra strength contains 0.09 g of medication. How much extra medication is in the extra-strength tablet?

78. A small can of Planters sunflower seeds weighs 177 g. If each can contains 6 servings, find the weight of one serving.

79. Doris Johnson has two open containers of rice. If she combines 1 lb 10 oz from one container with 3 lb 14 oz from the other container, how much total rice does she have?

80. Dru Mizel maintains the records of the amount of coal delivered to his department in the steel mill. In January, 3 tons 1500 lb were delivered. In February, 2 tons 1200 lb were delivered. Find the total amount delivered in these two months.

81. Carla Hamtini was amazed when she grew a 28-lb 10-oz zucchini in her garden, but later she learned that the heaviest zucchini ever grown weighed 64 lb 8 oz in Llanharry, Wales, by B. Lavery in 1990. How far below the record weight was Carla's zucchini? (*Source: Guinness World Records*)

82. The heaviest baby born in good health weighed an incredible 22 lb 8 oz. He was born in Italy in September 1955. How much heavier is this than a 7-lb 12-oz baby? (*Source: Guinness World Records*)

83. The smallest baby born in good health weighed only 8.6 ounces, less than a can of soda. She was born in Chicago in December 2004. How much lighter was she than an average baby, who weighs about 7 lb 8 ounces?

84. A large bottle of Hire's Root Beer weighs 1900 grams. If a carton contains 6 large bottles of root beer, find the weight in kilograms of 5 cartons.

85. Three milligrams of preservatives are added to a 0.5-kg box of dried fruit. How many milligrams of preservatives are in 3 cartons of dried fruit if each carton contains 16 boxes?

86. One box of Swiss Miss Cocoa Mix weighs 0.385 kg, but 39 grams of this weight is the packaging. Find the actual weight of the cocoa in 8 boxes.

87. A carton of 12 boxes of Quaker Oats Oatmeal weighs 6.432 kg. Each box includes 26 grams of packaging material. What is the actual weight of the oatmeal in the carton?

88. The supermarket prepares hamburger in 85-gram market packages. When Leo Gonzalas gets home, he divides the package in half before refrigerating the meat. How much will each package weigh?

89. The Shop 'n Bag supermarket chain ships hamburger meat by placing 10 packages of hamburger in a box, with each package weighing 3 lb 4 oz. How much will 4 boxes of hamburger weigh?

90. The Quaker Oats Company ships its 1-lb 2-oz boxes of oatmeal in cartons containing 12 boxes of oatmeal. How much will 3 such cartons weigh?

91. A carton of Del Monte Pineapple weighs 55 lb 4 oz, but 2 lb 8 oz of this weight is due to packaging. Find the actual weight of the pineapple in 4 cartons.

92. The Hormel Corporation ships cartons of canned ham weighing 43 lb 2 oz each. Of this weight, 3 lb 4 oz is due to packaging. Find the actual weight of the ham found in 3 cartons.

Review

Write each fraction as a decimal. See Section 5.5.

93. $\dfrac{4}{25}$

94. $\dfrac{3}{5}$

95. $\dfrac{7}{8}$

96. $\dfrac{3}{16}$

Concept Extensions

Determine whether the measurement in each statement is reasonable.

97. The doctor prescribed a pill containing 2 kg of medication.

98. A full-grown cat weighs approximately 15 g.

99. A bag of flour weighs 4.5 kg.

100. A staple weighs 15 mg.

101. A professor weighs less than 150 g.

102. A car weighs 2000 mg.

103. Use a unit other than centigram and write a mass that is equivalent to 25 centigrams. (*Hint:* There are many possibilities.)

104. Use a unit other than pound and write a weight that is equivalent to 4000 pounds. (*Hint:* There are many possibilities.)

True or false? See the second Concept Check in this section.

105. A kilogram is larger than a gram.

106. A decigram is larger than a milligram.

107. Why is the decimal point moved to the right when grams are converted to milligrams?

108. To change 8 pounds to ounces, multiply by 16. Why is this the correct procedure?

Capacity

Objectives

A Define U.S. Units of Capacity and Convert from One Unit to Another. ▶

B Perform Arithmetic Operations on U.S. Units of Capacity. ▶

C Define Metric Units of Capacity and Convert from One Unit to Another. ▶

D Perform Arithmetic Operations on Metric Units of Capacity. ▶

Objective **A** Defining and Converting U.S. System Units of Capacity ▶

Units of **capacity** are generally used to measure liquids. The number of gallons of gasoline needed to fill a gas tank in a car, the number of cups of water needed in a bread recipe, and the number of quarts of milk sold each day at a supermarket are all examples of using units of capacity. The following summary shows equivalencies between units of capacity:

> ### U.S. Units of Capacity
>
> $$8 \text{ fluid ounces (fl oz)} = 1 \text{ cup (c)}$$
> $$2 \text{ cups} = 1 \text{ pint (pt)}$$
> $$2 \text{ pints} = 1 \text{ quart (qt)}$$
> $$4 \text{ quarts} = 1 \text{ gallon (gal)}$$

Just as with units of length and weight, we can form unit fractions to convert between different units of capacity. For instance,

$$\frac{2 \text{ c}}{1 \text{ pt}} = \frac{1 \text{ pt}}{2 \text{ c}} = 1 \quad \text{and} \quad \frac{2 \text{ pt}}{1 \text{ qt}} = \frac{1 \text{ qt}}{2 \text{ pt}} = 1$$

Practice 1

Convert 43 pints to quarts.

Example 1 Convert 9 quarts to gallons.

Solution: We multiply by the unit fraction $\dfrac{1 \text{ gal}}{4 \text{ qt}}$.

$$9 \text{ qt} = \frac{9 \text{ qt}}{1} \cdot 1$$

$$= \frac{9 \text{ qt}}{1} \cdot \frac{1 \text{ gal}}{4 \text{ qt}}$$

$$= \frac{9 \text{ gal}}{4}$$

$$= 2\frac{1}{4} \text{ gal}$$

Thus, 9 quarts is the same as $2\dfrac{1}{4}$ gallons, as shown in the diagram:

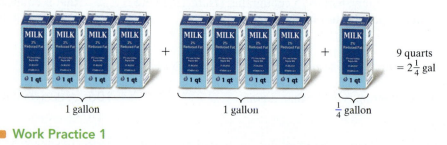

1 gallon + 1 gallon + $\frac{1}{4}$ gallon

9 quarts = $2\frac{1}{4}$ gal

■ **Work Practice 1**

Answer

1. $21\frac{1}{2}$ qt

Example 2   Convert 14 cups to quarts.

Solution: Our equivalency table contains no direct conversion from cups to quarts. However, from this table we know that

$$1 \text{ qt} = 2 \text{ pt} = \frac{2 \text{ pt}}{1} \cdot 1 = \frac{2 \text{ pt}}{1} \cdot \frac{2 \text{ c}}{1 \text{ pt}} = 4 \text{ c}$$

so 1 qt = 4 c. Now we have the unit fraction $\dfrac{1 \text{ qt}}{4 \text{ c}}$. Thus,

$$14 \text{ c} = \frac{14 \text{ c}}{1} \cdot 1 = \frac{14 \text{ c}}{1} \cdot \frac{1 \text{ qt}}{4 \text{ c}} = \frac{14 \text{ qt}}{4} = \frac{7}{2} \text{ qt} \quad \text{or} \quad 3\frac{1}{2} \text{ qt}$$

1 quart + 1 quart + 1 quart + $\frac{1}{2}$ quart   14 cups = $3\frac{1}{2}$ qt

■ **Work Practice 2**

✔**Concept Check** If 50 cups is converted to quarts, will the equivalent number of quarts be less than or greater than 50? Explain.

Objective B Performing Operations on U.S. System Units of Capacity ▶

As is true of units of length and weight, units of capacity can be added, subtracted, multiplied, and divided.

Example 3   Subtract 3 qt from 4 gal 2 qt.

Solution: To subtract, we line up similar units.

        4 gal 2 qt
    −   3 qt

We cannot subtract 3 qt from 2 qt. We need to borrow 1 gallon from the 4 gallons, convert it to 4 quarts, and then combine it with the 2 quarts.

    3 gal + (1 gal) 4 qt

$$\begin{array}{rcl} 4 \text{ gal } 2 \text{ qt} & = & 3 \text{ gal } 6 \text{ qt} \\ -\qquad 3 \text{ qt} & = & -\qquad 3 \text{ qt} \\ \hline & & 3 \text{ gal } 3 \text{ qt} \end{array}$$

To check, see that the sum of 3 gal 3 qt and 3 qt is 4 gal 2 qt.

■ **Work Practice 3**

Example 4   Finding the Amount of Water in an Aquarium

An aquarium contains 6 gal 3 qt of water. If 2 gal 2 qt of water is added, what is the total amount of water in the aquarium?

Solution:
beginning water	→	6 gal 3 qt
+ water added	→	+ 2 gal 2 qt
total water	→	8 gal 5 qt

(Continued on next page)

(Continued on next page)

Practice 2
Convert 26 quarts to cups.

Practice 3
Subtract 2 qt from 1 gal 1 qt.

Practice 4
A large oil drum contains 15 gal 3 qt of oil. How much will be in the drum if an additional 4 gal 3 qt of oil is poured into it?

Answers
2. 104 c   **3.** 3 qt   **4.** 20 gal 2 qt

✔**Concept Check Answer**
less than 50

Since 5 qt = 1 gal 1 qt, we have

$$\overbrace{= 8\ \text{gal}}^{8\ \text{gal}} + \overbrace{1\ \text{gal}\ 1\ \text{qt}}^{5\ \text{qt}}$$
$$= 9\ \text{gal}\ 1\ \text{qt}$$

The total amount of water is 9 gal 1 qt.

■ **Work Practice 4**

Objective C Defining and Converting Metric System Units of Capacity ▶

Thus far, we know that the basic unit of length in the metric system is the meter and that the basic unit of mass in the metric system is the gram. What is the basic unit of capacity? The **liter.** By definition, a **liter** is the capacity or volume of a cube measuring 10 centimeters on each side.

10 cm

10 cm

10 cm

The following examples may help you get a feeling for metric capacities:

- One liter of liquid is slightly more than one quart.

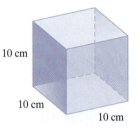

1 liter   1 quart

- Many soft drinks are packaged in 2-liter bottles.

2 liters

The metric system was designed to be a consistent system. Once again, the prefixes for metric units of capacity are the same as for metric units of length and mass, as summarized in the following table:

Metric Units of Capacity
1 **kilo**liter (kl) = 1000 liters (L)
1 **hecto**liter (hl) = 100 L
1 **deka**liter (dal) = 10 L
1 liter (L) = 1 L
1 **deci**liter (dl) = 1/10 L or 0.1 L
1 **centi**liter (cl) = 1/100 L or 0.01 L
1 **milli**liter (ml) = 1/1000 L or 0.001 L

The **milliliter** and the **liter** are the two most commonly used metric units of capacity.

Converting from one unit of capacity to another involves multiplying by powers of 10 or moving the decimal point to the left or to the right. Listing units of capacity in order from largest to smallest helps to keep track of how many places to move the decimal point when converting.

Let's convert 2.6 liters to milliliters. To convert from liters to milliliters, we move along the chart 3 units to the right.

kl   hl   dal   **L**   dl   cl   **ml**

3 units to the right

This means that we move the decimal point 3 places to the right to convert from liters to milliliters.

$$2.600 \, L = 2600 . \, ml$$

This same conversion can be done with unit fractions.

$$2.6 \, L = \frac{2.6 \, L}{1} \cdot 1$$

$$= \frac{2.6 \, L}{1} \cdot \frac{1000 \, ml}{1 \, L}$$

$$= 2.6 \cdot 1000 \, ml$$

$$= 2600 \, ml \qquad \text{To multiply by 1000, move the decimal point 3 places to the right.}$$

To visualize the result, study the diagram below:

MILK 2% Reduced Fat 1 L   MILK 2% Reduced Fat 1 L   MILK 2% Reduced Fat 0.6 L   2.6 L = 2600 ml

1000 ml   1000 ml   600 ml

Thus, 2.6 L = 2600 ml.

Example 5   Convert 3210 ml to liters.

Solution: Let's use the unit fraction method first.

Unit fraction

$$3210 \, ml = \frac{3210 \, ml}{1} \cdot 1 = 3210 \, ml \cdot \frac{1 \, L}{1000 \, ml} = 3.21 \, L$$

Now let's list the unit measures in order from left to right and move from milliliters to liters.

kl   hl   dal   L   dl   cl   ml

3 units to the left

3210 ml = 3.210 L, the same results as before and

3 places to the left   shown below in the diagram.

1000 ml   1000 ml   1000 ml

MILK 2% Reduced Fat 1 L   MILK 2% Reduced Fat 1 L   MILK 2% Reduced Fat 1 L   MILK 0.210 L   210 ml   3210 ml = 3.210 L

1 L   1 L   1 L   0.210 L

Work Practice 5

Practice 5

Convert 2100 ml to liters.

Practice 6

Convert 2.13 dal to liters.

Example 6 Convert 0.185 dl to milliliters.

Solution: We list the unit measures in order from left to right and move from deciliters to milliliters.

kl   hl   dal   L   dl   cl   ml
2 units to the right

$$0.185 \text{ dl} = 18.5 \text{ ml}$$
2 places to the right

■ **Work Practice 6**

Objective D Performing Operations on Metric System Units of Capacity

As was true for length and weight, arithmetic operations involving metric units of capacity can also be performed. Make sure that the metric units of capacity are the same before adding, subtracting, multiplying, or dividing.

Practice 7

Add 1250 ml to 2.9 L.

Example 7 Add 2400 ml to 8.9 L.

Solution: We must convert both to liters or both to milliliters before adding the capacities together.

2400 ml = 2.4 L	or	8.9 L = 8900ml

$$
\begin{array}{r}
2.4 \text{ L} \\
+ \ 8.9 \text{ L} \\
\hline
11.3 \text{ L}
\end{array}
\qquad
\begin{array}{r}
2400 \text{ ml} \\
+ \ 8900 \text{ ml} \\
\hline
11{,}300 \text{ ml}
\end{array}
$$

The total is 11.3 L or 11,300 ml. They both represent the same capacity.

■ **Work Practice 7**

✓**Concept Check** How could you estimate the following operation? Subtract 950 ml from 7.5 L.

Practice 8

If 28.6 L of water can be pumped every minute, how much water can be pumped in 85 minutes?

Example 8 Finding the Amount of Medication a Person Has Received

A patient hooked up to an IV unit in the hospital is to receive 12.5 ml of medication every hour. How much medication does the patient receive in 3.5 hours?

Solution: We multiply 12.5 ml by 3.5.

medication per hour	→	12.5 ml
× hours	→	× 3.5
total medication		625
		3750
		43.75 ml

The patient receives 43.75 ml of medication.

■ **Work Practice 8**

Answers
6. 21.3 L   7. 4150 ml or 4.15 L
8. 2431 L

✓**Concept Check Answer**
950 ml = 0.95 L; round 0.95 to 1;
7.5 − 1 = 6.5 L

Vocabulary, Readiness & Video Check

Use the choices below to fill in each blank. Some choices may be used more than once.

cups	pints	liter
quarts	fluid ounces	capacity

1. Units of _____ are generally used to measure liquids.

2. The basic unit of capacity in the metric system is the _____.

3. One cup equals 8 _____.

4. One quart equals 2 _____.

5. One pint equals 2 _____.

6. One quart equals 4 _____.

7. One gallon equals 4 _____ .

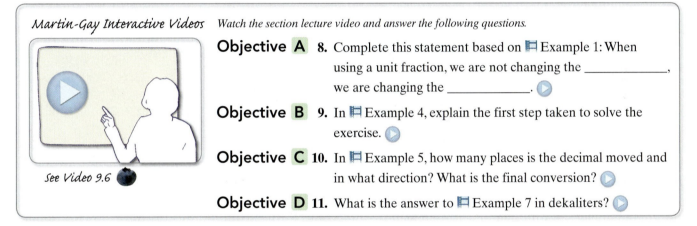

Martin-Gay Interactive Videos

See Video 9.6

Watch the section lecture video and answer the following questions.

Objective A **8.** Complete this statement based on ▭ Example 1: When using a unit fraction, we are not changing the _____, we are changing the _____. ▶

Objective B **9.** In ▭ Example 4, explain the first step taken to solve the exercise. ▶

Objective C **10.** In ▭ Example 5, how many places is the decimal moved and in what direction? What is the final conversion? ▶

Objective D **11.** What is the answer to ▭ Example 7 in dekaliters? ▶

9.6 Exercise Set MyMathLab® ▶

Objective A *Convert each measurement as indicated. See Examples 1 and 2.*

1. 32 fluid ounces to cups

2. 16 quarts to gallons

3. 8 quarts to pints

4. 9 pints to quarts

▶ **5.** 14 quarts to gallons

6. 11 cups to pints

7. 80 fluid ounces to pints

8. 18 pints to gallons

9. 2 quarts to cups

10. 3 pints to fluid ounces

11. 120 fluid ounces to quarts

12. 20 cups to gallons

▶ **13.** 42 cups to quarts

14. 7 quarts to cups

15. $4\frac{1}{2}$ pints to cups

16. $6\frac{1}{2}$ gallons to quarts

17. 5 gal 3 qt to quarts

18. 4 gal 1 qt to quarts

19. $\frac{1}{2}$ cup to pints

20. $\frac{1}{2}$ pint to quarts

▶ **21.** 58 qt = _____ gal _____ qt

22. 70 qt = _____ gal _____ qt

23. 39 pt = _____ gal _____ qt _____ pt

24. 29 pt = _____ gal _____ qt _____ pt

25. $2\frac{3}{4}$ gallons to pints

26. $3\frac{1}{4}$ quarts to cups

Objective B   *Perform each indicated operation. See Examples 3 and 4.*

27. 5 gal 3 qt + 7 gal 3 qt

28. 2 gal 2 qt + 9 gal 3 qt

29. 1 c 5 fl oz + 2 c 7 fl oz

30. 2 c 3 fl oz + 2 c 6 fl oz

▶ **31.** 3 gal − 1 gal 3 qt

32. 2 pt − 1 pt 1 c

33. 3 gal 1 qt − 1 qt 1 pt

34. 3 qt 1 c − 1 c 4 fl oz

35. 8 gal 2 qt × 2

36. 6 gal 1 pt × 2

37. 9 gal 2 qt ÷ 2

38. 5 gal 6 fl oz ÷ 2

Objective C   *Convert as indicated. See Examples 5 and 6.*

39. 5 L to milliliters

40. 8 L to milliliters

41. 0.16 L to kiloliters

42. 0.127 L to kiloliters

▶ **43.** 5600 ml to liters

44. 1500 ml to liters

45. 3.2 L to centiliters

46. 1.7 L to centiliters

47. 410 L to kiloliters

48. 250 L to kiloliters

49. 64 ml to liters

50. 39 ml to liters

▶ **51.** 0.16 kl to liters

52. 0.48 kl to liters

53. 3.6 L to milliliters

54. 1.9 L to milliliters

Objective D   *Perform each indicated operation. Remember to insert units when writing your answers. See Examples 7 and 8.*

55. 3.4 L + 15.9 L

56. 18.5 L + 4.6 L

▶ **57.** 2700 ml + 1.8 L

58. 4.6 L + 1600 ml

59. 8.6 L − 190 ml

60. 4.8 L − 283 ml

61. 17,500 ml − 0.9 L

62. 6850 ml − 0.3 L

63. 480 ml × 8

64. 290 ml × 6

65. 81.2 L ÷ 0.5

66. 5.4 L ÷ 3.6

Objectives A B C D Mixed Practice *Solve. Remember to insert units when writing your answers. For Exercises 67 through 70, complete the chart.*

	Capacity	Cups	Gallons	Quarts	Pints
67.	An average-size bath of water		21		
68.	A dairy cow's daily milk yield				38
69.	Your kidneys filter about this amount of blood every minute	4			
70.	The amount of water needed in a punch recipe	2			

71. Mike Schaferkotter drank 410 ml of Mountain Dew from a 2-liter bottle. How much Mountain Dew remains in the bottle?

72. The Werners' Volvo has a 54.5-L gas tank. Only 3.8 liters of gasoline still remain in the tank. How much is needed to fill it?

73. Margie Phitts added 354 ml of Prestone dry gas to the 18.6 L of gasoline in her car's tank. Find the total amount of gasoline in the tank.

74. Chris Peckaitis wishes to share a 2-L bottle of Coca-Cola equally with 7 of his friends. How much will each person get?

75. A garden tool engine requires a 30-to-1 gas-to-oil mixture. This means that $\frac{1}{30}$ of a gallon of oil should be mixed with 1 gallon of gas. Convert $\frac{1}{30}$ gallon to fluid ounces. Round to the nearest tenth.

76. Henning's Supermarket sells homemade soup in 1 qt 1 pt containers. How much soup is contained in three such containers?

77. Can 5 pt 1 c of fruit punch and 2 pt 1 c of ginger ale be poured into a 1-gal container without it overflowing?

78. Three cups of prepared Jell-O are poured into 6 dessert dishes. How many fluid ounces of Jell-O are in each dish?

79. Stanley Fisher paid $14 to fill his car with 44.3 liters of gasoline. Find the price per liter of gasoline to the nearest thousandth of a dollar.

80. A student carelessly misread the scale on a cylinder in the chemistry lab and added 40 cl of water to a mixture instead of 40 ml. Find the excess amount of water.

Review

Write each fraction in simplest form. See Section 4.2.

81. $\frac{20}{25}$ **82.** $\frac{75}{100}$ **83.** $\frac{27}{45}$ **84.** $\frac{56}{60}$ **85.** $\frac{72}{80}$ **86.** $\frac{18}{20}$

Concept Extensions

Determine whether the measurement in each statement is reasonable.

87. Clair took a dose of 2 L of cough medicine to cure her cough.

88. John drank 250 ml of milk for lunch.

89. Jeannie likes to relax in a tub filled with 3000 ml of hot water.

90. Sarah pumped 20 L of gasoline into her car yesterday.

Solve. See the Concept Checks in this section.

91. If 70 pints are converted to gallons, will the equivalent number of gallons be less than or greater than 70? Explain why.

92. If 30 gallons are converted to quarts, will the equivalent number of quarts be less than or greater than 30? Explain why.

93. Explain how to estimate the following operation: Add 986 ml to 6.9 L.

94. Explain how to borrow in order to subtract 1 gal 2 qt from 3 gal 1 qt.

95. Find the number of fluid ounces in 1 gallon.

96. Find the number of fluid ounces in 1.5 gallons.

A cubic centimeter (cc) is the amount of space that a volume of 1 ml occupies. Because of this, we will say that 1 cc = 1 ml.

A common syringe is one with a capacity of 3 cc. Use the diagram and give the measurement indicated by each arrow.

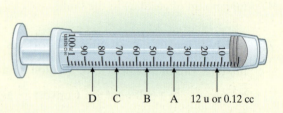

97. B

98. A

99. D

100. C

In order to measure small dosages, such as for insulin, u-100 syringes are used. For these syringes, 1 cc has been divided into 100 equal units (u). Use the diagram and give the measurement indicated by each arrow in units (u) and then in cubic centimeters. Use 100 u = 1 cc.

101. B

102. A

103. D

104. C

9.7 **Temperature and Conversions Between the U.S. and Metric Systems**

Objectives

A Convert Between the U.S. and Metric Systems.

B Convert Temperatures from Degrees Celsius to Degrees Fahrenheit.

C Convert Temperatures from Degrees Fahrenheit to Degrees Celsius.

Objective A Converting Between the U.S. and Metric Systems

The metric system probably had its beginnings in France in the 1600s, but it was the Metric Act of 1866 that made the use of this system legal (although not mandatory) in the United States. Other laws have followed that allow for a slow, but deliberate, transfer to the modernized metric system. In April 2001, for example, the U.S. Stock Exchanges completed their change to decimal trading instead of fractions. By the end of 2009, all products sold in Europe (with some exceptions) were required to have only metric units on their labels. (*Source:* U.S. Metric Association and National Institute of Standards and Technology)

You may be surprised at the number of everyday items we use that are already manufactured in metric units. We easily recognize 1 L and 2 L soda bottles, but what about the following?

- Pencil leads (0.5 mm or 0.7 mm)
- Camera film (35 mm)
- Sporting events (5-km or 10-km races)

- Medicines (500-mg capsules)

- Labels on retail goods (dual-labeled since 1994)

Since the United States has not completely converted to the metric system, we need to practice converting from one system to the other. Below is a table of mostly approximate conversions.

Length:		Capacity:		Weight (mass):	
Metric	U.S. System	Metric	U.S. System	Metric	U.S. System
1 m ≈ 1.09 yd		1 L ≈ 1.06 qt		1 kg ≈ 2.20 lb	
1 m ≈ 3.28 ft		1 L ≈ 0.26 gal		1 g ≈ 0.04 oz	
1 km ≈ 0.62 mi		3.79 L ≈ 1 gal		0.45 kg ≈ 1 lb	
2.54 cm = 1 in.		0.95 L ≈ 1 qt		28.35 g ≈ 1 oz	
0.30 m ≈ 1 ft		29.57 ml ≈ 1 fl oz			
1.61 km ≈ 1 mi					

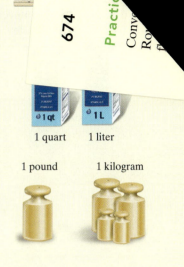

1 quart   1 liter

1 pound   1 kilogram

There are many ways to perform these metric-to-U.S. conversions. We will do so by using unit fractions.

Example 1 Compact Discs

Standard-sized compact discs are 12 centimeters in diameter. Convert this length to inches. Round the result to two decimal places.

Solution: From our length conversion table, we know that 2.54 cm = 1 in. This fact gives us two unit fractions: $\frac{2.54 \text{ cm}}{1 \text{ in.}}$ and $\frac{1 \text{ in.}}{2.54 \text{ cm}}$. We use the unit fraction with cm in the denominator so that these units divide out.

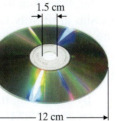

1.5 cm

12 cm

$$12 \text{ cm} = \frac{12 \text{ cm}}{1} \cdot 1 = \frac{12 \cancel{\text{ cm}}}{1} \cdot \underset{\leftarrow \text{ Original units}}{\overset{\leftarrow \text{ Units to convert to}}{\frac{1 \text{ in.}}{2.54 \cancel{\text{ cm}}}}}$$

$$= \frac{12 \text{ in.}}{2.54}$$

$$\approx 4.72 \text{ in.} \quad \text{Divide.}$$

Thus, the diameter of a standard compact disc is exactly 12 cm or approximately 4.72 inches. For a dimension this size, you can use a ruler to check. Another method is to approximate. Our result, 4.72 in., is close to 5 inches. Since 1 in. is about 2.5 cm, then 5 in. is about 5(2.5 cm) = 12.5 cm, which is close to 12 cm.

■ **Work Practice 1**

Example 2 Liver

The liver is your largest internal organ. It weighs about 3.5 pounds in a grown man. Convert this weight to kilograms. Round to the nearest tenth. (*Source: Some Body!* by Dr. Pete Rowan)

Solution: $3.5 \text{ lb} \approx \frac{3.5 \cancel{\text{ lb}}}{1} \cdot \overset{\text{Unit fraction}}{\frac{0.45 \text{ kg}}{1 \cancel{\text{ lb}}}} = 3.5(0.45 \text{ kg}) \approx 1.6 \text{ kg}$

Thus 3.5 pounds is approximately 1.6 kilograms. From the table of conversions, we know that 1 kg ≈ 2.2 lb. So that means 0.5 kg ≈ 1.1 lb and after adding, we have 1.5 kg ≈ 3.3 lb. Our result is reasonable.

■ **Work Practice 2**

Practice 1

The center hole of a standard-sized compact disc is 1.5 centimeters in diameter. Convert this length to inches. Round the result to 2 decimal places.

Practice 2

A full-grown human heart weighs about 8 ounces. Convert this weight to grams. If necessary, round your result to the nearest tenth of a gram.

Answers

1. 0.59 in.   **2.** 226.8 g

Example 3 Postage Stamp

Australia converted to the metric system in 1973. In that year, four postage stamps were issued to publicize this conversion. One such stamp is shown. Let's check the mathematics on the stamp by converting 7 fluid ounces to milliliters. Round to the nearest hundred.

Solution:
$$7 \text{ fl oz} \approx \frac{7 \text{ fl oz}}{1} \cdot \frac{29.57 \text{ ml}}{1 \text{ fl oz}} = 7(29.57 \text{ ml}) = 206.99 \text{ ml}$$

(Unit fraction)

Rounded to the nearest hundred, 7 fl oz ≈ 200 ml.

■ **Work Practice 3**

Now that we have practiced converting between two measurement systems, let's practice converting between two temperature scales.

Temperature When Gabriel Fahrenheit and Anders Celsius independently established units for temperature scales, each based his unit on the heat of water the moment it boils compared to the moment it freezes. One degree Celsius is $\frac{1}{100}$ of the difference in heat. One degree Fahrenheit is $\frac{1}{180}$ of the difference in heat. Celsius arbitrarily labeled the temperature at the freezing point at 0°C, making the boiling point 100°C; Fahrenheit labeled the freezing point 32°F, making the boiling point 212°F. Water boils at 212°F or 100°C.

By comparing the two scales in the figure, we see that a 20°C day is as warm as a 68°F day. Similarly, a sweltering 104°F day in the Mojave desert corresponds to a 40°C day.

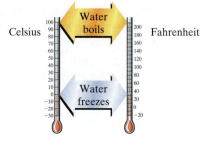

✓**Concept Check** Which of the following statements is correct? Explain.

a. 6°C is below the freezing point of water.

b. 6°F is below the freezing point of water.

Objective B Converting Degrees Celsius to Degrees Fahrenheit

To convert from Celsius temperatures to Fahrenheit temperatures, see the box below. In this box, we use the symbol F to represent degrees Fahrenheit and the symbol C to represent degrees Celsius.

Converting Celsius to Fahrenheit

$$F = \frac{9}{5}C + 32 \qquad \text{or} \qquad F = 1.8C + 32$$

(To convert to Fahrenheit temperature, multiply the Celsius temperature by $\frac{9}{5}$ or 1.8, and then add 32.)

---

e 3

rt 237 ml to fluid ounces. und to the nearest whole uid ounce.

Answer

3. 8 fl oz

✓**Concept Check Answer**

b

Example 4 Convert 15°C to degrees Fahrenheit.

Solution:

$$F = \frac{9}{5}C + 32$$

$$= \frac{9}{5} \cdot 15 + 32 \quad \text{Replace C with 15.}$$

$$= 27 + 32 \quad \text{Simplify.}$$

$$= 59 \quad \text{Add.}$$

Thus, 15°C is equivalent to 59°F.

■ **Work Practice 4**

Practice 4

Convert 60°C to degrees Fahrenheit.

Example 5 Convert 29°C to degrees Fahrenheit.

Solution:

$$F = 1.8\,C + 32$$

$$= 1.8 \cdot 29 + 32 \quad \text{Replace C with 29.}$$

$$= 52.2 + 32 \quad \text{Multiply 1.8 by 29.}$$

$$= 84.2 \quad \text{Add.}$$

Therefore, 29°C is the same as 84.2°F.

■ **Work Practice 5**

Practice 5

Convert 32°C to degrees Fahrenheit.

Objective C Converting Degrees Fahrenheit to Degrees Celsius ▶

To convert from Fahrenheit temperatures to Celsius temperatures, see the box below. The symbol C represents degrees Celsius and the symbol F represents degrees Fahrenheit.

> **Converting Fahrenheit to Celsius**
>
> $$C = \frac{5}{9}(F - 32)$$
>
> (To convert to Celsius temperature, subtract 32 from the Fahrenheit temperature, and then multiply by $\frac{5}{9}$.)

Example 6 Convert 59°F to degrees Celsius.

Solution: We evaluate the formula $C = \frac{5}{9}(F - 32)$ when F is 59.

$$C = \frac{5}{9}(F - 32)$$

$$= \frac{5}{9} \cdot (59 - 32) \quad \text{Replace F with 59.}$$

$$= \frac{5}{9} \cdot (27) \quad \text{Subtract inside parentheses.}$$

$$= 15 \quad \text{Multiply.}$$

Therefore, 59°F is the same temperature as 15°C.

■ **Work Practice 6**

Practice 6

Convert 68°F to degrees Celsius.

Answers

4. 140°F **5.** 89.6°F **6.** 20°C

Practice 7

Convert 113°F to degrees Celsius. If necessary, round to the nearest tenth of a degree.

Example 7 Convert 114°F to degrees Celsius. If necessary, round to the nearest tenth of a degree.

Solution: $C = \dfrac{5}{9}(F - 32)$

$$= \dfrac{5}{9}(114 - 32) \qquad \text{Replace F with 114.}$$

$$= \dfrac{5}{9} \cdot (82) \qquad \text{Subtract inside parentheses.}$$

$$\approx 45.6 \qquad \text{Multiply.}$$

Therefore, 114°F is approximately 45.6°C.

■ **Work Practice 7**

Practice 8

During a bout with the flu, Albert's temperature reaches 102.8°F. What is his temperature measured in degrees Celsius? Round to the nearest tenth of a degree.

Example 8 Body Temperature

Normal body temperature is 98.6°F. What is this temperature in degrees Celsius?

Solution: We evaluate the formula $C = \dfrac{5}{9}(F - 32)$ when F is 98.6.

$$C = \dfrac{5}{9}(F - 32)$$

$$= \dfrac{5}{9}(98.6 - 32) \qquad \text{Replace F with 98.6.}$$

$$= \dfrac{5}{9} \cdot (66.6) \qquad \text{Subtract inside parentheses.}$$

$$= 37 \qquad \text{Multiply.}$$

Therefore, normal body temperature is 37°C.

■ **Work Practice 8**

✓Concept Check Clarissa must convert 40°F to degrees Celsius. What is wrong with her work shown below?

$$F = 1.8 \cdot C + 32$$
$$F = 1.8 \cdot 40 + 32$$
$$F = 72 + 32$$
$$F = 104$$

Answers

7. 45°C **8.** 39.3°C

✓Concept Check Answer

She used the conversion for Celsius to Fahrenheit instead of Fahrenheit to Celsius.

Vocabulary, Readiness & Video Check

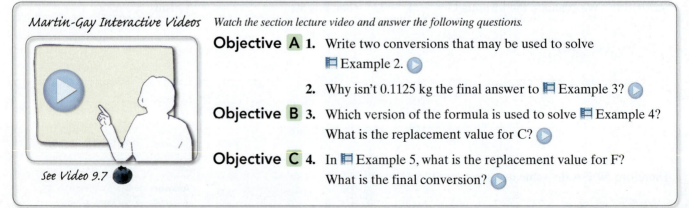

Martin-Gay Interactive Videos

See Video 9.7 ●

Watch the section lecture video and answer the following questions.

Objective A **1.** Write two conversions that may be used to solve ▣ Example 2. ▶

2. Why isn't 0.1125 kg the final answer to ▣ Example 3? ▶

Objective B **3.** Which version of the formula is used to solve ▣ Example 4? What is the replacement value for C? ▶

Objective C **4.** In ▣ Example 5, what is the replacement value for F? What is the final conversion? ▶

9.7 Exercise Set MyMathLab®

Note: Because approximations are used, your answers may vary slightly from the answers given in the back of the book.

Objective A *Convert as indicated. If necessary, round answers to two decimal places. See Examples 1 through 3.*

1. 756 milliliters to fluid ounces

2. 18 liters to quarts

3. 86 inches to centimeters

4. 86 miles to kilometers

5. 1000 grams to ounces

6. 100 kilograms to pounds

7. 93 kilometers to miles

8. 9.8 meters to feet

9. 14.5 liters to gallons

10. 150 milliliters to fluid ounces

11. 30 pounds to kilograms

12. 15 ounces to grams

Fill in the chart. Give exact answers or round to one decimal place. See Examples 1 through 3.

		Meters	Yards	Centimeters	Feet	Inches
13.	The height of a woman				5	
14.	Statue of Liberty length of nose	1.37				
15.	Leaning Tower of Pisa		60			
16.	Blue whale		36			

Solve. If necessary, round answers to two decimal places. See Examples 1 through 3.

17. The balance beam for female gymnasts is 10 centimeters wide. Convert this width to inches.

18. In men's gymnastics, the rings are 250 centimeters from the floor. Convert this height to inches, then to feet.

19. In many states, the maximum speed limit for recreational vehicles is 50 miles per hour. Convert this to kilometers per hour.

20. In some states, the speed limit is 70 miles per hour. Convert this to kilometers per hour.

21. Ibuprofen comes in 200-milligram tablets. Convert this to ounces. (Round your answer to this exercise to 3 decimal places.)

22. Vitamin C tablets come in 500-milligram caplets. Convert this to ounces.

The 70-meter-diameter antenna is the largest and most sensitive Deep Space Network antenna. See the Chapter Opener and answer Exercises 23–26.

70-Meter Antenna

23. Convert 70 meters to feet.

24. The Deep Space Network sites also have a 26-meter antenna. Convert 26 meters to feet.

25. The 70-meter-diameter antenna can track a spacecraft traveling more than 16 billion kilometers from Earth. Convert this distance to miles.

26. The dish reflector and the mount atop the concrete pedestal of the 70-meter antenna weigh nearly 2.7 million kilograms. Convert this number to tons.

27. A stone is a unit in the British customary system. Use the conversion 14 pounds = 1 stone to check the equivalencies in this 1973 Australian stamp. Is 100 kilograms approximately 15 stone 10 pounds?

28. Convert 5 feet 11 inches to centimeters and check the conversion on this 1973 Australian stamp. Is it correct?

29. The Monarch butterfly migrates annually between the northern United States and central Mexico. The trip is about 4500 km long. Convert this to miles.

30. There is a species of African termite that builds nests up to 18 ft high. Convert this to meters.

31. A $3\frac{1}{2}$-inch diskette is not really $3\frac{1}{2}$ inches. To find its actual width, convert this measurement to centimeters, then to millimeters. Round the result to the nearest ten.

32. The average two-year-old is 84 centimeters tall. Convert this to feet and inches.

33. For an average adult, the weight of the right lung is greater than the weight of the left lung. If the right lung weighs 1.5 pounds and the left lung weighs 1.25 pounds, find the difference in grams. (*Source: Some Body!*)

34. The skin of an average adult weighs 9 pounds and is the heaviest organ. Find the weight in grams. (*Source: Some Body!*)

35. A fast sneeze has been clocked at about 167 kilometers per hour. Convert this to miles per hour. Round to the nearest whole.

36. A Boeing 747 has a cruising speed of about 980 kilometers per hour. Convert this to miles per hour. Round to the nearest whole.

37. The General Sherman giant sequoia tree has a diameter of about 8 meters at its base. Convert this to feet. (*Source: Fantastic Book of Comparisons*)

38. The largest crater on the near side of the moon is Billy Crater. It has a diameter of 303 kilometers. Convert this to miles. (*Source: Fantastic Book of Comparisons*)

39. The total length of the track on a CD is about 4.5 kilometers. Convert this to miles. Round to the nearest whole mile.

40. The distance between Mackinaw City, Michigan, and Cheyenne, Wyoming, is 2079 kilometers. Convert this to miles. Round to the nearest whole mile.

41. A doctor orders a dosage of 5 ml of medicine every 4 hours for 1 week. How many fluid ounces of medicine should be purchased? Round up to the next whole fluid ounce.

42. A doctor orders a dosage of 12 ml of medicine every 6 hours for 10 days. How many fluid ounces of medicine should be purchased? Round up to the next whole fluid ounce.

Without actually converting, choose the most reasonable answer.

43. This math book has a height of about_____.
   a. 28 mm   **b.** 28 cm
   c. 28 m   **d.** 28 km

44. A mile is _____ a kilometer.
   a. shorter than   **b.** longer than
   c. the same length as

45. A liter has _____ capacity than a quart.
   a. less   **b.** greater
   c. the same

46. A foot is _____ a meter.
   a. shorter than   **b.** longer than
   c. the same length as

47. A kilogram weighs _____ a pound.
   a. the same as   **b.** less than
   c. greater than

48. A football field is 100 yards, which is about_____.
   a. 9 m   **b.** 90 m
   c. 900 m   **d.** 9000 m

49. An $8\frac{1}{2}$-ounce glass of water has a capacity of about _____.
   a. 250 L   **b.** 25 L
   c. 2.5 L   **d.** 250 ml

50. A 5-gallon gasoline can has a capacity of about _____.
   a. 19 L   **b.** 1.9 L
   c. 19 ml   **d.** 1.9 ml

51. The weight of an average man is about _____.
   a. 700 kg   **b.** 7 kg
   c. 0.7 kg   **d.** 70 kg

52. The weight of a pill is about _____.
   a. 200 kg   **b.** 20 kg
   c. 2 kg   **d.** 200 mg

Objectives B C **Mixed Practice** *Convert as indicated. When necessary, round to the nearest tenth of a degree.*
See Examples 4 through 8.

53. 77°F to degrees Celsius

54. 86°F to degrees Celsius

55. 104°F to degrees Celsius

56. 140°F to degrees Celsius

57. 50°C to degrees Fahrenheit

58. 80°C to degrees Fahrenheit

59. 115°C to degrees Fahrenheit

60. 225°C to degrees Fahrenheit

61. 20°F to degrees Celsius

62. 26°F to degrees Celsius

63. 142.1°F to degrees Celsius

64. 43.4°F to degrees Celsius

65. 92°C to degrees Fahrenheit

66. 75°C to degrees Fahrenheit

67. 12.4°C to degrees Fahrenheit

68. 48.6°C to degrees Fahrenheit

69. The hottest temperature ever recorded in the United States, in Death Valley, was 134°F. Convert this temperature to degrees Celsius. (*Source:* National Climatic Data Center)

70. The hottest temperature ever recorded in the United States in January was 95°F in Los Angeles. Convert this temperature to degrees Celsius. (*Source:* National Climatic Data Center)

71. A weather forecaster in Caracas predicts a high temperature of 27°C. Find this measurement in degrees Fahrenheit.

72. While driving to work, Alan Olda notices a temperature of 18°C flash on the local bank's temperature display. Find the corresponding temperature in degrees Fahrenheit.

73. At Mack Trucks' headquarters, the room temperature is to be set at 70°F, but the thermostat is calibrated in degrees Celsius. Find the temperature to be set.

74. The computer room at Merck, Sharp, and Dohm is normally cooled to 66°F. Find the corresponding temperature in degrees Celsius.

75. In a European cookbook, a recipe requires the ingredients for caramels to be heated to 118°C, but the cook has access only to a Fahrenheit thermometer. Find the temperature in degrees Fahrenheit that should be used to make the caramels.

76. The ingredients for divinity should be heated to 127°C, but the candy thermometer that Myung Kim has is calibrated to degrees Fahrenheit. Find how hot he should heat the ingredients.

77. The temperature of Earth's core is estimated to be 4000°C. Find the corresponding temperature in degrees Fahrenheit.

78. In 2012, the average temperature of Earth's surface was 58.3°F. Convert this temperature to degrees Celsius. (*Source:* NASA)

Review

Perform the indicated operations. See Section 1.7.

79. $6 \cdot 4 + 5 \div 1$ **80.** $10 \div 2 + 9(8)$ **81.** $3[(1 + 5) \cdot (8 - 6)]$ **82.** $5[(18 - 8) - 9]$

Concept Extensions

Determine whether the measurement in each statement is reasonable.

83. A 72°F room feels comfortable.

84. Water heated to 110°F will boil.

85. Josiah has a fever if a thermometer shows his temperature to be 40°F.

86. An air temperature of 20°F on a Vermont ski slope can be expected in the winter.

87. When the temperature is 30°C outside, an overcoat is needed.

88. An air-conditioned room at 60°C feels quite chilly.

89. Barbara has a fever when a thermometer records her temperature at 40°C.

90. Water cooled to 32°C will freeze.

Body surface area (BSA) is often used to calculate dosages for some drugs. BSA is calculated in square meters using a person's weight and height.

$$\text{BSA} = \sqrt{\frac{(\text{weight in kg}) \times (\text{height in cm})}{3600}}$$

For Exercises 91 through 96, calculate the BSA for each person. Round to the nearest hundredth. You will need to use the square root key on your calculator.

91. An adult whose height is 182 cm and weight is 90 kg

92. An adult whose height is 157 cm and weight is 63 kg

93. A child whose height is 40 in. and weight is 50 kg (*Hint:* Don't forget to first convert inches to centimeters.)

94. A child whose height is 26 in. and weight is 13 kg (*Hint:* Don't forget to first convert inches to centimeters.)

95. An adult whose height is 60 in. and weight is 150 lb

96. An adult whose height is 69 in. and weight is 172 lb

97. In February 2010, at the Brookhaven National Laboratory in Long Island, NY, the highest temperature produced in a laboratory was achieved. This temperature was 7,200,000,000°F. Convert this temperature to degrees Celsius. Round your answer to the nearest million degrees. (*Source: Guinness World Records*)

98. The hottest-burning substance known is carbon subnitride. Its flame at one atmospheric pressure reaches 9010°F. Convert this temperature to degrees Celsius. (*Source: Guinness World Records*)

99. In your own words, describe how to convert from degrees Celsius to degrees Fahrenheit.

100. In your own words, describe how to convert from degrees Fahrenheit to degrees Celsius.

Chapter 9 Group Activity

Map Reading

Sections 9.1, 9.4, and 9.7

Materials:

- ruler
- string
- calculator

This activity may be completed by working in groups or individually.

Investigate the route you would take from Santa Rosa, New Mexico, to San Antonio, New Mexico. Use the map in the figure to answer the following questions. You may find that using string to match the roads on the map is useful when measuring distances.

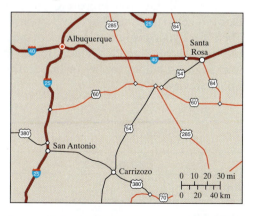

1. How many miles is it from Santa Rosa to San Antonio via Interstate 40 and Interstate 25? Convert this distance to kilometers.

2. How many miles is it from Santa Rosa to San Antonio via U.S. 54 and U.S. 380? Convert this distance to kilometers.

3. Assume that the speed limit on Interstates 40 and 25 is 65 miles per hour. How long would the trip take if you took this route and traveled 65 miles per hour the entire trip?

4. At what average speed would you have to travel on the U.S. routes to make the trip from Santa Rosa to San Antonio in the same amount of time that it would take on the interstate routes? Do you think this speed is reasonable on this route? Explain your reasoning.

5. Discuss in general the factors that might affect your decision between the different routes.

6. Explain which route you would choose in this case and why.

Chapter 9 Vocabulary Check

Fill in each blank with one of the words or phrases listed below.

transversal	line segment	obtuse	straight	adjacent	right	volume	area
acute	perimeter	vertical	supplementary	ray	angle	line	complementary
vertex	mass	unit fractions	gram	weight	meter	liter	surface area

1. _____ is a measure of the pull of gravity.

2. _____ is a measure of the amount of substance in an object. This measure does not change.

3. The basic unit of length in the metric system is the _____.

4. To convert from one unit of length to another, _____ may be used.

5. The _____ is the basic unit of mass in the metric system.

6. The _____ is the basic unit of capacity in the metric system.

7. A(n) _____ is a piece of a line with two endpoints.

8. Two angles that have a sum of 90° are called _____ angles.

9. A(n) _____ is a set of points extending indefinitely in two directions.

10. The _____ of a polygon is the distance around the polygon.

11. A(n) _____ is made up of two rays that share the same endpoint. The common endpoint is called the _____.

12. _____ measures the amount of surface of a region.

13. A(n) _____ is a part of a line with one endpoint. A ray extends indefinitely in one direction.

14. A line that intersects two or more lines at different points is called a(n) _____.

15. An angle that measures 180° is called a(n) _____ angle.

16. The measure of the space of a solid is called its _____.

17. When two lines intersect, four angles are formed. Two of these angles that are opposite each other are called _____ angles.

18. Two of the angles from Exercise **17** that share a common side are called _____ angles.

19. An angle whose measure is between 90° and 180° is called a(n) _____ angle.

20. An angle that measures 90° is called a(n) _____ angle.

21. An angle whose measure is between 0° and 90° is called a(n) _____ angle.

22. Two angles that have a sum of 180° are called _____ angles.

23. The _____ of a polyhedron is the sum of the areas of the faces of the polyhedron.

Helpful Hint

▶ Are you preparing for your test? Don't forget to take the Chapter 9 Test on page 692. Then check your answers at the back of the text and use the Chapter Test Prep Videos to see the fully worked-out solutions to any of the exercises you want to review.

9 Chapter Highlights

Definitions and Concepts	Examples
Section 9.1 Lines and Angles	
A **line** is a set of points extending indefinitely in two directions. A line has no width or height, but it does have length. We name a line by any two of its points.	Line AB or $\overleftrightarrow{AB}$
A **line segment** is a piece of a line with two endpoints.	Line segment AB or $\overline{AB}$

(continued)

Definitions and Concepts	F

Section 9.1   Lines and Angles   *(continued)*

A **ray** is a part of a line with one endpoint. A ray extends indefinitely in one direction.

An **angle** is made up of two rays that share the same endpoint. The common endpoint is called the **vertex.**

Ray AB or $\overrightarrow{AB}$

684

To convert
a **unit fra**

B
Vertex
C

Section 9.2   Perimeter

Perimeter Formulas

Rectangle: $P = 2l + 2w$

Square: $P = 4s$

Triangle: $P = a + b + c$

Circumference of a Circle: $C = 2\pi r$ or $C = \pi d$

where $\pi \approx 3.14$ or $\pi \approx \dfrac{22}{7}$

Find the perimeter of the rectangle.

28 m

15 m

$$P = 2l + 2w$$
$$= 2 \cdot 28 \text{ meters} + 2 \cdot 15 \text{ meters}$$
$$= 56 \text{ meters} + 30 \text{ meters}$$
$$= 86 \text{ meters}$$

The perimeter is 86 meters.

Section 9.3   Area, Volume, and Surface Area

Area Formulas

Rectangle: $A = lw$

Square: $A = s^2$

Triangle: $A = \dfrac{1}{2}bh$

Parallelogram: $A = bh$

Trapezoid: $A = \dfrac{1}{2}(b + B)h$

Circle: $A = \pi r^2$

Volume Formulas

Rectangular Solid: $V = lwh$

Cube: $V = s^3$

Sphere: $V = \dfrac{4}{3}\pi r^3$

Right Circular Cylinder: $V = \pi r^2 h$

Cone: $V = \dfrac{1}{3}\pi r^2 h$

Square-Based Pyramid: $V = \dfrac{1}{3}s^2 h$

Surface Area Formulas: See page 629.

Find the area of the square.

8 cm

$$A = s^2$$
$$= (8 \text{ centimeters})^2$$
$$= 64 \text{ square centimeters}$$

The area of the square is 64 square centimeters.

Find the volume of the sphere. Use $\dfrac{22}{7}$ for π.

4 in.

$$V = \dfrac{4}{3}\pi r^3$$
$$\approx \dfrac{4}{3} \cdot \dfrac{22}{7} \cdot (4 \text{ inches})^3$$
$$= \dfrac{4 \cdot 22 \cdot 64}{3 \cdot 7} \text{ cubic inches}$$
$$= \dfrac{5632}{21} \quad \text{or} \quad 268\dfrac{4}{21} \text{ cubic inches}$$

Definitions and Concepts	Examples

Section 9.4   Linear Measurement

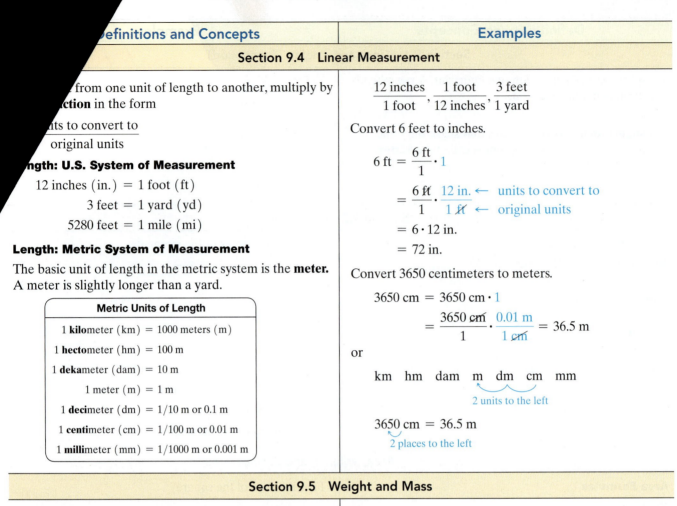

from one unit of length to another, multiply by ...ction in the form

$$\frac{\text{units to convert to}}{\text{original units}}$$

Length: U.S. System of Measurement

12 inches (in.) = 1 foot (ft)

3 feet = 1 yard (yd)

5280 feet = 1 mile (mi)

Length: Metric System of Measurement

The basic unit of length in the metric system is the **meter.** A meter is slightly longer than a yard.

Metric Units of Length
1 **kilo**meter (km) = 1000 meters (m)
1 **hecto**meter (hm) = 100 m
1 **deka**meter (dam) = 10 m
1 meter (m) = 1 m
1 **deci**meter (dm) = 1/10 m or 0.1 m
1 **centi**meter (cm) = 1/100 m or 0.01 m
1 **milli**meter (mm) = 1/1000 m or 0.001 m

$$\frac{12 \text{ inches}}{1 \text{ foot}}, \frac{1 \text{ foot}}{12 \text{ inches}}, \frac{3 \text{ feet}}{1 \text{ yard}}$$

Convert 6 feet to inches.

$$6 \text{ ft} = \frac{6 \text{ ft}}{1} \cdot 1$$

$$= \frac{6 \text{ ft}}{1} \cdot \frac{12 \text{ in.}}{1 \text{ ft}} \leftarrow \text{units to convert to} \\ \leftarrow \text{original units}$$

$$= 6 \cdot 12 \text{ in.}$$

$$= 72 \text{ in.}$$

Convert 3650 centimeters to meters.

$$3650 \text{ cm} = 3650 \text{ cm} \cdot 1$$

$$= \frac{3650 \text{ cm}}{1} \cdot \frac{0.01 \text{ m}}{1 \text{ cm}} = 36.5 \text{ m}$$

or

km   hm   dam   m   dm   cm   mm

2 units to the left

3650 cm = 36.5 m

2 places to the left

Section 9.5   Weight and Mass

Weight is really a measure of the pull of gravity. **Mass** is a measure of the amount of substance in an object and does not change.

Weight: U.S. System of Measurement

16 ounces (oz) = 1 pound (lb)

2000 pounds = 1 ton

Mass: Metric System of Measurement

The **gram** is the basic unit of mass in the metric system. It is the mass of water contained in a cube 1 centimeter on each side. A paper clip weighs about 1 gram.

Metric Units of Mass
1 **kilo**gram (kg) = 1000 grams (g)
1 **hecto**gram (hg) = 100 g
1 **deka**gram (dag) = 10 g
1 gram (g) = 1 g
1 **deci**gram (dg) = 1/10 g or 0.1 g
1 **centi**gram (cg) = 1/100 g or 0.01 g
1 **milli**gram (mg) = 1/1000 g or 0.001 g

Convert 5 pounds to ounces.

$$5 \text{ lb} = 5 \text{ lb} \cdot 1 = \frac{5 \text{ lb}}{1} \cdot \frac{16 \text{ oz}}{1 \text{ lb}} = 80 \text{ oz}$$

Convert 260 grams to kilograms.

$$260 \text{ g} = \frac{260 \text{ g}}{1} \cdot 1 = \frac{260 \text{ g}}{1} \cdot \frac{1 \text{ kg}}{1000 \text{ g}} = 0.26 \text{ kg}$$

or

kg   hg   dag   g   dg   cg   mg

3 units to the left

260 g = 0.26 kg

3 places to the left

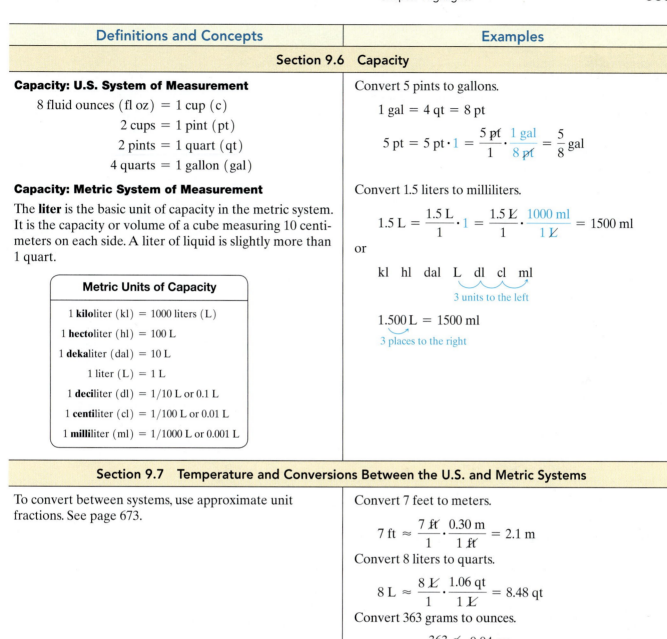

Definitions and Concepts	Examples

Section 9.6 Capacity

Capacity: U.S. System of Measurement

8 fluid ounces (fl oz) = 1 cup (c)

2 cups = 1 pint (pt)

2 pints = 1 quart (qt)

4 quarts = 1 gallon (gal)

Convert 5 pints to gallons.

1 gal = 4 qt = 8 pt

$$5 \text{ pt} = 5 \text{ pt} \cdot 1 = \frac{5 \text{ pt}}{1} \cdot \frac{1 \text{ gal}}{8 \text{ pt}} = \frac{5}{8} \text{ gal}$$

Capacity: Metric System of Measurement

The **liter** is the basic unit of capacity in the metric system. It is the capacity or volume of a cube measuring 10 centimeters on each side. A liter of liquid is slightly more than 1 quart.

Convert 1.5 liters to milliliters.

$$1.5 \text{ L} = \frac{1.5 \text{ L}}{1} \cdot 1 = \frac{1.5 \text{ L}}{1} \cdot \frac{1000 \text{ ml}}{1 \text{ L}} = 1500 \text{ ml}$$

or

kl hl dal L dl cl ml

3 units to the left

1.500 L = 1500 ml

3 places to the right

Metric Units of Capacity

1 **kilo**liter (kl) = 1000 liters (L)

1 **hecto**liter (hl) = 100 L

1 **deka**liter (dal) = 10 L

1 liter (L) = 1 L

1 **deci**liter (dl) = 1/10 L or 0.1 L

1 **centi**liter (cl) = 1/100 L or 0.01 L

1 **milli**liter (ml) = 1/1000 L or 0.001 L

Section 9.7   Temperature and Conversions Between the U.S. and Metric Systems

To convert between systems, use approximate unit fractions. See page 673.

Convert 7 feet to meters.

$$7 \text{ ft} \approx \frac{7 \text{ ft}}{1} \cdot \frac{0.30 \text{ m}}{1 \text{ ft}} = 2.1 \text{ m}$$

Convert 8 liters to quarts.

$$8 \text{ L} \approx \frac{8 \text{ L}}{1} \cdot \frac{1.06 \text{ qt}}{1 \text{ L}} = 8.48 \text{ qt}$$

Convert 363 grams to ounces.

$$363 \text{ g} \approx \frac{363 \text{ g}}{1} \cdot \frac{0.04 \text{ oz}}{1 \text{ g}} = 14.52 \text{ oz}$$

Celsius to Fahrenheit

$$F = \frac{9}{5}C + 32 \quad \text{or} \quad F = 1.8C + 32$$

Convert 35°C to degrees Fahrenheit.

$$F = \frac{9}{5} \cdot 35 + 32 = 63 + 32 = 95$$

35°C = 95°F

Fahrenheit to Celsius

$$C = \frac{5}{9}(F - 32)$$

Convert 50°F to degrees Celsius.

$$C = \frac{5}{9} \cdot (50 - 32) = \frac{5}{9} \cdot (18) = 10$$

50°F = 10°C

(9.1) *Classify each angle as acute, right, obtuse, or straight.*

1.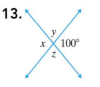

2.

3. *C*

4. *D*

5. Find the complement of a 25° angle.

6. Find the supplement of a 105° angle.

Find the measure of angle x in each figure.

7. 32° *x*

8. *x* 82°

9. 105° *x* 15°

10. 20° *x* 45°

11. Identify the pairs of supplementary angles.

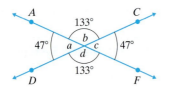

12. Identify the pairs of complementary angles.

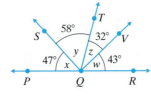

Find the measures of angles x, y, and z in each figure.

13. *y* *x* 100° *z*

14. *z* *x* *y* 25°

15. Given that $m \parallel n$.

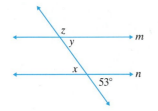

16. Given that $m \parallel n$.

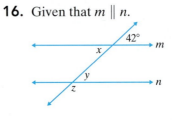

(9.2) *Find the perimeter of each figure.*

17. 23 m

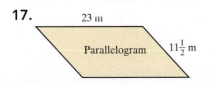

Parallelogram $11\frac{1}{2}$ m

18. 11 cm 7.6 cm

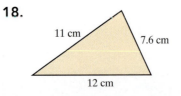

12 cm

686

19.
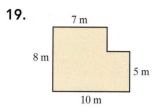

20.
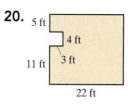

Solve.

21. Find the perimeter of a rectangular sign that measures 6 feet by 10 feet.

22. Find the perimeter of a town square that measures 110 feet on a side.

Find the circumference of each circle. Use $\pi \approx 3.14$.

23.

1.7 in.

24.

5 yd

(9.3) *Find the area of each figure. For the circles, find the exact area and then use $\pi \approx 3.14$ to approximate the area.*

25.
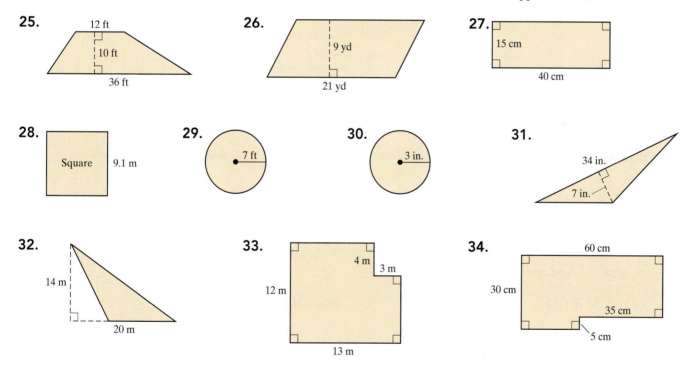

12 ft

10 ft

36 ft

26.

9 yd

21 yd

27.

15 cm

40 cm

28.

Square   9.1 m

29.

7 ft

30.

3 in.

31.

34 in.

7 in.

32.

14 m

20 m

33.

4 m

3 m

12 m

13 m

34.

60 cm

30 cm

35 cm

5 cm

35. The amount of sealer necessary to seal a driveway depends on the area. Find the area of a rectangular driveway 36 feet by 12 feet.

36. Find how much carpet is necessary to cover the floor of the room shown.

10 feet   13 feet

Find the volume and surface area of the solids in Exercises 37 and 38. For Exercises 39 and 40, give an exact volume and an approximation.

37.

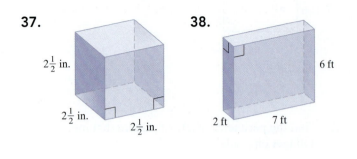

$2\frac{1}{2}$ in.

$2\frac{1}{2}$ in.

$2\frac{1}{2}$ in.

38.

6 ft

2 ft   7 ft

39. Use $\pi \approx 3.14$.

50 cm

20 cm

40. Use $\pi \approx \frac{22}{7}$.

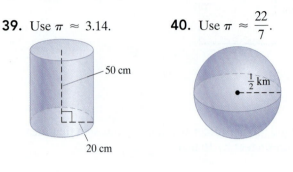

$\frac{1}{2}$ km

41. Find the volume of a pyramid with a square base 2 feet on a side and a height of 2 feet.

42. Approximate the volume of a tin can 8 inches high and 3.5 inches in radius. Use 3.14 for π.

43. A chest has 3 drawers. If each drawer has inside measurements of $2\frac{1}{2}$ feet by $1\frac{1}{2}$ feet by $\frac{2}{3}$ foot, find the total volume of the 3 drawers.

44. A cylindrical canister for a shop vacuum is 2 feet tall and 1 foot in *diameter*. Find its exact volume.

(9.4) *Convert.*

45. 108 in. to feet

46. 72 ft to yards

47. 1.5 mi to feet

48. $\frac{1}{2}$ yd to inches

49. 52 ft = _____ yd _____ ft

50. 46 in. = _____ ft _____ in.

51. 42 m to centimeters

52. 82 cm to millimeters

53. 12.18 mm to meters

54. 2.31 m to kilometers

Perform each indicated operation.

55. 4 yd 2 ft + 16 yd 2 ft

56. 7 ft 4 in. ÷ 2

57. 8 cm + 15 mm

58. 4 m − 126 cm

Solve.

59. A bolt of cloth contains 333 yd 1 ft of cotton ticking. Find the amount of material that remains after 163 yd 2 ft is removed from the bolt.

60. The student activities club is sponsoring a walk for hunger, and all students who participate will receive a sash with the name of the school to wear on the walk. If each sash requires 5 ft 2 in. of material and there are 50 students participating in the walk, how much material will the student activities club need?

61. The trip from El Paso, TX, to Ontario, CA, is about 1235 km each way. Four friends agree to share the driving equally. How far must each drive on this round-trip vacation?

Ontario
California
1235 km
El Paso
Texas

△ **62.** The college has ordered that NO SMOKING signs be placed above the doorway of each classroom. Each sign is 0.8 m long and 30 cm wide. Find the area of each sign. (*Hint:* Recall that the area of a rectangle = width · length.)

0.8 meter

30 centimeters

(9.5) *Convert.*

63. 66 oz to pounds

64. 2.3 tons to pounds

65. 52 oz = _____ lb _____ oz

66. 10,300 lb = _____ tons _____ lb

67. 27 mg to grams

68. 40 kg to grams

69. 2.1 hg to dekagrams

70. 0.03 mg to decigrams

Perform each indicated operation.

71. 6 lb 5 oz − 2 lb 12 oz

72. 8 lb 6 oz × 4

73. 4.3 mg × 5

74. 4.8 kg − 4200 g

Solve.

75. Donshay Berry ordered 1 lb 12 oz of soft-center candies and 2 lb 8 oz of chewy-center candies for his party. Find the total weight of the candy ordered.

76. Four local townships jointly purchase 38 tons 300 lb of cinders to spread on their roads during an ice storm. Determine the weight of the cinders each township receives if they share the purchase equally.

(9.6) *Convert.*

77. 28 pints to quarts

78. 40 fluid ounces to cups

79. 3 qt 1 pt to pints

80. 18 quarts to cups

81. 9 pt = _____ qt _____ pt

82. 15 qt = _____ gal _____ qt

83. 3.8 L to milliliters

84. 14 hl to kiloliters

85. 30.6 L to centiliters

86. 2.45 ml to liters

Perform each indicated operation.

87. 1 qt 1 pt + 3 qt 1 pt

88. 3 gal 2 qt × 2

89. 0.946 L − 210 ml

90. 6.1 L + 9400 ml

Solve.

91. Carlos Perez prepared 4 gal 2 qt of iced tea for a block party. During the first 30 minutes of the party, 1 gal 3 qt of the tea is consumed. How much iced tea remains?

92. A recipe for soup stock calls for 1 c 4 fl oz of beef broth. How much should be used if the recipe is cut in half?

93. Each bottle of Kiwi liquid shoe polish holds 85 ml of the polish. Find the number of liters of shoe polish contained in 8 boxes if each box contains 16 bottles.

94. Ivan Miller wants to pour three separate containers of saline solution into a single vat with a capacity of 10 liters. Will 6 liters of solution in the first container combined with 1300 milliliters in the second container and 2.6 liters in the third container fit into the larger vat?

(9.7) *Note: Because approximations are used in this section, your answers may vary slightly from the answers given in the back of the book.*

Convert as indicated. If necessary, round to two decimal places.

95. 7 meters to feet

96. 11.5 yards to meters

97. 17.5 liters to gallons

98. 7.8 liters to quarts

99. 15 ounces to grams

100. 23 pounds to kilograms

101. A compact disc is 1.2 mm thick. Find the height (in inches) of 50 discs.

102. If a person weighs 82 kilograms, how many pounds is this?

Convert. Round to the nearest tenth of a degree, if necessary.

103. 42°C to degrees Fahrenheit

104. 160°C to degrees Fahrenheit

105. 41.3°F to degrees Celsius

106. 80°F to degrees Celsius

Solve. Round to the nearest tenth of a degree, if necessary.

107. A sharp dip in the jet stream caused the temperature in New Orleans to drop to 35°F. Find the corresponding temperature in degrees Celsius.

108. A recipe for meat loaf calls for a 165°C oven. Find the setting used if the oven has a Fahrenheit thermometer.

Mixed Review

Find the following.

109. Find the supplement of a 72° angle.

110. Find the complement of a 1° angle.

Copyright 2015 Pearson Education Inc.

Find the measure of angle x in each figure.

111.

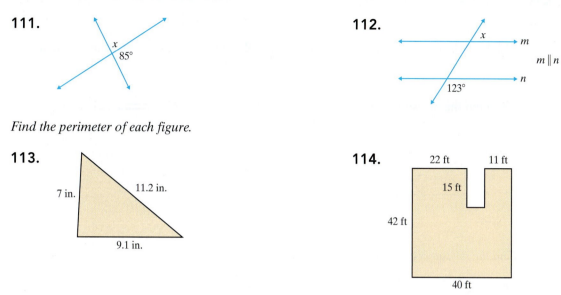

112.

113.

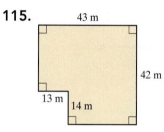

7 in.   11.2 in.

9.1 in.

114.

22 ft   11 ft

15 ft

42 ft

40 ft

Find the perimeter of each figure.

Find the area of each figure. For the circle, find the exact area and then use $\pi \approx 3.14$ *to approximate the area.*

115.

43 m

42 m

13 m

14 m

116.

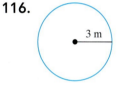

3 m

Find the volume of each solid.

117. Give an approximation using $\dfrac{22}{7}$ for π.

$5\frac{1}{4}$ in.

12 in.

118. Find the surface area also.

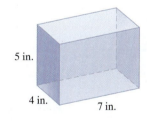

5 in.

4 in.   7 in.

Convert the following.

119. 6.25 ft to inches

120. 8200 lb = _____ tons _____ lb

121. 5 m to centimeters

122. 286 mm to kilometers

123. 1400 mg to grams

124. 6.75 gallons to quarts

125. 86°C to degrees Fahrenheit

126. 51.8°F to degrees Celsius

Perform the indicated operations and simplify.

127. 9.3 km − 183 m

128. 35 L + 700 ml

129. 3 gal 3 qt + 4 gal 2 qt

130. 3.2 kg × 4

Chapter 9   Test

Step-by-step test solutions are found on the Chapter Test Prep Videos. Where available: **MyMathLab®** or **You Tube**

Answers

△ **1.** Find the complement of a 78° angle.   △ **2.** Find the supplement of a 124° angle.

△ **3.** Find the measure of ∠x.

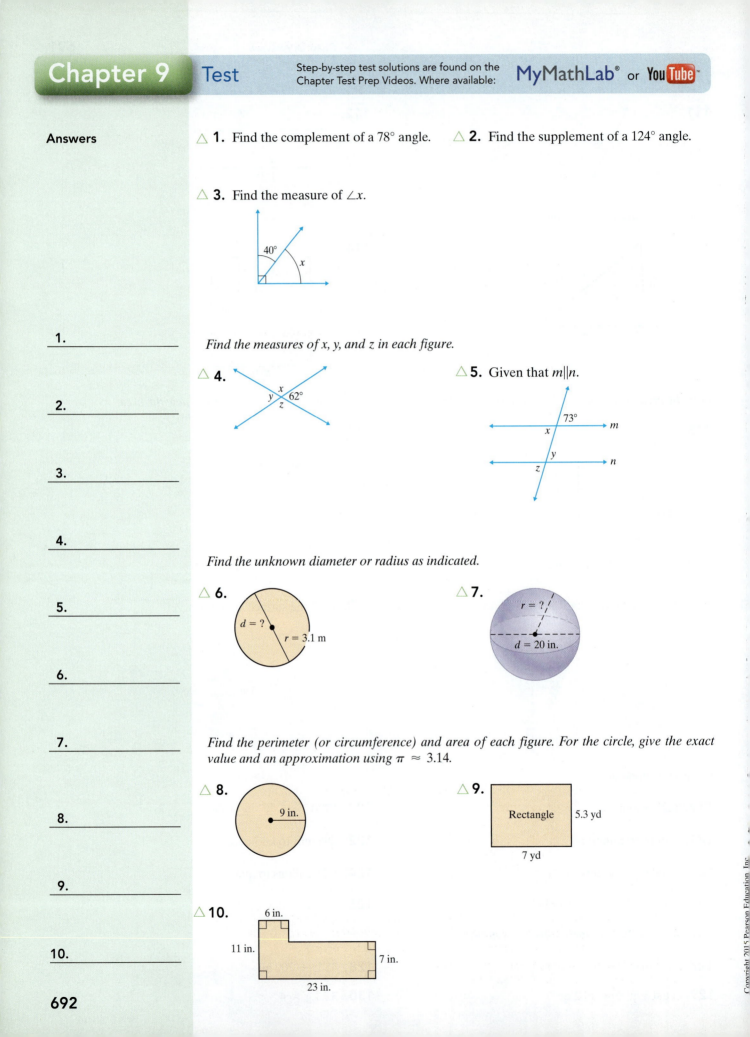

40°

x

1. _____

Find the measures of x, y, and z in each figure.

△ **4.**

x
y   62°
z

△ **5.** Given that m∥n.

73°
x   m
y
z   n

2. _____

3. _____

Find the unknown diameter or radius as indicated.

△ **6.**

d = ?
r = 3.1 m

△ **7.**

r = ?
d = 20 in.

4. _____

5. _____

6. _____

7. _____

Find the perimeter (or circumference) and area of each figure. For the circle, give the exact value and an approximation using π ≈ 3.14.

△ **8.**

9 in.

△ **9.**

Rectangle   5.3 yd

7 yd

8. _____

9. _____

△ **10.**

6 in.

11 in.

7 in.

23 in.

10. _____

Find the volume of each solid. For the cylinder, use $\pi \approx \frac{22}{7}$.

△**11.**

5 in.

2 in.

△**12.**

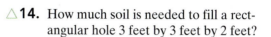

2 ft

3 ft   5 ft

Solve.

△**13.** Find the perimeter of a square photo with a side length of 4 inches.

△**14.** How much soil is needed to fill a rectangular hole 3 feet by 3 feet by 2 feet?

15. Find how much baseboard is needed to go around a rectangular room that measures 18 feet by 13 feet. If baseboard costs $1.87 per foot, also calculate the total cost needed for materials.

Convert.

16. 280 in. = _____ ft _____ in.

17. $2\frac{1}{2}$ gal to quarts

18. 30 oz to pounds

19. 2.8 tons to pounds

20. 38 pt to gallons

21. 40 mg to grams

22. 2.4 kg to grams

23. 3.6 cm to millimeters

24. 4.3 dg to grams

25. 0.83 L to milliliters

Perform each indicated operation.

26. 3 qt 1 pt + 2 qt 1 pt

27. 8 lb 6 oz − 4 lb 9 oz

28. 2 ft 9 in. × 3

29. 5 gal 2 qt ÷ 2

30. 8 cm − 14 mm

31. 1.8 km + 456 m

Convert. Round to the nearest tenth of a degree, if necessary.

32. 84°F to degrees Celsius

33. 12.6°C to degrees Fahrenheit

34. The sugar maples in front of Bette MacMillan's house are 8.4 meters tall. Because they interfere with the phone lines, the telephone company plans to remove the top third of the trees. How tall will the maples be after they are shortened?

35. A total of 15 gal 1 qt of oil has been removed from a 20-gallon drum. How much oil still remains in the container?

36. The engineer in charge of bridge construction said that the span of a certain bridge would be 88 m. But the actual construction required it to be 340 cm longer. Find the span of the bridge, in meters.

37. If 2 ft 9 in. of material is used to manufacture one scarf, how much material is needed for 6 scarves?

11. _____

12. _____

13. _____

14. _____

15. _____

16. _____

17. _____

18. _____

19. _____

20. _____

21. _____

22. _____

23. _____

24. _____

25. _____

26. _____

27. _____

28. _____

29. _____

30. _____

31. _____

32. _____

33. _____

34. _____

35. _____

36. _____

37. _____

Answers

1. _____

2. _____

3. a. _____

   b. _____

4. a. _____

   b. _____

5. _____

6. _____

7. _____

8. _____

9. _____

10. _____

11. _____

12. _____

13. _____

14. _____

15. _____

16. _____

1. Solve: $3a - 6 = a + 4$

2. Solve: $2x + 1 = 3x - 5$

3. Evaluate:

   a. $\left(\dfrac{2}{5}\right)^4$ b. $\left(-\dfrac{1}{4}\right)^2$

4. Evaluate:

   a. $\left(-\dfrac{1}{3}\right)^3$ b. $\left(\dfrac{3}{7}\right)^2$

5. Add: $2\dfrac{4}{5} + 5 + 1\dfrac{1}{2}$

6. Add: $2\dfrac{1}{3} + 4\dfrac{2}{5} + 3$

7. Simplify by combining like terms:
   $11.1x - 6.3 + 8.9x - 4.6$

8. Simplify by combining like terms:
   $2.5y + 3.7 - 1.3y - 1.9$

9. Simplify: $\dfrac{5.68 + (0.9)^2 \div 100}{0.2}$

10. Simplify: $\dfrac{0.12 + 0.96}{0.5}$

11. Insert $<$, $>$, or $=$ to form a true statement.

    $0.\overline{7}$ $\dfrac{7}{9}$

12. Insert $<$, $>$, or $=$ to form a true statement.

    0.43 $\dfrac{2}{5}$

13. Solve: $0.5y + 2.3 = 1.65$

14. Solve: $0.4x - 9.3 = 2.7$

△ 15. An inner-city park is in the shape of a square that measures 300 feet on a side. Find the length of the diagonal of the park, rounded to the nearest whole foot.

△ 16. A rectangular field is 200 feet by 125 feet. Find the length of the diagonal of the field, rounded to the nearest whole foot.

△ **17.** Given the rectangle shown:

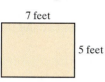

7 feet

5 feet

    a. Find the ratio of its width to its length.

    b. Find the ratio of its length to its perimeter.

△ **18.** A square is 9 inches by 9 inches.

9 inches

Square   9 inches

    a. Find the ratio of a side to its perimeter.

    b. Find the ratio of its perimeter to its area.

19. Write the rate as a fraction in simplest form: $2160 for 12 weeks.

20. Write the rate as a fraction in simplest form: 8 chaperones for 40 students

21. Solve for x: $\dfrac{1.6}{1.1} = \dfrac{x}{0.3}$

Round the solution to the nearest hundredth.

22. Solve for x: $\dfrac{2.4}{3.5} = \dfrac{0.7}{x}$

Round the solution to the nearest hundredth.

23. The standard dose of an antibiotic is 4 cc (cubic centimeters) for every 25 pounds (lb) of body weight. At this rate, find the standard dose for a 140-lb woman.

24. A recipe that makes 2 pie crusts calls for 3 cups of flour. How much flour is needed to make 5 pie crusts?

25. In a survey of 100 people, 17 people drive blue cars. What percent of people drive blue cars?

26. Of 100 shoppers surveyed at a mall, 38 paid for their purchases using only cash. What percent of shoppers used only cash to pay for their purchases?

27. 13 is $6\frac{1}{2}\%$ of what number?

28. 54 is $4\frac{1}{2}\%$ of what number?

29. What number is 30% of 9?

30. What number is 42% of 30?

31. The number of applications for a mathematics scholarship at one university increased from 34 to 45 in one year. What is the percent increase? Round to the nearest whole percent.

32. The price of a gallon of paint rose from $15 to $19. Find the percent increase, rounded to the nearest whole percent.

17. a. _____

    b. _____

18. a. _____

    b. _____

19. _____

20. _____

21. _____

22. _____

23. _____

24. _____

25. _____

26. _____

27. _____

28. _____

29. _____

30. _____

31. _____

32. _____

33. Find the sales tax and the total price on the purchase of an $85.50 atlas in a city where the sales tax rate is 7.5%.

34. A sofa has a purchase price of $375. If the sales tax rate is 8%, find the amount of sales tax and the total cost of the sofa.

35. Find the ordered pair corresponding to each point plotted on the rectangular coordinate system.

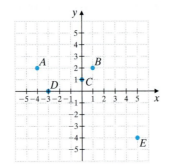

36. Find the ordered pair corresponding to each point plotted on the rectangular coordinate system.

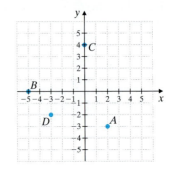

37. Graph $y = 4$.

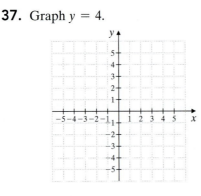

38. Graph $y = -2$.

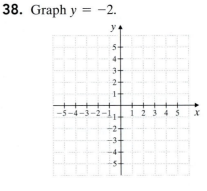

39. Find the median of the list of numbers: 25, 54, 56, 57, 60, 71, 98

40. Find the median of the list of scores: 60, 95, 89, 72, 83

41. Find the probability of choosing a red marble from a box containing 1 red, 1 yellow, and 2 blue marbles.

42. Find the probability of choosing a nickel at random in a coin purse that contains 2 pennies, 2 nickels, and 3 quarters.

43. Find the complement of a 48° angle.

44. Find the supplement of a 137° angle.

45. Convert 8 feet to inches.

46. Convert 7 yards to feet.

47. Find the area of a circle with a radius of 3 feet. Find the exact area, then an approximation using 3.14 for π.

48. Find the area of a circle with a radius of 2 miles. Find the exact area, then an approximation using 3.14 for π.

49. Convert 59°F to degrees Celsius.

50. Convert 86°F to degrees Celsius.

33. _____

34. _____

35. _____

36. _____

37. _____

38. _____

39. _____

40. _____

41. _____

42. _____

43. _____

44. _____

45. _____

46. _____

47. _____

48. _____

49. _____

50. _____

Exponents and Polynomials

Tablet **Hybrid** **Netbook**

Ultrabook **Notebook and Laptop**

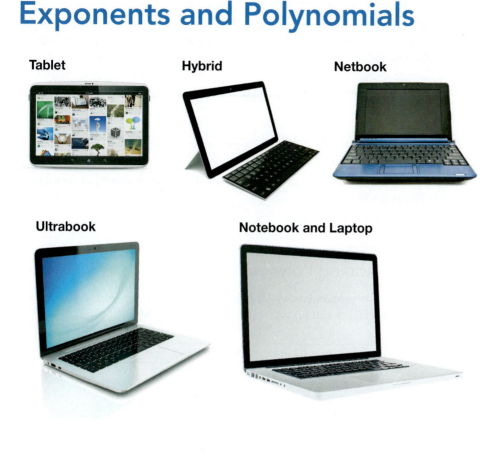

Just a few years ago, we had our cell phones to make mobile phone calls, and our large desktop computer or laptop for computing. Now, the lines are really starting to blur and we have new words such as *palmtop* (now obsolete), *tablet* (or *tablet PC*), *hybrid PC*, *netbook*, *ultrabook*, *notebook*, and *laptop*, just to name a few. What is different about these? In short, all these devices open and close except for a tablet. A tablet PC is a type of mobile computer that may have a touchscreen or a pen-enabled interface. Most companies that study the computer market say that tablets are the only growing portion of the PC market. It is predicted that more than half of Internet users will have a tablet of some kind by 2015. In the Chapter 10 Integrated Review, Exercise 31, we study the growth in sales of tablet PCs.

Recall that an exponent is a shorthand way of representing repeated multiplication. In this chapter, we learn more about exponents and a special type of expression containing exponents, called a *polynomial*. Studying polynomials is a major part of algebra. Polynomials are also useful for modeling many real-world situations. This chapter serves as an introduction to polynomials and some operations that can be performed on them.

Tablet PC Sales in the U.S.

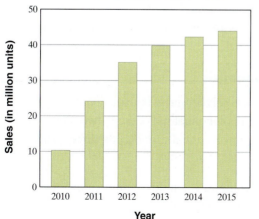

Source: Forrester Research (some years are projections)

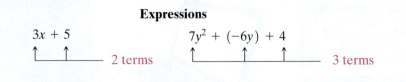
Objectives

A Add Polynomials.

B Subtract Polynomials.

C Evaluate Polynomials at Given Replacement Values.

Before we add and subtract polynomials, let's first review some definitions presented in Section 3.1. Recall that the *addends* of an algebraic expression are the *terms* of the expression.

Expressions

$$3x + 5$$

2 terms

$$7y^2 + (-6y) + 4$$

3 terms

Also, recall that *like terms* can be added or subtracted by using the distributive property. For example,

$$7x + 3x = (7 + 3)x = 10x$$

Objective A Adding Polynomials

Some terms are also **monomials.** A term is a monomial if the term contains only whole number exponents and no variable in the denominator.

Monomials	Not Monomials	
$3x^2$	$\dfrac{2}{y}$	Variable in denominator
$-\dfrac{1}{2}a^2bc^3$	$-2x^{-5}$	Not a whole number exponent
7		

A monomial or a sum and/or difference of monomials is called a **polynomial.**

Polynomial

A **polynomial** is a monomial or a sum and/or difference of monomials.

Examples of Polynomials

$$5x^3 - 6x^2 + 2x + 10, \quad -1.2y^3 + 0.7y, \quad z, \quad \frac{1}{3}r - \frac{1}{2}, \quad 0$$

Some polynomials are given special names depending on their number of terms.

Types of Polynomials

A **monomial** is a polynomial with exactly one term.
A **binomial** is a polynomial with exactly two terms.
A **trinomial** is a polynomial with exactly three terms.

The next page contains examples of monomials, binomials, and trinomials. Each of these examples is also a polynomial.

Polynomials			
Monomials	**Binomials**	**Trinomials**	**More than Three Terms**
z	$x + 2$	$x^2 - 2x + 1$	$5x^3 - 6x^2 + 2x - 10$
4	$\dfrac{1}{3}r - \dfrac{1}{2}$	$y^5 + 3y^2 - 1.7$	$t^7 - t^5 + t^3 - t + 1$
$0.2x^2$	$-1.2y^3 + 0.7y$	$-a^3 + 2a^2 - 5a$	$z^8 - z^4 + 3z^2 - 2z$
↑	↑	↑	
1 term	2 terms	3 terms	

To add polynomials, we use the commutative and associative properties to rearrange and group like terms. Then, we combine like terms.

> **Adding Polynomials**
>
> To add polynomials, combine like terms.

Example 1   Add: $(3x - 1) + (-6x + 2)$

Solution:

$$
\begin{aligned}
(3x - 1) + (-6x + 2) &= (3x - 6x) + (-1 + 2) &&\text{Group like terms.} \\
&= (-3x) + (1) &&\text{Combine like terms.} \\
&= -3x + 1
\end{aligned}
$$

■ **Work Practice 1**

Example 2   Add: $(9y^2 - 6y) + (7y^2 + 10y + 2)$

Solution:

$$
\begin{aligned}
(9y^2 - 6y) + (7y^2 + 10y + 2) &= 9y^2 + 7y^2 - 6y + 10y + 2 &&\text{Group like terms.} \\
&= 16y^2 + 4y + 2
\end{aligned}
$$

■ **Work Practice 2**

Example 3   Find the sum of $(-y^2 + 2y + 1.7)$ and $(12y^2 - 6y - 3.6)$.

Solution: Recall that "sum" means addition.

$$
\begin{aligned}
&(-y^2 + 2y + 1.7) + (12y^2 - 6y - 3.6) \\
&= \underbrace{-y^2 + 12y^2} + \underbrace{2y - 6y} + \underbrace{1.7 - 3.6} &&\text{Group like terms.} \\
&= 11y^2 - 4y - 1.9 &&\text{Combine like terms.}
\end{aligned}
$$

■ **Work Practice 3**

Polynomials can also be added vertically. To do this, line up like terms underneath one another. Let's vertically add the polynomials in Example 3.

Example 4   Find the sum of $(-y^2 + 2y + 1.7)$ and $(12y^2 - 6y - 3.6)$. Use a vertical format.

Solution: Line up like terms underneath one another.

$$
\begin{array}{r}
-y^2 + 2y + 1.7 \\
+12y^2 - 6y - 3.6 \\
\hline
11y^2 - 4y - 1.9
\end{array}
$$

■ **Work Practice 4**

Practice 1

Add:
$(3y + 7) + (-9y - 14)$

Practice 2

Add:
$(x^2 - 4x - 3) + (5x^2 - 6x)$

Practice 3

Find the sum of
$(-z^2 - 4.2z + 11)$ and
$(9z^2 - 1.9z + 6.3)$.

Practice 4

Add vertically:
$(x^2 - x + 1.1)$
$+ (-8x^2 - x - 6.7)$

Answers

1. $-6y - 7$ **2.** $6x^2 - 10x - 3$
3. $8z^2 - 6.1z + 17.3$
4. $-7x^2 - 2x - 5.6$

Notice that we are finding the same sum in Example 4 as in Example 3. Of course, the results are the same.

Objective B Subtracting Polynomials ▶

To subtract one polynomial from another, recall how we subtract numbers. Recall from Section 2.3 that to subtract a number, we add its opposite: $a - b = a + (-b)$.

For example,

$$7 - 10 = 7 + (-10)$$
$$= -3$$

To subtract a polynomial, we also add its opposite. Just as the opposite of 3 is -3, the opposite of $(2x^2 - 5x + 1)$ is $-(2x^2 - 5x + 1)$. Let's practice simplifying the opposite of a polynomial.

Practice 5

Simplify: $-(3y^2 + y - 2)$

Example 5   Simplify: $-(2x^2 - 5x + 1)$

Solution: Rewrite $-(2x^2 - 5x + 1)$ as $-1(2x^2 - 5x + 1)$ and use the distributive property.

$$-(2x^2 - 5x + 1) = -1(2x^2 - 5x + 1)$$
$$= -1(2x^2) + (-1)(-5x) + (-1)(1)$$
$$= -2x^2 + 5x - 1$$

■ Work Practice 5

Notice the result of Example 5.

$$-(2x^2 - 5x + 1) = -2x^2 + 5x - 1$$

This means that **the opposite of a polynomial can be found by changing the signs of the terms of the polynomial.** This leads to the following.

Subtracting Polynomials

To subtract polynomials, change the signs of the terms of the polynomial being subtracted, then add.

Practice 6

Subtract:
$(9b + 8) - (11b - 20)$

Example 6   Subtract: $(5a + 7) - (2a - 10)$

Solution:

$$(5a + 7) - (2a - 10) = (5a + 7) + (-2a + 10) \quad \text{Add the opposite of } 2a - 10.$$
$$= 5a - 2a + 7 + 10 \quad \text{Group like terms.}$$
$$= 3a + 17$$

■ Work Practice 6

Practice 7

Subtract:
$(11x^2 + 7x + 2) - (15x^2 + 4x)$

Example 7   Subtract: $(8x^2 - 4x + 1) - (10x^2 + 4)$

Solution:

$$(8x^2 - 4x + 1) - (10x^2 + 4) = (8x^2 - 4x + 1) + (-10x^2 - 4) \quad \begin{array}{l}\text{Add the opposite} \\ \text{of } 10x^2 + 4.\end{array}$$
$$= 8x^2 - 10x^2 - 4x + 1 - 4 \quad \text{Group like terms.}$$
$$= -2x^2 - 4x - 3$$

■ Work Practice 7

Answers

5. $-3y^2 - y + 2$ **6.** $-2b + 28$
7. $-4x^2 + 3x + 2$

Example 8 Subtract $(-6z^2 - 2z + 13)$ from $(4z^2 - 20z)$.

Solution: Be careful when arranging the polynomials in this example.

$$(4z^2 - 20z) - (-6z^2 - 2z + 13) = (4z^2 - 20z) + (6z^2 + 2z - 13)$$
$$= 4z^2 + 6z^2 - 20z + 2z - 13 \quad \text{Group like terms.}$$
$$= 10z^2 - 18z - 13$$

■ **Work Practice 8**

✓**Concept Check** Find and explain the error in the following subtraction.

$$(3x^2 + 4) - (x^2 - 3x) = (3x^2 + 4) + (-x^2 - 3x)$$
$$= 3x^2 - x^2 - 3x + 4$$
$$= 2x^2 - 3x + 4$$

Just as with adding polynomials, we can subtract polynomials using a vertical format. Let's subtract the polynomials in Example 8 using a vertical format.

Example 9 Subtract $(-6z^2 - 2z + 13)$ from $(4z^2 - 20z)$. Use a vertical format.

Solution: Line up like terms underneath one another.

$$
\begin{array}{r}
4z^2 - 20z \\
-(-6z^2 - 2z + 13)
\end{array}
\quad \text{can be written as} \quad
\begin{array}{r}
4z^2 - 20z \\
+6z^2 + 2z - 13 \\
\hline
10z^2 - 18z - 13
\end{array}
$$

■ **Work Practice 9**

Notice that the answers to Examples 8 and 9 are the same regardless of which format is used.

Objective C Evaluating Polynomials ▶

Polynomials have different values depending on the replacement values for the variables.

Example 10 Find the value of the polynomial $3t^3 - 2t + 5$ when $t = 1$.

Solution: Replace t with 1 and simplify.

$$3t^3 - 2t + 5 = 3(1)^3 - 2(1) + 5 \quad \text{Let } t = 1.$$
$$= 3(1) - 2(1) + 5 \quad (1)^3 = 1.$$
$$= 3 - 2 + 5$$
$$= 6$$

The value of $3t^3 - 2t + 5$ when $t = 1$ is 6.

■ **Work Practice 10**

Many real-world applications can be modeled by polynomials.

Practice 8
Subtract $(-7y^2 + y - 4)$ from $(-3y^2 + 5y)$.

Practice 9
Subtract $(3x^2 - 12x)$ from $(-4x^2 + 20x + 17)$. Use a vertical format.

Practice 10
Find the value of the polynomial $2y^3 + y^2 - 6$ when $y = 3$.

Answers
8. $4y^2 + 4y + 4$
9. $-7x^2 + 32x + 17$ **10.** 57

✓**Concept Check Answer**
$(3x^2 + 4) - (x^2 - 3x)$
$= (3x^2 + 4) + (-x^2 + 3x)$
$= 3x^2 - x^2 + 3x + 4$
$= 2x^2 + 3x + 4$

Practice 11

An object is dropped from the top of a 530-foot cliff. Its height in feet at time t seconds is given by the polynomial $-16t^2 + 530$. Find the height of the object when $t = 1$ second and when $t = 4$ seconds.

Answer

11. 514 feet; 274 feet

Example 11 Finding the Height of an Object

An object is dropped from the top of an 800-foot-tall building. Its height at time t seconds is given by the polynomial $-16t^2 + 800$. Find the height of the object when $t = 1$ second and when $t = 3$ seconds.

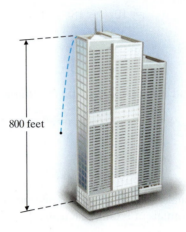

800 feet

Solution: To find each height, we evaluate the polynomial when $t = 1$ and when $t = 3$.

$$-16t^2 + 800 = -16(1)^2 + 800$$
$$= -16 + 800$$
$$= 784$$

The height of the object at 1 second is 784 feet.

$$-16t^2 + 800 = -16(3)^2 + 800$$
$$= -16(9) + 800$$
$$= -144 + 800$$
$$= 656$$

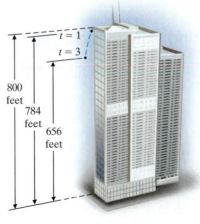

$t = 1$
$t = 3$
800 feet
784 feet
656 feet

The height of the object at 3 seconds is 656 feet.

■ **Work Practice 11**

Vocabulary, Readiness & Video Check

Use the choices below to fill in each blank.

trinomial	binomial	add
terms	monomial	subtract

1. The addends of an algebraic expression are the _____ of the expression.
2. A polynomial with exactly one term is called a _____.
3. A polynomial with exactly two terms is called a _____.
4. A polynomial with exactly three terms is called a _____.
5. To _____ polynomials, combine like terms.
6. To _____ polynomials, change the signs of the terms of the polynomial being subtracted; then add.

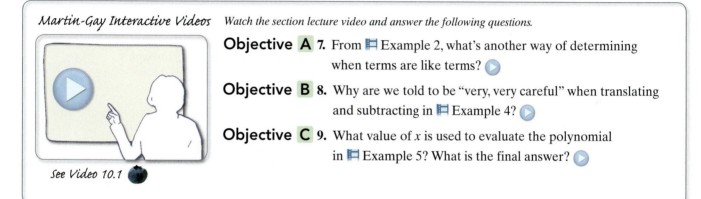

Martin-Gay Interactive Videos　Watch the section lecture video and answer the following questions.

Objective A **7.** From ▥ Example 2, what's another way of determining when terms are like terms? ▶

Objective B **8.** Why are we told to be "very, very careful" when translating and subtracting in ▥ Example 4? ▶

Objective C **9.** What value of x is used to evaluate the polynomial in ▥ Example 5? What is the final answer? ▶

See Video 10.1 ●

10.1 Exercise Set MyMathLab® ▶

Objective A *Add the polynomials. See Examples 1 through 4.*

1. $(2x + 3) + (-7x - 27)$

2. $(9y - 16) + (-43y + 16)$

▶ **3.** $(-3z^2 + 5z - 5) + (-8z^2 - 8z + 4)$

4. $(8a^2 + 5a - 9) + (5a^2 - 11a + 6)$

5. $(12y - 20) + (9y^2 + 13y - 20)$

6. $(5x^2 - 6) + (-3x^2 + 17x - 2)$

7. $(4.3a^4 + 5) + (-8.6a^4 - 2a^2 + 4)$

8. $(-12.7z^3 - 14z) + (-8.9z^3 + 12z + 2)$

Objective B *Simplify. See Example 5.*

9. $-(9x - 16)$

10. $-(4y - 12)$

11. $-(-3z^2 + z - 7)$

12. $-(-2x^2 - x + 1)$

Subtract the polynomials. See Examples 6 through 9.

13. $(8a - 5) - (3a + 8)$

14. $(3b + 5) - (-2b + 9)$

15. $(3x^2 - 2x + 1) - (5x^2 - 6x)$

16. $(-9z^2 + 6z + 2) - (3z^2 + 1)$

17. $(10y^2 - 7) - (20y^3 - 2y^2 - 3)$

18. $(11x^3 + 15x - 9) - (-x^3 + 10x^2 - 9)$

19. Subtract $(9x^2 + 3x - 4)$ from $(2x + 12)$.

20. Subtract $(4a^2 + 6a + 1)$ from $(-7a + 7)$.

21. Subtract $(5y^2 + 4y - 6)$ from $(13y^2 - 6y - 14)$.

22. Subtract $(16x^2 - x + 1)$ from $(12x^2 - 3x - 12)$.

Objectives A B Mixed Practice *Perform each indicated operation. See Examples 1 through 9.*

23. $(25x - 5) + (-20x - 7)$

24. $(14x + 2) + (-7x - 1)$

25. $(4y + 4) - (3y + 8)$

26. $(6z - 3) - (8z + 5)$

27. $(3x^2 + 3x - 4) + (-8x^2 + 9)$

28. $(-2a^2 - 5a) + (6a^2 - 2a + 9)$

29. $(5x + 4.5) + (-x - 8.6)$

30. $(20x - 0.8) + (x + 1.2)$

31. $(a - 5) - (-3a + 2)$

32. $(t + 9) - (-2t + 6)$

33. $(21y - 4.6) - (36y - 8.2)$

34. $(8.6x + 4) - (9.7x - 93)$

35. $(18t^2 - 4t + 2) - (-t^2 + 7t - 1)$

36. $(35x^2 + x - 5) - (17x^2 - x + 5)$

37. $(2b^3 + 5b^2 - 5b - 8) + (8b^2 + 9b + 6)$

38. $(3z^2 - 8z + 5) + (-3z^3 - 5z^2 - 2z - 4)$

39. Add $(6x^2 - 7)$ and $(-11x^2 - 11x + 20)$.

40. Add $(-2x^2 + 3x)$ and $(9x^2 - x + 14)$.

41. Subtract $\left(3z - \dfrac{3}{7}\right)$ from $\left(3z + \dfrac{6}{7}\right)$.

42. Subtract $\left(8y^2 - \dfrac{7}{10}y\right)$ from $\left(-5y^2 + \dfrac{3}{10}y\right)$.

Objective C *Find the value of each polynomial when $x = 2$. See Examples 10 and 11.*

43. $-2x + 9$

44. $-5x - 7$

45. $x^2 - 6x + 3$

46. $5x^2 + 4x - 100$

47. $\dfrac{3x^2}{2} - 14$

48. $\dfrac{7x^3}{14} - x + 5$

Find the value of each polynomial when x = 5. See Examples 10 and 11.

49. $2x + 10$

50. $-5x - 6$

51. x^2

52. x^3

53. $2x^2 + 4x - 20$

54. $4x^2 - 5x + 10$

Solve. See Example 11.

The distance in feet traveled by a free-falling object in t seconds is given by the polynomial $16t^2$. Use this polynomial for Exercises 55 and 56.

55. Find the distance traveled by an object that falls for 6 seconds.

56. It takes 8 seconds for a hard hat to fall from the top of a building. How high is the building?

Office Supplies, Inc. manufactures office products. The company determines that the total cost for manufacturing x file cabinets is given by the polynomial $3000 + 20x$. Use this polynomial for Exercises 57 and 58.

57. Find the total cost to manufacture 10 file cabinets.

58. Find the total cost to manufacture 100 file cabinets.

Devils Tower National Monument in Wyoming became America's first national monument in 1906. This rock formation has a height of 867 feet and is a popular climbing site. It was also used as the alien spacecraft landing site in the 1977 movie Close Encounters of the Third Kind.

59. One of the climbers of Devils Tower accidentally drops a piece of climbing chalk when he reaches the summit. The chalk's height above the ground, in feet, can be modeled by the equation $h = 867 - 16t^2$, where t stands for the number of seconds after the chalk is dropped. How far above the ground would the chalk be in 4 seconds?

60. At 7 seconds, the chalk in Exercise **59** passes another climber, who is closer to the base. How far above the ground is this climber?

An object is dropped from the deck of the Royal Gorge Bridge, which stretches across Royal Gorge at a height of 1053 feet above the Arkansas River. The height of the object above the river at t seconds is given by the polynomial $1053 - 16t^2$. Use this polynomial for Exercises 61 and 62. (Source: Royal Gorge Bridge Co.)

61. How far above the river is an object when $t = 3$ seconds?

62. How far above the river is an object when $t = 6$ seconds?

Solve.

63. The number of individuals using the Internet is still increasing every year. The number of individuals in Africa using the Internet (in millions) in a year can be modeled by $1.1x^2 - 3.3x + 5$, where x stands for the number of years since 2000. If the rate of growth continues as it has, how many individuals using the Internet should we expect in Africa in the year 2020? (*Source:* CTIA—The Wireless Association)

64. The continuously growing number of cellular subscribers means that the number of cell phone antennas needs to be constantly increased to carry the additional cellular traffic. The number of cell phone antennas (in thousands) located in the United States can be modeled by $0.3x^2 + 12x + 118$, where x stands for the number of years since 2000. If cellular services continue to grow at this rate, how many cell phone antennas should we expect in the United States in the year 2020? (*Source:* CTIA—The Wireless Association)

Review

Evaluate. See Sections 1.7 and 2.4.

65. 3^4

66. $(-2)^5$

67. $(-5)^2$

68. 4^3

Write using exponential notation. See Section 1.7.

69. $x \cdot x \cdot x$

70. $y \cdot y \cdot y \cdot y \cdot y$

71. $2 \cdot 2 \cdot a \cdot a \cdot a \cdot a$

72. $5 \cdot 5 \cdot 5 \cdot b \cdot b$

Concept Extensions

Find the perimeter of each figure.

△**73.**

(5x − 10) inches, (2x + 1) inches, (x + 11) inches

△**74.**

(x² − 6) meters, (x + 1) meters, (3x − 10) meters, (5x² + 2x) meters

Given the lengths in the figure below, we find the unknown length by subtracting. Use the information to find the unknown lengths in Exercises 75 and 76.

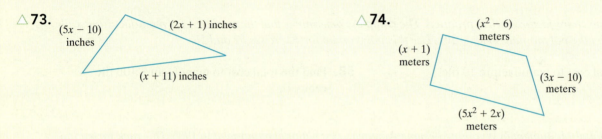

8 units

3 units   ?
(8 − 3) units

75.

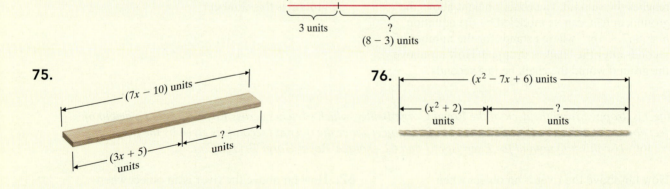

(7x − 10) units

(3x + 5) units   ? units

76.

(x² − 7x + 6) units

(x² + 2) units   ? units

Fill in the blanks.

77. $(3x^2 + \underline{\hspace{1cm}} x - \underline{\hspace{1cm}}) + (\underline{\hspace{1cm}} x^2 - 6x + 2) = 5x^2 + 14x - 4$

78. $(\underline{\hspace{1cm}} y^2 + 4y - 3) + (8y^2 - \underline{\hspace{1cm}} y + \underline{\hspace{1cm}}) = 9y^2 + 2y + 7$

79. Find the value of $7a^4 - 6a^2 + 2a - 1$ when $a = 1.2$.

80. Find the value of $3b^3 + 4b^2 - 100$ when $b = -2.5$.

81. For Exercises **61** and **62**, the polynomial $1053 - 16t^2$ was used to give the height of an object above the river at t seconds. Find the height when $t = 8$ seconds and $t = 9$ seconds. Explain what happened and why.

10.2 Multiplication Properties of Exponents ▶

Objective A Using the Product Property ▶

Recall from Section 1.8 that an exponent has the same meaning whether the base is a number or a variable. For example,

$$5^3 = \underbrace{5 \cdot 5 \cdot 5}_{\text{3 factors of 5}} \quad \text{and} \quad x^3 = \underbrace{x \cdot x \cdot x}_{\text{3 factors of } x}$$

We can use this definition of an exponent to discover properties that will help us to simplify products and powers of exponential expressions.

For example, let's use the definition of an exponent to find the product of x^3 and x^4.

$$x^3 \cdot x^4 = (x \cdot x \cdot x)(x \cdot x \cdot x \cdot x)$$
$$= \underbrace{x \cdot x \cdot x \cdot x \cdot x \cdot x \cdot x}_{\text{7 factors of } x}$$
$$= x^7$$

Notice that the result is the same if we add the exponents.

$$x^3 \cdot x^4 = x^{3+4} = x^7$$

This suggests the following product rule or property for exponents.

> **Product Property for Exponents**
>
> If m and n are positive integers and a is a real number, then
>
> $$a^m \cdot a^n = a^{m+n}$$

In other words, to multiply two exponential expressions with the same base, keep the base and add the exponents.

Example 1 Multiply: $y^7 \cdot y^2$

Solution: $y^7 \cdot y^2 = y^{7+2}$ Use the product property for exponents.
$\qquad\qquad = y^9$ Simplify.

■ Work Practice 1

Example 2 Multiply: $3x^5 \cdot 6x^3$

Solution: $3x^5 \cdot 6x^3 = (3 \cdot 6)(x^5 \cdot x^3)$ Apply the commutative and associative properties.
$\qquad\qquad\quad = 18x^{5+3}$ Use the product property for exponents.
$\qquad\qquad\quad = 18x^8$ Simplify.

■ Work Practice 2

Example 3 Multiply: $(-2a^4b^{10})(9a^5b^3)$

Solution: Use properties of multiplication to group numbers and like variables together.

$$(-2a^4b^{10})(9a^5b^3) = (-2 \cdot 9)(a^4 \cdot a^5)(b^{10} \cdot b^3)$$
$$= -18a^{4+5}b^{10+3}$$
$$= -18a^9b^{13}$$

■ Work Practice 3

Objectives

A Use the Product Property for Exponents. ▶

B Use the Power Property for Exponents. ▶

C Use the Power of a Product Property for Exponents. ▶

Practice 1

Multiply: $z^5 \cdot z^6$

Practice 2

Multiply: $8y^5 \cdot 4y^9$

Practice 3

Multiply: $(-4r^6s^2)(-3r^2s^5)$

Answers
1. z^{11} 2. $32y^{14}$ 3. $12r^8s^7$

Practice 4
Multiply: $11y^5 \cdot 3y^2 \cdot y$. (Recall that $y = y^1$.)

Example 4   Multiply: $2x^3 \cdot 3x \cdot 5x^6$

Solution: First notice the factor $3x$. Since there is one factor of x in $3x$, it can also be written as $3x^1$.

$$2x^3 \cdot 3x^1 \cdot 5x^6 = (2 \cdot 3 \cdot 5)\left(x^3 \cdot x^1 \cdot x^6\right)$$
$$= 30x^{10}$$

Helpful Hint   Don't forget that if an exponent is not written, it is assumed to be 1.

■ **Work Practice 4**

Helpful Hint

These examples will remind you of the difference between adding and multiplying terms.

Addition	Multiplication
$5x^3 + 3x^3 = (5 + 3)x^3 = 8x^3$	$\left(5x^3\right)\left(3x^3\right) = 5 \cdot 3 \cdot x^3 \cdot x^3 = 15x^{3+3} = 15x^6$
$7x + 4x^2 = 7x + 4x^2$	$\left(7x\right)\left(4x^2\right) = 7 \cdot 4 \cdot x \cdot x^2 = 28x^{1+2} = 28x^3$

Objective B   Using the Power Property ▶

Next suppose that we want to simplify an exponential expression raised to a power. To see how we simplify $\left(x^2\right)^3$, we again use the definition of an exponent.

$$\left(x^2\right)^3 = \underbrace{\left(x^2\right) \cdot \left(x^2\right) \cdot \left(x^2\right)}_{3 \text{ factors of } x^2} \qquad \text{Apply the definition of an exponent.}$$

$$= x^{2+2+2} \qquad \text{Use the product property for exponents.}$$

$$= x^6 \qquad \text{Simplify.}$$

Notice the result is exactly the same if we multiply the exponents.

$$\left(x^2\right)^3 = x^{2 \cdot 3} = x^6$$

This suggests the following power rule or property for exponents.

Power Property for Exponents

If m and n are positive integers and a is a real number, then

$$\left(a^m\right)^n = a^{m \cdot n}$$

In other words, to raise a power to a power, keep the base and multiply the exponents.

Helpful Hint

Take a moment to make sure that you understand when to apply the product rule and when to apply the power rule.

Product Property → Add Exponents	Power Property → Multiply Exponents
$x^5 \cdot x^7 = x^{5+7} = x^{12}$	$\left(x^5\right)^7 = x^{5 \cdot 7} = x^{35}$
$y^6 \cdot y^2 = y^{6+2} = y^8$	$\left(y^6\right)^2 = y^{6 \cdot 2} = y^{12}$

Answer

4. $33y^8$

Example 5 Simplify: $(y^8)^2$

Solution: $(y^8)^2 = y^{8 \cdot 2}$ Use the power property.

$= y^{16}$

■ Work Practice 5

Practice 5
Simplify: $(z^3)^6$

Example 6 Simplify: $(a^3)^4 \cdot (a^2)^9$

Solution: $(a^3)^4 \cdot (a^2)^9 = a^{12} \cdot a^{18}$ Use the power property.

$= a^{12+18}$ Use the product property.

$= a^{30}$ Simplify.

■ Work Practice 6

Practice 6
Simplify: $(z^4)^5 \cdot (z^3)^7$

Objective C Using the Power of a Product Property ▶

Next, let's simplify the power of a product.

$(xy)^3 = xy \cdot xy \cdot xy$ Apply the definition of an exponent.

$= (x \cdot x \cdot x)(y \cdot y \cdot y)$ Group like bases.

$= x^3 y^3$ Simplify.

Notice that the power of a product can be written as the product of powers. This leads to the following power of a product rule or property.

Power of a Product Property for Exponents

If n is a positive integer and a and b are real numbers, then

$(ab)^n = a^n b^n$

In other words, to raise a product to a power, raise each factor to the power.

✔**Concept Check** Which property is needed to simplify $(x^6)^3$? Explain.

a. Product property for exponents

b. Power property for exponents

c. Power of a product property for exponents

Example 7 Simplify: $(5t)^3$

Solution: $(5t)^3 = 5^3 t^3$ Apply the power of a product property.

$= 125t^3$ Write 5^3 as 125.

■ Work Practice 7

Practice 7
Simplify: $(3b)^4$

Example 8 Simplify: $(2a^5 b^3)^3$

Solution: $(2a^5 b^3)^3 = 2^3 (a^5)^3 (b^3)^3$ Apply the power of a product property.

$= 8a^{15} b^9$ Apply the power property.

■ Work Practice 8

Practice 8
Simplify: $(4x^2 y^6)^3$

Answers

5. z^{18} **6.** z^{41} **7.** $81b^4$ **8.** $64x^6 y^{18}$

✔**Concept Check Answer**

b

Practice 9

Simplify: $(2x^2y^4)^4(3x^6y^9)^2$

Answer

9. $144x^{20}y^{34}$

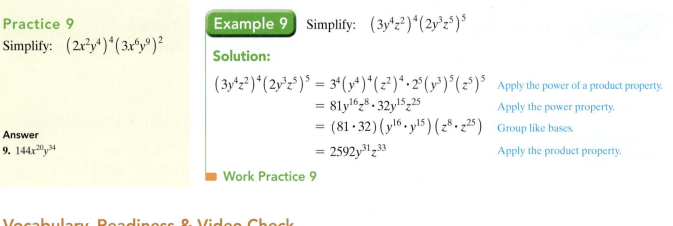

Example 9 Simplify: $(3y^4z^2)^4(2y^3z^5)^5$

Solution:

$$(3y^4z^2)^4(2y^3z^5)^5 = 3^4(y^4)^4(z^2)^4 \cdot 2^5(y^3)^5(z^5)^5 \quad \text{Apply the power of a product property.}$$
$$= 81y^{16}z^8 \cdot 32y^{15}z^{25} \quad \text{Apply the power property.}$$
$$= (81 \cdot 32)(y^{16} \cdot y^{15})(z^8 \cdot z^{25}) \quad \text{Group like bases.}$$
$$= 2592y^{31}z^{33} \quad \text{Apply the product property.}$$

■ **Work Practice 9**

Vocabulary, Readiness & Video Check

Use the choices below to fill in each blank. Not all choices will be used.

add	multiply	exponent	$6x^2$
subtract	divide	$36x$	$36x^2$

1. In $7x^2$, the 2 is called the _____.

2. To simplify $x^4 \cdot x^3$, we _____ the exponents.

3. To simplify $(x^4)^3$, we _____ the exponents.

4. The expression $(6x)^2$ simplifies to _____.

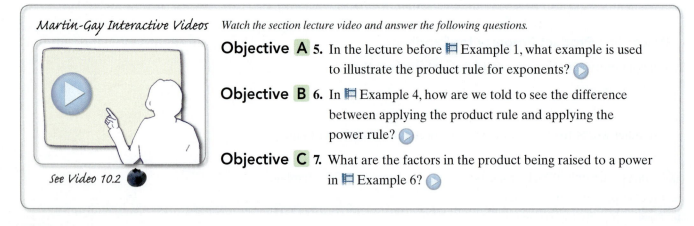

Martin-Gay Interactive Videos   *Watch the section lecture video and answer the following questions.*

Objective A **5.** In the lecture before ⊞ Example 1, what example is used to illustrate the product rule for exponents? ▶

Objective B **6.** In ⊞ Example 4, how are we told to see the difference between applying the product rule and applying the power rule? ▶

Objective C **7.** What are the factors in the product being raised to a power in ⊞ Example 6? ▶

See Video 10.2

10.2 Exercise Set MyMathLab® ▶

Objective A *Multiply. See Examples 1 through 4.*

▶ **1.** $x^5 \cdot x^9$

2. $y^4 \cdot y^7$

3. $a^3 \cdot a$

4. $b \cdot b^4$

5. $3z^3 \cdot 5z^2$

6. $8r^2 \cdot 2r^{15}$

7. $-4x \cdot 10x$

8. $-9y \cdot 3y$

▶ **9.** $2x \cdot 3x \cdot 7x$

10. $4y \cdot 3y \cdot 5y$

11. $a \cdot 4a^{11} \cdot 3a^5$

12. $b \cdot 7b^{10} \cdot 5b^8$

13. $(-5x^2y^3)(-5x^4y)$

14. $(-2xy^4)(-6x^3y^7)$

▶ **15.** $(7ab)(4a^4b^5)$

16. $(3a^3b^6)(12a^2b^9)$

Objectives B C **Mixed Practice** *Simplify. See Examples 5 through 9.*

▶ **17.** $(x^5)^3$

18. $(y^4)^7$

19. $(z^3)^{10}$

20. $(a^9)^3$

▶ **21.** $(b^7)^6 \cdot (b^2)^{10}$

22. $(x^2)^9 \cdot (x^5)^3$

▶ **23.** $(3a)^4$

24. $(2y)^5$

25. $(a^{11}b^8)^3$

26. $(x^7y^4)^8$

27. $(10x^5y^3)^3$

28. $(8a^5b^7)^2$

29. $(-3y)(2y^7)^3$

30. $(-2x)(5x^2)^4$

▶ **31.** $(4xy)^3(2x^3y^5)^2$

32. $(2xy)^4(3x^4y^3)^3$

Review

Multiply. See Section 3.1.

33. $7(x - 3)$

34. $4(y + 2)$

35. $-2(3a + 2b)$

36. $-3(8r + 3s)$

37. $9(x + 2y - 3)$

38. $5(a + 7b - 3)$

Concept Extensions

Find the area of each figure.

△ **39.**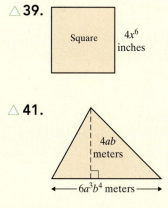

Square $4x^6$ inches

△ **40.**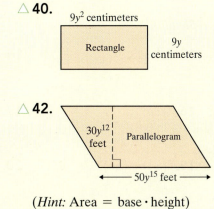

$9y^2$ centimeters

Rectangle $9y$ centimeters

△ **41.**

4ab meters

6a³b⁴ meters

△ **42.**

$30y^{12}$ feet Parallelogram

$50y^{15}$ feet

(*Hint:* Area = base · height)

Multiply and simplify.

🖩 **43.** $(14a^7b^6)^3(9a^6b^3)^4$

🖩 **44.** $(5x^{14}y^6)^7(3x^{20}y^{19})^5$

🖩 **45.** $(8.1x^{10})^5$

🖩 **46.** $(4.6a^{14})^4$

47. $(x^{90}y^{72})^3$

48. $(a^{20}b^{10}c^5)^5(a^9b^{12})^3$

✏ **49.** In your own words, explain why $x^2 \cdot x^3 = x^5$ and $(x^2)^3 = x^6$.

Operations on Polynomials

Answers

1. _____

2. _____

3. _____

4. _____

5. _____

6. _____

7. _____

8. _____

9. _____

10. _____

11. _____

12. _____

13. _____

14. _____

15. _____

16. _____

17. _____

18. _____

19. _____

20. _____

21. _____

22. _____

23. _____

24. _____

25. _____

26. _____

27. _____

28. _____

29. _____

30. _____

31. _____

Add or subtract the polynomials as indicated.

1. $(3x + 5) + (-x - 8)$

2. $(15y - 7) + (5y - 4)$

3. $(7x + 1) - (-3x - 2)$

4. $(14y - 6) - (19y - 2)$

5. $(a^4 + 5a) - (3a^4 - 3a^2 - 4a)$

6. $(2a^3 - 6a^2 + 11) - (6a^3 + 6a^2 + 11)$

7. $(4.5x^2 + 8.1x) + (2.8x^2 - 12.3x - 5.3)$ **8.** $(1.2y^2 - 3.6y) + (0.6y^2 + 1.2y - 5.6)$

9. Subtract $(2x - 6)$ from $(8x + 1)$.

10. Subtract $(3x^2 - x + 2)$ from $(5x^2 + 2x - 10)$.

Find the value of each polynomial when $x = 3$.

11. $2x - 7$

12. $x^2 + 5x + 2$

Simplify.

13. $x^9 \cdot x^{11}$

14. $x^5 \cdot x^5$

15. $y^3 \cdot y$

16. $a \cdot a^{10}$

17. $(x^7)^{11}$

18. $(x^6)^6$

19. $(x^3)^4 \cdot (x^5)^6$

20. $(y^2)^9 \cdot (y^3)^3$

21. $(5x)^3$

22. $(2y)^5$

23. $(-6xy^2)(2xy^5)$

24. $(-4a^2b^3)(-3ab)$

25. $(y^{11}z^{13})^3$

26. $(a^5b^{12})^4$

27. $(10x^2y)^2(3y)$

28. $(8y^3z)^2(2z^5)$

29. $(2a^5b)^4(3a^9b^4)^2$

30. $(5x^4y^6)^3(x^2y^2)^5$

31. The sales of tablet PCs in the United States are still increasing, but are predicted to slow down and even decrease as the market becomes saturated. The number of tablet PCs sold (in millions of units) in each year can be modeled by $-1.7x^2 + 49x - 308$, where x is the number of years since 2000. If the growth continues as it has, how many tablet PCs should we expect to be sold in the year 2018? (See the Chapter 10 Opener.)

10.3 Multiplying Polynomials

Objective A Multiplying a Monomial and a Polynomial

Recall from Section 10.1 that a polynomial that consists of one term is called a **monomial.** For example, $5x$ is a monomial. To multiply a monomial and any polynomial, we use the distributive property

$$a(b + c) = a \cdot b + a \cdot c$$

and apply properties of exponents.

Objectives

A Multiply a Monomial and Any Polynomial.

B Multiply Two Binomials.

C Square a Binomial.

D Use the FOIL Order to Multiply Binomials.

E Multiply Any Two Polynomials.

Example 1 Multiply: $5x(3x^2 + 2)$

Solution:

$$5x(3x^2 + 2) = 5x \cdot 3x^2 + 5x \cdot 2 \quad \text{Apply the distributive property.}$$
$$= 15x^3 + 10x$$

■ Work Practice 1

Practice 1

Multiply: $4y(8y^2 + 5)$

Example 2 Multiply: $2z(4z^2 + 6z - 9)$

Solution:

$$2z(4z^2 + 6z - 9) = 2z \cdot 4z^2 + 2z \cdot 6z + 2z(-9)$$
$$= 8z^3 + 12z^2 - 18z$$

■ Work Practice 2

Practice 2

Multiply: $3r(8r^2 - r + 11)$

To visualize multiplication by a monomial, let's look at two ways we can represent the area of the same rectangle.

Method 1: The width of the rectangle is x and its length is $x + 3$. One way to calculate the area of the rectangle is

area = width · length
$$= x(x + 3)$$

Method 2: Another way to calculate the area of the rectangle is to find the sum of the areas of the smaller figures.

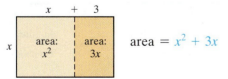

area = $x^2 + 3x$

Since the areas must be equal, we have that

$$x(x + 3) = x^2 + 3x \quad \text{As expected from the distributive property.}$$

Answers
1. $32y^3 + 20y$ 2. $24r^3 - 3r^2 + 33r$

Objective B Multiplying Binomials ▶

Recall also from Section 10.1 that a polynomial that consists of exactly two terms is called a **binomial.** To multiply two binomials, we use a version of the distributive property:

$$(b + c)a = b \cdot a + c \cdot a$$

Practice 3

Multiply: $(b + 3)(b + 5)$

> **Example 3**   Multiply: $(x + 2)(x + 3)$
>
> **Solution:**
>
> $$\begin{aligned}(x + 2)(x + 3) &= x(x + 3) + 2(x + 3) &&\text{Apply the distributive property.}\\ &= x \cdot x + x \cdot 3 + 2 \cdot x + 2 \cdot 3 &&\text{Apply the distributive property.}\\ &= x^2 + 3x + 2x + 6 &&\text{Multiply.}\\ &= x^2 + 5x + 6 &&\text{Combine like terms.}\end{aligned}$$
>
> ■ Work Practice 3

Practice 4

Multiply: $(7x - 1)(5x + 4)$

> **Example 4**   Multiply: $(4y + 9)(3y - 2)$
>
> **Solution:**
>
> $$\begin{aligned}(4y + 9)(3y - 2) &= 4y(3y - 2) + 9(3y - 2) &&\text{Apply he distributive property.}\\ &= 4y \cdot 3y + 4y(-2) + 9 \cdot 3y + 9(-2) &&\text{Apply the distributive property.}\\ &= 12y^2 - 8y + 27y - 18 &&\text{Multiply.}\\ &= 12y^2 + 19y - 18 &&\text{Combine like terms.}\end{aligned}$$
>
> ■ Work Practice 4

Objective C Squaring a Binomial ▶

Raising a binomial to the power of 2 is also called squaring a binomial. To square a binomial, we use the definition of an exponent, and then multiply.

Practice 5

Multiply: $(6y - 1)^2$

> **Example 5**   Multiply: $(2x + 1)^2$
>
> **Solution:**
>
> $$\begin{aligned}(2x + 1)^2 &= (2x + 1)(2x + 1) &&\text{Apply the definition of an exponent.}\\ &= 2x(2x + 1) + 1(2x + 1) &&\text{Apply the distributive property.}\\ &= 2x \cdot 2x + 2x \cdot 1 + 1 \cdot 2x + 1 \cdot 1 &&\text{Apply the distributive property.}\\ &= 4x^2 + 2x + 2x + 1 &&\text{Multiply.}\\ &= 4x^2 + 4x + 1 &&\text{Combine like terms.}\end{aligned}$$
>
> ■ Work Practice 5

Answers

3. $b^2 + 8b + 15$ **4.** $35x^2 + 23x - 4$
5. $36y^2 - 12y + 1$

✔ **Concept Check Answer**
$$\begin{aligned}(x + 5)^2 &= (x + 5)(x + 5)\\ &= x^2 + 10x + 25\end{aligned}$$

✔ **Concept Check**   Correct and explain the error:

$$(x + 5)^2 = x^2 + 25$$

Objective D Using the FOIL Order to Multiply Binomials

Recall from Example 3 that

$$(x + 2)(x + 3) = x \cdot x + x \cdot 3 + 2 \cdot x + 2 \cdot 3$$
$$= x^2 + 5x + 6$$

One way to remember the products $x \cdot x$, $x \cdot 3$, $2 \cdot x$, and $2 \cdot 3$ is to use a special order for multiplying binomials, called the FOIL order. Of course, the product is the same no matter what order or method you choose to use.

FOIL stands for the products of the First terms, Outer terms, Inner terms, and then Last terms. For example,

$$(x + 2)(x + 3) = \overset{F}{x \cdot x} + \overset{O}{x \cdot 3} + \overset{I}{2 \cdot x} + \overset{L}{2 \cdot 3} = x^2 + 3x + 2x + 6$$
$$= x^2 + 5x + 6$$

> **Helpful Hint**
> The product is the same no matter what order or method you choose to use.

Examples Use the FOIL order to multiply.

6. $(3x - 6)(2x + 1) = \overset{F}{3x \cdot 2x} + \overset{O}{3x \cdot 1} + \overset{I}{(-6)(2x)} + \overset{L}{(-6)(1)}$

$\qquad\qquad\qquad\qquad = 6x^2 + 3x - 12x - 6$ Multiply.

$\qquad\qquad\qquad\qquad = 6x^2 - 9x - 6$ Combine like terms.

7. $(3x - 5)^2 = (3x - 5)(3x - 5)$

$\qquad\qquad = \overset{F}{3x \cdot 3x} + \overset{O}{3x(-5)} + \overset{I}{(-5)(3x)} + \overset{L}{(-5)(-5)}$

$\qquad\qquad = 9x^2 - 15x - 15x + 25$ Multiply.

$\qquad\qquad = 9x^2 - 30x + 25$ Combine like terms.

■ **Work Practice 6–7**

Practice 6–7

Use the FOIL order to multiply.
6. $(10x - 7)(2x + 3)$
7. $(3x + 2)^2$

> **Helpful Hint**
> Remember that the FOIL order can only be used to multiply **two binomials**.

Objective E Multiplying Polynomials

In Section 10.1, we learned that a polynomial that consists of exactly three terms is called a **trinomial.** Next, we multiply a binomial by a trinomial.

Example 8 Multiply: $(3a + 2)(a^2 - 6a + 3)$

Solution: Use the distributive property to multiply $3a$ by the trinomial $(a^2 - 6a + 3)$ and then 2 by the trinomial.

$(3a + 2)(a^2 - 6a + 3) = 3a(a^2 - 6a + 3) + 2(a^2 - 6a + 3)$ Apply the distributive property.

$\qquad = 3a \cdot a^2 + 3a(-6a) + 3a \cdot 3 +$ Apply the distributive
$\qquad\qquad 2 \cdot a^2 + 2(-6a) + 2 \cdot 3$ property.

$\qquad = 3a^3 - 18a^2 + 9a + 2a^2 - 12a + 6$ Multiply.

$\qquad = 3a^3 - 16a^2 - 3a + 6$ Combine like terms.

■ **Work Practice 8**

Practice 8

Multiply:
$(2x + 5)(x^2 + 4x - 1)$

Answers
6. $20x^2 + 16x - 21$
7. $9x^2 + 12x + 4$
8. $2x^3 + 13x^2 + 18x - 5$

In general, we have the following.

> **To Multiply Two Polynomials**
>
> Multiply each term of the first polynomial by each term of the second polynomial, and then combine like terms.

A convenient method of multiplying polynomials is to use a vertical format similar to multiplying real numbers.

✓**Concept Check** True or false? When a trinomial is multiplied by a trinomial, the result will have at most nine terms. Explain.

Practice 9

Multiply $(x^2 + 3x - 2)$ and $(3x + 4)$ vertically.

Example 9 Find the product of $(a^2 - 6a + 3)$ and $(3a + 2)$ vertically.

Solution:

$$
\begin{array}{r}
a^2 - 6a + 3 \\
\times \quad\quad 3a + 2 \\
\hline
2a^2 - 12a + 6 \\
3a^3 - 18a^2 + 9a \\
\hline
3a^3 - 16a^2 - 3a + 6 \\
\end{array}
$$

Multiply $a^2 - 6a + 3$ by 2.

Multiply $a^2 - 6a + 3$ by $3a$. Line up like terms.

Combine like terms.

Notice that this example is the same as Example 8, and that of course the products are the same.

■ **Work Practice 9**

Answer

9. $3x^3 + 13x^2 + 6x - 8$

✓**Concept Check Answer**

true

Vocabulary, Readiness & Video Check

Martin-Gay Interactive Videos

See Video 10.3

Watch the section lecture video and answer the following questions.

Objective A **1.** How is the distributive property used to solve ▤ Example 1? ▶

Objective B **2.** In ▤ Example 2, how many times is the distributive property used? List each distribution using specific steps from the example. ▶

Objective C **3.** In ▤ Example 3, why is the power rule for exponents mentioned? ▶

Objective D **4.** From ▤ Examples 4 and 5, what's the only type of multiplication for which we can apply the FOIL order of multiplying? ▶

Objective E **5.** Can the FOIL order of multiplying be used to solve ▤ Example 6? Why or why not? ▶

10.3 Exercise Set MyMathLab®

Objective A *Multiply. See Examples 1 and 2.*

1. $3x(9x^2 - 3)$

2. $4y(10y^3 + 2y)$

3. $-3a(2a^2 - 3a - 5)$

4. $-4b(-2b^2 - 5b + 8)$

▶ 5. $7x^2(6x^2 - 5x + 7)$

6. $6z^2(-3z^2 - z + 4)$

Objectives B C D Mixed Practice *Multiply. See Examples 3 through 7.*

7. $(x + 3)(x + 10)$

8. $(y + 5)(y + 9)$

▶ 9. $(2x - 6)(x + 4)$

10. $(7z + 1)(z - 6)$

11. $(6a + 4)^2$

12. $(8b - 3)^2$

Objective E *Multiply. See Examples 8 and 9.*

13. $(a + 6)(a^2 - 6a + 3)$

14. $(y + 4)(y^2 + 8y - 2)$

15. $(4x - 5)(2x^2 + 3x - 10)$

16. $(9z - 2)(2z^2 + z + 1)$

17. $(x^3 + 2x + x^2)(3x + 1 + x^2)$

18. $(y^2 - 2y + 5)(y^3 + 2 + y)$

Objectives A B C D E Mixed Practice *Multiply. See Examples 1 through 9.*

19. $10r(-3r + 2)$

20. $5x(4x^2 + 5)$

21. $-2y^2(3y + y^2 - 6)$

22. $3z^3(4z^4 - 2z + z^3)$

23. $(x + 2)(x + 12)$

24. $(y + 7)(y - 7)$

▶ 25. $(2a + 3)(2a - 3)$

26. $(6s + 1)(3s - 1)$

27. $(x + 5)^2$

28. $(x + 3)^2$

▶ 29. $\left(b + \dfrac{3}{5}\right)\left(b + \dfrac{4}{5}\right)$

30. $\left(a - \dfrac{7}{10}\right)\left(a + \dfrac{3}{10}\right)$

31. $(6x + 1)(x^2 + 4x + 1)$

32. $(9y - 1)(y^2 + 3y - 5)$

▶ 33. $(7x + 5)^2$

34. $(5x + 9)^2$

35. $(2x - 1)^2$

36. $(4a - 3)^2$

37. $(2x^2 - 3)(4x^3 + 2x - 3)$ **38.** $(3y^2 + 2)(5y^2 - y + 2)$ **39.** $(x^3 + x^2 + x)(x^2 + x + 1)$

40. $(a^4 + a^2 + 1)(a^4 + a^2 - 1)$ **41.** $(2z^2 - z + 1)(5z^2 + z - 2)$ **42.** $(2b^2 - 4b + 3)(b^2 - b + 2)$

Review

Write each number as a product of prime numbers. See Section 4.2.

43. 50   **44.** 48   **45.** 72

46. 36   **47.** 200   **48.** 300

Concept Extensions

Find the area of each figure.

△ **49.**

$(y - 6)$ feet

$(y^2 + 3y + 2)$ feet

△ **50.**

Square $(2x + 11)$ centimeters

Find the area of the shaded figure. To do so, subtract the area of the smaller square from the area of the larger geometric figure.

△ **51.**

Square

$(x^2 - 1)$ meters

x meters

△ **52.**

$(3x + 5)$ miles

$2x$ miles

$(3x - 5)$ miles

53. Suppose that a classmate asked you why $(2x + 1)^2$ is **not** $4x^2 + 1$. Write down your response to this classmate.

Introduction to Factoring Polynomials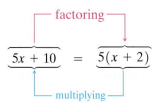

Recall from Section 2.4 that when an integer is written as the product of two or more integers, each of these integers is called a *factor* of the product. This is true of polynomials also. When a polynomial is written as the product of two or more other polynomials, each of these polynomials is called a factor of the product.

factor · factor = product
$-2 \cdot 4 = -8$
$x^3 \cdot x^7 = x^{10}$
$5(x + 2) = 5x + 10$

The process of writing a polynomial as a product is called **factoring.** Notice that factoring is the reverse process of multiplying.

$$\underbrace{5x + 10}_{\text{multiplying}} \overset{\text{factoring}}{=} \underbrace{5(x + 2)}$$

Objective A Finding the GCF of a List of Integers

Before we factor polynomials, let's practice finding the greatest common factor of a list of integers. The **greatest common factor (GCF)** of a list of integers is the largest integer that is a factor of all the integers in the list. For example,

the GCF of 30 and 18 is 6

because 6 is the largest integer that is a factor of both 30 and 18.
If the GCF cannot be found by inspection, the following steps can be used.

> **To Find the GCF of a List of Integers**
>
> **Step 1:** Write each number as a product of prime numbers.
>
> **Step 2:** Identify the common prime factors.
>
> **Step 3:** The product of all common prime factors found in Step 2 is the greatest common factor. If there are no common prime factors, the greatest common factor is 1.

✓**Concept Check** Which of the following is the prime factorization of 36?
a. $4 \cdot 9$ **b.** $2 \cdot 2 \cdot 3 \cdot 3$ **c.** $6 \cdot 6$

Recall from Section 4.2 that a prime number is a whole number other than 1 whose only factors are 1 and itself.

Example 1 Find the GCF of 12 and 20.

Solution:

Step 1: Write each number as a product of primes.

$12 = 2 \cdot 2 \cdot 3$
$20 = 2 \cdot 2 \cdot 5$

(Continued on next page)

Objectives

A Find the Greatest Common Factor of a List of Integers.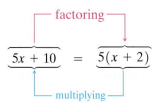

B Find the Greatest Common Factor of a List of Terms.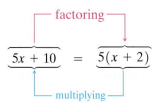

C Factor the Greatest Common Factor from the Terms of a Polynomial.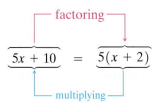

Practice 1

Find the GCF of 42 and 28.

Answer
1. 14

✓**Concept Check Answer**
b

Step 2: $12 = \boxed{2} \cdot \boxed{2} \cdot 3$
$20 = \boxed{2} \cdot \boxed{2} \cdot 5$

$\downarrow \quad \downarrow$

$2 \cdot 2$ Identify the common factors.

Step 3: The GCF is $2 \cdot 2 = 4$.

■ Work Practice 1

Objective B Finding the GCF of a List of Terms ▶

How do we find the GCF of a list of variables raised to powers? For example, what is the GCF of y^3, y^5, and y^{10}? Notice that each variable term contains a factor of y^3 and no higher power of y is a factor of each term.

$$y^3 = y^3$$
$$y^5 = y^3 \cdot y^2 \quad \text{Recall the product property for exponents.}$$
$$y^{10} = y^3 \cdot y^7$$

The GCF of y^3, y^5, and y^{10} is y^3. From this example, we can see that **the GCF of a list of variables raised to powers is the variable raised to the smallest exponent in the list.**

Practice 2

Find the GCF of z^7, z^8, and z.

| Example 2 | Find the GCF of x^{11}, x^4, and x^6.

Solution: The GCF is x^4 since 4 is the smallest exponent to which x is raised.

■ Work Practice 2

In general, **the GCF of a list of terms is the product of all common factors.**

Practice 3

Find the GCF of $6a^4$, $3a^5$, and $15a^2$.

| Example 3 | Find the GCF of $4x^3$, $12x$, and $10x^5$.

Solution: The GCF of 4, 12, and 10 is 2.

The GCF of x^3, x^1, and x^5 is x^1.

Thus, the GCF of $4x^3$, $12x$, and $10x^5$ is $2x^1$ or $2x$.

■ Work Practice 3

Helpful Hint

If you ever have trouble finding the GCF, remember that you can always use the method below.

Example 3:

$$4x^3 = \boxed{2} \cdot 2 \cdot \boxed{x} \cdot x \cdot x$$
$$12x = \boxed{2} \cdot 2 \cdot 3 \cdot \boxed{x}$$
$$10x^5 = \boxed{2} \cdot 5 \cdot \boxed{x} \cdot x \cdot x \cdot x \cdot x$$
$$\text{GCF} = 2 \cdot x \quad \text{or} \quad 2x$$

Objective C Factoring Out the GCF ▶

Next, we practice factoring a polynomial by factoring the GCF from its terms. To do so, we write each term of the polynomial as a product of the GCF and another factor, and then apply the distributive property.

Answers

2. z **3.** $3a^2$

Example 4   Factor: $7x^3 + 14x^2$

Solution: The GCF of $7x^3$ and $14x^2$ is $7x^2$.

$$7x^3 + 14x^2 = 7x^2 \cdot x^1 + 7x^2 \cdot 2$$
$$= 7x^2(x + 2) \qquad \text{Apply the distributive property.}$$

■ **Work Practice 4**

Notice in Example 4 that we factored $7x^3 + 14x^2$ by writing it as the product $7x^2(x + 2)$. Also notice that to check factoring, we multiply

$$7x^2(x + 2) = 7x^2 \cdot x + 7x^2 \cdot 2$$
$$= 7x^3 + 14x^2$$

which is the original binomial.

Example 5   Factor: $6x^2 - 24x + 6$

Solution: The GCF of the terms is 6.

$$6x^2 - 24x + 6 = 6 \cdot x^2 - 6 \cdot 4x + 6 \cdot 1$$
$$= 6(x^2 - 4x + 1)$$

■ **Work Practice 5**

> **Helpful Hint**
>
> A common mistake in the example above is to forget to write down the term of 1. Remember to mentally check by multiplying.
>
> $$6(x^2 - 4x) = 6x^2 - 24x \qquad \text{Not the original trinomial}$$
>
> $$6(x^2 - 4x + 1) = 6x^2 - 24x + 6 \qquad \text{The original trinomial}$$

Example 6   Factor: $-2a + 20b - 4b^2$

Solution:

$$-2a + 20b - 4b^2 = 2 \cdot -a + 2 \cdot 10b - 2 \cdot 2b^2$$
$$= 2(-a + 10b - 2b^2)$$

When the coefficient of the first term is a negative number, we often factor out a negative common factor.

$$-2a + 20b - 4b^2 = (-2)(a) + (-2)(-10b) + (-2)(2b^2)$$
$$= -2(a - 10b + 2b^2)$$

Both $2(-a + 10b - 2b^2)$ and $-2(a - 10b + 2b^2)$ are factorizations of $-2a + 20b - 4b^2$.

■ **Work Practice 6**

✓**Concept Check**   Check both factorizations given in Example 6.

Practice 4

Factor: $10y^7 + 5y^9$

Practice 5

Factor: $4z^2 - 12z + 2$

> **Helpful Hint**   Don't forget to include the term 1.

Practice 6

Factor: $-3y^2 - 9y + 15x^2$

Answers

4. $5y^7(2 + y^2)$ **5.** $2(2z^2 - 6z + 1)$
6. $-3(y^2 + 3y - 5x^2)$ or
    $3(-y^2 - 3y + 5x^2)$

✓**Concept Check Answer**

answers may vary

Vocabulary, Readiness & Video Check

Use the choices below to fill in each blank. Not all choices will be used.

factoring	smallest	factor
product	largest	greatest common factor (GCF)

1. In $-3 \cdot x^4 = -3x^4$, the -3 and the x^4 are each called a _____ and $-3x^4$ is called a _____.

2. The _____ of a list of integers is the largest integer that is a factor of all integers in the list.

3. The GCF of a list of variables raised to powers is the variable raised to the _____ exponent in the list.

4. _____ is the process of writing an expression as a product.

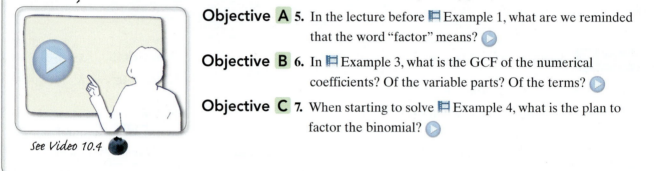

Martin-Gay Interactive Videos   Watch the section lecture video and answer the following question.

Objective A **5.** In the lecture before ▣ Example 1, what are we reminded that the word "factor" means? ▶

Objective B **6.** In ▣ Example 3, what is the GCF of the numerical coefficients? Of the variable parts? Of the terms? ▶

Objective C **7.** When starting to solve ▣ Example 4, what is the plan to factor the binomial? ▶

See Video 10.4

10.4   Exercise Set   MyMathLab® ▶

Objective A *Find the greatest common factor of each list of numbers. See Example 1.*

1. 48 and 15

2. 36 and 20

3. 60 and 72

4. 96 and 45

▶ **5.** 12, 20, and 36

6. 18, 24, and 60

7. 8, 32, and 100

8. 30, 50, and 200

Objective B *Find the greatest common factor of each list of terms. See Examples 2 and 3.*

▶ **9.** y^7, y^2, y^{10}

10. x^3, x, x^5

11. a^5, a^5, a^5

12. b^6, b^6, b^4

13. x^3y^2, xy^2, x^4y^2

14. a^5b^3, a^5b^2, a^5b

15. $3x^4, 5x^7, 10x$

16. $9z^6, 4z^5, 2z^3$

▶ **17.** $2z^3, 14z^5, 18z^3$

18. $6y^7, 9y^6, 15y^5$

Objective C *Factor. Check by multiplying. See Examples 4 through 6.*

▶ **19.** $3y^2 + 18y$

20. $2x^2 + 18x$

21. $10a^6 - 5a^8$

22. $21y^5 + y^{10}$

23. $4x^3 + 12x^2 + 20x$

24. $9b^3 - 54b^2 + 9b$

25. $z^7 - 6z^5$

26. $y^{10} + 4y^5$

▶ **27.** $-35 + 14y - 7y^2$

28. $-20x + 4x^2 - 2$

29. $12a^5 - 36a^6$

30. $25z^3 - 20z^2$

Review

Solve. See Sections 7.1–7.3.

31. Find 30% of 120.

32. Find 45% of 265.

33. Write 80% as a fraction in simplified form.

34. Write 65% as a fraction in simplified form.

35. Write $\dfrac{3}{8}$ as a percent.

36. Write $\dfrac{3}{4}$ as a percent.

Concept Extensions

△**37.** The area of the largest rectangle below is $x(x + 2)$.

    a. Find another expression for the area by writing the sum of the areas of the smaller rectangles.

    b. Explain how $x(x + 2)$ and the answer to part **a** are related.

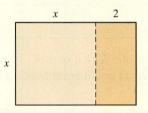

△**38.** **a.** Write an expression for the area of the largest rectangle in two different ways.

    b. Explain how the two answers to part **a** are related.

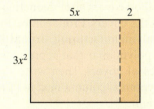

39. In your own words, define the greatest common factor of a list of numbers.

40. Suppose that a classmate asks you why $4x^2 + 6x + 2$ does not factor as $2(2x^2 + 3x)$. Write down your response to this classmate.

41. For the expression $(xy + z)^x$, let $x = 2$ and $z = 7$. Then multiply and simplify.

42. For the expression $(xy + z)^x$, let $x = 2$ and $z = -7$. Then multiply and simplify.

43. Explain two ways in which $(2y + 7)^2$ and your answer to Exercise **41** are related. Use the words *multiply* and *factor* in your explanations.

44. Explain two ways in which $(2y - 7)^2$ and your answer to Exercise **42** are related. Use the words *multiply* and *factor* in your explanations.

Chapter 10 Group Activity

Business Analysis

This activity may be completed by working in groups or individually.

Suppose you own a small business that manufactures specialized iPod covers. You need to decide how many covers to make. The more covers you make, the lower the price you will have to charge to sell them all. Naturally, each cover you make costs you money because you must buy the materials to make each cover. The following table summarizes some factors you must consider in deciding how many covers to make, along with algebraic representations of those factors.

	Description	Algebraic Expression
Number of Covers	**Unknown**	x
Total manufacturing expenses	This is the total amount that it will cost to manufacture all the iPod covers. It will cost $100 to buy special equipment to manufacture the covers in addition to materials costing $0.50 per cover.	$100 + 0.50x$
Price charged per cover	For each additional cover produced, the price that must be charged per cover decreases from $40 by an additional $0.05.	$40 - 0.05x$

1. *Revenue* is the amount of money collected from selling the iPod covers. Revenue can be found by multiplying the price charged per cover by the number of covers sold. Use the algebraic expressions given in the table above to find a polynomial that represents the revenue from sales of covers. Then write this polynomial in the Polynomial column next to "Revenue" in the table to the right.

2. *Profit* is the amount of money you make from selling the iPod covers after deducting the expenses for making the covers. Profit can be found by subtracting total manufacturing expenses from revenue. Find a polynomial that represents the profit from the sales of covers. Then write this polynomial in the Polynomial column next to "Profit" in the table to the right.

3. Complete the following table by evaluating each polynomial for each of the numbers of covers given in the table.

		Number of Covers, x				
	Polynomial	**200**	**300**	**400**	**500**	**600**
Revenue						
Total manufacturing expenses	$100 + 0.50x$					
Profit						

4. Study the table. Which number of covers will give you the largest profit from making and selling iPod covers?

Chapter 10 Vocabulary Check

Fill in each blank with one of the words or phrases listed below.

trinomial	monomial	greatest common factor	binomial
exponent	factoring	polynomials	FOIL

1. _____ is the process of writing an expression as a product.

2. The _____ of a list of terms is the product of all common factors.

3. The _____ method may be used when multiplying two binomials.

4. A polynomial with exactly 3 terms is called a(n) _____.

5. A polynomial with exactly 2 terms is called a(n) _____.

6. A polynomial with exactly 1 term is called a(n) _____.

7. Monomials, binomials, and trinomials are all examples of _____.

8. In $5x^3$, the 3 is called a(n) _____.

Helpful Hint ▶ Are you preparing for your test? Don't forget to take the Chapter 10 Test on page 729. Then check your answers at the back of the text and use the Chapter Test Prep Videos to see the fully worked-out solutions to any of the exercises you want to review.

10 Chapter Highlights

Definitions and Concepts	Examples
Section 10.1 Adding and Subtracting Polynomials	

A **monomial** is a term that contains whole number exponents and no variable in the denominator.

Monomial: $-2x^2y^3$

A **polynomial** is a monomial or a sum or difference of monomials.

Polynomials:
$$5x^2 - 6x + 2, \quad -\frac{9}{10}y, \quad 7$$

A **binomial** is a polynomial with two terms.

Binomial: $5x - y$

A **trinomial** is a polynomial with three terms.

Trinomial: $7z^3 + 0.5z + 1$

To add polynomials, combine like terms.

Add: $(7z^2 - 6z + 2) + (5z^2 - 4z + 5)$
$(7z^2 - 6z + 2) + (5z^2 - 4z + 5)$
$= 7z^2 + 5z^2 - 6z - 4z + 2 + 5$ Group like terms.
$= 12z^2 - 10z + 7$ Combine like terms.

To subtract polynomials, change the signs of the terms being subtracted, then add.

Subtract: $(20x - 6) - (30x - 6)$
$(20x - 6) - (30x - 6)$
$= (20x - 6) + (-30x + 6)$
$= 20x - 30x - 6 + 6$ Group like terms.
$= -10x$ Combine like terms.

| **Section 10.2   Multiplication Properties of Exponents** | |

Product property for exponents
$a^m \cdot a^n = a^{m+n}$

$x^3 \cdot x^{11} = x^{3+11} = x^{14}$

Power property for exponents
$(a^m)^n = a^{m \cdot n}$

$(y^5)^3 = y^{5 \cdot 3} = y^{15}$

Power of a product property for exponents
$(ab)^n = a^n b^n$

$(2z^5)^4 = 2^4(z^5)^4 = 16z^{20}$

Definitions and Concepts	Examples

Section 10.3   Multiplying Polynomials

To multiply two polynomials, multiply each term of the first polynomial by each term of the second polynomial, and then combine like terms.

$(x + 2)(x^2 + 5x - 1)$

$$= x(x^2 + 5x - 1) + 2(x^2 + 5x - 1)$$
$$= x \cdot x^2 + x \cdot 5x + x(-1) + 2 \cdot x^2 + 2 \cdot 5x + 2(-1)$$
$$= x^3 + 5x^2 - x + 2x^2 + 10x - 2$$
$$= x^3 + 7x^2 + 9x - 2$$

Section 10.4   Introduction to Factoring Polynomials

To Find the Greatest Common Factor of a List of Integers

Step 1: Write each number as a product of prime numbers.

Step 2: Identify the common prime factors.

Step 3: The product of all common prime factors found in Step 2 is the greatest common factor. If there are no common prime factors, the greatest common factor is 1.

The **GCF of a list of variables** raised to powers is the variable raised to the smallest exponent in the list.

The **GCF of a list of terms** is the product of all common factors.

To factor the GCF from the terms of a polynomial, write each term as a product of the GCF and another factor, then apply the distributive property.

Find the GCF of 18 and 30.
$$18 = 2 \cdot 3 \cdot 3$$
$$30 = 2 \cdot 3 \cdot 5$$

The GCF is $2 \cdot 3$ or 6.

The GCF of x^6, x^8, and x^3 is x^3.

Find the GCF of $6y^3$, $12y$, and $4y^7$.
The GCF of 6, 12, and 4 is 2.
The GCF of y^3, y, and y^7 is y.
The GCF of $6y^3$, $12y$, and $4y^7$ is $2y$.

Factor $4y^6 + 6y^5$.
The GCF of $4y^6$ and $6y^5$ is $2y^5$.
$$4y^6 + 6y^5 = 2y^5 \cdot 2y + 2y^5 \cdot 3$$
$$= 2y^5(2y + 3)$$

Chapter 10   Review

(10.1) *Perform each indicated operation.*

1. $(2b + 7) + (8b - 10)$

2. $(7s - 6) + (14s - 9)$

3. $(3x + 0.2) - (4x - 2.6)$

4. $(10y - 6) - (11y + 6)$

5. $(4z^2 + 6z - 1) + (5z - 5)$

6. $(17a^3 + 11a^2 + a) + (14a^2 - a)$

7. $\left(9y^2 - y + \frac{1}{2}\right) - \left(20y^2 - \frac{1}{4}\right)$

8. Subtract $(x - 2)$ from $(x^2 - 6x + 1)$.

Find the value of each polynomial when $x = 3$.

9. $5x^2$

10. $2 - 7x$

△**11.** Find the perimeter of the given rectangle.

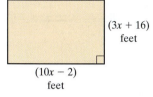

$(3x + 16)$ feet

$(10x - 2)$ feet

12. Find the perimeter of a square whose side length is $(4x^2 + 1)$ meters.

(10.2) *Multiply and simplify.*

13. $x^{10} \cdot x^{14}$

14. $y \cdot y^6$

15. $4z^2 \cdot 6z^5$

16. $(-3x^2y)(5xy^4)$

17. $(a^5)^7$

18. $(x^2)^4 \cdot (x^{10})^2$

19. $(9b)^2$

20. $(a^4b^2c)^5$

21. $(7x)(2x^5)^3$

22. $(3x^6y^5)^3(2x^6y^5)^2$

△**23.** Find the area of the square.

$9a^7$ miles

24. Find the area of a rectangle whose length is $3x^4$ inches and whose width is $9x$ inches.

(10.3) *Multiply.*

25. $2a(5a^2 - 6)$

26. $-3y^2(y^2 - 2y + 1)$

27. $(x + 2)(x + 6)$

28. $(3x - 1)(5x - 9)$

29. $(y - 5)^2$

30. $(7a + 1)^2$

31. $(x + 1)(x^2 - 2x + 3)$

32. $(4y^2 - 3)(2y^2 + y + 1)$

33. $(3z^2 + 2z + 1)(z^2 + z + 1)$

△**34.** Find the area of the given rectangle.

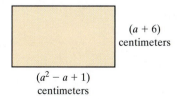

$(a + 6)$ centimeters

$(a^2 - a + 1)$ centimeters

(10.4) *Find the greatest common factor (GCF) of each list.*

35. 20 and 35

36. 12 and 32

37. 24, 30, and 60

38. 10, 20, and 25

39. x^3, x^2, x^{10}

40. y^{10}, y^7, y^7

41. xy^2, xy, x^3y^3

42. a^5b^4, a^6b^3, a^7b^2

43. $5a^3, 10a, 20a^4$

44. $12y^2z, 20y^2z, 24y^5z$

Factor out the GCF.

45. $2x^2 + 12x$

46. $6a^2 - 12a$

47. $6y^4 - y^6$

48. $7x^2 - 14x + 7$

49. $5a^7 - a^4 + a^3$

50. $10y^6 - 10y$

Mixed Review

Perform the indicated operations.

51. $(z^2 - 5z + 8) + (6z - 4)$

52. $(8y - 5) - (12y - 3)$

53. $x^5 \cdot x^{16}$

54. $y^8 \cdot y$

55. $(a^3b^5c)^6$

56. $(9x^2) \cdot (3x^2)^2$

57. $3a(4a^3 - 5)$

58. $(x + 4)(x + 5)$

59. $(3x + 4)^2$

60. $(6z + 5)(z - 2)$

Find the greatest common factor (GCF) of each list.

61. 28, 32, and 40

62. $5z^5, 12z^8, 3z^4$

Factor out the GCF.

63. $z^9 - 4z^7$

64. $x^{12} + 6x^5$

65. $15a^4 + 45a^5$

66. $16z^5 - 24z^8$

Add or subtract as indicated.

1. $(11x - 3) + (4x - 1)$

2. $(11x - 3) - (4x - 1)$

3. $(1.3y^2 + 5y) + (2.1y^2 - 3y - 3)$

4. Subtract $(8a^2 + a)$ from $(6a^2 + 2a + 1)$.

5. Find the value of $x^2 - 6x + 1$ when $x = 8$.

Multiply and simplify.

6. $y^3 \cdot y^{11}$

7. $(y^3)^{11}$

8. $(2x^2)^4$

9. $(6a^3)(-2a^7)$

10. $(p^6)^7(p^2)^6$

11. $(3a^4b)^2(2ba^4)^3$

12. $5x(2x^2 + 1.3)$

13. $-2y(y^3 + 6y^2 - 4)$

14. $(x - 3)(x + 2)$

15. $(5x + 2)^2$

16. $(a + 2)(a^2 - 2a + 4)$

△**17.** Find the area and the perimeter of the parallelogram. (*Hint:* $A = b \cdot h$.)

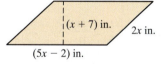

$(x + 7)$ in. $2x$ in.

$(5x - 2)$ in.

Find the greatest common factor of each list.

18. 45 and 60

19. $6y^3, 9y^5, 18y^4$

Factor out the GCF.

20. $3y^2 - 15y$

21. $10a^2 + 12a$

22. $6x^2 - 12x - 30$

23. $7x^6 - 6x^4 + x^3$

Answers

1. _____

2. _____

3. _____

4. _____

5. _____

6. _____

7. _____

8. _____

9. _____

10. _____

11. _____

12. _____

13. _____

14. _____

15. _____

16. _____

17. _____

18. _____

19. _____

20. _____

21. _____

22. _____

23. _____

Answers

1. _____

2. _____

3. _____

4. _____

5. _____

6. _____

7. _____

8. _____

9. _____

10. _____

11. _____

12. _____

13. _____

14. _____

15. _____

16. _____

17. _____

18. _____

19. _____

20. _____

△ **1.** The state of Colorado is in the shape of a rectangle whose length is 380 miles and whose width is 280 miles. Find its area.

2. In a pecan orchard, there are 21 trees in each row and 7 rows of trees. How many pecan trees are there?

3. Add: $1 + (-10) + (-8) + 9$

4. Add: $-2 + (-7) + 3 + (-4)$

Subtract.

5. $8 - 15$

6. $4 - 7$

7. $-4 - (-5)$

8. $3 - (-2)$

9. Solve: $x = -60 + 4 + 10$

10. Solve: $x = -12 + 3 + 7$

11. Solve: $17 - 7x + 3 = -3x + 21 - 3x$

12. $20 - 6x + 4 = -2x + 18 + 2x$

13. Add: $\dfrac{2x}{15} + \dfrac{3x}{10}$

14. Subtract: $\dfrac{5}{7y} - \dfrac{9}{14y}$

15. Round 736.2359 to the nearest tenth.

16. Round 328.174 to the nearest tenth.

17. Add: $23.85 + 1.604$

18. Add: $12.762 + 4.29$

19. Is -9 a solution of the equation $3.7y = -3.33$?

20. Is 6 a solution of the equation $2.8x = 16.8$?

Divide.

21. $\dfrac{786.1}{1000}$

22. $\dfrac{818}{1000}$

23. $\dfrac{0.12}{10}$

24. $\dfrac{5.03}{100}$

25. Evaluate $-2x + 5$ for $x = 3.8$.

26. Evaluate $6x - 1$ for $x = -2.1$.

27. Write $\dfrac{22}{7}$ as a decimal. Round to the nearest hundredth.

28. Write $\dfrac{37}{19}$ as a decimal. Round to the nearest thousandth.

29. Find: $\sqrt{\dfrac{1}{36}}$

30. Find: $\sqrt{\dfrac{4}{25}}$

31. Mel Wagstaff is a 6-foot-tall park ranger who needs to know the height of a particular tree. He measures the shadow of the tree to be 69 feet long when his own shadow is 9 feet long. Find the height of the tree.

32. Phoebe, a very intelligent dog, wants to estimate the height of a fire hydrant. She notices that when her shadow is 2 feet long, the shadow of the hydrant is 6 feet long. Find the height of the hydrant if Phoebe is 1 foot tall.

33. Translate to an equation: 1.2 is 30% of what number?

34. Translate to an equation: 9 is 45% of what number?

35. What percent of 50 is 8?

36. What percent of 16 is 4?

37. Mr. Percy, the principal at Slidell High School, counted 31 freshmen absent during a particular day. If this is 4% of the total number of freshmen, how many freshmen are there at Slidell High School?

38. Two percent of the apples in a shipment are rotten. If there are 29 rotten apples, how many apples are in the shipment?

21. _____

22. _____

23. _____

24. _____

25. _____

26. _____

27. _____

28. _____

29. _____

30. _____

31. _____

32. _____

33. _____

34. _____

35. _____

36. _____

37. _____

38. _____

39. _____

40. _____

41. _____

42. _____

43. _____

44. _____

45. _____

46. _____

47. _____

48. _____

49. _____

50. _____

51. _____

52. _____

53. _____

54. _____

39. A recent college graduate borrowed $2400 at 10% simple interest for 8 months to buy a used Toyota Corolla. Find the simple interest he paid.

40. Find the amount of simple interest earned on a $1000 CD for 10 months at an interest rate of 3%.

41. Using the circle graph shown, determine the percent of visitors who come to the United States from Mexico or Canada.

42. Using the circle graph for Exercise **41**, find the percent of visitors who come to the United States from Europe and Asia.

Visitors to U.S. by Region

Mexico 21%

Other 7%

Europe 19%

Asia 12%

South America 7%

Canada 34%

Source: Office of Travel and Tourism Industries, 2012

△ **43.** Find the perimeter of a rectangle with a length of 11 inches and a width of 3 inches.

△ **44.** Find the perimeter of a triangular yard whose sides are 6 feet, 8 feet, and 11 feet.

△ **45.** Find the area of the parallelogram.

1.5 mi

3.4 mi

△ **46.** Find the area of the triangle.

8 inches

17 inches

47. Subtract 3 tons 1350 lb from 8 tons 1000 lb.

48. Multiply 5 tons 700 lb by 3.

49. Convert 3210 ml to liters.

50. Convert 4321 cl to liters.

51. Add: $(3x - 1) + (-6x + 2)$

52. Subtract: $(7a + 4) - (3a - 8)$

53. Multiply: $(x + 2)(x + 3)$

54. Multiply: $(2x + 5)(x + 7)$

Tables

A.1 Tables of Geometric Figures ▶

Plane Figures Have Length and Width but No Thickness or Depth		
Name	Description	Figure
Polygon	Union of three or more coplanar line segments that intersect with each other only at each endpoint, with each endpoint shared by two segments.	
Triangle	Polygon with three sides (sum of measures of three angles is 180°).	
Scalene Triangle	Triangle with no sides of equal length.	
Isosceles Triangle	Triangle with two sides of equal length.	
Equilateral Triangle	Triangle with all sides of equal length.	
Right Triangle	Triangle that contains a right angle.	leg, hypotenuse, leg
Quadrilateral	Polygon with four sides (sum of measures of four angles is 360°).	
Trapezoid	Quadrilateral with exactly one pair of opposite sides parallel.	base, leg, parallel sides, leg, base
Isosceles Trapezoid	Trapezoid with legs of equal length.	
Parallelogram	Quadrilateral with both pairs of opposite sides parallel.	
Rhombus	Parallelogram with all sides of equal length.	
Rectangle	Parallelogram with four right angles.	

(Continued)

Plane Figures Have Length and Width but No Thickness or Depth (*continued*)		
Name	Description	Figure
Square	Rectangle with all sides of equal length.	
Circle	All points in a plane the same distance from a fixed point called the **center.**	

Solid Figures Have Length, Width, and Height or Depth		
Name	Description	Figure
Rectangular Solid	A solid with six sides, all of which are rectangles.	
Cube	A rectangular solid whose six sides are squares.	
Sphere	All points the same distance from a fixed point called the **center.**	
Right Circular Cylinder	A cylinder having two circular bases that are perpendicular to its altitude.	
Right Circular Cone	A cone with a circular base that is perpendicular to its altitude.	

Percent	Decimal	Fraction
1%	0.01	$\frac{1}{100}$
5%	0.05	$\frac{1}{20}$
10%	0.1	$\frac{1}{10}$
12.5% or $12\frac{1}{2}$%	0.125	$\frac{1}{8}$
$16.\overline{6}$% or $16\frac{2}{3}$%	$0.1\overline{6}$	$\frac{1}{6}$
20%	0.2	$\frac{1}{5}$
25%	0.25	$\frac{1}{4}$
30%	0.3	$\frac{3}{10}$
$33.\overline{3}$% or $33\frac{1}{3}$%	$0.\overline{3}$	$\frac{1}{3}$
37.5% or $37\frac{1}{2}$%	0.375	$\frac{3}{8}$
40%	0.4	$\frac{2}{5}$
50%	0.5	$\frac{1}{2}$
60%	0.6	$\frac{3}{5}$
62.5% or $62\frac{1}{2}$%	0.625	$\frac{5}{8}$
$66.\overline{6}$% or $66\frac{2}{3}$%	$0.\overline{6}$	$\frac{2}{3}$
70%	0.7	$\frac{7}{10}$
75%	0.75	$\frac{3}{4}$
80%	0.8	$\frac{4}{5}$
$83.\overline{3}$% or $83\frac{1}{3}$%	$0.8\overline{3}$	$\frac{5}{6}$
87.5% or $87\frac{1}{2}$%	0.875	$\frac{7}{8}$
90%	0.9	$\frac{9}{10}$
100%	1.0	1
110%	1.1	$1\frac{1}{10}$
125%	1.25	$1\frac{1}{4}$
$133.\overline{3}$% or $133\frac{1}{3}$%	$1.\overline{3}$	$1\frac{1}{3}$
150%	1.5	$1\frac{1}{2}$
$166.\overline{6}$% or $166\frac{2}{3}$%	$1.\overline{6}$	$1\frac{2}{3}$
175%	1.75	$1\frac{3}{4}$
200%	2.0	2

A.3 Table on Finding Common Percents of a Number

Common Percent Equivalences*	Shortcut Method for Finding Percent	Examples
$1\% = 0.01 \left(\text{or } \frac{1}{100}\right)$	To find 1% of a number, multiply by 0.01. To do so, move the decimal point 2 places to the left.	1% of 210 is 2.10 or 2.1. 1% of 1500 is 15. 1% of 8.6 is 0.086.
$10\% = 0.1 \left(\text{or } \frac{1}{10}\right)$	To find 10% of a number, multiply by 0.1, or move the decimal point of the number 1 place to the left.	10% of 140 is 14. 10% of 30 is 3. 10% of 17.6 is 1.76.
$25\% = \frac{1}{4}$	To find 25% of a number, find $\frac{1}{4}$ of the number, or divide the number by 4.	25% of 20 is $\frac{20}{4}$ or 5. 25% of 8 is 2. 25% of 10 is $\frac{10}{4}$ or $2\frac{1}{2}$.
$50\% = \frac{1}{2}$	To find 50% of a number, find $\frac{1}{2}$ of the number, or divide the number by 2.	50% of 64 is $\frac{64}{2}$ or 32. 50% of 1000 is 500. 50% of 9 is $\frac{9}{2}$ or $4\frac{1}{2}$.
$100\% = 1$	To find 100% of a number, multiply the number by 1. In other words, 100% of a number is the number.	100% of 98 is 98. 100% of 1407 is 1407. 100% of 18.4 is 18.4.
$200\% = 2$	To find 200% of a number, multiply the number by 2.	200% of 31 is $31 \cdot 2$ or 62. 200% of 750 is 1500. 200% of 6.5 is 13.

*See Appendix A.2.

A.4 Table of Squares and Square Roots

n	n^2	$\sqrt{n}$	n	n^2	$\sqrt{n}$
1	1	1.000	51	2601	7.141
2	4	1.414	52	2704	7.211
3	9	1.732	53	2809	7.280
4	16	2.000	54	2916	7.348
5	25	2.236	55	3025	7.416
6	36	2.449	56	3136	7.483
7	49	2.646	57	3249	7.550
8	64	2.828	58	3364	7.616
9	81	3.000	59	3481	7.681
10	100	3.162	60	3600	7.746
11	121	3.317	61	3721	7.810
12	144	3.464	62	3844	7.874
13	169	3.606	63	3969	7.937
14	196	3.742	64	4096	8.000
15	225	3.873	65	4225	8.062
16	256	4.000	66	4356	8.124
17	289	4.123	67	4489	8.185
18	324	4.243	68	4624	8.246
19	361	4.359	69	4761	8.307
20	400	4.472	70	4900	8.367
21	441	4.583	71	5041	8.426
22	484	4.690	72	5184	8.485
23	529	4.796	73	5329	8.544
24	576	4.899	74	5476	8.602
25	625	5.000	75	5625	8.660
26	676	5.099	76	5776	8.718
27	729	5.196	77	5929	8.775
28	784	5.292	78	6084	8.832
29	841	5.385	79	6241	8.888
30	900	5.477	80	6400	8.944
31	961	5.568	81	6561	9.000
32	1024	5.657	82	6724	9.055
33	1089	5.745	83	6889	9.110
34	1156	5.831	84	7056	9.165
35	1225	5.916	85	7225	9.220
36	1296	6.000	86	7396	9.274
37	1369	6.083	87	7569	9.327
38	1444	6.164	88	7744	9.381
39	1521	6.245	89	7921	9.434
40	1600	6.325	90	8100	9.487
41	1681	6.403	91	8281	9.539
42	1764	6.481	92	8464	9.592
43	1849	6.557	93	8649	9.644
44	1936	6.633	94	8836	9.695
45	2025	6.708	95	9025	9.747
46	2116	6.782	96	9216	9.798
47	2209	6.856	97	9409	9.849
48	2304	6.928	98	9604	9.899
49	2401	7.000	99	9801	9.950
50	2500	7.071	100	10,000	10.000

Quotient Rule and Negative Exponents

Objectives

A Use the Quotient Rule for Exponents, and Define a Number Raised to the 0 Power. ▶

B Simplify Expressions Containing Negative Exponents. ▶

C Use the Rules and Definitions for Exponents to Simplify Exponential Expressions. ▶

Objective A Using the Quotient Rule and Defining the Zero Exponent ▶

Let's study a pattern for simplifying exponential expressions that involves quotients.

$$\frac{x^5}{x^3} = \frac{x \cdot x \cdot x \cdot x \cdot x}{x \cdot x \cdot x}$$

$$= \frac{x \cdot x \cdot x \cdot x \cdot x}{x \cdot x \cdot x}$$

$$= 1 \cdot 1 \cdot 1 \cdot x \cdot x$$

$$= x \cdot x$$

$$= x^2$$

Notice that the result is exactly the same if we subtract exponents of the common bases.

$$\frac{x^5}{x^3} = x^{5-3} = x^2$$

The following rule states this result in a general way.

Quotient Rule for Exponents

If m and n are positive integers and a is a real number, then

$$\frac{a^m}{a^n} = a^{m-n}, \quad a \neq 0$$

For example,

$$\frac{x^6}{x^2} = x^{6-2} = x^4, \quad x \neq 0$$

In other words, to divide one exponential expression by another with a common base, we keep the base and subtract the exponents.

Practice 1–3

Simplify each quotient.

1. $\dfrac{y^{10}}{y^6}$ **2.** $\dfrac{5^{11}}{5^8}$ **3.** $\dfrac{12a^4b^{11}}{ab}$

Examples Simplify each quotient.

1. $\dfrac{x^5}{x^2} = x^{5-2} = x^3$ Use the quotient rule.

2. $\dfrac{4^7}{4^3} = 4^{7-3} = 4^4 = 256$ Use the quotient rule.

3. $\dfrac{2x^5y^2}{xy} = 2 \cdot \dfrac{x^5}{x^1} \cdot \dfrac{y^2}{y^1}$

$$= 2 \cdot \left(x^{5-1}\right) \cdot \left(y^{2-1}\right) \quad \text{Use the quotient rule.}$$

$$= 2x^4y^1 \quad \text{or} \quad 2x^4y$$

■ Work Practice 1–3

Answers

1. y^4 **2.** 125 **3.** $12a^3b^{10}$

Let's now give meaning to an expression such as x^0. To do so, we will simplify $\dfrac{x^3}{x^3}$ in two ways and compare the results.

$$\dfrac{x^3}{x^3} = x^{3-3} = x^0 \qquad \text{Apply the quotient rule.}$$

$$\dfrac{x^3}{x^3} = \dfrac{x \cdot x \cdot x}{x \cdot x \cdot x} = 1 \qquad \text{Apply the fundamental property for fractions.}$$

Since $\dfrac{x^3}{x^3} = x^0$ and $\dfrac{x^3}{x^3} = 1$, we define that $x^0 = 1$ as long as x is not 0.

> **Zero Exponent**
>
> $a^0 = 1$, as long as a is not 0. $\qquad$ For example, $5^0 = 1$.

In other words, a base raised to the 0 power is 1, as long as the base is not 0.

Examples Simplify each expression.

4. $3^0 = 1$

5. $(-4)^0 = 1$

6. $-4^0 = -1 \cdot 4^0 = -1 \cdot 1 = -1$

7. $5x^0 = 5 \cdot x^0 = 5 \cdot 1 = 5$

🟧 **Work Practice 4–7**

Objective B Simplifying Expressions Containing Negative Exponents ▶

Our work with exponential expressions so far has been limited to exponents that are positive integers or 0. Here we will also give meaning to an expression like x^{-3}.

Suppose that we wish to simplify the expression $\dfrac{x^2}{x^5}$. If we use the quotient rule for exponents, we subtract exponents:

$$\dfrac{x^2}{x^5} = x^{2-5} = x^{-3}, \quad x \neq 0$$

But what does x^{-3} mean? Let's simplify $\dfrac{x^2}{x^5}$ using the definition of a^n.

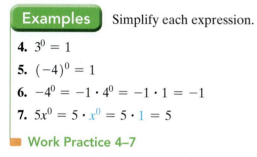

$$\dfrac{x^2}{x^5} = \dfrac{x \cdot x}{x \cdot x \cdot x \cdot x \cdot x}$$

$$= \dfrac{x \cdot x}{x \cdot x \cdot x \cdot x \cdot x} \qquad \begin{array}{l}\text{Divide numerator and denominator by common factors by applying} \\ \text{the fundamental principle for fractions.}\end{array}$$

$$= \dfrac{1}{x^3}$$

If the quotient rule is to hold true for negative exponents, then x^{-3} must equal $\dfrac{1}{x^3}$.

Practice 4–7

Simplify each expression.

4. 6^0 **5.** $(-8)^0$ **6.** -8^0

7. $7y^0$

Answers

4. 1 **5.** 1 **6.** −1 **7.** 7

From this example, we state the definition for negative exponents.

Negative Exponents

If a is a real number other than 0 and n is an integer, then

$$a^{-n} = \frac{1}{a^n}$$

For example,

$$x^{-3} = \frac{1}{x^3}$$

In other words, another way to write a^{-n} is to take its reciprocal and change the sign of its exponent.

Practice 8–10

Simplify by writing each expression with positive exponents only.

8. 5^{-2} **9.** $5x^{-2}$ **10.** $4^{-1} + 3^{-1}$

Examples   Simplify by writing each expression with positive exponents only.

8. $3^{-2} = \frac{1}{3^2} = \frac{1}{9}$ Use the definition of negative exponents.

9. $2x^{-3} = 2^1 \cdot \frac{1}{x^3} = \frac{2^1}{x^3}$ or $\frac{2}{x^3}$ Use the definition of negative exponents.

10. $2^{-1} + 4^{-1} = \frac{1}{2} + \frac{1}{4} = \frac{2}{4} + \frac{1}{4} = \frac{3}{4}$

Work Practice 8–10

Helpful Hint

Don't forget that since there are no parentheses, only x is the base for the exponent -3.

Helpful Hint

A negative exponent *does not affect* the sign of its base.

Remember: Another way to write a^{-n} is to take its reciprocal and change the sign of its exponent: $a^{-n} = \frac{1}{a^n}$. For example,

$$x^{-2} = \frac{1}{x^2},$$

$$\frac{1}{y^{-4}} = \frac{1}{\frac{1}{y^4}} = y^4,$$

$$2^{-3} = \frac{1}{2^3} \text{ or } \frac{1}{8}$$

$$\frac{1}{5^{-2}} = 5^2 \text{ or } 25$$

From the preceding Helpful Hint, we know that $x^{-2} = \frac{1}{x^2}$ and $\frac{1}{y^{-4}} = y^4$. We can use this to include another statement in our definition of negative exponents.

Negative Exponents

If a is a real number other than 0 and n is an integer, then

$$a^{-n} = \frac{1}{a^n} \quad \text{and} \quad \frac{1}{a^{-n}} = a^n$$

Answers

8. $\frac{1}{25}$ **9.** $\frac{5}{x^2}$ **10.** $\frac{7}{12}$

Examples Simplify each expression. Write each result using positive exponents only.

11. $\left(\dfrac{2}{x}\right)^{-3} = \dfrac{2^{-3}}{x^{-3}} = \dfrac{2^{-3}}{1} \cdot \dfrac{1}{x^{-3}} = \dfrac{1}{2^3} \cdot \dfrac{x^3}{1} = \dfrac{x^3}{2^3} = \dfrac{x^3}{8}$ Use the negative exponents rule.

12. $\dfrac{y}{y^{-2}} = \dfrac{y^1}{y^{-2}} = y^{1-(-2)} = y^3$ Use the quotient rule.

13. $\dfrac{x^{-5}}{x^7} = x^{-5-7} = x^{-12} = \dfrac{1}{x^{12}}$

■ **Work Practice 11–13**

Practice 11–13

Simplify each expression. Write each result using positive exponents only.

11. $\left(\dfrac{6}{7}\right)^{-2}$ **12.** $\dfrac{x}{x^{-4}}$ **13.** $\dfrac{y^{-4}}{y^6}$

Objective C Simplifying Exponential Expressions

All the previously stated rules for exponents apply for negative exponents also. Here is a summary of the rules and definitions for exponents. Notice that there is a power of a quotient rule.

Summary of Exponent Rules

If m and n are integers and a, b, and c are real numbers, then

Product rule for exponents:	$a^m \cdot a^n = a^{m+n}$
Power rule for exponents:	$\left(a^m\right)^n = a^{m \cdot n}$
Power of a product:	$(ab)^n = a^n b^n$
Power of a quotient:	$\left(\dfrac{a}{c}\right)^n = \dfrac{a^n}{c^n}, \quad c \neq 0$
Quotient rule for exponents:	$\dfrac{a^m}{a^n} = a^{m-n}, \quad a \neq 0$
Zero exponent:	$a^0 = 1, \quad a \neq 0$
Negative exponent:	$a^{-n} = \dfrac{1}{a^n}, \quad a \neq 0$

Examples Simplify each expression. Write each result using positive exponents only.

14. $x^{-3} \cdot x^2 \cdot x^{-7} = x^{-3+2} \cdot x^{-7}$ Use the product rule.

$= x^{-1} \cdot x^{-7}$

$= x^{-1+(-7)}$ Use the product rule.

$= x^{-8}$

$= \dfrac{1}{x^8}$ Use the definition of negative exponents.

15. $\left(x^4 y^{-9}\right)\left(x^{-1} y^{11}\right) = x^{4+(-1)} \cdot y^{-9+11}$ Use the product rule.

$= x^3 y^2$ Simplify.

16. $\left(4m^2 n^5\right)\left(5m^6 n^{-8}\right) = 4 \cdot 5 \cdot m^{2+6} \cdot n^{5+(-8)}$ Use the product rule.

$= 20 \cdot m^8 \cdot n^{-3}$ Simplify.

$= \dfrac{20m^8}{n^3}$ Use the definition of a negative exponent.

■ **Work Practice 14–16**

Practice 14–16

Simplify each expression. Write each result using positive exponents only.

14. $y^{-6} \cdot y^3 \cdot y^{-4}$

15. $\left(a^6 b^{-4}\right)\left(a^{-3} b^8\right)$

16. $\left(3y^9 z^{10}\right)\left(2y^3 z^{-12}\right)$

Answers

11. $\dfrac{49}{36}$ **12.** x^5 **13.** $\dfrac{1}{y^{10}}$

14. $\dfrac{1}{y^7}$ **15.** $a^3 b^4$ **16.** $\dfrac{6y^{12}}{z^2}$

B   Exercise Set   MyMathLab®

Objective **A**   *Use the quotient rule and simplify each expression. See Examples 1 through 3.*

1. $\dfrac{x^3}{x}$ **2.** $\dfrac{y^{10}}{y^9}$ **3.** $\dfrac{9^8}{9^6}$ **4.** $\dfrac{5^7}{5^4}$

5. $\dfrac{p^7 q^{20}}{p q^{15}}$ **6.** $\dfrac{x^8 y^6}{x y^5}$ **7.** $\dfrac{7 x^3 y^6}{14 x^2 y^3}$ **8.** $\dfrac{9 a^4 b^7}{27 a b^2}$

Simplify each expression. See Examples 4 through 7.

9. 7^0 **10.** 23^0 **11.** $2x^0$ **12.** $4y^0$

13. -7^0 **14.** -2^0 **15.** $(-7)^0$ **16.** $(-2)^0$

Objective **B**   *Simplify each expression. Write each result using positive exponents only. See Examples 8 through 13.*

17. 4^{-3} **18.** 6^{-2} **19.** $7x^{-3}$ **20.** $5y^{-4}$ **21.** $3^{-1} + 2^{-1}$ **22.** $4^{-1} + 4^{-2}$

23. $\dfrac{1}{p^{-3}}$ **24.** $\dfrac{1}{q^{-5}}$ **25.** $\dfrac{x^{-2}}{x}$ **26.** $\dfrac{y}{y^{-3}}$ **27.** $\dfrac{z^{-4}}{z^{-7}}$ **28.** $\dfrac{x^{-4}}{x^{-1}}$

29. $3^{-2} + 3^{-1}$ **30.** $4^{-2} - 4^{-3}$ **31.** $\left(\dfrac{5}{y}\right)^{-2}$ **32.** $\left(\dfrac{3}{x}\right)^{-3}$ **33.** $\dfrac{1}{p^{-4}}$ **34.** $\dfrac{1}{y^{-6}}$

Objective **C**   *Simplify each expression. Write each result using positive exponents only. See Examples 14 through 16.*

35. $a^2 \cdot a^{-9} \cdot a^{13}$ **36.** $z^4 \cdot z^{-5} \cdot z^3$ **37.** $\left(x^8 y^{-6}\right)\left(x^{-2} y^{12}\right)$ **38.** $\left(a^{-20} b^8\right)\left(a^{22} b^{-4}\right)$

39. $x^{-7} \cdot x^{-8} \cdot x^4$ **40.** $y^{-6} \cdot y^{-3} \cdot y^2$ **41.** $\left(5x^{-7}\right)\left(3x^4\right)$ **42.** $\left(4x^9\right)\left(6x^{-13}\right)$

43. $y^5 \cdot y^{-7} \cdot y^{-10}$ **44.** $x^8 \cdot x^{-11} \cdot x^{-2}$ **45.** $\left(8m^5 n^{-1}\right)\left(7m^2 n^{-4}\right)$ **46.** $\left(2x^{10} y^{-3}\right)\left(9x^4 y^{-7}\right)$

Objectives **A** **B** **C** **Mixed Practice** *Simplify each expression. Write each result using positive exponents only. See Examples 1 through 16.*

47. $\dfrac{x^{15}}{x^8}$ **48.** $\dfrac{y^{19}}{y^{10}}$ **49.** $\dfrac{a^9 b^{14}}{ab}$ **50.** $\dfrac{x^{11} y^7}{xy}$ **51.** $\dfrac{x^3}{x^9}$

52. $\dfrac{z^4}{z^{12}}$ **53.** $3z^0$ **54.** $5y^0$ **55.** 5^{-3} **56.** 7^{-2}

57. $8x^{-9}$ **58.** $9y^{-7}$ **59.** $5^{-1} + 10^{-1}$ **60.** $7^{-1} + 14^{-1}$ **61.** $\dfrac{z^{-8}}{z^{-1}}$

62. $\dfrac{r^{-15}}{r^{-4}}$ **63.** $x^{-7} \cdot x^5 \cdot x^{-7}$ **64.** $y^{-9} \cdot y^6 \cdot y^{-9}$ **65.** $\left(a^{-2} b^3\right)\left(a^{10} b^{-11}\right)$

66. $\left(x^{-4} y^5\right)\left(x^{13} y^{-14}\right)$ **67.** $\left(3x^{20} y^{-1}\right)\left(10x^{-11} y^{-5}\right)$ **68.** $\left(4m^{16} n^{-3}\right)\left(11m^{-6} n^{-3}\right)$

Scientific Notation

Objective A Writing Numbers in Scientific Notation ▶

Both very large and very small numbers frequently occur in many fields of science. For example, the distance between the Sun and the dwarf planet Pluto is approximately 5,906,000,000 kilometers, and the mass of a proton is approximately 0.000000000000000000000000165 gram. It can be tedious to write these numbers in this standard decimal notation, so **scientific notation** is used as a convenient shorthand for expressing very large and very small numbers.

Objectives

A Write Numbers in Scientific Notation. ▶

B Convert Numbers in Scientific Notation to Standard Form. ▶

C Perform Operations on Numbers Written in Scientific Notation. ▶

5,906,000,000 kilometers

Pluto

Scientific Notation

A positive number is written in scientific notation if it is written as the product of a number a, where $1 \leq a < 10$, and an integer power r of 10: $a \times 10^r$.

The following numbers are written in scientific notation. The $\times$ sign for multiplication is used as part of the notation.

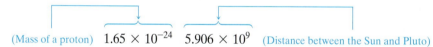

(Mass of a proton) 1.65×10^{-24} 5.906×10^9 (Distance between the Sun and Pluto)

The following steps are useful when writing numbers in scientific notation.

To Write a Number in Scientific Notation

Step 1: Move the decimal point in the original number so that the new number has a value between 1 and 10.

Step 2: Count the number of decimal places the decimal point is moved in Step 1. If the original number is 10 or greater, the count is positive. If the original number is less than 1, the count is negative.

Step 3: Multiply the new number in Step 1 by 10 raised to an exponent equal to the count found in Step 2.

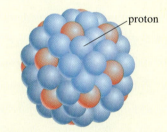

Mass of proton is approximately 0.000000000000000000000000165 gram

Practice 1

Write each number in scientific notation.

a. 760,000

b. 0.00035

Example 1   Write each number in scientific notation.

a. 367,000,000   b. 0.000003

Solution: Move each decimal point until the number is between 1 and 10.

a. 367,000,000. $= 3.67 \times 10^8$ The original number is 10 or greater, so the exponent is positive.

　　　　8 places

b. 0.000003 $= 3.0 \times 10^{-6}$ The original number is less than 1, so the exponent is negative.

　　　6 places

■ Work Practice 1

Objective B Converting Numbers to Standard Form ▶

A number written in scientific notation can be rewritten in standard form. For example, to write 8.63×10^3 in standard form, recall that $10^3 = 1000$.

$$8.63 \times 10^3 = 8.63(1000) = 8630$$

Notice that the exponent on the 10 is positive 3, and we moved the decimal point 3 places to the right.

To write 7.29×10^{-3} in standard form, recall that $10^{-3} = \dfrac{1}{10^3} = \dfrac{1}{1000}$.

$$7.29 \times 10^{-3} = 7.29\left(\dfrac{1}{1000}\right) = \dfrac{7.29}{1000} = 0.00729$$

The exponent on the 10 is negative 3, and we moved the decimal to the left 3 places.

In general, **to write a scientific notation number in standard form,** move the decimal point the same number of places as the exponent on 10. If the exponent is positive, move the decimal point to the right; if the exponent is negative, move the decimal point to the left.

Practice 2

Write the numbers in standard notation, without exponents.

a. 9.062×10^{-4}

b. 8.002×10^6

Example 2   Write each number in standard notation, without exponents.

a. 1.02×10^5 b. 7.358×10^{-3}

Solution:

a. Move the decimal point 5 places to the right.

$$1.02 \times 10^5 = 102,000.$$

b. Move the decimal point 3 places to the left.

$$7.358 \times 10^{-3} = 0.007358$$

■ Work Practice 2

Objective C Performing Operations on Numbers Written in Scientific Notation ▶

Performing operations on numbers written in scientific notation makes use of the rules and definitions for exponents.

Answers

1. a. 7.6×10^5 b. 3.5×10^{-4}

2. a. 0.0009062 b. 8,002,000

Example 3 Perform each indicated operation. Write each result in standard decimal notation.

a. $(8 \times 10^{-6})(7 \times 10^3)$

b. $\dfrac{12 \times 10^2}{6 \times 10^{-3}}$

Solution:

a. $(8 \times 10^{-6})(7 \times 10^3) = 8 \cdot 7 \cdot 10^{-6} \cdot 10^3$
$$= 56 \times 10^{-3}$$
$$= 0.056$$

b. $\dfrac{12 \times 10^2}{6 \times 10^{-3}} = \dfrac{12}{6} \times 10^{2-(-3)} = 2 \times 10^5 = 200{,}000$

■ **Work Practice 3**

Practice 3

Perform each indicated operation. Write each result in standard decimal notation.

a. $(8 \times 10^7)(3 \times 10^{-9})$

b. $\dfrac{8 \times 10^4}{2 \times 10^{-3}}$

Answers
3. a. 0.24   **b.** 40,000,000

C **Exercise Set** MyMathLab® ▶

Objective A *Write each number in scientific notation. See Example 1.*

1. 78,000

2. 9,300,000,000

3. 0.00000167

4. 0.00000017

5. 0.00635

6. 0.00194

7. 1,160,000

8. 700,000

9. When it is completed in 2022, the Thirty Meter Telescope is expected to be the world's largest optical telescope. Located in an observatory complex at the summit of Mauna Kea in Hawaii, the elevation of the Thirty Meter Telescope will be roughly 4200 meters above sea level. Write 4200 in scientific notation.

10. The Thirty Meter Telescope (see Exercise **9**) will have the ability to view objects 13,000,000,000 light-years away. Write 13,000,000,000 in scientific notation.

Objective B *Write each number in standard notation. See Example 2.*

11. 8.673×10^{-10}

12. 9.056×10^{-4}

13. 3.3×10^{-2}

14. 4.8×10^{-6}

15. 2.032×10^4

16. 9.07×10^{10}

17. Each second, the Sun converts 7.0×10^8 tons of hydrogen into helium and energy in the form of gamma rays. Write this number in standard notation. (*Source:* Students for the Exploration and Development of Space)

18. In chemistry, Avogadro's number is the number of atoms in one mole of an element. Avogadro's number is $6.02214199 \times 10^{23}$. Write this number in standard notation. (*Source:* National Institute of Standards and Technology)

Objectives A B Mixed Practice *See Examples 1 and 2. The bar graph below shows estimates of the top six national debts as of December 31, 2012. If a number is written in standard form, write it in scientific notation. If a number is written in scientific notation, write it in standard form.*

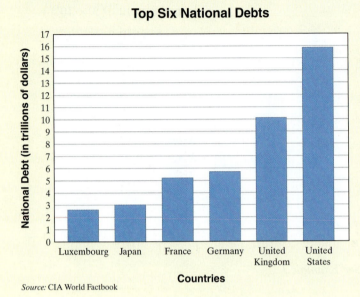

Top Six National Debts

Source: CIA World Factbook

19. Germany's national debt as of the end of 2012 was $5,700,000,000,000.

20. Luxembourg's national debt as of the end of 2012 was $2,600,000,000,000.

21. The United Kingdom's national debt as of the end of 2012 was 1.01×10^{13}.

22. France's national debt as of the end of 2012 was 5.2×10^{12}.

23. Use the bar graph to estimate the national debt of Japan and then express it in both standard and scientific notation.

24. Use the bar graph to estimate the national debt of the United States and then express it in both standard and scientific notation.

Objectives C *Evaluate each expression using exponential rules. Write each result in standard notation. See Example 3.*

25. $\left(1.2 \times 10^{-3}\right)\left(3 \times 10^{-2}\right)$

26. $\left(2.5 \times 10^{6}\right)\left(2 \times 10^{-6}\right)$

27. $\left(4 \times 10^{-10}\right)\left(7 \times 10^{-9}\right)$

28. $\left(5 \times 10^{6}\right)\left(4 \times 10^{-8}\right)$

29. $\dfrac{8 \times 10^{-1}}{16 \times 10^{5}}$

30. $\dfrac{25 \times 10^{-4}}{5 \times 10^{-9}}$

31. $\dfrac{1.4 \times 10^{-2}}{7 \times 10^{-8}}$

32. $\dfrac{0.4 \times 10^{5}}{0.2 \times 10^{11}}$

33. Although the actual amount varies by season and time of day, the average volume of water that flows over Niagara Falls (the American and Canadian falls combined) each second is 7.5×10^{5} gallons. How much water flows over Niagara Falls in an hour? Write the result in scientific notation. (*Hint:* 1 hour equals 3600 seconds.) (*Source:* niagara-fallslive.com)

34. A beam of light travels 9.460×10^{12} kilometers per year. How far does light travel in 10,000 years? Write the result in scientific notation.

Geometric Formulas

Rectangle

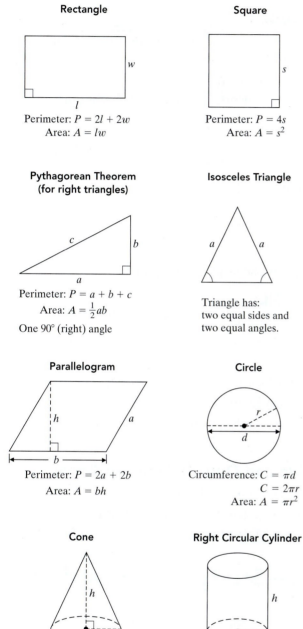

Perimeter: $P = 2l + 2w$
Area: $A = lw$

Square

Perimeter: $P = 4s$
Area: $A = s^2$

Triangle

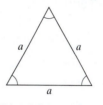

Perimeter: $P = a + b + c$
Area: $A = \frac{1}{2}bh$

Sum of Angles of Triangle

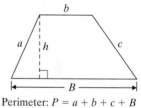

$A + B + C = 180°$

The sum of the measures of the three angles is 180.

Pythagorean Theorem (for right triangles)

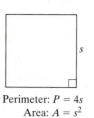

Perimeter: $P = a + b + c$
Area: $A = \frac{1}{2}ab$
One 90° (right) angle

Isosceles Triangle

Triangle has:
two equal sides and
two equal angles.

Equilateral Triangle

Triangle has:
three equal sides and
three equal angles.
Measure of each angle is 60°.

Trapezoid

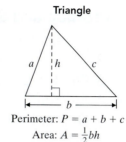

Perimeter: $P = a + b + c + B$
Area: $A = \frac{1}{2}h(B + b)$

Parallelogram

Perimeter: $P = 2a + 2b$
Area: $A = bh$

Circle

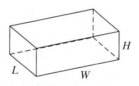

Circumference: $C = \pi d$
$C = 2\pi r$
Area: $A = \pi r^2$

Rectangular Solid

Volume: $V = LWH$
Surface Area:
$S = 2LW + 2HL + 2HW$

Cube

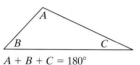

Volume: $V = s^3$
Surface Area: $S = 6s^2$

Cone

Volume: $V = \frac{1}{3}\pi r^2 h$
Lateral Surface Area:
$S = \pi r \sqrt{r^2 + h^2}$

Right Circular Cylinder

Volume: $V = \pi r^2 h$
Surface Area: $S = 2\pi r^2 + 2\pi rh$

Sphere

Volume: $V = \frac{4}{3}\pi r^3$
Surface Area: $S = 4\pi r^2$

Square-Based Pyramid

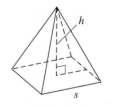

Volume: $V = a \cdot s^2 \cdot h$

Contents of Student Resources

Study Skills Builders

Bigger Picture—Study Guide Outline

Practice Final Exam

Answers to Selected Exercises

Solutions to Selected Exercises

Student Resources

Study Skills Builders

Attitude and Study Tips

Study Skills Builder 1

Have You Decided to Complete This Course Successfully?

Ask yourself if one of your current goals is to complete this course successfully.

If it is not a goal of yours, ask yourself why. One common reason is fear of failure. Amazingly enough, fear of failure alone can be strong enough to keep many of us from doing our best in any endeavor.

Another common reason is that you simply haven't taken the time to think about or write down your goals for this course. To help accomplish this, answer the questions below.

Exercises

1. Write down your goal(s) for this course.

2. Now list steps you will take to make sure your goal(s) in Exercise **1** are accomplished.

3. Rate your commitment to this course with a number between 1 and 5. Use the diagram below to help.

High Commitment		Average Commitment		Not Committed at All
5	4	3	2	1

4. If you have rated your personal commitment level (from the exercise above) as a 1, 2, or 3, list the reasons why this is so. Then determine whether it is possible to increase your commitment level to a 4 or 5.

Good luck, and don't forget that a positive attitude will make a big difference.

Study Skills Builder 2

Tips for Studying for an Exam

To prepare for an exam, try the following study techniques:

- Start the study process days before your exam.

- Make sure that you are up-to-date on your assignments.

- If there is a topic that you are unsure of, use one of the many resources that are available to you. For example,

  See your instructor.
  View a lecture video on the topic.
  Visit a learning resource center on campus.
  Read the textbook material and examples on the topic.

- Reread your notes and carefully review the Chapter Highlights at the end of any chapter.

- Work the review exercises at the end of the chapter.

- Find a quiet place to take the Chapter Test found at the end of the chapter. Do not use any resources when taking this sample test. This way, you will have a clear indication of how prepared you are for your exam. Check your answers and use the Chapter Test Prep Videos to make sure that you correct any missed exercises.

Good luck, and keep a positive attitude.

Exercises

Let's see how you did on your last exam.

1. How many days before your last exam did you start studying for that exam?

2. Were you up-to-date on your assignments at that time or did you need to catch up on assignments?

3. List the most helpful text supplement (if you used one).

4. List the most helpful campus supplement (if you used one).

5. List your process for preparing for a mathematics test.

6. Was this process helpful? In other words, were you satisfied with your performance on your exam?

7. If not, what changes can you make in your process that will make it more helpful to you?

Study Skills Builder 3

What to Do the Day of an Exam

Your first exam may be soon. On the day of an exam, don't forget to try the following:

- Allow yourself plenty of time to arrive.

- Read the directions on the test carefully.

- Read each problem carefully as you take your test. Make sure that you answer the question asked.

- Watch your time and pace yourself so that you may attempt each problem on your test.

- Check your work and answers.

- ***Do not turn your test in early.*** If you have extra time, spend it double-checking your work.

Good luck!

Exercises

Answer the following questions based on your most recent mathematics exam, whenever that was.

1. How soon before class did you arrive?

2. Did you read the directions on the test carefully?

3. Did you make sure you answered the question asked for each problem on the exam?

4. Were you able to attempt each problem on your exam?

5. If your answer to Exercise **4** is no, list reasons why.

6. Did you have extra time on your exam?

7. If your answer to Exercise **6** is yes, describe how you spent that extra time.

Study Skills Builder 4

Are You Satisfied with Your Performance on a Particular Quiz or Exam?

If not, don't forget to analyze your quiz or exam and look for common errors. Were most of your errors a result of:

- *Carelessness?* Did you turn in your quiz or exam before the allotted time expired? If so, resolve to use any extra time to check your work.

- *Running out of time?* Try completing any questions that you are unsure of last and delay checking your work until all questions have been answered.

- *Not understanding a concept?* If so, review that concept and correct your work so that you make sure you understand the concept before the next quiz or the final exam.

- *Test conditions?* When studying for a quiz or exam, make sure you place yourself in conditions similar to test conditions. For example, before your next quiz or exam, take a sample test without the aid of your notes or text.

(For a sample test, see your instructor or use the Chapter Test at the end of each chapter.)

Exercises

1. Have you corrected all your previous quizzes and exams?

2. List any errors you have found common to two or more of your graded papers.

3. Is one of your common errors not understanding a concept? If so, are you making sure you understand all the concepts for the next quiz or exam?

4. Is one of your common errors making careless mistakes? If so, are you now taking all the time allotted to check over your work so that you can minimize the number of careless mistakes?

5. Are you satisfied with your grades thus far on quizzes and tests?

6. If your answer to Exercise **5** is no, are there any more suggestions you can make to your instructor or yourself to help? If so, list them here and share them with your instructor.

Study Skills Builder 5

How Are You Doing?

If you haven't done so yet, take a few moments and think about how you are doing in this course. Are you working toward your goal of successfully completing this course? Is your performance on homework, quizzes, and tests satisfactory? If not, you might want to see your instructor to see if he/she has any suggestions on how you can improve your performance. Reread Section 1.1 for ideas on places to get help with your mathematics course.

Exercises

Answer the following.

1. List any textbook supplements you are using to help you through this course.

2. List any campus resources you are using to help you through this course.

3. Write a short paragraph describing how you are doing in your mathematics course.

4. If improvement is needed, list ways that you can work toward improving your situation as described in Exercise **3**.

Study Skills Builder 6

Are You Preparing for Your Final Exam?

To prepare for your final exam, try the following study techniques:

- Review the material that you will be responsible for on your exam. This includes material from your textbook, your notebook, and any handouts from your instructor.
- Review any formulas that you may need to memorize.
- Check to see if your instructor or mathematics department will be conducting a final exam review.
- Check with your instructor to see whether final exams from previous semesters/quarters are available to students for review.

- Use your previously taken exams as a practice final exam. To do so, rewrite the test questions in mixed order on blank sheets of paper. This will help you prepare for exam conditions.
- If you are unsure of a few concepts, see your instructor or visit a learning lab for assistance. Also, view the video segment of any troublesome sections.
- If you need further exercises to work, try the Cumulative Reviews at the end of the chapters.

Once again, good luck! I hope you are enjoying this textbook and your mathematics course.

Organizing Your Work

Study Skills Builder 7

Learning New Terms

Many of the terms used in this text may be new to you. It will be helpful to make a list of new mathematical terms and symbols as you encounter them and to review them frequently. Placing these new terms (including page references) on 3×5 index cards might help you later when you're preparing for a quiz.

Exercises

1. Name one way you might place a word and its definition on a 3×5 card.

2. How do new terms stand out in this text so that they can be found?

Study Skills Builder 8

Are You Organized?

Have you ever had trouble finding a completed assignment? When it's time to study for a test, are your notes neat and organized? Have you ever had trouble reading your own mathematics handwriting? (Be honest—I have.)

When any of these things happen, it's time to get organized. Here are a few suggestions:

- Write your notes and complete your homework assignments in a notebook with pockets (spiral or ring binder).

- Take class notes in this notebook, and then follow the notes with your completed homework assignment.

- When you receive graded papers or handouts, place them in the notebook pocket so that you will not lose them.

- Mark (possibly with an exclamation point) any note(s) that seem extra important to you.

- Mark (possibly with a question mark) any notes or homework that you are having trouble with.

- See your instructor or a math tutor for help with the concepts or exercises that you are having trouble understanding.

- If you are having trouble reading your own handwriting, *slow down* and write your mathematics work clearly!

Exercises

1. Have you been completing your assignments on time?

2. Have you been correcting any exercises you may be having difficulty with?

3. If you are having trouble understanding a mathematical concept or correcting any homework exercises, have you visited your instructor, a tutor, or your campus math lab?

4. Are you taking lecture notes in your mathematics course? (By the way, these notes should include worked-out examples solved by your instructor.)

5. Is your mathematics course material (handouts, graded papers, lecture notes) organized?

6. If your answer to Exercise 5 is no, take a moment and review your course material. List at least two ways that you might better organize it.

Study Skills Builder 9

Organizing a Notebook

It's never too late to get organized. If you need ideas about organizing a notebook for your mathematics course, try some of these:

- Use a spiral or ring binder notebook with pockets and use it for mathematics only.
- Start each page by writing the book's section number you are working on at the top.
- When your instructor is lecturing, take notes. *Always* include any examples your instructor works for you.
- Place your worked-out homework exercises in your notebook immediately after the lecture notes from that section. This way, a section's worth of material is together.
- Homework exercises: Attempt and check all assigned homework.
- Place graded quizzes in the pockets of your notebook or a special section of your binder.

Exercises

Check your notebook organization by answering the following questions.

1. Do you have a spiral or ring binder notebook for your mathematics course only?

2. Have you ever had to flip through several sheets of notes and work in your mathematics notebook to determine what section's work you are in?

3. Are you now writing the textbook's section number at the top of each notebook page?

4. Have you ever lost or had trouble finding a graded quiz or test?

5. Are you now placing all your graded work in a dedicated place in your notebook?

6. Are you attempting all of your homework and placing all of your work in your notebook?

7. Are you checking and correcting your homework in your notebook? If not, why not?

8. Are you writing in your notebook the examples your instructor works for you in class?

Study Skills Builder 10

How Are Your Homework Assignments Going?

It is very important in mathematics to keep up with homework. Why? Many concepts build on each other. Often your understanding of a day's concepts depends on an understanding of the previous day's material.

Remember that completing your homework assignment involves a lot more than attempting a few of the problems assigned.

To complete a homework assignment, remember these four things:

- Attempt all of it.
- Check it.
- Correct it.
- If needed, ask questions about it.

Exercises

Take a moment and review your completed homework assignments. Answer the questions below based on this review.

1. Approximate the fraction of your homework you have attempted.

2. Approximate the fraction of your homework you have checked (if possible).

3. If you are able to check your homework, have you corrected it when errors have been found?

4. When working homework, if you do not understand a concept, what do you do?

MyMathLab and MathXL

Study Skills Builder 11

Tips for Turning in Your Homework on Time

It is very important to keep up with your mathematics homework assignments. Why? Many concepts in mathematics build upon each other.

Remember these 4 tips to help ensure your work is completed on time:

- Know the assignments and due dates set by your instructor.
- Do not wait until the last minute to submit your homework.
- Set a goal to submit your homework 6–8 hours before the scheduled due date in case you have unexpected technology trouble.
- Schedule enough time to complete each assignment.

Following the tips above will also help you avoid potentially losing points for late or missed assignments.

Exercises

Take a moment to consider your work on your homework assignments to date and answer the following questions:

1. What percentage of your assignments have you turned in on time?

2. Why might it be a good idea to submit your homework 6–8 hours before the scheduled deadline?

3. If you have missed submitting any homework by the due date, list some of the reasons why this occurred.

4. What steps do you plan to take in the future to ensure your homework is submitted on time?

Study Skills Builder 12

Tips for Doing Your Homework Online

Practice is one of the main keys to success in any mathematics course. Did you know that MyMathLab/MathXL provides you with **immediate feedback** for each exercise? If you are incorrect, you are given hints to work the exercise correctly. You have **unlimited practice opportunities** and can rework any exercises you have trouble with until you master them, and submit homework assignments unlimited times before the deadline.

Remember these success tips when doing your homework online:

- Attempt all assigned exercises.
- Write down (neatly) your step-by-step work for each exercise before entering your answer.
- Use the immediate feedback provided by the program to help you check and correct your work for each exercise.
- Rework any exercises you have trouble with until you master them.
- Work through your homework assignment as many times as necessary until you are satisfied.

Exercises

Take a moment to think about your homework assignments to date and answer the following:

1. Have you attempted all assigned exercises?

2. Of the exercises attempted, have you also written out your work before entering your answer—so that you can check it?

3. Are you familiar with how to enter answers using the MathXL player so that you avoid answer-entry type errors?

4. List some ways the immediate feedback and practice supports have helped you with your homework. If you have not used these supports, how do you plan to use them with the success tips above on your next assignment?

Study Skills Builder 13

Organizing Your Work

Have you ever used any readily available paper (such as the back of a flyer, another course assignment, Post-its, etc.) to work out your homework exercises before entering the answer in MathXL? To save time, have you ever entered answers directly into MathXL without working the exercises on paper? When it's time to study, have you ever been unable to find your completed work or read and follow your own mathematics handwriting?

When any of these things happen, it's time to get organized. Here are some suggestions:

- Write your step-by-step work for each homework exercise (neatly) on lined, loose-leaf paper and keep this in a 3-ring binder.

- Refer to your step-by-step work when you receive feedback that your answer is incorrect in MathXL. Double-check using the steps and hints provided by the program and correct your work accordingly.

- Keep your written homework with your class notes for that section.

- Identify any exercises you are having trouble with and ask questions about them.

- Keep all graded quizzes and tests in this binder as well to study later.

If you follow the suggestions above, you and your instructor or tutor will be able to follow your steps and correct any mistakes. You will also have a written copy of your work to refer to later to ask questions and study for tests.

Exercises

1. Why is it important that you write out your step-by-step work for homework exercises and keep a hard copy of all work submitted online?

2. If you have gotten an incorrect answer, are you able to follow your steps and find your error?

3. If you were asked today to review your previous homework assignments and 1st test, could you find them? If not, list some ways you might better organize your work.

Study Skills Builder 14

Getting Help with Your Homework Assignments

There are many resources available to you through MathXL to help you work through any homework exercises you may have trouble with. It is important that you know what these resources are and know when and how to use them.

Let's review the features found on the right side of the screen in the homework exercises:

- **Help Me Solve This**—provides step-by-step help for the exercise you are working. You must work an additional exercise of the same type (without this help) before you can get credit for having worked it correctly.

- **View an Example**—allows you to view a correctly worked exercise similar to the one you are having trouble with. You can then go back to your original exercise and work it on your own.

- **E-Book**—allows you to read examples from your text and find similar exercises.

- **Video**—your text author, Elayn Martin-Gay, works an exercise similar to the one you need help with. **Not all exercises have an accompanying video clip.

- **Ask My Instructor**—allows you to e-mail your instructor for help with an exercise.

Exercises

1. How does the "Help Me Solve This" feature work?

2. If the "View an Example" feature is used, is it necessary to work an additional problem before continuing the assignment?

3. When might be a good time to use the "Video" feature? Do all exercises have an accompanying video clip?

4. Which of the features above have you used? List those you found the most helpful to you.

5. If you haven't used the features discussed, list those you plan to try on your next homework assignment.

Study Skills Builder 15

Tips for Preparing for an Exam

Did you know that you can rework your previous homework assignments in MyMathLab and MathXL? This is a great way to prepare for tests. To do this, open a previous homework assignment and click "similar exercise." This will generate new exercises similar to the homework you have submitted. You can then rework the exercises and assignments until you feel confident that you understand them.

To prepare for an exam, follow these tips:

- Review your written work for your previous homework assignments along with your class notes.
- Identify any exercises or topics that you have questions on or have difficulty understanding.
- Rework your previous assignments in MyMathLab and MathXL until you fully understand them and can do them without help.
- Get help for any topics you feel unsure of or for which you have questions.

Exercises

1. Are your current homework assignments up to date and is your written work for them organized in a binder or notebook? If the answer is no, it's time to get organized. For tips on this, see Study Skills Builder 13—Organizing Your Work.

2. How many days in advance of an exam do you usually start studying?

3. List some ways you think that working previous homework assignments can help you prepare for your test.

4. List 2–3 resources you can use to get help for any topics you are unsure of or have questions on.

Good luck!

Study Skills Builder 16

How Well Do You Know the Resources Available to You in MyMathLab?

There are many helpful resources available to you in MyMathLab. Let's take a moment to locate and explore a few of them now. Go into your MyMathLab course, and visit the Multimedia Library, Tools for Success, and E-Book.

Let's see what you found.

Exercises

1. List the resources available to you in the Multimedia Library.

2. List the resources available to you in the Tools for Success folder.

3. Where did you find the English/Spanish Audio Glossary?

4. Can you view videos from the E-Book?

5. Did you find any resources you did not know about? If so, which ones?

6. Which resources have you used most often or find most helpful?

Additional Help Inside and Outside Your Textbook

Study Skills Builder 17

How Well Do You Know Your Textbook?

The questions below will help determine whether you are familiar with your textbook. For additional information, see Section 1.1 in this text.

1. What does the ▶ icon mean?

2. What does the ✎ icon mean?

3. What does the △ icon mean?

4. Where can you find a review for each chapter? What answers to this review can be found in the back of your text?

5. Each chapter contains an overview of the chapter along with examples. What is this feature called?

6. Each chapter contains a review of vocabulary. What is this feature called?

7. There are practice exercises that are contained in this text. What are they and how can they be used?

8. This text contains a student section in the back entitled Student Resources. List the contents of this section and how they might be helpful.

9. What exercise answers are available in this text? Where are they located?

Study Skills Builder 18

Are You Familiar with Your Textbook Supplements?

Below is a review of some of the student supplements available for additional study. Check to see if you are using the ones most helpful to you.

- Chapter Test Prep Videos. These videos provide video clip solutions to the Chapter Test exercises in this text. You will find them extremely useful when studying for tests or exams.
- Interactive DVD Lecture Series. These are keyed to each section of the text. The material is presented by me, Elayn Martin-Gay, and I have placed a ▶ by the exercises in the text that I have worked on the video.
- The *Student Solutions Manual*. This contains worked-out solutions to odd-numbered exercises as well as every exercise in the Integrated Reviews, Chapter Reviews, Chapter Tests, and Cumulative Reviews.
- Pearson Tutor Center. Mathematics questions may be phoned, faxed, or e-mailed to this center.
- MyMathLab is a text-specific online course. MathXL is an online homework, tutorial, and assessment system.

Take a moment and determine whether these are available to you.

As usual, your instructor is your best source of information.

Exercises

Let's see how you are doing with textbook supplements.

1. Name one way the Lecture Videos can be helpful to you.

2. Name one way the Chapter Test Prep Videos can help you prepare for a chapter test.

3. List any textbook supplements that you have found useful.

4. Have you located and visited a learning resource lab located on your campus?

5. List the textbook supplements that are currently housed in your campus' learning resource lab.

Study Skills Builder 19

Are You Getting All the Mathematics Help That You Need?

Remember that, in addition to your instructor, there are many places to get help with your mathematics course. For example:

- This text has an accompanying video lesson for every section.
- The back of the book contains answers to odd-numbered exercises and selected solutions.
- A *Student Solutions Manual* is available that contains worked-out solutions to odd-numbered exercises as well as solutions to every exercise in the Integrated Reviews, Chapter Reviews, Chapter Tests, and Cumulative Reviews.
- Don't forget to check with your instructor for other local resources available to you, such as a tutor center.

Exercises

1. List items you find helpful in the text and all student supplements to this text.

2. List all the campus help that is available to you for this course.

3. List any help (besides the textbook) from Exercises **1** and **2** above that you are using.

4. List any help (besides the textbook) that you feel you should try.

5. Write a goal for yourself that includes trying anything you listed in Exercise **4** during the next week.

Bigger Picture—
Study Guide Outline

I. Operations on Sets of Numbers

A. Whole Numbers

1. **Add or Subtract:**

$$\begin{array}{r} 14 \\ + 39 \\ \hline 53 \end{array} \qquad \begin{array}{r} 300 \\ - 27 \\ \hline 273 \end{array}$$

2. **Multiply or Divide:**

$$\begin{array}{r} 238 \\ \times \ 47 \\ \hline 1666 \\ 9520 \\ \hline 11{,}186 \end{array}$$

$$\begin{array}{r} 127 \ \text{R2} \\ 7\overline{)891} \\ -7 \\ \hline 19 \\ -14 \\ \hline 51 \\ -49 \\ \hline 2 \end{array}$$

3. **Exponent:** 4 factors of 3

$$3^4 = \overbrace{3 \cdot 3 \cdot 3 \cdot 3}^{} = 81$$

4. **Order of Operations:**

$$24 \div 3 \cdot 2 - (2 + 8) = 24 \div 3 \cdot 2 - (10) \quad \text{Simplify within parentheses.}$$
$$= 8 \cdot 2 - 10 \quad \text{Multiply or divide from left to right.}$$
$$= 16 - 10 \quad \text{Multiply or divide from left to right.}$$
$$= 6 \quad \text{Add or subtract from left to right.}$$

5. **Square Root:**

$\sqrt{25} = 5$ because $5 \cdot 5 = 25$ and 5 is a positive number.

B. Integers

1. **Add:** $-5 + (-2) = -7$ Adding like signs
Add absolute values. Attach the common sign.

$-5 + 2 = -3$ Adding unlike signs
Subtract absolute values. Attach the sign of the number with the larger absolute value.

2. **Subtract:** Add the first number to the opposite of the second number.

$$7 - 10 = 7 + (-10) = -3$$

3. **Multiply or Divide:** Multiply or divide as usual. If the signs of the two numbers are the same, the answer is positive. If the signs of the two numbers are different, the answer is negative.

$$-5 \cdot 5 = -25, \quad \frac{-32}{-8} = 4$$

C. Fractions

1. **Simplify:** Factor the numerator and denominator. Then divide out factors of 1 by dividing out common factors in the numerator and denominator.

Simplify: $\dfrac{20}{28} = \dfrac{4 \cdot 5}{4 \cdot 7} = \dfrac{5}{7}$

2. **Multiply:** Numerator times numerator over denominator times denominator.

$$\frac{5}{9} \cdot \frac{2}{7} = \frac{10}{63}$$

3. **Divide:** First fraction times the reciprocal of the second fraction.

$$\frac{2}{11} \div \frac{3}{4} = \frac{2}{11} \cdot \frac{4}{3} = \frac{8}{33}$$

4. **Add or Subtract:** Must have same denominators. If not, find the LCD, and write each fraction as an equivalent fraction with the LCD as denominator.

$$\frac{2}{5} + \frac{1}{15} = \frac{2}{5} \cdot \frac{3}{3} + \frac{1}{15} = \frac{6}{15} + \frac{1}{15} = \frac{7}{15}$$

D. Decimals

1. **Add or Subtract:** Line up decimal points.

$$\begin{array}{r} 1.27 \\ +\ 0.6 \\ \hline 1.87 \end{array}$$

2. **Multiply:**

$$\begin{array}{r} 2.56 \\ \times\ 3.2 \\ \hline 512 \\ 7680 \\ \hline 8.192 \end{array}$$

2 decimal places

1 decimal place

$2 + 1 = 3$

3 decimal places

3. **Divide:**

$$\begin{array}{r} 0.7 \\ 8\overline{)5.6} \end{array} \qquad \begin{array}{r} 1.\ 31 \\ 0.6\overline{)0.786} \end{array}$$

II. Solving Equations

A. Equations in General: Simplify both sides of the equation by removing parentheses and combining any like terms. Then use the addition property to write variable terms on one side and constants (numbers) on the other side. Then use the multiplication property to solve for the variable by dividing both sides of the equation by the coefficient of the variable.

$$\text{Solve: } 2(x - 5) = 80$$

$$2x - 10 = 80 \qquad \text{Use the distributive property.}$$

$$2x - 10 + 10 = 80 + 10 \qquad \text{Add 10 to both sides.}$$

$$2x = 90 \qquad \text{Simplify.}$$

$$\frac{2x}{2} = \frac{90}{2} \qquad \text{Divide both sides by 2.}$$

$$x = 45 \qquad \text{Simplify.}$$

B. Proportions: Set cross products equal to each other. Then solve.

$$\frac{14}{3} = \frac{2}{n} \text{ or } 14 \cdot n = 3 \cdot 2 \text{ or } 14 \cdot n = 6 \text{ or } n = \frac{6}{14} = \frac{3}{7}$$

C. Percent Problems

1. **Solved by Equations:** Remember that "of" means multiplication and "is" means equals.

   "12% of some number is 6" translates to

$$12\% \cdot n = 6 \text{ or } 0.12 \cdot n = 6 \text{ or } n = \frac{6}{0.12} \text{ or } n = 50$$

2. **Solved by Proportions:** Remember that percent, p, is identified by % or "percent"; base, b, usually appears after "of"; and amount, a, is the part compared to the whole.

   "12% of some number is 6" translates to

$$\frac{6}{b} = \frac{12}{100} \text{ or } 6 \cdot 100 = b \cdot 12 \text{ or } \frac{600}{12} = b \text{ or } 50 = b$$

Practice Final Exam

Simplify by performing the indicated operations.

1. $2^3 \cdot 5^2$

2. $16 + 9 \div 3 \cdot 4 - 7$

3. $18 - 24$

4. $5 \cdot (-20)$

5. $\sqrt{49}$

6. $(-5)^3 - 24 \div (-3)$

7. $0 \div 49$

8. $62 \div 0$

9. $-\dfrac{8}{15y} - \dfrac{2}{15y}$

10. $\dfrac{11}{12} - \dfrac{3}{8} + \dfrac{5}{24}$

11. $\dfrac{3a}{8} \cdot \dfrac{16}{6a^3}$

12. $-\dfrac{16}{3} \div -\dfrac{3}{12}$

13. $\dfrac{19}{-2\frac{3}{11}}$

14. $\dfrac{0.23 + 1.63}{-0.3}$

15. 10.2×4.01

16. Write 0.6% as a decimal.

17. Write 6.1 as a percent.

18. Write $\dfrac{3}{8}$ as a percent.

19. Write 0.345 as a fraction.

20. Write $-\dfrac{13}{26}$ as a decimal.

21. Round 34.8923 to the nearest tenth.

Evaluate each expression for the given replacement values.

22. $5(x^3 - 2)$ for $x = 2$

23. $10 - y^2$ for $y = -3$

24. $x \div y$ for $x = \dfrac{1}{2}$ and $y = 3\dfrac{7}{8}$

25. Simplify: $-(3z + 2) - 5z - 18$

△ 26. Write an expression that represents the perimeter of the equilateral triangle. Then simplify the expression.

$(5x + 5)$ inches

Answer blanks:

1. _____
2. _____
3. _____
4. _____
5. _____
6. _____
7. _____
8. _____
9. _____
10. _____
11. _____
12. _____
13. _____
14. _____
15. _____
16. _____
17. _____
18. _____
19. _____
20. _____
21. _____
22. _____
23. _____
24. _____
25. _____
26. _____

Solve each equation.

27. $\dfrac{n}{-7} = 4$

28. $-4x + 7 = 15$

29. $-4(x - 11) - 34 = 10 - 12$

30. $\dfrac{x}{5} + x = -\dfrac{24}{5}$

31. $2(x + 5.7) = 6x - 3.4$

32. $\dfrac{8}{x} = \dfrac{11}{6}$

△ **33.** Find the perimeter and area.

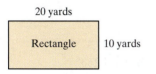

20 yards

Rectangle   10 yards

34. Find the average of $-12, -13, 0,$ and 9.

Solve.

35. The difference of three times a number and five times the same number is 4. Find the number.

36. During a 258-mile trip, a car used $10\dfrac{3}{4}$ gallons of gas. How many miles would we expect the car to travel on 1 gallon of gas?

37. In a 10-kilometer race, there are 112 more men entered than women. Find the number of female runners if the total number of runners in the race is 600.

38. The standard dose of medicine for a dog is 10 grams for every 15 pounds of body weight. What is the standard dose for a dog that weighs 80 pounds?

39. A \$120 framed picture is on sale for 15% off. Find the discount and the sale price.

Graph each linear equation.

40. $y + x = -4$

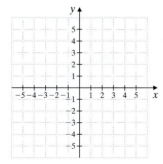

41. $y = 3x - 5$

27. _____

28. _____

29. _____

30. _____

31. _____

32. _____

33. _____

34. _____

35. _____

36. _____

37. _____

38. _____

39. _____

40. _____

41. _____

42. _____

43. _____

44. _____

45. _____

46. _____

47. _____

48. _____

49. _____

50. _____

51. _____

52. _____

53. _____

54. _____

42. $y = -4$

43. Add: $(11x - 3) + (4x - 1)$

44. Subtract $\left(8a^2 + a\right)$ from $\left(6a^2 + 2a + 1\right)$.

Multiply and simplify.

45. $\left(6a^3\right)\left(-2a^7\right)$

46. $\left(3a^4b\right)^2\left(2ba^4\right)^3$

47. $(x - 3)(x + 2)$

48. Factor out the GCF: $3y^2 - 15y$

49. Find the complement of a 78° angle.

50. Given that $m \parallel b$, find the measures of $x, y,$ and z.

51. Find the perimeter and area.

52. Find the circumference and area. Give the exact values and then approximations using $\pi \approx 3.14$.

Convert.

53. $2\dfrac{1}{2}$ gallons to quarts

54. 2.4 kilograms to grams

Answers to Selected Exercises

Chapter 1 The Whole Numbers

Section 1.2

Vocabulary, Readiness & Video Check **1.** whole **3.** words **5.** period **7.** hundreds **9.** 80,000

Exercise Set 1.2 **1.** tens **3.** thousands **5.** hundred-thousands **7.** millions **9.** three hundred fifty-four **11.** eight thousand, two hundred seventy-nine **13.** twenty-six thousand, nine hundred ninety **15.** two million, three hundred eighty-eight thousand **17.** twenty-four million, three hundred fifty thousand, one hundred eighty-five **19.** three hundred twenty-one thousand, eight hundred **21.** two thousand, seven hundred twenty **23.** sixteen million, three hundred thousand **25.** fourteen thousand, four hundred thirty-three **27.** eighteen thousand, twenty-seven **29.** 6587 **31.** 59,800 **33.** 13,601,011 **35.** 7,000,017 **37.** 260,997 **39.** 395 **41.** 2080 **43.** 200,300,000 **45.** 565 **47.** 200 + 9 **49.** 3000 + 400 + 70 **51.** 80,000 + 700 + 70 + 4 **53.** 60,000 + 6000 + 40 + 9 **55.** 30,000,000 + 9,000,000 + 600,000 + 80,000 **57.** 1786 **59.** Mount Baker **61.** Glacier Peak **63.** German shepherd **65.** Labrador retriever; seventy-five **67.** 70 pounds **69.** 9861 **71.** no; one hundred five **73.** answers may vary **75.** 1,000,000,000,000

Section 1.3

Calculator Explorations **1.** 134 **3.** 340 **5.** 2834 **7.** 770 **9.** 109 **11.** 8978

Vocabulary, Readiness & Video Check **1.** number **3.** 0 **5.** minuend; subtrahend; difference **7.** order; commutative **9.** place; right; left **11.** triangle; 3

Exercise Set 1.3 **1.** 36 **3.** 292 **5.** 49 **7.** 5399 **9.** 209,078 **11.** 25 **13.** 212 **15.** 11,926 **17.** 16,717 **19.** 35,901 **21.** 632,389 **23.** 600 **25.** 25 **27.** 288 **29.** 168 **31.** 5723 **33.** 504 **35.** 79 **37.** 32,711 **39.** 5041 **41.** 31,213 **43.** 1034 **45.** 9 **47.** 8518 **49.** 22,876 **51.** 25 ft **53.** 24 in. **55.** 29 in. **57.** 44 m **59.** 2093 **61.** 266 **63.** 20 **65.** 544 **67.** 72 **69.** 88 **71.** 3170 thousand **73.** $619 **75.** 264,000 sq mi **77.** 283,000 sq mi **79.** 340 ft **81.** 264 pages **83.** 31,300,000 **85.** 100 dB **87.** 58 dB **89.** 3444 **91.** 124 ft **93.** California **95.** 529 stores **97.** Pennsylvania and New York **99.** 5894 mi **101.** minuend: 48; subtrahend: 1 **103.** minuend: 70; subtrahend: 7 **105.** answers may vary **107.** correct **109.** incorrect; 530 **111.** incorrect; 685 **113.** correct **115.**

$$\begin{array}{r} 5269 \\ -\ 2385 \\ \hline 28\underline{8}4 \end{array}$$

117. answers may vary **119.** no; 1089 more pages

Section 1.4

Vocabulary, Readiness & Video Check **1.** graph **3.** 70; 60 **5.** 3 is the place we're rounding to (tens), and the digit to the right of this place is 5 or greater, so we need to add 1 to the 3. **7.** Each circled digit is to the right of the place value being rounded to and is used to determine whether or not we add 1 to the digit in the place value being rounded to.

Exercise Set 1.4 **1.** 420 **3.** 640 **5.** 2800 **7.** 500 **9.** 21,000 **11.** 34,000 **13.** 328,500 **15.** 36,000 **17.** 39,990 **19.** 30,000,000 **21.** 5280; 5300; 5000 **23.** 9440; 9400; 9000 **25.** 14,880; 14,900; 15,000 **27.** 311,000 miles **29.** 60,100 days **31.** $190,000,000,000 **33.** $3,200,000 **35.** US: 332,000,000; India: 894,000,000 **37.** 130 **39.** 80 **41.** 5700 **43.** 300 **45.** 11,400 **47.** incorrect **49.** correct **51.** correct **53.** $3400 **55.** 900 mi **57.** 6000 ft **59.** Springfield is larger by approximately 40,000. **61.** The increase was 55,000. **63.** 66,700,000; 67,000,000; 70,000,000 **65.** 57,800,000; 58,000,000; 60,000,000 **67.** 5723, for example **69. a.** 8550 **b.** 8649 **71.** answers may vary **73.** 140 m

Section 1.5

Calculator Explorations **1.** 3456 **3.** 15,322 **5.** 272,291

Vocabulary, Readiness & Video Check **1.** 0 **3.** product; factor **5.** grouping; associative **7.** length **9.** distributive **11.** Area is measured in square units, and here we have meters times meters, or square meters; the correct answer is 63 *square* meters, or the correct units are square meters.

Exercise Set 1.5 **1.** 24 **3.** 0 **5.** 0 **7.** 87 **9.** 6 · 3 + 6 · 8 **11.** 4 · 3 + 4 · 9 **13.** 20 · 14 + 20 · 6 **15.** 512 **17.** 3678 **19.** 1662 **21.** 6444 **23.** 1157 **25.** 24,418 **27.** 24,786 **29.** 15,600 **31.** 0 **33.** 6400 **35.** 48,126 **37.** 142,506 **39.** 2,369,826 **41.** 64,790 **43.** 3,949,935 **45.** area: 63 sq m; perimeter: 32 m **47.** area: 680 sq ft; perimeter: 114 ft **49.** 240,000 **51.** 300,000 **53.** c **55.** c **57.** 880 **59.** 4200 **61.** 4480 **63.** 375 cal **65.** $3290 **67. a.** 20 **b.** 100 **c.** 2000 lb **69.** 8800 sq ft **71.** 56,000 sq ft **73.** 5828 pixels **75.** 2100 characters **77.** 1280 cal **79.** $10, $60; $10, $200; $12, $36; $12, $36: total cost $372 **81.** 1,440,000 tea bags **83.** 135 **85.** 2144 **87.** 23 **89.** 15 **91.** 5 · 6 or 6 · 5 **93. a.** 5 + 5 + 5 or 3 + 3 + 3 + 3 + 3 **b.** answers may vary **95.**

$$\begin{array}{r} 203 \\ \times\ \ 14 \\ \hline 812 \\ 2030 \\ \hline 2842 \end{array}$$

97.

$$\begin{array}{r} 42 \\ \times 9\underline{3} \end{array}$$

99. answers may vary **101.** 506 windows

Section 1.6

Calculator Explorations **1.** 53 **3.** 62 **5.** 261 **7.** 0

Vocabulary, Readiness & Video Check **1.** quotient; dividend; divisor **3.** 1 **5.** undefined **7.** 0 **9.** $202 \cdot 102 + 15 = 20{,}619$
11. addition and division

Exercise Set 1.6 **1.** 6 **3.** 12 **5.** 0 **7.** 31 **9.** 1 **11.** 8 **13.** undefined **15.** 1 **17.** 0 **19.** 9 **21.** 29 **23.** 74 **25.** 338 **27.** undefined
29. 9 **31.** 25 **33.** 68 R 3 **35.** 236 R 5 **37.** 38 R 1 **39.** 326 R 4 **41.** 13 **43.** 49 **45.** 97 R 8 **47.** 209 R 11 **49.** 506 **51.** 202 R 7
53. 54 **55.** 99 R 100 **57.** 202 R 15 **59.** 579 R 72 **61.** 17 **63.** 511 R 3 **65.** 2132 R 32 **67.** 6080 **69.** 23 R 2 **71.** 5 R 25
73. 20 R 2 **75.** 33 students **77.** 165 lb **79.** 310 yd **81.** 89 bridges **83.** 11 light poles **85.** 5 mi **87.** 1760 yd **89.** 20 **91.** 387
93. 79 **95.** 74° **97.** 9278 **99.** 15,288 **101.** 679 **103.** undefined **105.** 9 R 12 **107.** c **109.** b **111.** $180,845,200 **113.** increase;
answers may vary **115.** no; answers may vary **117.** 12 ft **119.** answers may vary **121.** 5 R 1

Integrated Review **1.** 194 **2.** 6555 **3.** 4524 **4.** 562 **5.** 67 **6.** undefined **7.** 1 **8.** 5 **9.** 0 **10.** 0 **11.** 0 **12.** 3 **13.** 63
14. 9 **15.** 138 **16.** 276 **17.** 1169 **18.** 9826 **19.** 182 R 4 **20.** 79,317 **21.** 1099 R 2 **22.** 111 R 1 **23.** 663 R 24 **24.** 1076 R 60
25. 1037 **26.** 9899 **27.** 30,603 **28.** 47,500 **29.** 71 **30.** 558 **31.** 6 R 8 **32.** 53 **33.** 183 **34.** 231 **35.** 9740; 9700; 10,000
36. 1430; 1400; 1000 **37.** 20,800; 20,800; 21,000 **38.** 432,200; 432,200; 432,000 **39.** perimeter: 24 ft; area: 36 sq ft **40.** perimeter:
42 in.; area: 98 sq in. **41.** 28 mi **42.** 26 m **43.** 24 **44.** 124 **45.** Lake Pontchartrain Bridge; 2175 ft **46.** $5562

Section 1.7

Calculator Explorations **1.** 4096 **3.** 3125 **5.** 2048 **7.** 2526 **9.** 4295 **11.** 8

Vocabulary, Readiness & Video Check **1.** base; exponent **3.** addition **5.** division **7.** 1 **9.** The area of a rectangle is
length $\cdot$ width. A square is a special rectangle where length = width. Thus, the area of a square is side $\cdot$ side or $(\text{side})^2$.

Exercise Set 1.7 **1.** 4^3 **3.** 7^6 **5.** 12^3 **7.** $6^2 \cdot 5^3$ **9.** $9 \cdot 8^2$ **11.** $3 \cdot 2^4$ **13.** $3 \cdot 2^4 \cdot 5^5$ **15.** 64 **17.** 125 **19.** 32 **21.** 1 **23.** 7
25. 128 **27.** 256 **29.** 256 **31.** 729 **33.** 144 **35.** 100 **37.** 20 **39.** 729 **41.** 192 **43.** 162 **45.** 21 **47.** 7 **49.** 5 **51.** 16
53. 46 **55.** 8 **57.** 64 **59.** 83 **61.** 2 **63.** 48 **65.** 4 **67.** undefined **69.** 59 **71.** 52 **73.** 44 **75.** 12 **77.** 21 **79.** 3 **81.** 43
83. 8 **85.** 16 **87.** area: 49 sq m; perimeter: 28 m **89.** area: 529 sq mi; perimeter: 92 mi **91.** true **93.** false **95.** $(2 + 3) \cdot 6 - 2$
97. $24 \div (3 \cdot 2) + 2 \cdot 5$ **99.** 1260 ft **101.** 6,384,814 **103.** answers may vary

Section 1.8

Vocabulary, Readiness & Video Check **1.** expression **3.** expression; variables **5.** equation **7.** multiplication **9.** decreased by

Exercise Set 1.8 **1.** 28; 14; 147; 3 **3.** 152; 152; 0; undefined **5.** 57; 55; 56; 56 **7.** 9 **9.** 8 **11.** 6 **13.** 5 **15.** 117 **17.** 94 **19.** 5
21. 34 **23.** 20 **25.** 4 **27.** 4 **29.** 0 **31.** 33 **33.** 125 **35.** 121 **37.** 100 **39.** 60 **41.** 4 **43.** 16; 64; 144; 256 **45.** yes **47.** no
49. no **51.** yes **53.** no **55.** yes **57.** 12 **59.** 6 **61.** 4 **63.** none **65.** 11 **67.** $x + 8$ **69.** $x + 8$ **71.** $20 - x$ **73.** $512x$
75. $8 \div x$ or $\dfrac{8}{x}$ **77.** $5x + (17 + x)$ **79.** $5x$ **81.** $11 - x$ **83.** $x - 5$ **85.** $6 \div x$ or $\dfrac{6}{x}$ **87.** $50 - 8x$ **89.** 274,657 **91.** 777
93. $5x$; answers may vary **95.** As t gets larger $16t^2$ gets larger.

Chapter 1 Vocabulary Check **1.** whole numbers **2.** perimeter **3.** place value **4.** exponent **5.** area **6.** digits **7.** variable
8. equation **9.** expression **10.** solution **11.** set **12.** sum **13.** divisor **14.** dividend **15.** quotient **16.** factor **17.** product
18. minuend **19.** subtrahend **20.** difference **21.** addend

Chapter 1 Review **1.** tens **2.** ten-millions **3.** seven thousand, six hundred forty **4.** forty-six million, two hundred
thousand, one hundred twenty **5.** $3000 + 100 + 50 + 8$ **6.** $400{,}000{,}000 + 3{,}000{,}000 + 200{,}000 + 20{,}000 + 5000$
7. 81,900 **8.** 6,304,000,000 **9.** 467,000,000 **10.** 145,000,000 **11.** Oceania/Australia **12.** Asia **13.** 67 **14.** 67 **15.** 65 **16.** 304
17. 449 **18.** 840 **19.** 3914 **20.** 7908 **21.** 4211 **22.** 1967 **23.** 1334 **24.** 886 **25.** 17,897 **26.** 34,658 **27.** 7523 mi
28. $197,699 **29.** 216 ft **30.** 66 km **31.** 82 million or 82,000,000 **32.** 4 million or 4,000,000 **33.** May **34.** August **35.** $110
36. $240 **37.** 40 **38.** 50 **39.** 880 **40.** 500 **41.** 3800 **42.** 58,000 **43.** 40,000,000 **44.** 800,000 **45.** 7300 **46.** 4100 **47.** 2700 mi
48. Europe: 821,000,000; Latin America/Caribbean: 594,000,000; difference: 227,000,000 **49.** 2208 **50.** 1396 **51.** 2280 **52.** 2898
53. 560 **54.** 900 **55.** 0 **56.** 0 **57.** 16,994 **58.** 8954 **59.** 113,634 **60.** 44,763 **61.** 411,426 **62.** 636,314 **63.** 1500 **64.** 4920
65. $898 **66.** $122,240 **67.** 91 sq mi **68.** 500 sq cm **69.** 7 **70.** 4 **71.** 5 R 2 **72.** 4 R 2 **73.** undefined **74.** 0 **75.** 33 R 2
76. 19 R 7 **77.** 24 R 2 **78.** 35 R 15 **79.** 506 R 10 **80.** 907 R 40 **81.** 2793 R 140 **82.** 2012 R 60 **83.** 18 R 2 **84.** 21 R 2
85. 27 boxes **86.** 13 miles **87.** 51 **88.** 59 **89.** 64 **90.** 125 **91.** 405 **92.** 400 **93.** 16 **94.** 10 **95.** 15 **96.** 7 **97.** 12
98. 9 **99.** 42 **100.** 33 **101.** 9 **102.** 2 **103.** 6 **104.** 29 **105.** 40 **106.** 72 **107.** 5 **108.** 7 **109.** 49 sq m **110.** 9 sq in.

111. 5 **112.** 17 **113.** undefined **114.** 0 **115.** 121 **116.** 2 **117.** 4 **118.** 20 **119.** $x - 5$ **120.** $x + 7$ **121.** $10 \div x$ or $\dfrac{10}{x}$

122. $5x$ **123.** yes **124.** no **125.** no **126.** yes **127.** 11 **128.** 175 **129.** 14 **130.** none **131.** 417 **132.** 682 **133.** 2196
134. 2516 **135.** 1101 **136.** 1411 **137.** 458 R 8 **138.** 237 R 1 **139.** 70,848 **140.** 95,832 **141.** 1644 **142.** 8481 **143.** 840
144. 300,000 **145.** 12 **146.** 6 **147.** no **148.** yes **149.** 53 full boxes with 18 left over **150.** $86

Chapter 1 Test **1.** eighty-two thousand, four hundred twenty-six **2.** 402,550 **3.** 141 **4.** 113 **5.** 14,880 **6.** 766 R 42 **7.** 200 **8.** 98 **9.** 0 **10.** undefined **11.** 33 **12.** 21 **13.** 48 **14.** 36 **15.** 5,698,000 **16.** 82 **17.** 52,000 **18.** 13,700 **19.** 1600 **20.** 92 **21.** 122 **22.** 1605 **23.** 7 R 2 **24.** $17 **25.** $126 **26.** 360 cal **27.** $7905 **28.** 20 cm; 25 sq cm **29.** 60 yd; 200 sq yd **30.** 30 **31.** 1 **32. a.** $x \div 17$ or $\frac{x}{17}$ **b.** $2x - 20$ **33.** yes **34.** 10

Chapter 2 Integers and Introduction to Solving Equations
Section 2.1

Vocabulary, Readiness & Video Check **1.** integers **3.** inequality symbols **5.** is less than; is greater than **7.** absolute value **9.** number of feet a miner works underground **11.** negative **13.** opposite of

Exercise Set 2.1 **1.** -1235 **3.** $+14,433$ **5.** $+120$ **7.** $-11,810$ **9.** -3140 million **11.** $-160, -147$; Guillermo **13.** -2

15. **17.** **19.**

21. **23.** $>$ **25.** $<$ **27.** $>$ **29.** $<$ **31.** 5 **33.** 8 **35.** 0 **37.** 55 **39.** -5 **41.** 4 **43.** -23 **45.** 85 **47.** 7 **49.** -20 **51.** -3 **53.** 43 **55.** 15 **57.** 33 **59.** 6 **61.** -2 **63.** 32 **65.** -7 **67.** $<$ **69.** $<$ **71.** $=$ **73.** $<$ **75.** $>$ **77.** $<$ **79.** $>$ **81.** $<$ **83.** 31; -31 **85.** 28; 28 **87.** Caspian Sea **89.** Lake Superior **91.** iodine **93.** oxygen **95.** 13 **97.** 35 **99.** 360 **101.** $-|-8|, -|3|, 2^2, -(-5)$ **103.** $-|-6|, -|1|, |-1|, -(-6)$ **105.** $-10, -|-9|, -(-2), |-12|, 5^2$ **107.** a, d **109.** 8 **111.** false **113.** true **115.** false **117.** answers may vary **119.** no; answers may vary

Section 2.2

Calculator Explorations **1.** -159 **3.** 44 **5.** $-894,855$

Vocabulary, Readiness & Video Check **1.** 0 **3.** a **5.** Negative; the numbers have different signs and the sign of the sum is the same as the sign of the number with the larger absolute value, -6. **7.** The diver's current depth is 231 feet below the surface.

Exercise Set 2.2

1. **3.** **5.** **7.** 67

9. -10 **11.** 0 **13.** 4 **15.** -6 **17.** -2 **19.** -9 **21.** -24 **23.** -840 **25.** 7 **27.** -3 **29.** -30 **31.** 40 **33.** -20 **35.** -125 **37.** -7 **39.** -246 **41.** 16 **43.** 13 **45.** -28 **47.** -11 **49.** 20 **51.** -34 **53.** -1 **55.** 0 **57.** -42 **59.** -70 **61.** 3 **63.** -21 **65.** $-6 + 25; 19$ **67.** $-31 + (-9) + 30; -10$ **69.** $0 + (-215) + (-16) = -231; 231$ ft below the surface **71.** Dufner: -7; Furyk: -2 **73.** $41,733,000,000 **75.** $67,655,000,000 **77.** $2°C$ **79.** $13,609 **81.** $-2°F$ **83.** -7535 m **85.** 44 **87.** 141 **89.** answers may vary **91.** -3 **93.** -22 **95.** true **97.** false **99.** answers may vary

Section 2.3

Vocabulary, Readiness & Video Check **1.** b **3.** d **5.** additive inverse **7.** to follow the order of operations

Exercise Set 2.3 **1.** 0 **3.** 3 **5.** -5 **7.** 22 **9.** 3 **11.** -20 **13.** -12 **15.** -13 **17.** 508 **19.** -14 **21.** -4 **23.** -12 **25.** $-25 - 17; -42$ **27.** $-22 - (-3); -19$ **29.** $2 - (-12); 14$ **31.** -56 **33.** -5 **35.** -145 **37.** -37 **39.** 3 **41.** 1 **43.** -1 **45.** -31 **47.** 44 **49.** -32 **51.** -9 **53.** 14 **55.** -11 **57.** 31 **59.** 12 **61.** 20 **63.** $12°F$ **65.** $4°F$ **67.** $263°F$ **69.** 14 strokes **71.** $-10°C$ **73.** 154 ft **75.** 69 ft **77.** 652 ft **79.** 144 ft **81.** $1197°F$ **83.** $-$34 billion **85.** $-5 + x$ **87.** $-20 - x$ **89.** 5 **91.** 1058 **93.** answers may vary **95.** 16 **97.** -20 **99.** -4 **101.** 0 **103.** -14 **105.** false **107.** answers may vary

Section 2.4

Vocabulary, Readiness & Video Check **1.** negative **3.** positive **5.** 0 **7.** undefined **9.** multiplication **11.** The phrase "lost four yards" in the example translates to the negative number -4.

Exercise Set 2.4 **1.** 12 **3.** -36 **5.** -81 **7.** 0 **9.** 48 **11.** -12 **13.** 80 **15.** 0 **17.** -15 **19.** -9 **21.** -27 **23.** -36 **25.** -64 **27.** -8 **29.** -5 **31.** 7 **33.** 0 **35.** undefined **37.** -14 **39.** 0 **41.** -15 **43.** -63 **45.** 42 **47.** -24 **49.** 49 **51.** -5 **53.** -9 **55.** -6 **57.** 120 **59.** -1080 **61.** undefined **63.** -6 **65.** -7 **67.** 3 **69.** -1 **71.** -32 **73.** 180 **75.** 1 **77.** -30 **79.** -1104 **81.** -2870 **83.** -56 **85.** -18 **87.** 35 **89.** -1 **91.** undefined **93.** 6 **95.** 16; 4 **97.** 0; 0 **99.** $-54 \div 9; -6$ **101.** $-42(-6); 252$ **103.** $-71 \cdot x$ or $-71x$ **105.** $-16 - x$ **107.** $-29 + x$ **109.** $\frac{x}{-33}$ or $x \div (-33)$ **111.** $3 \cdot (-4) = -12$; a loss of 12 yd **113.** $5 \cdot (-20) = -100$; a depth of 100 feet **115.** $-210°C$ **117.** $-189°C$ **119.** $-$11 million per month **121. a.** $-26,932$ movie screens **b.** -6733 movie screens per year **123.** 109 **125.** 8 **127.** -19 **129.** -28 **131.** -8 **133.** negative **135.** $(-5)^{17}, (-2)^{17}, (-2)^{12}, (-5)^{12}$ **137.** answers may vary

Integrated Review   **1.** -50; $+122$ or 122 **2.**   **3.** $>$ **4.** $<$ **5.** $<$ **6.** $>$ **7.** 3   **8.** 9   **9.** -4 **10.** 5   **11.** -11 **12.** 3   **13.** -64 **14.** 0   **15.** 12   **16.** -20 **17.** -48 **18.** -9 **19.** 10   **20.** -2 **21.** 106   **22.** -3 **23.** 0   **24.** 4   **25.** 42   **26.** 6   **27.** 19   **28.** -900 **29.** -12 **30.** -19 **31.** undefined   **32.** 0   **33.** $-12 - (-8)$; -4 **34.** $-17 + (-27)$; -44 **35.** $-5(-25)$; 125   **36.** $-100 \div (-5)$; 20   **37.** $\dfrac{x}{-17}$ or $x \div (-17)$ **38.** $-3 + x$ **39.** $x - (-18)$ **40.** $-7 \cdot x$ or $-7x$ **41.** 9   **42.** -15 **43.** 27   **44.** 33   **45.** -15 **46.** -4

Section 2.5

Calculator Explorations   **1.** 48   **3.** -258

Vocabulary, Readiness & Video Check   **1.** division   **3.** average   **5.** subtraction   **7.** A fraction bar means divided by and it is a grouping symbol.   **9.** Finding the average is a good application of both order of operations and adding and dividing integers.

Exercise Set 2.5   **1.** -125 **3.** -64 **5.** 32   **7.** -8 **9.** -11 **11.** -43 **13.** -8 **15.** 17   **17.** -1 **19.** 4   **21.** -3 **23.** 16   **25.** 13   **27.** -77 **29.** 80   **31.** 256   **33.** 53   **35.** 4   **37.** -64 **39.** 4   **41.** 16   **43.** -27 **45.** 34   **47.** 65   **49.** -7 **51.** 36   **53.** -117 **55.** 30   **57.** -3 **59.** -30 **61.** 1   **63.** -12 **65.** 0   **67.** -20 **69.** 9   **71.** -16 **73.** -128 **75.** 1   **77.** -50 **79.** -2 **81.** -19 **83.** 18   **85.** -1 **87.** no; answers may vary   **89.** 4050   **91.** 45   **93.** 32 in.   **95.** 30 ft   **97.** $2 \cdot (7 - 5) \cdot 3$ **99.** $-6 \cdot (10 - 4)$ **101.** answers may vary   **103.** answers may vary   **105.** 20,736   **107.** 8900   **109.** 9

Section 2.6

Vocabulary, Readiness & Video Check   **1.** expression   **3.** equation; expression   **5.** solution   **7.** addition   **9.** an equal sign   **11.** original; true

Exercise Set 2.6   **1.** yes   **3.** no   **5.** yes   **7.** yes   **9.** 18   **11.** -12 **13.** 9   **15.** -17 **17.** 4   **19.** -4 **21.** -14 **23.** -17 **25.** 0   **27.** 1   **29.** -7 **31.** -50 **33.** -25 **35.** 36   **37.** 21   **39.** 12   **41.** -80 **43.** -2 **45.** $x - (-2)$ **47.** $-6 \cdot x$ or $-6x$ **49.** $-15 + x$ **51.** $-8 \div x$ or $\dfrac{-8}{x}$ **53.** 41,574   **55.** -409 **57.** answers may vary   **59.** answers may vary

Chapter 2 Vocabulary Check   **1.** opposites   **2.** absolute value   **3.** integers   **4.** negative   **5.** positive   **6.** inequality symbols   **7.** solution   **8.** average   **9.** expression   **10.** equation   **11.** is less than; is greater than   **12.** addition   **13.** multiplication

Chapter 2 Review   **1.** -1572 **2.** $+11,239$ **3.**   **4.**   **5.** 11   **6.** 0   **7.** -8 **8.** 9   **9.** -16 **10.** 2   **11.** $>$ **12.** $<$ **13.** $>$ **14.** $>$ **15.** 18   **16.** -42 **17.** false   **18.** true   **19.** true   **20.** true   **21.** 2   **22.** 3   **23.** -5 **24.** -10 **25.** Elevator D   **26.** Elevator B   **27.** 2   **28.** 14   **29.** 4   **30.** 17   **31.** -23 **32.** -22 **33.** -21 **34.** -70 **35.** 0   **36.** 0   **37.** -151 **38.** -606 **39.** $-20°C$ **40.** -150 ft   **41.** -21 **42.** 12   **43.** 8   **44.** -16 **45.** -24 **46.** -10 **47.** 20   **48.** 8   **49.** 0   **50.** -32 **51.** 0   **52.** 7   **53.** -10 **54.** -27 **55.** 692 ft   **56.** -25 **57.** -14 or 14 ft below ground   **58.** 82 ft   **59.** true   **60.** false   **61.** 21   **62.** -18 **63.** -64 **64.** 60   **65.** 25   **66.** -1 **67.** 0   **68.** 24   **69.** -5 **70.** 3   **71.** 0   **72.** undefined   **73.** -20 **74.** -9 **75.** 38   **76.** -5 **77.** $(-5)(2) = -10$ **78.** $(-50)(4) = -200$ **79.** $-1024 \div 4 = -256$ **80.** $-45 \div 9 = -5$ **81.** 49   **82.** -49 **83.** 0   **84.** -8 **85.** -16 **86.** 35   **87.** -32 **88.** -8 **89.** 7   **90.** -14 **91.** 39   **92.** -117 **93.** -2 **94.** -12 **95.** -3 **96.** -35 **97.** -5 **98.** 5   **99.** -1 **100.** -7 **101.** no   **102.** yes   **103.** -13 **104.** -20 **105.** -3 **106.** -9 **107.** -13 **108.** -31 **109.** 44   **110.** -26 **111.** -19 **112.** 38   **113.** 6   **114.** -5 **115.** -15 **116.** -19 **117.** 48   **118.** -21 **119.** 21   **120.** -5 **121.** $-27°C$ **122.** $6°C$ **123.** 13,118 ft   **124.** $-\$9$ **125.** 2   **126.** 3   **127.** -5 **128.** -25 **129.** -20 **130.** 17   **131.** -21 **132.** -17 **133.** 12   **134.** -9 **135.** -200 **136.** 3

Chapter 2 Test   **1.** 3   **2.** -6 **3.** -100 **4.** 4   **5.** -30 **6.** 12   **7.** 65   **8.** 5   **9.** 12   **10.** -6 **11.** 50   **12.** -2 **13.** -11 **14.** -46 **15.** -117 **16.** 3456   **17.** 28   **18.** -213 **19.** -2 **20.** 2   **21.** -5 **22.** -32 **23.** -17 **24.** 1   **25.** -1 **26.** 88 ft below sea level   **27.** 45   **28.** 31,642   **29.** 3820 ft below sea level   **30.** -4 **31. a.** $17 \cdot x$ or $17x$ **b.** $20 - x$ **32.** 5   **33.** -28 **34.** -20 **35.** -4

Cumulative Review   **1.** hundred-thousands; Sec. 1.2, Ex. 1   **2.** hundreds; Sec. 1.2   **3.** thousands; Sec. 1.2, Ex. 2   **4.** thousands; Sec. 1.2   **5.** ten-millions; Sec. 1.2, Ex. 3   **6.** hundred-thousands; Sec. 1.2   **7. a.** $<$ **b.** $>$ **c.** $>$; Sec. 2.1, Ex. 3   **8. a.** $>$ **b.** $>$ **c.** $<$; Sec. 2.1   **9.** 39; Sec. 1.3, Ex. 3   **10.** 39; Sec. 1.3   **11.** 7321; Sec. 1.3, Ex. 6   **12.** 3013; Sec. 1.3   **13.** 36,184 mi; Sec. 1.3, Ex. 11   **14.** $\$525$; Sec. 1.3   **15.** 570; Sec. 1.4, Ex. 1   **16.** 600; Sec. 1.4   **17.** 1800; Sec. 1.4, Ex. 5   **18.** 5000; Sec. 1.4   **19. a.** $5 \cdot 6 + 5 \cdot 5$ **b.** $20 \cdot 4 + 20 \cdot 7$ **c.** $2 \cdot 7 + 2 \cdot 9$; Sec. 1.5, Ex. 2   **20. a.** $5 \cdot 2 + 5 \cdot 12$ **b.** $9 \cdot 3 + 9 \cdot 6$ **c.** $4 \cdot 8 + 4 \cdot 1$; Sec. 1.5   **21.** 78,875; Sec. 1.5, Ex. 5   **22.** 31,096; Sec. 1.5   **23. a.** 6   **b.** 8   **c.** 7; Sec. 1.6, Ex. 1   **24. a.** 7   **b.** 8   **c.** 12; Sec. 1.6   **25.** 741; Sec. 1.6, Ex. 4   **26.** 456; Sec. 1.6   **27.** 12 cards each; 10 cards left over; Sec. 1.6, Ex. 11   **28.** $\$9$; Sec. 1.6   **29.** 81; Sec. 1.7, Ex. 5   **30.** 125; Sec. 1.7   **31.** 6; Sec. 1.7, Ex. 6   **32.** 4; Sec. 1.7   **33.** 180; Sec. 1.7, Ex. 8   **34.** 56; Sec. 1.7   **35.** 2; Sec. 1.7, Ex. 13   **36.** 5; Sec. 1.7   **37.** 14; Sec. 1.8, Ex. 1   **38.** 14; Sec. 1.8   **39. a.** 9   **b.** 8   **c.** 0; Sec. 2.1, Ex. 4   **40. a.** 4   **b.** 7; Sec. 2.1   **41.** 23; Sec. 2.2, Ex. 7   **42.** 5; Sec. 2.2   **43.** 22; Sec. 2.3, Ex. 12   **44.** 5; Sec. 2.3   **45.** -21; Sec. 2.4, Ex. 1   **46.** -10; Sec. 2.4   **47.** 0; Sec. 2.4, Ex. 3   **48.** -54; Sec. 2.4   **49.** -16; Sec. 2.5, Ex. 8   **50.** -27; Sec. 2.5

Chapter 3 Solving Equations and Problem Solving

Section 3.1

Vocabulary, Readiness & Video Check **1.** expression; term **3.** combine like terms **5.** variable; constant **7.** associative **9.** numerical coefficient **11.** distributive property **13.** addition; multiplication; $P = $ perimeter, $A = $ area

Exercise Set 3.1 **1.** $8x$ **3.** $-n$ **5.** $-2c$ **7.** $-6x$ **9.** $12a - 5$ **11.** $42x$ **13.** $-33y$ **15.** $72a$ **17.** $2y + 6$ **19.** $3a - 18$ **21.** $-12x - 28$ **23.** $2x + 1$ **25.** $15c + 3$ **27.** $-21n + 20$ **29.** $7w + 15$ **31.** $11x - 8$ **33.** $-2y + 16$ **35.** $-2y$ **37.** $-7z$ **39.** $8d - 3c$ **41.** $6y - 14$ **43.** $-q$ **45.** $2x + 22$ **47.** $-3x - 35$ **49.** $-3z - 15$ **51.** $-6x + 6$ **53.** $3x - 30$ **55.** $-r + 8$ **57.** $-7n + 3$ **59.** $9z - 14$ **61.** -6 **63.** $-4x + 10$ **65.** $2x + 20$ **67.** $7a + 12$ **69.** $3y + 5$ **71.** $(14y + 22)$ m **73.** $(11a + 12)$ ft **75.** $(-25x + 55)$ in. **77.** $36y$ sq in. **79.** $(32x - 64)$ sq km **81.** $(60y + 20)$ sq mi **83.** 4700 sq ft **85.** 64 ft **87.** $(3x + 6)$ ft **89.** -3 **91.** 8 **93.** 0 **95.** incorrect; $15x - 10$ **97.** incorrect; $2xy$ **99.** correct **101.** incorrect; $4y - 12 + 11$ or $4y - 1$ **103.** distributive **105.** associative **107.** $(20x + 16)$ sq mi **109.** $4824q + 12{,}274$ **111.** answers may vary **113.** answers may vary

Section 3.2

Vocabulary, Readiness & Video Check **1.** equivalent **3.** simplifying **5.** addition **7.** Simplify the left side of the equation by combining like terms. **9.** addition property of equality.

Exercise Set 3.2 **1.** 6 **3.** 8 **5.** -4 **7.** 6 **9.** -1 **11.** 18 **13.** -8 **15.** -50 **17.** 3 **19.** -22 **21.** -6 **23.** 24 **25.** -30 **27.** 12 **29.** 4 **31.** 3 **33.** -3 **35.** 1 **37.** 5 **39.** -11 **41.** -4 **43.** -3 **45.** -1 **47.** 0 **49.** 3 **51.** -6 **53.** -35 **55.** 10 **57.** -2 **59.** 0 **61.** 28 **63.** -5 **65.** -28 **67.** -28 **69.** 5 **71.** -4 **73.** $-7 + x$ **75.** $x - 11$ **77.** $-13x$ **79.** $\dfrac{x}{-12}$ or $-\dfrac{x}{12}$ **81.** $-11x + 5$ **83.** $-10 - 7x$ **85.** $4x + 7$ **87.** $2x - 17$ **89.** $-6(x + 15)$ **91.** $\dfrac{45}{-5x}$ or $-\dfrac{45}{5x}$ **93.** $\dfrac{17}{x} + (-15)$ or $\dfrac{17}{x} - 15$ **95.** California **97.** \$88 billion **99.** answers may vary **101.** no; answers may vary **103.** answers may vary **105.** 67,896 **107.** -48 **109.** 42

Integrated Review **1.** expression **2.** equation **3.** equation **4.** expression **5.** simplify **6.** solve **7.** $8x$ **8.** $-4y$ **9.** $-2a - 2$ **10.** $5a - 26$ **11.** $-8x - 14$ **12.** $-6x + 30$ **13.** $5y - 10$ **14.** $15x - 31$ **15.** $(12x - 6)$ sq m **16.** $(2x + 9)$ ft **17.** -4 **18.** -3 **19.** -10 **20.** 6 **21.** -15 **22.** -120 **23.** -5 **24.** -13 **25.** -24 **26.** -54 **27.** 12 **28.** -42 **29.** 2 **30.** 2 **31.** -3 **32.** 5 **33.** -5 **34.** 6 **35.** $x - 10$ **36.** $-20 + x$ **37.** $10x$ **38.** $\dfrac{10}{x}$ **39.** $-2x + 5$ **40.** $-4(x - 1)$

Section 3.3

Calculator Explorations **1.** yes **3.** no **5.** yes

Vocabulary, Readiness & Video Check **1.** $3x - 9 + x - 16$; $5(2x + 6) - 1 = 39$ **3.** addition **5.** distributive **7.** the addition property of equality; to make sure we get an equivalent equation **9.** gives; amounts to

Exercise Set 3.3 **1.** -12 **3.** -3 **5.** 1 **7.** -45 **9.** -9 **11.** 6 **13.** 8 **15.** 5 **17.** 0 **19.** -5 **21.** -22 **23.** 6 **25.** -11 **27.** -7 **29.** -5 **31.** 270 **33.** 5 **35.** 3 **37.** 9 **39.** -6 **41.** 11 **43.** 3 **45.** 4 **47.** -4 **49.** 3 **51.** -1 **53.** -4 **55.** -5 **57.** 0 **59.** 4 **61.** 1 **63.** -30 **65.** $-42 + 16 = -26$ **67.** $-5(-29) = 145$ **69.** $3(-14 - 2) = -48$ **71.** $\dfrac{100}{2(50)} = 1$ **73.** 122,000,000 returns **75.** 37,000,000 returns **77.** 33 **79.** -37 **81.** b **83.** a **85.** $6x - 10 = 5x - 7$; $6x = 5x + 3$; $x = 3$ **87.** 0 **89.** -4 **91.** no; answers may vary

Section 3.4

Vocabulary, Readiness & Video Check **1.** The phrase is "a number subtracted from -20" so -20 goes first and we subtract the number from that. **3.** The original application asks for the fastest speeds of a pheasant and a falcon. The value of x is the speed in mph for a pheasant, so the falcon's speed still needs to be found.

Exercise Set 3.4 **1.** $-5 + x = -7$ **3.** $3x = 27$ **5.** $-20 - x = 104$ **7.** $2x = 108$ **9.** $5(-3 + x) = -20$ **11.** $9 + 3x = 33$; 8 **13.** $3 + 4 + x = 16$; 9 **15.** $x - 3 = \dfrac{10}{5}$; 5 **17.** $30 - x = 3(x + 6)$; 3 **19.** $5x - 40 = x + 8$; 12 **21.** $3(x - 5) = \dfrac{108}{12}$; 8 **23.** $4x = 30 - 2x$; 5 **25.** California: 55 votes; Florida: 27 votes **27.** falcon: 185 mph; pheasant: 37 mph **29.** *The New York Times*: 1439 thousand; *The Los Angeles Times*: 1055 thousand **31.** Coca-Cola: \$72 billion; Disney: \$29 billion **33.** Xbox: \$330; games \$110 **35.** 5470 miles **37.** Beaver Stadium: 107,282; Michigan Stadium: 106,201 **39.** China: 140 million; Spain: 70 million **41.** US: 8086 cars per day; Germany: 16,172 cars per day **43.** \$225 **45.** 93 **47.** USA: 310,640 thousand; China: 195,140 thousand **49.** 590 **51.** 1000 **53.** 3000 **55.** yes; answers may vary **57.** \$216,200 **59.** \$549

Chapter 3 Vocabulary Check   1. simplified; combined.   **2.** like   **3.** variable   **4.** algebraic expression   **5.** terms   **6.** numerical coefficient   **7.** evaluating the expression   **8.** constant   **9.** equation   **10.** solution   **11.** distributive   **12.** multiplication   **13.** addition

Chapter 3 Review   1. $10y - 15$ **2.** $-6y - 10$ **3.** $-6a - 7$ **4.** $-8y + 2$ **5.** $2x + 10$ **6.** $-3y - 24$ **7.** $11x - 12$ **8.** $-4m - 12$ **9.** $-5a + 4$ **10.** $12y - 9$ **11.** $16y - 5$ **12.** $x - 2$ **13.** $(4x + 6)$ yd   **14.** $20y$ m   **15.** $(6x - 3)$ sq yd   **16.** $(45x + 8)$ sq cm   **17.** -2 **18.** 10   **19.** -7 **20.** 5   **21.** -12 **22.** 45   **23.** -6 **24.** -1 **25.** -25 **26.** -8 **27.** -2 **28.** -2 **29.** -8 **30.** -45 **31.** 5   **32.** -5 **33.** -63 **34.** -15 **35.** 5   **36.** 12   **37.** -6 **38.** 4   **39.** $-5x$ **40.** $x - 3$ **41.** $-5 + x$ **42.** $\dfrac{-2}{x}$ or $-\dfrac{2}{x}$ **43.** $-5x - 50$ **44.** $2x + 11$ **45.** $\dfrac{70}{x + 6}$ **46.** $2(x - 13)$ **47.** 21   **48.** -10 **49.** 2   **50.** 2   **51.** 11   **52.** -5 **53.** -15 **54.** 10   **55.** -2 **56.** -6 **57.** -1 **58.** 1   **59.** 0   **60.** 20   **61.** $20 - (-8) = 28$ **62.** $-2 - 19 = -21$ **63.** $\dfrac{-75}{5 + 20} = -3$ **64.** $5[2 + (-6)] = -20$ **65.** $2x - 8 = 40$ **66.** $6x = x + 20$ **67.** $\dfrac{x}{2} - 12 = 10$ **68.** $x - 3 = \dfrac{x}{4}$ **69.** 5   **70.** -16 **71.** 2386 votes   **72.** 84 DVDs   **73.** $-11x$ **74.** $-35x$ **75.** $22x - 19$ **76.** $-9x - 32$ **77.** -1 **78.** -25 **79.** 13   **80.** -6 **81.** -22 **82.** -6 **83.** -15 **84.** 18   **85.** -5 **86.** 11   **87.** 2   **88.** -1 **89.** 0   **90.** -6 **91.** 5   **92.** 1   **93.** Hawaii: 4371 mi; Delaware: 6302 mi   **94.** North Dakota 86,843 mi; South Dakota: 82,354 mi

Chapter 3 Test   1. $-5x + 5$ **2.** $-6y - 14$ **3.** $-8z - 20$ **4.** $(15x + 15)$ in.   **5.** $(12x - 4)$ sq m   **6.** -6 **7.** -6 **8.** 24   **9.** -2 **10.** 6   **11.** 3   **12.** -2 **13.** 0   **14.** 4   **15.** $-23 + x$ **16.** $-2 - 3x$ **17.** $2 \cdot 5 + (-15) = -5$ **18.** $3x + 6 = -30$ **19.** -2 **20.** 8 free throws   **21.** 244 women

Cumulative Review Chapters 1–3   1. three hundred eight million, sixty-three thousand, five hundred fifty-seven; Sec. 1.2, Ex. 7   **2.** two hundred seventy-six thousand, four; Sec. 1.2   **3.** 13 in.; Sec. 1.3, Ex. 9   **4.** 18 in.; Sec. 1.3   **5.** 726; Sec. 1.3, Ex. 8   **6.** 9585; Sec. 1.3   **7.** 249,000; Sec. 1.4, Ex. 3   **8.** 844,000; Sec. 1.4   **9.** 200; Sec. 1.5, Ex. 3a   **10.** 29,230; Sec. 1.5   **11.** 208; Sec. 1.6, Ex. 5   **12.** 86; Sec. 1.6   **13.** 7; Sec. 1.7, Ex. 9   **14.** 35; Sec. 1.7   **15.** 26; Sec. 1.8, Ex. 4   **16.** 10; Sec. 1.8   **17.** 20 is a solution; 26 and 40 are not solutions.; Sec. 1.8, Ex. 7   **18. a.** $<$ **b.** $>$; Sec. 2.1   **19.** 3; Sec. 2.2, Ex. 1   **20.** -7; Sec. 2.2   **21.** -25; Sec. 2.2, Ex. 5   **22.** -4; Sec. 2.2   **23.** 23; Sec. 2.2, Ex. 7   **24.** 17; Sec. 2.2   **25.** -14; Sec. 2.3, Ex. 2   **26.** -5; Sec. 2.3   **27.** 11; Sec. 2.3, Ex. 3   **28.** 29; Sec. 2.3   **29.** -4; Sec. 2.3, Ex. 4   **30.** -3; Sec. 2.3   **31.** -2; Sec. 2.4, Ex. 10   **32.** 6; Sec. 2.4   **33.** 5; Sec. 2.4, Ex. 11   **34.** -13; Sec. 2.4   **35.** -16; Sec. 2.4, Ex. 12   **36.** -10; Sec. 2.4   **37.** 9; Sec. 2.5, Ex. 1   **38.** -32; Sec. 2.5   **39.** -9; Sec. 2.5, Ex. 2   **40.** 25; Sec. 2.5   **41.** $6y + 2$; Sec. 3.1, Ex. 2   **42.** $3x + 9$; Sec. 3.1   **43.** not a solution; Sec. 2.6, Ex. 1   **44.** solution; Sec. 2.6   **45.** 3; Sec. 2.6, Ex. 7   **46.** -5; Sec. 2.6   **47.** 12; Sec. 3.2, Ex. 7   **48.** -2; Sec. 3.2   **49.** software: \$420; computer system: \$1680; Sec. 3.4, Ex. 4   **50.** 11; Sec. 3.4

Chapter 4 Fractions and Mixed Numbers

Section 4.1

Vocabulary, Readiness & Video Check   1. fraction; denominator; numerator   **3.** improper; proper; mixed number   **5.** equal; improper   **7.** how many equal parts to divide each whole number into   **9.** addition; +

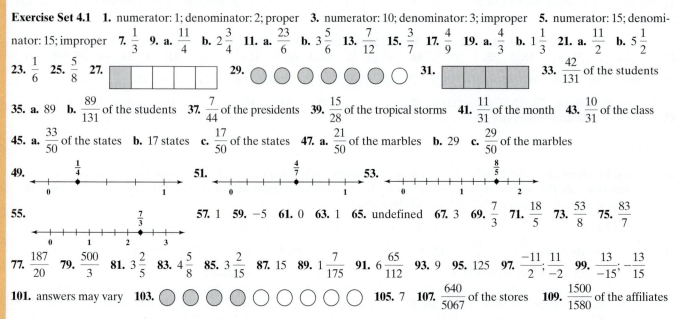

Exercise Set 4.1   1. numerator: 1; denominator: 2; proper   **3.** numerator: 10; denominator: 3; improper   **5.** numerator: 15; denominator: 15; improper   **7.** $\dfrac{1}{3}$ **9. a.** $\dfrac{11}{4}$ **b.** $2\dfrac{3}{4}$ **11. a.** $\dfrac{23}{6}$ **b.** $3\dfrac{5}{6}$ **13.** $\dfrac{7}{12}$ **15.** $\dfrac{3}{7}$ **17.** $\dfrac{4}{9}$ **19. a.** $\dfrac{4}{3}$ **b.** $1\dfrac{1}{3}$ **21. a.** $\dfrac{11}{2}$ **b.** $5\dfrac{1}{2}$ **23.** $\dfrac{1}{6}$ **25.** $\dfrac{5}{8}$ **27.**   **29.**   **31.**   **33.** $\dfrac{42}{131}$ of the students

35. a. 89   **b.** $\dfrac{89}{131}$ of the students   **37.** $\dfrac{7}{44}$ of the presidents   **39.** $\dfrac{15}{28}$ of the tropical storms   **41.** $\dfrac{11}{31}$ of the month   **43.** $\dfrac{10}{31}$ of the class

45. a. $\dfrac{33}{50}$ of the states   **b.** 17 states   **c.** $\dfrac{17}{50}$ of the states   **47. a.** $\dfrac{21}{50}$ of the marbles   **b.** 29   **c.** $\dfrac{29}{50}$ of the marbles

49.   **51.**   **53.**

55.   **57.** 1   **59.** -5 **61.** 0   **63.** 1   **65.** undefined   **67.** 3   **69.** $\dfrac{7}{3}$ **71.** $\dfrac{18}{5}$ **73.** $\dfrac{53}{8}$ **75.** $\dfrac{83}{7}$

77. $\dfrac{187}{20}$ **79.** $\dfrac{500}{3}$ **81.** $3\dfrac{2}{5}$ **83.** $4\dfrac{5}{8}$ **85.** $3\dfrac{2}{15}$ **87.** 15   **89.** $1\dfrac{7}{175}$ **91.** $6\dfrac{65}{112}$ **93.** 9   **95.** 125   **97.** $\dfrac{-11}{2}; \dfrac{11}{-2}$ **99.** $\dfrac{13}{-15}; -\dfrac{13}{15}$

101. answers may vary   **103.**   **105.** 7   **107.** $\dfrac{640}{5067}$ of the stores   **109.** $\dfrac{1500}{1580}$ of the affiliates

Section 4.2

Calculator Explorations **1.** $\dfrac{4}{7}$ **3.** $\dfrac{20}{27}$ **5.** $\dfrac{8}{15}$ **7.** $\dfrac{2}{9}$

Vocabulary, Readiness & Video Check **1.** prime factorization **3.** prime **5.** equivalent **7.** Check that every factor is a prime number and check that the product of the factors is the original number **9.** You can simplify the two fractions and then compare them. $\dfrac{3}{9}$ and $\dfrac{6}{18}$ both simplify to $\dfrac{1}{3}$ so the original fractions are equivalent.

Exercise Set 4.2 **1.** $2^2 \cdot 5$ **3.** $2^4 \cdot 3$ **5.** 3^4 **7.** $2 \cdot 3^4$ **9.** $2 \cdot 5 \cdot 11$ **11.** $5 \cdot 17$ **13.** $2^4 \cdot 3 \cdot 5$ **15.** $2^2 \cdot 3^2 \cdot 23$ **17.** $\dfrac{1}{4}$ **19.** $\dfrac{2x}{21}$ **21.** $\dfrac{7}{8}$ **23.** $\dfrac{2}{3}$ **25.** $\dfrac{7}{10}$ **27.** $-\dfrac{7}{9}$ **29.** $\dfrac{5x}{6}$ **31.** $\dfrac{27}{64}$ **33.** $\dfrac{5x}{8}$ **35.** $-\dfrac{5}{8}$ **37.** $\dfrac{3x^2y}{2}$ **39.** $\dfrac{3}{4}$ **41.** $\dfrac{5}{8z}$ **43.** $\dfrac{3}{14}$ **45.** $-\dfrac{11}{17y}$ **47.** $\dfrac{7}{8}$ **49.** $14a^2$ **51.** equivalent **53.** not equivalent **55.** equivalent **57.** equivalent **59.** not equivalent **61.** not equivalent **63.** $\dfrac{1}{4}$ of a shift **65.** $\dfrac{1}{2}$ mi **67. a.** $\dfrac{8}{25}$ **b.** 34 states **c.** $\dfrac{17}{25}$ **69.** $\dfrac{5}{12}$ of the wall **71. a.** 8 **b.** $\dfrac{4}{25}$ **73.** $\dfrac{13}{232}$ of U.S astronauts **75.** -3 **77.** -14 **79.** answers may vary **81.** $\dfrac{3}{5}$ **83.** $\dfrac{9}{25}$ **85.** $\dfrac{1}{25}$ **87.** $2^2 \cdot 3^5 \cdot 5 \cdot 7$ **89.** answers may vary **91.** no; answers may vary **93.** $\dfrac{3}{50}$ **95.** answers may vary **97.** $\dfrac{2}{25}$ **99.** answers may vary **101.** 786, 222, 900, 1470 **103.** 6; answers may vary

Section 4.3

Vocabulary, Readiness & Video Check **1.** $\dfrac{a \cdot c}{b \cdot d}$ **3.** $\dfrac{2 \cdot 2 \cdot 2}{7}; \dfrac{2}{7} \cdot \dfrac{2}{7} \cdot \dfrac{2}{7}$ **5.** $\dfrac{a \cdot d}{b \cdot c}$ **7.** We have a negative fraction times a positive fraction and a negative number times a positive number is a negative number. **9.** numerator; denominator **11.** radius $= \dfrac{1}{2} \cdot$ diameter

Exercise Set 4.3 **1.** $\dfrac{18}{77}$ **3.** $-\dfrac{5}{28}$ **5.** $\dfrac{1}{15}$ **7.** $\dfrac{18x}{55}$ **9.** $\dfrac{3a^2}{4}$ **11.** $\dfrac{x^2}{y}$ **13.** 0 **15.** $-\dfrac{17}{25}$ **17.** $\dfrac{1}{56}$ **19.** $\dfrac{1}{125}$ **21.** $\dfrac{4}{9}$ **23.** $-\dfrac{4}{27}$ **25.** $\dfrac{4}{5}$ **27.** $-\dfrac{1}{6}$ **29.** $-\dfrac{16}{9x}$ **31.** $\dfrac{121y}{60}$ **33.** $-\dfrac{1}{6}$ **35.** $\dfrac{x}{25}$ **37.** $\dfrac{10}{27}$ **39.** $\dfrac{18x^2}{35}$ **41.** $\dfrac{10}{9}$ **43.** $-\dfrac{1}{4}$ **45.** $\dfrac{9}{16}$ **47.** xy^2 **49.** $\dfrac{77}{2}$ **51.** $-\dfrac{36}{x}$ **53.** $\dfrac{3}{49}$ **55.** $-\dfrac{19y}{7}$ **57.** $\dfrac{4}{11}$ **59.** $\dfrac{8}{3}$ **61.** $\dfrac{15x}{4}$ **63.** $\dfrac{8}{9}$ **65.** $\dfrac{1}{60}$ **67.** b **69.** $\dfrac{2}{5}$ **71.** $-\dfrac{5}{3}$ **73. a.** $\dfrac{1}{3}$ **b.** $\dfrac{12}{25}$ **75. a.** $-\dfrac{36}{55}$ **b.** $-\dfrac{44}{45}$ **77.** yes **79.** no **81.** 50 **83.** 20 **85.** 112 **87.** 108 million **89.** 868 mi **91.** $\dfrac{3}{16}$ in. **93.** $1838 **95.** 50 libraries **97.** $\dfrac{1}{14}$ sq ft **99.** 3840 mi **101.** 2400 mi **103.** 201 **105.** 196 **107.** answers may vary **109.** 5 **111.** 39,239,250 people **113.** 8505

Section 4.4

Vocabulary, Readiness & Video Check **1.** like; unlike **3.** $-\dfrac{a}{b}$ **5.** least common denominator (LCD) **7.** numerators; denominator **9.** $P = \dfrac{5}{12} + \dfrac{7}{12} + \dfrac{5}{12} + \dfrac{7}{12}; 2$ meters **11.** Multiplying by 1 does not change the value of the fraction.

Exercise Set 4.4 **1.** $\dfrac{7}{11}$ **3.** $\dfrac{2}{3}$ **5.** $-\dfrac{1}{4}$ **7.** $-\dfrac{1}{2}$ **9.** $\dfrac{2}{3x}$ **11.** $-\dfrac{x}{9}$ **13.** $\dfrac{6}{11}$ **15.** $\dfrac{3}{4}$ **17.** $-\dfrac{3}{y}$ **19.** $-\dfrac{19}{33}$ **21.** $-\dfrac{1}{3}$ **23.** $\dfrac{7a - 3}{4}$ **25.** $\dfrac{9}{10}$ **27.** $-\dfrac{13x}{14}$ **29.** $\dfrac{9x + 1}{15}$ **31.** $-\dfrac{x}{2}$ **33.** $-\dfrac{2}{3}$ **35.** $\dfrac{3x}{4}$ **37.** $\dfrac{5}{4}$ **39.** $\dfrac{2}{5}$ **41.** 1 in. **43.** 2 m **45.** $\dfrac{7}{10}$ mi **47.** $\dfrac{13}{50}$ **49.** $\dfrac{7}{25}$ **51.** $\dfrac{1}{50}$ **53.** 45 **55.** 72 **57.** 150 **59.** $24x$ **61.** 126 **63.** 168 **65.** 14 **67.** 20 **69.** 25 **71.** $56x$ **73.** $8y$ **75.** $20a$ **77.** $\dfrac{54}{100}, \dfrac{50}{100}, \dfrac{46}{100}, \dfrac{50}{100}, \dfrac{15}{100}, \dfrac{65}{100}, \dfrac{45}{100}, \dfrac{52}{100}, \dfrac{60}{100}, \dfrac{61}{100}, \dfrac{48}{100}, \dfrac{50}{100}$ **79.** drugs, health and beauty aids **81.** 9 **83.** 125 **85.** 49 **87.** 24 **89.** $\dfrac{2}{7} + \dfrac{9}{7} = \dfrac{11}{7}$ **91.** answers may vary **93.** 1; answers may vary **95.** $814x$ **97.** answers may vary **99.** a, b, and d

Section 4.5

Calculator Explorations   **1.** $\dfrac{37}{80}$ **3.** $\dfrac{95}{72}$ **5.** $\dfrac{394}{323}$

Vocabulary, Readiness & Video Check   **1.** equivalent; least common denominator   **3.** $\dfrac{4}{24}; \dfrac{15}{24}; \dfrac{19}{24}$ **5.** expression; equation

7. They are unlike terms and so cannot be combined.   **9.** $\dfrac{3}{2} + \dfrac{1}{3}$; 6

Exercise Set 4.5   **1.** $\dfrac{5}{6}$ **3.** $\dfrac{1}{6}$ **5.** $-\dfrac{4}{33}$ **7.** $-\dfrac{3}{14}$ **9.** $\dfrac{3x}{5}$ **11.** $\dfrac{24-y}{12}$ **13.** $\dfrac{11}{36}$ **15.** $-\dfrac{44}{7}$ **17.** $\dfrac{89a}{99}$ **19.** $\dfrac{4y-1}{6}$ **21.** $\dfrac{x+6}{2x}$ **23.** $-\dfrac{8}{33}$

25. $\dfrac{3}{14}$ **27.** $\dfrac{11y-10}{35}$ **29.** $-\dfrac{11}{36}$ **31.** $\dfrac{1}{20}$ **33.** $\dfrac{33}{56}$ **35.** $\dfrac{17}{16}$ **37.** $-\dfrac{2}{9}$ **39.** $-\dfrac{11}{30}$ **41.** $\dfrac{15+11y}{33}$ **43.** $-\dfrac{53}{42}$ **45.** $\dfrac{7x}{8}$ **47.** $\dfrac{11}{18}$ **49.** $\dfrac{44a}{39}$

51. $-\dfrac{11}{60}$ **53.** $\dfrac{5y+9}{9y}$ **55.** $\dfrac{19}{20}$ **57.** $\dfrac{56}{45}$ **59.** $\dfrac{40+9x}{72x}$ **61.** $-\dfrac{5}{24}$ **63.** $\dfrac{37x-20}{56}$ **65.** $<$ **67.** $>$ **69.** $>$ **71.** $\dfrac{13}{12}$ **73.** $\dfrac{1}{4}$ **75.** $\dfrac{11}{6}$

77. $\dfrac{34}{15}$ or $2\dfrac{4}{15}$ cm   **79.** $\dfrac{17}{10}$ or $1\dfrac{7}{10}$ m   **81.** $x + \dfrac{1}{2}$ **83.** $-\dfrac{3}{8} - x$ **85.** $\dfrac{7}{100}$ mph   **87.** $\dfrac{5}{8}$ in.   **89.** $\dfrac{49}{100}$ of students   **91.** $\dfrac{47}{32}$ in.   **93.** $\dfrac{19}{25}$

95. $\dfrac{1}{25}$ **97.** $\dfrac{79}{100}$ **99.** -20 **101.** -6 **103.** $\dfrac{3}{5} + \dfrac{4}{5} = \dfrac{7}{5}$ **105.** $\dfrac{223}{540}$ **107.** $\dfrac{49}{44}$ **109.** answers may vary   **111.** standard mail

Integrated Review   **1.** $\dfrac{3}{7}$ **2.** $\dfrac{5}{4}$ or $1\dfrac{1}{4}$ **3.** $\dfrac{73}{85}$ **4.**   **5.** -1 **6.** 17   **7.** 0

8. undefined   **9.** $5 \cdot 13$ **10.** $2 \cdot 5 \cdot 7$ **11.** $3^2 \cdot 5 \cdot 7$ **12.** $3^2 \cdot 7^2$ **13.** $\dfrac{1}{7}$ **14.** $\dfrac{6}{5}$ **15.** $-\dfrac{14}{15}$ **16.** $-\dfrac{9}{10}$ **17.** $\dfrac{2x}{5}$ **18.** $\dfrac{3}{8y}$ **19.** $\dfrac{11z^2}{14}$

20. $\dfrac{7}{11ab^2}$ **21.** not equivalent   **22.** equivalent   **23. a.** $\dfrac{1}{25}$ **b.** 48   **c.** $\dfrac{24}{25}$ **24. a.** $\dfrac{2}{5}$ **b.** 54,000   **c.** $\dfrac{3}{5}$ **25.** 30   **26.** 14   **27.** 90

28. $\dfrac{28}{36}$ **29.** $\dfrac{55}{75}$ **30.** $\dfrac{40}{48}$ **31.** $\dfrac{4}{5}$ **32.** $-\dfrac{2}{5}$ **33.** $\dfrac{3}{25}$ **34.** $\dfrac{1}{3}$ **35.** $\dfrac{4}{5}$ **36.** $\dfrac{5}{9}$ **37.** $-\dfrac{1}{6y}$ **38.** $\dfrac{3x}{2}$ **39.** $\dfrac{1}{18}$ **40.** $\dfrac{4}{21}$ **41.** $-\dfrac{7}{48z}$ **42.** $-\dfrac{9}{50}$

43. $\dfrac{37}{40}$ **44.** $\dfrac{11}{36}$ **45.** $\dfrac{11}{18}$ **46.** $\dfrac{25y+12}{50}$ **47.** $\dfrac{2}{3} \cdot x$ or $\dfrac{2}{3}x$ **48.** $x \div \left(-\dfrac{1}{5}\right)$ **49.** $-\dfrac{8}{9} - x$ **50.** $\dfrac{6}{11} + x$ **51.** 1020   **52.** 24 lots   **53.** $\dfrac{3}{4}$ ft

Section 4.6

Vocabulary, Readiness & Video Check   **1.** complex   **3.** division   **5.** addition   **7.** distributive property   **9.** Since x is squared and the replacement value is negative, we use parentheses to make sure the whole value of x is squared. Without parentheses, the exponent would not apply to the negative sign.

Exercise Set 4.6   **1.** $\dfrac{1}{6}$ **3.** $\dfrac{7}{3}$ **5.** $\dfrac{x}{6}$ **7.** $\dfrac{23}{22}$ **9.** $\dfrac{2x}{13}$ **11.** $\dfrac{17}{60}$ **13.** $\dfrac{5}{8}$ **15.** $\dfrac{35}{9}$ **17.** $-\dfrac{17}{45}$ **19.** $\dfrac{11}{8}$ **21.** $\dfrac{29}{10}$ **23.** $\dfrac{27}{32}$ **25.** $\dfrac{1}{100}$ **27.** $\dfrac{9}{64}$

29. $\dfrac{7}{6}$ **31.** $-\dfrac{2}{5}$ **33.** $-\dfrac{2}{9}$ **35.** $\dfrac{11}{9}$ **37.** $\dfrac{5a}{2}$ **39.** $\dfrac{7}{2}$ **41.** $\dfrac{9}{20}$ **43.** $-\dfrac{13}{2}$ **45.** $\dfrac{9}{25}$ **47.** $-\dfrac{5}{32}$ **49.** 1   **51.** $-\dfrac{2}{5}$ **53.** $-\dfrac{11}{40}$ **55.** $\dfrac{x+6}{16}$

57. $\dfrac{7}{2}$ or $3\dfrac{1}{2}$ **59.** $\dfrac{49}{6}$ or $8\dfrac{1}{6}$ **61.** no; answers may very   **63.** $\dfrac{5}{8}$ **65.** $\dfrac{11}{56}$ **67.** halfway between a and b **69.** false   **71.** true

73. true   **75.** addition: answers may vary   **77.** subtraction, multiplication, addition, division   **79.** division, multiplication, subtraction, addition   **81.** $-\dfrac{77}{16}$ **83.** $-\dfrac{55}{16}$

Section 4.7

Calculator Explorations   **1.** $\dfrac{280}{11}$ **3.** $\dfrac{3776}{35}$ **5.** $26\dfrac{1}{14}$ **7.** $92\dfrac{3}{10}$

Vocabulary, Readiness & Video Check   **1.** mixed number   **3.** round   **5.** The denominator of the mixed number we're graphing, $-3\dfrac{4}{5}$, is 5.   **7.** The fractional part of a mixed number should always be a proper fraction.   **9.** We're adding two mixed numbers with unlike signs, so the answer has the sign of the mixed number with the larger absolute value, which in this case is negative.

Exercise Set 4.7

1.   **3.**   **5.** b   **7.** a   **9.** $\dfrac{8}{21}$ **11.** $4\dfrac{3}{8}$ **13.** $7\dfrac{7}{10}$; 8   **15.** $23\dfrac{31}{35}$; 24

17. $12\frac{1}{2}$ **19.** $5\frac{1}{2}$ **21.** $18\frac{2}{3}$ **23.** a **25.** b **27.** $6\frac{2}{3}$; 7 **29.** $13\frac{11}{14}$; 14 **31.** $17\frac{7}{25}$ **33.** $47\frac{53}{84}$ **35.** $25\frac{5}{14}$ **37.** $13\frac{13}{24}$ **39.** $2\frac{3}{5}$; 3

41. $7\frac{5}{14}$; 7 **43.** $\frac{24}{25}$ **45.** $3\frac{5}{9}$ **47.** $15\frac{3}{4}$ **49.** 4 **51.** $5\frac{11}{14}$ **53.** $6\frac{2}{9}$ **55.** $\frac{25}{33}$ **57.** $35\frac{13}{18}$ **59.** $2\frac{1}{2}$ **61.** $72\frac{19}{30}$ **63.** $\frac{11}{14}$ **65.** $5\frac{4}{7}$

67. $13\frac{16}{33}$ **69.** $-5\frac{2}{7}-x$ **71.** $1\frac{9}{10}\cdot x$ **73.** $3\frac{3}{16}$ mi **75.** $9\frac{2}{5}$ in. **77.** $7\frac{13}{20}$ in. **79.** $3\frac{1}{2}$ sq yd **81.** $\frac{15}{16}$ sq in. **83.** $21\frac{5}{24}$ m

85. no; she will be short $\frac{1}{2}$ ft **87.** $4\frac{2}{3}$ m **89.** $9\frac{3}{4}$ min **91.** $1\frac{4}{5}$ min **93.** $-10\frac{3}{25}$ **95.** $-24\frac{7}{8}$ **97.** $-13\frac{59}{60}$ **99.** $4\frac{2}{7}$ **101.** $-1\frac{23}{24}$

103. $\frac{73}{1000}$ **105.** $1x$ or x **107.** $1a$ or a **109.** a, b, c **111.** Incorrect; to divide mixed numbers, first write each mixed number as an improper fraction. **113.** answers may vary **115.** answers may vary **117.** answers may vary

Section 4.8

Vocabulary, Readiness & Video Check **1.** 6 **3.** 15 **5.** addition property of equality **7.** We multiply by 12 because it is the LCD of all fractions in the equation. The equation no longer contains fractions.

Exercise Set 4.8 **1.** $-\frac{2}{3}$ **3.** $\frac{1}{13}$ **5.** $\frac{4}{5}$ **7.** $\frac{11}{12}$ **9.** $-\frac{7}{10}$ **11.** $\frac{11}{16}$ **13.** $\frac{11}{18}$ **15.** $\frac{2}{7}$ **17.** 12 **19.** -27 **21.** $\frac{27}{8}$ **23.** $\frac{1}{21}$ **25.** $\frac{2}{11}$

27. $-\frac{3}{10}$ **29.** 1 **31.** 10 **33.** -1 **35.** -15 **37.** $\frac{3x-28}{21}$ **39.** $\frac{y+10}{2}$ **41.** $\frac{7x}{15}$ **43.** $\frac{4}{3}$ **45.** 2 **47.** $\frac{21}{10}$ **49.** $-\frac{1}{14}$ **51.** -3

53. $-\frac{1}{9}$ **55.** 50 **57.** $-\frac{3}{7}$ **59.** 4 **61.** $\frac{3}{5}$ **63.** $-\frac{1}{24}$ **65.** $-\frac{5}{14}$ **67.** 4 **69.** -36 **71.** 57,200 **73.** 330 **75.** answers may vary

77. $\frac{112}{11}$ **79.** area: $\frac{3}{16}$ sq in.: perimeter: 2 in.

Chapter 4 Vocabulary Check **1.** reciprocals **2.** composite number **3.** equivalent **4.** improper fraction **5.** prime number **6.** simplest form **7.** proper fraction **8.** mixed number **9.** numerator; denominator **10.** prime factorization **11.** undefined **12.** 0 **13.** like **14.** least common denominator **15.** complex fraction **16.** cross products

Chapter 4 Review **1.** $\frac{2}{6}$ **2.** $\frac{4}{7}$ **3.** $\frac{7}{3}$ or $2\frac{1}{3}$ **4.** $\frac{13}{4}$ or $3\frac{1}{4}$ **5.** $\frac{11}{12}$ **6. a.** 108 **b.** $\frac{108}{131}$ **7.** -1 **8.** 1 **9.** 0 **10.** undefined

11. number line: $\frac{7}{9}$ plotted between 0 and 1 **12.** number line: $\frac{4}{7}$ plotted between 0 and 1 **13.** number line: $\frac{5}{4}$ plotted between 1 and 2

14. number line: $\frac{7}{5}$ plotted between 1 and 2 **15.** $3\frac{3}{4}$ **16.** 3 **17.** $\frac{11}{5}$ **18.** $\frac{35}{9}$ **19.** $\frac{3}{7}$ **20.** $\frac{5}{9}$ **21.** $-\frac{1}{3x}$ **22.** $-\frac{y^2}{2}$ **23.** $\frac{29}{32c}$ **24.** $\frac{18z}{23}$ **25.** $\frac{5x}{3y^2}$

26. $\frac{7b}{5c^2}$ **27.** $\frac{2}{3}$ of a foot **28.** $\frac{3}{5}$ of the cars **29.** no **30.** yes **31.** $\frac{3}{10}$ **32.** $-\frac{5}{14}$ **33.** $\frac{9}{x^2}$ **34.** $\frac{y}{2}$ **35.** $-\frac{1}{27}$ **36.** $-\frac{25}{144}$ **37.** -2

38. $\frac{15}{4}$ **39.** $\frac{27}{2}$ **40.** $-\frac{5}{6y}$ **41.** $\frac{12}{7}$ **42.** $-\frac{63}{10}$ **43.** $\frac{77}{48}$ sq ft **44.** $\frac{4}{9}$ sq m **45.** $\frac{10}{11}$ **46.** $\frac{2}{3}$ **47.** $-\frac{1}{3}$ **48.** $\frac{4x}{5}$ **49.** $\frac{4y-3}{21}$ **50.** $-\frac{1}{15}$

51. $3x$ **52.** 24 **53.** 20 **54.** 35 **55.** $49a$ **56.** $45b$ **57.** 40 **58.** 10 **59.** $\frac{3}{4}$ of his homework **60.** $\frac{3}{2}$ mi **61.** $\frac{11}{18}$ **62.** $\frac{7}{26}$

63. $-\frac{1}{12}$ **64.** $-\frac{5}{12}$ **65.** $\frac{25x+2}{55}$ **66.** $\frac{4+3b}{15}$ **67.** $\frac{7y}{36}$ **68.** $\frac{11x}{18}$ **69.** $\frac{4y+45}{9y}$ **70.** $-\frac{15}{14}$ **71.** $\frac{91}{150}$ **72.** $\frac{5}{18}$ **73.** $\frac{19}{9}$ m **74.** $\frac{3}{2}$ ft

75. $\frac{21}{50}$ of the donors **76.** $\frac{1}{4}$ yd **77.** $\frac{4x}{7}$ **78.** $\frac{3y}{11}$ **79.** -2 **80.** $-7y$ **81.** $\frac{15}{4}$ **82.** $-\frac{5}{24}$ **83.** $\frac{8}{13}$ **84.** $-\frac{1}{27}$ **85.** $\frac{29}{110}$ **86.** $-\frac{1}{7}$

87. $20\frac{7}{24}$ **88.** $2\frac{51}{55}$; 3 **89.** $5\frac{1}{5}$; 6 **90.** $5\frac{1}{4}$ **91.** 22 mi **92.** $36\frac{2}{3}$ g **93.** each measurement is $4\frac{1}{4}$ in. **94.** $\frac{7}{10}$ yd **95.** $-27\frac{5}{14}$

96. $-\frac{33}{40}$ **97.** $1\frac{5}{27}$ **98.** $-3\frac{15}{16}$ **99.** $\frac{5}{6}$ **100.** $-\frac{9}{10}$ **101.** -10 **102.** -6 **103.** 15 **104.** 1 **105.** 5 **106.** -4 **107.** $\frac{1}{4}$ **108.** $\frac{y^2}{2x}$

109. $\frac{1}{5}$ **110.** $-\frac{7}{12x}$ **111.** $\frac{11x}{12}$ **112.** $-\frac{23}{55}$ **113.** $-6\frac{2}{5}$ **114.** $12\frac{3}{8}$; 13 **115.** $2\frac{19}{35}$; 2 **116.** $\frac{22}{7}$ or $3\frac{1}{7}$ **117.** $-\frac{1}{12}$ **118.** $-\frac{9}{14}$

119. $-\frac{4}{9}$ **120.** -19 **121.** $44\frac{1}{2}$ yd **122.** $40\frac{1}{2}$ sq ft

Chapter 4 Test **1.** $\frac{7}{16}$ **2.** $\frac{23}{3}$ **3.** $18\frac{3}{4}$ **4.** $\frac{4}{35}$ **5.** $-\frac{3x}{5}$ **6.** not equivalent **7.** equivalent **8.** $2^2\cdot3\cdot7$ **9.** $3^2\cdot5\cdot11$ **10.** $\frac{4}{3}$

11. $-\frac{4}{3}$ **12.** $\frac{8x}{9}$ **13.** $\frac{x-21}{7x}$ **14.** y^2 **15.** $\frac{16}{45}$ **16.** $\frac{9a+4}{10}$ **17.** $-\frac{2}{3y}$ **18.** $\frac{1}{a^2}$ **19.** $\frac{3}{4}$ **20.** $14\frac{1}{40}$ **21.** $16\frac{8}{11}$ **22.** $\frac{64}{3}$ or $21\frac{1}{3}$ **23.** $22\frac{1}{2}$

24. $-\frac{5}{3}$ or $-1\frac{2}{3}$ **25.** $\frac{9}{16}$ **26.** $\frac{3}{8}$ **27.** $\frac{11}{12}$ **28.** $\frac{3}{4x}$ **29.** $\frac{76}{21}$ or $3\frac{13}{21}$ **30.** -2 **31.** -4 **32.** 1 **33.** $\frac{5}{2}$ **34.** $\frac{4}{31}$ **35.** $3\frac{3}{4}$ ft   **36.** $\frac{23}{50}$

37. $\frac{13}{50}$ **38.** \$2820   **39.** perimeter: $3\frac{1}{3}$ ft; area: $\frac{2}{3}$ sq ft   **40.** 24 mi

Cumulative Review Chapters 1–4   **1.** five hundred forty-six; Sec. 1.2. Ex. 5   **2.** one hundred fifteen; Sec. 1.2   **3.** twenty-seven thousand, thirty-four; Sec 1.2, Ex. 6   **4.** six thousand, five hundred seventy-three; Sec. 1.2   **5.** 759; Sec. 1.3, Ex. 1   **6.** 631; Sec. 1.3   **7.** 514; Sec. 1.3, Ex. 7   **8.** 933; Sec. 1.3   **9.** 278,000; Sec. 1.4, Ex. 2   **10.** 1440; Sec. 1.4   **11.** 57,600 megabytes; Sec. 1.5, Ex. 7   **12.** 1305 mi; Sec. 1.5   **13.** 7089 R 5; Sec. 1.6, Ex. 7   **14.** 379 R 10; Sec. 1.6   **15.** 7^3; Sec. 1.7, Ex. 1   **16.** 7^2; Sec. 1.7   **17.** $3^4 \cdot 9^3$; Sec. 1.7, Ex. 4   **18.** $9^4 \cdot 5^2$; Sec 1.7   **19.** 6; Sec. 1.8, Ex. 2   **20.** 52; Sec. 1.8   **21.** -7188; Sec. 2.1, Ex. 1   **22.** -21; Sec. 2.1   **23.** -4; Sec. 2.2, Ex. 3   **24.** 5; Sec. 2.2   **25.** 3; Sec. 2.3, Ex. 9   **26.** 10; Sec. 2.3   **27.** 25; Sec. 2.4, Ex. 8   **28.** -16; Sec. 2.4   **29.** -16; Sec. 2.5, Ex. 8   **30.** 25; Sec. 2.5   **31.** $6y + 2$; Sec. 3.1, Ex. 2   **32.** $6x + 9$; Sec. 3.1   **33.** -14; Sec. 3.2, Ex. 4   **34.** -18; Sec. 3.2   **35.** -1; Sec. 3.3, Ex. 2   **36.** -11; Sec. 3.3   **37.** $\frac{2}{5}$; Sec. 4.1, Ex. 3   **38.** $2^2 \cdot 3 \cdot 13$; Sec. 4.2   **39. a.** $\frac{38}{9}$ **b.** $\frac{19}{11}$; Sec. 4.1, Ex. 20   **40.** $7\frac{4}{5}$; Sec. 4.1   **41.** $\frac{7x}{11}$; Sec. 4.2, Ex. 5   **42.** $\frac{2}{3y}$; Sec. 4.2   **43.** $2\frac{11}{12}$; Sec. 4.7, Ex. 2   **44.** $2\frac{2}{3}$; Sec. 4.7   **45.** $\frac{5}{12}$; Sec. 4.3, Ex. 11   **46.** $\frac{11}{56}$; Sec. 4.7

Chapter 5 Decimals

Section 5.1

Vocabulary, Readiness & Video Check   **1.** words; standard form   **3.** decimals   **5.** tenths; tens   **7.** as "and"   **9.** Reading a decimal correctly gives you the correct place value, which tells you the denominator of your equivalent fraction.   **11.** When rounding, we look to the digit to the right of the place value we're rounding to. In this case we look to the hundredths-place digit, which is 7.

Exercise Set 5.1   **1.** five and sixty-two hundredths   **3.** sixteen and twenty-three hundredths   **5.** negative two hundred five thousandths   **7.** one hundred sixty-seven and nine thousandths   **9.** three thousand and four hundredths   **11.** one hundred five and six tenths   **13.** two and forty-three hundredths

15.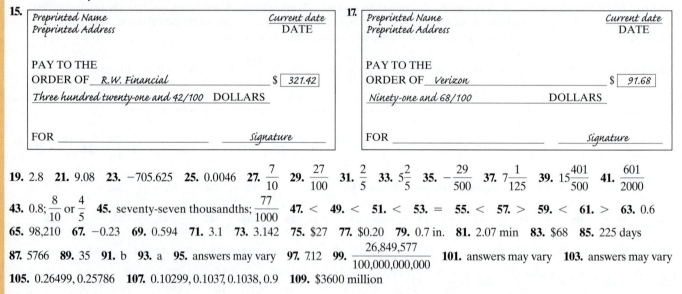

19. 2.8   **21.** 9.08   **23.** -705.625 **25.** 0.0046   **27.** $\frac{7}{10}$ **29.** $\frac{27}{100}$ **31.** $\frac{2}{5}$ **33.** $5\frac{2}{5}$ **35.** $-\frac{29}{500}$ **37.** $7\frac{1}{125}$ **39.** $15\frac{401}{500}$ **41.** $\frac{601}{2000}$

43. $0.8; \frac{8}{10}$ or $\frac{4}{5}$ **45.** seventy-seven thousandths; $\frac{77}{1000}$ **47.** $<$ **49.** $<$ **51.** $<$ **53.** $=$ **55.** $<$ **57.** $>$ **59.** $<$ **61.** $>$ **63.** 0.6

65. 98,210   **67.** -0.23 **69.** 0.594   **71.** 3.1   **73.** 3.142   **75.** \$27   **77.** \$0.20   **79.** 0.7 in.   **81.** 2.07 min   **83.** \$68   **85.** 225 days

87. 5766   **89.** 35   **91.** b   **93.** a   **95.** answers may vary   **97.** 7.12   **99.** $\frac{26,849,577}{100,000,000,000}$ **101.** answers may vary   **103.** answers may vary

105. 0.26499, 0.25786   **107.** 0.10299, 0.1037, 0.1038, 0.9   **109.** \$3600 million

Section 5.2

Calculator Explorations   **1.** 328.742   **3.** 5.2414   **5.** 865.392

Vocabulary, Readiness & Video Check   **1.** last   **3.** like   **5.** false   **7.** Lining up the decimal points also lines up place values, so we only add or subtract digits in the same place values.   **9.** So the subtraction can be written vertically with decimal points lined up.   **11.** perimeter

Exercise Set 5.2   **1.** 7.7   **3.** 10.35   **5.** 27.0578   **7.** -8.57 **9.** 10.33   **11.** 465.56;

$$\begin{array}{r} 230 \\ + \ 230 \\ \hline 460 \end{array}$$

13. 115.123;

$$\begin{array}{r} 100 \\ 6 \\ + \quad 9 \\ \hline 115 \end{array}$$

15. 50.409   **17.** 4.4

19. 15.3   **21.** 598.23   **23.** 1.83; $6 - 4 = 2$ **25.** 876.6; $\begin{array}{r} 1000 \\ -\ 100 \\ \hline 900 \end{array}$ **27.** 194.4   **29.** -6.32 **31.** -6.15 **33.** 3.1   **35.** 2.9988   **37.** 16.3

39. 3.1   **41.** -5.62 **43.** 363.36   **45.** -549.8 **47.** 861.6   **49.** 115.123   **51.** 0.088   **53.** -180.44 **55.** -1.1 **57.** 3.81   **59.** 3.39
61. 1.61   **63.** no   **65.** yes   **67.** no   **69.** $6.9x + 6.9$ **71.** $3.47y - 10.97$ **73.** $7.52   **75.** $-$0.42   **77.** 28.56 m   **79.** 14.36 in.
81. 195.8 mph   **83.** 11.8 texts per day   **85.** $2042.5 million   **87.** 326.3 in.   **89.** 67.44 ft   **91.** 13.462 mph   **93.** Switzerland   **95.** 7.94 lb

97.

Country	Pounds of Chocolate per Person
Switzerland	26.24
Ireland	21.83
UK	20.94
Austria	19.40
Belgium	18.30

99. 138   **101.** $\frac{4}{9}$ **103.** incorrect; $\begin{array}{r} 9.200 \\ 8.630 \\ +\ +\ 4.005 \\ \hline 21.835 \end{array}$ **105.** 6.08 in.   **107.** $1.20

109. 1 nickel, 1 dime, and 2 pennies; 3 nickels and 2 pennies; 1 dime and 7 pennies; 2 nickels and 7 pennies   **111.** answers may vary
113. answers may vary   **115.** $22.181x - 22.984$

Section 5.3

Vocabulary, Readiness & Video Check   1. sum   **3.** right; zeros   **5.** circumference   **7.** Whether we placed the decimal point correctly in our product   **9.** $3(5.7) - (-0.2)$ **11.** This is an application problem and needs units attached. The complete answer is 24.8 grams.

Exercise Set 5.3   1. 1.36   **3.** 0.6   **5.** -17.595 **7.** 55.008   **9.** 28.56; $7 \times 4 = 28$ **11.** 0.1041   **13.** 8.23854; 1   **15.** 11.2746   **17.** $\begin{array}{r} 65 \\ \times\ 8 \\ \hline 8 \end{array}$

19. 0.83   **21.** -7093 **23.** 70   **25.** 0.0983   **27.** 0.02523   **29.** 0.0492   **31.** 14,790   **33.** 1.29   **35.** -9.3762 **37.** 0.5623   **39.** 36.024
41. 1,500,000,000   **43.** 49,800,000   **45.** -0.6 **47.** 17.3   **49.** no   **51.** yes   **53.** 10π cm ≈ 31.4 cm   **55.** 18.2π yd ≈ 57.148 yd
57. $715.20   **59.** 24.8 g   **61.** 11.201 sq in.   **63.** 250π ft ≈ 785 ft   **65.** 135π m ≈ 423.9 m   **67.** 64.9605 in.   **69. a.** 62.8 m and 125.6 m
b. yes   **71.** $730   **73.** 786.9 Canadian dollars   **75.** 1024.67 New Zealand dollars   **77.** 486   **79.** -9 **81.** 3.64   **83.** 3.56
85. -0.1105 **87.** 3,831,600 mi   **89.** answers may vary   **91.** answers may vary

Section 5.4

Calculator Explorations   1. not reasonable   **3.** reasonable

Vocabulary, Readiness & Video Check   1. quotient; divisor; dividend   **3.** left; zeros   **5.** a whole number   **7.** We just need to know how to move the decimal point. 1000 has three zeros, so we move the decimal point in the decimal number three places to the left.
9. We want the answer rounded to the nearest tenth, so we go to one extra place value, to the hundredths place, in order to round.

Exercise Set 5.4   1. 4.6   **3.** 0.094   **5.** 300   **7.** 7.3   **9.** 6.6; $6\overline{)36}^{\,6}$ **11.** 0.413   **13.** -600 **15.** 7   **17.** 4.8   **19.** 2100   **21.** 5.8   **23.** 5.5
25. 9.8; $7\overline{)70}^{\,10}$ **27.** 9.6   **29.** 45   **31.** 54.592   **33.** 0.0055   **35.** 23.87   **37.** 114.0   **39.** 0.83397   **41.** 2.687   **43.** -0.0129 **45.** 12.6   **47.** 1.31
49. 0.045625   **51.** 0.413   **53.** -8 **55.** -7.2 **57.** 1400   **59.** 30   **61.** $-58,000$ **63.** -0.69 **65.** 0.024   **67.** 65   **69.** -5.65 **71.** -7.0625
73. yes   **75.** no   **77.** 11 qt   **79.** 5.1 m   **81.** 11.4 boxes   **83.** 24 tsp   **85.** 8 days   **87.** 146.6 mi per week   **89.** 345.5 thousand books
per hr   **91.** $\frac{21}{50}$ **93.** $-\frac{1}{10}$ **95.** 4.26   **97.** 1.578   **99.** -26.66 **101.** 904.29   **103.** c   **105.** b   **107.** 85.5   **109.** 8.6 ft
111. answers may vary   **113.** $65.2 - 82.6$ knots   **115.** 27.3 m

Integrated Review   1. 2.57   **2.** 4.05   **3.** 8.9   **4.** 3.5   **5.** 0.16   **6.** 0.24   **7.** 0.27   **8.** 0.52   **9.** -4.8 **10.** 6.09   **11.** 75.56   **12.** 289.12
13. -24.974 **14.** -43.875 **15.** -8.6 **16.** 5.4   **17.** -280 **18.** 1600   **19.** 224.938   **20.** 145.079   **21.** 0.56   **22.** -0.63 **23.** 27.6092
24. 145.6312   **25.** 5.4   **26.** -17.74 **27.** -414.44 **28.** -1295.03 **29.** -34 **30.** -28 **31.** 116.81   **32.** 18.79   **33.** 156.2   **34.** 1.562
35. 25.62   **36.** 5.62   **37.** Exact: 204.1 mi; Estimate: 200 mi   **38.** $0.81   **39.** $8.8 billion or $8,800,000,000

Section 5.5

Vocabulary, Readiness & Video Check   1. false   **3.** false   **5.** We place a bar over just the repeating digits and only 6 repeats in our decimal answer.   **7.** The fraction bar serves as a grouping symbol   **9.** $4(0.3) - (-2.4)$

Exercise Set 5.5   **1.** 0.2   **3.** 0.68   **5.** 0.75   **7.** −0.08   **9.** 2.25   **11.** $0.91\overline{6}$ **13.** 0.425   **15.** 0.45   **17.** $-0.\overline{3}$ **19.** 0.4375   **21.** $0.\overline{63}$
23. 5.85   **25.** 0.624   **27.** −0.33   **29.** 0.44   **31.** 0.6   **33.** 0.62   **35.** 0.86   **37.** 0.02   **39.** <   **41.** =   **43.** <   **45.** <   **47.** <   **49.** >
51. <   **53.** <   **55.** 0.32, 0.34, 0.35   **57.** 0.49, 0.491, 0.498   **59.** $5.23, \frac{42}{8}, 5.34$ **61.** $0.612, \frac{5}{8}, 0.649$ **63.** 0.59   **65.** −3   **67.** 5.29
69. 9.24   **71.** 0.2025   **73.** −1.29   **75.** −15.4   **77.** −3.7   **79.** 25.65 sq in.   **81.** 0.248 sq yd   **83.** 5.76   **85.** 5.7   **87.** 3.6   **89.** $\frac{77}{50}$
91. $\frac{5}{2}$ **93.** = 1   **95.** > 1   **97.** < 1   **99.** 0.057   **101.** 6300 stations   **103.** answers may vary

Section 5.6

Vocabulary, Readiness & Video Check   **1.** So that we are no longer working with decimals.

Exercise Set 5.6   **1.** 5.9   **3.** −0.43   **5.** 10.2   **7.** −4.5   **9.** 4   **11.** 0.45   **13.** 4.2   **15.** −4   **17.** 1.8   **19.** 10   **21.** 7.6   **23.** 60
25. −0.07   **27.** 20   **29.** 0.0148   **31.** −8.13   **33.** 1.5   **35.** −1   **37.** −7   **39.** 7   **41.** 53.2   **43.** $3x - 16$ **45.** $\frac{3}{5x}$ **47.** $\frac{13x}{21}$ **49.** 3.7
51. $6x - 0.61$ **53.** $-2y + 6.8$ **55.** 9.1   **57.** −3   **59.** $-4z + 16.67$ **61.** 15.7   **63.** 5.85   **65.** $-2.1z - 10.1$ **67.** answers may vary
69. answers may vary   **71.** 7.683   **73.** 4.683

Section 5.7

Vocabulary, Readiness & Video Check   **1.** average   **3.** mean (or average)   **5.** grade point average   **7.** Place the data numbers in numerical order (or verify that they already are)

Exercise Set 5.7   **1.** mean: 21; median: 23; no mode   **3.** mean: 8.1; median: 8.2; mode: 8.2   **5.** mean: 0.5; median: 0.5; mode: 0.2 and 0.5
7. mean: 370.9; median: 313.5; no mode   **9.** 1911.6 ft   **11.** 1601 ft   **13.** answers may vary   **15.** 2.79   **17.** 3.64   **19.** 6.8   **21.** 6.9
23. 88.5   **25.** 73   **27.** 70 and 71   **29.** 9 rates   **31.** $\frac{1}{3}$ **33.** $\frac{3}{5y}$ **35.** $\frac{11}{15}$ **37.** 35, 35, 37, 43   **39.** yes; answers may vary

Chapter 5 Vocabulary Check   **1.** decimal   **2.** numerator; denominator   **3.** vertically   **4.** and   **5.** sum   **6.** mode
7. circumference   **8.** median; mean   **9.** mean   **10.** standard form

Chapter 5 Review   **1.** tenths   **2.** hundred-thousandths   **3.** negative twenty-three and forty-five hundredths   **4.** three hundred
forty-five hundred-thousandths   **5.** one hundred nine and twenty-three hundredths   **6.** two hundred and thirty-two millionths
7. 8.06   **8.** −503.102   **9.** 16,025.0014   **10.** 14.011   **11.** $\frac{4}{25}$ **12.** $-12\frac{23}{1000}$ **13.** 0.00231   **14.** 25.25   **15.** >   **16.** =   **17.** <   **18.** >
19. 0.6   **20.** 0.94   **21.** −42.90   **22.** 16.349   **23.** 887,000,000   **24.** 600,000   **25.** 18.1   **26.** 5.1   **27.** −7.28   **28.** −12.04   **29.** 320.312
30. 148.74236   **31.** 1.7   **32.** 2.49   **33.** −1324.5   **34.** −10.136   **35.** 65.02   **36.** 199.99802   **37.** 52.6 mi   **38.** −5.7   **39.** 22.2 in.
40. 38.9 ft   **41.** 72   **42.** 9345   **43.** −78.246   **44.** 73,246.446   **45.** 14π m $\approx$ 43.96 m   **46.** 20π in. $\approx$ 62.8 in.   **47.** 0.0877
48. 15.825   **49.** 70   **50.** −0.21   **51.** 8.059   **52.** 30.4   **53.** 0.02365   **54.** −9.3   **55.** 7.3 m   **56.** 45 months   **57.** 0.8   **58.** −0.923
59. $2.\overline{3}$ or 2.333   **60.** $0.21\overline{6}$ or 0.217   **61.** =   **62.** <   **63.** <   **64.** <   **65.** 0.832, 0.837, 0.839   **66.** $\frac{5}{8}, 0.626, 0.685$ **67.** $0.42, \frac{3}{7}, 0.43$
68. $\frac{19}{12}, 1.63, \frac{18}{11}$ **69.** −11.94   **70.** 3.89   **71.** 7.26   **72.** 0.81   **73.** 55   **74.** −129   **75.** 6.9 sq ft   **76.** 5.46 sq in.   **77.** 0.3   **78.** 92.81
79. 8.6   **80.** −80   **81.** 1.98   **82.** −1.5   **83.** −20   **84.** 1   **85.** mean: 17.8; median: 14; no mode   **86.** mean: 58.1; median: 60; mode:
45 and 86   **87.** mean: 24,500; median: 20,000; mode: 20,000   **88.** mean: 447.3; median: 420; mode: 400   **89.** 3.25   **90.** 2.57   **91.** two
hundred and thirty-two ten-thousandths   **92.** −16.09   **93.** $\frac{847}{10,000}$ **94.** $0.75, \frac{6}{7}, \frac{8}{9}$ **95.** −0.07   **96.** 0.1125   **97.** 51.057   **98.** >
99. <   **100.** 86.91   **101.** 3.115   **102.** $123.00   **103.** $3646.00   **104.** −1.7   **105.** 5.26   **106.** −12.76   **107.** −14.907   **108.** 8.128
109. −7.245   **110.** 4900   **111.** 23.904   **112.** 9600 sq ft   **113.** yes   **114.** 0.1024   **115.** 3.6   **116.** mean: 74.4; median: 73; mode: none
117. mean: 619.17; median: 647.5; mode: 327

Chapter 5 Test   **1.** forty-five and ninety-two thousandths   **2.** 3000.059   **3.** 17.595   **4.** −51.20 or −51.2   **5.** −20.42   **6.** 40.902   **7.** 0.037
8. 34.9   **9.** 0.862   **10.** <   **11.** <   **12.** $\frac{69}{200}$ **13.** $-24\frac{73}{100}$ **14.** −0.5   **15.** 0.941   **16.** 1.93   **17.** −6.2   **18.** $0.5x - 13.4$ **19.** −3
20. 3.7   **21.** mean: 38.4; median: 42; no mode   **22.** mean: 12.625; median: 12.5; mode: 12 and 16   **23.** 3.07   **24.** 4,583,000,000
25. 2.31 sq mi   **26.** 18π mi $\approx$ 56.52 mi   **27. a.** 9904 sq ft   **b.** 198.08 oz   **28.** 54 mi

Cumulative Review Chapters 1–5   **1.** seventy-two; Sec. 1.2, Ex. 4   **2.** one hundred seven; Sec. 1.2   **3.** five hundred forty-six;
Sec. 1.2, Ex. 5   **4.** five thousand, twenty-six; Sec. 1.2   **5.** 759; Sec. 1.3, Ex. 1   **6.** 19 in.; Sec. 1.3   **7.** 514; Sec. 1.3, Ex. 7   **8.** 121 R 1;
Sec. 1.6   **9.** 278,000; Sec. 1.4, Ex. 2   **10.** $2 \cdot 3 \cdot 5$; Sec. 4.2   **11.** 20,296; Sec. 1.5, Ex. 4   **12.** 0; Sec. 1.5   **13. a.** 7   **b.** 12
c. 1   **d.** 1   **e.** 20   **f.** 1; Sec. 1.6, Ex. 2   **14.** 25; Sec. 1.6   **15.** 7; Sec. 1.7, Ex. 9   **16.** 49; Sec. 1.7   **17.** 81; Sec. 1.7, Ex. 5   **18.** 125; Sec. 1.7
19. 81; Sec. 1.7, Ex. 7   **20.** 1000; Sec. 1.7   **21.** 2; Sec. 1.8, Ex. 3   **22.** 6; Sec. 1.8   **23. a.** −13   **b.** 2   **c.** 0; Sec. 2.1, Ex. 5

24. a. 7   **b.** -4 **c.** 1; Sec. 2.1   **25.** -23; Sec. 2.2, Ex. 4   **26.** -22; Sec. 2.2   **27.** 180; Sec. 1.7, Ex. 8   **28.** 32; Sec. 1.7   **29.** -49; Sec. 2.4, Ex. 9   **30.** -32; Sec. 2.4   **31.** 25; Sec. 2.4, Ex. 8   **32.** -9; Sec. 2.4   **33.** $\frac{4}{3}$; $1\frac{1}{3}$; Sec. 4.1, Ex. 10   **34.** $\frac{7}{4}$; $1\frac{3}{4}$; Sec. 4.1   **35.** $\frac{11}{4}$; $2\frac{3}{4}$; Sec. 4.1, Ex. 11   **36.** $\frac{14}{3}$; $4\frac{2}{3}$; Sec. 4.1   **37.** $2^2 \cdot 3^2 \cdot 7$; Sec. 4.2, Ex. 3   **38.** 62; Sec. 1.3   **39.** $-\frac{36}{13}$; Sec. 4.2, Ex. 8   **40.** $\frac{79}{8}$; Sec. 4.1   **41.** equivalent; Sec. 4.2, Ex. 10   **42.** >; Sec. 4.5   **43.** $\frac{10}{33}$; Sec. 4.3, Ex. 1   **44.** $1\frac{1}{2}$; Sec. 4.7   **45.** $\frac{1}{8}$; Sec. 4.3, Ex. 2   **46.** 37; Sec. 4.7   **47.** -24; Sec. 3.2, Ex. 3   **48.** -8; Sec. 3.2   **49.** 829.6561; Sec. 5.2, Ex. 2   **50.** 230.8628; Sec. 5.2   **51.** 18.408; Sec. 5.3, Ex. 1   **52.** 28.251; Sec. 5.3

Chapter 6 Ratio, Proportion, and Triangle Applications
Section 6.1

Vocabulary, Readiness & Video Check   1. unit   **3.** division   **5.** numerator; denominator   **7.** false   **9.** The units are different in Example 5 (shrubs and feet); they were the same in Example 4 (days).   **11.** When shopping for the best buy

Exercise Set 6.1   1. $\frac{2}{3}$ **3.** $\frac{77}{100}$ **5.** $\frac{463}{821}$ **7.** $\frac{3}{8}$ **9.** $\frac{8}{25}$ **11.** $\frac{12}{7}$ **13.** $\frac{2}{7}$ **15.** $\frac{4}{1}$ **17.** $\frac{10}{29}$ **19.** $\frac{25}{144}$ **21.** $\frac{5}{4}$ **23.** $\frac{15}{1}$ **25.** $\frac{2}{3}$ **27.** $\frac{10}{21}$ **29.** $\frac{17}{40}$ **31.** $\frac{191}{141}$ **33.** $\frac{1\ \text{shrub}}{3\ \text{ft}}$ **35.** $\frac{3\ \text{returns}}{20\ \text{sales}}$ **37.** $\frac{3\ \text{laser printers}}{14\ \text{computers}}$ **39.** $\frac{9\ \text{gal}}{2\ \text{acres}}$ **41.** 110 cal/oz   **43.** 75 riders/car   **45.** 90 wingbeats/sec   **47.** \$50,000/yr   **49.** 315,750 voters/senator   **51.** 300 good/defective   **53.** \$5,924,700/player   **55. a.** 31.25 computer boards/hr   **b.** 33.5 computer boards/hr   **c.** Suellen   **57. a.** ≈ 27.6 miles/gal   **b.** ≈ 29.2 miles/gal   **c.** the truck   **59.** \$11.50 per compact disc   **61.** \$0.17 per banana   **63.** 8 oz: \$0.411 per oz; 12 oz: \$0.399 per oz; 12 oz   **65.** 16 oz: \$0.118 per oz; 6 oz: \$0.115 per oz; 6 oz   **67.** 12 oz: \$0.191 per oz; 8 oz: \$0.186 per oz; 8 oz   **69.** 100: \$0.006 per napkin; 180: \$0.005 per napkin; 180 napkins   **71.** 2.3   **73.** 0.15   **75.** no; answers may vary   **77.** no; $\frac{2}{5}$ **79.** yes, the machine should be repaired   **81.** 257; 19.2   **83.** 347; 21.6   **85.** 1.5 steps/foot   **87. a.** $\frac{11}{25}$ **b.** $\frac{11}{14}$ **c.** no; answers may vary   **89.** answers may vary   **91.** no; answers may vary

Section 6.2

Vocabulary, Readiness & Video Check   1. proportion; ratio   **3.** true   **5.** equals or $=$ **7.** It is a ratio equal to a ratio

Exercise Set 6.2   1. $\frac{10\ \text{diamonds}}{6\ \text{opals}} = \frac{5\ \text{diamonds}}{3\ \text{opals}}$ **3.** $\frac{20\ \text{students}}{5\ \text{microscopes}} = \frac{4\ \text{students}}{1\ \text{microscope}}$ **5.** $\frac{6\ \text{eagles}}{58\ \text{sparrows}} = \frac{3\ \text{eagles}}{29\ \text{sparrows}}$ **7.** $\frac{2\frac{1}{4}\ \text{cups flour}}{24\ \text{cookies}} = \frac{6\frac{3}{4}\ \text{cups flour}}{72\ \text{cookies}}$ **9.** $\frac{22\ \text{vanilla wafers}}{1\ \text{cup cookie crumbs}} = \frac{55\ \text{vanilla wafers}}{2.5\ \text{cups cookie crumbs}}$ **11.** true   **13.** false   **15.** true   **17.** true   **19.** false   **21.** true   **23.** true   **25.** false   **27.** true   **29.** $\frac{10}{15} = \frac{4}{6}$; true   **31.** $\frac{11}{4} = \frac{5}{2}$; false   **33.** $\frac{0.15}{3} = \frac{0.35}{7}$; true   **35.** $\frac{\frac{2}{3}}{\frac{1}{5}} = \frac{\frac{2}{5}}{\frac{1}{9}}$; false   **37.** 3   **39.** -9 **41.** 5   **43.** 3.2   **45.** 38.4   **47.** 25   **49.** 0.0025   **51.** 1   **53.** $\frac{9}{20}$ **55.** 12   **57.** $\frac{3}{4}$ **59.** $\frac{35}{18}$ **61.** 14.9   **63.** 0.07   **65.** 1.9   **67.** 3.163   **69.** $<$ **71.** $>$ **73.** $\frac{3}{5}$ **75.** $\frac{2x}{7}$ **77.** $\frac{9}{3} = \frac{15}{5}$; $\frac{5}{15} = \frac{3}{9}$; $\frac{15}{9} = \frac{5}{3}$ **79.** $\frac{6}{1} = \frac{18}{3}$; $\frac{3}{18} = \frac{1}{6}$; $\frac{18}{6} = \frac{3}{1}$ **81.** possible answers: $\frac{d}{b} = \frac{c}{a}$; $\frac{a}{c} = \frac{b}{d}$; $\frac{b}{a} = \frac{d}{c}$ **83.** answers may vary   **85.** 0   **87.** 1400   **89.** 252.5

Integrated Review   1. $\frac{9}{10}$ **2.** $\frac{9}{25}$ **3.** $\frac{47}{50}$ **4.** $\frac{8}{23}$ **5.** $\frac{173}{139}$ **6.** $\frac{6}{7}$ **7.** $\frac{7}{26}$ **8.** $\frac{20}{33}$ **9.** $\frac{2}{3}$ **10.** $\frac{1}{8}$ **11.** $\frac{2}{3}$ **12. a.** 44   **b.** $\frac{23}{103}$ **13.** $\frac{1\ \text{professor}}{5\ \text{graduate assistants}}$ **14.** $\frac{3\ \text{lights}}{10\ \text{ft}}$ **15.** $\frac{2\ \text{Senators}}{1\ \text{state}}$ **16.** $\frac{1\ \text{teacher}}{28\ \text{students}}$ **17.** $\frac{3\ \text{inches}}{1\ \text{second}}$ **18.** $\frac{\$8}{1\ \text{hour}}$ **19.** $\frac{19\ \text{households with computers}}{25\ \text{households}}$ **20.** $\frac{269\ \text{electoral votes}}{25\ \text{states}}$ **21.** 140 ft/sec   **22.** 65 mi/hr   **23.** 21 employees/fax line   **24.** 17 phone calls/teenager   **25.** 26 mi/gal   **26.** 16 teachers/computer   **27.** 6.5 books/student   **28.** 165 lb/adult   **29.** 8 lb: \$0.27 per lb; 18 lb: \$0.277 per lb; 8 lb   **30.** 100: \$0.020 per plate; 500: \$0.018 per plate; 500 paper plates   **31.** 3 packs: \$0.797 per pack; 8 packs: \$0.749 per pack; 8 packs   **32.** 4: \$1.173 per battery; 10: \$1.489 per battery; 4 batteries   **33.** no   **34.** yes   **35.** 24   **36.** 32.5   **37.** $2.\overline{72}$ or $2\frac{8}{11}$ **38.** 18

Section 6.3

Vocabulary, Readiness & Video Check   **1.** ones

Exercise Set 6.3   **1.** 360 baskets   **3.** 165 min   **5.** 630 applications   **7.** 23 ft   **9.** 270 sq ft   **11.** 25 gal   **13.** 450 km   **15.** 16 bags   **17.** 15 hits   **19.** 27 people   **21.** 18 applications   **23.** 5 weeks   **25.** $10\frac{2}{3}$ servings   **27.** 37.5 seconds   **29. a.** 18 tsp   **b.** 6 tbsp   **31.** 6 people   **33.** 112 ft; 11-in. difference   **35.** 102.9 mg   **37.** 1248 ft; coincidentally, this is the actual height of the Empire State Building   **39.** 434 emergency room visits   **41.** 28 workers   **43.** 2.4 c   **45. a.** 0.1 gal   **b.** 13 fl oz   **47. a.** 2062.5 mg   **b.** no   **49.** $2^3 \cdot 5^2$ **51.** 2^5 **53.** 0.8 ml   **55.** 1.25 ml   **57.** $11 \approx 12$ or 1 dozen; $1.5 \times 8 = 12$; 12 cups of milk   **59.** $4\frac{2}{3}$ ft   **61.** answers may vary

Section 6.4

Calculator Explorations   **1.** 32   **3.** 3.873   **5.** 9.849

Vocabulary, Readiness & Video Check   **1.** $10; -10$ **3.** radical   **5.** perfect squares   **7.** $c^2; b^2$ **9.** The square roots of 49 are 7 and -7 since $7^2 = 49$ and $(-7)^2 = 49$. The radical sign means the positive square root only, so $\sqrt{49} = 7$.   **11.** The hypotenuse is the side across from the right angle.

Exercise Set 6.4   **1.** 2   **3.** 11   **5.** $\frac{1}{9}$ **7.** $\frac{4}{8} = \frac{1}{2}$ **9.** 1.732   **11.** 3.873   **13.** 5.568   **15.** 5.099   **17.** 6, 7   **19.** 10, 11   **21.** 16   **23.** 9.592   **25.** $\frac{7}{12}$ **27.** 8.426   **29.** 13 in.   **31.** 6.633 cm   **33.** 52.802 m   **35.** 117 mm   **37.** 5   **39.** 12   **41.** 17.205   **43.** 44.822   **45.** 42.426   **47.** 1.732   **49.** 8.5   **51.** 141.42 yd   **53.** 25.0 ft   **55.** 340 ft   **57.** $\frac{5}{6}$ **59.** $\frac{x}{30}$ **61.** $\frac{21}{13y}$ **63.** $\frac{9x}{64}$ **65.** 6   **67.** 10   **69.** answers may vary   **71.** yes   **73.** $\sqrt{80} - 6 \approx 2.94$ in.

Section 6.5

Vocabulary, Readiness & Video Check   **1.** false   **3.** true   **5.** false   **7.** The ratios of corresponding sides are the same.

Exercise Set 6.5   **1.** congruent; SSS   **3.** not congruent   **5.** congruent; ASA   **7.** congruent; SAS   **9.** $\frac{2}{1}$ **11.** $\frac{3}{2}$ **13.** 4.5   **15.** 6   **17.** 5   **19.** 13.5   **21.** 17.5   **23.** 10   **25.** 28.125   **27.** 10   **29.** 520 ft   **31.** 500 ft   **33.** 14.4 ft   **35.** 52 neon tetras   **37.** 381 ft   **39.** 4.01   **41.** -1.23 **43.** $3\frac{8}{9}$ in.; no   **45.** 8.4   **47.** answers may vary   **49.** 200 ft, 300 ft, 425 ft

Chapter 6 Vocabulary Check   **1.** ratio   **2.** proportion   **3.** unit rate   **4.** unit price   **5.** rate   **6.** cross products   **7.** equal   **8.** not equal   **9.** Congruent   **10.** Similar   **11.** leg   **12.** leg   **13.** hypotenuse   **14.** right   **15.** Pythagorean

Chapter 6 Review   **1.** $\frac{23}{37}$ **2.** $\frac{11}{13}$ **3.** $\frac{17}{35}$ **4.** $\frac{18}{35}$ **5.** $\frac{9}{4}$ **6.** $\frac{2}{13}$ **7.** $\frac{5 \text{ people}}{2 \text{ pets}}$ **8.** $\frac{5 \text{ pages}}{2 \text{ min}}$ **9.** 52 mi/hr   **10.** 15 ft/sec   **11.** $1.74/$ diskette   **12.** 13 bushels/tree   **13.** 8 oz: $0.124 per oz; 12 oz: $0.141 per oz; 8-oz size   **14.** 18 oz: $0.083; 28 oz: $0.085; 18-oz size   **15.** $\frac{24 \text{ uniforms}}{8 \text{ players}} = \frac{3 \text{ uniforms}}{1 \text{ player}}$ **16.** $\frac{12 \text{ tires}}{3 \text{ cars}} = \frac{4 \text{ tires}}{1 \text{ car}}$ **17.** no   **18.** yes   **19.** 5   **20.** 21   **21.** -5.625 **22.** 60   **23.** 0.94   **24.** $1\frac{1}{8}$ **25.** 0.63   **26.** 30.9   **27.** 14   **28.** 35   **29.** 8 bags   **30.** 16 bags   **31.** $213\frac{1}{3}$ mi   **32.** 9.6 in.   **33.** 8   **34.** 12   **35.** 3.464   **36.** 3.873   **37.** 0   **38.** 1   **39.** 7.071   **40.** 8.062   **41.** $\frac{2}{5}$ **42.** $\frac{1}{10}$ **43.** 13   **44.** 29   **45.** 10.7   **46.** 55.1   **47.** 28.28 cm   **48.** 88.2 ft   **49.** congruent; ASA   **50.** not congruent   **51.** $13\frac{1}{3}$ **52.** 17.4   **53.** 33 ft   **54.** $x = \frac{5}{6}$ in.; $y = 2\frac{1}{6}$ in.   **55.** $\frac{3}{5}$ **56.** $\frac{1}{27}$ **57.** $\frac{1 \text{ teacher}}{9 \text{ students}}$ **58.** $\frac{1 \text{ nurse}}{4 \text{ patients}}$ **59.** 34 mi/hr   **60.** 2 gal/cow   **61.** $\frac{10}{481}$ **62.** 166 steps/min   **63.** 4 oz: $1.235 per oz; 8 oz: $1.248 per oz; 4-oz size   **64.** 12 oz: $0.054 per oz; 64 oz: $0.047 per oz; 64-oz size   **65.** $\frac{2 \text{ cups cookie dough}}{30 \text{ cookies}} = \frac{4 \text{ cups cookie dough}}{60 \text{ cookies}}$ **66.** $\frac{5 \text{ nickels}}{3 \text{ dollars}} = \frac{20 \text{ nickels}}{12 \text{ dollars}}$ **67.** 1.6   **68.** 25   **69.** 3.75   **70.** $\frac{2}{5}$ **71.** 6   **72.** $\frac{4}{9}$ **73.** 10.247   **74.** 5.657   **75.** 86.6   **76.** 20.8   **77.** 12   **78.** $6\frac{1}{2}$

Chapter 6 Test   **1.** $\frac{9}{13}$ **2.** $\frac{3 \text{ in.}}{10 \text{ days}}$ **3.** $\frac{43}{50}$ **4.** $\frac{47}{78}$ **5.** $\frac{293}{93}$ **6.** 81.25 km/hr   **7.** 28 students/teacher   **8.** 16 in./oz min   **9.** 8 oz: $0.149 per oz; 12 oz: $0.158 per oz; 8-oz size   **10.** 16 oz: $0.093 per oz; 24 oz: $0.100 per oz; 16-oz size   **11.** true   **12.** false   **13.** 5   **14.** $4\frac{4}{11}$ **15.** $\frac{7}{3}$ **16.** 8   **17.** $49\frac{1}{2}$ ft   **18.** $3\frac{3}{4}$ hr   **19.** $53\frac{1}{3}$ g   **20.** 7   **21.** 12.530   **22.** $\frac{8}{10} = \frac{4}{5}$ **23.** 5.66 cm   **24.** 7.5   **25.** 69 ft

Cumulative Review Chapters 1–6   1. a. 3   **b.** 15   **c.** 0   **d.** 70; Sec. 1.3, Ex. 5   **2. a.** 0   **b.** 20   **c.** 0   **d.** 20; Sec. 1.5   **3.** 249,000; Sec. 1.4, Ex. 3   **4.** 249,000; Sec. 1.4   **5. a.** 200   **b.** 1230; Sec. 1.5, Ex. 3   **6.** 373 R 24; Sec. 1.6   **7.** −8; Sec. 2.2, Ex. 15   **8.** −84; Sec. 2.4

9. $2^4 \cdot 5$; Sec. 4.2, Ex. 2   **10.** 8; Sec. 1.7   **11.** $\frac{3}{5}$; Sec. 4.2, Ex. 4   **12.** 243; Sec. 1.7   **13.** $\frac{2}{5}$; Sec. 4.3, Ex. 6   **14.** $15\frac{3}{8}$; Sec. 4.7   **15.** 2;

Sec. 4.4, Ex. 3   **16.** $\frac{4}{5}$; Sec. 4.4   **17.** 14; Sec. 4.4, Ex. 12   **18.** $\frac{49}{50}$; Sec. 4.5   **19.** $\frac{15}{20}$; Sec. 4.4, Ex. 17   **20.** yes; Sec. 4.2   **21.** $-\frac{8}{33}$; Sec. 4.5,

Ex. 4   **22.** $7\frac{47}{72}$; Sec. 4.7   **23.** $\frac{1}{6}$ hr; Sec. 4.5, Ex. 11   **24.** 27; Sec. 1.7   **25.** $7\frac{17}{24}$; Sec. 4.7, Ex. 9   **26.** $\frac{16}{27}$; Sec. 4.6   **27.** <; Sec. 4.5, Ex. 7

28. −8; Sec. 3.2   **29.** −3; Sec. 3.2, Ex. 1   **30.** 0; Sec. 3.3   **31.** 5; Sec. 3.3, Ex. 1   **32.** 5; Sec. 3.3   **33.** 2; Sec. 3.3, Ex. 4   **34.** 0.075; Sec. 5.1

35. 736.2; Sec. 5.1, Ex. 15   **36.** 736.236; Sec. 5.1   **37.** 25.454; Sec. 5.2, Ex. 1   **38.** 681.24; Sec. 5.2   **39.** 0.8496; Sec. 5.3, Ex. 2   **40.** 0.375; Sec. 5.5   **41.** −0.052; Sec. 5.4, Ex. 3   **42.** $\frac{79}{10}$; Sec. 5.5   **43.** −3.7; Sec. 5.5, Ex. 12   **44.** 3; Sec. 6.2   **45.** $\frac{4}{9}, \frac{9}{20}$, 0.456; Sec. 5.5, Ex. 10

46. 140 m/sec; Sec. 6.1   **47.** $\frac{3}{2}$; Sec. 6.1, Ex. 2   **48.** $\frac{1}{3}$; Sec. 6.1   **49.** $\frac{50}{63}$; Sec. 6.1, Ex. 3   **50.** $\frac{1}{10}$; Sec. 6.1

Chapter 7 Percent
Section 7.1

Vocabulary, Readiness & Video Check   1. Percent   **3.** percent   **5.** 0.01   **7.** Percent means "per 100."   **9.** 1

Exercise Set 7.1   1. 96%   **3.** football; 37%   **5.** 50%   **7.** 0.41   **9.** 0.06   **11.** 1.00 or 1   **13.** 0.736   **15.** 0.028   **17.** 0.006   **19.** 3.00 or 3

21. 0.3258   **23.** $\frac{2}{25}$ **25.** $\frac{1}{25}$ **27.** $\frac{9}{200}$ **29.** $\frac{7}{4}$ or $1\frac{3}{4}$ **31.** $\frac{1}{16}$ **33.** $\frac{31}{300}$ **35.** $\frac{179}{800}$ **37.** 22%   **39.** 0.6%   **41.** 530%   **43.** 5.6%

45. 22.28%   **47.** 300%   **49.** 70%   **51.** 70%   **53.** 80%   **55.** 68%   **57.** $37\frac{1}{2}$%   **59.** $33\frac{1}{3}$%   **61.** 450%   **63.** 190%   **65.** 81.82%

67. 26.67%   **69.** 0.6; $\frac{3}{5}$; $23\frac{1}{2}$%; $\frac{47}{200}$; 80%; 0.8; $0.333\overline{3}$; $\frac{1}{3}$; 87.5%; 0.875; 0.075; $\frac{3}{40}$ **71.** 2; 2; 280%; $2\frac{4}{5}$; 7.05; $7\frac{1}{20}$; 454%; 4.54

73. 0.38; $\frac{19}{50}$ **75.** 0.358; $\frac{179}{500}$ **77.** 0.91; $\frac{91}{100}$ **79.** 0.005; $\frac{1}{200}$ **81.** 0.142; $\frac{71}{500}$ **83.** 78.1%   **85.** 0.7%   **87.** 0.40 or 0.4   **89.** $\frac{11}{36}$

91. $1\frac{5}{6}$ **93. a.** 52.9%   **b.** 52.86%   **95.** b, d   **97.** 4%   **99.** 75%   **101.** greater   **103.** 0.266; 26.6%   **105.** biomedical engineers

107. 0.39   **109.** answers may vary

Section 7.2

Vocabulary, Readiness & Video Check   1. is   **3.** amount; base; percent   **5.** greater   **7.** "of" translates to multiplication; "is" (or something equivalent) translates to an equal sign; "what" or "unknown" translates to our variable

Exercise Set 7.2   1. $18\% \cdot 81 = x$ **3.** $20\% \cdot x = 105$ **5.** $0.6 = 40\% \cdot x$ **7.** $x \cdot 80 = 3.8$ **9.** $x = 9\% \cdot 43$ **11.** $x \cdot 250 = 150$ **13.** 3.5
15. 28.7   **17.** 10   **19.** 600   **21.** 110%   **23.** 34%   **25.** 1   **27.** 645   **29.** 500   **31.** 5.16%   **33.** 25.2   **35.** 35%   **37.** 35   **39.** 0.624

41. 0.5%   **43.** 145   **45.** 63%   **47.** 4%   **49.** 30   **51.** $3\frac{7}{11}$ **53.** $\frac{17}{12} = \frac{x}{20}$ **55.** $\frac{8}{9} = \frac{14}{x}$ **57.** c   **59.** b   **61.** answers may vary

63. b   **65.** c   **67.** c   **69.** a   **71.** a   **73.** answers may vary   **75.** 686.625   **77.** 12,285

Section 7.3

Vocabulary, Readiness & Video Check   1. amount; base; percent   **3.** amount   **5.** 45 follows the word "of" so it is the base

Exercise Set 7.3   1. $\frac{a}{45} = \frac{98}{100}$ **3.** $\frac{a}{150} = \frac{4}{100}$ **5.** $\frac{14.3}{b} = \frac{26}{100}$ **7.** $\frac{84}{b} = \frac{35}{100}$ **9.** $\frac{70}{400} = \frac{p}{100}$ **11.** $\frac{8.2}{82} = \frac{p}{100}$ **13.** 26   **15.** 18.9
17. 600   **19.** 10   **21.** 120%   **23.** 28%   **25.** 37   **27.** 1.68   **29.** 1000   **31.** 210%   **33.** 55.18   **33.** 45%   **37.** 75   **39.** 0.864   **41.** 0.5%

43. 140   **45.** 9.6   **47.** 113%   **49.** $-\frac{7}{8}$ **51.** $3\frac{2}{15}$ **53.** 0.7   **55.** 2.19   **57.** answers may vary   **59.** no; $a = 16$ **61.** yes

63. answers may vary   **65.** 12,011.2   **67.** 7270.6

Integrated Review   1. 94%   **2.** 17%   **3.** 37.5%   **4.** 350%   **5.** 470%   **6.** 800%   **7.** 45%   **8.** 106%   **9.** 675%   **10.** 325%

11. 2%   **12.** 6%   **13.** 0.71   **14.** 0.31   **15.** 0.03   **16.** 0.04   **17.** 2.24   **18.** 7   **19.** 0.029   **20.** 0.066   **21.** 0.07; $\frac{7}{100}$ **22.** 0.05; $\frac{1}{20}$

23. 0.068; $\frac{17}{250}$ **24.** 0.1125; $\frac{9}{80}$ **25.** 0.74; $\frac{37}{50}$ **26.** 0.45; $\frac{9}{20}$ **27.** 0.163; $\frac{49}{300}$ **28.** 0.127; $\frac{19}{150}$ **29.** 13.5   **30.** 100   **31.** 350   **32.** 120%

33. 28%   **34.** 76   **35.** 34   **36.** 130%   **37.** 46%   **38.** 37.8   **39.** 150   **40.** 62

Section 7.4

Vocabulary, Readiness & Video Check   **1.** The price of the home is $175,000.

Exercise Set 7.4   **1.** 1600 bolts   **3.** 8.8 pounds   **5.** 14%   **7.** 13,600 screens   **9.** 17%   **11.** 496 chairs; 5704 chairs   **13.** 108,680 physician assistants   **15.** 1,049,841   **17.** 30%   **19.** 50%   **21.** 12.5%   **23.** 29.2%   **25.** $175,000   **27.** 31.2 hr   **29.** $867.87; $20,153.87   **31.** 35 ft   **33.** increase: $1328; tuition in 2013–2014: $10,007   **35.** increase: 164,082 associate degrees; 2020–2021: 1,107,082 associate degrees   **37.** 30; 60%   **39.** 52; 80%   **41.** 2; 25%   **43.** 120; 75%   **45.** 44%   **47.** 137.5%   **49.** 69.5%   **51.** 12.9%   **53.** 8.3%   **55.** 5.4%   **57.** 141.4%   **59.** 62.3%   **61.** 4.56   **63.** 11.18   **65.** $\frac{1}{24}$ **67.** $\frac{28}{39}$ **69.** The increased number is double the original number.   **71.** answers may vary   **73.** percent increase $= \frac{30}{150} = 20\%$ **75.** False; the percents are different.

Section 7.5

Vocabulary, Readiness & Video Check   **1.** sales tax   **3.** commission   **5.** sale price   **7.** We write the commission rate as a percent.

Exercise Set 7.5   **1.** $7.50   **3.** $858.93   **5.** 7%   **7. a.** $120   **b.** $130.20   **9.** $117; $1917   **11.** $485   **13.** 6%   **15.** $16.10; $246.10   **17.** $53,176.04   **19.** 14%   **21.** $4888.50   **23.** $185,500   **25.** $8.90; $80.10   **27.** $98.25; $98.25   **29.** $143.50; $266.50   **31.** $3255; $18,445   **33.** $45; $255   **35.** $27.45; $332.45   **37.** $3.08; $59.08   **39.** $7074   **41.** 8%   **43.** 1200   **45.** 132   **47.** 16   **49.** d   **51.** $4.00; $6.00; $8.00   **53.** $7.20; $10.80; $14.40   **55.** a discount of 60% is better; answers may vary   **57.** $26,838.45

Section 7.6

Calculator Explorations   **1.** $936.31   **3.** $9674.77   **5.** $634.49

Vocabulary, Readiness & Video Check   **1.** simple   **3.** Compound   **5.** Total amount   **7.** principal

Exercise Set 7.6   **1.** $32   **3.** $73.60   **5.** $750   **7.** $33.75   **9.** $700   **11.** $101,562.50; $264,062.50   **13.** $5562.50   **15.** $14,280   **17.** $46,815.37   **19.** $2327.14   **21.** $58,163.65   **23.** $2915.75   **25.** $2938.66   **27.** $2971.89   **29.** 32 yd   **31.** 35 m   **33.** $\frac{9x}{20}$ **35.** $-\frac{131}{225}$ **37.** answers may vary   **39.** answers may vary

Chapter 7 Vocabulary Check   **1.** of   **2.** is   **3.** Percent   **4.** Compound interest   **5.** $\frac{\text{amount}}{\text{base}}$ **6.** 100%   **7.** 0.01   **8.** $\frac{1}{100}$ **9.** base; amount   **10.** Percent of decrease   **11.** Percent of increase   **12.** Sales tax   **13.** Total price   **14.** Commission   **15.** Amount of discount   **16.** Sale price

Chapter 7 Review   **1.** 37%   **2.** 77%   **3.** 0.26   **4.** 0.75   **5.** 0.035   **6.** 0.015   **7.** 2.75   **8.** 4.00 or 4   **9.** 0.4785   **10.** 0.8534   **11.** 160%   **12.** 5.5%   **13.** 7.6%   **14.** 8.5%   **15.** 71%   **16.** 65%   **17.** 600%   **18.** 900%   **19.** $\frac{7}{100}$ **20.** $\frac{3}{20}$ **21.** $\frac{1}{4}$ **22.** $\frac{17}{200}$ **23.** $\frac{51}{500}$ **24.** $\frac{1}{6}$ **25.** $\frac{1}{3}$ **26.** $1\frac{1}{10}$ **27.** 40%   **28.** 70%   **29.** $58\frac{1}{3}\%$ **30.** $166\frac{2}{3}\%$ **31.** 125%   **32.** 60%   **33.** 6.25%   **34.** 62.5%   **35.** 100,000   **36.** 8000   **37.** 23%   **38.** 114.5   **39.** 108.8   **40.** 150%   **41.** 418   **42.** 300   **43.** 159.6   **44.** 180%   **45.** 110%   **46.** 165   **47.** 66%   **48.** 16%   **49.** 20.9%   **50.** 106.25%   **51.** $13.23   **52.** $206,400   **53.** $273.75   **54.** $2.17   **55.** $5000   **56.** $300.38   **57.** discount: $900; sale price: $2100   **58.** discount: $9; sale price: $81   **59.** $160   **60.** $325   **61.** $30,104.61   **62.** $17,506.54   **63.** $180.61   **64.** $33,830.10   **65.** 0.038   **66.** 1.245   **67.** 54%   **68.** 9520%   **69.** $\frac{47}{100}$ **70.** $\frac{7}{125}$ **71.** 12.5%   **72.** 120%   **73.** 268.75   **74.** 110%   **75.** 708.48   **76.** 134%   **77.** 300%   **78.** 38.4   **79.** 560   **80.** 325%   **81.** 26%   **82.** $6786.50   **83.** $617.70   **84.** $3.45   **85.** 12.5%   **86.** $1491   **87.** $11,687.50

Chapter 7 Test   **1.** 0.85   **2.** 5   **3.** 0.006   **4.** 5.6%   **5.** 610%   **6.** 35%   **7.** $1\frac{1}{5}$ **8.** $\frac{77}{200}$ **9.** $\frac{1}{500}$ **10.** 55%   **11.** 37.5%   **12.** 175%   **13.** 20%   **14.** $\frac{16}{25}$ **15.** 33.6   **16.** 1250   **17.** 75%   **18.** 38.4 lb   **19.** $56,750   **20.** $383.21   **21.** 5%   **22.** discount: $18; sale price: $102   **23.** $395   **24.** 9%   **25.** $647.50   **26.** $2005.63   **27.** $427   **28.** 4.5%

Cumulative Review Chapters 1–7   **1.** 20,296; Sec. 1.5, Ex. 4   **2.** 31,084; Sec. 1.5   **3.** −10; Sec. 2.3, Ex. 8   **4.** 10; Sec. 2.3   **5.** 1; Sec. 2.6, Ex. 2   **6.** −1; Sec. 2.6   **7.** 2; Sec. 3.3, Ex. 4   **8.** 5; Sec. 3.3   **9.** $\frac{21}{7}$; Sec. 4.4, Ex. 20   **10.** $\frac{40}{5}$; Sec. 4.4   **11.** $-\frac{10}{27}$; Sec. 4.2, Ex. 6   **12.** $\frac{5y}{16}$; Sec. 4.2   **13.** $\frac{7}{10}$; Sec. 4.3, Ex. 13   **14.** $-\frac{4}{7}$; Sec. 4.3   **15.** $-\frac{1}{2}$; Sec. 4.4, Ex. 9   **16.** $\frac{1}{5}$; Sec. 4.4   **17.** $\frac{1}{28}$; Sec. 4.5, Ex. 5   **18.** $\frac{16}{45}$; Sec. 4.5   **19.** $\frac{3}{2}$; Sec. 4.6, Ex. 2   **20.** $\frac{50}{9}$; Sec. 4.6   **21.** 3; Sec. 4.8, Ex. 9   **22.** 4; Sec. 4.8   **23. a.** $\frac{38}{9}$ **b.** $\frac{19}{11}$; Sec. 4.1, Ex. 20   **24. a.** $\frac{17}{5}$ **b.** $\frac{44}{7}$; Sec. 4.1   **25.** $\frac{1}{8}$; Sec. 5.1, Ex. 9   **26.** $\frac{17}{20}$; Sec. 5.1   **27.** $-105\frac{83}{1000}$; Sec. 5.1, Ex. 11   **28.** $17\frac{3}{200}$; Sec. 5.1   **29.** 67.69; Sec. 5.2, Ex. 6   **30.** 27.94; Sec. 5.2   **31.** 76.8; Sec. 5.3, Ex. 5   **32.** 1248.3; Sec. 5.3   **33.** −76,300; Sec. 5.3, Ex. 7   **34.** −8537.5; Sec. 5.3   **35.** 50; Sec. 5.4, Ex. 10

36. no; Sec. 5.4 **37.** 80.5; Sec. 5.7, Ex. 4 **38.** 48; Sec. 5.7 **39.** $\frac{50}{63}$; Sec. 6.1, Ex. 3 **40.** $\frac{29}{38}$; Sec. 6.1 **41.** $0.21/oz; Sec. 6.1, Ex. 11 **42.** $2.25 per sq ft; Sec. 6.1 **43.** no; Sec. 6.2, Ex. 3 **44.** yes; Sec. 6.2 **45.** 17.5 mi; Sec. 6.3, Ex. 1 **46.** 35; Sec. 6.3 **47.** $\frac{19}{1000}$; Sec. 7.1, Ex. 9 **48.** $\frac{23}{1000}$; Sec. 7.1 **49.** $\frac{1}{3}$; Sec. 7.1, Ex. 11 **50.** $1\frac{2}{25}$; Sec. 7.1

Chapter 8 Graphing and Introduction to Statistics

Section 8.1

Vocabulary, Readiness & Video Check **1.** bar **3.** line **5.** Count the number of symbols and multiply this number by how much each symbol stands for (from the key). **7.** bar graph

Exercise Set 8.1 **1.** Kansas **3.** 5.5 million or 5,500,000 acres **5.** Texas **7.** Montana (or Oklahoma) and Washington **9.** 78,000 **11.** 2006 **13.** 30,000 **15.** 86,000 wildfires/year **17.** September **19.** 77 **21.** $\frac{2}{77}$ **23.** Tokyo, Japan; about 34.7 million or 34,700,000 **25.** New York; 21.6 million or 21,600,000 **27.** approximately 2 million

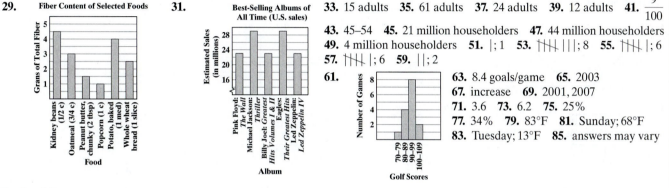

29. **31.** **33.** 15 adults **35.** 61 adults **37.** 24 adults **39.** 12 adults **41.** $\frac{9}{100}$ **43.** 45–54 **45.** 21 million householders **47.** 44 million householders **49.** 4 million householders **51.** |; 1 **53.** ||||| |||; 8 **55.** ||||| |; 6 **57.** ||||| |; 6 **59.** ||; 2 **61.** **63.** 8.4 goals/game **65.** 2003 **67.** increase **69.** 2001, 2007 **71.** 3.6 **73.** 6.2 **75.** 25% **77.** 34% **79.** 83°F **81.** Sunday; 68°F **83.** Tuesday; 13°F **85.** answers may vary

Section 8.2

Vocabulary, Readiness & Video Check **1.** circle **3.** 360 **5.** 100%

Exercise Set 8.2 **1.** parent or guardian's home **3.** $\frac{9}{35}$ **5.** $\frac{9}{16}$ **7.** Asia **9.** 37% **11.** 17,100,000 sq mi **13.** 2,850,000 sq mi **15.** 55% **17.** nonfiction **19.** 31,400 books **21.** 27,632 books **23.** 25,120 books

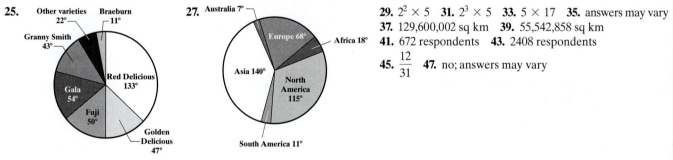

25. **27.** **29.** $2^2 \times 5$ **31.** $2^3 \times 5$ **33.** 5×17 **35.** answers may vary **37.** 129,600,002 sq km **39.** 55,542,858 sq km **41.** 672 respondents **43.** 2408 respondents **45.** $\frac{12}{31}$ **47.** no; answers may vary

Section 8.3

Vocabulary, Readiness & Video Check **1.** $x; y$ **3.** four **5.** plotting **7.** plane **9.** right; coordinate **11.** −7

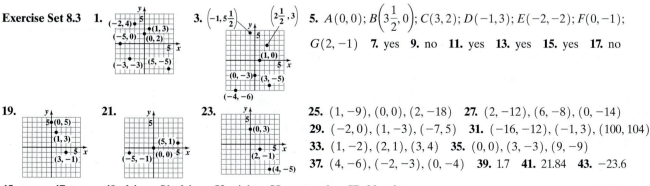

Exercise Set 8.3 **1.** **3.** $\left(-1, 5\frac{1}{2}\right)$, $\left(2\frac{1}{2}, 3\right)$ **5.** $A(0, 0)$; $B\left(3\frac{1}{2}, 0\right)$; $C(3, 2)$; $D(-1, 3)$; $E(-2, -2)$; $F(0, -1)$; $G(2, -1)$ **7.** yes **9.** no **11.** yes **13.** yes **15.** yes **17.** no **19.** **21.** **23.** **25.** $(1, -9), (0, 0), (2, -18)$ **27.** $(2, -12), (6, -8), (0, -14)$ **29.** $(-2, 0), (1, -3), (-7, 5)$ **31.** $(-16, -12), (-1, 3), (100, 104)$ **33.** $(1, -2), (2, 1), (3, 4)$ **35.** $(0, 0), (3, -3), (9, -9)$ **37.** $(4, -6), (-2, -3), (0, -4)$ **39.** 1.7 **41.** 21.84 **43.** −23.6 **45.** true **47.** true **49.** false **51.** false **53.** right **55.** rectangle **57.** 20 units

Integrated Review   **1.** 700,000   **2.** 725,000   **3.** registered nurses   **4.** food preparation and service workers   **5.** Oroville Dam; 755 ft   **6.** New Bullards Bar Dam; 635 ft   **7.** 15 ft   **8.** 4 dams   **9.** Thursday and Saturday; 100°F   **10.** Monday; 82°F   **11.** Sunday, Monday, and Tuesday   **12.** Wednesday, Thursday, Friday, and Saturday   **13.** 70 qt containers   **14.** 52 qt containers   **15.** 2 qt containers   **16.** 6 qt containers   **17.** ||; 2   **18.** |; 1   **19.** |||; 3   **20.** ||||| |; 6   **21.** |||||; 5   **22.**   **23.**   **24.** no   **25.** yes   **26.** $(0, -6), (6, 0), (2, -4)$

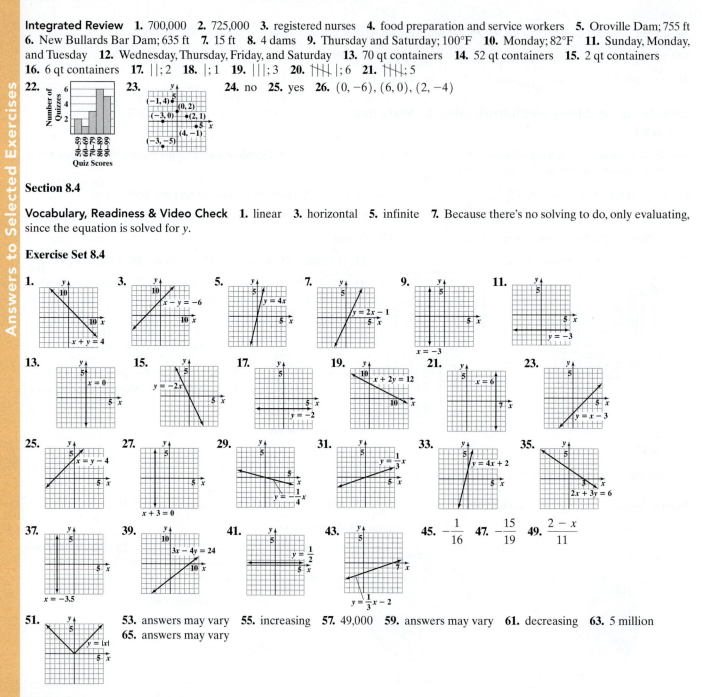

Section 8.4

Vocabulary, Readiness & Video Check   **1.** linear   **3.** horizontal   **5.** infinite   **7.** Because there's no solving to do, only evaluating, since the equation is solved for y.

Exercise Set 8.4

45. $-\dfrac{1}{16}$ **47.** $-\dfrac{15}{19}$ **49.** $\dfrac{2 - x}{11}$

53. answers may vary   **55.** increasing   **57.** 49,000   **59.** answers may vary   **61.** decreasing   **63.** 5 million   **65.** answers may vary

Section 8.5

Vocabulary, Readiness & Video Check   **1.** outcome   **3.** probability   **5.** 0   **7.** The number of outcomes equal the ending number of branches drawn

Exercise Set 8.5

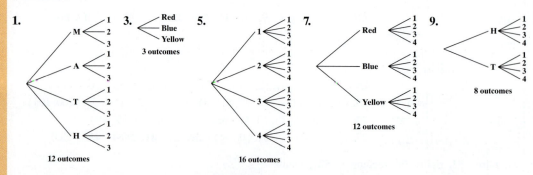

11. $\frac{1}{6}$ **13.** $\frac{1}{3}$ **15.** $\frac{1}{2}$ **17.** $\frac{2}{3}$ **19.** $\frac{1}{3}$ **21.** 1 **23.** $\frac{2}{3}$ **25.** $\frac{1}{7}$ **27.** $\frac{2}{7}$ **29.** $\frac{4}{7}$ **31.** $\frac{19}{100}$ **33.** $\frac{1}{20}$ **35.** $\frac{5}{6}$ **37.** $\frac{1}{6}$ **39.** $6\frac{2}{3}$ **41.** $\frac{1}{52}$
43. $\frac{1}{13}$ **45.** $\frac{1}{4}$ **47.** $\frac{1}{2}$ **49.** $\frac{5}{36}$ **51.** 0 **53.** answers may vary

Chapter 8 Vocabulary Check   1. bar **2.** outcomes **3.** pictograph **4.** line **5.** $x; y$ **6.** tree diagram **7.** experiment **8.** circle
9. probability **10.** histogram; class interval; class frequency **11.** origin **12.** quadrants **13.** plotting **14.** linear

Chapter 8 Review   1. 2,250,000 homes **2.** 4,750,000 homes **3.** South **4.** Northeast **5.** South, West **6.** Northeast, Midwest
7. 30% **8.** 2012 **9.** 1990, 2000, 2010, 2012 **10.** answers may vary **11.** 962 (exact number) **12.** 920 **13.** 930 **14.** 815 **15.** 25
16. 147 (exact number) **17.** 1 employee **18.** 4 employees **19.** 18 employees **20.** 9 employees **21.** ⊮⊮; 5 **22.** |||; 3 **23.** ||||; 4

24.   **25.** mortgage payment **26.** utilities **27.** $1225 **28.** $700 **29.** $\frac{39}{160}$ **30.** $\frac{7}{40}$ **31.** 31 **32.** 6 **33.** 5 **34.** 16

35. $(0, 0), (6, -1), (-6, 1)$ **36.** $(0, -2), (-1, -5), (2, 4)$ **37.** $(-1, -3), (-4, 0), (-5, 1)$ **38.** $(4, 1), (0, -3), (6, 3)$

39. $(1, 3), (-2, -6), (0, 0)$ **40.** $(1, -5), (6, 0), (2, -4)$ **41.** **42.** **43.**

44. **45.** **46.** **47.** **48.** **49.**

50. **51.** **52.** **53.** **54.**

55. **56.** $\frac{1}{6}$ **57.** $\frac{1}{6}$ **58.** $\frac{1}{5}$ **59.** $\frac{1}{5}$ **60.** $\frac{3}{5}$ **61.** $\frac{2}{5}$ **62.** $\frac{1}{4}$ **63.** $\frac{3}{8}$ **64.** $\frac{1}{4}$ **65.** $\frac{1}{8}$ **66.**

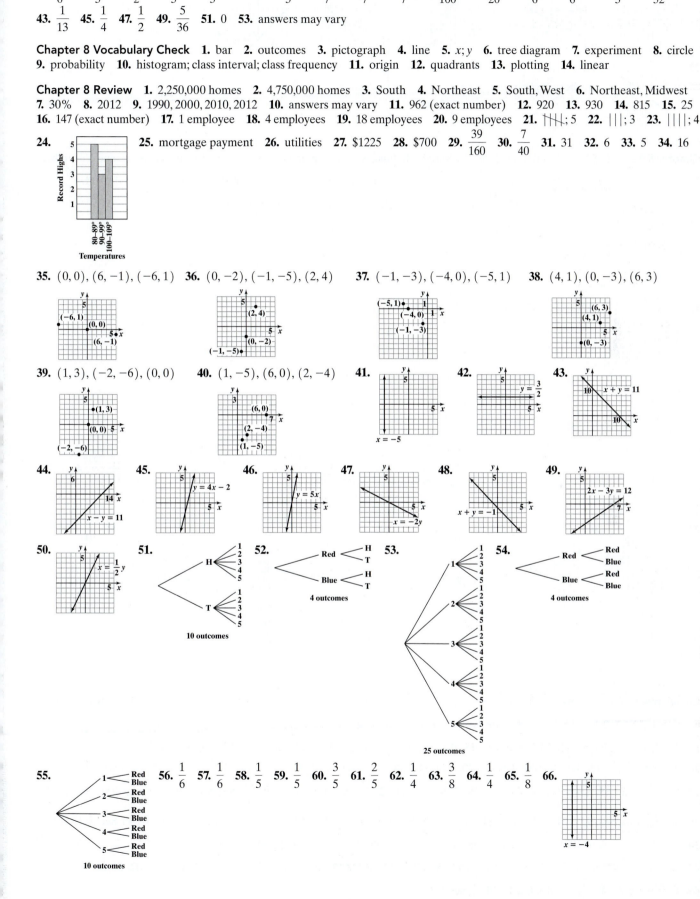

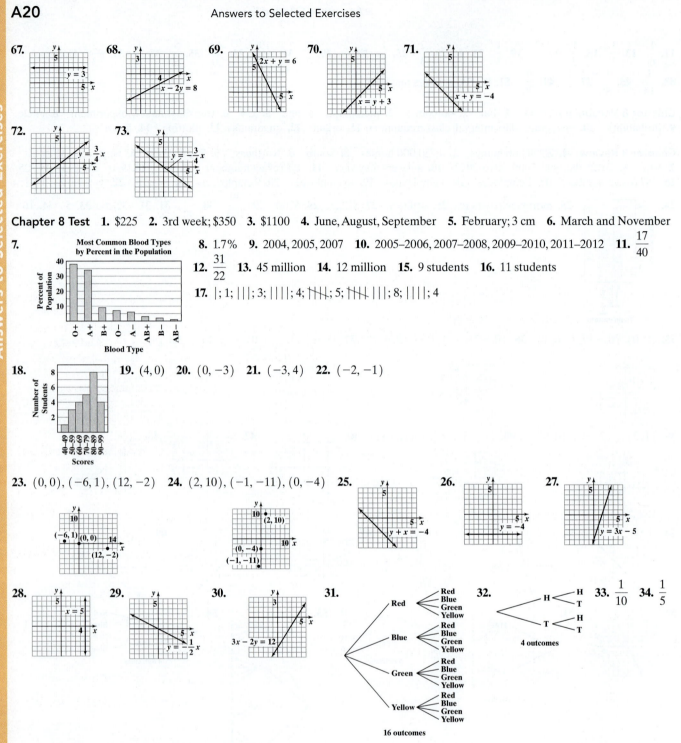

Chapter 8 Test **1.** \$225 **2.** 3rd week; \$350 **3.** \$1100 **4.** June, August, September **5.** February; 3 cm **6.** March and November

7.

Most Common Blood Types by Percent in the Population

8. 1.7% **9.** 2004, 2005, 2007 **10.** 2005–2006, 2007–2008, 2009–2010, 2011–2012 **11.** $\frac{17}{40}$

12. $\frac{31}{22}$ **13.** 45 million **14.** 12 million **15.** 9 students **16.** 11 students

17. |; 1; | | |; 3; | | | |; 4; ||||; 5; |||| | | |; 8; | | | |; 4

18.

19. $(4, 0)$ **20.** $(0, -3)$ **21.** $(-3, 4)$ **22.** $(-2, -1)$

23. $(0, 0), (-6, 1), (12, -2)$ **24.** $(2, 10), (-1, -11), (0, -4)$ **25.** **26.** **27.**

28. **29.** **30.** **31.** **32.** 4 outcomes **33.** $\frac{1}{10}$ **34.** $\frac{1}{5}$

16 outcomes

Cumulative Review Chapters 1–8 **1.** 47; Sec. 1.7, Ex. 12 **2.** 180; Sec. 1.7 **3.** -12; Sec. 2.3, Ex. 11 **4.** 9; Sec. 2.3 **5.** -3; Sec. 3.2, Ex. 2

6. -6; Sec. 3.2 **7.** 2; Sec. 4.8, Ex. 7 **8.** -10; Sec. 4.8 **9.** $7\frac{17}{24}$; Sec. 4.7, Ex. 9 **10.** $8\frac{3}{20}$; Sec. 4.7 **11.** $5\frac{9}{10}$; Sec. 5.1, Ex. 8 **12.** $2\frac{4}{5}$; Sec. 5.1

13. 3.432; Sec. 5.2, Ex. 5 **14.** 7.327; Sec. 5.2 **15.** 0.8496; Sec. 5.3, Ex. 2 **16.** 0.0294; Sec. 5.3 **17.** -0.052; Sec. 5.4, Ex. 3

18. 0.136; Sec. 5.4 **19.** 4.09; Sec. 5.5, Ex. 13 **20.** 7.29; Sec. 5.5 **21.** 0.25; Sec. 5.5, Ex. 1 **22.** 0.375; Sec. 5.5 **23.** 0.7; Sec. 5.6, Ex. 5

24. 1.68; Sec. 5.6 **25.** 8.944; Sec. 6.4, Ex. 7b **26.** 7.746; Sec. 6.4 **27.** $\frac{12}{29}$; Sec. 6.1, Ex. 1 **28.** $\frac{7}{15}$; Sec. 6.1 **29.** 22.5 mi/gal; Sec. 6.1, Ex. 10

30. \$0.53 per oz; Sec. 6.1 **31.** -2; Sec. 6.2, Ex. 6 **32.** 16; Sec. 6.2 **33.** $\frac{12}{19}$; Sec. 6.5, Ex. 2 **34.** $\frac{4}{9}$; Sec. 6.5 **35.** 0.046; Sec. 7.1, Ex. 4

36. 0.32; Sec. 7.1 **37.** 0.0074; Sec. 7.1, Ex. 6 **38.** 0.027; Sec. 7.1 **39.** 21; Sec. 7.2, Ex. 7 **40.** 14.4; Sec. 7.2 **41.** 52; Sec. 7.3, Ex. 9

42. 38; Sec. 7.3 **43.** 8.5%; Sec. 7.5, Ex. 2 **44.** 6.5%; Sec. 7.5 **45.** \$6772.12; Sec. 7.6, Ex. 5 **46.** \$144.05; Sec. 7.6 **47.** 57; Sec. 5.7, Ex. 3

48. 48.5; Sec. 5.7 **49.** $\frac{1}{3}$; Sec. 8.5, Ex. 4 **50.** $\frac{1}{2}$; Sec. 8.5

Chapter 9 Geometry and Measurement

Section 9.1

Vocabulary, Readiness & Video Check **1.** plane **3.** Space **5.** ray **7.** straight **9.** acute **11.** Parallel; intersecting **13.** degrees **15.** vertical **17.** $\angle WUV, \angle VUW, \angle U, \angle x$ **19.** $180° - 17° = 163°$

Exercise Set 9.1 **1.** line; line CD or line l or $\overleftrightarrow{CD}$ **3.** line segment; line segment MN or $\overline{MN}$ **5.** angle; $\angle GHI$ or $\angle IHG$ or $\angle H$ **7.** ray; ray UW or $\overrightarrow{UW}$ **9.** $\angle CPR, \angle RPC$ **11.** $\angle TPM, \angle MPT$ **13.** straight **15.** right **17.** obtuse **19.** acute **21.** 67° **23.** 163° **25.** 32° **27.** 30° **29.** $\angle MNP$ and $\angle RNO$; $\angle PNQ$ and $\angle QNR$ **31.** $\angle SPT$ and $\angle TPQ$; $\angle SPR$ and $\angle RPQ$; $\angle SPT$ and $\angle SPR$; $\angle TPQ$ and $\angle QPR$ **33.** 27° **35.** 132° **37.** $m\angle x = 30°$; $m\angle y = 150°$; $m\angle z = 30°$ **39.** $m\angle x = 77°$; $m\angle y = 103°$; $m\angle z = 77°$ **41.** $m\angle x = 100°$; $m\angle y = 80°$; $m\angle z = 100°$ **43.** $m\angle x = 134°$; $m\angle y = 46°$; $m\angle z = 134°$ **45.** $\angle ABC$ or $\angle CBA$ **47.** $\angle DBE$ or $\angle EBD$ **49.** 15° **51.** 50° **53.** 65° **55.** 95° **57.** $\frac{9}{8}$ or $1\frac{1}{8}$ **59.** $\frac{7}{32}$ **61.** $\frac{5}{6}$ **63.** $1\frac{1}{3}$ **65.** 360° **67.** 54.8° **69.** false; answers may vary **71.** true **73.** $m\angle a = 60°$; $m\angle b = 50°$; $m\angle c = 110°$; $m\angle d = 70°$; $m\angle e = 120°$ **75.** no; answers may vary **77.** 45°; 45°

Section 9.2

Vocabulary, Readiness & Video Check **1.** perimeter **3.** π **5.** $\frac{22}{7}$ (or 3.14); $3.14\left(\text{or } \frac{22}{7}\right)$ **7.** Opposite sides of a rectangle have the same measure, so we can just find the sum of the measures of all four sides.

Exercise Set 9.2 **1.** 64 ft **3.** 120 cm **5.** 21 in. **7.** 48 ft **9.** 42 in. **11.** 155 cm **13.** 21 ft **15.** 624 ft **17.** 346 yd **19.** 22 ft **21.** $55 **23. a.** 8 **b.** 72 in. **25.** 28 in. **27.** $36.12 **29.** 96 m **31.** 66 ft **33.** 74 cm **35.** 17π cm; 53.38 cm **37.** 16π mi; 50.24 mi **39.** 26π m; 81.64 m **41.** 15π ft; 47.1 ft **43.** 12,560 ft **45.** 30.7 mi **47.** 14π cm $\approx$ 43.96 cm **49.** 40 mm **51.** 84 ft **53.** 23 **55.** 1 **57.** 6 **59.** 10 **61. a.** width: 30 yd; length: 40 yd **b.** 140 yd **63.** b **65. a.** 62.8 m; 125.6 m **b.** yes **67.** answers may vary **69.** 27.4 m **71.** 75.4 m

Section 9.3

Vocabulary, Readiness & Video Check **1.** surface area **3.** Area **5.** square **7.** We don't have a formula for an L-shaped figure, so we divide it into two rectangles, use the formula to find the area of each, and then add these two areas.

Exercise Set 9.3 **1.** 7 sq m **3.** $9\frac{3}{4}$ sq yd **5.** 15 sq yd **7.** 2.25π sq in. $\approx$ 7.065 sq in. **9.** 36.75 sq ft **11.** 28 sq m **13.** 22 sq yd **15.** $36\frac{3}{4}$ sq ft **17.** $22\frac{1}{2}$ sq in. **19.** 25 sq cm **21.** 86 sq mi **23.** 24 sq cm **25.** 36π sq in. $\approx 113\frac{1}{7}$ sq in. **27.** $V = 72$ cu in.; $SA = 108$ sq in. **29.** $V = 512$ cu cm; $SA = 384$ sq cm **31.** $V = 4\pi$ cu yd $\approx 12\frac{4}{7}$ cu yd; $SA = (2\pi\sqrt{13} + 4\pi)$ sq yd ≈ 35.20 sq yd **33.** $V = \frac{500}{3}\pi$ cu in. $\approx 523\frac{17}{21}$ cu in.; $SA = 100\pi$ sq in. $\approx 314\frac{2}{7}$ sq in. **35.** $V = 9\pi$ cu in. $\approx 28\frac{2}{7}$ cu in. **37.** $V = 75$ cu cm **39.** $2\frac{10}{27}$ cu in. **41.** $V = 8.4$ cu ft; $SA = 26$ sq ft **43.** 113,625 sq ft **45.** 168 sq ft **47.** 960 cu cm **49.** 9200 sq ft **51.** $V = \frac{1372}{3}\pi$ cu in. or $457\frac{1}{3}\pi$ cu in.; $SA = 196\pi$ sq in. **53. a.** 381 sq ft **b.** 4 squares **55.** $V = 5.25\pi$ cu in. **57.** 4π sq ft $\approx$ 12.56 sq ft **59.** $V = 4.5\pi$ cu m; 14.13 cu m **61.** 168 sq ft **63.** $10\frac{5}{6}$ cu in. **65.** 8.8 cu in. **67.** 25 **69.** 9 **71.** 5 **73.** 20 **75.** perimeter **77.** area **79.** area **81.** perimeter **83.** 12-in. pizza **85.** 2093.33 cu m **87.** no; answers may vary **89.** 7.74 sq in. **91.** 298.5 sq m **93.** no; answers may vary

Integrated Review **1.** 153°; 63° **2.** $m\angle x = 75°$; $m\angle y = 105°$; $m\angle z = 75°$ **3.** $m\angle x = 128°$; $m\angle y = 52°$; $m\angle z = 128°$ **4.** $m\angle x = 52°$ **5.** 4.6 in. **6.** $4\frac{1}{4}$ in. **7.** 20 m; 25 sq m **8.** 12 ft; 6 sq ft **9.** 10π cm $\approx$ 31.4 cm; 25π sq cm $\approx$ 78.5 sq cm **10.** 32 mi; 44 sq mi **11.** 54 cm; 143 sq cm **12.** 62 ft; 238 sq ft **13.** $V = 64$ cu in.; $SA = 96$ sq in. **14.** $V = 30.6$ cu ft; $SA = 63$ sq ft **15.** $V = 400$ cu cm **16.** $V = 4\frac{1}{2}\pi$ cu mi $\approx 14\frac{1}{7}$ cu mi

Section 9.4

Vocabulary, Readiness & Video Check **1.** meter **3.** yard **5.** feet **7.** feet **9.** Both mean addition; $5\frac{2}{5} = 5 + \frac{2}{5}$ and 5 ft 2 in. = 5 ft + 2 in. **11.** Since the metric system is based on base 10, we just need to move the decimal point to convert from one unit to another.

Exercise Set 9.4   1. 5 ft   **3.** 36 ft   **5.** 8 mi   **7.** 102 in.   **9.** $3\frac{1}{3}$ yd   **11.** 33,792 ft   **13.** 4.5 yd   **15.** 0.25 ft   **17.** 13 yd 1 ft   **19.** 7 ft 1 in.
21. 1 mi 4720 ft   **23.** 62 in.   **25.** 26 ft   **27.** 84 in.   **29.** 11 ft 2 in.   **31.** 22 yd 1 ft   **33.** 6 ft 5 in.   **35.** 7 ft 6 in.   **37.** 14 ft 4 in.
39. 83 yd 1 ft   **41.** 6000 cm   **43.** 4 cm   **45.** 0.5 km   **47.** 1.7 m   **49.** 15 m   **51.** 42,000 cm   **53.** 7000 m   **55.** 83 mm   **57.** 0.201 dm
59. 40 mm   **61.** 8.94 m   **63.** 2.94 m or 2940 mm   **65.** 1.29 cm or 12.9 mm   **67.** 12.64 km or 12,640 m   **69.** 54.9 m   **71.** 1.55 km
73. $348\frac{2}{3}$; 12,552   **75.** $11\frac{2}{3}$; 420   **77.** 5000; 0.005; 500   **79.** 0.065; 65; 0.000065   **81.** 342,000; 342,000,000; 34,200,000   **83.** 10 ft 6 in.
85. 5100 ft   **87.** 5.0 times   **89.** 26.7 mm   **91.** 15 ft 9 in.   **93.** 3.35 m   **95.** $121\frac{1}{3}$ yd   **97.** $\frac{21}{100}$ **99.** 0.13   **101.** 0.25   **103.** no   **105.** yes
107. no   **109.** Estimate: 13 yd   **111.** answers may vary; for example, $1\frac{1}{3}$ yd or 48 in.   **113.** answers may vary   **115.** 334.89 sq m

Section 9.5

Vocabulary, Readiness & Video Check   1. Mass   **3.** gram   **5.** 2000   **7.** We can't subtract 9 oz from 4 oz, so we borrow
1 lb (= 16 oz) from 12 lb to add to the 4 oz; 12 lb 4 oz becomes 11 lb 20 oz.   **9.** 18.50 dg

Exercise Set 9.5   1. 32 oz   **3.** 10,000 lb   **5.** 9 tons   **7.** $3\frac{3}{4}$ lb   **9.** $1\frac{3}{4}$ tons   **11.** 204 oz   **13.** 9800 lb   **15.** 76 oz   **17.** 1.5 tons
19. $\frac{1}{20}$ lb   **21.** 92 oz   **23.** 161 oz   **25.** 5 lb 9 oz   **27.** 53 lb 10 oz   **29.** 8 tons 750 lb   **31.** 3 tons 175 lb   **33.** 8 lb 11 oz
35. 31 lb 2 oz   **37.** 1 ton 700 lb   **39.** 0.5 kg   **41.** 4000 mg   **43.** 25,000 g   **45.** 0.048 g   **47.** 0.0063 kg   **49.** 15,140 mg   **51.** 6250 g
53. 350,000 cg   **55.** 13.5 mg   **57.** 5.815 g or 5815 mg   **59.** 1850 mg or 1.85 g   **61.** 1360 g or 1.36 kg   **63.** 13.52 kg   **65.** 2.125 kg
67. 200,000; 3,200,000   **69.** $\frac{269}{400}$ or 0.6725; 21,520   **71.** 0.5; 0.0005; 50   **73.** 21,000; 21,000,000; 2,100,000   **75.** 8.064 kg   **77.** 30 mg
79. 5 lb 8 oz   **81.** 35 lb 14 oz   **83.** 6 lb 15.4 oz   **85.** 144 mg   **87.** 6.12 kg   **89.** 130 lb   **91.** 211 lb   **93.** 0.16   **95.** 0.875   **97.** no
99. yes   **101.** no   **103.** answers may vary; for example, 250 mg or 0.25 g   **105.** true   **107.** answers may vary

Section 9.6

Vocabulary, Readiness & Video Check   1. capacity   **3.** fluid ounces   **5.** cups   **7.** quarts   **9.** We can't subtract 3 qt from 0 qt, so
we borrow 1 gal (= 4 qt) from 3 gal to get 2 gal 4 qt.   **11.** 0.45 dal

Exercise Set 9.6   1. 4 c   **3.** 16 pt   **5.** $3\frac{1}{2}$ gal   **7.** 5 pt   **9.** 8 c   **11.** $3\frac{3}{4}$ qt   **13.** $10\frac{1}{2}$ qt   **15.** 9 c   **17.** 23 qt   **19.** $\frac{1}{4}$ pt   **21.** 14 gal 2 qt
23. 4 gal 3 qt 1 pt   **25.** 22 pt   **27.** 13 gal 2 qt   **29.** 4 c 4 fl oz   **31.** 1 gal 1 qt   **33.** 2 gal 3 qt 1 pt   **35.** 17 gal   **37.** 4 gal 3 qt
39. 5000 ml   **41.** 0.00016 kl   **43.** 5.6 L   **45.** 320 cl   **47.** 0.41 kl   **49.** 0.064 L   **51.** 160 L   **53.** 3600 ml   **55.** 19.3 L   **57.** 4.5 L or
4500 ml   **59.** 8410 ml or 8.41 L   **61.** 16,600 ml or 16.6 L   **63.** 3840 ml   **65.** 162.4 L   **67.** 336; 84; 168   **69.** $\frac{1}{4}$; 1; 2   **71.** 1.59 L
73. 18.954 L   **75.** 4.3 fl oz   **77.** yes   **79.** $0.316   **81.** $\frac{4}{5}$ **83.** $\frac{3}{5}$ **85.** $\frac{9}{10}$ **87.** no   **89.** no   **91.** less than; answers may vary
93. answers may vary   **95.** 128 fl oz   **97.** 1.5 cc   **99.** 2.7 cc   **101.** 54 u or 0.54 cc   **103.** 86 u or 0.86 cc

Section 9.7

Vocabulary, Readiness & Video Check   1. 1 L ≈ 0.26 gal or 3.79 L ≈ 1 gal   **3.** F = 1.8C + 32; 27

Exercise Set 9.7   1. 25.57 fl oz   **3.** 218.44 cm   **5.** 40 oz   **7.** 57.66 mi   **9.** 3.77 gal   **11.** 13.5 kg   **13.** 1.5; $1\frac{2}{3}$; 150; 60   **15.** 55; 5500;
180; 2160   **17.** 3.94 in.   **19.** 80.5 kph   **21.** 0.008 oz   **23.** 229.6 ft   **25.** 9.92 billion mi   **27.** yes   **29.** 2790 mi   **31.** 90 mm   **33.** 112.5 g
35. 104 mph   **37.** 26.24 ft   **39.** 3 mi   **41.** 8 fl oz   **43.** b   **45.** b   **47.** c   **49.** d   **51.** d   **53.** 25°C   **55.** 40°C   **57.** 122°F   **59.** 239°F
61. −6.7°C   **63.** 61.2°C   **65.** 197.6°F   **67.** 54.3°F   **69.** 56.7°C   **71.** 80.6°F   **73.** 21.1°C   **75.** 244.4°F   **77.** 7232°F   **79.** 29   **81.** 36
83. yes   **85.** no   **87.** no   **89.** yes   **91.** 2.13 sq m   **93.** 1.19 sq m   **95.** 1.69 sq m   **97.** 4,000,000,000°C   **99.** answers may vary

Chapter 9 Vocabulary Check   1. Weight   **2.** Mass   **3.** meter   **4.** unit fractions   **5.** gram   **6.** liter   **7.** line segment   **8.** complementary   **9.** line   **10.** perimeter   **11.** angle; vertex   **12.** Area   **13.** ray   **14.** transversal   **15.** straight   **16.** volume   **17.** vertical
18. adjacent   **19.** obtuse   **20.** right   **21.** acute   **22.** supplementary   **23.** surface area

Chapter 9 Review   1. right   **2.** straight   **3.** acute   **4.** obtuse   **5.** 65°   **6.** 75°   **7.** 58°   **8.** 98°   **9.** 90°   **10.** 25°   **11.** $\angle a$ and $\angle b$;
$\angle b$ and $\angle c$; $\angle c$ and $\angle d$; $\angle d$ and $\angle a$ **12.** $\angle x$ and $\angle w$; $\angle y$ and $\angle z$ **13.** $m\angle x = 100°, m\angle y = 80°; m\angle z = 80°$ **14.** $m\angle x = 155°$;
$m\angle y = 155°; m\angle z = 25°$ **15.** $m\angle x = 53°; m\angle y = 53°; m\angle z = 127°$ **16.** $m\angle x = 42°; m\angle y = 42°; m\angle z = 138°$ **17.** 69 m
18. 30.6 cm   **19.** 36 m   **20.** 90 ft   **21.** 32 ft   **22.** 440 ft   **23.** 5.338 in.   **24.** 31.4 yd   **25.** 240 sq ft   **26.** 189 sq yd   **27.** 600 sq cm
28. 82.81 sq m   **29.** 49π sq ft ≈ 153.86 sq ft   **30.** 9π sq in. ≈ 28.26 sq in.   **31.** 119 sq in.   **32.** 140 sq m   **33.** 144 sq m   **34.** 1625 sq cm
35. 432 sq ft   **36.** 130 sq ft   **37.** $V = 15\frac{5}{8}$ cu in.; $SA = 37\frac{1}{2}$ sq in.   **38.** $V = 84$ cu ft; $SA = 136$ sq ft

39. $V = 20,000\pi$ cu cm $\approx 62,800$ cu cm **40.** $V = \frac{1}{6}\pi$ cu km $\approx \frac{11}{21}$ cu km **41.** $2\frac{2}{3}$ cu ft **42.** 307.72 cu in. **43.** $7\frac{1}{2}$ cu ft

44. 0.5π cu ft or $\frac{1}{2}\pi$ cu ft **45.** 9 ft **46.** 24 yd **47.** 7920 ft **48.** 18 in. **49.** 17 yd 1 ft **50.** 3 ft 10 in. **51.** 4200 cm **52.** 820 mm

53. 0.01218 m **54.** 0.00231 km **55.** 21 yd 1 ft **56.** 3 ft 8 in. **57.** 9.5 cm or 95 mm **58.** 2.74 m or 274 cm **59.** 169 yd 2 ft

60. 258 ft 4 in. **61.** 617.5 km **62.** 0.24 sq m **63.** $4\frac{1}{8}$ lb **64.** 4600 lb **65.** 3 lb 4 oz **66.** 5 tons 300 lb **67.** 0.027 g **68.** 40,000 g

69. 21 dag **70.** 0.0003 dg **71.** 3 lb 9 oz **72.** 33 lb 8 oz **73.** 21.5 mg **74.** 0.6 kg or 600 g **75.** 4 lb 4 oz **76.** 9 tons 1075 lb
77. 14 qt **78.** 5 c **79.** 7 pt **80.** 72 c **81.** 4 qt 1 pt **82.** 3 gal 3 qt **83.** 3800 ml **84.** 1.4 kl **85.** 3060 cl **86.** 0.00245 L
87. 1 gal 1 qt **88.** 7 gal **89.** 736 ml or 0.736 L **90.** 15.5 L or 15,500 ml **91.** 2 gal 3 qt **92.** 6 fl oz **93.** 10.88 L **94.** yes
95. 22.96 ft **96.** 10.55 m **97.** 4.55 gal **98.** 8.27 qt **99.** 425.25 g **100.** 10.35 kg **101.** 2.36 in. **102.** 180.4 lb **103.** 107.6°F
104. 320°F **105.** 5.2°C **106.** 26.7°C **107.** 1.7°C **108.** 329°F **109.** 108° **110.** 89° **111.** 95° **112.** 57° **113.** 27.3 in.

114. 194 ft **115.** 1624 sq m **116.** 9π sq m ≈ 28.26 sq m **117.** $346\frac{1}{2}$ cu in. **118.** $V = 140$ cu in.; $SA = 166$ sq in. **119.** 75 in.

120. 4 tons 200 lb **121.** 500 cm **122.** 0.000286 km **123.** 1.4 g **124.** 27 qt **125.** 186.8°F **126.** 11°C **127.** 9117 m or 9.117 km
128. 35.7 L or 35,700 ml **129.** 8 gal 1 qt **130.** 12.8 kg

Chapter 9 Test **1.** 12° **2.** 56° **3.** 50° **4.** $m\angle x = 118°; m\angle y = 62°; m\angle z = 118°$ **5.** $m\angle x = 73°; m\angle y = 73°;$
$m\angle z = 73°$ **6.** 6.2 m **7.** 10 in. **8.** circumference $= 18\pi$ in. ≈ 56.52 in.; area $= 81\pi$ sq in. ≈ 254.34 sq in.

9. perimeter $= 24.6$ yd; area $= 37.1$ sq yd **10.** perimeter $= 68$ in.; area $= 185$ sq in. **11.** $62\frac{6}{7}$ cu in. **12.** 30 cu ft **13.** 16 in.

14. 18 cu ft **15.** 62 ft; \$115.94 **16.** 23 ft 4 in. **17.** 10 qt **18.** $1\frac{7}{8}$ lb **19.** 5600 lb **20.** $4\frac{3}{4}$ gal **21.** 0.04 g **22.** 2400 g **23.** 36 mm

24. 0.43 g **25.** 830 ml **26.** 1 gal 2 qt **27.** 3 lb 13 oz **28.** 8 ft 3 in. **29.** 2 gal 3 qt **30.** 66 mm or 6.6 cm **31.** 2.256 km or 2256 m
32. 28.9°C **33.** 54.7°F **34.** 5.6 m **35.** 4 gal 3 qt **36.** 91.4 m **37.** 16 ft 6 in.

Cumulative Review Chapters 1–9 **1.** 5; Sec. 3.3, Ex. 1 **2.** 6; Sec. 3.3 **3. a.** $\frac{16}{625}$ **b.** $\frac{1}{16}$; Sec. 4.3, Ex. 10 **4. a.** $-\frac{1}{27}$ **b.** $\frac{9}{49}$;
Sec. 4.3 **5.** $9\frac{3}{10}$; Sec. 4.7, Ex. 11 **6.** $9\frac{11}{15}$; Sec. 4.7 **7.** $20x - 10.9$; Sec. 5.2, Ex. 13 **8.** $1.2y + 1.8$; Sec. 5.2 **9.** 28.4405; Sec. 5.5, Ex. 14
10. 2.16; Sec. 5.5 **11.** =; Sec. 5.5, Ex. 9 **12.** >; Sec. 5.5 **13.** -1.3; Sec. 5.6, Ex. 6 **14.** 30; Sec. 5.6 **15.** 424 ft; Sec. 6.4, Ex. 12
16. 236 ft; Sec. 6.4 **17. a.** $\frac{5}{7}$ **b.** $\frac{7}{24}$; Sec. 6.1, Ex. 6 **18. a.** $\frac{1}{4}$ **b.** $\frac{4}{9}$; Sec. 6.1 **19.** $\frac{180 \text{ dollars}}{1 \text{ week}}$; Sec. 6.1, Ex. 7 **20.** $\frac{1 \text{ chaperone}}{5 \text{ students}}$;
Sec. 6.1 **21.** 0.44; Sec. 6.2; Ex. 10 **22.** 1.02; Sec. 6.2 **23.** 22.4 cc; Sec. 6.3, Ex. 2 **24.** 7.5 cups; Sec. 6.3 **25.** 17%; Sec. 7.1, Ex. 1
26. 38%; Sec. 7.1 **27.** 200; Sec. 7.2, Ex. 10 **28.** 1200; Sec. 7.2 **29.** 2.7; Sec. 7.3, Ex. 7 **30.** 12.6; Sec. 7.3 **31.** 32%; Sec. 7.4, Ex. 5
32. 27%; Sec. 7.4 **33.** sales tax: \$6.41; total price: \$91.91; Sec. 7.5, Ex. 1 **34.** sales tax: \$30; total price: \$405; Sec. 7.5
35. $A(-4, 2), B(1, 2), C(0, 1), D(-3, 0), E(5, -4)$; Sec. 8.3, Ex. 2 **36.** $A(2, -3), B(-5, 0), C(0, 4), D(-3, -2)$; Sec. 8.3

37. ; Sec. 8.4, Ex. 4 **38.** ; Sec. 8.4 **39.** 57; Sec. 5.7, Ex. 3 **40.** 83; Sec. 5.7 **41.** $\frac{1}{4}$; Sec. 8.5, Ex. 5

42. $\frac{2}{7}$; Sec. 8.5 **43.** 42°; Sec. 9.1, Ex. 4 **44.** 43°; Sec. 9.1 **45.** 96 in.; Sec. 9.4, Ex. 1 **46.** 21 feet; Sec. 9.4 **47.** 9π sq ft ≈ 28.26 sq ft;
Sec. 9.3, Ex. 4 **48.** 4π sq mi ≈ 12.56 sq mi; Sec. 9.3 **49.** 15°C; Sec. 9.7, Ex. 6 **50.** 30°C; Sec. 9.7

Chapter 10 Exponents and Polynomials
Section 10.1

Vocabulary, Readiness & Video Check **1.** terms **3.** binomial **5.** add **7.** Terms where everything is the same except for the
numerical coefficient. **9.** $2; -5$

Exercise Set 10.1 **1.** $-5x - 24$ **3.** $-11z^2 - 3z - 1$ **5.** $9y^2 + 25y - 40$ **7.** $-4.3a^4 - 2a^2 + 9$ **9.** $-9x + 16$ **11.** $3z^2 - z + 7$
13. $5a - 13$ **15.** $-2x^2 + 4x + 1$ **17.** $-20y^3 + 12y^2 - 4$ **19.** $-9x^2 - x + 16$ **21.** $8y^2 - 10y - 8$ **23.** $5x - 12$
25. $y - 4$ **27.** $-5x^2 + 3x + 5$ **29.** $4x - 4.1$ **31.** $4a - 7$ **33.** $-15y + 3.6$ **35.** $19t^2 - 11t + 3$ **37.** $2b^3 + 13b^2 + 4b - 2$
39. $-5x^2 - 11x + 13$ **41.** $\frac{9}{7}$ **43.** 5 **45.** -5 **47.** -8 **49.** 20 **51.** 25 **53.** 50 **55.** 576 ft **57.** \$3200 **59.** 611 ft **61.** 909 ft
63. 379 million **65.** 81 **67.** 25 **69.** x^3 **71.** $2^2 a^4$ **73.** $(8x + 2)$ in. **75.** $(4x - 15)$ units **77.** 20; 6; 2 **79.** 7.2752
81. 29 ft; -243 ft; answers may vary

Section 10.2

Vocabulary, Readiness & Video Check **1.** exponent **3.** multiply **5.** $x^2 \cdot x^3 = x^5$ **7.** 3 and a

Exercise Set 10.2 **1.** x^{14} **3.** a^4 **5.** $15z^5$ **7.** $-40x^2$ **9.** $42x^3$ **11.** $12a^{17}$ **13.** $25x^6y^4$ **15.** $28a^5b^6$ **17.** x^{15} **19.** z^{30} **21.** b^{62} **23.** $81a^4$ **25.** $a^{33}b^{24}$ **27.** $1000x^{15}y^9$ **29.** $-24y^{22}$ **31.** $256x^9y^{13}$ **33.** $7x - 21$ **35.** $-6a - 4b$ **37.** $9x + 18y - 27$ **39.** $16x^{12}$ sq in. **41.** $12a^4b^5$ sq m **43.** $18{,}003{,}384a^{45}b^{30}$ **45.** $34{,}867.84401x^{50}$ **47.** $x^{270}y^{216}$ **49.** answers may vary

Integrated Review **1.** $2x - 3$ **2.** $20y - 11$ **3.** $10x + 3$ **4.** $-5y - 4$ **5.** $-2a^4 + 3a^2 + 9a$ **6.** $-4a^3 - 12a^2$ **7.** $7.3x^2 - 4.2x - 5.3$ **8.** $1.8y^2 - 2.4y - 5.6$ **9.** $6x + 7$ **10.** $2x^2 + 3x - 12$ **11.** -1 **12.** 26 **13.** x^{20} **14.** x^{10} **15.** y^4 **16.** a^{11} **17.** x^{77} **18.** x^{36} **19.** x^{42} **20.** y^{27} **21.** $125x^3$ **22.** $32y^5$ **23.** $-12x^2y^7$ **24.** $12a^3b^4$ **25.** $y^{33}z^{39}$ **26.** $a^{20}b^{48}$ **27.** $300x^4y^3$ **28.** $128y^6z^7$ **29.** $144a^{38}b^{12}$ **30.** $125x^{22}y^{28}$ **31.** 23.2 million units

Section 10.3

Vocabulary, Readiness & Video Check **1.** The monomial is multiplied by each term in the trinomial. **3.** To make the point that the power rule applies only to products and not to sums, so we cannot apply the power rule to a binomial squared. **5.** No; it is a binomial times a trinomial, and FOIL can only be used to multiply a binomial times a binomial.

Exercise Set 10.3 **1.** $27x^3 - 9x$ **3.** $-6a^3 + 9a^2 + 15a$ **5.** $42x^4 - 35x^3 + 49x^2$ **7.** $x^2 + 13x + 30$ **9.** $2x^2 + 2x - 24$ **11.** $36a^2 + 48a + 16$ **13.** $a^3 - 33a + 18$ **15.** $8x^3 + 2x^2 - 55x + 50$ **17.** $x^5 + 4x^4 + 6x^3 + 7x^2 + 2x$ **19.** $-30r^2 + 20r$ **21.** $-6y^3 - 2y^4 + 12y^2$ **23.** $x^2 + 14x + 24$ **25.** $4a^2 - 9$ **27.** $x^2 + 10x + 25$ **29.** $b^2 + \frac{7}{5}b + \frac{12}{25}$ **31.** $6x^3 + 25x^2 + 10x + 1$ **33.** $49x^2 + 70x + 25$ **35.** $4x^2 - 4x + 1$ **37.** $8x^5 - 8x^3 - 6x^2 - 6x + 9$ **39.** $x^5 + 2x^4 + 3x^3 + 2x^2 + x$ **41.** $10z^4 - 3z^3 + 3z - 2$ **43.** $2 \cdot 5^2$ **45.** $2^3 \cdot 3^2$ **47.** $2^3 \cdot 5^2$ **49.** $\left(y^3 - 3y^2 - 16y - 12\right)$ sq ft **51.** $\left(x^4 - 3x^2 + 1\right)$ sq m **53.** answers may vary

Section 10.4

Vocabulary, Readiness & Video Check **1.** factor; product **3.** smallest **5.** Factor means to write as a product. **7.** Find the GCF of the terms of the binomial and then factor it out.

Exercise Set 10.4 **1.** 3 **3.** 12 **5.** 4 **7.** 4 **9.** y^2 **11.** a^5 **13.** xy^2 **15.** x **17.** $2z^3$ **19.** $3y(y + 6)$ **21.** $5a^6\left(2 - a^2\right)$ **23.** $4x\left(x^2 + 3x + 5\right)$ **25.** $z^5\left(z^2 - 6\right)$ **27.** $-7\left(5 - 2y + y^2\right)$ or $7\left(-5 + 2y - y^2\right)$ **29.** $12a^5(1 - 3a)$ **31.** 36 **33.** $\frac{4}{5}$ **35.** 37.5% **37. a.** $x^2 + 2x$ **b.** answers may vary **39.** answers may vary **41.** $4y^2 + 28y + 49$ **43.** answers may vary

Chapter 10 Vocabulary Check **1.** Factoring **2.** greatest common factor **3.** FOIL **4.** trinomial **5.** binomial **6.** monomial **7.** polynomials **8.** exponent

Chapter 10 Review **1.** $10b - 3$ **2.** $21s - 15$ **3.** $-x + 2.8$ **4.** $-y - 12$ **5.** $4z^2 + 11z - 6$ **6.** $17a^3 + 25a^2$ **7.** $-11y^2 - y + \frac{3}{4}$ **8.** $x^2 - 7x + 3$ **9.** 45 **10.** -19 **11.** $(26x + 28)$ ft **12.** $(16x^2 + 4)$ m **13.** x^{24} **14.** y^7 **15.** $24z^7$ **16.** $-15x^3y^5$ **17.** a^{35} **18.** x^{28} **19.** $81b^2$ **20.** $a^{20}b^{10}c^5$ **21.** $56x^{16}$ **22.** $108x^{30}y^{25}$ **23.** $81a^{14}$ sq mi **24.** $27x^5$ sq in. **25.** $10a^3 - 12a$ **26.** $-3y^4 + 6y^3 - 3y^2$ **27.** $x^2 + 8x + 12$ **28.** $15x^2 - 32x + 9$ **29.** $y^2 - 10y + 25$ **30.** $49a^2 + 14a + 1$ **31.** $x^3 - x^2 + x + 3$ **32.** $8y^4 + 4y^3 - 2y^2 - 3y - 3$ **33.** $3z^4 + 5z^3 + 6z^2 + 3z + 1$ **34.** $\left(a^3 + 5a^2 - 5a + 6\right)$ sq cm **35.** 5 **36.** 4 **37.** 6 **38.** 5 **39.** x^2 **40.** y^7 **41.** xy **42.** a^5b^2 **43.** $5a$ **44.** $4y^2z$ **45.** $2x(x + 6)$ **46.** $6a(a - 2)$ **47.** $y^4\left(6 - y^2\right)$ **48.** $7\left(x^2 - 2x + 1\right)$ **49.** $a^3\left(5a^4 - a + 1\right)$ **50.** $10y\left(y^5 - 1\right)$ **51.** $z^2 + z + 4$ **52.** $-4y - 2$ **53.** x^{21} **54.** y^9 **55.** $a^{18}b^{30}c^6$ **56.** $81x^6$ **57.** $12a^4 - 15a$ **58.** $x^2 + 9x + 20$ **59.** $9x^2 + 24x + 16$ **60.** $6z^2 - 7z - 10$ **61.** 4 **62.** z^4 **63.** $z^7\left(z^2 - 4\right)$ **64.** $x^5\left(x^7 + 6\right)$ **65.** $15a^4(1 + 3a)$ **66.** $8z^5\left(2 - 3z^3\right)$

Chapter 10 Test **1.** $15x - 4$ **2.** $7x - 2$ **3.** $3.4y^2 + 2y - 3$ **4.** $-2a^2 + a + 1$ **5.** 17 **6.** y^{14} **7.** y^{33} **8.** $16x^8$ **9.** $-12a^{10}$ **10.** p^{54} **11.** $72a^{20}b^5$ **12.** $10x^3 + 6.5x$ **13.** $-2y^4 - 12y^3 + 8y$ **14.** $x^2 - x - 6$ **15.** $25x^2 + 20x + 4$ **16.** $a^3 + 8$ **17.** perimeter: $(14x - 4)$ in.; area: $(5x^2 + 33x - 14)$ sq in. **18.** 15 **19.** $3y^3$ **20.** $3y(y - 5)$ **21.** $2a(5a + 6)$ **22.** $6\left(x^2 - 2x - 5\right)$ **23.** $x^3\left(7x^3 - 6x + 1\right)$

Cumulative Review Chapters 1–10 **1.** 106,400 sq mi; Sec. 1.5, Ex. 6 **2.** 147 trees; Sec. 1.5 **3.** -8; Sec. 2.2, Ex. 15 **4.** -10; Sec. 2.2 **5.** -7; Sec. 2.3, Ex. 6 **6.** -3; Sec. 2.3 **7.** 1; Sec. 2.3, Ex. 7 **8.** 5; Sec. 2.3 **9.** 4; Sec. 2.6, Ex. 4 **10.** -2; Sec. 2.6 **11.** -1; Sec. 3.3, Ex. 2 **12.** 1; Sec. 3.3 **13.** $\frac{13x}{30}$; Sec. 4.5, Ex. 2 **14.** $\frac{1}{14y}$; Sec. 4.5 **15.** 736.2; Sec. 5.1, Ex. 15 **16.** 328.2; Sec. 5.1 **17.** 25.454; Sec. 5.2, Ex. 1 **18.** 17.052; Sec. 5.2 **19.** no; Sec. 5.3, Ex. 13 **20.** yes; Sec. 5.3 **21.** 0.7861; Sec. 5.4, Ex. 8 **22.** 0.818; Sec. 5.4 **23.** 0.012; Sec. 5.4, Ex. 9 **24.** 0.0503; Sec. 5.4 **25.** -2.6; Sec. 5.5, Ex. 16 **26.** -13.6; Sec. 5.5 **27.** 3.14; Sec. 5.5, Ex. 4 **28.** 1.947; Sec. 5.5 **29.** $\frac{1}{6}$; Sec. 6.4, Ex. 5 **30.** $\frac{2}{5}$; Sec. 6.4 **31.** 46 ft; Sec. 6.5, Ex. 4 **32.** 3 ft; Sec. 6.5 **33.** $1.2 = 30\% \cdot x$; Sec. 7.2, Ex. 2 **34.** $9 = 45\% \cdot x$; Sec. 7.2 **35.** 16%; Sec. 7.3, Ex. 10 **36.** 25%; Sec. 7.3 **37.** 775 freshmen; Sec. 7.4, Ex. 3 **38.** 1450 apples; Sec. 7.4 **39.** $160; Sec. 7.6, Ex. 2 **40.** $25; Sec. 7.6 **41.** 55%; Sec. 8.2, Ex. 2 **42.** 31%; Sec. 8.2 **43.** 28 in.; Sec. 9.2, Ex. 2 **44.** 25 ft; Sec. 9.2 **45.** 5.1 sq mi; Sec. 9.3, Ex. 2 **46.** 68 sq in.; Sec. 9.3 **47.** 4 tons 1650 lb; Sec. 9.5, Ex. 4 **48.** 16 tons 100 lb; Sec. 9.5 **49.** 3.21 L; Sec. 9.6, Ex. 5 **50.** 43.21 L; Sec. 9.6 **51.** $-3x + 1$; Sec. 10.1, Ex. 1 **52.** $4a + 12$; Sec. 10.1 **53.** $x^2 + 5x + 6$; Sec. 10.3, Ex. 3 **54.** $2x^2 + 19x + 35$; Sec. 10.3

Appendices

Appendix B Exercise Set 1   **1.** x^2 **3.** 81 **5.** p^6q^5 **7.** $\dfrac{xy^3}{2}$ **9.** 1 **11.** 2 **13.** -1 **15.** 1 **17.** $\dfrac{1}{64}$ **19.** $\dfrac{7}{x^3}$ **21.** $\dfrac{5}{6}$ **23.** p^3 **25.** $\dfrac{1}{x^3}$

27. z^3 **29.** $\dfrac{4}{9}$ **31.** $\dfrac{y^2}{25}$ **33.** p^4 **35.** a^6 **37.** x^6y^6 **39.** $\dfrac{1}{x^{11}}$ **41.** $\dfrac{15}{x^3}$ **43.** $\dfrac{1}{y^{12}}$ **45.** $\dfrac{56m^7}{n^5}$ **47.** x^7 **49.** a^8b^{13} **51.** $\dfrac{1}{x^6}$ **53.** 3

55. $\dfrac{1}{125}$ **57.** $\dfrac{8}{x^9}$ **59.** $\dfrac{3}{10}$ **61.** $\dfrac{1}{z^7}$ **63.** $\dfrac{1}{x^9}$ **65.** $\dfrac{a^8}{b^8}$ **67.** $\dfrac{30x^9}{y^6}$

Appendix C Exercise Set   **1.** 7.8×10^4 **3.** 1.67×10^{-6} **5.** 6.35×10^{-3} **7.** 1.16×10^6 **9.** 4.2×10^3

11. 0.0000000008673 **13.** 0.033 **15.** $20,320$ **17.** $700,000,000$ **19.** 5.7×10^{12} **21.** $10,100,000,000,000$ **23.** $3,000,000,000,000$; 3×10^{12} **25.** 0.000036 **27.** 0.0000000000000000028 **29.** 0.0000005 **31.** $200,000$ **33.** 2.7×10^9 gal

Practice Final Exam   **1.** 200 **2.** 21 **3.** -6 **4.** -100 **5.** 7 **6.** -117 **7.** 0 **8.** undefined   **9.** $-\dfrac{2}{3y}$ **10.** $\dfrac{3}{4}$ **11.** $\dfrac{1}{a^2}$

12. $\dfrac{64}{3}$ or $21\dfrac{1}{3}$ **13.** $16\dfrac{8}{11}$ **14.** -6.2 **15.** 40.902 **16.** 0.006 **17.** 610% **18.** 37.5% **19.** $\dfrac{69}{200}$ **20.** -0.5 **21.** 34.9

22. 30 **23.** 1 **24.** $\dfrac{4}{31}$ **25.** $-8z - 20$ **26.** $(15x + 15)$ in.   **27.** -28 **28.** -2 **29.** 3 **30.** -4 **31.** 3.7 **32.** $\dfrac{48}{11}$ or $4\dfrac{4}{11}$

33. 60 yd; 200 sq yd   **34.** -4 **35.** -2 **36.** 24 mi   **37.** 244 women   **38.** $53\dfrac{1}{3}$g   **39.** discount: \$18; sale price: \$102

40.   **41.**   **42.**

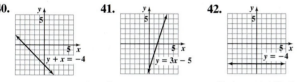

43. $15x - 4$ **44.** $-2a^2 + a + 1$ **45.** $-12a^{10}$ **46.** $72a^{20}b^5$ **47.** $x^2 - x - 6$ **48.** $3y(y - 5)$ **49.** $12°$
50. $m\angle x = 73°$; $m\angle y = 73°$; $m\angle z = 73°$ **51.** perimeter $= 68$ in.; area $= 185$ sq in.   **52.** circumference $= 18\pi$ in. ≈ 56.52 in.; area $= 81\pi$ sq in. ≈ 254.34 sq in.   **53.** 10 qt   **54.** 2400 g

Solutions to Selected Exercises

Exercise Set 1.2

1. The place value of the 5 in 657 is tens.

5. The place value of the 5 in 43,526,000 is hundred-thousands.

9. 354 is written as three hundred fifty-four.

13. 26,990 is written as twenty-six thousand, nine hundred ninety.

17. 24,350,185 is written as twenty-four million, three hundred fifty thousand, one hundred eighty-five.

21. 2720 is written as two thousand, seven hundred twenty.

25. 14,433 is written as fourteen thousand, four hundred thirty-three.

29. Six thousand, five hundred eighty-seven in standard form is 6587.

33. Thirteen million, six hundred one thousand, eleven in standard form is 13,601,011.

37. Two hundred sixty thousand, nine hundred ninety-seven in standard form is 260,997.

41. Two thousand eighty in standard form is 2080.

45. Five hundred sixty-five in standard form is 565.

49. $3470 = 3000 + 400 + 70$

53. $66,049 = 60,000 + 6000 + 40 + 9$

57. Mount Shasta erupted in 1786.

61. The earliest date in the table is 1750, which is an eruption of Glacier Peak.

65. Labrador retrievers are the most popular; 75 is written as seventy-five

69. The largest number is 9861.

73. answers may vary

Exercise Set 1.3

1.
```
   14
 + 22
 ----
   36
```

5.
```
   12
   13
 + 24
 ----
   49
```

9.
```
  1 11
   22,781
 + 186,297
 --------
  209,078
```

13.
```
  22
   81
   17
   23
   79
 + 12
 ----
  212
```

17.
```
 111
  6 820
  4 271
 +5 626
 ------
 16,717
```

21.
```
 122 21
 121,742
  57,279
  26,586
+426,782
--------
 632,389
```

25.
```
   62
 - 37
 ----
   25
```
Check:
```
    1
   25
 + 37
 ----
   62
```

29.
```
  600
 -432
 ----
  168
```
Check:
```
   11
  168
 +432
 ----
  600
```

33.
```
  533
 - 29
 ----
  504
```
Check:
```
    1
  504
 + 29
 ----
  533
```

37.
```
  50,000
 -17,289
 -------
  32,711
```
Check:
```
  11 11
  32,711
 +17,289
 -------
  50,000
```

41.
```
  51,111
 -19,898
 -------
  31,213
```
Check:
```
  11 11
  31,213
 +19,898
 -------
  51,111
```

45.
```
   76
 - 67
 ----
    9
```

49.
```
 11 1
 10,962
  4851
 +7063
 ------
 22,876
```

53. Opposite sides of a rectangle have the same length.
$4 + 8 + 4 + 8 = 12 + 12 = 24$
The perimeter is 24 inches.

57. The unknown vertical side has length
$12 - 5 = 7$ meters.
The unknown horizontal side has length
$10 - 5 = 5$ meters.
$10 + 12 + 5 + 7 + 5 + 5 = 44$
The perimeter is 44 meters.

61. "Find the total" indicates addition.
```
  13
  76
  39
   8
  17
 +126
 ----
  266
```
The total of 76, 39, 8, 17, and 126 is 266.

65. "Increased by" indicates addition.

$$\begin{array}{r} \overset{1}{4}52 \\ +\ 92 \\ \hline 544 \end{array}$$

452 increased by 92 is 544.

69. "Subtracted from" indicates subtraction.

$$\begin{array}{r} 100 \\ -\ 12 \\ \hline 88 \end{array}$$

12 subtracted from 100 is 88.

73. Subtract the cost of the DVD player from the amount in her savings account.

$$\begin{array}{r} 914 \\ -\ 295 \\ \hline 619 \end{array}$$

She will have $619 left.

77.

$$\begin{array}{r} 530{,}000 \\ -\ 247{,}000 \\ \hline 283{,}000 \end{array}$$

The Missouri sub-basin drains 283,000 square miles more than the Arkansas Red-White sub-basin.

81.

$$\begin{array}{r} 503 \\ -\ 239 \\ \hline 264 \end{array}$$

She must read 264 more pages.

85. Live rock music has a decibel level of 100 dB.

89.

$$\begin{array}{r} 2410 \\ +1034 \\ \hline 3444 \end{array}$$

There were 3444 stores worldwide.

93. California has the most Target stores.

97. Pennsylvania and New York:

$$\begin{array}{r} \overset{11}{6}3 \\ +67 \\ \hline 130 \end{array}$$

Michigan and Ohio:

$$\begin{array}{r} \overset{11}{5}9 \\ +64 \\ \hline 123 \end{array}$$

Pennsylvania and New York have more Target stores.

101. The minuend is 48 and the subtrahend is 1.

105. answers may vary

109.

$$\begin{array}{r} \overset{22}{1}4 \\ 173 \\ 86 \\ +\ 257 \\ \hline 530 \end{array}$$

The given sum is incorrect, the correct sum is 530.

113. The given difference is correct.

117. answers may vary

Exercise Set 1.4

1. To round 423 to the nearest ten, observe that the digit in the ones place is 3. Since this digit is less than 5, we do not add 1 to the digit in the tens place. The number 423 rounded to the nearest ten is 420.

5. To round 2791 to the nearest hundred, observe that the digit in the tens place is 9. Since this digit is at least 5, we add 1 to the digit in the hundreds place. The number 2791 rounded to the nearest hundred is 2800.

9. To round 21,094 to the nearest thousand, observe that the digit in the hundreds place is 0. Since this digit is less than 5, we do not add 1 to the digit in the thousands place. The number 21,094 rounded to the nearest thousand is 21,000.

13. To round 328,495 to the nearest hundred, observe that the digit in the tens place is 9. Since this digit is at least 5, we add 1 to the digit in the hundreds place. The number 328,495 rounded to the nearest hundred is 328,500.

17. To round 39,994 to the nearest ten, observe that the digit in the ones place is 4. Since this digit is less than 5, we do not add 1 to the digit in the tens place. The number 39,994 rounded to the nearest ten is 39,990.

21. Estimate 5281 to a given place value by rounding it to that place value. 5281 rounded to the tens place is 5280, to the hundreds place is 5300, and to the thousands place is 5000.

25. Estimate 14,876 to a given place value by rounding it to that place value. 14,876 rounded to the tens place is 14,880, to the hundreds place is 14,900, and to the thousands place is 15,000.

29. To round 60,149 to the nearest hundred, observe that the digit in the tens place is 4. Since this digit is less than 5, we do not add 1 to the digit in the hundreds place. Therefore, 60,149 days rounded to the nearest thousand is 60,100 days.

33. To round 3,213,479 to the nearest hundred-thousand, observe that the digit in the ten-thousands place is 1. Since this digit is less than 5, we do not add 1 to the digit in the hundred-thousands place. Therefore, $3,213,470 rounded to the nearest hundred-thousand is $3,200,000.

37.

$$\begin{array}{rll} 39 & \text{rounds to} & 40 \\ 45 & \text{rounds to} & 50 \\ 22 & \text{rounds to} & 20 \\ +\ 17 & \text{rounds to} & +\ 20 \\ \hline & & 130 \end{array}$$

41.

$$\begin{array}{rll} 1913 & \text{rounds to} & 1900 \\ 1886 & \text{rounds to} & 1900 \\ +\ 1925 & \text{rounds to} & +\ 1900 \\ \hline & & 5700 \end{array}$$

45.

$$\begin{array}{rll} 3995 & \text{rounds to} & 4000 \\ 2549 & \text{rounds to} & 2500 \\ +\ 4944 & \text{rounds to} & +\ 4900 \\ \hline & & 11{,}400 \end{array}$$

49. $229 + 443 + 606$ is approximately $230 + 440 + 610 = 1280$.

The answer of 1278 is correct.

53.

$$\begin{array}{rll} 899 & \text{rounds to} & 900 \\ 1499 & \text{rounds to} & 1500 \\ +\ 999 & \text{rounds to} & +\ 1000 \\ \hline & & 3400 \end{array}$$

The total cost is approximately $3400.

57.

$$\begin{array}{rll} 20{,}320 & \text{rounds to} & 20{,}000 \\ -\ 14{,}410 & \text{rounds to} & -\ 14{,}000 \\ \hline & & 6000 \end{array}$$

The difference in elevation is approximately 6000 feet.

61.

$$
\begin{array}{ll}
1{,}128{,}030 & \text{rounds to} \\
-\ 1{,}073{,}440 & \text{rounds to}
\end{array}
\quad
\begin{array}{r}
1{,}128{,}000 \\
-\ 1{,}073{,}000 \\
\hline
55{,}000
\end{array}
$$

The increase in enrollment was approximately 55,000 children.

65. 578 hundred-thousand is 57,800,000 in standard form
57,800,000 rounded to the nearest million is 58,000,000.
57,800,000 rounded to the nearest ten-million is
60,000,000.

69. a. The smallest possible number that rounds to 8600 is 8550.

   b. The largest possible number that rounds to 8600 is 8649.

73. 54 rounds to 50

17 rounds to 20

$50 + 20 + 50 + 20 = 140$

The perimeter is approximately 140 meters.

Exercise Set 1.5

1. $1 \cdot 24 = 24$

5. $8 \cdot 0 \cdot 9 = 0$

9. $6(3 + 8) = 6 \cdot 3 + 6 \cdot 8$

13. $20(14 + 6) = 20 \cdot 14 + 20 \cdot 6$

17.
$$
\begin{array}{r}
613 \\
\times\ \ \ 6 \\
\hline
3678
\end{array}
$$

21.
$$
\begin{array}{r}
1074 \\
\times\ \ \ 6 \\
\hline
6444
\end{array}
$$

25.
$$
\begin{array}{r}
421 \\
\times\ \ 58 \\
\hline
3\ 368 \\
21\ 050 \\
\hline
24{,}418
\end{array}
$$

29.
$$
\begin{array}{r}
780 \\
\times\ \ 20 \\
\hline
15{,}600
\end{array}
$$

33. $(640)(1)(10) = (640)(10) = 6400$

37.
$$
\begin{array}{r}
609 \\
\times\ \ 234 \\
\hline
2\ 436 \\
18\ 270 \\
121\ 800 \\
\hline
142{,}506
\end{array}
$$

41.
$$
\begin{array}{r}
589 \\
\times\ \ 110 \\
\hline
5\ 890 \\
58\ 900 \\
\hline
64{,}790
\end{array}
$$

45. Area $=$ (length)(width)

$\phantom{\text{Area}} = (9\ \text{meters})(7\ \text{meters})$

$\phantom{\text{Area}} = 63$ square meters

Perimeter $= (9 + 7 + 9 + 7)$ meters

$\phantom{\text{Perimeter}} = 32$ meters

49.
$$
\begin{array}{ll}
576 & \text{rounds to} \\
\times\ 354 & \text{rounds to}
\end{array}
\quad
\begin{array}{r}
600 \\
\times\ \ \ 400 \\
\hline
240{,}000
\end{array}
$$

53. 38×42 is approximately 40×40, which is 1600. The best estimate is c.

57.
$$
\begin{aligned}
80 \times 11 &= (8 \times 10) \times 11 \\
&= 8 \times (10 \times 11) \\
&= 8 \times 110 \\
&= 880
\end{aligned}
$$

61.
$$
\begin{array}{r}
2240 \\
\times\ \ \ 2 \\
\hline
4480
\end{array}
$$

65.
$$
\begin{array}{r}
94 \\
\times\ 35 \\
\hline
470 \\
2820 \\
\hline
3290
\end{array}
$$

The total cost is $3290.

69. Area $=$ (length)(width)

$\phantom{\text{Area}} = (110\ \text{feet})(80\ \text{feet})$

$\phantom{\text{Area}} = 8800$ square feet

The area is 8800 square feet.

73.
$$
\begin{array}{r}
94 \\
\times\ 62 \\
\hline
188 \\
5640 \\
\hline
5828
\end{array}
$$

There are 5828 pixels on the screen.

77.
$$
\begin{array}{r}
160 \\
\times\ \ 8 \\
\hline
1280
\end{array}
$$

There are 1280 calories in 8 ounces.

81. There are 60 minutes in one hour, so there are
24×60 minutes in one day.

$$
\begin{aligned}
24 \times 60 \times 1000 &= 24 \times 6 \times 10 \times 1000 \\
&= 144 \times 10{,}000 \\
&= 1{,}440{,}000
\end{aligned}
$$

They produce 1,440,000 tea bags in one day.

85.
$$
\begin{array}{r}
134 \\
\times\ 16 \\
\hline
804 \\
1340 \\
\hline
2144
\end{array}
$$

89.
$$
\begin{array}{r}
19 \\
-\ \ 4 \\
\hline
15
\end{array}
$$

The difference of 19 and 4 is 15.

93. a. $3 \cdot 5 = 5 + 5 + 5 = 3 + 3 + 3 + 3 + 3$

   b. answers may vary

97. $42 \times 3 = 126$

$42 \times 9 = 378$

The problem is
$$
\begin{array}{r}
42 \\
\times\ 93
\end{array}
$$

101. On a side with 7 windows per row, there are
$7 \times 23 = 161$ windows. On a side with 4 windows per row, there are $4 \times 23 = 92$ windows.

$161 + 161 + 92 + 92 = 506$

There are 506 windows on the building.

Exercise Set 1.6

1. $54 \div 9 = 6$

5. $0 \div 8 = 0$

9. $\dfrac{18}{18} = 1$

13. $26 \div 0$ is undefined.

17. $0 \div 14 = 0$

21.
$$\begin{array}{r} 29 \\ 3\overline{)87} \\ -6 \\ \hline 27 \\ -27 \\ \hline 0 \end{array}$$
Check: $3 \cdot 29 = 87$

25.
$$\begin{array}{r} 338 \\ 3\overline{)1014} \\ -9 \\ \hline 11 \\ -9 \\ \hline 24 \\ -24 \\ \hline 0 \end{array}$$
Check: $3 \cdot 338 = 1014$

29.
$$\begin{array}{r} 9 \\ 7\overline{)63} \\ -63 \\ \hline 0 \end{array}$$
Check: $7 \cdot 9 = 63$

33.
$$\begin{array}{r} 68 \text{ R } 3 \\ 7\overline{)479} \\ -42 \\ \hline 59 \\ -56 \\ \hline 3 \end{array}$$
Check: $7 \cdot 68 + 3 = 479$

37.
$$\begin{array}{r} 38 \text{ R } 1 \\ 8\overline{)305} \\ -24 \\ \hline 65 \\ -64 \\ \hline 1 \end{array}$$
Check: $8 \cdot 38 + 1 = 305$

41.
$$\begin{array}{r} 13 \\ 55\overline{)715} \\ -55 \\ \hline 165 \\ -165 \\ \hline 0 \end{array}$$
Check: $55 \cdot 13 = 715$

45.
$$\begin{array}{r} 97 \text{ R } 8 \\ 97\overline{)9417} \\ -873 \\ \hline 687 \\ -679 \\ \hline 8 \end{array}$$
Check: $97 \cdot 97 + 8 = 9417$

49.
$$\begin{array}{r} 506 \\ 13\overline{)6578} \\ -65 \\ \hline 07 \\ -0 \\ \hline 78 \\ -78 \\ \hline 0 \end{array}$$
Check: $13 \cdot 506 = 6578$

53.
$$\begin{array}{r} 54 \\ 236\overline{)12744} \\ -1180 \\ \hline 944 \\ -944 \\ \hline 0 \end{array}$$
Check: $236 \cdot 54 = 12{,}744$

57.
$$\begin{array}{r} 202 \text{ R } 15 \\ 102\overline{)20619} \\ -204 \\ \hline 21 \\ -0 \\ \hline 219 \\ -204 \\ \hline 15 \end{array}$$
Check: $102 \cdot 202 + 15 = 20{,}619$

61.
$$\begin{array}{r} 17 \\ 7\overline{)119} \\ -7 \\ \hline 49 \\ -49 \\ \hline 0 \end{array}$$

65.
$$\begin{array}{r} 2132 \text{ R } 32 \\ 40\overline{)85312} \\ -80 \\ \hline 53 \\ -40 \\ \hline 131 \\ -120 \\ \hline 112 \\ -80 \\ \hline 32 \end{array}$$

69.
$$\begin{array}{r} 23 \text{ R } 2 \\ 5\overline{)117} \\ -10 \\ \hline 17 \\ -15 \\ \hline 2 \end{array}$$
The quotient is 23 R 2.

73.
$$\begin{array}{r} 20 \text{ R } 2 \\ 3\overline{)62} \\ -6 \\ \hline 02 \\ -0 \\ \hline 2 \end{array}$$
The quotient is 20 R 2.

77.

$$
\begin{array}{r}
165 \\
318\overline{)52470} \\
-318 \\
\hline
2067 \\
-1908 \\
\hline
1590 \\
-1590 \\
\hline
0
\end{array}
$$

The person weighs 165 pounds on Earth.

81.

$$
\begin{array}{r}
88\ \text{R}\ 1 \\
3\overline{)265} \\
-24 \\
\hline
25 \\
-24 \\
\hline
1
\end{array}
$$

There are 88 bridges every 3 miles over the 265 miles, plus the first bridge for a total of 89 bridges.

85.

$$
\begin{array}{r}
5 \\
5280\overline{)26400} \\
-26400 \\
\hline
0
\end{array}
$$

Broad Peak is 5 miles tall.

89.

$$
\begin{array}{r}
\overset{2}{10} \\
24 \\
35 \\
22 \\
17 \\
+12 \\
\hline
120
\end{array}
\qquad
\begin{array}{r}
20 \\
6\overline{)120} \\
-12 \\
\hline
0
\end{array}
$$

$$\text{Average} = \frac{120}{6} = 20$$

93.

$$
\begin{array}{r}
\overset{2}{86} \\
79 \\
81 \\
69 \\
+80 \\
\hline
395
\end{array}
\qquad
\begin{array}{r}
79 \\
5\overline{)395} \\
-35 \\
\hline
45 \\
-45 \\
\hline
0
\end{array}
$$

$$\text{Average} = \frac{395}{5} = 79$$

97.

$$
\begin{array}{r}
\overset{1\ 1\ 1}{82} \\
463 \\
29 \\
+8704 \\
\hline
9278
\end{array}
$$

101.

$$
\begin{array}{r}
722 \\
-43 \\
\hline
679
\end{array}
$$

105.

$$
\begin{array}{r}
9\ \text{R}\ 12 \\
24\overline{)228} \\
-216 \\
\hline
12
\end{array}
$$

109. 200 divided by 20 is $200 \div 20$, which is choice b.

113. The average will increase; answers may vary.

117. Since Area $=$ length $\cdot$ width, length $= \dfrac{\text{Area}}{\text{width}}$.

$$\text{length} = \frac{60\ \text{square feet}}{5\ \text{feet}} = \frac{60}{5}\ \text{feet} = 12\ \text{feet}$$

The length is 12 feet.

121.

$$
\begin{array}{r}
26 \\
-5 \\
\hline
21 \\
-5 \\
\hline
16 \\
-5 \\
\hline
11 \\
-5 \\
\hline
6 \\
-5 \\
\hline
1
\end{array}
$$

Thus $26 \div 5 = 5\ \text{R}\ 1$.

Exercise Set 1.7

1. $4 \cdot 4 \cdot 4 = 4^3$

5. $12 \cdot 12 \cdot 12 = 12^3$

9. $9 \cdot 8 \cdot 8 = 9 \cdot 8^2$

13. $3 \cdot 2 \cdot 2 \cdot 2 \cdot 2 \cdot 5 \cdot 5 \cdot 5 \cdot 5 \cdot 5 = 3 \cdot 2^4 \cdot 5^5$

17. $5^3 = 5 \cdot 5 \cdot 5 = 125$

21. $1^{10} = 1 \cdot 1 \cdot 1 \cdot 1 \cdot 1 \cdot 1 \cdot 1 \cdot 1 \cdot 1 \cdot 1 = 1$

25. $2^7 = 2 \cdot 2 \cdot 2 \cdot 2 \cdot 2 \cdot 2 \cdot 2 = 128$

29. $4^4 = 4 \cdot 4 \cdot 4 \cdot 4 = 256$

33. $12^2 = 12 \cdot 12 = 144$

37. $20^1 = 20$

41. $3 \cdot 2^6 = 3 \cdot 2 \cdot 2 \cdot 2 \cdot 2 \cdot 2 \cdot 2 = 192$

45. $15 + 3 \cdot 2 = 15 + 6 = 21$

49. $32 \div 4 - 3 = 8 - 3 = 5$

53. $6 \cdot 5 + 8 \cdot 2 = 30 + 16 = 46$

57.
$$
\begin{aligned}
(7 + 5^2) \div 4 \cdot 2^3 &= (7 + 25) \div 4 \cdot 2^3 \\
&= 32 \div 4 \cdot 2^3 \\
&= 32 \div 4 \cdot 8 \\
&= 8 \cdot 8 \\
&= 64
\end{aligned}
$$

61. $\dfrac{18 + 6}{2^4 - 2^2} = \dfrac{24}{16 - 4} = \dfrac{24}{12} = 2$

65. $\dfrac{7(9 - 6) + 3}{3^2 - 3} = \dfrac{7(3) + 3}{9 - 3} = \dfrac{21 + 3}{6} = \dfrac{24}{6} = 4$

69.
$$
\begin{aligned}
2^4 \cdot 4 - (25 \div 5) &= 2^4 \cdot 4 - 5 \\
&= 16 \cdot 4 - 5 \\
&= 64 - 5 \\
&= 59
\end{aligned}
$$

73.
$$
\begin{aligned}
(7 \cdot 5) + [9 \div (3 \div 3)] &= (7 \cdot 5) + [9 \div (1)] \\
&= 35 + 9 \\
&= 44
\end{aligned}
$$

77. $\dfrac{9^2 + 2^2 - 1^2}{8 \div 2 \cdot 3 \cdot 1 \div 3} = \dfrac{81 + 4 - 1}{4 \cdot 3 \cdot 1 \div 3}$

$\qquad\qquad = \dfrac{85 - 1}{12 \cdot 1 \div 3}$

$\qquad\qquad = \dfrac{84}{12 \div 3}$

$\qquad\qquad = \dfrac{84}{4}$

$\qquad\qquad = 21$

81. $9 \div 3 + 5^2 \cdot 2 - 10 = 9 \div 3 + 25 \cdot 2 - 10$

$\qquad\qquad\qquad\qquad = 3 + 50 - 10$

$\qquad\qquad\qquad\qquad = 53 - 10$

$\qquad\qquad\qquad\qquad = 43$

85. $7^2 - \{18 - [40 \div (5 \cdot 1) + 2] + 5^2\}$

$\quad = 7^2 - \{18 - [40 \div (5) + 2] + 5^2\}$

$\quad = 7^2 - \{18 - [8 + 2] + 5^2\}$

$\quad = 7^2 - \{18 - [10] + 5^2\}$

$\quad = 7^2 - \{18 - 10 + 25\}$

$\quad = 7^2 - \{8 + 25\}$

$\quad = 7^2 - \{33\}$

$\quad = 49 - 33$

$\quad = 16$

89. Area of a square $= (\text{side})^2$

$\qquad\qquad\qquad\quad = (23 \text{ miles})^2$

$\qquad\qquad\qquad\quad = 529 \text{ square miles}$

Perimeter $= (23 + 23 + 23 + 23) \text{ meters}$

$\qquad\quad = 92 \text{ meters}$

93. $2^5 = 2 \cdot 2 \cdot 2 \cdot 2 \cdot 2$

The statement is false.

97. $24 \div (3 \cdot 2) + 2 \cdot 5 = 24 \div 6 + 2 \cdot 5 = 4 + 10 = 14$

101. $(7 + 2^4)^5 - (3^5 - 2^4)^2 = (7 + 16)^5 - (243 - 16)^2$

$\qquad\qquad\qquad\qquad\qquad = 23^5 - 227^2$

$\qquad\qquad\qquad\qquad\qquad = 6{,}436{,}343 - 51{,}529$

$\qquad\qquad\qquad\qquad\qquad = 6{,}384{,}814$

Exercise Set 1.8

1.

a	b	$a + b$	$a - b$	$a \cdot b$	$a \div b$
21	7	$21 + 7 = 28$	$21 - 7 = 14$	$21 \cdot 7 = 147$	$21 \div 7 = 3$

5.

a	b	$a + b$	$a - b$	$a \cdot b$	$a \div b$
56	1	$56 + 1 = 57$	$56 - 1 = 55$	$56 \cdot 1 = 56$	$56 \div 1 = 56$

9. $3xz - 5x = 3(2)(3) - 5(2) = 18 - 10 = 8$

13. $4x - z = 4(2) - 3 = 8 - 3 = 5$

17. $2xy^2 - 6 = 2(2)(5)^2 - 6$

$\qquad\qquad = 2 \cdot 2 \cdot 25 - 6$

$\qquad\qquad = 100 - 6$

$\qquad\qquad = 94$

21. $x^5 + (y - z) = 2^5 + (5 - 3)$

$\qquad\qquad\qquad = 2^5 + 2$

$\qquad\qquad\qquad = 32 + 2$

$\qquad\qquad\qquad = 34$

25. $\dfrac{2y - 2}{x} = \dfrac{2(5) - 2}{2} = \dfrac{10 - 2}{2} = \dfrac{8}{2} = 4$

29. $\dfrac{5x}{y} - \dfrac{10}{y} = \dfrac{5(2)}{5} - \dfrac{10}{5} = \dfrac{10}{5} - 2 = 2 - 2 = 0$

33. $(4y - 5z)^3 = (4 \cdot 5 - 5 \cdot 3)^3$

$\qquad\qquad\quad = (20 - 15)^3$

$\qquad\qquad\quad = (5)^3$

$\qquad\qquad\quad = 125$

37. $2y(4z - x) = 2 \cdot 5(4 \cdot 3 - 2)$

$\qquad\qquad\quad = 2 \cdot 5(12 - 2)$

$\qquad\qquad\quad = 2 \cdot 5(10)$

$\qquad\qquad\quad = 10(10)$

$\qquad\qquad\quad = 100$

41. $\dfrac{7x + 2y}{3x} = \dfrac{7(2) + 2(5)}{3(2)} = \dfrac{14 + 10}{6} = \dfrac{24}{6} = 4$

45. Let n be 10.

$\quad n - 8 = 2$

$\quad 10 - 8 \overset{?}{=} 2$

$\qquad\quad 2 = 2$ True

Yes, 10 is a solution.

49. Let n be 7.

$\quad 3n - 5 = 10$

$\quad 3(7) - 5 \overset{?}{=} 10$

$\quad 21 - 5 \overset{?}{=} 10$

$\qquad\quad 16 = 10$ False

No, 7 is not a solution.

53. Let x be 0.

$\quad 5x + 3 = 4x + 13$

$\quad 5(0) + 3 \overset{?}{=} 4(0) + 13$

$\quad 0 + 3 \overset{?}{=} 0 + 13$

$\qquad\quad 3 = 13$ False

No, 0 is not a solution.

57. $n - 2 = 10$

Let n be 10.

$\quad 10 - 2 \overset{?}{=} 10$

$\qquad\quad 8 = 10$ False

Let n be 12.

$\quad 12 - 2 \overset{?}{=} 10$

$\qquad 10 = 10$ True

Let n be 14.

$\quad 14 - 2 \overset{?}{=} 10$

$\qquad 12 = 10$ False

12 is a solution.

61. $6n + 2 = 26$

Let n be 0.

$\quad 6(0) + 2 \overset{?}{=} 26$

$\qquad 0 + 2 \overset{?}{=} 26$

$\qquad\qquad 2 = 26$ False

Let n be 2.

$\quad 6(2) + 2 \overset{?}{=} 26$

$\quad 12 + 2 \overset{?}{=} 26$

$\qquad\quad 14 = 26$ False

Let n be 4.

$\quad 6(4) + 2 \overset{?}{=} 26$

$\quad 24 + 2 \overset{?}{=} 26$

$\qquad\quad 26 = 26$ True

4 is a solution.

65. $7x - 9 = 5x + 13$

Let x be 3.

$\quad 7(3) - 9 \overset{?}{=} 5(3) + 13$

$\quad 21 - 9 \overset{?}{=} 15 + 13$

$\qquad\quad 12 = 28$ False

Let x be 7.
$$7(7) - 9 \overset{?}{=} 5(7) + 13$$
$$49 - 9 \overset{?}{=} 35 + 13$$
$$40 = 48 \quad \text{False}$$

Let x be 11.
$$7(11) - 9 \overset{?}{=} 5(11) + 13$$
$$77 - 9 \overset{?}{=} 55 + 13$$
$$68 = 68 \quad \text{True}$$

11 is a solution.

69. The total of a number and 8 is $x + 8$.

73. The product of 512 and a number is $512x$.

77. The sum of seventeen and a number added to the product of five and the number is $5x + (17 + x)$.

81. A number subtracted from 11 is $11 - x$.

85. 6 divided by a number is $6 \div x$ or $\dfrac{6}{x}$.

89. $x^4 - y^2 = 23^4 - 72^2$
$$= 279{,}841 - 5184$$
$$= 274{,}657$$

93. $5x$ is the largest; answers may vary.

Chapter 1 Test

1. 82,426 in words is eighty-two thousand, four hundred twenty-six.

5.
$$\begin{array}{r} 496 \\ \times\ 30 \\ \hline 14{,}880 \end{array}$$

9. $0 \div 49 = 0$

13. $6^1 \cdot 2^3 = 6 \cdot 2 \cdot 2 \cdot 2 = 48$

17. To round 52,369 to the nearest thousand, observe that the digit in the hundreds place is 3. Since this digit is less than 5, we do not add 1 to the digit in the thousands place. The number 52,369 rounded to the nearest thousand is 52,000.

21.
$$\begin{array}{r} \overset{1}{15} \\ +\ 107 \\ \hline 122 \end{array}$$

25.
$$\begin{array}{r} 725 \\ -599 \\ \hline 126 \end{array}$$
The higher-priced one is $126 more.

29. Perimeter $= (20 + 10 + 20 + 10)$ yards $= 60$ yards
$$\begin{aligned} \text{Area} &= (\text{length})(\text{width}) \\ &= (20 \text{ yards})(10 \text{ yards}) \\ &= 200 \text{ square yards} \end{aligned}$$

33. Let n be 6.
$$5n - 11 = 19$$
$$5(6) - 11 \overset{?}{=} 19$$
$$30 - 11 \overset{?}{=} 19$$
$$19 = 19 \quad \text{True}$$
6 is a solution.

Chapter 2

Exercise Set 2.1

1. If 0 represents ground level, then 1235 feet underground is -1235.

5. If 0 represents zero degrees Fahrenheit, then 120 degrees above zero is $+120$.

9. If 0 represents a loss of $0, then a loss of $3140 million is -3140 million.

13. If 0 represents a decrease of 0%, then a 2% decrease is -2.

17. ⟵—+—◆—+—◆—+—◆—+—◆—+—◆—⟶
    $-5\ -4\ -3\ -2\ -1\ \ 0\ \ 1\ \ 2\ \ 3\ \ 4$

21. ⟵—+—◆—+—◆—+—+—+—+—◆—⟶
    $-8\ -7\ -6\ -5\ -4\ -3\ -2\ -1\ \ 0\ \ 1$

25. $-7 < -5$ since -7 is to the left of -5 on a number line.

29. $-26 < 26$ since -26 is to the left of 26 on a number line.

33. $|-8| = 8$ since -8 is 8 units from 0 on a number line.

37. $|-55| = 55$ since -55 is 55 units from 0 on a number line.

41. The opposite of negative 4 is 4. $-(-4) = 4$

45. The opposite of negative 85 is 85. $-(-85) = 85$

49. $-|20| = -20$

53. $-(-43) = 43$

57. $-(-33) = 33$

61. $-|-x| = -|-2| = -2$

65. $-|x| = -|7| = -7$

69. $|-8| = 8$
$|-11| = 11$
Since $8 < 11$, $|-8| < |-11|$.

73. $-|-12| = -12$
$-(-12) = 12$
Since $-12 < 12$, $-|-12| < -(-12)$.

77. $|0| = 0$
$|-9| = 9$
Since $0 < 9$, $|0| < |-9|$.

81. $-(-12) = 12$
$-(-18) = 18$
Since $12 < 18$, $-(-12) < -(-18)$.

85. If the opposite of a number is -28, then the number is 28, and its absolute value is 28.

89. The tallest bar on the graph corresponds to Lake Superior, so Lake Superior has the highest elevation.

93. The number on the graph closest to $-200°C$ is $-186°C$, which corresponds to the element oxygen.

97.
$$\begin{array}{r} 15 \\ +20 \\ \hline 35 \end{array}$$

101. $2^2 = 4$, $-|3| = -3$, $-(-5) = 5$, and $-|-8| = -8$, so the numbers in order from least to greatest are $-|-8|$, $-|3|$, 2^2, $-(-5)$.

105. $-(-2) = 2$, $5^2 = 25$, $-10 = -10$, $-|-9| = -9$, and $|-12| = 12$, so the numbers in order from least to greatest are -10, $-|-9|$, $-(-2)$, $|-12|$, 5^2.

109. $-(-|-8|) = -(-8) = 8$

113. True; a positive number will always be to the right of a negative number on a number line.

117. answers may vary

Exercise Set 2.2

1.

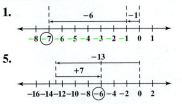

5.

9. $|-8| + |-2| = 8 + 2 = 10$

The common sign is negative, so $-8 + (-2) = -10$.

13. $|6| - |-2| = 6 - 2 = 4$

$6 > 2$, so the answer is positive.

$6 + (-2) = 4$

17. $|-5| - |3| = 5 - 3 = 2$

$5 > 3$, so the answer is negative.

$3 + (-5) = -2$

21. $|-12| + |-12| = 12 + 12 = 24$

The common sign is negative, so $-12 + (-12) = -24$.

25. $|12| - |-5| = 12 - 5 = 7$

$12 > 5$, so the answer is positive.

$12 + (-5) = 7$

29. $|-56| - |26| = 56 - 26 = 30$

$56 > 26$, so the answer is negative.

$-56 + 26 = -30$

33. $|-144| - |124| = 144 - 124 = 20$

$144 > 124$, so the answer is negative.

$124 + (-144) = -20$

37. $-4 + 2 + (-5) = -2 + (-5) = -7$

41. $12 + (-4) + (-4) + 12 = 8 + (-4) + 12$

$$= 4 + 12$$

$$= 16$$

45. $-6 + (-15) + (-7) = -21 + (-7) = -28$

49. $5 + (-2) + 17 = 3 + 17 = 20$

53. $3 + 14 + (-18) = 17 + (-18) = -1$

57. $-13 + 8 + (-10) + (-27) = -5 + (-10) + (-27)$

$$= -15 + (-27)$$

$$= -42$$

61. $3x + y = 3(2) + (-3) = 6 + (-3) = 3$

65. The sum of -6 and 25 is $-6 + 25 = 19$.

69. $0 + (-215) + (-16) = -215 + (-16) = -231$

The diver's final depth is 231 feet below the surface.

73. The bar for 2012 has a height of 41,733, so the net income in 2012 was $41,733,000,000.

77. $-10 + 12 = 2$

The temperature at 11 p.m. was 2°C.

81. $-23 + 21 = -2$

Florida's all-time record low temperature is $-2°F$.

85. $44 - 0 = 44$

89. answers may vary

93. $-10 + (-12) = -22$

97. False; for example, $4 + (-2) = 2 > 0$.

Exercise Set 2.3

1. $-8 - (-8) = -8 + 8 = 0$

5. $3 - 8 = 3 + (-8) = -5$

9. $-4 - (-7) = -4 + 7 = 3$

13. $3 - 15 = 3 + (-15) = -12$

17. $478 - (-30) = 478 + 30 = 508$

21. $-7 - (-3) = -7 + 3 = -4$

25. $-25 - 17 = -25 + (-17) = -42$

29. $2 - (-12) = 2 + 12 = 14$

33. $8 - 13 = 8 + (-13) = -5$

37. $30 - 67 = 30 + (-67) = -37$

41. $13 - 5 - 7 = 13 + (-5) + (-7) = 8 + (-7) = 1$

45. $-11 + (-6) - 14 = -11 + (-6) + (-14)$

$$= -17 + (-14)$$

$$= -31$$

49. $-(-5) - 21 + (-16) = 5 + (-21) + (-16)$

$$= -16 + (-16)$$

$$= -32$$

53. $-3 + 4 - (-23) - 10 = -3 + 4 + 23 + (-10)$

$$= 1 + 23 + (-10)$$

$$= 24 + (-10)$$

$$= 14$$

57. $x - y = 8 - (-23) = 8 + 23 = 31$

61. $2x - y = 2(1) - (-18) = 2 + 18 = 20$

65. The two months with the lowest temperatures are January, $-8°F$, and December, $-4°F$.

$-4 - (-8) = -4 + 8 = 4$

The difference is 4°F.

69. $5 - (-9) = 5 + 9 = 14$

The difference in scores was 14 strokes

73. $-282 - (-436) = -282 + 436 = 154$

The difference in elevation is 154 feet.

77. $600 - (-52) = 600 + 52 = 652$

The difference in elevation is 652 feet.

81. $867 - (-330) = 867 + 330 = 1197$

The difference in temperatures is 1197°F.

85. The sum of -5 and a number is $-5 + x$.

89. $\dfrac{100}{20} = 5$

93. answers may vary

97. $10 - 30 = 10 + (-30) = -20$

101. $|-5| - |5| = 5 - 5 = 0$

105. $|-8 - 3| = |-8 + (-3)| = |-11| = 11$

$8 - 3 = 8 + (-3) = 5$

Since $11 \neq 5$, the statement is false.

Exercise Set 2.4

1. $-6(-2) = 12$

5. $9(-9) = -81$

9. $6(-2)(-4) = -12(-4) = 48$

13. $-4(4)(-5) = -16(-5) = 80$

17. $-5(3)(-1)(-1) = -15(-1)(-1) = 15(-1) = -15$

21. $(-3)^3 = (-3)(-3)(-3) = 9(-3) = -27$

25. $(-4)^3 = (-4)(-4)(-4) = 16(-4) = -64$

29. $\dfrac{-30}{6} = -5$

33. $\dfrac{0}{-21} = 0$

37. $\dfrac{56}{-4} = -14$

41. $-5(3) = -15$

45. $-7(-6) = 42$

49. $(-7)^2 = (-7)(-7) = 49$

53. $-\dfrac{72}{8} = -9$

57. $4(-10)(-3) = -40(-3) = 120$

61. $\dfrac{-25}{0}$ is undefined.

65. $280 \div (-40) = \dfrac{280}{-40} = -7$

69. $-1^4 = -(1 \cdot 1 \cdot 1 \cdot 1) = -1$

73. $-2(3)(5)(-6) = -6(5)(-6) = -30(-6) = 180$

77. $-2(-3)(-5) = 6(-5) = -30$

81.
$$
\begin{array}{r}
35 \\
\times\ 82 \\
\hline
70 \\
2800 \\
\hline
2870
\end{array}
$$
$35 \cdot (-82) = -2870$

85. $ab = 9(-2) = -18$

89. $\dfrac{x}{y} = \dfrac{5}{-5} = -1$

93. $\dfrac{x}{y} = \dfrac{-36}{-6} = 6$

97. $xy = 0(-8) = 0$

$\dfrac{x}{y} = \dfrac{0}{-8} = 0$

101.
$$
\begin{array}{r}
42 \\
\times\ 6 \\
\hline
252
\end{array}
$$
$-42(-6) = 252$

105. Subtract a number from -16 is $-16 - x$.

109. Divide a number by -33 is $\dfrac{x}{-33}$ or $x \div (-33)$.

113. Each move of 20 feet down is represented by -20.
$5 \cdot (-20) = -100$
The diver is at a depth of 100 feet.

117. $-3 \cdot 63 = -189$
The melting point of argon is $-189°C$.

121. a. $33{,}319 - 6387 = 26{,}932$

There were 26,932 fewer analog movie screens in 2012. This is a change of $-26{,}932$ movie screens.

b. This is a period of 4 years.
$\dfrac{-26{,}932}{4} = -6733$

The average change was -6733 movie screens per year.

125. $12 \div 4 - 2 + 7 = 3 - 2 + 7 = 1 + 7 = 8$

129. $-8 - 20 = -8 + (-20) = -28$

133. The product of an odd number of negative numbers is negative, so the product of seven negative numbers is negative.

137. answers may vary

Exercise Set 2.5

1. $(-5)^3 = (-5)(-5)(-5) = 25(-5) = -125$

5. $8 \cdot 2^2 = 8 \cdot 4 = 32$

9. $7 + 3(-6) = 7 + (-18) = -11$

13. $-10 + 4 \div 2 = -10 + 2 = -8$

17. $\dfrac{16 - 13}{-3} = \dfrac{3}{-3} = -1$

21. $5(-3) - (-12) = -15 - (-12) = -15 + 12 = -3$

25. $8 \cdot 6 - 3 \cdot 5 + (-20) = 48 - 3 \cdot 5 + (-20)$
$= 48 - 15 + (-20)$
$= 33 + (-20)$
$= 13$

29. $|7 + 3| \cdot 2^3 = |10| \cdot 2^3 = 10 \cdot 2^3 = 10 \cdot 8 = 80$

33. $7^2 - (4 - 2^3) = 7^2 - (4 - 8)$
$= 7^2 - (-4)$
$= 49 - (-4)$
$= 49 + 4$
$= 53$

37. $-(-2)^6 = -64$

41. $|8 - 24| \cdot (-2) \div (-2) = |-16| \cdot (-2) \div (-2)$
$= 16 \cdot (-2) \div (-2)$
$= -32 \div (-2)$
$= 16$

45. $5(5 - 2) + (-5)^2 - 6 = 5(3) + (-5)^2 - 6$
$= 5(3) + 25 - 6$
$= 15 + 25 - 6$
$= 40 - 6$
$= 34$

49. $(-36 \div 6) - (4 \div 4) = -6 - 1 = -7$

53. $2(8 - 10)^2 - 5(1 - 6)^2 = 2(-2)^2 - 5(-5)^2$
$= 2(4) - 5(25)$
$= 8 - 125$
$= -117$

57. $\dfrac{(-7)(-3) - (4)(3)}{3[7 \div (3 - 10)]} = \dfrac{21 - 12}{3[7 \div (-7)]}$
$= \dfrac{9}{3(-1)}$
$= \dfrac{9}{-3}$
$= -3$

61. $x + y + z = -2 + 4 + (-1) = 2 + (-1) = 1$

65. $x^2 - y = (-2)^2 - 4 = 4 - 4 = 0$

69. $x^2 = (-3)^2 = 9$

73. $2z^3 = 2(-4)^3 = 2(-64) = -128$

77. $2x^3 - z = 2(-3)^3 - (-4)$
$= 2(-27) - (-4)$
$= -54 - (-4)$
$= -54 + 4$
$= -50$

81. average $= \dfrac{-17 + (-26) + (-20) + (-13)}{4}$
$= \dfrac{-76}{4}$
$= -19$

85. average $= \dfrac{-5 + (-1) + 0 + 2}{4}$
$= \dfrac{-4}{4}$
$= -1$

The average of the scores is -1.

89.
$$
\begin{array}{r}
45 \\
\times\,90 \\
\hline
4050
\end{array}
$$

93. $8 + 8 + 8 + 8 = 32$

The perimeter is 32 inches.

97. $2 \cdot (7 - 5) \cdot 3 = 2 \cdot 2 \cdot 3 = 4 \cdot 3 = 12$

101. answers may vary

105. $(-12)^4 = (-12)(-12)(-12)(-12) = 20{,}736$

109. $(xy + z)^x = [2(-5) + 7]^2$
$$= [-10 + 7]^2$$
$$= [-3]^2$$
$$= 9$$

Exercise Set 2.6

1. $x - 8 = -2$
$6 - 8 \overset{?}{=} -2$
$\quad -2 = -2 \quad$ True

Since $-2 = -2$ is true, 6 is a solution of the equation.

5. $-9f = 64 - f$
$-9(-8) \overset{?}{=} 64 - (-8)$
$\quad 72 \overset{?}{=} 64 + 8$
$\quad 72 = 72 \quad$ True

Since $72 = 72$ is true, -8 is a solution of the equation.

9. $a + 5 = 23$
$a + 5 - 5 = 23 - 5$
$\quad a = 18$
Check: $\quad a + 5 = 23$
$\quad 18 + 5 \overset{?}{=} 23$
$\quad\quad 23 = 23 \quad$ True

The solution is 18.

13. $7 = y - 2$
$7 + 2 = y - 2 + 2$
$\quad 9 = y$
Check: $\quad 7 = y - 2$
$\quad 7 \overset{?}{=} 9 - 2$
$\quad\quad 7 = 7 \quad$ True

The solution is 9.

17. $5x = 20$
$\dfrac{5x}{5} = \dfrac{20}{5}$
$\dfrac{5}{5} \cdot x = \dfrac{20}{5}$
$\quad x = 4$
Check: $5x = 20$
$\quad 5(4) \overset{?}{=} 20$
$\quad\quad 20 = 20 \quad$ True

The solution is 4.

21. $\dfrac{n}{7} = -2$

$7 \cdot \dfrac{n}{7} = 7 \cdot (-2)$

$\dfrac{7}{7} \cdot n = 7 \cdot (-2)$

$\quad n = -14$

Check: $\dfrac{n}{7} = -2$

$\quad \dfrac{-14}{7} \overset{?}{=} -2$

$\quad\quad -2 = -2 \quad$ True

The solution is -14.

25. $-4y = 0$
$\dfrac{-4y}{-4} = \dfrac{0}{-4}$
$\dfrac{-4}{-4} \cdot y = \dfrac{0}{-4}$
$\quad y = 0$
Check: $\quad -4y = 0$
$\quad -4(0) \overset{?}{=} 0$
$\quad\quad\quad 0 = 0 \quad$ True

The solution is 0.

29. $5x = -35$
$\dfrac{5x}{5} = \dfrac{-35}{5}$
$\dfrac{5}{5} \cdot x = \dfrac{-35}{5}$
$\quad x = -7$

The solution is -7.

33. $-15 = y + 10$
$-15 - 10 = y + 10 - 10$
$\quad -25 = y$

The solution is -25.

37. $n = -10 + 31$
$n = 21$

The solution is 21.

41. $\dfrac{n}{4} = -20$

$4 \cdot \dfrac{n}{4} = 4 \cdot (-20)$

$\dfrac{4}{4} \cdot n = 4 \cdot (-20)$

$\quad n = -80$

The solution is -80.

45. A number decreased by -2 is $x - (-2)$.

49. The sum of -15 and a number is $-15 + x$.

53. $n - 42{,}860 = -1286$
$n - 42{,}860 + 42{,}860 = -1286 + 42{,}860$
$\quad n = 41{,}574$

The solution is 41,574.

57. answers may vary

Chapter 2 Test

1. $-5 + 8 = 3$

5. $-18 + (-12) = -30$

9. $|-25| + (-13) = 25 + (-13) = 12$

13. $-8 + 9 \div (-3) = -8 + (-3) = -11$

17. $-(-7)^2 \div 7 \cdot (-4) = -49 \div 7 \cdot (-4) = -7 \cdot (-4) = 28$

21. $\dfrac{|25 - 30|^2}{2(-6) + 7} = \dfrac{|-5|^2}{-12 + 7} = \dfrac{(5)^2}{-5} = \dfrac{25}{-5} = -5$

25. $\dfrac{3z}{2y} = \dfrac{3(2)}{2(-3)} = \dfrac{6}{-6} = -1$

29. Subtract the depth of the lake from the elevation of the surface.

$$1495 - 5315 = 1495 + (-5315) = -3820$$

The deepest point on the lake is 3820 feet below sea level.

33.
$$\frac{n}{-7} = 4$$
$$-7 \cdot \frac{n}{-7} = -7 \cdot 4$$
$$\frac{-7}{-7} \cdot n = -7 \cdot 4$$
$$n = -28$$

The solution is -28.

Chapter 3

Exercise Set 3.1

1. $3x + 5x = (3 + 5)x = 8x$

5. $4c + c - 7c = (4 + 1 - 7)c = -2c$

9. $3a + 2a + 7a - 5 = (3 + 2 + 7)a - 5 = 12a - 5$

13. $-3(11y) = (-3 \cdot 11)y = -33y$

17. $2(y + 3) = 2 \cdot y + 2 \cdot 3 = 2y + 6$

21. $-4(3x + 7) = -4 \cdot 3x + (-4) \cdot 7 = -12x - 28$

25. $8 + 5(3c - 1) = 8 + 5 \cdot 3c - 5 \cdot 1$
$$= 8 + 15c - 5$$
$$= 15c + 8 - 5$$
$$= 15c + 3$$

29. $3 + 6(w + 2) + w = 3 + 6 \cdot w + 6 \cdot 2 + w$
$$= 3 + 6w + 12 + w$$
$$= 6w + w + 3 + 12$$
$$= 7w + 15$$

33. $-(2y - 6) + 10 = -1(2y - 6) + 10$
$$= -1 \cdot 2y - (-1) \cdot 6 + 10$$
$$= -2y + 6 + 10$$
$$= -2y + 16$$

37. $z - 8z = (1 - 8)z = -7z$

41. $2y - 6 + 4y - 8 = 2y + 4y - 6 - 8 = 6y - 14$

45. $2(x + 1) + 20 = 2 \cdot x + 2 \cdot 1 + 20$
$$= 2x + 2 + 20$$
$$= 2x + 22$$

49. $-5(z + 3) + 2z = -5 \cdot z + (-5) \cdot 3 + 2z$
$$= -5z - 15 + 2z$$
$$= -5z + 2z - 15$$
$$= -3z - 15$$

53. $-7(x + 5) + 5(2x + 1)$
$$= -7 \cdot x + (-7) \cdot 5 + 5 \cdot 2x + 5 \cdot 1$$
$$= -7x - 35 + 10x + 5$$
$$= -7x + 10x - 35 + 5$$
$$= 3x - 30$$

57. $-3(n - 1) - 4n = -3 \cdot n - (-3) \cdot 1 - 4n$
$$= -3n + 3 - 4n$$
$$= -3n - 4n + 3$$
$$= -7n + 3$$

61. $6(2x - 1) - 12x = 6 \cdot 2x - 6 \cdot 1 - 12x$
$$= 12x - 6 - 12x$$
$$= 12x - 12x - 6$$
$$= -6$$

65. $-(4x - 10) + 2(3x + 5)$
$$= -1(4x - 10) + 2 \cdot 3x + 2 \cdot 5$$
$$= -1 \cdot 4x - (-1) \cdot 10 + 6x + 10$$
$$= -4x + 10 + 6x + 10$$
$$= -4x + 6x + 10 + 10$$
$$= 2x + 20$$

69. $5y - 2(y - 1) + 3 = 5y - 2 \cdot y - (-2) \cdot 1 + 3$
$$= 5y - 2y + 2 + 3$$
$$= 3y + 5$$

73. $2a + 2a + 6 + 5a + 6 + 2a$
$$= 2a + 2a + 5a + 2a + 6 + 6$$
$$= (2 + 2 + 5 + 2)a + 6 + 6$$
$$= 11a + 12$$

The perimeter is $(11a + 12)$ feet.

77. Area $= (\text{length}) \cdot (\text{width})$
$$= (4y) \cdot (9)$$
$$= (4 \cdot 9)y$$
$$= 36y$$

The area is $36y$ square inches.

81. Area $= (\text{length}) \cdot (\text{width})$
$$= (3y + 1) \cdot (20)$$
$$= 3y \cdot 20 + 1 \cdot 20$$
$$= (3 \cdot 20)y + 20$$
$$= 60y + 20$$

The area is $(60y + 20)$ square miles.

85. Perimeter $= 2 \cdot (\text{length}) + 2 \cdot (\text{width})$
$$= 2 \cdot (18) + 2 \cdot (14)$$
$$= 36 + 28$$
$$= 64$$

The perimeter is 64 feet.

89. $-13 + 10 = -3$

93. $-4 + 4 = 0$

97. $2(xy) = 2 \cdot x \cdot y$
$$= 2xy$$

The expressions are not equivalent.

101. $4(y - 3) + 11 = 4 \cdot y - 4 \cdot 3 + 11$
$$= 4y - 12 + 11$$
$$= 4y - 1$$

The expressions are not equivalent.

105. The order of the terms is not changed, but the grouping is. This is the associative property of addition.

109. $9684q - 686 - 4860q + 12{,}960$
$$= 9684q - 4860q - 686 + 12{,}960$$
$$= 4824q + 12{,}274$$

113. answers may vary

Exercise Set 3.2

1.
$$x - 3 = -1 + 4$$
$$x - 3 = 3$$
$$x - 3 + 3 = 3 + 3$$
$$x = 6$$

5. $2w - 12w = 40$

$\quad\quad -10w = 40$

$\quad\quad \dfrac{-10w}{-10} = \dfrac{40}{-10}$

$\quad\quad\quad\quad w = -4$

9. $2z = 12 - 14$

$\quad 2z = -2$

$\quad \dfrac{2z}{2} = \dfrac{-2}{2}$

$\quad\quad z = -1$

13. $-3x - 3x = 50 - 2$

$\quad\quad\quad -6x = 48$

$\quad\quad\quad \dfrac{-6x}{-6} = \dfrac{48}{-6}$

$\quad\quad\quad\quad x = -8$

17. $7x + 7 - 6x = 10$

$\quad 7x - 6x + 7 = 10$

$\quad\quad\quad x + 7 = 10$

$\quad x + 7 - 7 = 10 - 7$

$\quad\quad\quad\quad x = 3$

21. $\quad 2(5x - 3) = 11x$

$\quad 2 \cdot 5x - 2 \cdot 3 = 11x$

$\quad\quad 10x - 6 = 11x$

$\quad 10x - 10x - 6 = 11x - 10x$

$\quad\quad\quad\quad -6 = x$

25. $\quad\quad 21y = 5(4y - 6)$

$\quad\quad 21y = 5 \cdot 4y - 5 \cdot 6$

$\quad\quad 21y = 20y - 30$

$\quad 21y - 20y = 20y - 20y - 30$

$\quad\quad\quad\quad y = -30$

29. $\quad\quad 2x - 8 = 0$

$\quad 2x - 8 + 8 = 0 + 8$

$\quad\quad\quad 2x = 8$

$\quad\quad\quad \dfrac{2x}{2} = \dfrac{8}{2}$

$\quad\quad\quad\quad x = 4$

33. $\quad\quad -7 = 2x - 1$

$\quad -7 + 1 = 2x - 1 + 1$

$\quad\quad\quad -6 = 2x$

$\quad\quad\quad \dfrac{-6}{2} = \dfrac{2x}{2}$

$\quad\quad\quad -3 = x$

37. $\quad\quad 11(x - 6) = -4 - 7$

$\quad 11 \cdot x - 11 \cdot 6 = -11$

$\quad\quad\quad 11x - 66 = -11$

$\quad 11x - 66 + 66 = -11 + 66$

$\quad\quad\quad\quad 11x = 55$

$\quad\quad\quad\quad \dfrac{11x}{11} = \dfrac{55}{11}$

$\quad\quad\quad\quad\quad x = 5$

41. $\quad\quad y - 20 = 6y$

$\quad y - y - 20 = 6y - y$

$\quad\quad\quad -20 = 5y$

$\quad\quad\quad \dfrac{-20}{5} = \dfrac{5y}{5}$

$\quad\quad\quad -4 = y$

45. $-2 - 3 = -4 + x$

$\quad\quad -5 = -4 + x$

$\quad -5 + 4 = -4 + 4 + x$

$\quad\quad\quad -1 = x$

49. $3w - 12w = -27$

$\quad\quad -9w = -27$

$\quad\quad \dfrac{-9w}{-9} = \dfrac{-27}{-9}$

$\quad\quad\quad w = 3$

53. $18 - 11 = \dfrac{x}{-5}$

$\quad\quad\quad 7 = \dfrac{x}{-5}$

$\quad -5 \cdot 7 = -5 \cdot \dfrac{x}{-5}$

$\quad -5 \cdot 7 = \dfrac{-5}{-5} \cdot x$

$\quad\quad -35 = x$

57. $\quad 10 = 7t - 12t$

$\quad 10 = -5t$

$\quad \dfrac{10}{-5} = \dfrac{-5t}{-5}$

$\quad -2 = t$

61. $\quad\quad 50y = 7(7y + 4)$

$\quad\quad 50y = 7 \cdot 7y + 7 \cdot 4$

$\quad\quad 50y = 49y + 28$

$\quad 50y - 49y = 49y - 49y + 28$

$\quad\quad\quad\quad y = 28$

65. $7x + 14 - 6x = -4 - 10$

$\quad 7x - 6x + 14 = -14$

$\quad\quad\quad x + 14 = -14$

$\quad x + 14 - 14 = -14 - 14$

$\quad\quad\quad\quad x = -28$

69. $23x + 8 - 25x = 7 - 9$

$\quad 23x - 25x + 8 = -2$

$\quad\quad -2x + 8 = -2$

$\quad -2x + 8 - 8 = -2 - 8$

$\quad\quad\quad -2x = -10$

$\quad\quad\quad \dfrac{-2x}{-2} = \dfrac{-10}{-2}$

$\quad\quad\quad\quad x = 5$

73. "The sum of -7 and a number" is $-7 + x$.

77. "The product of -13 and a number" is $-13x$.

81. "The product of -11 and a number, increased by 5" is $-11x + 5$.

85. "Seven added to the product of 4 and a number" is $4x + 7$.

89. "The product of -6 and the sum of a number and 15" is $-6(x + 15)$.

93. "The quotient of seventeen and a number, increased by -15" is $\dfrac{17}{x} + (-15)$ or $\dfrac{17}{x} - 15$.

97. From the graph, travelers spent $67 billion in Florida and $21 billion in Georgia.

$67 + 21 = 88$

The combined spending for Florida and Georgia was $88 billion.

101. no; answers may vary

105.
$$\frac{y}{72} = -86 - (-1029)$$
$$\frac{y}{72} = -86 + 1029$$
$$\frac{y}{72} = 943$$
$$72 \cdot \frac{y}{72} = 72 \cdot 943$$
$$\frac{72}{72} \cdot y = 72 \cdot 943$$
$$y = 67{,}896$$

109.
$$|-13| + 3^2 = 100y - |-20| - 99y$$
$$13 + 3^2 = 100y - 20 - 99y$$
$$13 + 9 = 100y - 99y - 20$$
$$22 = y - 20$$
$$22 + 20 = y - 20 + 20$$
$$42 = y$$

Exercise Set 3.3

1.
$$3x - 7 = 4x + 5$$
$$3x - 3x - 7 = 4x - 3x + 5$$
$$-7 = x + 5$$
$$-7 - 5 = x + 5 - 5$$
$$-12 = x$$

5.
$$19 - 3x = 14 + 2x$$
$$19 - 3x + 3x = 14 + 2x + 3x$$
$$19 = 14 + 5x$$
$$19 - 14 = 14 - 14 + 5x$$
$$5 = 5x$$
$$\frac{5}{5} = \frac{5x}{5}$$
$$1 = x$$

9.
$$x + 20 + 2x = -10 - 2x - 15$$
$$x + 2x + 20 = -10 - 15 - 2x$$
$$3x + 20 = -25 - 2x$$
$$3x + 2x + 20 = -25 - 2x + 2x$$
$$5x + 20 = -25$$
$$5x + 20 - 20 = -25 - 20$$
$$5x = -45$$
$$\frac{5x}{5} = \frac{-45}{5}$$
$$x = -9$$

13.
$$35 - 17 = 3(x - 2)$$
$$18 = 3x - 6$$
$$18 + 6 = 3x - 6 + 6$$
$$24 = 3x$$
$$\frac{24}{3} = \frac{3x}{3}$$
$$8 = x$$

17.
$$2(y - 3) = y - 6$$
$$2y - 6 = y - 6$$
$$2y - y - 6 = y - y - 6$$
$$y - 6 = -6$$
$$y - 6 + 6 = -6 + 6$$
$$y = 0$$

21.
$$2t - 1 = 3(t + 7)$$
$$2t - 1 = 3t + 21$$
$$2t - 2t - 1 = 3t - 2t + 21$$
$$-1 = t + 21$$
$$-1 - 21 = t + 21 - 21$$
$$-22 = t$$

25.
$$-4x = 44$$
$$\frac{-4x}{-4} = \frac{44}{-4}$$
$$x = -11$$

29.
$$8 - b = 13$$
$$8 - 8 - b = 13 - 8$$
$$-b = 5$$
$$\frac{-b}{-1} = \frac{5}{-1}$$
$$b = -5$$

33.
$$3r + 4 = 19$$
$$3r + 4 - 4 = 19 - 4$$
$$3r = 15$$
$$\frac{3r}{3} = \frac{15}{3}$$
$$r = 5$$

37.
$$8y - 13y = -20 - 25$$
$$-5y = -45$$
$$\frac{-5y}{-5} = \frac{-45}{-5}$$
$$y = 9$$

41.
$$-4 + 12 = 16x - 3 - 15x$$
$$8 = x - 3$$
$$8 + 3 = x - 3 + 3$$
$$11 = x$$

45.
$$4x + 3 = 2x + 11$$
$$4x - 2x + 3 = 2x - 2x + 11$$
$$2x + 3 = 11$$
$$2x + 3 - 3 = 11 - 3$$
$$2x = 8$$
$$\frac{2x}{2} = \frac{8}{2}$$
$$x = 4$$

49.
$$-8n + 1 = -6n - 5$$
$$-8n + 8n + 1 = -6n + 8n - 5$$
$$1 = 2n - 5$$
$$1 + 5 = 2n - 5 + 5$$
$$6 = 2n$$
$$\frac{6}{2} = \frac{2n}{2}$$
$$3 = n$$

53.
$$9a + 29 + 7 = 0$$
$$9a + 36 = 0$$
$$9a + 36 - 36 = 0 - 36$$
$$9a = -36$$
$$\frac{9a}{9} = \frac{-36}{9}$$
$$a = -4$$

57.
$$12 + 5t = 6(t + 2)$$
$$12 + 5t = 6t + 12$$
$$12 + 5t - 5t = 6t - 5t + 12$$
$$12 = t + 12$$
$$12 - 12 = t + 12 - 12$$
$$0 = t$$

61. $10 + 5(z - 2) = 4z + 1$
$$10 + 5z - 10 = 4z + 1$$
$$5z = 4z + 1$$
$$5z - 4z = 4z - 4z + 1$$
$$z = 1$$

65. "The sum of -42 and 16 is -26" translates to
$-42 + 16 = -26$.

69. "Three times the difference of -14 and 2 amounts to -48"
translates to $3(-14 - 2) = -48$.

73. The 2012 bar is labelled 122. Thus 122,000,000 returns
were filed electronically in 2012.

77. $x^3 - 2xy = 3^3 - 2(3)(-1)$
$$= 27 - 2(3)(-1)$$
$$= 27 - (-6)$$
$$= 27 + 6$$
$$= 33$$

81. The first step in solving $2x - 5 = -7$ is to add 5 to both
sides, which is choice b.

85. The error is in the second line.
$$2(3x - 5) = 5x - 7$$
$$6x - 10 = 5x - 7$$
$$6x - 10 + 10 = 5x - 7 + 10$$
$$6x = 5x + 3$$
$$6x - 5x = 5x + 3 - 5x$$
$$x = 3$$

89.
$$2^3(x + 4) = 3^2(x + 4)$$
$$8(x + 4) = 9(x + 4)$$
$$8x + 32 = 9x + 36$$
$$8x + 32 - 36 = 9x + 36 - 36$$
$$8x - 4 = 9x$$
$$8x - 4 - 8x = 9x - 8x$$
$$-4 = x$$

Exercise Set 3.4

1. "A number added to -5 is -7" is $-5 + x = -7$.

5. "A number subtracted from -20 amounts to 104" is
$-20 - x = 104$.

9. "The product of 5 and the sum of -3 and a number is -20"
is $5(-3 + x) = -20$.

13. "The sum of 3, 4, and a number amounts to 16" is
$$3 + 4 + x = 16$$
$$7 + x = 16$$
$$x + 7 - 7 = 16 - 7$$
$$x = 9$$

17. "Thirty less a number is equal to the product of 3 and the
sum of the number and 6" is
$$30 - x = 3(x + 6)$$
$$30 - x = 3x + 18$$
$$30 - 30 - x = 3x + 18 - 30$$
$$-x = 3x - 12$$
$$-x - 3x = 3x - 3x - 12$$
$$-4x = -12$$
$$\frac{-4x}{-4} = \frac{-12}{-4}$$
$$x = 3$$

21. "Three times the difference of some number and 5
amounts to the quotient of 108 and 12" is
$$3(x - 5) = \frac{108}{12}$$
$$3x - 15 = 9$$
$$3x - 15 + 15 = 9 + 15$$
$$3x = 24$$
$$\frac{3x}{3} = \frac{24}{3}$$
$$x = 8$$

25. Let x be the number of electoral votes that California has.
Since Florida has 28 fewer electoral votes, Florida has
$x - 28$ electoral votes. Since the two states have a total of
82 electoral votes, the sum of x and $x - 28$ is 82.
$$x + x - 28 = 82$$
$$2x - 28 = 82$$
$$2x - 28 + 28 = 82 + 28$$
$$2x = 110$$
$$\frac{2x}{2} = \frac{110}{2}$$
$$x = 55$$

Thus, California has 55 electoral votes and Florida has
$55 - 28 = 27$ electroral votes.

29. Let x be the Sunday circulation, in thousands, of *The Los
Angeles Times*. Since the Sunday circulation of *The New
York Times* is 384 thousand more than the Sunday circulation of *The Los Angeles Times*, the Sunday circulation
of *The New York Times* is $x + 384$. Since the combined
Sunday circulation is 2494 thousand, the sum of x and
$x + 384$ is 2494.
$$x + x + 384 = 2494$$
$$2x + 384 = 2494$$
$$2x + 384 - 384 = 2494 - 384$$
$$2x = 2110$$
$$\frac{2x}{2} = \frac{2110}{2}$$
$$x = 1055$$

Thus the Sunday circulation of *The Los Angeles Times*
is 1055 thousand and the Sunday circulation of *The New
York Times* is $1055 + 384 = 1439$ thousand.

33. Let x be the cost of the games. Since the cost of the
Xbox 360 is 3 times as much as the cost of the games, the

cost of the Xbox 360 is $3x$. Since the total cost of the Xbox 360 and the games is $440, the sum of x and $3x$ is 440.

$$x + 3x = 440$$
$$4x = 440$$
$$\frac{4x}{4} = \frac{440}{4}$$
$$x = 110$$
$$3x = 3(110) = 330$$

The cost of the games is $110 and the cost of the Xbox 360 is $330.

37. Let x be the capacity of Michigan Stadium. Since the capacity of Beaver Stadium is 1081 more than the capacity of Michigan Stadium, the capacity of Beaver Stadium is $x + 1081$. Since the combined capacity of the two stadiums is 213,483, the sum of x and $x + 1081$ is 213,483.

$$x + x + 1081 = 213,483$$
$$2x + 1081 = 213,483$$
$$2x + 1081 - 1081 = 213,483 - 1081$$
$$2x = 212,402$$
$$\frac{2x}{2} = \frac{212,402}{2}$$
$$x = 106,201$$

The capacity of Michigan Stadium is 106,201 and the capacity of Beaver Stadium is $106,201 + 1081 = 107,282$.

41. Let x be the number of cars produced each day in the United States. Since the number of cars produced each day in Germany is twice the number produced each day in the United States, the number of cars produced each day in Germany is $2x$. Since the total number of cars produced by these countries each day is 24,258, the sum of x and $2x$ is 24,258.

$$x + 2x = 24,258$$
$$3x = 24,258$$
$$\frac{3x}{3} = \frac{24,258}{3}$$
$$x = 8086$$
$$2x = 2(8086) = 16,172$$

The number of cars produced in the United States is 8086 and the number of cars produced in Germany is 16,172.

45. Let x be the number of points scored by the Louisville Cardinals. Since the Connecticut Huskies scored 33 points more than the Louisville Cardinals, the number of points scored by the Connecticut Huskies was $x + 33$. Since the total number of points scored by both teams was 153, the sum of x and $x + 33$ is 153.

$$x + x + 33 = 153$$
$$2x + 33 = 153$$
$$2x + 33 - 33 = 153 - 33$$
$$2x = 120$$
$$\frac{2x}{2} = \frac{120}{2}$$
$$x = 60$$
$$x + 33 = 60 + 33 = 93$$

The Connecticut Huskies scored 93 points.

49. 586 rounded to the nearest ten is 590

53. 2986 rounded to the nearest thousand is 3000.

57. Use $P = A + C$, where $P = 230,000$ and $C = 13,800$.
$$P = A + C$$
$$230,000 = A + 13,800$$
$$230,000 - 13,800 = A + 13,800 - 13,800$$
$$216,200 = A$$
The seller received $216,200.

Chapter 3 Test

1. $7x - 5 - 12x + 10 = 7x - 12x - 5 + 10$
$$= (7 - 12)x - 5 + 10$$
$$= -5x + 5$$

5. Area $= (\text{length}) \cdot (\text{width})$
$$= 4 \cdot (3x - 1)$$
$$= 4 \cdot 3x - 4 \cdot 1$$
$$= 12x - 4$$
The area is $(12x - 4)$ square meters.

9. $-4x + 7 = 15$
$$-4x + 7 - 7 = 15 - 7$$
$$-4x = 8$$
$$\frac{-4x}{-4} = \frac{8}{-4}$$
$$x = -2$$

13. $4(5x + 3) = 2(7x + 6)$
$$20x + 12 = 14x + 12$$
$$20x + 12 - 14x = 14x + 12 - 14x$$
$$6x + 12 = 12$$
$$6x + 12 - 12 = 12 - 12$$
$$6x = 0$$
$$\frac{6x}{6} = \frac{0}{6}$$
$$x = 0$$

17. "The sum of twice 5 and -15 is -5" translates to $2 \cdot 5 + (-15) = -5$.

21. Let x be the number of women runners entered in the race. Since the number of men entered in the race is 112 more than the number of women, the number of men is $x + 112$. Since the total number of runners in the race is 600, the sum of x and $x + 112$ is 600.

$$x + x + 112 = 600$$
$$2x + 112 = 600$$
$$2x + 112 - 112 = 600 - 112$$
$$2x = 488$$
$$\frac{2x}{2} = \frac{488}{2}$$
$$x = 244$$

244 women entered the race.

Chapter 4

Exercise Set 4.1

1. In the fraction $\frac{1}{2}$, the numerator is 1 and the denominator is 2. Since $1 < 2$, the fraction is proper.

5. In the fraction $\frac{15}{15}$, the numerator is 15 and the denominator is 15. Since $15 \geq 15$, the fraction is improper.

9. Each part is $\frac{1}{4}$ of a whole and there are 11 parts shaded, or 2 wholes and 3 more parts.
  a. $\frac{11}{4}$ **b.** $2\frac{3}{4}$

13. 7 out of 12 equal parts are shaded: $\frac{7}{12}$

17. 4 out of 9 equal parts are shaded: $\frac{4}{9}$

21. Each part is $\frac{1}{2}$ of a whole and there are 11 parts shaded, or 5 wholes and 1 more part.
  a. $\frac{11}{2}$ **b.** $5\frac{1}{2}$

25. 5 of 8 equal parts are shaded: $\frac{5}{8}$

29.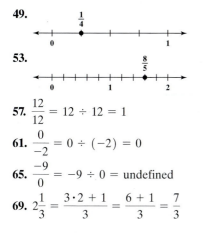

33. freshmen $\rightarrow \dfrac{42}{131}$
students $\rightarrow$
$\frac{42}{131}$ of the students are freshmen.

37. born in Ohio $\rightarrow \dfrac{7}{44}$
U.S. presidents $\rightarrow$
$\frac{7}{44}$ of U.S. presidents were born in Ohio.

41. 11 of 31 days of March is $\frac{11}{31}$ of the month.

45. There are 50 states total. 33 states contain federal Indian reservations.
  a. $\frac{33}{50}$ of the states contain federal Indian reservations.
  b. $50 - 33 = 17$
    17 states do not contain federal Indian reservations.
  c. $\frac{17}{50}$ of the states do not contain federal Indian reservations.

49.

53.

57. $\frac{12}{12} = 12 \div 12 = 1$

61. $\frac{0}{-2} = 0 \div (-2) = 0$

65. $\frac{-9}{0} = -9 \div 0 =$ undefined

69. $2\frac{1}{3} = \frac{3 \cdot 2 + 1}{3} = \frac{6 + 1}{3} = \frac{7}{3}$

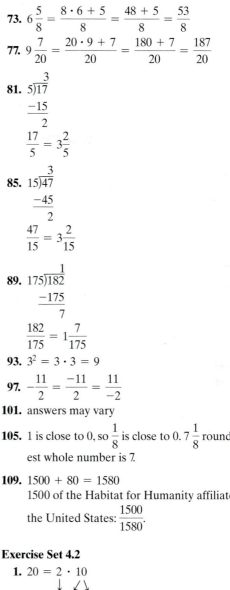

73. $6\frac{5}{8} = \frac{8 \cdot 6 + 5}{8} = \frac{48 + 5}{8} = \frac{53}{8}$

77. $9\frac{7}{20} = \frac{20 \cdot 9 + 7}{20} = \frac{180 + 7}{20} = \frac{187}{20}$

81. $5\overline{)17}$ $\frac{-15}{2}$ $\frac{17}{5} = 3\frac{2}{5}$

85. $15\overline{)47}$ $\frac{-45}{2}$ $\frac{47}{15} = 3\frac{2}{15}$

89. $175\overline{)182}$ $\frac{-175}{7}$ $\frac{182}{175} = 1\frac{7}{175}$

93. $3^2 = 3 \cdot 3 = 9$

97. $-\frac{11}{2} = \frac{-11}{2} = \frac{11}{-2}$

101. answers may vary

105. 1 is close to 0, so $\frac{1}{8}$ is close to 0. $7\frac{1}{8}$ rounded to the nearest whole number is 7.

109. $1500 + 80 = 1580$
1500 of the Habitat for Humanity affiliates are located in the United States: $\frac{1500}{1580}$.

Exercise Set 4.2

1. $20 = 2 \cdot 10$
$2 \cdot 2 \cdot 5 = 2^2 \cdot 5$

5. $81 = 9 \cdot 9$
$3 \cdot 3 \cdot 3 \cdot 3 = 3^4$

9. $110 = 2 \cdot 55$
$2 \cdot 5 \cdot 11 = 2 \cdot 5 \cdot 11$

13. $240 = 2 \cdot 120$
$2 \cdot 2 \cdot 60$
$2 \cdot 2 \cdot 2 \cdot 30$
$2 \cdot 2 \cdot 2 \cdot 2 \cdot 15$
$2 \cdot 2 \cdot 2 \cdot 2 \cdot 3 \cdot 5 = 2^4 \cdot 3 \cdot 5$

In exercises 17 through 49, common factors are divided out to simplify.

17. $\frac{3}{12} = \frac{3 \cdot 1}{3 \cdot 4} = \frac{1}{4}$

21. $\dfrac{14}{16} = \dfrac{2 \cdot 7}{2 \cdot 8} = \dfrac{7}{8}$

25. $\dfrac{35a}{50a} = \dfrac{5 \cdot 7 \cdot a}{5 \cdot 10 \cdot a} = \dfrac{7}{10}$

29. $\dfrac{30x^2}{36x} = \dfrac{5 \cdot 6 \cdot x \cdot x}{6 \cdot 6 \cdot x} = \dfrac{5 \cdot x}{6} = \dfrac{5x}{6}$

33. $\dfrac{25xy}{40y} = \dfrac{5 \cdot 5 \cdot x \cdot y}{5 \cdot 8 \cdot y} = \dfrac{5 \cdot x}{8} = \dfrac{5x}{8}$

37. $\dfrac{36x^3y^2}{24xy} = \dfrac{3 \cdot 12 \cdot x \cdot x \cdot x \cdot y \cdot y}{2 \cdot 12 \cdot x \cdot y} = \dfrac{3 \cdot x^2 \cdot y}{2} = \dfrac{3x^2y}{2}$

41. $\dfrac{40xy}{64xyz} = \dfrac{5 \cdot 8 \cdot x \cdot y}{8 \cdot 8 \cdot x \cdot y \cdot z} = \dfrac{5}{8 \cdot z} = \dfrac{5}{8z}$

45. $-\dfrac{55}{85y} = -\dfrac{5 \cdot 11}{5 \cdot 17 \cdot y} = -\dfrac{11}{17 \cdot y} = -\dfrac{11}{17y}$

49. $\dfrac{224a^3b^4c^2}{16ab^4c^2} = \dfrac{14 \cdot 16 \cdot a \cdot a \cdot a \cdot b \cdot b \cdot b \cdot b \cdot c \cdot c}{1 \cdot 16 \cdot a \cdot b \cdot b \cdot b \cdot b \cdot c \cdot c}$

$\qquad = \dfrac{14 \cdot a^2}{1}$

$\qquad = 14a^2$

53. Not equivalent, since the cross products are not equal: $7 \cdot 8 = 56$ and $5 \cdot 11 = 55$.

57. Equivalent, since the cross products are equal: $3 \cdot 18 = 54$ and $9 \cdot 6 = 54$

61. Not equivalent, since the cross products are not equal: $8 \cdot 24 = 192$ and $12 \cdot 18 = 216$

65. $\dfrac{2640 \text{ feet}}{5280 \text{ feet}} = \dfrac{2640 \cdot 1}{2640 \cdot 2} = \dfrac{1}{2}$

2640 feet represents $\dfrac{1}{2}$ of a mile.

69. $\dfrac{10 \text{ inches}}{24 \text{ inches}} = \dfrac{2 \cdot 5}{2 \cdot 12} = \dfrac{5}{12}$

$\dfrac{5}{12}$ of the wall is concrete.

73. $\dfrac{26 \text{ individuals}}{464 \text{ individuals}} = \dfrac{2 \cdot 13}{2 \cdot 232} = \dfrac{13}{232}$

$\dfrac{13}{232}$ of U.S. astronauts were born in Texas.

77. $2y = 2(-7) = -14$

81. $\dfrac{3975}{6625} = \dfrac{3 \cdot 1325}{5 \cdot 1325} = \dfrac{3}{5}$

85. $3 + 1 = 4$ blood donors have an AB blood type.

$\dfrac{4 \text{ donors}}{100 \text{ donors}} = \dfrac{4 \cdot 1}{4 \cdot 25} = \dfrac{1}{25}$

$\dfrac{1}{25}$ of blood donors have an AB blood type.

89. answers may vary

93. The piece representing education is labeled $\dfrac{6}{100}$.

$\dfrac{6}{100} = \dfrac{2 \cdot 3}{2 \cdot 50} = \dfrac{3}{50}$

$\dfrac{3}{50}$ of entering college freshmen plan to major in education.

97. The piece representing Memorials is labeled $\dfrac{8}{100}$.

$\dfrac{8}{100} = \dfrac{2 \cdot 4}{25 \cdot 4} = \dfrac{2}{25}$

$\dfrac{2}{25}$ of National Park Service areas are National Memorials.

101. 786, 22, 222, 900, and 1470 are divisible by 2 because each number ends with an even digit. 8691, 786, 2235, 105, 222, 900, and 1470 are divisible by 3 because the sum of each number's digits is divisible by 3.

786, 222, 900, and 1470 are divisible by both 2 and 3.

Exercise Set 4.3

In exercises 1 through 113, common factors are divided out as needed to simplify.

1. $\dfrac{6}{11} \cdot \dfrac{3}{7} = \dfrac{6 \cdot 3}{11 \cdot 7} = \dfrac{18}{77}$

5. $-\dfrac{1}{2} \cdot -\dfrac{2}{15} = \dfrac{1 \cdot 2}{2 \cdot 15} = \dfrac{1}{15}$

9. $3a^2 \cdot \dfrac{1}{4} = \dfrac{3a^2}{1} \cdot \dfrac{1}{4} = \dfrac{3a^2 \cdot 1}{1 \cdot 4} = \dfrac{3a^2}{4}$

13. $0 \cdot \dfrac{8}{9} = 0$

17. $\dfrac{11}{20} \cdot \dfrac{1}{7} \cdot \dfrac{5}{22} = \dfrac{11 \cdot 1 \cdot 5}{20 \cdot 7 \cdot 22}$

$\qquad = \dfrac{11 \cdot 1 \cdot 5}{5 \cdot 4 \cdot 7 \cdot 11 \cdot 2}$

$\qquad = \dfrac{1}{4 \cdot 7 \cdot 2}$

$\qquad = \dfrac{1}{56}$

21. $\left(-\dfrac{2}{3}\right)^2 = -\dfrac{2}{3} \cdot -\dfrac{2}{3} = \dfrac{2 \cdot 2}{3 \cdot 3} = \dfrac{4}{9}$

25. $\dfrac{2}{3} \div \dfrac{5}{6} = \dfrac{2}{3} \cdot \dfrac{6}{5} = \dfrac{2 \cdot 6}{3 \cdot 5} = \dfrac{2 \cdot 2 \cdot 3}{3 \cdot 5} = \dfrac{2 \cdot 2}{5} = \dfrac{4}{5}$

29. $-\dfrac{8}{9} \div \dfrac{x}{2} = -\dfrac{8}{9} \cdot \dfrac{2}{x} = -\dfrac{8 \cdot 2}{9 \cdot x} = -\dfrac{16}{9x}$

33. $-\dfrac{2}{3} \div 4 = -\dfrac{2}{3} \div \dfrac{4}{1}$

$\qquad = -\dfrac{2}{3} \cdot \dfrac{1}{4}$

$\qquad = -\dfrac{2 \cdot 1}{3 \cdot 4}$

$\qquad = -\dfrac{2 \cdot 1}{3 \cdot 2 \cdot 2}$

$\qquad = -\dfrac{1}{3 \cdot 2}$

$\qquad = -\dfrac{1}{6}$

37. $\dfrac{2}{3} \cdot \dfrac{5}{9} = \dfrac{2 \cdot 5}{3 \cdot 9} = \dfrac{10}{27}$

41. $\dfrac{16}{27y} \div \dfrac{8}{15y} = \dfrac{16}{27y} \cdot \dfrac{15y}{8}$

$\qquad = \dfrac{16 \cdot 15y}{27y \cdot 8}$

$\qquad = \dfrac{8 \cdot 2 \cdot 3 \cdot 5 \cdot y}{3 \cdot 9 \cdot y \cdot 8}$

$\qquad = \dfrac{2 \cdot 5}{9}$

$\qquad = \dfrac{10}{9}$

45. $\left(-\dfrac{3}{4}\right)^2 = -\dfrac{3}{4} \cdot -\dfrac{3}{4} = \dfrac{3 \cdot 3}{4 \cdot 4} = \dfrac{9}{16}$

49. $7 \div \dfrac{2}{11} = \dfrac{7}{1} \div \dfrac{2}{11} = \dfrac{7}{1} \cdot \dfrac{11}{2} = \dfrac{7 \cdot 11}{1 \cdot 2} = \dfrac{77}{2}$

53. $\left(\dfrac{2}{7} \div \dfrac{7}{2} \right) \cdot \dfrac{3}{4} = \left(\dfrac{2}{7} \cdot \dfrac{2}{7} \right) \cdot \dfrac{3}{4}$

$\qquad\qquad\qquad\quad = \dfrac{4}{49} \cdot \dfrac{3}{4}$

$\qquad\qquad\qquad\quad = \dfrac{3}{49}$

57. $-\dfrac{2}{3} \cdot -\dfrac{6}{11} = \dfrac{2 \cdot 6}{3 \cdot 11} = \dfrac{2 \cdot 2 \cdot 3}{3 \cdot 11} = \dfrac{2 \cdot 2}{11} = \dfrac{4}{11}$

61. $\dfrac{21x^2}{10y} \div \dfrac{14x}{25y} = \dfrac{21x^2}{10y} \cdot \dfrac{25y}{14x}$

$\qquad\qquad\qquad = \dfrac{21x^2 \cdot 25y}{10y \cdot 14x}$

$\qquad\qquad\qquad = \dfrac{3 \cdot 7 \cdot x \cdot x \cdot 5 \cdot 5 \cdot y}{2 \cdot 5 \cdot y \cdot 2 \cdot 7 \cdot x}$

$\qquad\qquad\qquad = \dfrac{3 \cdot x \cdot 5}{2 \cdot 2}$

$\qquad\qquad\qquad = \dfrac{15x}{4}$

65. $\dfrac{a^3}{2} \div 30a^3 = \dfrac{a^3}{2} \div \dfrac{30a^3}{1}$

$\qquad\qquad\quad = \dfrac{a^3}{2} \cdot \dfrac{1}{30a^3}$

$\qquad\qquad\quad = \dfrac{a^3 \cdot 1}{2 \cdot 30a^3}$

$\qquad\qquad\quad = \dfrac{a \cdot a \cdot a \cdot 1}{2 \cdot 30 \cdot a \cdot a \cdot a}$

$\qquad\qquad\quad = \dfrac{1}{2 \cdot 30}$

$\qquad\qquad\quad = \dfrac{1}{60}$

69. $\left(\dfrac{1}{2} \cdot \dfrac{2}{3} \right) \div \dfrac{5}{6} = \left(\dfrac{1 \cdot 2}{2 \cdot 3} \right) \div \dfrac{5}{6}$

$\qquad\qquad\qquad\quad = \dfrac{1}{3} \div \dfrac{5}{6}$

$\qquad\qquad\qquad\quad = \dfrac{1}{3} \cdot \dfrac{6}{5}$

$\qquad\qquad\qquad\quad = \dfrac{1 \cdot 6}{3 \cdot 5}$

$\qquad\qquad\qquad\quad = \dfrac{1 \cdot 2 \cdot 3}{3 \cdot 5}$

$\qquad\qquad\qquad\quad = \dfrac{1 \cdot 2}{5}$

$\qquad\qquad\qquad\quad = \dfrac{2}{5}$

73. a. $xy = \dfrac{2}{5} \cdot \dfrac{5}{6} = \dfrac{2 \cdot 5}{5 \cdot 6} = \dfrac{2 \cdot 5}{5 \cdot 2 \cdot 3} = \dfrac{1}{3}$

b. $x \div y = \dfrac{2}{5} \div \dfrac{5}{6} = \dfrac{2}{5} \cdot \dfrac{6}{5} = \dfrac{2 \cdot 6}{5 \cdot 5} = \dfrac{12}{25}$

77. $\qquad\qquad 3x = -\dfrac{5}{6}$

$\qquad 3\left(-\dfrac{5}{18} \right) \overset{?}{=} -\dfrac{5}{6}$

$\qquad \dfrac{3}{1} \cdot -\dfrac{5}{18} \overset{?}{=} -\dfrac{5}{6}$

$\qquad -\dfrac{3 \cdot 5}{1 \cdot 18} \overset{?}{=} -\dfrac{5}{6}$

$\qquad -\dfrac{3 \cdot 5}{3 \cdot 6} \overset{?}{=} -\dfrac{5}{6}$

$\qquad\qquad -\dfrac{5}{6} = -\dfrac{5}{6}$ True

Yes, $-\dfrac{5}{18}$ is a solution.

81. $\dfrac{1}{4}$ of $200 = \dfrac{1}{4} \cdot 200$

$\qquad\qquad\quad = \dfrac{1}{4} \cdot \dfrac{200}{1}$

$\qquad\qquad\quad = \dfrac{1 \cdot 200}{4 \cdot 1}$

$\qquad\qquad\quad = \dfrac{1 \cdot 4 \cdot 50}{4 \cdot 1}$

$\qquad\qquad\quad = \dfrac{50}{1}$

$\qquad\qquad\quad = 50$

85. $\dfrac{7}{50}$ of $800 = \dfrac{7}{50} \cdot 800$

$\qquad\qquad\quad = \dfrac{7}{50} \cdot \dfrac{800}{1}$

$\qquad\qquad\quad = \dfrac{7 \cdot 800}{50 \cdot 1}$

$\qquad\qquad\quad = \dfrac{7 \cdot 50 \cdot 16}{50 \cdot 1}$

$\qquad\qquad\quad = \dfrac{7 \cdot 16}{1}$

$\qquad\qquad\quad = 112$

112 of the students would be expected to major in business.

89. $\dfrac{2}{5}$ of $2170 = \dfrac{2}{5} \cdot 2170$

$\qquad\qquad\quad = \dfrac{2}{5} \cdot \dfrac{2170}{1}$

$\qquad\qquad\quad = \dfrac{2 \cdot 2170}{5 \cdot 1}$

$\qquad\qquad\quad = \dfrac{2 \cdot 5 \cdot 434}{5 \cdot 1}$

$\qquad\qquad\quad = \dfrac{2 \cdot 434}{1}$

$\qquad\qquad\quad = 868$

He has hiked 868 miles.

93. $\dfrac{2}{3}$ of $2757 = \dfrac{2}{3} \cdot 2757$

$\qquad\qquad\quad = \dfrac{2}{3} \cdot \dfrac{2757}{1}$

$\qquad\qquad\quad = \dfrac{2 \cdot 2757}{3 \cdot 1}$

$\qquad\qquad\quad = \dfrac{2 \cdot 3 \cdot 919}{3 \cdot 1}$

$\qquad\qquad\quad = \dfrac{2 \cdot 919}{1}$

$\qquad\qquad\quad = 1838$ The sale price is $1838.

A44

97. Area = length · width = $\frac{5}{14} \cdot \frac{1}{5} = \frac{5 \cdot 1}{14 \cdot 5} = \frac{1}{14}$

The area is $\frac{1}{14}$ square foot.

101. $\frac{1}{5} \cdot 12{,}000 = \frac{1}{5} \cdot \frac{12{,}000}{1}$

$= \frac{1 \cdot 5 \cdot 2400}{5 \cdot 1}$

$= \frac{1 \cdot 2400}{1}$

$= 2400$

The family drove 2400 miles for family business.

105. $\begin{array}{r} 968 \\ -772 \\ \hline 196 \end{array}$

109. $\frac{42}{25} \cdot \frac{125}{36} \div \frac{7}{6} = \frac{42}{25} \cdot \frac{125}{36} \cdot \frac{6}{7}$

$= \frac{42 \cdot 125 \cdot 6}{25 \cdot 36 \cdot 7}$

$= \frac{6 \cdot 7 \cdot 5 \cdot 25 \cdot 6}{25 \cdot 6 \cdot 6 \cdot 7}$

$= \frac{5}{1}$

$= 5$

113. $\frac{63}{200}$ of $27{,}000 = \frac{63}{200} \cdot 27{,}000$

$= \frac{63}{200} \cdot \frac{27{,}000}{1}$

$= \frac{63 \cdot 200 \cdot 135}{200 \cdot 1}$

$= \frac{63 \cdot 135}{1}$

$= 8505$

The National Park Service is charged with maintaining 8505 monuments and statues.

Exercise Set 4.4

In exercises 1 through 93, common factors are divided out as needed to simplify.

1. $\frac{5}{11} + \frac{2}{11} = \frac{5+2}{11} = \frac{7}{11}$

5. $-\frac{6}{20} + \frac{1}{20} = \frac{-6+1}{20} = \frac{-5}{20} = -\frac{1 \cdot 5}{4 \cdot 5} = -\frac{1}{4}$

9. $\frac{2}{9x} + \frac{4}{9x} = \frac{2+4}{9x} = \frac{6}{9x} = \frac{2 \cdot 3}{3 \cdot 3 \cdot x} = \frac{2}{3x}$

13. $\frac{10}{11} - \frac{4}{11} = \frac{10-4}{11} = \frac{6}{11}$

17. $\frac{1}{y} - \frac{4}{y} = \frac{1-4}{y} = \frac{-3}{y} = -\frac{3}{y}$

21. $\frac{20}{21} - \frac{10}{21} - \frac{17}{21} = \frac{20-10-17}{21}$

$= \frac{-7}{21}$

$= -\frac{1 \cdot 7}{3 \cdot 7} = -\frac{1}{3}$

25. $-\frac{9}{100} + \frac{99}{100} = \frac{-9+99}{100} = \frac{90}{100} = \frac{9 \cdot 10}{10 \cdot 10} = \frac{9}{10}$

29. $\frac{9x}{15} + \frac{1}{15} = \frac{9x+1}{15}$

33. $\frac{9}{12} - \frac{7}{12} - \frac{10}{12} = \frac{9-7-10}{12} = \frac{-8}{12} = -\frac{2 \cdot 4}{3 \cdot 4} = -\frac{2}{3}$

37. $x + y = \frac{3}{4} + \frac{2}{4} = \frac{3+2}{4} = \frac{5}{4}$

41. $\frac{4}{20} + \frac{7}{20} + \frac{9}{20} = \frac{4+7+9}{20} = \frac{20}{20} = 1$

The perimeter is 1 inch.

45. To find the remaining amount of track to be inspected, subtract the $\frac{5}{20}$ mile that has already been inspected from the $\frac{19}{20}$ mile total that must be inspected.

$\frac{19}{20} - \frac{5}{20} = \frac{19-5}{20} = \frac{14}{20} = \frac{2 \cdot 7}{2 \cdot 10} = \frac{7}{10}$

$\frac{7}{10}$ of a mile of track remains to be inspected.

49. North America takes up $\frac{16}{100}$ of the world's land area, while South America takes up $\frac{12}{100}$ of the land area.

$\frac{16}{100} + \frac{12}{100} = \frac{16+12}{100} = \frac{28}{100} = \frac{4 \cdot 7}{4 \cdot 25} = \frac{7}{25}$

$\frac{7}{25}$ of the world's land area is within North America and South America.

53. Multiples of 15:

$15 \cdot 1 = 15$, not a multiple of 9

$15 \cdot 2 = 30$, not a multiple of 9

$15 \cdot 3 = 45$, a multiple of 9

LCD: 45

57. $6 = \boxed{2} \cdot 3$

$15 = \boxed{3} \cdot 5$

$25 = \boxed{5 \cdot 5}$

LCD $= 2 \cdot 3 \cdot 5 \cdot 5 = 150$

61. $18 = \boxed{2 \cdot 3 \cdot 3}$

$21 = 3 \cdot \boxed{7}$

LCD $= 2 \cdot 3 \cdot 3 \cdot 7 = 126$

65. $\frac{2}{3} = \frac{2 \cdot 7}{3 \cdot 7} = \frac{14}{21}$

69. $\frac{1}{2} = \frac{1 \cdot 25}{2 \cdot 25} = \frac{25}{50}$

73. $\frac{2y}{3} = \frac{2y \cdot 4}{3 \cdot 4} = \frac{8y}{12}$

77. books and magazines: $\frac{27}{50} = \frac{27 \cdot 2}{50 \cdot 2} = \frac{54}{100}$

clothing and accessories: $\frac{1}{2} = \frac{1 \cdot 50}{2 \cdot 50} = \frac{50}{100}$

computer hardware: $\frac{23}{50} = \frac{23 \cdot 2}{50 \cdot 2} = \frac{46}{100}$

computer software: $\frac{1}{2} = \frac{1 \cdot 50}{2 \cdot 50} = \frac{50}{100}$

drugs, health and beauty aids: $\frac{3}{20} = \frac{3 \cdot 5}{20 \cdot 5} = \frac{15}{100}$

electronics and appliances: $\frac{13}{20} = \frac{13 \cdot 5}{20 \cdot 5} = \frac{65}{100}$

Solutions to Selected Exercises

food, beer, and wine: $\dfrac{9}{20} = \dfrac{9 \cdot 5}{20 \cdot 5} = \dfrac{45}{100}$

home furnishings: $\dfrac{13}{25} = \dfrac{13 \cdot 4}{25 \cdot 4} = \dfrac{52}{100}$

music and videos: $\dfrac{3}{5} = \dfrac{3 \cdot 20}{5 \cdot 20} = \dfrac{60}{100}$

office equipment and supplies: $\dfrac{61}{100} = \dfrac{61}{100}$

sporting goods: $\dfrac{12}{25} = \dfrac{12 \cdot 4}{25 \cdot 4} = \dfrac{48}{100}$

toys, hobbies, and games: $\dfrac{1}{2} = \dfrac{1 \cdot 50}{2 \cdot 50} = \dfrac{50}{100}$

81. $3^2 = 3 \cdot 3 = 9$

85. $7^2 = 7 \cdot 7 = 49$

89. $\dfrac{2}{7} + \dfrac{9}{7} = \dfrac{2+9}{7} = \dfrac{11}{7}$

93. $\dfrac{16}{100} + \dfrac{12}{100} + \dfrac{7}{100} + \dfrac{20}{100} + \dfrac{30}{100} + \dfrac{6}{100} + \dfrac{9}{100}$

$= \dfrac{16 + 12 + 7 + 20 + 30 + 6 + 9}{100}$

$= \dfrac{100}{100}$

$= 1$

answers may vary

97. answers may vary

Exercise Set 4.5

1. The LCD of 3 and 6 is 6.

$\dfrac{2}{3} + \dfrac{1}{6} = \dfrac{2 \cdot 2}{3 \cdot 2} + \dfrac{1}{6} = \dfrac{4}{6} + \dfrac{1}{6} = \dfrac{5}{6}$

5. The LCD of 11 and 33 is 33.

$-\dfrac{2}{11} + \dfrac{2}{33} = -\dfrac{2 \cdot 3}{11 \cdot 3} + \dfrac{2}{33} = -\dfrac{6}{33} + \dfrac{2}{33} = -\dfrac{4}{33}$

9. The LCD of 35 and 7 is 35.

$\dfrac{11x}{35} + \dfrac{2x}{7} = \dfrac{11x}{35} + \dfrac{2x \cdot 5}{7 \cdot 5}$

$= \dfrac{11x}{35} + \dfrac{10x}{35}$

$= \dfrac{21x}{35}$

$= \dfrac{3 \cdot 7 \cdot x}{5 \cdot 7}$

$= \dfrac{3x}{5}$

13. The LCD of 12 and 9 is 36.

$\dfrac{5}{12} - \dfrac{1}{9} = \dfrac{5 \cdot 3}{12 \cdot 3} - \dfrac{1 \cdot 4}{9 \cdot 4} = \dfrac{15}{36} - \dfrac{4}{36} = \dfrac{11}{36}$

17. The LCD of 11 and 9 is 99.

$\dfrac{5a}{11} + \dfrac{4a}{9} = \dfrac{5a \cdot 9}{11 \cdot 9} + \dfrac{4a \cdot 11}{9 \cdot 11}$

$= \dfrac{45a}{99} + \dfrac{44a}{99}$

$= \dfrac{89a}{99}$

21. The LCD of 2 and x is $2x$.

$\dfrac{1}{2} + \dfrac{3}{x} = \dfrac{1 \cdot x}{2 \cdot x} + \dfrac{3 \cdot 2}{x \cdot 2} = \dfrac{x}{2x} + \dfrac{6}{2x} = \dfrac{x+6}{2x}$

25. The LCD of 14 and 7 is 14.

$\dfrac{9}{14} - \dfrac{3}{7} = \dfrac{9}{14} - \dfrac{3 \cdot 2}{7 \cdot 2} = \dfrac{9}{14} - \dfrac{6}{14} = \dfrac{3}{14}$

29. The LCD of 9 and 12 is 36.

$\dfrac{1}{9} - \dfrac{5}{12} = \dfrac{1 \cdot 4}{9 \cdot 4} - \dfrac{5 \cdot 3}{12 \cdot 3} = \dfrac{4}{36} - \dfrac{15}{36} = -\dfrac{11}{36}$

33. The LCD of 7 and 8 is 56.

$\dfrac{5}{7} - \dfrac{1}{8} = \dfrac{5 \cdot 8}{7 \cdot 8} - \dfrac{1 \cdot 7}{8 \cdot 7} = \dfrac{40}{56} - \dfrac{7}{56} = \dfrac{33}{56}$

37. $\dfrac{3}{9} - \dfrac{5}{9} = \dfrac{3-5}{9} = \dfrac{-2}{9} = -\dfrac{2}{9}$

41. The LCD of 11 and 3 is 33.

$\dfrac{5}{11} + \dfrac{y}{3} = \dfrac{5 \cdot 3}{11 \cdot 3} + \dfrac{y \cdot 11}{3 \cdot 11}$

$= \dfrac{15}{33} + \dfrac{11y}{33}$

$= \dfrac{15 + 11y}{33}$

45. The LCD of 2, 4, and 16 is 16.

$\dfrac{x}{2} + \dfrac{x}{4} + \dfrac{2x}{16} = \dfrac{x \cdot 8}{2 \cdot 8} + \dfrac{x \cdot 4}{4 \cdot 4} + \dfrac{2x}{16}$

$= \dfrac{8x}{16} + \dfrac{4x}{16} + \dfrac{2x}{16}$

$= \dfrac{14x}{16}$

$= \dfrac{2 \cdot 7x}{2 \cdot 8}$

$= \dfrac{7x}{8}$

49. The LCD of 3 and 13 is 39.

$\dfrac{2a}{3} + \dfrac{6a}{13} = \dfrac{2a \cdot 13}{3 \cdot 13} + \dfrac{6a \cdot 3}{13 \cdot 3}$

$= \dfrac{26a}{39} + \dfrac{18a}{39}$

$= \dfrac{44a}{39}$

53. The LCD of 9 and y is $9y$.

$\dfrac{5}{9} + \dfrac{1}{y} = \dfrac{5 \cdot y}{9 \cdot y} + \dfrac{1 \cdot 9}{y \cdot 9}$

$= \dfrac{5y}{9y} + \dfrac{9}{9y}$

$= \dfrac{5y + 9}{9y}$

57. The LCD of 5 and 9 is 45.

$\dfrac{4}{5} + \dfrac{4}{9} = \dfrac{4 \cdot 9}{5 \cdot 9} + \dfrac{4 \cdot 5}{9 \cdot 5} = \dfrac{36}{45} + \dfrac{20}{45} = \dfrac{56}{45}$

61. The LCD of 12, 24, and 6 is 24.

$-\dfrac{9}{12} + \dfrac{17}{24} - \dfrac{1}{6} = -\dfrac{9 \cdot 2}{12 \cdot 2} + \dfrac{17}{24} - \dfrac{1 \cdot 4}{6 \cdot 4}$

$= -\dfrac{18}{24} + \dfrac{17}{24} - \dfrac{4}{24}$

$= -\dfrac{5}{24}$

65. The LCD of 7 and 10 is 70. Write each fraction as an equivalent fraction with a denominator of 70.

$$\frac{2}{7} = \frac{2 \cdot 10}{7 \cdot 10} = \frac{20}{70}$$

$$\frac{3}{10} = \frac{3 \cdot 7}{10 \cdot 7} = \frac{21}{70}$$

Since $20 < 21, \frac{20}{70} < \frac{21}{70}$, so $\frac{2}{7} < \frac{3}{10}$.

69. The LCD of 4 and 14 is 28. Write each fraction as an equivalent fraction with a denominator of 28.

$$-\frac{3}{4} = -\frac{3 \cdot 7}{4 \cdot 7} = -\frac{21}{28}$$

$$-\frac{11}{14} = -\frac{11 \cdot 2}{14 \cdot 2} = -\frac{22}{28}$$

Since $-21 > -22, -\frac{21}{28} > -\frac{22}{28}$, so $-\frac{3}{4} > -\frac{11}{14}$.

73. $xy = \frac{1}{3} \cdot \frac{3}{4} = \frac{1 \cdot 3}{3 \cdot 4} = \frac{1}{4}$

77. The LCD of 3 and 5 is 15.

$$\frac{4}{5} + \frac{1}{3} + \frac{4}{5} + \frac{1}{3} = \frac{4}{5} \cdot \frac{3}{3} + \frac{1}{3} \cdot \frac{5}{5} + \frac{4}{5} \cdot \frac{3}{3} + \frac{1}{3} \cdot \frac{5}{5}$$

$$= \frac{12}{15} + \frac{5}{15} + \frac{12}{15} + \frac{5}{15}$$

$$= \frac{34}{15}$$

The perimeter is $\frac{34}{15}$ or $2\frac{4}{15}$ centimeters.

81. "The sum of a number and $\frac{1}{2}$" translates as $x + \frac{1}{2}$.

85. The LCD of 10 and 100 is 100.

$$\frac{17}{100} - \frac{1}{10} = \frac{17}{100} - \frac{1}{10} \cdot \frac{10}{10} = \frac{17}{100} - \frac{10}{100} = \frac{7}{100}$$

A sloth can travel $\frac{7}{100}$ mph faster in the trees.

89. The LCD of 20 and 25 is 100.

$$\frac{13}{20} - \frac{4}{25} = \frac{13}{20} \cdot \frac{5}{5} - \frac{4}{25} \cdot \frac{4}{4} = \frac{65}{100} - \frac{16}{100} = \frac{49}{100}$$

Math or science is the favorite subject for $\frac{49}{100}$ of these students.

93. The LCD of 50 and 2 is 50.

$$\frac{13}{50} + \frac{1}{2} = \frac{13}{50} + \frac{1}{2} \cdot \frac{25}{25}$$

$$= \frac{13}{50} + \frac{25}{50}$$

$$= \frac{38}{50}$$

$$= \frac{2 \cdot 19}{2 \cdot 25}$$

$$= \frac{19}{25}$$

The Pacific and Atlantic Oceans account for $\frac{19}{25}$ of the world's water surface area.

97. $1 - \frac{21}{100} = \frac{100}{100} - \frac{21}{100} = \frac{79}{100}$

$\frac{79}{100}$ of the recreation areas maintained by the National Park Service are not National Monuments.

101. $(8 - 6) \cdot (4 - 7) = 2 \cdot (-3) = -6$

105. The LCD of 3, 4, and 540 is 540.

$$\frac{2}{3} - \frac{1}{4} - \frac{2}{540} = \frac{2}{3} \cdot \frac{180}{180} - \frac{1}{4} \cdot \frac{135}{135} - \frac{2}{540}$$

$$= \frac{360}{540} - \frac{135}{540} - \frac{2}{540}$$

$$= \frac{225}{540} - \frac{2}{540}$$

$$= \frac{223}{540}$$

109. answers may vary

Exercise Set 4.6

In exercises 1 through 53, common factors are divided out as needed to simplify.

1. $\dfrac{\frac{1}{8}}{\frac{3}{4}} = \frac{1}{8} \div \frac{3}{4} = \frac{1}{8} \cdot \frac{4}{3} = \frac{1 \cdot 4}{8 \cdot 3} = \frac{1 \cdot 4}{2 \cdot 4 \cdot 3} = \frac{1}{6}$

5. $\dfrac{\frac{2x}{27}}{\frac{4}{9}} = \frac{2x}{27} \div \frac{4}{9} = \frac{2x}{27} \cdot \frac{9}{4} = \frac{2x \cdot 9}{27 \cdot 4} = \frac{2 \cdot x \cdot 9}{3 \cdot 9 \cdot 2 \cdot 2} = \frac{x}{6}$

9. $\dfrac{\frac{3x}{4}}{5 - \frac{1}{8}} = \dfrac{8 \cdot \left(\frac{3x}{4}\right)}{8 \cdot \left(5 - \frac{1}{8}\right)}$

$$= \dfrac{\frac{8}{1} \cdot \frac{3x}{4}}{8 \cdot 5 - 8 \cdot \frac{1}{8}}$$

$$= \dfrac{\frac{2 \cdot 4 \cdot 3x}{1 \cdot 4}}{40 - 1}$$

$$= \frac{6x}{39}$$

$$= \frac{2 \cdot 3 \cdot x}{3 \cdot 13}$$

$$= \frac{2x}{13}$$

13. $\dfrac{5}{6} \div \dfrac{1}{3} \cdot \dfrac{1}{4} = \dfrac{5}{6} \cdot \dfrac{3}{1} \cdot \dfrac{1}{4}$

$$= \frac{5 \cdot 3 \cdot 1}{2 \cdot 3 \cdot 1 \cdot 4}$$

$$= \frac{5}{2 \cdot 4}$$

$$= \frac{5}{8}$$

17. $\left(\dfrac{2}{9} + \dfrac{4}{9}\right)\left(\dfrac{1}{3} - \dfrac{9}{10}\right) = \left(\dfrac{6}{9}\right)\left(\dfrac{1}{3} \cdot \dfrac{10}{10} - \dfrac{9}{10} \cdot \dfrac{3}{3}\right)$

$$= \left(\frac{6}{9}\right)\left(\frac{10}{30} - \frac{27}{30}\right)$$

$$= \left(\frac{6}{9}\right)\left(\frac{-17}{30}\right)$$

$$= -\frac{6 \cdot 17}{9 \cdot 30}$$

$$= -\frac{3 \cdot 2 \cdot 17}{3 \cdot 3 \cdot 2 \cdot 15}$$

$$= -\frac{17}{3 \cdot 15}$$

$$= -\frac{17}{45}$$

21. $2 \cdot \left(\dfrac{1}{4} + \dfrac{1}{5}\right) + 2 = 2 \cdot \left(\dfrac{1}{4} \cdot \dfrac{5}{5} + \dfrac{1}{5} \cdot \dfrac{4}{4}\right) + 2$

$= 2 \cdot \left(\dfrac{5}{20} + \dfrac{4}{20}\right) + 2$

$= 2 \cdot \left(\dfrac{9}{20}\right) + 2$

$= \left(\dfrac{2}{1}\right)\left(\dfrac{9}{20}\right) + 2$

$= \dfrac{2 \cdot 9}{1 \cdot 20} + 2$

$= \dfrac{2 \cdot 9}{1 \cdot 2 \cdot 10} + 2$

$= \dfrac{9}{10} + 2$

$= \dfrac{9}{10} + \dfrac{2}{1} \cdot \dfrac{10}{10}$

$= \dfrac{9}{10} + \dfrac{20}{10}$

$= \dfrac{29}{10}$

25. $\left(\dfrac{2}{5} - \dfrac{3}{10}\right)^2 = \left(\dfrac{2}{5} \cdot \dfrac{2}{2} - \dfrac{3}{10}\right)^2$

$= \left(\dfrac{4}{10} - \dfrac{3}{10}\right)^2$

$= \left(\dfrac{1}{10}\right)^2$

$= \dfrac{1}{10} \cdot \dfrac{1}{10}$

$= \dfrac{1}{100}$

29. $5y - z = 5\left(\dfrac{2}{5}\right) - \dfrac{5}{6}$

$= 2 - \dfrac{5}{6}$

$= \dfrac{2}{1} \cdot \dfrac{6}{6} - \dfrac{5}{6}$

$= \dfrac{12}{6} - \dfrac{5}{6}$

$= \dfrac{7}{6}$

33. $x^2 - yz = \left(-\dfrac{1}{3}\right)^2 - \left(\dfrac{2}{5}\right)\left(\dfrac{5}{6}\right)$

$= \left(-\dfrac{1}{3}\right)\left(-\dfrac{1}{3}\right) - \left(\dfrac{2}{5}\right)\left(\dfrac{5}{6}\right)$

$= \dfrac{1}{9} - \dfrac{2 \cdot 5}{5 \cdot 2 \cdot 3}$

$= \dfrac{1}{9} - \dfrac{1}{3}$

$= \dfrac{1}{9} - \dfrac{1}{3} \cdot \dfrac{3}{3}$

$= \dfrac{1}{9} - \dfrac{3}{9}$

$= -\dfrac{2}{9}$

37. $\dfrac{\frac{5a}{24}}{\frac{1}{12}} = \dfrac{5a}{24} \div \dfrac{1}{12} = \dfrac{5a}{24} \cdot \dfrac{12}{1} = \dfrac{5 \cdot a \cdot 12}{12 \cdot 2 \cdot 1} = \dfrac{5a}{2}$

41. $\left(-\dfrac{1}{2}\right)^2 + \dfrac{1}{5} = \dfrac{1}{4} + \dfrac{1}{5} = \dfrac{1 \cdot 5}{4 \cdot 5} + \dfrac{1 \cdot 4}{5 \cdot 4} = \dfrac{5}{20} + \dfrac{4}{20} = \dfrac{9}{20}$

45. $\left(1 - \dfrac{2}{5}\right)^2 = \left(\dfrac{5}{5} - \dfrac{2}{5}\right)^2 = \left(\dfrac{3}{5}\right)^2 = \dfrac{3}{5} \cdot \dfrac{3}{5} = \dfrac{9}{25}$

49. $\left(-\dfrac{2}{9} - \dfrac{7}{9}\right)^4 = \left(-\dfrac{9}{9}\right)^4$

$= (-1)^4$

$= (-1)(-1)(-1)(-1)$

$= 1$

53. $\left(\dfrac{3}{4} \div \dfrac{6}{5}\right) - \left(\dfrac{3}{4} \cdot \dfrac{6}{5}\right) = \left(\dfrac{3}{4} \cdot \dfrac{5}{6}\right) - \left(\dfrac{3}{4} \cdot \dfrac{6}{5}\right)$

$= \dfrac{3 \cdot 5}{4 \cdot 2 \cdot 3} - \dfrac{3 \cdot 2 \cdot 3}{2 \cdot 2 \cdot 5}$

$= \dfrac{5}{4 \cdot 2} - \dfrac{3 \cdot 3}{2 \cdot 5}$

$= \dfrac{5}{8} - \dfrac{9}{10}$

$= \dfrac{5 \cdot 5}{8 \cdot 5} - \dfrac{9 \cdot 4}{10 \cdot 4}$

$= \dfrac{25}{40} - \dfrac{36}{40}$

$= -\dfrac{11}{40}$

57. $3 + \dfrac{1}{2} = \dfrac{3}{1} \cdot \dfrac{2}{2} + \dfrac{1}{2} = \dfrac{6}{2} + \dfrac{1}{2} = \dfrac{7}{2}$ or $3\dfrac{1}{2}$

61. no; answers may vary

65. $\dfrac{\frac{1}{4} + \frac{2}{14}}{2} = \dfrac{28\left(\frac{1}{4} + \frac{2}{14}\right)}{28(2)}$

$= \dfrac{28 \cdot \frac{1}{4} + 28 \cdot \frac{2}{14}}{28 \cdot 2}$

$= \dfrac{7 + 4}{56}$

$= \dfrac{11}{56}$

69. False; the average cannot be greater than the greatest number.

73. true

77. subtraction, multiplication, addition, division

81. $\dfrac{2 + x}{y} = \dfrac{2 + \frac{3}{4}}{-\frac{4}{7}}$

$= \dfrac{28\left(2 + \frac{3}{4}\right)}{28\left(-\frac{4}{7}\right)}$

$= \dfrac{28 \cdot 2 + 28 \cdot \frac{3}{4}}{28 \cdot \left(-\frac{4}{7}\right)}$

$= \dfrac{56 + 21}{-16}$

$= -\dfrac{77}{16}$

Exercise Set 4.7

1.

5. $2\dfrac{11}{12}$ rounds to 3.

$1\dfrac{1}{4}$ rounds to 1.

$3 \cdot 1 = 3$

The best estimate is b.

9. $2\frac{2}{3} \cdot \frac{1}{7} = \frac{8}{3} \cdot \frac{1}{7} = \frac{8}{21}$

13. Exact: $2\frac{1}{5} \cdot 3\frac{1}{2} = \frac{11}{5} \cdot \frac{7}{2} = \frac{77}{10}$ or $7\frac{7}{10}$

Estimate: $2\frac{1}{5}$ rounds to 2, $3\frac{1}{2}$ rounds to 4.

$2 \cdot 4 = 8$ so the answer is reasonable.

17. $5 \cdot 2\frac{1}{2} = \frac{5}{1} \cdot \frac{5}{2} = \frac{25}{2}$ or $12\frac{1}{2}$

21. $2\frac{2}{3} \div \frac{1}{7} = \frac{8}{3} \cdot \frac{7}{1} = \frac{56}{3}$ or $18\frac{2}{3}$

25. $8\frac{1}{3}$ rounds to 8.

$1\frac{1}{2}$ rounds to 2.

$8 + 2 = 10$

The best estimate is b.

29. Exact: $10\frac{3}{14} + 3\frac{4}{7} = 10\frac{3}{14} + 3\frac{8}{14} = 13\frac{11}{14}$

Estimate: $10\frac{3}{14}$ rounds to 10, $3\frac{4}{7}$ rounds to 4.

$10 + 4 = 14$ so the answer is reasonable.

33. $12\frac{3}{14} \qquad 12\frac{18}{84}$
$10 \qquad\quad\ \ 10$
$+25\frac{5}{12} \qquad +25\frac{35}{84}$
$\rule{2cm}{0.4pt} \qquad \rule{2cm}{0.4pt}$
$\qquad\qquad\quad\ 47\frac{53}{84}$

37. $3\frac{5}{8} \qquad 3\frac{15}{24}$
$2\frac{1}{6} \qquad 2\frac{4}{24}$
$+7\frac{3}{4} \qquad +7\frac{18}{24}$
$\rule{2cm}{0.4pt} \qquad \rule{2cm}{0.4pt}$
$\qquad\quad 12\frac{37}{24} = 12 + 1\frac{13}{24} = 13\frac{13}{24}$

41. $10\frac{13}{14} \qquad 10\frac{13}{14}$
$-3\frac{4}{7} \qquad -3\frac{8}{14}$
$\rule{2cm}{0.4pt} \qquad \rule{2cm}{0.4pt}$
Exact: $\qquad 7\frac{5}{14}$

Estimate: $10\frac{13}{14}$ rounds to 11, $3\frac{4}{7}$ rounds to 4.

$11 - 4 = 7$ so the answer is reasonable.

45. $6 \qquad\quad 5\frac{9}{9}$
$-2\frac{4}{9} \qquad -2\frac{4}{9}$
$\rule{2cm}{0.4pt} \qquad \rule{2cm}{0.4pt}$
$\qquad\qquad\ 3\frac{5}{9}$

49. $2\frac{3}{4}$
$+1\frac{1}{4}$
$\rule{2cm}{0.4pt}$
$3\frac{4}{4} = 3 + 1 = 4$

53. $3\frac{1}{9} \cdot 2 = \frac{28}{9} \cdot \frac{2}{1} = \frac{56}{9}$ or $6\frac{2}{9}$

57. $22\frac{4}{9} + 13\frac{5}{18} = 22\frac{8}{18} + 13\frac{5}{18} = 35\frac{13}{18}$

61. $15\frac{1}{5} \qquad 15\frac{6}{30}$
$20\frac{3}{10} \qquad 20\frac{9}{30}$
$+37\frac{2}{15} \qquad +37\frac{4}{30}$
$\rule{2cm}{0.4pt} \qquad \rule{2cm}{0.4pt}$
$\qquad\qquad\ 72\frac{19}{30}$

65. $4\frac{2}{7} \cdot 1\frac{3}{10} = \frac{30}{7} \cdot \frac{13}{10}$
$\qquad = \frac{30 \cdot 13}{7 \cdot 10}$
$\qquad = \frac{3 \cdot 10 \cdot 13}{7 \cdot 10}$
$\qquad = \frac{39}{7}$ or $5\frac{4}{7}$

69. "$-5\frac{2}{7}$ decreased by a number" translates as $-5\frac{2}{7} - x$.

73. $12\frac{3}{4} \div 4 = \frac{51}{4} \div \frac{4}{1} = \frac{51}{4} \cdot \frac{1}{4} = \frac{51}{16}$ or $3\frac{3}{16}$

The patient walked $3\frac{3}{16}$ miles per day.

77. $11\frac{1}{4} \qquad 11\frac{5}{20} \qquad 10\frac{25}{20}$
$-3\frac{3}{5} \qquad -3\frac{12}{20} \qquad -3\frac{12}{30}$
$\rule{1.5cm}{0.4pt} \qquad \rule{1.5cm}{0.4pt} \qquad \rule{1.5cm}{0.4pt}$
$\qquad\qquad\qquad\qquad\qquad 7\frac{13}{20}$

Tucson gets an average of $7\frac{13}{20}$ inches more rain than Yuma.

81. $\frac{3}{4} \cdot 1\frac{1}{4} = \frac{3}{4} \cdot \frac{5}{4} = \frac{3 \cdot 5}{4 \cdot 4} = \frac{15}{16}$

The area is $\frac{15}{16}$ square inch.

85. $15\frac{2}{3} - \left(3\frac{1}{4} + 2\frac{1}{2}\right) = 15\frac{2}{3} - \left(3\frac{1}{4} + 2\frac{2}{4}\right)$
$\qquad\qquad\qquad\qquad = 15\frac{2}{3} - 5\frac{3}{4}$

$15\frac{2}{3} \qquad 15\frac{8}{12} \qquad 14\frac{20}{12}$
$-5\frac{3}{4} \qquad -5\frac{9}{12} \qquad -5\frac{9}{12}$
$\rule{1.5cm}{0.4pt} \qquad \rule{1.5cm}{0.4pt} \qquad \rule{1.5cm}{0.4pt}$
$\qquad\qquad\qquad\qquad\qquad 9\frac{11}{12}$

No; the remaining pipe is $9\frac{11}{12}$ feet, which is $\frac{1}{12}$ foot short.

89. $2\frac{2}{3} \qquad 2\frac{40}{60}$
$4\frac{7}{15} \qquad 4\frac{28}{60}$
$+2\frac{37}{60} \qquad +2\frac{37}{60}$
$\rule{1.5cm}{0.4pt} \qquad \rule{1.5cm}{0.4pt}$
$\qquad\quad 8\frac{105}{60} = 8\frac{7}{4} = 8 + 1\frac{3}{4} = 9\frac{3}{4}$

The total duration of the eclipses is $9\frac{3}{4}$ minutes.

93. $-4\dfrac{2}{5} \cdot 2\dfrac{3}{10} = -\dfrac{22}{5} \cdot \dfrac{23}{10}$

$\qquad\qquad = -\dfrac{2 \cdot 11 \cdot 23}{5 \cdot 2 \cdot 5}$

$\qquad\qquad = -\dfrac{253}{25}$ or $-10\dfrac{3}{25}$

97. $-31\dfrac{2}{15} + 17\dfrac{3}{20} = -31\dfrac{8}{60} + 17\dfrac{9}{60}$

$\qquad\qquad\qquad = -30\dfrac{68}{60} + 17\dfrac{9}{60}$

$\qquad\qquad\qquad = -13\dfrac{59}{60}$

101. $11\dfrac{7}{8} - 13\dfrac{5}{6} = 11\dfrac{21}{24} - 13\dfrac{20}{24}$

$\qquad\qquad\quad = -\left(13\dfrac{20}{24} - 11\dfrac{21}{24}\right)$

$\qquad\qquad\quad = -\left(12\dfrac{44}{24} - 11\dfrac{21}{24}\right)$

$\qquad\qquad\quad = -1\dfrac{23}{24}$

105. $\dfrac{1}{3}(3x) = \left(\dfrac{1}{3} \cdot 3\right)x = 1 \cdot x = x$

109. a. $9\dfrac{5}{5} = 9 + 1 = 10$

b. $9\dfrac{100}{100} = 9 + 1 = 10$

c. $6\dfrac{44}{11} = 6 + 4 = 10$

d. $8\dfrac{13}{13} = 8 + 1 = 9$

a, b, and c are equivalent to 10.

113. answers may vary

117. answers may vary

Exercise Set 4.8

1. $\qquad x + \dfrac{1}{3} = -\dfrac{1}{3}$

$\quad x + \dfrac{1}{3} - \dfrac{1}{3} = -\dfrac{1}{3} - \dfrac{1}{3}$

$\qquad\qquad\quad x = -\dfrac{2}{3}$

Check: $\quad x + \dfrac{1}{3} = -\dfrac{1}{3}$

$\qquad -\dfrac{2}{3} + \dfrac{1}{3} \overset{?}{=} -\dfrac{1}{3}$

$\qquad\qquad -\dfrac{1}{3} = -\dfrac{1}{3}$ True

The solution is $-\dfrac{2}{3}$.

5. $3x - \dfrac{1}{5} - 2x = \dfrac{1}{5} + \dfrac{2}{5}$

$\qquad\quad x - \dfrac{1}{5} = \dfrac{3}{5}$

$\quad x - \dfrac{1}{5} + \dfrac{1}{5} = \dfrac{3}{5} + \dfrac{1}{5}$

$\qquad\qquad\quad x = \dfrac{4}{5}$

Check: $\quad 3x - \dfrac{1}{5} - 2x = \dfrac{1}{5} + \dfrac{2}{5}$

$\quad 3 \cdot \dfrac{4}{5} - \dfrac{1}{5} - 2 \cdot \dfrac{4}{5} \overset{?}{=} \dfrac{1}{5} + \dfrac{2}{5}$

$\qquad \dfrac{12}{5} - \dfrac{1}{5} - \dfrac{8}{5} \overset{?}{=} \dfrac{1}{5} + \dfrac{2}{5}$

$\qquad\qquad\qquad \dfrac{3}{5} = \dfrac{3}{5}$ True

The solution is $\dfrac{4}{5}$.

9. $\qquad\qquad \dfrac{2}{5} + y = -\dfrac{3}{10}$

$\qquad\quad \dfrac{2}{5} + y - \dfrac{2}{5} = -\dfrac{3}{10} - \dfrac{2}{5}$

$\qquad\qquad\qquad y = -\dfrac{3}{10} - \dfrac{2}{5} \cdot \dfrac{2}{2}$

$\qquad\qquad\qquad y = -\dfrac{3}{10} - \dfrac{4}{10}$

$\qquad\qquad\qquad y = -\dfrac{7}{10}$

Check: $\qquad \dfrac{2}{5} + y = -\dfrac{3}{10}$

$\qquad\quad \dfrac{2}{5} + \left(-\dfrac{7}{10}\right) \overset{?}{=} -\dfrac{3}{10}$

$\qquad \dfrac{2}{5} \cdot \dfrac{2}{2} + \left(-\dfrac{7}{10}\right) \overset{?}{=} -\dfrac{3}{10}$

$\qquad\quad \dfrac{4}{10} + \left(-\dfrac{7}{10}\right) \overset{?}{=} -\dfrac{3}{10}$

$\qquad\qquad\qquad -\dfrac{3}{10} = -\dfrac{3}{10}$ True

The solution is $-\dfrac{7}{10}$.

13. $\qquad\qquad -\dfrac{2}{9} = x - \dfrac{5}{6}$

$\qquad\quad -\dfrac{2}{9} + \dfrac{5}{6} = x - \dfrac{5}{6} + \dfrac{5}{6}$

$\quad -\dfrac{2}{9} \cdot \dfrac{2}{2} + \dfrac{5}{6} \cdot \dfrac{3}{3} = x$

$\qquad\quad -\dfrac{4}{18} + \dfrac{15}{18} = x$

$\qquad\qquad\qquad \dfrac{11}{18} = x$

Check: $\qquad -\dfrac{2}{9} = x - \dfrac{5}{6}$

$\qquad\quad -\dfrac{2}{9} \overset{?}{=} \dfrac{11}{18} - \dfrac{5}{6}$

$\quad -\dfrac{2}{9} \cdot \dfrac{2}{2} \overset{?}{=} \dfrac{11}{18} - \dfrac{5}{6} \cdot \dfrac{3}{3}$

$\qquad -\dfrac{4}{18} \overset{?}{=} \dfrac{11}{18} - \dfrac{15}{18}$

$\qquad -\dfrac{4}{18} = -\dfrac{4}{18}$ True

The solution is $\dfrac{11}{18}$.

17. $\qquad \dfrac{1}{4}x = 3$

$\quad 4 \cdot \dfrac{1}{4}x = 4 \cdot 3$

$\qquad\qquad x = 12$

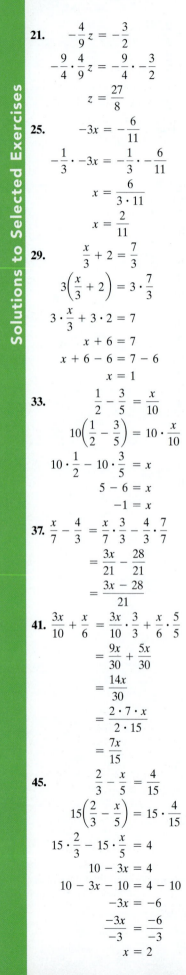

21.
$$-\frac{4}{9}z = -\frac{3}{2}$$
$$-\frac{9}{4}\cdot\frac{4}{9}z = -\frac{9}{4}\cdot-\frac{3}{2}$$
$$z = \frac{27}{8}$$

25.
$$-3x = -\frac{6}{11}$$
$$-\frac{1}{3}\cdot-3x = -\frac{1}{3}\cdot-\frac{6}{11}$$
$$x = \frac{6}{3\cdot11}$$
$$x = \frac{2}{11}$$

29.
$$\frac{x}{3} + 2 = \frac{7}{3}$$
$$3\left(\frac{x}{3} + 2\right) = 3\cdot\frac{7}{3}$$
$$3\cdot\frac{x}{3} + 3\cdot2 = 7$$
$$x + 6 = 7$$
$$x + 6 - 6 = 7 - 6$$
$$x = 1$$

33.
$$\frac{1}{2} - \frac{3}{5} = \frac{x}{10}$$
$$10\left(\frac{1}{2} - \frac{3}{5}\right) = 10\cdot\frac{x}{10}$$
$$10\cdot\frac{1}{2} - 10\cdot\frac{3}{5} = x$$
$$5 - 6 = x$$
$$-1 = x$$

37.
$$\frac{x}{7} - \frac{4}{3} = \frac{x}{7}\cdot\frac{3}{3} - \frac{4}{3}\cdot\frac{7}{7}$$
$$= \frac{3x}{21} - \frac{28}{21}$$
$$= \frac{3x - 28}{21}$$

41.
$$\frac{3x}{10} + \frac{x}{6} = \frac{3x}{10}\cdot\frac{3}{3} + \frac{x}{6}\cdot\frac{5}{5}$$
$$= \frac{9x}{30} + \frac{5x}{30}$$
$$= \frac{14x}{30}$$
$$= \frac{2\cdot7\cdot x}{2\cdot15}$$
$$= \frac{7x}{15}$$

45.
$$\frac{2}{3} - \frac{x}{5} = \frac{4}{15}$$
$$15\left(\frac{2}{3} - \frac{x}{5}\right) = 15\cdot\frac{4}{15}$$
$$15\cdot\frac{2}{3} - 15\cdot\frac{x}{5} = 4$$
$$10 - 3x = 4$$
$$10 - 3x - 10 = 4 - 10$$
$$-3x = -6$$
$$\frac{-3x}{-3} = \frac{-6}{-3}$$
$$x = 2$$

49.
$$-3m - 5m = \frac{4}{7}$$
$$-8m = \frac{4}{7}$$
$$-\frac{1}{8}\cdot-8m = -\frac{1}{8}\cdot\frac{4}{7}$$
$$m = -\frac{1\cdot4}{2\cdot4\cdot7}$$
$$m = -\frac{1}{14}$$

53.
$$\frac{5}{9} - \frac{2}{3} = \frac{5}{9} - \frac{2}{3}\cdot\frac{3}{3} = \frac{5}{9} - \frac{6}{9} = -\frac{1}{9}$$

57.
$$\frac{5}{7}y = -\frac{15}{49}$$
$$\frac{7}{5}\cdot\frac{5}{7}y = \frac{7}{5}\cdot-\frac{15}{49}$$
$$y = -\frac{7\cdot15}{5\cdot49}$$
$$y = -\frac{7\cdot3\cdot5}{5\cdot7\cdot7}$$
$$y = -\frac{3}{7}$$

61.
$$-\frac{5}{8}y = \frac{3}{16} - \frac{9}{16}$$
$$16\left(-\frac{5}{8}y\right) = 16\left(\frac{3}{16} - \frac{9}{16}\right)$$
$$-10y = 16\cdot\frac{3}{16} - 16\cdot\frac{9}{16}$$
$$-10y = 3 - 9$$
$$-10y = -6$$
$$\frac{-10y}{-10} = \frac{-6}{-10}$$
$$y = \frac{3}{5}$$

65.
$$\frac{7}{6}x = \frac{1}{4} - \frac{2}{3}$$
$$12\cdot\frac{7}{6}x = 12\left(\frac{1}{4} - \frac{2}{3}\right)$$
$$14x = 12\cdot\frac{1}{4} - 12\cdot\frac{2}{3}$$
$$14x = 3 - 8$$
$$14x = -5$$
$$\frac{14x}{14} = \frac{-5}{14}$$
$$x = -\frac{5}{14}$$

69.
$$\frac{x}{3} + 2 = \frac{x}{2} + 8$$
$$6\left(\frac{x}{3} + 2\right) = 6\left(\frac{x}{2} + 8\right)$$
$$6\cdot\frac{x}{3} + 6\cdot2 = 6\cdot\frac{x}{2} + 6\cdot8$$
$$2x + 12 = 3x + 48$$
$$2x + 12 - 2x = 3x + 48 - 2x$$
$$12 = x + 48$$
$$12 - 48 = x + 48 - 48$$
$$-36 = x$$

73. 327 rounded to the nearest ten is 330.

77.
$$\frac{14}{11} + \frac{3x}{8} = \frac{x}{2}$$
$$88\left(\frac{14}{11} + \frac{3x}{8}\right) = 88 \cdot \frac{x}{2}$$
$$88 \cdot \frac{14}{11} + 88 \cdot \frac{3x}{8} = 44x$$
$$112 + 33x = 44x$$
$$112 + 33x - 33x = 44x - 33x$$
$$112 = 11x$$
$$\frac{112}{11} = \frac{11x}{11}$$
$$\frac{112}{11} = x$$

Chapter 4 Test

1. 7 of the 16 equal parts are shaded: $\frac{7}{16}$

5. $-\frac{42x}{70} = -\frac{3 \cdot 14 \cdot x}{5 \cdot 14} = -\frac{3x}{5}$

9. $495 = 3 \cdot 165 = 3 \cdot 3 \cdot 55 = 3 \cdot 3 \cdot 5 \cdot 11 = 3^2 \cdot 5 \cdot 11$

13. The LCD of 7 and x is $7x$.
$$\frac{1}{7} - \frac{3}{x} = \frac{1}{7} \cdot \frac{x}{x} - \frac{3}{x} \cdot \frac{7}{7} = \frac{x}{7x} - \frac{21}{7x} = \frac{x - 21}{7x}$$

17. $-\frac{8}{15y} - \frac{2}{15y} = \frac{-8 - 2}{15y} = \frac{-10}{15y} = -\frac{2 \cdot 5}{3 \cdot 5 \cdot y} = -\frac{2}{3y}$

21.
$$19 \qquad 18\frac{11}{11}$$
$$\underline{-2\frac{3}{11}} \qquad \underline{-2\frac{3}{11}}$$
$$16\frac{8}{11}$$

25. $\frac{1}{2} \div \frac{2}{3} \cdot \frac{3}{4} = \frac{1}{2} \cdot \frac{3}{2} \cdot \frac{3}{4} = \frac{1 \cdot 3 \cdot 3}{2 \cdot 2 \cdot 4} = \frac{9}{16}$

29. $\frac{5 + \frac{3}{7}}{2 - \frac{1}{2}} = \frac{14\left(5 + \frac{3}{7}\right)}{14\left(2 - \frac{1}{2}\right)} = \frac{14 \cdot 5 + 14 \cdot \frac{3}{7}}{14 \cdot 2 - 14 \cdot \frac{1}{2}} = \frac{70 + 6}{28 - 7} = \frac{76}{21}$ or $3\frac{13}{21}$

33. $-5x = -5\left(-\frac{1}{2}\right) = \frac{5}{1} \cdot \frac{1}{2} = \frac{5 \cdot 1}{1 \cdot 2} = \frac{5}{2}$

37. Education: $\frac{1}{50}$

Transportation: $\frac{1}{5}$

Clothing: $\frac{1}{25}$

$$\frac{1}{50} + \frac{1}{5} + \frac{1}{25} = \frac{1}{50} + \frac{1}{5} \cdot \frac{10}{10} + \frac{1}{25} \cdot \frac{2}{2}$$
$$= \frac{1}{50} + \frac{10}{50} + \frac{2}{50}$$
$$= \frac{13}{50}$$

$\frac{13}{50}$ of spending goes for education, transportation, and clothing.

Chapter 5

Exercise Set 5.1

1. 5.62 in words is five and sixty-two hundredths.

5. -0.205 in words is negative two hundred five thousandths.

9. 3000.04 in words is three thousand and four hundredths.

13. 2.43 in words is two and forty-three hundredths.

17. The check should be paid to "Verizon," for the amount of "91.68," which is written in words as "Ninety-one and $\frac{68}{100}$."

21. Nine and eight hundredths is 9.08.

25. Forty-six ten-thousandths is 0.0046.

29. $0.27 = \frac{27}{100}$

33. $5.4 = 5\frac{4}{10} = 5\frac{2}{5}$

37. $7.008 = 7\frac{8}{1000} = 7\frac{1}{125}$

41. $0.3005 = \frac{3005}{10,000} = \frac{601}{2000}$

45. In words, 0.077 is seventy-seven thousandths. As a fraction, $0.077 = \frac{77}{1000}$.

49. 0.57 0.54
$$\uparrow \quad \uparrow$$
$$7 > 4$$
so $0.57 > 0.54$
Thus $-0.57 < -0.54$.

53. 0.54900 0.549
$$\uparrow \qquad \uparrow$$
$$9 \quad = \quad 9$$
so $0.54900 = 0.549$

57. 1.062 1.07
$$\uparrow \qquad \uparrow$$
$$6 \quad < \quad 7$$
so $1.062 < 1.07$
Thus, $-1.062 > -1.07$.

61. 0.023 0.024
$$\uparrow \qquad \uparrow$$
$$3 \quad < \quad 4$$
so $0.023 < 0.024$
Thus, $-0.023 > -0.024$.

65. To round 98,207.23 to the nearest ten, observe that the digit in the ones place is 7. Since this digit is at least 5, we add 1 to the digit in the tens place. The number 98,207.23 rounded to the nearest ten is 98,210.

69. To round 0.5942 to the nearest thousandth, observe that the digit in the ten-thousandths place is 2. Since this digit is less than 5, we do not add 1 to the digit in the thousandths place. The number 0.5942 rounded to the nearest thousandth is 0.594.

73. To round $\pi \approx 3.14159265$ to the nearest thousandth, observe that the digit in the ten-thousandths place is 5. Since this digit is at least 5, we add 1 to the digit in the thousandths place. The number $\pi \approx 3.14159265$ rounded to the nearest thousandth is 3.142.

77. To round 0.1992 to the nearest hundredth, observe that the digit in the thousandths place is 9. Since this digit is at least 5, we add 1 to the digit in the hundredths place. The number 0.1992 rounded to the nearest hundredth is 0.2. The amount is $0.20.

81. To round 2.0677 to the nearest hundredth, observe that the digit in the thousandths place is 7. Since this digit is at least 5, we add 1 to the digit in the hundredths place. The number 2.0677 rounded to the nearest hundredth is 2.07. The time is 2.07 minutes.

85. To round 224.695 to the nearest one, observe that the digit in the tenths place is 6. Since this digit is at least 5, we add 1 to the digit in the ones place. The number 224.695 rounded to the nearest one is 225. This is 225 days.

89.
$$\begin{array}{r} 82 \\ -\ 47 \\ \hline 35 \end{array}$$

93. To round 2849.1738 to the nearest hundredth, observe that the digit in the thousandths place is 3. Since this digit is less than 5, we do not add 1 to the digit in the hundredths place. 2849.1738 rounded to the nearest hundredth is 2849.17, which is choice a.

97. $7\dfrac{12}{100} = 7.12$

101. answers may vary

105. 0.26499 and 0.25786 rounded to the nearest hundredths are 0.26. 0.26559 rounds to 0.27 and 0.25186 rounds to 0.25.

109. Round to the nearest hundred million, then add.
$$\begin{array}{r} 800 \\ 700 \\ 600 \\ 500 \\ 500 \\ +\ 500 \\ \hline 3600 \end{array}$$
The total amount of money is estimated as $3600 million.

Exercise Set 5.2

1.
$$\begin{array}{r} 5.6 \\ +\ 2.1 \\ \hline 7.7 \end{array}$$

5.
$$\begin{array}{r} \overset{1\ 1}{24.6000} \\ 2.3900 \\ +\ 0.0678 \\ \hline 27.0578 \end{array}$$

9. $18.56 + (-8.23)$

Subtract the absolute values.
$$\begin{array}{r} 18.56 \\ -\ 8.23 \\ \hline 10.33 \end{array}$$
Attach the sign of the larger absolute value.
$18.56 + (-8.23) = 10.33$

13. Exact:
$$\begin{array}{r} \overset{1\ \ 11}{100.009} \\ 6.080 \\ +\ 9.034 \\ \hline 115.123 \end{array}$$
Estimate:
$$\begin{array}{r} 100 \\ 6 \\ +\ 9 \\ \hline 115 \end{array}$$

17.
$$\begin{array}{r} 12.6 \\ -\ 8.2 \\ \hline 4.4 \end{array}$$
Check:
$$\begin{array}{r} 4.4 \\ +\ 8.2 \\ \hline 12.6 \end{array}$$

21.
$$\begin{array}{r} 654.90 \\ -\ 56.67 \\ \hline 598.23 \end{array}$$
Check:
$$\begin{array}{r} \overset{11\ \ \ 1}{598.23} \\ +\ 56.67 \\ \hline 654.90 \end{array}$$

25. Exact:
$$\begin{array}{r} 1000.0 \\ -\ 123.4 \\ \hline 876.6 \end{array}$$
Check:
$$\begin{array}{r} \overset{1111}{876.6} \\ +\ 123.4 \\ \hline 1000.0 \end{array}$$
Estimate:
$$\begin{array}{r} 1000 \\ -\ 100 \\ \hline 900 \end{array}$$

29. $-1.12 - 5.2 = -1.12 + (-5.2)$

Add the absolute values.
$$\begin{array}{r} 1.12 \\ +\ 5.20 \\ \hline 6.32 \end{array}$$
Attach the common sign.
$-1.12 - 5.2 = -6.32$

33. $-2.6 - (-5.7) = -2.6 + 5.7$

Subtract the absolute values.
$$\begin{array}{r} 5.7 \\ -\ 2.6 \\ \hline 3.1 \end{array}$$
Attach the sign of the larger absolute value.
$-2.6 - (-5.7) = 3.1$

37.
$$\begin{array}{r} 23.0 \\ -\ 6.7 \\ \hline 16.3 \end{array}$$
Check:
$$\begin{array}{r} \overset{1\ 1}{16.3} \\ +\ 6.7 \\ \hline 23.0 \end{array}$$

41. $-6.06 + 0.44$

Subtract the absolute values.
$$\begin{array}{r} 6.06 \\ -\ 0.44 \\ \hline 5.62 \end{array}$$
Attach the sign of the larger absolute value.
$-6.06 + 0.44 = -5.62$

45. $50.2 - 600 = 50.2 + (-600)$

Subtract the absolute values.
$$\begin{array}{r} 600.0 \\ -\ 50.2 \\ \hline 549.8 \end{array}$$
Attach the sign of the larger absolute value.
$50.2 - 600 = -549.8$

49.
$$\begin{array}{r} \overset{1\ \ 11}{100.009} \\ 6.080 \\ +\ 9.034 \\ \hline 115.123 \end{array}$$

53. $-102.4 - 78.04 = -102.4 + (-78.04)$

Add the absolute values.
$$\begin{array}{r} 102.40 \\ +\ 78.04 \\ \hline 180.44 \end{array}$$
Attach the common sign.
$-102.4 - 78.04 = -180.44$

57. $x + z = 3.6 + 0.21 = 3.81$

61. $y - x + z = 5 - 3.6 + 0.21$

$= 5.00 - 3.60 + 0.21$

$= 1.40 + 0.21$

$= 1.61$

65. $27.4 + y = 16$

$27.4 + (-11.4) \overset{?}{=} 16$

$16 = 16$ True

Yes, -11.4 is a solution.

69. $30.7x + 17.6 - 23.8x - 10.7$

$= 30.7x - 23.8x + 17.6 - 10.7$

$= 6.9x + 6.9$

73. 40.00
$\underline{-\ 32.48}$
  7.52

Her change was $7.52.

77. Perimeter $= 7.14 + 7.14 + 7.14 + 7.14 = 28.56$ meters

81. The phrase "How much faster" indicates that we should subtract the average wind speed from the record speed.

  231.0
$\underline{-\ \ 35.2}$
  195.8

The highest wind speed is 195.8 miles per hour faster than the average wind speed.

85. To find the total, we add.

  $\overset{2\,1\,1\,1}{760.5}$
  658.7
$\underline{+\ 623.3}$
  2042.5

The total ticket sales were $2042.5 million.

89. Add the lengths of the sides to get the perimeter.

  $\overset{1\,1}{12.40}$
  29.34
$\underline{+\ 25.70}$
  67.44

67.44 feet of border material is needed.

93. The tallest bar indicates the greatest chocolate consumption per person, so Switzerland has the greatest chocolate consumption per person.

97.

Country	Pounds of Chocolate per Person
Switzerland	26.24
Ireland	21.83
UK	20.94
Austria	19.40
Belgium	18.30

101. $\left(\dfrac{2}{3}\right)^2 = \dfrac{2}{3} \cdot \dfrac{2}{3} = \dfrac{2 \cdot 2}{3 \cdot 3} = \dfrac{4}{9}$

105. $10.68 - (2.3 + 2.3) = 10.68 - 4.60 = 6.08$

The unknown length is 6.08 inches

109. 1 nickel, 1 dime, and 2 pennies:

$0.05 + 0.10 + 0.01 + 0.01 = 0.17$

3 nickels and 2 pennies:

$0.05 + 0.05 + 0.05 + 0.01 + 0.01 = 0.17$

1 dime and 7 pennies:

$0.10 + 0.01 + 0.01 + 0.01 + 0.01 + 0.01 + 0.01$
$+ 0.01 = 0.17$

2 nickels and 7 pennies:

$0.05 + 0.05 + 0.01 + 0.01 + 0.01 +$

$0.01 + 0.01 + 0.01 + 0.01 = 0.17$

113. answers may vary

Exercise Set 5.3

1. 0.17 2 decimal places
$\underline{\times\ 8}$ 0 decimal places
1.36 $2 + 0 = 2$ decimal places

5. The product $(-2.3)(7.65)$ is negative.
  7.65 2 decimal places
$\underline{\times\ 2.3}$ 1 decimal place
  2 295
$\underline{15\ 300}$
-17.595 $2 + 1 = 3$ decimal places and include the negative sign

9. Exact: 6.8 Estimate: 7
$\underline{\times\ 4.2}$ $\underline{\times\ 4}$
  1 36 28
$\underline{27\ 20}$
  28.56

13. Exact: 1.0047 Estimate: 1
$\underline{\times\ 8.2}$ $\underline{\times\ 8}$
  20094 8
$\underline{803760}$
  8.23854

17. $6.5 \times 10 = 65$

21. $(-7.093)(1000) = -7093$

25. $(-9.83)(-0.01) = 0.0983$

29. 0.123
$\underline{\times\ \ 8.4}$
  0.492

33. 430
$\underline{860}$
  1.290

    or 1.29

37. $562.3 \times 0.001 = 0.5623$

41. 1.5 billion $= 1.5 \times 1$ billion

$= 1.5 \times 1,000,000,000$

$= 1,500,000,000$

The cost at launch was $1,500,000,000.

45. $xy = 3(-0.2) = -0.6$

49. $0.6x = 4.92$

$0.6(14.2) \overset{?}{=} 4.92$

$8.52 \overset{?}{=} 4.92$ False

No, 14.2 is not a solution.

53. $C = \pi d$ is $\pi(10 \text{ cm}) = 10\pi$ cm
$C \approx 10(3.14)$ cm $= 31.4$ cm

57. Multiply his hourly wage by the number of hours worked.

$$\begin{array}{r} 17.88 \\ \times\quad 40 \\ \hline 715.20 \end{array}$$

His pay for last week was \$715.20.

61. Area $=$ length $\cdot$ width

$$\begin{array}{r} 4.87 \\ \times\ 2.3 \\ \hline 1461 \\ 9740 \\ \hline 11.201 \end{array}$$

The face is 11.201 square inches.

65. $C = \pi \cdot d$
$C = \pi \cdot 135 = 135\pi$

$$\begin{array}{r} 135 \\ \times\ 3.14 \\ \hline 5\ 40 \\ 13\ 50 \\ 405\ 00 \\ \hline 423.90 \end{array}$$

He travels 135π meters or approximately 423.9 meters.

69. a. Circumference $= 2 \cdot \pi \cdot$ radius
Smaller circle:
$C = 2 \cdot \pi \cdot 10 = 20\pi$
$C \approx 20(3.14) = 62.8$
The circumference of the smaller circle is approximately 62.8 meters.
Larger circle:
$C = 2 \cdot \pi \cdot 20 = 40\pi$
$C \approx 40(3.14) = 125.6$
The circumference of the larger circle is approximately 125.6 meters.

b. Yes, the circumference gets doubled when the radius is doubled.

73.
$$\begin{array}{r} 1.04920 \\ \times\qquad 750 \\ \hline 524600 \\ 7344400 \\ \hline 786.9000 \end{array}$$

750 U.S. dollars is equivalent to 786.9 Canadian dollars.

77.
$$\begin{array}{r} 486 \\ 6)\overline{2916} \\ -24 \\ \hline 51 \\ -48 \\ \hline 36 \\ -36 \\ \hline 0 \end{array}$$

81.
$$\begin{array}{r} 3.60 \\ +\ 0.04 \\ \hline 3.64 \end{array}$$

85. The product of a negative number and a positive number is a negative number.

$$\begin{array}{r} 0.221 \\ \times\ 0.5 \\ \hline 0.1105 \end{array}$$

The product is -0.1105.

89. answers may vary

Exercise Set 5.4

1.
$$\begin{array}{r} 4.6 \\ 6)\overline{27.6} \\ -24 \\ \hline 3\ 6 \\ -3\ 6 \\ \hline 0 \end{array}$$

5. $0.06)\overline{18}$ becomes
$$\begin{array}{r} 300 \\ 6)\overline{1800} \\ -18 \\ \hline 0 \end{array}$$

9. Exact: $5.5)\overline{36.3}$ becomes
$$\begin{array}{r} 6.6 \\ 55)\overline{363.0} \\ -330 \\ \hline 33\ 0 \\ -33\ 0 \\ \hline 0 \end{array}$$

Estimate: $6)\overline{36}$ with quotient 6

13. A positive number divided by a negative number is a negative number.

$0.06)\overline{36}$ becomes
$$\begin{array}{r} 600 \\ 6)\overline{3600} \\ -36 \\ \hline 0 \end{array}$$
$36 \div (-0.06) = -600$

17. $0.27)\overline{1.296}$ becomes
$$\begin{array}{r} 4.8 \\ 27)\overline{129.6} \\ -108 \\ \hline 21\ 6 \\ -21\ 6 \\ \hline 0 \end{array}$$

21. $0.82)\overline{4.756}$ because
$$\begin{array}{r} 5.8 \\ 82)\overline{475.6} \\ -410 \\ \hline 65\ 6 \\ -65\ 6 \\ \hline 0 \end{array}$$

25. Exact: $7.2)\overline{70.56}$ becomes
$$\begin{array}{r} 9.8 \\ 72)\overline{705.6} \\ -648 \\ \hline 57\ 6 \\ -57\ 6 \\ \hline 0 \end{array}$$

Estimate: $7)\overline{70}$ with quotient 10

29. $0.027\overline{)1.215}$ becomes
$$
\begin{array}{r}
45 \\
27\overline{)\,1215} \\
-108 \\
\hline
135 \\
-135 \\
\hline
0
\end{array}
$$

33. $3.78\overline{)0.02079}$ becomes
$$
\begin{array}{r}
0.0055 \\
378\overline{)\,2.0790} \\
-1\,890 \\
\hline
1890 \\
-1890 \\
\hline
0
\end{array}
$$

37. $0.6\overline{)68.39}$ becomes
$$
\begin{array}{r}
113.98 \approx 114.0 \\
6\overline{)\,683.90} \\
-6 \\
\hline
08 \\
-6 \\
\hline
23 \\
-18 \\
\hline
5\,9 \\
-5\,4 \\
\hline
50 \\
-48 \\
\hline
2
\end{array}
$$

41. $\dfrac{26.87}{10} = 2.687$

45.
$$
\begin{array}{r}
12.6 \\
7\overline{)\,88.2} \\
-7 \\
\hline
18 \\
-14 \\
\hline
4\,2 \\
-4\,2 \\
\hline
0
\end{array}
$$

49. $\dfrac{456.25}{10,000} = 0.045625$

53. $0.6\overline{)4.8}$ becomes
$$
\begin{array}{r}
8 \\
6\overline{)\,48} \\
-48 \\
\hline
0
\end{array}
$$
$4.8 \div (-0.6) = -8$

57. $0.03\overline{)42}$ becomes
$$
\begin{array}{r}
1400 \\
3\overline{)\,4200} \\
-3 \\
\hline
12 \\
-12 \\
\hline
0
\end{array}
$$

61. $0.0015\overline{)87}$ becomes
$$
\begin{array}{r}
58,000 \\
15\overline{)\,870,000} \\
-75 \\
\hline
120 \\
-120 \\
\hline
0
\end{array}
$$
$87 \div (-0.0015) = -58,000$

65. $-2.4 \div (-100) = \dfrac{-2.4}{-100} = 0.024$

69. $z \div y = 4.52 \div (-0.8)$

$0.8\overline{)4.52}$ becomes
$$
\begin{array}{r}
5.65 \\
8\overline{)\,45.20} \\
-40 \\
\hline
5\,2 \\
-4\,8 \\
\hline
40 \\
-40 \\
\hline
0
\end{array}
$$
$z \div y = 4.52 \div (-0.8) = -5.65$

73. $\dfrac{x}{4} = 3.04$

$\dfrac{12.16}{4} \overset{?}{=} 3.04$

$3.04 = 3.04$ True

Yes, 12.16 is a solution.

77. Divide the square feet by the square feet per quart.
$$
\begin{array}{r}
10.5 \approx 11 \\
52\overline{)\,546.0} \\
-52 \\
\hline
26 \\
-0 \\
\hline
26\,0 \\
-26\,0 \\
\hline
0
\end{array}
$$
Since only whole quarts are sold 11 quarts are needed.

81. Divide the number of crayons by 64.
$$
\begin{array}{r}
11.40 \text{ rounded to the nearest tenth is 11.4 boxes.} \\
64\overline{)\,730.00} \\
-64 \\
\hline
90 \\
-64 \\
\hline
26\,0 \\
-25\,6 \\
\hline
40
\end{array}
$$

85. From Exercise 83, we know that there are 24 teaspoons in 4 fluid ounces. Thus, there are 48 half teaspoons (0.5 tsp) or doses in 4 fluid ounces. To see how long the medicine will last, if a dose is taken every 4 hours, there are $24 \div 4 = 6$ doses taken per day. 48 (doses) $\div$ 6 (per day) = 8 days. The medicine will last 8 days.

89. Divide the number of books sold by the number of hours.
$$
\begin{array}{r}
345.5 \\
24\overline{)\,8292.0} \\
-72 \\
\hline
109 \\
-96 \\
\hline
132 \\
-120 \\
\hline
12\,0 \\
-12\,0 \\
\hline
0
\end{array}
$$
There were 345.5 thousand books sold per hour.

93. $\dfrac{3}{5} - \dfrac{7}{10} = \dfrac{3}{5} \cdot \dfrac{2}{2} - \dfrac{7}{10} = \dfrac{6}{10} - \dfrac{7}{10} = -\dfrac{1}{10}$

97.
```
  1.278
+0.300
 1.578
```

101.
```
 1000.00
-  95.71
  904.29
```

105. $78.6 \div 97$ is approximately $78.6 \div 100 = 0.786$, which is choice b.

109. Area $=$ (length)(width)

$4.5\overline{)38.7}$ becomes
```
        8.6
45)  387.0
    -360
      27 0
     -27 0
         0
```
The length is 8.6 feet.

113. $1.15\overline{)75}$ becomes
```
            65.21   ≈ 65.2
115)  7500.00
     -690
       600
      -575
       250
      -230
       200
      -115
        85
```

$1.15\overline{)95}$ becomes
```
            82.60   ≈ 82.6
115)  9500.00
     -920
       300
      -230
       700
      -690
       100
       -0
       100
```
The range of wind speeds is 65.2–82.6 knots.

Exercise Set 5.5

1.
```
      0.2
5)  1.0
   -1.0
      0
```
$\dfrac{1}{5} = 0.2$

5.
```
      0.75
4)  3.00
   -2 8
      20
     -20
       0
```
$\dfrac{3}{4} = 0.75$

9.
```
      2.25
4)  9.00
   -8
    1 0
    -8
     20
    -20
      0
```
$\dfrac{9}{4} = 2.25$

13.
```
        0.425
40)  17.000
    -16 0
      1 00
      -80
       200
      -200
         0
```
$\dfrac{17}{40} = 0.425$

17.
```
      0.333...
3)  1.000
   -9
    10
    -9
    10
    -9
     1
```
$-\dfrac{1}{3} = -0.\overline{3}$

21.
```
       0.636363...
11)  7.000000
    -6 6
       40
      -33
       70
      -66
       40
      -33
       70
      -66
       40
      -33
        7
```
$\dfrac{7}{11} = 0.\overline{63}$

25.
```
        0.624
125)  78.000
     -75 0
       3 00
      -2 50
        500
       -500
          0
```
$\dfrac{78}{125} = 0.624$

29. $\dfrac{7}{16} = 0.4375 \approx 0.44$

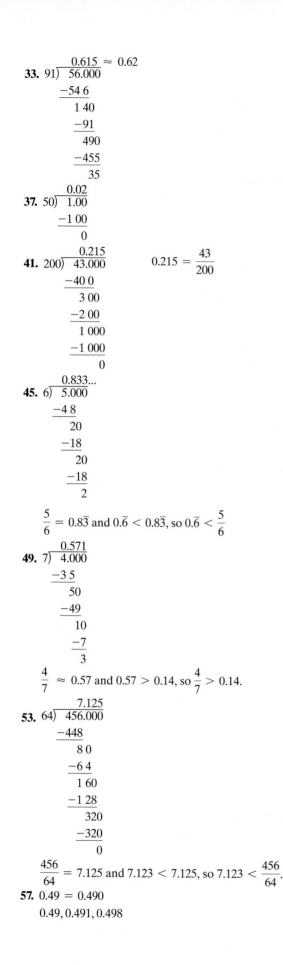

33.
$$\begin{array}{r} 0.615 \approx 0.62 \\ 91\overline{)\ 56.000} \\ -54\ 6 \\ \hline 1\ 40 \\ -91 \\ \hline 490 \\ -455 \\ \hline 35 \end{array}$$

37.
$$\begin{array}{r} 0.02 \\ 50\overline{)\ 1.00} \\ -1\ 00 \\ \hline 0 \end{array}$$

41.
$$\begin{array}{r} 0.215 \\ 200\overline{)\ 43.000} \\ -40\ 0 \\ \hline 3\ 00 \\ -2\ 00 \\ \hline 1\ 000 \\ -1\ 000 \\ \hline 0 \end{array}$$
$\qquad 0.215 = \dfrac{43}{200}$

45.
$$\begin{array}{r} 0.833... \\ 6\overline{)\ 5.000} \\ -4\ 8 \\ \hline 20 \\ -18 \\ \hline 20 \\ -18 \\ \hline 2 \end{array}$$

$\dfrac{5}{6} = 0.8\overline{3}$ and $0.\overline{6} < 0.8\overline{3}$, so $0.\overline{6} < \dfrac{5}{6}$

49.
$$\begin{array}{r} 0.571 \\ 7\overline{)\ 4.000} \\ -3\ 5 \\ \hline 50 \\ -49 \\ \hline 10 \\ -7 \\ \hline 3 \end{array}$$

$\dfrac{4}{7} \approx 0.57$ and $0.57 > 0.14$, so $\dfrac{4}{7} > 0.14$.

53.
$$\begin{array}{r} 7.125 \\ 64\overline{)\ 456.000} \\ -448 \\ \hline 8\ 0 \\ -6\ 4 \\ \hline 1\ 60 \\ -1\ 28 \\ \hline 320 \\ -320 \\ \hline 0 \end{array}$$

$\dfrac{456}{64} = 7.125$ and $7.123 < 7.125$, so $7.123 < \dfrac{456}{64}$.

57. $0.49 = 0.490$
$0.49, 0.491, 0.498$

61. $\dfrac{5}{8} = 0.625$
$0.612 = 0.612$
$0.649 = 0.649$
$0.612, \dfrac{5}{8}, 0.649$

65. $\dfrac{1 + 0.8}{-0.6} = \dfrac{1.8}{-0.6} = \dfrac{18}{-6} = -3$

69. $(5.6 - 2.3)(2.4 + 0.4) = (3.3)(2.8) = 9.24$

73. $\dfrac{7 + 0.74}{-6} = \dfrac{7.74}{-6} = -1.29$

77. $\dfrac{1}{4}(-9.6 - 5.2) = \dfrac{1}{4}(-14.8) = -3.7$

81. Area $= l \cdot w = (0.62)\left(\dfrac{2}{5}\right) = (0.62)(0.4) = 0.248$
The area is 0.248 square yard.

85. $x - y = 6 - 0.3 = 5.7$

89. $\dfrac{9}{10} + \dfrac{16}{25} = \dfrac{9}{10} \cdot \dfrac{5}{5} + \dfrac{16}{25} \cdot \dfrac{2}{2} = \dfrac{45}{50} + \dfrac{32}{50} = \dfrac{77}{50}$

93. $1.0 = 1$

97. $\dfrac{99}{100} = 0.99$
$\dfrac{99}{100} < 1$

101.

2020	rounds to	2000
1503	rounds to	1500
657	rounds to	700
597	rounds to	600
816	rounds to	800
+ 692	rounds to	+ 700
		6300

The total number of stations is estimated to be 6300.

Exercise Set 5.6

1.
$$\begin{aligned} x + 1.2 &= 7.1 \\ x + 1.2 - 1.2 &= 7.1 - 1.2 \\ x &= 5.9 \end{aligned}$$

5.
$$\begin{aligned} 6.2 &= y - 4 \\ 6.2 + 4 &= y - 4 + 4 \\ 10.2 &= y \end{aligned}$$

9.
$$\begin{aligned} -3.5x + 2.8 &= -11.2 \\ -3.5x + 2.8 - 2.8 &= -11.2 - 2.8 \\ -3.5x &= -14 \\ \dfrac{-3.5x}{-3.5} &= \dfrac{-14}{-3.5} \\ x &= 4 \end{aligned}$$

13.
$$\begin{aligned} 2(x - 1.3) &= 5.8 \\ 2x - 2.6 &= 5.8 \\ 2x - 2.6 + 2.6 &= 5.8 + 2.6 \\ 2x &= 8.4 \\ \dfrac{2x}{2} &= \dfrac{8.4}{2} \\ x &= 4.2 \end{aligned}$$

17.
$$7x - 10.8 = x$$
$$7x - 10.8 - 7x = x - 7x$$
$$-10.8 = -6x$$
$$\frac{-10.8}{-6} = \frac{-6x}{-6}$$
$$1.8 = x$$

21.
$$y - 3.6 = 4$$
$$y - 3.6 + 3.6 = 4 + 3.6$$
$$y = 7.6$$

25.
$$6.5 = 10x + 7.2$$
$$6.5 - 7.2 = 10x + 7.2 - 7.2$$
$$-0.7 = 10x$$
$$\frac{-0.7}{10} = \frac{10x}{10}$$
$$-0.07 = x$$

29.
$$200x - 0.67 = 100x + 0.81$$
$$200x - 0.67 + 0.67 = 100x + 0.81 + 0.67$$
$$200x = 100x + 1.48$$
$$200x - 100x = 100x - 100x + 1.48$$
$$100x = 1.48$$
$$\frac{100x}{100} = \frac{1.48}{100}$$
$$x = 0.0148$$

33.
$$8x - 5 = 10x - 8$$
$$8x - 5 + 8 = 10x - 8 + 8$$
$$8x + 3 = 10x$$
$$8x + 3 - 8x = 10x - 8x$$
$$3 = 2x$$
$$\frac{3}{2} = \frac{2x}{2}$$
$$1.5 = x$$

37. $-0.9x + 2.65 = -0.5x + 5.45$

Multiply each term by 100.
$$-90x + 265 = -50x + 545$$
$$-90x + 265 + 90x = -50x + 545 + 90x$$
$$265 = 40x + 545$$
$$265 - 545 = 40x + 545 - 545$$
$$-280 = 40x$$
$$\frac{-280}{40} = \frac{40x}{40}$$
$$-7 = x$$

41. $0.7x + 13.8 = x - 2.16$

Multiply each term by 100.
$$70x + 1380 = 100x - 216$$
$$70x + 1380 + 216 = 100x - 216 + 216$$
$$70x + 1596 = 100x$$
$$70x + 1596 - 70x = 100x - 70x$$
$$1596 = 30x$$
$$\frac{1596}{30} = \frac{30x}{30}$$
$$53.2 = x$$

45. $\dfrac{6x}{5} \cdot \dfrac{1}{2x^2} = \dfrac{6x \cdot 1}{5 \cdot 2x^2} = \dfrac{2 \cdot 3 \cdot x}{5 \cdot 2 \cdot x \cdot x} = \dfrac{3}{5x}$

49.
$$b + 4.6 = 8.3$$
$$b + 4.6 - 4.6 = 8.3 - 4.6$$
$$b = 3.7$$

53.
$$5y - 1.2 - 7y + 8 = 5y - 7y - 1.2 + 8$$
$$= -2y + 6.8$$

57.
$$4.7x + 8.3 = -5.8$$
$$4.7x + 8.3 - 8.3 = -5.8 - 8.3$$
$$4.7x = -14.1$$
$$\frac{4.7x}{4.7} = \frac{-14.1}{4.7}$$
$$x = -3$$

61.
$$5(x - 3.14) = 4x$$
$$5 \cdot x - 5 \cdot 3.14 = 4x$$
$$5x - 15.7 = 4x$$
$$5x - 5x - 15.7 = 4x - 5x$$
$$-15.7 = -x$$
$$15.7 = x$$

65.
$$9.6z - 3.2 - 11.7z - 6.9 = 9.6z - 11.7z - 3.2 - 6.9$$
$$= -2.1z - 10.1$$

69. answers may vary

73.
$$1.95y + 6.834 = 7.65y - 19.8591$$
$$1.95y + 6.834 - 6.834 = 7.65y - 19.8591 - 6.834$$
$$1.95y = 7.65y - 26.6931$$
$$1.95y - 7.65y = 7.65y - 7.65y - 26.6931$$
$$-5.7y = -26.6931$$
$$\frac{-5.7y}{-5.7} = \frac{-26.6931}{-5.7}$$
$$y = 4.683$$

Exercise Set 5.7

1. Mean: $\dfrac{15 + 23 + 24 + 18 + 25}{5} = \dfrac{105}{5} = 21$

Median: Write the numbers in order: $15, 18, 23, 24, 25$

The middle number is 23.

Mode: There is no mode, since each number occurs once.

5. Mean:
$$\frac{0.5 + 0.2 + 0.2 + 0.6 + 0.3 + 1.3 + 0.8 + 0.1 + 0.5}{9}$$
$$= \frac{4.5}{9}$$
$$= 0.5$$

Median: Write the numbers in order:

$0.1, 0.2, 0.2, 0.3, 0.5, 0.5, 0.6, 0.8, 1.3$

The middle number is 0.5.

Mode: Since 0.2 and 0.5 occur twice, there are two modes, 0.2 and 0.5.

9. Mean:
$$\frac{2717 + 1972 + 1667 + 1614 + 1588}{5} = \frac{9558}{5}$$
$$= 1911.6$$

The mean height of the five tallest buildings is 1911.6 feet.

13. answers may vary

17. $\text{GPA} = \dfrac{4 \cdot 3 + 4 \cdot 3 + 4 \cdot 4 + 3 \cdot 3 + 2 \cdot 1}{3 + 3 + 4 + 3 + 1}$

$= \dfrac{51}{14}$

≈ 3.64

21. Mode: 6.9 since this number appears twice.

25. Mean: $\dfrac{\text{sum of 15 pulse rates}}{15} = \dfrac{1095}{15} = 73$

29. There are 9 rates lower than the mean. They are 66, 68, 71, 64, 71, 70, 65, 70, and 72.

33. $\dfrac{18}{30y} = \dfrac{3 \cdot 6}{5 \cdot 6 \cdot y} = \dfrac{3}{5y}$

37. Since the mode is 35, 35 must occur at least twice in the set.

Since there is an odd number of numbers in the set, the median, 37, is in the set.

Let n be the remaining unknown number.

Mean: $\dfrac{35 + 35 + 37 + 40 + n}{5} = 38$

$\dfrac{147 + n}{5} = 38$

$5 \cdot \dfrac{147 + n}{5} = 5 \cdot 38$

$147 + n = 190$

$147 - 147 + n = 190 - 147$

$n = 43$

The missing numbers are 35, 35, 37, and 43.

Chapter 5 Test

1. 45.092 in words is forty-five and ninety-two thousandths.

5. Subtract the absolute values.

```
  30.25
−  9.83
───────
  20.42
```

Attach the sign of the larger absolute value.
$9.83 - 30.25 = -20.42$

9. 0.8623 rounded to the nearest thousandth is 0.862.

13. $-24.73 = -24\dfrac{73}{100}$

17. $\dfrac{0.23 + 1.63}{-0.3} = \dfrac{1.86}{-0.3} = -6.2$

21. Mean: $\dfrac{26 + 32 + 42 + 43 + 49}{5} = \dfrac{192}{5} = 38.4$

Median: The numbers are listed in order. The middle number is 42.

Mode: There is no mode since each number occurs only once.

25. Area $= \dfrac{1}{2}(4.2 \text{ miles})(1.1 \text{ miles})$

$= 0.5(4.2)(1.1)$ square miles

$= 2.31$ square miles

Chapter 6

Exercise Set 6.1

1. The ratio of 16 to 24 is $\dfrac{16}{24} = \dfrac{8 \cdot 2}{8 \cdot 3} = \dfrac{2}{3}$.

5. The ratio of 4.63 to 8.21 is

$\dfrac{4.63}{8.21} = \dfrac{4.63 \cdot 100}{8.21 \cdot 100} = \dfrac{463}{821}$.

9. The ratio of \$32 to \$100 is $\dfrac{32}{100} = \dfrac{4 \cdot 8}{4 \cdot 25} = \dfrac{8}{25}$.

13. The ratio of $3\dfrac{1}{2}$ to $12\dfrac{1}{4}$ is

$\dfrac{3\frac{1}{2}}{12\frac{1}{4}} = 3\dfrac{1}{2} \div 12\dfrac{1}{4}$

$= \dfrac{7}{2} \div \dfrac{49}{4}$

$= \dfrac{7}{2} \cdot \dfrac{4}{49}$

$= \dfrac{7 \cdot 2 \cdot 2}{2 \cdot 7 \cdot 7}$

$= \dfrac{2}{7}$

17. The ratio of the average Fin Whale weight to the average Blue Whale weight is

$\dfrac{50 \text{ tons}}{145 \text{ tons}} = \dfrac{5 \cdot 10}{5 \cdot 29} = \dfrac{10}{29}$.

21. The ratio of women to men is

$\dfrac{125}{100} = \dfrac{5 \cdot 25}{4 \cdot 25} = \dfrac{5}{4}$.

25. The ratio of the mini DVD diameter to the standard DVD diameter is

$\dfrac{8 \text{ centimeters}}{12 \text{ centimeters}} = \dfrac{2 \cdot 4}{3 \cdot 4} = \dfrac{2}{3}$.

29. Perimeter $= 8 + 15 + 17 = 40$

The ratio of the longest side to the perimeter is

$\dfrac{17 \text{ feet}}{40 \text{ feet}} = \dfrac{17}{40}$.

33. The rate of 5 shrubs every 15 feet is

$\dfrac{5 \text{ shrubs}}{15 \text{ feet}} = \dfrac{1 \text{ shrub}}{3 \text{ feet}}$.

37. The rate of 6 laser printers for 28 computers is

$\dfrac{6 \text{ laser printers}}{28 \text{ computers}} = \dfrac{3 \text{ laser printers}}{14 \text{ computers}}$.

41.
```
      110
  3)330
   −3
   ──
    03
   − 3
   ──
     0
```

330 calories in a 3-ounce serving is $\dfrac{110 \text{ calories}}{1 \text{ ounce}}$ or 110 calories/ounce.

45.
```
       90
  60)5400
    −540
    ────
       0
```

5400 wingbeats per 60 seconds is $\dfrac{90 \text{ wingbeats}}{1 \text{ second}}$ or 90 wingbeats/second.

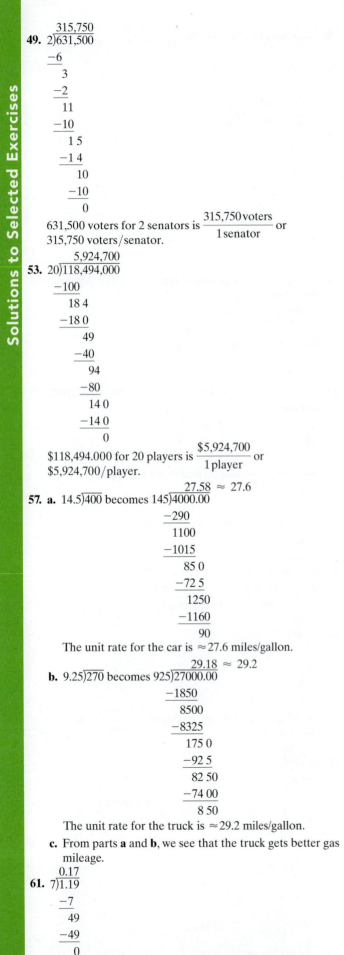

49.
$$
\begin{array}{r}
315{,}750 \\
2\overline{)631{,}500} \\
\underline{-6} \\
3 \\
\underline{-2} \\
11 \\
\underline{-10} \\
1\,5 \\
\underline{-1\,4} \\
10 \\
\underline{-10} \\
0
\end{array}
$$

631,500 voters for 2 senators is $\dfrac{315{,}750 \,\text{voters}}{1\,\text{senator}}$ or 315,750 voters/senator.

53.
$$
\begin{array}{r}
5{,}924{,}700 \\
20\overline{)118{,}494{,}000} \\
\underline{-100} \\
18\,4 \\
\underline{-18\,0} \\
49 \\
\underline{-40} \\
94 \\
\underline{-80} \\
14\,0 \\
\underline{-14\,0} \\
0
\end{array}
$$

$118,494.000 for 20 players is $\dfrac{\$5{,}924{,}700}{1\,\text{player}}$ or $5,924,700/player.

57. a. $14.5\overline{)400}$ becomes $145\overline{)4000.00}$
$$
\begin{array}{r}
27.58 \approx 27.6 \\
\underline{-290} \\
1100 \\
\underline{-1015} \\
85\,0 \\
\underline{-72\,5} \\
1250 \\
\underline{-1160} \\
90
\end{array}
$$

The unit rate for the car is ≈ 27.6 miles/gallon.

b. $9.25\overline{)270}$ becomes $925\overline{)27000.00}$
$$
\begin{array}{r}
29.18 \approx 29.2 \\
\underline{-1850} \\
8500 \\
\underline{-8325} \\
175\,0 \\
\underline{-92\,5} \\
82\,50 \\
\underline{-74\,00} \\
8\,50
\end{array}
$$

The unit rate for the truck is ≈ 29.2 miles/gallon.

c. From parts **a** and **b**, we see that the truck gets better gas mileage.

61.
$$
\begin{array}{r}
0.17 \\
7\overline{)1.19} \\
\underline{-7} \\
49 \\
\underline{-49} \\
0
\end{array}
$$

The unit price is $0.17 per banana.

65.
$$
\begin{array}{r}
0.1181 \approx 0.118 \\
16\overline{)1.8900} \\
\underline{-16} \\
29 \\
\underline{-16} \\
130 \\
\underline{-128} \\
20 \\
\underline{-16} \\
4
\end{array}
$$

The 16-ounce size costs $0.118 per ounce.

$$
\begin{array}{r}
0.115 \\
6\overline{)0.690} \\
\underline{-6} \\
9 \\
\underline{-6} \\
30 \\
\underline{-30} \\
0
\end{array}
$$

The 6-ounce size costs $0.115 per ounce.

The 6-ounce size is the better buy.

69.
$$
\begin{array}{r}
0.0059 \approx 0.006 \\
100\overline{)0.5900} \\
\underline{-500} \\
900 \\
\underline{-900} \\
0
\end{array}
$$

The 100-count size costs $0.006 per napkin.

$$
\begin{array}{r}
0.0051 \approx 0.005 \\
180\overline{)0.9300} \\
\underline{-900} \\
300 \\
\underline{-180} \\
120
\end{array}
$$

The 180-count size costs $0.005 per napkin.

The 180-count size is the better buy.

73. $3.7\overline{)0.555}$ becomes $37\overline{)5.55}$
$$
\begin{array}{r}
0.15 \\
\underline{-3\,7} \\
1\,85 \\
\underline{-1\,85} \\
0
\end{array}
$$

77. no; $\dfrac{6\ \text{inches}}{15\ \text{inches}} = \dfrac{2 \cdot 3}{5 \cdot 3} = \dfrac{2}{5}$

81. $29{,}543 - 29{,}286 = 257$

$13.4\overline{)257}$ becomes $134\overline{)2570.00}$
$$
\begin{array}{r}
19.17 \approx 19.2 \\
\underline{-134} \\
1230 \\
\underline{-1206} \\
24\,0 \\
\underline{-13\,4} \\
10\,60 \\
\underline{-9\,38} \\
1\,22
\end{array}
$$

There were 257 miles driven and the miles per gallon was 19.2.

85.
$$7759\overline{)11{,}674.00} \quad \frac{1.50}{} \approx 1.5$$

$$\frac{7\,759}{3\,915\,0}$$

$$\frac{-3\,879\,5}{35\,50}$$

The unit rate is 1.5 steps/foot.

89. answers may vary

Exercise Set 6.2

1. $\dfrac{10\ \text{diamonds}}{6\ \text{opals}} = \dfrac{5\ \text{diamonds}}{3\ \text{opals}}$

5. $\dfrac{6\ \text{eagles}}{58\ \text{sparrows}} = \dfrac{3\ \text{eagles}}{29\ \text{sparrows}}$

9. $\dfrac{22\ \text{vanilla wafers}}{1\ \text{cup cookie crumbs}} = \dfrac{55\ \text{vanilla wafers}}{2.5\ \text{cups cookie crumbs}}$

13. $\dfrac{5}{8} \overset{?}{=} \dfrac{4}{7}$

$5 \cdot 7 \overset{?}{=} 8 \cdot 4$

$35 \neq 32$

The proportion is false.

17. $\dfrac{5}{8} \overset{?}{=} \dfrac{625}{1000}$

$5 \cdot 1000 \overset{?}{=} 8 \cdot 625$

$5000 = 5000$

The proportion is true.

21. $\dfrac{4.2}{8.4} \overset{?}{=} \dfrac{5}{10}$

$4.2 \cdot 10 \overset{?}{=} 8.4 \cdot 5$

$42 = 42$

The proportion is true.

25. $\dfrac{2\frac{2}{5}}{\frac{2}{3}} \overset{?}{=} \dfrac{\frac{10}{9}}{\frac{1}{4}}$

$2\frac{2}{5} \cdot \frac{1}{4} \overset{?}{=} \frac{2}{3} \cdot \frac{10}{9}$

$\dfrac{12}{5} \cdot \dfrac{1}{4} \overset{?}{=} \dfrac{20}{27}$

$\dfrac{12}{20} \neq \dfrac{20}{27}$

The proportion is false.

29. $\dfrac{10}{15} \overset{?}{=} \dfrac{4}{6}$

$10 \cdot 6 \overset{?}{=} 15 \cdot 4$

$60 = 60$

The proportion $\dfrac{10}{15} = \dfrac{4}{6}$ is true.

33. $\dfrac{0.15}{3} \overset{?}{=} \dfrac{0.35}{7}$

$0.15 \cdot 7 \overset{?}{=} 3 \cdot 0.35$

$1.05 = 1.05$

The proportion $\dfrac{0.15}{3} = \dfrac{0.35}{7}$ is true.

37. $\dfrac{x}{5} = \dfrac{6}{10}$

$x \cdot 10 = 5 \cdot 6$

$10x = 30$

$\dfrac{10x}{10} = \dfrac{30}{10}$

$x = 3$

41. $\dfrac{30}{10} = \dfrac{15}{y}$

$30 \cdot y = 10 \cdot 15$

$30y = 150$

$\dfrac{30y}{30} = \dfrac{150}{30}$

$y = 5$

45. $\dfrac{24}{x} = \dfrac{60}{96}$

$24 \cdot 96 = x \cdot 60$

$2304 = 60x$

$\dfrac{2304}{60} = \dfrac{60x}{60}$

$38.4 = x$

49. $\dfrac{n}{0.6} = \dfrac{0.05}{12}$

$n \cdot 12 = 0.6 \cdot 0.05$

$12n = 0.030$

$\dfrac{12n}{12} = \dfrac{0.03}{12}$

$n = 0.0025$

53. $\dfrac{\frac{1}{3}}{\frac{3}{8}} = \dfrac{\frac{2}{5}}{n}$

$\dfrac{1}{3} \cdot n = \dfrac{3}{8} \cdot \dfrac{2}{5}$

$\dfrac{n}{3} = \dfrac{3}{20}$

$3 \cdot \dfrac{n}{3} = 3 \cdot \dfrac{3}{20}$

$n = \dfrac{9}{20}$

57. $\dfrac{n}{1\frac{1}{5}} = \dfrac{4\frac{1}{6}}{6\frac{2}{3}}$

$n \cdot 6\frac{2}{3} = 1\frac{1}{5} \cdot 4\frac{1}{6}$

$n \cdot \dfrac{20}{3} = \dfrac{6}{5} \cdot \dfrac{25}{6}$

$\dfrac{20}{3}n = 5$

$\dfrac{3}{20} \cdot \dfrac{20}{3}n = \dfrac{3}{20} \cdot 5$

$n = \dfrac{3}{4}$

61.
$$\frac{3.2}{0.3} = \frac{x}{1.4}$$
$$3.2 \cdot 1.4 = 0.3 \cdot x$$
$$4.48 = 0.3x$$
$$\frac{4.48}{0.3} = \frac{0.3x}{0.3}$$
$$14.9 \approx x$$

65.
$$\frac{7}{18} = \frac{x}{5}$$
$$7 \cdot 5 = 18 \cdot x$$
$$35 = 18x$$
$$\frac{35}{18} = \frac{18x}{18}$$
$$1.9 \approx x$$

69. 8.01　　8.1
$$\uparrow \qquad \uparrow$$
$$0 \; < \; 1$$
$$8.01 < 8.1$$

73. $\dfrac{75}{125} = \dfrac{3 \cdot 25}{5 \cdot 25} = \dfrac{3}{5}$

77.
$$\frac{9}{15} = \frac{3}{5}$$
$$\frac{9}{3} = \frac{15}{5}$$
$$\frac{5}{15} = \frac{3}{9}$$
$$\frac{15}{9} = \frac{5}{3}$$

81. $\dfrac{a}{b} = \dfrac{c}{d}$

Possible answers include:
$$\frac{d}{b} = \frac{c}{a}$$
$$\frac{a}{c} = \frac{b}{d}$$
$$\frac{b}{a} = \frac{d}{c}$$

85.
$$\frac{x}{7} = \frac{0}{8}$$
$$x \cdot 8 = 7 \cdot 0$$
$$8x = 0$$
$$\frac{8x}{8} = \frac{0}{8}$$
$$x = 0$$

89.
$$\frac{222}{1515} = \frac{37}{y}$$
$$222 \cdot y = 1515 \cdot 37$$
$$222y = 56{,}055$$
$$\frac{222y}{222} = \frac{56{,}055}{222}$$
$$y = 252.5$$

Exercise Set 6.3

1. Let x be the number of field goals (baskets) made.

baskets $\to \dfrac{45}{100} = \dfrac{x}{800} \gets$ baskets
attempts $\to \phantom{\dfrac{45}{100}} \phantom{\dfrac{x}{800}} \gets$ attempts
$$45 \cdot 800 = 100 \cdot x$$
$$36{,}000 = 100x$$
$$\frac{36{,}000}{100} = \frac{100x}{100}$$
$$360 = x$$

He made 360 baskets.

5. Let x be the number of applications received.

accepted $\to \dfrac{2}{7} = \dfrac{180}{x} \gets$ accepted
applied $\to \phantom{\dfrac{2}{7}} \phantom{\dfrac{180}{x}} \gets$ applied
$$2 \cdot x = 7 \cdot 180$$
$$2x = 1260$$
$$\frac{2x}{2} = \frac{1260}{2}$$
$$x = 630$$

The school received 630 applications.

9. Let x be the number of square feet required.

floor space $\to \dfrac{9}{1} = \dfrac{x}{30} \gets$ floor space
students $\to \phantom{\dfrac{9}{1}} \phantom{\dfrac{x}{30}} \gets$ students
$$9 \cdot 30 = 1 \cdot x$$
$$270 = x$$

30 students require 270 square feet of floor space.

13. Let x be the distance between Milan and Rome.

kilometers $\to \dfrac{30}{1} = \dfrac{x}{15} \gets$ kilometers
cm on map $\to \phantom{\dfrac{30}{1}} \phantom{\dfrac{x}{15}} \gets$ cm on map
$$30 \cdot 15 = 1 \cdot x$$
$$450 = x$$

Milan and Rome are 450 kilometers apart.

17. Let x be the number of hits the player is expected to get.

hits $\to \dfrac{3}{8} = \dfrac{x}{40} \gets$ hits
at bats $\to \phantom{\dfrac{3}{8}} \phantom{\dfrac{x}{40}} \gets$ at bats
$$3 \cdot 40 = 8 \cdot x$$
$$120 = 8x$$
$$\frac{120}{8} = \frac{8x}{8}$$
$$15 = x$$

The player would be expected to get 15 hits.

21. Let x be the number of applications she should expect.

applications $\to \dfrac{4}{3} = \dfrac{x}{14} \gets$ applications
ounces $\to \phantom{\dfrac{4}{3}} \phantom{\dfrac{x}{14}} \gets$ ounces
$$4 \cdot 14 = 3 \cdot x$$
$$56 = 3x$$
$$\frac{56}{3} = \frac{3x}{3}$$
$$18\frac{2}{3} = x$$

She should expect 18 applications from the 14-ounce bottle.

89. $\dfrac{3}{4} - \dfrac{1}{2} \cdot \dfrac{8}{9} = \dfrac{3}{4} - \dfrac{1 \cdot 8}{2 \cdot 9}$

$\phantom{\dfrac{3}{4} - \dfrac{1}{2} \cdot \dfrac{8}{9}} = \dfrac{3}{4} - \dfrac{1 \cdot 2 \cdot 4}{2 \cdot 9}$

$\phantom{\dfrac{3}{4} - \dfrac{1}{2} \cdot \dfrac{8}{9}} = \dfrac{3}{4} - \dfrac{4}{9}$

$\phantom{\dfrac{3}{4} - \dfrac{1}{2} \cdot \dfrac{8}{9}} = \dfrac{3}{4} \cdot \dfrac{9}{9} - \dfrac{4}{9} \cdot \dfrac{4}{4}$

$\phantom{\dfrac{3}{4} - \dfrac{1}{2} \cdot \dfrac{8}{9}} = \dfrac{27}{36} - \dfrac{16}{36}$

$\phantom{\dfrac{3}{4} - \dfrac{1}{2} \cdot \dfrac{8}{9}} = \dfrac{27 - 16}{36}$

$\phantom{\dfrac{3}{4} - \dfrac{1}{2} \cdot \dfrac{8}{9}} = \dfrac{11}{36}$

93. a. 52.8647% rounded to the nearest tenth percent is 52.9%.

    b. 52.8647% rounded to the nearest hundredth percent is 52.86%.

97. $45\% + 40\% + 11\% = 96\%$

$100\% - 96\% = 4\%$

4% of the U.S. population have AB blood type.

101. A fraction written as a percent is greater than 100% when the numerator is <u>greater</u> than the denominator.

105. The longest bar corresponds to biomedical engineers, so that is predicted to be the fastest growing occupation.

109. answers may vary

Exercise Set 7.2

1. 18% of 81 is what number?

$18\% \cdot 81 = x$

5. 0.6 is 40% of what number?

$0.6 = 40\% \cdot x$

9. what number is 9% of 43?

$x = 9\% \cdot 43$

13. $10\% \cdot 35 = x$

$0.10 \cdot 35 = x$

$3.5 = x$

10% of 35 is 3.5.

17. $1.2 = 12\% \cdot x$

$1.2 = 0.12 \cdot x$

$\dfrac{1.2}{0.12} = \dfrac{0.12x}{0.12}$

$10 = x$

1.2 is 12% of 10.

21. $x \cdot 80 = 88$

$\dfrac{x \cdot 80}{80} = \dfrac{88}{80}$

$x = 1.1$

$x = 110\%$

88 is 110% of 80.

25. $0.1 = 10\% \cdot x$

$0.1 = 0.10 \cdot x$

$\dfrac{0.1}{0.1} = \dfrac{0.1x}{0.1}$

$1 = x$

0.1 is 10% of 1.

29. $82.5 = 16\dfrac{1}{2}\% \cdot x$

$82.5 = 0.165 \cdot x$

$\dfrac{82.5}{0.165} = \dfrac{0.165x}{0.165}$

$500 = x$

82.5 is $16\dfrac{1}{2}\%$ of 500.

33. $x = 42\% \cdot 60$

$x = 0.42 \cdot 60$

$x = 25.2$

25.2 is 42% of 60.

37. $120\% \cdot x = 42$

$1.20 \cdot x = 42$

$\dfrac{1.2x}{1.2} = \dfrac{42}{1.2}$

$x = 35$

120% of 35 is 42.

41. $x \cdot 600 = 3$

$\dfrac{x \cdot 600}{600} = \dfrac{3}{600}$

$x = 0.005$

$x = 0.5\%$

0.5% of 600 is 3.

45. $1575 = x \cdot 2500$

$\dfrac{1575}{2500} = \dfrac{x \cdot 2500}{2500}$

$0.63 = x$

$63\% = x$

1575 is 63% of 2500.

49. $\dfrac{27}{x} = \dfrac{9}{10}$

$27 \cdot 10 = x \cdot 9$

$270 = 9x$

$\dfrac{270}{9} = \dfrac{9x}{9}$

$30 = x$

53. $\dfrac{17}{12} = \dfrac{x}{20}$

57. $5 \cdot n = 32$

$\dfrac{5 \cdot n}{5} = \dfrac{32}{5}$

$n = \dfrac{32}{5}$

Choice c is correct.

61. answers may vary

65. Since 85 is less than 120, the percent is less than 100%; c.

69. Since 100% is 1, 100% of 45 is equal to 45; a.

73. answers may vary

77. $22,113 = 180\% \cdot x$

$22,113 = 1.80 \cdot x$

$\dfrac{22,113}{1.8} = \dfrac{1.8 \cdot x}{1.8}$

$12,285 = x$

22,113 is 180% of 12,285.

Exercise Set 7.3

1.   98% of 45 is <u>what number?</u>

  ↓   ↓   ↓

  percent   base   amount = a

$\dfrac{a}{45} = \dfrac{98}{100}$

5.   14.3 is   26% of <u>what number?</u>

   ↓   ↓   ↓

  amount percent   base = b

$\dfrac{14.3}{b} = \dfrac{26}{100}$

9. <u>what percent?</u> of 400 is 70?

        ↓   ↓   ↓

   percent = p   base amount

$\dfrac{70}{400} = \dfrac{p}{100}$

13. $\dfrac{a}{65} = \dfrac{40}{100}$ or $\dfrac{a}{65} = \dfrac{2}{5}$

$a \cdot 5 = 65 \cdot 2$

$5a = 130$

$\dfrac{5a}{5} = \dfrac{130}{5}$

$a = 26$

40% of 65 is 26.

17. $\dfrac{90}{b} = \dfrac{15}{100}$ or $\dfrac{90}{b} = \dfrac{3}{20}$

$90 \cdot 20 = b \cdot 3$

$1800 = 3b$

$\dfrac{1800}{3} = \dfrac{3b}{3}$

$600 = b$

15% of 600 is 90.

21. $\dfrac{42}{35} = \dfrac{p}{100}$ or $\dfrac{6}{5} = \dfrac{p}{100}$

$6 \cdot 100 = 5 \cdot p$

$600 = 5p$

$\dfrac{600}{5} = \dfrac{5p}{5}$

$120 = p$

42 is 120% of 35.

25. $\dfrac{3.7}{b} = \dfrac{10}{100}$ or $\dfrac{3.7}{b} = \dfrac{1}{10}$

$3.7 \cdot 10 - b \cdot 1$

$37 = b$

3.7 is 10% of 37.

29. $\dfrac{160}{b} = \dfrac{16}{100}$ or $\dfrac{160}{b} = \dfrac{4}{25}$

$160 \cdot 25 = b \cdot 4$

$4000 = 4b$

$\dfrac{4000}{4} = \dfrac{4b}{4}$

$1000 = b$

160 is 16% of 1000.

33. $\dfrac{a}{62} = \dfrac{89}{100}$

$a \cdot 100 = 62 \cdot 89$

$100a = 5518$

$\dfrac{100a}{100} = \dfrac{5518}{100}$

$a = 55.18$

55.18 is 89% of 62.

37. $\dfrac{105}{b} = \dfrac{140}{100}$ or $\dfrac{105}{b} = \dfrac{7}{5}$

$105 \cdot 5 = b \cdot 7$

$525 = 7b$

$\dfrac{525}{7} = \dfrac{7b}{7}$

$75 = b$

140% of 75 is 105.

41. $\dfrac{4}{800} = \dfrac{p}{100}$ or $\dfrac{1}{200} = \dfrac{p}{100}$

$1 \cdot 100 = 200 \cdot p$

$100 = 200p$

$\dfrac{100}{200} = \dfrac{200p}{200}$

$0.5 = p$

0.5% of 800 is 4.

45. $\dfrac{a}{48} = \dfrac{20}{100}$ or $\dfrac{a}{48} = \dfrac{1}{5}$

$a \cdot 5 = 48 \cdot 1$

$5a = 48$

$\dfrac{5a}{5} = \dfrac{48}{5}$

$a = 9.6$

20% of 48 is 9.6.

49. $-\dfrac{11}{16} + \left(-\dfrac{3}{16}\right) = \dfrac{-11 - 3}{16} = \dfrac{-14}{16} = -\dfrac{2 \cdot 7}{2 \cdot 8} = -\dfrac{7}{8}$

53. $\overset{1}{0.41}$

$\underline{+0.29}$

0.70

57. answers may vary

61. $\dfrac{p}{100} = \dfrac{13}{52}$

$\dfrac{25}{100} \overset{?}{=} \dfrac{13}{52}$

$\dfrac{1}{4} = \dfrac{1}{4}$ True

Yes, the percent is 25.

65. $\dfrac{a}{53{,}862} = \dfrac{22.3}{100}$

$a \cdot 100 = 22.3 \cdot 53{,}862$

$100a = 1{,}201{,}122.6$

$\dfrac{100a}{100} = \dfrac{1{,}201{,}122.6}{100}$

$a \approx 12{,}011.2$

Exercise Set 7.4

1. 24 is 1.5% of what number?

Method 1:

$24 = 1.5\% \cdot x$

$24 = 0.015x$

$\dfrac{24}{0.015} = \dfrac{0.015x}{0.015}$

$1600 = x$

1600 bolts were inspected.

Method 2:

$\dfrac{24}{b} = \dfrac{1.5}{100}$

$24 \cdot 100 = 1.5 \cdot b$

$2400 = 1.5b$

$\dfrac{2400}{1.5} = \dfrac{1.5b}{1.5}$

$1600 = b$

1600 bolts were inspected.

5. 378 is what percent of 2700?

Method 1:

$378 = x \cdot 2700$

$\dfrac{378}{2700} = \dfrac{x \cdot 2700}{2700}$

$0.14 = x$

$14\% = x$

The student spent 14% of last semester's college costs on books.

Method 2:

$\dfrac{378}{2700} = \dfrac{p}{100}$

$378 \cdot 100 = 2700 \cdot p$

$37{,}800 = 2700p$

$\dfrac{37{,}800}{2700} = \dfrac{2700p}{2700}$

$14 = p$

The student spent 14% of last semester's college costs on books.

9. 160,650 is what percent of 945,000?

Method 1:

$160{,}650 = x \cdot 945{,}000$

$\dfrac{160{,}650}{945{,}000} = \dfrac{x \cdot 945{,}000}{945{,}000}$

$0.17 = x$

$17\% = x$

17% of restaurants in America are pizza restaurants.

Method 2:

$\dfrac{160{,}650}{945{,}000} = \dfrac{p}{100}$

$160{,}650 \cdot 100 = 945{,}000 \cdot p$

$16{,}065{,}000 = 945{,}000p$

$\dfrac{16{,}065{,}000}{945{,}000} = \dfrac{945{,}000p}{945{,}000}$

$17 = p$

17% restaurants in America are pizza restaurants.

13. What number is 30% of 83,600?

Method 1:

$x = 30\% \cdot 83{,}600$

$x = 0.30 \cdot 83{,}600$

$x = 25{,}080$

The number of people employed as physician assistants is expected to be $83{,}600 + 25{,}080 = 108{,}680$.

Method 2:

$\dfrac{a}{83{,}600} = \dfrac{30}{100}$

$a \cdot 100 = 83{,}600 \cdot 30$

$100a = 2{,}508{,}000$

$\dfrac{100a}{100} = \dfrac{2{,}508{,}000}{100}$

$a = 25{,}080$

The number of people employed as physician assistants is expected to be $83{,}600 + 25{,}080 = 108{,}680$.

17. 41 is what percent of 135?

Method 1:

$41 = x \cdot 135$

$\dfrac{41}{135} = \dfrac{x \cdot 135}{135}$

$0.30 \approx x$

$30\% \approx x$

30% of the ski runs at Keystone ski area are rated intermediate.

Method 2:

$\dfrac{41}{135} = \dfrac{p}{100}$

$41 \cdot 100 = 135 \cdot p$

$4100 = 135p$

$\dfrac{4100}{135} = \dfrac{135p}{135}$

$0.30 \approx p$

30% of the ski runs at Keystone ski area are rated intermediate.

21. 10 is what percent of 80?

Method 1:

$10 = x \cdot 80$

$\dfrac{10}{80} = \dfrac{x \cdot 80}{80}$

$0.125 = x$

$12.5\% = x$

12.5% of the total calories come from fat.

Method 2:

$$\frac{10}{80} = \frac{p}{100}$$

$$10 \cdot 100 = 80 \cdot p$$

$$1000 = 80p$$

$$\frac{1000}{80} = \frac{80p}{80}$$

$$12.5 = p$$

12.5% of the total calories come from fat.

25. 26,250 is 15% of what number?

Method 1:

$$26{,}250 = 15\% \cdot x$$

$$26{,}250 = 0.15 \cdot x$$

$$\frac{26{,}250}{0.15} = \frac{0.15 \cdot x}{0.15}$$

$$175{,}000 = x$$

The price of the home was $175,000.

Method 2:

$$\frac{26{,}250}{b} = \frac{15}{100}$$

$$26{,}250 \cdot 100 = b \cdot 15$$

$$2{,}625{,}000 = 15b$$

$$\frac{2{,}625{,}000}{15} = \frac{15b}{15}$$

$$175{,}000 = b$$

The price of the home was $175,000.

29. What number is 4.5% of 19,286?

Method 1:

$$x = 4.5\% \cdot 19{,}286$$

$$x = 0.045 \cdot 19{,}286$$

$$x = 867.87$$

The price of the car will increase by $867.87. The new price of that model will be $19,286 + $867.87 = $20,153.87.

Method 2:

$$\frac{a}{19{,}286} = \frac{4.5}{100}$$

$$a \cdot 100 = 4.5 \cdot 19{,}286$$

$$100a = 86{,}787$$

$$a = \frac{86{,}787}{100}$$

$$a = 867.87$$

The price of the car will increase by $867.87. The new price of that model will be $19,286 + $867.87 = $20,153.87.

33. 15.3% of $8679 is what number?

Method 1:

$$15.3\%\,(8679) = x$$

$$0.153\,(8679) = x$$

$$1327.887 = x$$

$$1328 \approx x$$

The increase in tuition is $1328. The tuition for the 2013–2014 school year is $1328 + $8679 = $10,007.

Method 2:

$$\frac{a}{8679} = \frac{15.3}{100}$$

$$a \cdot 100 = 15.3 \cdot 8679$$

$$100a = 132788.7$$

$$a = \frac{132788.7}{100}$$

$$a = 1327.887$$

$$a \approx 1328$$

The increase in tuition is $1328. The tuition for the 2013–2014 school year is $1328 + $8679 = $10,007.

37.

Original Amount	New Amount	Amount of Increase	Percent Increase
50	80	$80 - 50 = 30$	$\dfrac{30}{50} = 0.6 = 60\%$

41.

Original Amount	New Amount	Amount of Decrease	Percent Decrease
8	6	$8 - 6 = 2$	$\dfrac{2}{8} = 0.25 = 25\%$

45. percent decrease $= \dfrac{\text{amount of decrease}}{\text{original amount}}$

$$= \frac{150 - 84}{150}$$

$$= \frac{66}{150}$$

$$= 0.44$$

The decrease in calories is 44%.

49. percent increase $= \dfrac{\text{amount of increase}}{\text{original amount}}$

$$= \frac{301{,}779 - 178{,}025}{178{,}025}$$

$$= \frac{123{,}754}{178{,}025}$$

$$\approx 0.695$$

The increase in cell sites was 69.5%.

53. percent decrease $= \dfrac{\text{amount of decrease}}{\text{original amount}}$

$$= \frac{5813 - 5331}{5813}$$

$$= \frac{482}{5813}$$

$$\approx 0.083$$

The decrease in indoor cinema sites was 8.3%.

57. percent increase $= \dfrac{\text{amount of increase}}{\text{original amount}}$

$$= \frac{420 - 174}{174}$$

$$= \frac{246}{174}$$

$$\approx 1.414$$

The increase in the size of farms in the United States was 141.4%.

61.
$$\begin{array}{r} 0.12 \\ \times\ 38 \\ \hline 96 \\ 360 \\ \hline 4.56 \end{array}$$

65. $-\dfrac{3}{8} + \dfrac{5}{12} = -\dfrac{3}{8}\cdot\dfrac{3}{3} + \dfrac{5}{12}\cdot\dfrac{2}{2}$

$\qquad\qquad = -\dfrac{9}{24} + \dfrac{10}{24}$

$\qquad\qquad = \dfrac{1}{24}$

69. The increased number is double the original number.

73. percent increase $= \dfrac{\text{amount of increase}}{\text{original amount}}$

$\qquad\qquad = \dfrac{180 - 150}{150}$

$\qquad\qquad = \dfrac{30}{150}$

$\qquad\qquad = 0.20$

The increase in population was 20%.

Exercise Set 7.5

1. sales tax $= 5\% \cdot \$150 = 0.05 \cdot \$150 = \$7.50$

The sales tax is $7.50.

5. $\$335.30 = r \cdot \$4790$

$\qquad \dfrac{335.30}{4790} = r$

$\qquad\quad 0.07 = r$

The sales tax rate is 7%.

9. sales tax $= 6.5\% \cdot \$1800 = 0.065 \cdot \$1800 = \$117$

total price $= \$1800 + \$117 = \$1917$

The sales tax is $117 and the total price of the bracelet is $1917.

13. $\$98.70 = r \cdot \$1645$

$\qquad \dfrac{98.70}{1645} = r$

$\qquad\quad 0.06 = r$

The sales tax rate is 6%.

17. commission $= 4\% \cdot \$1,329,401 = 0.04 \cdot \$1,329,401$

$\qquad\qquad\quad = \$53,176.04$

Her commission was $53,176.04.

21. commission $= 1.5\% \cdot \$325,900 = 0.015 \cdot \$325,900$

$\qquad\qquad\quad = \$4888.50$

His commission will be $4888.50.

	Original Price	Discount Rate	Amount of Discount	Sale Price
25.	$89	10%	$10\% \cdot \$89$ $= \$8.90$	$\$89 - \$8.90$ $= \$80.10$
29.	$410	35%	$35\% \cdot \$410$ $= \$143.50$	$\$410 - \$143.50$ $= \$266.50$

33. discount $= 15\% \cdot \$300 = 0.15 \cdot \$300 = \$45$

sale price $= \$300 - \$45 = \$255$

The discount is $45 and the sale price is $255.

37.

Purchase Price	Tax Rate	Sales Tax	Total Price
$56	5.5%	$5.5\% \cdot \$56 = \$3.08$	$\$56 + \$3.08$ $= \$59.08$

41.

Sale	Commission Rate	Commission
$17,900	$\dfrac{\$1432}{\$17,900} = 0.08 = 8\%$	$1432

45. $400 \cdot \dfrac{3}{100} \cdot 11 = 12 \cdot 11 = 132$

49. Round $68 to $70 and 9.5% to 10%.

$10\% \cdot \$70 = 0.10 \cdot \$70 = \$7$

$\$70 + \$7 = \$77$

The best estimate of the total price is $77; d.

53.

Bill Amount	10%	15%	20%
$\$72.17 \approx \$72.00$	$7.20	$\$7.20 + \dfrac{1}{2}(\$7.20)$ $= \$7.20 + \$3.60$ $= \$10.80$	$2(\$7.20)$ $= \$14.40$

57. $7.5\% \cdot \$24,966 = 0.075 \cdot \$24,966 = \$1872.45$

$\$24,966 + \$1872.45 = \$26,838.45$

The total price of the necklace is $26,838.45.

Exercise Set 7.6

1. simple interest $=$ principal $\cdot$ rate $\cdot$ time

$\qquad\qquad\qquad = (\$200)(8\%)(2)$

$\qquad\qquad\qquad = (\$200)(0.08)(2) = \$32$

5. simple interest $=$ principal $\cdot$ rate $\cdot$ time

$\qquad\qquad\qquad = (\$5000)(10\%)\left(1\dfrac{1}{2}\right)$

$\qquad\qquad\qquad = (\$5000)(0.10)(1.5) = \$750$

9. simple interest $=$ principal $\cdot$ rate $\cdot$ time

$\qquad\qquad\qquad = (\$2500)(16\%)\left(\dfrac{21}{12}\right)$

$\qquad\qquad\qquad = (\$2500)(0.16)(1.75) = \$700$

13. simple interest $=$ principal $\cdot$ rate $\cdot$ time

$\qquad\qquad\qquad = \$5000(9\%)\left(\dfrac{15}{12}\right)$

$\qquad\qquad\qquad = \$5000(0.09)(1.25) = \$562.50$

Total $= \$5000 + \$562.50 = \$5562.50$

17. $A = P\left(1 + \dfrac{r}{n}\right)^{n\cdot t}$

$\quad = 6150\left(1 + \dfrac{0.14}{2}\right)^{2\cdot 15}$

$\quad = 6150(1.07)^{30}$

$\quad \approx 46,815.37$

The total amount is $46,815.37.

21. $A = P\left(1 + \dfrac{r}{n}\right)^{n\cdot t}$

$\quad = 10,000\left(1 + \dfrac{0.09}{2}\right)^{2\cdot 20}$

$\quad = 10,000(1.045)^{40}$

$\quad \approx 58,163.65$

The total amount is $58,163.65.

25. $A = P\left(1 + \dfrac{r}{n}\right)^{n \cdot t}$

$\qquad = 2000\left(1 + \dfrac{0.08}{1}\right)^{1 \cdot 5}$

$\qquad = 2000(1.08)^5$

$\qquad \approx 2938.66$

The total amount is $2938.66

29. perimeter $= 10 + 6 + 10 + 6 = 32$

The perimeter is 32 yards.

33. $\dfrac{x}{4} + \dfrac{x}{5} = \dfrac{x}{4} \cdot \dfrac{5}{5} + \dfrac{x}{5} \cdot \dfrac{4}{4} = \dfrac{5x}{20} + \dfrac{4x}{20} = \dfrac{9x}{20}$

37. answers may vary

Chapter 7 Test

1. $85\% = 85(0.01) = 0.85$

5. $6.1 = 6.1(100\%) = 610\%$

9. $0.2\% = 0.2 \cdot \dfrac{1}{100} = \dfrac{0.2}{100} = \dfrac{2}{1000} = \dfrac{1}{500}$

13. $\dfrac{1}{5} = \dfrac{1}{5} \cdot \dfrac{100\%}{1} = \dfrac{100}{5}\% = 20\%$

17. *Method 1*:

$\qquad 567 = x \cdot 756$

$\qquad \dfrac{567}{756} = \dfrac{x \cdot 756}{756}$

$\qquad 0.75 = x$

$\qquad 75\% = x$

$\qquad$ 567 is 75% of 756.

Method 2:

$\qquad \dfrac{567}{756} = \dfrac{p}{100}$

$\qquad 567 \cdot 100 = 756 \cdot p$

$\qquad 56{,}700 = 756p$

$\qquad \dfrac{56{,}700}{756} = \dfrac{756p}{756}$

$\qquad 75 = p$

$\qquad$ 567 is 75% of 756.

21. percent increase $= \dfrac{\text{amount of increase}}{\text{original amount}}$

$\qquad = \dfrac{26{,}460 - 25{,}200}{25{,}200}$

$\qquad = \dfrac{1260}{25{,}200}$

$\qquad = 0.05$

The increase in population was 5%.

25. simple interest $= \text{principal} \cdot \text{rate} \cdot \text{time}$

$\qquad = (\$2000)(9.25\%)\left(3\dfrac{1}{2}\right)$

$\qquad = (\$2000)(0.0925)(3.5)$

$\qquad = \$647.50$

Chapter 8

Exercise Set 8.1

1. Kansas has the greatest number of wheat icons, so the greatest acreage of wheat was planted in the state of Kansas.

5. Texas is represented by 6 wheat icons, so the state of Texas plants about 6,000,000 acres of wheat.

9. The year 2008 has 6.5 flames and each flame represents 12,000 wildfires, so there were approximately $6.5(12{,}000) = 78{,}000$ wildfires in 2008.

13. 2006 has 8 flames, and 2012 has 5.5 flames, which is 2.5 less. Thus, the decrease in the number of wildfires from 2006 to 2012 was about $2.5(12{,}000) = 30{,}000$.

17. The longest bar corresponds to September, so the month in which most hurricanes made landfall is September.

21. Two of the 77 hurricanes that made landfall in August did so in 2008. The fraction is $\dfrac{2}{77}$.

25. The only bar corresponding to a city in the United States is the bar for New York. The population is approximately 21.6 million or 21,600,000.

29.

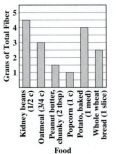

Fiber Content of Selected Foods

33. The height of the bar for 100–149 miles per week is 15, so 15 of the adults drive 100–149 miles per week.

37. 15 of the adults drive 100–149 miles per week and 9 of the adults drive 150–199 miles per week, so $15 + 9 = 24$ of the adults drive 100–199 miles per week.

41. 9 of the 100 adults surveyed drive 150–199 miles per week, so the ratio is $\dfrac{9}{100}$.

45. According to the bar graph, approximately 21 million householders were 55–64 years old.

49. According to the graph, 25 million householders were 45–54 years old and 21 million householders were 55–64 years old. $25 - 21 = 4$ million more householders were 45–54 than were 55–64 years old.

53.

Class Interval (Scores)	Tally	Class Frequency (Number of Games)
90–99	||||| |||	8

57.

Class Interval (Account Balances)	Tally	Class Frequency (Number of People)
$200–$299	|||||		6

61.

65. The highest point on the graph corresponds to 2003, so the average number of goals per game was the greatest in 2003.

69. The dots for 2001 and 2007 are below the 8-level, so the average number of goals per game was less than 8 in 2001 and 2007.

73. 10% of 62 is $0.10 \cdot 62 = 6.2$

77. $\dfrac{17}{50} = \dfrac{17}{50} \cdot 100\% = \dfrac{17 \cdot 2 \cdot 50}{50}\% = 34\%$

81. The lowest point on the graph of low temperatures corresponds to Sunday. The low temperature on Sunday was 68°F.

85. answers may vary

Exercise Set 8.2

1. The largest sector corresponds to the category "parent or guardian's home," so most of the students live in a parent or guardian's home.

5. 180 of the students live in campus housing while 320 live in a parent or guardian's home.
$$\dfrac{180}{320} = \dfrac{9}{16}$$
The ratio is $\dfrac{9}{16}$.

9. $30\% + 7\% = 37\%$
37% of the land on Earth is accounted for by Europe and Asia.

13. Australia accounts for 5% of the land on Earth.
5% of $57{,}000{,}000 = 0.05 \cdot 57{,}000{,}000$
$\qquad\qquad\qquad = 2{,}850{,}000$
Australia is 2,850,000 square miles.

17. The second-largest sector corresponds to nonfiction, so the second-largest category of books is nonfiction.

21. Children's fiction accounts for 22% of the books.
22% of $125{,}600 = 0.22 \cdot 125{,}600$
$\qquad\qquad\qquad = 27{,}632$
The library has 27,632 children's fiction books.

25.

Type of Apple	Percent	Degrees in Sector
Red Delicious	37%	37% of 360° = 0.37(360°) ≈ 133°
Golden Delicious	13%	13% of 360° = 0.13(360°) ≈ 47°
Fuji	14%	14% of 360° = 0.14(360°) ≈ 50°
Gala	15%	15% of 360° = 0.15(360°) = 54°
Granny Smith	12%	12% of 360° = 0.12(360°) ≈ 43°
Other varieties	6%	6% of 360° = 0.06(360°) ≈ 22°
Braeburn	3%	3% of 360° = 0.03(360°) ≈ 11°

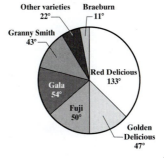

29. $20 = 2 \times 10 = 2 \times 2 \times 5 = 2^2 \times 5$

33. $85 = 5 \times 17$

37. Pacific Ocean:
$49\% \cdot 264{,}489{,}800 = 0.49 \cdot 264{,}489{,}800$
$\qquad\qquad\qquad\qquad = 129{,}600{,}002$ square kilometers

41. $24\% \cdot 2800 = 0.24 \cdot 2800 = 672$
672 respondents said that they spend $0 online each month.

45. $\dfrac{\text{number of respondents who spend } \$0}{\text{number of respondents who spend } \$1\text{–}\$100} = \dfrac{672}{1736}$
$$= \dfrac{12 \cdot 56}{31 \cdot 56}$$
$$= \dfrac{12}{31}$$

Exercise Set 8.3

1.

5. Point A has coordinates $(0, 0)$.
Point B has coordinates $\left(3\dfrac{1}{2}, 0\right)$.
Point C has coordinates $(3, 2)$.
Point D has coordinates $(-1, 3)$.
Point E has coordinates $(-2, -2)$.
Point F has coordinates $(0, -1)$.
Point G has coordinates $(2, -1)$.

9. $x - y = 3$
$1 - 2 \overset{?}{=} 3$
$\quad -1 \overset{?}{=} 3$ False
No, $(1, 2)$ is not a solution of $x - y = 3$.

13. $x = -3y$
$6 \overset{?}{=} -3(-2)$
$6 = 6$ True
Yes, $(6, -2)$ is a solution of $x = -3y$.

17. $x - 5y = -1$
$3 - 5(1) \overset{?}{=} -1$
$\quad 3 - 5 \overset{?}{=} -1$
$\qquad -2 \overset{?}{=} -1$ False
No, $(3, 1)$ is not a solution of $x - 5y = -1$.

21.

25. $y = -9x$
$y = -9(1)$
$y = -9$
The solution is $(1, -9)$.
$y = -9x$
$y = -9(0)$
$y = 0$
The solution is $(0, 0)$.
$y = -9x$
$-18 = -9x$
$\dfrac{-18}{-9} = \dfrac{-9x}{-9}$
$2 = x$
The solution is $(2, -18)$.

29. $x + y = -2$
$-2 + y = -2$
$-2 + 2 + y = -2 + 2$
$y = 0$
The solution is $(-2, 0)$.
$x + y = -2$
$1 + y = -2$
$1 - 1 + y = -2 - 1$
$y = -3$
The solution is $(1, -3)$.
$x + y = -2$
$x + 5 = -2$
$x + 5 - 5 = -2 - 5$
$x = -7$
The solution is $(-7, 5)$.

33. $y = 3x - 5$
$y = 3 \cdot 1 - 5$
$y = 3 - 5$
$y = -2$
The solution is $(1, -2)$.
$y = 3x - 5$
$y = 3 \cdot 2 - 5$
$y = 6 - 5$
$y = 1$
The solution is $(2, 1)$.
$y = 3x - 5$
$4 = 3x - 5$
$4 + 5 = 3x - 5 + 5$
$9 = 3x$
$\dfrac{9}{3} = \dfrac{3x}{3}$
$3 = x$
The solution is $(3, 4)$.

37. $x + 2y = -8$
$4 + 2y = -8$
$4 - 4 + 2y = -8 - 4$
$2y = -12$
$\dfrac{2y}{2} = \dfrac{-12}{2}$
$y = -6$

The solution is $(4, -6)$.
$x + 2y = -8$
$x + 2(-3) = -8$
$x - 6 = -8$
$x - 6 + 6 = -8 + 6$
$x = -2$
The solution is $(-2, -3)$.
$x + 2y = -8$
$0 + 2y = -8$
$2y = -8$
$\dfrac{2y}{2} = \dfrac{-8}{2}$
$y = -4$
The solution is $(0, -4)$.

41. $5.6 \cdot 3.9 = 21.84$

45. To plot (a, b), start at the origin and move a units to the right and b units up. The point will be in quadrant I. Thus, (a, b) is in quadrant I is a true statement.

49. To plot $(0, -b)$, start at the origin and move 0 units to the right and b units down. The point will be on the y-axis. Thus, $(0, -b)$ lies on the x-axis is a false statement.

53. To plot $(4, -3)$, start at the origin and move 4 units to the right, so the point $(4, -3)$ is plotted to the right of the y-axis.

57. $P = 2l + 2w$
Let $l = 6$ and $w = 4$.
$P = 2 \cdot 6 + 2 \cdot 4 = 12 + 8 = 20$
The perimeter is 20 units.

Exercise Set 8.4

1. $x + y = 4$
Find any 3 ordered pair solutions.
Let $x = 0$.
$x + y = 4$
$0 + y = 4$
$y = 4$
$(0, 4)$
Let $x = 2$.
$x + y = 4$
$2 + y = 4$
$2 - 2 + y = 4 - 2$
$y = 2$
$(2, 2)$
Let $x = 4$.
$x + y = 4$
$4 + y = 4$
$4 - 4 + y = 4 - 4$
$y = 0$
$(4, 0)$
Plot $(0, 4), (2, 2),$ and $(4, 0)$. Then draw the line through them.

5. $y = 4x$
Find any 3 ordered-pair solutions.
Let $x = 0$.
$y = 4x$
$y = 4(0)$
$y = 0$
$(0, 0)$

Let $x = 1$.
$y = 4x$
$y = 4(1)$
$y = 4$
$(1, 4)$
Let $x = -1$.
$y = 4x$
$y = 4(-1)$
$y = -4$
$(-1, -4)$
Plot $(0, 0)$, $(1, 4)$, and $(-1, -4)$. Then draw the line through them.

9. $x = -3$

No matter what y-value we choose, x is always -3.

x	y
-3	-4
-3	0
-3	4

13. $x = 0$

No matter what y-value we choose, x is always 0.

x	y
0	-3
0	0
0	3

17. $y = -2$

No matter what x-value we choose, y is always -2.

x	y
-4	-2
0	-2
4	-2

21. $x = 6$

No matter what y-value we choose, x is always 6.

x	y
6	-5
6	0
6	5

25. $x = y - 4$

Find any 3 ordered pair solutions.
Let $y = 0$.
$x = y - 4$
$x = 0 - 4$
$x = -4$
$(-4, 0)$
Let $y = 4$.
$x = y - 4$
$x = 4 - 4$
$x = 0$
$(0, 4)$

Let $y = 6$.
$x = y - 4$
$x = 6 - 4$
$x = 2$
$(2, 6)$
Plot $(-4, 0)$, $(0, 4)$, and $(2, 6)$. Then draw the line through them.

29. $y = -\dfrac{1}{4}x$

Find any 3 ordered pair solutions.
Let $x = -4$.
$y = -\dfrac{1}{4}x$
$y = -\dfrac{1}{4}(-4)$
$y = 1$
$(-4, 1)$
Let $x = 0$.
$y = -\dfrac{1}{4}x$
$y = -\dfrac{1}{4} \cdot 0$
$y = 0$
$(0, 0)$
Let $x = 4$.
$y = -\dfrac{1}{4}x$
$y = -\dfrac{1}{4} \cdot 4$
$y = -1$
$(4, -1)$

Plot $(-4, 1)$, $(0, 0)$, and $(4, -1)$. Then draw the line through them.

33. $y = 4x + 2$

Find any 3 ordered pair solutions.
Let $x = -1$.
$y = 4x + 2$
$y = 4(-1) + 2$
$y = -4 + 2$
$y = -2$
$(-1, -2)$
Let $x = 0$.
$y = 4x + 2$
$y = 4(0) + 2$
$y = 0 + 2$
$y = 2$
$(0, 2)$

Let $x = 1$.
$y = 4x + 2$
$y = 4(1) + 2$
$y = 4 + 2$
$y = 6$
$(1, 6)$
Plot $(-1, -2), (0, 2),$ and $(1, 6)$. Then draw the line through them.

37. $x = -3.5$

No matter what y-value we choose, x is always -3.5.

x	y
-3.5	-3
-3.5	0
-3.5	3

$x = -3.5$

41. $y = \dfrac{1}{2}$

No matter what x-value we choose, y is always $\dfrac{1}{2}$.

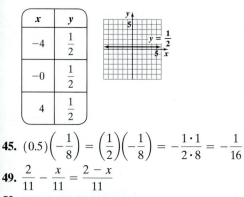

x	y
-4	$\dfrac{1}{2}$
-0	$\dfrac{1}{2}$
4	$\dfrac{1}{2}$

$y = \dfrac{1}{2}$

45. $(0.5)\left(-\dfrac{1}{8}\right) = \left(\dfrac{1}{2}\right)\left(-\dfrac{1}{8}\right) = -\dfrac{1 \cdot 1}{2 \cdot 8} = -\dfrac{1}{16}$

49. $\dfrac{2}{11} - \dfrac{x}{11} = \dfrac{2 - x}{11}$

53. answers may vary

57. The HEV line appears to be at 49,000 above May 2013. Approximately 49,000 HEVs were sold during May 2013.

61. Because the line corresponding to passenger cars from 2000 to 2004 is decreasing from left to right, passenger car sales were decreasing from 2000 to 2004.

65. answers may vary

Exercise Set 8.5

1.

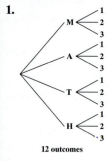

12 outcomes

5.

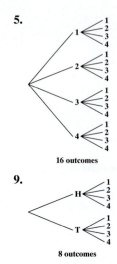

16 outcomes

9.

8 outcomes

13. A 1 or a 6 are two of the six possible outcomes. The probability is $\dfrac{2}{6} = \dfrac{1}{3}$.

17. Four of the six possible outcomes are numbers greater than 2. The probability is $\dfrac{4}{6} = \dfrac{2}{3}$.

21. A 1, a 2, or a 3 are three of three possible outcomes. The probability is $\dfrac{3}{3} = 1$.

25. One of the seven marbles is red. The probability is $\dfrac{1}{7}$.

29. Four of the seven marbles are either green or red. The probability is $\dfrac{4}{7}$.

33. The blood pressure did not change for 10 of the 200 people. The probability is $\dfrac{10}{200} = \dfrac{1}{20}$.

37. $\dfrac{1}{2} \cdot \dfrac{1}{3} = \dfrac{1 \cdot 1}{2 \cdot 3} = \dfrac{1}{6}$

41. One of the 52 cards is the king of hearts. The probability is $\dfrac{1}{52}$.

45. Thirteen of the 52 cards are hearts. The probability is $\dfrac{13}{52} = \dfrac{1}{4}$.

49.

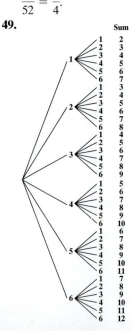

Five of the 36 sums are 6. The probability is $\dfrac{5}{36}$.

53. answers may vary

Chapter 8 Test

1. There are $4\dfrac{1}{2}$ dollar symbols for the second

week. Each dollar symbol corresponds to $50.
$$4\dfrac{1}{2} \cdot \$50 = \dfrac{9}{2} \cdot \$50 = \dfrac{\$450}{2} = \$225$$
$225 was collected during the second week.

5. The shortest bar corresponds to February. The normal monthly precipitation in February in Chicago is 3 centimeters.

9. The line graph is above the 3 level for 2004, 2005, and 2007. Thus the inflation rate was greater than 3% in 2004, 2005, and 2007.

13. People in their twenties are expected to account for 13.7% of the U.S. population in 2015.
13.7% of 326 million $= 0.137 \times 326$ million
$\qquad\qquad\qquad\qquad = 44.662$ million
45 million people are projected to be in their twenties in 2015.

17.

Class Interval (Scores)	Tally	Class Frequency (Number of Students)
40–49	|	1
50–59	|||	3
60–69	||||	4
70–79	⊮	5
80–89	⊮ |||	8
90–99	||||	4

21. Point C has coordinates $(-3, 4)$

25. $y + x = -4$

Find any 3 ordered pair solutions.
Let $x = 0$
$y + x = -4$
$y + 0 = -4$
$\quad y = -4$
$(0, -4)$
Let $y = 0$.
$y + x = -4$
$0 + x = -4$
$\quad x = -4$
$(-4, 0)$
Let $x = -2$.
$\quad y + x = -4$
$y + (-2) = -4$
$\quad y - 2 = -4$
$y - 2 + 2 = -4 + 2$
$\qquad\quad y = -2$
$(-2, -2)$
Plot $(0, -4), (-4, 0)$, and $(-2, -2)$. Then draw the line through them.

29. $y = -\dfrac{1}{2}x$

Find any 3 ordered pair solutions.
Let $x = 0$
$y = -\dfrac{1}{2}x$
$y = -\dfrac{1}{2} \cdot 0$
$y = 0$
$(0, 0)$
Let $x = -2$.
$y = -\dfrac{1}{2}x$
$y = -\dfrac{1}{2} \cdot -2$
$y = 1$
$(-2, 1)$
Let $x = 2$.
$y = -\dfrac{1}{2}x$
$y = -\dfrac{1}{2} \cdot 2$
$y = -1$
$(2, -1)$

Plot $(0, 0), (-2, 1)$, and $(2, -1)$. Then draw the line through them.

33. One of the ten possible outcomes is a 6. The probability is $\dfrac{1}{10}$.

Chapter 9

Exercise Set 9.1

1. The figure extends indefinitely in two directions. It is line CD, line l, or $\overleftrightarrow{CD}$.

5. The figure has two rays with a common endpoint. It is an angle, which can be named $\angle GHI$, $\angle IHG$, or $\angle H$.

9. Two other ways to name $\angle x$ are $\angle CPR$ and $\angle RPC$.

13. $\angle S$ is a straight angle.

17. $\angle Q$ measures between 90° and 180°. It is an obtuse angle.

21. The complement of an angle that measures 23° is an angle that measures $90° - 23° = 67°$.

25. The complement of an angle that measures 58° is an angle that measures $90° - 58° = 32°$.

29. $52° + 38° = 90°$, so $\angle PNQ$ and $\angle QNR$ are complementary. $60° + 30° = 90°$, so $\angle MNP$ and $\angle RNO$ are complementary.

33. $m\angle x = 74° - 47° = 27°$

37. $\angle x$ and the angle marked 150° are supplementary, so $m\angle x = 180° - 150° = 30°$. $\angle y$ and the angle marked 150° are vertical angles, so $m\angle y = 150°$. $\angle z$ and $\angle x$ are vertical angles so $m\angle z = m\angle x = 30°$.

41. $\angle x$ and the angle marked 80° are supplementary, so $m\angle x = 180° - 80° = 100°$. $\angle y$ and the angle marked 80° are alternate interior angles, so $m\angle y = 80°$. $\angle x$ and $\angle z$ are corresponding angles, so $m\angle z = m\angle x = 100°$.

45. $\angle x$ can also be named $\angle ABC$ or $\angle CBA$.

49. $m\angle ABC = 15°$

53. $m\angle DBA = m\angle DBC + m\angle CBA$
$$= 50° + 15°$$
$$= 65°$$

57. $\dfrac{7}{8} + \dfrac{1}{4} = \dfrac{7}{8} + \dfrac{2}{8} = \dfrac{9}{8}$ or $1\dfrac{1}{8}$

61. $3\dfrac{1}{3} - 2\dfrac{1}{2} = \dfrac{10}{3} - \dfrac{5}{2}$
$$= \dfrac{20}{6} - \dfrac{15}{6}$$
$$= \dfrac{5}{6}$$

65. Since there are 360° in a full revolution, there are 360° around the earth at the equator.

69. False; since 100° is greater than 90°, it is not possible to find the complement of a 100° angle.

73. Since lines m and n are parallel, $\angle a$ and the angle labeled 60° are alternative interior angles, so $m\angle a = 60°$.

Since $\angle a$, $\angle b$, and the angle labeled 70° form a straight angle, $m\angle a + m\angle b + 70° = 180°$.

$m\angle a + m\angle b + 70° = 180°$
$60° + m\angle b + 70° = 180°$
$130° + m\angle b = 180°$
$130° - 130° + m\angle b = 180° - 130°$
$m\angle b = 50°$

Since lines m and n are parallel, $\angle d$ and the angle labeled 70° are alternative interior angles, so $m\angle d = 70°$.

Since $\angle c$ and $\angle d$ are supplementary angles, $m\angle c = 180° - m\angle d = 180° - 70° = 110°$.

Since $\angle e$ and the angle labeled 60° are supplementary angles, $m\angle e = 180° - 60° = 120°$.

77. Let each of the two equal angles have measure $x°$. Since the angles are complementary, the sum of their measures is 90°.

$x° + x° = 90°$
$2x° = 90°$
$\dfrac{2x°}{2} = \dfrac{90°}{2}$
$x° = 45°$

The two angles measure 45° and 45°.

Exercise Set 9.2

1. $P = 2 \cdot l + 2 \cdot w$
$$= 2 \cdot 17 \text{ ft} + 2 \cdot 15 \text{ ft}$$
$$= 34 \text{ ft} + 30 \text{ ft}$$
$$= 64 \text{ ft}$$
The perimeter is 64 feet.

5. $P = a + b + c$
$$= 5 \text{ in.} + 7 \text{ in.} + 9 \text{ in.}$$
$$= 21 \text{ in.}$$
The perimeter is 21 inches.

9. All sides of a regular triangle have the same length. $P = a + b + c = 14 \text{ in.} + 14 \text{ in.} + 14 \text{ in.} = 42 \text{ in.}$ The perimeter is 42 inches.

13. Sum the lengths of the sides.
$P = 5 \text{ ft} + 3 \text{ ft} + 2 \text{ ft} + 7 \text{ ft} + 4 \text{ ft}$
$$= 21 \text{ ft}$$
The perimeter is 21 feet.

17. $P = 2 \cdot l + 2 \cdot w$
$$= 2 \cdot 120 \text{ yd} + 2 \cdot 53 \text{ yd}$$
$$= 240 \text{ yd} + 106 \text{ yd}$$
$$= 346 \text{ yd}$$
The perimeter of the football field is 346 yards.

21. The amount of stripping needed is 22 feet.
22 feet $\cdot$ \$2.50 per foot $=$ \$55
The total cost of the stripping is \$55.

25. $P = 4 \cdot s = 4 \cdot 7 \text{ in.} = 28 \text{ in.}$
The perimeter is 28 inches.

29. The unmarked vertical side must have length $28 \text{ m} - 20 \text{ m} = 8 \text{ m}$.
The unmarked horizontal side must have length $20 \text{ m} - 17 \text{ m} = 3 \text{ m}$.
Sum the lengths of the sides.
$P = 17 \text{ m} + 8 \text{ m} + 3 \text{ m} + 20 \text{ m} + 20 \text{ m} + 28 \text{ m}$
$$= 96 \text{ m}$$
The perimeter is 96 meters.

33. The unmarked vertical side must have length $5 \text{ cm} + 14 \text{ cm} = 19 \text{ cm}$.
The unmarked horizontal side must have length $18 \text{ cm} - 9 \text{ cm} = 9 \text{ cm}$.
Sum the lengths of the sides.
$P = 18 \text{ cm} + 19 \text{ cm} + 9 \text{ cm} + 14 \text{ cm} + 9 \text{ cm} + 5 \text{ cm}$
$$= 74 \text{ cm}$$
The perimeter is 74 centimeters.

37. $C = 2 \cdot \pi \cdot r$
$$= 2 \cdot \pi \cdot 8 \text{ mi}$$
$$= 16\pi \text{ mi}$$
$$\approx 50.24 \text{ mi}$$
The circumference is exactly 16π miles, which is approximately 50.24 miles.

41. $\pi \cdot d = \pi \cdot 15 \text{ ft} = 15\pi \text{ ft} \approx 47.1 \text{ ft}$
He needs 15π feet of netting or 47.1 feet.

45. Sum the lengths of the sides.
$P = 9 \text{ mi} + 6 \text{ mi} + 11 \text{ mi} + 4.7 \text{ mi}$
$$= 30.7 \text{ mi}$$
The perimeter is 30.7 miles.

49. The sides of a regular pentagon all have the same length. Sum the lengths of the sides.
$P = 8 \text{ mm} + 8 \text{ mm} + 8 \text{ mm} + 8 \text{ mm} + 8 \text{ mm}$
$$= 40 \text{ mm}$$
The perimeter is 40 millimeters.

53. $5 + 6 \cdot 3 = 5 + 18 = 23$

57. $72 \div (2 \cdot 6) = 72 \div 12 = 6$

61. a. The first age category that 8-year-old children fit into is "Under 9." Thus the minimum width is 30 yards, and the minimum length is 40 yards.

b. $P = 2 \cdot l + 2 \cdot w$
$= 2 \cdot 40 \text{ yd} + 2 \cdot 30 \text{ yd}$
$= 80 \text{ yd} + 60 \text{ yd}$
$= 140 \text{ yd}$

The perimeter of the field is 140 yards.

65. a. Smaller circle:
$C = 2 \cdot \pi \cdot r$
$= 2 \cdot \pi \cdot 10 \text{ m}$
$= 20\pi \text{ m}$
$\approx 62.8 \text{ m}$

Larger circle:
$C = 2 \cdot \pi \cdot r$
$= 2 \cdot \pi \cdot 20 \text{ m}$
$= 40\pi \text{ m}$
$\approx 125.6 \text{ m}$

b. Yes, when the radius of a circle is doubled, the circumference is also doubled.

69. The three linear sides each have length 6 meters. The length of the curved side is half of the circumference of a circle with diameter 6 meters, or
$\frac{1}{2}\pi \cdot d = \frac{1}{2}\pi \cdot 6 = 3\pi \approx 9.4 \text{ meters}$
$6 \text{ m} + 6 \text{ m} + 6 \text{ m} + 9.4 \text{ m} = 27.4 \text{ m}$

The perimeter of the window is 27.4 meters.

Exercise Set 9.3

1. $A = l \cdot w = 3.5 \text{ m} \cdot 2 \text{ m} = 7 \text{ sq m}$

5. $A = \frac{1}{2} \cdot b \cdot h = \frac{1}{2} \cdot 6 \text{ yd} \cdot 5 \text{ yd} = 15 \text{ sq yd}$

9. $A = b \cdot h = 7 \text{ ft} \cdot 5.25 \text{ ft} = 36.75 \text{ sq ft}$

13. $A = \frac{1}{2}(b + B) \cdot h$
$= \frac{1}{2}(7 \text{ yd} + 4 \text{ yd}) \cdot 4 \text{ yd}$
$= \frac{1}{2}(11 \text{ yd}) \cdot 4 \text{ yd}$
$= 22 \text{ sq yd}$

17. $A = b \cdot h$
$= 5 \text{ in.} \cdot 4\frac{1}{2} \text{ in.}$
$= 5 \text{ in.} \cdot \frac{9}{2} \text{ in.}$
$= \frac{45}{2} \text{ sq in.}$
$= 22\frac{1}{2} \text{ sq in.}$

21.

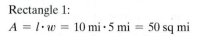

Rectangle 1:
$A = l \cdot w = 10 \text{ mi} \cdot 5 \text{ mi} = 50 \text{ sq mi}$

Rectangle 2:
$A = l \cdot w = 12 \text{ mi} \cdot 3 \text{ mi} = 36 \text{ sq mi}$
The area of the figure is
$50 \text{ sq mi} + 36 \text{ sq mi} = 86 \text{ sq mi}.$

25. $A = \pi r^2 = \pi(6 \text{ in.})^2 = 36\pi \text{ sq in.} \approx 113\frac{1}{7} \text{ sq in.}$

29. $V = s^3 = (8 \text{ cm})^3 = 512 \text{ cu cm}$
$SA = 6s^2 = 6(8 \text{ cm})^2 = 6 \cdot 64 \text{ sq cm} = 384 \text{ sq cm}$

33. $r = \frac{1}{2} \cdot d = \frac{1}{2} \cdot 10 \text{ in.} = 5 \text{ in.}$
$V = \frac{4}{3}\pi r^3$
$= \frac{4}{3}\pi(5 \text{ in.})^3$
$= \frac{4}{3}\pi \cdot 125 \text{ cu in.}$
$= \frac{500}{3}\pi \text{ cu in.}$
$\approx \frac{500}{3} \cdot \frac{22}{7} \text{ cu in.} = 523\frac{17}{21} \text{ cu in.}$
$SA = 4\pi r^2$
$= 4\pi(5 \text{ in.})^2$
$= 4\pi \cdot 25 \text{ sq in.}$
$= 100\pi \text{ sq in.}$
$\approx 100 \cdot \frac{22}{7} \text{ sq in.} = 314\frac{2}{7} \text{ sq in.}$

37. $V = \frac{1}{3}s^2 h$
$= \frac{1}{3}(5 \text{ cm})^2(9 \text{ cm})$
$= \frac{1}{3} \cdot 25 \text{ sq cm} \cdot 9 \text{ cm}$
$= 75 \text{ cu cm}$

41. $V = lwh = (2 \text{ ft})(1.4 \text{ ft})(3 \text{ ft}) = 8.4 \text{ cu ft}$
$SA = 2lh + 2wh + 2lw$
$= 2(2 \text{ ft})(3 \text{ ft}) + 2(1.4 \text{ ft})(3 \text{ ft}) + 2(2 \text{ ft})(1.4 \text{ ft})$
$= 12 \text{ sq ft} + 8.4 \text{ sq ft} + 5.6 \text{ sq ft}$
$= 26 \text{ sq ft}$

45. $A = l \cdot w = 7 \text{ ft} \cdot 6 \text{ ft} = 42 \text{ sq ft}$
$4 \cdot 42 \text{ sq ft} = 168 \text{ sq ft}$
Four panels have an area of 168 square feet.

49. $A = \frac{1}{2}(b + B) \cdot h$
$= \frac{1}{2}(140 \text{ ft} + 90 \text{ ft}) \cdot 80 \text{ ft}$
$= \frac{1}{2} \cdot 230 \text{ ft} \cdot 80 \text{ ft}$
$= 9200 \text{ sq ft}$

53. a. $A = \frac{1}{2}(b + B) \cdot h$
$= \frac{1}{2}(25 \text{ ft} + 36 \text{ ft}) \cdot 12\frac{1}{2} \text{ ft}$
$= \frac{1}{2} \cdot 61 \text{ ft} \cdot 12\frac{1}{2} \text{ ft}$
$= 381\frac{1}{4} \text{ sq ft}$

To the nearest square foot, the area is 381 square feet.

b. Since each square covers 100 square feet, 4 squares of shingles need to be purchased.

57. $r = \frac{1}{2} \cdot d = \frac{1}{2} \cdot 4 \text{ ft} = 2 \text{ ft}$

$A = \pi r^2 = \pi (2 \text{ ft})^2 = \pi \cdot 4 \text{ sq ft}$

The area of the pizza is 4π square feet, or approximately $4 \cdot 3.14 = 12.56$ square feet.

61. $A = l \cdot w = 16 \text{ ft} \cdot 10\frac{1}{2} \text{ ft}$

$\qquad = 16 \text{ ft} \cdot \frac{21}{2} \text{ ft}$

$\qquad = 168 \text{ sq ft}$

The area of the wall is 168 square feet.

65. $V = lwh = (2 \text{ in.})(2 \text{ in.})(2.2 \text{ in.}) = 8.8 \text{ cu in.}$

The volume of the Space Cube is 8.8 cubic inches.

69. $3^2 = 3 \cdot 3 = 9$

73. $4^2 + 2^2 = 4 \cdot 4 + 2 \cdot 2 = 16 + 4 = 20$

77. Carpet covers the entire floor of a room, so the situation involves area.

81. A wallpaper border goes around the edge of a room, so the situation involves perimeter.

85. $r = \frac{1}{2} \cdot d = \frac{1}{2} \cdot 20 \text{ m} = 10 \text{ m}$

$V = \frac{1}{2} \cdot \frac{4}{3} \pi r^3$

$\quad = \frac{2}{3} \pi (10 \text{ m})^3$

$\quad = \frac{2}{3} \pi \cdot 1000 \text{ cu m}$

$\quad = \frac{2000}{3} \pi \text{ cu m}$

$\quad \approx \frac{2000}{3} \cdot 3.14 \text{ cu m}$

$\quad \approx 2093.33 \text{ cu m}$

The volume of the dome is about 2093.33 cubic meters.

89. The area of the shaded region is the area of the square minus the area of the circle.

Square: $A = s^2 = (6 \text{ in.})^2 = 36 \text{ sq in.}$

Circle:

$r = \frac{1}{2} \cdot d = \frac{1}{2}(6 \text{ in.}) = 3 \text{ in.}$

$A = \pi \cdot r^2 = \pi (3 \text{ in.})^2 = 9\pi \text{ sq in.} \approx 28.26 \text{ sq in.}$

$36 \text{ sq in.} - 28.26 \text{ sq in.} = 7.74 \text{ sq in.}$

The shaded region has an area of approximately 7.74 square inches.

93. no; answers may vary

Exercise Set 9.4

1. $60 \text{ in.} = \frac{60 \text{ in.}}{1} \cdot \frac{1 \text{ ft}}{12 \text{ in.}} = \frac{60}{12} \text{ ft} = 5 \text{ ft}$

5. $42{,}240 \text{ ft} = \frac{42{,}240 \text{ ft}}{1} \cdot \frac{1 \text{ mi}}{5280 \text{ ft}}$

$\qquad = \frac{42{,}240}{5280} \text{ mi}$

$\qquad = 8 \text{ mi}$

9. $10 \text{ ft} = \frac{10 \text{ ft}}{1} \cdot \frac{1 \text{ yd}}{3 \text{ ft}} = \frac{10}{3} \text{ yd} = 3\frac{1}{3} \text{ yd}$

13. $162 \text{ in.} = \frac{162 \text{ in.}}{1} \cdot \frac{1 \text{ ft}}{12 \text{ in.}} \cdot \frac{1 \text{ yd}}{3 \text{ ft}}$

$\qquad = \frac{162}{36} \text{ yd}$

$\qquad = 4.5 \text{ yd}$

17. $40 \text{ ft} = \frac{40 \text{ ft}}{1} \cdot \frac{1 \text{ yd}}{3 \text{ ft}} = \frac{40}{3} \text{ yd}$

$\qquad \begin{array}{r} 13 \text{ yd } 1 \text{ ft} \\ 3\overline{)40} \\ \underline{-3} \\ 10 \\ \underline{-9} \\ 1 \end{array}$

$40 \text{ ft} = 13 \text{ yd } 1 \text{ ft}$

21. $10{,}000 \text{ ft} = \frac{10{,}000 \text{ ft}}{1} \cdot \frac{1 \text{ mi}}{5280 \text{ ft}} = \frac{10{,}000}{5280} \text{ mi}$

$\qquad \begin{array}{r} 1 \text{ mi } 4720 \text{ ft} \\ 5280\overline{)10{,}000} \\ \underline{-5280} \\ 4720 \end{array}$

$10{,}000 \text{ ft} = 1 \text{ mi } 4720 \text{ ft}$

25. $8 \text{ yd } 2 \text{ ft} = \frac{8 \text{ yd}}{1} \cdot \frac{3 \text{ ft}}{1 \text{ yd}} + 2 \text{ ft}$

$\qquad = 24 \text{ ft} + 2 \text{ ft}$

$\qquad = 26 \text{ ft}$

29. $3 \text{ ft } 10 \text{ in.} + 7 \text{ ft } 4 \text{ in.} = 10 \text{ ft } 14 \text{ in.}$

$\qquad\qquad\qquad\qquad\qquad = 10 \text{ ft} + 1 \text{ ft } 2 \text{ in.}$

$\qquad\qquad\qquad\qquad\qquad = 11 \text{ ft } 2 \text{ in.}$

33. $\begin{array}{r} 22 \text{ ft } 8 \text{ in.} \\ \underline{-16 \text{ ft } 3 \text{ in.}} \\ 6 \text{ ft } 5 \text{ in.} \end{array}$

37. $28 \text{ ft } 8 \text{ in.} \div 2 = 14 \text{ ft } 4 \text{ in.}$

41. $60 \text{ m} = \frac{60 \text{ m}}{1} \cdot \frac{100 \text{ cm}}{1 \text{ m}} = 6000 \text{ cm}$

45. $500 \text{ m} = \frac{500 \text{ m}}{1} \cdot \frac{1 \text{ km}}{1000 \text{ cm}} = \frac{500}{1000} \text{ km} = 0.5 \text{ km}$

49. $1500 \text{ cm} = \frac{1500 \text{ cm}}{1} \cdot \frac{1 \text{ m}}{100 \text{ cm}} = \frac{1500}{100} \text{ m} = 15 \text{ m}$

53. $7 \text{ km} = \frac{7 \text{ km}}{1} \cdot \frac{1000 \text{ m}}{1 \text{ km}} = 7000 \text{ m}$

57. $20.1 \text{ mm} = \frac{20.1 \text{ mm}}{1} \cdot \frac{1 \text{ dm}}{100 \text{ mm}}$

$\qquad = \frac{20.1}{100} \text{ dm}$

$\qquad = 0.201 \text{ dm}$

61. $\begin{array}{r} 8.6 \text{ m} \\ \underline{+0.34 \text{ m}} \\ 8.94 \text{ m} \end{array}$

65. $\begin{array}{r} 24.8 \text{ mm} \\ \underline{-1.19 \text{ cm}} \end{array}$ $\begin{array}{r} 24.8 \text{ mm} \\ \underline{-11.9 \text{ mm}} \\ 12.9 \text{ mm} \end{array}$ or $\begin{array}{r} 2.48 \text{ cm} \\ \underline{-1.19 \text{ cm}} \\ 1.29 \text{ cm} \end{array}$

69. $18.3 \text{ m} \times 3 = 54.9 \text{ m}$

73.

	Yards	Feet	Inches
Chrysler Building in New York City	$348\frac{2}{3}$	1046	12,552

		Meters	Millimeters	Kilometers	Centimeters
77.	Length of elephant	5	5000	0.005	500
81.	Distance from London to Paris	342,000	342,000,000	342	34,200,000

85.
$$\begin{array}{r} 6000 \text{ ft} \\ -900 \text{ ft} \\ \hline 5100 \text{ ft} \end{array}$$

The Grand Canyon of the Colorado River is 5100 feet deeper than the Grand Canyon of the Yellowstone River.

89.
$$\begin{array}{r} 80 \text{ mm} \qquad 80.0 \text{ mm} \\ -5.33 \text{ cm} \quad -53.3 \text{ mm} \\ \hline 26.7 \text{ mm} \end{array}$$

The ice must be 26.7 mm thicker before skating is allowed.

93.
$$\begin{array}{r} 3.35 \\ 20)\overline{67.00} \\ -60 \\ \hline 70 \\ -60 \\ \hline 100 \\ -100 \\ \hline 0 \end{array}$$

Each piece will be 3.35 meters long.

97. $0.21 = \dfrac{21}{100}$

101. $\dfrac{1}{4} = \dfrac{1}{4} \cdot \dfrac{25}{25} = \dfrac{25}{100} = 0.25$

105. Yes, glass for a drinking glass being 2 millimeters thick is reasonable.

109. 5 yd 2 in. is close to 5 yd. 7 yd 30 in. is close to 7 yd 36 in. = 8 yd. Estimate: 5 yd + 8 yd = 13 yd.

113. answers may vary

Exercise Set 9.5

1. $2 \text{ lb} = \dfrac{2 \text{ lb}}{1} \cdot \dfrac{16 \text{ oz}}{1 \text{ lb}} = 2 \cdot 16 \text{ oz} = 32 \text{ oz}$

5. $18,000 \text{ lb} = \dfrac{18,000 \text{ lb}}{1} \cdot \dfrac{1 \text{ ton}}{2000 \text{ lb}}$
$= \dfrac{18,000}{2000} \text{ tons}$
$= 9 \text{ tons}$

9. $3500 \text{ lb} = \dfrac{3500 \text{ lb}}{1} \cdot \dfrac{1 \text{ ton}}{2000 \text{ lb}}$
$= \dfrac{3500}{2000} \text{ tons}$
$= \dfrac{7}{4} \text{ tons}$
$= 1\dfrac{3}{4} \text{ tons}$

13. $4.9 \text{ tons} = \dfrac{4.9 \text{ tons}}{1} \cdot \dfrac{2000 \text{ lb}}{1 \text{ ton}}$
$= 4.9 \cdot 2000 \text{ lb}$
$= 9800 \text{ lb}$

17. $2950 \text{ lb} = \dfrac{2950 \text{ lb}}{1} \cdot \dfrac{1 \text{ ton}}{2000 \text{ lb}}$
$= \dfrac{2950}{2000} \text{ tons}$
$= \dfrac{59}{40} \text{ tons}$
$\approx 1.5 \text{ tons}$

21. $5\dfrac{3}{4} \text{ lb} = \dfrac{23}{4} \text{ lb}$
$= \dfrac{\frac{23}{4} \text{ lb}}{1} \cdot \dfrac{16 \text{ oz}}{1 \text{ lb}}$
$= \dfrac{23}{4} \cdot 16 \text{ oz}$
$= 23 \cdot 4 \text{ oz}$
$= 92 \text{ oz}$

25. $89 \text{ oz} = \dfrac{89 \text{ oz}}{1} \cdot \dfrac{1 \text{ lb}}{16 \text{ oz}} = \dfrac{89}{16} \text{ lb}$

$$\begin{array}{r} 5 \text{ lb } 9 \text{ oz} \\ 16)\overline{89} \\ -80 \\ \hline 9 \end{array}$$

$89 \text{ oz} = 5 \text{ lb } 9 \text{ oz}$

29. $3 \text{ tons } 1820 \text{ lb} + 4 \text{ tons } 930 \text{ lb}$
$= 7 \text{ tons } 2750 \text{ lb}$
$= 7 \text{ tons} + 1 \text{ ton } 750 \text{ lb}$
$= 8 \text{ tons } 750 \text{ lb}$

33.
$$\begin{array}{r} 12 \text{ lb } 4 \text{ oz} \qquad 11 \text{ lb } 20 \text{ oz} \\ -3 \text{ lb } 9 \text{ oz} \quad -3 \text{ lb } 9 \text{ oz} \\ \hline 8 \text{ lb } 11 \text{ oz} \end{array}$$

37. $6 \text{ tons } 1500 \text{ lb} \div 5 = \dfrac{6}{5} \text{ tons } 300 \text{ lb}$
$= 1\dfrac{1}{5} \text{ tons } 300 \text{ lb}$
$= 1 \text{ ton} + \dfrac{2000 \text{ lb}}{5} + 300 \text{ lb}$
$= 1 \text{ ton} + 400 \text{ lb} + 300 \text{ lb}$
$= 1 \text{ ton } 700 \text{ lb}$

41. $4 \text{ g} = \dfrac{4 \text{ g}}{1} \cdot \dfrac{1000 \text{ mg}}{1 \text{ g}} = 4 \cdot 1000 \text{ mg} = 4000 \text{ mg}$

45. $48 \text{ mg} = \dfrac{48 \text{ mg}}{1} \cdot \dfrac{1 \text{ g}}{1000 \text{ mg}} = \dfrac{48}{1000} \text{ g} = 0.048 \text{ g}$

49. $15.14 \text{ g} = \dfrac{15.14 \text{ g}}{1} \cdot \dfrac{1000 \text{ mg}}{1 \text{ g}}$
$= 15.14 \cdot 1000 \text{ mg}$
$= 15,140 \text{ mg}$

53. $35 \text{ hg} = \dfrac{35 \text{ hg}}{1} \cdot \dfrac{10,000 \text{ cg}}{1 \text{ hg}}$
$= 35 \cdot 10,000 \text{ cg}$
$= 350,000 \text{ cg}$

57. $205 \text{ mg} + 5.61 \text{ g} = 0.205 \text{ g} + 5.61 \text{ g} = 5.815 \text{ g}$
or
$205 \text{ mg} + 5.61 \text{ g} = 205 \text{ mg} + 5610 \text{ mg}$
$= 5815 \text{ mg}$

61. $1.61 \text{ kg} - 250 \text{ g} = 1.61 \text{ kg} - 0.250 \text{ kg} = 1.36 \text{ kg}$

    or

    $1.61 \text{ kg} - 250 \text{ g} = 1610 \text{ g} - 250 \text{ g} = 1360 \text{ g}$

65. $17 \text{ kg} \div 8 = \dfrac{17}{8} \text{ kg}$

$$
\begin{array}{r}
2.125 \\
8\overline{)17.000} \\
\underline{-16} \\
10 \\
\underline{-8} \\
20 \\
\underline{-16} \\
40 \\
\underline{-40} \\
0
\end{array}
$$

    $17 \text{ kg} \div 8 = 2.125 \text{ kg}$

69.

Object	Tons	Pounds	Ounces
A 12-inch cube of osmium	$\dfrac{269}{400}$ or 0.6725	1345	21,520

73.

Object	Grams	Kilograms	Milligrams	Centigrams
A six-year-old boy	21,000	21	21,000,000	2,100,000

77. $0.09 \text{ g} = \dfrac{0.09 \text{ g}}{1} \cdot \dfrac{1000 \text{ mg}}{1 \text{ g}} = 0.09 \cdot 1000 \text{ mg} = 90 \text{ mg}$

    $90 \text{ mg} - 60 \text{ mg} = 30 \text{ mg}$

    The extra-strength tablet contains 30 mg more medication.

81.
$$
\begin{array}{ll}
64 \text{ lb} \; 8 \text{ oz} & 63 \text{ lb} \; 24 \text{ oz} \\
\underline{-28 \text{ lb} \; 10 \text{ oz}} & \underline{-28 \text{ lb} \; 10 \text{ oz}} \\
& 35 \text{ lb} \; 14 \text{ oz}
\end{array}
$$

    Carla's zucchini was 35 lb 14 oz lighter than the record weight.

85. $3 \times 16 = 48$

    3 cartons contain 48 boxes of fruit.

    $3 \text{ mg} \times 48 = 144 \text{ mg}$

    3 cartons contain 144 mg of preservatives.

89. $3 \text{ lb} 4 \text{ oz} \times 10 = 30 \text{ lb} 40 \text{ oz}$

$$
\begin{aligned}
&= 30 \text{ lb} + 2 \text{ lb} 8 \text{ oz} \\
&= 32 \text{ lb} 8 \text{ oz}
\end{aligned}
$$

    Each box weighs 32 lb 8 oz.

    $32 \text{ lb} 8 \text{ oz} \times 4 = 128 \text{ lb} 32 \text{ oz}$

$$
\begin{aligned}
&= 128 \text{ lb} + 2 \text{ lb} \\
&= 130 \text{ lb}
\end{aligned}
$$

    4 boxes of meat weigh 130 lb.

93. $\dfrac{4}{25} = \dfrac{4}{25} \cdot \dfrac{4}{4} = \dfrac{16}{100} = 0.16$

97. No, a pill containing 2 kg of medication is not reasonable.

101. No, a professor weighing less than 150 g is not reasonable.

105. True, a kilogram is 1000 grams.

Exercise Set 9.6

1. $32 \text{ fl oz} = \dfrac{32 \text{ fl oz}}{1} \cdot \dfrac{1 \text{ c}}{8 \text{ fl oz}} = \dfrac{32}{8} \text{ c} = 4 \text{ c}$

5. $14 \text{ qt} = \dfrac{14 \text{ qt}}{1} \cdot \dfrac{1 \text{ gal}}{4 \text{ qt}} = \dfrac{14}{4} \text{ gal} = 3\dfrac{1}{2} \text{ gal}$

9. $2 \text{ qt} = \dfrac{2 \text{ qt}}{1} \cdot \dfrac{2 \text{ pt}}{1 \text{ qt}} \cdot \dfrac{2 \text{ c}}{1 \text{ pt}} = 2 \cdot 2 \cdot 2 \cdot \text{c} = 8 \text{ c}$

13. $42 \text{ c} = \dfrac{42 \text{ c}}{1} \cdot \dfrac{1 \text{ qt}}{4 \text{ c}} = \dfrac{42}{4} \text{ qt} = 10\dfrac{1}{2} \text{ qt}$

17. $5 \text{ gal} 3 \text{ qt} = \dfrac{5 \text{ gal}}{1} \cdot \dfrac{4 \text{ qt}}{1 \text{ gal}} + 3 \text{ qt}$

$$
\begin{aligned}
&= 5 \cdot 4 \text{ qt} + 3 \text{ qt} \\
&= 20 \text{ qt} + 3 \text{ qt} \\
&= 23 \text{ qt}
\end{aligned}
$$

21. $58 \text{ qt} = 56 \text{ qt} + 2 \text{ qt}$

$$
\begin{aligned}
&= \dfrac{56 \text{ qt}}{1} \cdot \dfrac{1 \text{ gal}}{4 \text{ qt}} + 2 \text{ qt} \\
&= \dfrac{56}{4} \text{ gal} + 2 \text{ qt} \\
&= 14 \text{ gal} 2 \text{ qt}
\end{aligned}
$$

25. $2\dfrac{3}{4} \text{ gal} = \dfrac{11}{4} \text{ gal}$

$$
\begin{aligned}
&= \dfrac{\frac{11}{4} \text{ gal}}{1} \cdot \dfrac{4 \text{ qt}}{1 \text{ gal}} \cdot \dfrac{2 \text{ pt}}{1 \text{ qt}} \\
&= \dfrac{11}{4} \cdot 4 \cdot 2 \text{ pt} \\
&= 22 \text{ pt}
\end{aligned}
$$

29. $1 \text{ c} 5 \text{ fl oz} + 2 \text{ c} 7 \text{ fl oz} = 3 \text{ c} 12 \text{ fl oz}$

$$
\begin{aligned}
&= 3 \text{ c} + 1 \text{ c} 4 \text{ fl oz} \\
&= 4 \text{ c} 4 \text{ fl oz}
\end{aligned}
$$

33.
$$
\begin{array}{lll}
3 \text{ gal} 1 \text{ qt} & 2 \text{ gal} 5 \text{ qt} & 2 \text{ gal} 4 \text{ qt} 2 \text{ pt} \\
\underline{-\quad\quad 1 \text{ qt} 1 \text{ pt}} & \underline{-\quad\quad 1 \text{ qt} 1 \text{ pt}} & \underline{-\quad\quad 1 \text{ qt} 1 \text{ pt}} \\
& & 2 \text{ gal} 3 \text{ qt} 1 \text{ pt}
\end{array}
$$

37. $9 \text{ gal} 2 \text{ qt} \div 2 = (8 \text{ gal} 4 \text{ qt} + 2 \text{ qt}) \div 2$

$$
\begin{aligned}
&= 8 \text{ gal} 6 \text{ qt} \div 2 \\
&= 4 \text{ gal} 3 \text{ qt}
\end{aligned}
$$

41. $0.16 \text{ L} = \dfrac{0.16 \text{ L}}{1} \cdot \dfrac{1 \text{ kl}}{1000 \text{ L}} = \dfrac{0.16}{1000} \text{ kl} = 0.00016 \text{ kl}$

45. $3.2 \text{ L} = \dfrac{3.2 \text{ L}}{1} \cdot \dfrac{100 \text{ cl}}{1 \text{ L}} = 3.2 \cdot 100 \text{ cl} = 320 \text{ cl}$

49. $64 \text{ ml} = \dfrac{64 \text{ ml}}{1} \cdot \dfrac{1 \text{ L}}{1000 \text{ ml}} = \dfrac{64}{1000} \text{ L} = 0.064 \text{ L}$

53. $3.6 \text{ L} = \dfrac{3.6 \text{ L}}{1} \cdot \dfrac{1000 \text{ ml}}{1 \text{ L}} = 3.6 \cdot 1000 \text{ ml} = 3600 \text{ ml}$

57. $2700 \text{ ml} + 1.8 \text{ L} = 2.7 \text{ L} + 1.8 \text{ L} = 4.5 \text{ L}$

    or

    $2700 \text{ ml} + 1.8 \text{ L} = 2700 \text{ ml} + 1800 \text{ ml} = 4500 \text{ ml}$

61. $17,500 \text{ ml} - 0.9 \text{ L} = 17,500 \text{ ml} - 900 \text{ ml}$

$$
= 16,600 \text{ ml}
$$

    or

    $17,500 \text{ ml} - 0.9 \text{ L} = 17.5 \text{ L} - 0.9 \text{ L} = 16.6 \text{ L}$

65. $81.2 \text{ L} \div 0.5 = 81.2 \text{ L} \div \dfrac{1}{2}$

$$
\begin{aligned}
&= 81.2 \text{ L} \cdot 2 \\
&= 162.4 \text{ L}
\end{aligned}
$$

69.

Capacity	Cups	Gallons	Quarts	Pints
Your kidneys filter about this amount of blood every minute	4	$\frac{1}{4}$	1	2

73. 354 ml + 18.6 L = 0.354 L + 18.6 L = 18.954 L

There were 18.954 liters of gasoline in her tank.

77. 5 pt 1 c + 2 pt 1 c = 7 pt 2 c

$$= 7 \text{ pt} + 1 \text{ pt}$$
$$= 8 \text{ pt}$$
$$= \frac{8 \text{ pt}}{1} \cdot \frac{1 \text{ qt}}{2 \text{ pt}}$$
$$= \frac{8}{2} \text{ qt}$$
$$= 4 \text{ qt}$$
$$= \frac{4 \text{ qt}}{1} \cdot \frac{1 \text{ gal}}{4 \text{ qt}}$$
$$= \frac{4}{4} \text{ gal}$$
$$= 1 \text{ gal}$$

Yes, the liquid can be poured into the container without causing it to overflow.

81. $\frac{20}{25} = \frac{4 \cdot 5}{5 \cdot 5} = \frac{4}{5}$

85. $\frac{72}{80} = \frac{8 \cdot 9}{8 \cdot 10} = \frac{9}{10}$

89. No, a tub filled with 3000 ml of hot water is not reasonable.

93. answers may vary

97. B indicates 1.5 cc.

101. B indicates 54 u or 0.54 cc.

Exercise Set 9.7

1. $756 \text{ ml} \approx \frac{756 \text{ ml}}{1} \cdot \frac{1 \text{ fl oz}}{29.57 \text{ ml}} \approx 25.57 \text{ fl oz}$

5. $1000 \text{ g} \approx \frac{1000 \text{ g}}{1} \cdot \frac{0.04 \text{ oz}}{1 \text{ g}} \approx 40 \text{ oz}$

9. $14.5 \text{ L} \approx \frac{14.5 \text{ L}}{1} \cdot \frac{0.26 \text{ gal}}{1 \text{ L}} \approx 3.77 \text{ gal}$

13.

	Meters	Yards	Centimeters	Feet	Inches
The height of a woman	1.5	$1\frac{2}{3}$	150	5	60

17. $10 \text{ cm} = \frac{10 \text{ cm}}{1} \cdot \frac{1 \text{ in.}}{2.54 \text{ cm}} \approx 3.94 \text{ in.}$

The balance beam is approximately 3.94 inches wide.

21. $200 \text{ mg} = 0.2 \text{ g} \approx \frac{0.2 \text{ g}}{1} \cdot \frac{0.04 \text{ oz}}{1 \text{ g}} \approx 0.008 \text{ oz}$

25. $16 \text{ billion km} \approx \frac{16 \text{ billion km}}{1} \cdot \frac{0.62 \text{ mi}}{1 \text{ km}} \approx 9.92 \text{ billion mi}$

The antenna can track a spacecraft that is 9.92 billion miles from Earth.

29. $4500 \text{ km} \approx \frac{4500 \text{ km}}{1} \cdot \frac{0.62 \text{ mi}}{1 \text{ km}} \approx 2790 \text{ mi}$

The trip is about 2790 miles.

33. 1.5 lb − 1.25 lb = 0.25 lb

$0.25 \text{ lb} \approx \frac{0.25 \text{ lb}}{1} \cdot \frac{0.45 \text{ kg}}{1 \text{ lb}} \cdot \frac{1000 \text{ g}}{1 \text{ kg}} \approx 112.5 \text{ g}$

The difference is approximately 112.5 grams.

37. $8 \text{ m} \approx \frac{8 \text{ m}}{1} \cdot \frac{3.28 \text{ ft}}{1 \text{ m}} \approx 26.24 \text{ ft}$

The base diameter is approximately 26.24 feet.

41. One dose every 4 hours results in $\frac{24}{4} = 6$ doses per day and $6 \times 7 = 42$ doses per week.

$5 \text{ ml} \times 42 = 210 \text{ ml}$

$210 \text{ ml} \approx \frac{210 \text{ ml}}{1} \cdot \frac{1 \text{ fl oz}}{29.57 \text{ ml}} \approx 7.1 \text{ fl oz}$

8 fluid ounces of medicine should be purchased.

45. A liter has greater capacity than a quart; b.

49. An $8\frac{1}{2}$-ounce glass of water has a capacity of about $250 \text{ ml} \left(\frac{1}{4} \text{ L}\right)$; d.

53. $C = \frac{5}{9}(F - 32)$

$$= \frac{5}{9}(77 - 32)$$
$$= \frac{5}{9}(45)$$
$$= 25$$

77°F is 25°C.

57. $F = \frac{9}{5}C + 32$

$$= \frac{9}{5}(50) + 32$$
$$= 90 + 32$$
$$= 122$$

50°C is 122°F.

61. $C = \frac{5}{9}(F - 32)$

$$= \frac{5}{9}(20 - 32)$$
$$= \frac{5}{9}(-12)$$
$$\approx -6.7$$

20°F is −6.7°C.

65. $F = 1.8C + 32$

$$= 1.8(92) + 32$$
$$= 165.6 + 32$$
$$= 197.6$$

92°C is 197.6°F.

69. $C = \frac{5}{9}(F - 32)$

$$= \frac{5}{9}(134 - 32)$$
$$= \frac{5}{9}(102)$$
$$\approx 56.7$$

134°F is 56.7°C.

73. $C = \dfrac{5}{9}(F - 32)$

$ = \dfrac{5}{9}(70 - 32)$

$ = \dfrac{5}{9}(38)$

$ \approx 21.1$

70°F is 21.1°C.

77. $F = 1.8C + 32$

$ = 1.8(4000) + 32$

$ = 7200 + 32$

$ = 7232$

4000°C is 7232°F.

81. $3[(1 + 5) \cdot (8 - 6)] = 3(6 \cdot 2) = 3(12) = 36$

85. No, a fever of 40°F is not reasonable.

89. Yes, a fever of 40°C is reasonable.

93. $40 \text{ in.} = \dfrac{40 \text{ in.}}{1} \cdot \dfrac{2.54 \text{ cm}}{1 \text{ in.}} = 101.6 \text{ cm}$

$\text{BSA} = \sqrt{\dfrac{50 \times 101.6}{3600}} \approx 1.19$

The BSA is approximately 1.19 sq m.

97. $C = \dfrac{5}{9}(F - 32)$

$ = \dfrac{5}{9}(7{,}200{,}000{,}000 - 32)$

$ = \dfrac{5}{9}(7{,}199{,}999{,}968)$

$ \approx 4{,}000{,}000{,}000$

7,200,000,000°F is approximately 4,000,000,000°C.

Chapter 9 Test

1. The complement of an angle that measures 78° is an angle that measures $90° - 78° = 12°$.

5. $\angle x$ and the angle marked 73° are vertical angles, so $m\angle x = 73°$. $\angle x$ and $\angle y$ are alternate interior angles, so $m\angle y = m\angle x = 73°$. $\angle x$ and $\angle z$ are corresponding angles, so $m\angle z = m\angle x = 73°$.

9. $P = 2 \cdot l + 2 \cdot w$

$ = 2(7 \text{ yd}) + 2(5.3 \text{ yd})$

$ = 14 \text{ yd} + 10.6 \text{ yd}$

$ = 24.6 \text{ yd}$

$A = l \cdot w = 7 \text{ yd} \cdot 5.3 \text{ yd} = 37.1 \text{ sq yd}$

13. $P = 4 \cdot s = 4 \cdot 4 \text{ in.} = 16 \text{ in.}$

The perimeter of the photo is 16 inches.

17. $2\dfrac{1}{2} \text{ gal} = \dfrac{2\frac{1}{2} \text{ gal}}{1} \cdot \dfrac{4 \text{ qt}}{1 \text{ gal}} = 10 \text{ qt}$

21. $40 \text{ mg} = \dfrac{40 \text{ mg}}{1} \cdot \dfrac{1 \text{ g}}{1000 \text{ mg}} = 0.04 \text{ g}$

25. $0.83 \text{ L} = \dfrac{0.83 \text{ L}}{1} \cdot \dfrac{1000 \text{ ml}}{1 \text{ L}} = 830 \text{ ml}$

29. $5 \text{ gal } 2 \text{ qt} \div 2 = 4 \text{ gal } 6 \text{ qt} \div 2$

$\phantom{5 \text{ gal } 2 \text{ qt} \div 2} = \dfrac{4}{2} \text{ gal} \dfrac{6}{2} \text{ qt}$

$\phantom{5 \text{ gal } 2 \text{ qt} \div 2} = 2 \text{ gal } 3 \text{ qt}$

33. $F = 1.8\,C + 32 = 1.8(12.6) + 32 = 22.68 + 32 \approx 54.7$

12.6°C is 54.7°F

37.
$$\begin{array}{r} 2 \text{ ft} \quad 9 \text{ in.} \\ \times \qquad\quad 6 \\ \hline 12 \text{ ft} \ \ 54 \text{ in.} \end{array} = 12 \text{ ft} + 4 \text{ ft } 6 \text{ in.} = 16 \text{ ft } 6 \text{ in.}$$

Thus, 16 ft 6 in. of material is needed.

Chapter 10

Exercise Set 10.1

1. $(2x + 3) + (-7x - 27) = (2x - 7x) + (3 - 27)$

$ = -5x + (-24)$

$ = -5x - 24$

5. $(12y - 20) + (9y^2 + 13y - 20)$

$ = 9y^2 + (12y + 13y) + (-20 - 20)$

$ = 9y^2 + 25y - 40$

9. $-(9x - 16) = -1(9x - 16)$

$ = -1(9x) + (-1)(-16)$

$ = -9x + 16$

13. $(8a - 5) - (3a + 8) = (8a - 5) + (-3a - 8)$

$ = 8a - 3a - 5 - 8$

$ = 5a - 13$

17. $(10y^2 - 7) - (20y^3 - 2y^2 - 3)$

$ = (10y^2 - 7) + (-20y^3 + 2y^2 + 3)$

$ = (-20y^3) + (10y^2 + 2y^2) + (-7 + 3)$

$ = -20y^3 + 12y^2 - 4$

21.
$$\begin{array}{r} 13y^2 - 6y - 14 \\ -\left(5y^2 + 4y - 6\right) \\ \hline \end{array} \qquad \begin{array}{r} 13y^2 - \ \ 6y - 14 \\ + -5y^2 - \ \ 4y + \ \ 6 \\ \hline 8y^2 - 10y - \ \ 8 \end{array}$$

25. $(4y + 4) - (3y + 8) = (4y + 4) + (-3y - 8)$

$ = (4y - 3y) + (4 - 8)$

$ = y - 4$

29. $(5x + 4.5) + (-x - 8.6) = (5x - x) + (4.5 - 8.6)$

$ = 4x - 4.1$

33. $(21y - 4.6) - (36y - 8.2)$

$ = (21y - 4.6) + (-36y + 8.2)$

$ = (21y - 36y) + (-4.6 + 8.2)$

$ = -15y + 3.6$

37. $(2b^3 + 5b^2 - 5b - 8) + (8b^2 + 9b + 6)$

$ = 2b^3 + (5b^2 + 8b^2) + (-5b + 9b) + (-8 + 6)$

$ = 2b^3 + 13b^2 + 4b - 2$

41. $\left(3z + \dfrac{6}{7}\right) - \left(3z - \dfrac{3}{7}\right) = \left(3z + \dfrac{6}{7}\right) + \left(-3z + \dfrac{3}{7}\right)$

$ = (3z - 3z) + \left(\dfrac{6}{7} + \dfrac{3}{7}\right)$

$ = \dfrac{9}{7}$

45. $x^2 - 6x + 3 = 2^2 - 6(2) + 3$

$ = 4 - 6(2) + 3$

$ = 4 - 12 + 3$

$ = -5$

49. $2x + 10 = 2(5) + 10 = 10 + 10 = 20$

53. $2x^2 + 4x - 20 = 2(5)^2 + 4(5) - 20$

$ = 2(25) + 4(5) - 20$

$ = 50 + 20 - 20$

$ = 50$

57. Let $x = 10$.

$3000 + 20x = 3000 + 20(10) = 3000 + 200 = 3200$

It costs \$3200 to manufacture 10 file cabinets.

61. Let $t = 3$.

$$1053 - 16t^2 = 1053 - 16(3)^2$$
$$= 1053 - 16(9)$$
$$= 1053 - 144$$
$$= 909$$

After 3 seconds, the object is 909 feet above the river.

65. $3^4 = 3 \cdot 3 \cdot 3 \cdot 3 = 81$

69. $x \cdot x \cdot x = x^3$

73. $P = (5x - 10) + (2x + 1) + (x + 11)$
$$= 5x + 2x + x - 10 + 1 + 11$$
$$= 8x + 2$$

The perimeter is $(8x + 2)$ inches.

77.
$$
\begin{array}{r}
3x^2 + \underline{}\,x - \underline{} \\
+ \underline{}\,x^2 - \;6x + \;2 \\
\hline
5x^2 + 14x - \;4
\end{array}
$$

Since $3x^2 + 2x^2 = 5x^2, 20x - 6x = 14x$ and $-6 + 2 = -4$, the missing numbers are $20, 6,$ and 2.

$(3x^2 + \underline{20}x - \underline{6}) + (2x^2 - 6x + 2) = 5x^2 + 14x - 4$

81. Let $t = 8$.

$$1053 - 16t^2 = 1053 - 16(8)^2$$
$$= 1053 - 16(64)$$
$$= 1053 - 1024$$
$$= 29$$

The height after 8 seconds is 29 feet.

Let $t = 9$.

$$1053 - 16t^2 = 1053 - 16(9)^2$$
$$= 1053 - 16(81)$$
$$= 1053 - 1296$$
$$= -243$$

The height after 9 seconds is -243 feet. answers may vary

Exercise Set 10.2

1. $x^5 \cdot x^9 = x^{5+9} = x^{14}$

5. $3z^3 \cdot 5z^2 = (3 \cdot 5)(z^3 \cdot z^2) = 15z^5$

9. $2x \cdot 3x \cdot 7x = (2 \cdot 3 \cdot 7)(x \cdot x \cdot x) = 42x^3$

13. $(-5x^2y^3)(-5x^4y) = (-5)(-5)(x^2 \cdot x^4)(y^3 \cdot y^1)$
$$= 25x^6y^4$$

17. $(x^5)^3 = x^{5 \cdot 3} = x^{15}$

21. $(b^7)^6 \cdot (b^2)^{10} = b^{7 \cdot 6} \cdot b^{2 \cdot 10}$
$$= b^{42} \cdot b^{20}$$
$$= b^{42+20}$$
$$= b^{62}$$

25. $(a^{11}b^8)^3 = a^{11 \cdot 3}b^{8 \cdot 3} = a^{33}b^{24}$

29. $(-3y)(2y^7)^3 = (-3y) \cdot 2^3(y^7)^3$
$$= (-3y) \cdot 8y^{21}$$
$$= (-3)(8)(y^1 \cdot y^{21})$$
$$= -24y^{22}$$

33. $7(x - 3) = 7x - 21$

37. $9(x + 2y - 3) = 9x + 18y - 27$

41. Area $= \dfrac{1}{2}bh$

$$= \dfrac{1}{2} \cdot (6a^3b^4) \cdot (4ab)$$

$$= \left(\dfrac{1}{2} \cdot 6 \cdot 4\right)(a^3 \cdot a)(b^4 \cdot b)$$

$$= 12a^4b^5$$

The area is $12a^4b^5$ square meters.

45. $(8.1x^{10})^5 = 8.1^5(x^{10})^5 = 34{,}867.84401x^{50}$

49. answers may vary

Exercise Set 10.3

1. $3x(9x^2 - 3) = 3x \cdot 9x^2 + 3x \cdot (-3)$
$$= (3 \cdot 9)(x \cdot x^2) + (3)(-3)(x)$$
$$= 27x^3 + (-9x)$$
$$= 27x^3 - 9x$$

5. $7x^2(6x^2 - 5x + 7)$
$$= (7x^2)(6x^2) + (7x^2)(-5x) + (7x^2)(7)$$
$$= (7 \cdot 6)(x^2 \cdot x^2) + (7)(-5)(x^2 \cdot x) + (7 \cdot 7)x^2$$
$$= 42x^4 - 35x^3 + 49x^2$$

9. $(2x - 6)(x + 4)$
$$= 2x(x + 4) - 6(x + 4)$$
$$= 2x \cdot x + 2x \cdot 4 - 6 \cdot x - 6 \cdot 4$$
$$= 2x^2 + 8x - 6x - 24$$
$$= 2x^2 + 2x - 24$$

13. $(a + 6)(a^2 - 6a + 3)$
$$= a(a^2 - 6a + 3) + 6(a^2 - 6a + 3)$$
$$= a \cdot a^2 + a(-6a) + a \cdot 3 + 6 \cdot a^2 + 6(-6a) + 6 \cdot 3$$
$$= a^3 - 6a^2 + 3a + 6a^2 - 36a + 18$$
$$= a^3 - 33a + 18$$

17. $(x^3 + 2x + x^2)(3x + 1 + x^2)$
$$= x^3(3x + 1 + x^2) + 2x(3x + 1 + x^2) + x^2(3x + 1 + x^2)$$
$$= x^3 \cdot 3x + x^3 \cdot 1 + x^3 \cdot x^2 + 2x \cdot 3x + 2x \cdot 1 + 2x \cdot x^2$$
$$\quad + x^2 \cdot 3x + x^2 \cdot 1 + x^2 \cdot x^2$$
$$= 3x^4 + x^3 + x^5 + 6x^2 + 2x + 2x^3 + 3x^3 + x^2 + x^4$$
$$= x^5 + 4x^4 + 6x^3 + 7x^2 + 2x$$

21. $-2y^2(3y + y^2 - 6)$
$$= -2y^2 \cdot 3y + (-2y^2) \cdot y^2 + (-2y^2)(-6)$$
$$= -6y^3 - 2y^4 + 12y^2$$

25. $(2a + 3)(2a - 3) = 2a(2a - 3) + 3(2a - 3)$
$$= 2a \cdot 2a + 2a(-3) + 3 \cdot 2a + 3(-3)$$
$$= 4a^2 - 6a + 6a - 9$$
$$= 4a^2 - 9$$

29. $\left(b + \dfrac{3}{5}\right)\left(b + \dfrac{4}{5}\right) = b\left(b + \dfrac{4}{5}\right) + \dfrac{3}{5}\left(b + \dfrac{4}{5}\right)$

$$= b^2 + \dfrac{4}{5}b + \dfrac{3}{5}b + \dfrac{3}{5} \cdot \dfrac{4}{5}$$

$$= b^2 + \dfrac{7}{5}b + \dfrac{12}{25}$$

33. $(7x + 5)^2 = (7x + 5)(7x + 5)$
$$= 7x(7x + 5) + 5(7x + 5)$$
$$= 49x^2 + 35x + 35x + 25$$
$$= 49x^2 + 70x + 25$$

37. $(2x^2 - 3)(4x^3 + 2x - 3)$
$$= 2x^2(4x^3 + 2x - 3) - 3(4x^3 + 2x - 3)$$
$$= 8x^5 + 4x^3 - 6x^2 - 12x^3 - 6x + 9$$
$$= 8x^5 - 8x^3 - 6x^2 - 6x + 9$$

41.
$$
\begin{array}{r}
2z^2 - \;z + 1 \\
\times\; 5z^2 + \;z - 2 \\
\hline
-4z^2 + 2z - 2 \\
2z^3 - \;z^2 + \;z \\
10z^4 - 5z^3 + 5z^2 \\
\hline
10z^4 - 3z^3 \qquad\quad + 3z - 2
\end{array}
$$

45. $72 = 2 \cdot 2 \cdot 2 \cdot 3 \cdot 3 = 2^3 \cdot 3^2$

49. $(y - 6)(y^2 + 3y + 2)$
$= y(y^2 + 3y + 2) - 6(y^2 + 3y + 2)$
$= y^3 + 3y^2 + 2y - 6y^2 - 18y - 12$
$= y^3 - 3y^2 - 16y - 12$
The area is $(y^3 - 3y^2 - 16y - 12)$ square feet.

53. answers may vary

Exercise Set 10.4

1. $48 = 2 \cdot 2 \cdot 2 \cdot 2 \cdot 3$
$15 = 3 \cdot 5$
$GCF = 3$

5. $12 = 2 \cdot 2 \cdot 3$
$20 = 2 \cdot 2 \cdot 5$
$36 = 2 \cdot 2 \cdot 3 \cdot 3$
$GCF = 2 \cdot 2 = 4$

9. $y^7 = y^2 \cdot y^5$
$y^2 = y^2$
$y^{10} = y^2 \cdot y^8$
$GCF = y^2$

13. $x^3 y^2 = x \cdot x^2 \cdot y^2$
$xy^2 = x \cdot y^2$
$x^4 y^2 = x \cdot x^3 \cdot y^2$
$GCF = x \cdot y^2 = xy^2$

17. $2 = 2$
$14 = 2 \cdot 7$
$18 = 2 \cdot 3 \cdot 3$
$GCF = 2$
$z^3 = z^3$
$z^5 = z^3 \cdot z^2$
$z^3 = z^3$
$GCF = z^3$
$GCF = 2z^3$

21. $10a^6 = 5a^6 \cdot 2$
$5a^8 = 5a^6 \cdot a^2$
$GCF = 5a^6$
$10a^6 - 5a^8 = 5a^6 \cdot 2 - 5a^6 \cdot a^2$
$\qquad = 5a^6(2 - a^2)$

25. $z^7 = z^5 \cdot z^2$
$6z^5 = z^5 \cdot 6$
$GCF = z^5$
$z^7 - 6z^5 = z^5 \cdot z^2 - z^5 \cdot 6 = z^5(z^2 - 6)$

29. $12a^5 = 12a^5$
$36a^6 = 12a^5 \cdot 3a$
$GCF = 12a^5$
$12a^5 - 36a^6 = 12a^5 \cdot 1 - 12a^5 \cdot 3a$
$\qquad = 12a^5(1 - 3a)$

33. $80\% = \dfrac{80}{100} = \dfrac{4}{5}$

37. a. area on the left: $x \cdot x = x^2$
area on the right: $2 \cdot x = 2x$
total area: $x^2 + 2x$
b. answers may vary; notice that $x^2 + 2x = x(x + 2)$

41. Let $x = 2$ and $z = 7$.
$(xy + z)^2 = (2y + 7)^2$
$\qquad = (2y + 7)(2y + 7)$
$\qquad = (2y)(2y) + 2y(7) + 7(2y) + 7(7)$
$\qquad = 4y^2 + 14y + 14y + 49$
$\qquad = 4y^2 + 28y + 49$

Chapter 10 Test

1. $(11x - 3) + (4x - 1) = (11x + 4x) + (-3 - 1)$
$\qquad = 15x + (-4)$
$\qquad = 15x - 4$

5. Let $x = 8$.
$x^2 - 6x + 1 = 8^2 - 6(8) + 1$
$\qquad = 64 - 6(8) + 1$
$\qquad = 64 - 48 + 1$
$\qquad = 17$

9. $(6a^3)(-2a^7) = (6)(-2)(a^3 \cdot a^7) = -12a^{10}$

13. $-2y(y^3 + 6y^2 - 4)$
$\qquad = -2y \cdot y^3 - 2y \cdot 6y^2 - 2y \cdot (-4)$
$\qquad = -2y^4 - 12y^3 + 8y$

17. Area:
$(x + 7)(5x - 2) = x(5x - 2) + 7(5x - 2)$
$\qquad = 5x^2 - 2x + 35x - 14$
$\qquad = 5x^2 + 33x - 14$
The area is $(5x^2 + 33x - 14)$ square inches.
Perimeter:
$2(2x) + 2(5x - 2) = 4x + 10x - 4 = 14x - 4$
The perimeter is $(14x - 4)$ inches.

21. $10a^2 = 2a \cdot 5a$
$12a = 2a \cdot 6$
$GCF = 2a$
$10a^2 + 12a = 2a \cdot 5a + 2a \cdot 6 = 2a(5a + 6)$

Appendices

Appendix B Exercise Set

1. $\dfrac{x^3}{x} = \dfrac{x^3}{x^1} = x^{3-1} = x^2$

5. $\dfrac{p^7 q^{20}}{pq^{15}} = \dfrac{p^7}{p^1} \cdot \dfrac{q^{20}}{q^{15}} = p^{7-1} \cdot q^{20-15} = p^6 q^5$

9. $7^0 = 1$

13. $-7^0 = -(7^0) = -1$

17. $4^{-3} = \dfrac{1}{4^3} = \dfrac{1}{64}$

21. $3^{-1} + 2^{-1} = \dfrac{1}{3} + \dfrac{1}{2} = \dfrac{1}{3} \cdot \dfrac{2}{2} + \dfrac{1}{2} \cdot \dfrac{3}{3} = \dfrac{2}{6} + \dfrac{3}{6} = \dfrac{5}{6}$

25. $\dfrac{x^{-2}}{x} = \dfrac{x^{-2}}{x^1} = x^{-2-1} = x^{-3} = \dfrac{1}{x^3}$

29. $3^{-2} + 3^{-1} = \dfrac{1}{3^2} + \dfrac{1}{3^1} = \dfrac{1}{9} + \dfrac{1}{3} = \dfrac{1}{9} + \dfrac{3}{9} = \dfrac{4}{9}$

33. $\dfrac{1}{p^{-4}} = p^4$

37. $(x^8 y^{-6})(x^{-2} y^{12}) = x^{8+(-2)} \cdot y^{-6+12}$
$\qquad = x^6 y^6$

41. $(5x^{-7})(3x^4) = 5 \cdot 3 \cdot x^{-7+4}$
$\qquad = 15 \cdot x^{-3}$
$\qquad = 15 \cdot \dfrac{1}{x^3}$
$\qquad = \dfrac{15}{x^3}$

45. $\left(8m^5n^{-1}\right)\left(7m^2n^{-4}\right) = 8\cdot 7\cdot m^{5+2}n^{-1+(-4)}$
$= 56\cdot m^7\cdot n^{-5}$
$= 56\cdot m^7\cdot \dfrac{1}{n^5}$
$= \dfrac{56m^7}{n^5}$

49. $\dfrac{a^9b^{14}}{ab} = \dfrac{a^9}{a^1}\cdot\dfrac{b^{14}}{b^1} = a^{9-1}\cdot b^{14-1} = a^8b^{13}$

53. $3z^0 = 3\cdot z^0 = 3\cdot 1 = 3$

57. $8x^{-9} = 8\cdot x^{-9} = 8\cdot\dfrac{1}{x^9} = \dfrac{8}{x^9}$

61. $\dfrac{z^{-8}}{z^{-1}} = z^{-8-(-1)} = z^{-8+1} = z^{-7} = \dfrac{1}{z^7}$

65. $\left(a^{-2}b^3\right)\left(a^{10}b^{-11}\right) = a^{-2+10}\cdot b^{3+(-11)}$
$= a^8\cdot b^{-8}$
$= a^8\cdot\dfrac{1}{b^8}$
$= \dfrac{a^8}{b^8}$

Appendix C Exercise Set

1. $78{,}000 = 7.8\times 10^4$

5. $0.00635 = 6.35\times 10^{-3}$

9. $4200 = 4.2\times 10^3$

13. $3.3\times 10^{-2} = 0.033$

17. $7.0\times 10^8 = 700{,}000{,}000$

21. $1.01\times 10^{13} = 10{,}100{,}000{,}000{,}000$

25. $\left(1.2\times 10^{-3}\right)\left(3\times 10^{-2}\right) = 1.2\cdot 3\times 10^{-3+(-2)}$
$= 3.6\times 10^{-5}$
$= 0.000036$

29. $\dfrac{8\times 10^{-1}}{16\times 10^5} = \dfrac{8}{16}\cdot\dfrac{10^{-1}}{10^5}$
$= 0.5\times 10^{-1-5}$
$= 0.5\times 10^{-6}$
$= 0.0000005$

33. $\left(7.5\times 10^5\right)(3600) = \left(7.5\times 10^5\right)\left(3.6\times 10^3\right)$
$= 7.5\cdot 3.6\times 10^{5+3}$
$= 27\times 10^8$
$= 2.7\times 10^1\times 10^8$
$= 2.7\times 10^9$

On average, 2.7×10^9 gallons of water flow over Niagara Falls each hour.

Practice Final Exam

1. $2^3\cdot 5^2 = 2\cdot 2\cdot 2\cdot 5\cdot 5 = 200$

5. $\sqrt{49} = 7$ because $7^2 = 49$

9. $-\dfrac{8}{15y} - \dfrac{2}{15y} = \dfrac{-8-2}{15y} = \dfrac{-10}{15y} = -\dfrac{2\cdot 5}{3\cdot 5\cdot y} = -\dfrac{2}{3y}$

13. $\begin{array}{r} 19 \\ -2\,\dfrac{3}{11} \\ \hline 11 \end{array}$ $\begin{array}{r} 18\dfrac{11}{11} \\ -2\dfrac{3}{11} \\ \hline 16\dfrac{8}{11} \end{array}$

17. $6.1 = 6.1(100\%) = 610\%$

21. 34.8923 rounded to the nearest tenth is 34.9.

25. $-(3z + 2) - 5z - 18 = -1(3z + 2) - 5z - 18$
$= -1\cdot 3z + (-1)\cdot 2 - 5z - 18$
$= -3z - 2 - 5z - 18$
$= -3z - 5z - 2 - 18$
$= -8z - 20$

29. $-4(x - 11) - 34 = 10 - 12$
$-4x + 44 - 34 = 10 - 12$
$-4x + 10 = -2$
$-4x + 10 - 10 = -2 - 10$
$-4x = -12$
$\dfrac{-4x}{-4} = \dfrac{-12}{-4}$
$x = 3$

33. Perimeter $= (20 + 10 + 20 + 10)\,\text{yards} = 60\,\text{yards}$
Area $= (\text{length})(\text{width})$
$= (20\,\text{yards})(10\,\text{yards})$
$= 200$ square yards

37. Let x be the number of women runners entered in the race. Since the number of men entered in the race is 112 more than the number of women, the number of men is $x + 112$. Since the total number of runners in the race is 600, the sum of x and $x + 112$ is 600.
$x + x + 112 = 600$
$2x + 112 = 600$
$2x + 112 - 112 = 600 - 112$
$2x = 448$
$\dfrac{2x}{2} = \dfrac{488}{2}$
$x = 244.$

244 women entered the race.

41. $y = 3x - 5$
Find 3 ordered pair solutions.
Let $x = 0$.
$y = 3x - 5$
$y = 3\cdot 0 - 5$
$y = 0 - 5$
$y = -5$
$(0, -5)$
Let $x = 1$.
$y = 3x - 5$
$y = 3\cdot 1 - 5$
$y = 3 - 5$
$y = -2$
$(1, -2)$
Let $x = 2$.
$y = 3x - 5$
$y = 3\cdot 2 - 5$
$y = 6 - 5$
$y = 1$
$(2, 1)$

Plot $(0, -5)$, $(1, -2)$, and $(2, 1)$. Then draw the line through them.

45. $\left(6a^3\right)\left(-2a^7\right) = (6)(-2)\left(a^3 \cdot a^7\right) = -12a^{10}$

49. The complement of an angle that measures $78°$ is an angle that measures $90° - 78° = 12°$.

53. $2\dfrac{1}{2}$ gal $= \dfrac{2\dfrac{1}{2} \text{ gal}}{1} \cdot \dfrac{4 \text{ qt}}{1 \text{ gal}} = 10$ qt

Subject Index

Photo Credits

Chapter 1 **p. 1** (tr) Rachel Youdelman/Pearson Education, Inc.; (mr) Rachel Youdelman/Pearson Education, Inc.; (bl) Rachel Youdelman/Pearson Education, Inc. **p. 6** Maciej Noskowski/Vetta/Getty Images **p. 14** NASA **p. 15** Jeanne Provost/Fotolia **p. 17** Rachel Youdelman/Pearson Education, Inc. **p. 23** Mediagram/Shutterstock **p. 24** Kikkerdirk/Fotolia **p. 35** Palessimages/Fotolia **p. 37** Auttapon Moonsawad/Fotolia **p. 46** Greg Henry/Shutterstock **p. 50** Photographee.eu/Fotolia

Chapter 2 **p. 97** Stephen Alvarez/National Geographic/Getty Images **p. 103** (l) Snehit/Fotolia; (r) Lunamarina/ Fotolia **p. 111** Paul Maguire/Fotolia **p. 126** Samott/Fotolia **p. 130** James Steidl/Fotolia **p. 150** *Melancolia* (1514), Albrecht Durer. Engraving, 9.5 × 7.5 in./Superstock **p. 153** (r) Tusharkoley/ Fotolia; (l) Chris Turner/Shutterstock **p. 155** Itsallgood/Fotolia

Chapter 3 **p. 162** Imago Stock & People/Newscom **p. 194** Shock/Fotolia **p. 196** (t) Falconhy/Fotolia; (b) David Benton/Shutterstock **p. 198** (l) Gretchen Owen/Fotolia; (r) HelleM/Fotolia **p. 208** Greg Roden/ Rough Guides/DK Images

Chapter 4 **p. 211** William87/Fotolia **p. 237** Keith Brofsky/Photodisc/Getty Images **p. 245** Pefkos/Fotolia **p. 264** Andres Rodriguez/Fotolia **p. 302** James Thew/Fotolia

Chapter 5 **p. 328** Dan Race/Fotolia **p. 330** Borsheim's Jewelry Store/AP Images **p. 337** SeanPavonePhoto/ Fotolia **p. 339** Grant V. Faint/Photodisc/Getty Images **p. 350** Courtesy of Apple **p. 361** Giemmephoto/ Fotolia **p. 371** WavebreakMediaMicro/Fotolia

Chapter 6 **p. 410** Studio306fotolia/Fotolia **p. 415** Rachel Youdelman/Pearson Education, Inc. **p. 420** (r) Monkey Business/Fotolia; (l) Stefan Huwiler/ Imagebroker/Alamy **p. 454** Stocktrek Images/ Getty Images **p. 459** (l) Rachel Youdelman/Pearson Education, Inc.; (r) Rachel Youdelman/Pearson Education, Inc.

Chapter 7 **p. 467** Fuse/Getty Images **p. 468** Pressmaster/Fotolia **p. 473** Clay Gay **p. 497** Michael Jung/ Fotolia **p. 498** Remik44992/Fotolia **p. 502** (l) Monkey Business/Fotolia; (r) Elayn Martin-Gay **p. 504** Kubais/Fotolia **p. 505** Rafa Irusta/Fotolia **p. 506** (l) Kletr/Fotolia; (r) Andreamuscatello/Fotolia **p. 508** Narvf/Fotolia **p. 512** (tr) Sashkin/Fotolia; (bl) Terex/Fotolia; (br) LuckyPhoto/Fotolia

Chapter 8 **p. 531** (l) Michael Shake/Fotolia; (r) Danr13/Fotolia

Chapter 9 **p. 613** (r) Donyanedomam/Fotolia; (l) Asa Gauen/Alamy **p. 622** Frankix/Fotolia **p. 637** (l) Jörg Hackemann/Fotolia; (r) Shuttoz/Fotolia **p. 642** daphot75/Fotolia **p. 662** (l) Vuktopua/ Fotolia; (r) Bill/Fotolia **p. 677** (r) iStockphoto; (l) Huaxiadragon/Fotolia **p. 678** Dave/Fotolia **p. 693** Worker/Shutterstock

Chapter 10 **p. 697** (tl) Bloomua/Fotolia; (tc) Chesky/Fotolia; (tr) Tarasov_vl/Fotolia; (bl) Koya979/ Fotolia; (br) Lateci/Fotolia

Cover Tamara Newman